MultiSIM® Software Offer!

Circuit Simulation **SOFTWARE**

AT STUDENT-FRIENDLY PRICES!

WHILE SUPPLIES LAST

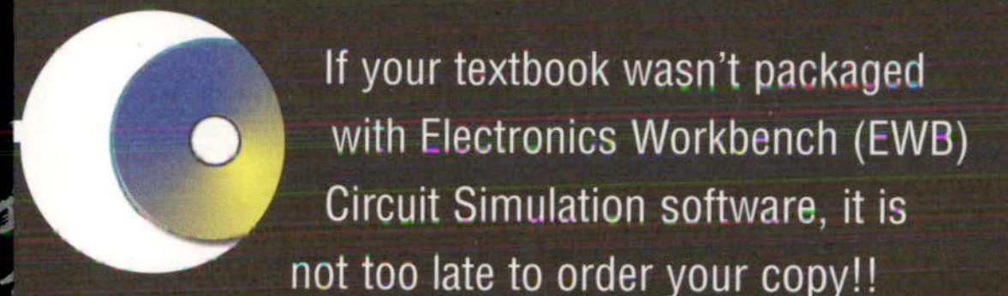

If your textbook wasn't packaged with Electronics Workbench (EWB) Circuit Simulation software, it is not too late to order your copy!!

Your purchase of this text entitles you to preferred pricing for the EWB Student Suite and Student Edition Lite software offered only by Prentice Hall.

The EWB Student Suite software provides you with a powerful tool used in electronics labs and in the electronics industry.

Benefits of owning your own copy:

- Save money!
- Save time by being more productive away from school!
- Complete lab assignments from home!
- Use it for more than one course!
- Save and Print your work and use it on campus in the lab!
- Add a line to your resume!

Prentice Hall offers pricing that you cannot receive from anyone else!!

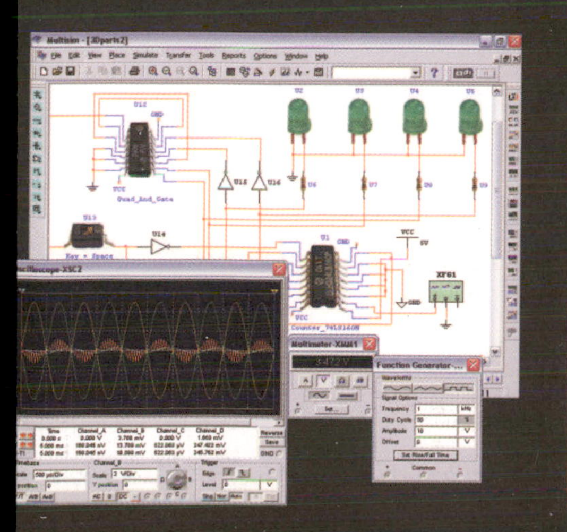

OPTION 1: EWB STUDENT SUITE

Student Suite provides electronics and electrical engineering students with a complete electronics lab in a computer. It is for use on a home PC and augments the Electronics Workbench labs at school. This package contains MultiSIM for circuit simulation and UltiBOARD for circuit board layout in 3D.

OPTION 2: EWB Student Edition Lite

Student Edition Lite provides students with the essentials needed to complete basic circuit simulation. Student Edition Lite contains the MultiSIM portion of EWB Student Suite and is a cost-effective solution for anyone looking for basic circuit simulation.

For ordering, pricing, and product specifications and comparisons, visit www.prenhall.com/ewb

*"**The Benefits of using simulation software (Multisim) before the actual connection of the circuit and signal measurement is substantial.** Students will have a good understanding of how the circuit will operate, and what the signals should look like."*

— Massaud M. Rabisee, Eastern Kentucky University

*"**Electronics Workbench (Multisim) is a terrific aid to help students learn circuit design.** It provides a learning experience that is very similar to the actual lab experience, allows a far greater variety of circuits to be studied and, of course, is far more convenient."*

— Dr. Adel Sedra, Dean of Engineering, University of Waterloo & Author

Ordering Electronics Workbench software from Prentice Hall is easy.

Visit www.prenhall.com/ewb or Email us at phewb@prenhall.com

Digital Systems

Principles and Applications

TENTH EDITION

Digital Systems

Principles and Applications

Ronald J. Tocci
Monroe Community College

Neal S. Widmer
Purdue University

Gregory L. Moss
Purdue University

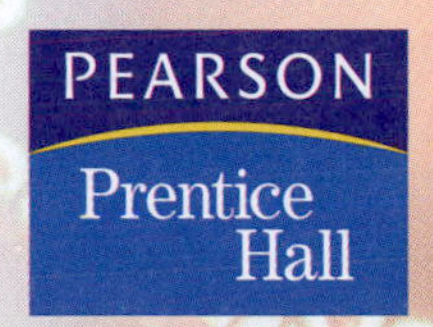

Upper Saddle River, New Jersey
Columbus, Ohio

Library of Congress Cataloging-in-Publication Data

Tocci, Ronald J.
Digital systems : principles and applications / Ronald J. Tocci, Neal S. Widmer, Gregory L. Moss.—10th ed.
p. cm.
Includes bibliographical references and index.
ISBN 0-13-172579-3
1. Digital electronics—Textbooks. I. Widmer, Neal S. II. Moss, Gregory L. III. Title.
TK7868.D5T62 2007
621.381—dc22

2005035835

Director of Development: Vern Anthony
Editorial Assistant: Lara Dimmick
Production Editor: Stephen C. Robb
Production Coordination: Peggy Hood, TechBooks/GTS
Design Coordinator: Diane Y. Ernsberger
Cover Designer: Jason Moore
Cover Art: Getty One
Production Manager: Matt Ottenweller
Marketing Manager: Ben Leonard

This book was set in TimesEuropa Roman by *TechBooks/GTS* York, PA Campus. It was printed and bound by Courier Kendallville, Inc. The cover was printed by Phoenix Color Corp.

Pearson Education Ltd.
Pearson Education Singapore, Pte. Ltd.
Pearson Education Canada, Ltd.
Pearson Education—Japan
Pearson Education Australia Pty. Limited
Pearson Education North Asia Ltd.
Pearson Educación de Mexico, S.A. de C.V.
Pearson Education Malaysia, Pte. Ltd.

10 9 8 7 6 5 4 3 2 1
ISBN: 0-13-172579-3

To you, Cap, for loving me for so long; and for the million and one ways you brighten the lives of everyone you touch.

—RJT

To my wife, Kris, and our children, John, Brad, Blake, Matt, and Katie: the lenders of their rights to my time and attention that this revision might be accomplished.

—NSW

To my family, Marita, David, and Ryan.

—GLM

PREFACE

This book is a comprehensive study of the principles and techniques of modern digital systems. It teaches the fundamental principles of digital systems and covers thoroughly both traditional and modern methods of applying digital design and development techniques, including how to manage a systems-level project. The book is intended for use in two- and four-year programs in technology, engineering, and computer science. Although a background in basic electronics is helpful, most of the material requires no electronics training. Portions of the text that use electronics concepts can be skipped without adversely affecting the comprehension of the logic principles.

General Improvements

The tenth edition of *Digital Systems* reflects the authors' views of the direction of modern digital electronics. In industry today, we see the importance of getting a product to market very quickly. The use of modern design tools, CPLDs, and FPGAs allows engineers to progress from concept to functional silicon very quickly. Microcontrollers have taken over many applications that once were implemented by digital circuits, and DSP has been used to replace many analog circuits. It is amazing that microcontrollers, DSP, and all the necessary glue logic can now be consolidated onto a single FPGA using a hardware description language with advanced development tools. Today's students must be exposed to these modern tools, even in an introductory course. It is every educator's responsibility to find the best way to prepare graduates for the work they will encounter in their professional lives.

The standard SSI and MSI parts that have served as "bricks and mortar" in the building of digital systems for nearly 40 years are now nearing obsolescence. Many of the techniques that have been taught over that time have focused on optimizing circuits that are built from these outmoded devices. The topics that are uniquely suited to applying the old technology *but do not contribute to an understanding of the new technology* must be removed from the curriculum. From an educational standpoint, however, these small ICs do offer a way to study simple digital circuits, and the wiring of circuits using breadboards is a valuable pedagogic exercise. They help to solidify concepts such as binary inputs and outputs, physical device operation, and practical limitations, using a very simple platform. Consequently, we have chosen to continue to introduce the conceptual descriptions of digital circuits and to offer examples using conventional standard logic parts. For instructors who continue to teach the fundamentals using SSI and MSI circuits, this edition retains those qualities that have made the text so widely accepted in the past. Many hardware design tools even provide an easy-to-use design entry technique that will employ the functionality of conventional standard parts with the flexibility of programmable logic devices. A digital design can be described using a schematic drawing with pre-created building blocks that are equivalent to conventional standard parts, which can be compiled and then programmed directly into a target PLD with the added capability of easily simulating the design within the same development tool.

We believe that graduates will actually apply the concepts presented in this book using higher-level description methods and more complex programmable devices. The major shift in the field is a greater need to understand the description methods, rather than focusing on the architecture of an actual device. Software tools have evolved to the point where there is little need for concern about the inner workings of the hardware but much more need to focus on what goes in, what comes out, and how the designer can describe what the device is supposed to do. We also believe that graduates will be involved with projects using state-of-the-art design tools and hardware solutions.

This book offers a strategic advantage for teaching the vital new topic of hardware description languages to beginners in the digital field. VHDL is undisputedly an industry standard language at this time, but it is also very complex and has a steep learning curve. Beginning students are often discouraged by the rigorous requirements of various data types, and they struggle with understanding edge-triggered events in VHDL. Fortunately, Altera offers AHDL, a less demanding language that uses the same basic concepts as VHDL but is much easier for beginners to master. So, instructors can opt to use AHDL to teach introductory students or VHDL for more advanced classes. This edition offers more than 40 AHDL examples, more than 40 VHDL examples, and many examples of simulation testing. All of these design files are available on the enclosed CD-ROM.

Altera's latest software development system is Quartus II. The MAX+PLUS II software that has been used for many years is still popular in industry and is supported by Altera. Its main drawback is that it does not program the latest devices. The material in this text does not attempt to teach a particular hardware platform or the details of using a software development system. New revisions of software tools appear so frequently that a textbook cannot remain current if it tries to describe all of the details. We have tried to show what this tool can do, rather than train the reader how to use it. However, tutorials have been included on the accompanying CD-ROM that make it easy to learn either software package. The AHDL and VHDL examples are compatible with either Quartus or MAX+PLUS systems. The timing simulations were developed using MAX+PLUS but can also be done with Quartus.

Many laboratory hardware options are available to users of this book. A number of CPLD and FPGA development boards are available for students to use in the laboratory. There are several earlier generation boards similar to Altera's UP2 that contain MAX7000 family CPLDs. A more recent example of an available board is the UP3 board from Altera's university program (see Figure P-1), which contains a larger FPGA from the Cyclone family. An even

FIGURE P-1 Altera's UP3 development board.

newer board from Altera is called the DE2 board (see Figure P-2), which has a powerful new 672-pin Cyclone II FPGA and a number of basic features such as switches, LEDs, and displays as well as many additional features for more advanced projects. More development boards are entering the market every year, and many are becoming very affordable. These boards, along with powerful educational software, offer an excellent way to teach and demonstrate the practical implementation of the concepts presented in this text.

The most significant improvements in the tenth edition are found in Chapter 7. Although asynchronous (ripple) counters provide a good introduction to sequential circuits, the real world uses synchronous counter circuits. Chapter 7 and subsequent examples have been rewritten to emphasize synchronous counter ICs and include techniques for analysis, cascading, and using HDL to describe them. A section has also been added to improve the coverage of state machines and the HDL features used to describe them. Other improvements include analysis techniques for combinational circuits, expanded coverage of 555 timer applications, and better coverage of signed binary numbers.

FIGURE P-2 Altera's DE2 development board.

Our approach to HDL and PLDs gives instructors several options:

1. The HDL material can be skipped entirely without affecting the continuity of the text.
2. HDL can be taught as a separate topic by skipping the material initially and then going back to the last sections of Chapters 3, 4, 5, 6, 7, and 9 and then covering Chapter 10.
3. HDL and the use of PLDs can be covered as the course unfolds—chapter by chapter—and woven into the fabric of the lecture/lab experience.

Among all specific hardware description languages, VHDL is clearly the industry standard and is most likely to be used by graduates in their careers. We have always felt that it is a bold proposition, however, to try to teach VHDL in an introductory course. The nature of the syntax, the subtle distinctions in object types, and the higher levels of abstraction can pose obstacles for a beginner. For this reason, we have included Altera's AHDL as the recommended introductory language for freshman courses. We have also included VHDL as the recommended language for more advanced classes or introductory courses offered to more mature students. We do not recommend trying to cover both languages in the same course. Sections of the text that cover the specifics of a language are clearly designated with a color bar in the margin. The HDL code figures are set in a color to match the color-coded text explanation. The reader can focus only on the language of his or her choice and skip the other. Obviously, we have attempted to appeal to the diverse interests of our market, but we believe we have created a book that can be used in multiple courses and will serve as an excellent reference after graduation.

Chapter Organization

It is a rare instructor who uses the chapters of a textbook in the sequence in which they are presented. This book was written so that, for the most part, each chapter builds on previous material, but it is possible to alter the chapter sequence somewhat. The first part of Chapter 6 (arithmetic operations) can be covered right after Chapter 2 (number systems), although this will lead to a long interval before the arithmetic circuits of Chapter 6 are encountered. Much of the material in Chapter 8 (IC characteristics) can be covered earlier (e.g., after Chapter 4 or 5) without creating any serious problems.

This book can be used either in a one-term course or in a two-term sequence. In a one-term course, limits on available class hours might require omitting some topics. Obviously, the choice of deletions will depend on factors such as program or course objectives and student background. A list of sections and chapters that can be deleted with minimal disruption follows:

- Chapter 1: All
- Chapter 2: Section 6
- Chapter 3: Sections 15–20
- Chapter 4: Sections 7, 10–13
- Chapter 5: Sections 3, 23–27
- Chapter 6: Sections 5–7, 11, 13, 16–23
- Chapter 7: Sections 9–14, 21–24
- Chapter 8: Sections 10, 14–19

- Chapter 9: Sections 5, 9, 15–20
- Chapter 10: All
- Chapter 11: Sections 7, 14–17
- Chapter 12: Sections 17–21
- Chapter 13: All

PROBLEM SETS This edition includes six categories of problems: basic (B), challenging (C), troubleshooting (T), new (N), design (D), and HDL (H). Undesignated problems are considered to be of intermediate difficulty, between basic and challenging. Problems for which solutions are printed in the back of the text or on the enclosed CD-ROM are marked with an asterisk (see Figure P-3).

PROJECT MANAGEMENT AND SYSTEM-LEVEL DESIGN Several real-world examples are included in Chapter 10 to describe the techniques used to manage projects. These applications are generally familiar to most students studying electronics, and the primary example of a digital clock is familiar to everyone. Many texts talk about top-down design, but this text demonstrates the key features of this approach and how to use the modern tools to accomplish it.

DATA SHEETS The CD-ROM containing Texas Instruments data sheets that accompanied the ninth edition has been removed. The information that was included on this CD-ROM is now readily available online.

SIMULATION FILES This edition also includes simulation files that can be loaded into Electronics Workbench Multisim®. The circuit schematics of many of the figures throughout the text have been captured as input files for this popular simulation tool. Each file has some way of demonstrating the operation of the circuit or reinforcing a concept. In many cases, instruments are attached to the circuit and input sequences are applied to demonstrate the concept presented in one of the figures of the text. These circuits can then be modified as desired to expand on topics or create assignments and tutorials

FIGURE P-3 Letters denote categories of problems, and asterisks indicate that corresponding solutions are provided at the end of the text.

PROBLEMS

SECTION 9-1

B 9-1. Refer to Figure 9-3. Determine the levels at each decoder output for the following sets of input conditions.

(a)*All inputs LOW

(b)*All inputs LOW except E_3 = HIGH

(c) All inputs HIGH except $\overline{E}_1 = \overline{E}_2$ = LOW

(d) All inputs HIGH

B 9-2.*What is the number of inputs and outputs of a decoder that accepts 64 different input combinations?

*Answers to problems marked with an asterisk can be found in the back of the text.

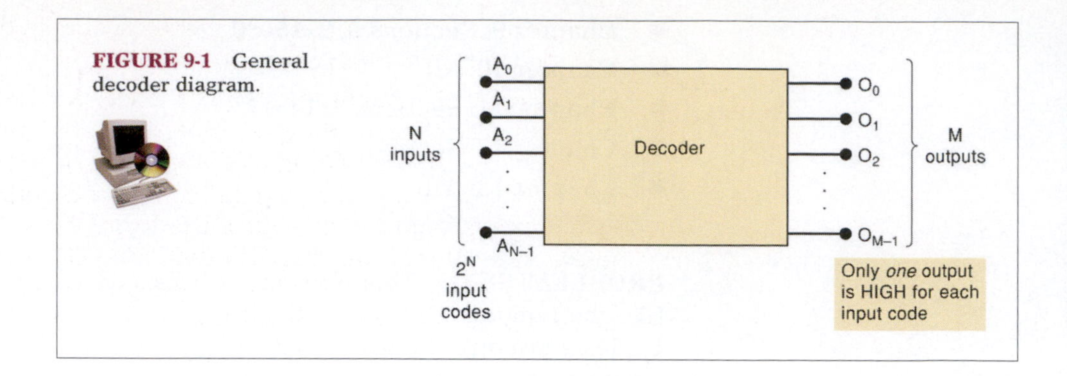

FIGURE P-4 The icon denotes a corresponding simulation file on the CD-ROM.

for students. All figures in the text that have a corresponding simulation file on the CD-ROM are identified by the icon shown in Figure P-4.

IC TECHNOLOGY This new edition continues the practice begun with the last three editions of giving more prominence to CMOS as the principal IC technology in small- and medium-scale integration applications. This depth of coverage has been accomplished while retaining the substantial coverage of TTL logic.

Specific Changes

The major changes in the topical coverage are listed here.

- **Chapter 1**. Many explanations covering digital/analog issues have been updated and improved.
- **Chapter 2**. The octal number system has been removed and the Gray code has been added. A complete standard ASCII code table has been included, along with new examples that relate ASCII characters, hex representation, and computer object code transfer files. New material on framing ASCII characters for asynchronous data transfer has also been added.
- **Chapter 3**. Along with some new practical examples of logic functions, the major improvement in Chapter 3 is a new analysis technique using tables that evaluate intermediate points in the logic circuit.
- **Chapter 4**. Very few changes were necessary in Chapter 4.
- **Chapter 5**. A new section covers digital pulses and associated definitions such as pulse width, period, rise time, and fall time. The terminology used for latch circuit inputs has been changed from Clear to Reset in order to be compatible with Altera component descriptions. The definition of a master/slave flip-flop has been removed as well. The discussion of Schmitt trigger applications has been improved to emphasize their role in eliminating the effects of noise. The inner workings of the 555 timer are now explained, and some improved timing circuits are proposed that make the device more versatile. The HDL coverage of SR and D latches has been rewritten to use a more intuitive behavioral description, and the coverage of counters has been modified to focus on structural techniques to interconnect flip-flop blocks.
- **Chapter 6**. Signed numbers are covered in more detail in this edition, particularly regarding sign extension in 2's complement numbers and arithmetic overflow. A new calculator hint simplifies negation of binary numbers represented in hex. A number circle model is used to compare

signed and unsigned number formats and help students to visualize add/subtract operation using both.

- **Chapter 7.** This chapter has been heavily revised to emphasize synchronous counter circuits. Simple ripple counters are still introduced to provide a basic understanding of the concept of counting and asynchronous cascading. After examining the limitations of ripple counters in Section 2, synchronous counters are introduced in Section 3 and used in all subsequent examples throughout the text. The IC counters presented are the 74160, '161, '162, and '163. These common devices offer an excellent assortment of features that teach the difference between synchronous and asynchronous control inputs and cascading techniques. The 74190 and '191 are used as an example of a synchronous up/down counter IC, further reinforcing the techniques required for synchronous cascading. A new section is devoted to analysis techniques for synchronous circuits using JK and D flip-flops. Synchronous design techniques now also include the use of D flip-flop registers that best represent the way sequential circuits are implemented in modern PLD technology. The HDL sections have been improved to demonstrate the implementation of synchronous/asynchronous loading, clearing, and cascading. A new emphasis is placed on simulation and testing of HDL modules. State machines are now presented as a topic, the traditional Mealy and Moore models are defined, and a new traffic light control system is presented as an example. Minor improvements have been made in the second half of Chapter 7 also. All of the problems at the end of Chapter 7 have been rewritten to reinforce the concepts.
- **Chapter 8.** This chapter remains a very technical description of the technology available in standard logic families and digital components. The mixed-voltage interfacing sections have been improved to cover low-voltage technology. The latest Texas Instruments life-cycle curve shows the history and current position of various logic series between introduction and obsolescence. Low-voltage differential signaling (LVDS) is introduced as well.
- **Chapter 9.** The many different building blocks of digital systems are still covered in this chapter and demonstrated using HDL. Many other HDL techniques, such as tristate outputs and various HDL control structures, are also introduced. A 74ALS148 is described as another example of an encoder. The examples of systems that use counters have all been updated to synchronous operation. The serial transmission system using MUX and DEMUX is particularly improved. The technique of using a MUX to implement SOP expressions has been explained in a more structured way as an independent study exercise in the end-of-the-chapter problems.
- **Chapter 10.** Chapter 10, which was new to the ninth edition, has remained essentially unchanged.
- **Chapter 11.** The material on bipolar DACs has been improved, and an example of using DACs as a digital amplitude control for analog waveforms is presented. The more common A/D converter accuracy specification in the form of +/− LSB is explained in this edition.
- **Chapter 12.** Minor improvements were made to this chapter to consolidate and compress some of the material on older technologies of memory such as UV EPROM. Flash technology is still introduced using a first-generation example, but the more recent improvements, as well as some of the applications of flash technology in modern consumer devices, are described.
- **Chapter 13.** This chapter, which was new to the ninth edition, has been updated to introduce the new Cyclone family of PLDs.

Retained Features

This edition retains all of the features that made the previous editions so widely accepted. It utilizes a block diagram approach to teach the basic logic operations without confusing the reader with the details of internal operation. All but the most basic electrical characteristics of the logic ICs are withheld until the reader has a firm understanding of logic principles. In Chapter 8, the reader is introduced to the internal IC circuitry. At that point, the reader can interpret a logic block's input and output characteristics and "fit" it properly into a complete system.

The treatment of each new topic or device typically follows these steps: the principle of operation is introduced; thoroughly explained examples and applications are presented, often using actual ICs; short review questions are posed at the end of the section; and finally, in-depth problems are available at the end of the chapter. These problems, ranging from simple to complex, provide instructors with a wide choice of student assignments. These problems are often intended to reinforce the material without simply repeating the principles. They require students to demonstrate comprehension of the principles by applying them to different situations. This approach also helps students to develop confidence and expand their knowledge of the material.

The material on PLDs and HDLs is distributed throughout the text, with examples that emphasize key features in each application. These topics appear at the end of each chapter, making it easy to relate each topic to the general discussion earlier in the chapter or to address the general discussion separately from the PLD/HDL coverage.

The extensive troubleshooting coverage is spread over Chapters 4 through 12 and includes presentation of troubleshooting principles and techniques, case studies, 25 troubleshooting examples, and 60 *real* troubleshooting problems. When supplemented with hands-on lab exercises, this material can help foster the development of good troubleshooting skills.

The tenth edition offers more than 200 worked-out examples, more than 400 review questions, and more than 450 chapter problems/exercises. Some of these problems are applications that show how the logic devices presented in the chapter are used in a typical microcomputer system. Answers to a majority of the problems immediately follow the Glossary. The Glossary provides concise definitions of all terms in the text that have been highlighted in boldface type.

An IC index is provided at the back of the book to help readers locate easily material on any IC cited or used in the text. The back endsheets provide tables of the most often used Boolean algebra theorems, logic gate summaries, and flip-flop truth tables for quick reference when doing problems or working in the lab.

Supplements

An extensive complement of teaching and learning tools has been developed to accompany this textbook. Each component provides a unique function, and each can be used independently or in conjunction with the others.

CD-ROM A CD-ROM is packaged with each copy of the text. It contains the following material:

- **MAX+PLUS® II Educational Version software from Altera.** This is a fully functional, professional-quality, integrated development environment for digital systems that has been used for many years and is still supported by Altera. Students can use it to write, compile, and simulate their designs at home before going to the lab. They can use the same software to program and test an Altera CPLD.

- **Quartus II Web Version software from Altera.** This is the latest development system software from Altera, which offers more advanced features and supports new PLD devices such as the Cyclone family of FPGAs, found on many of the newest educational boards.
- **Tutorials.** Gregory Moss has developed tutorials that have been used successfully for several years to teach introductory students how to use Altera MAX+PLUS II software. These tutorials are available in PDF and PPT (Microsoft® PowerPoint® presentation) formats and have been adapted to teach Quartus II as well. With the help of these tutorials, anyone can learn to modify and test all the examples presented in this text, as well as develop his or her own designs.
- **Design files from the textbook figures.** More than 40 design files in each language are presented in figures throughout the text. Students can load these into the Altera software and test them.
- **Solutions to selected problems: HDL design files.** A few of the end-of-chapter problem solutions are available to students. (All of the HDL solutions are available to instructors in the *Instructor's Resource Manual.)* Solutions for Chapter 7 problems include some large graphic and HDL files that are not published in the back of the book but are available on the enclosed CD-ROM.
- **Circuits from the text rendered in Multisim®.** Students can open and work interactively with approximately 100 circuits to increase their understanding of concepts and prepare for laboratory activities. The Multisim circuit files are provided for use by anyone who has Multisim software. Anyone who does not have Multisim software and wishes to purchase it in order to use the circuit files may do so by ordering it from www.prenhall.com/ewb.
- **Supplemental material introducing microprocessors and microcontrollers.** For the flexibility to serve the diverse needs of the many different schools, an introduction to this topic is presented as a convenient bridge between a digital systems course and an introduction to microprocessors/microcontrollers course.

STUDENT RESOURCES

- ***Lab Manual: A Design Approach.*** This lab manual, written by Gregory Moss, contains topical units with lab projects that emphasize simulation and design. It utilizes the Altera MAX+PLUS II or Quartus II software in its programmable logic exercises and features both schematic capture and hardware description language techniques. The new edition contains many new projects and examples. (ISBN 0-13-188138-8)
- ***Lab Manual: A Troubleshooting Approach.*** This manual, written by Jim DeLoach and Frank Ambrosio, is presented with an analysis and troubleshooting approach and is fully updated for this edition of the text. (ISBN 0-13-188136-1)
- **Companion Website (www.prenhall.com/tocci).** This site offers students a free online study guide with which they can review the material learned in the text and check their understanding of key topics.

INSTRUCTOR RESOURCES

- ***Instructor's Resource Manual.*** This manual contains worked-out solutions for all end-of-chapter problems in this textbook. (ISBN 0-13-172665-X)
- ***Lab Solutions Manual.*** Worked-out lab results for both lab manuals are featured in this manual. (ISBN 0-13-172664-1)
- **PowerPoint® presentations.** Figures from the text, in addition to Lecture Notes for each chapter, are available on CD-ROM. (ISBN 0-13-172667-6)
- **TestGen.** A computerized test bank is available on CD-ROM. (ISBN 0-13-172666-8)

To access supplementary materials online, instructors need to request an instructor access code. Go to **www.prenhall.com,** click the **Instructor Resource Center** link, and then click **Register Today** for an instructor access code. Within 48 hours after registering, you will receive a confirming e-mail including an instructor access code. When you have received your code, go to the site and log on for full instructions on downloading the materials you wish to use.

ACKNOWLEDGMENTS

We are grateful to all those who evaluated the ninth edition and provided answers to an extensive questionnaire: Ali Khabari, Wentworth Institute of Technology; Al Knebel, Monroe Community College; Rex Fisher, Brigham Young University; Alan Niemi, LeTourneau University; and Roger Sash, University of Nebraska. Their comments, critiques, and suggestions were given serious consideration and were invaluable in determining the final form of the tenth edition.

We also are greatly indebted to Professor Frank Ambrosio, Monroe Community College, for his usual high-quality work on the indexes and the *Instructor's Resource Manual;* and Professor Thomas L. Robertson, Purdue University, for providing his magnetic levitation system as an example; and Professors Russ Aubrey and Gene Harding, Purdue University, for their technical review of topics and many suggestions for improvements. We appreciate the cooperation of Mike Phipps and the Altera Corporation for their support in granting permission to use their software package and their figures from technical publications.

A writing project of this magnitude requires conscientious and professional editorial support, and Prentice Hall came through again in typical fashion. We thank the staffs at Prentice Hall and TechBooks/GTS for their help to make this publication a success.

And finally, we want to let our wives and our children know how much we appreciate their support and their understanding. We hope that we can eventually make up for all the hours we spent away from them while we worked on this revision.

Ronald J. Tocci
Neal S. Widmer
Gregory L. Moss

BRIEF CONTENTS

CONTENTS

Chapter 3 Describing Logic Circuits 54

Chapter 4 Combinational Logic Circuits 118

Chapter 5 Flip-Flops and Related Devices 208

Chapter 6 Digital Arithmetic: Operations and Circuits 296

Chapter 11 Interfacing with the Analog World 718

Chapter 12 Memory Devices 784

Digital Systems

Principles and Applications

CHAPTER 1

INTRODUCTORY CONCEPTS

OUTLINE

OBJECTIVES

Upon completion of this chapter, you will be able to:

- Distinguish between analog and digital representations.
- Cite the advantages and drawbacks of digital techniques compared with analog.
- Understand the need for analog-to-digital converters (ADCs) and digital-to-analog converters (DACs).
- Recognize the basic characteristics of the binary number system.
- Convert a binary number to its decimal equivalent.
- Count in the binary number system.
- Identify typical digital signals.
- Identify a timing diagram.
- State the differences between parallel and serial transmission.
- Describe the property of memory.
- Describe the major parts of a digital computer and understand their functions.
- Distinguish among microcomputers, microprocessors, and microcontrollers.

INTRODUCTION

In today's world, the term *digital* has become part of our everyday vocabulary because of the dramatic way that digital circuits and digital techniques have become so widely used in almost all areas of life: computers, automation, robots, medical science and technology, transportation, telecommunications, entertainment, space exploration, and on and on. You are about to begin an exciting educational journey in which you will discover the fundamental principles, concepts, and operations that are common to all digital systems, from the simplest on/off switch to the most complex computer. If this book is successful, you should gain a deep understanding of how all digital systems work, and you should be able to apply this understanding to the analysis and troubleshooting of any digital system.

We start by introducing some underlying concepts that are a vital part of digital technology; these concepts will be expanded on as they are needed later in the book. We also introduce some of the terminology that is necessary when embarking on a new field of study, and add to this list of important terms in every chapter.

1-1 NUMERICAL REPRESENTATIONS

In science, technology, business, and, in fact, most other fields of endeavor, we are constantly dealing with *quantities.* Quantities are measured, monitored, recorded, manipulated arithmetically, observed, or in some other way utilized in most physical systems. It is important when dealing with various quantities that we be able to represent their values efficiently and accurately. There are basically two ways of representing the numerical value of quantities: **analog** and **digital**.

Analog Representations

In **analog representation** a quantity is represented by a continuously variable, proportional indicator. An example is an automobile speedometer from the classic muscle cars of the 1960s and 1970s. The deflection of the needle is proportional to the speed of the car and follows any changes that occur as the vehicle speeds up or slows down. On older cars, a flexible mechanical shaft connected the transmission to the speedometer on the dash board. It is interesting to note that on newer cars, the analog representation is usually preferred even though speed is now measured digitally.

Thermometers before the digital revolution used analog representation to measure temperature, and many are still in use today. Mercury thermometers use a column of mercury whose height is proportional to temperature. These devices are being phased out of the market because of environmental concerns, but nonetheless they are an excellent example of analog representation. Another example is an outdoor thermometer on which the position of the pointer rotates around a dial as a metal coil expands and contracts with temperature changes. The position of the pointer is proportional to the temperature. Regardless of how small the change in temperature, there will be a proportional change in the indication.

In these two examples the physical quantities (speed and temperature) are being coupled to an indicator by purely mechanical means. In electrical analog systems, the physical quantity that is being measured or processed is converted to a proportional voltage or current (electrical signal). This voltage or current is then used by the system for display, processing, or control purposes.

Sound is an example of a physical quantity that can be represented by an electrical analog signal. A microphone is a device that generates an output voltage that is proportional to the amplitude of the sound waves that strike it. Variations in the sound waves will produce variations in the microphone's output voltage. Tape recordings can then store sound waves by using the output voltage of the microphone to proportionally change the magnetic field on the tape.

Analog quantities such as those cited above have an important characteristic, no matter how they are represented: *they can vary over a continuous range of values.* The automobile speed can have *any* value between zero and, say, 100 mph. Similarly, the microphone output might have any value within a range of zero to 10 mV (e.g., 1 mV, 2.3724 mV, 9.9999 mV).

Digital Representations

In **digital representation** the quantities are represented not by continuously variable indicators but by symbols called *digits.* As an example, consider the digital clock, which provides the time of day in the form of decimal digits that represent hours and minutes (and sometimes seconds). As we know, the time of day changes continuously, but the digital clock reading does not change continuously; rather, it changes in steps of one per minute (or per second). In other words, this digital representation of the time of day changes in *discrete* steps, as compared with the representation of time provided by an analog ac line-powered wall clock, where the dial reading changes continuously.

The major difference between analog and digital quantities, then, can be simply stated as follows:

analog ≡ continuous
digital ≡ discrete (step by step)

Because of the discrete nature of digital representations, there is no ambiguity when reading the value of a digital quantity, whereas the value of an analog quantity is often open to interpretation. In practice, when we take a measurement of an analog quantity, we always "round" to a convenient level of precision. In other words, we digitize the quantity. The digital representation is the result of assigning a number of limited precision to a continuously variable quantity. For example, when you take your temperature with a mercury (analog) thermometer, the mercury column is usually between two graduation lines, but you would pick the nearest line and assign it a number of, say, 98.6°F.

EXAMPLE 1-1

Which of the following involve analog quantities and which involve digital quantities?

(a) Ten-position switch
(b) Current flowing from an electrical outlet
(c) Temperature of a room
(d) Sand grains on the beach
(e) Automobile fuel gauge

Solution

(a) Digital
(b) Analog
(c) Analog
(d) Digital, since the number of grains can be only certain discrete (integer) values and not every possible value over a continuous range
(e) Analog, if needle type; digital, if numerical readout or bar graph display

REVIEW QUESTION *

1. Concisely describe the major difference between analog and digital quantities.

1-2 DIGITAL AND ANALOG SYSTEMS

A **digital system** is a combination of devices designed to manipulate logical information or physical quantities that are represented in digital form; that is, the quantities can take on only discrete values. These devices are most

*Answers to review questions are found at the end of the chapter in which they occur.

often electronic, but they can also be mechanical, magnetic, or pneumatic. Some of the more familiar digital systems include digital computers and calculators, digital audio and video equipment, and the telephone system—the world's largest digital system.

An **analog system** contains devices that manipulate physical quantities that are represented in analog form. In an analog system, the quantities can vary over a continuous range of values. For example, the amplitude of the output signal to the speaker in a radio receiver can have any value between zero and its maximum limit. Other common analog systems are audio amplifiers, magnetic tape recording and playback equipment, and a simple light dimmer switch.

Advantages of Digital Techniques

An increasing majority of applications in electronics, as well as in most other technologies, use digital techniques to perform operations that were once performed using analog methods. The chief reasons for the shift to digital technology are:

1. *Digital systems are generally easier to design.* The circuits used in digital systems are *switching circuits,* where *exact* values of voltage or current are not important, only the range (HIGH or LOW) in which they fall.
2. *Information storage is easy.* This is accomplished by special devices and circuits that can latch onto digital information and hold it for as long as necessary, and mass storage techniques that can store billions of bits of information in a relatively small physical space. Analog storage capabilities are, by contrast, extremely limited.
3. *Accuracy and precision are easier to maintain throughout the system.* Once a signal is digitized, the information it contains does not deteriorate as it is processed. In analog systems, the voltage and current signals tend to be distorted by the effects of temperature, humidity, and component tolerance variations in the circuits that process the signal.
4. *Operation can be programmed.* It is fairly easy to design digital systems whose operation is controlled by a set of stored instructions called a *program.* Analog systems can also be *programmed,* but the variety and the complexity of the available operations are severely limited.
5. *Digital circuits are less affected by noise.* Spurious fluctuations in voltage (noise) are not as critical in digital systems because the exact value of a voltage is not important, as long as the noise is not large enough to prevent us from distinguishing a HIGH from a LOW.
6. *More digital circuitry can be fabricated on IC chips.* It is true that analog circuitry has also benefited from the tremendous development of IC technology, but its relative complexity and its use of devices that cannot be economically integrated (high-value capacitors, precision resistors, inductors, transformers) have prevented analog systems from achieving the same high degree of integration.

Limitations of Digital Techniques

There are really very few drawbacks when using digital techniques. The two biggest problems are:

The real world is analog.
Processing digitized signals takes time.

Most physical quantities are analog in nature, and these quantities are often the inputs and outputs that are being monitored, operated on, and controlled by a system. Some examples are temperature, pressure, position, velocity, liquid level, flow rate, and so on. We are in the habit of expressing these quantities *digitally*, such as when we say that the temperature is 64° (63.8° when we want to be more precise), but we are really making a digital approximation to an inherently analog quantity.

To take advantage of digital techniques when dealing with analog inputs and outputs, four steps must be followed:

1. Convert the physical variable to an electrical signal (analog).
2. Convert the electrical (analog) signal into digital form.
3. Process (operate on) the digital information.
4. Convert the digital outputs back to real-world analog form.

An entire book could be written about step 1 alone. There are many kinds of devices that convert various physical variables into electrical analog signals (sensors). These are used to measure things that are found in our "real" analog world. On your car alone, there are sensors for fluid level (gas tank), temperature (climate control and engine), velocity (speedometer), acceleration (airbag collision detection), pressure (oil, manifold), and flow rate (fuel), to name just a few.

To illustrate a typical system that uses this approach Figure 1-1 describes a precision temperature regulation system. A user pushes up or down buttons to set the desired temperature in 0.1° increments (digital representation). A temperature sensor in the heated space converts the measured temperature to a proportional voltage. This analog voltage is converted to a digital quantity by an **analog-to-digital converter (ADC)**. This value is then compared to the desired value and used to determine a digital value of how much heat is needed. The digital value is converted to an analog quantity (voltage) by a **digital-to-analog converter (DAC)**. This voltage is applied to a heating element, which will produce heat that is related to the voltage applied and will affect the temperature of the space.

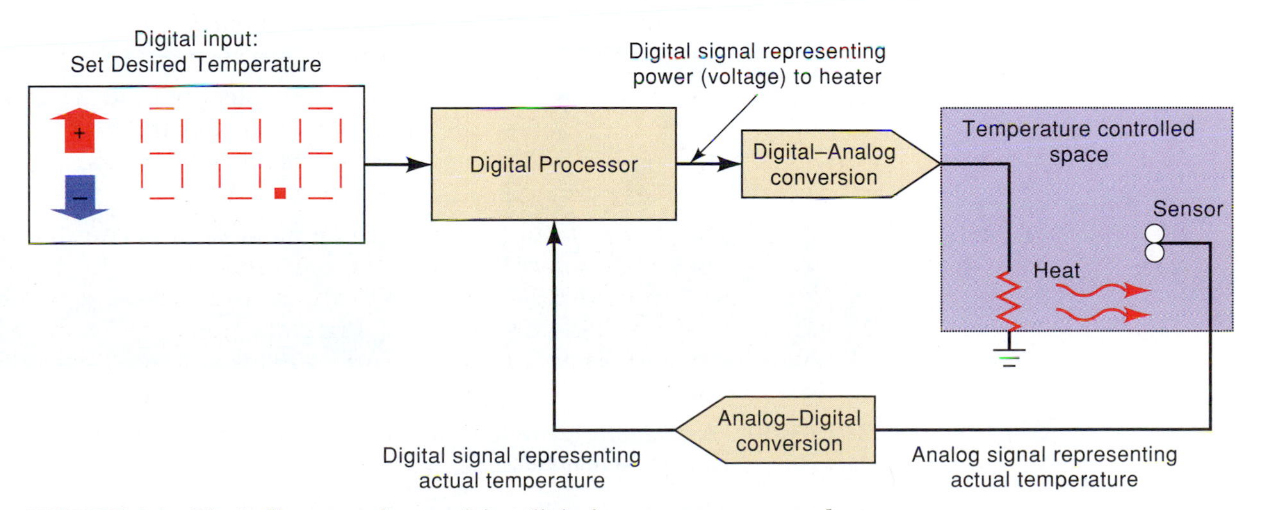

FIGURE 1-1 Block diagram of a precision digital temperature control system.

Another good example where conversion between analog and digital takes place is in the recording of audio. Compact disks (CDs) have replaced cassette tapes because they provide a much better means for recording and playing back music. The process works something like this: (1) sounds from instruments and human voices produce an analog voltage signal in a microphone; (2) this analog signal is converted to a digital format using an analog-to-digital conversion process; (3) the digital information is stored on the CD's surface; (4) during playback, the CD player takes the digital information from the CD surface and converts it into an analog signal that is then amplified and fed to a speaker, where it can be picked up by the human ear.

The second drawback to digital systems is that processing these digitized signals (lists of numbers) takes time. And we also need to convert between the analog and digital forms of information, which can add complexity and expense to a system. The more precise the numbers need to be, the longer it takes to process them. In many applications, these factors are outweighed by the numerous advantages of using digital techniques, and so the conversion between analog and digital quantities has become quite commonplace in the current technology.

There are situations, however, where use of analog techniques is simpler or more economical. For example, several years ago, a colleague (Tom Robertson) decided to create a control system demonstration for tour groups. He planned to suspend a metallic object in a magnetic field, as shown in Figure 1-2. An electromagnet was made by winding a coil of wire and controlling the amount of current through the coil. The position of the metal object was measured by passing an infrared light beam across the magnetic field. As the object drew closer to the magnetic coil, it began to block the light beam. By measuring small changes in the light level, the magnetic field could be controlled to keep the metal object hovering and stationary, with no strings attached. All attempts at using a microcomputer to measure these very small changes, run the control calculations, and drive the magnet proved to be too slow, even when using the fastest, most powerful PC available at the time. His final solution used just a couple of op-amps and a few dollars' worth of other components: a totally analog approach. Today we have access to processors fast enough and measurement techniques precise enough to accomplish this feat, but the simplest solution is still analog.

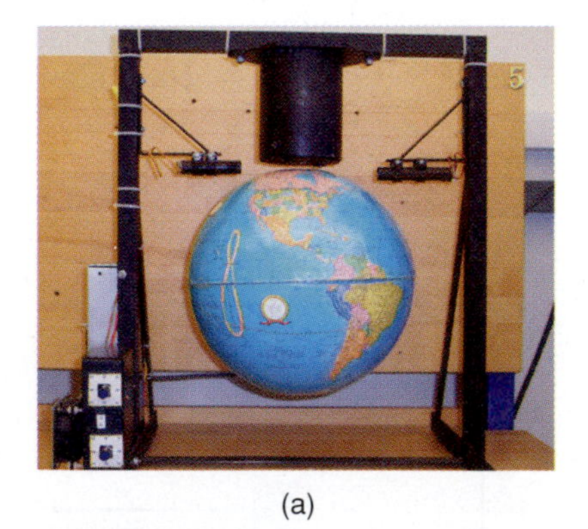

(a)

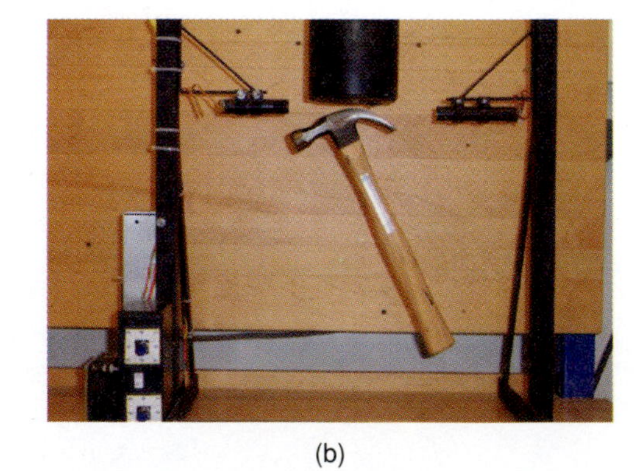

(b)

FIGURE 1-2 A magnetic levitation system suspending: (a) a globe with a steel plate inserted and (b) a hammer.

It is common to see both digital and analog techniques employed within the same system to be able to profit from the advantages of each. In these *hybrid* systems, one of the most important parts of the design phase involves determining what parts of the system are to be analog and what parts are to be digital. The trend in most systems is to digitize the signal as early as possible and convert it back to analog as late as possible as the signals flow through the system.

The Future Is Digital

The advances in digital technology over the past three decades have been nothing short of phenomenal, and there is every reason to believe that more is coming. Think of the everyday items that have changed from analog format to digital in your lifetime. An indoor/outdoor wireless digital thermometer can be purchased for less then $10.00. Cars have gone from having very few electronic controls to being predominantly digitally controlled vehicles. Digital audio has moved us to the compact disk and MP3 player. Digital video brought the DVD. Digital home video and still cameras; digital recording with systems like TiVo; digital cellular phones; and digital imaging in x-ray, magnetic resonance imaging (MRI), and ultrasound systems in hospitals are just a few of the applications that have been taken over by the digital revolution. As soon as the infrastructure is in place, telephone and television systems will go digital. The growth rate in the digital realm continues to be staggering. Maybe your automobile is equipped with a system such as GM's On Star, which turns your dashboard into a hub for wireless communication, information, and navigation. You may already be using voice commands to send or retrieve e-mail, call for a traffic report, check on the car's maintenance needs, or just switch radio stations or CDs—all without taking your hands off the wheel or your eyes off the road. Cars can report their exact location in case of emergency or mechanical breakdown. In the coming years wireless communication will continue to expand coverage to provide connectivity wherever you are. Telephones will be able to receive, sort, and maybe respond to incoming calls like a well-trained secretary. The digital television revolution will provide not only higher definition of the picture, but also much more flexibility in programming. You will be able to select the programs that you want to view and load them into your television's memory, allowing you to pause or replay scenes at your convenience, very much like viewing a DVD today. As virtual reality continues to improve, you will be able to interact with the subject matter you are studying. This may not sound exciting when studying electronics, but imagine studying history from the standpoint of being a participant, or learning proper techniques for everything from athletics to surgery through simulations based on your actual performance.

Digital technology will continue its high-speed incursion into current areas of our lives as well as break new ground in ways we may never have considered. These applications (and many more) are based on the principles presented in this text. The software tools to develop complex systems are constantly being upgraded and are available to anyone over the Web. We will study the technical underpinnings necessary to communicate with any of these tools, and prepare you for a fascinating and rewarding career.

REVIEW QUESTIONS

1. What are the advantages of digital techniques over analog?
2. What is the chief limitation to the use of digital techniques?

1-3 DIGITAL NUMBER SYSTEMS

Many number systems are in use in digital technology. The most common are the decimal, binary, octal, and hexadecimal systems. The decimal system is clearly the most familiar to us because it is a tool that we use every day. Examining some of its characteristics will help us to understand the other systems better.

Decimal System

The **decimal system** is composed of *10* numerals or symbols. These 10 symbols are 0, 1, 2, 3, 4, 5, 6, 7, 8, 9; using these symbols as *digits* of a number, we can express any quantity. The decimal system, also called the *base-10* system because it has 10 digits, has evolved naturally as a result of the fact that people have 10 fingers. In fact, the word *digit* is derived from the Latin word for "finger."

The decimal system is a *positional-value system* in which the value of a digit depends on its position. For example, consider the decimal number 453. We know that the digit 4 actually represents 4 *hundreds,* the 5 represents 5 *tens,* and the 3 represents 3 *units.* In essence, the 4 carries the most weight of the three digits; it is referred to as the *most significant digit (MSD)*. The 3 carries the least weight and is called the *least significant digit (LSD)*.

Consider another example, 27.35. This number is actually equal to 2 tens plus 7 units plus 3 tenths plus 5 hundredths, or $2 \times 10 + 7 \times 1 + 3 \times 0.1 + 5 \times 0.01$. The decimal point is used to separate the integer and fractional parts of the number.

More rigorously, the various positions relative to the decimal point carry weights that can be expressed as powers of 10. This is illustrated in Figure 1-3, where the number 2745.214 is represented. The decimal point separates the positive powers of 10 from the negative powers. The number 2745.214 is thus equal to

$$(2 \times 10^{+3}) + (7 \times 10^{+2}) + (4 \times 10^{1}) + (5 \times 10^{0}) + (2 \times 10^{-1}) + (1 \times 10^{-2}) + (4 \times 10^{-3})$$

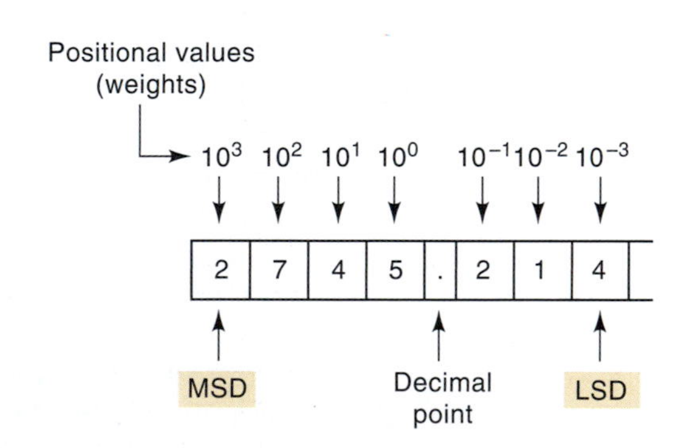

FIGURE 1-3 Decimal position values as powers of 10.

In general, any number is simply the sum of the products of each digit value and its positional value.

Decimal Counting

When counting in the decimal system, we start with 0 in the units position and take each symbol (digit) in progression until we reach 9. Then we add a 1 to the next higher position and start over with 0 in the first position (see

Figure 1-4). This process continues until the count of 99 is reached. Then we add a 1 to the third position and start over with 0s in the first two positions. The same pattern is followed continuously as high as we wish to count.

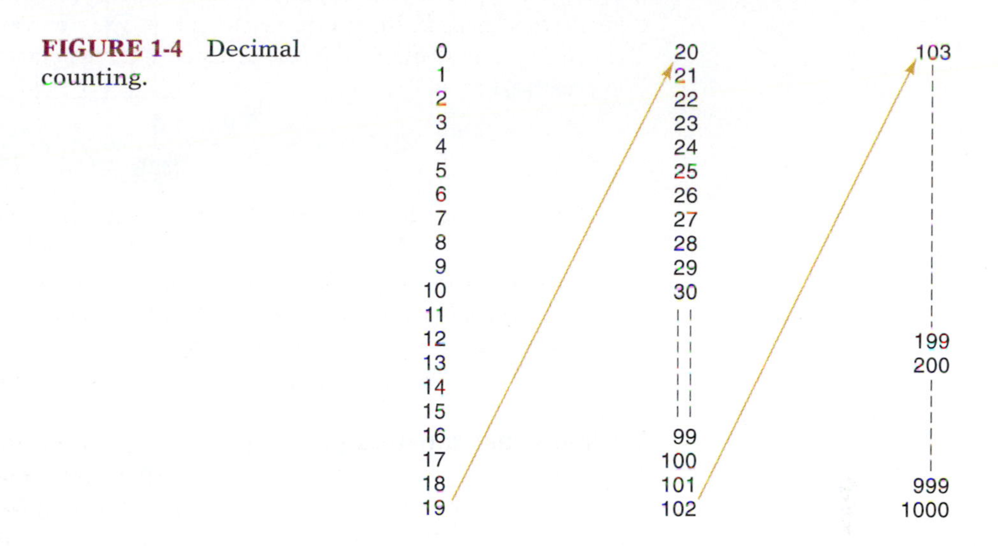

FIGURE 1-4 Decimal counting.

It is important to note that in decimal counting, the units position (LSD) changes upward with each step in the count, the tens position changes upward every 10 steps in the count, the hundreds position changes upward every 100 steps in the count, and so on.

Another characteristic of the decimal system is that using only two decimal places, we can count through $10^2 = 100$ different numbers (0 to 99).* With three places we can count through 1000 numbers (0 to 999), and so on. In general, with N places or digits, we can count through 10^N different numbers, starting with and including zero. The largest number will always be $10^N - 1$.

Binary System

Unfortunately, the decimal number system does not lend itself to convenient implementation in digital systems. For example, it is very difficult to design electronic equipment so that it can work with 10 different voltage levels (each one representing one decimal character, 0 through 9). On the other hand, it is very easy to design simple, accurate electronic circuits that operate with only two voltage levels. For this reason, almost every digital system uses the binary (base-2) number system as the basic number system of its operations. Other number systems are often used to interpret or represent binary quantities for the convenience of the people who work with and use these digital systems.

In the **binary system** there are only two symbols or possible digit values, 0 and 1. Even so, this base-2 system can be used to represent any quantity that can be represented in decimal or other number systems. In general though, it will take a greater number of binary digits to express a given quantity.

All of the statements made earlier concerning the decimal system are equally applicable to the binary system. The binary system is also a positional-value system, wherein each binary digit has its own value or weight expressed as a power of 2. This is illustrated in Figure 1-5. Here, places to the left of the

*Zero is counted as a number.

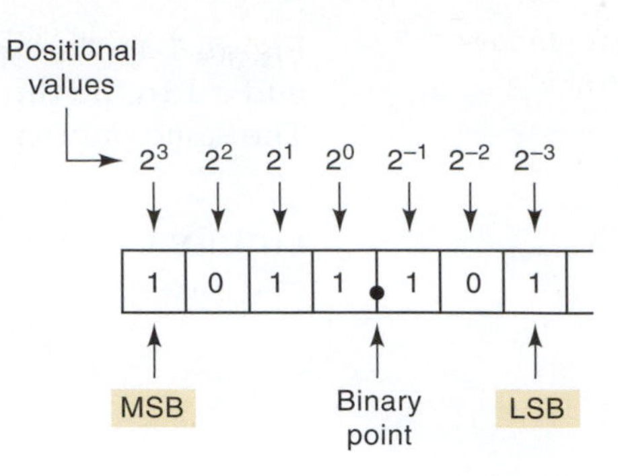

FIGURE 1-5 Binary position values as powers of 2.

binary point (counterpart of the decimal point) are positive powers of 2, and places to the right are negative powers of 2. The number 1011.101 is shown represented in the figure. To find its equivalent in the decimal system, we simply take the sum of the products of each digit value (0 or 1) and its positional value:

$$\begin{aligned} 1011.101_2 &= (1 \times 2^3) + (0 \times 2^2) + (1 \times 2^1) + (1 \times 2^0) \\ &\quad + (1 \times 2^{-1}) + (0 \times 2^{-2}) + (1 \times 2^{-3}) \\ &= 8 + 0 + 2 + 1 + 0.5 + 0 + 0.125 \\ &= 11.625_{10} \end{aligned}$$

Notice in the preceding operation that subscripts (2 and 10) were used to indicate the base in which the particular number is expressed. This convention is used to avoid confusion whenever more than one number system is being employed.

In the binary system, the term *binary digit* is often abbreviated to the term **bit**, which we will use from now on. Thus, in the number expressed in Figure 1-5 there are four bits to the left of the binary point, representing the integer part of the number, and three bits to the right of the binary point, representing the fractional part. The most significant bit (MSB) is the leftmost bit (largest weight). The least significant bit (LSB) is the rightmost bit (smallest weight). These are indicated in Figure 1-5. Here, the MSB has a weight of 2^3; the LSB has a weight of 2^{-3}.

Binary Counting

When we deal with binary numbers, we will usually be restricted to a specific number of bits. This restriction is based on the circuitry used to represent these binary numbers. Let's use four-bit binary numbers to illustrate the method for counting in binary.

The sequence (shown in Figure 1-6) begins with all bits at 0; this is called the *zero count*. For each successive count, the units (2^0) position *toggles*; that is, it changes from one binary value to the other. Each time the units bit changes from a 1 to a 0, the twos (2^1) position will toggle (change states). Each time the twos position changes from 1 to 0, the fours (2^2) position will toggle (change states). Likewise, each time the fours position goes from 1 to 0, the eights (2^3) position toggles. This same process would be continued for the higher-order bit positions if the binary number had more than four bits.

The binary counting sequence has an important characteristic, as shown in Figure 1-6. The units bit (LSB) changes either from 0 to 1 or 1 to 0 with *each* count. The second bit (twos position) stays at 0 for two counts, then at 1 for two counts, then at 0 for two counts, and so on. The third bit (fours position) stays at 0 for four counts, then at 1 for four counts, and so on. The fourth bit (eights position) stays at 0 for eight counts, then at 1 for eight counts. If we wanted to

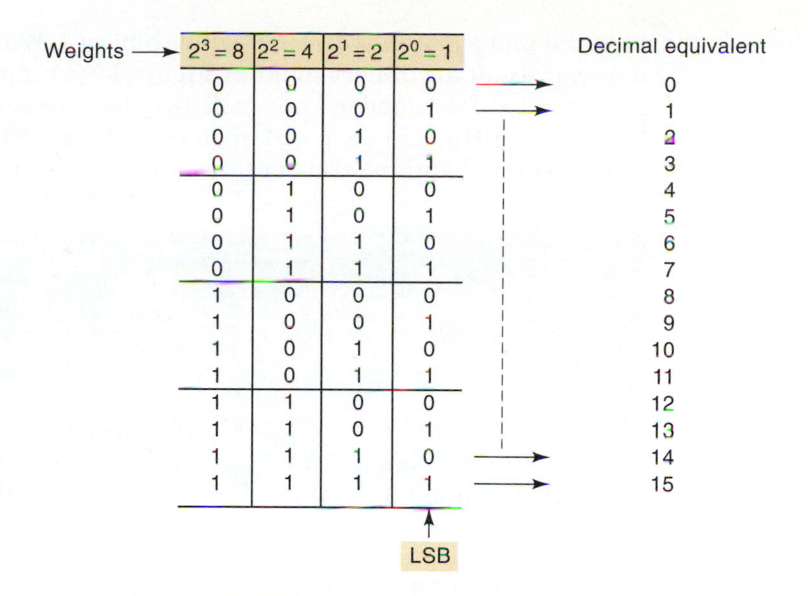

$2^3 = 8$	$2^2 = 4$	$2^1 = 2$	$2^0 = 1$	Decimal equivalent
0	0	0	0	0
0	0	0	1	1
0	0	1	0	2
0	0	1	1	3
0	1	0	0	4
0	1	0	1	5
0	1	1	0	6
0	1	1	1	7
1	0	0	0	8
1	0	0	1	9
1	0	1	0	10
1	0	1	1	11
1	1	0	0	12
1	1	0	1	13
1	1	1	0	14
1	1	1	1	15

FIGURE 1-6 Binary counting sequence.

count further, we would add more places, and this pattern would continue with 0s and 1s alternating in groups of 2^{N-1}. For example, using a fifth binary place, the fifth bit would alternate sixteen 0s, then sixteen 1s, and so on.

As we saw for the decimal system, it is also true for the binary system that by using N bits or places, we can go through 2^N counts. For example, with two bits we can go through $2^2 = 4$ counts (00_2 through 11_2); with four bits we can go through $2^4 = 16$ counts (0000_2 through 1111_2); and so on. The last count will always be all 1s and is equal to 2^N-1 in the decimal system. For example, using four bits, the last count is $1111_2 = 2^4-1 = 15_{10}$.

EXAMPLE 1-2

What is the largest number that can be represented using eight bits?

Solution

$2^N-1 = 2^8-1 = 255_{10} = 11111111_2$.

This has been a brief introduction of the binary number system and its relation to the decimal system. We will spend much more time on these two systems and several others in the next chapter.

REVIEW QUESTIONS

1. What is the decimal equivalent of 1101011_2?
2. What is the next binary number following 10111_2 in the counting sequence?
3. What is the largest decimal value that can be represented using 12 bits?

1-4 REPRESENTING BINARY QUANTITIES

In digital systems, the information being processed is usually present in binary form. Binary quantities can be represented by any device that has only two operating states or possible conditions. For example, a switch has only two states: open or closed. We can arbitrarily let an open switch represent binary 0 and a closed switch represent binary 1. With this assignment we can now represent any binary number. Figure 1-7(a) shows a binary code number for a garage door opener. The small switches are set to form the binary number 1000101010. The door will open only if a matching pattern of bits is set in the receiver and the transmitter.

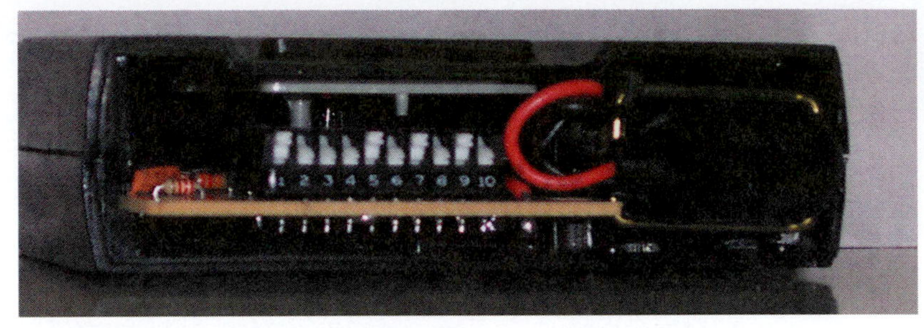

(a)

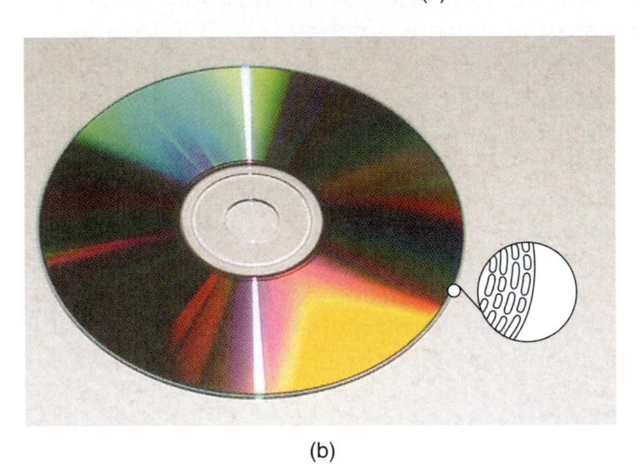

(b)

FIGURE 1-7 (a) Binary code settings for a garage door opener. (b) Digital audio on a CD.

Another example is shown in Figure 1-7(b), where binary numbers are stored on a CD. The inner surface (under a transparent plastic layer) is coated with a highly reflective aluminum layer. Holes are burned through this reflective coating to form "pits" that do not reflect light the same as the unburned areas. The areas where the pits are burned are considered "1" and the reflective areas are "0."

There are numerous other devices that have only two operating states or can be operated in two extreme conditions. Among these are: light bulb (bright or dark), diode (conducting or nonconducting), electromagnet (energized or deenergized), transistor (cut off or saturated), photocell (illuminated or dark), thermostat (open or closed), mechanical clutch (engaged or disengaged), and spot on a magnetic disk (magnetized or demagnetized).

In electronic digital systems, binary information is represented by voltages (or currents) that are present at the inputs and outputs of the various circuits. Typically, the binary 0 and 1 are represented by two nominal voltage levels. For example, zero volts (0 V) might represent binary 0, and +5 V might represent binary 1. In actuality, because of circuit variations, the 0 and 1 would be represented by voltage ranges. This is illustrated in Figure 1-8(a), where any voltage between 0 and 0.8 V represents a 0 and any voltage between 2 and 5 V represents a 1. All input and output signals will normally fall within one of these ranges, except during transitions from one level to another.

We can now see another significant difference between digital and analog systems. In digital systems, the exact value of a voltage *is not* important;

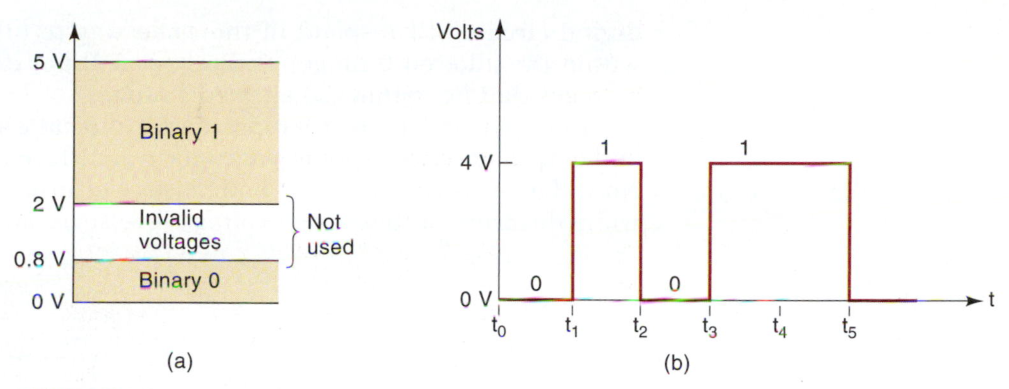

FIGURE 1-8 (a) Typical voltage assignments in digital system; (b) typical digital signal timing diagram.

for example, for the voltage assignments of Figure 1-8(a), a voltage of 3.6 V means the same as a voltage of 4.3 V. In analog systems, the exact value of a voltage *is* important. For instance, if the analog voltage is proportional to the temperature measured by a transducer, the 3.6 V would represent a different temperature than would 4.3 V. In other words, the voltage value carries significant information. This characteristic means that the design of accurate analog circuitry is generally more difficult than that of digital circuitry because of the way in which exact voltage values are affected by variations in component values, temperature, and noise (random voltage fluctuations).

Digital Signals and Timing Diagrams

Figure 1-8(b) shows a typical digital signal and how it varies over time. It is actually a graph of voltage versus time (*t*) and is called a **timing diagram**. The horizontal time scale is marked off at regular intervals beginning at t_0 and proceeding to t_1, t_2, and so on. For the example timing diagram shown here, the signal starts at 0 V (a binary 0) at time t_0 and remains there until time t_1. At t_1, the signal makes a rapid transition (jump) up to 4 V (a binary 1). At t_2, it jumps back down to 0 V. Similar transitions occur at t_3 and t_5. Note that the signal does not change at t_4 but stays at 4 V from t_3 to t_5.

The transitions on this timing diagram are drawn as vertical lines, and so they appear to be instantaneous, when in reality they are not. In many situations, however, the transition times are so short compared to the times between transitions that we can show them on the diagram as vertical lines. We will encounter situations later where it will be necessary to show the transitions more accurately on an expanded time scale.

Timing diagrams are used extensively to show how digital signals change with time, and especially to show the relationship between two or more digital signals in the same circuit or system. By displaying one or more digital signals on an *oscilloscope* or *logic analyzer,* we can compare the signals to their expected timing diagrams. This is a very important part of the testing and troubleshooting procedures used in digital systems.

1-5 DIGITAL CIRCUITS/LOGIC CIRCUITS

Digital circuits are designed to produce output voltages that fall within the prescribed 0 and 1 voltage ranges such as those defined in Figure 1-8. Likewise, digital circuits are designed to respond predictably to input voltages that are within the defined 0 and 1 ranges. What this means is that a digital circuit will respond in the same way to all input voltages that fall within the allowed 0 range; similarly, it will not distinguish between input voltages that lie within the allowed 1 range.

To illustrate, Figure 1-9 represents a typical digital circuit with input v_i and output v_o. The output is shown for two different input signal waveforms. Note that v_o is the same for both cases because the two input waveforms, while differing in their exact voltage levels, are at the same binary levels.

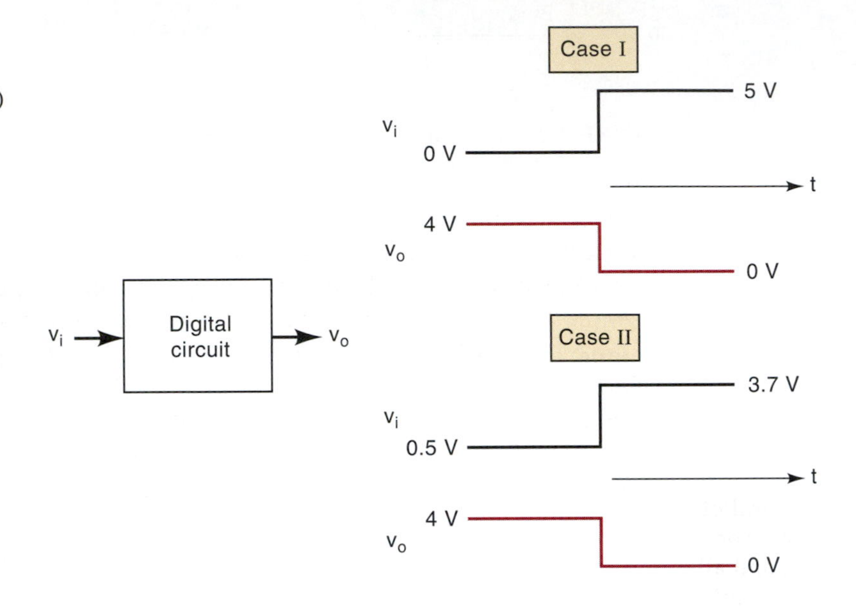

FIGURE 1-9 A digital circuit responds to an input's binary level (0 or 1) and not to its actual voltage.

Logic Circuits

The manner in which a digital circuit responds to an input is referred to as the circuit's *logic*. Each type of digital circuit obeys a certain set of logic rules. For this reason, digital circuits are also called **logic circuits**. We will use both terms interchangeably throughout the text. In Chapter 3, we will see more clearly what is meant by a circuit's "logic."

We will be studying all the types of logic circuits that are currently used in digital systems. Initially, our attention will be focused only on the logical operation that these circuits perform—that is, the relationship between the circuit inputs and outputs. We will defer any discussion of the internal circuit operation of these logic circuits until after we have developed an understanding of their logical operation.

Digital Integrated Circuits

Almost all of the digital circuits used in modern digital systems are integrated circuits (ICs). The wide variety of available logic ICs has made it possible to construct complex digital systems that are smaller and more reliable than their discrete-component counterparts.

Several integrated-circuit fabrication technologies are used to produce digital ICs, the most common being CMOS, TTL, NMOS, and ECL. Each differs in the type of circuitry used to provide the desired logic operation. For example, TTL (transistor-transistor logic) uses the bipolar transistor as its main circuit element, while CMOS (complementary metal-oxide-semiconductor) uses the enhancement-mode MOSFET as its principal circuit element. We will learn about the various IC technologies, their characteristics, and their relative advantages and disadvantages after we master the basic logic circuit types.

REVIEW QUESTIONS

1. *True or false:* The exact value of an input voltage is critical for a digital circuit.
2. Can a digital circuit produce the same output voltage for different input voltage values?
3. A digital circuit is also referred to as a ________ circuit.
4. A graph that shows how one or more digital signals change with time is called a ________.

1-6 PARALLEL AND SERIAL TRANSMISSION

One of the most common operations that occur in any digital system is the transmission of information from one place to another. The information can be transmitted over a distance as small as a fraction of an inch on the same circuit board, or over a distance of many miles when an operator at a computer terminal is communicating with a computer in another city. The information that is transmitted is in binary form and is generally represented as voltages at the outputs of a sending circuit that are connected to the inputs of a receiving circuit. Figure 1-10 illustrates the two basic methods for digital information transmission: **parallel** and **serial**.

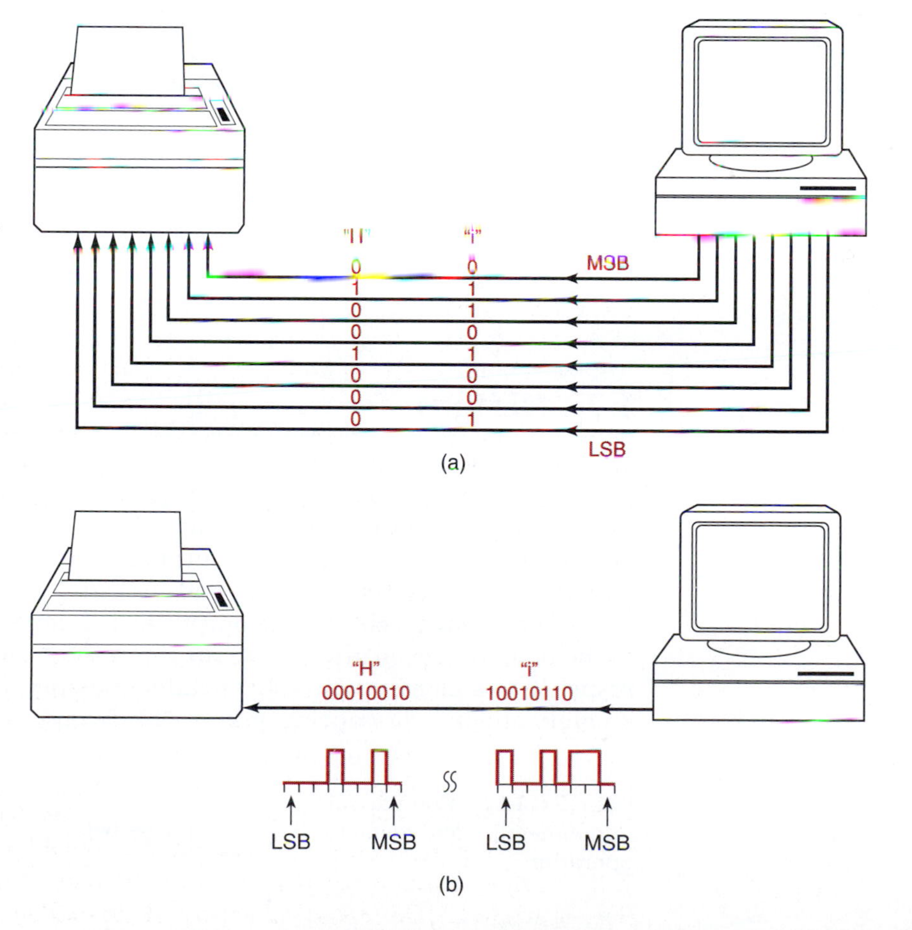

FIGURE 1-10 (a) Parallel transmission uses one connecting line per bit, and all bits are transmitted simultaneously; (b) serial transmission uses only one signal line, and the individual bits are transmitted serially (one at a time).

Figure 1-10(a) demonstrates parallel transmission of data from a computer to a printer using the parallel printer port (LPT1) of the computer. In this scenario, assume we are trying to print the word "Hi" on the printer. The binary code for "H" is 01001000 and the binary code for "i" is 01101001. Each character (the "H" and the "i") are made up of eight bits. Using parallel transmission, all eight bits are sent simultaneously over eight wires. The "H" is sent first, followed by the "i."

Figure 1-10(b) demonstrates serial transmission such as is employed when using a serial COM port on your computer to send data to a modem, or when using a USB (Universal Serial Bus) port to send data to a printer. Although the details of the data format and speed of transmission are quite different between a COM port and a USB port, the actual data are sent in the same way: one bit at a time over a single wire. The bits are shown in the diagram as though they were actually moving down the wire in the order shown. The least significant bit of "H" is sent first and the most significant bit of "i" is sent last. Of course, in reality, only one bit can be on the wire at any point in time and time is usually drawn on a graph starting at the left and advancing to the right. This produces a graph of logic bits versus time of the serial transmission called a timing diagram. Notice that in this presentation, the least significant bit is shown on the left because it was sent first.

The principal trade-off between parallel and serial representations is one of speed versus circuit simplicity. The transmission of binary data from one part of a digital system to another can be done more quickly using parallel representation because all the bits are transmitted simultaneously, while serial representation transmits one bit at a time. On the other hand, parallel requires more signal lines connected between the sender and the receiver of the binary data than does serial. In other words, parallel is faster, and serial requires fewer signal lines. This comparison between parallel and serial methods for representing binary information will be encountered many times in discussions throughout the text.

REVIEW QUESTION

1. Describe the relative advantages of parallel and serial transmission of binary data.

1-7 MEMORY

When an input signal is applied to most devices or circuits, the output somehow changes in response to the input, and when the input signal is removed, the output returns to its original state. These circuits do not exhibit the property of *memory* because their outputs revert back to normal. In digital circuitry certain types of devices and circuits do have memory. When an input is applied to such a circuit, the output will change its state, but it will remain in the new state even after the input is removed. This property of retaining its response to a momentary input is called **memory**. Figure 1-11 illustrates nonmemory and memory operations.

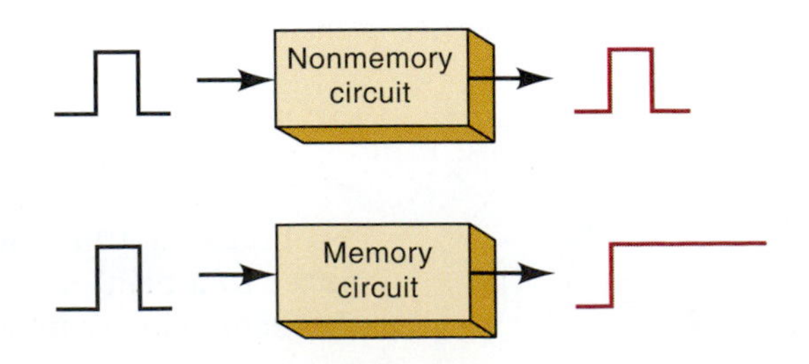

FIGURE 1-11 Comparison of nonmemory and memory operation.

Memory devices and circuits play an important role in digital systems because they provide a means for storing binary numbers either temporarily or permanently, with the ability to change the stored information at any time. As we shall see, the various memory elements include magnetic and optical types and those that utilize electronic latching circuits (called *latches* and *flip-flops*).

1-8 DIGITAL COMPUTERS

Digital techniques have found their way into innumerable areas of technology, but the area of automatic **digital computers** is by far the most notable and most extensive. Although digital computers affect some part of all of our lives, it is doubtful that many of us know exactly what a computer does. In simplest terms, *a computer is a system of hardware that performs arithmetic operations, manipulates data (usually in binary form), and makes decisions.*

For the most part, human beings can do whatever computers can do, but computers can do it with much greater speed and accuracy, in spite of the fact that computers perform all their calculations and operations one step at a time. For example, a human being can take a list of 10 numbers and find their sum all in one operation by listing the numbers one over the other and adding them column by column. A computer, on the other hand, can add numbers only two at a time, so that adding this same list of numbers will take nine actual addition steps. Of course, the fact that the computer requires only a few nanoseconds per step makes up for this apparent inefficiency.

A computer is faster and more accurate than people are, but unlike most of us, it must be given a complete set of instructions that tell it *exactly* what to do at each step of its operation. This set of instructions, called a **program**, is prepared by one or more persons for each job the computer is to do. Programs are placed in the computer's memory unit in binary-coded form, with each instruction having a unique code. The computer takes these instruction codes from memory *one at a time* and performs the operation called for by the code.

Major Parts of a Computer

There are several types of computer systems, but each can be broken down into the same functional units. Each unit performs specific functions, and all units function together to carry out the instructions given in the program. Figure 1-12 shows the five major functional parts of a digital computer and

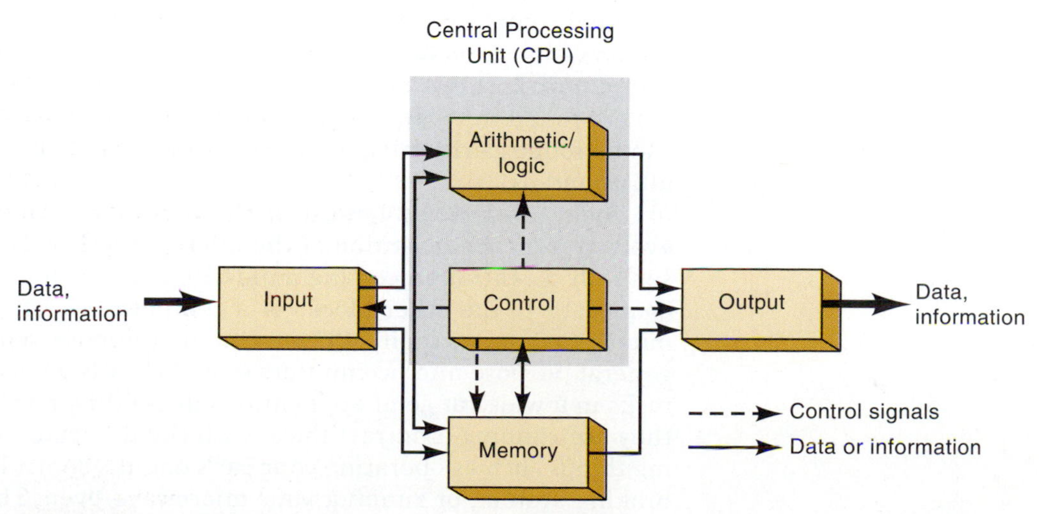

FIGURE 1-12 Functional diagram of a digital computer.

their interaction. The solid lines with arrows represent the flow of data and information. The dashed lines with arrows represent the flow of timing and control signals.

The major functions of each unit are:

1. **Input unit**. Through this unit, a complete set of instructions and data is fed into the computer system and into the memory unit, to be stored until needed. The information typically enters the input unit from a keyboard or a disk.
2. **Memory unit**. The memory stores the instructions and data received from the input unit. It stores the results of arithmetic operations received from the arithmetic unit. It also supplies information to the output unit.
3. **Control unit**. This unit takes instructions from the memory unit one at a time and interprets them. It then sends appropriate signals to all the other units to cause the specific instruction to be executed.
4. **Arithmetic/logic unit**. All arithmetic calculations and logical decisions are performed in this unit, which can then send results to the memory unit to be stored.
5. **Output unit**. This unit takes data from the memory unit and prints out, displays, or otherwise presents the information to the operator (or process, in the case of a process control computer).

Central Processing Unit (CPU)

As the diagram in Figure 1-12 shows, the control and arithmetic/logic units are often considered as one unit, called the **central processing unit (CPU)**. The CPU contains all of the circuitry for fetching and interpreting instructions and for controlling and performing the various operations called for by the instructions.

TYPES OF COMPUTERS All computers are made up of the basic units described above, but they can differ as to physical size, operating speed, memory capacity, and computational power, as well as other characteristics. Computer systems are configured in many and various ways today, with many common characteristics and distinguishing differences. Large computer systems that are permanently installed in multiple cabinets are used by corporations and universities for information technology support. Desktop personal computers are used in our homes and offices to run useful application programs that enhance our lives and provide communication with other computers. Portable computers are found in PDAs and specialized computers are found in video game systems. The most prevalent form of computers can be found performing dedicated routine tasks in appliances and systems all around us.

Today, all but the largest of these systems utilize technology that has evolved from the invention of the **microprocessor**. The microprocessor is essentially a central processing unit (CPU) in an integrated circuit that can be connected to the other blocks of a computer system. Computers that use a microprocessor as their CPU are usually referred to as **microcomputers**. The general-purpose microcomputers (e.g., PCs, PDAs, etc.) perform a variety of tasks in a wide range of applications depending on the software (programs) they are running. Contrast these with the dedicated computers that are doing things such as operating your car's engine, controlling your car's antilock braking system, or running your microwave oven. These computers cannot be programmed by the user, but simply perform their intended control task: they are referred to as **microcontrollers**. Since these microcontrollers are an integral part of a bigger system and serve a dedicated purpose, they also are called *embedded controllers*. Microcontrollers generally have all the elements of a complete computer (CPU, memory, and input/output ports), all contained on a single integrated circuit. You can find them embedded in your kitchen appliances, entertainment equipment, photocopiers, automatic teller machines, automated manufacturing equipment, medical instrumentation, and much, much more.

So you see, even people who don't own a PC or use one at work or school are using microcomputers every day because so many modern consumer electronic devices, appliances, office equipment, and much more are built around embedded microcontrollers. If you work, play, or go to school in this digital age, there's no escaping it: you'll use a microcomputer somewhere.

REVIEW QUESTIONS

1. Explain how a digital circuit that has memory differs from one that does not.
2. Name the five major functional units of a computer.
3. Which two units make up the CPU?
4. An IC chip that contains a CPU is called a _____.

SUMMARY

1. The two basic ways of representing the numerical value of physical quantities are analog (continuous) and digital (discrete).
2. Most quantities in the real world are analog, but digital techniques are generally superior to analog techniques, and most of the predicted advances will be in the digital realm.
3. The binary number system (0 and 1) is the basic system used in digital technology.
4. Digital or logic circuits operate on voltages that fall in prescribed ranges that represent either a binary 0 or a binary 1.
5. The two basic ways to transfer digital information are parallel—all bits simultaneously—and serial—one bit at a time.
6. The main parts of all computers are the input, control, memory, arithmetic/logic, and output units.
7. The combination of the arithmetic/logic unit and the control unit makes up the CPU (central processing unit).
8. A microcomputer usually has a CPU that is on a single chip called a *microprocessor*.
9. A microcontroller is a microcomputer especially designed for dedicated (not general-purpose) control applications.

IMPORTANT TERMS*

analog representation
digital representation
digital system
analog system
analog-to-digital converter (ADC)
digital-to-analog converter (DAC)
decimal system

*These terms can be found in **boldface** type in the chapter and are defined in the Glossary at the end of the book. This applies to all chapters.

binary system
bit
timing diagram
digital circuits/logic circuits
parallel transmission
serial transmission
memory
digital computer
program
input unit
memory unit
control unit
arithmetic/logic unit
output unit
central processing unit (CPU)
microprocessor
microcomputer
microcontroller

PROBLEMS

SECTION 1-2

1-1.* Which of the following are analog quantities, and which are digital?
  (a) Number of atoms in a sample of material
  (b) Altitude of an aircraft
  (c) Pressure in a bicycle tire
  (d) Current through a speaker
  (e) Timer setting on a microwave oven

1-2. Which of the following are analog quantities, and which are digital?
  (a) Width of a piece of lumber
  (b) The amount of time before the oven buzzer goes off
  (c) The time of day displayed on a quartz watch
  (d) Altitude above sea level measured on a staircase
  (e) Altitude above sea level measured on a ramp

SECTION 1-3

1-3.* Convert the following binary numbers to their equivalent decimal values.
  (a) 11001_2
  (b) 1001.1001_2
  (c) 10011011001.10110_2

1-4. Convert the following binary numbers to decimal.
  (a) 10011_2
  (b) 1100.0101
  (c) 10011100100.10010

1-5.* Using three bits, show the binary counting sequence from 000 to 111.

1-6. Using six bits, show the binary counting sequence from 000000 to 111111.

1-7.* What is the maximum number that we can count up to using 10 bits?

1-8. What is the maximum number that we can count up to using 14 bits?

1-9.* How many bits are needed to count up to a maximum of 511?

1-10. How many bits are needed to count up to a maximum of 63?

SECTION 1-4

1-11.* Draw the timing diagram for a digital signal that continuously alternates between 0.2 V (binary 0) for 2 ms and 4.4 V (binary 1) for 4 ms.

---

*Answers to problems marked with an asterisk can be found in the back of the text.

1-12. Draw the timing diagram for a signal that alternates between 0.3 V (binary 0) for 5 ms and 3.9 V (binary 1) for 2 ms.

SECTION 1-6

1-13.* Suppose that the decimal integer values from 0 to 15 are to be transmitted in binary.

(a) How many lines will be needed if parallel representation is used?

(b) How many will be needed if serial representation is used?

SECTIONS 1-7 AND 1-8

1-14. How is a microprocessor different from a microcomputer?

1-15. How is a microcontroller different from a microcomputer?

ANSWERS TO SECTION REVIEW QUESTIONS

SECTION 1-1

1. Analog quantities can take on *any* value over a continuous range; digital quantities can take on only *discrete* values.

SECTION 1-2

1. Easier to design; easier to store information; greater accuracy and precision; programmability; less affected by noise; higher degree of integration
2. Real-world physical quantities are analog. Digital processing takes time.

SECTION 1-3

1. 107_{10} 2. 11000_2 3. 4095_{10}

SECTION 1-5

1. False 2. Yes, provided that the two input voltages are within the same logic level range 3. Logic 4. Timing diagram

SECTION 1-6

1. Parallel is faster; serial requires only one signal line.

SECTION 1-8

1. One that has memory will have its output changed and *remain* changed in response to a momentary change in the input signal. 2. Input, output, memory, arithmetic/logic, control 3. Control and arithmetic/logic 4. Microprocessor

CHAPTER 2

NUMBER SYSTEMS AND CODES

OUTLINE

OBJECTIVES

Upon completion of this chapter, you will be able to:

- Convert a number from one number system (decimal, binary, hexadecimal) to its equivalent in one of the other number systems.
- Cite the advantages of the hexadecimal number system.
- Count in hexadecimal.
- Represent decimal numbers using the BCD code; cite the pros and cons of using BCD.
- Understand the difference between BCD and straight binary.
- Understand the purpose of alphanumeric codes such as the ASCII code.
- Explain the parity method for error detection.
- Determine the parity bit to be attached to a digital data string.

INTRODUCTION

The binary number system is the most important one in digital systems, but several others are also important. The decimal system is important because it is universally used to represent quantities outside a digital system. This means that there will be situations where decimal values must be converted to binary values before they are entered into the digital system. For example, when you punch a decimal number into your hand calculator (or computer), the circuitry inside the machine converts the decimal number to a binary value.

Likewise, there will be situations where the binary values at the outputs of a digital system must be converted to decimal values for presentation to the outside world. For example, your calculator (or computer) uses binary numbers to calculate answers to a problem and then converts the answers to decimal digits before displaying them.

As you will see, it is not easy to simply look at a large binary number and convert it to its equivalent decimal value. It is very tedious to enter a long sequence of 1s and 0s on a keypad, or to write large binary numbers on a piece of paper. It is especially difficult to try to convey a binary quantity while speaking to someone. The hexadecimal (base-16) number system has become a very standard way of communicating numeric values in digital systems. The great advantage is that hexadecimal numbers can be converted easily to and from binary.

Other methods of representing decimal quantities with binary-encoded digits have been devised that are not truly number systems but offer the ease of conversion between the binary code and the decimal number system. This is referred to as binary-coded decimal. Quantities and patterns of bits might be represented by any of these methods in any given system and throughout the written material that supports the system, so it is very important that you are able to interpret values in any system and convert between any of these numeric representations. Other codes that use 1s and 0s to represent things such as alphanumeric characters will be covered because they are so common in digital systems.

2-1 BINARY-TO-DECIMAL CONVERSIONS

As explained in Chapter 1, the binary number system is a positional system where each binary digit (bit) carries a certain weight based on its position relative to the LSB. Any binary number can be converted to its decimal equivalent simply by summing together the weights of the various positions in the binary number that contain a 1. To illustrate, let's change 11011_2 to its decimal equivalent.

$$\begin{aligned} &1 \quad 1 \quad 0 \quad 1 \quad 1_2 \\ &2^4 + 2^3 + 0 + 2^1 + 2^0 = 16 + 8 + 2 + 1 \\ &\qquad\qquad\qquad\qquad = 27_{10} \end{aligned}$$

Let's try another example with a greater number of bits:

$$\begin{aligned} &1 \quad 0 \quad 1 \quad 1 \quad 0 \quad 1 \quad 0 \quad 1_2 = \\ &2^7 + 0 + 2^5 + 2^4 + 0 + 2^2 + 0 + 2^0 = 181_{10} \end{aligned}$$

Note that the procedure is to find the weights (i.e., powers of 2) for each bit position that contains a 1, and then to add them up. Also note that the MSB has a weight of 2^7 even though it is the eighth bit; this is because the LSB is the first bit and has a weight of 2^0.

REVIEW QUESTIONS

1. Convert 100011011011_2 to its decimal equivalent.
2. What is the weight of the MSB of a 16-bit number?

2-2 DECIMAL-TO-BINARY CONVERSIONS

There are two ways to convert a decimal *whole* number to its equivalent binary-system representation. The first method is the reverse of the process described in Section 2-1. The decimal number is simply expressed as a sum of powers of 2, and then 1s and 0s are written in the appropriate bit positions. To illustrate:

$$\begin{aligned} 45_{10} = 32 + 8 + 4 + 1 &= 2^5 + 0 + 2^3 + 2^2 + 0 + 2^0 \\ &= 1 \quad 0 \quad 1 \quad 1 \quad 0 \quad 1_2 \end{aligned}$$

Note that a 0 is placed in the 2^1 and 2^4 positions, since all positions must be accounted for. Another example is the following:

$$\begin{aligned} 76_{10} = 64 + 8 + 4 &= 2^6 + 0 + 0 + 2^3 + 2^2 + 0 + 0 \\ &= 1 \quad 0 \quad 0 \quad 1 \quad 1 \quad 0 \quad 0_2 \end{aligned}$$

Repeated Division

Another method for converting decimal integers uses repeated division by 2. The conversion, illustrated below for 25_{10}, requires repeatedly dividing the decimal number by 2 and writing down the remainder after each division until a quotient of 0 is obtained. Note that the binary result is obtained by writing the first remainder as the LSB and the last remainder as the MSB. This process, diagrammed in the flowchart of Figure 2-1, can also be used to convert from decimal to any other number system, as we shall see.

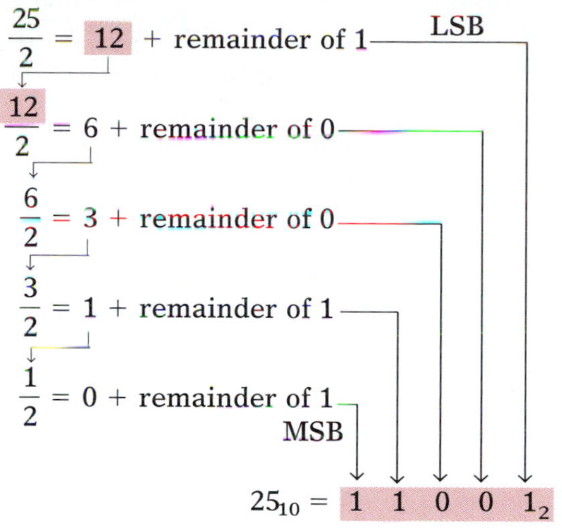

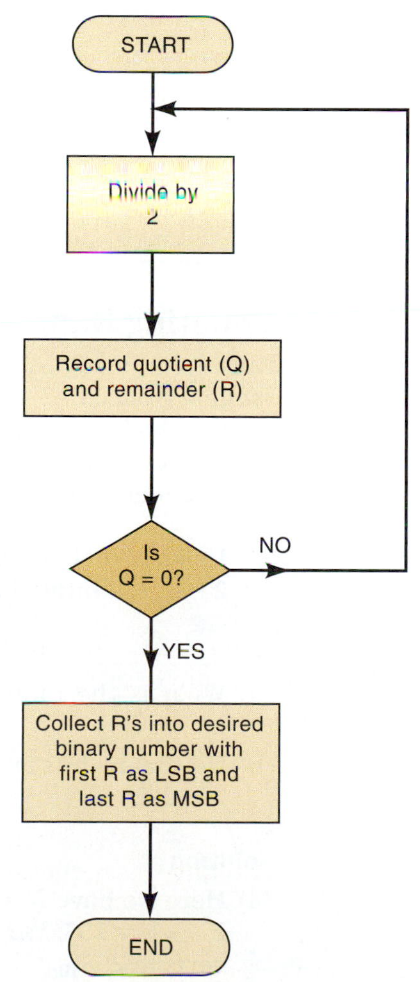

FIGURE 2-1 Flowchart for repeated-division method of decimal-to-binary conversion of integers. The same process can be used to convert a decimal integer to any other number system.

CALCULATOR HINT:

If you use a calculator to perform the divisions by 2, you can tell whether the remainder is 0 or 1 by whether or not the result has a fractional part. For instance, 25/2 would produce 12.5. Since there is a fractional part (the .5), the remainder is a 1. If there were no fractional part, such as 12/2 = 6, then the remainder would be 0. The following example illustrates this.

EXAMPLE 2-1

Convert 37_{10} to binary. Try to do it on your own before you look at the solution.

Solution

$$\frac{37}{2} = 18.5 \longrightarrow \text{remainder of 1 (LSB)}$$

$$\frac{18}{2} = 9.0 \longrightarrow 0$$

$$\frac{9}{2} = 4.5 \longrightarrow 1$$

$$\frac{4}{2} = 2.0 \longrightarrow 0$$

$$\frac{2}{2} = 1.0 \longrightarrow 0$$

$$\frac{1}{2} = 0.5 \longrightarrow 1 \text{ (MSB)}$$

Thus, $37^{10} = \mathbf{100101_2}$.

Counting Range

Recall that using N bits, we can count through 2^N different decimal numbers ranging from 0 to $2^N - 1$. For example, for $N = 4$, we can count from 0000_2 to 1111_2, which is 0_{10} to 15_{10}, for a total of 16 different numbers. Here, the largest decimal value is $2^4 - 1 = 15$, and there are 2^4 different numbers.

In general, then, we can state:

Using N bits, we can represent decimal numbers ranging from 0 to $2^N - 1$, a total of 2^N different numbers.

EXAMPLE 2-2

(a) What is the total range of decimal values that can be represented in eight bits?

(b) How many bits are needed to represent decimal values ranging from 0 to 12,500?

Solution

(a) Here we have $N = 8$. Thus, we can represent decimal numbers from 0 to $2^8 - 1 = \mathbf{255}$. We can verify this by checking to see that 11111111_2 converts to 255_{10}.

(b) With 13 bits, we can count from decimal 0 to $2^{13} - 1 = 8191$. With 14 bits, we can count from 0 to $2^{14} - 1 = 16{,}383$. Clearly, 13 bits aren't enough, but 14 bits will get us up beyond 12,500. Thus, the required number of bits is **14**.

REVIEW QUESTIONS

1. Convert 83_{10} to binary using both methods.
2. Convert 729_{10} to binary using both methods. Check your answer by converting back to decimal.
3. How many bits are required to count up to decimal 1 million?

2-3 HEXADECIMAL NUMBER SYSTEM

The **hexadecimal number system** uses base 16. Thus, it has 16 possible digit symbols. It uses the digits 0 through 9 plus the letters A, B, C, D, E, and F as the 16 digit symbols. The digit positions are weighted as powers of 16 as shown below, rather than as powers of 10 as in the decimal system.

16^4	16^3	16^2	16^1	16^0	16^{-1}	16^{-2}	16^{-3}	16^{-4}

Hexadecimal point

Table 2-1 shows the relationships among hexadecimal, decimal, and binary. Note that each hexadecimal digit represents a group of four binary digits. It is important to remember that hex (abbreviation for "hexadecimal") digits A through F are equivalent to the decimal values 10 through 15.

TABLE 2-1

Hexadecimal	Decimal	Binary
0	0	0000
1	1	0001
2	2	0010
3	3	0011
4	4	0100
5	5	0101
6	6	0110
7	7	0111
8	8	1000
9	9	1001
A	10	1010
B	11	1011
C	12	1100
D	13	1101
E	14	1110
F	15	1111

Hex-to-Decimal Conversion

A hex number can be converted to its decimal equivalent by using the fact that each hex digit position has a weight that is a power of 16. The LSD has a weight of $16^0 = 1$; the next higher digit position has a weight of $16^1 = 16$; the next has a weight of $16^2 = 256$; and so on. The conversion process is demonstrated in the examples below.

CALCULATOR HINT:

You can use the y^x calculator function to evaluate the powers of 16.

$$
\begin{aligned}
356_{16} &= 3 \times 16^2 + 5 \times 16^1 + 6 \times 16^0 \\
&= 768 + 80 + 6 \\
&= 854_{10}
\end{aligned}
$$

$$
\begin{aligned}
2AF_{16} &= 2 \times 16^2 + 10 \times 16^1 + 15 \times 16^0 \\
&= 512 + 160 + 15 \\
&= 687_{10}
\end{aligned}
$$

Note that in the second example, the value 10 was substituted for A and the value 15 for F in the conversion to decimal.

For practice, verify that $1BC2_{16}$ is equal to 7106_{10}.

Decimal-to-Hex Conversion

Recall that we did decimal-to-binary conversion using repeated division by 2. Likewise, decimal-to-hex conversion can be done using repeated division by 16 (Figure 2-1). The following example contains two illustrations of this conversion.

EXAMPLE 2-3

(a) Convert 423_{10} to hex.

Solution

$$\frac{423}{16} = 26 + \text{remainder of } 7$$

$$\frac{26}{16} = 1 + \text{remainder of } 10$$

$$\frac{1}{16} = 0 + \text{remainder of } 1$$

$$423_{10} = 1A7_{16}$$

(b) Convert 214_{10} to hex.

Solution

$$\frac{214}{16} = 13 + \text{remainder of } 6$$

$$\frac{13}{16} = 0 + \text{remainder of } 13$$

$$214_{10} = D6_{16}$$

Again note that the remainders of the division processes form the digits of the hex number. Also note that any remainders that are greater than 9 are represented by the letters A through F.

CALCULATOR HINT:

If a calculator is used to perform the divisions in the conversion process, the results will include a decimal fraction instead of a remainder. The remainder can be obtained by multiplying the fraction by 16. To illustrate, in Example 2-3(b), the calculator would have produced

$$\frac{214}{16} = 13.375$$

The remainder becomes $(0.375) \times 16 = 6$.

Hex-to-Binary Conversion

The hexadecimal number system is used primarily as a "shorthand" method for representing binary numbers. It is a relatively simple matter to convert a hex number to binary. *Each* hex digit is converted to its four-bit binary equivalent (Table 2-1). This is illustrated below for $9F2_{16}$.

$$\begin{aligned} 9F2_{16} &= \quad 9 \qquad\quad F \qquad\quad 2 \\ &\qquad\ \downarrow \qquad\quad \downarrow \qquad\quad \downarrow \\ &= 1\ 0\ 0\ 1 \quad 1\ 1\ 1\ 1 \quad 0\ 0\ 1\ 0 \\ &= 100111110010_2 \end{aligned}$$

For practice, verify that $BA6_{16} = 101110100110_2$.

Binary-to-Hex Conversion

Conversion from binary to hex is just the reverse of the process above. The binary number is grouped into groups of *four* bits, and each group is converted to its equivalent hex digit. Zeros (shown shaded) are added, as needed, to complete a four-bit group.

$$1\,1\,1\,0\,1\,0\,0\,1\,1\,0_2 = \underbrace{0\,0\,1\,1}_{3}\ \underbrace{1\,0\,1\,0}_{A}\ \underbrace{0\,1\,1\,0}_{6} = 3A6_{16}$$

To perform these conversions between hex and binary, it is necessary to know the four-bit binary numbers (0000 through 1111) and their equivalent hex digits. Once these are mastered, the conversions can be performed quickly without the need for any calculations. This is why hex is so useful in representing large binary numbers.

For practice, verify that $101011111_2 = 15F_{16}$.

Counting in Hexadecimal

When counting in hex, each digit position can be incremented (increased by 1) from 0 to F. Once a digit position reaches the value F, it is reset to 0, and the next digit position is incremented. This is illustrated in the following hex counting sequences:

(a) 38, 39, 3A, 3B, 3C, 3D, 3E, 3F, 40, 41, 42

(b) 6F8, 6F9, 6FA, 6FB, 6FC, 6FD, 6FE, 6FF, 700

Note that when there is a 9 in a digit position, it becomes an A when it is incremented.

With N hex digit positions, we can count from decimal 0 to $16^N - 1$, for a total of 16^N different values. For example, with three hex digits, we can count from 000_{16} to FFF_{16}, which is 0_{10} to 4095_{10}, for a total of $4096 = 16^3$ different values.

Usefulness of Hex

Hex is often used in a digital system as sort of a "shorthand" way to represent strings of bits. In computer work, strings as long as 64 bits are not uncommon. These binary strings do not always represent a numerical value, but—as you will find out—can be some type of code that conveys nonnumerical information. When dealing with a large number of bits, it is more convenient and less error-prone to write the binary numbers in hex and, as we have seen, it is relatively easy to convert back and forth between binary and hex. To illustrate the advantage of hex representation of a binary string, suppose you had in front of you a printout of the contents of 50 memory locations, each of which was a 16-bit number, and you were checking it against a list. Would you rather check 50 numbers like this one: 0110111001100111, or 50 numbers like this one: 6E67? And which one would you be more apt to read incorrectly? It is important to keep in mind, though, that digital circuits all work in binary. Hex is simply used as a convenience for the humans involved. You should memorize the 4-bit binary pattern for each hexadecimal digit. Only then will you realize the usefulness of this tool in digital systems.

EXAMPLE 2-4

Convert decimal 378 to a 16-bit binary number by first converting to hexadecimal.

Solution

$$\frac{378}{16} = 23 + \text{remainder of } 10_{10} = A_{16}$$

$$\frac{23}{16} = 1 + \text{remainder of } 7$$

$$\frac{1}{16} = 0 + \text{remainder of } 1$$

Thus, $378_{10} = 17A_{16}$. This hex value can be converted easily to binary 000101111010. Finally, we can express 378_{10} as a 16-bit number by adding four leading 0s:

$$378_{10} = 0000\ \ 0001\ \ 0111\ \ 1010_2$$

EXAMPLE 2-5

Convert $B2F_{16}$ to decimal.

Solution

$$\begin{aligned} B2F_{16} &= B \times 16^2 + 2 \times 16^1 + F \times 16^0 \\ &= 11 \times 256 + 2 \times 16 + 15 \\ &= 2863_{10} \end{aligned}$$

Summary of Conversions

Right now, your head is probably spinning as you try to keep straight all of these different conversions from one number system to another. You probably realize that many of these conversions can be done *automatically* on your calculator just by pressing a key, but it is important for you to master these conversions so that you understand the process. Besides, what happens if your calculator battery dies at a crucial time and you have no handy replacement? The following summary should help you, but nothing beats practice, practice, practice!

1. When converting from binary [or hex] to decimal, use the method of taking the weighted sum of each digit position.
2. When converting from decimal to binary [or hex], use the method of repeatedly dividing by 2 [or 16] and collecting remainders (Figure 2-1).
3. When converting from binary to hex, group the bits in groups of four, and convert each group into the correct hex digit.
4. When converting from hex to binary, convert each digit into its four-bit equivalent.

REVIEW QUESTIONS

1. Convert $24CE_{16}$ to decimal.
2. Convert 3117_{10} to hex, then from hex to binary.
3. Convert 1001011110110101_2 to hex.
4. Write the next four numbers in this hex counting sequence: E9A, E9B, E9C, E9D, _____, _____, _____, _____.
5. Convert 3527 to $binary_{16}$.
6. What range of decimal values can be represented by a four-digit hex number?

2-4 BCD CODE

When numbers, letters, or words are represented by a special group of symbols, we say that they are being encoded, and the group of symbols is called a *code*. Probably one of the most familiar codes is the Morse code, where a series of dots and dashes represents letters of the alphabet.

We have seen that any decimal number can be represented by an equivalent binary number. The group of 0s and 1s in the binary number can be thought of as a code representing the decimal number. When a decimal number is represented by its equivalent binary number, we call it **straight binary coding**.

Digital systems all use some form of binary numbers for their internal operation, but the external world is decimal in nature. This means that conversions between the decimal and binary systems are being performed often. We have seen that the conversions between decimal and binary can become long and complicated for large numbers. For this reason, a means of encoding decimal numbers that combines some features of both the decimal and the binary systems is used in certain situations.

Binary-Coded-Decimal Code

If *each* digit of a decimal number is represented by its binary equivalent, the result is a code called **binary-coded-decimal** (hereafter abbreviated BCD). Since a decimal digit can be as large as 9, four bits are required to code each digit (the binary code for 9 is 1001).

To illustrate the BCD code, take a decimal number such as 874. Each *digit* is changed to its binary equivalent as follows:

8	7	4	(decimal)
↓	↓	↓	
1000	0111	0100	(BCD)

As another example, let us change 943 to its BCD-code representation:

9	4	3	(decimal)
↓	↓	↓	
1001	0100	0011	(BCD)

Once again, each decimal digit is changed to its straight binary equivalent. Note that four bits are *always* used for each digit.

The BCD code, then, represents each digit of the decimal number by a four-bit binary number. Clearly only the four-bit binary numbers from 0000 through 1001 are used. The BCD code does not use the numbers 1010, 1011, 1100, 1101, 1110, and 1111. In other words, only 10 of the 16 possible four-bit binary code groups are used. If any of the "forbidden" four-bit numbers ever occurs in a machine using the BCD code, it is usually an indication that an error has occurred.

EXAMPLE 2-6

Convert 0110100000111001 (BCD) to its decimal equivalent.

Solution

Divide the BCD number into four-bit groups and convert each to decimal.

0110	1000	0011	1001
6	8	3	9

EXAMPLE 2-7

Convert the BCD number 011111000001 to its decimal equivalent.

Solution

0111	1100	0001
7	↓	1

The forbidden code group indicates an error in the BCD number

Comparison of BCD and Binary

It is important to realize that BCD is not another number system like binary, decimal, and hexadecimal. In fact, it is the decimal system with each digit encoded in its binary equivalent. It is also important to understand that a BCD number is *not* the same as a straight binary number. A straight binary number takes the *complete* decimal number and represents it in binary; the BCD code converts *each* decimal *digit* to binary individually. To illustrate, take the number 137 and compare its straight binary and BCD codes:

$$137_{10} = 10001001_2 \quad \text{(binary)}$$
$$137_{10} = 0001\ 0011\ 0111 \quad \text{(BCD)}$$

The BCD code requires 12 bits, while the straight binary code requires only eight bits to represent 137. BCD requires more bits than straight binary to represent decimal numbers of more than one digit because BCD does not use all possible four-bit groups, as pointed out earlier, and is therefore somewhat inefficient.

The main advantage of the BCD code is the relative ease of converting to and from decimal. Only the four-bit code groups for the decimal digits 0 through 9 need to be remembered. This ease of conversion is especially important from a hardware standpoint because in a digital system, it is the logic circuits that perform the conversions to and from decimal.

REVIEW QUESTIONS

1. Represent the decimal value 178 by its straight binary equivalent. Then encode the same decimal number using BCD.
2. How many bits are required to represent an eight-digit decimal number in BCD?
3. What is an advantage of encoding a decimal number in BCD rather than in straight binary? What is a disadvantage?

2-5 THE GRAY CODE

Digital systems operate at very fast speeds and respond to changes that occur in the digital inputs. Just as in life, when multiple input conditions are changing at the same time, the situation can be misinterpreted and cause an erroneous reaction. When you look at the bits in a binary count sequence, it is clear that there are often several bits that must change states at the same time. For example, consider when the three-bit binary number for 3 changes to 4: all three bits must change state.

In order to reduce the likelihood of a digital circuit misinterpreting a changing input, the **Gray code** has been developed as a way to represent a sequence of numbers. The unique aspect of the Gray code is that only one bit ever changes between two successive numbers in the sequence. Table 2-2 shows the translation between three-bit binary and Gray code values. To convert binary to Gray, simply start on the most significant bit and use it as the Gray MSB as shown in Figure 2-2(a). Now compare the MSB binary with the next binary bit (B1). If they are the same, then G1 = 0. If they are different, then G1 = 1. G0 can be found by comparing B1 with B0.

TABLE 2-2 Three-bit binary and Gray code equivalents.

B_2	B_1	B_0	G_2	G_1	G_0
0	0	0	0	0	0
0	0	1	0	0	1
0	1	0	0	1	1
0	1	1	0	1	0
1	0	0	1	1	0
1	0	1	1	1	1
1	1	0	1	0	1
1	1	1	1	0	0

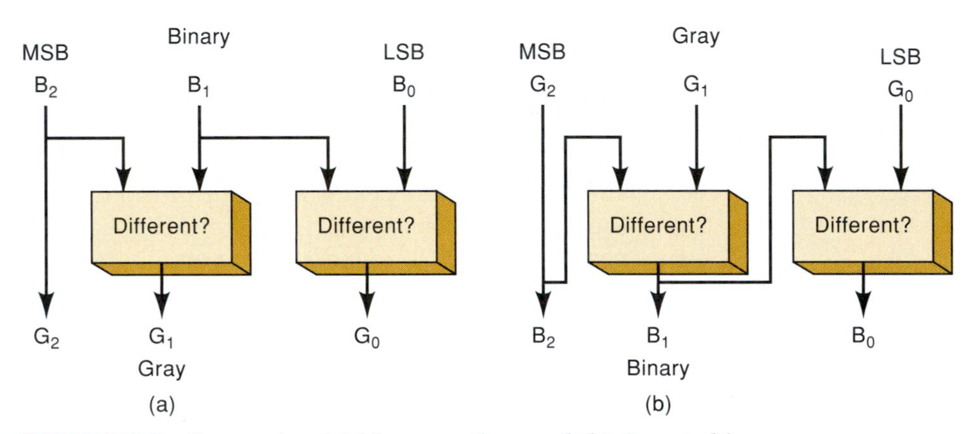

FIGURE 2-2 Converting (a) binary to Gray and (b) Gray to binary.

Conversion from Gray code back into binary is shown in Figure 2-2(b). Note that the MSB in Gray is always the same as the MSB in binary. The next binary bit is found by comparing the *binary* bit to the left with the *corresponding Gray code bit*. Similar bits produce a 0 and differing bits produce a 1. The most common application of the Gray code is in shaft position encoders as shown in Figure 2-3. These devices produce a binary value that represents the position of a rotating mechanical shaft. A practical shaft encoder would use many more bits than just three and divide the rotation into many more segments than eight, so that it could detect much smaller increments of rotation.

FIGURE 2-3 An eight-position, three-bit shaft encoder.

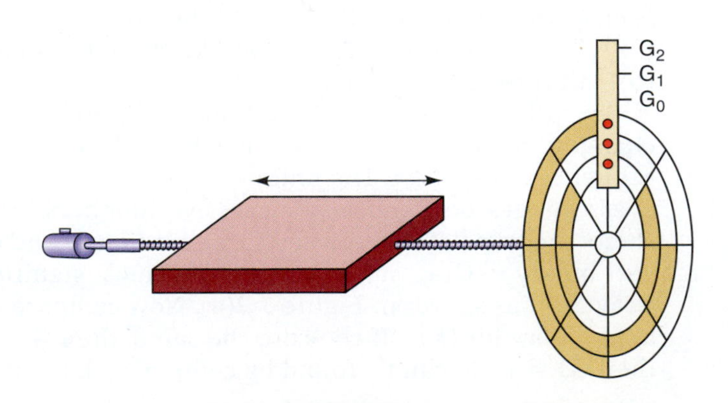

REVIEW QUESTIONS

1. Convert the number 0101 (binary) to its Gray code equivalent.
2. Convert 0101 (Gray code) to its binary number equivalent.

2-6 PUTTING IT ALL TOGETHER

Table 2-3 gives the representation of the decimal numbers 1 through 15 in the binary and hex number systems and also in the BCD and Gray codes. Examine it carefully and make sure you understand how it was obtained. Especially note how the BCD representation always uses four bits for each decimal digit.

TABLE 2-3

Decimal	Binary	Hexadecimal	BCD	GRAY
0	0	0	0000	0000
1	1	1	0001	0001
2	10	2	0010	0011
3	11	3	0011	0010
4	100	4	0100	0110
5	101	5	0101	0111
6	110	6	0110	0101
7	111	7	0111	0100
8	1000	8	1000	1100
9	1001	9	1001	1101
10	1010	A	0001 0000	1111
11	1011	B	0001 0001	1110
12	1100	C	0001 0010	1010
13	1101	D	0001 0011	1011
14	1110	E	0001 0100	1001
15	1111	F	0001 0101	1000

2-7 THE BYTE, NIBBLE, AND WORD

Bytes

Most microcomputers handle and store binary data and information in groups of eight bits, so a special name is given to a string of eight bits: it is called a **byte.** A byte always consists of eight bits, and it can represent any of numerous types of data or information. The following examples will illustrate.

EXAMPLE 2-8

How many bytes are in a 32-bit string (a string of 32 bits)?

Solution

32/8 = 4, so there are **four** bytes in a 32-bit string.

EXAMPLE 2-9

What is the largest decimal value that can be represented in binary using two bytes?

Solution

Two bytes is 16 bits, so the largest binary value will be equivalent to decimal $2^{16} - 1 = 65,535$.

EXAMPLE 2-10

How many bytes are needed to represent the decimal value 846,569 in BCD?

Solution

Each decimal digit converts to a four-bit BCD code. Thus, a six-digit decimal number requires 24 bits. These 24 bits are equal to **three** bytes. This is diagrammed below.

```
   8 4 6 5 6 9              (decimal)
1000 0100 0110 0101 0110 1001 (BCD)
  byte1     byte2     byte3
```

Nibbles

Binary numbers are often broken down into groups of four bits, as we have seen with BCD codes and hexadecimal number conversions. In the early days of digital systems, a term caught on to describe a group of four bits. Because it is half as big as a byte, it was named a **nibble**. The following examples illustrate the use of this term.

EXAMPLE 2-11

How many nibbles are in a byte?

Solution

2

EXAMPLE 2-12

What is the hex value of the least significant nibble of the binary number 1001 0101?

Solution

1001 0101

The least significant nibble is 0101 = 5.

Words

Bits, nibbles, and bytes are terms that represent a fixed number of binary digits. As systems have grown over the years, their capacity (appetite?) for handling binary data has also grown. A **word** is a group of bits that represents a certain unit of information. The size of the word depends on the size of the data pathway in the system that uses the information. The **word size** can be defined as the number of bits in the binary word that a digital system operates on. For example, the computer in your microwave oven can probably handle only one byte at a time. It has a word size of eight bits. On the other hand, the personal computer on your desk can handle eight bytes at a time, so it has a word size of 64 bits.

REVIEW QUESTIONS

1. How many bytes are needed to represent 235_{10} in binary?
2. What is the largest decimal value that can be represented in BCD using two bytes?
3. How many hex digits can a nibble represent?
4. How many nibbles are in one BCD digit?

2-8 ALPHANUMERIC CODES

In addition to numerical data, a computer must be able to handle nonnumerical information. In other words, a computer should recognize codes that represent letters of the alphabet, punctuation marks, and other special characters as well as numbers. These codes are called **alphanumeric codes**. A complete alphanumeric code would include the 26 lowercase letters, 26 uppercase letters, 10 numeric digits, 7 punctuation marks, and anywhere from 20 to 40 other characters, such as +, /, #, %, *, and so on. We can say that an alphanumeric code represents all of the various characters and functions that are found on a computer keyboard.

ASCII Code

The most widely used alphanumeric code is the **American Standard Code for Information Interchange (ASCII)**. The ASCII (pronounced "askee") code is a seven-bit code, and so it has $2^7 = 128$ possible code groups. This is more than enough to represent all of the standard keyboard characters as well as the control functions such as the (RETURN) and (LINEFEED) functions. Table 2-4 shows a listing of the standard seven-bit ASCII code. The table gives the hexadecimal and decimal equivalents. The seven-bit binary code for each character can be obtained by converting the hex value to binary.

EXAMPLE 2-13

Use Table 2-4 to find the seven-bit ASCII code for the backslash character (\).

Solution

The hex value given in Table 2-4 is 5C. Translating each hex digit into four-bit binary produces 0101 1100. The lower seven bits represent the ASCII code for \, or 1011100.

TABLE 2-4 Standard ASCII codes.

Character	HEX	Decimal	Character	HEX	Decimal	Character	HEX	Decimal	Character	HEX	Decimal
NUL (null)	0	0	Space	20	32	@	40	64	`	60	96
Start Heading	1	1	!	21	33	A	41	65	a	61	97
Start Text	2	2	“	22	34	B	42	66	b	62	98
End Text	3	3	#	23	35	C	43	67	c	63	99
End Transmit.	4	4	$	24	36	D	44	68	d	64	100
Enquiry	5	5	%	25	37	E	45	69	e	65	101
Acknowlege	6	6	&	26	38	F	46	70	f	66	102
Bell	7	7	'	27	39	G	47	71	g	67	103
Backspace	8	8	(	28	40	H	48	72	h	68	104
Horiz. Tab	9	9	)	29	41	I	49	73	i	69	105
Line Feed	A	10	*	2A	42	J	4A	74	j	6A	106
Vert. Tab	B	11	+	2B	43	K	4B	75	k	6B	107
Form Feed	C	12	,	2C	44	L	4C	76	l	6C	108
Carriage Return	D	13	-	2D	45	M	4D	77	m	6D	109
Shift Out	E	14	.	2E	46	N	4E	78	n	6E	110
Shift In	F	15	/	2F	47	O	4F	79	o	6F	111
Data Link Esc	10	16	0	30	48	P	50	80	p	70	112
Direct Control 1	11	17	1	31	49	Q	51	81	q	71	113
Direct Control 2	12	18	2	32	50	R	52	82	r	72	114
Direct Control 3	13	19	3	33	51	S	53	83	s	73	115
Direct Control 4	14	20	4	34	52	T	54	84	t	74	116
Negative ACK	15	21	5	35	53	U	55	85	u	75	117
Synch Idle	16	22	6	36	54	V	56	86	v	76	118
End Trans Block	17	23	7	37	55	W	57	87	w	77	119
Cancel	18	24	8	38	56	X	58	88	x	78	120
End of Medium	19	25	9	39	57	Y	59	89	y	79	121
Substitue	1A	26	:	3A	58	Z	5A	90	z	7A	122
Escape	1B	27	;	3B	59	[	5B	91	{	7B	123
Form separator	1C	28	<	3C	60	\	5C	92	|	7C	124
Group separator	1D	29	=	3D	61	]	5D	93	}	7D	125
Record Separator	1E	30	>	3E	62	^	5E	94	~	7E	126
Unit Separator	1F	31	?	3F	63	_	5F	95	Delete	7F	127

The ASCII code is used for the transfer of alphanumeric information between a computer and the external devices such as a printer or another computer. A computer also uses ASCII internally to store the information that an operator types in at the computer's keyboard. The following example illustrates this.

EXAMPLE 2-14

An operator is typing in a C language program at the keyboard of a certain microcomputer. The computer converts each keystroke into its ASCII code and stores the code as a byte in memory. Determine the binary strings that will be entered into memory when the operator types in the following C statement:

```
if (x>3)
```

Solution

Locate each character (including the space) in Table 2-4 and record its ASCII code.

i	69	0110	1001
f	66	0110	0110
space	20	0010	0000
(	28	0010	1000
x	78	0111	1000
>	3E	0011	1110
3	33	0011	0011
)	29	0010	1001

Note that a 0 was added to the leftmost bit of each ASCII code because the codes must be stored as bytes (eight bits). This adding of an extra bit is called *padding with 0s.*

REVIEW QUESTIONS

1. Encode the following message in ASCII code using the hex representation: "COST = $72."
2. The following padded ASCII-coded message is stored in successive memory locations in a computer:

   01010011 01010100 01001111 01010000

What is the message?

2-9 PARITY METHOD FOR ERROR DETECTION

The movement of binary data and codes from one location to another is the most frequent operation performed in digital systems. Here are just a few examples:

- The transmission of digitized voice over a microwave link
- The storage of data in and retrieval of data from external memory devices such as magnetic and optical disk
- The transmission of digital data from a computer to a remote computer over telephone lines (i.e., using a modem). This is one of the major ways of sending and receiving information on the Internet.

Whenever information is transmitted from one device (the transmitter) to another device (the receiver), there is a possibility that errors can occur such that the receiver does not receive the identical information that was sent by the transmitter. The major cause of any transmission errors is *electrical noise,* which consists of spurious fluctuations in voltage or current that are present in all electronic systems to varying degrees. Figure 2-4 is a simple illustration of a type of transmission error.

The transmitter sends a relatively noise-free serial digital signal over a signal line to a receiver. However, by the time the signal reaches the receiver,

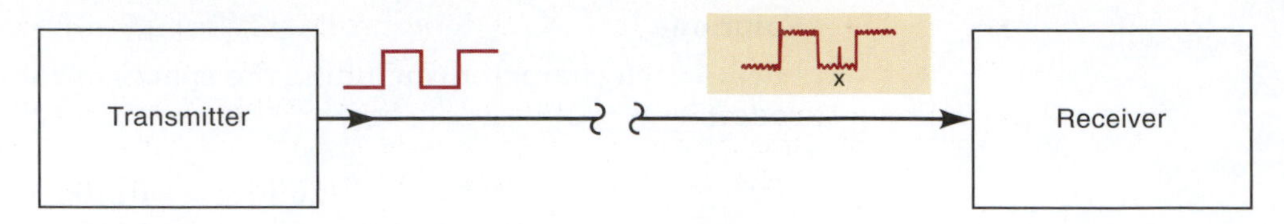

FIGURE 2-4 Example of noise causing an error in the transmission of digital data.

it contains a certain degree of noise superimposed on the original signal. Occasionally, the noise is large enough in amplitude that it will alter the logic level of the signal, as it does at point x. When this occurs, the receiver may incorrectly interpret that bit as a logic 1, which is not what the transmitter has sent.

Most modern digital equipment is designed to be relatively error-free, and the probability of errors such as the one shown in Figure 2-4 is very low. However, we must realize that digital systems often transmit thousands, even millions, of bits per second, so that even a very low rate of occurrence of errors can produce an occasional error that might prove to be bothersome, if not disastrous. For this reason, many digital systems employ some method for detection (and sometimes correction) of errors. One of the simplest and most widely used schemes for error detection is the **parity method**.

Parity Bit

A **parity bit** is an extra bit that is attached to a code group that is being transferred from one location to another. The parity bit is made either 0 or 1, depending on the number of 1s that are contained in the code group. Two different methods are used.

In the *even-parity* method, the value of the parity bit is chosen so that the total number of 1s in the code group (including the parity bit) is an *even* number. For example, suppose that the group is 1000011. This is the ASCII character "C." The code group has *three* 1s. Therefore, we will add a parity bit of 1 to make the total number of 1s an even number. The *new* code group, *including the parity bit,* thus becomes

1 1 0 0 0 0 1 1
↑ added parity bit*

If the code group contains an even number of 1s to begin with, the parity bit is given a value of 0. For example, if the code group were 1000001 (the ASCII code for "A"), the assigned parity bit would be 0, so that the new code, *including the parity bit,* would be 01000001.

The *odd-parity* method is used in exactly the same way except that the parity bit is chosen so the total number of 1s (including the parity bit) is an *odd* number. For example, for the code group 1000001, the assigned parity bit would be a 1. For the code group 1000011, the parity bit would be a 0.

Regardless of whether even parity or odd parity is used, the parity bit becomes an actual part of the code word. For example, adding a parity bit to

*The parity bit can be placed at either end of the code group, but it is usually placed to the left of the MSB.

the seven-bit ASCII code produces an eight-bit code. Thus, the parity bit is treated just like any other bit in the code.

The parity bit is issued to detect any *single-bit* errors that occur during the transmission of a code from one location to another. For example, suppose that the character "A" is being transmitted and *odd* parity is being used. The transmitted code would be

1 1 0 0 0 0 0 1

When the receiver circuit receives this code, it will check that the code contains an odd number of 1s (including the parity bit). If so, the receiver will assume that the code has been correctly received. Now, suppose that because of some noise or malfunction the receiver actually receives the following code:

1 1 0 0 0 0 0 0

The receiver will find that this code has an *even* number of 1s. This tells the receiver that there must be an error in the code because presumably the transmitter and receiver have agreed to use odd parity. There is no way, however, that the receiver can tell which bit is in error because it does not know what the code is supposed to be.

It should be apparent that this parity method would not work if *two* bits were in error, because two errors would not change the "oddness" or "evenness" of the number of 1s in the code. In practice, the parity method is used only in situations where the probability of a single error is very low and the probability of double errors is essentially zero.

When the parity method is being used, the transmitter and the receiver must have agreement, in advance, as to whether odd or even parity is being used. There is no advantage of one over the other, although even parity seems to be used more often. The transmitter must attach an appropriate parity bit to each unit of information that it transmits. For example, if the transmitter is sending ASCII-coded data, it will attach a parity bit to each seven-bit ASCII code group. When the receiver examines the data that it has received from the transmitter, it checks each code group to see that the total number of 1s (including the parity bit) is consistent with the agreed-upon type of parity. This is often called *checking the parity* of the data. In the event that it detects an error, the receiver may send a message back to the transmitter asking it to retransmit the last set of data. The exact procedure that is followed when an error is detected depends on the particular system.

EXAMPLE 2-15

Computers often communicate with other remote computers over telephone lines. For example, this is how dial-up communication over the internet takes place. When one computer is transmitting a message to another, the information is usually encoded in ASCII. What actual bit strings would a computer transmit to send the message HELLO, using ASCII with even parity?

Solution

First, look up the ASCII codes for each character in the message. Then for each code, count the number of 1s. If it is an even number, attach a 0 as the

MSB. If it is an odd number, attach a 1. Thus, the resulting eight-bit codes (bytes) will all have an even number of 1s (including parity).

attached even-parity bits
↓

		parity	bits
H	=	0	1 0 0 1 0 0 0
E	=	1	1 0 0 0 1 0 1
L	=	1	1 0 0 1 1 0 0
L	=	1	1 0 0 1 1 0 0
O	=	1	1 0 0 1 1 1 1

REVIEW QUESTIONS

1. Attach an odd-parity bit to the ASCII code for the $ symbol, and express the result in hexadecimal.
2. Attach an even-parity bit to the BCD code for decimal 69.
3. Why can't the parity method detect a double error in transmitted data?

2-10 APPLICATIONS

Here are several applications that will serve as a review of some of the concepts covered in this chapter. These applications should give a sense of how the various number systems and codes are used in the digital world. More applications are presented in the end-of-chapter problems.

APPLICATION 2-1

A typical CD-ROM can store 650 megabytes of digital data. Since mega $= 2^{20}$, how many bits of data can a CD-ROM hold?

Solution

Remember that a byte is eight bits. Therefore, 650 megabytes is $650 \times 2^{20} \times 8 =$ **5,452,595,200 bits**.

APPLICATION 2-2

In order to program many microcontrollers, the binary instructions are stored in a file on a personal computer in a special way known as Intel Hex Format. The hexadecimal information is encoded into ASCII characters so it can be displayed easily on the PC screen, printed, and easily transmitted one character at a time over a standard PC's serial COM port. One line of an Intel Hex Format file is shown below:

:10002000F7CFFFCF1FEF2FEF2A95F1F71A95D9F7EA

The first character sent is the ASCII code for a colon, followed by a 1. Each has an even parity bit appended as the most significant bit. A test instrument captures the binary bit pattern as it goes across the cable to the microcontroller.

(a) What should the binary bit pattern (including parity) look like? (MSB – LSB)

(b) The value 10, following the colon, represents the total hexadecimal number of bytes that are to be loaded into the micro's memory. What is the decimal number of bytes being loaded?

(c) The number 0020 is a four-digit hex value representing the address where the first byte is to be stored. What is the biggest address possible? How many bits would it take to represent this address?

(d) The value of the first data byte is F7. What is the value (in binary) of the least significant nibble of this byte?

FFFF 1111 1111 1111 1111 16 bits

Solution

(a) ASCII codes are 3A (for :) and 31 (for 1) 00111010 10110001
even parity bit

(b) 10 hex $= 1 \times 16 + 0 \times 1 =$ 16 decimal bytes

(c) FFFF is the biggest possible value. Each hex digit is 4 bits, so we need 16 bits.

(d) The least significant nibble (4 bits) is represented by hex 7. In binary this would be 0111.

APPLICATION 2-3

A small process-control computer uses hexadecimal codes to represent its 16-bit memory addresses.

(a) How many hex digits are required?

(b) What is the range of addresses in hex?

(c) How many memory locations are there?

Solution

(a) Since 4 bits convert to a single hex digit, $16/4 =$ **4** hex digits are needed.

(b) The binary range is 0000000000000000_2 to 1111111111111111_2. In hex, this becomes 0000_{16} to $FFFF_{16}$.

(c) With 4 hex digits, the total number of addresses is $16^4 =$ **65,536.**

APPLICATION 2-4

Numbers are entered into a microcontroller-based system in BCD, but stored in straight binary. As a programmer, you must decide whether you need a one-byte or two-byte storage location.

(a) How many bytes do you need if the system takes a two-digit decimal entry?

(b) What if you needed to be able to enter three digits?

Solution

(a) With two digits you can enter values up to 99 ($1001\ 1001_{BCD}$). In binary this value is 01100011, which will fit into an eight-bit memory location. Thus you can use a single byte.

(b) Three digits can represent up to 999 (1001 1001 1001). In binary this value is 1111100111 (10 bits). Thus you cannot use a single byte; you need two bytes.

APPLICATION 2-5

When ASCII characters must be transmitted between two independent systems (such as between a computer and a modem), there must be a way of telling the receiver when a new character is coming in. There is often a need to detect errors in the transmission as well. The method of transfer is called asynchronous data communication. The normal resting state of the transmission line is logic 1. When the transmitter sends an ASCII character, it must be "framed" so the receiver knows where the data begins and ends. The first bit must always be a start bit (logic 0). Next the ASCII code is sent LSB first and MSB last. After the MSB, a parity bit is appended to check for transmission errors. Finally, the transmission is ended by sending a stop bit (logic 1). A typical asynchronous transmission of a seven-bit ASCII code for the pound sign # (23 Hex) with even parity is shown in Figure 2-5.

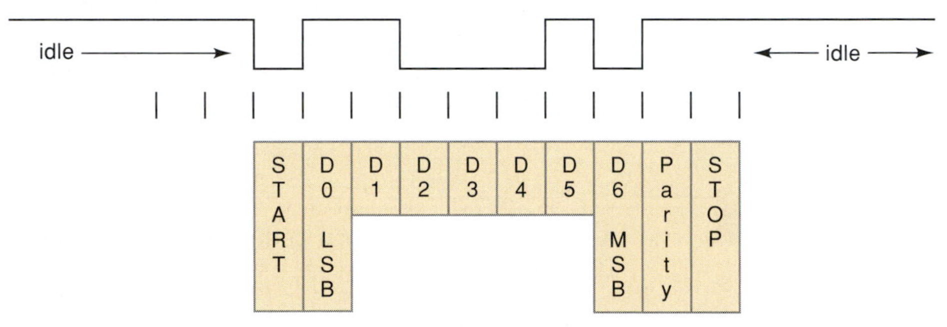

FIGURE 2-5 Asynchronous serial data with even parity.

SUMMARY

1. The hexadecimal number system is used in digital systems and computers as an efficient way of representing binary quantities.
2. In conversions between hex and binary, each hex digit corresponds to four bits.
3. The repeated-division method is used to convert decimal numbers to binary or hexadecimal.
4. Using an N-bit binary number, we can represent decimal values from 0 to $2^N - 1$.
5. The BCD code for a decimal number is formed by converting each digit of the decimal number to its four-bit binary equivalent.
6. The Gray code defines a sequence of bit patterns in which only one bit changes between successive patterns in the sequence.
7. A byte is a string of eight bits. A nibble is four bits. The word size depends on the system.
8. An alphanumeric code is one that uses groups of bits to represent all of the various characters and functions that are part of a typical computer's keyboard. The ASCII code is the most widely used alphanumeric code.
9. The parity method for error detection attaches a special parity bit to each transmitted group of bits.

IMPORTANT TERMS

hexadecimal number system
straight binary coding
binary-coded-decimal (BCD) code
Gray code
byte
nibble
word
word size
alphanumeric code
American Standard Code for Information Interchange (ASCII)
parity method
parity bit

PROBLEMS

SECTIONS 2-1 AND 2-2

2-1. Convert these binary numbers to decimal.

(a)*10110 (b) 10010101 (c)*100100001001 (d) 01101011 (e)*11111111 (f) 01101111 (g)*1111010111 (h) 11011111

2-2. Convert the following decimal values to binary.

(a)*37 (b) 13 (c)*189 (d) 1000 (e)*77 (f) 390 (g)*205 (h) 2133 (i)* 511

2-3. What is the largest decimal value that can be represented by (a)* an eight-bit binary number? (b) A 16-bit number?

SECTION 2-4

2-4. Convert each hex number to its decimal equivalent.

(a)*743 (b) 36 (c)*37FD (d) 2000 (e)* 165 (f) ABCD (g)*7FF (h) 1204

2-5. Convert each of the following decimal numbers to hex.

(a)*59 (b) 372 (c)*919 (d) 1024 (e)*771 (f) 2313 (g)*65,536 (h) 255

2-6. Convert each of the hex values from Problem 2-4 to binary.

2-7. Convert the binary numbers in Problem 2-1 to hex.

2-8. List the hex numbers in sequence from 195_{16} to 280_{16}.

2-9. When a large decimal number is to be converted to binary, it is sometimes easier to convert it first to hex, and then from hex to binary. Try this procedure for 2133_{10} and compare it with the procedure used in Problem 2-2(h).

2-10. How many hex digits are required to represent decimal numbers up to 20,000?

2-11. Convert these hex values to decimal.

(a)*92 (b) 1A6 (c)*37FD (d) ABCD (e)* 000F (f) 55 (g)*2C0 (h) 7FF

*Answers to problems marked with an asterisk can be found in the back of the text.

2-12. Convert these decimal values to hex.

(a)*75 (d) 24 (g)*25,619
(b) 314 (e)*7245 (h) 4095
(c)*2048 (f) 498

2-13. Take each four-bit binary number in the order they are written and write the equivalent hex digit without performing a calculation by hand or by calculator.

(a) 1001 (e) 1111 (i) 1011 (m) 0001
(b) 1101 (f) 0010 (j) 1100 (n) 0101
(c) 1000 (g) 1010 (k) 0011 (o) 0111
(d) 0000 (h) 1001 (l) 0100 (p) 0110

2-14. Take each hex digit and write its four-bit binary value without performing any calculations by hand or by calculator.

(a) 6 (e) 4 (i) 9 (m) 0
(b) 7 (f) 3 (j) A (n) 8
(c) 5 (g) C (k) 2 (o) D
(d) 1 (h) B (l) F (p) 9

2-15.* Convert the binary numbers in Problem 2-1 to hexadecimal.

2-16.* Convert the hex values in Problem 2-11 to binary.

2-17.* List the hex numbers in sequence from 280 to 2A0.

2-18. How many hex digits are required to represent decimal numbers up to 1 million?

SECTION 2-5

2-19. Encode these decimal numbers in BCD.

(a)*47 (d) 6727 (g)*89,627
(b) 962 (e)*13 (h) 1024
(c)*187 (f) 529

2-20. How many bits are required to represent the decimal numbers in the range from 0 to 999 using (a) straight binary code? (b) Using BCD code?

2-21. The following numbers are in BCD. Convert them to decimal.

(a)*1001011101010010 (d) 0111011101110101
(b) 000110000100 (e)*010010010010
(c)*011010010101 (f) 010101010101

SECTION 2-7

2-22.*(a) How many bits are contained in eight bytes?
(b) What is the largest hex number that can be represented in four bytes?
(c) What is the largest BCD-encoded decimal value that can be represented in three bytes?

2-23. (a) Refer to Table 2-4. What is the most significant nibble of the ASCII code for the letter X?
(b) How many nibbles can be stored in a 16-bit word?
(c) How many bytes does it take to make up a 24-bit word?

SECTIONS 2-8 AND 2-9

2-24. Represent the statement "X = 3 × Y" in ASCII code. Attach an odd-parity bit.

2-25.* Attach an *even*-parity bit to each of the ASCII codes for Problem 2-24, and give the results in hex.

2-26. The following bytes (shown in hex) represent a person's name as it would be stored in a computer's memory. Each byte is a padded ASCII code. Determine the name of each person.

(a)* 42 45 4E 20 53 4D 49 54 48

(b) 4A 6F 65 20 47 72 65 65 6E

2-27. Convert the following decimal numbers to BCD code and then attach an *odd* parity bit.

(a)* 74	(c)* 8884	(e)* 165
(b) 38	(d) 275	(f) 9201

2-28.* In a certain digital system, the decimal numbers from 000 through 999 are represented in BCD code. An *odd*-parity bit is also included at the end of each code group. Examine each of the code groups below, and assume that each one has just been transferred from one location to another. Some of the groups contain errors. Assume that *no more than* two errors have occurred for each group. Determine which of the code groups have a single error and which of them *definitely* have a double error. (*Hint:* Remember that this is a BCD code.)

(a) 1001010110000

(b) 0100011101100

(c) 0111110000011

(d) 1000011000101

2-29. Suppose that the receiver received the following data from the transmitter of Example 2-16:

```
0 1 0 0 1 0 0 0
1 1 0 0 0 1 0 1
1 1 0 0 1 1 0 0
1 1 0 0 1 0 0 0
1 1 0 0 1 1 0 0
```

What errors can the receiver determine in these received data?

DRILL QUESTIONS

2-30.* Perform each of the following conversions. For some of them, you may want to try several methods to see which one works best for you. For example, a binary-to-decimal conversion may be done directly, or it may be done as a binary-to-hex conversion followed by a hex-to-decimal conversion.

(a) 1417_{10} = ______$_2$

(b) 255_{10} = ______$_2$

(c) 11010001_2 = ______$_{10}$

(d) 1110101000100111_2 = ______$_{10}$

(e) 2497_{10} = _____ $_{16}$
(f) 511_{10} = _____ (BCD)
(g) 235_{16} = _____ $_{10}$
(h) 4316_{10} = _____ $_{16}$
(i) $7A9_{16}$ = _____ $_{10}$
(j) $3E1C_{16}$ = _____ $_{10}$
(k) 1600_{10} = _____ $_{16}$
(l) $38{,}187_{10}$ = _____ $_{16}$
(m) 865_{10} = _____ (BCD)
(n) 100101000111 (BCD) = _____ $_{10}$
(o) 465_{16} = _____ $_{2}$
(p) $B34_{16}$ = _____ $_{2}$
(q) 01110100 (BCD) = _____ $_{2}$
(r) 111010_2 = _____ (BCD)

2-31.*Represent the decimal value 37 in each of the following ways.
(a) straight binary
(b) BCD
(c) hex
(d) ASCII (i.e., treat each digit as a character)

2-32.*Fill in the blanks with the correct word or words.
(a) Conversion from decimal to _____ requires repeated division by 16.
(b) Conversion from decimal to binary requires repeated division by _____.
(c) In the BCD code, each _____ is converted to its four-bit binary equivalent.
(d) The _____ code has the characteristic that only one bit changes in going from one step to the next.
(e) A transmitter attaches a _____ to a code group to allow the receiver to detect _____.
(f) The _____ code is the most common alphanumeric code used in computer systems.
(g) _____ is often used as a convenient way to represent large binary numbers.
(h) A string of eight bits is called a _____.

2-33. Write the binary number that results when each of the following numbers is incremented by one.
(a)*0111 (b) 010011 (c) 1011

2-34. Decrement each binary number.
(a)*1110 (b) 101000 (c) 1110

2-35. Write the number that results when each of the following is incremented.
(a)*7779_{16} (c)*$0FFF_{16}$ (e)*$9FF_{16}$
(b) 9999_{16} (d) 2000_{16} (f) $100A_{16}$

2-36.*Repeat Problem 2-35 for the decrement operation.

CHALLENGING EXERCISES

2-37.* In a microcomputer, the *addresses* of memory locations are binary numbers that identify each memory circuit where a byte is stored. The number of bits that make up an address depends on how many memory locations there are. Since the number of bits can be very large, the addresses are often specified in hex instead of binary.

(a) If a microcomputer uses a 20-bit address, how many different memory locations are there?

(b) How many hex digits are needed to represent the address of a memory location?

(c) What is the hex address of the 256th memory location? (*Note:* The first address is always 0.)

2-38. In an audio CD, the audio voltage signal is typically sampled about 44,000 times per second, and the value of each sample is recorded on the CD surface as a binary number. In other words, each recorded binary number represents a single voltage point on the audio signal waveform.

(a) If the binary numbers are six bits in length, how many different voltage values can be represented by a single binary number? Repeat for eight bits and ten bits.

(b) If ten-bit numbers are used, how many bits will be recorded on the CD in 1 second?

(c) If a CD can typically store 5 billion bits, how many seconds of audio can be recorded when ten-bit numbers are used?

2-39.* A black-and-white digital camera lays a fine grid over an image and then measures and records a binary number representing the level of gray it sees in each cell of the grid. For example, if four-bit numbers are used, the value of black is set to 0000 and the value of white to 1111, and any level of gray is somewhere between 0000 and 1111. If six-bit numbers are used, black is 000000, white is 111111, and all grays are between the two.

Suppose we wanted to distinguish among 254 different levels of gray within each cell of the grid. How many bits would we need to use to represent these levels?

2-40. A 3-Megapixel digital camera stores an eight-bit number for the brightness of each of the primary colors (red, green, blue) found in each picture element (pixel). If every bit is stored (no data compression), how many pictures can be stored on a 128-Megabyte memory card? (Note: In digital systems, Mega means 2^{20}.)

2-41. Construct a table showing the binary, hex, and BCD representations of all decimal numbers from 0 to 15. Compare your table with Table 2-3.

ANSWERS TO SECTION REVIEW QUESTIONS

SECTION 2-1

1. 2267 2. 32768

SECTION 2-2

1. 1010011 2. 1011011001 3. 20 bits

SECTION 2-3

1. 9422 2. C2D; 110000101101 3. 97B5 4. E9E, E9F, EA0, EA1
5. 11010100100111 6. 0 to 65,535

SECTION 2-4

1. 10110010_2; 000101111000 (BCD) 2. 32 3. Advantage: ease of conversion. Disadvantage: BCD requires more bits.

SECTION 2-5

1. 0111 2. 0110

SECTION 2-7

1. One 2. 9999 3. One 4. One

SECTION 2-8

1. 43, 4F, 53, 54, 20, 3D, 20, 24, 37, 32 2. STOP

SECTION 2-9

1. A4 2. 001101001 3. Two errors in the data would not change the oddness or evenness of the number of 1s in the data.

CHAPTER 3

DESCRIBING LOGIC CIRCUITS

OUTLINE

OBJECTIVES

Upon completion of this chapter, you will be able to:

- Perform the three basic logic operations.
- Describe the operation of and construct the truth tables for the AND, NAND, OR, and NOR gates, and the NOT (INVERTER) circuit.
- Draw timing diagrams for the various logic-circuit gates.
- Write the Boolean expression for the logic gates and combinations of logic gates.
- Implement logic circuits using basic AND, OR, and NOT gates.
- Appreciate the potential of Boolean algebra to simplify complex logic circuits.
- Use DeMorgan's theorems to simplify logic expressions.
- Use either of the universal gates (NAND or NOR) to implement a circuit represented by a Boolean expression.
- Explain the advantages of constructing a logic-circuit diagram using the alternate gate symbols versus the standard logic-gate symbols.
- Describe the concept of active-LOW and active-HIGH logic signals.
- Draw and interpret the IEEE/ANSI standard logic-gate symbols.
- Use several methods to describe the operation of logic circuits.
- Interpret simple circuits defined by a hardware description language (HDL).
- Explain the difference between an HDL and a computer programming language.
- Create an HDL file for a simple logic gate.
- Create an HDL file for combinational circuits with intermediate variables.

INTRODUCTION

Chapters 1 and 2 introduced the concepts of logic levels and logic circuits. In logic, only two possible conditions exist for any input or output: true and false. The binary number system uses only two digits, 1 and 0, so it is perfect for representing logical relationships. Digital logic circuits use predefined voltage ranges to represent these binary states. Using these concepts, we can create circuits made of little more than processed beach sand and wire that make consistent, intelligent, logical decisions. It is vitally important that we have a method to describe the logical decisions made by these circuits. In other words, we must describe how they operate. In this chapter, we will discover many ways to describe their operation. Each description method is important because all these methods commonly appear in technical literature and system documentation and are used in conjunction with modern design and development tools.

Life is full of examples of circumstances that are in one state or another. For example, a creature is either alive or dead, a light is either on or off, a door is locked or unlocked, and it is either raining or it is not. In 1854, a mathematician named George Boole wrote *An Investigation of the Laws of Thought,* in which he described the way we make logical decisions based on true or false circumstances. The methods he described are referred to today as Boolean logic, and the system of using symbols and operators to describe these decisions is called Boolean algebra. In the same way we use symbols such as x and y to represent unknown numerical values in regular algebra, Boolean algebra uses symbols to represent a logical expression that has one of two possible values: true or false. The logical expression might be *door is closed, button is pressed,* or *fuel is low*. Writing these expressions is very tedious, and so we tend to substitute symbols such as A, B, and C.

The main purpose of these logical expressions is to describe the relationship between a logic circuit's output (the decision) and its inputs (the circumstances). In this chapter, we will study the most basic logic circuits—*logic gates*—which are the fundamental building blocks from which all other logic circuits and digital systems are constructed. We will see how the operation of the different logic gates and the more complex circuits formed from combinations of logic gates can be described and analyzed using Boolean algebra. We will also get a glimpse of how Boolean algebra can be used to simplify a circuit's Boolean expression so that the circuit can be rebuilt using fewer logic gates and/or fewer connections. Much more will be done with circuit simplification in Chapter 4.

Boolean algebra is not only used as a tool for analysis and simplification of logic systems. It can also be used as a tool to create a logic circuit that will produce the desired input/output relationship. This process is often called synthesis of logic circuits as opposed to analysis. Other techniques have been used in the analysis, synthesis, and documentation of logic systems and circuits including truth tables, schematic symbols, timing diagrams, and—last but by no means least—language. To categorize these methods, we could say that Boolean algebra is a mathematic tool, truth tables are data organizational tools, schematic symbols are drawing tools, timing diagrams are graphing tools, and language is the universal description tool.

Today, any of these tools can be used to provide input to computers. The computers can be used to simplify and translate between these various forms of description and ultimately provide an output in the form necessary to implement a digital system. To take advantage of the powerful benefits of computer software, we must first fully understand the acceptable ways for describing these systems in terms the computer can understand. This chapter will lay the groundwork for further study of these vital tools for synthesis and analysis of digital systems.

Clearly the tools described here are invaluable tools in describing, analyzing, designing, and implementing digital circuits. The student who expects to work in the digital field must work hard at understanding and becoming comfortable with Boolean algebra (believe us, it's much, much easier than conventional algebra) and all the other tools. Do *all* of the examples, exercises, and problems, even the ones your instructor doesn't assign. When those run out, make up your own. The time you spend will be well worth it because you will see your skills improve and your confidence grow.

3-1 BOOLEAN CONSTANTS AND VARIABLES

Boolean algebra differs in a major way from ordinary algebra because Boolean constants and variables are allowed to have only two possible values, 0 or 1. A Boolean variable is a quantity that may, at different times, be equal to either 0 or 1. Boolean variables are often used to represent the voltage level present on a wire or at the input/output terminals of a circuit. For example, in a certain digital system, the Boolean value of 0 might be assigned to any voltage in the range from 0 to 0.8 V, while the Boolean value of 1 might be assigned to any voltage in the range 2 to 5 V.*

Thus, Boolean 0 and 1 do not represent actual numbers but instead represent the state of a voltage variable, or what is called its **logic level**. A voltage in a digital circuit is said to be at the logic 0 level or the logic 1 level, depending on its actual numerical value. In digital logic, several other terms are used synonymously with 0 and 1. Some of the more common ones are shown in Table 3-1. We will use the 0/1 and LOW/HIGH designations most of the time.

TABLE 3-1

Logic 0	Logic 1
False	True
Off	On
Low	High
No	Yes
Open switch	Closed switch

As we said in the introduction, **Boolean algebra** is a means for expressing the relationship between a logic circuit's inputs and outputs. The inputs are considered logic variables whose logic levels at any time determine the output levels. In all our work to follow, we shall use letter symbols to represent logic variables. For example, the letter A might represent a certain digital circuit input or output, and at any time we must have either $A = 0$ or $A = 1$: if not one, then the other.

Because only two values are possible, Boolean algebra is relatively easy to work with compared with ordinary algebra. In Boolean algebra, there are no fractions, decimals, negative numbers, square roots, cube roots, logarithms, imaginary numbers, and so on. In fact, in Boolean algebra there are only *three* basic operations: *OR*, *AND*, and *NOT*.

These basic operations are called *logic operations*. Digital circuits called *logic gates* can be constructed from diodes, transistors, and resistors connected so that the circuit output is the result of a basic logic operation *(OR, AND, NOT)* performed on the inputs. We will be using Boolean algebra first to describe and analyze these basic logic gates, then later to analyze and design combinations of logic gates connected as logic circuits.

3-2 TRUTH TABLES

A **truth table** is a means for describing how a logic circuit's output depends on the logic levels present at the circuit's inputs. Figure 3-1(a) illustrates a truth table for one type of two-input logic circuit. The table lists all possible

*Voltages between 0.8 and 2 V are undefined (neither 0 nor 1) and should not occur under normal circumstances.

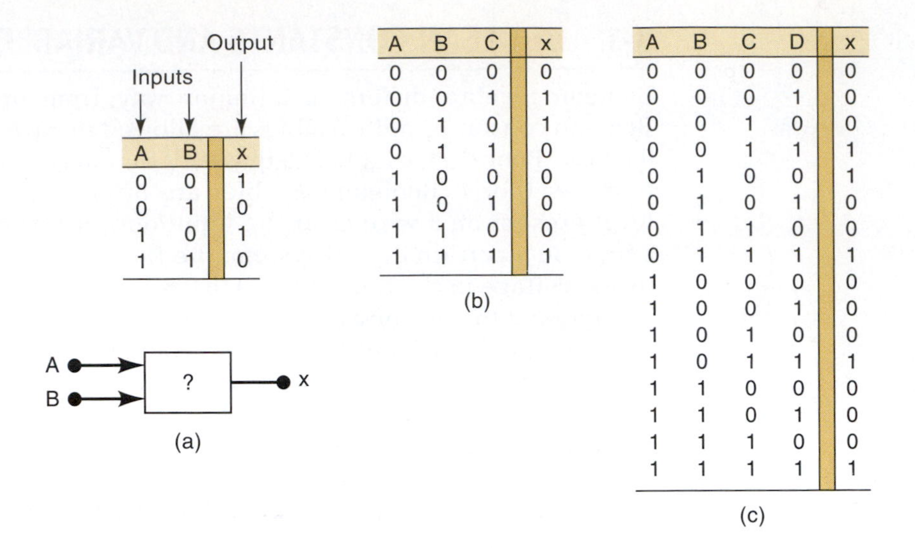

(a)

A	B	x
0	0	1
0	1	0
1	0	1
1	1	0

(b)

A	B	C	x
0	0	0	0
0	0	1	1
0	1	0	1
0	1	1	0
1	0	0	0
1	0	1	0
1	1	0	0
1	1	1	1

(c)

A	B	C	D	x
0	0	0	0	0
0	0	0	1	0
0	0	1	0	0
0	0	1	1	1
0	1	0	0	1
0	1	0	1	0
0	1	1	0	0
0	1	1	1	1
1	0	0	0	0
1	0	0	1	0
1	0	1	0	0
1	0	1	1	1
1	1	0	0	0
1	1	0	1	0
1	1	1	0	0
1	1	1	1	1

FIGURE 3-1 Example truth tables for (a) two-input, (b) three-input, and (c) four-input circuits.

combinations of logic levels present at inputs A and B, along with the corresponding output level x. The first entry in the table shows that when A and B are both at the 0 level, the output x is at the 1 level or, equivalently, in the 1 state. The second entry shows that when input B is changed to the 1 state, so that $A = 0$ and $B = 1$, the output x becomes a 0. In a similar way, the table shows what happens to the output state for any set of input conditions.

Figures 3-1(b) and (c) show samples of truth tables for three- and four-input logic circuits. Again, each table lists all possible combinations of input logic levels on the left, with the resultant logic level for output x on the right. Of course, the actual values for x will depend on the type of logic circuit.

Note that there are 4 table entries for the two-input truth table, 8 entries for a three-input truth table, and 16 entries for the four-input truth table. The number of input combinations will equal 2^N for an N-input truth table. Also note that the list of all possible input combinations follows the binary counting sequence, and so it is an easy matter to write down all of the combinations without missing any.

REVIEW QUESTIONS

1. What is the output state of the four-input circuit represented in Figure 3-1(c) when all inputs except B are 1?
2. Repeat question 1 for the following input conditions: $A = 1, B = 0, C = 1, D = 0$.
3. How many table entries are needed for a five-input circuit?

3-3 OR OPERATION WITH OR GATES

The **OR operation** is the first of the three basic Boolean operations to be learned. An example can be found in the kitchen oven. The light inside the oven should turn on if either the *oven light switch is on* OR if the *door is opened*. The letter A could be used to represent the *oven light switch is on* and B could represent *door is opened*. The letter x could represent the *light is on*. The truth table in Figure 3-2(a) shows what happens when two logic inputs, A and B, are combined using the OR operation to produce the output x. The table shows that x is a logic 1 for every combination of input levels where one *or* more inputs are 1. The only case where x is a 0 is when both inputs are 0.

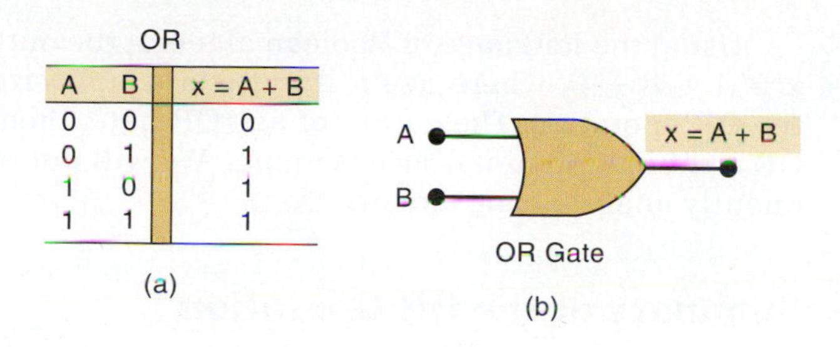

OR

A	B	x = A + B
0	0	0
0	1	1
1	0	1
1	1	1

FIGURE 3-2 (a) Truth table defining the OR operation; (b) circuit symbol for a two-input OR gate.

The Boolean expression for the OR operation is

$$x = A + B$$

In this expression, the + sign does not stand for ordinary addition; it stands for the OR operation. The OR operation is similar to ordinary addition except for the case where A and B are both 1; the OR operation produces $1 + 1 = 1$, not $1 + 1 = 2$. In Boolean algebra, 1 is as high as we go, so we can never have a result greater than 1. The same holds true for combining three inputs using the OR operation. Here we have $x = A + B + C$. If we consider the case where all three inputs are 1, we have

$$x = 1 + 1 + 1 = 1$$

The expression $x = A + B$ is read as "x equals A OR B," which means that x will be 1 when A or B or both are 1. Likewise, the expression $x = A + B + C$ is read as "x equals A OR B OR C," which means that x will be 1 when A or B or C or any combination of them are 1. To describe this circuit in the English language we could say that *x is true (1)* ***WHEN*** *A is true (1)* **OR** *B is true (1)* **OR** *C is true (1)*.

OR Gate

In digital circuitry, an **OR gate*** is a circuit that has two or more inputs and whose output is equal to the OR combination of the inputs. Figure 3-2(b) is the logic symbol for a two-input OR gate. The inputs A and B are logic voltage levels, and the output x is a logic voltage level whose value is the result of the OR operation on A and B; that is, $x = A + B$. In other words, the OR gate operates so that its output is HIGH (logic 1) if either input *A or B or both* are at a logic 1 level. The OR gate output will be LOW (logic 0) only if all its inputs are at logic 0.

This same idea can be extended to more than two inputs. Figure 3-3 shows a three-input OR gate and its truth table. Examination of this truth table shows again that the output will be 1 for every case where one or more inputs are 1. This general principle is the same for OR gates with any number of inputs.

FIGURE 3-3 Symbol and truth table for a three-input OR gate.

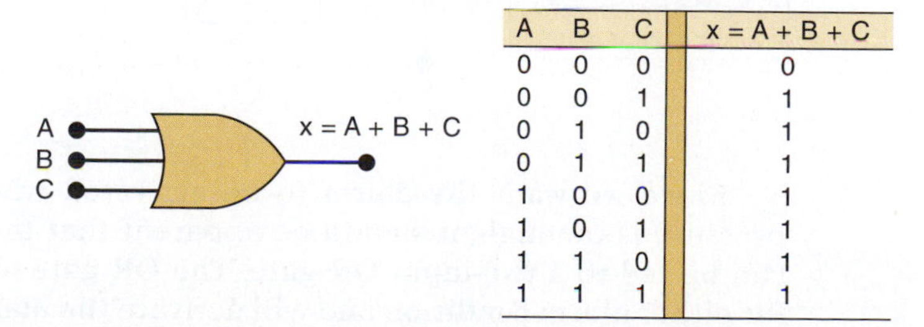

A	B	C	x = A + B + C
0	0	0	0
0	0	1	1
0	1	0	1
0	1	1	1
1	0	0	1
1	0	1	1
1	1	0	1
1	1	1	1

*The term *gate* comes from the inhibit/enable operation discussed in Chapter 4.

Using the language of Boolean algebra, the output x can be expressed as $x = A + B + C$, where again it must be emphasized that the + represents the OR operation. The output of any OR gate, then, can be expressed as the OR combination of its various inputs. We will put this to use when we subsequently analyze logic circuits.

Summary of the OR Operation

The important points to remember concerning the OR operation and OR gates are:

1. The OR operation produces a result (output) of 1 whenever *any* input is a 1. Otherwise the output is 0.
2. An OR gate is a logic circuit that performs an OR operation on the circuit's inputs.
3. The expression $x = A + B$ is read as "x equals A OR B."

EXAMPLE 3-1

In many industrial control systems, it is required to activate an output function whenever any one of several inputs is activated. For example, in a chemical process it may be desired that an alarm be activated whenever the process temperature exceeds a maximum value *or* whenever the pressure goes above a certain limit. Figure 3-4 is a block diagram of this situation. The temperature transducer circuit produces an output voltage proportional to the process temperature. This voltage, V_T, is compared with a temperature reference voltage, V_{TR}, in a voltage comparator circuit. The comparator output, T_H, is normally a low voltage (logic 0), but it switches to a high voltage (logic 1) when V_T exceeds V_{TR}, indicating that the process temperature is too high. A similar arrangement is used for the pressure measurement, so that its associated comparator output, P_H, goes from LOW to HIGH when the pressure is too high.

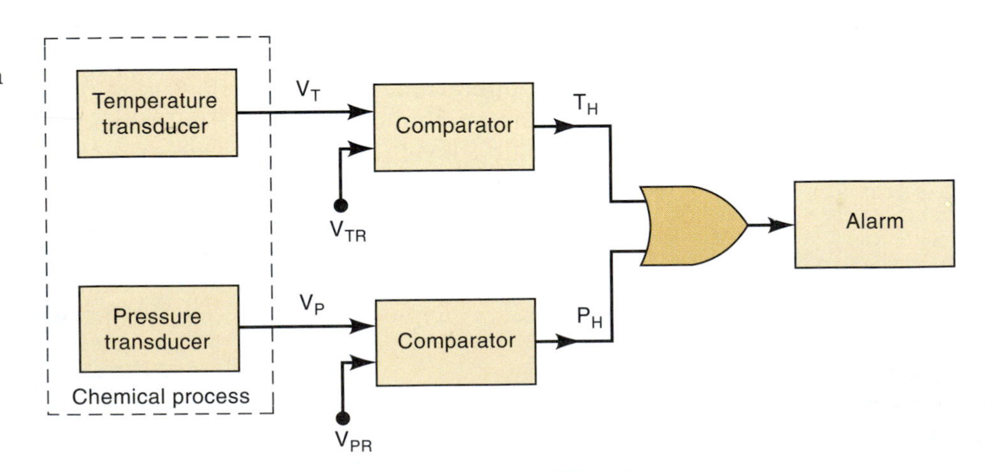

FIGURE 3-4 Example of the use of an OR gate in an alarm system.

Since we want the alarm to be activated when either temperature *or* pressure is too high, it should be apparent that the two comparator outputs can be fed to a two-input OR gate. The OR gate output thus goes HIGH (1) for either alarm condition and will activate the alarm. This same idea can obviously be extended to situations with more than two process variables.

EXAMPLE 3-2

Determine the OR gate output in Figure 3-5. The OR gate inputs A and B are varying according to the timing diagrams shown. For example, A starts out LOW at time t_0, goes HIGH at t_1, back to LOW at t_3, and so on.

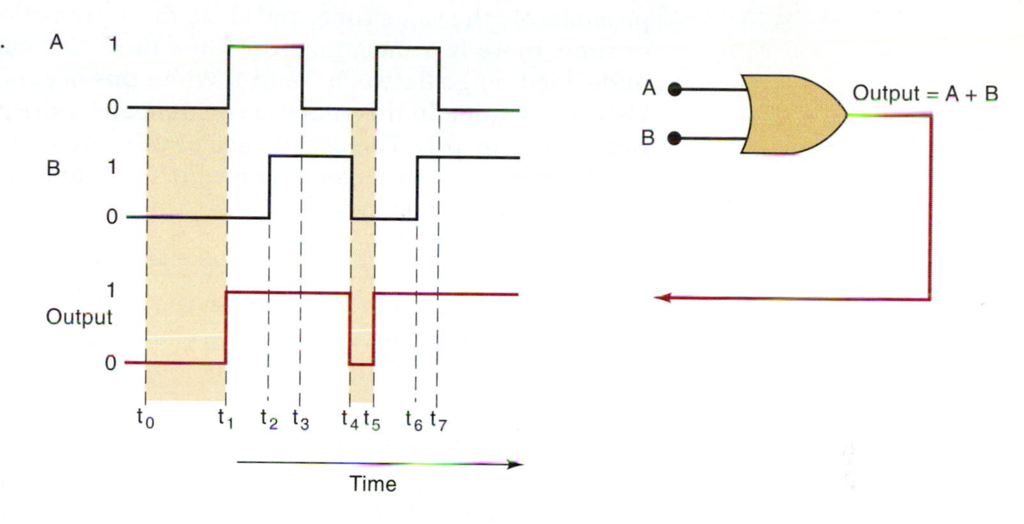

FIGURE 3-5 Example 3-2.

Solution

The OR gate output will be HIGH whenever *any* input is HIGH. Between time t_0 and t_1, both inputs are LOW, so OUTPUT = LOW. At t_1, input A goes HIGH while B remains LOW. This causes OUTPUT to go HIGH at t_1 and stay HIGH until t_4 because, during this interval, one or both inputs are HIGH. At t_4, input B goes from 1 to 0 so that now both inputs are LOW, and this drives OUTPUT back to LOW. At t_5, A goes HIGH, sending OUTPUT back HIGH, where it stays for the rest of the shown time span.

EXAMPLE 3-3A

For the situation depicted in Figure 3-6, determine the waveform at the OR gate output.

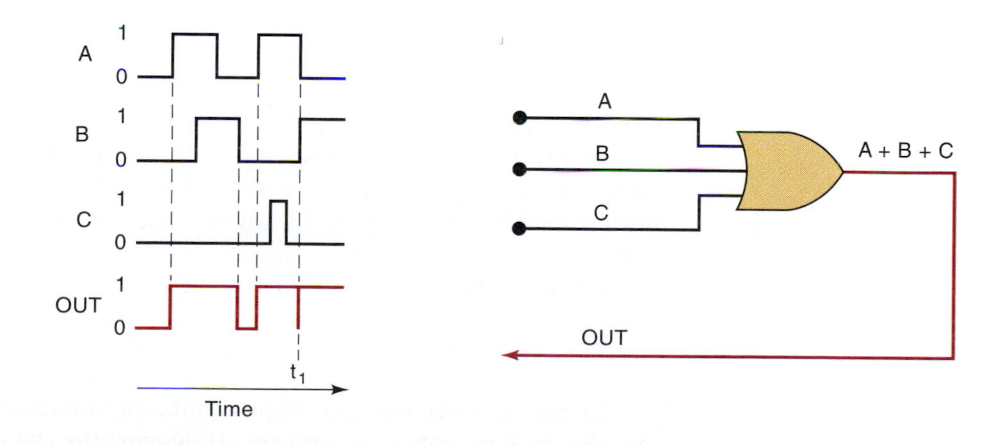

FIGURE 3-6 Examples 3-3A and B.

Solution

The three OR gate inputs A, B, and C are varying, as shown by their waveform diagrams. The OR gate output is determined by realizing that it will be

HIGH whenever *any* of the three inputs is at a HIGH level. Using this reasoning, the OR output waveform is as shown in the figure. Particular attention should be paid to what occurs at time t_1. The diagram shows that, at that instant of time, input A is going from HIGH to LOW while input B is going from LOW to HIGH. Since these inputs are making their transitions at approximately the same time, and since these transitions take a certain amount of time, there is a short interval when these OR gate inputs are both in the undefined range between 0 and 1. When this occurs, the OR gate output also becomes a value in this range, as evidenced by the glitch or spike on the output waveform at t_1. The occurrence of this glitch and its size (amplitude and width) depend on the speed with which the input transitions occur.

EXAMPLE 3-3B

What would happen to the glitch in the output in Figure 3-6 if input C sat in the HIGH state while A and B were changing at time t_1?

Solution

With the C input HIGH at t_1, the OR gate output will remain HIGH, regardless of what is occurring at the other inputs, because any HIGH input will keep an OR gate output HIGH. Therefore, the glitch will not appear in the output.

REVIEW QUESTIONS

1. What is the only set of input conditions that will produce a LOW output for any OR gate?
2. Write the Boolean expression for a six-input OR gate.
3. If the A input in Figure 3-6 is permanently kept at the 1 level, what will the resultant output waveform be?

3-4 AND OPERATION WITH AND GATES

The **AND operation** is the second basic Boolean operation. As an example of the use of AND logic, consider a typical clothes dryer. It is drying clothes (heating, tumbling) only if the *timer is set above zero* AND the *door is closed*. Let's assign A to represent *timer is set*, B to represent *door is closed*, and x can represent the *heater and motor are on*. The truth table in Figure 3-7(a) shows what happens when two logic inputs, A and B, are combined using the AND operation to produce output x. The table shows that x is a logic 1 only when both A and B are at the logic 1 level. For any case where one of the inputs is 0, the output is 0.

The Boolean expression for the AND operation is

$$x = A \cdot B$$

In this expression, the $\cdot$ sign stands for the Boolean AND operation and not the multiplication operation. However, the AND operation on Boolean variables operates the same as ordinary multiplication, as examination of the truth table shows, so we can think of them as being the same. This characteristic can be helpful when evaluating logic expressions that contain AND operations.

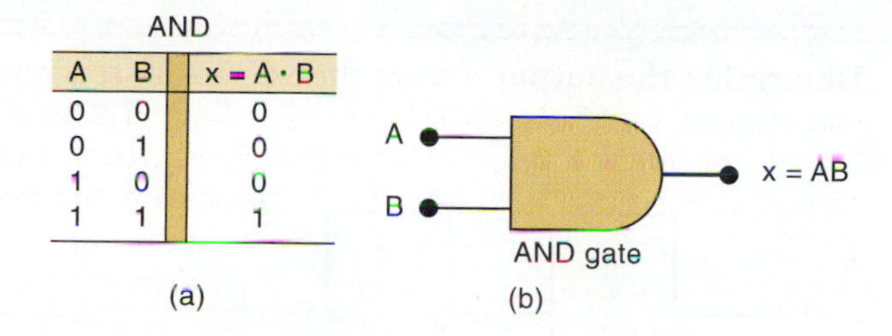

AND

A	B	x = A · B
0	0	0
0	1	0
1	0	0
1	1	1

(a)

FIGURE 3-7 (a) Truth table for the AND operation; (b) AND gate symbol.

The expression $x = A \cdot B$ is read as "x equals A AND B," which means that x will be 1 only when A and B are both 1. The $\cdot$ sign is usually omitted so that the expression simply becomes $x = AB$. For the case when three inputs are ANDed, we have $x = A \cdot B \cdot C = ABC$. This is read as "$x$ equals A AND B AND C," which means that x will be 1 only when A and B and C are all 1.

AND Gate

The logic symbol for a two-input **AND gate** is shown in Figure 3-7(b). The AND gate output is equal to the AND product of the logic inputs; that is, $x = AB$. In other words, the AND gate is a circuit that operates so that its output is HIGH only when all its inputs are HIGH. For all other cases, the AND gate output is LOW.

This same operation is characteristic of AND gates with more than two inputs. For example, a three-input AND gate and its accompanying truth table are shown in Figure 3-8. Once again, note that the gate output is 1 only for the case where $A = B = C = 1$. The expression for the output is $x = ABC$. For a four-input AND gate, the output is $x = ABCD$, and so on.

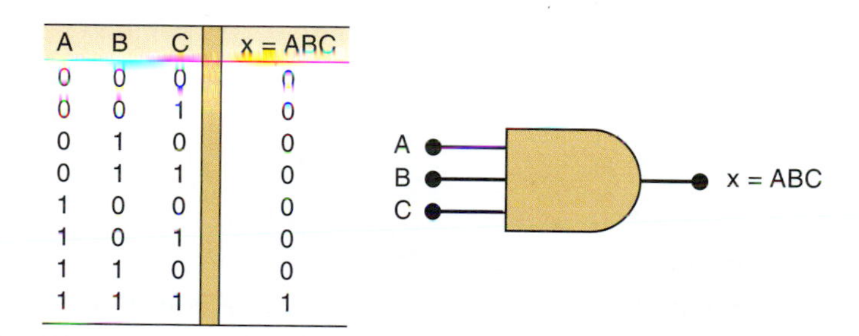

A	B	C	x = ABC
0	0	0	0
0	0	1	0
0	1	0	0
0	1	1	0
1	0	0	0
1	0	1	0
1	1	0	0
1	1	1	1

FIGURE 3-8 Truth table and symbol for a three-input AND gate.

Note the difference between the symbols for the AND gate and the OR gate. Whenever you see the AND symbol on a logic-circuit diagram, it tells you that the output will go HIGH *only* when *all* inputs are HIGH. Whenever you see the OR symbol, it means that the output will go HIGH when *any* input is HIGH.

Summary of the AND Operation

1. The AND operation is performed the same as ordinary multiplication of 1s and 0s.
2. An AND gate is a logic circuit that performs the AND operation on the circuit's inputs.
3. An AND gate output will be 1 *only* for the case when *all* inputs are 1; for all other cases, the output will be 0.
4. The expression $x = AB$ is read as "x equals A AND B."

EXAMPLE 3-4

Determine the output x from the AND gate in Figure 3-9 for the given input waveforms.

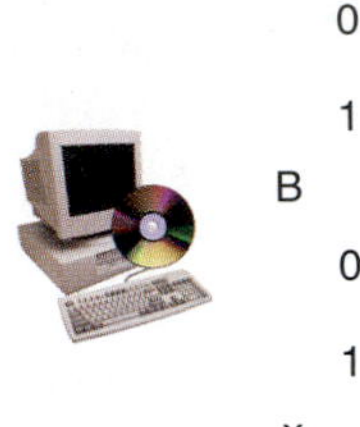

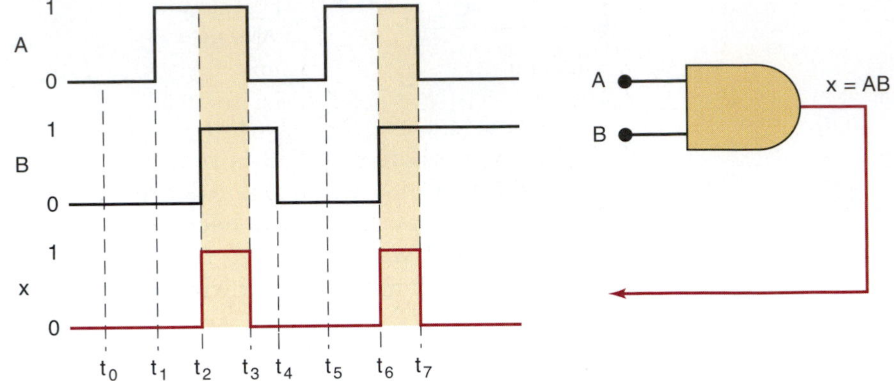

FIGURE 3-9 Example 3-4.

Solution

The output of an AND gate is determined by realizing that it will be HIGH only when all inputs are HIGH at the same time. For the input waveforms given, this condition is met only during intervals t_2-t_3 and t_6-t_7. At all other times, one or more of the inputs are 0, thereby producing a LOW output. Note that input level changes that occur while the other input is LOW have no effect on the output.

EXAMPLE 3-5A

Determine the output waveform for the AND gate shown in Figure 3-10.

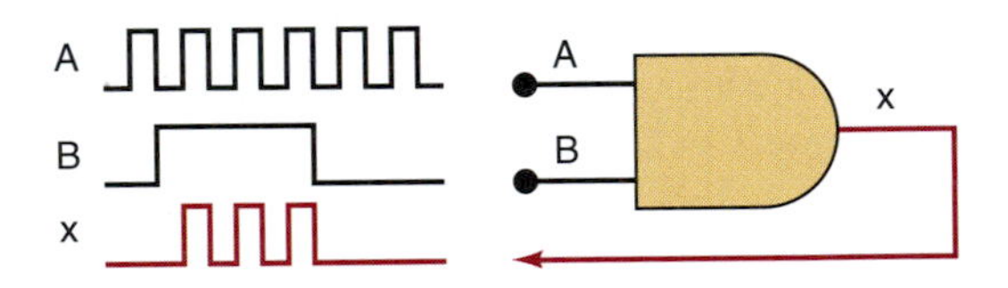

FIGURE 3-10 Examples 3-5A and B.

Solution

The output x will be at 1 only when A and B are both HIGH at the same time. Using this fact, we can determine the x waveform as shown in the figure.

Notice that the x waveform is 0 whenever B is 0, regardless of the signal at A. Also notice that whenever B is 1, the x waveform is the same as A. Thus, we can think of the B input as a *control* input whose logic level determines whether or not the A waveform gets through to the x output. In this situation, the AND gate is used as an *inhibit circuit*. We can say that $B = 0$ is the inhibit condition producing a 0 output. Conversely, $B = 1$ is the *enable* condition, which enables A to reach the output. This inhibit operation is an important application of AND gates, which will be encountered later.

EXAMPLE 3-5B

What will happen to the x output waveform in Figure 3-10 if the B input is kept at the 0 level?

Solution

With B kept LOW, the x output will also stay LOW. This can be reasoned in two different ways. First, with $B = 0$ we have $x = A \cdot B = A \cdot 0 = 0$ because anything multiplied (ANDed) by 0 will be 0. Another way to look at it is that an AND gate requires that all inputs be HIGH for the output to be HIGH, and this cannot happen if B is kept LOW.

REVIEW QUESTIONS

1. What is the only input combination that will produce a HIGH at the output of a five-input AND gate?
2. What logic level should be applied to the second input of a two-input AND gate if the logic signal at the first input is to be inhibited (prevented) from reaching the output?
3. *True or false:* An AND gate output will always differ from an OR gate output for the same input conditions.

3-5 NOT OPERATION

The **NOT operation** is unlike the OR and AND operations because it can be performed on a single input variable. For example, if the variable A is subjected to the NOT operation, the result x can be expressed as

$$x = \overline{A}$$

where the overbar represents the NOT operation. This expression is read as "x equals NOT A" or "x equals the *inverse* of A" or "x equals the *complement* of A." Each of these is in common usage, and all indicate that the logic value of $x = \overline{A}$ *is opposite to* the logic value of A. The truth table in Figure 3-11(a) clarifies this for the two cases $A = 0$ and $A = 1$. That is,

$$0 = \overline{1} \quad \text{because 0 is not 1}$$

and

$$1 = \overline{0} \quad \text{because 1 is not 0}$$

The NOT operation is also referred to as **inversion** or **complementation**, and these terms will be used interchangeably throughout the book. Although we will always use the overbar indicator to represent inversion, it is important to mention that another indicator for inversion is the prime symbol ('). That is,

$$A' = \overline{A}$$

Both should be recognized as indicating the inversion operation.

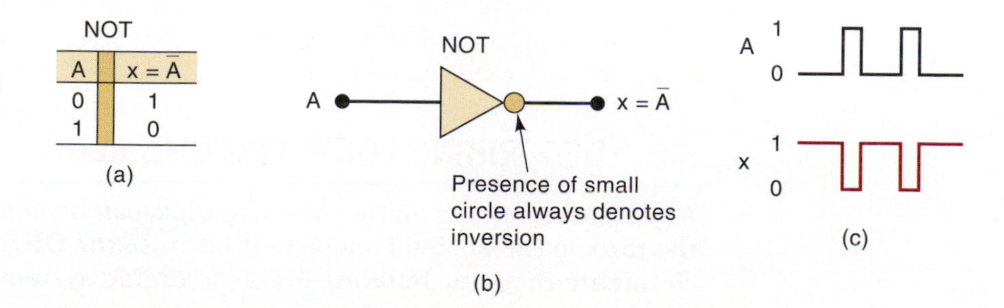

FIGURE 3-11 (a) Truth table; (b) symbol for the INVERTER (NOT circuit); (c) sample waveforms.

NOT Circuit (INVERTER)

Figure 3-11(b) shows the symbol for a **NOT circuit**, which is more commonly called an **INVERTER**. This circuit *always* has only a single input, and its output logic level is always opposite to the logic level of this input. Figure 3-11(c) shows how the INVERTER affects an input signal. It inverts (complements) the input signal at all points on the waveform so that whenever the input = 0, output = 1, and vice versa.

APPLICATION 3-1

Figure 3-12 shows a typical application of the NOT gate. The push button is wired to produce a logic 1 (true) when it is pressed. Sometimes we want to know if the push button is not being pressed, and so this circuit provides an expression that is true when the button is not pressed.

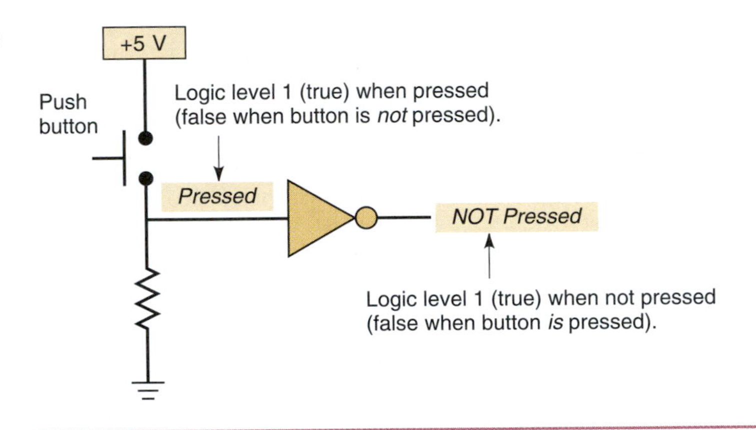

FIGURE 3-12 A NOT gate indicating a button is *not* pressed when its output is true.

Summary of Boolean Operations

The rules for the OR, AND, and NOT operations may be summarized as follows:

OR	*AND*	*NOT*
$0 + 0 = 0$	$0 \cdot 0 = 0$	$\overline{0} = 1$
$0 + 1 = 1$	$0 \cdot 1 = 0$	$\overline{1} = 0$
$1 + 0 = 1$	$1 \cdot 0 = 0$	
$1 + 1 = 1$	$1 \cdot 1 = 1$	

REVIEW QUESTIONS

1. The output of the INVERTER of Figure 3-11 is connected to the input of a second INVERTER. Determine the output level of the second INVERTER for each level of input A.
2. The output of the AND gate in Figure 3-7 is connected to the input of an INVERTER. Write the truth table showing the INVERTER output, y, for each combination of inputs A and B.

3-6 DESCRIBING LOGIC CIRCUITS ALGEBRAICALLY

Any logic circuit, no matter how complex, can be described completely using the three basic Boolean operations because the OR gate, AND gate, and NOT circuit are the basic building blocks of digital systems. For example, consider

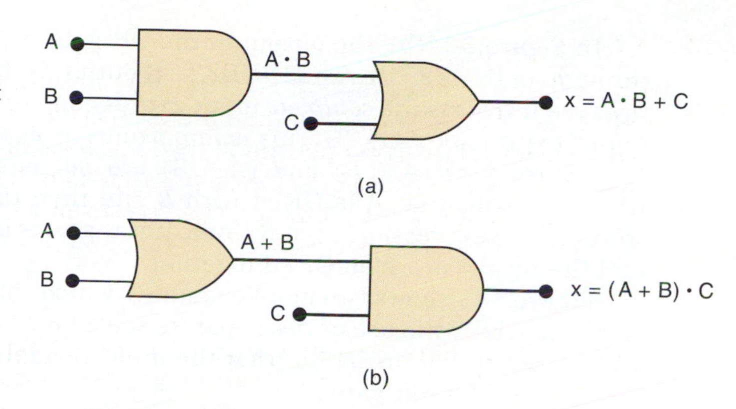

FIGURE 3-13 (a) Logic circuit with its Boolean expression; (b) logic circuit whose expression requires parentheses.

the circuit in Figure 3-13(a). This circuit has three inputs, A, B, and C, and a single output, x. Utilizing the Boolean expression for each gate, we can easily determine the expression for the output.

The expression for the AND gate output is written $A \cdot B$. This AND output is connected as an input to the OR gate along with C, another input. The OR gate operates on its inputs so that its output is the OR sum of the inputs. Thus, we can express the OR output as $x = A \cdot B + C$. (This final expression could also be written as $x = C + A \cdot B$ because it does not matter which term of the OR sum is written first.)

Operator Precedence

Occasionally, there may be confusion about which operation in an expression is performed first. The expression $A \cdot B + C$ can be interpreted in two different ways: (1) $A \cdot B$ is ORed with C, or (2) A is ANDed with the term $B + C$. To avoid this confusion, it will be understood that if an expression contains both AND and OR operations, the AND operations are performed first, unless there are *parentheses* in the expression, in which case the operation inside the parentheses is to be performed first. This is the same rule that is used in ordinary algebra to determine the order of operations.

To illustrate further, consider the circuit in Figure 3-13(b). The expression for the OR gate output is simply $A + B$. This output serves as an input to the AND gate along with another input, C. Thus, we express the output of the AND gate as $x = (A + B) \cdot C$. Note the use of parentheses here to indicate that A and B are ORed *first*, before their OR sum is ANDed with C. Without the parentheses it would be interpreted *incorrectly*, because $A + B \cdot C$ means that A is ORed with the product $B \cdot C$.

Circuits Containing INVERTERs

Whenever an INVERTER is present in a logic-circuit diagram, its output expression is simply equal to the input expression with a bar over it. Figure 3-14 shows two examples using INVERTERs. In Figure 3-14(a), input A is fed through an INVERTER, whose output is therefore $\overline{A}$. The INVERTER output is fed to an OR gate together with B, so that the *OR* output is equal to $\overline{A} + B$. Note that the bar is over the A alone, indicating that A is first inverted and then ORed with B.

FIGURE 3-14 Circuits using INVERTERs.

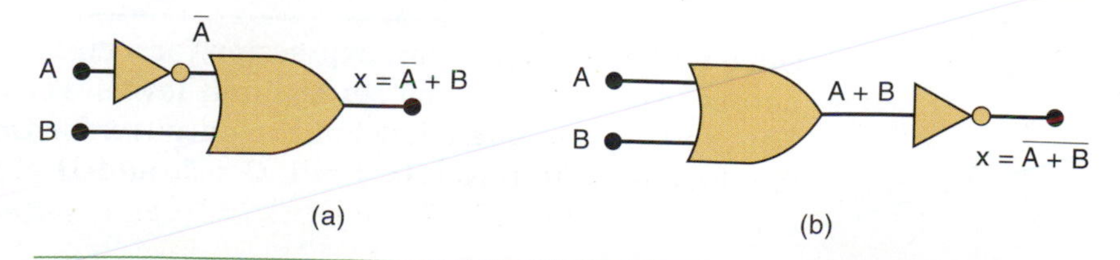

In Figure 3-14(b), the output of the OR gate is equal to $A + B$ and is fed through an INVERTER. The INVERTER output is therefore equal to $(\overline{A + B})$ because it inverts the *complete* input expression. Note that the bar covers the entire expression $(A + B)$. This is important because, as will be shown later, the expressions $(\overline{A + B})$ and $(\overline{A} + \overline{B})$ are *not* equivalent. The expression $(\overline{A + B})$ means that A is ORed with B and then their OR sum is inverted, whereas the expression $(\overline{A} + \overline{B})$ indicates that A is inverted and B is inverted and the results are then ORed together.

Figure 3-15 shows two more examples, which should be studied carefully. Note especially the use of *two* separate sets of parentheses in Figure 3-15(b). Also notice in Figure 3-15(a) that the input variable A is connected as an input to two different gates.

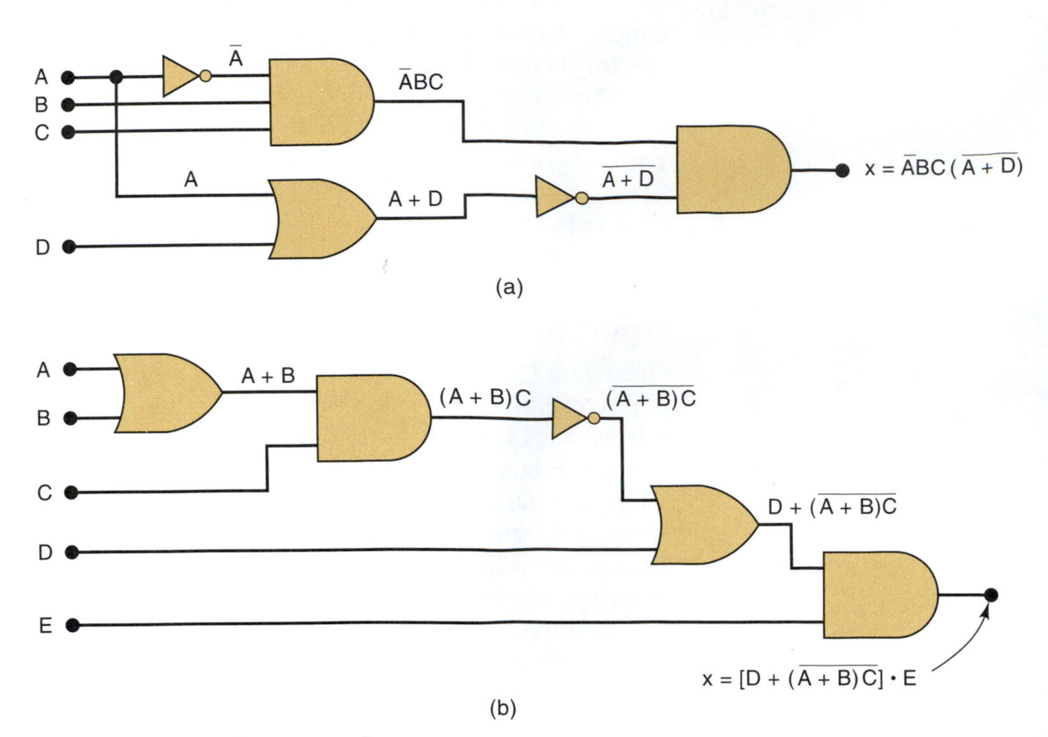

FIGURE 3-15 More examples.

REVIEW QUESTIONS

1. In Figure 3-15(a), change each AND gate to an OR gate, and change the OR gate to an AND gate. Then write the expression for output x.
2. In Figure 3-15(b), change each AND gate to an OR gate, and each OR gate to an AND gate. Then write the expression for x.

3-7 EVALUATING LOGIC-CIRCUIT OUTPUTS

Once we have the Boolean expression for a circuit output, we can obtain the output logic level for any set of input levels. For example, suppose that we want to know the logic level of the output x for the circuit in Figure 3-15(a) for the case where $A = 0, B = 1, C = 1$, and $D = 1$. As in ordinary algebra, the value of x can be found by "plugging" the values of the variables into the expression and performing the indicated operations as follows:

$$\begin{aligned} x &= \bar{A}BC\overline{(A + D)} \\ &= \bar{0}\cdot 1\cdot 1\cdot\overline{(0 + 1)} \\ &= 1\cdot 1\cdot 1\cdot\overline{(0 + 1)} \\ &= 1\cdot 1\cdot 1\cdot(\bar{1}) \\ &= 1\cdot 1\cdot 1\cdot 0 \\ &= 0 \end{aligned}$$

As another illustration, let us evaluate the output of the circuit in Figure 3-15(b) for $A = 0, B = 0, C = 1, D = 1$, and $E = 1$.

$$\begin{aligned} x &= [D + \overline{(A + B)C}]\cdot E \\ &= [1 + \overline{(0 + 0)\cdot 1}]\cdot 1 \\ &= [1 + \overline{0\cdot 1}]\cdot 1 \\ &= [1 + \bar{0}]\cdot 1 \\ &= [1 + 1]\cdot 1 \\ &= 1\cdot 1 \\ &= 1 \end{aligned}$$

In general, the following rules must always be followed when evaluating a Boolean expression:

1. First, perform all inversions of single terms; that is, $\bar{0} = 1$ or $\bar{1} = 0$.
2. Then perform all operations within parentheses.
3. Perform an AND operation before an OR operation unless parentheses indicate otherwise.
4. If an expression has a bar over it, perform the operations inside the expression first and then invert the result.

For practice, determine the outputs of both circuits in Figure 3-15 for the case where all inputs are 1. The answers are $x = 0$ and $x = 1$, respectively.

Analysis Using a Table

Whenever you have a combinational logic circuit and you want to know how it works, the best way to analyze it is to use a truth table. The advantages of this method are:

It allows you to analyze one gate or logic combination at a time.

It allows you to easily double-check your work.

When you are done, you have a table that is of tremendous benefit in troubleshooting the logic circuit.

Recall that a truth table lists all the possible input combinations in numerical order. For each possible input combination, we can determine the logic state at every point (node) in the logic circuit including the output. For example refer to Figure 3-16(a). There are several intermediate nodes in this circuit that are neither inputs nor outputs to the circuit. They are simply connections between one gate's output and another gate's input. In this diagram they have been labeled u, v, and w. The first step after listing all the input combinations is to create a column in the truth table for each intermediate signal (node) as shown in Figure 3-16(b). Node u has been filled in as the complement of A.

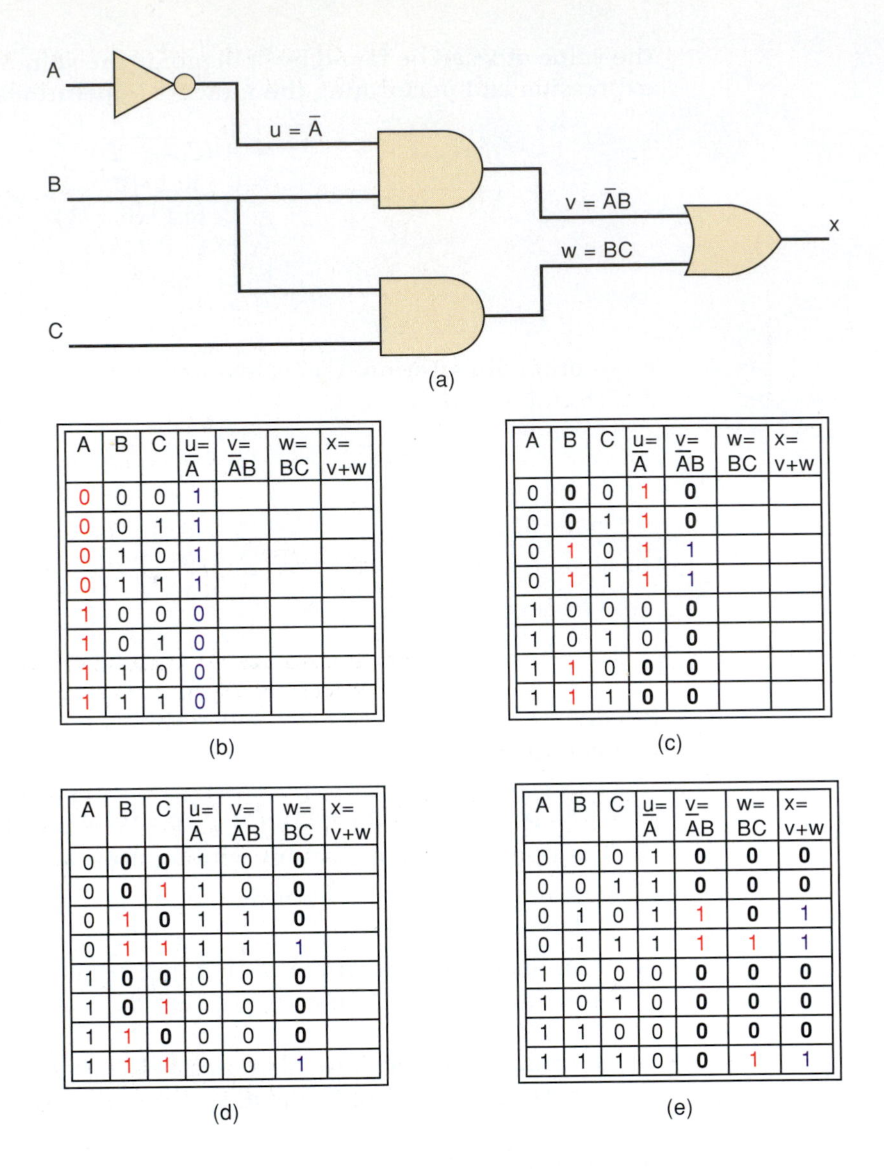

(b)

A	B	C	u= $\overline{A}$	v= $\overline{A}B$	w= BC	x= v+w
0	0	0	1			
0	0	1	1			
0	1	0	1			
0	1	1	1			
1	0	0	0			
1	0	1	0			
1	1	0	0			
1	1	1	0			

(c)

A	B	C	u= $\overline{A}$	v= $\overline{A}B$	w= BC	x= v+w
0	0	0	1	0		
0	0	1	1	0		
0	1	0	1	1		
0	1	1	1	1		
1	0	0	0	0		
1	0	1	0	0		
1	1	0	0	0		
1	1	1	0	0		

(d)

A	B	C	u= $\overline{A}$	v= $\overline{A}B$	w= BC	x= v+w
0	0	0	1	0	0	
0	0	1	1	0	0	
0	1	0	1	1	0	
0	1	1	1	1	1	
1	0	0	0	0	0	
1	0	1	0	0	0	
1	1	0	0	0	0	
1	1	1	0	0	1	

(e)

A	B	C	u= $\overline{A}$	v= $\overline{A}B$	w= BC	x= v+w
0	0	0	1	0	0	0
0	0	1	1	0	0	0
0	1	0	1	1	0	1
0	1	1	1	1	1	1
1	0	0	0	0	0	0
1	0	1	0	0	0	0
1	1	0	0	0	0	0
1	1	1	0	0	1	1

FIGURE 3-16 Analysis of a logic circuit using truth tables.

The next step is to fill in the values for column v as shown in Figure 3-16(c). From the diagram we can see that $v = \overline{A}B$. The node v should be HIGH when $\overline{A}$ (node u) is HIGH AND B is HIGH. This occurs whenever A is LOW AND B is HIGH. The third step is to predict the values at node w which is the logical product of BC. This column is HIGH whenever B is HIGH AND C is HIGH as shown in Figure 3-16(d). The final step is to logically combine columns v and w to predict the output x. Since $x = v + w$, the x output will be HIGH when v is HIGH OR w is HIGH as shown in Figure 3-16(e).

If you built this circuit and it was not producing the correct output for x under all conditions, this table could be used to find the trouble. The general procedure is to test the circuit under each combination of inputs. If any input combination produces an incorrect output (i.e., a fault), compare the actual logic state of each intermediate node in the circuit with the correct theoretical value in the table while applying that input condition. If the logic state for an intermediate node is *correct*, the problem must be farther to the right of that node. If the logic state for an intermediate node is *incorrect*, the problem must be to the left of that node (or that node is shorted to something). Detailed troubleshooting procedures and possible circuit faults will be covered more extensively in Chapter 4.

EXAMPLE 3-6

Analyze the operation of Figure 3-15(a) by creating a table showing the logic state at each node of the circuit.

Solution

Fill in the column for t by entering a 1 for all entries where $A = 0$ and $B = 1$ and $C = 1$.

Fill in the column for u by entering a 1 for all entries where $A = 1$ or $D = 1$.

Fill in the column for v by complementing all entries in column u.

Fill in the column for x by entering a 1 for all entries where $t = 1$ and $v = 1$.

A	B	C	D	$t = \overline{A}BC$	$u = A + D$	$v = \overline{A + D}$	$x = tv$
0	0	0	0	0	0	1	0
0	0	0	1	0	1	0	0
0	0	1	0	0	0	1	0
0	0	1	1	0	1	0	0
0	1	0	0	0	0	1	0
0	1	0	1	0	1	0	0
0	1	1	0	1	0	1	1
0	1	1	1	1	1	0	0
1	0	0	0	0	1	0	0
1	0	0	1	0	1	0	0
1	0	1	0	0	1	0	0
1	0	1	1	0	1	0	0
1	1	0	0	0	1	0	0
1	1	0	1	0	1	0	0
1	1	1	0	0	1	0	0
1	1	1	1	0	1	0	0

REVIEW QUESTIONS

1. Use the expression for x to determine the output of the circuit in Figure 3-15(a) for the conditions $A = 0, B = 1, C = 1$, and $D = 0$.
2. Use the expression for x to determine the output of the circuit in Figure 3-15(b) for the conditions $A = B = E = 1, C = D = 0$.
3. Determine the answers to Questions 1 and 2 by finding the logic levels present at each gate output using a table as in Figure 3-16.

3-8 IMPLEMENTING CIRCUITS FROM BOOLEAN EXPRESSIONS

When the operation of a circuit is defined by a Boolean expression, we can draw a logic-circuit diagram directly from that expression. For example, if we needed a circuit that was defined by $x = A \cdot B \cdot C$, we would immediately know that all that was needed was a three-input AND gate. If we needed a circuit that was defined by $x = A + \overline{B}$, we would use a two-input OR gate with an INVERTER on one of the inputs. The same reasoning used for these simple cases can be extended to more complex circuits.

Suppose that we wanted to construct a circuit whose output is $y = AC + B\overline{C} + \overline{A}BC$. This Boolean expression contains three terms (AC, $B\overline{C}$, $\overline{A}BC$), which are ORed together. This tells us that a three-input OR gate is required with inputs that are equal to AC, $B\overline{C}$, and $\overline{A}BC$. This is illustrated in Figure 3-17(a), where a three-input OR gate is drawn with inputs labeled as AC, $B\overline{C}$, and $\overline{A}BC$.

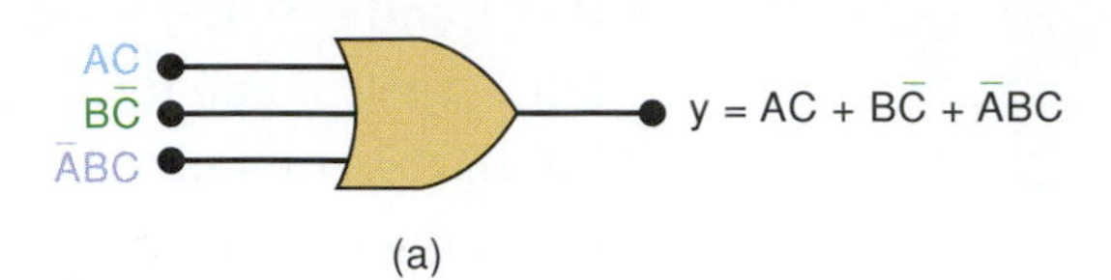

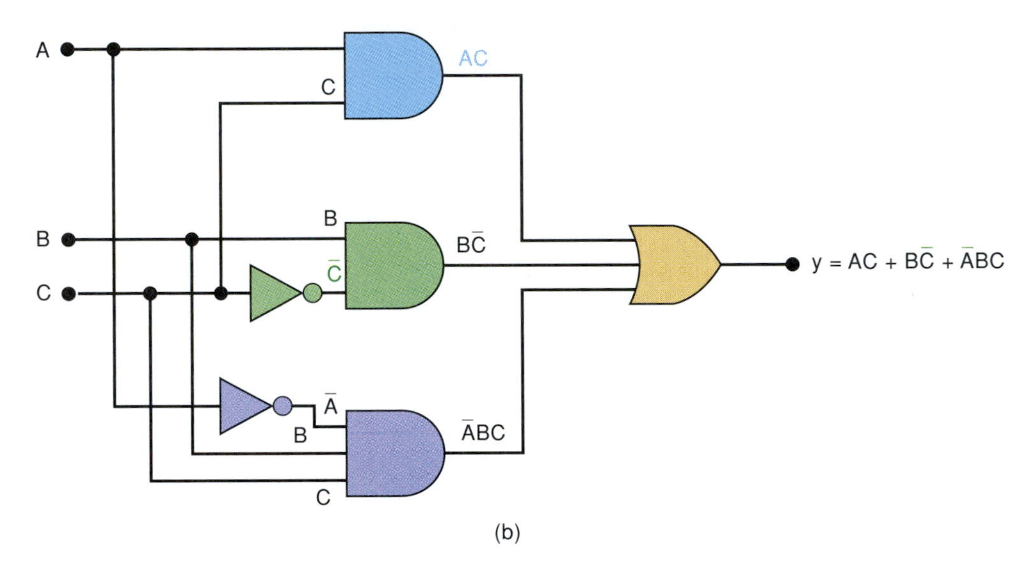

FIGURE 3-17 Constructing a logic circuit from a Boolean expression.

Each OR gate input is an AND product term, which means that an AND gate with appropriate inputs can be used to generate each of these terms. This is shown in Figure 3-17(b), which is the final circuit diagram. Note the use of INVERTERs to produce the $\overline{A}$ and $\overline{C}$ terms required in the expression.

This same general approach can always be followed, although we shall find that there are some clever, more efficient techniques that can be employed. For now, however, this straightforward method will be used to minimize the number of new items that are to be learned.

EXAMPLE 3-7

Draw the circuit diagram to implement the expression $x = (A + B)(\overline{B} + C)$.

Solution

This expression shows that the terms $A + B$ and $\overline{B} + C$ are inputs to an AND gate, and each of these two terms is generated from a separate OR gate. The result is drawn in Figure 3-18.

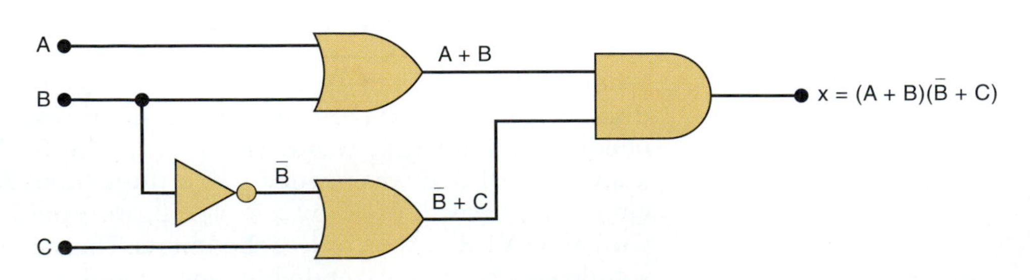

FIGURE 3-18 Example 3-7.

REVIEW QUESTIONS

1. Draw the circuit diagram that implements the expression $x = \overline{A}BC(\overline{A + D})$ using gates with no more than three inputs.
2. Draw the circuit diagram for the expression $y = AC + B\overline{C} + \overline{A}BC$.
3. Draw the circuit diagram for $x = [D + \overline{(A + B)C}]\cdot E$.

3-9 NOR GATES AND NAND GATES

Two other types of logic gates, NOR gates and NAND gates, are widely used in digital circuits. These gates actually combine the basic AND, OR, and NOT operations, so it is a relatively simple matter to write their Boolean expressions.

NOR Gate

The symbol for a two-input **NOR gate** is shown in Figure 3-19(a). It is the same as the OR gate symbol except that it has a small circle on the output. The small circle represents the inversion operation. Thus, the NOR gate operates like an OR gate followed by an INVERTER, so that the circuits in Figure 3-19(a) and (b) are equivalent, and the output expression for the NOR gate is $x = \overline{A + B}$.

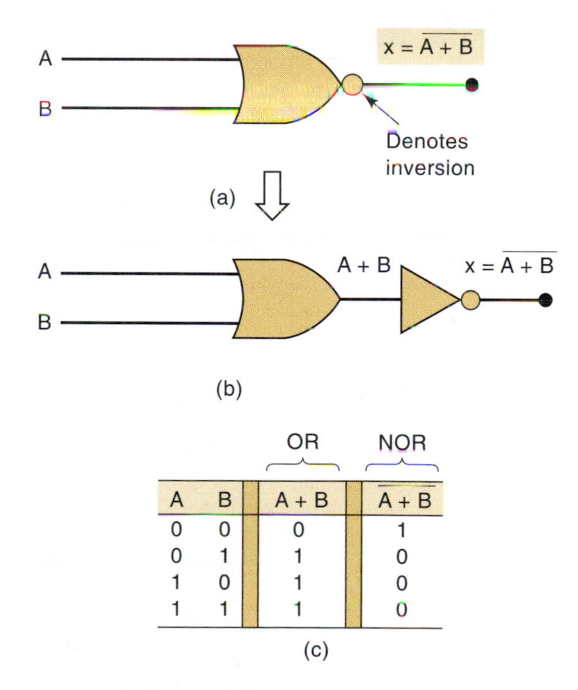

A	B	A + B	$\overline{A + B}$
0	0	0	1
0	1	1	0
1	0	1	0
1	1	1	0

FIGURE 3-19 (a) NOR symbol; (b) equivalent circuit; (c) truth table.

The truth table in Figure 3-19(c) shows that the NOR gate output is the exact inverse of the OR gate output for all possible input conditions. An OR gate output goes HIGH when any input is HIGH; the NOR gate output goes LOW when any input is HIGH. This same operation can be extended to NOR gates with more than two inputs.

EXAMPLE 3-8

Determine the waveform at the output of a NOR gate for the input waveforms shown in Figure 3-20.

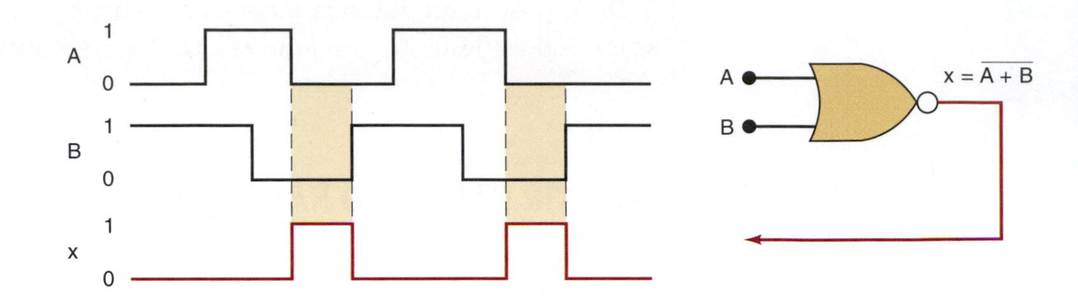

FIGURE 3-20 Example 3-8.

Solution

One way to determine the NOR output waveform is to find first the OR output waveform and then invert it (change all 1s to 0s, and vice versa). Another way utilizes the fact that a NOR gate output will be HIGH *only* when all inputs are LOW. Thus, you can examine the input waveforms, find those time intervals where they are all LOW, and make the NOR output HIGH for those intervals. The NOR output will be LOW for all other time intervals. The resultant output waveform is shown in the figure.

EXAMPLE 3-9

Determine the Boolean expression for a three-input NOR gate followed by an INVERTER.

Solution

Refer to Figure 3-21, where the circuit diagram is shown. The expression at the NOR output is $(\overline{A + B + C})$, which is then fed through an INVERTER to produce

$$x = (\overline{\overline{A + B + C}})$$

The presence of the double inversion signs indicates that the quantity $(A + B + C)$ has been inverted and then inverted again. It should be clear that this simply results in the expression $(A + B + C)$ being unchanged. That is,

$$x = (\overline{\overline{A + B + C}}) = (A + B + C)$$

Whenever two inversion bars are over the same variable or quantity, they cancel each other out, as in the example above. However, in cases such as $\overline{\overline{A} + \overline{B}}$ the inversion bars do not cancel. This is because the smaller inversion bars invert the single variables A and B, while the wide bar inverts the quantity $(\overline{A} + \overline{B})$. Thus, $\overline{\overline{A} + \overline{B}} \neq A + B$. Similarly, $\overline{\overline{A}\,\overline{B}} \neq AB$.

FIGURE 3-21 Example 3-9.

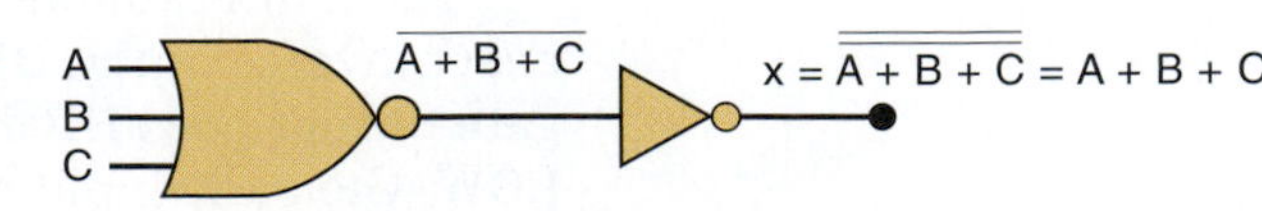

NAND Gate

The symbol for a two-input **NAND gate** is shown in Figure 3-22(a). It is the same as the AND gate symbol except for the small circle on its output. Once again, this small circle denotes the inversion operation. Thus, the NAND operates like an AND gate followed by an INVERTER, so that the circuits of Figure 3-22(a) and (b) are equivalent, and the output expression for the NAND gate is $x = \overline{AB}$.

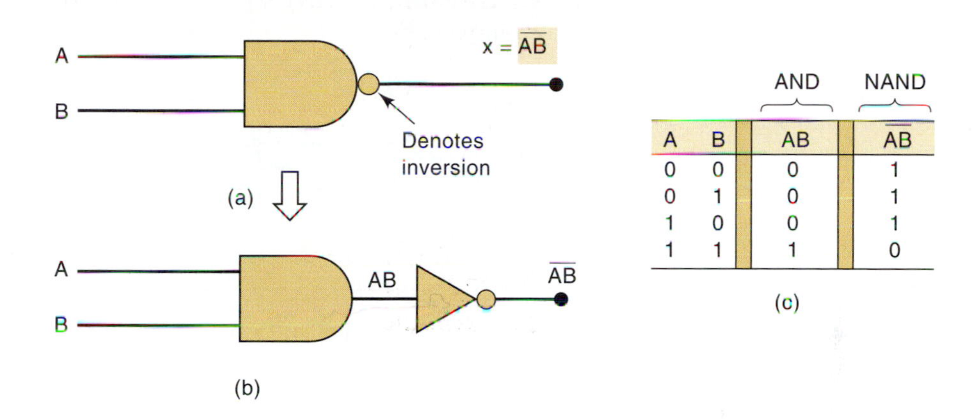

A	B	AB	$\overline{AB}$
0	0	0	1
0	1	0	1
1	0	0	1
1	1	1	0

FIGURE 3-22 (a) NAND symbol; (b) equivalent circuit; (c) truth table.

The truth table in Figure 3-22(c) shows that the NAND gate output is the exact inverse of the AND gate for all possible input conditions. The AND output goes HIGH only when all inputs are HIGH, while the NAND output goes LOW only when all inputs are HIGH. This same characteristic is true of NAND gates having more than two inputs.

EXAMPLE 3-10

Determine the output waveform of a NAND gate having the inputs shown in Figure 3-23.

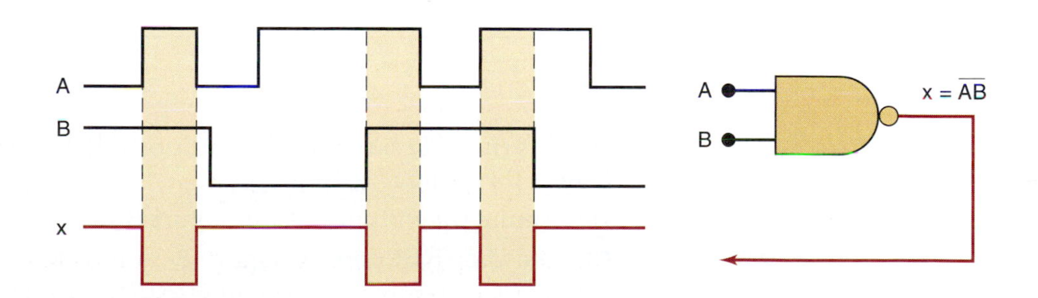

FIGURE 3-23 Example 3-10.

Solution

One way is to draw first the output waveform for an AND gate and then invert it. Another way utilizes the fact that a NAND output will be LOW only when all inputs are HIGH. Thus, you can find those time intervals during which the inputs are all HIGH, and make the NAND output LOW for those intervals. The output will be HIGH at all other times.

EXAMPLE 3-11

Implement the logic circuit that has the expression $x = \overline{AB \cdot (\overline{C + D})}$ using only NOR and NAND gates.

Solution

The $(\overline{C + D})$ term is the expression for the output of a NOR gate. This term is ANDed with A and B, and the result is inverted; this, of course, is the NAND operation. Thus, the circuit is implemented as shown in Figure 3-24. Note that the NAND gate first ANDs the A, B, and $(\overline{C + D})$ terms, and then it inverts the *complete* result.

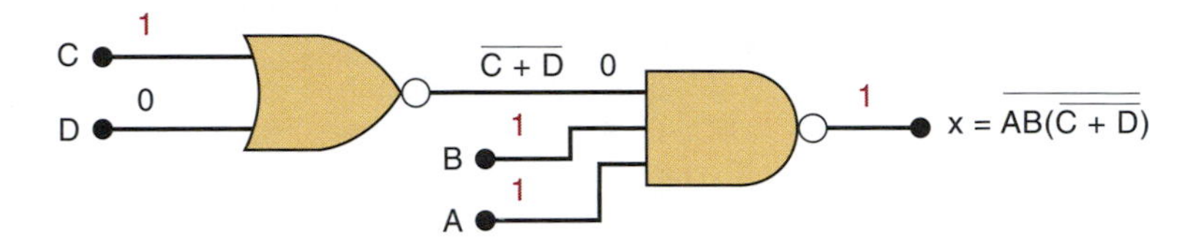

FIGURE 3-24 Examples 3-11 and 3-12.

EXAMPLE 3-12

Determine the output level in Figure 3-24 for $A = B = C = 1$ and $D = 0$.

Solution

In the first method we use the expression for x.

$$\begin{aligned} x &= \overline{AB(\overline{C + D})} \\ &= \overline{1 \cdot 1 \cdot (\overline{1 + 0})} \\ &= \overline{1 \cdot 1 \cdot (\overline{1})} \\ &= \overline{1 \cdot 1 \cdot 0} \\ &= \overline{0} = 1 \end{aligned}$$

In the second method, we write down the input logic levels on the circuit diagram (shown in color in Figure 3-24) and follow these levels through each gate to the final output. The NOR gate has inputs of 1 and 0 to produce an output of 0 (an OR would have produced an output of 1). The NAND gate thus has input levels of 0, 1, and 1 to produce an output of 1 (an AND would have produced an output of 0).

REVIEW QUESTIONS

1. What is the only set of input conditions that will produce a HIGH output from a three-input NOR gate?
2. Determine the output level in Figure 3-24 for $A = B = 1, C = D = 0$.
3. Change the NOR gate of Figure 3-24 to a NAND gate, and change the NAND to a NOR. What is the new expression for x?

3-10 BOOLEAN THEOREMS

We have seen how Boolean algebra can be used to help analyze a logic circuit and express its operation mathematically. We will continue our study of Boolean algebra by investigating the various **Boolean theorems** (rules) that can help us to simplify logic expressions and logic circuits. The first group of theorems is given in Figure 3-25. In each theorem, x is a logic variable that

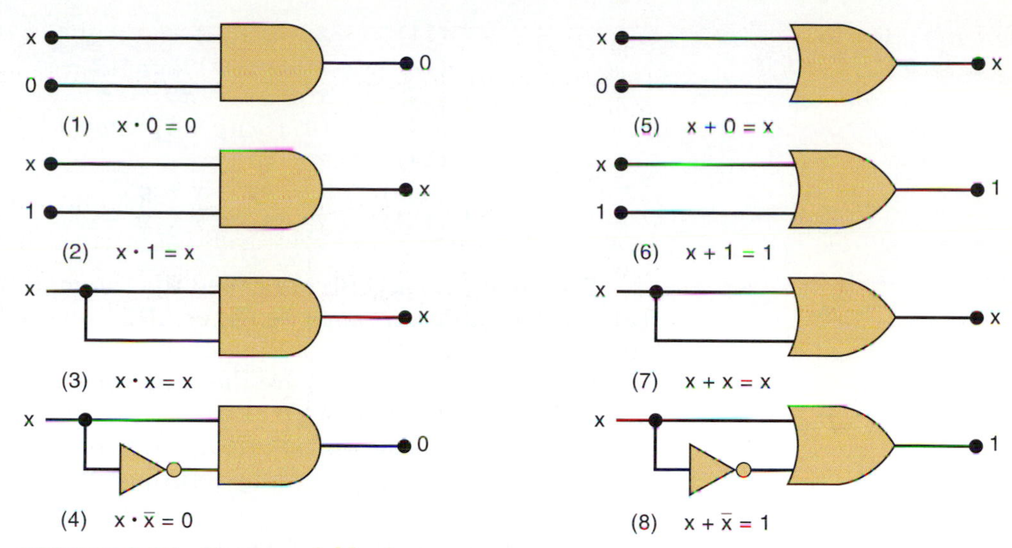

FIGURE 3-25 Single-variable theorems.

can be either a 0 or a 1. Each theorem is accompanied by a logic-circuit diagram that demonstrates its validity.

Theorem (1) states that if any variable is ANDed with 0, the result must be 0. This is easy to remember because the AND operation is just like ordinary multiplication, where we know that anything multiplied by 0 is 0. We also know that the output of an AND gate will be 0 whenever any input is 0, regardless of the level on the other input.

Theorem (2) is also obvious by comparison with ordinary multiplication.

Theorem (3) can be proved by trying each case. If $x = 0$, then $0 \cdot 0 = 0$; if $x = 1$, then $1 \cdot 1 = 1$. Thus, $x \cdot x = x$.

Theorem (4) can be proved in the same manner. However, it can also be reasoned that at any time either x or its inverse $\overline{x}$ must be at the 0 level, and so their AND product always must be 0.

Theorem (5) is straightforward, since 0 *added* to anything does not affect its value, either in regular addition or in OR addition.

Theorem (6) states that if any variable is ORed with 1, the result will always be 1. We check this for both values of x: $0 + 1 = 1$ and $1 + 1 = 1$. Equivalently, we can remember that an OR gate output will be 1 when *any* input is 1, regardless of the value of the other input.

Theorem (7) can be proved by checking for both values of x: $0 + 0 = 0$ and $1 + 1 = 1$.

Theorem (8) can be proved similarly, or we can just reason that at any time either x or $\overline{x}$ must be at the 1 level so that we are always ORing a 0 and a 1, which always results in 1.

Before introducing any more theorems, we should point out that when theorems (1) through (8) are applied, the variable x may actually represent an expression containing more than one variable. For example, if we have $A\overline{B}(\overline{A\overline{B}})$, we can invoke theorem (4) by letting $x = A\overline{B}$. Thus, we can say that $A\overline{B}(\overline{A\overline{B}}) = 0$. The same idea can be applied to the use of any of these theorems.

Multivariable Theorems

The theorems presented below involve more than one variable:

$$(9) \qquad x + y = y + x$$

$$(10) \qquad x \cdot y = y \cdot x$$

(11)	$x + (y + z) = (x + y) + z = x + y + z$
(12)	$x(yz) = (xy)z = xyz$
(13*a*)	$x(y + z) = xy + xz$
(13*b*)	$(w + x)(y + z) = wy + xy + wz + xz$
(14)	$x + xy = x$
(15*a*)	$x + \overline{x}y = x + y$
(15*b*)	$\overline{x} + xy = \overline{x} + y$

Theorems (9) and (10) are called the *commutative laws.* These laws indicate that the order in which we OR or AND two variables is unimportant; the result is the same.

Theorems (11) and (12) are the *associative laws,* which state that we can group the variables in an AND expression or OR expression any way we want.

Theorem (13) is the *distributive law,* which states that an expression can be expanded by multiplying term by term just the same as in ordinary algebra. This theorem also indicates that we can factor an expression. That is, if we have a sum of two (or more) terms, each of which contains a common variable, the common variable can be factored out just as in ordinary algebra. For example, if we have the expression $A\overline{B}C + \overline{A}\,\overline{B}\,\overline{C}$, we can factor out the $\overline{B}$ variable:

$$A\overline{B}C + \overline{A}\,\overline{B}\,\overline{C} = \overline{B}(AC + \overline{A}\,\overline{C})$$

As another example, consider the expression $ABC + ABD$. Here the two terms have the variables A and B in common, and so $A \cdot B$ can be factored out of both terms. That is,

$$ABC + ABD = AB(C + D)$$

Theorems (9) to (13) are easy to remember and use because they are identical to those of ordinary algebra. Theorems (14) and (15), on the other hand, do not have any counterparts in ordinary algebra. Each can be proved by trying all possible cases for x and y. This is illustrated (for theorem 14) by creating an analysis table for the equation $x + xy$ as follows:

x	y	xy	x + xy
0	0	0	0
0	1	0	0
1	0	0	1
1	1	1	1

Notice that the value of the entire expression $(x + xy)$ is always the same as x.

Theorem (14) can also be proved by factoring and using theorems (6) and (2) as follows:

$$\begin{aligned} x + xy &= x(1 + y) & \\ &= x \cdot 1 & \text{[using theorem (6)]} \\ &= x & \text{[using theorem (2)]} \end{aligned}$$

All of these Boolean theorems can be useful in simplifying a logic expression—that is, in reducing the number of terms in the expression. When this is done, the reduced expression will produce a circuit that is less complex than the one that the original expression would have produced. A good portion of the next chapter will be devoted to the process of circuit simplification. For now, the following examples will serve to illustrate how the Boolean theorems can be applied. **Note:** You can find all the Boolean theorems on the inside back cover.

EXAMPLE 3-13

Simplify the expression $y = A\overline{B}D + A\overline{B}\,\overline{D}$.

Solution

Factor out the common variables $A\overline{B}$ using theorem (13):

$$y = A\overline{B}(D + \overline{D})$$

Using theorem (8), the term in parentheses is equivalent to 1. Thus,

$$\begin{aligned} y &= A\overline{B} \cdot 1 \\ &= A\overline{B} \qquad \text{[using theorem (2)]} \end{aligned}$$

EXAMPLE 3-14

Simplify $z = (\overline{A} + B)(A + B)$.

Solution

The expression can be expanded by multiplying out the terms [theorem (13)]:

$$z = \overline{A} \cdot A + \overline{A} \cdot B + B \cdot A + B \cdot B$$

Invoking theorem (4), the term $\overline{A} \cdot A = 0$. Also, $B \cdot B = B$ [theorem (3)]:

$$z = 0 + \overline{A} \cdot B + B \cdot A + B = \overline{A}B + AB + B$$

Factoring out the variable B [theorem (13)], we have

$$z = B(\overline{A} + A + 1)$$

Finally, using theorems (2) and (6),

$$z = B$$

EXAMPLE 3-15

Simplify $x = ACD + \overline{A}BCD$.

Solution

Factoring out the common variables CD, we have

$$x = CD(A + \overline{A}B)$$

Utilizing theorem (15a), we can replace $A + \overline{A}B$ by $A + B$, so

$$\begin{aligned} x &= CD(A + B) \\ &= ACD + BCD \end{aligned}$$

REVIEW QUESTIONS

1. Use theorems (13) and (14) to simplify $y = A\overline{C} + AB\overline{C}$.
2. Use theorems (13) and (8) to simplify $y = \overline{A}\,\overline{B}C\overline{D} + \overline{A}\,\overline{B}\,\overline{C}\,\overline{D}$.
3. Use theorems (13) and (15b) to simplify $y = \overline{A}D + ABD$.

3-11 DEMORGAN'S THEOREMS

Two of the most important theorems of Boolean algebra were contributed by a great mathematician named DeMorgan. **DeMorgan's theorems** are extremely useful in simplifying expressions in which a product or sum of variables is inverted. The two theorems are:

$$(16) \quad (\overline{x + y}) = \overline{x} \cdot \overline{y}$$
$$(17) \quad (\overline{x \cdot y}) = \overline{x} + \overline{y}$$

Theorem (16) says that when the OR sum of two variables is inverted, this is the same as inverting each variable individually and then ANDing these inverted variables. Theorem (17) says that when the AND product of two variables is inverted, this is the same as inverting each variable individually and then ORing them. Each of DeMorgan's theorems can readily be proven by checking for all possible combinations of x and y. This will be left as an end-of-chapter exercise.

Although these theorems have been stated in terms of single variables x and y, they are equally valid for situations where x and/or y are expressions that contain more than one variable. For example, let's apply them to the expression $(\overline{A\overline{B} + C})$ as shown below:

$$(\overline{A\overline{B} + C}) = (\overline{A\overline{B}}) \cdot \overline{C}$$

Note that we used theorem (16) and treated $A\overline{B}$ as x and C as y. The result can be further simplified because we have a product $A\overline{B}$ that is inverted. Using theorem (17), the expression becomes

$$\overline{A\overline{B}} \cdot \overline{C} = (\overline{A} + \overline{\overline{B}}) \cdot \overline{C}$$

Notice that we can replace $\overline{\overline{B}}$ by B, so that we finally have

$$(\overline{A} + B) \cdot \overline{C} = \overline{A}\,\overline{C} + B\overline{C}$$

This final result contains only inverter signs that invert a single variable.

EXAMPLE 3-16

Simplify the expression $z = \overline{(\overline{A} + C) \cdot (B + \overline{D})}$ to one having only single variables inverted.

Solution

Using theorem (17), and treating $(\overline{A} + C)$ as x and $(B + \overline{D})$ as y, we have

$$z = (\overline{\overline{A} + C}) + (\overline{B + \overline{D}})$$

We can think of this as breaking the large inverter sign down the middle and changing the AND sign (·) to an OR sign (+). Now the term $\overline{(\overline{A} + C)}$ can be simplified by applying theorem (16). Likewise, $\overline{(B + \overline{D})}$ can be simplified:

$$
\begin{aligned}
z &= \overline{(\overline{A} + C)} + \overline{(B + \overline{D})} \\
&= (\overline{\overline{A}} \cdot \overline{C}) + \overline{B} \cdot \overline{\overline{D}}
\end{aligned}
$$

Here we have broken the larger inverter signs down the middle and replaced the (+) with a (·). Canceling out the double inversions, we have finally

$$z = \boldsymbol{A\overline{C} + \overline{B}D}$$

Example 3-16 points out that when using DeMorgan's theorems to reduce an expression, we may break an inverter sign at any point in the expression and change the operator sign at that point in the expression to its opposite (+ is changed to ·, and vice versa). This procedure is continued until the expression is reduced to one in which only single variables are inverted. Two more examples are given below.

Example 1

$$
\begin{aligned}
z &= \overline{A + \overline{B} \cdot C} \\
&= \overline{A} \cdot (\overline{\overline{B} \cdot C}) \\
&= \overline{A} \cdot (\overline{\overline{B}} + \overline{C}) \\
&= \overline{A} \cdot (B + \overline{C})
\end{aligned}
$$

Example 2

$$
\begin{aligned}
\omega &= \overline{(A + BC) \cdot (D + EF)} \\
&= \overline{(A + BC)} + \overline{(D + EF)} \\
&= (\overline{A} \cdot \overline{BC}) + (\overline{D} \cdot \overline{EF}) \\
&= [\overline{A} \cdot (\overline{B} + \overline{C})] + [\overline{D} \cdot (\overline{E} + \overline{F})] \\
&= \overline{A}\,\overline{B} + \overline{A}\,\overline{C} + \overline{D}\,\overline{E} + \overline{D}\,\overline{F}
\end{aligned}
$$

DeMorgan's theorems are easily extended to more than two variables. For example, it can be proved that

$$
\begin{aligned}
\overline{x + y + z} &= \overline{x} \cdot \overline{y} \cdot \overline{z} \\
\overline{x \cdot y \cdot z} &= \overline{x} + \overline{y} + \overline{z}
\end{aligned}
$$

Here, we see that the large inverter sign is broken at *two* points in the expression and the operator sign is changed to its opposite. This can be extended to any number of variables. Again, realize that the variables can themselves be expressions rather than single variables. Here is another example.

$$
\begin{aligned}
x &= \overline{\overline{AB} \cdot \overline{CD} \cdot \overline{EF}} \\
&= \overline{\overline{AB}} + \overline{\overline{CD}} + \overline{\overline{EF}} \\
&= AB + CD + EF
\end{aligned}
$$

Implications of DeMorgan's Theorems

Let us examine theorems (16) and (17) from the standpoint of logic circuits. First, consider theorem (16):

$$\overline{x + y} = \overline{x} \cdot \overline{y}$$

The left-hand side of the equation can be viewed as the output of a NOR gate whose inputs are x and y. The right-hand side of the equation, on the other

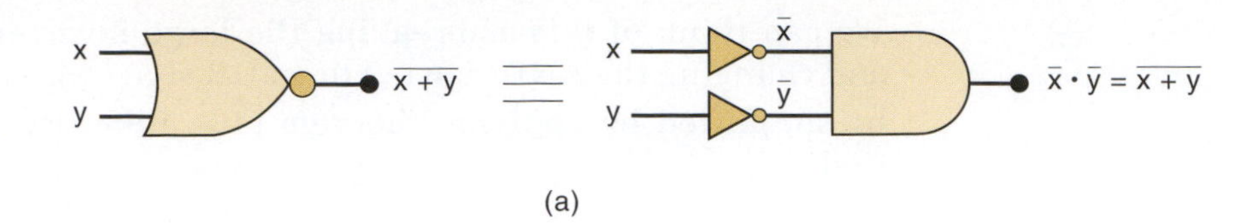

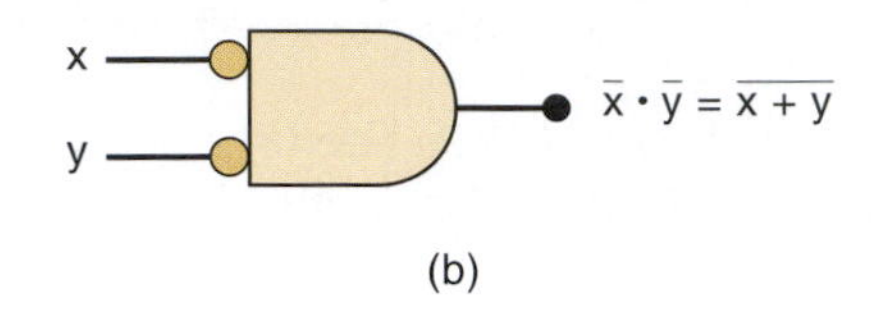

FIGURE 3-26 (a) Equivalent circuits implied by theorem (16); (b) alternative symbol for the NOR function.

hand, is the result of first inverting both x and y and then putting them through an AND gate. These two representations are equivalent and are illustrated in Figure 3-26(a). What this means is that an AND gate with INVERTERs on each of its inputs is equivalent to a NOR gate. In fact, both representations are used to represent the NOR function. When the AND gate with inverted inputs is used to represent the NOR function, it is usually drawn as shown in Figure 3-26(b), where the small circles on the inputs represent the inversion operation.

Now consider theorem (17):

$$\overline{x \cdot y} = \overline{x} + \overline{y}$$

The left side of the equation can be implemented by a NAND gate with inputs x and y. The right side can be implemented by first inverting inputs x and y and then putting them through an OR gate. These two equivalent representations are shown in Figure 3-27(a). The OR gate with INVERTERs on each of its inputs is equivalent to the NAND gate. In fact, both representations are used to represent the NAND function. When the OR gate with inverted inputs is used to represent the NAND function, it is usually drawn as shown in Figure 3-27(b), where the circles again represent inversion.

FIGURE 3-27 (a) Equivalent circuits implied by theorem (17); (b) alternative symbol for the NAND function.

$$x \cdot y \;\equiv\; \overline{x} + \overline{y} = \overline{xy}$$

(a)

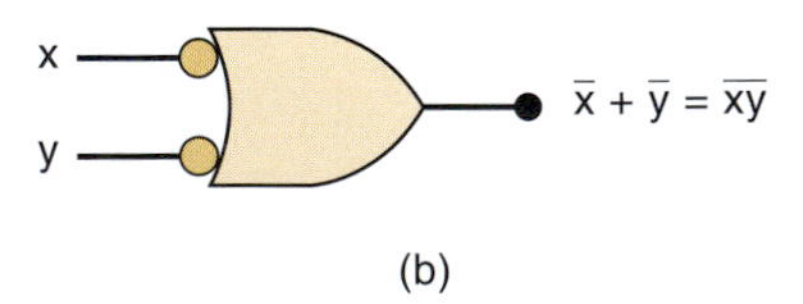

EXAMPLE 3-17

Determine the output expression for the circuit of Figure 3-28 and simplify it using DeMorgan's theorems.

FIGURE 3-28 Example 3-17.

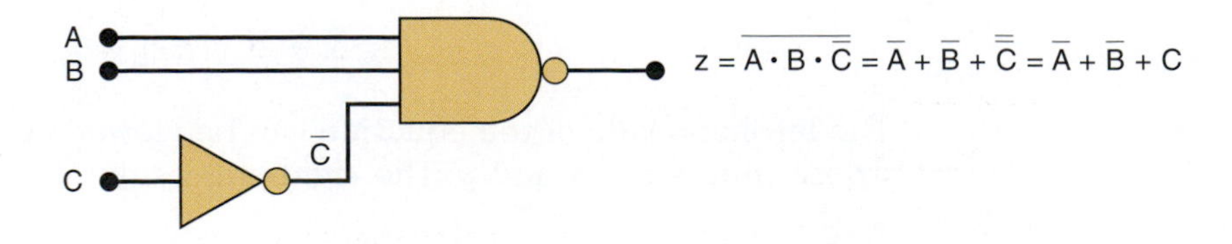

Solution

The expression for z is $z = \overline{A\overline{B}C}$. Use DeMorgan's theorem to break the large inversion sign:

$$z = \overline{A} + \overline{\overline{B}} + \overline{C}$$

Cancel the double inversions over C to obtain

$$z = \overline{A} + B + \overline{C}$$

REVIEW QUESTIONS

1. Use DeMorgan's theorems to convert the expression $z = \overline{(A + B) \cdot \overline{C}}$ to one that has only single-variable inversions.
2. Repeat question 1 for the expression $y = \overline{R\overline{S}T + \overline{Q}}$.
3. Implement a circuit having output expression $z = \overline{A}\,\overline{B}C$ using only a NOR gate and an INVERTER.
4. Use DeMorgan's theorems to convert $y = \overline{\overline{A} + \overline{B} + \overline{C}D}$ to an expression containing only single-variable inversions.

3-12 UNIVERSALITY OF NAND GATES AND NOR GATES

All Boolean expressions consist of various combinations of the basic operations of OR, AND, and INVERT. Therefore, any expression can be implemented using combinations of OR gates, AND gates, and INVERTERs. It is possible, however, to implement any logic expression using *only* NAND gates and no other type of gate. This is because NAND gates, in the proper combination, can be used to perform each of the Boolean operations OR, AND, and INVERT. This is demonstrated in Figure 3-29.

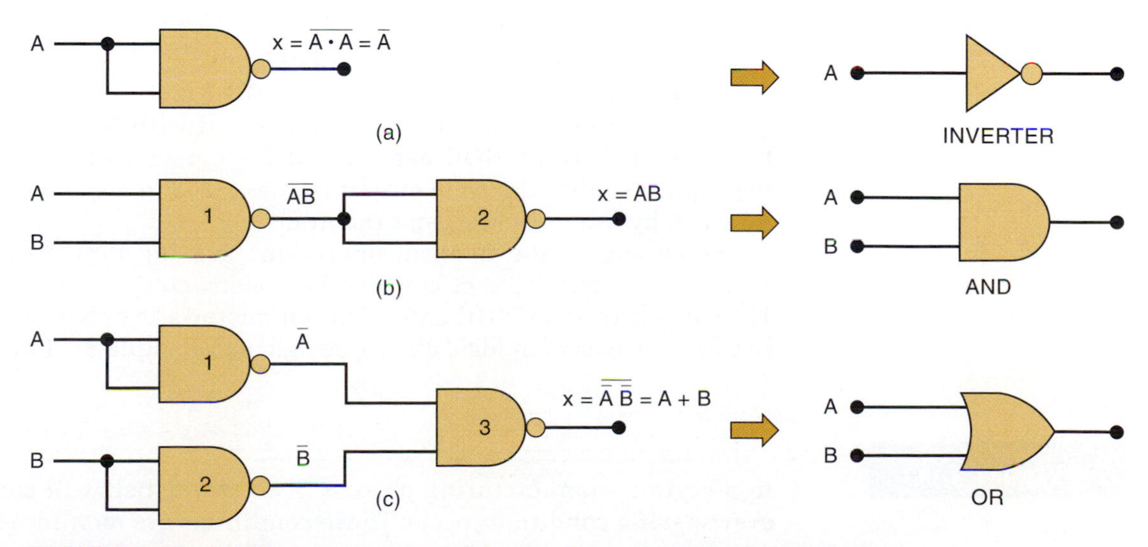

FIGURE 3-29 NAND gates can be used to implement any Boolean function.

First, in Figure 3-29(a), we have a two-input NAND gate whose inputs are purposely connected together so that the variable A is applied to both. In this configuration, the NAND simply acts as INVERTER because its output is $x = \overline{A \cdot A} = \overline{A}$.

In Figure 3-29(b), we have two NAND gates connected so that the AND operation is performed. NAND gate 2 is used as an INVERTER to change $\overline{AB}$ to $\overline{\overline{AB}} = AB$, which is the desired AND function.

The OR operation can be implemented using NAND gates connected as shown in Figure 3-29(c). Here NAND gates 1 and 2 are used as INVERTERs to invert the inputs, so that the final output is $x = \overline{\overline{A} \cdot \overline{B}}$, which can be simplified to $x = A + B$ using DeMorgan's theorem.

In a similar manner, it can be shown that NOR gates can be arranged to implement any of the Boolean operations. This is illustrated in Figure 3-30. Part (a) shows that a NOR gate with its inputs connected together behaves as an INVERTER because the output is $x = \overline{A + A} = \overline{A}$.

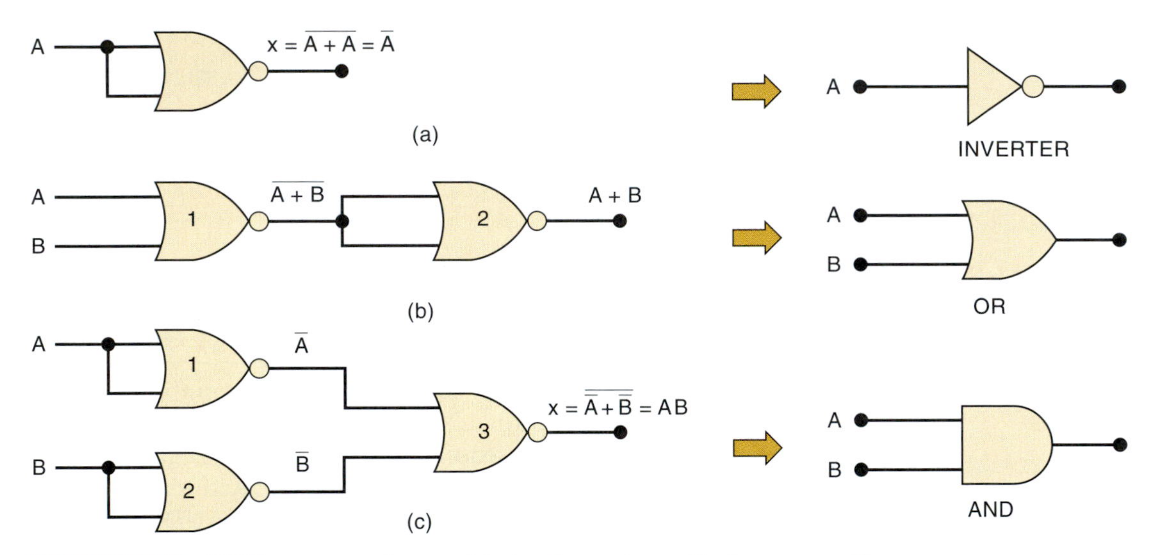

FIGURE 3-30 NOR gates can be used to implement any Boolean operation.

In Figure 3-30(b), two NOR gates are arranged so that the OR operation is performed. NOR gate 2 is used as an INVERTER to change $\overline{A + B}$ to $\overline{\overline{A + B}} = A + B$, which is the desired OR function.

The AND operation can be implemented with NOR gates as shown in Figure 3-30(c). Here, NOR gates 1 and 2 are used as INVERTERs to invert the inputs, so that the final output is $x = \overline{\overline{A} + \overline{B}}$, which can be simplified to $x = A \cdot B$ by use of DeMorgan's theorem.

Since any of the Boolean operations can be implemented using only NAND gates, any logic circuit can be constructed using only NAND gates. The same is true for NOR gates. This characteristic of NAND and NOR gates can be very useful in logic-circuit design, as Example 3-18 illustrates.

EXAMPLE 3-18

In a certain manufacturing process, a conveyor belt will shut down whenever specific conditions occur. These conditions are monitored and reflected by the states of four logic signals as follows: signal A will be HIGH whenever the conveyor belt speed is too fast; signal B will be HIGH whenever the collection bin at the end of the belt is full; signal C will be HIGH when the belt tension is too high; signal D will be HIGH when the manual override is off.

A logic circuit is needed to generate a signal x that will go HIGH whenever conditions A and B exist simultaneously or whenever conditions C and D exist simultaneously. Clearly, the logic expression for x will be $x = AB + CD$. The circuit is to be implemented with a minimum number of ICs. The TTL integrated circuits shown in Figure 3-31 are available. Each IC is a *quad*, which means that it contains *four* identical gates on one chip.

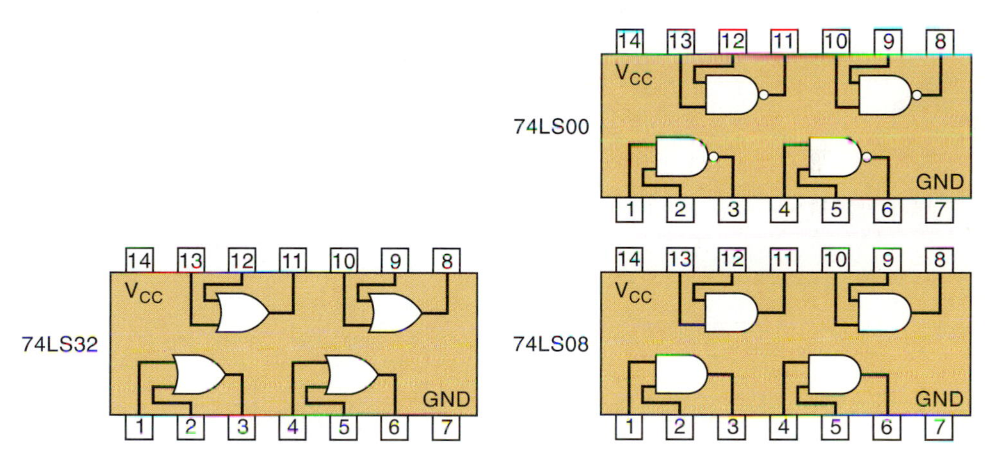

FIGURE 3-31 ICs available for Example 3-18.

Solution

The straightforward method for implementing the given expression uses two AND gates and an OR gate, as shown in Figure 3-32(a). This implementation uses two gates from the 74LS08 IC and a single gate from the 74LS32 IC. The numbers in parentheses at each input and output are the pin numbers of the respective IC. These are always shown on any logic-circuit wiring diagram. For our purposes, most logic diagrams will not show pin numbers unless they are needed in the description of circuit operation.

Another implementation can be accomplished by taking the circuit of Figure 3-32(a) and replacing each AND gate and OR gate by its equivalent NAND gate implementation from Figure 3-29. The result is shown in Figure 3-32(b).

At first glance, this new circuit looks as if it requires seven NAND gates. However, NAND gates 3 and 5 are connected as INVERTERs in series and can be eliminated from the circuit because they perform a double inversion of the signal out of NAND gate 1. Similarly, NAND gates 4 and 6 can be eliminated. The final circuit, after eliminating the double INVERTERs, is drawn in Figure 3-32(c).

This final circuit is more efficient than the one in Figure 3-32(a) because it uses three two-input NAND gates that can be implemented from one IC, the 74LS00.

FIGURE 3-32 Possible implementations for Example 3-18.

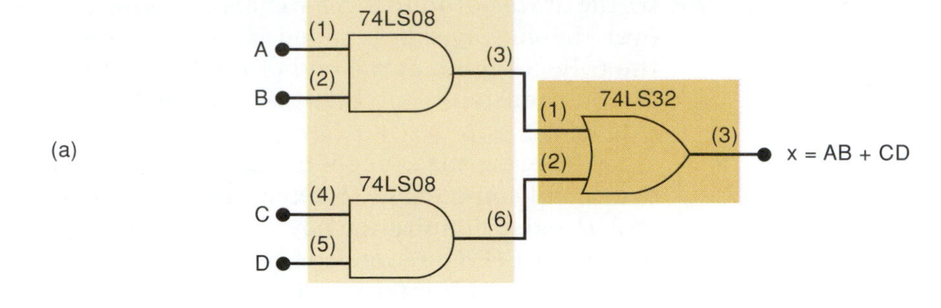

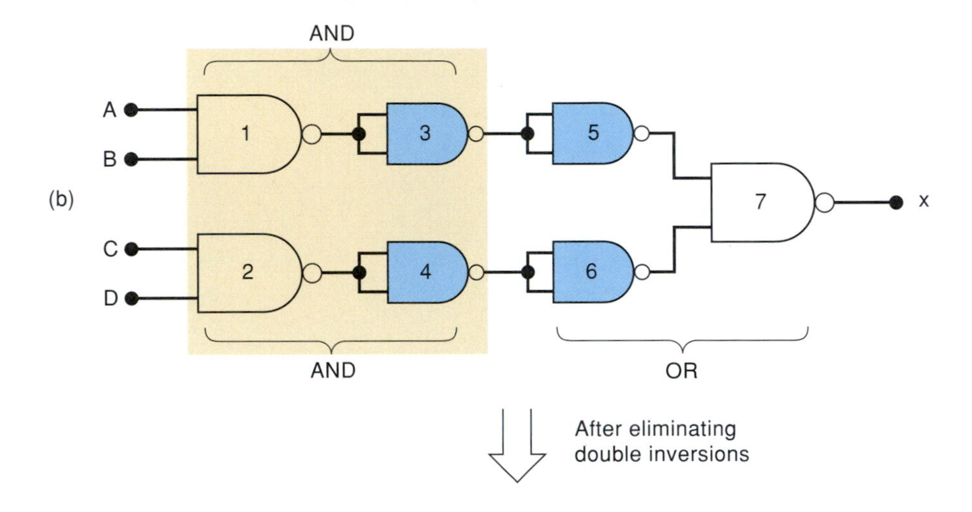

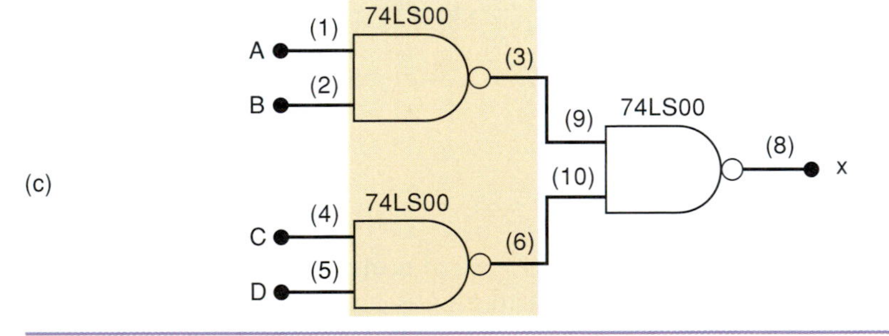

REVIEW QUESTIONS

1. How many different ways do we now have to implement the inversion operation in a logic circuit?
2. Implement the expression $x = (A + B)(C + D)$ using OR and AND gates. Then implement the expression using only NOR gates by converting each OR and AND gate to its NOR implementation from Figure 3-30. Which circuit is more efficient?
3. Write the output expression for the circuit of Figure 3-32(c), and use DeMorgan's theorems to show that it is equivalent to the expression for the circuit of Figure 3-32(a).

3-13 ALTERNATE LOGIC-GATE REPRESENTATIONS

We have introduced the five basic logic gates (AND, OR, INVERTER, NAND, and NOR) and the standard symbols used to represent them on logic-circuit diagrams. Although you may find that some circuit diagrams still use these standard symbols exclusively, it has become increasingly more common to find circuit diagrams that utilize **alternate logic symbols** *in addition* to the standard symbols.

Before discussing the reasons for using an alternate symbol for a logic gate, we will present the alternate symbols for each gate and show that they are equivalent to the standard symbols. Refer to Figure 3-33; the left side of the illustration shows the standard symbol for each logic gate, and the right side shows the alternate symbol. The alternate symbol for each gate is obtained from the standard symbol by doing the following:

1. Invert each input and output of the standard symbol. This is done by adding bubbles (small circles) on input and output lines that do not have bubbles and by removing bubbles that are already there.
2. Change the operation symbol from AND to OR, or from OR to AND. (In the special case of the INVERTER, the operation symbol is not changed.)

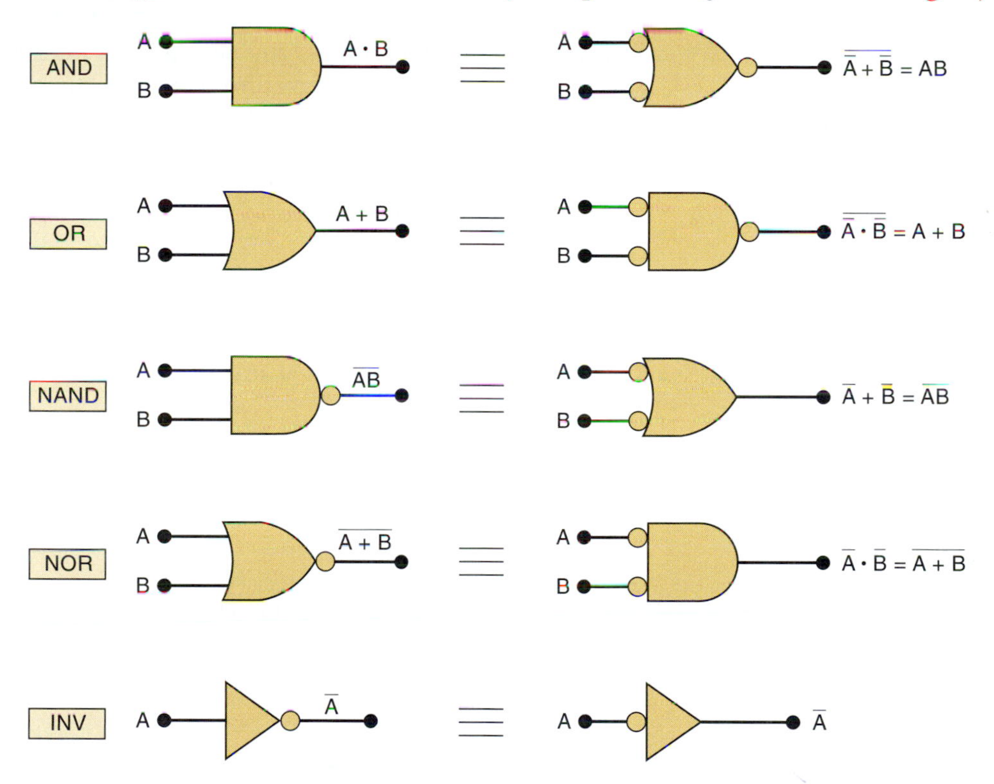

FIGURE 3-33 Standard and alternate symbols for various logic gates and inverter.

For example, the standard NAND symbol is an AND symbol with a bubble on its output. Following the steps outlined above, remove the bubble from the output, and add a bubble to each input. Then change the AND symbol to an OR symbol. The result is an OR symbol with bubbles on its inputs.

We can easily prove that this alternate symbol is equivalent to the standard symbol by using DeMorgan's theorems and recalling that the bubble represents an inversion operation. The output expression from the standard NAND symbol is $\overline{AB} = \overline{A} + \overline{B}$, which is the same as the output expression for the alternate symbol. This same procedure can be followed for each pair of symbols in Figure 3-33.

Several points should be stressed regarding the logic symbol equivalences:

1. The equivalences can be extended to gates with *any* number of inputs.
2. None of the standard symbols have bubbles on their inputs, and all the alternate symbols do.

3. The standard and alternate symbols for each gate represent the same physical circuit; *there is no difference in the circuits represented by the two symbols.*
4. NAND and NOR gates are inverting gates, and so both the standard and the alternate symbols for each will have a bubble on *either* the input or the output. AND and OR gates are *noninverting* gates, and so the alternate symbols for each will have bubbles on *both* inputs and output.

Logic-Symbol Interpretation

Each of the logic-gate symbols of Figure 3-33 provides a unique interpretation of how the gate operates. Before we can demonstrate these interpretations, we must first establish the concept of **active logic levels**.

When an input or output line on a logic circuit symbol has *no bubble* on it, that line is said to be **active-HIGH**. When an input or output line *does* have a *bubble* on it, that line is said to be **active-LOW**. The presence or absence of a bubble, then, determines the active-HIGH/active-LOW status of a circuit's inputs and output, and is used to interpret the circuit operation.

To illustrate, Figure 3-34(a) shows the standard symbol for a NAND gate. The standard symbol has a bubble on its output and no bubbles on its inputs. Thus, it has an active-LOW output and active-HIGH inputs. The logic operation represented by this symbol can therefore be interpreted as follows:

The output goes LOW only when *all* of the inputs are HIGH.

Note that this says that the output will go to its active state only when *all* of the inputs are in their active states. The word *all* is used because of the AND symbol.

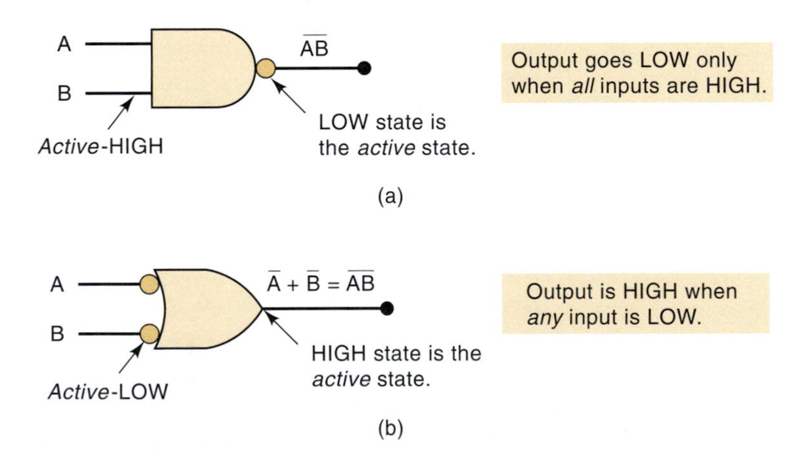

FIGURE 3-34 Interpretation of the two NAND gate symbols.

The alternate symbol for a NAND gate shown in Figure 3-34(b) has an active-HIGH output and active-LOW inputs, and so its operation can be stated as follows:

The output goes HIGH when *any input* is LOW.

This says that the output will be in its active state whenever *any* of the inputs is in its active state. The word *any* is used because of the OR symbol.

With a little thought, you can see that the two interpretations for the NAND symbols in Figure 3-34 are different ways of saying the same thing.

Summary

At this point you are probably wondering why there is a need to have two different symbols and interpretations for each logic gate. We hope the reasons will become clear after reading the next section. For now, let us summarize the important points concerning the logic-gate representations.

1. To obtain the alternate symbol for a logic gate, take the standard symbol and change its operation symbol (OR to AND, or AND to OR), and change the bubbles on both inputs and output (i.e., delete bubbles that are present, and add bubbles where there are none).
2. To interpret the logic-gate operation, first note which logic state, 0 or 1, is the active state for the inputs and which is the active state for the output. Then realize that the output's active state is produced by having *all* of the inputs in their active state (if an AND symbol is used) or by having *any* of the inputs in its active state (if an OR symbol is used).

EXAMPLE 3-19

Give the interpretation of the two OR gate symbols.

Solution

The results are shown in Figure 3-35. Note that the word *any* is used when the operation symbol is an OR symbol and the word *all* is used when it includes an AND symbol.

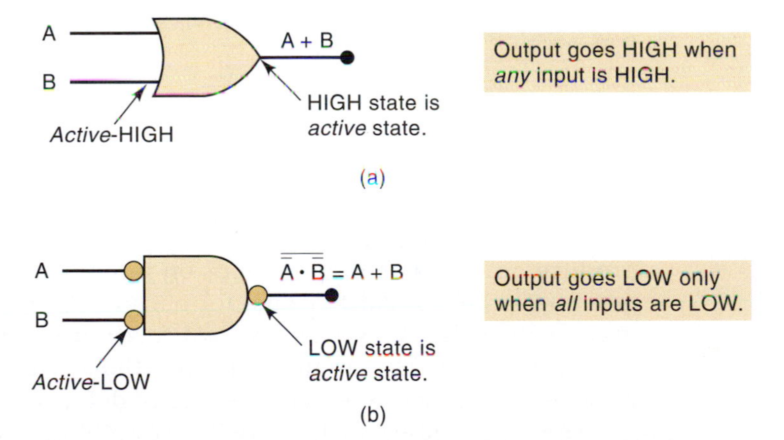

FIGURE 3-35 Interpretation of the two OR gate symbols.

REVIEW QUESTIONS

1. Write the interpretation of the operation performed by the standard NOR gate symbol in Figure 3-33.
2. Repeat question 1 for the alternate NOR gate symbol.
3. Repeat question 1 for the alternate AND gate symbol.
4. Repeat question 1 for the standard AND gate symbol.

3-14 WHICH GATE REPRESENTATION TO USE

Some logic-circuit designers and some textbooks use only the standard logic-gate symbols in their circuit schematics. While this practice is not incorrect, it does nothing to make the circuit operation easier to follow. Proper use of the alternate gate symbols in the circuit diagram can make the circuit operation

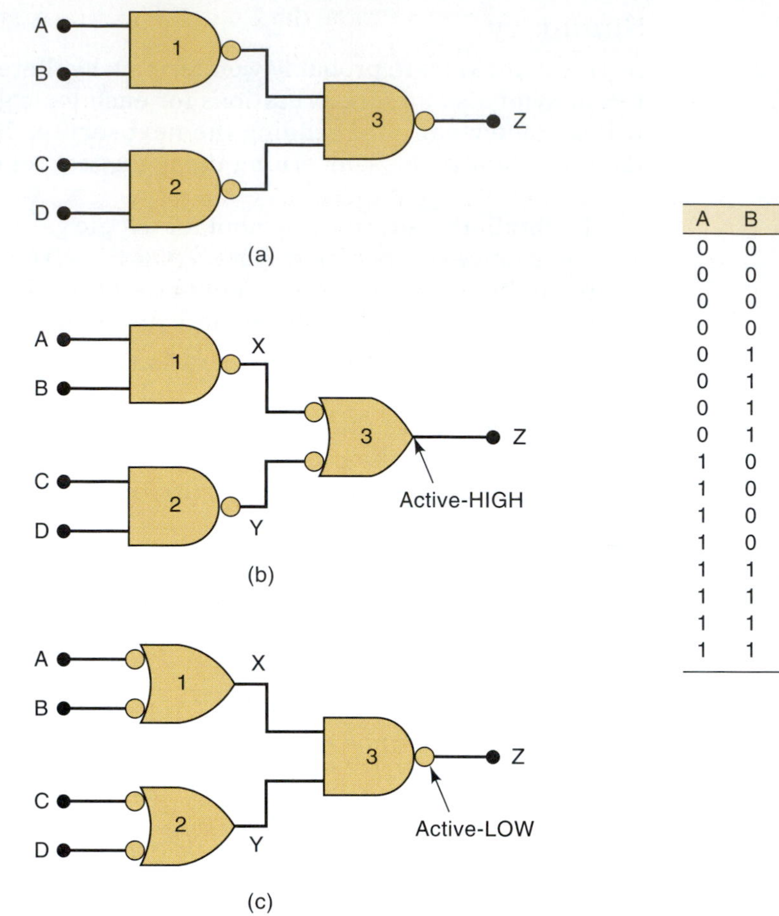

A	B	C	D	Z
0	0	0	0	0
0	0	0	1	0
0	0	1	0	0
0	0	1	1	1
0	1	0	0	0
0	1	0	1	0
0	1	1	0	0
0	1	1	1	1
1	0	0	0	0
1	0	0	1	0
1	0	1	0	0
1	0	1	1	1
1	1	0	0	1
1	1	0	1	1
1	1	1	0	1
1	1	1	1	1

(d)

FIGURE 3-36 (a) Original circuit using standard NAND symbols; (b) equivalent representation where output Z is active-HIGH; (c) equivalent representation where output Z is active-LOW; (d) truth table.

much clearer. This can be illustrated by considering the example shown in Figure 3-36.

The circuit in Figure 3-36(a) contains three NAND gates connected to produce an output Z that depends on inputs A, B, C, and D. The circuit diagram uses the standard symbol for each of the NAND gates. While this diagram is logically correct, it does not facilitate an understanding of how the circuit functions. The circuit representations given in Figures 3-36(b) and (c), however, can be analyzed more easily to determine the circuit operation.

The representation of Figure 3-36(b) is obtained from the original circuit diagram by replacing NAND gate 3 with its alternate symbol. In this diagram, output Z is taken from a NAND gate symbol that has an active-HIGH output. Thus, we can say that Z will go HIGH when either X or Y is LOW. Now, since X and Y each appear at the output of NAND symbols having active-LOW outputs, we can say that X will go LOW only if $A = B = 1$, and Y will go LOW only if $C = D = 1$. Putting this all together, we can describe the circuit operation as follows:

> **Output Z will go HIGH whenever either $A = B = 1$ or $C = D = 1$ (or both).**

This description can be translated to truth-table form by setting $Z = 1$ for those cases where $A = B = 1$ and for those cases where $C = D = 1$. For all other cases, Z is made a 0. The resultant truth table is shown in Figure 3-36(d).

The representation of Figure 3-36(c) is obtained from the original circuit diagram by replacing NAND gates 1 and 2 by their alternate symbols. In this equivalent representation, the Z output is taken from a NAND gate that has an active-LOW output. Thus, we can say that Z will go LOW only when $X = Y = 1$. Because X and Y are active-HIGH outputs, we can say that X will be HIGH when either A or B is LOW, and Y will be HIGH when either C or D is LOW. Putting this all together, we can describe the circuit operation as follows:

Output *Z* will go LOW only when *A* or *B* is LOW and *C* or *D* is LOW.

This description can be translated to truth-table form by making $Z = 0$ for all cases where at least one of the A or B inputs is LOW at the same time that at least one of the C or D inputs is LOW. For all other cases, Z is made a 1. The resultant truth table is the same as that obtained for the circuit diagram of Figure 3-36(b).

Which Circuit Diagram Should Be Used?

The answer to this question depends on the particular function being performed by the circuit output. If the circuit is being used to cause some action (e.g., turn on an LED or activate another logic circuit) when output Z goes to the 1 state, then we say that Z is to be active-HIGH, and the circuit diagram of Figure 3-36(b) should be used. On the other hand, if the circuit is being used to cause some action when Z goes to the 0 state, then Z is to be active-LOW, and the diagram of Figure 3-36(c) should be used.

Of course, there will be situations where *both* output states are used to produce different actions and either one can be considered to be the active state. For these cases, either circuit representation can be used.

Bubble Placement

Refer to the circuit representation of Figure 3-36(b) and note that the symbols for NAND gates 1 and 2 were chosen to have active-LOW outputs to match the active-LOW inputs of NAND gate 3. Refer to the circuit representation of Figure 3-36(c) and note that the symbols for NAND gates 1 and 2 were chosen to have active-HIGH outputs to match the active-HIGH inputs of NAND gate 3. This leads to the following general rule for preparing logic-circuit schematics:

Whenever possible, choose gate symbols so that bubble outputs are connected to bubble inputs, and nonbubble outputs to nonbubble inputs.

The following examples will show how this rule can be applied.

EXAMPLE 3-20

The logic circuit in Figure 3-37(a) is being used to activate an alarm when its output Z goes HIGH. Modify the circuit diagram so that it represents the circuit operation more effectively.

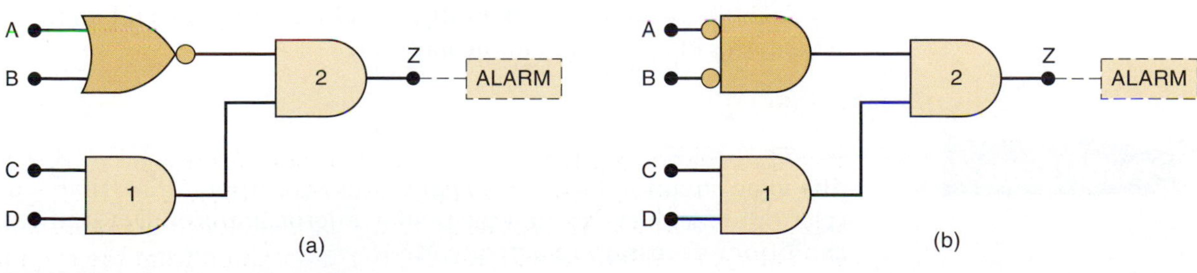

FIGURE 3-37 Example 3-20.

Solution

Because $Z = 1$ will activate the alarm, Z is to be active-HIGH. Thus, the AND gate 2 symbol does not have to be changed. The NOR gate symbol should be changed to the alternate symbol with a nonbubble (active-HIGH) output to match the nonbubble input of AND gate 2, as shown in Figure 3-37(b). Note that the circuit now has nonbubble outputs connected to the nonbubble inputs of gate 2.

EXAMPLE 3-21

When the output of the logic circuit in Figure 3-38(a) goes LOW, it activates another logic circuit. Modify the circuit diagram to represent the circuit operation more effectively.

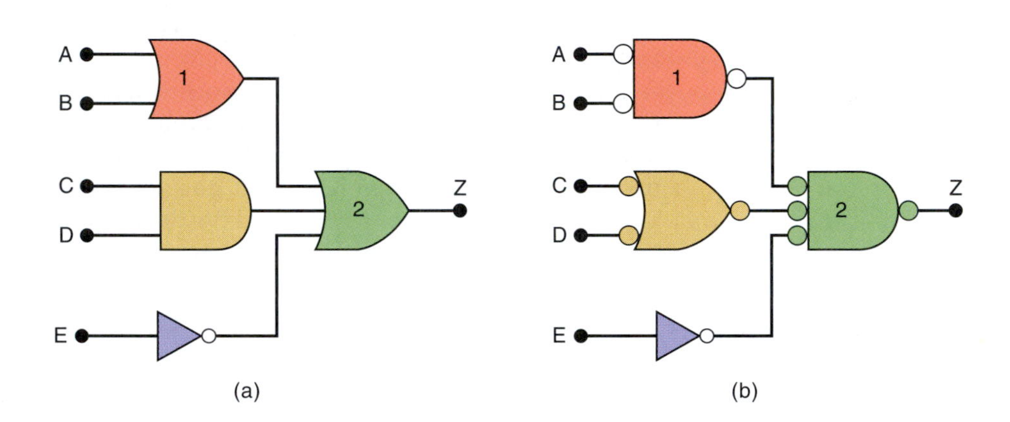

FIGURE 3-38 Example 3-21.

Solution

Because Z is to be active-LOW, the symbol for OR gate 2 must be changed to its alternate symbol, as shown in Figure 3-38(b). The new OR gate 2 symbol has bubble inputs, and so the AND gate and OR gate 1 symbols must be changed to bubbled outputs, as shown in Figure 3-38(b). The INVERTER already has a bubble output. Now the circuit has all bubble outputs connected to bubble inputs of gate 2.

Analyzing Circuits

When a logic-circuit schematic is drawn using the rules we followed in these examples, it is much easier for an engineer or technician (or student) to follow the signal flow through the circuit and to determine the input conditions that are needed to activate the output. This will be illustrated in the following examples—which, incidentally, use circuit diagrams taken from the logic schematics of an actual microcomputer.

EXAMPLE 3-22

The logic circuit in Figure 3-39 generates an output, *MEM*, that is used to activate the memory ICs in a particular microcomputer. Determine the input conditions necessary to activate *MEM*.

FIGURE 3-39 Example 3-22.

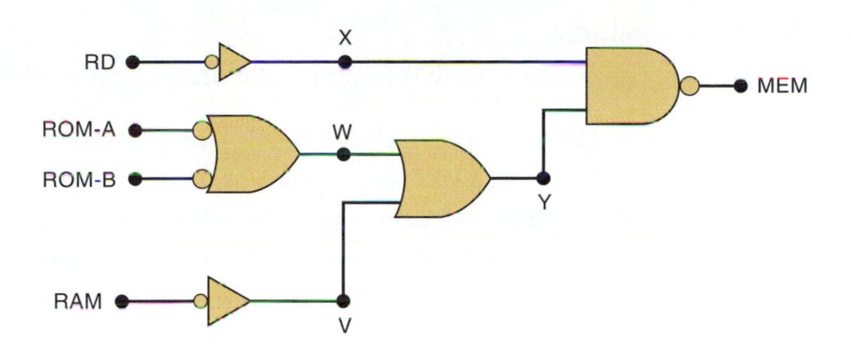

Solution

One way to do this would be to write the expression for *MEM* in terms of the inputs *RD*, *ROM-A*, *ROM-B*, and *RAM*, and to evaluate it for the 16 possible combinations of these inputs. While this method would work, it would require a lot more work than is necessary.

A more efficient method is to interpret the circuit diagram using the ideas we have been developing in the last two sections. These are the steps:

1. *MEM* is active-LOW, and it will go LOW only when *X* and *Y* are HIGH.
2. *X* will be HIGH only when *RD* = 0.
3. *Y* will be HIGH when either *W* or *V* is HIGH.
4. *V* will be HIGH when *RAM* = 0.
5. *W* will be HIGH when either *ROM-A* or *ROM-B* = 0.
6. Putting this all together, *MEM* will go LOW only when *RD* = 0 *and* at least one of the three inputs *ROM-A*, *ROM-B*, or *RAM* is LOW.

EXAMPLE 3-23

The logic circuit in Figure 3-40 is used to control the drive spindle motor for a floppy disk drive when the microcomputer is sending data to or receiving data from the disk. The circuit will turn on the motor when *DRIVE* = 1. Determine the input conditions necessary to turn on the motor.

FIGURE 3-40 Example 3-23.

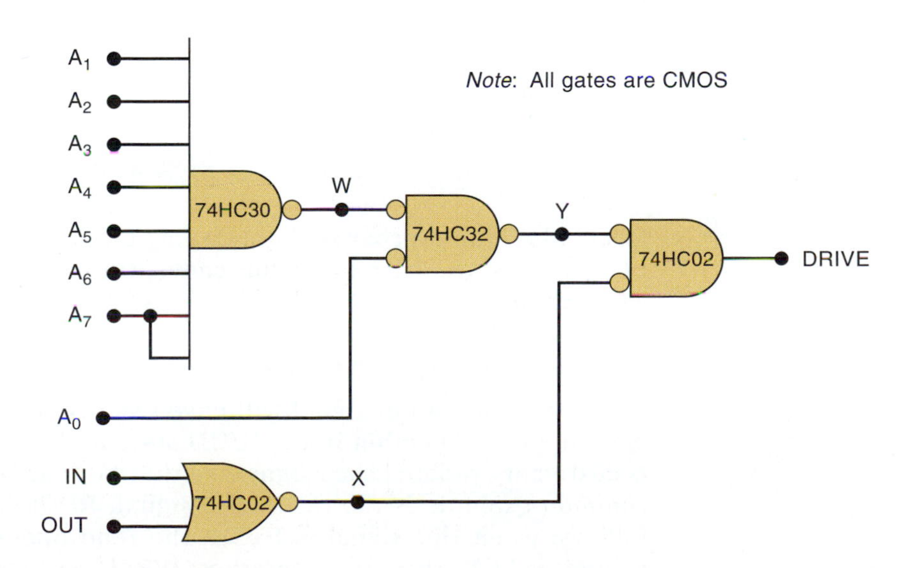

Solution

Once again, we will interpret the diagram in a step-by-step fashion:

1. *DRIVE* is active-HIGH, and it will go HIGH only when $X = Y = 0$.
2. X will be LOW when either *IN* or *OUT* is HIGH.
3. Y will be LOW only when $W = 0$ and $A_0 = 0$.
4. W will be LOW only when A_1 through A_7 are all HIGH.
5. Putting this all together, *DRIVE* will be HIGH when $A_1 = A_2 = A_3 = A_4 = A_5 = A_6 = A_7 = 1$ and $A_0 = 0$, and either *IN* or *OUT* or both are 1.

Note the strange symbol for the eight-input CMOS NAND gate (74HC30); also note that signal A_7 is connected to two of the NAND inputs.

Asserted Levels

We have been describing logic signals as being active-LOW or active-HIGH. For example, the output *MEM* in Figure 3-39 is active-LOW, and the output *DRIVE* in Figure 3-40 is active-HIGH because these are the output states that cause something to happen. Similarly, Figure 3-40 has active-HIGH inputs A_1 to A_7, and active-LOW input A_0.

When a logic signal is in its active state, it can be said to be **asserted**. For example, when we say that input A_0 is asserted, we are saying that it is in its active-LOW state. When a logic signal is not in its active state, it is said to be **unasserted**. Thus, when we say that *DRIVE* is unasserted, we mean that it is in its inactive state (low).

Clearly, the terms *asserted* and *unasserted* are synonymous with *active* and *inactive*, respectively:

asserted = active
unasserted = inactive

Both sets of terms are in common use in the digital field, so you should recognize both ways of describing a logic signal's active state.

Labeling Active-LOW Logic Signals

It has become common practice to use an overbar to label active-LOW signals. The overbar serves as another indication that the signal is active-LOW; of course, the absence of an overbar means that the signal is active-HIGH.

To illustrate, all of the signals in Figure 3-39 are active-LOW, and so they can be labeled as follows:

$$\overline{RD},\quad \overline{ROM\text{-}A},\quad \overline{ROM\text{-}B},\quad \overline{RAM},\quad \overline{MEM}$$

Remember, the overbar is simply a way to emphasize that these are active-LOW signals. We will employ this convention for labeling logic signals whenever appropriate.

Labeling Bistate Signals

Very often, an output signal will have two active states; that is, it will have one important function in the HIGH state and another in the LOW state. It is customary to label such signals so that both active states are apparent. A common example is the read/write signal, $RD/\overline{WR}$, which is interpreted as follows: when this signal is HIGH, the read operation (*RD*) is performed; when it is LOW, the write operation (*WR*) is performed.

REVIEW QUESTIONS

1. Use the method of Examples 3-22 and 3-23 to determine the input conditions needed to activate the output of the circuit in Figure 3-37(b).
2. Repeat question 1 for the circuit of Figure 3-38(b).
3. How many NAND gates are shown in Figure 3-39?
4. How many NOR gates are shown in Figure 3-40?
5. What will be the output level in Figure 3-38(b) when all of the inputs are asserted?
6. What inputs are required to assert the alarm output in Figure 3-37(b)?
7. Which of the following signals is active-LOW: RD, $\overline{W}$, $R/\overline{W}$?

3-15 IEEE/ANSI STANDARD LOGIC SYMBOLS

The logic symbols we have used so far in this chapter are the *traditional* standard symbols used in the digital industry for many, many years. These traditional symbols use a distinctive shape for each logic gate. A newer standard for logic symbols was developed in 1984; it is called the **IEEE/ANSI** Standard 91-1984 for logic symbols. The IEEE/ANSI standard uses rectangular symbols to represent all logic gates and circuits. A special *dependency notation* inside the rectangular symbol indicates how the device outputs depend on the device inputs. Figure 3-41 shows the IEEE/ANSI symbols alongside the traditional symbols for the basic logic gates. Note the following points:

1. The rectangular symbols use a small right triangle (◺) in place of the small bubble of the traditional symbols to indicate the inversion of the logic level. The presence or absence of the triangle also signifies whether an input or output is active-LOW or active-HIGH.

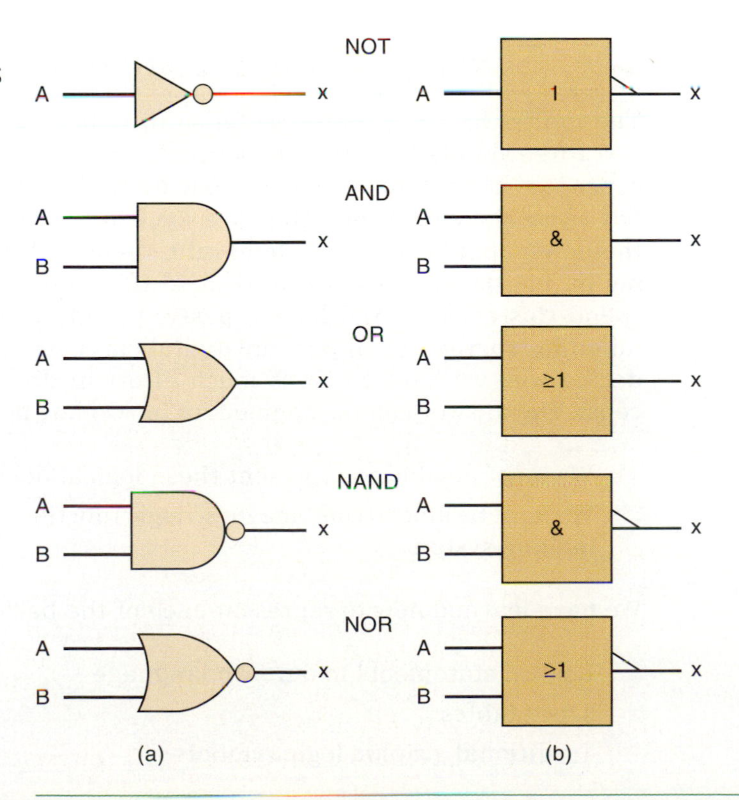

FIGURE 3-41 Standard logic symbols: (a) traditional; (b) IEEE/ANSI.

2. A special notation inside each rectangular symbol describes the logic relation between inputs and output. The "1" inside the INVERTER symbol denotes a device with only *one* input; the triangle on the output indicates that the output will go to its active-LOW state when that one input is in its active-HIGH state. The "&" inside the AND symbol means that the output will go to its active-HIGH state when all of the inputs are in their active-HIGH state. The "≥" inside the OR gate means that the output will go to its active state (HIGH) whenever *one or more* inputs are in their active state (HIGH).
3. The rectangular symbols for the NAND and the NOR are the same as those for the AND and the OR, respectively, with the addition of the small inversion triangle on the output.

Traditional or IEEE/ANSI?

The IEEE/ANSI standard has not yet been widely accepted for use in the digital field, although you will run across it in some newer equipment schematics. Most digital IC data books include both the traditional and IEEE/ANSI symbols, and it is possible that the newer standard might eventually become more widely used. We will employ the traditional symbols in most of the circuit diagrams throughout this book.

REVIEW QUESTIONS

1. Draw all of the basic logic gates using both the traditional symbols and the IEEE/ANSI symbols.
2. Draw the IEEE/ANSI symbol for a NOR gate with active-HIGH output.

3-16 SUMMARY OF METHODS TO DESCRIBE LOGIC CIRCUITS

The topics we have covered so far in this chapter have all centered around just three simple logic functions that we refer to as AND, OR, and NOT. The concept is not new to anyone because we all use these logical functions every day as we make decisions. Here are some logical examples. If it is raining OR the newspaper says that it could rain, then I will take my umbrella. If I get my paycheck today AND I make it to the bank, then I will have money to spend this evening. If I have a passing grade in lecture AND I have NOT failed in lab, then I will pass my digital class. At this point, you may be wondering why we have spent so much effort in describing such familiar concepts. The answer can be summed up in two key points:

1. We must be able to represent these logical decisions.
2. We must be able to combine these logic functions and implement a decision-making system.

We have learned how to represent each of the basic logic functions using:

Logical statements in our own language
Truth tables
Traditional graphic logic symbols

IEEE/ANSI standard logic symbols
Boolean algebra expressions
Timing diagrams

EXAMPLE 3-24

The following English expression describes the way a logic circuit needs to operate in order to drive a seatbelt warning indicator in a car.

If the driver is present AND the driver is NOT buckled up AND the ignition switch is on, THEN turn on the warning light.

Describe the circuit using Boolean algebra, schematic diagrams with logic symbols, truth tables, and timing diagrams.

Solution

See Figure 3-42.

Boolean expression

$\text{warning\_light} = \text{driver\_present} \bullet \overline{\text{buckled\_up}} \bullet \text{ignition\_on}$

(a)

Schematic diagram

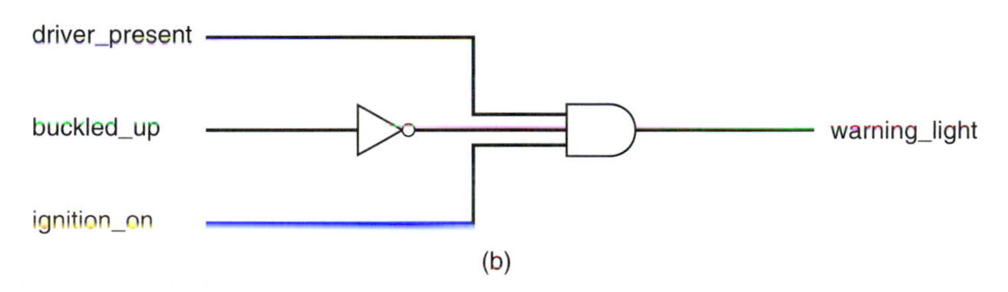

(b)

Truth table

driver_present	buckled_up	ignition_on	warning_light
0	0	0	0
0	0	1	0
0	1	0	0
0	1	1	0
1	0	0	0
1	0	1	1
1	1	0	0
1	1	1	0

(c)

Timing diagram

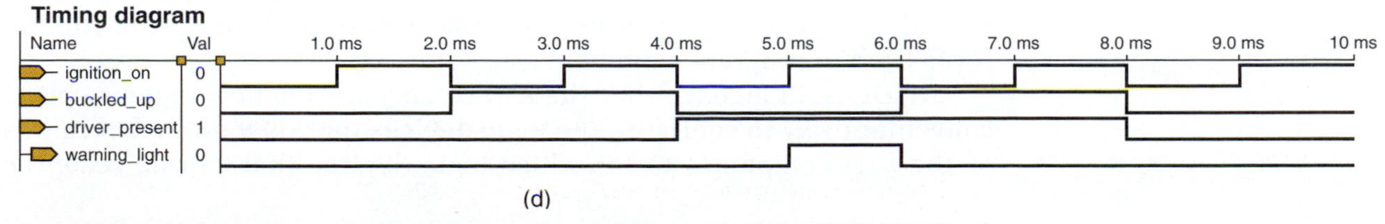

(d)

FIGURE 3-42 Methods of describing logic circuits: (a) Boolean expression; (b) schematic diagram; (c) truth table; (d) timing diagram.

Figure 3-42 shows four different ways of representing the logic circuit that was described in English as the problem statement of Example 3-24. There are many other ways in which we could represent the logic of this decision. As an example we could dream up an entirely new set of graphic symbols, or state the logical relationship in French or Japanese. Of course, we cannot cover all the possible ways of describing a logic circuit, but we must understand the most common methods to be able to communicate with others in this profession. Furthermore, certain situations are easier to describe using one method over another. In some cases, a picture is worth a thousand words, and in other cases words are concise enough and are more easily communicated to others. The important point here is that we need ways to describe and communicate the operation of digital systems.

REVIEW QUESTION

1. Name five ways to describe the operation of logic circuits.

3-17 DESCRIPTION LANGUAGES VERSUS PROGRAMMING LANGUAGES*

Recent trends in the field of digital systems are favoring text-based language description of digital circuits. You probably noticed that each description method in Figure 3-42 offers challenges to computer entry, whether it is due to overbars, symbols, formatting, or line-drawing issues. In this section, we will begin to learn some of the more advanced tools that professionals in the digital field use to describe the circuits that implement their ideas. These tools are referred to as **hardware description languages (HDLs)**. Even with the powerful computers we have today, it is not possible to describe a logic circuit in English prose and expect the computer to understand what you mean. Computers need a more rigidly defined language. We will focus on two languages in this text: **Altera hardware description language (AHDL)** and **very high speed integrated circuit (VHSIC) hardware description language (VHDL)**.

VHDL and AHDL

VHDL is not a new language. It was developed by the Department of Defense in the early 1980s as a concise way to document the designs in the very high speed integrated circuit (VHSIC) program. Appending HDL onto this acronym was too much, even for the military, and so the language was abbreviated to VHDL. Computer programs were developed to take the VHDL language files and simulate the operation of the circuits. With the growth of complex programmable logic devices in digital systems, VHDL has evolved into one of the primary high-level hardware description languages for designing and implementing digital circuits (synthesis). The language has been standardized by the IEEE, making it universally appealing for engineers as well as the makers of software tools that translate designs into the bit patterns used to program actual devices.

AHDL is a language that the Altera Corporation developed to provide a convenient way to configure the logic devices that they offer. Altera was one of the first companies to introduce logic devices that can be reconfigured

*All sections covering hardware description languages may by skipped without loss of continuity in the balance of Chapters 1–12.

electronically. These devices are called **programmable logic devices (PLDs)**. Unlike VHDL, this language is not intended to be used as a universal language for describing any logic circuit. It is intended to be used for programming complex digital systems into Altera PLDs in a language that is generally perceived to be easier to learn yet very similar to VHDL. It also has features that take full advantage of the architecture of Altera devices. All of the examples in this text will use the Altera MAX+PLUS II or Quartus II software to develop both AHDL and VHDL design files. You will see the advantage of using Altera's development system for both languages when you program an actual device. The Altera system makes circuit development very easy and contains all the necessary tools to translate from the HDL design file to a file ready to load into an Altera PLD. It also allows you to develop building blocks using schematic entry, AHDL, VHDL, and other methods and then interconnect them to form a complete system.

Other HDLs are available that are more suitable for programming simple programmable logic devices. You will find any of these languages easy to use after learning the basics of AHDL or VHDL as covered in this text.

Computer Programming Languages

It is important to distinguish between hardware description languages intended to describe the hardware configuration of a circuit and programming languages that represent a sequence of instructions intended to be carried out by a computer to accomplish some task. In both cases, we use a *language* to *program* a device. However, computers are complex digital systems that are made up of logic circuits. Computers operate by following a laundry list of tasks (i.e., instructions, or "the program"), each of which must be done in sequential order. The speed of operation is determined by how fast the computer can execute each instruction. For example, if a computer were to respond to four different inputs, it would require at least four separate instructions (sequential tasks) to detect and identify which input changed state. A digital logic circuit, on the other hand, is limited in its speed only by how quickly the circuitry can change the outputs in response to changes in the inputs. It is monitoring all inputs **concurrently** (at the same time) and responding to any changes.

The following analogy will help you understand the difference between computer operation and digital logic circuit operation and the role of language elements used to describe what the systems do. Consider the challenge of describing what is done to an Indy 500 car during a pit stop. If a single person performed all the necessary tasks one at a time, he or she would need to be very fast at each task. This is the way a computer operates: one task at a time but very quickly. Of course, at Indy, there is an entire pit crew that swarms the car, and each member of the crew does his or her task while the others do theirs. All crew members operate concurrently, like the elements of a digital circuit. Now consider how you would describe to someone else what is being done to the Indy car during the pit stop using (1) the individual-mechanic approach or (2) the pit-crew approach. Wouldn't the two English language descriptions of what is being done sound very similar? As we will see, the languages used to describe digital hardware (HDL) are very similar to languages that describe computer programs (e.g., BASIC, C, JAVA), even though the resulting implementation operates quite differently. Knowledge of any of these computer programming languages is not necessary to understand HDL. The important thing is that when you have learned both an HDL and a computer language, you must understand their different roles in digital systems.

EXAMPLE 3-25

Compare the operation of a computer and a logic circuit in performing the simple logical operation of $y = AB$.

Solution

The logic circuit is a simple AND gate. The output y will be HIGH within approximately 10 nanoseconds of the point when A and B are HIGH simultaneously. Within approximately 10 nanoseconds after either input goes LOW, the output y will be LOW.

The computer must run a program of instructions that makes decisions. Suppose each instruction takes 20 ns (that's pretty fast!). Each shape in the flowchart shown in Figure 3-43 represents one instruction. Clearly, it will take a minimum of two or three instructions (40–60 ns) to respond to changes in the inputs.

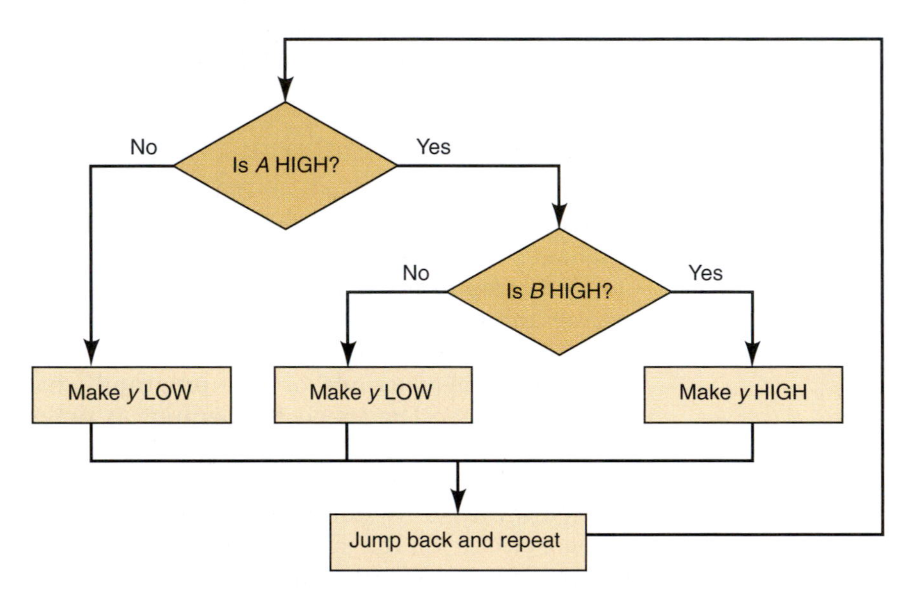

FIGURE 3-43 Decision process of a computer program.

REVIEW QUESTIONS

1. What does HDL stand for?
2. What is the purpose of an HDL?
3. What is the purpose of a computer programming language?
4. What is the key difference between HDL and computer programming languages?

3-18 IMPLEMENTING LOGIC CIRCUITS WITH PLDs

Many digital circuits today are implemented using programmable logic devices (PLDs). These devices are not like microcomputers or microcontrollers that "run" the program of instructions. Instead, they are configured electronically, and their internal circuits are "wired" together electronically to form a logic circuit. This programmable wiring can be thought of as thousands of connections that are either connected (1) or not connected (0). Figure 3-44 shows a small area of programmable connections. Each intersection between a row (horizontal wire) and a column (vertical wire) is a programmable connection. You can imagine how difficult it would be to try to

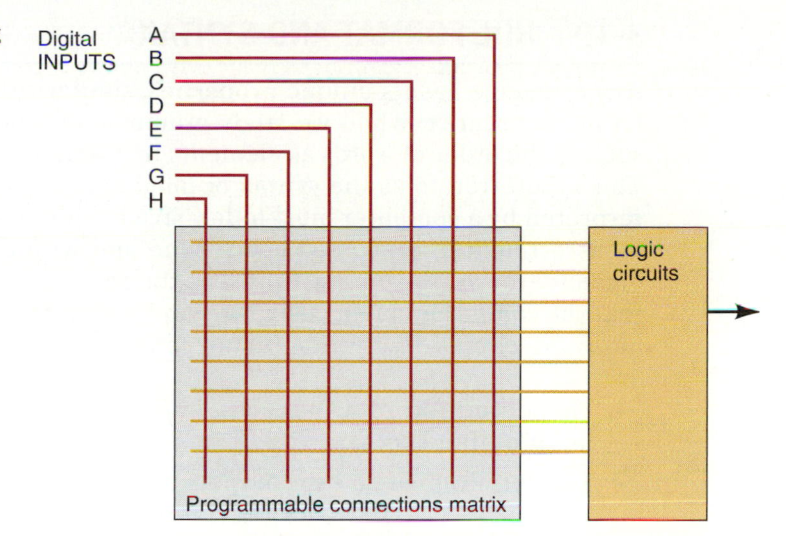

FIGURE 3-44 Configuring hardware connections with programmable logic devices.

configure these devices by placing 1s and 0s in a grid manually (which is how they did it back in the 1970s).

The role of the hardware description language is to provide a concise and convenient way for the designer to describe the operation of the circuit in a format that a personal computer can handle and store conveniently. The computer runs a special software application called a **compiler** to translate from the hardware description language into the grid of 1s and 0s that can be loaded into the PLD. If a person can master the higher-level hardware description language, it actually makes programming the PLDs much easier than trying to use Boolean algebra, schematic drawings, or truth tables. In much the same way that you learned the English language, we will start by expressing simple things and gradually learn the more complicated aspects of these languages. Our objective is to learn enough of HDL to be able to communicate with others and perform simple tasks. A full understanding of all the details of these languages is beyond the scope of this text and can really be mastered only by regular use.

In the sections throughout this book that cover the HDLs, we will present both AHDL and VHDL in a format that allows you to skip over one language and concentrate on the other without missing important information. Of course, this setup means there will be some redundant information presented if you choose to read about both languages. We feel this redundancy is worth the extra effort to provide you with the flexibility of focusing on either of the two languages or learning both by comparing and contrasting similar examples. The recommended way to use the text is to focus on one language. It is true that the easiest way to become bilingual, and fluent in both languages, is to be raised in an environment where both languages are spoken routinely. It is also very easy, however, to confuse details, so we will keep the specific examples separate and independent. We hope this format provides you with the opportunity to learn one language now and then use this book as a reference later in your career should you need to pick up the second language.

REVIEW QUESTIONS

1. What does PLD stand for?
2. How are the circuits reconfigured electronically in a PLD?
3. What does a compiler do?

3-19 HDL FORMAT AND SYNTAX

Any language has its unique properties, similarities to other languages, and its proper syntax. When we study grammar in school, we learn conventions such as the order of words as elements in a sentence and proper punctuation. This is referred to as the **syntax** of language. A language designed to be interpreted by a computer must follow strict rules of syntax. A computer is just an assortment of processed beach sand and wire that has no idea what you "meant" to say, so you must present the instructions using the exact syntax that the computer language expects and understands. The basic format of any hardware circuit description (in any language) involves two vital elements:

1. The definition of what goes into it and what comes out of it (i.e., input/output specs)
2. The definition of how the outputs respond to the inputs (i.e., its operation)

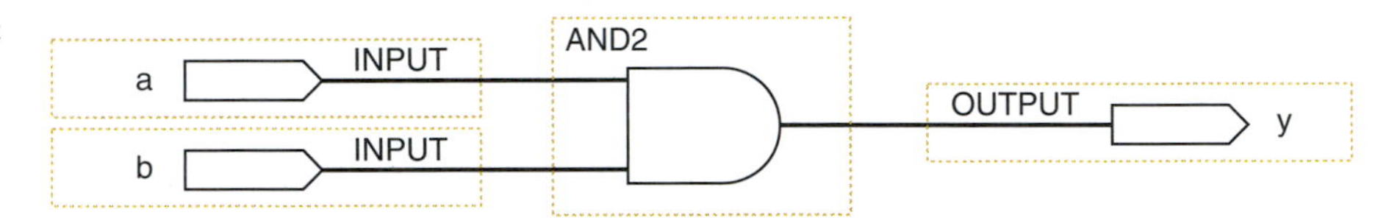

FIGURE 3-45 A schematic diagram description.

A circuit schematic diagram such as Figure 3-45 can be read and understood by a competent engineer or technician because both would understand the meaning of each symbol in the drawing. If you understand how each element works and how the elements are connected to each other, you can understand how the circuit operates. On the left side of the diagram is the set of inputs, and on the right is the set of outputs. The symbols in the middle define its operation. The text-based language must convey the same information. All HDLs use the format shown in Figure 3-46.

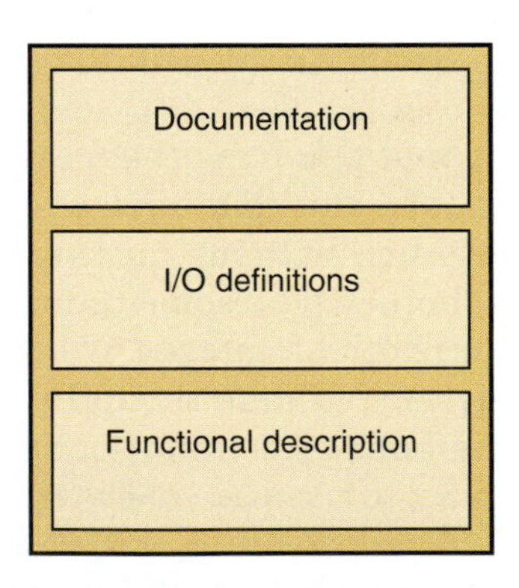

FIGURE 3-46 Format of HDL files.

In a text-based language, the circuit being described must be given a name. The inputs and outputs (sometimes called ports) must be assigned names and defined according to the nature of the port. Is it a single bit from a toggle switch? Or is it a four-bit number coming from a keypad? The text-based language must somehow convey the nature of these inputs and outputs. The **mode** of a port defines whether it is input, output, or both. The **type** refers to the number of bits and how those bits are grouped and interpreted. If the *type* of input is a single bit, then it can have only two possible values: 0 and 1. If the type of input is a four-bit binary number from a keypad, it can have any one of 16 different values (0000_2–1111_2). The type determines the range of possible values. The definition of the circuit's operation in a text-based language is contained in a set of statements that follow the circuit input/output (I/O) definition. The following two sections describe the very simple circuit of Figure 3-45 and illustrate the critical elements of AHDL and VHDL.

BOOLEAN DESCRIPTION USING AHDL

Refer to Figure 3-47. The keyword **SUBDESIGN** gives a name to the circuit block, which in this case is *and_gate.* The name of the file must also be and_gate.tdf. Notice that the keyword SUBDESIGN is capitalized. This is not required by the software, but use of a consistent style in capitalization makes the code much easier to read. The style guide that is provided with the Altera compiler for AHDL suggests the use of capital letters for the keywords in the language. Variables that are named by the designer should be lowercase.

FIGURE 3-47 Essential elements in AHDL.

```
SUBDESIGN and_gate
(
    a, b        :INPUT;
    y           :OUTPUT;
)
BEGIN
    y = a & b;
END;
```

The SUBDESIGN section defines the inputs and outputs of the logic circuit block. Something must enclose the circuit that we are trying to describe, much the same way that a block diagram encloses everything that makes up that part of the design. In AHDL, this input/output definition is enclosed in parentheses. The list of variables used for inputs to this block are separated by commas and followed by :INPUT;. In AHDL, the single-bit type is assumed unless the variable is designated as multiple bits. The single-output bit is declared with the mode :OUTPUT;. We will learn the proper way to describe other types of inputs, outputs, and variables as we need to use them.

The set of statements that describe the operation of the AHDL circuit are contained in the logic section between the keywords BEGIN and END. In this example, the operation of the hardware is described by a very simple Boolean algebra equation that states that the output (y) is assigned (=) the logic level produced by a AND b. This Boolean algebra equation is referred to as a **concurrent assignment statement**. Any statements (there is only one in this example) between BEGIN and END are evaluated constantly and concurrently. The order in which they are listed makes no difference. The basic Boolean operators are:

&	AND
#	OR
!	NOT
$	XOR

REVIEW QUESTIONS

1. What appears inside the parentheses ( ) after SUBDESIGN?
2. What appears between BEGIN and END?

VHDL

BOOLEAN DESCRIPTION USING VHDL

Refer to Figure 3-48. The keyword **ENTITY** gives a name to the circuit block, which in this case is and_gate. Notice that the keyword ENTITY is capitalized but and_gate is not. This is not required by the software, but use of a consistent style in capitalization makes the code much easier to read. The style guide provided with the Altera compiler for VHDL suggests using capital letters for the keywords in the language. Variables that are named by the designer should be lowercase.

FIGURE 3-48 Essential elements in VHDL.

```
ENTITY and_gate IS
PORT (    a, b     :IN BIT;
          y        :OUT BIT);
END and_gate;
ARCHITECTURE ckt OF and_gate IS
BEGIN
        y <= a AND b;
END ckt;
```

The ENTITY declaration can be thought of as a block description. Something must enclose the circuit we are trying to describe, much the same way a block diagram encloses everything that makes up that part of the design. In VHDL, the keyword PORT tells the compiler that we are defining inputs and outputs to this circuit block. The names used for inputs (separated by commas) are listed, ending with a colon and a description of the mode and type of input (:IN BIT;). In VHDL, the **BIT** description tells the compiler that each variable in the list is a single bit. We will learn the proper way to describe other types of inputs, outputs, and variables as we need to use them. The line containing END and_gate; terminates the ENTITY declaration.

The **ARCHITECTURE** declaration is used to describe the operation of everything inside the block. The designer makes up a name for this architectural description of the inner workings of the ENTITY block (*ckt* in this example). Every ENTITY must have at least one ARCHITECTURE associated with it. The words OF and IS are keywords in this declaration. The body of the architecture description is enclosed between the BEGIN and END keywords. END is followed by the name that has been assigned to this architecture. Within the body (between BEGIN and END) is the description of the block's operation. In this example, the operation of the hardware is described by a very simple Boolean algebra equation that states that the output (y) is assigned (<=) the logic level produced by a AND b. This is referred to as a **concurrent assignment statement**, which means that all the statements (there is only one in this example) between BEGIN and END are evaluated constantly and concurrently. The order in which they are listed makes no difference.

REVIEW QUESTIONS

1. What is the role of the ENTITY declaration?
2. Which key section defines the operation of the circuit?
3. What is the assignment operator used to give a value to a logic signal?

3-20 INTERMEDIATE SIGNALS

In many designs, there is a need to define signal points "inside" the circuit block. They are points in the circuit that are neither inputs nor outputs for the block but may be useful as a reference point. It may be a signal that needs to be connected to many other places within the block. In an analog or digital schematic diagram, they would be called test points or *nodes*. In an HDL, they are referred to as **buried nodes** or **local signals**. Figure 3-49 shows a very simple circuit that uses an intermediate signal named *m*. In the HDL, these nodes (signals) are not defined with the inputs and outputs but rather in the section that describes the operation of the block. The inputs and outputs are available to other circuit blocks in the system, but these local signals are recognized only within this block.

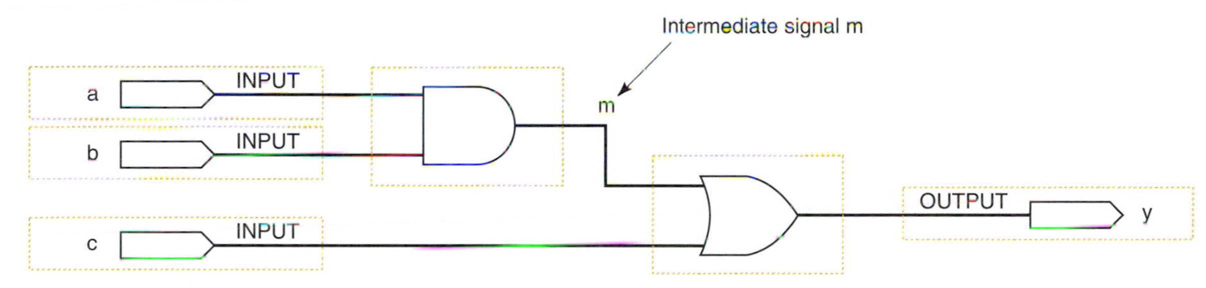

FIGURE 3-49 A logic circuit diagram with an intermediate variable.

In the example code that follows, notice the information at the top. The purpose of this information is strictly for documentation purposes. It is absolutely vital that the design is documented thoroughly. At a minimum, it should describe the project it is being used in, who wrote it, and the date. This information is often referred to as a header. We are keeping our headers brief to make this book a little lighter to carry to class, but remember: memory space is cheap and information is valuable. So don't be afraid to *document thoroughly!* There are also comments next to many of the statements in the code. These comments help the designer remember what she or he was trying to do and to help any other person to understand what was intended.

AHDL BURIED NODES

The AHDL code that describes the circuit in Figure 3-49 is shown in Figure 3-50. The **comments** in AHDL can be enclosed between % characters, as you can see in the figure between lines 1 and 4. This section of the code allows the designer to write many lines of information that will be ignored by computer programs using this file but can be read by any person trying to decipher the code. Notice that the comments at the end of lines 9, 10, 13, 15, and 16 are preceded by two dashes (--). The text following the dashes is for documentation only. Either type of comment symbol may be used, but percent signs must be used in pairs to open and close a comment. Double dashes indicate a comment that extends to the end of the line.

In AHDL, local signals are declared in the VARIABLE section, which is placed between the SUBDESIGN section and the logic section. The intermediate signal *m* is defined on line 11, following the keyword **VARIABLE**. The

AHDL

FIGURE 3-50
Intermediate variables in AHDL described in Figure 3-49.

```
1   % Intermediate variables in AHDL (Figure 3-49)
2     Digital Systems 10th ed
3     NS Widmer
4     MAY 23, 2005          %
5   SUBDESIGN fig3_50
6   (
7     a,b,c      :INPUT;   -- define inputs to block
8     y          :OUTPUT;  -- define block output
9   )
10  VARIABLE
11    m        :NODE;      -- name an intermediate signal
12  BEGIN
13    m = a & b;           -- generate buried product term
14    y = m # c;           -- generate sum on output
15  END;
```

keyword **NODE** designates the nature of the variable. Notice that a colon separates the variable name from its node designation. In the hardware description on line 13, the intermediate variable is assigned (connected to) a value (*m* = *a* & *b*;) and then m is used in the second statement on line 14 to assign (connect) a value to *y* (*y* = *m* # *c*;). Remember that the assignment statements are concurrent and, thus, the order in which they are given does not matter. For human readability, it may seem more logical to assign values to intermediate variables before they are used in other assignment statements, as shown here.

REVIEW QUESTIONS

1. What is the designation used for intermediate variables?
2. Where are these variables declared?
3. Does it matter whether the *m* or *y* equation comes first?
4. What character is used to limit a block of comments?
5. What characters are used to comment a single line?

VHDL

VHDL LOCAL SIGNALS

The VHDL code that describes the circuit in Figure 3-49 is shown in Figure 3-51. The **comments** in VHDL follow two dashes (--). Typing two successive dashes allows the designer to write information from that point to the end of the line. The information following the two successive dashes will be ignored by computer programs using this file, but can be read by any person trying to decipher the code.

The intermediate signal *m* is defined on line 13 following the keyword SIGNAL. The keyword BIT designates the type of the signal. Notice that a colon separates the signal name from its type designation. In the hardware description on line 16, the intermediate signal is assigned (connected to) a value

```
1    -- Intermediate variables in VHDL (Figure 3-49)
2    -- Digital Systems 10th ed
3    -- NS Widmer
4    -- MAY 23, 2005
5
6    ENTITY  fig3_51 IS
7    PORT( a, b, c     :IN BIT;      -- define inputs to block
8    y                 :OUT BIT);    -- define block output
9    END fig3_51;
10
11   ARCHITECTURE ckt OF fig3_51 IS
12
13      SIGNAL m       :BIT;         -- name an intermediate signal
14
15   BEGIN
16         m <= a AND b;             -- generate buried product term
17         y <= m OR c;              -- generate sum on output
18   END ckt;
```

FIGURE 3-51 Intermediate signals in VHDL described in Figure 3-49.

(*m*<=*a* AND *b*;) and then *m* is used in the statement on line 17 to assign (connect) a value to *y* (*y*<=*m* OR *c*;). Remember that the assignment statements are concurrent and, thus, the order in which they are given does not matter. For human readability, it may seem more logical to assign values to intermediate signals before they are used in other assignment statements, as shown here.

REVIEW QUESTIONS

1. What is the designation used for intermediate signals?
2. Where are these signals declared?
3. Does it matter whether the m or y equation comes first?
4. What characters are used to comment a single line?

SUMMARY

1. Boolean algebra is a mathematical tool used in the analysis and design of digital circuits.
2. The basic Boolean operations are the OR, AND, and NOT operations.
3. An OR gate produces a HIGH output when any input is HIGH. An AND gate produces a HIGH output only when all inputs are HIGH. A NOT circuit (INVERTER) produces an output that is the opposite logic level compared to the input.
4. A NOR gate is the same as an OR gate with its output connected to an INVERTER. A NAND gate is the same as an AND gate with its output connected to an INVERTER.

5. Boolean theorems and rules can be used to simplify the expression of a logic circuit and can lead to a simpler way of implementing the circuit.
6. NAND gates can be used to implement any of the basic Boolean operations. NOR gates can be used likewise.
7. Either standard or alternate symbols can be used for each logic gate, depending on whether the output is to be active-HIGH or active-LOW.
8. The IEEE/ANSI standard for logic symbols uses rectangular symbols for each logic device, with special notations inside the rectangles to show how the outputs depend on the inputs.
9. Hardware description languages have become an important method of describing digital circuits.
10. HDL code should always contain comments that document its vital characteristics so a person reading it later can understand what it does.
11. Every HDL circuit description contains a definition of the inputs and outputs, followed by a section that describes the circuit's operation.
12. In addition to inputs and outputs, intermediate connections that are buried within the circuit can be defined. These intermediate connections are called nodes or signals.

IMPORTANT TERMS

logic level
Boolean algebra
truth table
OR operation
OR gate
AND operation
AND gate
NOT operation
inversion (complementation)
NOT circuit (INVERTER)
NOR gate
NAND gate
Boolean theorems
DeMorgan's theorems
alternate logic symbols
active logic levels
active-HIGH
active-LOW
asserted
unasserted
IEEE/ANSI
hardware description languages (HDLs)
Altera hardware description language (AHDL)
very high speed integrated circuit (VHSIC) hardware description language (VHDL)
programmable logic devices (PLDs)
concurrent
compiler
syntax
mode
type
SUBDESIGN
concurrent assignment statement
ENTITY
BIT
ARCHITECTURE
buried nodes (local signals)
comments
VARIABLE
NODE

PROBLEMS

The color letters preceding some of the problems are used to indicate the nature or type of problem as follows:

B basic problem
T troubleshooting problem
D design or circuit-modification problem
N new concept or technique not covered in text
C challenging problem
H HDL problem

SECTION 3-3

B 3-1.* Draw the output waveform for the OR gate of Figure 3-52.

FIGURE 3-52

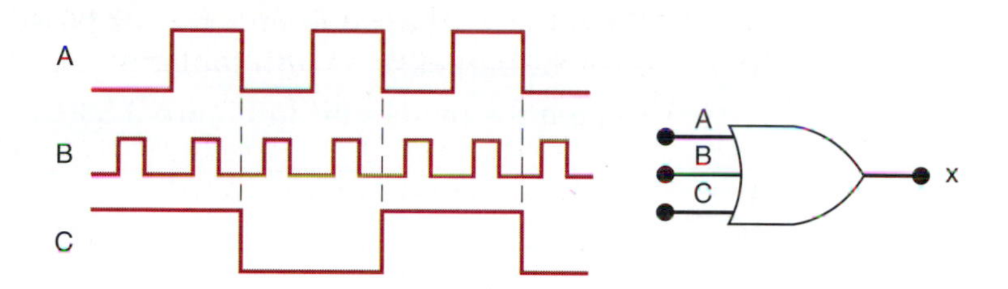

B 3-2. Suppose that the *A* input in Figure 3-52 is unintentionally shorted to ground (i.e., $A = 0$). Draw the resulting output waveform.

B 3-3.* Suppose that the *A* input in Figure 3-52 is unintentionally shorted to the +5 V supply line (i.e., $A = 1$). Draw the resulting output waveform.

C 3-4. Read the statements below concerning an OR gate. At first, they may appear to be valid, but after some thought you should realize that neither one is *always* true. Prove this by showing a specific example to refute each statement.

(a) If the output waveform from an OR gate is the same as the waveform at one of its inputs, the other input is being held permanently LOW.

(b) If the output waveform from an OR gate is always HIGH, one of its inputs is being held permanently HIGH.

B 3-5. How many different sets of input conditions will produce a HIGH output from a five-input OR gate?

SECTION 3-4

B 3-6. Change the OR gate in Figure 3-52 to an AND gate.

(a)* Draw the output waveform.

(b) Draw the output waveform if the *A* input is permanently shorted to ground.

(c) Draw the output waveform if *A* is permanently shorted to +5 V.

D 3-7.* Refer to Figure 3-4. Modify the circuit so that the alarm is to be activated only when the pressure and the temperature exceed their maximum limits at the same time.

B 3-8.* Change the OR gate in Figure 3-6 to an AND gate and draw the output waveform.

B 3-9. Suppose that you have an unknown two-input gate that is either an OR gate or an AND gate. What combination of input levels should you apply to the gate's inputs to determine which type of gate it is?

B 3-10. *True or false:* No matter how many inputs it has, an AND gate will produce a HIGH output for only one combination of input levels.

*Answers to problems marked with an asterisk can be found in the back of the text.

SECTIONS 3-5 TO 3-7

B 3-11. Apply the A waveform from Figure 3-23 to the input of an INVERTER. Draw the output waveform. Repeat for waveform B.

B 3-12. (a)* Write the Boolean expression for output x in Figure 3-53(a). Determine the value of x for all possible input conditions, and list the values in a truth table.

(b) Repeat for the circuit in Figure 3-53(b).

FIGURE 3-53

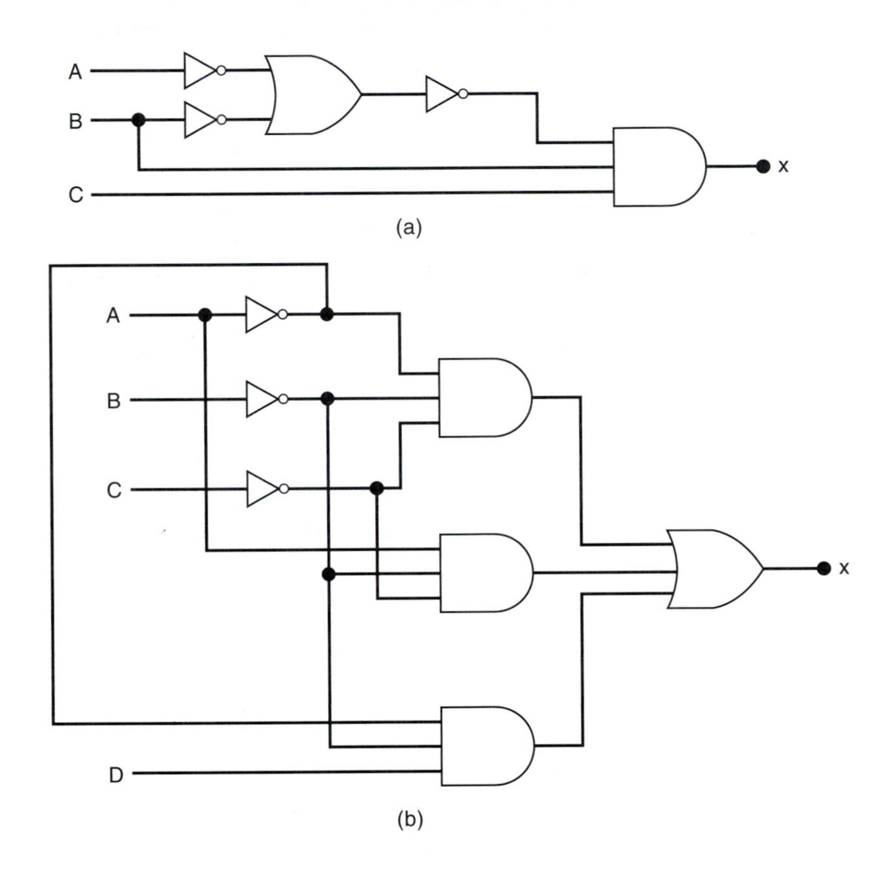

B 3-13.* Create a complete analysis table for the circuit of Figure 3-15(b) by finding the logic levels present at each gate output for each of the 32 possible input combinations.

B 3-14. (a)* Change each OR to an AND, and each AND to an OR, in Figure 3-15(b). Then write the expression for the output.

(b) Complete an analysis table.

B 3-15. Create a complete analysis table for the circuit of Figure 3-16 by finding the logic levels present at each gate output for each of the 16 possible combinations of input levels.

SECTION 3-8

B 3-16. For each of the following expressions, construct the corresponding logic circuit, using AND and OR gates and INVERTERs.

(a)* $x = \overline{AB(C + D}$)

(b)* $z = \overline{A + B + \overline{C}D\overline{E}}$) $+ \overline{B}C\overline{D}$

(c) $y = (\overline{M + N} + \overline{P}Q)$

(d) $x = \overline{W + P\overline{Q}}$

(e) $z = MN(P + \overline{N})$

(f) $x = (A + B)(\overline{A} + \overline{B})$

SECTION 3-9

B 3-17.* (a) Apply the input waveforms of Figure 3-54 to a NOR gate, and draw the output waveform.

(b) Repeat with C held permanently LOW.

(c) Repeat with C held HIGH.

FIGURE 3-54

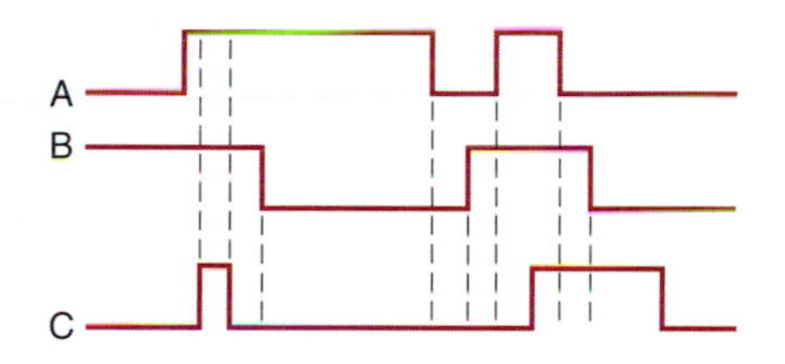

B 3-18. Repeat Problem 3-17 for a NAND gate.

C 3-19.* Write the expression for the output of Figure 3-55, and use it to determine the complete truth table. Then apply the waveforms of Figure 3-54 to the circuit inputs, and draw the resulting output waveform.

FIGURE 3-55

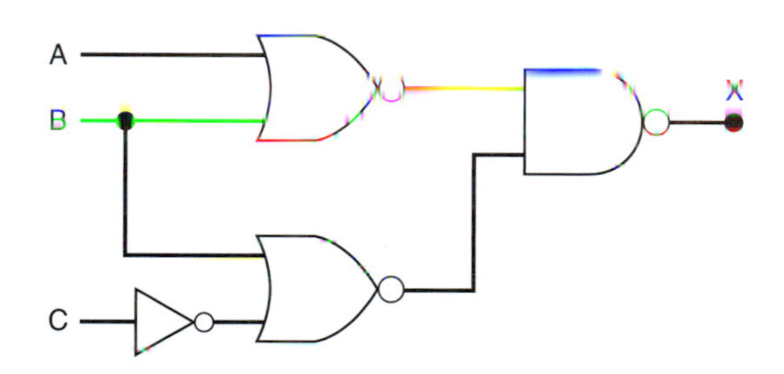

B 3-20. Determine the truth table for the circuit of Figure 3-24.

B 3-21. Modify the circuits that were constructed in Problem 3-16 so that NAND gates and NOR gates are used wherever appropriate.

SECTION 3-10

C 3-22. Prove theorems (15a) and (15b) by trying all possible cases.

B 3-23.* DRILL QUESTION

Complete each expression.

(a) $A + 1 =$ ________

(b) $A \cdot A =$ ________

(c) $B \cdot \overline{B} =$ ________

(d) $C + C =$ ________

(e) $x \cdot 0 =$ ________

(f) $D \cdot 1 =$ ________

(g) $D + 0 =$ ________

(h) $C + \overline{C} =$ ________

(i) $G + GF =$ ________

(j) $y + \overline{w}y =$ ________

C 3-24. (a)* Simplify the following expression using theorems (13b), (3), and (4):

$$x = (M + N)(\overline{M} + P)(\overline{N} + \overline{P})$$

(b) Simplify the following expression using theorems (13a), (8), and (6):

$$z = \overline{A}B\overline{C} + AB\overline{C} + B\overline{C}D$$

SECTIONS 3-11 AND 3-12

C 3-25. Prove DeMorgan's theorems by trying all possible cases.

B 3-26. Simplify each of the following expressions using DeMorgan's theorems.

(a)* $\overline{\overline{A}B\overline{C}}$	(d) $\overline{A + \overline{B}}$	(g)* $\overline{A\overline{(B + \overline{C})}D}$
(b) $\overline{\overline{A} + \overline{B}C}$	(e)* $\overline{\overline{AB}}$	(h) $\overline{(M + \overline{N})(\overline{M} + N)}$
(c)* $\overline{AB\overline{CD}}$	(f) $\overline{\overline{A} + \overline{C} + \overline{D}}$	(i) $\overline{\overline{\overline{AB}C}D}$

B 3-27.* Use DeMorgan's theorems to simplify the expression for the output of Figure 3-55.

C 3-28. Convert the circuit of Figure 3-53(b) to one using only NAND gates. Then write the output expression for the new circuit, simplify it using DeMorgan's theorems, and compare it with the expression for the original circuit.

C 3-29. Convert the circuit of Figure 3-53(a) to one using only NOR gates. Then write the expression for the new circuit, simplify it using DeMorgan's theorems, and compare it with the expression for the original circuit.

B 3-30. Show how a two-input NAND gate can be constructed from two-input NOR gates.

B 3-31. Show how a two-input NOR gate can be constructed from two-input NAND gates.

C 3-32. A jet aircraft employs a system for monitoring the rpm, pressure, and temperature values of its engines using sensors that operate as follows:

RPM sensor output $= 0$ only when speed < 4800 rpm
P sensor output $= 0$ only when pressure < 220 psi
T sensor output $= 0$ only when temperature < 200°F

Figure 3-56 shows the logic circuit that controls a cockpit warning light for certain combinations of engine conditions. Assume that a HIGH at output W activates the warning light.

(a)* Determine what engine conditions will give a warning to the pilot.

(b) Change this circuit to one using all NAND gates.

FIGURE 3-56

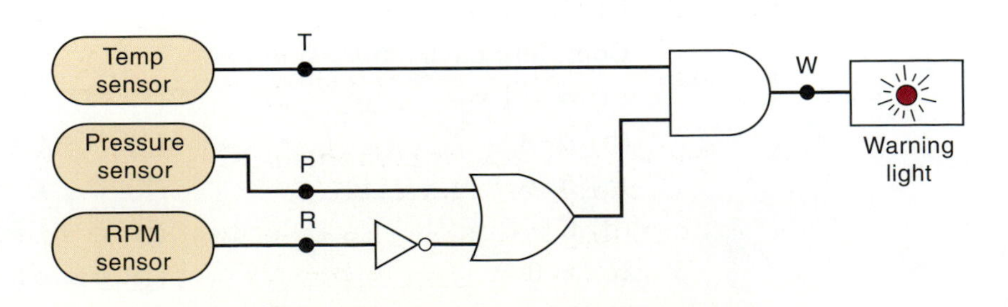

SECTIONS 3-13 AND 3-14

B 3-33. For each statement below, draw the appropriate logic-gate symbol—standard or alternate—for the given operation.

(a) A HIGH output occurs only when all three inputs are LOW.

(b) A LOW output occurs when any of the four inputs is LOW.

(c) A LOW output occurs only when all eight inputs are HIGH.

B 3-34. Draw the standard representations for each of the basic logic gates. Then draw the alternate representations.

C 3-35. The circuit of Figure 3-55 is supposed to be a simple digital combination lock whose output will generate an active-LOW $\overline{UNLOCK}$ signal for only one combination of inputs.

(a)* Modify the circuit diagram so that it represents more effectively the circuit operation.

(b) Use the new circuit diagram to determine the input combination that will activate the output. Do this by working back from the output using the information given by the gate symbols, as was done in Examples 3-22 and 3-23. Compare the results with the truth table obtained in Problem 3-19.

C 3-36. (a) Determine the input conditions needed to activate output Z in Figure 3-37(b). Do this by working back from the output, as was done in Examples 3-22 and 3-23.

(b) Assume that it is the LOW state of Z that is to activate the alarm. Change the circuit diagram to reflect this, and then use the revised diagram to determine the input conditions needed to activate the alarm.

D 3-37. Modify the circuit of Figure 3-40 so that $A_1 = 0$ is needed to produce $DRIVE = 1$ instead of $A_1 = 1$.

B 3-38.* Determine the input conditions needed to cause the output in Figure 3-57 to go to its active state.

FIGURE 3-57

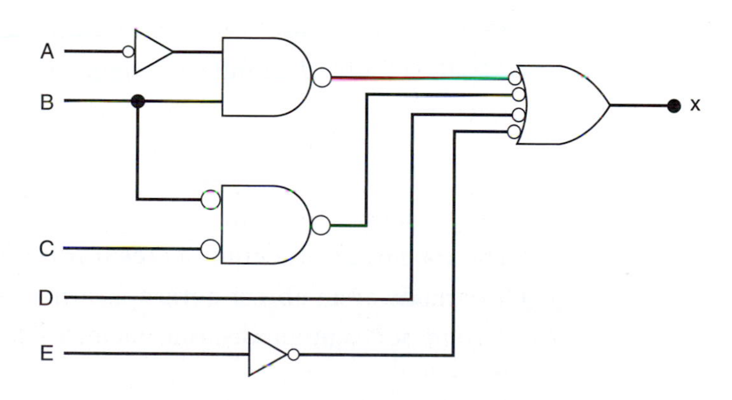

B 3-39.* What is the asserted state for the output of Figure 3-57? For the output of Figure 3-36(c)?

B 3-40. Use the results of Problem 3-38 to obtain the complete truth table for the circuit of Figure 3-57.

N 3-41.* Figure 3-58 shows an application of logic gates that simulates a two-way switch like the ones used in our homes to turn a light on or off from two different switches. Here the light is an LED that will be ON (conducting) when the NOR gate output is LOW. Note that this output is labeled $\overline{LIGHT}$ to indicate that it is active-LOW. Determine the input conditions needed to turn on the LED. Then verify that the circuit operates as a two-way switch using switches *A* and *B*. (In Chapter 4, you will learn how to design circuits like this one to produce a given relationship between inputs and outputs.)

FIGURE 3-58

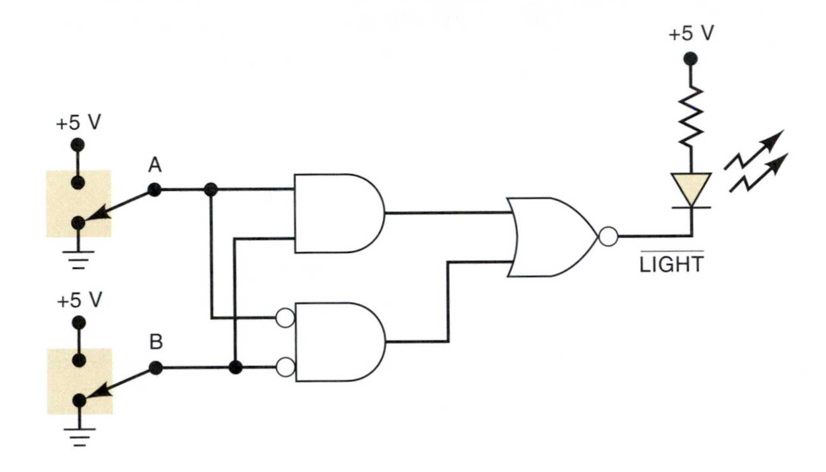

SECTION 3-15

B 3-42. Redraw the circuits of (a)* Figure 3-57 and (b) Figure 3-58 using the IEEE/ANSI symbols.

SECTION 3-17
HDL DRILL QUESTIONS

H 3-43.* *True or false:*

(a) VHDL is a computer programming language.

(b) VHDL can accomplish the same thing as AHDL.

(c) AHDL is an IEEE standard language.

(d) Each intersection in a switch matrix can be programmed as an open or short circuit between a row and column wire.

(e) The first item that appears at the top of an HDL listing is the functional description.

(f) The type of an object indicates if it is an input or an output.

(g) The mode of an object determines if it is an input or an output.

(h) Buried nodes are nodes that have been deleted and will never be used again.

(i) Local signals are another name for intermediate variables.

(j) The header is a block of comments that document vital information about the project.

SECTION 3-18

B 3-44. Redraw the programmable connection matrix from Figure 3-44. Label the output signals (horizontal lines) from the connection matrix (from top row to bottom row) as follows: AAABADHE. Draw an X in the appropriate intersections to short-circuit a row to a column and create these connections to the logic circuit.

H 3-45.*Write the HDL code in the language of your choice that will produce the following output functions:

$$X = A + B$$
$$Y = AB$$
$$Z = A + B + C$$

H 3-46. Write the HDL code in the language of your choice that will implement the logic circuit of Figure 3-39.

(a) Use a single Boolean equation.

(b) Use the intermediate variables V, W, X, and Y.

MICROCOMPUTER APPLICATION

C 3-47.*Refer to Figure 3-40 in Example 3-23. Inputs A_7 through A_0 are *address* inputs that are supplied to this circuit from outputs of the microprocessor chip in a microcomputer. The eight-bit address code A_7 to A_0 selects which device the microprocessor wants to activate. In Example 3-23, the required address code to activate the disk drive was A_7 through $A_0 = 11111110_2 = FE_{16}$.

Modify the circuit so that the microprocessor must supply an address code of $4A_{16}$ to activate the disk drive.

CHALLENGING EXERCISES

C 3-48. Show how $x = AB\overline{C}$ can be implemented with one two-input NOR and one two-input NAND gate.

C 3-49.*Implement $y = ABCD$ using only two-input NAND gates.

ANSWERS TO SECTION REVIEW QUESTIONS

SECTION 3-2

1. $x = 1$ 2. $x = 0$ 3. 32

SECTION 3-3

1. All inputs LOW 2. $x = A + B + C + D + E + F$ 3. Constant HIGH

SECTION 3-4

1. All five inputs = 1 2. A LOW input will keep the output LOW. 3. False; see truth table of each gate.

SECTION 3-5

1. Output of second INVERTER will be the same as input A. 2. y will be LOW only for $A = B = 1$.

SECTION 3-6

1. $x = \overline{A} + B + C + \overline{AD}$ 2. $x = D(\overline{AB + C}) + E$

SECTION 3-7

1. $x = 1$ 2. $x = 1$ 3. $x = 1$ for both.

SECTION 3-8

1. See Figure 3-15(a). 2. See Figure 3-17(b). 3. See Figure 3-15(b).

SECTION 3-9

1. All inputs LOW. 2. $x = 0$ 3. $x = \overline{A + B + \overline{CD}}$

SECTION 3-10

1. $y = A\overline{C}$ 2. $y = \overline{A}\,\overline{B}\,\overline{D}$ 3. $y = \overline{A}D + BD$

SECTION 3-11

1. $z = \overline{AB} + C$ 2. $y = (\overline{R} + S + \overline{T})Q$ 3. Same as Figure 3-28 except NAND is replaced by NOR. 4. $y = \overline{A}B(C + \overline{D})$

SECTION 3-12

1. Three. 2. NOR circuit is more efficient because it can be implemented with one 74LS02 IC. 3. $x = \overline{(\overline{AB})\,(\overline{CD})} = \overline{\overline{AB}} + \overline{\overline{(CD)}} + AB + CD$

SECTION 3-13

1. Output goes LOW when any input is HIGH. 2. Output goes HIGH only when all inputs are LOW. 3. Output goes LOW when any input is LOW. 4. Output goes HIGH only when all inputs are HIGH.

SECTION 3-14

1. Z will go HIGH when $A = B = 0$ and $C = D = 1$. 2. Z will go LOW when $A = B = 0$, $E = 1$, and either C or D or both are 0. 3. Two 4. Two 5. LOW 6. $A = B = 0$, $C = D = 1$ 7. $\overline{W}$

SECTION 3-15

1. See Figure 3-41. 2. Rectangle with & inside, and triangles on inputs.

SECTION 3-16

1. Boolean equation, truth table, logic diagram, timing diagram, language.

SECTION 3-17

1. Hardware description language 2. To describe a digital circuit and its operation. 3. To give a computer a sequential list of tasks. 4. HDL describes concurrent hardware circuits; computer instructions execute one at a time.

SECTION 3-18

1. Programmable logic device 2. By making and breaking connections in a switching matrix 3. It translates HDL code into a pattern of bits to configure the switching matix.

SECTION 3-19

AHDL

1. The input and output definitions. 2. The description of how it operates.

VHDL

1. To give a name to the circuit and define its inputs and outputs. 2. The ARCHITECTURE description. 3. <=

SECTION 3-20

AHDL

1. NODE 2. After the I/O definition and before BEGIN. 3. No 4. % 5. --

VHDL

1. SIGNAL 2. Inside ARCHITECTURE before BEGIN. 3. No 4. --

CHAPTER 4

COMBINATIONAL LOGIC CIRCUITS

OUTLINE

OBJECTIVES

Upon completion of this chapter, you will be able to:

- Convert a logic expression into a sum-of-products expression.
- Perform the necessary steps to reduce a sum-of-products expression to its simplest form.
- Use Boolean algebra and the Karnaugh map as tools to simplify and design logic circuits.
- Explain the operation of both exclusive-OR and exclusive-NOR circuits.
- Design simple logic circuits without the help of a truth table.
- Implement enable circuits.
- Cite the basic characteristics of TTL and CMOS digital ICs.
- Use the basic troubleshooting rules of digital systems.
- Deduce from observed results the faults of malfunctioning combinational logic circuits.
- Describe the fundamental idea of programmable logic devices (PLDs).
- Outline the steps involved in programming a PLD to perform a simple combinational logic function.
- Go to the Altera user manuals to acquire the information needed to do a simple programming experiment in the lab.
- Describe hierarchical design methods.
- Identify proper data types for single-bit, bit array, and numeric value variables.
- Describe logic circuits using HDL control structures IF/ELSE, IF/ELSIF, and CASE.
- Select the appropriate control structure for a given problem.

INTRODUCTION

In Chapter 3, we studied the operation of all the basic logic gates, and we used Boolean algebra to describe and analyze circuits that were made up of combinations of logic gates. These circuits can be classified as *combinational* logic circuits because, at any time, the logic level at the output depends on the combination of logic levels present at the inputs. A combinational circuit has no *memory* characteristic, so its output depends *only* on the current value of its inputs.

In this chapter, we will continue our study of combinational circuits. To start, we will go further into the simplification of logic circuits. Two methods will be used: one uses Boolean algebra theorems; the other uses a *mapping* technique. In addition, we will study simple techniques for designing combinational logic circuits to satisfy a given set of requirements. A complete study of logic-circuit design is not one of our objectives, but the methods we introduce will provide a good introduction to logic design.

A good portion of the chapter is devoted to the troubleshooting of combinational circuits. This first exposure to troubleshooting should begin to develop the type of analytical skills needed for successful troubleshooting. To make this material as practical as possible, we will first present some of the basic characteristics of logic-gate ICs in the TTL and CMOS logic families along with a description of the most common types of faults encountered in digital IC circuits.

In the last sections of this chapter, we will extend our knowledge of programmable logic devices and hardware description languages. The concept of programmable hardware connections will be reinforced, and we will provide more details regarding the role of the development system. You will learn the steps followed in the design and development of digital systems today. Enough information will be provided to allow you to choose the correct types of data objects for use in simple projects to be presented later in this text. Finally, several control structures will be explained, along with some instruction regarding their appropriate use.

4-1 SUM-OF-PRODUCTS FORM

The methods of logic-circuit simplification and design that we will study require the logic expression to be in a **sum-of-products (SOP)** form. Some examples of this form are:

1. $ABC + \overline{A}B\overline{C}$
2. $AB + \overline{A}B\overline{C} + \overline{C}\,\overline{D} + D$
3. $\overline{A}B + C\overline{D} + EF + GK + H\overline{L}$

Each of these sum-of-products expressions consists of two or more AND terms (products) that are ORed together. Each AND term consists of one or more variables *individually* appearing in either complemented or uncomplemented form. For example, in the sum-of-products expression $ABC + \overline{A}B\overline{C}$, the first AND product contains the variables A, B, and C in their uncomplemented (not inverted) form. The second AND term contains A and C in their complemented (inverted) form. Note that in a sum-of-products expression, one inversion sign *cannot* cover more than one variable in a term (e.g., we cannot have $\overline{ABC}$ or $\overline{RST}$).

Product-of-Sums

Another general form for logic expressions is sometimes used in logic-circuit design. Called the **product-of-sums (POS)** form, it consists of two or more OR terms (sums) that are ANDed together. Each OR term contains one or more variables in complemented or uncomplemented form. Here are some product-of-sum expressions:

1. $(A + \overline{B} + C)(A + C)$
2. $(A + \overline{B})(\overline{C} + D)F$
3. $(A + C)(B + \overline{D})(\overline{B} + C)(A + \overline{D} + \overline{E})$

The methods of circuit simplification and design that we will be using are based on the sum-of-products (SOP) form, so we will not be doing much with the product-of-sums (POS) form. It will, however, occur from time to time in some logic circuits that have a particular structure.

REVIEW QUESTIONS

1. Which of the following expressions is in SOP form?
   (a) $AB + CD + E$
   (b) $AB(C + D)$
   (c) $(A + B)(C + D + F)$
   (d) $\overline{MN} + PQ$
2. Repeat question 1 for the POS form.

4-2 SIMPLIFYING LOGIC CIRCUITS

Once the expression for a logic circuit has been obtained, we may be able to reduce it to a simpler form containing fewer terms or fewer variables in one or more terms. The new expression can then be used to implement a circuit that is equivalent to the original circuit but that contains fewer gates and connections.

To illustrate, the circuit of Figure 4-1(a) can be simplified to produce the circuit of Figure 4-1(b). Both circuits perform the same logic, so it should be obvious that the simpler circuit is more desirable because it contains fewer gates and will therefore be smaller and cheaper than the original. Furthermore, the circuit reliability will improve because there are fewer interconnections that can be potential circuit faults.

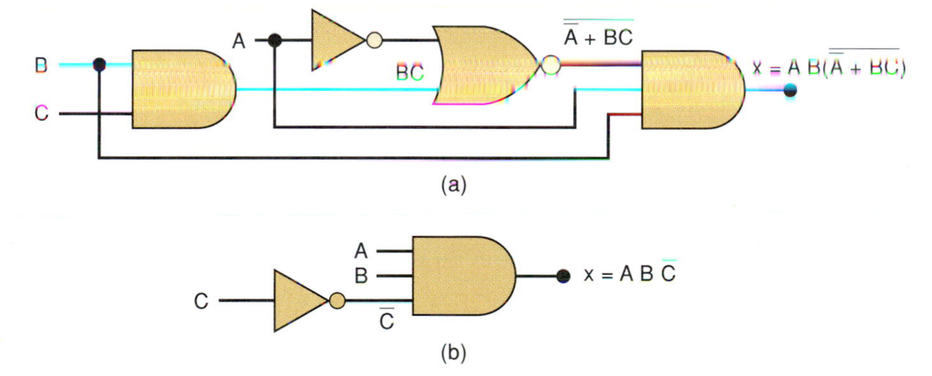

FIGURE 4-1 It is often possible to simplify a logic circuit such as that in part (a) to produce a more efficient implementation, shown in (b).

In subsequent sections, we will study two methods for simplifying logic circuits. One method will utilize the Boolean algebra theorems and, as we shall see, is greatly dependent on inspiration and experience. The other method (Karnaugh mapping) is a systematic, step-by-step approach. Some instructors may wish to skip over this latter method because it is somewhat mechanical and probably does not contribute to a better understanding of Boolean algebra. This can be done without affecting the continuity or clarity of the rest of the text.

4-3 ALGEBRAIC SIMPLIFICATION

We can use the Boolean algebra theorems that we studied in Chapter 3 to help us simplify the expression for a logic circuit. Unfortunately, it is not always obvious which theorems should be applied to produce the simplest result. Furthermore, there is no easy way to tell whether the simplified expression is in its simplest form or whether it could have been simplified further. Thus, algebraic simplification often becomes a process of trial and error. With experience, however, one can become adept at obtaining reasonably good results.

The examples that follow will illustrate many of the ways in which the Boolean theorems can be applied in trying to simplify an expression. You should notice that these examples contain two essential steps:

1. The original expression is put into SOP form by repeated application of DeMorgan's theorems and multiplication of terms.
2. Once the original expression is in SOP form, the product terms are checked for common factors, and factoring is performed wherever possible. The factoring should result in the elimination of one or more terms.

EXAMPLE 4-1

Simplify the logic circuit shown in Figure 4-2(a).

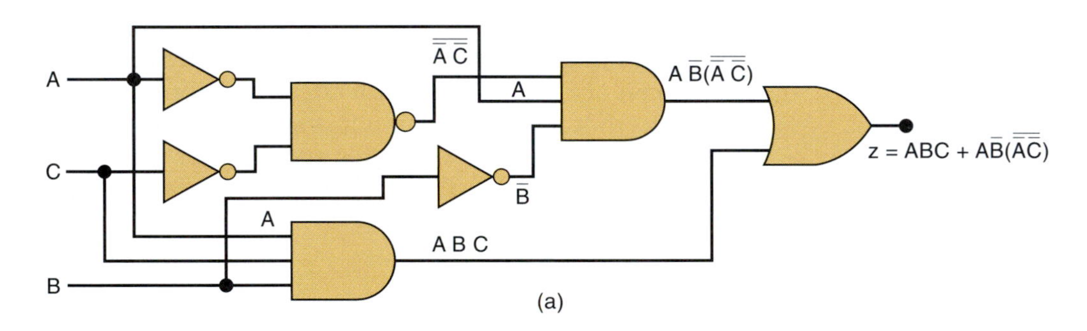

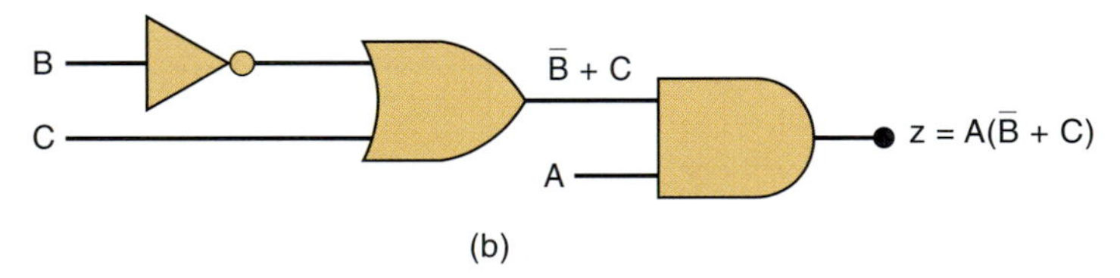

FIGURE 4-2 Example 4-1.

Solution

The first step is to determine the expression for the output using the method presented in Section 3-6. The result is

$$z = ABC + A\overline{B}\cdot(\overline{\overline{A}\,\overline{C}})$$

Once the expression is determined, it is usually a good idea to break down all large inverter signs using DeMorgan's theorems and then multiply out all terms.

$$\begin{aligned} z &= ABC + A\overline{B}(\overline{\overline{A}} + \overline{\overline{C}}) && \text{[theorem (17)]} \\ &= ABC + A\overline{B}(A + C) && \text{[cancel double inversions]} \\ &= ABC + A\overline{B}A + A\overline{B}C && \text{[multiply out]} \\ &= ABC + A\overline{B} + A\overline{B}C && [A \cdot A = A] \end{aligned}$$

With the expression now in SOP form, we should look for common variables among the various terms with the intention of factoring. The first and third terms above have AC in common, which can be factored out:

$$z = AC(B + \overline{B}) + A\overline{B}$$

Since $B + \overline{B} = 1$, then

$$\begin{aligned} z &= AC(1) + A\overline{B} \\ &= AC + A\overline{B} \end{aligned}$$

We can now factor out A, which results in

$$z = A(C + \overline{B})$$

This result can be simplified no further. Its circuit implementation is shown in Figure 4-2(b). It is obvious that the circuit in Figure 4-2(b) is a great deal simpler than the original circuit in Figure 4-2(a).

EXAMPLE 4-2

Simplify the expression $z = A\overline{B}\,\overline{C} + A\overline{B}C + ABC$.

Solution

The expression is already in SOP form.

Method 1: The first two terms in the expression have the product $A\overline{B}$ in common. Thus,

$$\begin{aligned} z &= A\overline{B}(\overline{C} + C) + ABC \\ &= A\overline{B}(1) + ABC \\ &= A\overline{B} + ABC \end{aligned}$$

We can factor the variable A from both terms:

$$z = A(\overline{B} + BC)$$

Invoking theorem (15b):

$$z = A(\overline{B} + C)$$

Method 2: The original expression is $z = A\overline{B}\,\overline{C} + A\overline{B}C + ABC$. The first two terms have $A\overline{B}$ in common. The last two terms have AC in common. How do we know whether to factor $A\overline{B}$ from the first two terms or AC from the last two terms? Actually, we can do both by using the $A\overline{B}C$ term *twice.* In other words, we can rewrite the expression as:

$$z = A\overline{B}\,\overline{C} + A\overline{B}C + A\overline{B}C + ABC$$

where we have added an extra term $A\overline{B}C$. This is valid and will not change the value of the expression because $A\overline{B}C + A\overline{B}C = A\overline{B}C$ [theorem (7)]. Now we can factor $A\overline{B}$ from the first two terms and AC from the last two terms:

$$\begin{aligned} z &= A\overline{B}(C + \overline{C}) + AC(\overline{B} + B) \\ &= A\overline{B}\cdot 1 + AC\cdot 1 \\ &= A\overline{B} + AC = A(\overline{B} + C) \end{aligned}$$

Of course, this is the same result obtained with method 1. This trick of using the same term twice can always be used. In fact, the same term can be used more than twice if necessary.

EXAMPLE 4-3

Simplify $z = \overline{A}C(\overline{\overline{A}BD}) + \overline{A}B\overline{C}\,\overline{D} + A\overline{B}C$.

Solution

First, use DeMorgan's theorem on the first term:

$$z = \overline{A}C(A + \overline{B} + \overline{D}) + \overline{A}B\overline{C}\,\overline{D} + A\overline{B}C \qquad \textbf{(step 1)}$$

Multiplying out yields

$$z = \overline{A}CA + \overline{A}C\overline{B} + \overline{A}C\overline{D} + \overline{A}B\overline{C}\,\overline{D} + A\overline{B}C \qquad \textbf{(2)}$$

Because $\overline{A}\cdot A = 0$, the first term is eliminated:

$$z = \overline{A}\,\overline{B}C + \overline{A}C\overline{D} + \overline{A}B\overline{C}\,\overline{D} + A\overline{B}C \qquad \textbf{(3)}$$

This is the desired SOP form. Now we must look for common factors among the various product terms. The idea is to check for the largest common factor between any two or more product terms. For example, the first and last terms have the common factor $\overline{B}C$, and the second and third terms share the common factor $\overline{A}\,\overline{D}$. We can factor these out as follows:

$$z = \overline{B}C(\overline{A} + A) + \overline{A}\,\overline{D}(C + B\overline{C}) \qquad \textbf{(4)}$$

Now, because $\overline{A} + A = 1$, and $C + B\overline{C} = C + B$ [theorem (15a)], we have

$$z = \overline{B}C + \overline{A}\,\overline{D}(B + C) \qquad \textbf{(5)}$$

This same result could have been reached with other choices for the factoring. For example, we could have factored C from the first, second, and fourth product terms in step 3 to obtain

$$z = C(\overline{A}\,\overline{B} + \overline{A}\,\overline{D} + A\overline{B}) + \overline{A}B\overline{C}\,\overline{D}$$

The expression inside the parentheses can be factored further:

$$z = C(\overline{B}[\overline{A} + A] + \overline{A}\,\overline{D}) + \overline{A}B\overline{C}\,\overline{D}$$

Because $\overline{A} + A = 1$, this becomes

$$z = C(\overline{B} + \overline{A}\,\overline{D}) + \overline{A}B\overline{C}\,\overline{D}$$

Multiplying out yields

$$z = \overline{B}C + \overline{A}C\overline{D} + \overline{A}B\overline{C}\,\overline{D}$$

Now we can factor $\overline{A}\,\overline{D}$ from the second and third terms to get

$$z = \overline{B}C + \overline{A}\,\overline{D}(C + B\overline{C})$$

If we use theorem (15a), the expression in parentheses becomes $B + C$. Thus, we finally have

$$z = \overline{B}C + \overline{A}\,\overline{D}(B + C)$$

This is the same result that we obtained earlier, but it took us many more steps. This illustrates why you should look for the largest common factors: it will generally lead to the final expression in the fewest steps.

Example 4-3 illustrates the frustration often encountered in Boolean simplification. Because we have arrived at the same equation (which appears irreducible) by two different methods, it might seem reasonable to conclude that this final equation is the simplest form. In fact, the simplest form of this equation is

$$z = \overline{A}B\overline{D} + \overline{B}C$$

But there is no apparent way to reduce step (5) to reach this simpler version. In this case, we missed an operation earlier in the process that could have led to the simpler form. The question is, "How could we have known that we missed a step?" Later in this chapter, we will examine a mapping technique that will always lead to the simplest SOP form.

EXAMPLE 4-4

Simplify the expression $x = (\overline{A} + B)(A + B + D)\overline{D}$.

Solution

The expression can be put into sum-of-products form by multiplying out all the terms. The result is

$$x = \overline{A}A\overline{D} + \overline{A}B\overline{D} + \overline{A}D\overline{D} + BA\overline{D} + BB\overline{D} + BD\overline{D}$$

The first term can be eliminated because $\overline{A}A = 0$. Likewise, the third and sixth terms can be eliminated because $D\overline{D} = 0$. The fifth term can be simplified to $B\overline{D}$ because $BB = B$. This gives us

$$x = \overline{A}B\overline{D} + AB\overline{D} + B\overline{D}$$

We can factor $B\overline{D}$ from each term to obtain

$$x = B\overline{D}(\overline{A} + A + 1)$$

Clearly, the term inside the parentheses is always 1, so we finally have

$$x = B\overline{D}$$

EXAMPLE 4-5

Simplify the circuit of Figure 4-3(a).

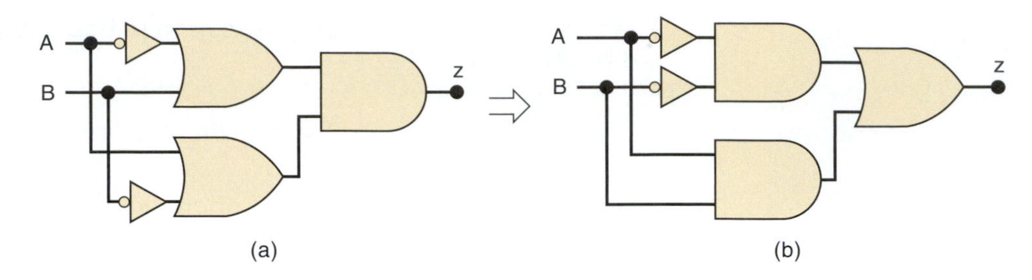

FIGURE 4-3 Example 4-5.

Solution

The expression for output z is

$$z = (\overline{A} + B)(A + \overline{B})$$

Multiplying out to get the sum-of-products form, we obtain

$$z = \overline{A}A + \overline{A}\,\overline{B} + BA + B\overline{B}$$

We can eliminate $\overline{A}A = 0$ and $B\overline{B} = 0$ to end up with

$$z = \overline{A}\,\overline{B} + AB$$

This expression is implemented in Figure 4-3(b), and if we compare it with the original circuit, we see that both circuits contain the same number of gates and connections. In this case, the simplification process produced an equivalent, but not simpler, circuit.

EXAMPLE 4-6

Simplify $x = A\overline{B}C + \overline{A}BD + \overline{C}\,\overline{D}$.

Solution

You can try, but you will not be able to simplify this expression any further.

REVIEW QUESTIONS

1. State which of the following expressions are *not* in the sum-of-products form:
   (a) $RS\overline{T} + \overline{R}S\overline{T} + \overline{T}$
   (b) $A\overline{DC} + \overline{A}DC$
   (c) $MN\overline{P} + (M + \overline{N})P$
   (d) $AB + \overline{A}B\overline{C} + A\,\overline{B}\,\overline{C}D$
2. Simplify the circuit in Figure 4-1(a) to arrive at the circuit of Figure 4-1(b).
3. Change each AND gate in Figure 4-1(a) to a NAND gate. Determine the new expression for x and simplify it.

4-4 DESIGNING COMBINATIONAL LOGIC CIRCUITS

When the desired output level of a logic circuit is given for all possible input conditions, the results can be conveniently displayed in a truth table. The Boolean expression for the required circuit can then be derived from the truth table. For example, consider Figure 4-4(a), where a truth table is shown for a circuit that has two inputs, A and B, and output x. The table shows that output x is to be at the 1 level *only* for the case where $A = 0$ and $B = 1$. It now remains to determine what logic circuit will produce this desired operation. It should be apparent that one possible solution is that shown in Figure 4-4(b). Here an AND gate is used with inputs $\overline{A}$ and B, so that $x = \overline{A} \cdot B$. Obviously x will be 1 *only if* both inputs to the AND gate are 1, namely, $\overline{A} = 1$ (which means that $A = 0$) and $B = 1$. For all other values of A and B, the output x will be 0.

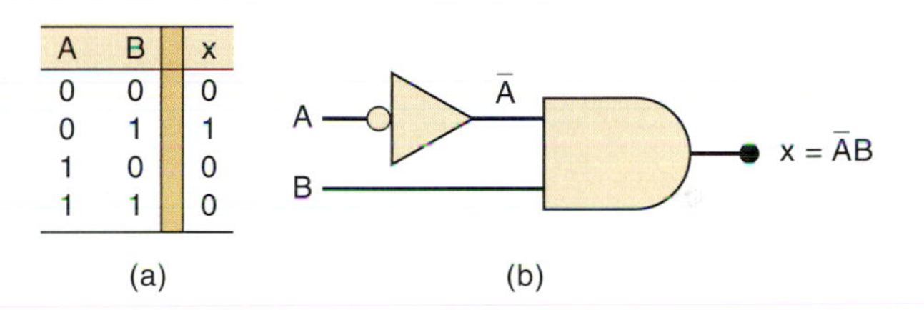

A	B	x
0	0	0
0	1	1
1	0	0
1	1	0

FIGURE 4-4 Circuit that produces a 1 output only for the $A = 0, B = 1$ condition.

A similar approach can be used for the other input conditions. For instance, if x were to be high only for the $A = 1, B = 0$ condition, the resulting circuit would be an AND gate with inputs A and $\overline{B}$. In other words, for any of the four possible input conditions, we can generate a high x output by using an AND gate with appropriate inputs to generate the required AND product. The four different cases are shown in Figure 4-5. Each of the AND gates shown generates an output that is 1 *only* for one given input condition and the output is 0 for all other conditions. It should be noted that the AND inputs are inverted or not inverted depending on the values that the variables have for the given condition. If the variable is 0 for the given condition, it is inverted before entering the AND gate.

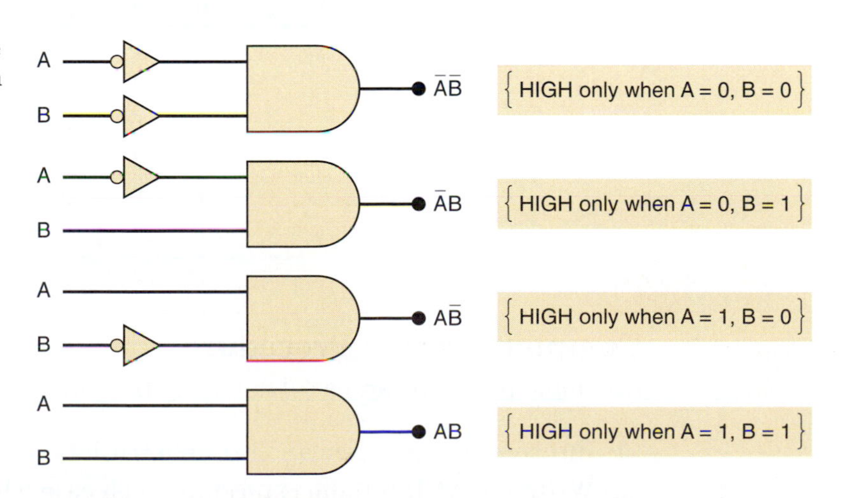

FIGURE 4-5 An AND gate with appropriate inputs can be used to produce a 1 output for a specific set of input levels.

Let us now consider the case shown in Figure 4-6(a), where we have a truth table that indicates that the output x is to be 1 for two different cases: $A = 0, B = 1$ and $A = 1, B = 0$. How can this be implemented? We know that the AND term $\overline{A} \cdot B$ will generate a 1 only for the $A = 0, B = 1$ condition, and the AND term $A \cdot \overline{B}$ will generate a 1 for the $A = 1, B = 0$ condition. Because x must be HIGH for *either* condition, it should be clear that these terms should be ORed together to produce the desired output, x. This implementation is shown in Figure 4-6(b), where the resulting expression for the output is $x = \overline{A}B + A\overline{B}$.

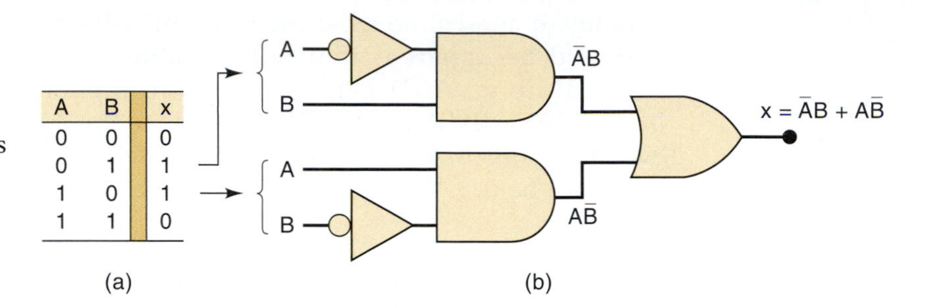

FIGURE 4-6 Each set of input conditions that is to produce a HIGH output is implemented by a separate AND gate. The AND outputs are ORed to produce the final output.

In this example, an AND term is generated for each case in the table where the output x is to be a 1. The AND gate outputs are then ORed together to produce the total output x, which will be 1 when either AND term is 1. This same procedure can be extended to examples with more than two inputs. Consider the truth table for a three-input circuit (Table 4-1). Here there are three cases where the output x is to be 1. The required AND term for each of these cases is shown. Again, note that for each case where a variable is 0, it appears inverted in the AND term. The sum-of-products expression for x is obtained by ORing the three AND terms.

$$x = \overline{A}B\overline{C} + \overline{A}BC + ABC$$

TABLE 4-1

A	B	C	x	
0	0	0	0	
0	0	1	0	
0	1	0	1	$\rightarrow \overline{A}B\overline{C}$
0	1	1	1	$\rightarrow \overline{A}BC$
1	0	0	0	
1	0	1	0	
1	1	0	0	
1	1	1	1	$\rightarrow ABC$

Complete Design Procedure

Any logic problem can be solved using the following step-by-step procedure.

1. Interpret the problem and set up a truth table to describe its operation.
2. Write the AND (product) term for each case where the output is 1.
3. Write the sum-of-products (SOP) expression for the output.
4. Simplify the output expression if possible.
5. Implement the circuit for the final, simplified expression.

The following example illustrates the complete design procedure.

EXAMPLE 4-7

Design a logic circuit that has three inputs, A, B, and C, and whose output will be HIGH only when a majority of the inputs are HIGH.

Solution

Step 1. Set up the truth table.

On the basis of the problem statement, the output x should be 1 whenever two or more inputs are 1; for all other cases, the output should be 0 (Table 4-2).

TABLE 4-2

A	B	C	x	
0	0	0	0	
0	0	1	0	
0	1	0	0	
0	1	1	1	$\rightarrow \overline{A}BC$
1	0	0	0	
1	0	1	1	$\rightarrow A\overline{B}C$
1	1	0	1	$\rightarrow AB\overline{C}$
1	1	1	1	$\rightarrow ABC$

Step 2. Write the AND term for each case where the output is a 1.

There are four such cases. The AND terms are shown next to the truth table (Table 4-2). Again note that each AND term contains each input variable in either inverted or noninverted form.

Step 3. Write the sum-of-products expression for the output.

$$x = \overline{A}BC + A\overline{B}C + AB\overline{C} + ABC$$

Step 4. Simplify the output expression.

This expression can be simplified in several ways. Perhaps the quickest way is to realize that the last term ABC has two variables in common with each of the other terms. Thus, we can use the ABC term to factor with each of the other terms. The expression is rewritten with the ABC term occurring three times (recall from Example 4-2 that this is legal in Boolean algebra):

$$x = \overline{A}BC + ABC + A\overline{B}C + ABC + AB\overline{C} + ABC$$

Factoring the appropriate pairs of terms, we have

$$x = BC(\overline{A} + A) + AC(\overline{B} + B) + AB(\overline{C} + C)$$

Each term in parentheses is equal to 1, so we have

$$x = BC + AC + AB$$

Step 5. Implement the circuit for the final expression.

This expression is implemented in Figure 4-7. Since the expression is in SOP form, the circuit consists of a group of AND gates working into a single OR gate.

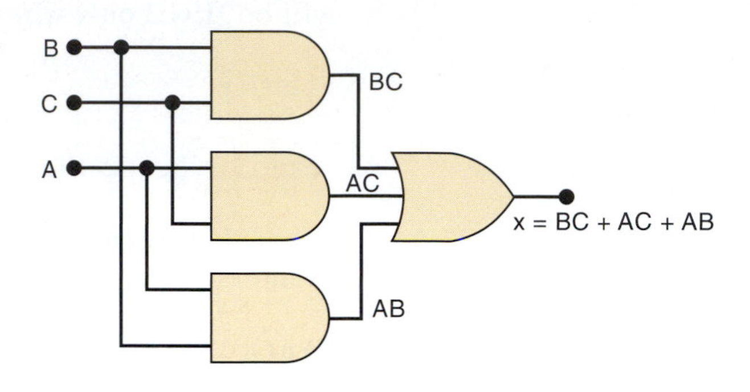

FIGURE 4-7 Example 4-7.

EXAMPLE 4-8

Refer to Figure 4-8(a), where an analog-to-digital converter is monitoring the dc voltage of a 12-V storage battery on an orbiting spaceship. The converter's output is a four-bit binary number, *ABCD*, corresponding to the battery voltage in steps of 1 V, with *A* as the MSB. The converter's binary outputs are fed to a logic circuit that is to produce a HIGH output as long as the binary value is greater than $0110_2 = 6_{10}$; that is, the battery voltage is greater than 6 V. Design this logic circuit.

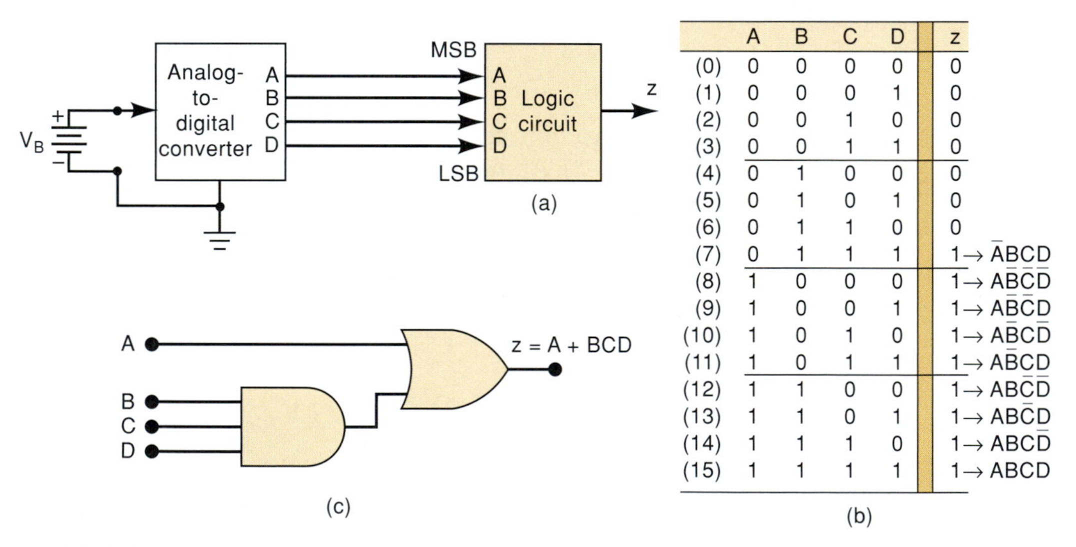

	A	B	C	D	z
(0)	0	0	0	0	0
(1)	0	0	0	1	0
(2)	0	0	1	0	0
(3)	0	0	1	1	0
(4)	0	1	0	0	0
(5)	0	1	0	1	0
(6)	0	1	1	0	0
(7)	0	1	1	1	1 → $\overline{A}BCD$
(8)	1	0	0	0	1 → $A\overline{B}\,\overline{C}\,\overline{D}$
(9)	1	0	0	1	1 → $A\overline{B}\,\overline{C}D$
(10)	1	0	1	0	1 → $A\overline{B}C\overline{D}$
(11)	1	0	1	1	1 → $A\overline{B}CD$
(12)	1	1	0	0	1 → $AB\overline{C}\,\overline{D}$
(13)	1	1	0	1	1 → $AB\overline{C}D$
(14)	1	1	1	0	1 → $ABC\overline{D}$
(15)	1	1	1	1	1 → $ABCD$

FIGURE 4-8 Example 4-8.

Solution

The truth table is shown in Figure 4-8(b). For each case in the truth table, we have indicated the decimal equivalent of the binary number represented by the *ABCD* combination.

The output *z* is set equal to 1 for all those cases where the binary number is greater than 0110. For all other cases, *z* is set equal to 0. This truth table gives us the following sum-of-products expression:

$$z = \overline{A}BCD + A\overline{B}\,\overline{C}\,\overline{D} + A\overline{B}\,\overline{C}D + A\overline{B}C\overline{D} + A\overline{B}CD + AB\overline{C}\,\overline{D} + AB\overline{C}D + ABC\overline{D} + ABCD$$

Simplification of this expression will be a formidable task, but with a little care it can be accomplished. The step-by-step process involves factoring and eliminating terms of the form $A + \bar{A}$:

$$
\begin{aligned}
z &= \bar{A}BCD + A\bar{B}\,\bar{C}(\bar{D} + D) + A\bar{B}C(\bar{D} + D) + AB\bar{C}(\bar{D} + D) + ABC(\bar{D} + D) \\
&= \bar{A}BCD + A\bar{B}\,\bar{C} + A\bar{B}C + AB\bar{C} + ABC \\
&= \bar{A}BCD + A\bar{B}(\bar{C} + C) + AB(\bar{C} + C) \\
&= \bar{A}BCD + A\bar{B} + AB \\
&= \bar{A}BCD + A(\bar{B} + B) \\
&= \bar{A}BCD + A
\end{aligned}
$$

This can be reduced further by invoking theorem (15a), which says that $x + \bar{x}y = x + y$. In this case $x = A$ and $y = BCD$. Thus,

$$z = \bar{A}BCD + A = BCD + A$$

This final expression is implemented in Figure 4-8(c).

As this example demonstrates, the algebraic simplification method can be quite lengthy when the original expression contains a large number of terms. This is a limitation that is not shared by the Karnaugh mapping method, as we will see later.

EXAMPLE 4-9

Refer to Figure 4-9(a). In a simple copy machine, a stop signal, S, is to be generated to stop the machine operation and energize an indicator light whenever either of the following conditions exists: (1) there is no paper in the paper feeder tray; or (2) the two microswitches in the paper path are

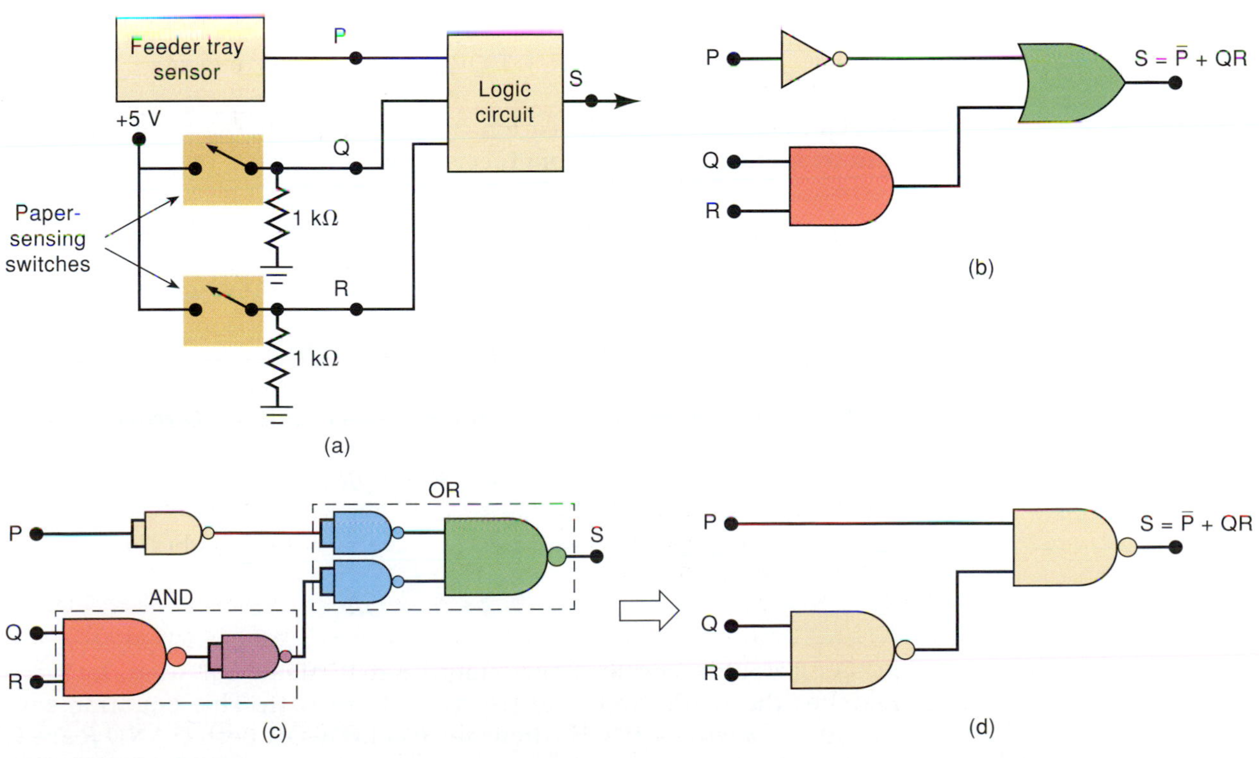

FIGURE 4-9 Example 4-9.

activated, indicating a jam in the paper path. The presence of paper in the feeder tray is indicated by a HIGH at logic signal P. Each of the microswitches produces a logic signal (Q and R) that goes HIGH whenever paper is passing over the switch to activate it. Design the logic circuit to produce a HIGH at output signal S for the stated conditions, and implement it using the 74HC00 CMOS quad two-input NAND chip.

Solution

We will use the five-step process used in Example 4-7. The truth table is shown in Table 4-3. The S output will be a logic 1 whenever $P = 0$ because this indicates no paper in the feeder tray. S will also be a 1 for the two cases where Q and R are both 1, indicating a paper jam. As the table shows, there are five different input conditions that produce a HIGH output. **(Step 1)**

TABLE 4-3

P	Q	R	S	
0	0	0	1	$\overline{P}\,\overline{Q}\,\overline{R}$
0	0	1	1	$\overline{P}\,\overline{Q}R$
0	1	0	1	$\overline{P}Q\overline{R}$
0	1	1	1	$\overline{P}QR$
1	0	0	0	
1	0	1	0	
1	1	0	0	
1	1	1	1	PQR

The AND terms for each of these cases are shown. **(Step 2)**
The sum-of-products expression becomes

$$S = \overline{P}\,\overline{Q}\,\overline{R} + \overline{P}\,\overline{Q}R + \overline{P}Q\overline{R} + \overline{P}QR + PQR \qquad \textbf{(Step 3)}$$

We can begin the simplification by factoring out $\overline{P}\,\overline{Q}$ from terms 1 and 2 and by factoring out $\overline{P}Q$ from terms 3 and 4:

$$S = \overline{P}\,\overline{Q}(\overline{R} + R) + \overline{P}Q(\overline{R} + R) + PQR \qquad \textbf{(Step 4)}$$

Now we can eliminate the $\overline{R} + R$ terms because they equal 1:

$$S = \overline{P}\,\overline{Q} + \overline{P}Q + PQR$$

Factoring $\overline{P}$ from terms 1 and 2 allows us to eliminate Q from these terms:

$$S = \overline{P} + PQR$$

Here we can apply theorem (15b) ($\overline{x} + xy = \overline{x} + y$) to obtain

$$S = \overline{P} + QR$$

As a double check of this simplified Boolean equation, let's see if it matches the truth table that we started out with. This equation says that the output S will be HIGH whenever P is LOW OR both Q AND R are HIGH. Look at Table 4-3 and observe that the output is HIGH for all four cases when P is LOW. S is also HIGH when Q AND R are both HIGH, regardless of the state of P. This agrees with the equation.

The AND/OR implementation for this circuit is shown in Figure 4-9(b).

(Step 5)

To implement this circuit using the 74HC00 quad two-input NAND chip, we must convert each gate and the INVERTER by their NAND-gate equivalents (per Section 3-12). This is shown in Figure 4-9(c). Clearly, we can eliminate the double inverters to produce the NAND-gate implementation shown in Figure 4-9(d).

The final wired-up circuit is obtained by connecting two of the NAND gates on the 74HC00 chip. This CMOS chip has the same gate configuration and pin numbers as the TTL 74LS00 chip of Figure 3-31. Figure 4-10 shows the wired-up circuit with pin numbers, including the +5 V and GROUND pins. It also includes an output driver transistor and LED to indicate the state of output S.

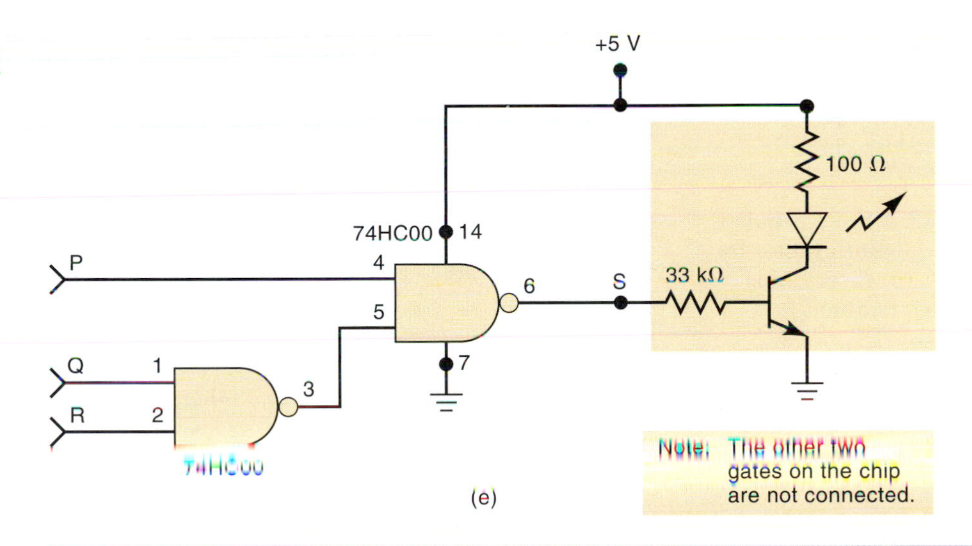

FIGURE 4-10 Circuit of Figure 4-9(d) implemented using 74HC00 NAND chip.

REVIEW QUESTIONS

1. Write the sum-of-products expression for a circuit with four inputs and an output that is to be HIGH only when input A is LOW at the same time that exactly two other inputs are LOW.
2. Implement the expression of question 1 using all four-input NAND gates. How many are required?

4-5 KARNAUGH MAP METHOD

The **Karnaugh map (K map)** is a graphical tool used to simplify a logic equation or to convert a truth table to its corresponding logic circuit in a simple, orderly process. Although a K map can be used for problems involving any number of input variables, its practical usefulness is limited to five or six variables. The following discussion will be limited to problems with up to four inputs because even five- and six-input problems are too involved and are best done by a computer program.

Karnaugh Map Format

The K map, like a truth table, is a means for showing the relationship between logic inputs and the desired output. Figure 4-11 shows three examples of K maps for two, three, and four variables, together with the corresponding truth tables. These examples illustrate the following important points:

1. The truth table gives the value of output X for each combination of input values. The K map gives the same information in a different format. Each case in the truth table corresponds to a square in the K map. For example, in Figure 4-11(a), the $A = 0, B = 0$ condition in the truth table corresponds to the $\overline{A}\,\overline{B}$ square in the K map. Because the truth table shows $X = 1$ for this case, a 1 is placed in the $\overline{A}\,\overline{B}$ square in the K map. Similarly, the $A = 1, B = 1$ condition in the truth table corresponds to the AB square of the K map. Because $X = 1$ for this case, a 1 is placed in the AB square. All other squares are filled with 0s. This same idea is used in the three- and four-variable maps shown in the figure.
2. The K-map squares are labeled so that horizontally adjacent squares differ only in one variable. For example, the upper left-hand square in the four-variable map is $\overline{A}\,\overline{B}\,\overline{C}\,\overline{D}$, while the square immediately to its right is $\overline{A}\,\overline{B}\,\overline{C}D$ (only the D variable is different). Similarly, vertically adjacent

FIGURE 4-11 Karnaugh maps and truth tables for (a) two, (b) three, and (c) four variables.

A	B	X
0	0	1 → $\overline{A}\overline{B}$
0	1	0
1	0	0
1	1	1 → AB

$$\{\, x = \overline{A}\overline{B} + AB \,\}$$

	$\overline{B}$	B
$\overline{A}$	1	0
A	0	1

(a)

A	B	C	X
0	0	0	1 → $\overline{A}\overline{B}\overline{C}$
0	0	1	1 → $\overline{A}\overline{B}C$
0	1	0	1 → $\overline{A}B\overline{C}$
0	1	1	0
1	0	0	0
1	0	1	0
1	1	0	1 → $AB\overline{C}$
1	1	1	0

$$\left\{\, X = \overline{A}\overline{B}\overline{C} + \overline{A}\overline{B}C + \overline{A}B\overline{C} + AB\overline{C} \,\right\}$$

	$\overline{C}$	C
$\overline{A}\overline{B}$	1	1
$\overline{A}B$	1	0
AB	1	0
$A\overline{B}$	0	0

(b)

A	B	C	D	X
0	0	0	0	0
0	0	0	1	1 → $\overline{A}\overline{B}\overline{C}D$
0	0	1	0	0
0	0	1	1	0
0	1	0	0	0
0	1	0	1	1 → $\overline{A}B\overline{C}D$
0	1	1	0	0
0	1	1	1	0
1	0	0	0	0
1	0	0	1	0
1	0	1	0	0
1	0	1	1	0
1	1	0	0	0
1	1	0	1	1 → $AB\overline{C}D$
1	1	1	0	0
1	1	1	1	1 → $ABCD$

$$\left\{\, X = \overline{A}\overline{B}\overline{C}D + \overline{A}B\overline{C}D + AB\overline{C}D + ABCD \,\right\}$$

	$\overline{C}\overline{D}$	$\overline{C}D$	CD	$C\overline{D}$
$\overline{A}\overline{B}$	0	1	0	0
$\overline{A}B$	0	1	0	0
AB	0	1	1	0
$A\overline{B}$	0	0	0	0

(c)

squares differ only in one variable. For example, the upper left-hand square is $\overline{A}\,\overline{B}\,\overline{C}\,\overline{D}$, while the square directly below it is $\overline{A}B\overline{C}\,\overline{D}$ (only the B variable is different).

Note that each square in the top row is considered to be adjacent to a corresponding square in the bottom row. For example, the $\overline{A}\,\overline{B}CD$ square in the top row is adjacent to the $A\overline{B}CD$ square in the bottom row because they differ only in the A variable. You can think of the top of the map as being wrapped around to touch the bottom of the map. Similarly, squares in the leftmost column are adjacent to corresponding squares in the rightmost column.

3. In order for vertically and horizontally adjacent squares to differ in only one variable, the top-to-bottom labeling must be done in the order shown: $\overline{A}\,\overline{B}$, $\overline{A}B$, AB, $A\overline{B}$. The same is true of the left-to-right labeling: $\overline{C}\,\overline{D}$, $\overline{C}D$, CD, $C\overline{D}$.
4. Once a K map has been filled with 0s and 1s, the sum-of-products expression for the output X can be obtained by ORing together those squares that contain a 1. In the three-variable map of Figure 4-11(b), the $\overline{A}\,\overline{B}\,\overline{C}$, $\overline{A}\,\overline{B}C$, $\overline{A}B\overline{C}$, and $AB\overline{C}$ squares contain a 1, so that $X = \overline{A}\,\overline{B}\,\overline{C} + \overline{A}\,\overline{B}C + \overline{A}B\overline{C} + AB\overline{C}$.

Looping

The expression for output X can be simplified by properly combining those squares in the K map that contain 1s. The process for combining these 1s is called **looping**.

Looping Groups of Two (Pairs)

Figure 4-12(a) is the K map for a particular three-variable truth table. This map contains a pair of 1s that are vertically adjacent to each other; the first

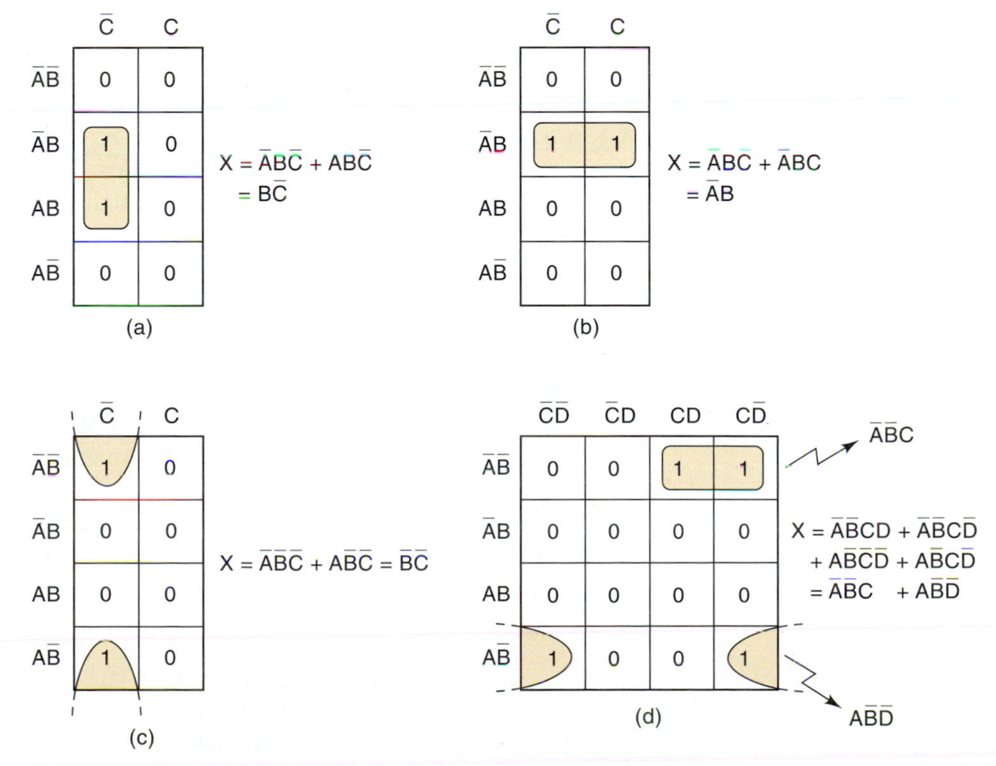

FIGURE 4-12 Examples of looping pairs of adjacent 1s.

represents $\overline{A}B\overline{C}$, and the second represents $AB\overline{C}$. Note that in these two terms only the A variable appears in both normal and complemented (inverted) form, while B and $\overline{C}$ remain unchanged. These two terms can be looped (combined) to give a resultant that eliminates the A variable because it appears in both uncomplemented and complemented forms. This is easily proved as follows:

$$\begin{aligned} X &= \overline{A}B\overline{C} + AB\overline{C} \\ &= B\overline{C}(\overline{A} + A) \\ &= B\overline{C}(1) = B\overline{C} \end{aligned}$$

This same principle holds true for any pair of vertically or horizontally adjacent 1s. Figure 4-12(b) shows an example of two horizontally adjacent 1s. These two can be looped and the C variable eliminated because it appears in both its uncomplemented and complemented forms to give a resultant of $X = \overline{A}B$.

Another example is shown in Figure 4-12(c). In a K map, the top row and bottom row of squares are considered to be adjacent. Thus, the two 1s in this map can be looped to provide a resultant of $\overline{A}\,\overline{B}\,\overline{C} + A\overline{B}\,\overline{C} = \overline{B}\,\overline{C}$.

Figure 4-12(d) shows a K map that has two pairs of 1s that can be looped. The two 1s in the top row are horizontally adjacent. The two 1s in the bottom row are also adjacent because, in a K map, the leftmost column and the rightmost column of squares are considered to be adjacent. When the top pair of 1s is looped, the D variable is eliminated (because it appears as both D and $\overline{D}$) to give the term $\overline{A}\,\overline{B}C$. Looping the bottom pair eliminates the C variable to give the term $A\overline{B}\,\overline{D}$. These two terms are ORed to give the final result for X.

To summarize:

Looping a pair of adjacent 1s in a K map eliminates the variable that appears in complemented and uncomplemented form.

Looping Groups of Four (Quads)

A K map may contain a group of four 1s that are adjacent to each other. This group is called a *quad.* Figure 4-13 shows several examples of quads. In Figure 4-13(a), the four 1s are vertically adjacent, and in Figure 4-13(b), they are horizontally adjacent. The K map in Figure 4-13(c) contains four 1s in a square, and they are considered adjacent to each other. The four 1s in Figure 4-13(d) are also adjacent, as are those in Figure 4-13(e), because, as pointed out earlier, the top and bottom rows are considered to be adjacent to each other, as are the leftmost and rightmost columns.

When a quad is looped, the resultant term will contain only the variables that do not change form for all the squares in the quad. For example, in Figure 4-13(a), the four squares that contain a 1 are $\overline{A}\,\overline{B}C$, $\overline{A}BC$, ABC, and $A\overline{B}C$. Examination of these terms reveals that only the variable C remains unchanged (both A and B appear in complemented and uncomplemented form). Thus, the resultant expression for X is simply $X = C$. This can be proved as follows:

$$\begin{aligned} X &= \overline{A}\,\overline{B}C + \overline{A}BC + ABC + A\overline{B}C \\ &= \overline{A}C(\overline{B} + B) + AC(B + \overline{B}) \\ &= \overline{A}C + AC \\ &= C(\overline{A} + A) = C \end{aligned}$$

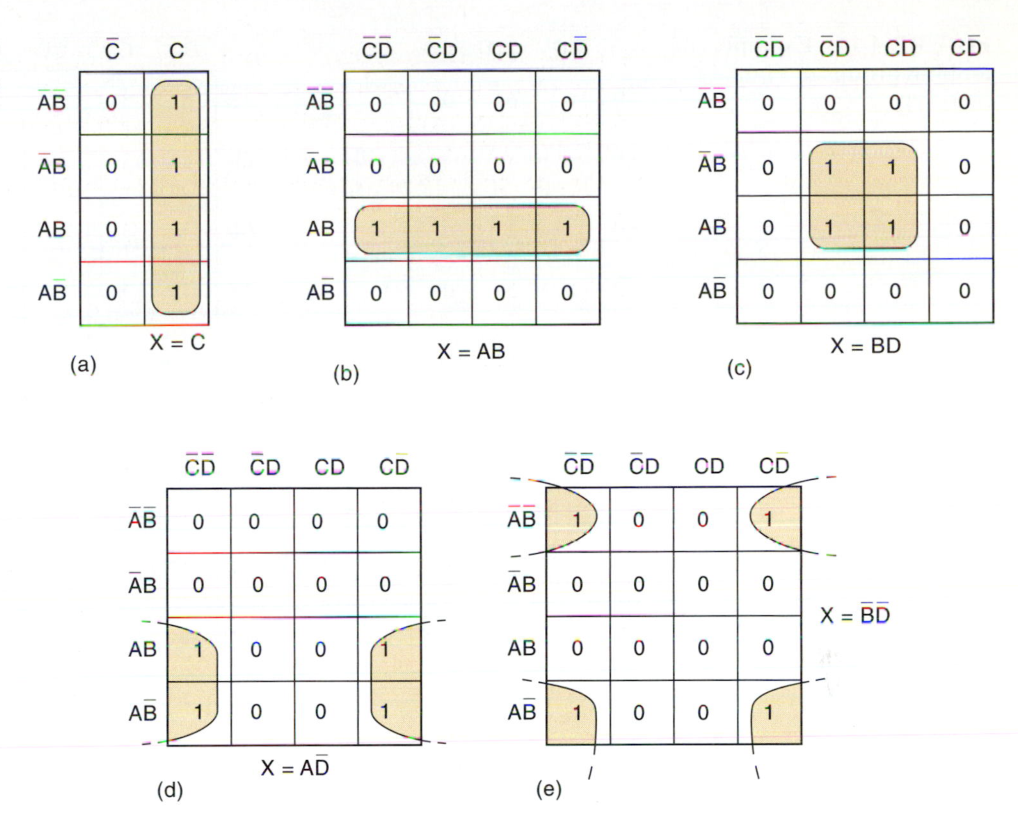

FIGURE 4-13 Examples of looping groups of four 1s (quads).

As another example, consider Figure 4-13(d), where the four squares containing 1s are $AB\overline{C}\,\overline{D}$, $A\overline{B}\,\overline{C}\,\overline{D}$, $ABC\overline{D}$, and $A\overline{B}C\overline{D}$. Examination of these terms indicates that only the variables A and $\overline{D}$ remain unchanged, so that the simplified expression for X is

$$X = A\overline{D}$$

This can be proved in the same manner that was used above. The reader should check each of the other cases in Figure 4-13 to verify the indicated expressions for X.

To summarize:

Looping a quad of adjacent 1s eliminates the two variables that appear in both complemented and uncomplemented form.

Looping Groups of Eight (Octets)

A group of eight 1s that are adjacent to one another is called an *octet.* Several examples of octets are shown in Figure 4-14. When an octet is looped in a four-variable map, three of the four variables are eliminated because only one variable remains unchanged. For example, examination of the eight looped squares in Figure 4-14(a) shows that only the variable B is in the same form for all eight squares: the other variables appear in complemented and uncomplemented form. Thus, for this map, $X = B$. The reader can verify the results for the other examples in Figure 4-14.

To summarize:

Looping an octet of adjacent 1s eliminates the three variables that appear in both complemented and uncomplemented form.

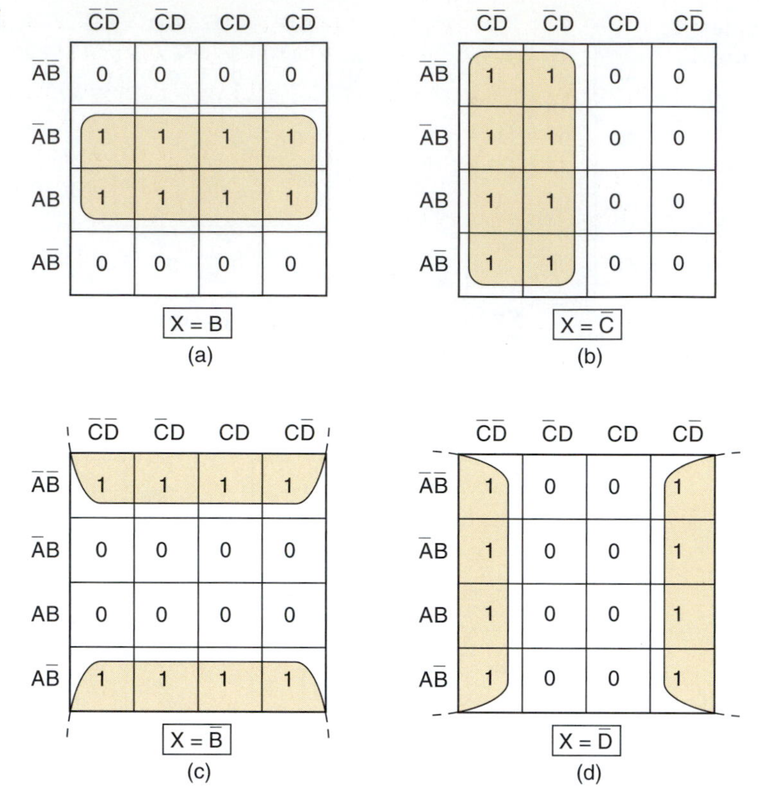

FIGURE 4-14 Examples of looping groups of eight 1s (octets).

Complete Simplification Process

We have seen how looping of pairs, quads, and octets on a K map can be used to obtain a simplified expression. We can summarize the rule for loops of *any* size:

> **When a variable appears in both complemented and uncomplemented form within a loop, that variable is eliminated from the expression. Variables that are the same for all squares of the loop must appear in the final expression.**

It should be clear that a larger loop of 1s eliminates more variables. To be exact, a loop of two eliminates one variable, a loop of four eliminates two variables, and a loop of eight eliminates three. This principle will now be used to obtain a simplified logic expression from a K map that contains any combination of 1s and 0s.

The procedure will first be outlined and then applied to several examples. The steps below are followed in using the K-map method for simplifying a Boolean expression:

Step 1 Construct the K map and place 1s in those squares corresponding to the 1s in the truth table. Place 0s in the other squares.

Step 2 Examine the map for adjacent 1s and loop those 1s that are *not* adjacent to any other 1s. These are called *isolated* 1s.

Step 3 Next, look for those 1s that are adjacent to only one other 1. Loop *any* pair containing such a 1.

Step 4 Loop any octet even if it contains some 1s that have already been looped.

Step 5 Loop any quad that contains one or more 1s that have not already been looped, *making sure to use the minimum number of loops.*

Step 6 Loop any pairs necessary to include any 1s that have not yet been looped, *making sure to use the minimum number of loops.*

Step 7 Form the OR sum of all the terms generated by each loop.

These steps will be followed exactly and referred to in the following examples. In each case, the resulting logic expression will be in its simplest sum-of-products form.

EXAMPLE 4-10

Figure 4-15(a) shows the K map for a four-variable problem. We will assume that the map was obtained from the problem truth table **(step 1)**. The squares are numbered for convenience in identifying each loop.

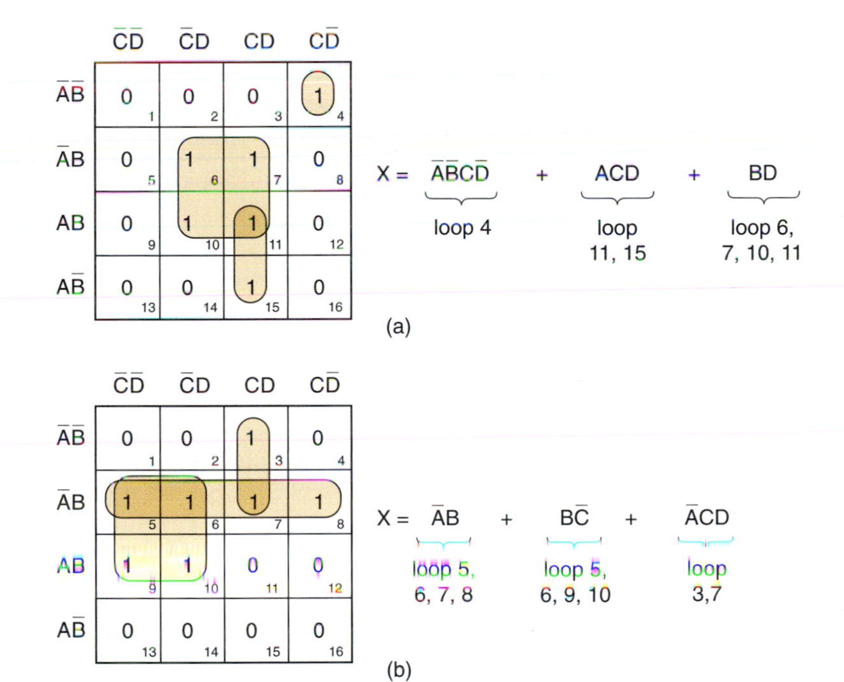

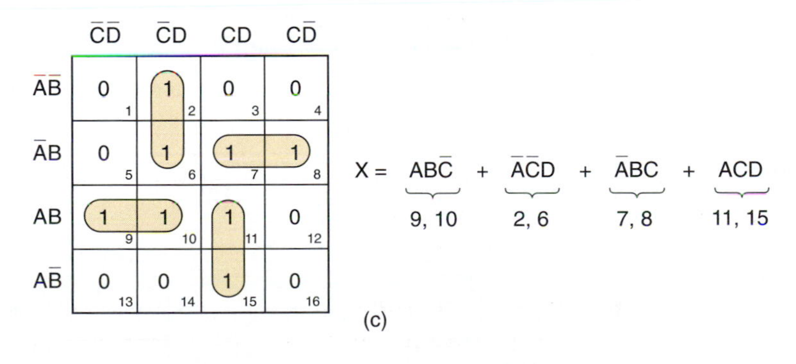

FIGURE 4-15 Examples 4-10 to 4-12.

Step 2 Square 4 is the only square containing a 1 that is not adjacent to any other 1. It is looped and is referred to as loop 4.

Step 3 Square 15 is adjacent *only* to square 11. This pair is looped and referred to as loop 11, 15.

Step 4 There are no octets.

Step 5 Squares 6, 7, 10, and 11 form a quad. This quad is looped (loop 6, 7, 10, 11). Note that square 11 is used again, even though it was part of loop 11, 15.

Step 6 All 1s have already been looped.

Step 7 Each loop generates a term in the expression for X. Loop 4 is simply $\overline{A}\,\overline{B}C\overline{D}$. Loop 11, 15 is ACD (the B variable is eliminated). Loop 6, 7, 10, 11 is BD (A and C are eliminated).

EXAMPLE 4-11

Consider the K map in Figure 4-15(b). Once again, we can assume that step 1 has already been performed.

Step 2 There are no isolated 1s.

Step 3 The 1 in square 3 is adjacent *only* to the 1 in square 7. Looping this pair (loop 3, 7) produces the term $\overline{A}CD$.

Step 4 There are no octets.

Step 5 There are two quads. Squares 5, 6, 7, and 8 form one quad. Looping this quad produces the term $\overline{A}B$. The second quad is made up of squares 5, 6, 9, and 10. This quad is looped because it contains two squares that have not been looped previously. Looping this quad produces $B\overline{C}$.

Step 6 All 1s have already been looped.

Step 7 The terms generated by the three loops are ORed together to obtain the expression for X.

EXAMPLE 4-12

Consider the K map in Figure 4-15(c).

Step 2 There are no isolated 1s.

Step 3 The 1 in square 2 is adjacent only to the 1 in square 6. This pair is looped to produce $\overline{A}\,\overline{C}D$. Similarly, square 9 is adjacent only to square 10. Looping this pair produces $AB\overline{C}$. Likewise, loop 7, 8 and loop 11, 15 produce the terms $\overline{A}BC$ and ACD, respectively.

Step 4 There are no octets.

Step 5 There is one quad formed by squares 6, 7, 10, and 11. This quad, however, is *not* looped because all the 1s in the quad have been included in other loops.

Step 6 All 1s have already been looped.

Step 7 The expression for X is shown in the figure.

EXAMPLE 4-13

Consider the K map in Figure 4-16(a).

FIGURE 4-16 The same K map with two equally good solutions.

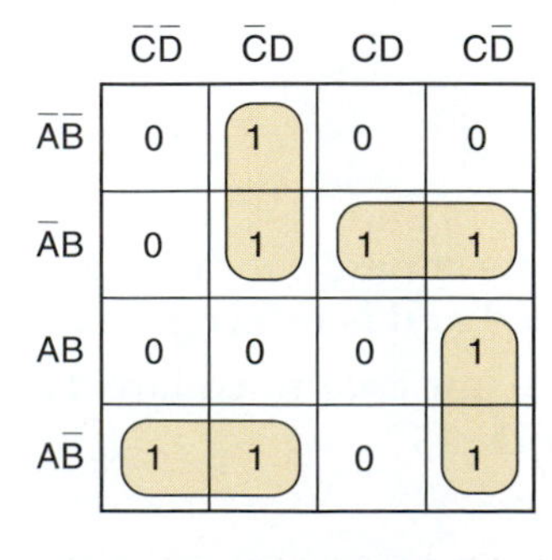

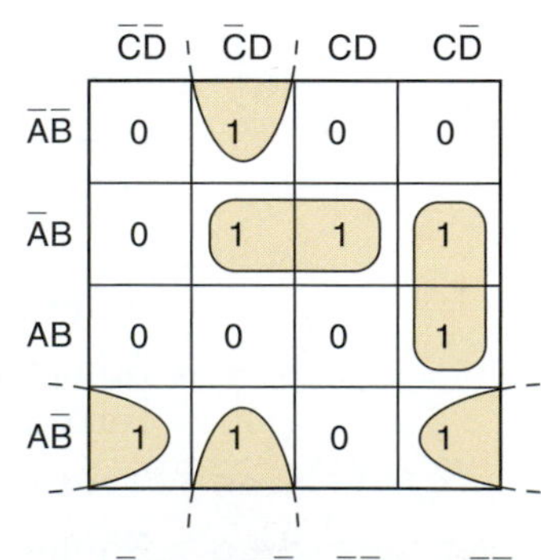

Step 2 There are no isolated 1s.

Step 3 There are no 1s that are adjacent to only one other 1.

Step 4 There are no octets.

Step 5 There are no quads.

Steps 6 and 7 There are many possible pairs. The looping must use the minimum number of loops to account for all the 1s. For this map, there are *two* possible loopings, which require only four looped pairs. Figure 4-16(a) shows one solution and its resultant expression. Figure 4-16(b) shows the other. Note that both expressions are of the same complexity, and so neither is better than the other.

Filling a K Map from an Output Expression

When the desired output is presented as a Boolean expression instead of a truth table, the K map can be filled by using the following steps:

1. Get the expression into SOP form if it is not already in that form.
2. For each product term in the SOP expression, place a 1 in each K-map square whose label contains the same combination of input variables. Place a 0 in all other squares.

The following example illustrates this procedure.

EXAMPLE 4-14

Use a K map to simplify $y = \overline{C}(\overline{A}\,\overline{B}\,\overline{D} + D) + A\overline{B}C + \overline{D}$.

Solution

1. Multiply out the first term to get $y = \overline{A}\,\overline{B}\,\overline{C}\,\overline{D} + \overline{C}D + A\overline{B}C + \overline{D}$, which is now in SOP form.
2. For the $\overline{A}\,\overline{B}\,\overline{C}\,\overline{D}$ term, simply put a 1 in the $\overline{A}\,\overline{B}\,\overline{C}\,\overline{D}$ square of the K map (Figure 4-17). For the $\overline{C}D$ term, place a 1 in all squares with $\overline{C}D$ in their labels, that is, $\overline{A}\,\overline{B}\,\overline{C}D$, $\overline{A}B\overline{C}D$, $AB\overline{C}D$, $A\overline{B}\,\overline{C}D$. For the $A\overline{B}C$ term, place a 1 in all squares that have an $A\overline{B}C$ in their labels, that is, $A\overline{B}C\overline{D}$, $A\overline{B}CD$. For the $\overline{D}$ term, place a 1 in all squares that have a $\overline{D}$ in their labels, that is, all squares in the leftmost and rightmost columns.

FIGURE 4-17 Example 4-14.

	$\overline{C}\,\overline{D}$	$\overline{C}D$	CD	$C\overline{D}$
$\overline{A}\,\overline{B}$	1	1	0	1
$\overline{A}B$	1	1	0	1
AB	1	1	0	1
$A\overline{B}$	1	1	1	1

$y = A\overline{B} + \overline{C} + \overline{D}$

The K map is now filled and can be looped for simplification. Verify that proper looping produces $y = A\overline{B} + \overline{C} + \overline{D}$.

Don't-Care Conditions

Some logic circuits can be designed so that there are certain input conditions for which there are no specified output levels, usually because these input conditions will never occur. In other words, there will be certain combinations of input levels where we "don't care" whether the output is HIGH or LOW. This is illustrated in the truth table of Figure 4-18(a).

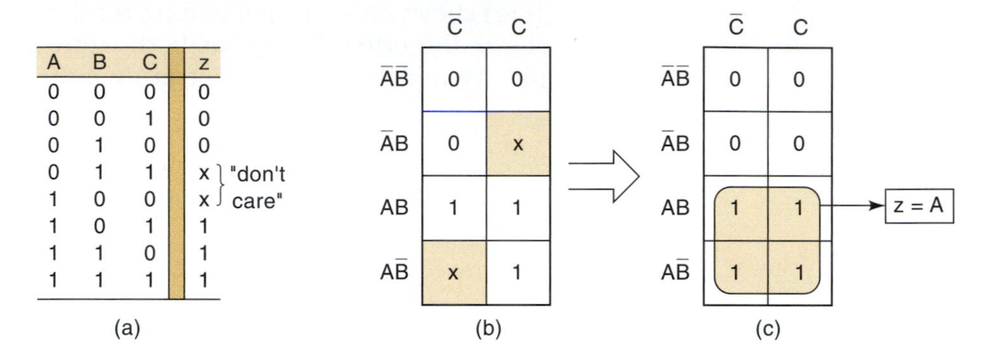

A	B	C	z	
0	0	0	0	
0	0	1	0	
0	1	0	0	
0	1	1	x	"don't care"
1	0	0	x	
1	0	1	1	
1	1	0	1	
1	1	1	1	

FIGURE 4-18 "Don't-care" conditions should be changed to 0 or 1 to produce K-map looping that yields the simplest expression.

Here the output z is not specified as either 0 or 1 for the conditions $A, B, C = 1, 0, 0$ and $A, B, C = 0, 1, 1$. Instead, an x is shown for these conditions. The x represents the **don't-care condition**. A don't-care condition can come about for several reasons, the most common being that in some situations certain input combinations can never occur, and so there is no specified output for these conditions.

A circuit designer is free to make the output for any don't-care condition either a 0 or a 1 to produce the simplest output expression. For example, the K map for this truth table is shown in Figure 4-18(b) with an x placed in the $A\overline{B}\,\overline{C}$ and $\overline{A}BC$ squares. The designer here would be wise to change the x in the $A\overline{B}\,\overline{C}$ square to a 1 and the x in the $\overline{A}BC$ square to a 0 because this would produce a quad that can be looped to produce $z = A$, as shown in Figure 4-18(c).

Whenever don't-care conditions occur, we must decide which x to change to 0 and which to 1 to produce the best K-map looping (i.e., the simplest expression). This decision is not always an easy one. Several end-of-chapter problems will provide practice in dealing with don't-care cases. Here's another example.

EXAMPLE 4-15

Let's design a logic circuit that controls an elevator door in a three-story building. The circuit in Figure 4-19(a) has four inputs. M is a logic signal that indicates when the elevator is moving ($M = 1$) or stopped ($M = 0$). $F1$, $F2$, and $F3$ are floor indicator signals that are normally LOW, and they go HIGH only when the elevator is positioned at the level of that particular floor. For example, when the elevator is lined up level with the second floor, $F2 = 1$ and $F1 = F3 = 0$. The circuit output is the *OPEN* signal, which is normally LOW and will go HIGH when the elevator door is to be opened.

We can fill in the truth table for the *OPEN* output [Figure 4-19(b)] as follows:

1. Because the elevator cannot be lined up with more than one floor at a time, only one of the floor inputs can be HIGH at any given time. This means that all those cases in the truth table where more than one floor

FIGURE 4-19 Example 4-15.

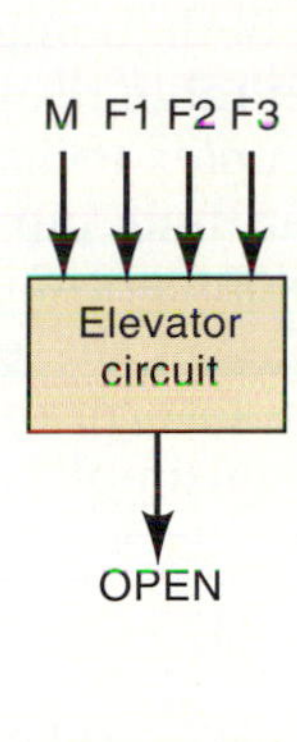

(a)

M	F1	F2	F3	OPEN
0	0	0	0	0
0	0	0	1	1
0	0	1	0	1
0	0	1	1	X
0	1	0	0	1
0	1	0	1	X
0	1	1	0	X
0	1	1	1	X
1	0	0	0	0
1	0	0	1	0
1	0	1	0	0
1	0	1	1	X
1	1	0	0	0
1	1	0	1	X
1	1	1	0	X
1	1	1	1	X

(b)

	$\overline{F2}\,\overline{F3}$	$\overline{F2}\,F3$	$F2\,F3$	$F2\,\overline{F3}$
$\overline{M}\,\overline{F1}$	0	1	X	1
$\overline{M}\,F1$	1	X	X	X
$M\,F1$	0	X	X	X
$M\,\overline{F1}$	0	0	X	0

(c)

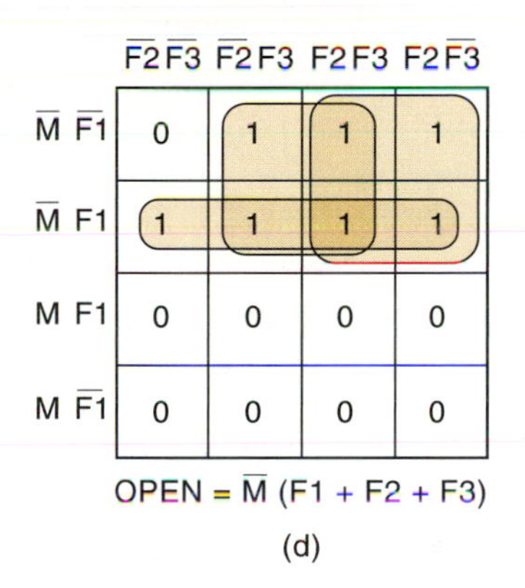

(d)

input is a 1 are don't-care conditions. We can place an x in the *OPEN* output column for those eight cases where more than one F input is 1.

2. Looking at the other eight cases, when $M = 1$ the elevator is moving, so *OPEN* must be a 0 because we do not want the elevator door to open. When $M = 0$ (elevator stopped) we want $OPEN = 1$ provided that one of the floor inputs is 1. When $M = 0$ and all floor inputs are 0, the elevator is stopped but is not properly lined up with any floor, so we want $OPEN = 0$ to keep the door closed.

The truth table is now complete and we can transfer its information to the K map in Figure 4-19(c). The map has only three 1s, but it has eight don't-cares. By changing four of these don't-care squares to 1s, we can produce quad loopings that contain the original 1s [Figure 4-19(d)]. This is the best we can do as far as minimizing the output expression. Verify that the loopings produce the *OPEN* output expression shown.

Summary

The K-map process has several advantages over the algebraic method. K mapping is a more orderly process with well-defined steps compared with the trial-and-error process sometimes used in algebraic simplification. K mapping usually requires fewer steps, especially for expressions containing many terms, and it always produces a minimum expression.

Nevertheless, some instructors prefer the algebraic method because it requires a thorough knowledge of Boolean algebra and is not simply a mechanical procedure. Each method has its advantages, and although most logic designers are adept at both, being proficient in one method is all that is necessary to produce acceptable results.

There are other, more complex techniques that designers use to minimize logic circuits with more than four inputs. These techniques are especially suited for circuits with large numbers of inputs where algebraic and K-mapping methods are not feasible. Most of these techniques can be translated into a computer program that will perform the minimization from input data that supply the truth table or the unsimplified expression.

REVIEW QUESTIONS

1. Use K mapping to obtain the expression of Example 4-7.
2. Use K mapping to obtain the expression of Example 4-8. This should emphasize the advantage of K mapping for expressions containing many terms.
3. Obtain the expression of Example 4-9 using a K map.
4. What is a don't-care condition?

4-6 EXCLUSIVE-OR AND EXCLUSIVE-NOR CIRCUITS

Two special logic circuits that occur quite often in digital systems are the *exclusive-OR* and *exclusive-NOR* circuits.

Exclusive-OR

Consider the logic circuit of Figure 4-20(a). The output expression of this circuit is

$$x = \overline{A}B + A\overline{B}$$

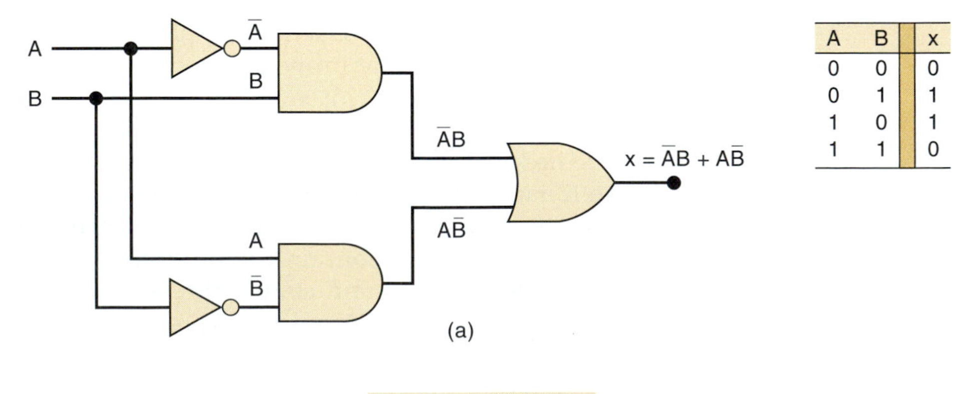

A	B	x
0	0	0
0	1	1
1	0	1
1	1	0

FIGURE 4-20 (a) Exclusive-OR circuit and truth table; (b) traditional XOR gate symbol; (c) IEEE/ANSI symbol for XOR gate.

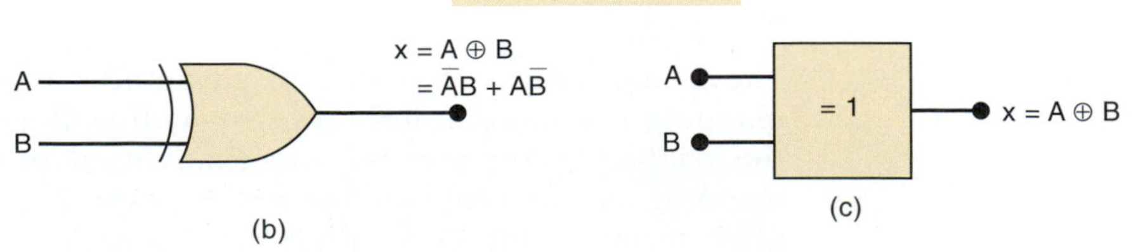

The accompanying truth table shows that $x = 1$ for two cases: $A = 0, B = 1$ (the $\overline{A}B$ term) and $A = 1, B = 0$ (the $A\overline{B}$ term). In other words:

This circuit produces a HIGH output whenever the two inputs are at opposite levels.

This is the **exclusive-OR** circuit, which will hereafter be abbreviated **XOR**.

This particular combination of logic gates occurs quite often and is very useful in certain applications. In fact, the XOR circuit has been given a symbol of its own, shown in Figure 4-20(b). This symbol is assumed to contain all of the logic contained in the XOR circuit and therefore has the same logic expression and truth table. This XOR circuit is commonly referred to as an XOR *gate*, and we consider it as another type of logic gate. The IEEE/ANSI symbol for an XOR gate is shown in Figure 4-20(c). The dependency notation (= 1) inside the block indicates that the output will be active-HIGH *only* when a single input is HIGH.

An XOR gate has only *two* inputs; there are no three-input or four-input XOR gates. The two inputs are combined so that $x = \overline{A}B + A\overline{B}$. A shorthand way that is sometimes used to indicate the XOR output expression is

$$x = A \oplus B$$

where the symbol $\oplus$ represents the XOR gate operation.

The characteristics of an XOR gate are summarized as follows:

1. It has only two inputs and its output is

$$x = \overline{A}B + A\overline{B} = A \oplus B$$

2. Its output is *HIGH* only when the two inputs are at *different* levels.

Several ICs are available that contain XOR gates. Those listed below are *quad* XOR chips containing four XOR gates.

74LS86	Quad XOR (TTL family)
74C86	Quad XOR (CMOS family)
74HC86	Quad XOR (high-speed CMOS)

Exclusive-NOR

The **exclusive-NOR** circuit (abbreviated **XNOR**) operates completely opposite to the XOR circuit. Figure 4-21(a) shows an XNOR circuit and its accompanying truth table. The output expression is

$$x = AB + \overline{A}\,\overline{B}$$

which indicates along with the truth table that x will be 1 for two cases: $A = B = 1$ (the AB term) and $A = B = 0$ (the $\overline{A}\,\overline{B}$ term). In other words:

The XNOR produces a HIGH output whenever the two inputs are at the same level.

It should be apparent that the output of the XNOR circuit is the exact inverse of the output of the XOR circuit. The traditional symbol for an XNOR

FIGURE 4-21 (a) Exclusive-NOR circuit; (b) traditional symbol for XNOR gate; (c) IEEE/ANSI symbol.

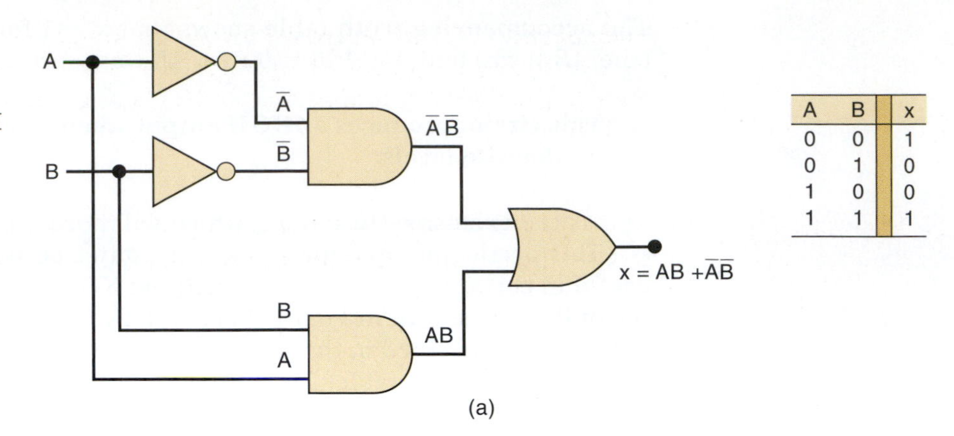

A	B	x
0	0	1
0	1	0
1	0	0
1	1	1

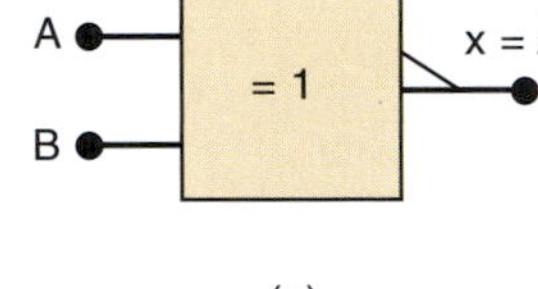

gate is obtained by simply adding a small circle at the output of the XOR symbol [Figure 4-21(b)]. The IEEE/ANSI symbol adds the small triangle on the output of the XOR symbol. Both symbols indicate an output that goes to its active-LOW state when *only one* input is HIGH.

The XNOR gate also has *only two* inputs, and it combines them so that its output is

$$x = AB + \overline{A}\,\overline{B}$$

A shorthand way to indicate the output expression of the XNOR is

$$x = \overline{A \oplus B}$$

which is simply the inverse of the XOR operation. The XNOR gate is summarized as follows:

1. It has only two inputs and its output is

$$x = AB + \overline{A}\,\overline{B} = \overline{A \oplus B}$$

2. Its output is HIGH only when the two inputs are at the *same* level.

Several ICs are available that contain XNOR gates. Those listed below are quad XNOR chips containing four XNOR gates.

74LS266	Quad XNOR (TTL family)
74C266	Quad XNOR (CMOS)
74HC266	Quad XNOR (high-speed CMOS)

Each of these XNOR chips, however, has special output circuitry that limits its use to special types of applications. Very often, a logic designer will obtain the XNOR function simply by connecting the output of an XOR to an INVERTER.

EXAMPLE 4-16

Determine the output waveform for the input waveforms given in Figure 4-22.

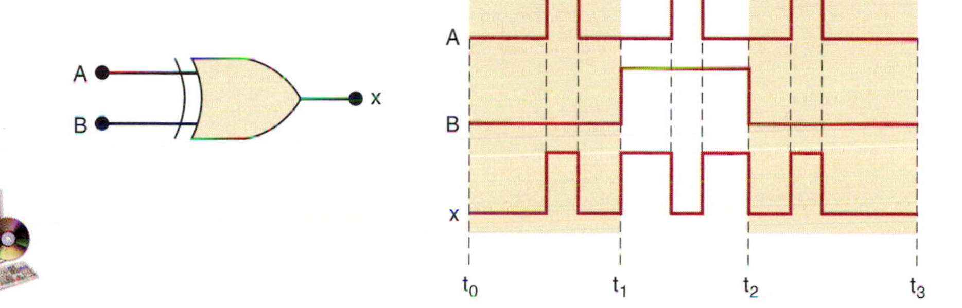

FIGURE 4-22 Example 4-16.

Solution

The output waveform is obtained using the fact that the XOR output will go HIGH only when its inputs are at different levels. The resulting output waveform reveals several interesting points:

1. The x waveform matches the A input waveform during those time intervals when $B = 0$. This occurs during the time intervals t_0 to t_1 and t_2 to t_3.
2. The x waveform is the *inverse* of the A input waveform during those time intervals when $B = 1$. This occurs during the interval t_1 to t_2.
3. These observations show that an XOR gate can be used as a *controlled INVERTER*; that is, one of its inputs can be used to control whether or not the signal at the other input will be inverted. This property will be useful in certain applications.

EXAMPLE 4-17

The notation x_1x_0 represents a two-bit binary number that can have any value (00, 01, 10, or 11); for example, when $x_1 = 1$ and $x_0 = 0$, the binary number is 10, and so on. Similarly, y_1y_0 represents another two-bit binary number. Design a logic circuit, using x_1, x_0, y_1, and y_0 inputs, whose output will be HIGH only when the two binary numbers x_1x_0 and y_1y_0 are *equal*.

Solution

The first step is to construct a truth table for the 16 input conditions (Table 4-4). The output z must be HIGH whenever the x_1x_0 values match the y_1y_0 values; that is, whenever $x_1 = y_1$ and $x_0 = y_0$. The table shows that there are four such cases. We could now continue with the normal procedure, which would be to obtain a sum-of-products expression for z, attempt to simplify it, and then implement the result. However, the nature of this problem makes it ideally suited for implementation using XNOR gates, and a little thought

TABLE 4-4

x_1	x_0	y_1	y_0	z (Output)
0	0	0	0	1
0	0	0	1	0
0	0	1	0	0
0	0	1	1	0
0	1	0	0	0
0	1	0	1	1
0	1	1	0	0
0	1	1	1	0
1	0	0	0	0
1	0	0	1	0
1	0	1	0	1
1	0	1	1	0
1	1	0	0	0
1	1	0	1	0
1	1	1	0	0
1	1	1	1	1

will produce a simple solution with minimum work. Refer to Figure 4-23; in this logic diagram, x_1 and y_1 are fed to one XNOR gate, and x_0 and y_0 are fed to another XNOR gate. The output of each XNOR will be HIGH only when its inputs are equal. Thus, for $x_0 = y_0$ and $x_1 = y_1$, both XNOR outputs will be HIGH. This is the condition we are looking for because it means that the two two-bit numbers are equal. The AND gate output will be HIGH only for this case, thereby producing the desired output.

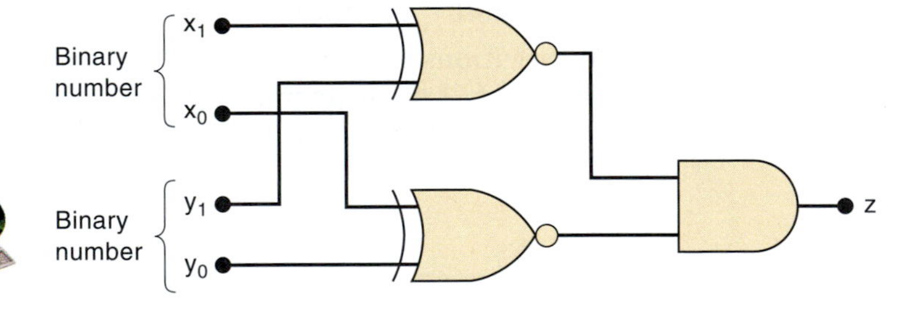

FIGURE 4-23 Circuit for detecting equality of two two-bit binary numbers.

EXAMPLE 4-18

When simplifying the expression for the output of a combinational logic circuit, you may encounter the XOR or XNOR operations as you are factoring. This will often lead to the use of XOR or XNOR gates in the implementation of the final circuit. To illustrate, simplify the circuit of Figure 4-24(a).

Solution

The unsimplified expression for the circuit is obtained as

$$z = ABCD + A\overline{B}\,\overline{C}D + \overline{A}\,\overline{D}$$

We can factor AD from the first two terms:

$$z = AD(BC + \overline{B}\,\overline{C}) + \overline{A}\,\overline{D}$$

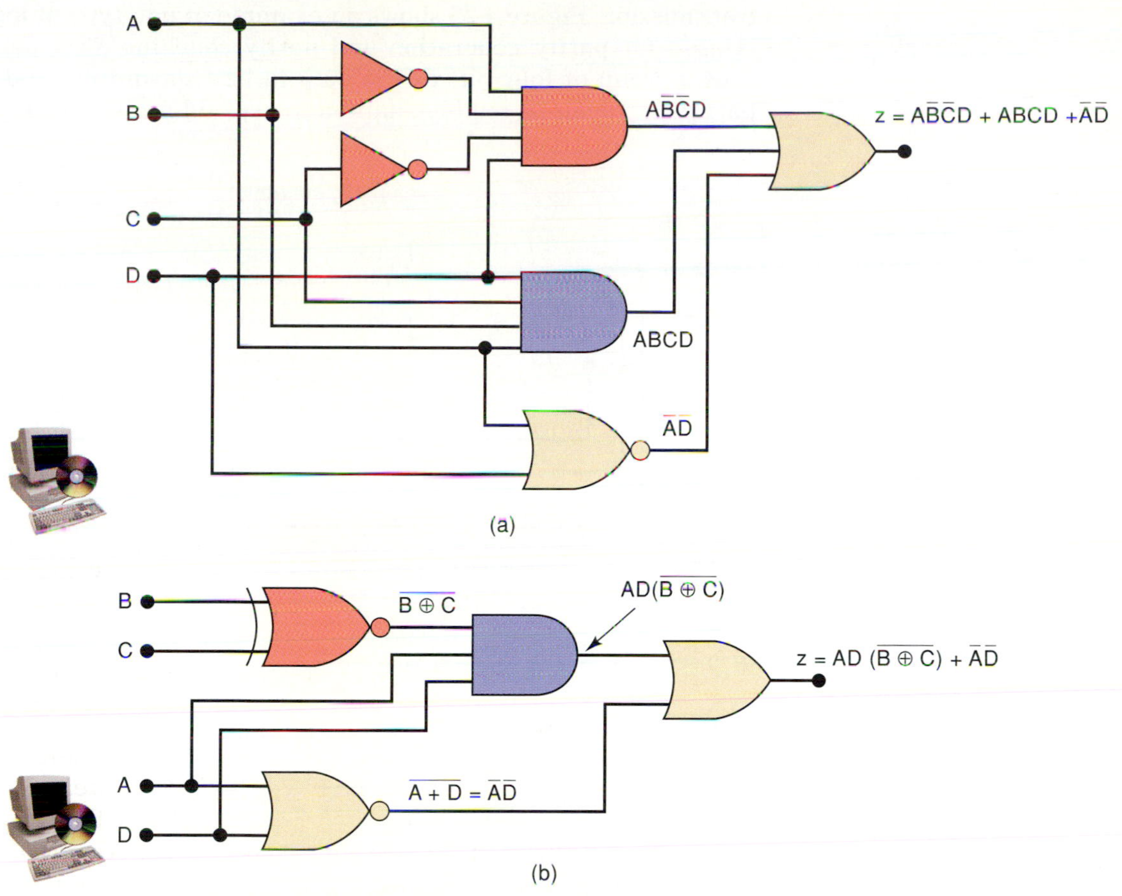

FIGURE 4-24 Example 4-18, showing how an XNOR gate may be used to simplify circuit implementation.

At first glance, you might think that the expression in parentheses can be replaced by 1. But that would be true only if it were $BC + \overline{B}\,\overline{C}$. You should recognize the expression in parentheses as the XNOR combination of B and C. This fact can be used to reimplement the circuit as shown in Figure 4-24(b). This circuit is much simpler than the original because it uses gates with fewer inputs and two INVERTERs have been eliminated.

REVIEW QUESTIONS

1. Use Boolean algebra to prove that the XNOR output expression is the exact inverse of the XOR output expression.
2. What is the output of an XNOR gate when a logic signal and its exact inverse are connected to its inputs?
3. A logic designer needs an INVERTER, and all that is available is one XOR gate from a 74HC86 chip. Does he need another chip?

4-7 PARITY GENERATOR AND CHECKER

In Chapter 2, we saw that a transmitter can attach a parity bit to a set of data bits before transmitting the data bits to a receiver. We also saw how this allows the receiver to detect any single-bit errors that may have occurred during the transmission. Figure 4-25 shows an example of one type of logic circuitry that is used for **parity generation** and **parity checking**. This particular example uses a group of four bits as the data to be transmitted, and it uses an even-parity bit. It can be readily adapted to use odd parity and any number of bits.

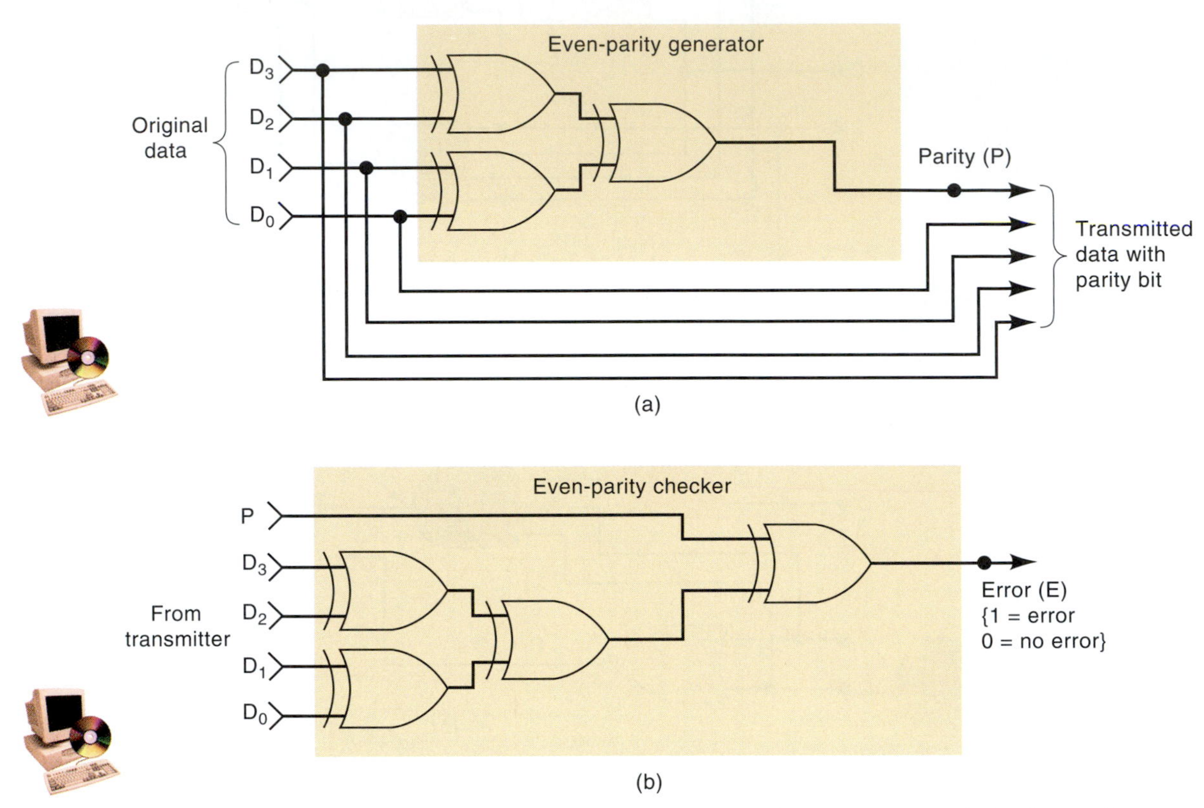

FIGURE 4-25 XOR gates used to implement (a) the parity generator and (b) the parity checker for an even-parity system.

In Figure 4-25(a), the set of data to be transmitted is applied to the parity-generator circuit, which produces the even-parity bit, P, at its output. This parity bit is transmitted to the receiver along with the original data bits, making a total of five bits. In Figure 4-25(b), these five bits (data + parity) enter the receiver's parity-checker circuit, which produces an error output, E, that indicates whether or not a single-bit error has occurred.

It should not be too surprising that both of these circuits employ XOR gates when we consider that a single XOR gate operates so that it produces a 1 output if an odd number of its inputs are 1, and a 0 output if an even number of its inputs are 1.

EXAMPLE 4-19

Determine the parity generator's output for each of the following sets of input data, $D_3D_2D_1D_0$: (a) 0111; (b) 1001; (c) 0000; (d) 0100. Refer to Figure 4-25(a).

Solution

For each case, apply the data levels to the parity-generator inputs and trace them through each gate to the P output. The results are: (a) 1; (b) 0; (c) 0; and (d) 1. Note that P is a 1 only when the original data contain an odd number of 1s. Thus, the total number of 1s sent to the receiver (data + parity) will be even.

EXAMPLE 4-20

Determine the parity checker's output [see Figure 4-25(b)] for each of the following sets of data from the transmitter:

	P	D_3	D_2	D_1	D_0
(a)	0	1	0	1	0
(b)	1	1	1	1	0
(c)	1	1	1	1	1
(d)	1	0	0	0	0

Solution

For each case, apply these levels to the parity-checker inputs and trace them through to the E output. The results are: (a) 0; (b) 0; (c) 1; (d) 1. Note that a 1 is produced at E only when an odd number of 1s appears in the inputs to the parity checker. This indicates that an error has occurred because even parity is being used.

4-8 ENABLE/DISABLE CIRCUITS

Each of the basic logic gates can be used to control the passage of an input logic signal through to the output. This is depicted in Figure 4-26, where a logic signal, A, is applied to one input of each of the basic logic gates. The

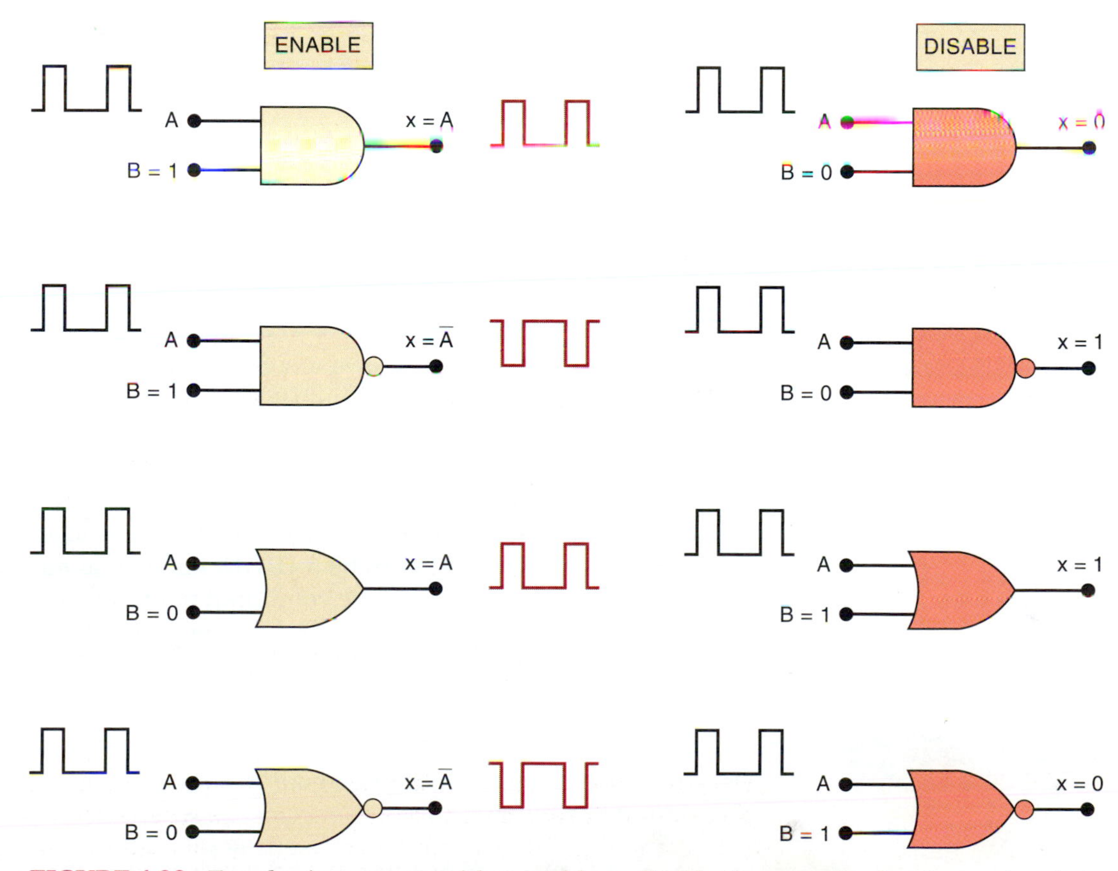

FIGURE 4-26 Four basic gates can either enable or disable the passage of an input signal, A, under control of the logic level at control input B.

other input of each gate is the control input, B. The logic level at this control input will determine whether the input signal is **enabled** to reach the output or **disabled** from reaching the output. This controlling action is why these circuits came to be called *gates.*

Examine Figure 4-26 and you should notice that when the noninverting gates (AND, OR) are enabled, the output will follow the A signal exactly. Conversely, when the inverting gates (NAND, NOR) are enabled, the output will be the exact inverse of the A signal.

Also notice in the figure that AND and NOR gates produce a constant LOW output when they are in the disabled condition. Conversely, the NAND and OR gates produce a constant HIGH output in the disabled condition.

There will be many situations in digital-circuit design where the passage of a logic signal is to be enabled or disabled, depending on conditions present at one or more control inputs. Several are shown in the following examples.

EXAMPLE 4-21

Design a logic circuit that will allow a signal to pass to the output only when control inputs B and C are both HIGH; otherwise, the output will stay LOW.

Solution

An AND gate should be used because the signal is to be passed without inversion, and the disable output condition is a LOW. Because the enable condition must occur only when $B = C = 1$, a three-input AND gate is used, as shown in Figure 4-27(a).

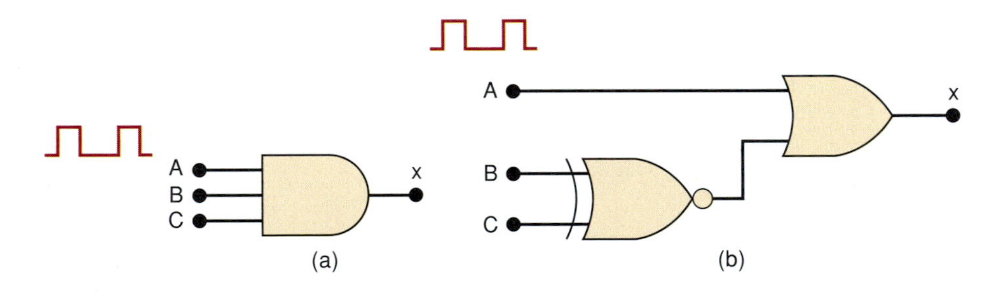

FIGURE 4-27 Examples 4-21 and 4-22.

EXAMPLE 4-22

Design a logic circuit that allows a signal to pass to the output only when one, but not both, of the control inputs are HIGH; otherwise, the output will stay HIGH.

Solution

The result is drawn in Figure 4-27(b). An OR gate is used because we want the output disable condition to be a HIGH, and we do not want to invert the signal. Control inputs B and C are combined in an XNOR gate. When B and C are different, the XNOR sends a LOW to enable the OR gate. When B and C are the same, the XNOR sends a HIGH to disable the OR gate.

EXAMPLE 4-23

Design a logic circuit with input signal A, control input B, and outputs X and Y to operate as follows:

1. When $B = 1$, output X will follow input A, and output Y will be 0.
2. When $B = 0$, output X will be 0, and output Y will follow input A.

Solution

The two outputs will be 0 when they are disabled and will follow the input signal when they are enabled. Thus, an AND gate should be used for each output. Because X is to be enabled when $B = 1$, its AND gate must be controlled by B, as shown in Figure 4-28. Because Y is to be enabled when $B = 0$, its AND gate is controlled by $\overline{B}$. The circuit in Figure 4-28 is called a *pulse-steering circuit* because it steers the input pulse to one output or the other, depending on B.

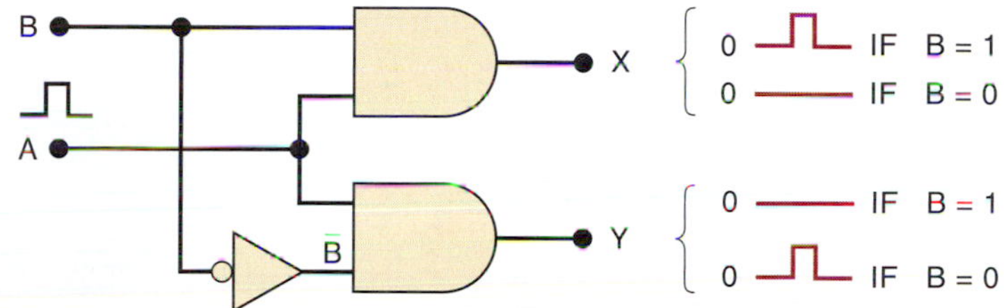

FIGURE 4-28 Example 4-23.

REVIEW QUESTIONS

1. Design a logic circuit with three inputs A, B, C and an output that goes LOW only when A is HIGH while B and C are different.
2. Which logic gates produce a 1 output in the disabled state?
3. Which logic gates pass the inverse of the input signal when they are enabled?

4-9 BASIC CHARACTERISTICS OF DIGITAL ICs

Digital ICs are a collection of resistors, diodes, and transistors fabricated on a single piece of semiconductor material (usually silicon) called a *substrate*, which is commonly referred to as a *chip*. The chip is enclosed in a protective plastic or ceramic package from which pins extend for connecting the IC to other devices. One of the more common types of package is the **dual-in-line package (DIP)**, shown in Figure 4-29(a), so called because it contains two parallel rows of pins. The pins are numbered counterclockwise when viewed from the top of the package with respect to an identifying notch or dot at one end of the package [see Figure 4-29(b)]. The DIP shown here is a 14-pin package that measures 0.75 in. by 0.25 in.; 16-, 20-, 24-, 28-, 40-, and 64-pin packages are also used.

Figure 4-29(c) shows that the actual silicon chip is much smaller than the DIP; typically, it might be a 0.05-in. square. The silicon chip is connected to the pins of the DIP by very fine wires (1-mil diameter).

The DIP is probably the most common digital IC package found in older digital equipment, but other types are becoming more and more popular. The IC shown in Figure 4-29(d) is only one of the many packages common to modern digital circuits. This particular package uses J-shaped leads that curl under the IC. We will take a look at some of these other types of IC packages in Chapter 8.

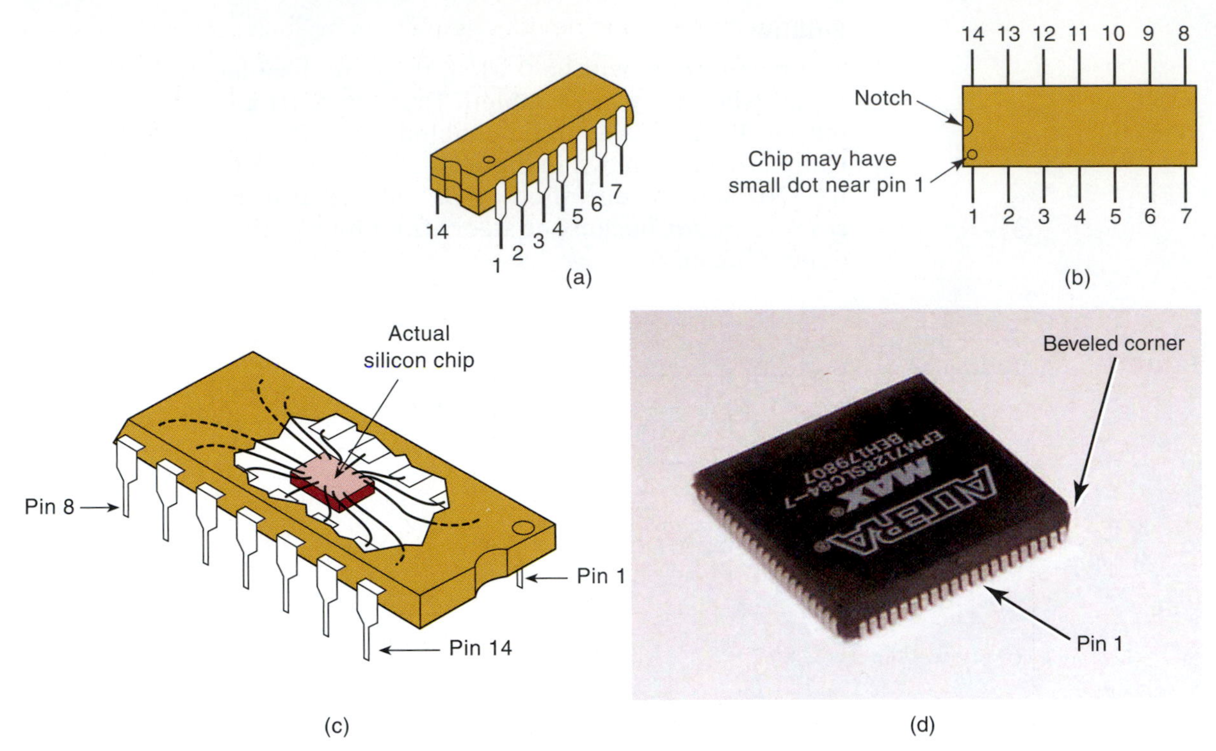

FIGURE 4-29 (a) Dual-in-line package (DIP); (b) top view; (c) actual silicon chip is much smaller than the protective package; (d) PLCC package.

Digital ICs are often categorized according to their circuit complexity as measured by the number of equivalent logic gates on the substrate. There are currently six levels of complexity that are commonly defined as shown in Table 4-5.

TABLE 4-5

Complexity	Gates per Chip
Small-scale integration (SSI)	Fewer than 12
Medium-scale integration (MSI)	12 to 99
Large-scale integration (LSI)	100 to 9999
Very large-scale integration (VLSI)	10,000 to 99,999
Ultra large-scale integration (ULSI)	100,000 to 999,999
Giga-scale integration (GSI)	1,000,000 or more

All of the specific ICs referred to in Chapter 3 and this chapter are **SSI** chips having a small number of gates. In modern digital systems, medium-scale integration **(MSI)** and large-scale integration devices **(LSI, VLSI, ULSI, GSI)** perform most of the functions that once required several circuit boards full of SSI devices. However, SSI chips are still used as the "interface," or "glue," between these more complex chips. The small-scale ICs also offer an excellent way to learn the basic building blocks of digital systems. Consequently, many laboratory-based courses use these ICs to build and test small projects.

The industrial world of digital electronics has now turned to programmable logic devices (PLDs) to implement a digital system of any significant size. Some simple PLDs are available in DIP packages, but the more complex programmable logic devices require many more pins than are available in DIPs. Larger integrated circuits that may need to be removed from a circuit and replaced are typically manufactured in a plastic leaded chip carrier (PLCC) package. Figure 4-29(d) shows the Altera EPM 7128SLC84 in a PLCC package, which is a very popular PLD used in many educational laboratories. The key features of this chip are more pins, closer spacing, and pins around the entire periphery. Notice that pin 1 is not "on the corner" like the DIP but rather at the middle of the top of the package.

Bipolar and Unipolar Digital ICs

Digital ICs can also be categorized according to the principal type of electronic component used in their circuitry. *Bipolar ICs* are made using the bipolar junction transistor (NPN and PNP) as their main circuit element. *Unipolar ICs* use the unipolar field-effect transistor (P-channel and N-channel MOSFETs) as their main element.

The **transistor-transistor logic (TTL)** family has been the major family of bipolar digital ICs for over 30 years. The standard 74 series was the first series of TTL ICs. It is no longer used in new designs, having been replaced by several higher-performance TTL series, but its basic circuit arrangement forms the foundation for all the TTL series ICs. This circuit arrangement is shown in Figure 4-30(a) for the standard TTL INVERTER. Notice that the circuit contains several bipolar transistors as the main circuit element.

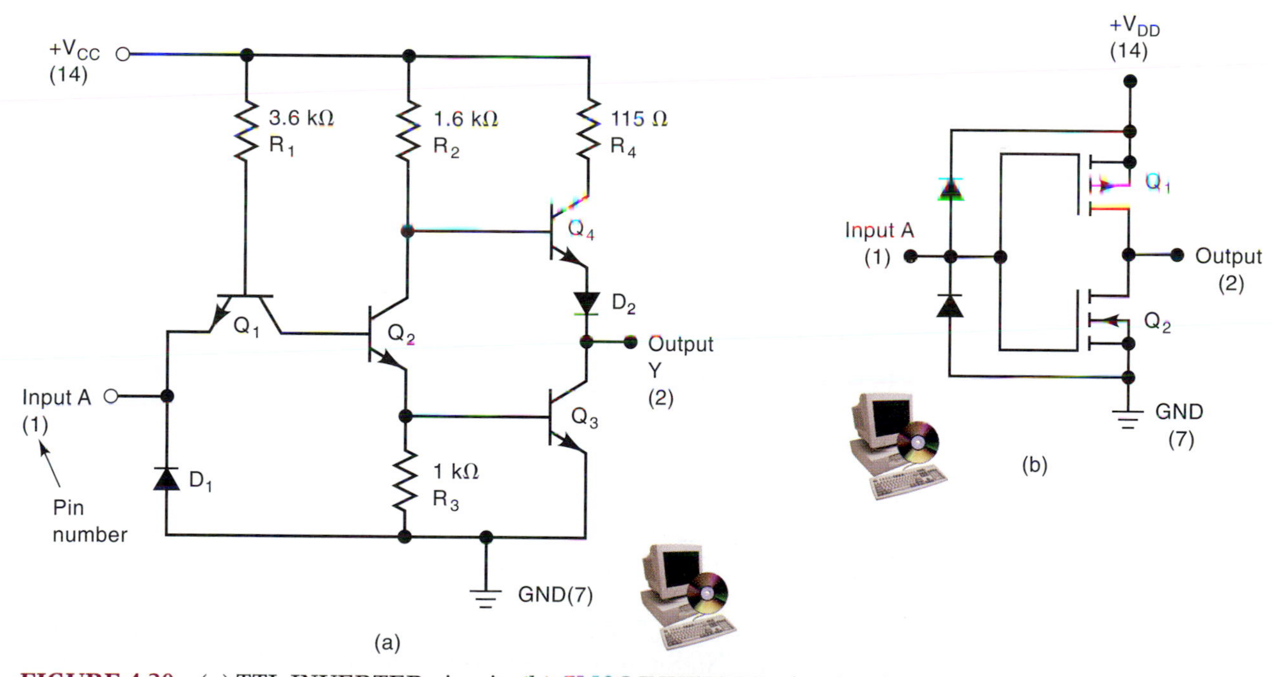

FIGURE 4-30 (a) TTL INVERTER circuit; (b) CMOS INVERTER circuit. Pin numbers are given in parentheses.

TTL had been the leading IC family in the SSI and MSI categories up until the last 12 or so years. Since then, its leading position has been challenged by the CMOS family, which has gradually displaced TTL from that position. The **complementary metal-oxide semiconductor (CMOS)** family belongs to the class of unipolar digital ICs because it uses P- and N-channel MOSFETs as the main circuit elements. Figure 4-30(b) is a standard CMOS INVERTER circuit. If we compare the TTL and CMOS circuits in Figure 4-30, it is apparent that the CMOS version uses fewer components. This is one of the main advantages of CMOS over TTL.

Because of the simplicity and compactness as well as some other superior attributes of CMOS, the modern large-scale ICs are manufactured primarily using CMOS technology. Teaching laboratories that use SSI and MSI devices often use TTL due to its durability, although some use CMOS as well. Chapter 8 will provide a comprehensive study of the circuitry and characteristics of TTL and CMOS. For now, we need to look at only a few of their basic characteristics so that we can talk about troubleshooting simple combinational circuits.

TTL Family

The TTL logic family actually consists of several subfamilies or series. Table 4-6 lists the name of each TTL series together with the prefix designation used to identify different ICs as being part of that series. For example, ICs that are part of the standard TTL series have an identification number that starts with 74. The 7402, 7438, and 74123 are all ICs in this series. Likewise, ICs that are part of the low-power Schottky TTL series have an identification number that starts with 74LS. The 74LS02, 74LS38, and 74LS123 are examples of devices in the 74LS series.

TABLE 4-6 Various series within the TTL logic family.

TTL Series	Prefix	Example IC
Standard TTL	74	7404 (hex INVERTER)
Schottky TTL	74S	74S04 (hex INVERTER)
Low-power Schottky TTL	74LS	74LS04 (hex INVERTER)
Advanced Schottky TTL	74AS	74AS04 (hex INVERTER)
Advanced low-power Schottky TTL	74ALS	74ALS04 (hex INVERTER)

The principal differences in the various TTL series have to do with their electrical characteristics such as power dissipation and switching speed. They do not differ in the pin layout or logic operations performed by the circuits on the chip. For example, the 7404, 74S04, 74LS04, 74AS04, and 74ALS04 are all hex-INVERTER ICs, each containing *six* INVERTERs on a single chip.

CMOS Family

Several CMOS series are available, and some of these are listed in Table 4-7. The 4000 series is the oldest CMOS series. This series contains many of the same logic functions as the TTL family but was not designed to be *pin-compatible* with TTL devices. For example, the 4001 quad NOR chip contains four two-input NOR gates, as does the TTL 7402 chip, but the gate inputs and outputs on the CMOS chip will not have the same pin numbers as the corresponding signals on the TTL chip.

The 74C, 74HC, 74HCT, 74AC, and 74ACT series are newer CMOS series. The first three are pin-compatible with correspondingly numbered TTL devices. For example, the 74C02, 74HC02, and 74HCT02 have the same pin layout as the 7402, 74LS02, and so on. The 74HC and 74HCT series operate at a higher speed than 74C devices. The 74HCT series is designed to be *electrically compatible* with TTL devices; that is, a 74HCT integrated circuit can be connected directly to TTL devices without any interfacing circuitry. The 74AC and 74ACT series are advanced-performance ICs. Neither is pin-compatible with

TABLE 4-7 Various series within the CMOS logic family.

CMOS Series	Prefix	Example IC
Metal-gate CMOS	40	4001 (quad NOR gates)
Metal-gate, pin-compatible with TTL	74C	74C02 (quad NOR gates)
Silicon-gate, pin-compatible with TTL, high-speed	74HC	74HC02 (quad NOR gates)
Silicon-gate, high-speed, pin-compatible and electrically compatible with TTL	74HCT	74HCT02 (quad NOR gates)
Advanced-performance CMOS, not pin-compatible or electrically compatible with TTL	74AC	74AC02 (quad NOR)
Advanced-performance CMOS, not pin-compatible with TTL, but electrically compatible with TTL	74ACT	74ACT02 (quad NOR)

TTL. The 74ACT devices are electrically compatible with TTL. We explore the various TTL and CMOS series in greater detail in Chapter 8.

Power and Ground

To use digital ICs, it is necessary to make the proper connections to the IC pins. The most important connections are *dc power* and *ground*. These are required for the circuits on the chip to operate correctly. In Figure 4-30, you can see that both the TTL and the CMOS circuits have a dc power supply voltage connected to one of their pins, and ground to another. The power supply pin is labeled V_{CC} for the TTL circuit, and V_{DD} for the CMOS circuit. Many of the newer CMOS integrated circuits that are designed to be compatible with TTL integrated circuits also use the V_{CC} designation as their power pin.

If either the power or the ground connection is not made to the IC, the logic gates on the chip will not respond properly to the logic inputs, and the gates will not produce the expected output logic levels.

Logic-Level Voltage Ranges

For TTL devices, V_{CC} is nominally +5 V. For CMOS integrated circuits, V_{DD} can range from +3 to +18 V, although +5 V is most often used when CMOS integrated circuits are used in the same circuit with TTL integrated circuits.

For standard TTL devices, the acceptable input voltage ranges for the logic 0 and logic 1 levels are defined as shown in Figure 4-31(a). A logic 0 is any voltage in the range from 0 to 0.8 V; a logic 1 is any voltage from 2 to 5 V. Voltages that are not in either of these ranges are said to be **indeterminate** and should not be used as inputs to any TTL device. The IC manufacturers cannot guarantee how a TTL circuit will respond to input levels that are in the indeterminate range (between 0.8 and 2.0 V).

The logic input voltage ranges for CMOS integrated circuits operating with V_{DD} = +5 V are shown in Figure 4-31(b). Voltages between 0 and 1.5 V are defined as a logic 0, and voltages from 3.5 to 5 V are defined as a logic 1. The indeterminate range includes voltages between 1.5 and 3.5 V.

Unconnected (Floating) Inputs

What happens when the input to a digital IC is left unconnected? An unconnected input is often called a **floating** input. The answer to this question will be different for TTL and CMOS.

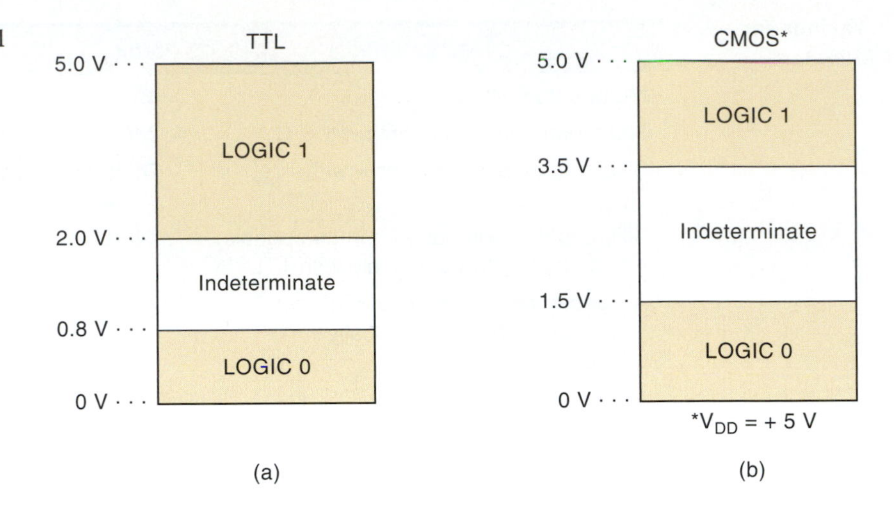

FIGURE 4-31 Logic-level input voltage ranges for (a) TTL and (b) CMOS digital ICs.

A floating TTL input acts just like a logic 1. In other words, the IC will respond as if the input had a logic HIGH level applied to it. This characteristic is often used when testing a TTL circuit. A lazy technician might leave certain inputs unconnected instead of connecting them to a logic HIGH. Although this is logically correct, it is not a recommended practice, especially in final circuit designs, because the floating TTL input is extremely susceptible to picking up noise signals that will probably adversely affect the device's operation.

A floating input on some TTL gates will measure a dc level of between 1.4 and 1.8 V when checked with a VOM or an oscilloscope. Even though this is in the indeterminate range for TTL, it will produce the same response as a logic 1. Being aware of this characteristic of a floating TTL input can be valuable when troubleshooting TTL circuits.

If a CMOS input is left floating, it may have disastrous results. The IC may become overheated and eventually destroy itself. For this reason all inputs to a CMOS integrated circuit must be connected to a LOW or a HIGH level or to the output of another IC. A floating CMOS input will not measure as a specific dc voltage but will fluctuate randomly as it picks up noise. Thus, it does not act as logic 1 or logic 0, and so its effect on the output is unpredictable. Sometimes the output will oscillate as a result of the noise picked up by the floating input.

Many of the more complex CMOS ICs have circuitry built into the inputs, which reduces the likelihood of any destructive reaction to an open input. With this circuitry, it is not necessary to ground each unused pin on these large ICs when experimenting. It is still good practice, however, to tie unused inputs to HIGH or LOW (whichever is appropriate) in the final circuit implementation.

Logic-Circuit Connection Diagrams

A connection diagram shows *all* electrical connections, pin numbers, IC numbers, component values, signal names, and power supply voltages. Figure 4-32 shows a typical connection diagram for a simple logic circuit. Examine it carefully and note the following important points:

1. The circuit uses logic gates from two different ICs. The two INVERTERs are part of a 74HC04 chip that has been given the designation Z1. The 74HC04 contains six INVERTERs; two of them are used in this circuit, and each is labeled as being part of chip Z1. Similarly, the two NAND gates are part of a 74HC00 chip that contains four NAND gates. All of

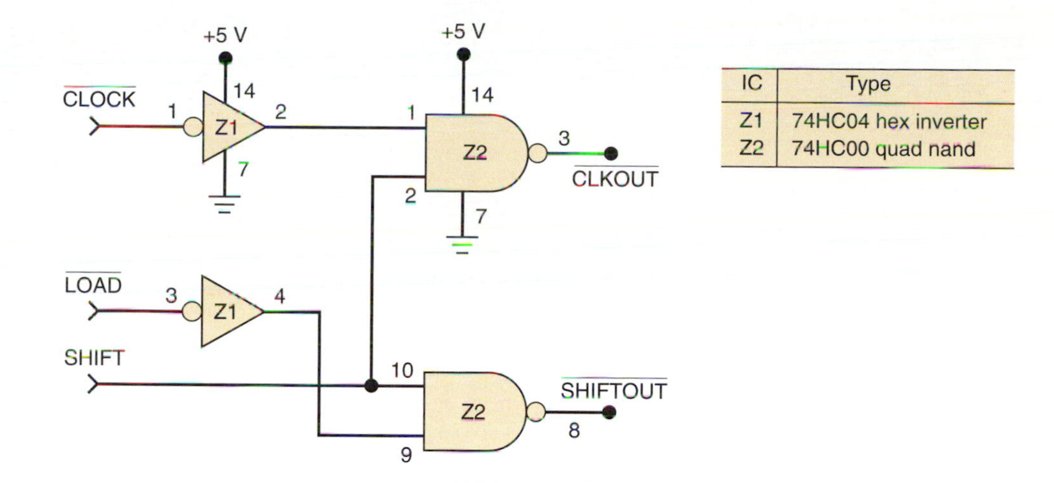

IC	Type
Z1	74HC04 hex inverter
Z2	74HC00 quad nand

FIGURE 4-32 Typical logic-circuit connection diagram.

the gates on this chip are designated with the label Z2. By numbering each gate as Z1, Z2, Z3, and so on, we can keep track of which gate is part of which chip. This is especially valuable in more complex circuits containing many ICs with several gates per chip.

2. Each gate input and output pin number is indicated on the diagram. These pin numbers and the IC labels are used to reference easily any point in the circuit. For example, Z1 pin 2 refers to the output pin of the top INVERTER. Similarly, we can say that Z1 pin 4 is connected to Z2 pin 9.
3. The power and ground connections to each IC (not each gate) are shown on the diagram. For example, Z1 pin 14 is connected to +5 V, and Z1 pin 7 is connected to ground. These connections provide power to *all* of the six INVERTERs that are part of Z1.
4. For the circuit contained in Figure 4-32, the signals that are inputs are on the left. The signals that are outputs are on the right. The bar over the signal name indicates that the signal is active when LOW. The bubbles are positioned on the diagram symbols also to indicate the active-LOW state. Each signal in this case is obviously a single bit.
5. Signals are defined graphically in Figure 4-32 as inputs and outputs, and the relationship between them (the operation of the circuit) is described graphically using interconnected logic symbols.

Manufacturers of electronic equipment generally supply detailed schematics that use a format similar to that in Figure 4-32. These connection diagrams are a virtual necessity when troubleshooting a faulty circuit. We have chosen to identify individual ICs as Z1, Z2, Z3, and so on. Other designations that are commonly used are IC1, IC2, IC3, and so on, and U1, U2, U3, and so on.

Personal computers with schematic diagram software can be used to draw logic circuits. Computer applications that can interpret these graphic symbols and signal connections and can translate them into logical relationships are often called schematic capture tools. The Altera MAX+PLUS development system for programmable logic allows the user to enter graphic design files (.gdf) using schematic capture techniques. Thus, designing the circuit is as easy as drawing the schematic diagram on the computer screen. Notice that in Figure 4-33 there are no pin numbers or chip designations on the logic symbols. The circuits will not be implemented using actual SSI or MSI chips, but rather the equivalent logic functionality will be "programmed" into a PLD. We will explain this further at a later point in this chapter.

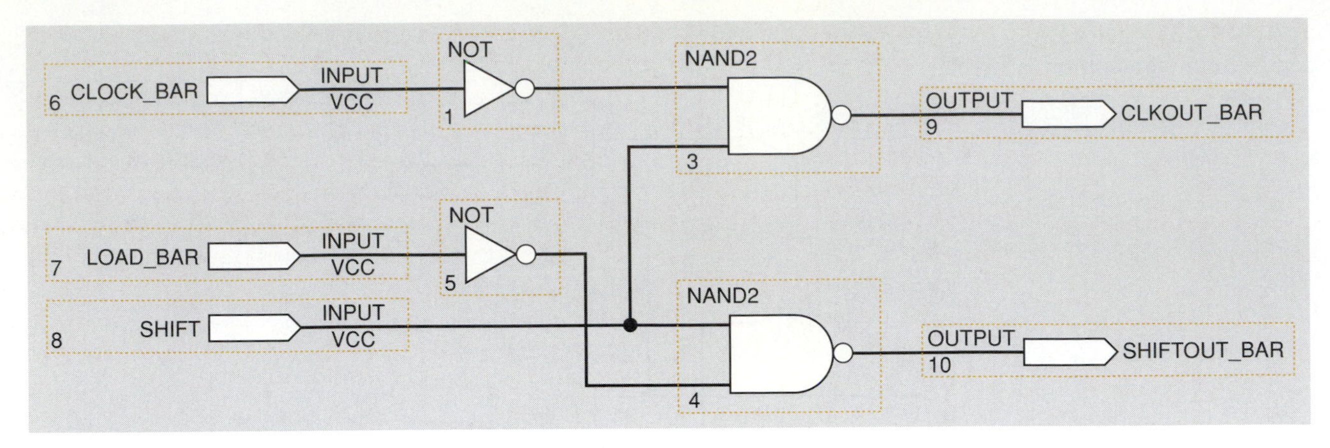

FIGURE 4-33 Logic diagram using schematic capture.

REVIEW QUESTIONS

1. What is the most common type of digital IC package?
2. Name the six common categories of digital ICs according to complexity.
3. *True or false:* A 74S74 chip will contain the same logic and pin layout as the 74LS74.
4. *True or false:* A 74HC74 chip will contain the same logic and pin layout as the 74AS74.
5. Which CMOS series are not pin-compatible with TTL?
6. What is the acceptable input voltage range of a logic 0 for TTL? What is it for a logic 1?
7. Repeat question 6 for CMOS operating at $V_{DD} = 5$ V.
8. How does a TTL integrated circuit respond to a floating input?
9. How does a CMOS integrated circuit respond to a floating input?
10. Which CMOS series can be connected directly to TTL with no interfacing circuitry?
11. What is the purpose of pin numbers on a logic circuit connection diagram?
12. What are the key similarities of graphic design files used for programmable logic and traditional logic circuit connection diagrams?

4-10 TROUBLESHOOTING DIGITAL SYSTEMS

There are three basic steps in fixing a digital circuit or system that has a fault (failure):

1. *Fault detection.* Observe the circuit/system operation and compare it with the expected correct operation.
2. *Fault isolation.* Perform tests and make measurements to isolate the fault.
3. *Fault correction.* Replace the faulty component, repair the faulty connection, remove the short, and so on.

Although these steps may seem relatively apparent and straightforward, the actual troubleshooting procedure that is followed is highly dependent on the type and complexity of the circuitry, and on the kinds of troubleshooting tools and documentation that are available.

Good troubleshooting techniques can be learned only in a laboratory environment through experimentation and actual troubleshooting of faulty circuits and systems. There is absolutely no better way to become an effective troubleshooter than to do as much troubleshooting as possible, and no amount of textbook reading can provide that kind of experience. We can, however, help you to develop the analytical skills that are the most essential part of effective troubleshooting. We will describe the types of faults that are common to systems that are made primarily from digital ICs and we will tell you how to recognize them. We will then present typical case studies to illustrate the analytical processes involved in troubleshooting. In addition, there will be end-of-chapter problems to provide you with the opportunity to go through these analytical processes to reach conclusions about faulty digital circuits.

For the troubleshooting discussions and exercises we will be doing in this book, it will be assumed that the troubleshooting technician has the basic troubleshooting tools available: *logic probe, oscilloscope, logic pulser.* Of course, the most important and effective troubleshooting tool is the technician's brain, and that's the tool we are hoping to develop by presenting troubleshooting principles and techniques, examples and problems, here and in the following chapters.

In the next three sections on troubleshooting, we will use only our brain and a **logic probe** such as the one illustrated in Figure 4-34. The logic probe has a pointy metal tip that is touched to the specific point you want to test. Here, it is shown probing (touching) pin 3 of an IC. It can also be touched to a printed circuit board trace, an uninsulated wire, a connector pin, a lead on a discrete component such as a transistor, or any other conducting point in a circuit. The logic level that is present at the probe tip will be indicated by the status of the indicator LEDs in the probe. The four possibilities are given in the table of Figure 4-34. Note that an *indeterminate* logic level produces no indicator light. This includes the condition where the probe tip is touched to a point in a circuit that is open or floating—that is, not connected to any source of voltage. This type of probe also offers a yellow LED to indicate the presence of a pulse train. Any transitions (LOW to HIGH or HIGH to LOW) will cause the yellow LED to flash on for a fraction of a second and then turn off. If the transitions are occurring frequently, the LED will continue to flash

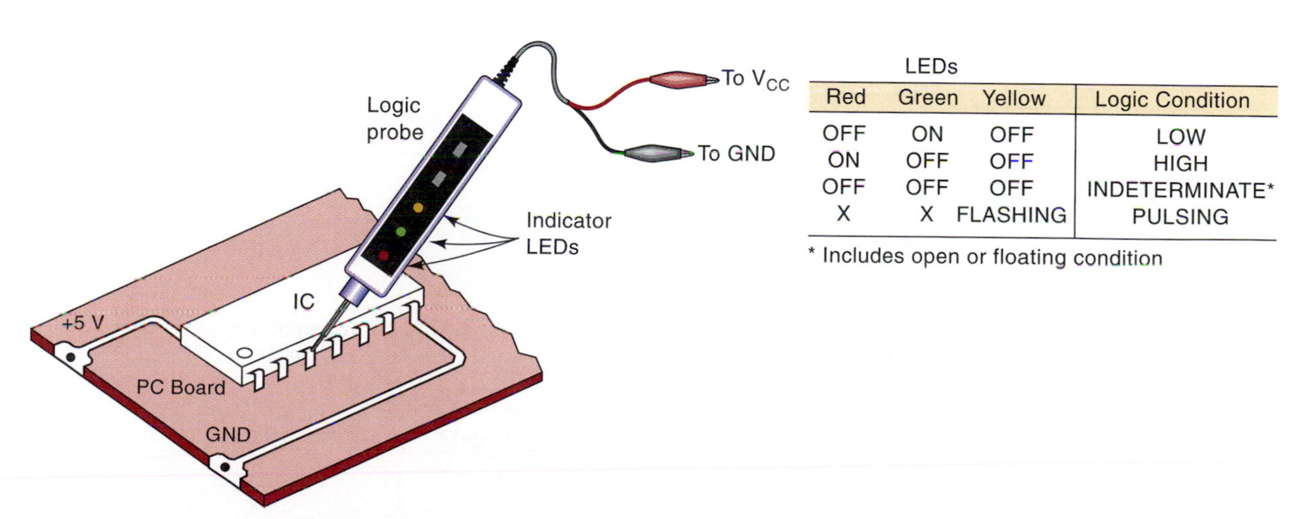

LEDs			
Red	Green	Yellow	Logic Condition
OFF	ON	OFF	LOW
ON	OFF	OFF	HIGH
OFF	OFF	OFF	INDETERMINATE*
X	X	FLASHING	PULSING

* Includes open or floating condition

FIGURE 4-34 A logic probe is used to monitor the logic level activity at an IC pin or any other accessible point in a circuit.

at around 3 Hz. By observing the green and red LEDs along with the flashing yellow, you can tell whether the signal is mostly HIGH or mostly LOW.

4-11 INTERNAL DIGITAL IC FAULTS

The most common internal failures of digital ICs are:

1. Malfunction in the internal circuitry
2. Inputs or outputs shorted to ground or V_{CC}
3. Inputs or outputs open-circuited
4. Short between two pins (other than ground or V_{CC})

We will now describe each of these types of failure.

Malfunction in Internal Circuitry

This is usually caused by one of the internal components failing completely or operating outside its specifications. When this happens, the IC outputs do not respond properly to the IC inputs. There is no way to predict what the outputs will do because it depends on what internal component has failed. Examples of this type of failure would be a base-emitter short in transistor Q_4 or an extremely large resistance value for R_2 in the TTL INVERTER of Figure 4-30(a). This type of internal IC failure is not as common as the other three.

Input Internally Shorted to Ground or Supply

This type of internal failure will cause an IC input to be stuck in the LOW or HIGH state. Figure 4-35(a) shows input pin 2 of a NAND gate shorted to ground within the IC. This will cause pin 2 always to be in the LOW state. If this input pin is being driven by a logic signal B, it will effectively short B to ground. Thus, this type of fault will affect the output of the device that is generating the B signal.

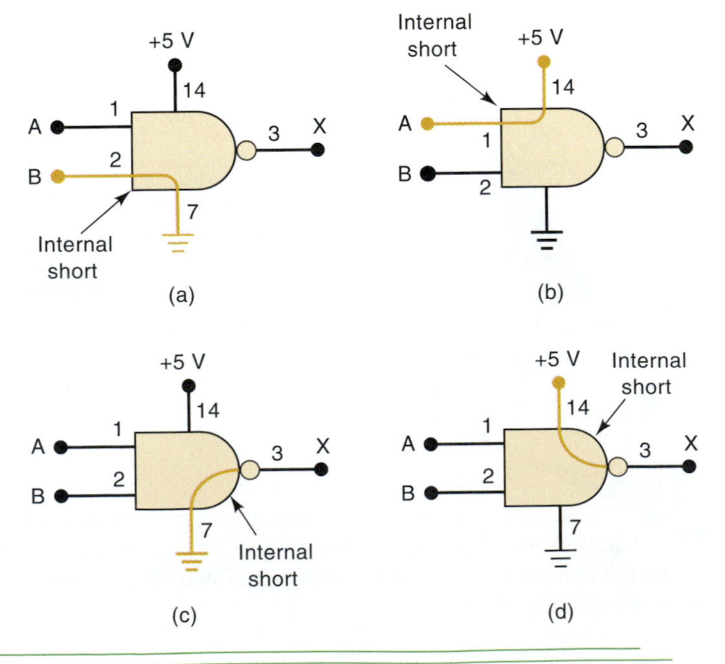

FIGURE 4-35 (a) IC input internally shorted to ground; (b) IC input internally shorted to supply voltage. These two types of failures force the input signal at the shorted pin to stay in the same state. (c) IC output internally shorted to ground; (d) output internally shorted to supply voltage. These two failures do not affect signals at the IC inputs.

Similarly, an IC input pin could be internally shorted to +5 V, as in Figure 4-35(b). This would keep that pin stuck in the HIGH state. If this input pin is being driven by a logic signal A, it will effectively short A to +5 V.

Output Internally Shorted to Ground or Supply

This type of internal failure will cause the output pin to be stuck in the LOW or HIGH state. Figure 4-35(c) shows pin 3 of the NAND gate shorted to ground within the IC. This output is stuck LOW, and it will not respond to the conditions applied to input pins 1 and 2; in other words, logic inputs A and B will have no effect on output X.

An IC output pin can also be shorted to +5 V within the IC, as shown in Figure 4-35(d). This forces the output pin 3 to be stuck HIGH regardless of the state of the signals at the input pins. Note that this type of failure has no effect on the logic signals at the IC inputs.

EXAMPLE 4-24

Refer to the circuit of Figure 4-36. A technician uses a logic probe to determine the conditions at the various IC pins. The results are recorded in the figure. Examine these results and determine if the circuit is working properly. If not, suggest some of the possible faults.

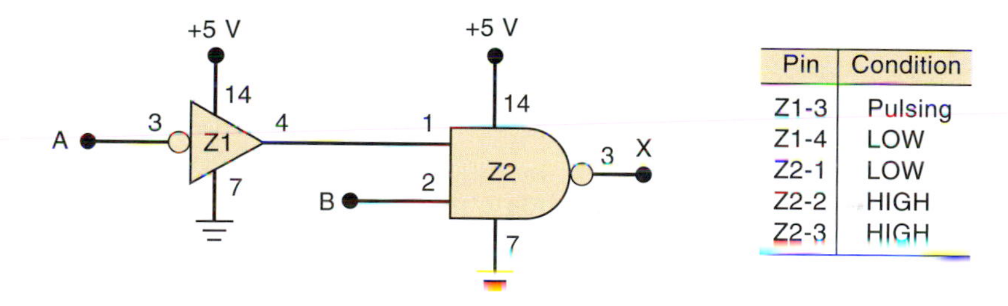

Pin	Condition
Z1-3	Pulsing
Z1-4	LOW
Z2-1	LOW
Z2-2	HIGH
Z2-3	HIGH

FIGURE 4-36 Example 4-24.

Solution

Output pin 4 of the INVERTER should be pulsing because its input is pulsing. The recorded results, however, show that pin 4 is stuck LOW. Because this is connected to Z2 pin 1, this keeps the NAND output HIGH. From our preceding discussion, we can list three possible faults that could produce this operation.

First, there could be an internal component failure in the INVERTER that prevents it from responding properly to its input. Second, pin 4 of the INVERTER could be shorted to ground internal to Z1, thereby keeping it stuck LOW. Third, pin 1 of Z2 could be shorted to ground internal to Z2. This would prevent the INVERTER output pin from changing.

In addition to these possible faults, there can be external shorts to ground anywhere in the conducting path between Z1 pin 4 and Z2 pin 1. We will see how to go about isolating the actual fault in a subsequent example.

Open-Circuited Input or Output

Sometimes the very fine conducting wire that connects an IC pin to the IC's internal circuitry will break, producing an open circuit. Figure 4-37 in Example 4-25 shows this for an input (pin 13) and an output (pin 6). If a signal is applied to pin 13, it will not reach the NAND-1 gate input and so will not have an effect on the NAND-1 output. The open gate input will be in the floating state. As stated earlier, TTL devices will respond as if this floating input is a logic 1, and CMOS devices will respond erratically and may even become damaged from overheating.

The open at the NAND-4 output prevents the signal from reaching IC pin 6, so there will be no stable voltage present at that pin. If this pin is connected to the input of another IC, it will produce a floating condition at that input.

EXAMPLE 4-25

What would a logic probe indicate at pin 13 and at pin 6 of Figure 4-37?

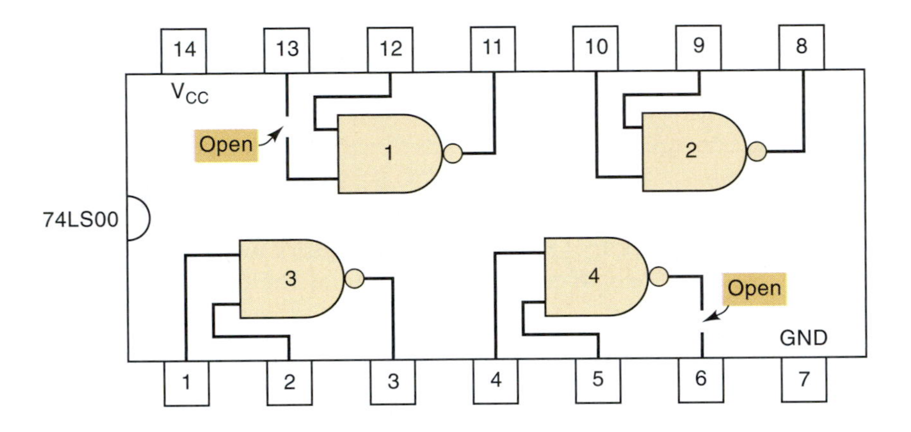

FIGURE 4-37 An IC with an internally open input will not respond to signals applied to that input pin. An internally open output will produce an unpredictable voltage at that output pin.

Solution

At pin 13, the logic probe will indicate the logic level of the external signal that is connected to pin 13 (which is not shown in the diagram). At pin 6, the logic probe will show no LED lit for an indeterminate logic level because the NAND output level never makes it to pin 6.

EXAMPLE 4-26

Refer to the circuit of Figure 4-38 and the recorded logic probe indications. What are some of the possible faults that could produce the recorded results? Assume that the ICs are TTL.

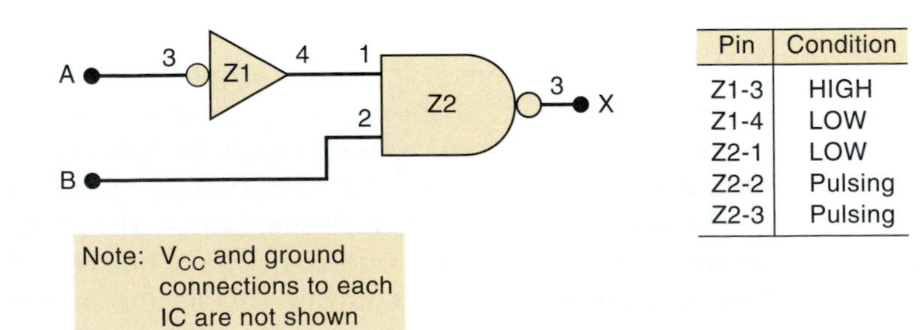

Pin	Condition
Z1-3	HIGH
Z1-4	LOW
Z2-1	LOW
Z2-2	Pulsing
Z2-3	Pulsing

FIGURE 4-38 Example 4-26.

Solution

Examination of the recorded results indicates that the INVERTER appears to be working properly, but the NAND output is inconsistent with its inputs. The NAND output should be HIGH because its input pin 1 is LOW. This LOW should prevent the NAND gate from responding to the pulses at pin 2. It is probable that this LOW is not reaching the internal NAND gate circuitry because of an internal open. Because the IC is TTL, this open circuit would produce the same effect as a logic HIGH at pin 1. If the IC had been CMOS, the internal open circuit at pin 1 might have produced an indeterminate output and possible overheating and destruction of the chip.

From our earlier statement regarding open TTL inputs, you might have expected that the voltage of pin 1 of Z2 would be 1.4 to 1.8 V and should have been registered as indeterminate by the logic probe. This would have been true if the open circuit had been *external* to the NAND chip. There is no open circuit between Z1 pin 4 and Z2 pin 1, and so the voltage at Z1 pin 4 is reaching Z2 pin 1, but it becomes disconnected *inside* the NAND chip.

Short Between Two Pins

An internal short between two pins of an IC will force the logic signals at those pins always to be identical. Whenever two signals that are supposed to be different show the same logic-level variations, there is a good possibility that the signals are shorted together.

Consider the circuit in Figure 4-39, where pins 5 and 6 of the NOR gate are internally shorted together. The short causes the two INVERTER output pins to be connected together so that the signals at Z1 pin 2 and Z1 pin 4 must be identical, even when the two INVERTER input signals are trying to produce different outputs. To illustrate, consider the input waveforms shown in the diagram. Even though these input waveforms are different, the waveforms at outputs Z1-2 and Z1-4 are the same.

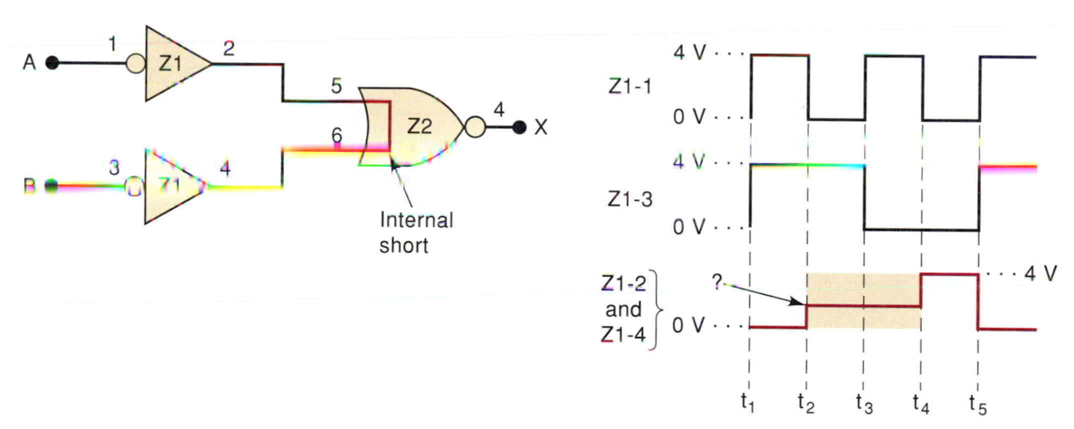

FIGURE 4-39 When two input pins are internally shorted, the signals driving these pins are forced to be identical, and usually a signal with three distinct levels results.

During the interval t_1 to t_2, both INVERTERs have a HIGH input and both are trying to produce a LOW output, so that their being shorted together makes no difference. During the interval t_4 to t_5, both INVERTERs have a LOW input and are trying to produce a HIGH output, so that again their being shorted has no effect. However, during the intervals t_2 to t_3 and t_3 to t_4, one INVERTER is trying to produce a HIGH output while the other is trying to produce a LOW output. This is called signal **contention** because the two signals are "fighting" each other. When this happens, the actual voltage level that appears at the shorted outputs will depend on the internal IC circuitry. For TTL devices, it will usually be a voltage in the high end of the logic 0 range (i.e., close to 0.8 V), although it may also be in the indeterminate range. For CMOS devices, it will often be a voltage in the indeterminate range.

Whenever you see a waveform like the Z1-2, Z1-4 signal in Figure 4-39 with three different levels, you should suspect that two output signals may be shorted together.

REVIEW QUESTIONS

1. List the different internal digital IC faults.
2. Which internal IC fault can produce signals that show three different voltage levels?
3. What would a logic probe indicate at Z1-2 and Z1-4 of Figure 4-39 if $A = 0$ and $B = 1$?
4. What is signal contention?

4-12 EXTERNAL FAULTS

We have seen how to recognize the effects of various faults internal to digital ICs. Many more things can go wrong external to the ICs; we will describe the most common ones in this section.

Open Signal Lines

This category includes any fault that produces a break or discontinuity in the conducting path such that a voltage level or signal is prevented from going from one point to another. Some of the causes of open signal lines are:

1. Broken wire
2. Poor solder connection; loose wire-wrap connection
3. Crack or cut trace on a printed circuit board (some of these are hairline cracks that are difficult to see without a magnifying glass)
4. Bent or broken pin on an IC
5. Faulty IC socket such that the IC pin does not make good contact with the socket

This type of circuit fault can often be detected by a careful visual inspection and then verified by disconnecting power from the circuit and checking for continuity (i.e., a low-resistance path) with an ohmmeter between the two points in question.

EXAMPLE 4-27

Consider the CMOS circuit of Figure 4-40 and the accompanying logic probe indications. What is the most probable circuit fault?

Solution

The indeterminate level at the NOR gate output is probably due to the indeterminate input at pin 2. Because there is a LOW at Z1-6, this LOW should also be at Z2-2. Clearly, the LOW from Z1-6 is not reaching Z2-2, and there must be an open circuit in the signal path between these two points. The location of this open circuit can be determined by starting at Z1-6 with the logic probe and tracing the LOW level along the signal path toward Z2-2 until it changes into an indeterminate level.

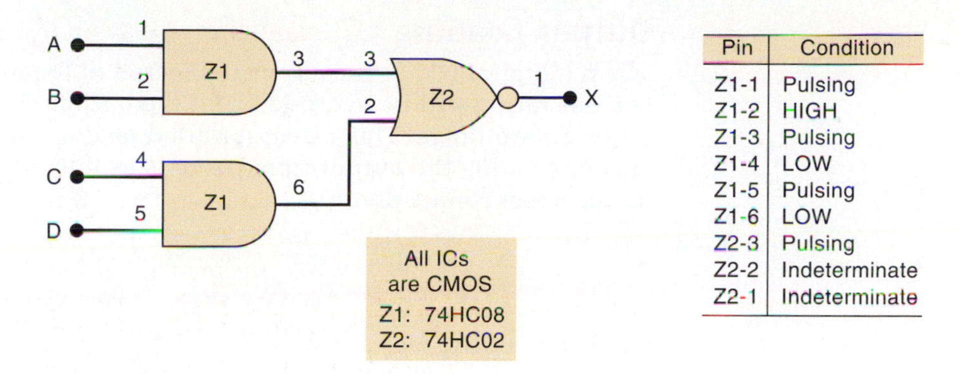

Pin	Condition
Z1-1	Pulsing
Z1-2	HIGH
Z1-3	Pulsing
Z1-4	LOW
Z1-5	Pulsing
Z1-6	LOW
Z2-3	Pulsing
Z2-2	Indeterminate
Z2-1	Indeterminate

FIGURE 4-40 Example 4-27.

Shorted Signal Lines

This type of fault has the same effect as an internal short between IC pins. It causes two signals to be exactly the same (signal contention). A signal line may be shorted to ground or V_{CC} rather than to another signal line. In those cases, the signal will be forced to the LOW or the HIGH state. The main causes for unexpected shorts between two points in a circuit are as follows:

1. *Sloppy wiring.* An example of this is stripping too much insulation from ends of wires that are in close proximity.
2. *Solder bridges.* These are splashes of solder that short two or more points together. They commonly occur between points that are very close together, such as adjacent pins on a chip.
3. *Incomplete etching.* The copper between adjacent conducting paths on a printed circuit board is not completely etched away.

Again, a careful visual inspection can very often uncover this type of fault, and an ohmmeter check can verify that the two points in the circuit are shorted together.

Faulty Power Supply

All digital systems have one or more dc power supplies that supply the V_{CC} and V_{DD} voltages required by the chips. A faulty power supply or one that is overloaded (supplying more than its rated amount of current) will provide poorly regulated supply voltages to the ICs, and the ICs either will not operate or will operate erratically.

A power supply may go out of regulation because of a fault in its internal circuitry, or because the circuits that it is powering are drawing more current than the supply is designed for. This can happen if a chip or a component has a fault that causes it to draw much more current than normal.

It is good troubleshooting practice to check the voltage levels at each power supply in the system to see that they are within their specified ranges. It is also a good idea to check them on an oscilloscope to verify that there is no significant amount of ac ripple on the dc levels and to verify that the voltage levels stay regulated during the system operation.

One of the most common signs of a faulty power supply is one or more chips operating erratically or not at all. Some ICs are more tolerant of power supply variations and may operate properly, while others do not. You should always check the power and ground levels at each IC that appears to be operating incorrectly.

Output Loading

When a digital IC has its output connected to too many IC inputs, its output current rating will be exceeded, and the output voltage can fall into the indeterminate range. This effect is called *loading* the output signal (actually it's overloading the output signal) and is usually the result of poor design or an incorrect connection.

REVIEW QUESTIONS

1. What are the most common types of external faults?
2. List some of the causes of signal-path open circuits.
3. What symptoms are caused by a faulty power supply?
4. How might loading affect an IC output voltage level?

4-13 TROUBLESHOOTING CASE STUDY

The following example will illustrate the analytical processes involved in troubleshooting digital circuits. Although the example is a fairly simple combinational logic circuit, the reasoning and the troubleshooting procedures used can be applied to the more complex digital circuits that we encounter in subsequent chapters.

EXAMPLE 4-28

Consider the circuit of Figure 4-41. Output Y is supposed to go HIGH for either of the following conditions:

1. $A = 1, B = 0$ regardless of the level on C
2. $A = 0, B = 1, C = 1$

You may wish to verify these results for yourself.

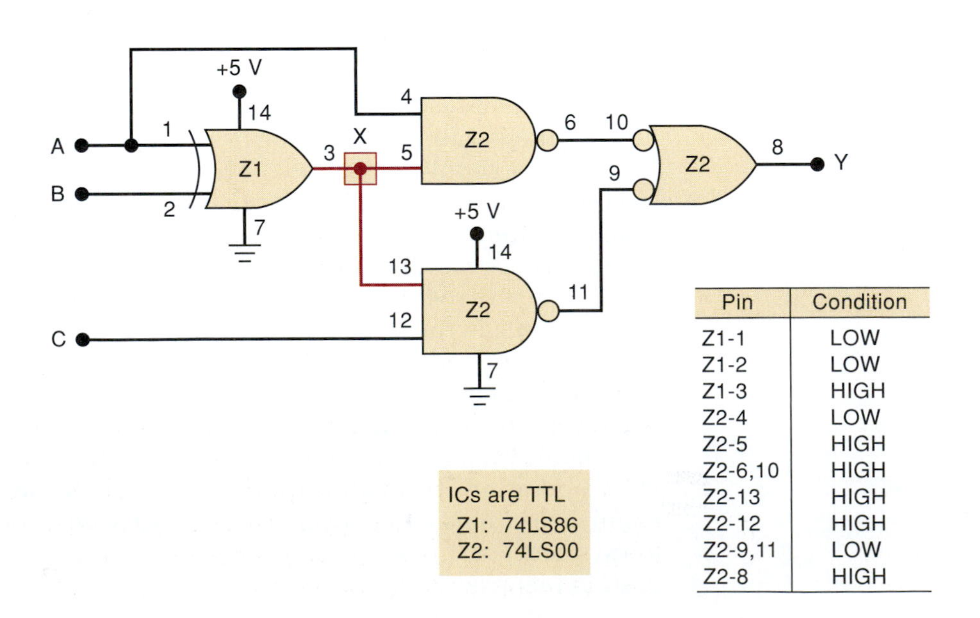

Pin	Condition
Z1-1	LOW
Z1-2	LOW
Z1-3	HIGH
Z2-4	LOW
Z2-5	HIGH
Z2-6,10	HIGH
Z2-13	HIGH
Z2-12	HIGH
Z2-9,11	LOW
Z2-8	HIGH

FIGURE 4-41 Example 4-28.

When the circuit is tested, the technician observes that output Y goes HIGH whenever A is HIGH or C is HIGH, regardless of the level at B. She takes logic probe measurements for the condition where $A = B = 0, C = 1$ and comes up with the indications recorded in Figure 4-41.

Examine the recorded levels and list the possible causes for the malfunction. Then develop a step-by-step procedure to determine the exact fault.

Solution

All of the NAND gate outputs are correct for the levels present at their inputs. The XOR gate, however, should be producing a LOW at output pin 3 because both of its inputs are at the same LOW level. It appears that Z1-3 is stuck HIGH, even though its inputs should produce a LOW. There are several possible causes for this:

1. An internal component failure in Z1 that prevents its output from going LOW
2. An external short to V_{CC} from any point along the conductors connected to node X (shaded in the diagram of the figure)
3. Pin 3 of Z1 internally shorted to V_{CC}
4. Pin 5 of Z2 internally shorted to V_{CC}
5. Pin 13 of Z2 internally shorted to V_{CC}

All of these possibilities except for the first one will short node X (and every IC pin connected to it) directly to V_{CC}.

The following procedure can be used to isolate the fault. This procedure is not the only approach that can be used and, as we stated earlier, the actual troubleshooting procedure that a technician uses is very dependent on what test equipment is available.

1. Check the V_{CC} and ground levels at the appropriate pins of Z1. Although it is unlikely that the absence of either of these might cause Z1-3 to stay HIGH, it is a good idea to make this check on any IC that is producing an incorrect output.
2. Turn off power to the circuit and use an ohmmeter to check for a short (resistance less than 1 Ω) between node X and any point connected to V_{CC} (such as Z1-14 or Z2-14). If no short is indicated, the last four possibilities in our list can be eliminated. This means that it is very likely that Z1 has an internal failure and should be replaced.
3. If step 2 shows that there is a short from node X to V_{CC}, perform a thorough visual examination of the circuit board and look for solder bridges, unetched copper slivers, uninsulated wires touching each other, and any other possible cause of an external short to V_{CC}. A likely spot for a solder bridge would be between adjacent pins 13 and 14 of Z2. Pin 14 is connected to V_{CC}, and pin 13 to node X. If an external short is found, remove it and perform an ohmmeter check to verify that node X is no longer shorted to V_{CC}.
4. If step 3 does not reveal an external short, the three possibilities that remain are internal shorts to V_{CC} at Z1-3, Z2-13, or Z2-5. One of these is shorting node X to V_{CC}.

To determine which of these IC pins is the culprit, we should disconnect each of them from node X *one at a time* and recheck for a short to V_{CC} after each disconnection. When the pin that is internally shorted to V_{CC} is disconnected, node X will no longer be shorted to V_{CC}.

The process of disconnecting each suspected pin from node *X* can be easy or difficult depending on how the circuit is constructed. If the ICs are in sockets, all you need to do is to pull the IC from its socket, bend out the suspected pin, and reinsert the IC into its socket. If the ICs are soldered into a printed circuit board, you will have to cut the trace that is connected to the pin and repair the cut trace when you are finished.

Example 4-28, although fairly simple, shows you the kinds of thinking that a troubleshooter must employ to isolate a fault. You will have the opportunity to begin developing your own troubleshooting skills by working on many end-of-chapter problems that have been designated with a **T** for troubleshooting.

4-14 PROGRAMMABLE LOGIC DEVICES*

In the previous sections, we briefly considered the class of ICs known as programmable logic devices. In Chapter 3, we introduced the concept of describing a circuit's operation using a hardware description language. In this section, we will explore these topics further and become prepared to use the tools of the trade to develop and implement digital systems using PLDs. Of course, it is impossible to understand all the complex details of how a PLD works before grasping the fundamentals of digital circuits. As we examine new fundamental concepts, we will expand our knowledge of the PLDs and the programming methods. The material is presented in such a way that anyone who is not interested in PLDs can easily skip over these sections without loss of continuity in the coverage of the basic principles.

Let's review the process we covered earlier of designing combinational digital circuits. The input devices are identified and assigned an algebraic name like *A*, *B*, *C*, or *LOAD*, *SHIFT*, *CLOCK*. Likewise, output devices are given names like X, Z, or *CLOCK_OUT*, *SHIFT_OUT*. Then a truth table is constructed that lists all the possible input combinations and identifies the required state of the outputs under each input condition. The truth table is one way of describing how the circuit is to operate. Another way to describe the circuit's operation is the Boolean expression. From this point the designer must find the simplest algebraic relationship and select digital ICs that can be wired together to implement the circuit. You have probably experienced that these last steps are the most tedious, time consuming, and prone to errors.

Programmable logic devices allow most of these tedious steps to be automated by a computer and PLD *development software*. Using programmable logic improves the efficiency of the design and development process. Consequently, most modern digital systems are implemented in this way. The job of the circuit designer is to identify inputs and outputs, specify the logical relationship in the most convenient manner, and select a programmable device that is capable of implementing the circuit at the lowest cost. The concept behind programmable logic devices is simple: put lots of logic gates in a single IC and control the interconnection of these gates electronically.

PLD Hardware

Recall from Chapter 3 that many digital circuits today are implemented using programmable logic devices (PLDs). These devices are configured electronically and their internal circuits are "wired" together electronically to

*All sections covering PLDs may be skipped without loss of continuity in the balance of Chapters 1–12.

form a logic circuit. This programmable wiring can be thought of as thousands of connections that are either connected (1) or not connected (0). It is very tedious to try to configure these devices by manually placing 1s and 0s in a grid, so the next logical question is, "How do we control the interconnection of gates in a PLD electronically?"

A common method of connecting one of many signals entering a network to one of many signal lines exiting the network is a switching matrix. Refer back to Figure 3-44, where this concept was introduced. A matrix is simply a grid of conductors (wires) arranged in rows and columns. Input signals are connected to the columns of the matrix, and the outputs are connected to the rows of the matrix. At each intersection of a row and a column is a switch that can electrically connect that row to that column. The switches that connect rows to columns can be mechanical switches, fusible links, electromagnetic switches (relays), or transistors. This is the general structure used in many applications and will be explored further when we study memory devices in Chapter 12.

PLDs also use a switch matrix that is often referred to as a programmable array. By deciding which intersections are connected and which ones are not, we can "program" the way the inputs are connected to the outputs of the array. In Figure 4-42, a programmable array is used to select the inputs for each AND gate. Notice that in this simple matrix, we can produce any logical product combination of variables A, B at any of the AND gate outputs. A matrix or programmable array such as the one shown in the figure can also be used to connect the AND outputs to OR gates. The details of various PLD architectures will be covered thoroughly in Chapter 13.

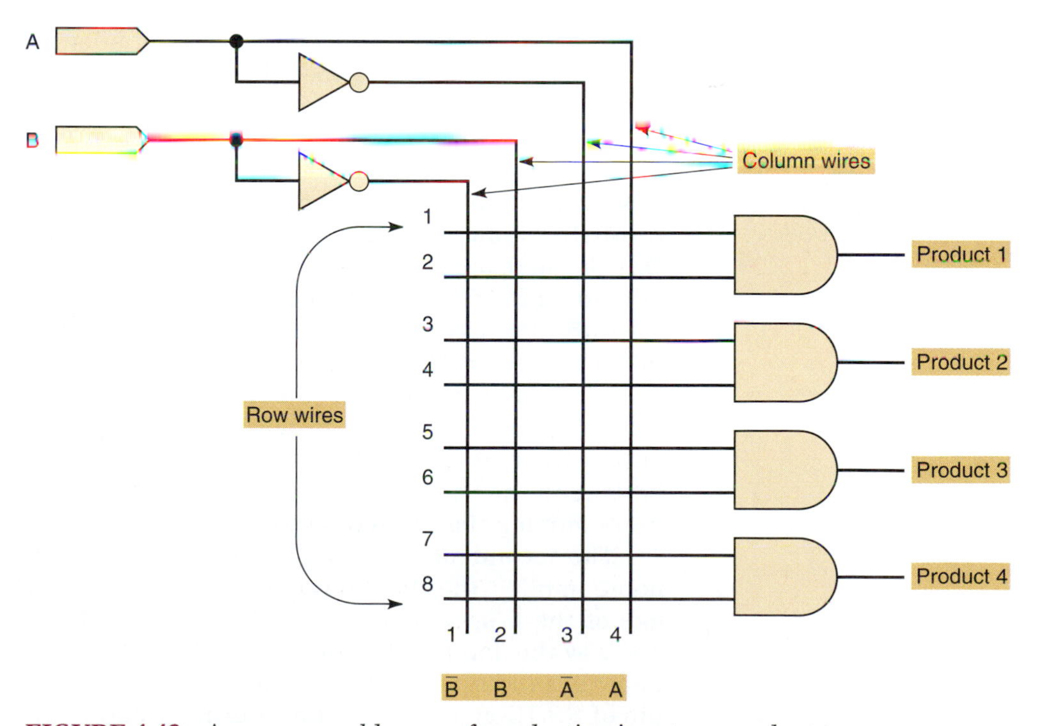

FIGURE 4-42 A programmable array for selecting inputs as product terms.

Programming a PLD

There are two ways to "program" a PLD IC. Programming means making the actual connections in the array. In other words, it means determining which of those connections are supposed to be open (0) and which are supposed to be closed (1). The first method involves removing the PLD IC chip from its circuit board. The chip is then placed in a special fixture called a **programmer**, shown in Figure 4-43. Most modern programmers are connected to a personal computer that is running software containing libraries of information about the many types of programmable devices available.

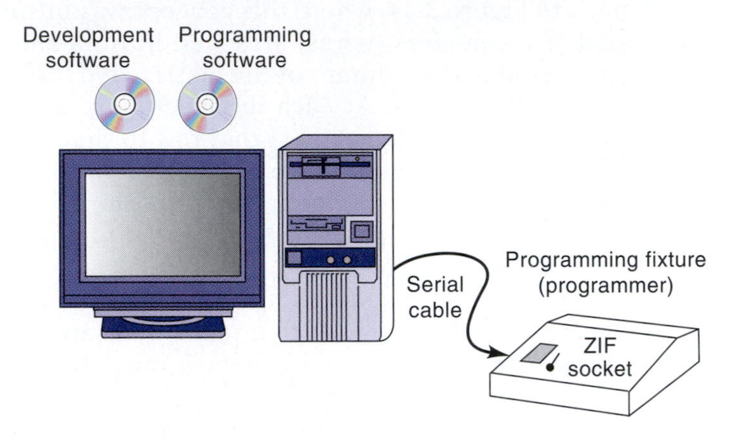

FIGURE 4-43 A PLD development system.

The programming software is invoked (called up and executed) on the PC to establish communication with the programmer. This software allows the user to set up the programmer for the type of device that is to be programmed, check if the device is blank, read the state of any programmable connection in the device, and provide instructions for the user to program a chip. Ultimately, the part is placed into a special socket that allows you to drop the chip in and then clamp the contacts onto the pins. This is called a **zero insertion force (ZIF)** socket. *Universal programmers* that can program any type of programmable device are available from numerous manufacturers.

Fortunately, as programmable parts began to proliferate, manufacturers saw the need to standardize pin assignments and programming methods. As a result, the Joint Electronic Device Engineering Council **(JEDEC)** was formed. One of the results was JEDEC standard 3, a format for transferring programming data for PLDs, independent of the PLD manufacturer or programming software. Pin assignments for various IC packages were also standardized, making universal programmers less complicated. Consequently, programming fixtures can program numerous types of PLDs. The software that allows the designer to specify a configuration for a PLD simply needs to produce an output file that conforms to the JEDEC standards. Then this JEDEC file can be loaded into any JEDEC-compatible PLD programmer that is capable of programming the desired type of PLD.

The second method is referred to as in system programming (ISP). As its name implies, the chip does not need to be removed from its circuit for storage of the programming information. A standard interface has been developed by the Joint Test Action Group **(JTAG)**. The interface was developed to allow ICs to be tested without actually connecting test equipment to every pin of the IC. It also allows for internal programming. Four pins on the IC are used as a portal to store data and retrieve information about the inner condition of the IC. Many ICs, including PLDs and microcontrollers, are manufactured today to include the JTAG interface. An interface cable connects the four JTAG pins on the IC to an output port (like the printer port) of a personal computer. Software running on the PC establishes contact with the IC and loads the information in the proper format.

Development Software

We have examined several methods of describing logic circuits now, including schematic capture, logic equations, truth tables, and HDL. We also described the fundamental methods of storing 1s and 0s into a PLD IC to connect the logic circuits in the desired way. The biggest challenge in getting a PLD programmed is converting from any form of description into the array of 1s and 0s. Fortunately, this task is accomplished quite easily by a computer running the development software. The development software that we will be referring to and using for examples is produced by Altera. This software allows the designer to enter a circuit description in any one of the many ways we have been discussing: graphic design files (schematics), AHDL, and VHDL. It also allows the use of another HDL, called Verilog, and the option of describing the circuit with timing diagrams. Circuit blocks described by any of these methods can also be "connected" together to implement a much larger digital system, as shown in Figure 4-44. Any logic diagram found in this text can be redrawn using the schematic entry tools in the Altera software to create a graphic design file. We will not focus on graphic design entry in this text because it is quite straightforward to pick up these skills in the laboratory. We will focus our examples on the methods that allow us to use HDL as an alternate means of describing a circuit. For more information on the Altera software, see the accompanying CD as well as user manuals from the Altera web site (http://www.altera.com).

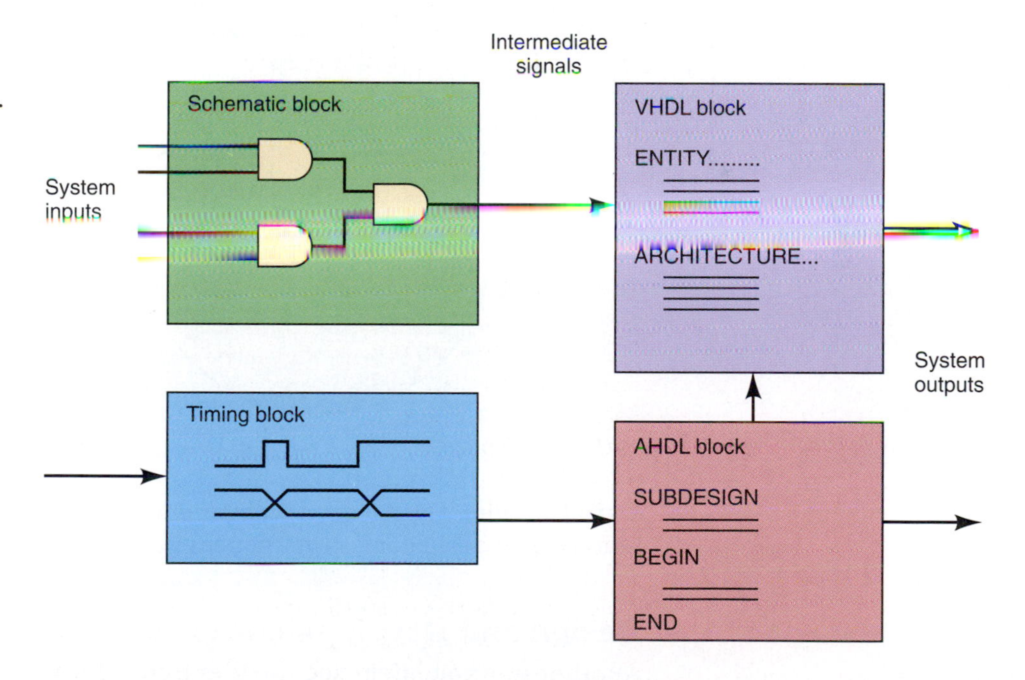

FIGURE 4-44 Combining blocks developed using different description methods.

This concept of using building blocks of circuits is called **hierarchical design**. Small, useful logic circuits can be defined in whatever manner is most convenient (graphic, HDL, timing, etc.) and then combined with other circuits to form a large section of a project. Sections can be combined and connected with other sections to form the whole system. Figure 4-45 shows the hierarchical structure of a CD player using a block diagram. The outer box encloses the entire system. The dashed lines identify each major subsection, and each subsection contains individual circuits. Although it is not shown in this diagram, each circuit may be made up of smaller building blocks of common

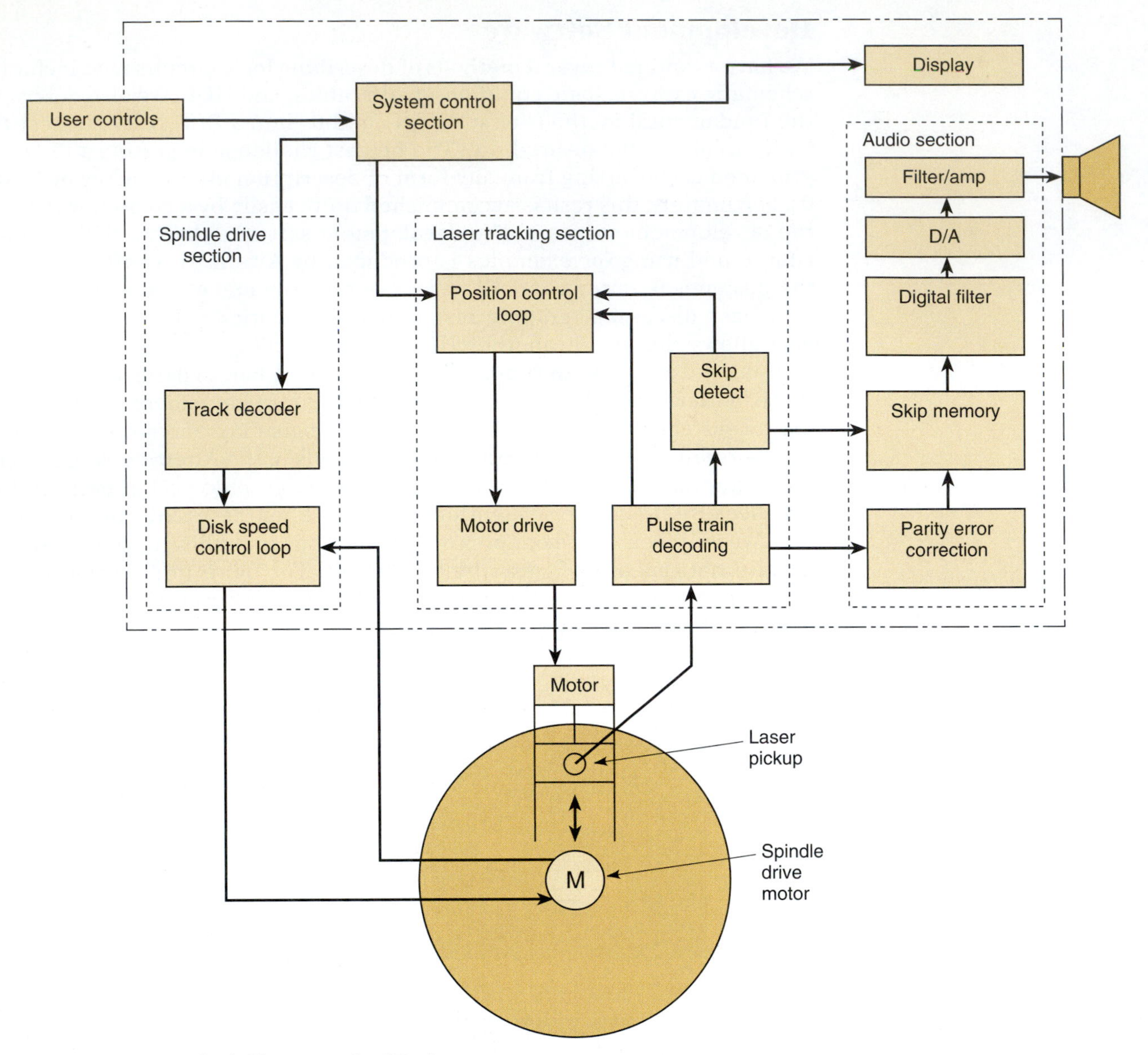

FIGURE 4-45 Block diagram of a CD player.

digital circuits. The Altera development software makes this type of modular, hierarchical design and development easy to accomplish.

Design and Development Process

Another way you might see the hierarchy of a system like the CD player just described is shown in Figure 4-46. The top level represents the entire system. It is made up of three subsections, each of which in turn is made up of the smaller circuits shown. Notice that this diagram does not show how the signals flow throughout the system but clearly identifies the various levels of the hierarchical structure of the project.

This type of diagram has led to the name for one of the most common methods of design: **top-down**. With this design approach, you start with the overall description of the entire system, such as the top box in Figure 4-46. Then you define several subsections that will make up the system. The subsections are further refined into individual circuits connected together.

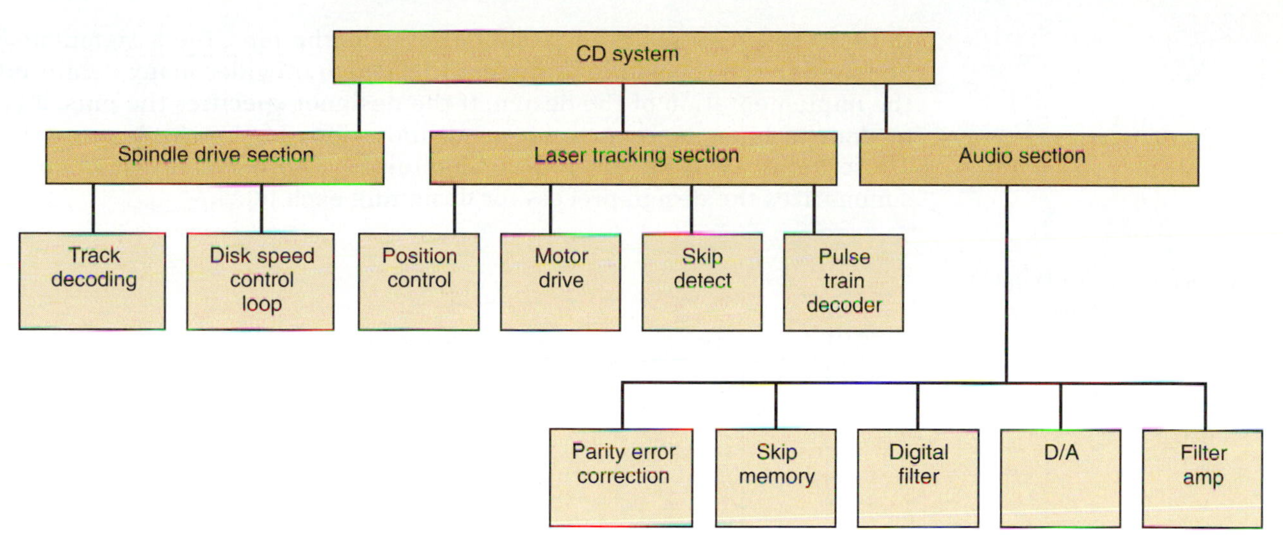

FIGURE 4-46 An organizational hierarchy chart.

Every one of these hierarchy levels has defined inputs, outputs, and behavior. Each can be tested individually before it is connected to the others.

After defining the blocks from the top down, the system is built from the bottom up. Each block in this system design has a design file that describes it. The lowest level blocks must be designed by opening a design file and writing a description of its operation. The designed block is then compiled using the development tools. The compiling process determines if you have made errors in your syntax. Until your syntax is correct, the computer cannot possibly translate your description into its proper form. After it has been compiled with no syntax errors, it should be tested to see if it operates correctly. Development systems offer simulator programs that run on the PC and simulate the way your circuit responds to inputs. A simulator is a computer program that calculates the correct output logic states based on a description of the logic circuit and the current inputs. A set of hypothetical inputs and their corresponding correct outputs are developed that will prove the block works as expected. These hypothetical inputs are often called **test vectors**. Thorough testing during simulation greatly increases the likelihood of the final system working reliably. Figure 4-47 shows the simulation file for the circuit described in Figure 3-13(a) of Chapter 3. Inputs a, b, and c were entered as test vectors, and the simulation produced output y.

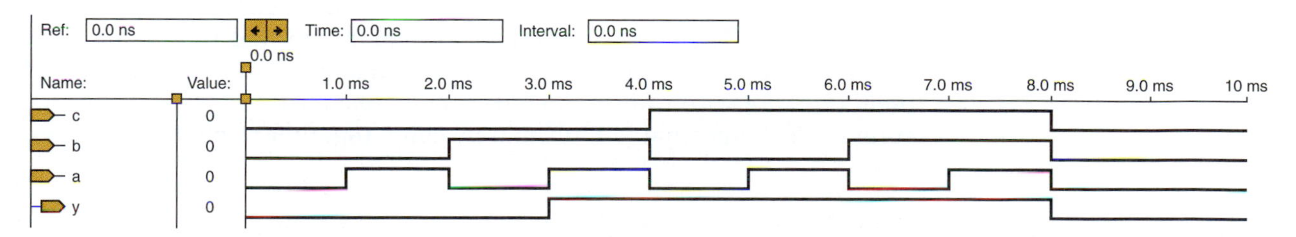

FIGURE 4-47 A timing simulation of a circuit described in HDL.

When the designer is satisfied that the design works, the design can be verified by actually programming a chip and testing. For a complex PLD, the designer can either let the development system assign pins and then lay out the final circuit board accordingly, or specify the pins for each signal using the software features. If the compiler assigns the pins, the assignments can be found in the report file or pin-out file, which provides many details about the implementation of the design. If the designer specifies the pins, it is important to know the constraints and limitations of the chip's architecture. These details will be covered in Chapter 13. The flowchart of Figure 4-48 summarizes the design process for designing each block.

FIGURE 4-48 PLD development cycle flowchart.

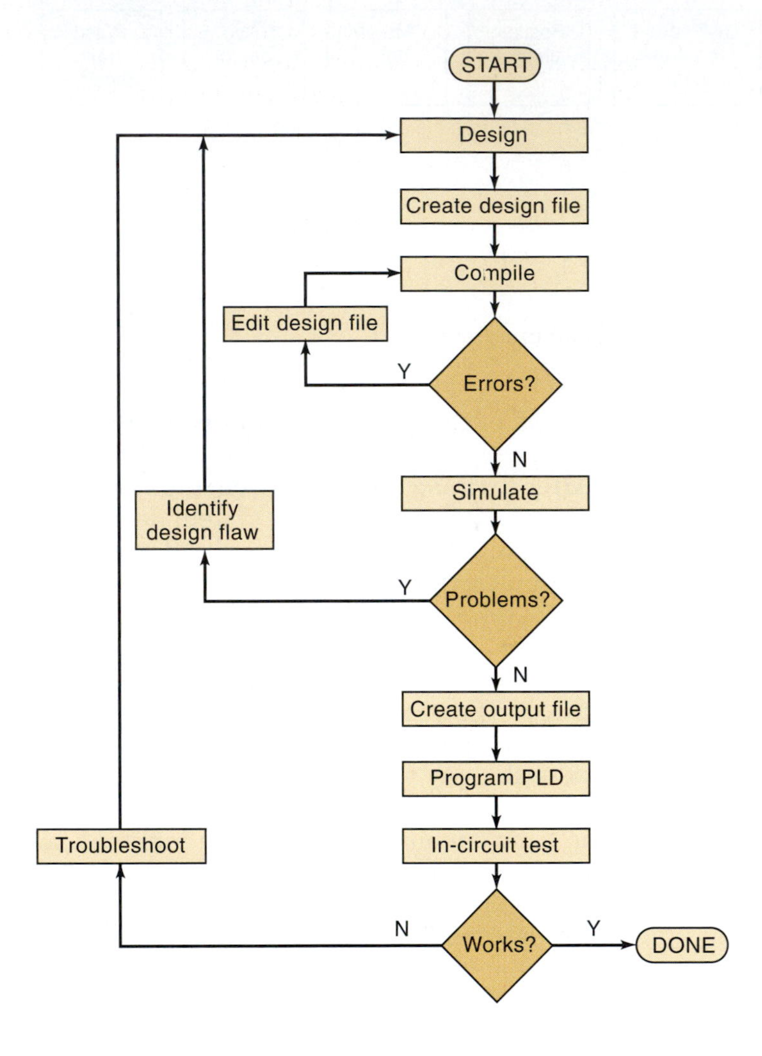

After each circuit in a subsection has been tested, all can be combined and the subsection can be tested following the same process that was used for the small circuits. Then the subsections are combined and the system is tested. This approach lends itself very well to a typical project environment, where a team of people are working together, each responsible for his or her own circuits and sections that will ultimately come together to make up the system.

REVIEW QUESTIONS

1. What is actually being "programmed" in a PLD?
2. What bits (column, row) in Figure 4-42 must be connected to make Product 1 = AB?
3. What bits (column, row) in Figure 4-42 must be connected to make Product 3 = $A\overline{B}$?

4-15 REPRESENTING DATA IN HDL

Numeric data can be represented in various ways. We have studied the use of the hexadecimal number system as a convenient way to communicate bit patterns. We naturally prefer to use the decimal number system for numeric data, but computers and digital systems can operate only on binary information, as we studied in previous chapters. When we write in HDL, we often need to use these various number formats, and the computer must be able to understand which number system we are using. So far in this text, we have used a subscript to indicate the number system. For example, 101_2 was binary, 101_{16} was hexadecimal, and 101_{10} was decimal. Every programming language and HDL has its own unique way of identifying the various number systems, generally done with a prefix to indicate the number system. In most languages, a number with no prefix is assumed to be decimal. When we read one of these number designations, we must think of it as a symbol that represents a binary bit pattern. These numeric values are referred to as scalars or **literals**. Table 4-8 summarizes the methods of specifying values in binary, hex, and decimal for AHDL and VHDL.

TABLE 4-8 Designating number systems in HDL.

Number System	AHDL	VHDL	Bit Pattern	Decimal Equivalent
Binary	B"101"	B"101"	101	5
Hexadecimal	H"101"	X"101"	100000001	257
Decimal	101	101	1100101	101

EXAMPLE 4-29

Express the following bit pattern's numeric value in binary, hex, and decimal using AHDL and VHDL notation:

11001

Solution

Binary is designated the same in both AHDL and VHDL: **B "11001"**.
Converting the binary to hex, we have 19_{16}.
In AHDL: **H "19"**
In VHDL: **X "19"**
Converting the binary to decimal, we have 25_{10}.
Decimal is designated the same in both AHDL and VHDL: 25.

Bit Arrays/Bit Vectors

In Chapter 3, we declared names for inputs to and outputs of a very simple logic circuit. These were defined as bits, or single binary digits. What if we want to represent an input, output, or signal that is made up of several bits? In an HDL, we must define the type of the signal and its range of acceptable values.

To understand the concepts used in HDLs, let's first consider some conventions for describing bits of binary words in common digital systems. Suppose we have an eight-bit number representing the current temperature, and the number is coming into our digital system through an input port that we have named P1, as shown in Figure 4-49. We can refer to the individual bits of this port as P1 bit 0 for the least significant bit, on up to P1 bit 7 for the most significant bit.

We can also describe this port by saying that it is named P1, with bits numbered 7 down to 0. The terms **bit array** and **bit vector** are often used to describe this type of data structure. It simply means that the overall data structure (eight-bit port) has a name (P1) and that each individual element (bit) has a unique **index** number (0–7) to describe that bit's position (and possibly its numeric weight) in the overall structure. The HDLs and computer programming languages take advantage of this notation. For example, the third bit from the right is designated as P1[2], and it can be connected to another signal bit by using an assignment operator.

EXAMPLE 4-30

Assume there is an eight-bit array named P1, as shown in Figure 4-49, and another four-bit array is named P5.

(a) Write the bit designation for the most significant bit of P1.
(b) Write the bit designation for the least significant bit of P5.
(c) Write an expression that causes the least significant bit of P5 to drive the most significant bit of P1.

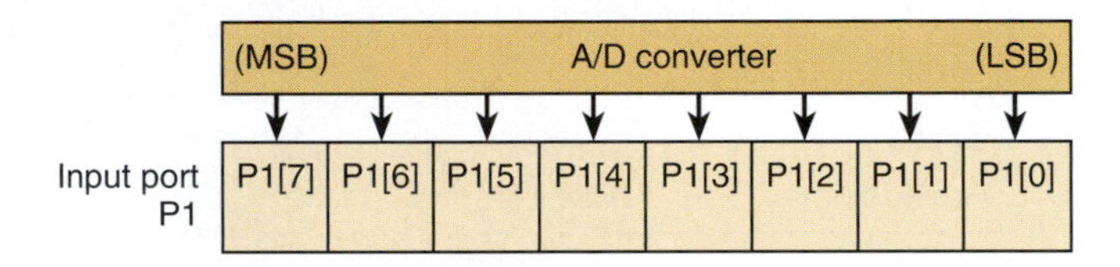

FIGURE 4-49 Bit array notation.

Solution

(a) The name of the port is P1 and the most significant bit is bit 7. The proper designation for P1 bit 7 is P1[7].
(b) The name of the port is P5 and the least significant bit is bit 0. The proper designation for P5 bit 0 is P5[0].
(c) The driving signal is placed on the right side of the assignment operator, and the driven signal is placed on the left: P1[7] = P5[0];.

AHDL

AHDL BIT ARRAY DECLARATIONS

In AHDL, port *p1* of Figure 4-49 is defined as an eight-bit input port, and the value on this port can be referred to using any number system, such as hex, binary, decimal, etc. The syntax for AHDL uses a name for the bit vector followed by the range of index designations, which are enclosed in square brackets. This declaration is included in the SUBDESIGN section. For example, to declare an eight-bit input port called *p1*, you would write

```
p1 [7..0] :INPUT; --define an 8-bit input port
```

EXAMPLE 4-31

Declare a four-bit input named *keypad* using AHDL.

Solution

```
keypad [3..0] :INPUT;
```

Intermediate variables can also be declared as an array of bits. As with single bits, they are declared just after the I/O declarations in SUBDESIGN. As an example, the eight-bit temperature port *p1* can be assigned (connected) to a node named *temp*, as follows:

```
VARIABLE temp [7..0] :NODE;
BEGIN
   temp[] = p1[];
END;
```

Notice that the input port *p1* has the data applied to it, and it is driving the signal wires named *temp*. Think of the term on the right of the equals sign as the source of the data and the term on the left as the destination. The empty brackets [] mean that each of the correspanding bits in the two arrays are being connected. Individual bits can also be "connected" by specifying the bits inside the brackets. For example, to connect only the least significant bit of p1 to the LSB of temp, the statement would be temp[0] = p1[0];.

VHDL BIT VECTOR DECLARATIONS

In VHDL, port *p1* of Figure 4-49 is defined as an eight-bit input port, and the value on this port can be referred to using only binary literals. The syntax for VHDL uses a name for the bit vector followed by the mode (:IN), the type (**BIT_VECTOR**), and the range of index designations, which are enclosed in parentheses. This declaration is included in the ENTITY section. For example, to declare an eight-bit input port called *p1*, you would write

```
PORT (p1 :IN BIT_VECTOR (7 DOWNTO 0);
```

EXAMPLE 4-32

Declare a four-bit input named *keypad* using VHDL.

Solution

```
PORT(keypad :IN BIT_VECTOR (3 DOWNTO 0);
```

Intermediate signals can also be declared as an array of bits. As with single bits, they are declared just inside the ARCHITECTURE definition. As an example, the eight-bit temperature on port *p1* can be assigned (connected) to a signal named *temp*, as follows:

```
SIGNAL      temp :BIT_VECTOR (7 DOWNTO 0);
BEGIN
    temp <= p1;
END;
```

Notice that the input port *p1* has the data applied to it, and it is driving the signal wires named *temp*. No elements in the bit vector are specified, which means that all the bits are being connected. Individual bits can also be "connected" using signal assignments and by specifying the bit numbers inside parentheses. For example, to connect only the least significant bit of *p1* to the LSB of *temp*, the statement would be temp(0) <= p1(0);.

VHDL is very particular regarding the definitions of each type of the data. The type "bit_vector" describes an array of individual bits. This is interpreted differently than an eight-bit binary number (called a scaler quantity), which has the type **integer**. Unfortunately, VHDL does not allow us to assign an integer value to a BIT_VECTOR signal directly. Data can be represented by any of the types shown in Table 4-9, but data assignments and other operations must be done between objects of the same type. For example, the compiler will not allow you to take a number from a keypad declared as an integer and connect it to four LEDs that are declared as BIT_VECTOR outputs. Notice in Table 4-9, under Possible Values, that individual BIT and STD_LOGIC data **objects** (e.g., signals, variables, inputs, and outputs) are designated by single quotes, whereas values assigned to BIT_VECTOR and STD_LOGIC_VECTOR types are strings of valid bit values enclosed in double quotes.

TABLE 4-9 Common VHDL data types.

Data Type	Sample Declaration	Possible Values	Use
BIT	y :OUT BIT;	'0' '1'	y <= '0';
STD_LOGIC	driver :STD_LOGIC	'0' '1' 'z' 'x' '-'	driver <= 'z';
BIT_VECTOR	bcd_data :BIT_VECTOR (3 DOWNTO 0);	"0101" "1001" "0000"	digit <= bcd_data;
STD_LOGIC_VECTOR	dbus :STD_LOGIC_VECTOR (3 DOWNTO 0);	"0Z1X"	IF rd = '0' THEN dbus <= "zzzz";
INTEGER	SIGNAL z:INTEGER RANGE −32 TO 31;	−32..−2,−1,0,1,2 . . . 31	IF z > 5 THEN . . .

VHDL also offers some standardized data types that are necessary when using logic functions that are contained in the **libraries**. As you might have guessed, libraries are simply collections of little pieces of VHDL code that can be used in your hardware descriptions without reinventing the wheel. These libraries offer convenient functions, called **macrofunctions**, like many of the standard TTL devices that are described throughout this text. Rather than writing a new description of a familiar TTL device, we can simply pull its macrofunction out of the library and use it in our system. Of course, you need to get signals into and out of these macrofunctions, and the types of the signals in your code must match the types in the functions (which someone else wrote). This means that everyone must use the same standard data types.

When VHDL was standardized through the IEEE, many data types were created at the same time. The two that we will use in this text are **STD_LOGIC**, which is equivalent to BIT type, and **STD_LOGIC_VECTOR**, which is equivalent to BIT_VECTOR. As you recall, BIT type can have values of only '0' and '1'. The standard logic types are defined in the IEEE library and have a broader range of possible values than their built-in counterparts. The possible values for a STD_LOGIC type or for any element in a STD_LOGIC_VECTOR are given in Table 4-10. The names of these categories will make much more sense after we study the characteristics of logic circuits in Chapter 8. For now, we will show examples using values of only '1' and '0'.

TABLE 4-10 STD_LOGIC values.

'1'	Logic 1 (just like BIT type)
'0'	Logic 0 (just like BIT type)
'z'	High impedance*
'-'	don't care (just like you used in your K maps)
'U'	Uninitialized
'X'	Unknown
'W'	Weak unknown
'L'	Weak '0'
'H'	Weak '1'

*We will study tristate logic in Chapter 8.

REVIEW QUESTIONS

1. How would you declare a six-bit input array named push_buttons in (a) AHDL or (b) VHDL?
2. What statement would you use to take the MSB from the array in question 1 and put it on a single-bit output port named z? Use (a) AHDL or (b) VHDL.
3. In VHDL, what is the IEEE standard type that is equivalent to the BIT type?
4. In VHDL, what is the IEEE standard type that is equivalent to the BIT_VECTOR type?

4-16 TRUTH TABLES USING HDL

We have learned that a truth table is another way of expressing the operation of a circuit block. It relates the output of the circuit to every possible combination of its inputs. As we saw in Section 4-4, a truth table is the starting point for a designer to define how the circuit should operate. Then a Boolean expression is derived from the truth table and simplified using K maps or Boolean algebra. Finally the circuit is implemented from the final Boolean equation. Wouldn't it be great if we could go from the truth table directly to the final circuit without all those steps? We can do exactly that by entering the truth table using HDL.

TRUTH TABLES USING AHDL

The code in Figure 4-50 uses AHDL to implement a circuit and uses a truth table to describe its operation. The truth table for this design was presented in Example 4-7. The key point of this example is the use of the TABLE keyword in AHDL. It allows the designer to specify the operation of the circuit just like you would fill out a truth table. On the first line after TABLE, the input variables (*a*,*b*,*c*) are listed exactly like you would create a column heading on a truth table. By including the three binary variables in parentheses, we tell the compiler that we want to use these three bits as a group and to refer to them as a three-bit binary number or bit pattern. The specific values for this bit pattern are listed below the group and are referred to as binary literals. The special operator (=>) is used in truth tables to separate the inputs from the output (*y*).

AHDL

FIGURE 4-50 AHDL design file for Figure 4-7

```
%     Figure 4-7 in AHDL
      Digital Systems 10th ed
      Neal Widmer
      MAY 23, 2005                    %
SUBDESIGN FIG4_50
(
      a,b,c :INPUT;             --define inputs to block
      y     :OUTPUT;            --define block output
)
BEGIN
      TABLE
            (a,b,c)             =>     y;    --column headings
            (0,0,0)             =>     0;
            (0,0,1)             =>     0;
            (0,1,0)             =>     0;
            (0,1,1)             =>     1;
            (1,0,0)             =>     0;
            (1,0,1)             =>     1;
            (1,1,0)             =>     1;
            (1,1,1)             =>     1;
      END TABLE;
END;
```

The TABLE in Figure 4-50 is intended to show the relationship between the HDL code and a truth table. A more common way of representing the input data heading is to use a variable bit array to represent the value on *a*, *b*, *c*. This method involves a declaration of the bit array on the line before BEGIN, such as:

```
VARIABLE in_bits[2..0]     :NODE;
```

Just before the TABLE keyword, the input bits can be assigned to the array, *inbits[ ]*:

```
in_bits[] = (a,b,c);
```

Grouping three independent bits in order like this is referred to as **concatenating**, and it is often done to connect individual bits to a bit array. The table heading on the input bit sets can be represented by *in_bits[ ]*, in this case. Note that as we list the possible combinations of the inputs, we have several options. We can make up a group of 1s and 0s in parentheses, as shown in Figure 4-50, or we can represent the same bit pattern using the equivalent binary, hex, or decimal number. It is up to the designer to decide which format is most appropriate depending on what the input variables represent.

TRUTH TABLES USING VHDL: SELECTED SIGNAL ASSIGNMENT

The code in Figure 4-51 uses VHDL to implement a circuit using a **selected signal assignment** to describe its operation. It allows the designer to specify the operation of the circuit, just like you would fill out a truth table. The truth table for this design was presented in Example 4-7. The primary point of this example is the use of the WITH signal_name SELECT statement in VHDL. A secondary point presented here shows how to put the data into a

```
--     Figure 4-7 in VHDL
--     Digital Systems 10th ed
--     Neal Widmer
--     MAY 23, 2005
ENTITY fig4_51 IS
PORT(
       a,b,c :IN BIT;              --declare individual input bits
       y     :OUT BIT);
END fig4_51;

ARCHITECTURE truth OF fig4_51 IS
       SIGNAL in_bits :BIT_VECTOR(2 DOWNTO 0);
       BEGIN
       in_bits <= a & b & c;     --concatenate input bits into bit_vector
             WITH in_bits SELECT
             y     <=     '0'  WHEN "000",    --Truth Table
                          '0'  WHEN "001",
                          '0'  WHEN "010",
                          '1'  WHEN "011",
                          '0'  WHEN "100",
                          '1'  WHEN "101",
                          '1'  WHEN "110",
                          '1'  WHEN "111";
END truth;
```

FIGURE 4-51 VHDL design file for Figure 4-7.

format that can be used conveniently with the selected signal assignment. Notice that the inputs are defined in the ENTITY declaration as three independent bits *a*, *b*, and *c*. Nothing in this declaration makes one of these more significant than another. The order in which they are listed does not matter. We want to compare the current value of these bits with each of the possible combinations that could be present. If we drew out a truth table, we would decide which bit to place on the left (MSB) and which to place on the right (LSB). This is accomplished in VHDL by **concatenating** (connecting in order) the bit variables to form a bit vector. The concatenation operator is "&". A signal is declared as a BIT_VECTOR to receive the ordered set of input bits and is used to compare the input's value with the string literals contained in quotes. The output (*y*) is assigned (<=) a bit value ('0' or '1') WHEN *in_bits* contains the value listed in double quotes.

VHDL is very strict in the way it allows us to assign and compare objects such as signals, variables, constants, and literals. The output *y* is a BIT, and so it must be assigned a value of '0' or '1'. The SIGNAL *in_bits* is a three-bit BIT_VECTOR, so it must be compared with a three-bit string literal value. VHDL will not allow *in_bits* (a BIT_VECTOR) to be compared with a hex number like X "5" or a decimal number like 3. These scalar quantities would be valid for assignment or comparison with integers.

EXAMPLE 4-33

Declare three signals in VHDL that are single bits named *too_hot, too_cold,* and *just_right.* Combine (concatenate) these three bits into a three-bit signal called *temp_status,* with hot on the left and cold on the right.

Solution

1. Declare signals first in Architecture.

```
SIGNAL too_hot, too_cold, just_right :BIT;
SIGNAL temp_status :BIT_VECTOR (2 DOWNTO 0);
```

2. Write concurrent assignment statements between BEGIN and END.

```
temp_status <= too_hot & just_right & too_cold;
```

REVIEW QUESTIONS

1. How would you concatenate three bits *x*, *y*, and *z* into a three-bit array named *omega*? Use AHDL or VHDL.
2. How are truth tables implemented in AHDL?
3. How are truth tables implemented in VHDL?

4-17 DECISION CONTROL STRUCTURES IN HDL

In this section, we will examine methods that allow us to tell the digital system how to make "logical" decisions in much the same way that we make decisions every day. In Chapter 3, we learned that concurrent assignment statements are evaluated such that the order in which they are written has no effect on the circuit being described. When using **decision control structures**, the order in which we ask the questions does matter. To summarize this concept in the terms used in HDL documentation, statements that can be written in any sequence are called **concurrent**, and statements that are evaluated in the sequence in which they are written are called **sequential**. The sequence of sequential statements affects the circuit's operation.

The examples we have considered so far involve several individual bits. Many digital systems require inputs that represent a numeric value. Refer again to Example 4-8, in which the purpose of the logic circuit is to monitor the battery voltage measured by an A/D converter. The digital value is represented by a four-bit number coming from the A/D into the logic circuit. These inputs are not independent binary variables but rather four binary digits of a number representing battery voltage. We need to give the data the correct type that will allow us to use it as a number.

IF/ELSE

Truth tables are great for listing all the possible combinations of independent variables, but there are better ways to handle numeric data. As an example, when a person leaves for school or work in the morning, she must make a logical decision about wearing a coat. Let's assume she decides this issue based only on the current temperature. How many of us would reason as follows?

I will wear a coat if the temperature is 0.
I will wear a coat if the temperature is 1.
I will wear a coat if the temperature is 2. . . .
I will wear a coat if the temperature is 55.

I will *not* wear a coat if the temperature is 56.

I will *not* wear a coat if the temperature is 57.

I will *not* wear a coat if the temperature is 58. . . .

I will *not* wear a coat if the temperature is 99.

This method is similar to the truth table approach of describing the decision. For every possible input, she decides what the output should be. Of course, what she would really do is decide as follows:

I will wear a coat if the temperature is less than 56 degrees.

Otherwise, I will *not* wear a coat.

An HDL gives us the power to describe logic circuits using this type of reasoning. First, we must describe the inputs as a *number within a given range,* and then we can write statements that decide what to do to the outputs based on the *value* of the incoming number. In most computer programming languages, as well as HDLs, these types of decisions are made using an IF/THEN/ELSE control structure. Whenever the decision is between doing something and doing nothing, an **IF/THEN** construct is used. The keyword IF is followed by a statement that is true or false. IF it is true, THEN do whatever is specified. In the event that the statement is false, no action is taken. Figure 4-52(a) shows graphically how this decision works. The diamond shape represents the decision being made by evaluating the statement contained within the diamond. Every decision has two possible outcomes: true or false. In this example, if the statement is false, no action is taken.

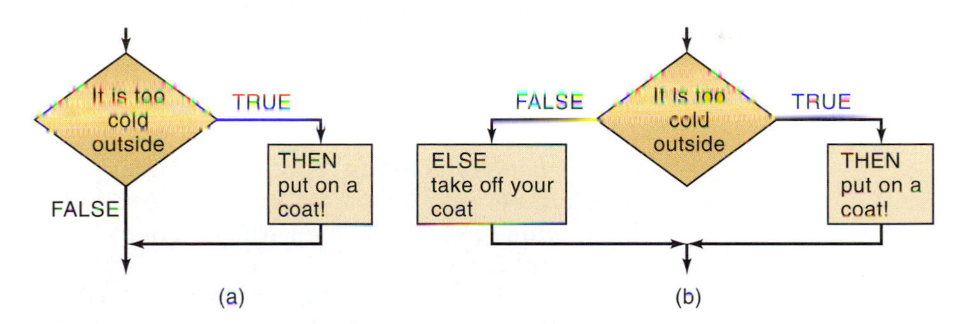

FIGURE 4-52 Logical flow of (a) IF/THEN and (b) IF/THEN/ELSE constructs.

In some cases it is not enough only to decide to act or not to act, but rather we must choose between two different actions. For example, in our analogy about the decision to wear a coat, if the person already has her coat on when making this decision, she will not be taking it off. The use of IF/THEN logic assumes that she is not wearing her coat initially.

When decisions demand two possible actions, the IF/THEN/**ELSE** control structure is used, as shown in Figure 4-52(b). Here again, the statement is evaluated as true or false. The difference is that, when the statement is false, a different action is performed. One of the two actions must occur with this construct. We can state it verbally as, "IF the statement is true, THEN do this. ELSE do that." In our coat analogy, this control structure would work, regardless of whether the person's coat was on or off initially.

Example 4-8 gave a simple example of a logic circuit that has as its input a numeric value representing battery voltage from an A/D converter. The inputs A, B, C, D are actually binary digits in a four-bit number, with A being the MSB and D being the LSB. Figure 4-53 shows the same circuit with the

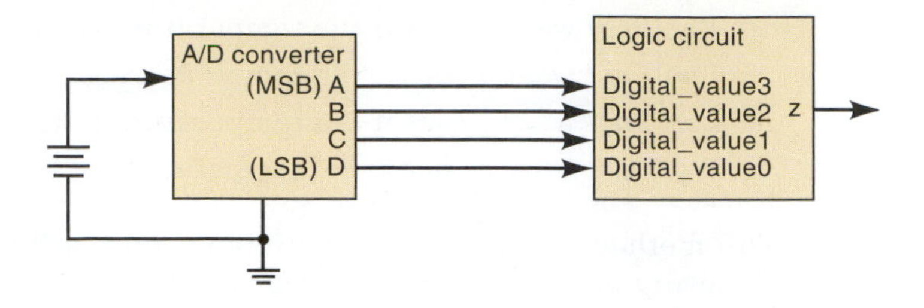

FIGURE 4-53 Logic circuit similar to Example 4-8.

inputs labeled as a four-bit number called *digital_value*. The relationship between bits is as follows:

A	*digital_value[3]*	digital value bit 3 (MSB)
B	*digital_value[2]*	digital value bit 2
C	*digital_value[1]*	digital value bit 1
D	*digital_value[0]*	digital value bit 0 (LSB)

The input can be treated as a decimal number between 0 and 15 if we specify the correct type of the input variable.

IF/THEN/ELSE USING AHDL

In AHDL, the inputs can be specified as a binary number made up of multiple bits by assigning a variable name followed by a list of the bit positions, as shown in Figure 4-54. The name is *digital_value*, and the bit positions range from 3 down to 0. Notice how simple the code becomes using this method along with an IF/ELSE construct. The IF is followed by a statement that refers to the value of the entire four-bit input variable and compares it with the number 6. Of course, 6 is a decimal form of a scalar quantity and *digital_value[ ]* actually represents a binary number. The compiler can interpret numbers in any system, so it creates a logic circuit that compares the binary value of *digital_value* with the binary number for 6 and decides if this statement is true or false. If it is true, THEN the next statement (z = VCC) is used to assign *z* a value. Notice that in AHDL, we must use VCC for a logic 1 and GND for a logic 0 when assigning a logic level to a single bit. When *digital_value* is 6 or less, it follows the statement after ELSE (z = GND). The END IF; terminates the control structure.

FIGURE 4-54 AHDL version.

```
SUBDESIGN FIG4_54
(
   digital_value[3..0]    :INPUT;  -- define inputs to block
   z                      :OUTPUT; -- define block output
)
BEGIN
   IF digital_value[] > 6 THEN
         z = VCC;                  -- output a 1
   ELSE  z = GND;                  -- output a 0
   END IF;
END;
```

IF/THEN/ELSE USING VHDL

In VHDL, the critical issue is the declaration of the type of inputs. Refer to Figure 4-55. The input is treated as a single variable called *digital_value*. Because its type is declared as INTEGER, the compiler knows to treat it as a number. By specifying a range of 0 to 15, the compiler knows it is a four-bit number. Notice that RANGE does not specify the index number of a bit vector but rather the limits of the numeric value of the integer. Integers are treated differently than bit arrays (BIT_VECTOR) in VHDL. An integer can be compared with other numbers using inequality operators. A BIT_VECTOR cannot be used with inequality operators.

FIGURE 4-55 VHDL version.

```
ENTITY fig4_55 IS
PORT( digital_value :IN INTEGER RANGE 0 TO 15; -- 4-bit input
      z             :OUT BIT);
END fig4_55;

ARCHITECTURE decision OF fig4_55 IS

BEGIN
   PROCESS (digital_value)
      BEGIN
         IF (digital_value > 6) THEN
            z <= '1';
         ELSE
            z <= '0';
      END IF;
END PROCESS ;
END decision;
```

To use the IF/THEN/ELSE control structure, VHDL requires that the code be put inside a "PROCESS." The statements that occur within a process are *sequential*, meaning that the order in which they are written affects the operation of the circuit. The keyword **PROCESS** is followed by a list of variables called a **sensitivity list**, which is a list of variables to which the process code must respond. Whenever *digital_value* changes, it causes the process code to be reevaluated. Even though we know *digital_value* is really a four-bit binary number, the compiler will evaluate it as a number between the equivalent decimal values of 0 and 15. IF the statement in parentheses is true, THEN the next statement is applied (z is assigned a value of logic 1). If this statement is not true, the logic follows the ELSE clause and assigns a value of 0 to z. The END IF; terminates the control structure, and the END PROCESS; terminates the evaluation of the sequential statements.

ELSIF

We often need to choose among many possible actions, depending on the situation. The IF construct chooses whether to perform a set of actions or not. The IF/ELSE construct selects one out of two possible actions. By combining

IF and ELSE decisions, we can create a control structure referred to as **ELSIF**, which chooses one of many possible outcomes. The decision structure is shown graphically in Figure 4-56.

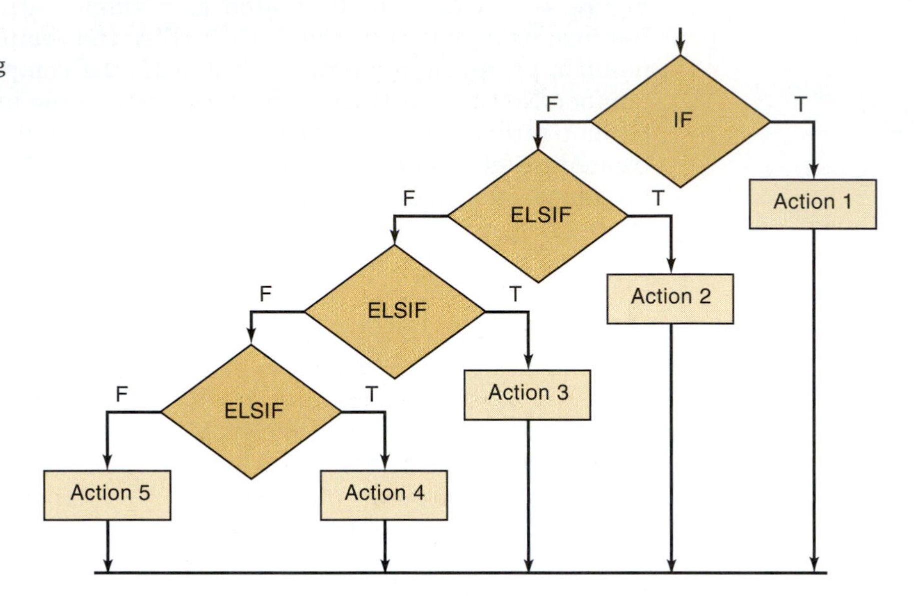

FIGURE 4-56 Flowchart for multiple decisions using IF/ELSIF.

Notice that as each condition is evaluated, it either performs an action if true or goes on to evaluate the next condition. Each action is associated with one condition, and there is no chance to select more than one action. Note also that the conditions used to decide the appropriate action can be any expression that evaluates as true or false. This fact allows the designer to use the inequality operators to choose an action based on a range of input values. As an example of this application, let's consider the temperature-measuring system that uses an A/D converter, as described in Figure 4-57. Suppose that we want to indicate when the temperature is in a certain range, which we will refer to as Too Cold, Just Right, and Too Hot.

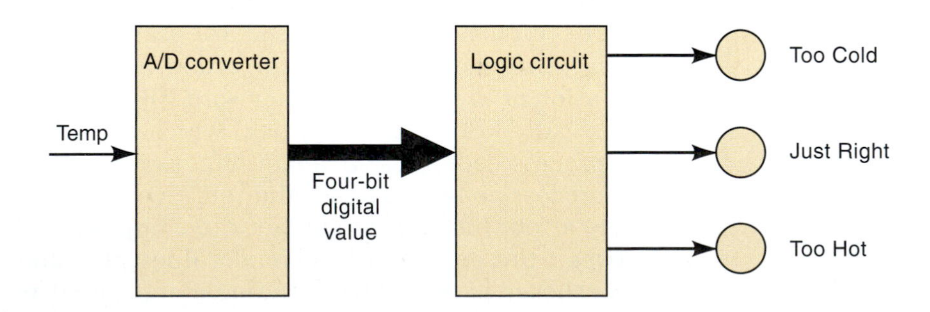

FIGURE 4-57 Temperature range indicator circuit.

The relationship between the digital values for temperature and the categories is

Digital Values	*Category*
0000–1000	Too Cold
1001–1010	Just Right
1011–1111	Too Hot

We can express the decision-making process for this logic circuit as follows:

> IF the digital value is less than or equal to 8, THEN light only the Too Cold indicator.
>
> ELSE IF the digital value is greater than 8 AND less than 11, THEN light only the Just Right indicator.
>
> ELSE light only the Too Hot indicator.

AHDL

ELSIF USING AHDL

The AHDL code in Figure 4-58 defines the inputs as a four-bit binary number. The outputs are three individual bits that drive the three range indicators. This example uses an intermediate variable (*status*) that allows us to assign a bit pattern representing the three conditions of *too_cold, just_right*, and *too_hot*. The sequential section of the code uses the IF, ELSIF, ELSE to identify the range in which the temperature lies and assigns the correct bit pattern to status. In the last statement, the bits of *status* are connected to the actual output port bits. These bits have been ordered in a group that relates to the bit patterns assigned to *status[ ]*. This could also have been written as three concurrent statements: too_cold = status[2]; just_right = status[1]; too_hot = status[0];.

```
SUBDESIGN fig4_58
(
   digital_value[3..0]            :INPUT;  --define inputs to block
   too_cold, just_right, too_hot :OUTPUT;--define outputs
)
VARIABLE
status[2..0]     :NODE;--holds state of too_cold, just_right, too_hot
BEGIN
   IF       digital_value[] <= 8 THEN status[] = b"100";
   ELSIF    digital_value[] > 8 AND digital_value[] < 11 THEN
         status[] = b"010";
   ELSE  status[] = b"001";
   END IF;
   (too_cold, just_right, too_hot) = status[]; -- update output bits
END;
```

FIGURE 4-58 Temperature range example in AHDL using ELSIF.

VHDL

ELSIF USING VHDL

The VHDL code in Figure 4-59 defines the inputs as a four-bit integer. The outputs are three individual bits that drive the three range indicators. This example uses an intermediate signal (*status*) that allows us to assign a bit pattern representing all three conditions of *too_cold, just_right,* and *too_hot.* The process section of the code uses the IF, ELSIF, and ELSE to identify the range in which the temperature lies and assigns the correct bit pattern to status. In the last three statements, each bit of *status* is connected to the correct output port bit.

```
ENTITY fig4_59  IS
PORT(digital_value:IN INTEGER RANGE 0 TO 15;    -- declare 4-bit input
      too_cold, just_right, too_hot :OUT BIT);
END fig4_59 ;

ARCHITECTURE howhot OF fig4_59  IS
SIGNAL status    :BIT_VECTOR (2 downto 0);
BEGIN
   PROCESS (digital_value)
      BEGIN
         IF (digital_value <= 8) THEN status <= "100";
         ELSIF (digital_value > 8 AND digital_value < 11) THEN
               status <= "010";
         ELSE  status <= "001";
         END IF;
      END PROCESS ;
   too_cold    <= status(2);      -- assign status bits to output
   just_right  <= status(1);
   too_hot     <= status(0);
END howhot;
```

FIGURE 4-59 Temperature range example in VHDL using ELSIF.

CASE

One more important control structure is useful for choosing actions based on current conditions. It is called by various names, depending on the programming language, but it nearly always involves the word **CASE**. This construct determines the value of an expression or object and then goes through a list of possible values (cases) for the expression or object being evaluated. Each case has a list of actions that should take place. A CASE construct is different from an IF/ELSIF because a case correlates one unique value of an object with a set of actions. Recall that an IF/ELSIF correlates a set of actions with a true statement. There can be only one match for a CASE statement. An IF/ELSIF can have more than one statement that is true, but will THEN perform the action associated with the first true statement it evaluates.

Another important point in the examples that follow is the need to combine several independent variables into a set of bits, called a bit vector. Recall that this action of linking several bits in a particular order is called *concatenation.* It allows us to consider the bit pattern as an ordered group.

AHDL

CASE USING AHDL

The AHDL example in Figure 4-60 demonstrates a case construct implementing the circuit of Figure 4-9 (see also Table 4-3). It uses individual bits as its inputs. In the first statement after BEGIN, these bits are concatenated and assigned to the intermediate variable called *status.* The CASE statement evaluates the variable *status* and finds the bit pattern (following the keyword WHEN) that matches the value of *status.* It then performs the action described following =>. In this example, it simply assigns logic 0 to the output for each of the three specified cases. All *other* cases result in a logic 1 on the output.

FIGURE 4-60 Figure 4-9 represented in AHDL.

```
SUBDESIGN fig4_60
(
   p, q, r          :INPUT;        -- define inputs to block
   s                :OUTPUT;       -- define outputs
)
VARIABLE
  status[2..0]      :NODE;
BEGIN
   status[]= (p, q, r);   -- link input bits in order
   CASE status[] IS
      WHEN b"100"       => s = GND;
      WHEN b"101"       => s = GND;
      WHEN b"110"       => s = GND;
      WHEN OTHERS       => s = VCC;
   END CASE;
END;
```

CASE USING VHDL

The VHDL example in Figure 4-61 demonstrates the case construct implementing the circuit of Figure 4-9 (see also Table 4-3). It uses individual bits as its inputs. In the first statement after BEGIN, these bits are concatenated and assigned to the intermediate variable called *status* using the & operator. The CASE statement evaluates the variable status and finds the bit pattern (following the keyword WHEN) that matches the value of *status.* It then performs the action described following =>. In this simple example, it merely assigns logic 0 to the output for each of the three specified cases. All *other* cases result in a logic 1 on the output.

FIGURE 4-61 Figure 4-9 represented in VHDL.

```
ENTITY fig4_61 IS
PORT( p, q, r       :IN bit;               --declare 3 bits input
      s             :OUT BIT);
END fig4_61;

ARCHITECTURE copy OF fig4_61 IS
SIGNAL status       :BIT_VECTOR (2 downto 0);
BEGIN
   status <= p & q & r;                    --link bits in order.
   PROCESS (status)
      BEGIN
         CASE status IS
            WHEN "100" =>  s <= '0';
            WHEN "101" =>  s <= '0';
            WHEN "110" =>  s <= '0';
            WHEN OTHERS => s <= '1';
         END CASE;
      END PROCESS;
END copy;
```

VHDL

EXAMPLE 4-34

A coin detector in a vending machine accepts quarters, dimes, and nickels and activates the corresponding digital signal (Q, D, N) only when the correct coin is present. It is physically impossible for multiple coins to be present at the same time. A digital circuit must use the Q, D, and N signals as inputs and produce a binary number representing the value of the coin as shown in Figure 4-62. Write the AHDL and VHDL code.

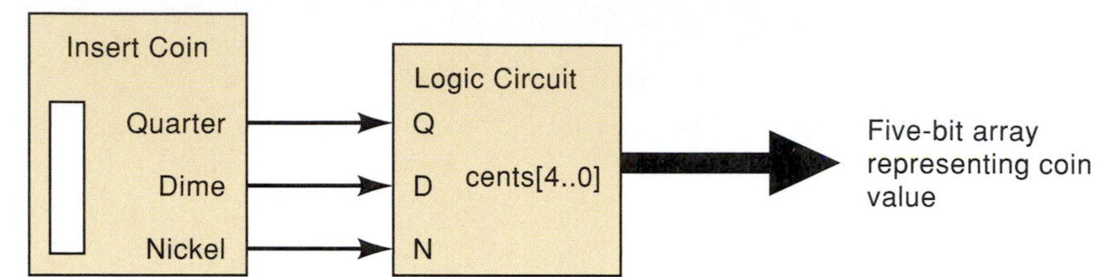

FIGURE 4-62 A coin detector circuit for a vending machine.

Solution

This is an ideal application of the CASE construct to describe the correct operation. The outputs must be declared as five-bit numbers in order to represent up to 25 cents. Figure 4-63 shows the AHDL solution and Figure 4-64 shows the VHDL solution.

AHDL

```
SUBDESIGN   fig4_63
(
   q, d, n         :INPUT;      -- define quarter, dime, nickel
   cents[4..0]     :OUTPUT;     -- define binary value of coins
)
BEGIN
   CASE (q, d, n) IS            -- group coins in an ordered set
      WHEN b"001" => cents[] = 5;
      WHEN b"010" => cents[] = 10;
      WHEN b"100" => cents[] = 25;
      WHEN others => cents[] = 0;
   END CASE;
END;
```

FIGURE 4-63 An AHDL coin detector.

VHDL

```
ENTITY   fig4_64 IS
PORT( q, d, n:IN BIT;                          -- quarter, dime, nickel
      cents  :OUT INTEGER RANGE 0 TO 25);  -- binary value of coins
END fig4_64;
ARCHITECTURE detector of fig4_64 IS
   SIGNAL   coins :BIT_VECTOR(2 DOWNTO 0);-- group the coin sensors
   BEGIN
      coins <= (q & d & n);                    -- assign sensors to group
      PROCESS (coins)
         BEGIN
            CASE (coins) IS
               WHEN "001"  => cents <= 5;
               WHEN "010"  => cents <= 10;
               WHEN "100"  => cents <= 25;
               WHEN others => cents <= 0;
            END CASE;
         END PROCESS;
   END detector;
```

FIGURE 4-64 A VHDL coin detector.

REVIEW QUESTIONS

1. Which control structure decides to do or not to do?
2. Which control structure decides to do this or to do that?
3. Which control structure(s) decides which one of several different actions to take?
4. Declare an input named *count* that can represent a numeric quantity as big as 205. Use AHDL or VHDL.

SUMMARY

1. The two general forms for logic expressions are the sum-of-products form and the product-of-sums form.
2. One approach to the design of a combinatorial logic circuit is to (1) construct its truth table, (2) convert the truth table to a sum-of-products expression, (3) simplify the expression using Boolean algebra or K mapping, (4) implement the final expression.
3. The K map is a graphical method for representing a circuit's truth table and generating a simplified expression for the circuit output.
4. An exclusive-OR circuit has the expression $x = A\overline{B} + \overline{A}B$. Its output x will be HIGH only when inputs A and B are at opposite logic levels.
5. An exclusive-NOR circuit has the expression $x = \overline{A}\,\overline{B} + AB$. Its output x will be HIGH only when inputs A and B are at the same logic level.
6. Each of the basic gates (AND, OR, NAND, NOR) can be used to enable or disable the passage of an input signal to its output.
7. The main digital IC families are the TTL and CMOS families. Digital ICs are available in a wide range of complexities (gates per chip), from the basic to the high-complexity logic functions.
8. To perform basic troubleshooting requires—at a minimum—an understanding of circuit operation, a knowledge of the types of possible faults, a complete logic-circuit connection diagram, and a logic probe.
9. A programmable logic device (PLD) is an IC that contains a large number of logic gates whose interconnections can be programmed by the user to generate the desired logic relationship between inputs and outputs.
10. To program a PLD, you need a development system that consists of a computer, PLD development software, and a programmer fixture that does the actual programming of the PLD chip.
11. The Altera system allows convenient hierarchical design techniques using any form of hardware description.
12. The type of data objects must be specified so that the HDL compiler knows the range of numbers to be represented.
13. Truth tables can be entered directly into the source file using the features of HDL.
14. Logical control structures such as IF, ELSE, and CASE can be used to describe the operation of a logic circuit, making the code and the problem's solution much more straightforward.

IMPORTANT TERMS

sum-of-products (SOP)
product-of-sums (POS)
Karnaugh map (K map)
looping
don't-care condition
exclusive-OR (XOR)
exclusive-NOR (XNOR)
parity generator
parity checker
enable/disable
dual-in-line package (DIP)
SSI, MSI, LSI, VLSI, ULSI, GSI
transistor-transistor logic (TTL)
complementary metal-oxide-semiconductor (CMOS)
indeterminate
floating
logic probe
contention
programmer
ZIF socket
JEDEC
JTAG
hierarchical design
top-down
test vectors
literals
bit array
bit vector
BIT_VECTOR
index
integer
objects
libraries
macrofunction
STD_LOGIC
STD_LOGIC_VECTOR
concatenate
selected signal assignment
decision control structure
concurrent
sequential
IF/THEN
ELSE
PROCESS
sensitivity list
ELSIF
CASE

PROBLEMS

SECTIONS 4-2 AND 4-3

B 4-1.* Simplify the following expressions using Boolean algebra.

(a) $x = ABC + \overline{A}C$

(b) $y = (Q + R)(\overline{Q} + \overline{R})$

(c) $w = ABC + A\overline{B}C + \overline{A}$

(d) $q = \overline{RST}(\overline{R + S + T})$

(e) $x = \overline{A}\,\overline{B}\,\overline{C} + \overline{A}BC + ABC + A\overline{B}\,\overline{C} + A\overline{B}C$

(f) $z = (B + \overline{C})(\overline{B} + C) + \overline{\overline{A} + B + \overline{C}}$

(g) $y = (\overline{C + D}) + \overline{A}C\overline{D} + A\overline{B}\,\overline{C} + \overline{A}\,\overline{B}CD + AC\overline{D}$

(h) $x = AB(\overline{\overline{C}D}) + \overline{A}BD + \overline{B}\,\overline{C}\,\overline{D}$

B 4-2. Simplify the circuit of Figure 4-65 using Boolean algebra.

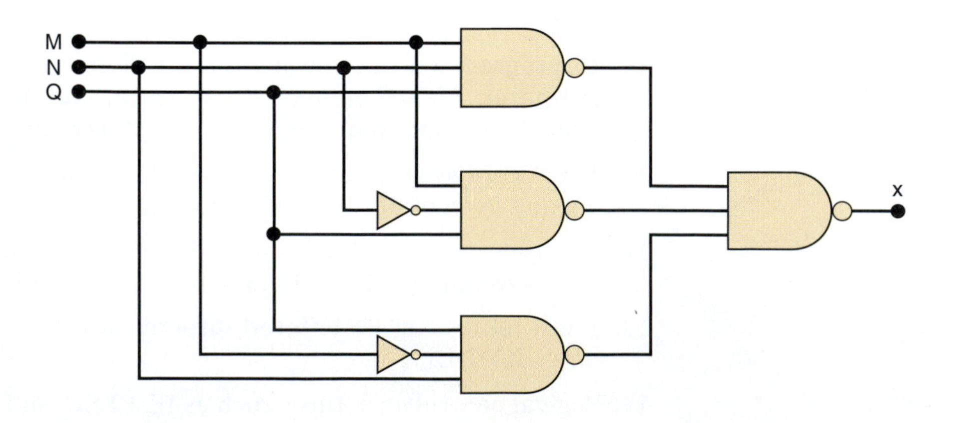

FIGURE 4-65 Problems 4-2 and 4-3.

*Answers to problems marked with an asterisk can be found in the back of the text.

B 4-3.* Change each gate in Problem 4-2 to a NOR gate, and simplify the circuit using Boolean algebra.

SECTION 4-4

B, D 4-4.* Design the logic circuit corresponding to the truth table shown in Table 4-11.

TABLE 4-11

A	B	C	x
0	0	0	1
0	0	1	0
0	1	0	1
0	1	1	1
1	0	0	1
1	0	1	0
1	1	0	0
1	1	1	1

B, D 4-5. Design a logic circuit whose output is HIGH *only* when a majority of inputs A, B, and C are LOW.

D 4-6. A manufacturing plant needs to have a horn sound to signal quitting time. The horn should be activated when either of the following conditions is met:

1. It's after 5 o'clock and all machines are shut down.
2. It's Friday, the production run for the day is complete, and all machines are shut down.

Design a logic circuit that will control the horn. (*Hint:* Use four logic input variables to represent the various conditions; for example, input A will be HIGH only when the time of day is 5 o'clock or later.)

D 4-7.* A four-bit binary number is represented as $A_3A_2A_1A_0$, where A_3, A_2, A_1, and A_0 represent the individual bits and A_0 is equal to the LSB. Design a logic circuit that will produce a HIGH output whenever the binary number is greater than 0010 and less than 1000.

D 4-8. Figure 4-66 shows a diagram for an automobile alarm circuit used to detect certain undesirable conditions. The three switches are used to

FIGURE 4-66 Problem 4-8.

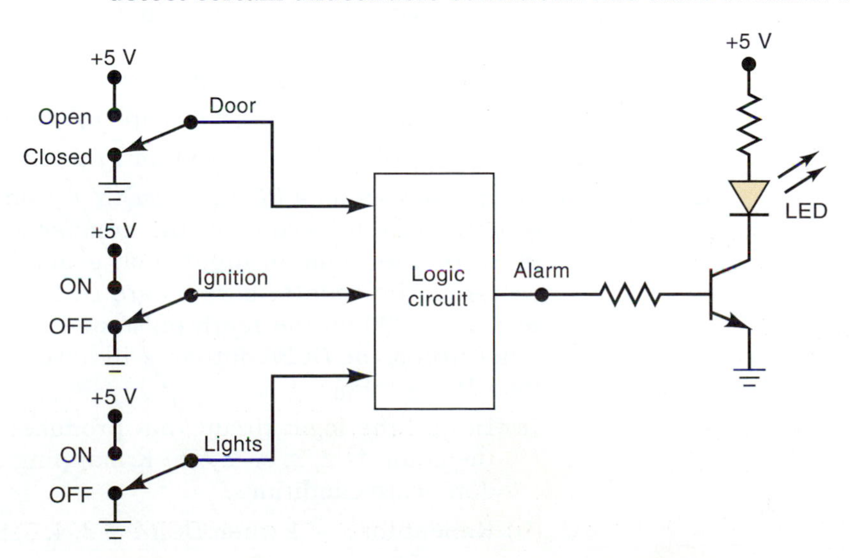

indicate the status of the door by the driver's seat, the ignition, and the headlights, respectively. Design the logic circuit with these three switches as inputs so that the alarm will be activated whenever either of the following conditions exists:

- The headlights are on while the ignition is off.
- The door is open while the ignition is on.

4-9.* Implement the circuit of Problem 4-4 using all NAND gates.

4-10. Implement the circuit of Problem 4-5 using all NAND gates.

SECTION 4-5

B 4-11. Determine the minimum expression for each K map in Figure 4-67. Pay particular attention to step 5 for the map in (a).

FIGURE 4-67 Problem 4-11.

	$\overline{C}\overline{D}$	$\overline{C}D$	CD	$C\overline{D}$
$\overline{A}\overline{B}$	1	1	1	1
$\overline{A}B$	1	1	0	0
AB	0	0	0	1
$A\overline{B}$	0	0	1	1

(a)*

	$\overline{C}\overline{D}$	$\overline{C}D$	CD	$C\overline{D}$
$\overline{A}\overline{B}$	1	0	1	1
$\overline{A}B$	1	0	0	1
AB	0	0	0	0
$A\overline{B}$	1	0	1	1

(b)

	$\overline{C}$	C
$\overline{A}\overline{B}$	1	1
$\overline{A}B$	0	0
AB	1	0
$A\overline{B}$	1	X

(c)

B 4-12. For the truth table below, create a 2 × 2 K map, group terms, and simplify. Then look at the truth table again to see if the expression is true for every entry in the table.

A	*B*	*y*
0	0	1
0	1	1
1	0	0
1	1	0

B 4-13. Starting with the truth table in Table 4-11, use a K map to find the simplest SOP equation.

B 4-14. Simplify the expression in (a)* Problem 4-1(e) using a K map. (b) Problem 4-1(g) using a K map. (c)* Problem 4-1(h) using a K map.

B 4-15.* Obtain the output expression for Problem 4-7 using a K map.

C, D 4-16. Figure 4-68 shows a *BCD counter* that produces a four-bit output representing the BCD code for the number of pulses that have been applied to the counter input. For example, after four pulses have occurred, the counter outputs are $DCBA = 0100_2 = 4_{10}$. The counter resets to 0000 on the tenth pulse and starts counting over again. In other words, the *DCBA* outputs will never represent a number greater than $1001_2 = 9_{10}$.

(a)* Design the logic circuit that produces a HIGH output whenever the count is 2, 3, or 9. Use K mapping and take advantage of the don't-care conditions.

(b) Repeat for $x = 1$ when $DCBA = 3, 4, 5, 8$.

FIGURE 4-68 Problem 4-16.

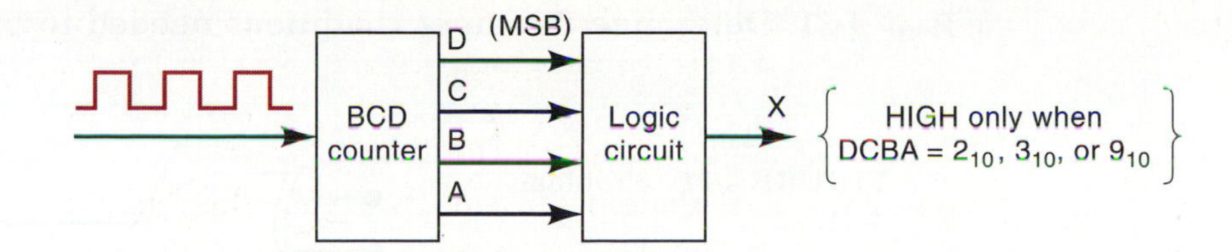

D 4-17.* Figure 4-69 shows four switches that are part of the control circuitry in a copy machine. The switches are at various points along the path of the copy paper as the paper passes through the machine. Each switch is normally open, and as the paper passes over a switch, the switch closes. It is impossible for switches SW1 and SW4 to be closed at the same time. Design the logic circuit to produce a HIGH output whenever *two or more* switches are closed at the same time. Use K mapping and take advantage of the don't-care conditions.

FIGURE 4-69 Problem 4-17.

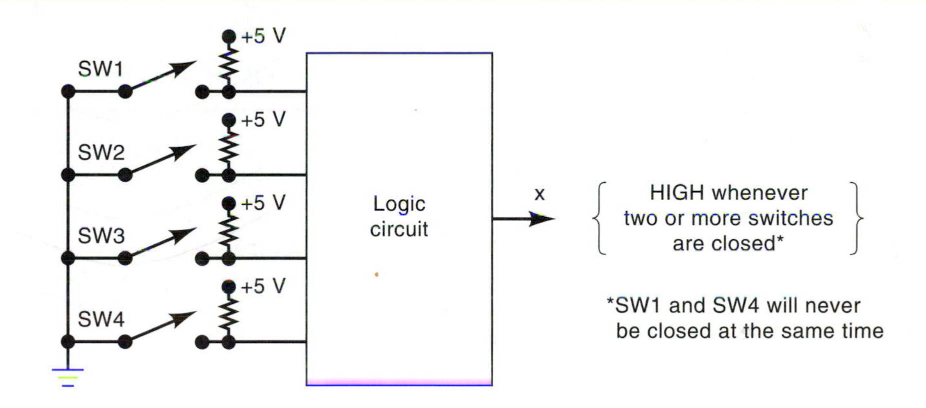

B 4-18. Example 4-3 demonstrated algebraic simplification. Step 3 resulted in the SOP equation $z = \overline{A}\,\overline{B}C + \overline{A}C\overline{D} + \overline{A}B\overline{C}\,\overline{D} + A\overline{B}C$. Use a K map to prove that this equation can be simplified further than the answer shown in the example.

C 4-19. Use Boolean algebra to arrive at the same result obtained by the K map method of Problem 4-18.

SECTION 4-6

B 4-20. (a) Determine the output waveform for the circuit of Figure 4-70.

(b) Repeat with the *B* input held LOW.

(c) Repeat with *B* held HIGH.

FIGURE 4-70 Problem 4-20.

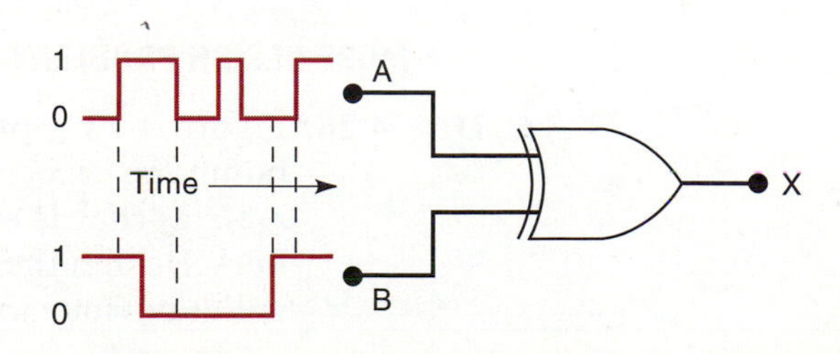

B 4-21.* Determine the input conditions needed to produce $x = 1$ in Figure 4-71.

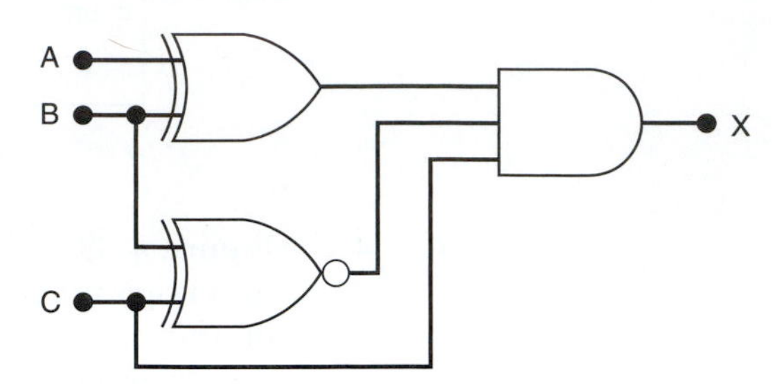

FIGURE 4-71 Problem 4-21.

B 4-22. Design a circuit that produces a HIGH out only when all three inputs are the same level.

(a) Use a truth table and K map to produce the SOP solution.

(b) Use two-input XOR and other gates to find a solution. (*Hint:* Recall the transitive property from algebra. . . if $a = b$ and $b = c$ then $a = c$.)

B 4-23.* A 7486 chip contains four XOR gates. Show how to make an XNOR gate using only a 7486 chip. *Hint:* See Example 4-16.

B 4-24.* Modify the circuit of Figure 4-23 to compare two four-bit numbers and produce a HIGH output when the two numbers match exactly.

B 4-25. Figure 4-72 represents a *relative-magnitude detector* that takes two three-bit binary numbers, $x_2x_1x_0$ and $y_2y_1y_0$, and determines whether they are equal and, if not, which one is larger. There are three outputs, defined as follows:

1. $M = 1$ only if the two input numbers are equal.
2. $N = 1$ only if $x_2x_1x_0$ is greater than $y_2y_1y_0$.
3. $P = 1$ only if $y_2y_1y_0$ is greater than $x_2x_1x_0$.

Design the logic circuitry for this detector. The circuit has *six* inputs and *three* outputs and is therefore much too complex to handle using the truth-table approach. Refer to Example 4-17 as a hint about how you might start to solve this problem.

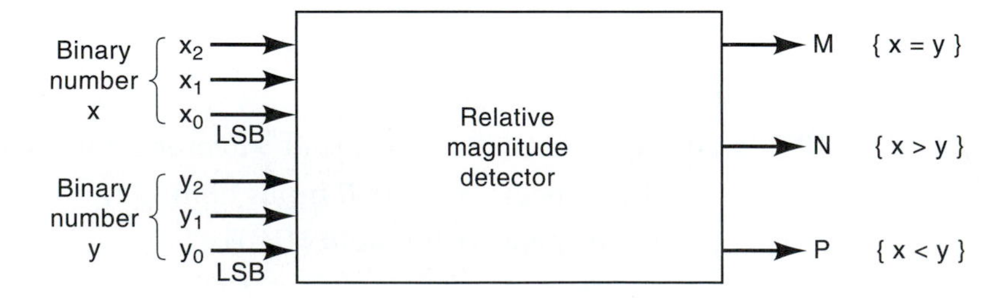

FIGURE 4-72 Problem 4-25.

MORE DESIGN PROBLEMS

C, D 4-26.* Figure 4-73 represents a multiplier circuit that takes two-bit binary numbers, x_1x_0 and y_1y_0, and produces an output binary number $z_3z_2z_1z_0$ that is equal to the arithmetic product of the two input numbers. Design the logic circuit for the multiplier. (*Hint:* The logic circuit will have four inputs and four outputs.)

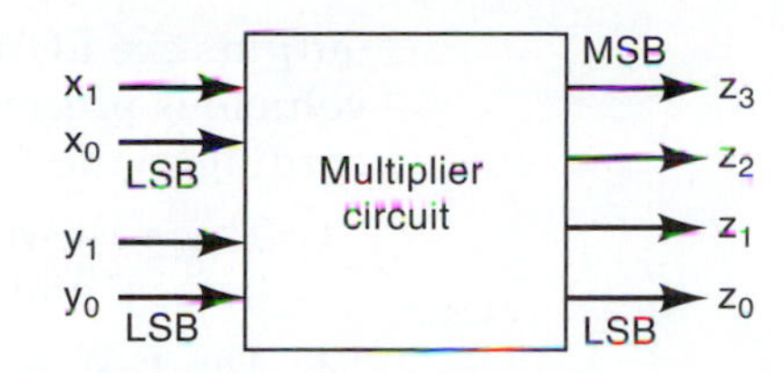

FIGURE 4-73 Problem 4-26.

D 4-27. A BCD code is being transmitted to a remote receiver. The bits are A_3, A_2, A_1, and A_0, with A_3 as the MSB. The receiver circuitry includes a *BCD error detector* circuit that examines the received code to see if it is a legal BCD code (i.e., ≤1001). Design this circuit to produce a HIGH for any error condition.

D 4-28.* Design a logic circuit whose output is HIGH whenever *A* and *B* are both HIGH as long as *C* and *D* are either both LOW or both HIGH. Try to do this without using a truth table. Then check your result by constructing a truth table from your circuit to see if it agrees with the problem statement.

D 4-29. Four large tanks at a chemical plant contain different liquids being heated. Liquid-level sensors are being used to detect whenever the level in tank *A* or tank *B* rises above a predetermined level. Temperature sensors in tanks *C* and *D* detect when the temperature in either of these tanks drops below a prescribed temperature limit. Assume that the liquid-level sensor outputs *A* and *B* are LOW when the level is satisfactory and HIGH when the level is too high. Also, the temperature-sensor outputs *C* and *D* are LOW when the temperature is satisfactory and HIGH when the temperature is too low. Design a logic circuit that will detect whenever the level in tank *A* or tank *B* is too high at the same time that the temperature in either tank *C* or tank *D* is too low.

C, D 4-30.* Figure 4-74 shows the intersection of a main highway with a secondary access road. Vehicle-detection sensors are placed along lanes *C* and *D* (main road) and lanes *A* and *B* (access road). These sensor

FIGURE 4-74 Problem 4-30.

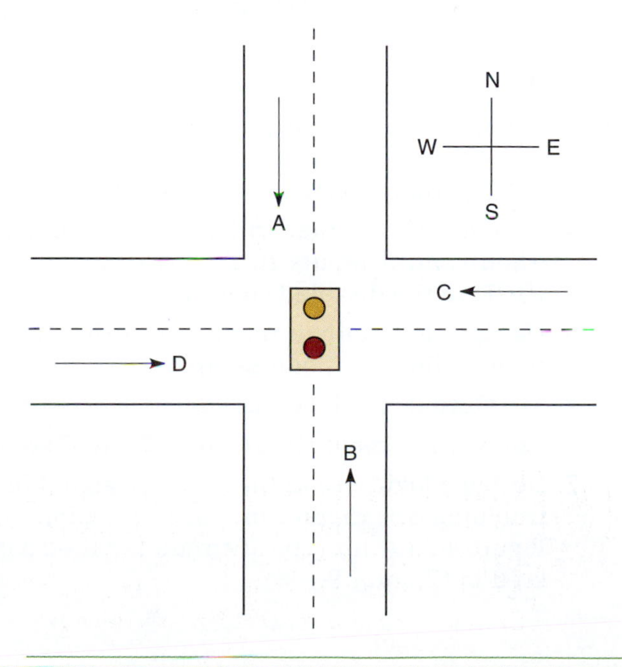

outputs are LOW (0) when no vehicle is present and HIGH (1) when a vehicle is present. The intersection traffic light is to be controlled according to the following logic:

1. The east-west (E-W) traffic light will be green whenever *both* lanes C and D are occupied.
2. The E-W light will be green whenever *either* C or D is occupied but lanes A and B are not *both* occupied.
3. The north-south (N-S) light will be green whenever *both* lanes A and B are occupied but C and D are not *both* occupied.
4. The N-S light will also be green when *either* A or B is occupied while C and D are *both* vacant.
5. The E-W light will be green when *no* vehicles are present.

Using the sensor outputs A, B, C, and D as inputs, design a logic circuit to control the traffic light. There should be two outputs, N-S and E-W, that go HIGH when the corresponding light is to be *green*. Simplify the circuit as much as possible and show *all* steps.

SECTION 4-7

D 4-31. Redesign the parity generator and checker of Figure 4-25 to (a) operate using odd parity. (*Hint:* What is the relationship between an odd-parity bit and an even-parity bit for the same set of data bits?) (b) Operate on eight data bits.

SECTION 4-8

B 4-32. (a) Under what conditions will an OR gate allow a logic signal to pass through to its output unchanged?

(b) Repeat (a) for an AND gate.

(c) Repeat for a NAND gate.

(d) Repeat for a NOR gate.

B 4-33.* (a) Can an INVERTER be used as an enable/disable circuit? Explain.

(b) Can an XOR gate be used as an enable/disable circuit? Explain.

D 4-34. Design a logic circuit that will allow input signal A to pass through to the output only when control input B is LOW while control input C is HIGH; otherwise, the output is LOW.

D 4-35.* Design a circuit that will *disable* the passage of an input signal only when control inputs B, C, and D are all HIGH; the output is to be HIGH in the disabled condition.

D 4-36. Design a logic circuit that controls the passage of a signal A according to the following requirements:

1. Output X will equal A when control inputs B and C are the same.
2. X will remain HIGH when B and C are different.

D 4-37. Design a logic circuit that has two signal inputs, A_1 and A_0, and a control input S so that it functions according to the requirements given in Figure 4-75. (This type of circuit is called a *multiplexer* and will be covered in Chapter 9.)

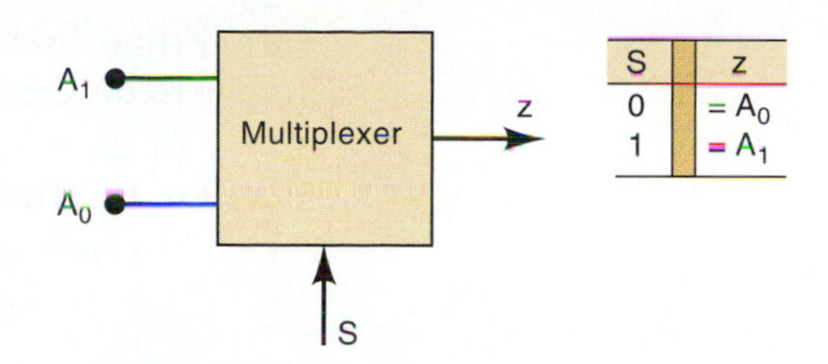

FIGURE 4-75 Problem 4-37.

D 4-38.* Use K mapping to design a circuit to meet the requirements of Example 4-17. Compare this circuit with the solution in Figure 4-23. This points out that the K-map method cannot take advantage of the XOR and XNOR gate logic. The designer must be able to determine when these gates are applicable.

SECTIONS 4-9 TO 4-13

T* 4-39. (a) A technician testing a logic circuit sees that the output of a particular INVERTER is stuck LOW while its input is pulsing. List as many possible reasons as you can for this faulty operation.

(b) Repeat part (a) for the case where the INVERTER output is stuck at an indeterminate logic level.

T 4-40.* The signals shown in Figure 4-76 are applied to the inputs of the circuit of Figure 4-32. Suppose that there is an internal open circuit at Z1-4.

(a) What will a logic probe indicate at Z1-4?

(b) What dc voltage reading would you expect at Z1-4? (Remember that the ICs are TTL.)

(c) Sketch what you think the $\overline{CLKOUT}$ and $\overline{SHIFTOUT}$ signals will look like.

(d) Instead of the open at Z1-4, suppose that pins 9 and 10 of Z2 are internally shorted. Sketch the probable signals at Z2-10, $\overline{CLOCKOUT}$, and $\overline{SHIFTOUT}$.

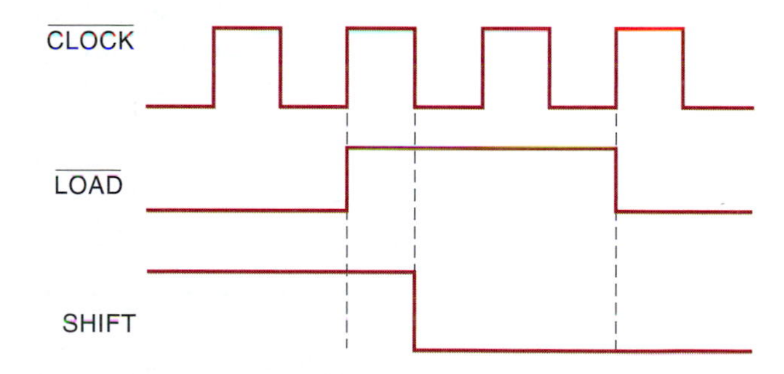

FIGURE 4-76 Problem 4-40.

T 4-41. Assume that the ICs in Figure 4-32 are CMOS. Describe how the circuit operation would be affected by an open circuit in the conductor connecting Z2-2 and Z2-10.

T 4-42. In Example 4-24, we listed three possible faults for the situation of Figure 4-36. What procedure would you follow to determine which of the faults is the actual one?

T 4-43.* Refer to the circuit of Figure 4-38. Assume that the devices are CMOS. Also assume that the logic probe indication at Z2-3 is "indeterminate"

*Recall that T indicates a troubleshooting exercise.

rather than "pulsing." List the possible faults, and write a procedure to follow to determine the actual fault.

T 4-44.* Refer to the logic circuit of Figure 4-41. Recall that output Y is supposed to be HIGH for either of the following conditions:

1. $A = 1, B = 0$, regardless of C
2. $A = 0, B = 1, C = 1$

When testing the circuit, the technician observes that Y goes HIGH only for the first condition but stays LOW for all other input conditions. Consider the following list of possible faults. For each one, write yes or no to indicate whether or not it could be the actual fault. Explain your reasoning for each no response.

(a) An internal short to ground at Z2-13
(b) An open circuit in the connection to Z2-13
(c) An internal short to V_{CC} at Z2-11
(d) An open circuit in the V_{CC} connection to Z2
(e) An internal open circuit at Z2-9
(f) An open in the connection from Z2-11 to Z2-9
(g) A solder bridge between pins 6 and 7 of Z2

T 4-45. Develop a procedure for isolating the fault that is causing the malfunction described in Problem 4-44.

T 4-46.* Assume that the gates in Figure 4-41 are all CMOS. When the technician tests the circuit, he finds that it operates correctly except for the following conditions:

1. $A = 1, B = 0, C = 0$
2. $A = 0, B = 1, C = 1$

For these conditions, the logic probe indicates indeterminate levels at Z2-6, Z2-11, and Z2-8. What do you think is the probable fault in the circuit? Explain your reasoning.

T 4-47. Figure 4-77 is a combinational logic circuit that operates an alarm in a car whenever the driver and/or passenger seats are occupied and the seatbelts are not fastened when the car is started. The active-HIGH signals *DRIV* and *PASS* indicate the presence of the driver and passenger, respectively, and are taken from pressure-actuated switches in the seats. The signal *IGN* is active-HIGH when the ignition switch is on. The signal $\overline{BELTD}$ is active-LOW and indicates that the driver's seatbelt is

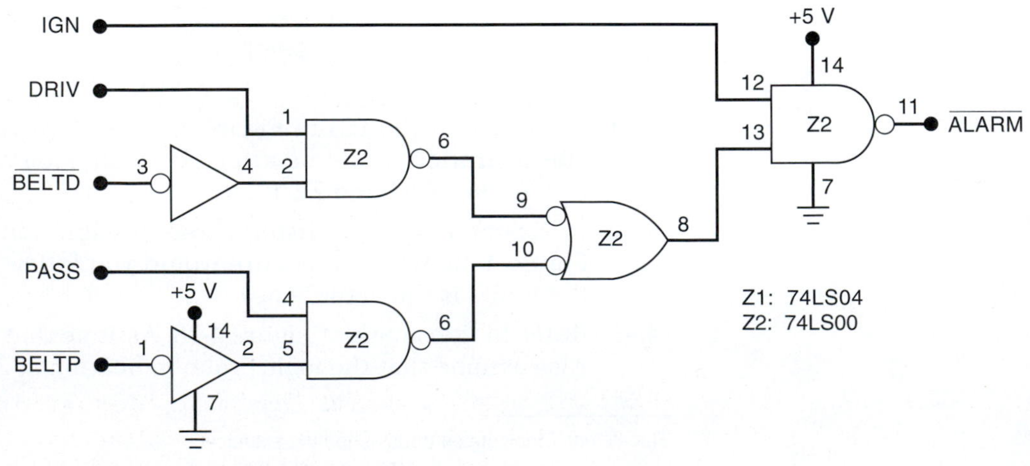

FIGURE 4-77 Problems 4-47, 4-48, and 4-49.

unfastened; $\overline{BELTP}$ is the corresponding signal for the passenger seatbelt. The alarm will be activated (LOW) whenever the car is started and either of the front seats is occupied and its seatbelt is not fastened.

(a) Verify that the circuit will function as described.

(b) Describe how this alarm system would operate if Z1-2 were internally shorted to ground.

(c) Describe how it would operate if there were an open connection from Z2-6 to Z2-10.

T 4-48.* Suppose that the system of Figure 4-77 is functioning so that the alarm is activated as soon as the driver and/or passenger are seated and the car is started, regardless of the status of the seatbelts. What are the possible faults? What procedure would you follow to find the actual fault?

T 4-49.* Suppose that the alarm system of Figure 4-77 is operating so that the alarm goes on continuously as soon as the car is started, regardless of the state of the other inputs. List the possible faults and write a procedure to isolate the fault.

DRILL QUESTIONS ON PLDs (50 THROUGH 55)

4-50.* *True or false:*

(a) Top-down design begins with an overall description of the entire system and it specifications.

(b) A JEDEC file can be used as the input file for a programmer.

(c) If an input file compiles with no errors, it means the PLD circuit will work correctly.

(d) A compiler can interpret code in spite of syntax errors.

(e) Test vectors are used to simulate and test a device.

H, B 4-51. What are the % characters used for in the AHDL design file?

H, B 4-52. How are comments indicated in a VHDL design file?

B 4-53. What is a ZIF socket?

B 4-54.* Name three entry modes used to input a circuit description into PLD development software.

B 4-55. What do JEDEC and HDL stand for?

SECTION 4-15

H, B 4-56. Declare the following data objects in AHDL or VHDL.

(a)* An array of eight output bits named *gadgets.*

(b) A single-output bit named *buzzer.*

(c) A 16-bit numeric input port named *altitude.*

(d) A single, intermediate bit within a hardware description file named *wire2.*

H, B 4-57. Express the following literal numbers in hex, binary, and decimal using the syntax of AHDL or VHDL.

(a)* 152_{10}

(b) 1001010100_2

(c) $3C4_{16}$

H, B 4-58.* The following similar I/O definition is given for AHDL and VHDL. Write four concurrent assignment statements that will connect the inputs to the outputs as shown in Figure 4-78.

FIGURE 4-78 Problem 4-58.

```
SUBDESIGN hw
(
    inbits[3..0]    :INPUT;
    outbits[3..0]   :OUTPUT;
)
```

```
ENTITY hw IS
PORT (
     inbits      :IN BIT_VECTOR (3 downto 0);
     outbits     :OUT BIT_VECTOR (3 downto 0)
     );
END hw;
```

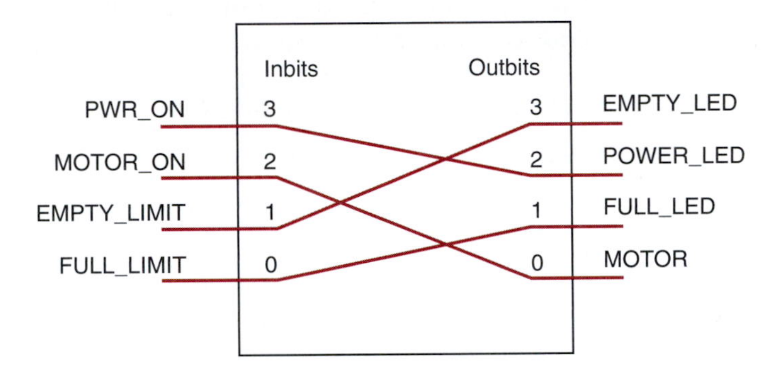

SECTION 4-16

H, D 4-59. Modify the AHDL truth table of Figure 4-50 to implement $AB + A\overline{C} + \overline{A}B$.

H, D 4-60.* Modify the AHDL design in Figure 4-54 so that $z = 1$ only when the digital value is less than 1010_2.

H, D 4-61. Modify the VHDL truth table of Figure 4-51 to implement $AB + A\overline{C} + \overline{A}B$.

H, D 4-62.* Modify the VHDL design in Figure 4-55 so that $z = 1$ only when the digital value is less than 1010_2.

H, B 4-63. Modify the code of (a) Figure 4-54 or (b) Figure 4-55 such that the output z is LOW only when digital_value is between 6 and 11 (inclusive).

H, D 4-64. Modify (a) the AHDL design in Figure 4-60 to implement Table 4-1. (b) the VHDL design in Figure 4-61 to implement Table 4-1.

H, D 4-65.* Write the hardware description design file Boolean equation to implement Example 4-9.

4-66. Write the hardware description design file Boolean equation to implement a four-bit parity generator as shown in Figure 4-25(a).

DRILL QUESTION

B 4-67. Define each of the following terms.

(a) Karnaugh map

(b) Sum-of-products form

(c) Parity generator

(d) Octet

(e) Enable circuit
(f) Don't-care condition
(g) Floating input
(h) Indeterminate voltage level
(i) Contention
(j) PLD
(k) TTL
(l) CMOS

MICROCOMPUTER APPLICATIONS

C 4-68. In a microcomputer, the microprocessor unit (MPU) is always communicating with one of the following: (1) random-access memory (RAM), which stores programs and data that can be readily changed; (2) read-only memory (ROM), which stores programs and data that never change; and (3) external input/output (I/O) devices such as keyboards, video displays, printers, and disk drives. As it is executing a program, the MPU will generate an address code that selects which type of device (RAM, ROM, or I/O) it wants to communicate with. Figure 4-79 shows a typical arrangement where the MPU outputs an eight-bit address code A_{15} through A_8. Actually, the MPU outputs a 16-bit address code, but the low-order bits A_7 through A_0 are not used in the device selection process. The address code is applied to a logic circuit that uses it to generate the device select signals: $\overline{RAM}$, $\overline{ROM}$, and $\overline{I/O}$.

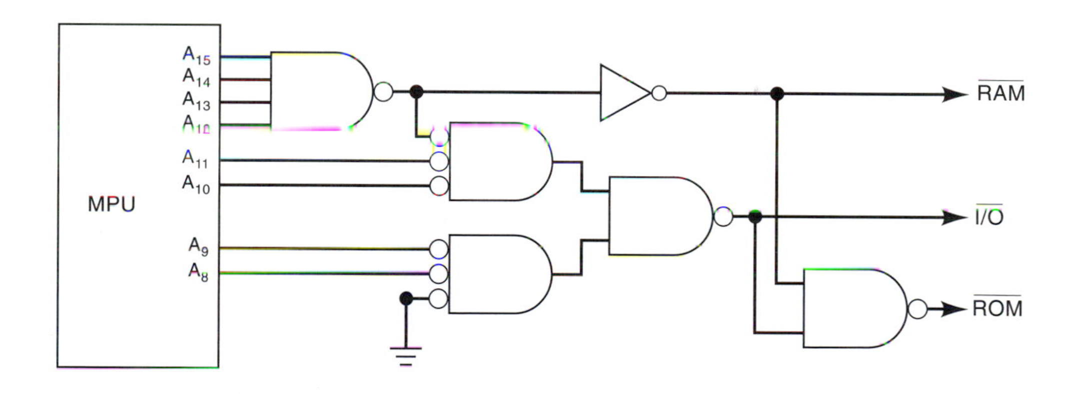

FIGURE 4-79 Problem 4-68.

Analyze this circuit and determine the following.

(a)* The range of addresses A_{15} through A_8 that will activate $\overline{\text{RAM}}$
(b) The range of addresses that activate $\overline{\text{I/O}}$
(c) The range of addresses that activate $\overline{\text{ROM}}$

Express the addresses in binary and hexadecimal. For example, the answer to (a) is A_{15} to $A_8 = 00000000_2$ to $11101111_2 = 00_{16}$ to EF_{16}.

C, D 4-69. In some microcomputers, the MPU can be *disabled* for short periods of time while another device controls the RAM, ROM, and I/O. During these intervals, a special control signal ($\overline{DMA}$) is activated by the MPU and is used to disable (deactivate) the device select logic so that the $\overline{\text{RAM}}$, $\overline{\text{ROM}}$, and $\overline{\text{I/O}}$ are all in their inactive state. Modify the circuit of Figure 4-79 so that $\overline{\text{RAM}}$, $\overline{\text{ROM}}$, and $\overline{\text{I/O}}$ will be deactivated whenever the signal $\overline{DMA}$ is active, regardless of the state of the address code.

ANSWERS TO SECTION REVIEW QUESTIONS

SECTION 4-1

1. Only (a)   2. Only (c)

SECTION 4-3

1. Expression (b) is not in sum-of-products form because of the inversion sign over both the C and D variables (i.e., the $A\overline{CD}$ term). Expression (c) is not in sum-of-products form because of the $(M + \overline{N})P$ term.   3. $x = \overline{A} + \overline{B} + \overline{C}$

SECTION 4-4

1. $x = \overline{A}\,\overline{B}\,\overline{C}D + \overline{A}\,\overline{B}C\overline{D} + \overline{A}B\overline{C}\,\overline{D}$ 2. Eight

SECTION 4-5

1. $x = AB + AC + BC$ 2. $x = A + BCD$ 3. $S = \overline{P} + QR$ 4. An input condition for which there is no specific required output condition; i.e., we are free to make it 0 or 1.

SECTION 4-6

2. A constant LOW   3. No; the available XOR gate can be used as an INVERTER by connecting one of its inputs to a constant HIGH (see Example 4-16).

SECTION 4-8

1. $x = \overline{A(B \oplus C)}$ 2. OR, NAND   3. NAND, NOR

SECTION 4-9

1. DIP   2. SSI, MSI, LSI, VLSI, ULSI, GSI   3. True   4. True   5. 40, 74AC, 74ACT series   6. 0 to 0.8 V; 2.0 to 5.0 V   7. 0 to 1.5 V; 3.5 to 5.0 V   8. As if the input were HIGH   9. Unpredictably; it may overheat and be destroyed.   10. 74HCT and 74ACT   11. They describe exactly how to interconnect the chips for laying out the circuit and troubleshooting.   12. Inputs and outputs are defined, and logical relationships are described.

SECTION 4-11

1. Open inputs or outputs; inputs or outputs shorted to V_{CC}; inputs or outputs shorted to ground; pins shorted together; internal circuit failures   2. Pins shorted together   3. For TTL, a LOW; for CMOS, indeterminate   4. Two or more outputs connected together

SECTION 4-12

1. Open signal lines; shorted signal lines; faulty power supply; output loading   2. Broken wires; poor solder connections; cracks or cuts in PC board; bent or broken IC pins; faulty IC sockets   3. ICs operating erratically or not at all   4. Logic level indeterminate

SECTION 4-14

1. Electrically controlled connections are being programmed as open or closed.   2. (4, 1) (2, 2) or (2, 1) (4, 2)   3. (4, 5) (1, 6) or (4, 6) (1, 5)

SECTION 4-15

1. (a) push_buttons[5..0] :INPUT;   (b) push_buttons :IN BIT_VECTOR (5 DOWNTO 0),   2. (a) z = push_buttons[5];   (b) z <= push_buttons(5);   3. STD_LOGIC   4. STD_LOGIC_VECTOR

SECTION 4-16

1. (AHDL) omega[] = (x, y, z); (VHDL) omega <= x & y & z; 2. Using the keyword TABLE 3. Using selected signal assignments

SECTION 4-17

1. IF/THEN 2. IF/THEN/ELSE 3. CASE or IF/ELSIF
4. (AHDL) count[7..0] :INPUT; (VHDL) count :IN INTEGER RANGE 0 TO 205

CHAPTER 5

FLIP-FLOPS AND RELATED DEVICES

OUTLINE

OBJECTIVES

Upon completion of this chapter, you will be able to:

- Construct and analyze the operation of a latch flip-flop made from NAND or NOR gates.
- Describe the difference between synchronous and asynchronous systems.
- Understand the operation of edge-triggered flip-flops.
- Analyze and apply the various flip-flop timing parameters specified by the manufacturers.
- Understand the major differences between parallel and serial data transfers.
- Draw the output timing waveforms of several types of flip-flops in response to a set of input signals.
- Recognize the various IEEE/ANSI flip-flop symbols.
- Use state transition diagrams to describe counter operation.
- Use flip-flops in synchronization circuits.
- Connect shift registers as data transfer circuits.
- Employ flip-flops as frequency-division and counting circuits.
- Understand the typical characteristics of Schmitt triggers.
- Apply two different types of one-shots in circuit design.
- Design a free-running oscillator using a 555 timer.
- Recognize and predict the effects of clock skew on synchronous circuits.
- Troubleshoot various types of flip-flop circuits.
- Write HDL code for latches.
- Use logic primitives, components, and libraries in HDL code.
- Build structural level circuits from components.

INTRODUCTION

The logic circuits considered thus far have been combinational circuits whose output levels at any instant of time are dependent on the levels present at the inputs at that time. Any prior input-level conditions have no effect on the present outputs because combinational logic circuits have no memory. Most digital systems consist of both combinational circuits and memory elements.

Figure 5-1 shows a block diagram of a general digital system that combines combinational logic gates with memory devices. The combinational portion accepts logic signals from external inputs and from the outputs of the memory elements. The combinational circuit operates on these inputs

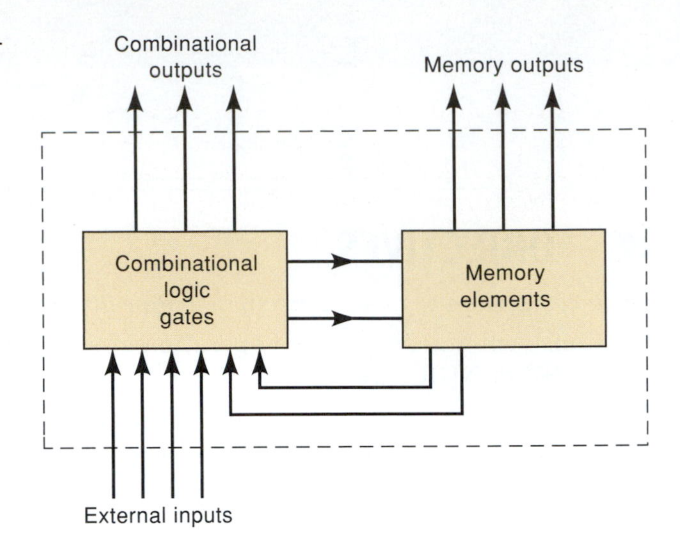

FIGURE 5-1 General digital system diagram.

to produce various outputs, some of which are used to determine the binary values to be stored in the memory elements. The outputs of some of the memory elements, in turn, go to the inputs of logic gates in the combinational circuits. This process indicates that the external outputs of a digital system are functions of both its external inputs and the information stored in its memory elements.

The most important memory element is the **flip-flop**, which is made up of an assembly of logic gates. Even though a logic gate, by itself, has no storage capability, several can be connected together in ways that permit information to be stored. Several different gate arrangements are used to produce these flip-flops (abbreviated FF).

Figure 5-2(a) is the general type of symbol used for a flip-flop. It shows two outputs, labeled Q and $\overline{Q}$, that are the inverse of each other. $Q/\overline{Q}$ are the most common designations used for a FF's outputs. From time to time, we will use other designations such as $X/\overline{X}$ and $A/\overline{A}$ for convenience in identifying different FFs in a logic circuit.

The Q output is called the *normal* FF output, and $\overline{Q}$ is the *inverted* FF output. Whenever we refer to the state of a FF, we are referring to the state of its normal (Q) output; it is understood that its inverted output ($\overline{Q}$) is in the opposite state. For example, if we say that a FF is in the HIGH (1) state, we mean that $Q = 1$; if we say that a FF is in the LOW (0) state, we mean that $Q = 0$. Of course, the $\overline{Q}$ state will always be the inverse of Q.

The two possible operating states for a FF are summarized in Figure 5-2(b). Note that the HIGH or 1 state ($Q = 1/\overline{Q} = 0$) is also referred to as the **SET** state. Whenever the inputs to a FF cause it to go to the $Q = 1$ state, we call this *setting* the FF; the FF has been set. In a similar way, the LOW or

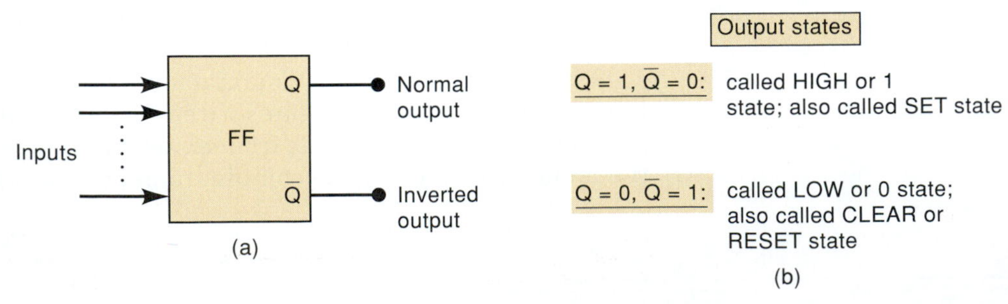

FIGURE 5-2 General flip-flop symbol and definition of its two possible output states.

0 state ($Q = 0/\overline{Q} = 1$) is also referred to as the **CLEAR** or **RESET** state. Whenever the inputs to a FF cause it to go to the $Q = 0$ state, we call this *clearing* or *resetting* the FF; the FF has been cleared (reset). As we shall see, many FFs will have a **SET** input and/or a **CLEAR (RESET)** input that is used to drive the FF into a specific output state.

As the symbol in Figure 5-2(a) implies, a FF can have one or more inputs. These inputs are used to cause the FF to switch back and forth ("flip-flop") between its possible output states. We will find out that most FF inputs need only to be momentarily activated (pulsed) in order to cause a change in the FF output state, and the output will remain in that new state even after the input pulse is over. This is the FF's *memory* characteristic.

The flip-flop is known by other names, including *latch* and *bistable multivibrator.* The term *latch* is used for certain types of flip-flops that we will describe. The term *bistable multivibrator* is the more technical name for a flip-flop, but it is too much of a mouthful to be used regularly.

5-1 NAND GATE LATCH

The most basic FF circuit can be constructed from either two NAND gates or two NOR gates. The NAND gate version, called a **NAND gate latch** or simply a **latch**, is shown in Figure 5-3(a). The two NAND gates are cross-coupled so that the output of NAND-1 is connected to one of the inputs of NAND-2, and vice versa. The gate outputs, labeled Q and $\overline{Q}$, respectively, are the latch outputs. Under normal conditions, these outputs will always be the inverse of each other. There are two latch inputs: the SET input is the input that *sets* Q to the 1 state; the RESET input is the input that *resets* Q to the 0 state.

The SET and RESET inputs are both normally resting in the HIGH state, and one of them will be pulsed LOW whenever we want to change the latch outputs. We begin our analysis by showing that there are two equally likely output states when SET = RESET = 1. One possibility is shown in Figure 5-3(a), where we have $Q = 0$ and $\overline{Q} = 1$. With $Q = 0$, the inputs to NAND-2 are 0 and 1, which produce $\overline{Q} = 1$. The 1 from $\overline{Q}$ causes NAND-1 to have a 1 at both inputs to produce a 0 output at Q. In effect, what we have is the LOW at the NAND-1 output producing a HIGH at the NAND-2 output, which, in turn, keeps the NAND-1 output LOW.

The second possibility is shown in Figure 5-3(b), where $Q = 1$ and $\overline{Q} = 0$. The HIGH from NAND-1 produces a LOW at the NAND-2 output, which, in turn, keeps the NAND-1 output HIGH. Thus, there are two possible output states when SET = RESET = 1; as we shall soon see, the one that actually exists will depend on what has occurred previously at the inputs.

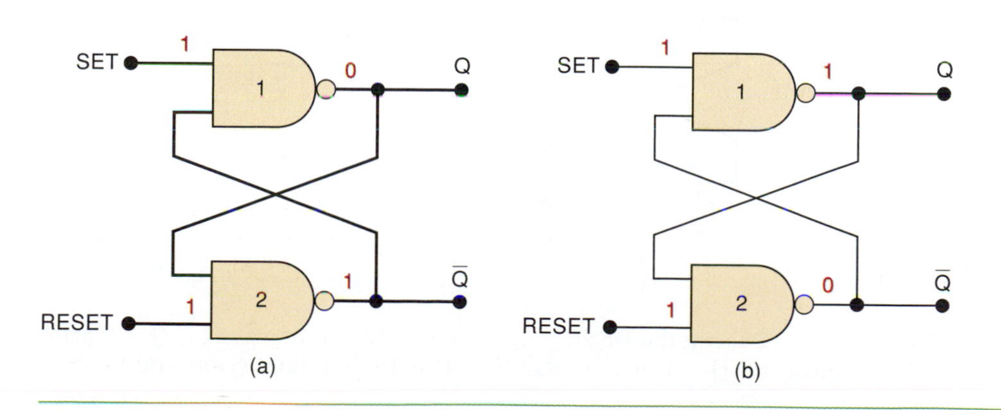

FIGURE 5-3 A NAND latch has two possible resting states when SET = RESET = 1.

Setting the Latch (FF)

Now let's investigate what happens when the SET input is momentarily pulsed LOW while RESET is kept HIGH. Figure 5-4(a) shows what happens when $Q = 0$ prior to the occurrence of the pulse. As SET is pulsed LOW at time t_0, Q will go HIGH, and this HIGH will force $\overline{Q}$ to go LOW so that NAND-1 now has two LOW inputs. Thus, when SET returns to the 1 state at t_1, the NAND-1 output *remains* HIGH, which, in turn, keeps the NAND-2 output LOW.

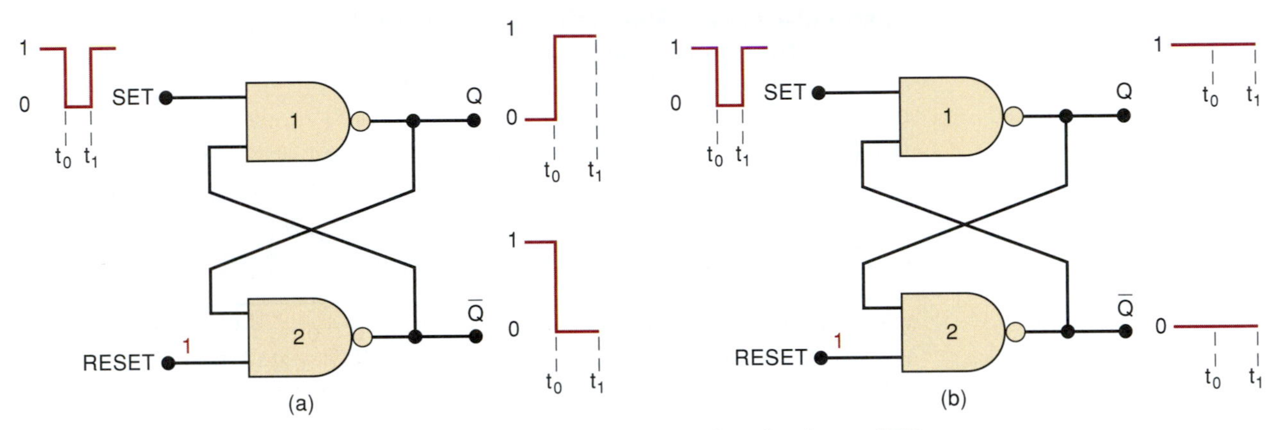

FIGURE 5-4 Pulsing the SET input to the 0 state when (a) $Q = 0$ prior to SET pulse; (b) $Q = 1$ prior to SET pulse. Note that, in both cases, Q ends up HIGH.

Figure 5-4(b) shows what happens when $Q = 1$ and $\overline{Q} = 0$ prior to the application of the SET pulse. Since $\overline{Q} = 0$ is already keeping the NAND-1 output HIGH, the LOW pulse at SET will not change anything. Thus, when SET returns HIGH, the latch outputs are still in the $Q = 1, \overline{Q} = 0$ state.

We can summarize Figure 5-4 by stating that a LOW pulse on the SET input will always cause the latch to end up in the $Q = 1$ state. This operation is called *setting* the latch or FF.

Resetting the Latch (FF)

Now let's consider what occurs when the RESET input is pulsed LOW while SET is kept HIGH. Figure 5-5(a) shows what happens when $Q = 0$ and $\overline{Q} = 1$

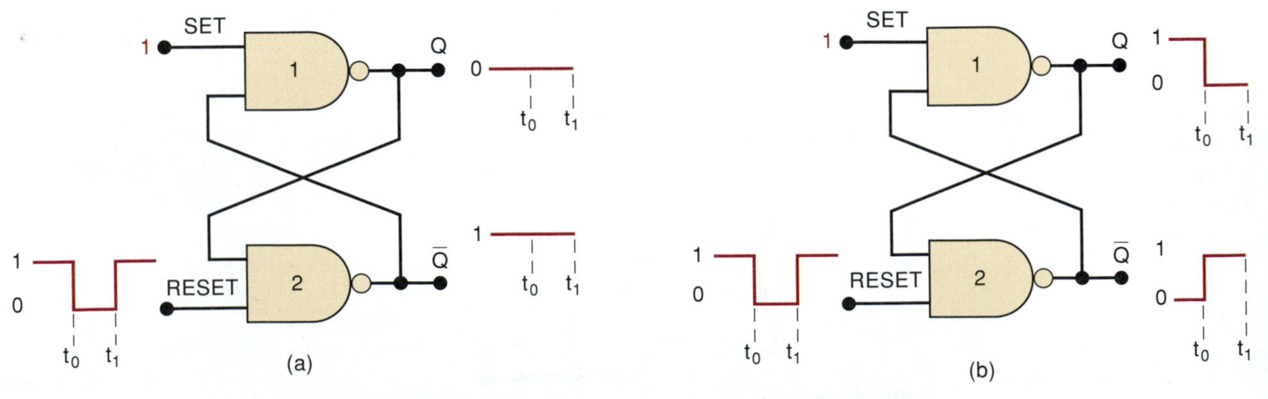

FIGURE 5-5 Pulsing the RESET input to the LOW state when (a) $Q = 0$ prior to RESET pulse; (b) $Q = 1$ prior to RESET pulse. In each case, Q ends up LOW.

prior to the application of the pulse. Since $Q = 0$ is already keeping the NAND-2 output HIGH, the LOW pulse at RESET will not have any effect. When RESET returns HIGH, the latch outputs are still $Q = 0$ and $\overline{Q} = 1$.

Figure 5-5(b) shows the situation where $Q = 1$ prior to the occurrence of the RESET pulse. As RESET is pulsed LOW at t_0, $\overline{Q}$ will go HIGH, and this HIGH forces Q to go LOW so that NAND-2 now has two LOW inputs. Thus, when RESET returns HIGH at t_1, the NAND-2 output *remains* HIGH, which, in turn, keeps the NAND-1 output LOW.

Figure 5-5 can be summarized by stating that a LOW pulse on the RESET input will always cause the latch to end up in the $Q = 0$ state. This operation is called *clearing* or *resetting* the latch.

Simultaneous Setting and Resetting

The last case to consider is the case where the SET and RESET inputs are simultaneously pulsed LOW. This will produce HIGH levels at both NAND outputs so that $Q = \overline{Q} = 1$. Clearly, this is an undesired condition because the two outputs are supposed to be inverses of each other. Furthermore, when the SET and RESET inputs return HIGH, the resulting output state will depend on which input returns HIGH first. Simultaneous transitions back to the 1 state will produce unpredictable results. For these reasons the SET = RESET = 0 condition is normally not used for the NAND latch.

Summary of NAND Latch

The operation described above can be conveniently placed in a function table (Figure 5-6) and is summarized as follows:

1. SET = RESET = 1. This condition is the normal resting state, and it has no effect on the output state. The Q and $\overline{Q}$ outputs will remain in whatever state they were in prior to this input condition.
2. SET = 0, RESET = 1. This will always cause the output to go to the $Q = 1$ state, where it will remain even after SET returns HIGH. This is called *setting* the latch.
3. SET = 1, RESET = 0. This will always produce the $Q = 0$ state, where the output will remain even after RESET returns HIGH. This is called *clearing* or *resetting* the latch.
4. SET = RESET = 0. This condition tries to set and clear the latch at the same time, and it produces $Q = \overline{Q} = 1$. If the inputs are returned to 1 simultaneously, the resulting state is unpredictable. This input condition should not be used.

FIGURE 5-6 (a) NAND latch; (b) function table.

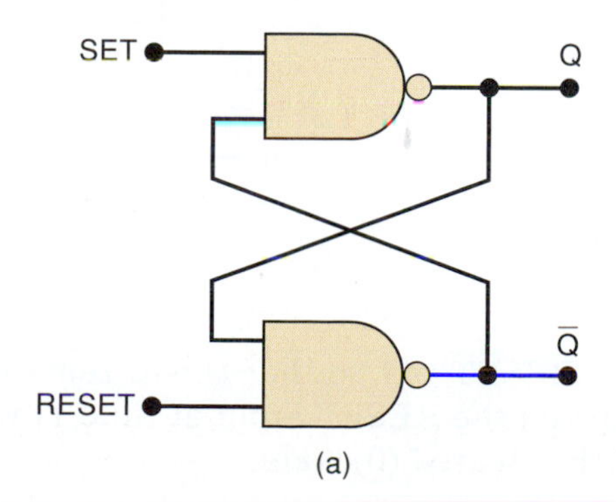

Set	Reset	Output
1	1	No change
0	1	Q = 1
1	0	Q = 0
0	0	Invalid*

*Produces $Q = \overline{Q} = 1$.

(b)

Alternate Representations

From the description of the NAND latch operation, it should be clear that the SET and RESET inputs are active-LOW. The SET input will set $Q = 1$ when SET goes LOW; the RESET input will clear $Q = 0$ when RESET goes LOW. For this reason, the NAND latch is often drawn using the alternate representation for each NAND gate, as shown in Figure 5-7(a). The bubbles on the inputs, as well as the labeling of the signals as $\overline{SET}$ and $\overline{RESET}$, indicate the active-LOW status of these inputs. (You may want to review Sections 3-13 and 3-14 on this topic.)

Figure 5-7(b) shows a simplified block representation that we will sometimes use. The S and R labels represent the SET and RESET inputs, and the bubbles indicate the active-LOW nature of these inputs. Whenever we use this block symbol, it represents a NAND latch.

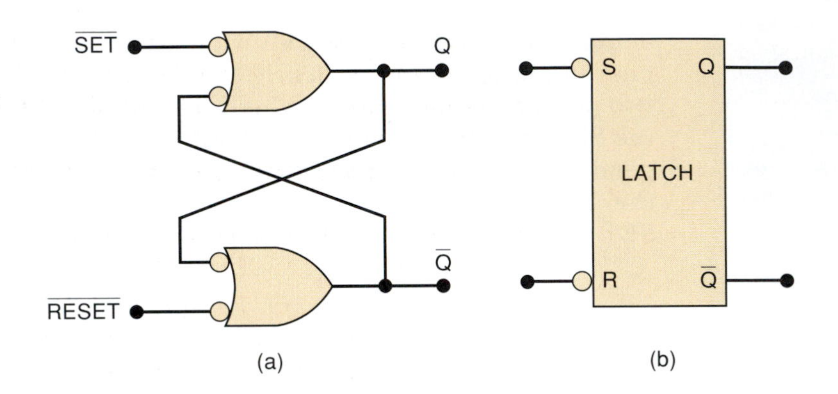

FIGURE 5-7 (a) NAND latch equivalent representation; (b) simplified block symbol.

Terminology

The action of *resetting* a FF or a latch is also called *clearing*, and both terms are used interchangeably in the digital field. In fact, a RESET input can also be called a CLEAR input, and a SET-RESET latch can be called a SET-CLEAR latch.

EXAMPLE 5-1

The waveforms of Figure 5-8 are applied to the inputs of the latch of Figure 5-7. Assume that initially $Q = 0$, and determine the Q waveform.

FIGURE 5-8 Example 5-1.

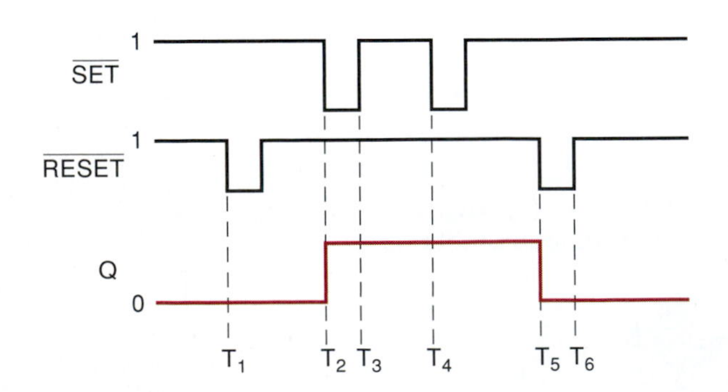

Solution

Initially, $\overline{SET} = \overline{RESET} = 1$ so that Q will remain in the 0 state. The LOW pulse that occurs on the $\overline{RESET}$ input at time T_1 will have no effect because Q is already in the cleared (0) state.

The only way that Q can go to the 1 state is by a LOW pulse on the $\overline{\text{SET}}$ input. This occurs at time T_2 when $\overline{\text{SET}}$ first goes LOW. When $\overline{\text{SET}}$ returns HIGH at T_3, Q will remain in its new HIGH state.

At time T_4 when $\overline{\text{SET}}$ goes LOW again, there will be no effect on Q because Q is already set to the 1 state.

The only way to bring Q back to the 0 state is by a LOW pulse on the $\overline{\text{RESET}}$ input. This occurs at time T_5. When $\overline{\text{RESET}}$ returns to 1 at time T_6, Q remains in the LOW state.

Example 5-1 shows that the latch output "remembers" the last input that was activated and will not change states until the opposite input is activated.

EXAMPLE 5-2

It is almost impossible to obtain a "clean" voltage transition from a mechanical switch because of the phenomenon of **contact bounce**. This is illustrated in Figure 5-9(a), where the action of moving the switch from contact position 1 to 2 produces several output voltage transitions as the switch bounces (makes and breaks contact with contact 2 several times) before coming to rest on contact 2.

The multiple transitions on the output signal generally last no longer than a few milliseconds, but they would be unacceptable in many applications. A NAND latch can be used to prevent the presence of contact bounce from affecting the output. Describe the operation of the "switch debouncing" circuit in Figure 5-9(b).

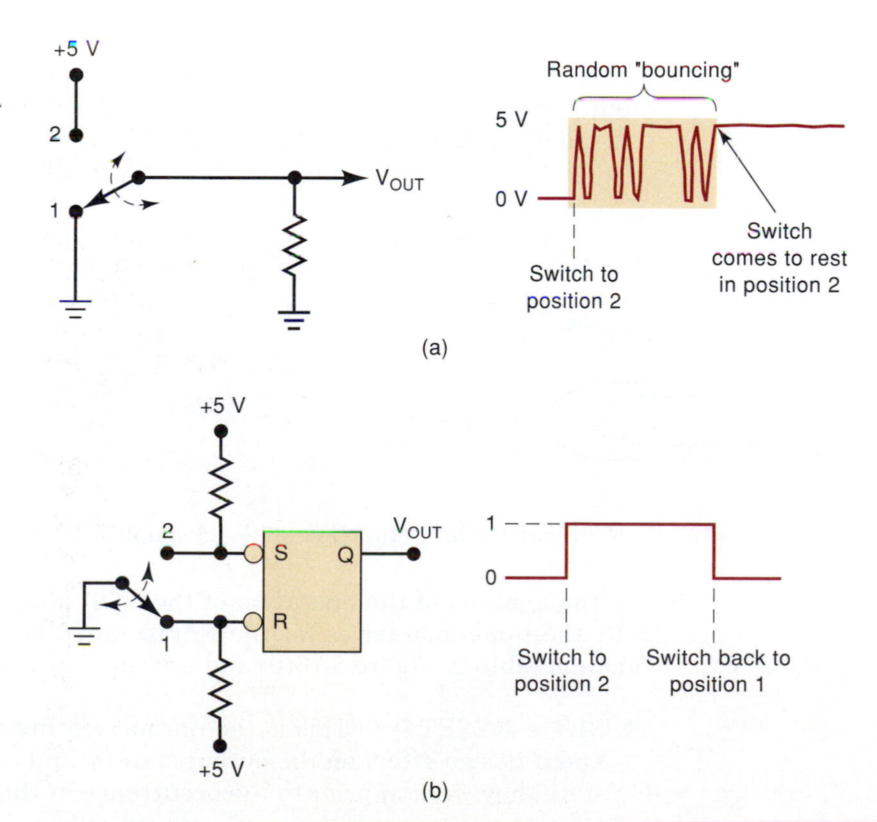

FIGURE 5-9 (a) Mechanical contact bounce will produce multiple transitions; (b) NAND latch used to debounce a mechanical switch.

Solution

Assume that the switch is resting in position 1 so that the $\overline{\text{RESET}}$ input is LOW and $Q = 0$. When the switch is moved to position 2, $\overline{\text{RESET}}$ will go HIGH, and a LOW will appear on the $\overline{\text{SET}}$ input as the switch first makes contact. This will set $Q = 1$ within a matter of a few nanoseconds (the response time of the NAND gate). Now if the switch bounces off contact 2, $\overline{\text{SET}}$ and $\overline{\text{RESET}}$ will both be HIGH, and Q will not be affected; it will stay HIGH. Thus, nothing will happen at Q as the switch bounces on and off contact 2 before finally coming to rest in position 2.

Likewise, when the switch is moved from position 2 back to position 1, it will place a LOW on the $\overline{\text{RESET}}$ input as it first makes contact. This clears Q to the LOW state, where it will remain even if the switch bounces on and off contact 1 several times before coming to rest.

Thus, the output at Q will consist of a single transition each time the switch is moved from one position to the other.

REVIEW QUESTIONS

1. What is the normal resting state of the $\overline{\text{SET}}$ and $\overline{\text{RESET}}$ inputs? What is the active state of each input?
2. What will be the states of Q and $\overline{Q}$ after a FF has been reset (cleared)?
3. *True or false:* The $\overline{\text{SET}}$ input can never be used to make $Q = 0$.
4. When power is first applied to any FF circuit, it is impossible to predict the initial states of Q and $\overline{Q}$. What can be done to ensure that a NAND latch always starts off in the $Q = 1$ state?

5-2 NOR GATE LATCH

Two cross-coupled NOR gates can be used as a **NOR gate latch**. The arrangement, shown in Figure 5-10(a), is similar to the NAND latch except that the Q and $\overline{Q}$ outputs have reversed positions.

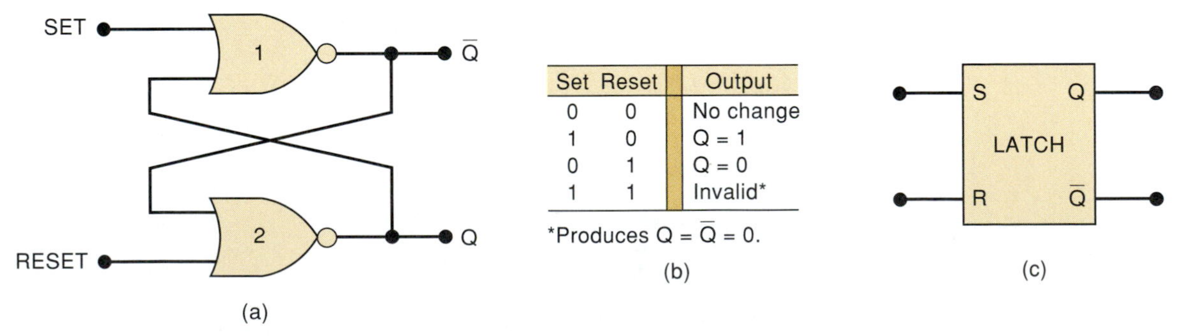

Set	Reset	Output
0	0	No change
1	0	Q = 1
0	1	Q = 0
1	1	Invalid*

*Produces $Q = \overline{Q} = 0$.

(b)

FIGURE 5-10 (a) NOR gate latch; (b) function table; (c) simplified block symbol.

The analysis of the operation of the NOR latch can be performed in exactly the same manner as for the NAND latch. The results are given in the function table in Figure 5-10(b) and are summarized as follows:

1. SET = RESET = 0. This is the normal resting state for the NOR latch, and it has no effect on the output state. Q and $\overline{Q}$ will remain in whatever state they were in prior to the occurrence of this input condition.

2. SET = 1, RESET = 0. This will always set $Q = 1$, where it will remain even after SET returns to 0.
3. SET = 0, RESET = 1. This will always clear $Q = 0$, where it will remain even after RESET returns to 0.
4. SET = 1, RESET = 1. This condition tries to set and reset the latch at the same time, and it produces $Q = \overline{Q} = 0$. If the inputs are returned to 0 simultaneously, the resulting output state is unpredictable. This input condition should not be used.

The NOR gate latch operates exactly like the NAND latch except that the SET and RESET inputs are active-HIGH rather than active-LOW, and the normal resting state is SET = RESET = 0. Q will be set HIGH by a HIGH pulse on the SET input, and it will be cleared LOW by a HIGH pulse on the RESET input. The simplified block symbol for the NOR latch in Figure 5-10(c) is shown with no bubbles on the S and R inputs; this indicates that these inputs are active-HIGH.

EXAMPLE 5-3

Assume that $Q = 0$ initially, and determine the Q waveform for the NOR latch inputs of Figure 5-11.

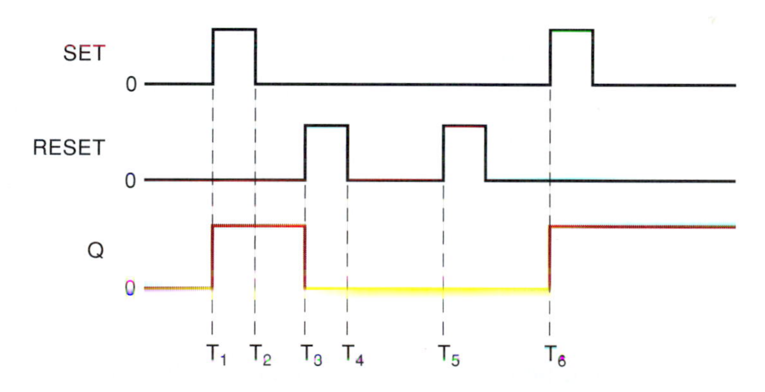

FIGURE 5-11 Example 5-3.

Solution

Initially, SET = RESET = 0, which has no effect on Q, and Q stays LOW. When SET goes HIGH at time T_1, Q will be set to 1 and will remain there even after SET returns to 0 at T_2.

At T_3 the RESET input goes HIGH and clears Q to the 0 state, where it remains even after RESET returns LOW at T_4.

The RESET pulse at T_5 has no effect on Q because Q is already LOW. The SET pulse at T_6 again sets Q back to 1, where it will stay.

Example 5-3 shows that the latch "remembers" the last input that was activated, and it will not change states until the opposite input is activated.

EXAMPLE 5-4

Figure 5-12 shows a simple circuit that can be used to detect the interruption of a light beam. The light is focused on a phototransistor that is connected in the common-emitter configuration to operate as a switch. Assume that the latch has previously been cleared to the 0 state by momentarily opening switch SW1, and describe what happens if the light beam is momentarily interrupted.

FIGURE 5-12
Example 5-4.

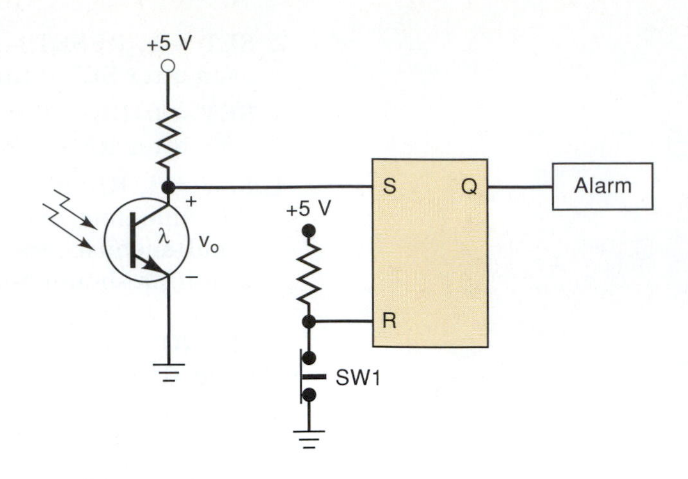

Solution

With light on the phototransistor, we can assume that it is fully conducting so that the resistance between the collector and the emitter is very small. Thus, v_0 will be close to 0 V. This places a LOW on the SET input of the latch so that SET = RESET = 0.

When the light beam is interrupted, the phototransistor turns off, and its collector-emitter resistance becomes very high (i.e., essentially an open circuit). This causes v_0 to rise to approximately 5 V; this activates the SET input, which sets Q HIGH and turns on the alarm.

Q will remain HIGH and the alarm will remain on even if v_0 returns to 0 V (i.e., the light beam was interrupted only momentarily) because SET and RESET will both be LOW, which will produce no change in Q.

In this application, the latch's memory characteristic is used to convert a momentary occurrence (beam interruption) into a constant output.

Flip-Flop State on Power-Up

When power is applied to a circuit, it is not possible to predict the starting state of a flip-flop's output if its SET and RESET inputs are in their inactive state (e.g., $S = R = 1$ for a NAND latch, $S = R = 0$ for a NOR latch). There is just as much chance that the starting state will be $Q = 0$ as $Q = 1$. It will depend on factors such as internal propagation delays, parasitic capacitance, and external loading. If a latch or FF must start off in a particular state to ensure the proper operation of a circuit, then it must be placed in that state by momentarily activating the SET or RESET input at the start of the circuit's operation. This is often achieved by application of a pulse to the appropriate input.

REVIEW QUESTIONS

1. What is the normal resting state of the NOR latch inputs? What is the active state?
2. When a latch is set, what are the states of Q and $\overline{Q}$?
3. What is the only way to cause the Q output of a NOR latch to change from 1 to 0?
4. If the NOR latch in Figure 5-12 were replaced by a NAND latch, why wouldn't the circuit work properly?

5-3 TROUBLESHOOTING CASE STUDY

The following two examples present an illustration of the kinds of reasoning used in troubleshooting a circuit containing a latch.

EXAMPLE 5-5

Analyze and describe the operation of the circuit in Figure 5-13.

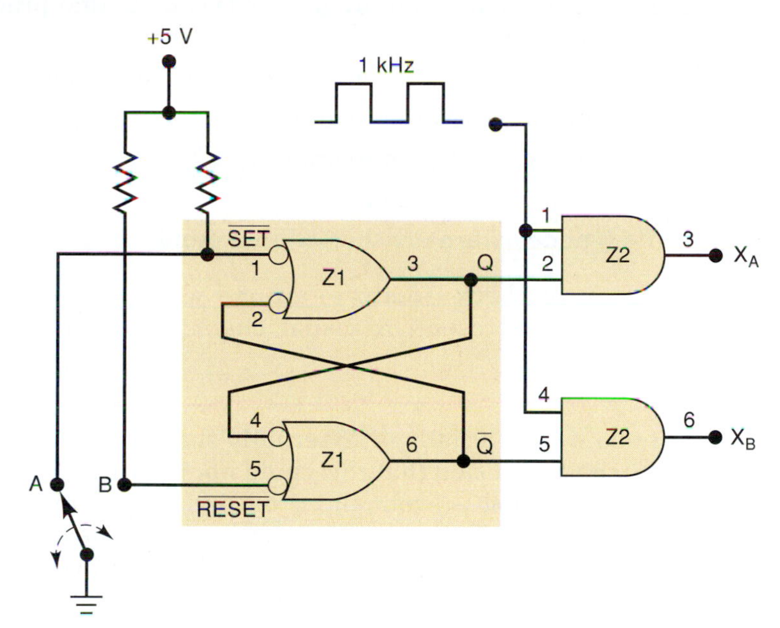

Switch position	X_A	X_B
A	Pulses	LOW
B	LOW	Pulses

FIGURE 5-13 Examples 5-5 and 5-6.

Solution

The switch is used to set or clear the NAND latch to produce clean, bounce-free signals at Q and $\overline{Q}$. These latch outputs control the passage of the 1-kHz pulse signal through to the AND outputs X_A and X_B.

When the switch moves to position A, the latch is set to $Q = 1$. This enables the 1-kHz pulses to pass through to X_A, while the LOW at $\overline{Q}$ keeps $X_B = 0$. When the switch moves to position B, the latch is cleared to $Q = 0$, which keeps $X_A = 0$, while the HIGH at $\overline{Q}$ enables the pulses to pass through to X_B.

EXAMPLE 5-6

A technician tests the circuit of Figure 5-13 and records the observations shown in Table 5-1. He notices that when the switch is in position B, the circuit functions correctly, but in position A the latch does not set to the $Q = 1$ state. What are the possible faults that could produce this malfunction?

Solution

There are several possibilities:

1. An internal open connection at Z1-1, which would prevent Q from responding to the $\overline{\text{SET}}$ input.

TABLE 5-1

Switch Position	$\overline{\text{SET}}$ (Z1-1)	$\overline{\text{RESET}}$ (Z1-5)	Q (Z1-3)	$\overline{Q}$ (Z1-6)	X_A (Z2-3)	X_B (Z2-6)
A	LOW	HIGH	LOW	HIGH	LOW	Pulses
B	HIGH	LOW	LOW	HIGH	LOW	Pulses

2. An internal component failure in NAND gate Z1 that prevents it from responding properly.
3. The Q output is stuck LOW, which could be caused by:
   (a) Z1-3 internally shorted to ground
   (b) Z1-4 internally shorted to ground
   (c) Z2-2 internally shorted to ground
   (d) The Q node externally shorted to ground

   An ohmmeter check from Q to ground will determine if any of these conditions are present. A visual check should reveal any external short.

What about $\overline{Q}$ internally or externally shorted to V_{CC}? A little thought will lead to the conclusion that this could not be the fault. If $\overline{Q}$ were shorted to V_{CC}, this would not prevent the Q output from going HIGH when $\overline{\text{SET}}$ goes LOW. Because Q *does not* go HIGH, this cannot be the fault. The reason that $\overline{Q}$ looks as if it is stuck HIGH is that Q is stuck LOW, and that keeps $\overline{Q}$ HIGH through the bottom NAND gate.

5-4 DIGITAL PULSES

As you can see from our discussion of SR latches, there are situations in digital systems when a signal switches from a normal inactive state to the opposite (active) state, thus causing something to happen in the circuit. Then the signal returns to its inactive state while the effect of the recently activated signal remains in the system. These signals are called **pulses**, and it is very important to understand the terminology associated with pulses and pulse waveforms. A pulse that performs its intended function when it goes HIGH is called a *positive* pulse, and a pulse that performs its intended function when it goes LOW is called a *negative* pulse. In actual circuits it takes time for a pulse waveform to change from one level to the other. These transition times are called the rise time (t_r) and the fall time (t_f) and are defined as the time it takes the voltage to change between 10% and 90% of the HIGH level voltage as shown on the positive pulse in Figure 5-14(a). The transition at the beginning of the pulse is called the leading edge and the transition at the end of the pulse is the trailing edge. The duration (width) of the pulse (t_w) is defined as the time between the points when the leading and trailing edges are at 50% of the HIGH level voltage. Figure 5-14(b) shows an active-LOW or negative pulse.

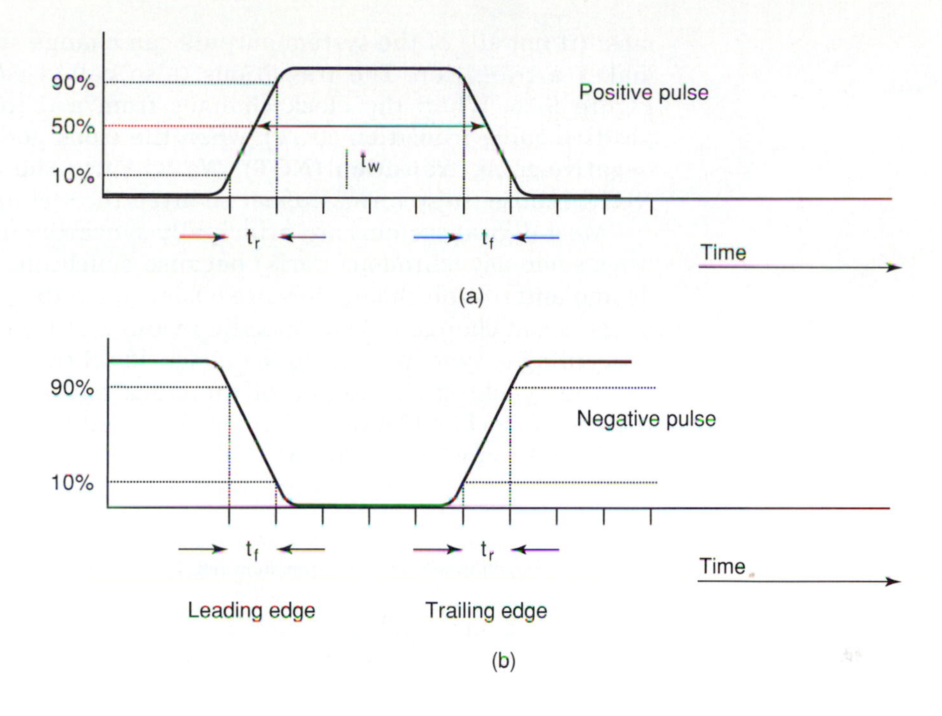

FIGURE 5-14 (a) A positive pulse and (b) a negative pulse.

EXAMPLE 5-7

When a microcontroller wants to access data in its external memory, it activates an active-LOW output pin called $\overline{RD}$ (read). The data book says that the $\overline{RD}$ pulse typically has a pulse width t_w of 50 ns, a rise time t_r of 15 ns, and a fall time t_f of 10 ns. Draw a scaled drawing of the $\overline{RD}$ pulse.

Solution

Figure 5-15 shows the drawing of the pulse. The $\overline{RD}$ pulse is active-LOW, so the leading edge is a falling edge measured by t_f and the trailing edge is the rising edge measured by t_r.

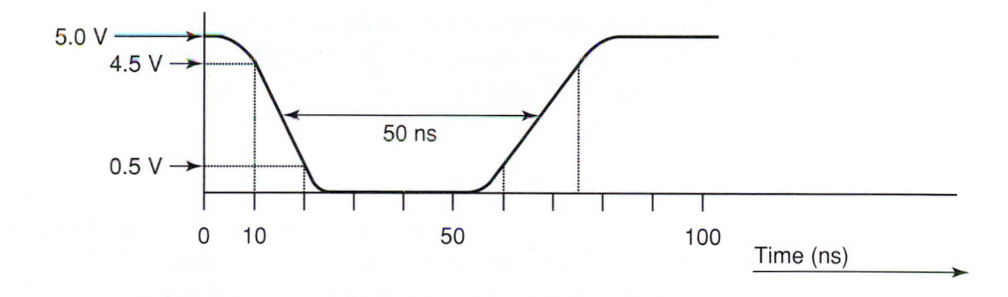

FIGURE 5-15 Example 5-7.

5-5 CLOCK SIGNALS AND CLOCKED FLIP-FLOPS

Digital systems can operate either *asynchronously* or *synchronously.* In asynchronous systems, the outputs of logic circuits can change state any time one or more of the inputs change. An asynchronous system is generally more difficult to design and troubleshoot than a synchronous system.

In synchronous systems, the exact times at which any output can change states are determined by a signal commonly called the **clock**. This clock signal is generally a rectangular pulse train or a square wave, as shown in Figure 5-16. The clock signal is distributed to all parts of the system, and most (if not all) of the system outputs can change state only when the clock makes a transition. The transitions (also called *edges*) are pointed out in Figure 5-16. When the clock changes from a 0 to a 1, this is called the **positive-going transition (PGT)**; when the clock goes from 1 to 0, this is the **negative-going transition (NGT)**. We will use the abbreviations PGT and NGT because these terms appear so often throughout the text.

Most digital systems are principally synchronous (although there are always some asynchronous parts) because synchronous circuits are easier to design and troubleshoot. They are easier to troubleshoot because the circuit outputs can change only at specific instants of time. In other words, almost everything is synchronized to the clock-signal transitions.

The synchronizing action of the clock signals is accomplished through the use of **clocked flip-flops** that are designed to change states on one or the other of the clock's transitions.

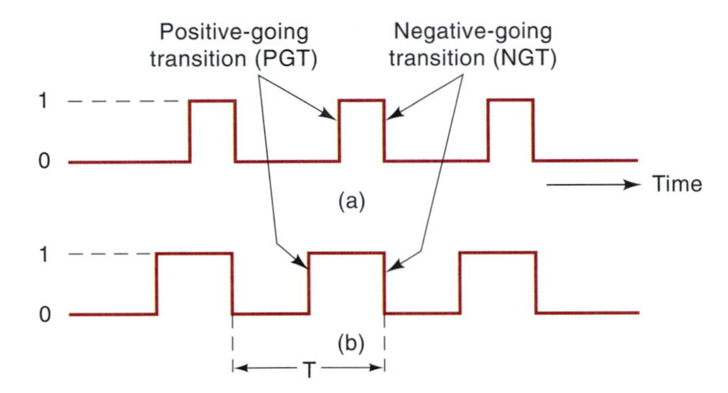

FIGURE 5-16 Clock signals.

The speed at which a synchronous digital system operates is dependent on how often the clock cycles occur. A clock cycle is measured from one PGT to the next PGT or from one NGT to the next NGT. The time it takes to complete one cycle (seconds/cycle) is called the **period (T)**, as shown in Figure 5-16(b). The speed of a digital system is normally referred to by the number of clock cycles that happen in 1 s (cycles/second), which is known as the **frequency (F)** of the clock. The standard unit for frequency is hertz. One hertz (1 Hz) = 1 cycle/second.

Clocked Flip-Flops

Several types of clocked FFs are used in a wide range of applications. Before we begin our study of the different clocked FFs, we will describe the principal ideas that are common to all of them.

1. Clocked FFs have a clock input that is typically labeled *CLK*, *CK*, or *CP*. We will normally use *CLK*, as shown in Figure 5-17. In most clocked FFs, the *CLK* input is **edge-triggered**, which means that it is activated by a signal transition; this is indicated by the presence of a small triangle on the *CLK* input. This contrasts with the latches, which are level-triggered.

   Figure 5-17(a) is a FF with a small triangle on its *CLK* input to indicate that this input is activated *only* when a positive-going transition (PGT) occurs; no other part of the input pulse will have an effect on the *CLK* input. In Figure 5-17(b), the FF symbol has a bubble as well as a triangle on its *CLK* input. This signifies that the *CLK* input is activated *only* when a negative-going transition occurs; no other part of the input pulse will have an effect on the *CLK* input.

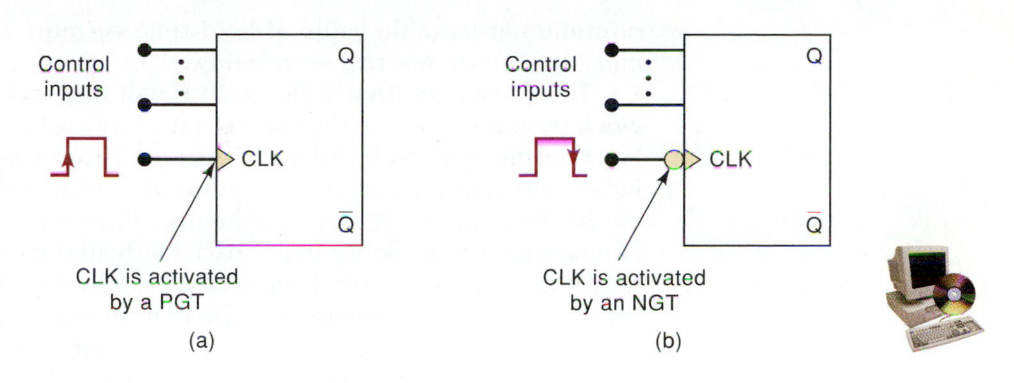

FIGURE 5-17 Clocked FFs have a clock input (*CLK*) that is active on either (a) the PGT or (b) the NGT. The control inputs determine the effect of the active clock transition.

2. Clocked FFs also have one or more **control inputs** that can have various names, depending on their operation. The control inputs will have no effect on *Q* until the active clock transition occurs. In other words, their effect is synchronized with the signal applied to *CLK*. For this reason they are called **synchronous control inputs**.

   For example, the control inputs of the FF in Figure 5-17(a) will have no effect on *Q* until the PGT of the clock signal occurs. Likewise, the control inputs in Figure 5-17(b) will have no effect until the NGT of the clock signal occurs.

3. In summary, we can say that the control inputs get the FF outputs ready to change, while the active transition at the *CLK* input actually *triggers* the change. The control inputs control the WHAT (i.e., what state the output will go to); the *CLK* input determines the WHEN.

Setup and Hold Times

Two timing requirements must be met if a clocked FF is to respond reliably to its control inputs when the active *CLK* transition occurs. These requirements are illustrated in Figure 5-18 for a FF that triggers on a PGT.

The **setup time, t_S**, is the time interval immediately preceding the active transition of the *CLK* signal during which the control input must be maintained at the proper level. IC manufacturers usually specify the minimum allowable setup time t_S(min). If this time requirement is not met, the FF may not respond reliably when the clock edge occurs.

The **hold time, t_H**, is the time interval immediately following the active transition of the *CLK* signal during which the synchronous control input must be maintained at the proper level. IC manufacturers usually specify the

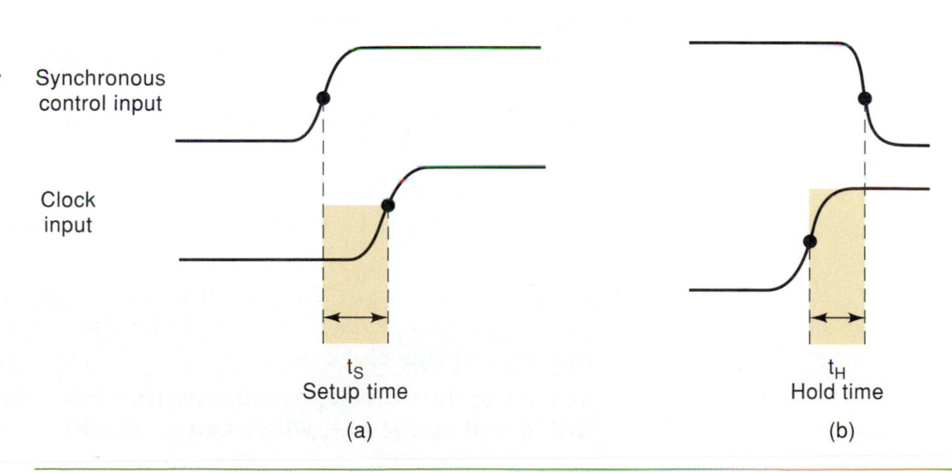

FIGURE 5-18 Control inputs must be held stable for (a) a time t_S prior to active clock transition and for (b) a time t_H after the active block transition.

minimum acceptable value of hold time t_H(min). If this requirement is not met, the FF will not trigger reliably.

Thus, to ensure that a clocked FF will respond properly when the active clock transition occurs, the control inputs must be stable (unchanging) for at least a time interval equal to t_S(min) *prior* to the clock transition, and for at least a time interval equal to t_H(min) *after* the clock transition.

IC flip-flops will have minimum allowable t_S and t_H values in the nanosecond range. Setup times are usually in the range of 5 to 50 ns, whereas hold times are generally from 0 to 10 ns. Notice that these times are measured between the 50 percent points on the transitions.

These timing requirements are very important in synchronous systems because, as we shall see, there will be many situations where the synchronous control inputs to a FF are changing at approximately the same time as the *CLK* input.

REVIEW QUESTIONS

1. What two types of inputs does a clocked FF have?
2. What is meant by the term *edge-triggered*?
3. *True or false:* The *CLK* input will affect the FF output only when the active transition of the control input occurs.
4. Define the setup time and hold time requirements of a clocked FF.

5-6 CLOCKED S-R FLIP-FLOP

Figure 5-19(a) shows the logic symbol for a **clocked S-R flip-flop** that is triggered by the positive-going edge of the clock signal. This means that the FF can change states *only* when a signal applied to its clock input makes a transition from 0 to 1. The *S* and *R* inputs control the state of the FF in the same manner as described earlier for the NOR gate latch, but the FF does not respond to these inputs until the occurrence of the PGT of the clock signal.

The function table in Figure 5-19(b) shows how the FF output will respond to the PGT at the *CLK* input for the various combinations of *S* and *R* inputs. This function table uses some new nomenclature. The up arrow (↑) indicates that a PGT is required at *CLK*; the label Q_0 indicates the level at Q prior to the PGT. This nomenclature is often used by IC manufacturers in their IC data manuals.

The waveforms in Figure 5-19(c) illustrate the operation of the clocked S-R flip-flop. If we assume that the setup and hold time requirements are being met in all cases, we can analyze these waveforms as follows:

1. Initially all inputs are 0 and the Q output is assumed to be 0; that is, $Q_0 = 0$.
2. When the PGT of the first clock pulse occurs (point *a*), the *S* and *R* inputs are both 0, so the FF is not affected and remains in the $Q = 0$ state (i.e., $Q = Q_0$).
3. At the occurrence of the PGT of the second clock pulse (point *c*), the *S* input is now high, with *R* still low. Thus, the FF sets to the 1 state at the rising edge of this clock pulse.
4. When the third clock pulse makes its positive transition (point *e*), it finds that $S = 0$ and $R = 1$, which causes the FF to clear to the 0 state.

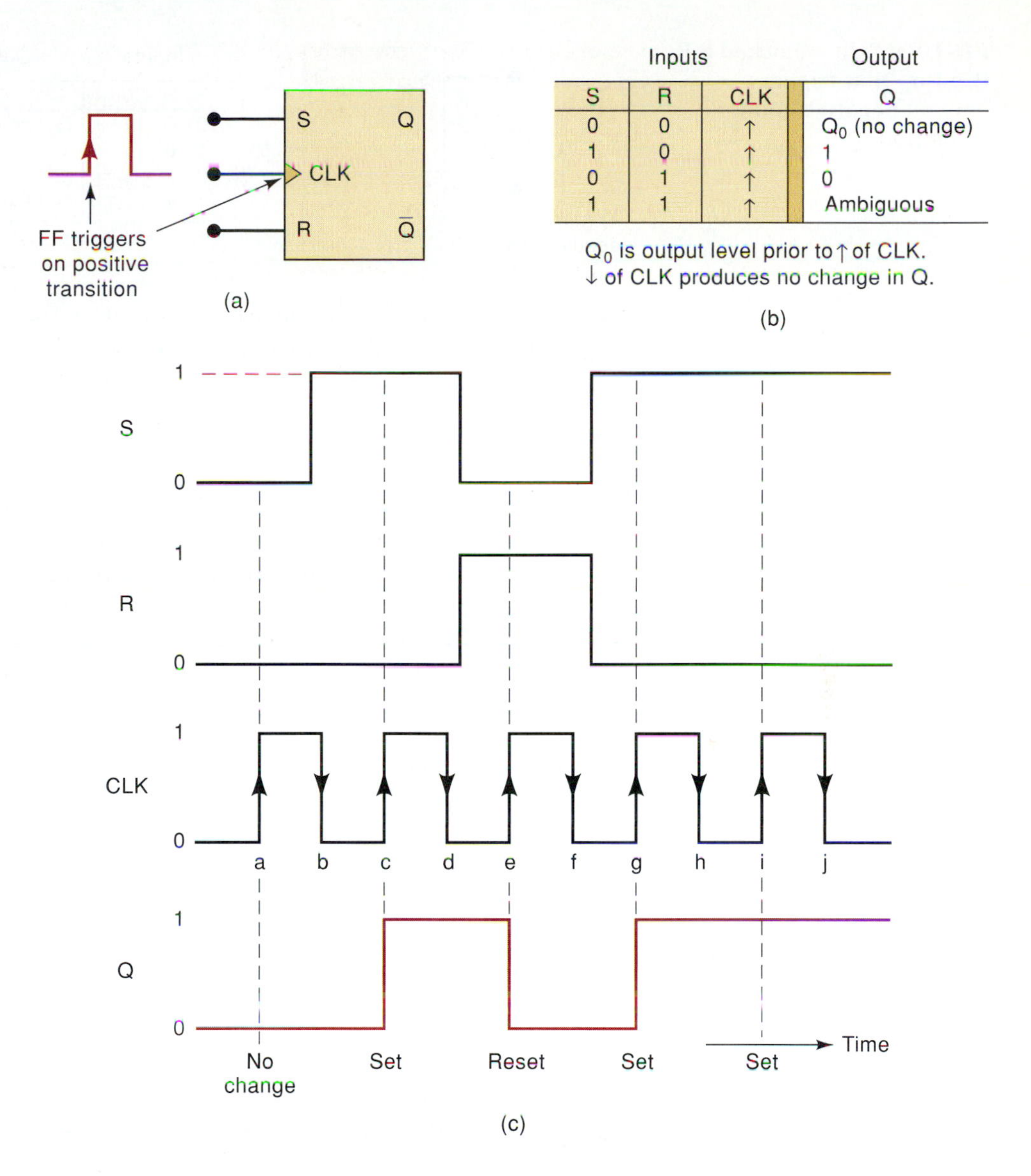

FIGURE 5-19 (a) Clocked S-R flip-flop that responds only to the positive-going edge of a clock pulse; (b) function table; (c) typical waveforms.

5. The fourth pulse sets the FF once again to the $Q = 1$ state (point g) because $S = 1$ and $R = 0$ when the positive edge occurs.
6. The fifth pulse also finds that $S = 1$ and $R = 0$ when it makes its positive-going transition. However, Q is already high, so it remains in that state.
7. The $S = R = 1$ condition should not be used because it results in an ambiguous condition.

It should be noted from these waveforms that the FF is not affected by the negative-going transitions of the clock pulses. Also, note that the S and R levels have no effect on the FF, except upon the occurrence of a positive-going transition of the clock signal. The S and R inputs are synchronous *control* inputs; they control which state the FF will go to when the clock pulse occurs. The *CLK* input is the **trigger** input that causes the FF to change states according to what the S and R inputs are when the active clock transition occurs.

Figure 5-20 shows the symbol and the function table for a clocked S-R flip-flop that triggers on the *negative*-going transition at its *CLK* input. The small circle and triangle on the *CLK* input indicates that this FF will trigger only when the *CLK* input goes from 1 to 0. This FF operates in the same

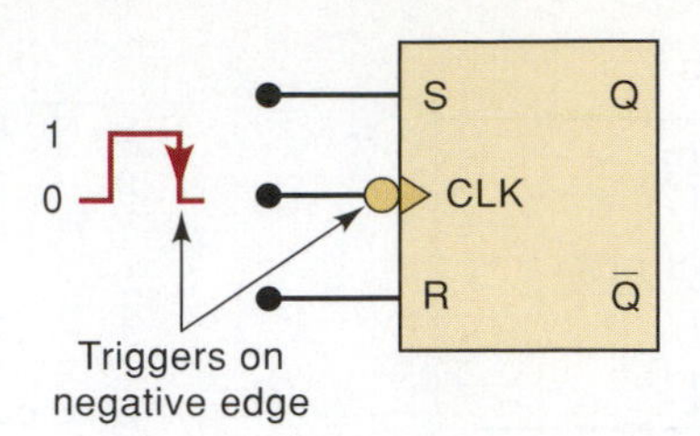

Inputs			Output
S	R	CLK	Q
0	0	↓	Q_0 (no change)
1	0	↓	1
0	1	↓	0
1	1	↓	Ambiguous

FIGURE 5-20 Clocked S-R flip-flop that triggers only on negative-going transitions.

manner as the positive-edge FF except that the output can change states only on the falling edge of the clock pulses (points *b*, *d*, *f*, *h*, and *j* in Figure 5-19). Both positive-edge and negative-edge triggering FFs are used in digital systems.

Internal Circuitry of the Edge-Triggered S-R Flip-Flop

A detailed analysis of the internal circuitry of a clocked FF is not necessary because all types are readily available as ICs. Although our main interest is in the FF's external operation, our understanding of this external operation can be aided by taking a look at a simplified version of the FF's internal circuitry. Figure 5-21 shows this for an edge-triggered S-R flip-flop.

The circuit contains three sections:

1. A basic NAND gate latch formed by NAND-3 and NAND-4
2. A **pulse-steering circuit** formed by NAND-1 and NAND-2
3. An **edge-detector circuit**

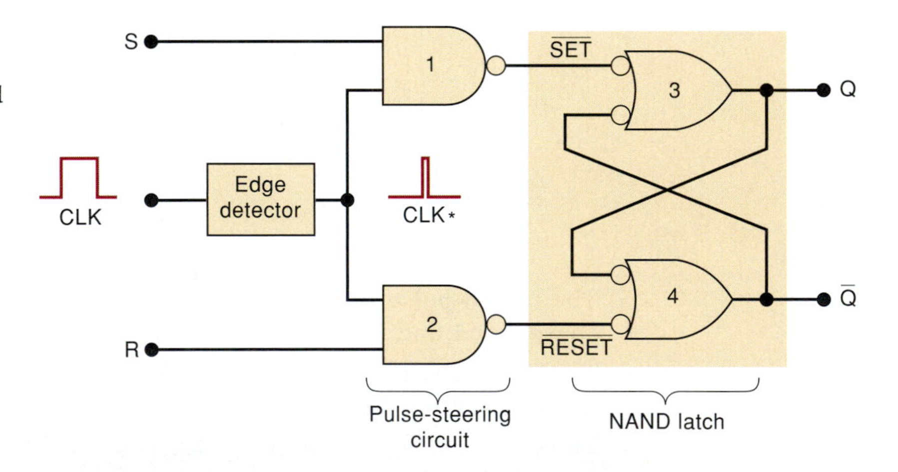

FIGURE 5-21 Simplified version of the internal circuitry for an edge-triggered S-R flip-flop.

As shown in Figure 5-21, the edge detector produces a narrow positive-going spike (*CLK**) that occurs coincident with the active transition of the *CLK* input pulse. The pulse-steering circuit "steers" the spike through to the SET or the RESET input of the latch in accordance with the levels present at *S* and *R*. For example, with $S = 1$ and $R = 0$, the *CLK** signal is inverted and passed through NAND-1 to produce a LOW pulse at the SET input of the latch that sets $Q = 1$. With $S = 0$, $R = 1$, the *CLK** signal is inverted and passed through NAND-2 to produce a low pulse at the RESET input of the latch that resets $Q = 0$.

Figure 5-22(a) shows how the *CLK** signal is generated for edge-triggered FFs that trigger on a PGT. The INVERTER produces a delay of a few nanoseconds so that the transitions of $\overline{CLK}$ occur a little bit after those of *CLK*. The AND

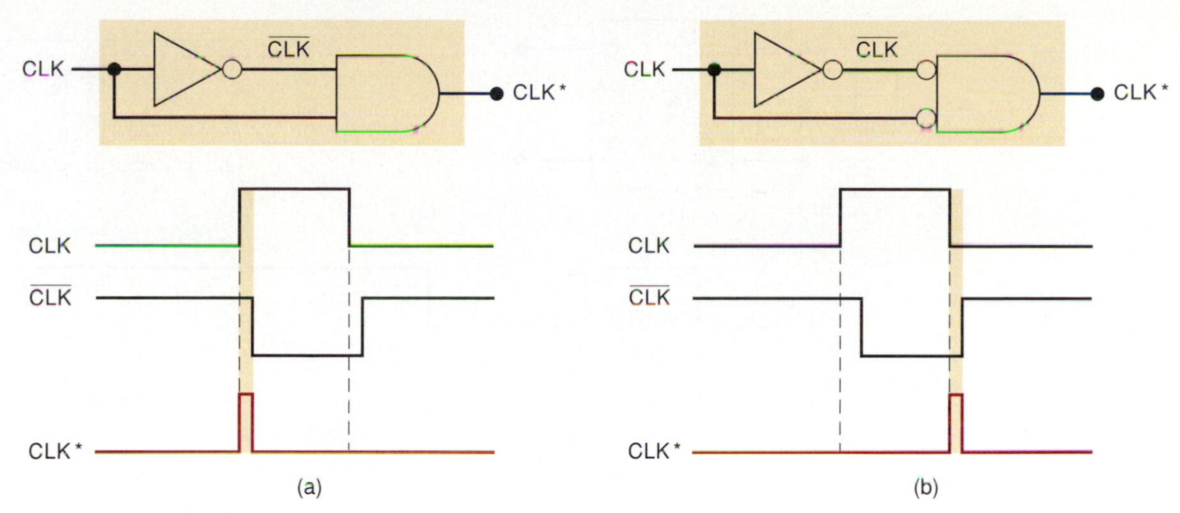

FIGURE 5-22 Implementation of edge-detector circuits used in edge-triggered flip-flops: (a) PGT; (b) NGT. The duration of the *CLK** pulses is typically 2–5 ns.

gate produces an output spike that is HIGH only for the few nanoseconds when *CLK* and $\overline{CLK}$ are both HIGH. The result is a narrow pulse at *CLK**, which occurs on the PGT of *CLK*. The arrangement of Figure 5-22(b) likewise produces *CLK** on the NGT of *CLK* for FFs that are to trigger on a NGT.

Because the *CLK** signal is HIGH for only a few nanoseconds, *Q* is affected by the levels at *S* and *R* only for a short time during and after the occurrence of the active edge of *CLK*. This is what gives the FF its edge-triggered property.

REVIEW QUESTIONS

1. Suppose that the waveforms of Figure 5-19(c) are applied to the inputs of the FF of Figure 5-20. What will happen to *Q* at point *b*? At point *f*? At point *h*?
2. Explain why the *S* and *R* inputs affect *Q* only during the active transition of *CLK*.

5-7 CLOCKED J-K FLIP-FLOP

Figure 5-23(a) shows a **clocked J-K flip-flop** that is triggered by the positive-going edge of the clock signal. The *J* and *K* inputs control the state of the FF in the same ways as the *S* and *R* inputs do for the clocked S-R flip-flop except for one major difference: *the* $J = K = 1$ *condition does not result in an ambiguous output.* For this 1, 1 condition, the FF will always go to its *opposite* state upon the positive transition of the clock signal. This is called the **toggle mode** of operation. In this mode, if both *J* and *K* are left HIGH, the FF will change states (toggle) for each PGT of the clock.

The function table in Figure 5-23(a) summarizes how the J-K flip-flop responds to the PGT for each combination of *J* and *K*. Notice that the function table is the same as for the clocked S-R flip-flop (Figure 5-19) except for the $J = K = 1$ condition. This condition results in $Q = \overline{Q}_0$, which means that the new value of *Q* will be the inverse of the value it had prior to the PGT; this is the toggle operation.

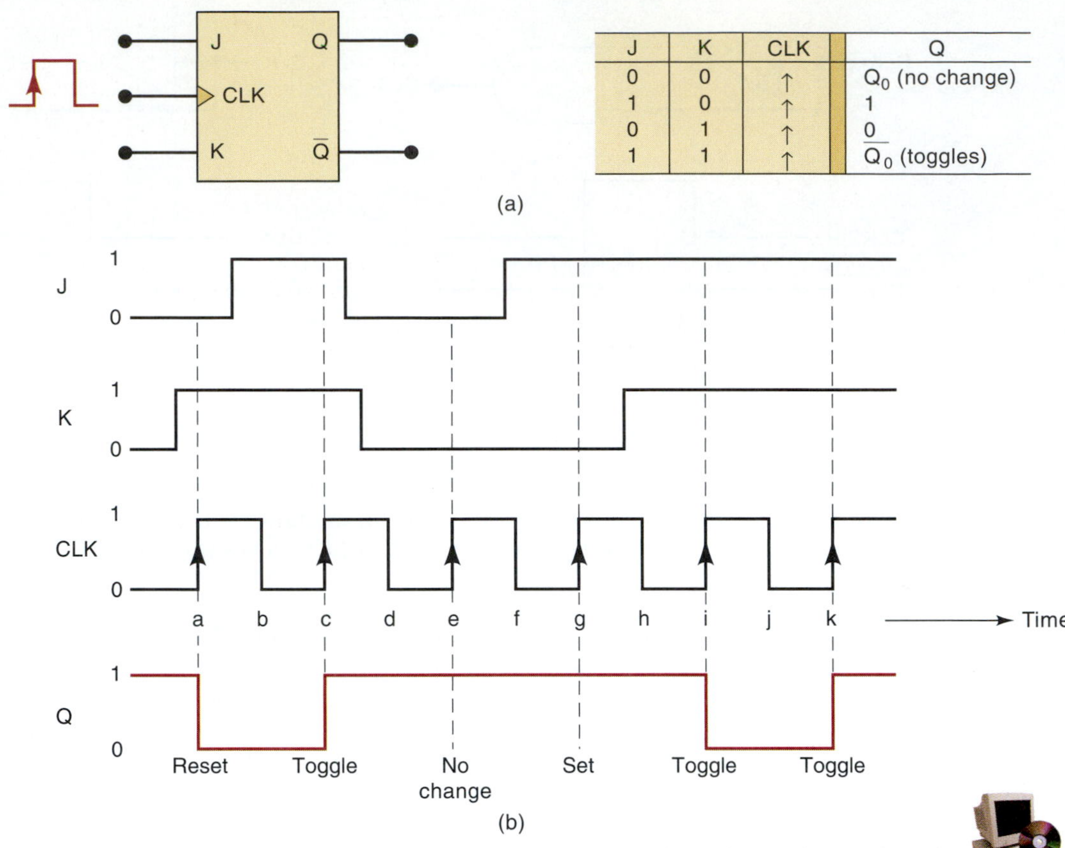

FIGURE 5-23 (a) Clocked J-K flip-flop that responds only to the positive edge of the clock; (b) waveforms.

The operation of this FF is illustrated by the waveforms in Figure 5-23(b). Once again, we assume that the setup and hold time requirements are being met.

1. Initially all inputs are 0, and the Q output is assumed to be 1; that is, $Q_0 = 1$.
2. When the positive-going edge of the first clock pulse occurs (point a), the $J = 0$, $K = 1$ condition exists. Thus, the FF will be reset to the $Q = 0$ state.
3. The second clock pulse finds $J = K = 1$ when it makes its positive transition (point c). This causes the FF to *toggle* to its opposite state, $Q = 1$.
4. At point e on the clock waveform, J and K are both 0, so that the FF does not change states on this transition.
5. At point g, $J = 1$ and $K = 0$. This is the condition that sets Q to the 1 state. However, it is already 1, and so it will remain there.
6. At point i, $J = K = 1$, and so the FF toggles to its opposite state. The same thing occurs at point k.

Note from these waveforms that the FF is not affected by the negative-going edge of the clock pulses. Also, the J and K input levels have no effect except upon the occurrence of the PGT of the clock signal. The J and K inputs by themselves cannot cause the FF to change states.

Figure 5-24 shows the symbol for a clocked J-K flip-flop that triggers on the negative-going clock-signal transitions. The small circle on the *CLK* input

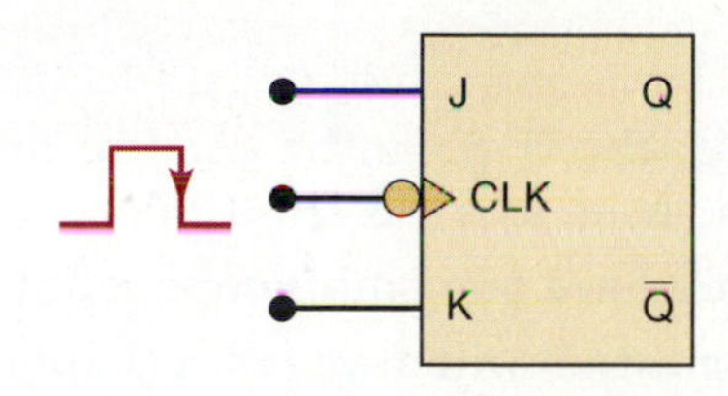

J	K	CLK	Q
0	0	↓	Q_0 (no change)
1	0	↓	1
0	1	↓	0
1	1	↓	$\overline{Q_0}$ (toggles)

FIGURE 5-24 J-K flip-flop that triggers only on negative-going transitions.

indicates that this FF will trigger when the *CLK* input goes from 1 to 0. This FF operates in the same manner as the positive-edge FF of Figure 5-23 except that the output can change states only on negative-going clock-signal transitions (points *b, d, f, h,* and *j*). Both polarities of edge-triggered J-K flip-flops are in common usage.

The J-K flip-flop is much more versatile than the S-R flip-flop because it has no ambiguous states. The $J = K = 1$ condition, which produces the toggling operation, finds extensive use in all types of binary counters. In essence, the J-K flip-flop can do anything the S-R flip-flop can do *plus* operate in the toggle mode.

Internal Circuitry of the Edge-Triggered J-K Flip-Flop

A simplified version of the internal circuitry of an edge-triggered J-K flip-flop is shown in Figure 5-25. It contains the same three sections as the edge-triggered S-R flip-flop (Figure 5-21). In fact, the only difference between the two circuits is that the Q and $\overline{Q}$ outputs are fed back to the pulse-steering NAND gates. This feedback connection is what gives the J-K flip-flop its toggle operation for the $J = K = 1$ condition.

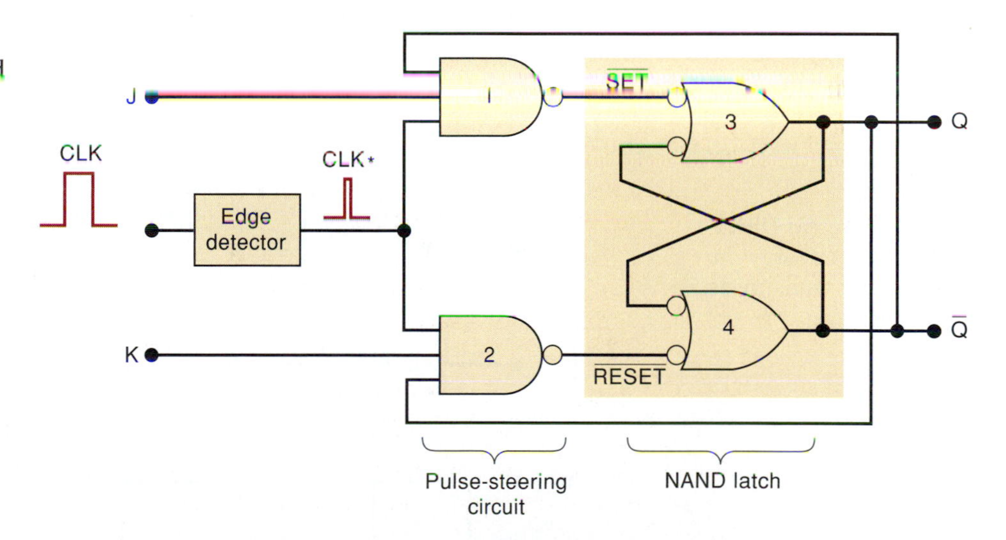

FIGURE 5-25 Internal circuit of the edge-triggered J-K flip-flop.

Let's examine this toggle condition more closely by assuming that $J = K = 1$ and that Q is sitting in the LOW state when a *CLK* pulse occurs. With $Q = 0$ and $\overline{Q} = 1$, NAND gate 1 will steer *CLK** (inverted) to the $\overline{\text{SET}}$ input of the NAND latch to produce $Q = 1$. If we assume that Q is HIGH when a *CLK* pulse occurs, NAND gate 2 will steer *CLK** (inverted) to the $\overline{\text{RESET}}$ input of the latch to produce $Q = 0$. Thus, Q always ends up in the opposite state.

In order for the toggle operation to work as described above, the *CLK** pulse must be very narrow. It must return to 0 before the Q and $\overline{Q}$ outputs toggle to their new values; otherwise, the new values of Q and $\overline{Q}$ will cause the *CLK** pulse to toggle the latch outputs again.

REVIEW QUESTIONS

1. *True or false:* A J-K flip-flop can be used as an S-R flip-flop, but an S-R flip-flop cannot be used as a J-K flip-flop.
2. Does a J-K flip-flop have any ambiguous input conditions?
3. What *J-K* input condition will always set *Q* upon the occurrence of the active *CLK* transition?

5-8 CLOCKED D FLIP-FLOP

Figure 5-26(a) shows the symbol and the function table for a **clocked D flip-flop** that triggers on a PGT. Unlike the S-R and J-K flip-flops, this flip-flop has only one synchronous control input, *D*, which stands for *data.* The operation of the D flip-flop is very simple: *Q* will go to the same state that is present on the *D* input when a PGT occurs at *CLK.* In other words, the level present at *D* will be *stored* in the flip-flop at the instant the PGT occurs. The waveforms in Figure 5-26(b) illustrate this operation.

Assume that *Q* is initially HIGH. When the first PGT occurs at point *a*, the *D* input is LOW; thus, *Q* will go to the 0 state. Even though the *D* input level changes between points *a* and *b*, it has no effect on *Q*; *Q* is storing the LOW that was on *D* at point *a*. When the PGT at *b* occurs, *Q* goes HIGH because *D* is HIGH at that time. *Q* stores this HIGH until the PGT at point *c* causes *Q* to go LOW because *D* is LOW at that time. In a similar manner, the *Q* output takes on the levels present at *D* when the PGTs occur at points *d*, *e*, *f*, and *g*. Note that *Q* stays HIGH at point *e* because *D* is still HIGH.

Again, it is important to remember that *Q* can change only when a PGT occurs. The *D* input has no effect between PGTs.

A negative-edge-triggered D flip-flop operates in the same way just described except that *Q* will take on the value of *D* when a NGT occurs at *CLK.* The symbol for the D flip-flop that triggers on NGTs will have a bubble on the *CLK* input.

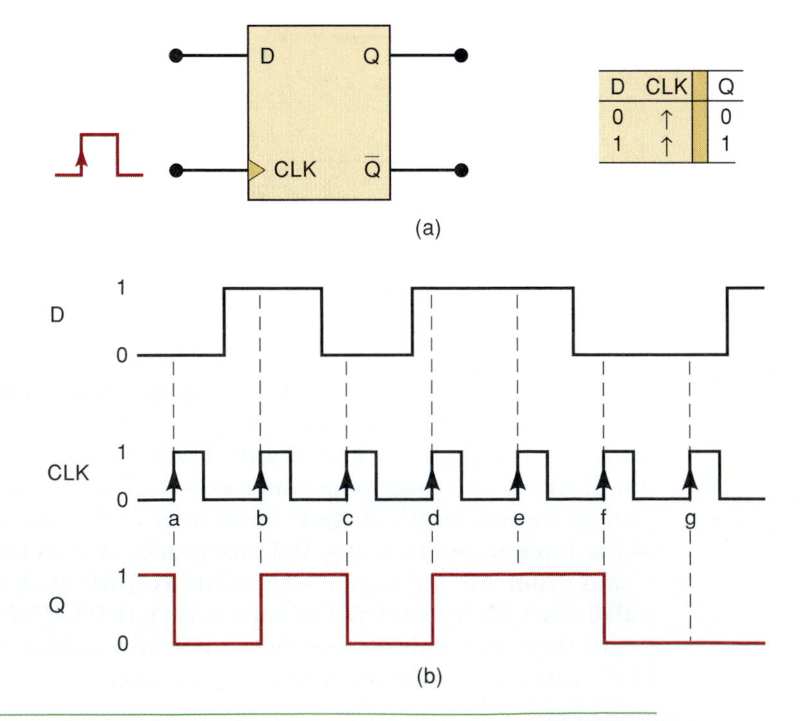

FIGURE 5-26 (a) D flip-flop that triggers only on positive-going transitions; (b) waveforms.

Implementation of the D Flip-Flop

An edge-triggered D flip-flop is easily implemented by adding a single INVERTER to the edge-triggered J-K flip-flop, as shown in Figure 5-27. If you try both values of D, you should see that Q takes on the level present at D when a PGT occurs. The same can be done to convert a S-R flip-flop to a D flip-flop.

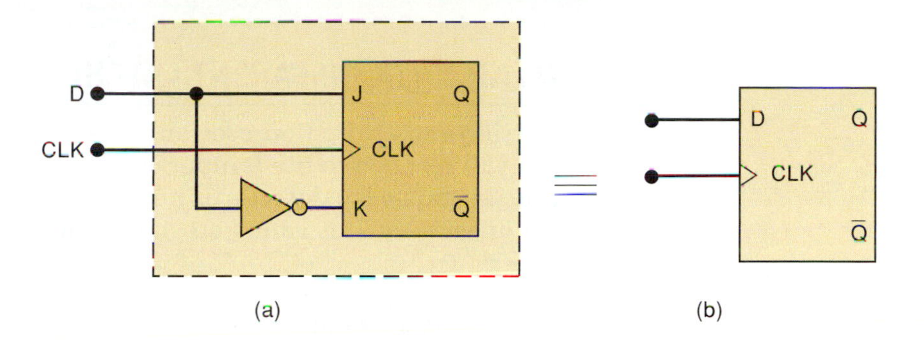

FIGURE 5-27 Edge-triggered D flip-flop implementation from a J-K flip-flop.

Parallel Data Transfer

At this point you may well be wondering about the usefulness of the D flip-flop because it appears that the Q output is the same as the D input. Not quite; remember, Q takes on the value of D only at certain time instances, and so it is not identical to D (e.g., see the waveforms in Figure 5-26).

In most applications of the D flip-flop, the Q output must take on the value at its D input only at precisely defined times. One example of this is illustrated in Figure 5-28. Outputs X, Y, Z from a logic circuit are to be transferred to FFs Q_1, Q_2, and Q_3 for storage. Using the D flip-flops, the levels present at X, Y, and Z will be transferred to Q_1, Q_2, and Q_3, respectively, upon application of a TRANSFER pulse to the common CLK inputs. The FFs can store these values for subsequent processing. This is an example of **parallel data transfer** of binary data; the three bits X, Y, and Z are all transferred *simultaneously*.

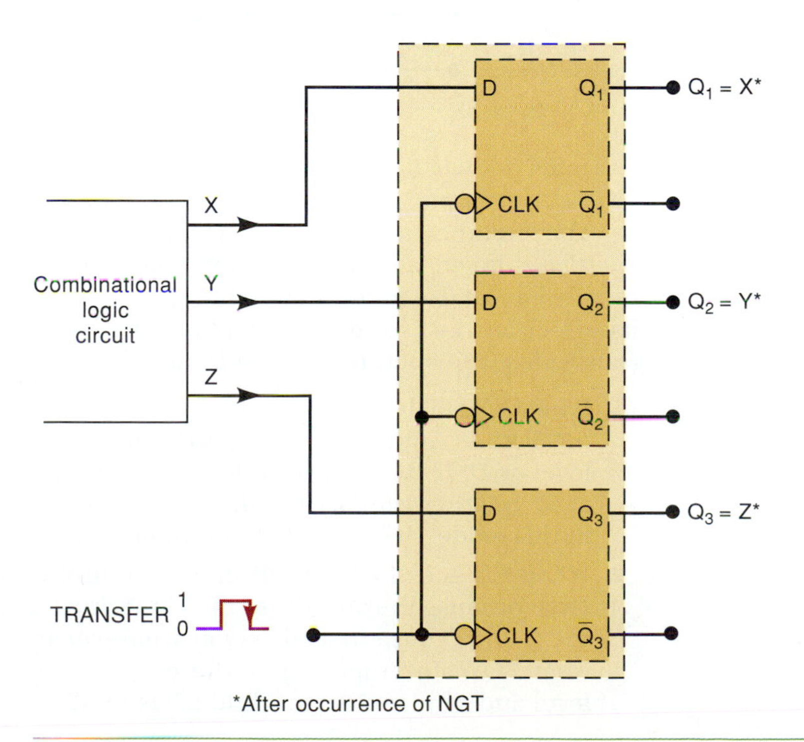

FIGURE 5-28 Parallel transfer of binary data using D flip-flops.

REVIEW QUESTIONS

1. What will happen to the *Q* waveform in Figure 5-26(b) if the *D* input is held permanently LOW?
2. *True or false:* The *Q* output will equal the level at the *D* input at all times.
3. Can J-K FFs be used for parallel data transfer?

5-9 *D* LATCH (TRANSPARENT LATCH)

The edge-triggered D flip-flop uses an edge-detector circuit to ensure that the output will respond to the *D* input *only* when the active transition of the clock occurs. If this edge detector is not used, the resultant circuit operates somewhat differently. It is called a ***D* latch** and has the arrangement shown in Figure 5-29(a).

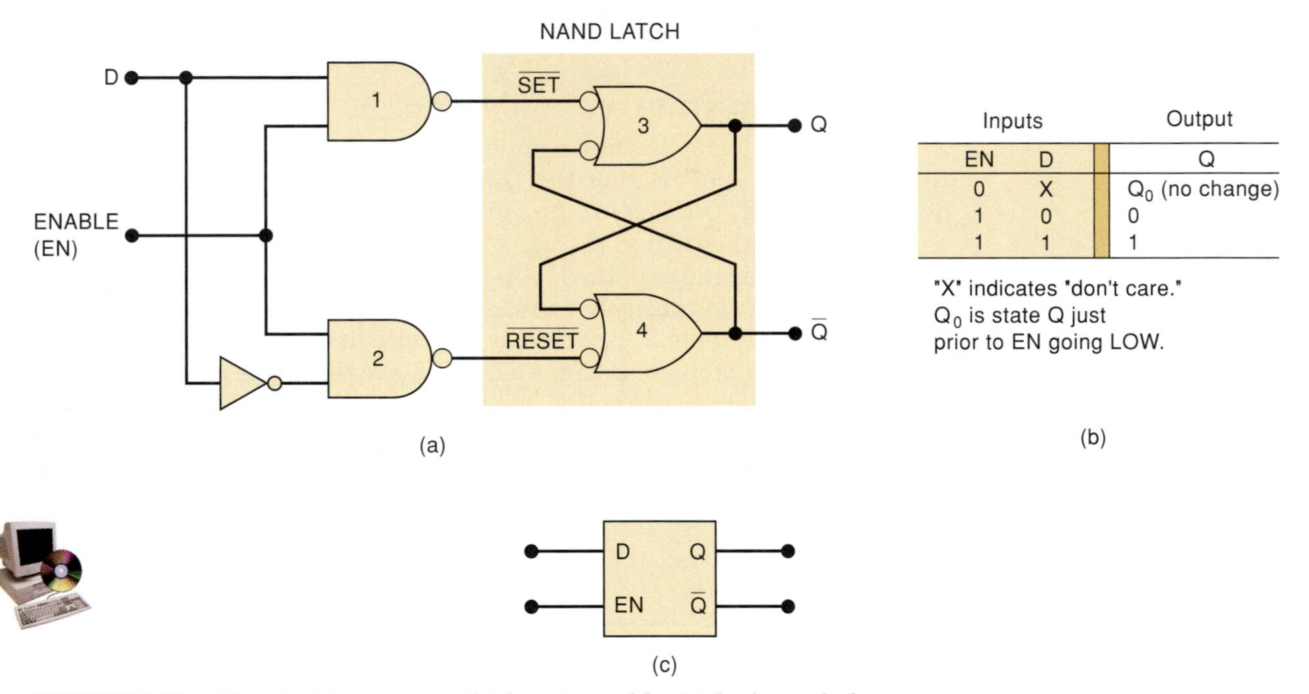

EN	D	Q
0	X	Q_0 (no change)
1	0	0
1	1	1

FIGURE 5-29 *D* latch: (a) structure; (b) function table; (c) logic symbol.

The circuit contains the NAND latch and the steering NAND gates 1 and 2 *without* the edge-detector circuit. The common input to the steering gates is called an *enable* input (abbreviated *EN*) rather than a clock input because its effect on the *Q* and $\overline{Q}$ outputs is not restricted to occurring only on its transitions. The operation of the *D* latch is described as follows:

1. When *EN* is HIGH, the *D* input will produce a LOW at either the $\overline{SET}$ or the $\overline{RESET}$ inputs of the NAND latch to cause *Q* to become the same level as *D*. If *D* changes while *EN* is HIGH, *Q* will follow the changes exactly. In other words, while $EN = 1$, the *Q* output will look exactly like *D*; in this mode, the *D* latch is said to be "transparent."
2. When *EN* goes LOW, the *D* input is inhibited from affecting the NAND latch because the outputs of both steering gates will be held HIGH. Thus, the *Q* and $\overline{Q}$ outputs will stay at whatever level they had just before *EN* went LOW. In other words, the outputs are "latched" to their current level and cannot change while *EN* is LOW even if *D* changes.

This operation is summarized in the function table in Figure 5-29(b). The logic symbol for the *D* latch is given in Figure 5-29(c). Note that even though the *EN* input operates much like the *CLK* input of an edge-triggered FF, there is no small triangle on the *EN* input. This is because the small triangle symbol is used strictly for inputs that can cause an output change only when a transition occurs. *The D latch is not edge-triggered.*

EXAMPLE 5-8

Determine the Q waveform for a *D* latch with the *EN* and *D* inputs of Figure 5-30. Assume that $Q = 0$ initially.

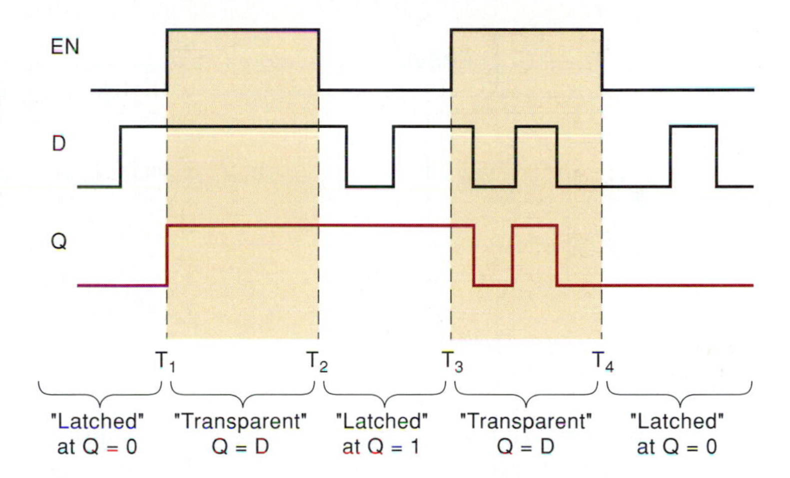

FIGURE 5-30 Waveforms for Example 5-8 showing the two modes of operation of the transparent *D* latch.

Solution

Prior to time T_1, *EN* is LOW, so that Q is "latched" at its current 0 level and cannot change even though *D* is changing. During the interval T_1 to T_2, *EN* is HIGH so that Q will follow the signal present at *D*. Thus, Q goes HIGH at T_1 and stays there because *D* is not changing. When *EN* returns LOW at T_2, Q will latch at the HIGH level that it has at T_2 and will remain there while *EN* is LOW.

At T_3 when *EN* goes HIGH again, Q will follow the changes in the *D* input until T_4 when *EN* returns LOW. During the interval T_3 to T_4, the *D* latch is "transparent" because the variations in *D* go through to the output Q. At T_4 when *EN* goes LOW, Q will latch at the 0 level because that is its level at T_4. After T_4 the variations in *D* will have no effect on Q because it is latched (i.e., $EN = 0$).

REVIEW QUESTIONS

1. Describe how a *D* latch operates differently from an edge-triggered D flip-flop.
2. *True or false:* A *D* latch is in its transparent mode when $EN = 0$.
3. *True or false:* In a *D* latch, the *D* input can affect Q only when $EN = 1$.

5-10 ASYNCHRONOUS INPUTS

For the clocked flip-flops that we have been studying, the *S*, *R*, *J*, *K*, and *D* inputs have been referred to as *control* inputs. These inputs are also called synchronous inputs because their effect on the FF output is synchronized with the *CLK* input. As we have seen, the synchronous control inputs must be used in conjunction with a clock signal to trigger the FF.

Most clocked FFs also have one or more **asynchronous inputs** that operate independently of the synchronous inputs and clock input. These asynchronous inputs can be used to set the FF to the 1 state or clear (reset) the FF to the 0 state *at any time, regardless of the conditions at the other inputs.* Stated in another way, the asynchronous inputs are **override inputs**, which can be used to override all the other inputs in order to place the FF in one state or the other.

Figure 5-31 shows a J-K flip-flop with two asynchronous inputs designated as $\overline{\text{PRESET}}$ and $\overline{\text{CLEAR}}$. These are active-LOW inputs, as indicated by the bubbles on the FF symbol. The accompanying function table summarizes how they affect the FF output. Let's examine the various cases.

FIGURE 5-31 Clocked J-K flip-flop with asynchronous inputs.

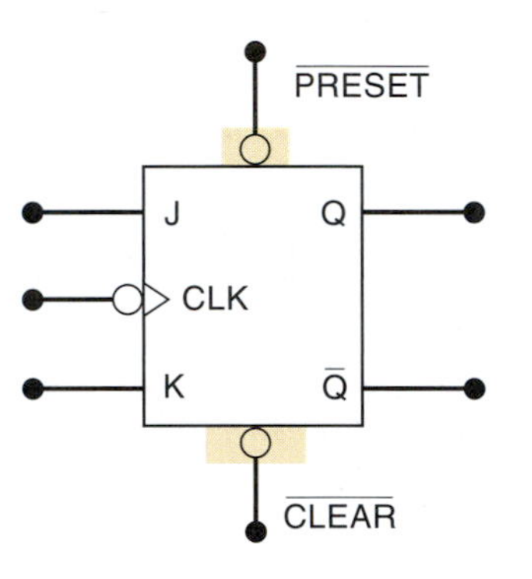

J	K	Clk	$\overline{\text{PRE}}$	$\overline{\text{CLR}}$	Q
0	0	↓	1	1	Q (no change)
0	1	↓	1	1	0 (Synch reset)
1	0	↓	1	1	1 (Synch set)
1	1	↓	1	1	$\overline{\text{Q}}$ (Synch toggle)
x	x	x	1	1	Q (no change)
x	x	x	1	0	0 (asynch clear)
x	x	x	0	1	1 (asynch preset)
x	x	x	0	0	(Invalid)

- $\overline{\text{PRESET}} = \overline{\text{CLEAR}} = 1$. The asynchronous inputs are inactive and the FF is free to respond to the *J*, *K*, and *CLK* inputs; in other words, the clocked operation can take place.
- $\overline{\text{PRESET}} = 0$; $\overline{\text{CLEAR}} = 1$. The $\overline{\text{PRESET}}$ is activated and *Q* is *immediately* set to 1 no matter what conditions are present at the *J*, *K*, and *CLK* inputs. The *CLK* input cannot affect the FF while $\overline{\text{PRESET}} = 0$.
- $\overline{\text{PRESET}} = 1$; $\overline{\text{CLEAR}} = 0$. The $\overline{\text{CLEAR}}$ is activated and *Q* is *immediately* cleared to 0 independent of the conditions on the *J*, *K*, or *CLK* inputs. The *CLK* input has no effect while $\overline{\text{CLEAR}} = 0$.
- $\overline{\text{PRESET}} = \overline{\text{CLEAR}} = 0$. This condition should not be used because it can result in an ambiguous response.

It is important to realize that these asynchronous inputs respond to dc levels. This means that if a constant 0 is held on the $\overline{\text{PRESET}}$ input, the FF will remain in the $Q = 1$ state regardless of what is occurring at the other inputs. Similarly, a constant LOW on the $\overline{\text{CLEAR}}$ input holds the FF in the $Q = 0$ state. Thus, the asynchronous inputs can be used to hold the FF in a particular state for any desired interval. Most often, however, the asynchronous inputs are used to set or clear the FF to the desired state by application of a momentary pulse.

Many clocked FFs that are available as ICs will have both of these asynchronous inputs; some will have only the $\overline{\text{CLEAR}}$ input. Some FFs will have asynchronous inputs that are active-HIGH rather than active-LOW. For these FFs the FF symbol would not have a bubble on the asynchronous inputs.

Designations for Asynchronous Inputs

IC manufacturers do not all agree on the nomenclature to use for these asynchronous inputs. The most common designations are *PRE* (short for PRESET) and *CLR* (short for CLEAR). These labels clearly distinguish them from the synchronous SET and RESET inputs. Other labels such as S_D (direct SET) and R_D (direct RESET) are also used. From now on, we will use the labels *PRE* and *CLR* to represent the asynchronous inputs because these seem to be the most commonly used labels. When these asynchronous inputs are active-LOW, as they generally are, we will use the overbar to indicate their active-LOW status, that is, $\overline{PRE}$ and $\overline{CLR}$.

Although most IC flip-flops have at least one or more asynchronous inputs, there are some circuit applications where they are not used. In such cases they are held permanently at their inactive level. Often, in our use of FFs throughout the remainder of the text, we will not show a FF's unused asynchronous inputs; it will be assumed that they are permanently connected to their inactive logic level.

EXAMPLE 5-9

Figure 5-32(a) shows the symbol for a J-K FF that responds to a NGT on its clock input and has active-LOW asynchronous inputs. Before proceeding with the example, take note of the way the inputs are labeled. First, note that the clock signal applied to the FF is labeled $\overline{CLK}$ (the overbar indicates that this signal is active on the NGT), whereas on the other side of the bubble (inside the block), it is labeled *CLK*. Likewise, the external active-LOW asynchronous

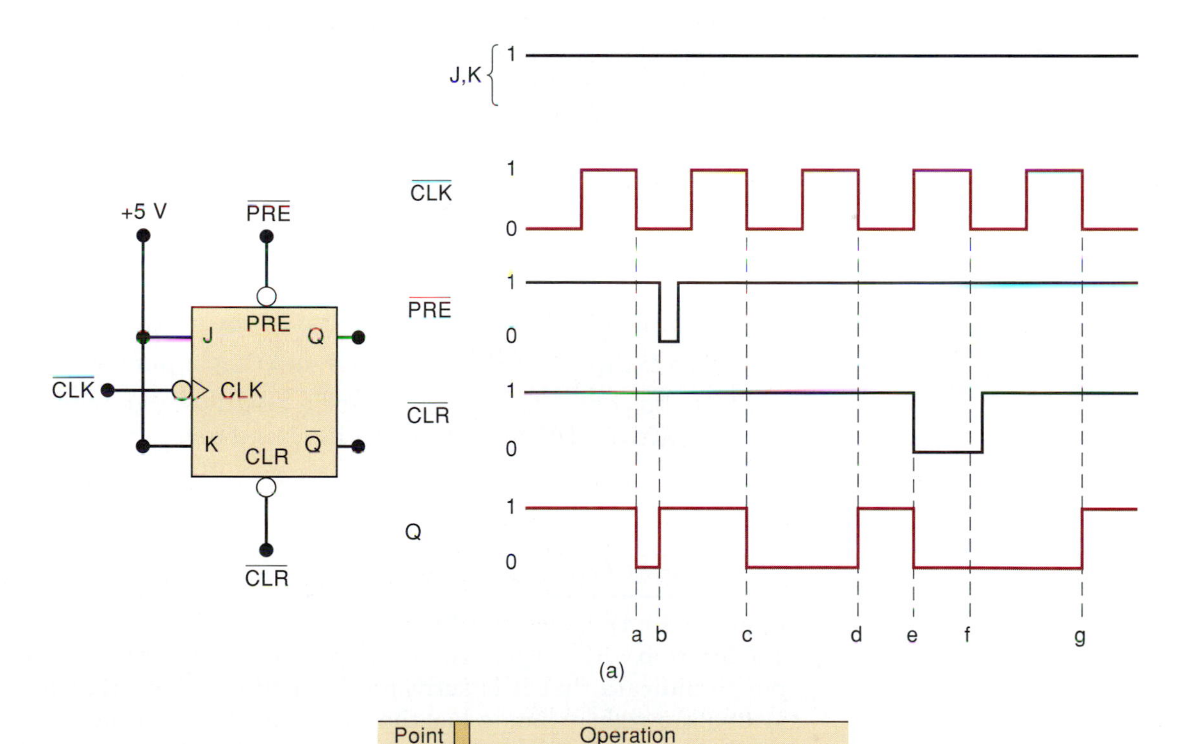

(a)

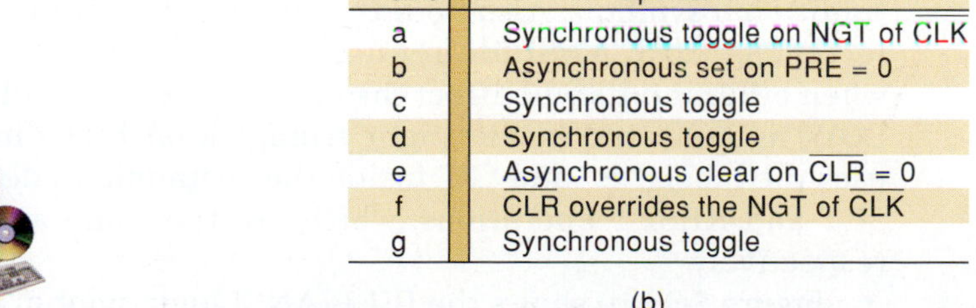

Point	Operation
a	Synchronous toggle on NGT of $\overline{CLK}$
b	Asynchronous set on $\overline{PRE}$ = 0
c	Synchronous toggle
d	Synchronous toggle
e	Asynchronous clear on $\overline{CLR}$ = 0
f	$\overline{CLR}$ overrides the NGT of $\overline{CLK}$
g	Synchronous toggle

(b)

FIGURE 5-32 Waveforms for Example 5-9 showing how a clocked flip-flop responds to asynchronous inputs.

inputs are labeled $\overline{PRE}$ and $\overline{CLR}$, whereas inside the block on the other side of the bubble, they are labeled PRE and CLR. The important thing to remember is that the presence of the bubble on an input means that the input responds to a logic LOW signal.

The J and K inputs are shown tied HIGH for this example. Determine the Q output in response to the input waveforms shown in Figure 5-32(a). Assume that Q is initially HIGH.

Solution

Initially, $\overline{PRE}$ and $\overline{CLR}$ are in their inactive HIGH state, so that they will have no effect on Q. Thus, when the first NGT of the $\overline{CLK}$ signal occurs at point a, Q will toggle to its opposite state; remember, $J = K = 1$ produces the toggle operation.

At point b, the $\overline{PRE}$ input is pulsed to its active-LOW state. This will *immediately* set $Q = 1$. Note that $\overline{PRE}$ produces $Q = 1$ without waiting for a NGT at $\overline{CLK}$. The asynchronous inputs operate independently of $\overline{CLK}$.

At point c, the NGT of $\overline{CLK}$ will again cause Q to toggle to its opposite state. Note that $\overline{PRE}$ has returned to its inactive state prior to point c. Likewise, the NGT of $\overline{CLK}$ at point d will toggle Q back HIGH.

At point e, the $\overline{CLR}$ input is pulsed to its active-LOW state and will *immediately* clear $Q = 0$. Again, it does this independently of $\overline{CLK}$.

The NGT of $\overline{CLK}$ at point f *will not toggle* Q because the $\overline{CLR}$ input is still active. The LOW at $\overline{CLR}$ overrides the $\overline{CLK}$ input and holds $Q = 0$.

When the NGT of $\overline{CLK}$ occurs at point g, it will toggle Q to the HIGH state because neither asynchronous input is active at that point.

These steps are summarized in Figure 5-32(b).

REVIEW QUESTIONS

1. How does the operation of an asynchronous input differ from that of a synchronous input?
2. Can a D flip-flop respond to its D and CLK inputs while $\overline{PRE} = 1$?
3. List the conditions necessary for a positive-edge-triggered J-K flip-flop with active-LOW asynchronous inputs to toggle to its opposite state.

5-11 IEEE/ANSI SYMBOLS

Figure 5-33(a) shows the IEEE/ANSI symbol for a negative-edge-triggered J-K flip-flop with asynchronous inputs. Note the right triangle on the CLK input to indicate that it is activated by a NGT. Recall that in the IEEE/ANSI symbols, a right triangle has the same meanings as the small bubble in the traditional symbols. Also note that the clock input is labeled "C" inside the rectangle. IEEE/ANSI always uses a "C" to denote any input that *controls* when other inputs will affect the output. The $\overline{PRE}$ and $\overline{CLR}$ inputs are active-LOW, as indicated by the right triangles on these inputs. IEEE/ANSI also uses the labels "S" and "R" inside the rectangle to denote the asynchronous SET and RESET operations, which are the same as PRESET and CLEAR, respectively.

Figure 5-33(b) shows the IEEE/ANSI logic symbol for an IC that is part of the 74LS series of TTL devices. The 74LS112 is a dual negative-edge-triggered J-K flip-flop with preset and clear capabilities. It contains two J-K flip-flops,

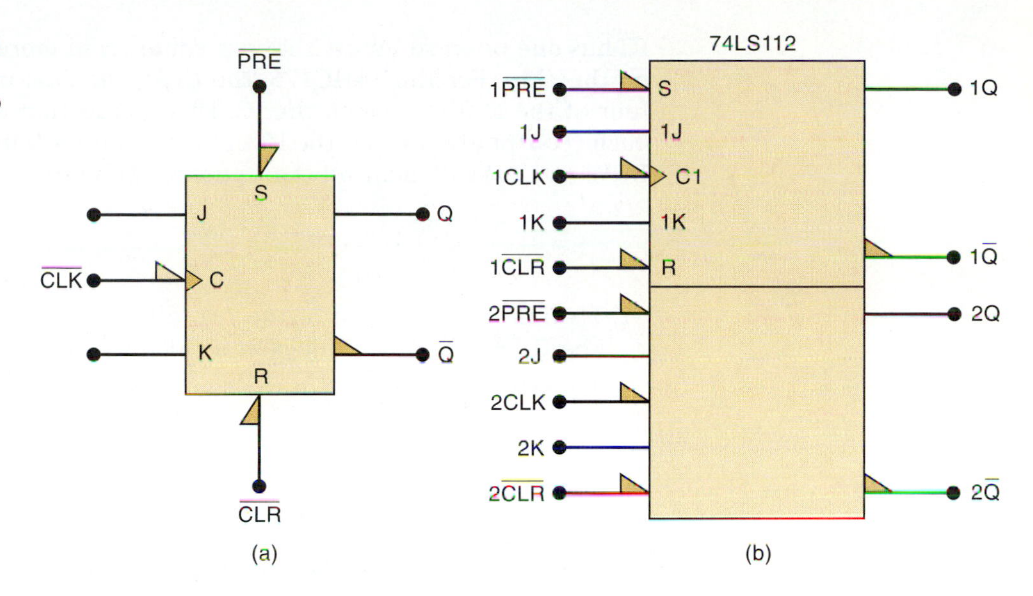

FIGURE 5-33 IEEE/ANSI symbols for (a) a single edge-triggered J-K flip-flop and (b) an actual IC (74LS112 dual negative-edge-triggered J-K flip-flop).

like the one symbolized in Figure 5-33(a). Note how the inputs and outputs are numbered. Also note that the input labels inside the rectangles are shown only for the top FF. It is understood that the inputs to the bottom FF are in the same arrangement as the top one. This same IC symbol applies to the CMOS 74HC112.

Figure 5-34(a) is the IEEE/ANSI symbol for a positive-edge-triggered D flip-flop with asynchronous inputs. There is no right triangle on the clock input because this FF is clocked by PGTs.

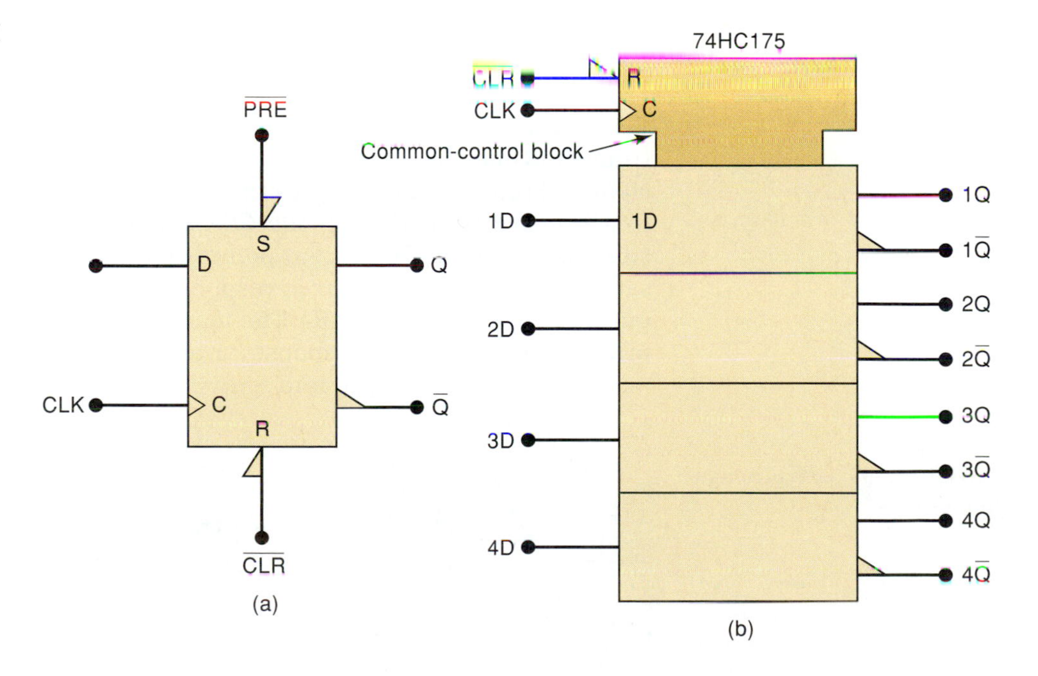

FIGURE 5-34 IEEE/ANSI symbols for (a) a single edge-triggered D flip-flop and (b) an actual IC (74HC175 quad flip-flop with common clock and clear).

Figure 5-34(b) is the IEEE/ANSI symbol for a 74HC175 IC, which contains four D flip-flops that share a common *CLK* input and a common $\overline{CLR}$ input. The FFs do not have a $\overline{PRE}$ input. This symbol contains a separate rectangle to represent each FF, and a special **common-control block**, which is the notched rectangle on top. The common-control block is used whenever an

IC has one or more inputs that are common to more than one of the circuits on the chip. For the 74HC175, the *CLK* and $\overline{CLR}$ inputs are common to all four of the D flip-flops on the IC. This means that a PGT on *CLK* will cause each *Q* output to take on the level present at its *D* input; it also means that a LOW on $\overline{CLR}$ will clear all *Q* outputs to the LOW state.

REVIEW QUESTIONS

1. Explain the meaning of the two different triangles that can be part of the IEEE/ANSI symbology at a clock input.
2. Describe the meaning of the common-control block.

5-12 FLIP-FLOP TIMING CONSIDERATIONS

Manufacturers of IC flip-flops will specify several important timing parameters and characteristics that must be considered before a FF is used in any circuit application. We will describe the most important of these and then give some actual examples of specific IC flip-flops from the TTL and CMOS logic families.

Setup and Hold Times

The setup and hold times have already been discussed, and you may recall from Section 5-5 that they represent requirements that must be met for reliable FF triggering. The manufacturer's IC data sheet will always specify the *minimum* values of t_S and t_H.

Propagation Delays

Whenever a signal is to change the state of a FF's output, there is a delay from the time the signal is applied to the time when the output makes its change. Figure 5-35 illustrates the **propagation delays** that occur in response to a positive transition on the *CLK* input. Note that these delays are measured between the 50 percent points on the input and output waveforms. The same types of delays occur in response to signals on a FF's asynchronous inputs (PRESET and CLEAR). The manufacturers' data sheets usually specify propagation delays in response to all inputs, and they usually specify the *maximum* values for t_{PLH} and t_{PHL}.

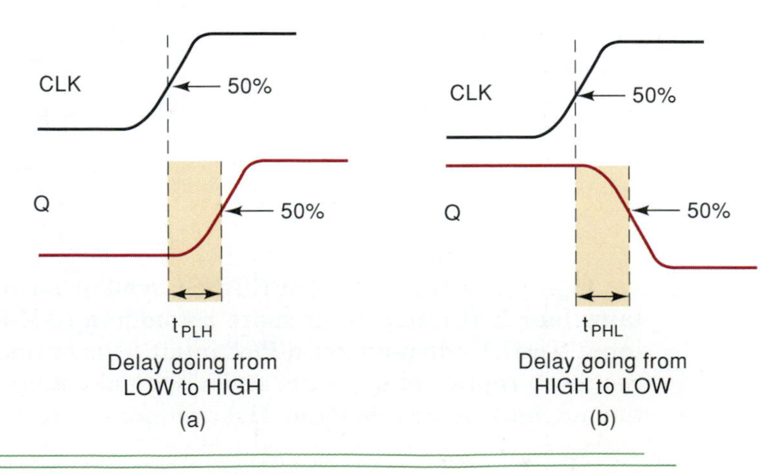

FIGURE 5-35 FF propagation delays.

Modern IC flip-flops have propagation delays that range from a few nanoseconds to around 100 ns. The values of t_{PLH} and t_{PHL} are generally not the same, and they increase in direct proportion to the number of loads being driven by the Q output. FF propagation delays play an important part in certain situations that we will encounter later.

Maximum Clocking Frequency, f_{MAX}

This is the highest frequency that may be applied to the *CLK* input of a FF and still have it trigger reliably. The f_{MAX} limit will vary from FF to FF, even with FFs having the same device number. For example, the manufacturer of the 7470 J-K flip-flop IC tests many of these FFs and may find that the f_{MAX} values fall in the range of 20 to 35 MHz. He will then specify the *minimum* f_{MAX} as 20 MHz. This may seem confusing, but a little thought should make it clear that what the manufacturer is saying is that he cannot guarantee that the 7470 FF that you put in your circuit will work above 20 MHz; most of them will, but some of them will not. If you operate them below 20 MHz, however, he guarantees that they will all work.

Clock Pulse HIGH and LOW Times

The manufacturer will also specify the *minimum* time duration that the *CLK* signal must remain LOW before it goes HIGH, sometimes called $t_W(L)$, and the *minimum* time that *CLK* must be kept HIGH before it returns LOW, sometimes called $t_W(H)$. These times are defined in Figure 5-36(a). Failure to meet these minimum time requirements can result in unreliable triggering. Note that these time values are measured between the halfway points on the signal transitions.

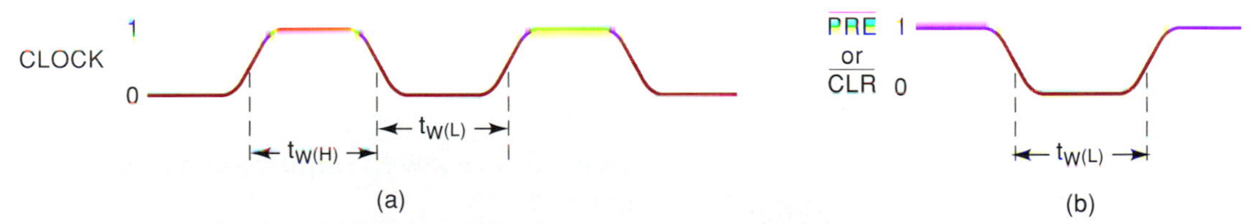

FIGURE 5-36 (a) Clock LOW and HIGH times; (b) asynchronous pulse width.

Asynchronous Active Pulse Width

The manufacturer will also specify the *minimum* time duration that a PRESET or CLEAR input must be kept in its active state in order to set or clear the FF reliably. Figure 5-36(b) shows $t_W(L)$ for active-LOW asynchronous inputs.

Clock Transition Times

For reliable triggering, the clock waveform transition times (rise and fall times) should be kept very short. If the clock signal takes too long to make the transitions from one level to the other, the FF may trigger erratically or not at all. Manufacturers usually do not list a maximum transition time requirement for each FF integrated circuit. Instead, it is usually given as a general requirement for all ICs within a given logic family. For example, the transition times should generally be $\leq$50 ns for TTL devices and $\leq$200 ns for CMOS. These requirements will vary among the different manufacturers and among the various subfamilies within the broad TTL and CMOS logic families.

Actual ICs

As practical examples of these timing parameters, let's take a look at several actual integrated-circuit FFs. In particular, we will look at the following ICs:

7474	Dual edge-triggered D flip-flop (standard TTL)
74LS112	Dual edge-triggered J-K flip-flop (low-power Schottky TTL)
74C74	Dual edge-triggered D flip-flop (metal-gate CMOS)
74HC112	Dual edge-triggered J-K flip-flop (high-speed CMOS)

Table 5-2 lists the various timing values for each of these FFs as they appear in the manufacturers' data books. All of the listed values are *minimum* values, except for the propagation delays, which are *maximum* values. Examination of Table 5-2 reveals two interesting points.

TABLE 5-2 Flip-flop timing values (in nanoseconds).

		TTL		CMOS	
		7474	74LS112	74C74	74HC112
t_S		20	20	60	25
t_H		5	0	0	0
t_{PHL}	from CLK to Q	40	24	200	31
t_{PLH}	from CLK to Q	25	16	200	31
t_{PHL}	from $\overline{CLR}$ to Q	40	24	225	41
t_{PLH}	from $\overline{PRE}$ to Q	25	16	225	41
$t_W(L)$	CLK LOW time	37	15	100	25
$t_W(H)$	CLK HIGH time	30	20	100	25
$t_W(L)$	at $\overline{PRE}$ or $\overline{CLR}$	30	15	60	25
f_{MAX}	in MHz	15	30	5	20

1. All of the FFs have very low t_H requirements; this is typical of most modern edge-triggered FFs.
2. The 74HC series of CMOS devices has timing values that are comparable to those of the TTL devices. The 74C series is much slower than the 74HC series.

EXAMPLE 5-10

From Table 5-2 determine the following.

(a) Assume that $Q = 0$. How long can it take for Q to go HIGH when a PGT occurs at the CLK input of a 7474?

(b) Assume that $Q = 1$. How long can it take for Q to go LOW in response to the $\overline{CLR}$ input of a 74HC112?

(c) What is the narrowest pulse that should be applied to the $\overline{CLR}$ input of the 74LS112 FF to clear Q reliably?

(d) Which FF in Table 5-2 requires that the control inputs remain stable *after* the occurrence of the active clock transition?

(e) For which FFs must the control inputs be held stable for a minimum time prior to the active clock transition?

Solution

(a) The PGT will cause Q to go from LOW to HIGH. The delay from CLK to Q is listed as $t_{PLH} = 25$ ns for the 7474.

(b) For the 74HC112, the time required for Q to go from HIGH to LOW in response to the $\overline{CLR}$ input is listed as $t_{PHL} = 41$ ns.

(c) For the 74LS112, the narrowest pulse at the $\overline{CLR}$ input is listed as $t_W(L) = 15$ ns.

(d) The 7474 is the only FF in Table 5-2 that has a nonzero hold time requirement.

(e) All of the FFs have a nonzero setup time requirement.

REVIEW QUESTIONS

1. Which FF timing parameters indicate the time it takes the Q output to respond to an input?
2. *True or false:* A FF that has an f_{MAX} rating of 25 MHz can be reliably triggered by any CLK pulse waveform with a frequency below 25 MHz.

5-13 POTENTIAL TIMING PROBLEM IN FF CIRCUITS

In many digital circuits, the output of one FF is connected either directly or through logic gates to the input of another FF, and both FFs are triggered by the same clock signal. This presents a potential timing problem. A typical situation is illustrated in Figure 5-37, where the output of Q_1 is connected to the J input of Q_2 and both FFs are clocked by the same signal at their CLK inputs.

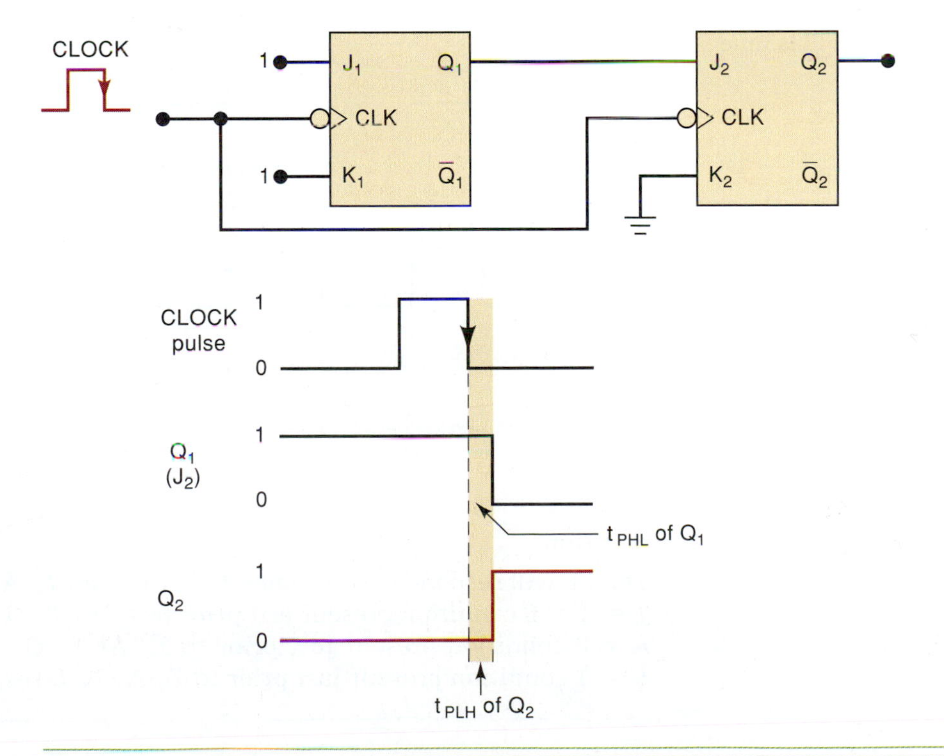

FIGURE 5-37 Q_2 will respond properly to the level present at Q_1 prior to the NGT of *CLK*, provided that Q_2's hold time requirement, t_H, is less than Q_1's propagation delay.

The potential timing problem is this: because Q_1 will change on the NGT of the clock pulse, the J_2 input of Q_2 will be changing as it receives the same NGT. This could lead to an unpredictable response at Q_2.

Let's assume that initially $Q_1 = 1$ and $Q_2 = 0$. Thus, the Q_1 FF has $J_1 = K_1 = 1$, and Q_2 has $J_2 = Q_1 = 1$, $K_2 = 0$ prior to the NGT of the clock pulse. When the NGT occurs, Q_1 will toggle to the LOW state, but it will not actually go LOW until after its propagation delay, t_{PHL}. The same NGT will reliably clock Q_2 to the HIGH state provided that t_{PHL} is greater than Q_2's hold time requirement, t_H. If this condition is not met, the response of Q_2 will be unpredictable.

Fortunately, all modern edge-triggered FFs have hold time requirements that are 5 ns or less; most have $t_H = 0$, which means that they have no hold time requirement. For these FFs, situations like that in Figure 5-37 will not be a problem.

Unless stated otherwise, in all of the FF circuits that we encounter throughout the text, we will assume that the FF's hold time requirement is short enough to respond reliably according to the following rule:

The FF output will go to a state determined by the logic levels present at its synchronous control inputs just prior to the active clock transition.

If we apply this rule to Figure 5-37, it says that Q_2 will go to a state determined by the $J_2 = 1$, $K_2 = 0$ condition that is present just prior to the NGT of the clock pulse. The fact that J_2 is changing in response to the same NGT has no effect.

EXAMPLE 5-11

Determine the Q output for a negative-edge-triggered J-K flip-flop for the input waveforms shown in Figure 5-38. Assume that $t_H = 0$ and that $Q = 0$ initially.

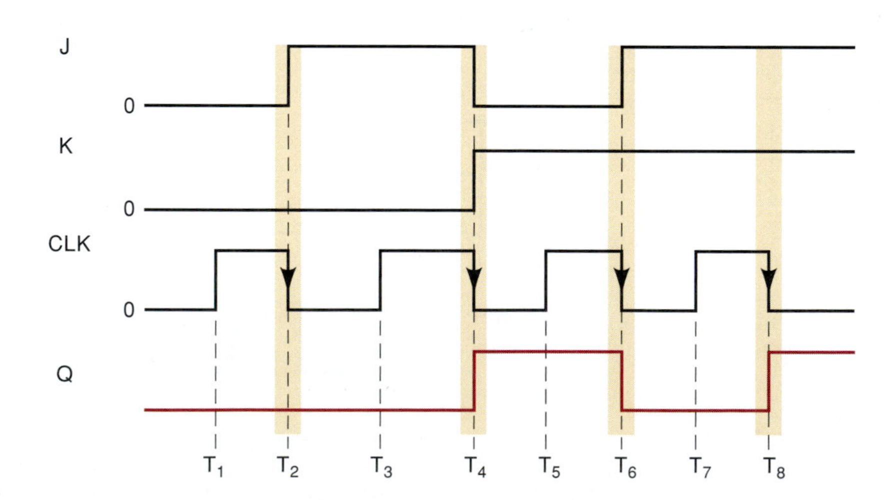

FIGURE 5-38 Example 5-11.

Solution

The FF will respond only at times T_2, T_4, T_6, and T_8. At T_2, Q will respond to the $J = K = 0$ condition present just prior to T_2. At T_4, Q will respond to the $J = 1$, $K = 0$ condition present just prior to T_4. At T_6, Q will respond to the $J = 0$, $K = 1$ condition present just prior to T_6. At T_8, Q responds to $J = K = 1$.

5-14 FLIP-FLOP APPLICATIONS

Edge-triggered (clocked) flip-flops are versatile devices that can be used in a wide variety of applications including counting, storing of binary data, transferring binary data from one location to another, and many more. Almost all of these applications utilize the FF's clocked operation. Many of them fall into the category of **sequential circuits**. A sequential circuit is one in which the outputs follow a predetermined sequence of states, with a new state occurring each time a clock pulse occurs. We will introduce some of the basic applications in the following sections, and we will expand on them in subsequent chapters.

5-15 FLIP-FLOP SYNCHRONIZATION

Most digital systems are principally synchronous in their operation because most of the signals will change states in synchronism with the clock transitions. In many cases, however, there will be an external signal that is not synchronized to the clock; in other words, it is asynchronous. Asynchronous signals often occur as a result of a human operator's actuating an input switch at some random time relative to the clock signal. This randomness can produce unpredictable and undesirable results. The following example illustrates how a FF can be used to synchronize the effect of an asynchronous input.

EXAMPLE 5-12

Figure 5-39(a) shows a situation where input signal A is generated from a debounced switch that is actuated by an operator (a debounced switch was first introduced in Example 5-2). A goes HIGH when the operator actuates the switch and goes LOW when the operator releases the switch. This A input is used to control the passage of the clock signal through the AND gate so that clock pulses appear at output X only as long as A is HIGH.

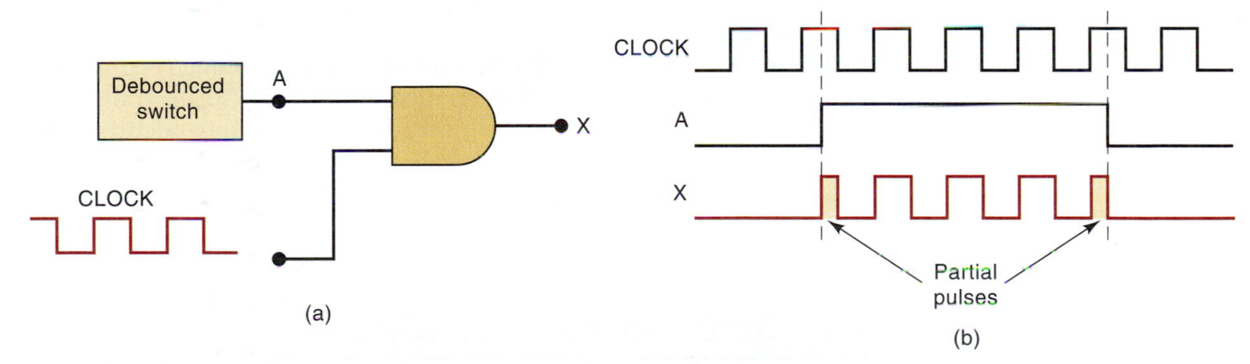

FIGURE 5-39 Asynchronous signal A can produce partial pulses at X.

The problem with this circuit is that A is asynchronous; it can change states at any time relative to the clock signal because the exact times when the operator actuates or releases the switch are essentially random. This can produce *partial* clock pulses at output X if either transition of A occurs while the clock signal is HIGH, as shown in the waveforms of Figure 5-39(b).

This type of output is often not acceptable, so a method for preventing the appearance of partial pulses at X must be developed. One solution is shown in Figure 5-40(a). Describe how this circuit solves the problem, and draw the X waveform for the same situation as in Figure 5-39(b).

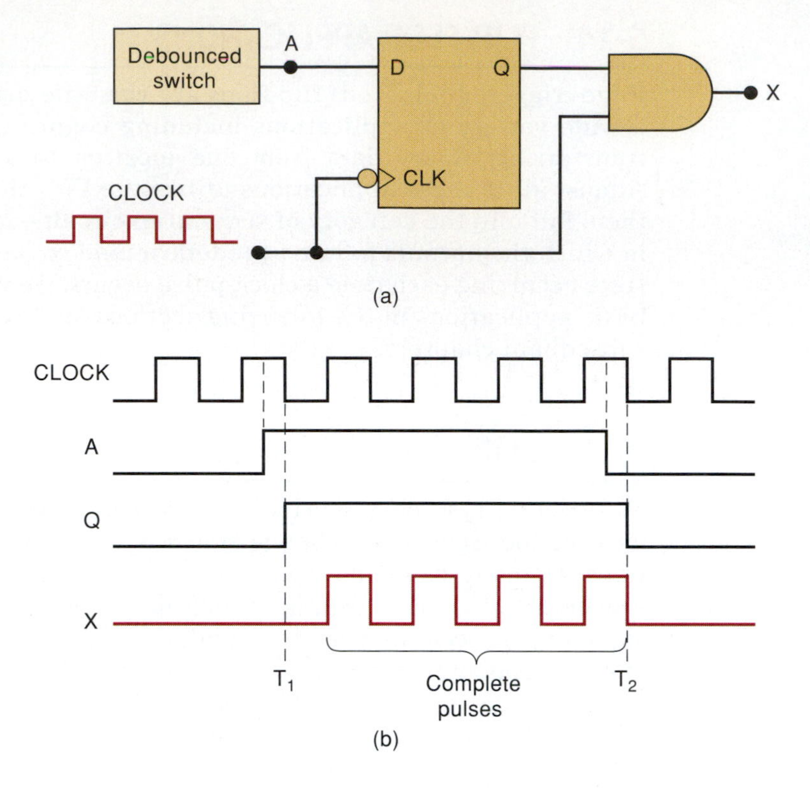

FIGURE 5-40 An edge-triggered D flip-flop is used to synchronize the enabling of the AND gate to the NGTs of the clock.

Solution

The A signal is connected to the D input of FF Q, which is clocked by the NGT of the clock signal. Thus, when A goes HIGH, Q will not go HIGH until the next NGT of the clock at time T_1. This HIGH at Q will enable the AND gate to pass subsequent *complete* clock pulses to X, as shown in Figure 5-40(b).

When A returns LOW, Q will not go LOW until the next NGT of the clock at T_2. Thus, the AND gate will not inhibit clock pulses until the clock pulse that ends at T_2 has been passed through to X. Therefore, output X contains only complete pulses.

There is a potential problem with this circuit. Since A could go HIGH at any moment, it may by random chance violate the setup time requirement of the flip-flop. In other words, the transition of A may occur so close to the clock edge that it causes an unstable response (glitch) from the Q output. Preventing this would require a more complex synchronizing circuit.

5-16 DETECTING AN INPUT SEQUENCE

In many situations, an output is to be activated only when the inputs are activated in a certain sequence. This cannot be accomplished using pure combinational logic but requires the storage characteristic of FFs.

For example, an AND gate can be used to determine when two inputs A and B are both HIGH, but its output will respond the same regardless of which input goes HIGH first. But suppose that we want to generate a HIGH output *only* if A goes HIGH and then B goes HIGH some time later. One way to accomplish this is shown in Figure 5-41(a).

The waveforms in Figure 5-41(b) and (c) show that Q will go HIGH only if A goes HIGH before B goes HIGH. This is because A must be HIGH in order for Q to go HIGH on the PGT of B.

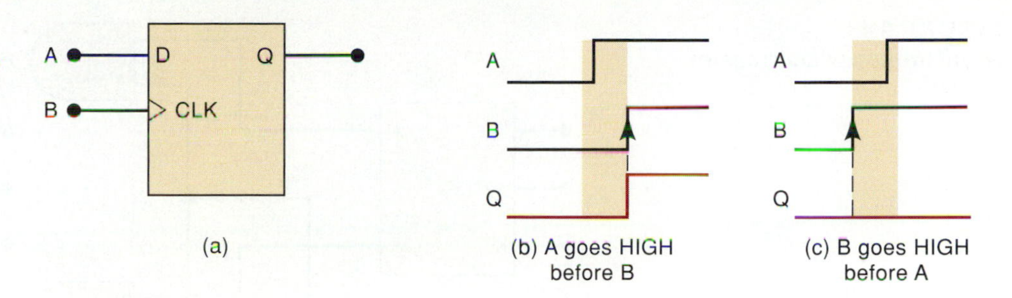

FIGURE 5-41 Clocked D flip-flop used to respond to a particular sequence of inputs.

In order for this circuit to work properly, A must go HIGH prior to B by at least an amount of time equal to the setup time requirement of the FF.

5-17 DATA STORAGE AND TRANSFER

By far the most common use of flip-flops is for the storage of data or information. The data may represent numerical values (e.g., binary numbers, BCD-coded decimal numbers) or any of a wide variety of types of data that have been encoded in binary. These data are generally stored in groups of FFs called **registers**.

The operation most often performed on data that are stored in a FF or a register is the **data transfer** operation. This involves the transfer of data from one FF or register to another. Figure 5-42 illustrates how data transfer can be accomplished between two FFs using clocked S-R, J-K, and D flip-flops. In each case, the logic value that is currently stored in FF A is transferred to FF B upon the NGT of the TRANSFER pulse. Thus, after this NGT, the B output will be the same as the A output.

The transfer operations in Figure 5-42 are examples of **synchronous transfer** because the synchronous control and CLK inputs are used to perform the transfer. A transfer operation can also be obtained using the asynchronous inputs of a FF. Figure 5-43 shows how an **asynchronous transfer** can be accomplished using the PRESET and CLEAR inputs of any type of FF.

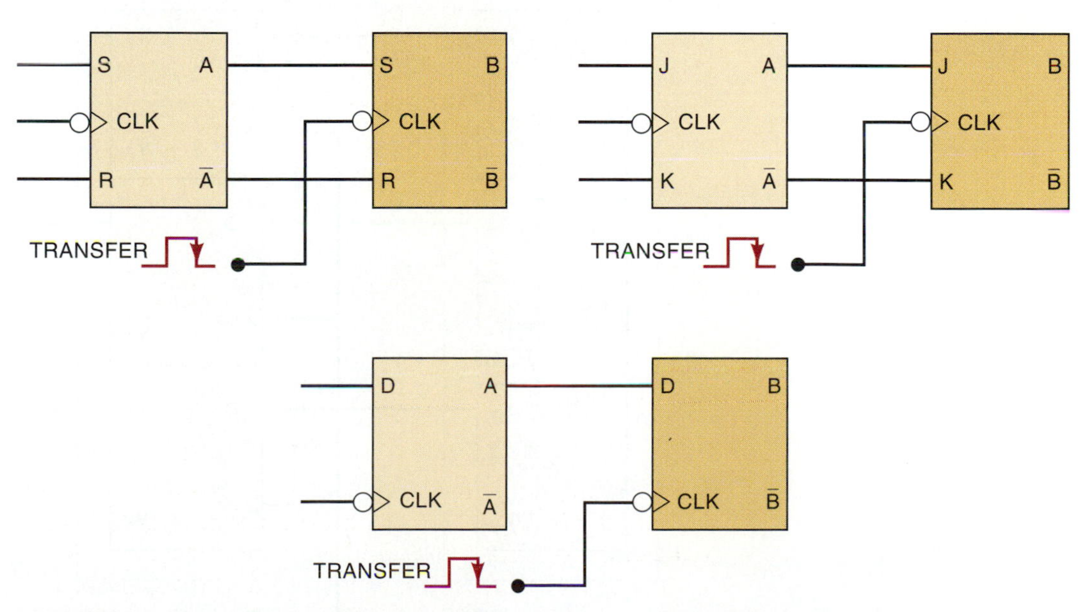

FIGURE 5-42 Synchronous data transfer operation performed by various types of clocked FFs.

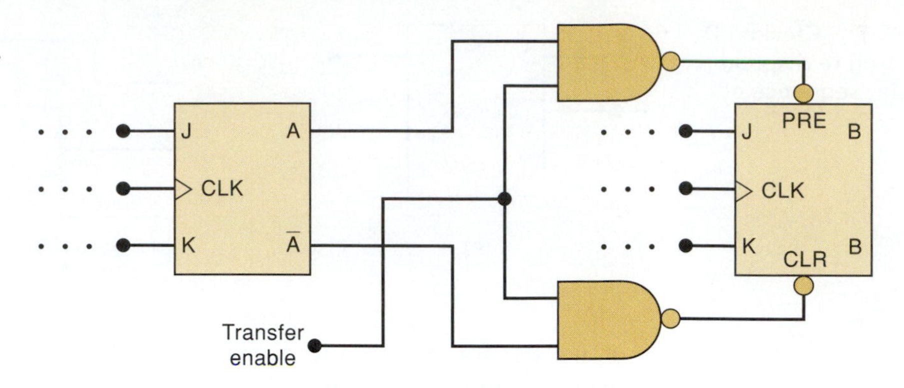

FIGURE 5-43 Asynchronous data transfer operation.

Here, the asynchronous inputs respond to LOW levels. When the TRANSFER ENABLE line is held LOW, the two NAND outputs are kept HIGH, with no effect on the FF outputs. When the TRANSFER ENABLE line is made HIGH, one of the NAND outputs will go LOW, depending on the state of the A and $\overline{A}$ outputs. This LOW will either set or clear FF B to the same state as FF A. This asynchronous transfer is done independently of the synchronous and CLK inputs of the FF. Asynchronous transfer is also called **jam transfer** because the data can be "jammed" into FF B even if its synchronous inputs are active.

Parallel Data Transfer

Figure 5-44 illustrates data transfer from one register to another using D-type FFs. Register X consists of FFs X_2, X_1, and X_0; register Y consists of FFs Y_2, Y_1, and Y_0. Upon application of the PGT of the TRANSFER pulse, the level stored in X_2 is transferred to Y_2, X_1 to Y_1, and X_0 to Y_0. The transfer of the

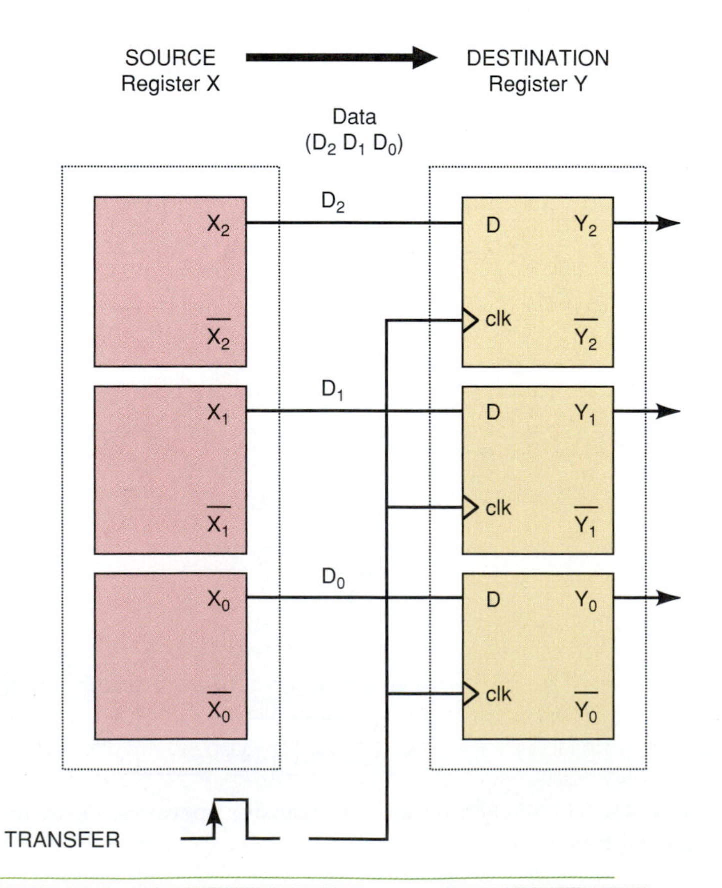

FIGURE 5-44 Parallel transfer of contents of register X into register Y.

contents of the X register into the Y register is a synchronous transfer. It is also referred to as a parallel transfer because the contents of X_2, X_1, and X_0 are transferred *simultaneously* into Y_2, Y_1, and Y_0, respectively. If a **serial data transfer** were performed, the contents of the X register would be transferred to the Y register one bit at a time. This will be examined in the next section.

It is important to understand that parallel transfer does not change the contents of the register that is the source of data. For example, in Figure 5-44, if $X_2X_1X_0 = 101$ and $Y_2Y_1Y_0 = 011$ prior to the occurrence of the TRANSFER pulse, then both registers will be holding 101 after the TRANSFER pulse.

REVIEW QUESTIONS

1. *True or false:* Asynchronous data transfer uses the *CLK* input.
2. Which type of FF is best suited for synchronous transfer because it requires the fewest interconnections from one FF to the other?
3. If J-K flip-flops were used in the registers of Figure 5-44, how many total interconnections would be required from register X to register Y?
4. *True or false:* Synchronous data transfer requires less circuitry than asynchronous transfer.

5-18 SERIAL DATA TRANSFER: SHIFT REGISTERS

Before we describe the serial data transfer operation, we must first examine the basic *shift-register* arrangement. A **shift register** is a group of FFs arranged so that the binary numbers stored in the FFs are shifted from one FF to the next for every clock pulse. You have undoubtedly seen shift registers in action in devices such as an electronic calculator, where the digits shown on the display shift over each time you key in a new digit. This is the same action taking place in a shift register.

Figure 5-45(a) shows one way to arrange J-K flip-flops to operate as a four-bit shift register. Note that the FFs are connected so that the output of X_3 transfers into X_2, X_2 into X_1, and X_1 into X_0. What this means is that upon the occurrence of the NGT of a shift pulse, each FF takes on the value stored previously in the FF on its left. Flip-flop X_3 takes on a value determined by the conditions present on its J and K inputs when the NGT occurs. For now, we will assume that X_3's J and K inputs are fed by the DATA IN waveform shown in Figure 5-45(b). We will also assume that all FFs are in the 0 state before shift pulses are applied.

The waveforms in Figure 5-45(b) show how the input data are shifted from left to right from FF to FF as shift pulses are applied. When the first NGT occurs at T_1, each of the FFs X_2, X_1, and X_0 will have the $J = 0$, $K = 1$ condition present at its inputs because of the state of the FF on its left. Flip-flop X_3 will have $J = 1$, $K = 0$ because of DATA IN. Thus, at T_1, only X_3 will go HIGH, while all the other FFs remain LOW. When the second NGT occurs at T_2, flip-flop X_3 will have $J = 0$, $K = 1$ because of DATA IN. Flip-flop X_2 will have $J = 1$, $K = 0$ because of the current HIGH at X_3. Flip-flops X_1 and X_0 will still have $J = 0$, $K = 1$. Thus, at T_2, only FF X_2 will go HIGH, FF X_3 will go LOW, and FFs X_1 and X_0 will remain LOW.

Similar reasoning can be used to determine how the waveforms change at T_3 and T_4. Note that on each NGT of the shift pulses, each FF output takes on the level that was present at the output of the FF on its left just *prior* to the NGT. Of course, X_3 takes on the level that was present at DATA IN just prior to the NGT.

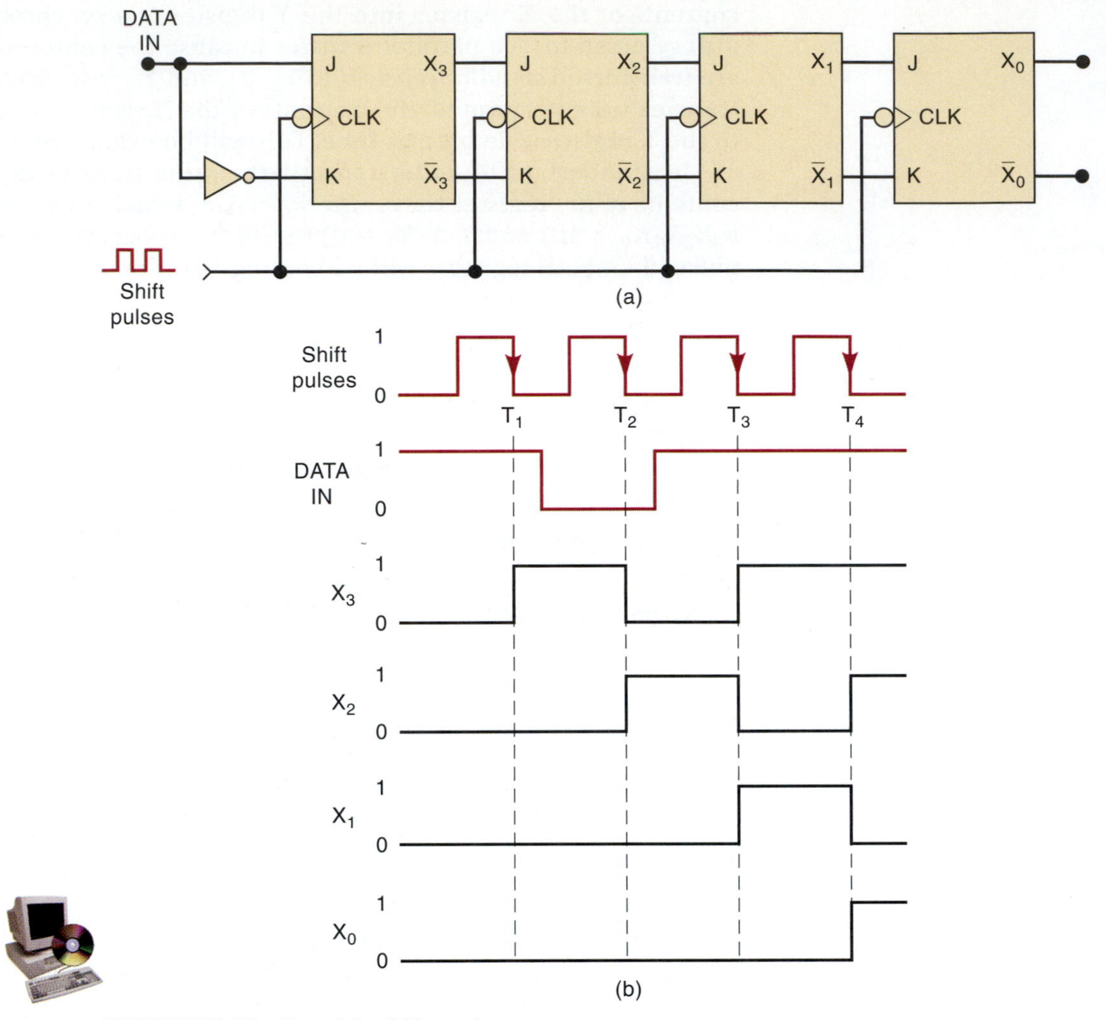

FIGURE 5-45 Four-bit shift register.

Hold Time Requirement

In this shift-register arrangement, it is necessary that the FFs have a very small hold time requirement because there are times when the *J*, *K* inputs are changing at about the same time as the *CLK* transition. For example, the X_3 output switches from 1 to 0 in response to the NGT at T_2, causing the *J*, *K* inputs of X_2 to change while its *CLK* input is changing. Actually, because of the propagation delay of X_3, the *J*, *K* inputs of X_2 won't change for a short time after the NGT. For this reason, a shift register should be implemented using edge-triggered FFs that have a t_H value less than one *CLK*-to-output propagation delay. This latter requirement is easily satisfied by most modern edge-triggered FFs.

Serial Transfer Between Registers

Figure 5-46(a) shows two three-bit shift registers connected so that the contents of the *X* register will be serially transferred (shifted) into register *Y*. We are using D flip-flops for each shift register because this requires fewer connections than J-K flip-flops. Notice how X_0, the last FF of register *X*, is connected to the *D* input of Y_2, the first FF of register *Y*. Thus, as the shift pulses are applied, the information transfer takes place as follows:

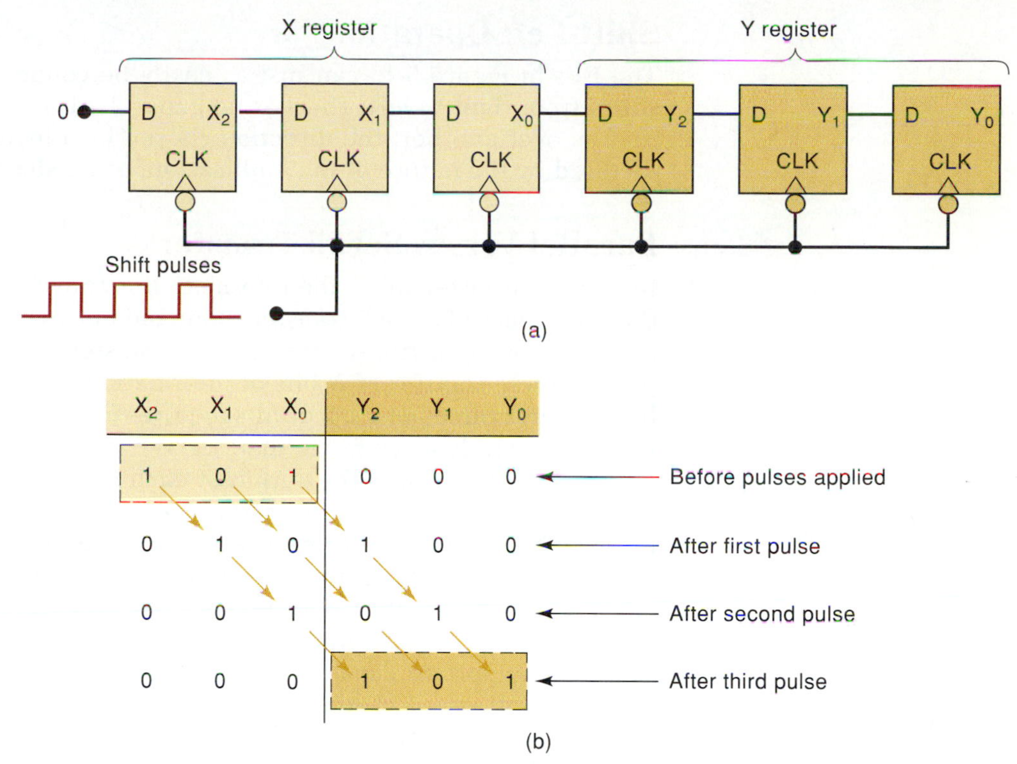

X_2	X_1	X_0	Y_2	Y_1	Y_0	
1	0	1	0	0	0	← Before pulses applied
0	1	0	1	0	0	← After first pulse
0	0	1	0	1	0	← After second pulse
0	0	0	1	0	1	← After third pulse

FIGURE 5-46 Serial transfer of information from *X* register into *Y* register.

$X_2 \rightarrow X_1 \rightarrow X_0 \rightarrow Y_2 \rightarrow Y_1 \rightarrow Y_0$. The X_2 FF will go to a state determined by its *D* input. For now, *D* will be held LOW, so that X_2 will go LOW on the first pulse and will remain there.

To illustrate, let us assume that before any shift pulses are applied, the contents of the *X* register are 101 (i.e., $X_2 = 1$, $X_1 = 0$, $X_0 = 1$) and the *Y* register is at 000. Refer to the table in Figure 5-46(b), which shows how the states of each FF change as shift pulses are applied. The following points should be noted:

1. On the NGT of each pulse, each FF takes on the value that was stored in the FF on its left prior to the occurrence of the pulse.
2. After *three* pulses, the 1 that was initially in X_2 is in Y_2, the 0 initially in X_1 is in Y_1, and the 1 initially in X_0 is in Y_0. In other words, the 101 stored in the *X* register has now been shifted into the *Y* register. The *X* register is at 000; it has lost its original data.
3. The complete transfer of the *three* bits of data requires *three* shift pulses.

EXAMPLE 5-13

Assume the same initial contents of the *X* and *Y* registers in Figure 5-46. What will be the contents of each FF after the occurrence of the sixth shift pulse?

Solution

If we continue the process shown in Figure 5-46(b) for three more shift pulses, we will find that all of the FFs will be in the 0 state after the sixth pulse. Another way to arrive at this result is to reason as follows: the constant 0 level at the *D* input of X_2 shifts in a new 0 with each pulse so that, after six pulses, the registers are filled up with 0s.

Shift-Left Operation

The FFs in Figure 5-46 can just as easily be connected so that information shifts from right to left. There is no general advantage of shifting in one direction over another; the direction chosen by a logic designer will often be dictated by the nature of the application, as we shall see.

Parallel Versus Serial Transfer

In parallel transfer, all of the information is transferred simultaneously upon the occurrence of a *single* transfer command pulse (Figure 5-44), no matter how many bits are being transferred. In serial transfer, as exemplified by Figure 5-46, the complete transfer of N bits of information requires N clock pulses (three bits requires three pulses, four bits requires four pulses, etc.). Parallel transfer, then, is obviously much faster than serial transfer using shift registers.

In parallel transfer, the output of each FF in register X is connected to a corresponding FF input in register Y. In serial transfer, only the last FF in register X is connected to register Y. In general, then, parallel transfer requires more interconnections between the sending register (X) and the receiving register (Y) than does serial transfer. This difference becomes more critical when a greater number of bits of information are being transferred. This is an important consideration when the sending and receiving registers are remote from each other because it determines how many lines (wires) are needed for the transmission of the information.

The choice of either parallel or serial transmission depends on the particular system application and specifications. Often, a combination of the two types is used to take advantage of the *speed* of parallel transfer and the *economy and simplicity* of serial transfer. More will be said later about information transfer.

REVIEW QUESTIONS

1. *True or false:* The fastest method for transferring data from one register to another is parallel transfer.
2. What is the major advantage of serial transfer over parallel transfer?
3. Refer to Figure 5-46. Assume that the initial contents of the registers are $X_2 = 0$, $X_1 = 1$, $X_0 = 0$, $Y_2 = 1$, $Y_1 = 1$, $Y_0 = 0$. Also assume that the D input of X_2 is held HIGH. Determine the value of each FF output after the occurrence of the fourth shift pulse.
4. In which form of data transfer does the source of the data not lose its data?

5-19 FREQUENCY DIVISION AND COUNTING

Refer to Figure 5-47(a). Each FF has its J and K inputs at the 1 level, so that it will change states (toggle) whenever the signal on its CLK input goes from HIGH to LOW. The clock pulses are applied only to the CLK input of FF Q_0. Output Q_0 is connected to the CLK input of FF Q_1, and output Q_1 is connected to the CLK input of FF Q_2. The waveforms in Figure 5-47(b) show how the FFs change states as the pulses are applied. The following important points should be noted:

1. Flip-flop Q_0 toggles on the negative-going transition of each input clock pulse. Thus, the Q_0 output waveform has a frequency that is exactly one-half of the clock pulse frequency.

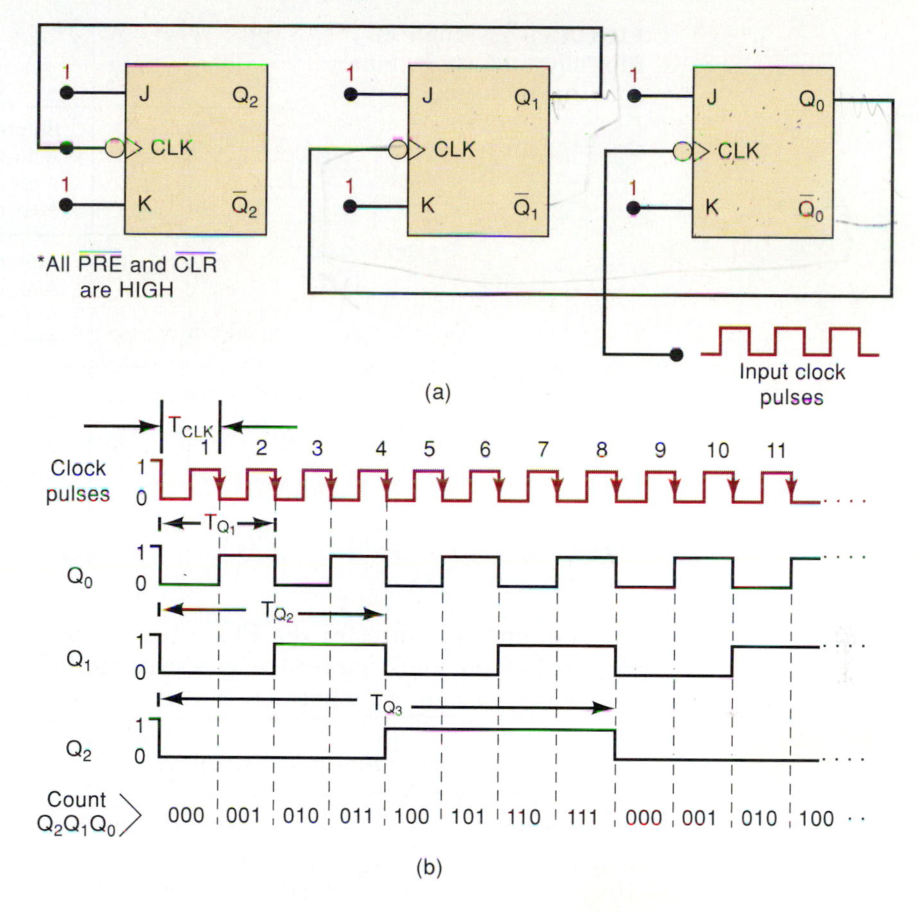

FIGURE 5-47 J-K flip-flops wired as a three-bit binary counter (MOD-8).

2. Flip-flop Q_1 toggles each time the Q_0 output goes from HIGH to LOW. The Q_1 waveform has a frequency equal to exactly one-half the frequency of the Q_0 output and therefore one-fourth of the clock frequency.
3. Flip-flop Q_2 toggles each time the Q_1 output goes from HIGH to LOW. Thus, the Q_2 waveform has one-half the frequency of Q_1 and therefore one-eighth of the clock frequency.
4. Each FF output is a square wave (50% duty cycle).

As described above, each FF divides the frequency of its input by 2. Thus, if we were to add a fourth FF to the chain, it would have a frequency equal to one-sixteenth of the clock frequency, and so on. Using the appropriate number of FFs, this circuit could divide a frequency by any power of 2. Specifically, using N flip-flops would produce an output frequency from the last FF, which is equal to $1/2^N$ of the input frequency.

This application of flip-flops is referred to as **frequency division**. Many applications require a frequency division. For example, your wristwatch is no doubt a "quartz" watch. The term *quartz watch* means that a quartz crystal is used to generate a very stable oscillator frequency. The natural resonant frequency of the quartz crystal in your watch is likely 1 MHz or more. In order to advance the "seconds" display once every second, the oscillator frequency is *divided* by a value that will produce a very stable and accurate 1 Hz output frequency.

Counting Operation

In addition to functioning as a frequency divider, the circuit of Figure 5-47 also operates as a **binary counter**. This can be demonstrated by examining

FIGURE 5-48 Table of flip-flop states shows binary counting sequence.

2^2 Q_2	2^1 Q_1	2^0 Q_0	
0	0	0	Before applying clock pulses
0	0	1	After pulse #1
0	1	0	After pulse #2
0	1	1	After pulse #3
1	0	0	After pulse #4
1	0	1	After pulse #5
1	1	0	After pulse #6
1	1	1	After pulse #7
0	0	0	After pulse #8 recycles to 000
0	0	1	After pulse #9
0	1	0	After pulse #10
0	1	1	After pulse #11
.	.	.	. . . .
.	.	.	. . . .
.	.	.	. . . .

the sequence of states of the FFs after the occurrence of each clock pulse. Figure 5-48 presents the results in a **state table**. Let the $Q_2Q_1Q_0$ values represent a binary number where Q_2 is in the 2^2 position, Q_1 is in the 2^1 position, and Q_0 is in the 2^0 position. The first eight $Q_2Q_1Q_0$ states in the table should be recognized as the binary counting sequence from 000 to 111. After the first NGT, the FFs are in the 001 state ($Q_2 = 0$, $Q_1 = 0$, $Q_0 = 1$), which represents 001_2 (equivalent to decimal 1); after the second NGT, the FFs represent 010_2, which is equivalent to 2_{10}; after three pulses, $011_2 = 3_{10}$; after four pulses, $100_2 = 4_{10}$; and so on, until after seven pulses, $111_2 = 7_{10}$. On the eighth NGT, the FFs return to the 000 state, and the binary sequence repeats itself for succeeding pulses.

Thus, for the first seven input pulses, the circuit functions as a binary counter in which the states of the FFs represent a binary number equivalent to the number of pulses that have occurred. This counter can count as high as $111_2 = 7_{10}$ before it returns to 000.

State Transition Diagram

Another way to show how the states of the FFs change with each applied clock pulse is to use a **state transition diagram**, as illustrated in Figure 5-49.

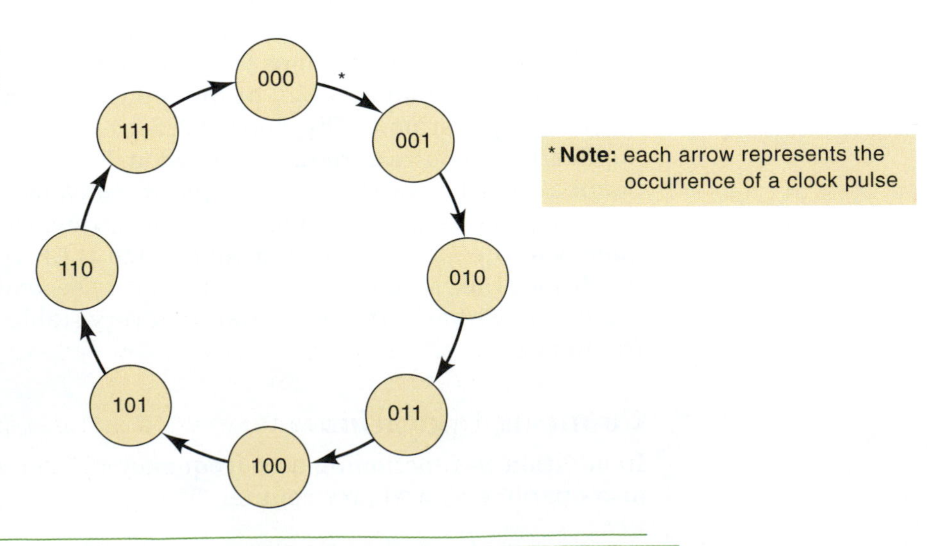

FIGURE 5-49 State transition diagram shows how the states of the counter flip-flops change with each applied clock pulse.

Each circle represents one possible state, as indicated by the binary number inside the circle. For example, the circle containing the number 100 represents the 100 state (i.e., $Q_2 = 1$, $Q_1 = Q_0 = 0$).

The arrows connecting one circle to another show how one state changes to another as a clock pulse is applied. By looking at a particular state circle, we can see which state precedes it and which state follows it. For example, looking at the 000 state, we see that this state is reached whenever the counter is in the 111 state and a clock pulse is applied. Likewise, we see that the 000 state is always followed by the 001 state.

We will use state transition diagrams to help describe, analyze, and design counters and other sequential circuits.

MOD Number

The counter of Figure 5-47 has $2^3 = 8$ different states (000 through 111). It would be referred to as a *MOD-8 counter,* where the **MOD number** indicates the number of states in the counting sequence. If a fourth FF were added, the sequence of states would count in binary from 0000 to 1111, a total of 16 states. This would be called a *MOD-16 counter.* In general, if N flip-flops are connected in the arrangement of Figure 5-47, the counter will have 2^N different states, and so it is a MOD-2^N counter. It would be capable of counting up to $2^N - 1$ before returning to its 0 state.

The MOD number of a counter also indicates the frequency division obtained from the last FF. For instance, a four-bit counter has four FFs, each representing one binary digit (bit), and so it is a MOD-2^4 = MOD-16 counter. It can therefore count up to 15 ($= 2^4 - 1$). It can also be used to divide the input pulse frequency by a factor of 16 (the MOD number).

We have looked only at the basic FF binary counter. We examine counters in much more detail in Chapter 7.

EXAMPLE 5-14

Assume that the MOD-8 counter in Figure 5-47 is in the 101 state. What will be the state (count) after 13 pulses have been applied?

Solution

Locate the 101 state on the state transition diagram. Proceed around the state diagram through eight state changes, and you should be back in the 101 state. Now continue through five more state changes (for a total of 13), and you should end up in the 010 state.

Notice that because this is a MOD-8 counter with eight states, it takes eight state transitions to make one complete excursion around the diagram back to the starting state.

EXAMPLE 5-15

Consider a counter circuit that contains six FFs wired in the arrangement of Figure 5-47 (i.e., $Q_5, Q_4, Q_3, Q_2, Q_1, Q_0$).

(a) Determine the counter's MOD number.

(b) Determine the frequency at the output of the last FF (Q_5) when the input clock frequency is 1 MHz.

(c) What is the range of counting states for this counter?

(d) Assume a starting state (count) of 000000. What will be the counter's state after 129 pulses?

Solution

(a) MOD number $= 2^6 = 64$.

(b) The frequency at the last FF will equal the input clock frequency divided by the MOD number. That is,

$$f(\text{at } Q_5) = \frac{1\text{ MHz}}{64} = 15.625\text{ kHz}$$

(c) The counter will count from 000000_2 to 111111_2 (0 to 63_{10}) for a total of 64 states. Note that the number of states is the same as the MOD number.

(d) Because this is a MOD-64 counter, every 64 clock pulses will bring the counter back to its starting state. Therefore, after 128 pulses, the count is back to 000000. The 129th pulse brings the counter to the 000001 counter.

REVIEW QUESTIONS

1. A 20-kHz clock signal is applied to a J-K flip-flop with $J = K = 1$. What is the frequency of the FF output waveform?
2. How many FFs are required for a counter that will count 0 to 255_{10}?
3. What is the MOD number of the counter in question 2?
4. What is the frequency of the output of the eighth FF when the input clock frequency is 512 kHz?
5. If this counter starts at 00000000, what will be its state after 520 pulses?

5-20 MICROCOMPUTER APPLICATION

Your study of digital systems is still in a relatively early stage, and you have not learned very much about microprocessors and microcomputers. However, you can get a basic idea of how FFs are employed in a typical microprocessor-controlled application without being concerned with all of the details you will need to know later.

Figure 5-50 shows a microprocessor unit (MPU) with its outputs used to transfer binary data to register X, which consists of four D flip-flops X_3, X_2, X_1, X_0. One set of MPU outputs is the *address code* made up of the eight

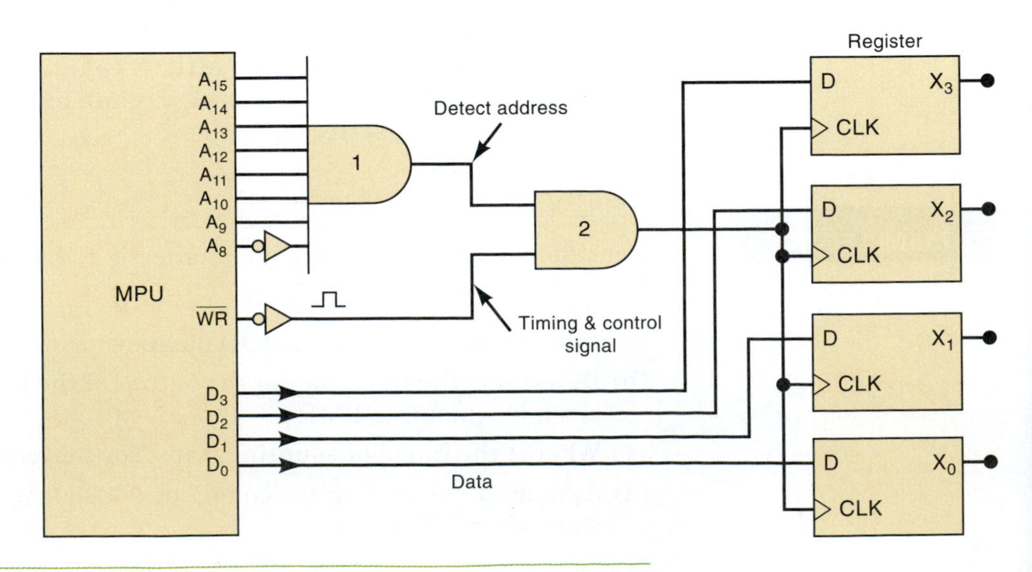

FIGURE 5-50 Example of a microprocessor transferring binary data to an external register.

outputs $A_{15}, A_{14}, A_{13}, A_{12}, A_{11}, A_{10}, A_9, A_8$. Most MPUs have at least 16 available address outputs, but they are not always all used. A second set of MPU outputs consists of the four *data lines* D_3, D_2, D_1, D_0. Most MPUs have at least eight available data lines. The other MPU output is a timing control signal $\overline{WR}$, which goes LOW when the MPU is ready to write.

Recall that the MPU is the central processing unit of a microcomputer, and its main function is to execute a program of instructions stored in the computer's memory. One of the instructions it might execute could be one that tells the MPU to transfer a binary number from a storage resister within the MPU to the external register X. This is called a *write* cycle. In executing this instruction, the MPU would perform the following steps:

1. Place the binary number onto its data output lines D_3 through D_0.
2. Place the proper address code on its output lines A_{15} through A_8 to select register X as the recipient of the data.
3. Once the data and address outputs are stabilized, the MPU generates the write pulse $\overline{WR}$ to clock the register and complete the parallel transfer of data into X.

There are many situations where an MPU, under the control of a program, will send data to an external register in order to control external events. For example, the individual FFs in the register can control the ON/OFF status of electromechanical devices such as solenoids, relays, motors, and so on (through appropriate interface circuits, of course). The data sent from the MPU to the register will determine which devices are ON and which are OFF. Another common example is when the register is used to hold a binary number for input to a digital-to-analog converter (DAC). The MPU sends the binary number to the register, and the DAC converts it to an analog voltage that may be used to control something such as the position of an electron beam on a CRT screen or the speed of a motor.

EXAMPLE 5-16

(a) What address code must the MPU generate in order for the data to be transferred into X?

(b) Assume that $X_3–X_0 = 0110$, $A_{15}–A_8 = 11111111$, and $D_3–D_0 = 1011$. What will be in X after a $\overline{WR}$ pulse occurs?

Solution

(a) In order for the data to be transferred into X, the clock pulse must pass through AND gate 2 into the *CLK* inputs of the FFs. This will happen only if the top input of AND gate 2 is HIGH. This means that all of the inputs to AND gate 1 must be HIGH; that is, A_{15} through A_9 must be 1, and A_8 must be 0. Thus, the presence of address code 11111110 is needed to allow data to be transferred into X.

(b) With $A_8 = 1$, the LOW from AND gate 1 will inhibit $\overline{WR}$ from getting through AND gate 2, and the FFs will not be clocked. Therefore, the contents of register X will not change from 0110.

REVIEW QUESTION

1. Show how the 74HC175 IC of Figure 5-34 can be used for the X register of Figure 5-50.

5-21 SCHMITT-TRIGGER DEVICES

A **Schmitt-trigger circuit** is not classified as a flip-flop, but it does exhibit a type of memory characteristic that makes it useful in certain special situations. One of those situations is shown in Figure 5-51(a). Here a standard INVERTER is being driven by a logic input that has relatively slow transition times. When these transition times exceed the maximum allowed values (this depends on the particular logic family), the outputs of logic gates and INVERTERs may produce oscillations as the input signal passes through the indeterminate range. The same input conditions can also produce erratic triggering of FFs.

A device that has a Schmitt-trigger type of input is designed to accept noisy slow-changing signals and produce an output that has oscillation-free transitions. The output will generally have very rapid transition times (typically 10 ns) that are independent of the input signal characteristics. Figure 5-51(b) shows a Schmitt-trigger INVERTER and its response to a slow-changing input.

If you examine the waveforms in Figure 5-51(b), you should note that the output does not change from HIGH to LOW until the input exceeds the *positive-going threshold* voltage, V_{T+}. Once the output goes LOW, it will remain there even when the input drops back below V_{T+} (this is its memory characteristic) until it drops all the way down below the *negative-going threshold* voltage, V_{T-}. The values of the two threshold voltages will vary from logic family to logic family, but V_{T-} will always be less than V_{T+}.

The Schmitt-trigger INVERTER, and all other devices with Schmitt-trigger inputs, uses the distinctive symbol shown in Figure 5-51(b) to indicate that they can reliably respond to slow-changing input signals. Logic designers use ICs with Schmitt-trigger inputs to convert slow-changing signals to clean, fast-changing signals that can drive standard IC inputs.

Several ICs are available with Schmitt-trigger inputs. The 7414, 74LS14, and 74HC14 are hex INVERTER ICs with Schmitt-trigger inputs. The 7413, 74LS13, and 74HC13 are dual four-input NANDs with Schmitt-trigger inputs.

REVIEW QUESTIONS

1. What could occur when a slow-changing signal is applied to a standard logic IC?
2. How does a Schmitt-trigger logic device operate differently from a standard logic device?

5-22 ONE-SHOT (MONOSTABLE MULTIVIBRATOR)

A digital circuit that is somewhat related to the FF is the **one-shot (OS)**. Like the FF, the OS has two outputs, Q and $\overline{Q}$, which are the inverse of each other. Unlike the FF, the OS has only one *stable* output state (normally $Q = 0$, $\overline{Q} = 1$), where it remains until it is triggered by an input signal. Once triggered, the OS outputs switch to the opposite state ($Q = 1, \overline{Q} = 0$). It remains in this **quasi-stable state** for a fixed period of time, t_p, which is usually determined by an RC time constant that results from the values of external components connected to the OS. After a time t_p, the OS outputs return to their resting state until triggered again.

Figure 5-52(a) shows the logic symbol for a OS. The value of t_p is often indicated somewhere on the OS symbol. In practice, t_p can vary from several nanoseconds to several tens of seconds. The exact value of t_p is variable and is determined by the values of external components R_T and C_T.

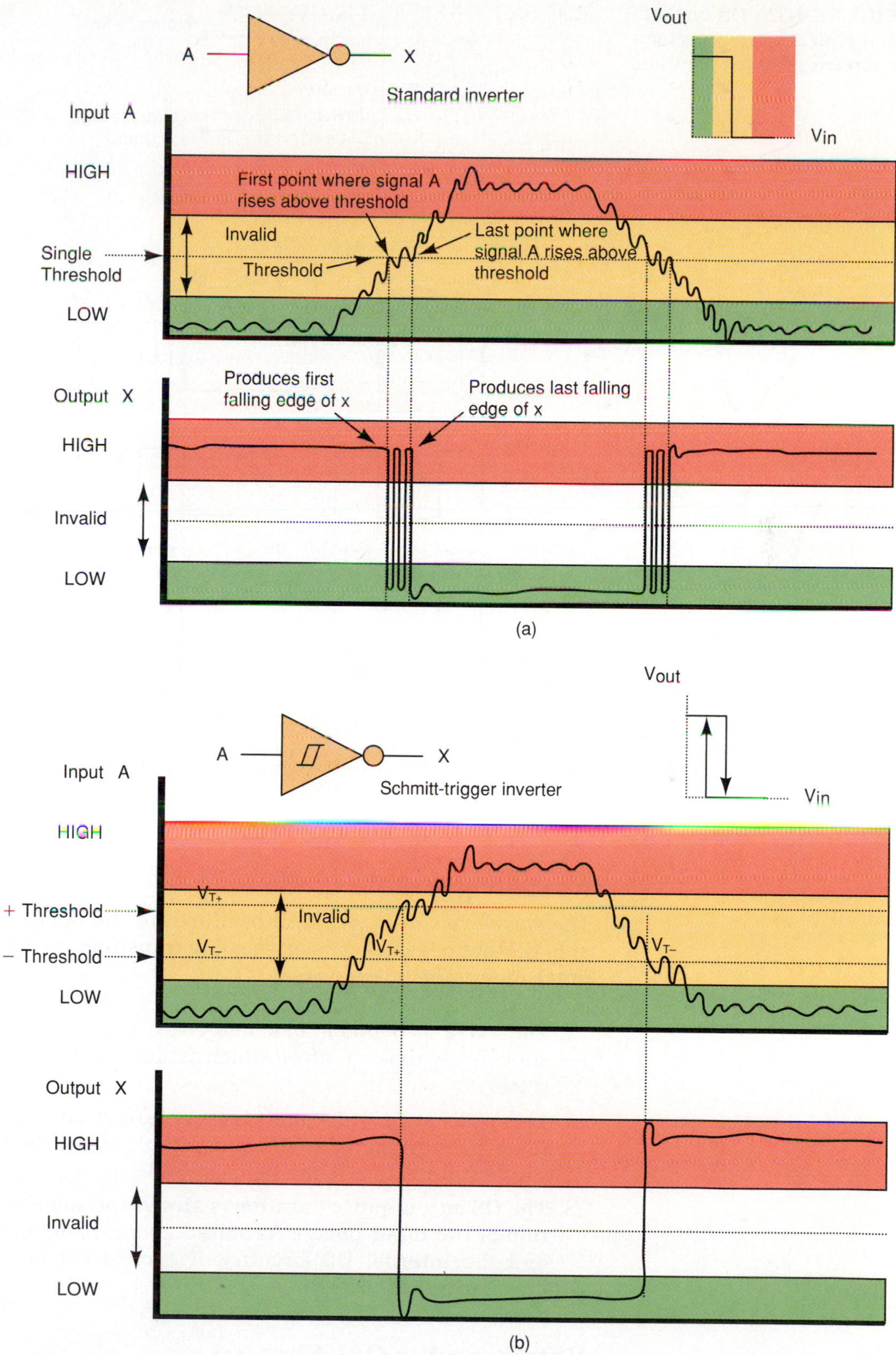

FIGURE 5-51 (a) Standard inverter response to slow noisy input, and (b) Schmitt-trigger response to slow noisy input.

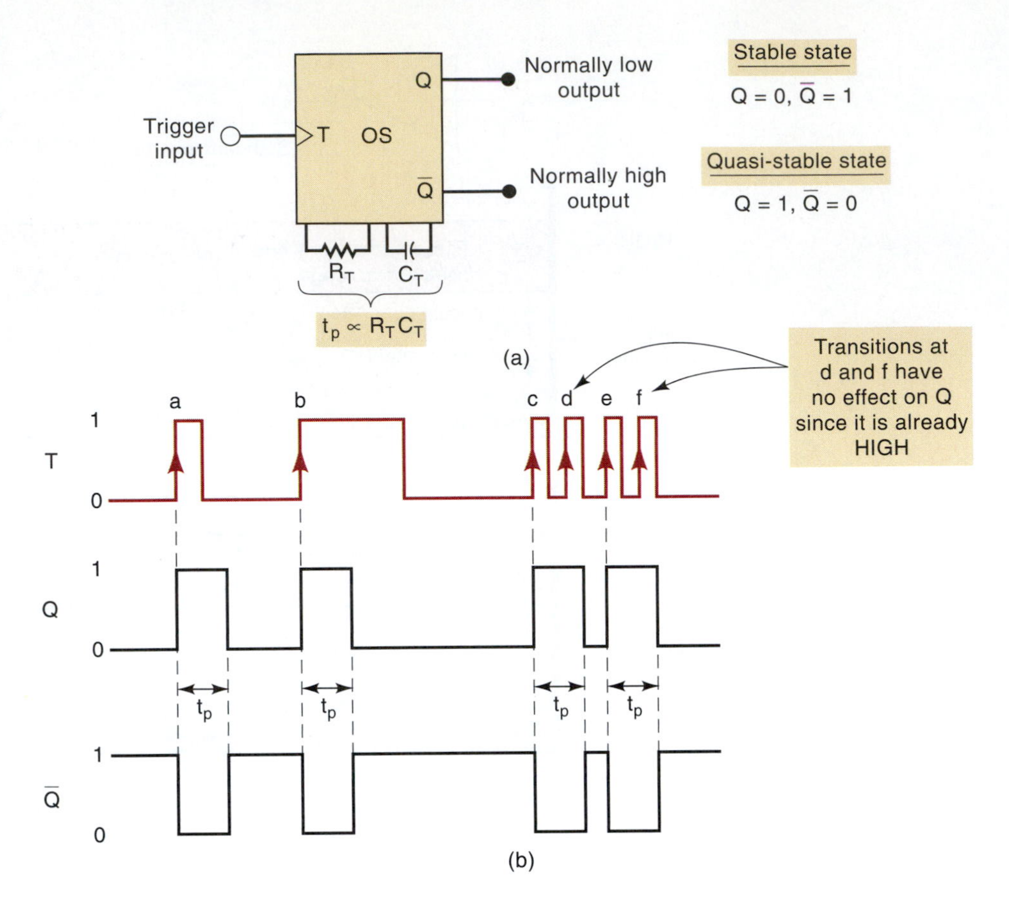

FIGURE 5-52 OS symbol and typical waveforms for nonretriggerable operation.

Two types of one-shots are available in IC form: the **nonretriggerable OS** and the **retriggerable OS**.

Nonretriggerable One-Shot

The waveforms in Figure 5-52(b) illustrate the operation of a nonretriggerable OS that triggers on positive-going transitions at its trigger (T) input. The important points to note are:

1. The PGTs at points a, b, c, and e will trigger the OS to its quasi-stable state for a time t_p, after which it automatically returns to the stable state.
2. The PGTs at points d and f have no effect on the OS because it has already been triggered to the quasi-stable state. The OS must return to the stable state before it can be triggered.
3. The OS output-pulse duration is always the same, regardless of the duration of the input pulses. As stated above, t_p depends only on R_T and C_T and the internal OS circuitry. A typical OS may have a t_p given by $t_p = 0.693\, R_T C_T$.

Retriggerable One-Shot

The retriggerable OS operates much like the nonretriggerable OS except for one major difference: *it can be retriggered while it is in the quasi-stable state, and it will begin a new t_p interval.* Figure 5-53(a) compares the response of both types of OS using a t_p of 2 ms. Let's examine these waveforms.

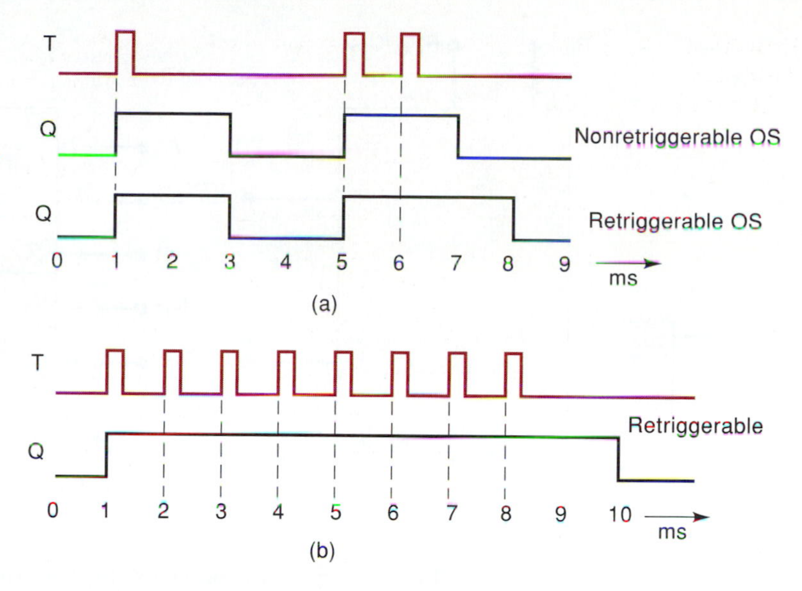

FIGURE 5-53 (a) Comparison of nonretriggerable and retriggerable OS responses for $t_p = 2$ ms. (b) Retriggerable OS begins a new t_p interval each time it receives a trigger pulse.

Both types of OS respond to the first trigger pulse at $t = 1$ ms by going HIGH for 2 ms and then returning LOW. The second trigger pulse at $t = 5$ ms triggers both one-shots to the HIGH state. The third trigger pulse at $t = 6$ ms has no effect on the nonretriggerable OS because it is already in its quasi-stable state. However, this trigger pulse will *retrigger* the retriggerable OS to begin a new $t_p = 2$ ms interval. Thus, it will stay HIGH for 2 ms *after* this third trigger pulse.

In effect, then, a retriggerable OS begins a new t_p interval each time a trigger pulse is applied, regardless of the current state of its Q output. In fact, trigger pulses can be applied at a rate fast enough that the OS will always be retriggered before the end of the t_p interval and Q will remain HIGH. This is shown in Figure 5-53(b), where eight pulses are applied every 1 ms. Q does not return LOW until 2 ms after the last trigger pulse.

Actual Devices

Several one-shot ICs are available in both the retriggerable and the nonretriggerable versions. The 74121 is a single nonretriggerable one-shot IC; the 74221, 74LS221, and 74HC221 are dual nonretriggerable one-shot ICs; the 74122 and 74LS122 are single retriggerable one-shot ICs; the 74123, 74LS123, and 74HC123 are dual retriggerable one-shot ICs.

Figure 5-54(a) shows the traditional symbol for the 74121 nonretriggerable one-shot IC. Note that it contains internal logic gates to allow inputs A_1, A_2, and B to trigger the OS in a variety of ways. The B input is a Schmitt-trigger type of input that is allowed to have slow transition times and still reliably trigger the OS. The pins labeled R_{INT}, R_{EXT}/C_{EXT}, and C_{EXT} are used to connect an external resistor and capacitor to achieve the desired output pulse duration. Figure 5-54(b) is the IEEE/ANSI symbol for the 74121 nonretriggerable OS. Note how this symbol represents the logic gates. Also notice the presence of a small pulse with 1 in front of it. This indicates that the device is a nonretriggerable OS. The IEEE/ANSI symbol for a retriggerable OS would not have the 1 in front of the pulse.

Monostable Multivibrator

Another name for the one-shot is *monostable multivibrator* because it has only one stable state. One-shots find limited application in most sequential

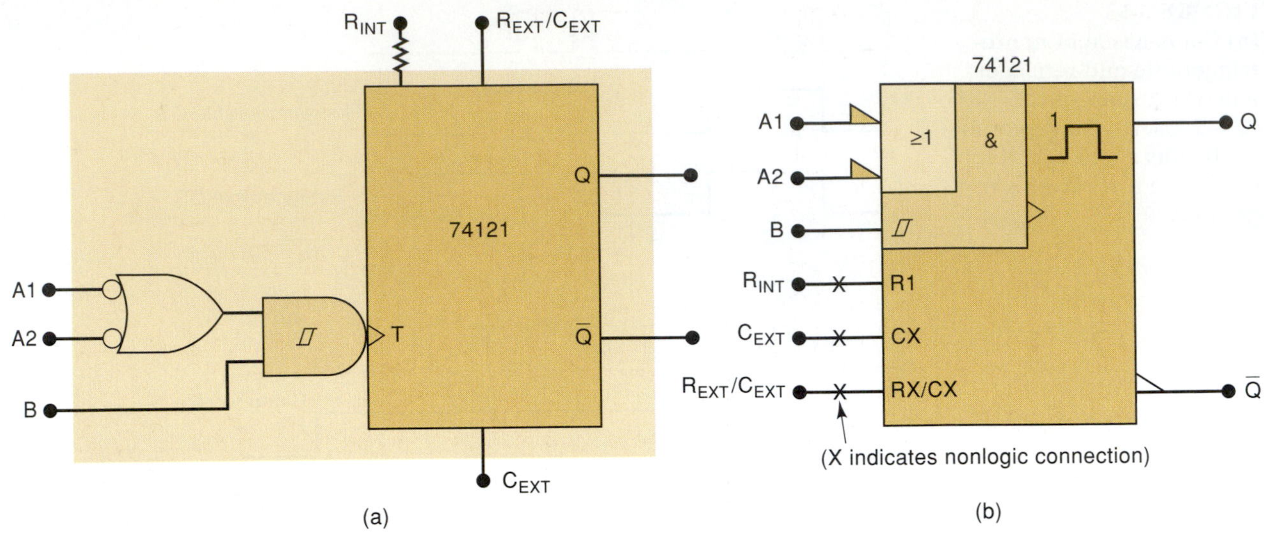

FIGURE 5-54 Logic symbols for the 74121 nonretriggerable one-shot: (a) traditional; (b) IEEE/ANSI.

clock-controlled systems, and experienced designers generally avoid using them because they are prone to false triggering by spurious noise. When they are used, it is usually in simple timing applications that utilize the predetermined t_p interval. Several of the end-of-chapter problems will illustrate how a OS is used.

REVIEW QUESTIONS

1. In the absence of a trigger pulse, what will be the state of a OS output?
2. *True or false:* When a nonretriggerable OS is pulsed while it is in its quasi-stable state, the output is not affected.
3. What determines the t_p value for a OS?
4. Describe how a retriggerable OS operates differently from a nonretriggerable OS.

5-23 CLOCK GENERATOR CIRCUITS

Flip-flops have two stable states; therefore, we can say that they are *bistable multivibrators.* One-shots have one stable state, and so we call them *monostable multivibrators.* A third type of multivibrator has no stable states; it is called an **astable** or **free-running multivibrator**. This type of logic circuit switches back and forth (oscillates) between two unstable output states. It is useful for generating clock signals for synchronous digital circuits.

Several types of astable multivibrators are in common use. We will present three of them without any attempt to analyze their operation. They are presented here so that you can construct a clock generator circuit if needed for a project or for testing digital circuits in the lab.

Schmitt-Trigger Oscillator

Figure 5-55 shows how a Schmitt-trigger INVERTER can be connected as an oscillator. The signal at V_{OUT} is an approximate square wave with a frequency that depends on the R and C values. The relationship between the frequency

FIGURE 5-55 Schmitt-trigger oscillator using a 7414 INVERTER. A 7413 Schmitt-trigger NAND may also be used.

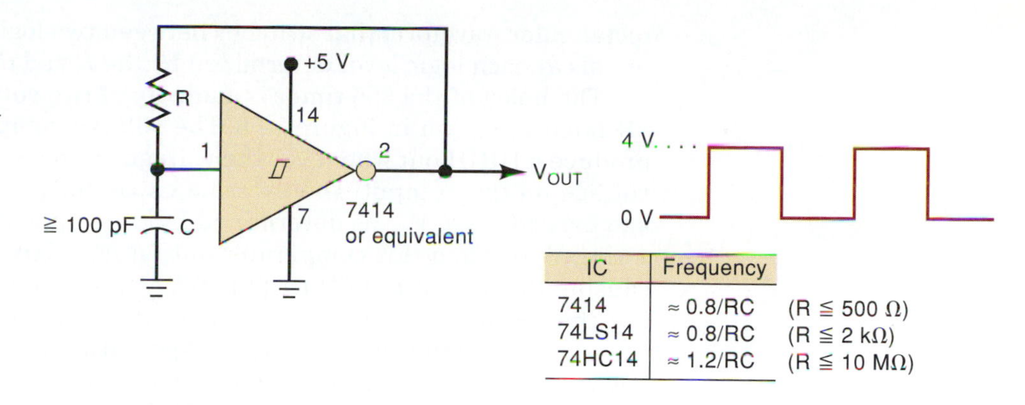

IC	Frequency	
7414	≈ 0.8/RC	(R ≦ 500 Ω)
74LS14	≈ 0.8/RC	(R ≦ 2 kΩ)
74HC14	≈ 1.2/RC	(R ≦ 10 MΩ)

and *RC* values is shown in Figure 5-55 for three different Schmitt-trigger INVERTERs. Note the maximum limits on the resistance value for each device. The circuit will fail to oscillate if *R* is not kept below these limits.

555 Timer Used as an Astable Multivibrator

The **555 timer** IC is a TTL-compatible device that can operate in several different modes. Figure 5-56 shows how external components can be connected to a 555 so that it operates as a free-running oscillator. Its output is a repetitive

FIGURE 5-56 555 timer IC used as an astable multivibrator.

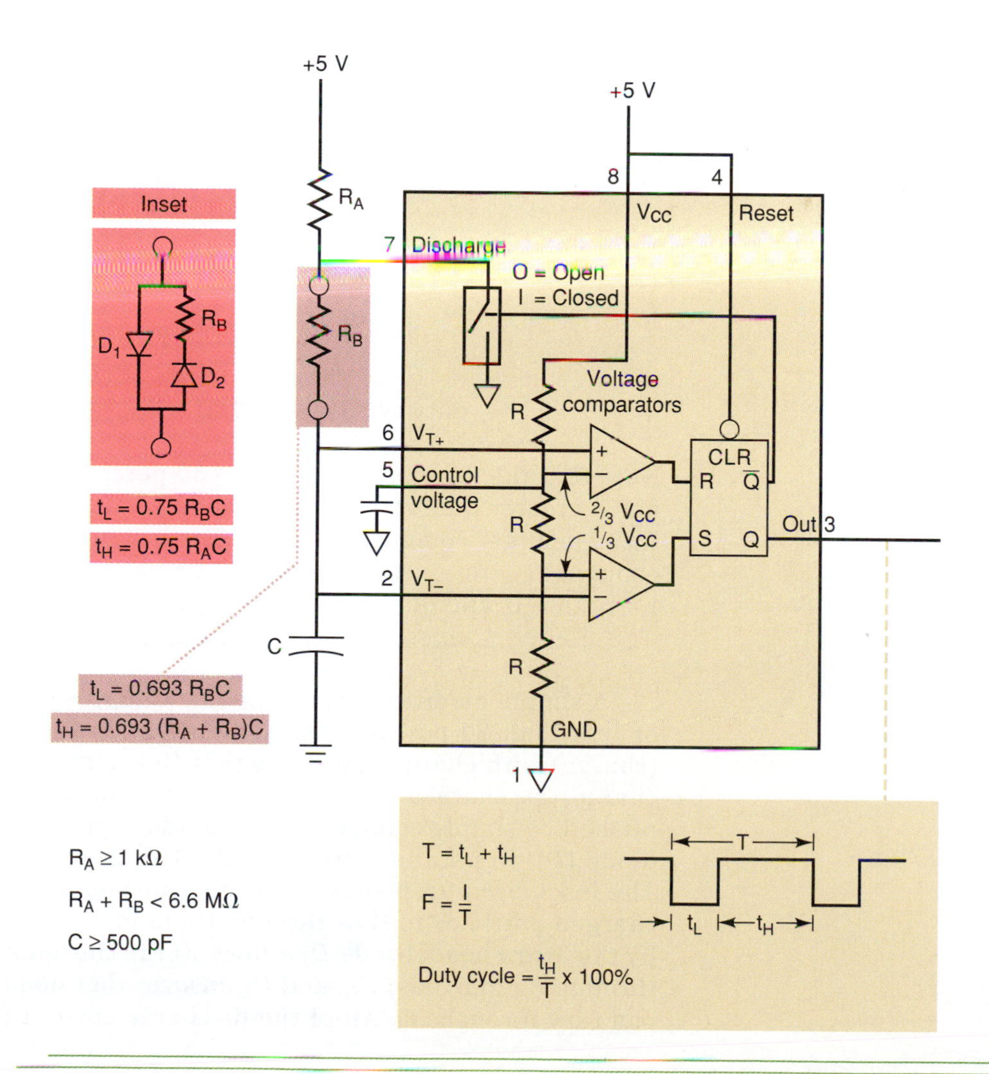

rectangular waveform that switches between two logic levels, with the time intervals at each logic level determined by the R and C values.

The heart of the 555 timer is made up of two voltage comparators and an SR latch as shown in Figure 5-56. The voltage comparators are devices that produce a HIGH out whenever the voltage on the + input is greater than the voltage on the − input. The external capacitor (C) charges up until its voltage exceeds $\frac{2}{3} \times V_{CC}$ as determined by the upper voltage comparator monitoring V_{T+}. When this comparator output goes HIGH, it resets the *SR* latch, causing the output pin (3) to go LOW. At the same time, $\overline{Q}$ goes HIGH, closing the discharge switch and causing the capacitor to begin to discharge through R_B. It will continue to discharge until the capacitor voltage drops below $\frac{1}{3} \times V_{CC}$ as determined by the lower-voltage comparator monitoring V_{T-}. When this comparator output goes HIGH, it sets the *SR* latch, causing the output pin to go HIGH, opening the discharge switch, and allowing the capacitor to start charging again as the cycle repeats.

The formulas for these time intervals, t_L and t_H, and the overall period of the oscillations, T, are given in the figure. The frequency of the oscillations is, of course, the reciprocal of T. As the formulas in the diagram indicate, the t_L and t_H intervals cannot be equal unless R_A is made zero. This cannot be done without producing excess current through the device. This means that it is impossible to produce a perfect 50 percent duty-cycle square wave output with this circuit. It is possible, however, to get very close to a 50 percent duty cycle by making $R_B >> R_A$ (while keeping R_A greater than $1\ \text{k}\Omega$), so that $t_L \approx t_H$.

EXAMPLE 5-17

Calculate the frequency and the duty cycle of the 555 astable multivibrator output for $C = 0.001\ \mu\text{F}$, $R_A = 2.2\ \text{k}\Omega$, and $R_B = 100\ \text{k}\Omega$.

Solution

$$\begin{aligned} t_L &= 0.693(100\ \text{k}\Omega)(0.001\ \mu\text{F}) = 69.3\ \mu\text{s} \\ t_H &= 0.693(102.2\ \text{k}\Omega)(0.001\ \mu\text{F}) = 70.7\ \mu\text{s} \\ T &= 69.3 + 70.7 = 140\ \mu\text{s} \\ f &= 1/140\ \mu\text{s} = 7.29\ \text{kHz} \\ \text{duty cycle} &= 70.7/140 = 50.5\% \end{aligned}$$

Note that the duty cycle is close to 50 percent (square wave) because R_B is much greater than R_A. It can be made even closer to 50 percent by making R_B even larger compared with R_A. For instance, you should verify that if we change R_A to $1\ \text{k}\Omega$ (its minimum allowed value), the results are $f = 7.18\ \text{kHz}$ and duty cycle = 50.3 percent.

A simple modification can be made to this circuit to allow a duty cycle of less than 50 percent. The strategy is to allow the capacitor to fill up (charge) with charged particles that flow only through R_A and empty (discharge) as charged particles flow only through R_B. This can be accomplished by simply connecting one diode (D_2) in series with R_B and another diode (D_1) in parallel with R_B and D_2 as shown in the inset of Figure 5-56. The inset circuit replaces R_B in the drawing. Diodes are devices that allow charged particles to flow through them in only one direction, as indicated by the arrow head. Diode D_1 allows all the charging current which has come through R_A to bypass R_B, and D_2 ensures that none of the charging current can flow through R_B. All of the discharge current flows through D_2 and R_B

when the discharge switch is closed. The equations for the time high and time low for this circuit are

$$t_L = 0.75\, R_B C$$
$$t_H = 0.75\, R_A C$$

Note: The constant 0.75 is correct only for $V_{CC} = 5$ V.

EXAMPLE 5-18

Using the diodes along with R_B as shown in Figure 5-56, calculate the values of R_A and R_B necessary to get a 1 kHz, 25 percent duty cycle waveform out of a 555. Assume C is a 0.1 μF capacitor.

Solution

$$T = \frac{1}{F} = \frac{1}{1000} = 0.001 \text{ s} = 1 \text{ ms}$$
$$t_H = 0.25 \times T = 0.25 \times 1 \text{ ms} = 250\ \mu\text{s}$$
$$R_A = \frac{250\ \mu\text{s}}{0.75 \times C} = \frac{250\ \mu\text{s}}{0.75 \times 0.1\ \mu\text{F}} = 3.3 \text{ k}\Omega$$
$$R_B = \frac{750\ \mu\text{s}}{0.75 \times C} = \frac{750\ \mu\text{s}}{0.75 \times 0.1\ \mu\text{F}} = 10 \text{ k}\Omega$$

Crystal-Controlled Clock Generators

The output frequencies of the signals from the clock-generating circuits described above depend on the values of resistors and capacitors, and thus they are not extremely accurate or stable. Even if variable resistors are used so that the desired frequency can be adjusted by "tweaking" the resistance values, changes in the R and C values will occur with changes in ambient temperature and with aging, thereby causing the adjusted frequency to drift. If frequency accuracy and stability are critical, another method of generating clock signals can be used: a **crystal-controlled clock generator**. It employs a highly stable and accurate component called a *quartz crystal*. A piece of quartz crystal can be cut to a specific size and shape to vibrate (resonate) at a precise frequency that is extremely stable with temperature and aging; frequencies from 10 kHz to 80 MHz are readily achievable. When a crystal is placed in certain circuit configurations, it can produce oscillations at an accurate and stable frequency equal to the crystal's resonant frequency. Crystal oscillators are available as IC packages.

Crystal-controlled clock generator circuits are used in all microprocessor-based systems and microcomputers, and in any application in which a clock signal is used to generate accurate timing intervals. We will see this in some of the applications we encounter in the following chapters.

REVIEW QUESTIONS

1. Determine the approximate frequency of a Schmitt-trigger oscillator that uses a 74HC14 with $R = 10$ kΩ and $C = 0.005\ \mu$F.
2. Determine the approximate frequency and duty cycle of the 555 oscillator for $R_A = R_B = 2.2$ kΩ and $C = 2000$ pF.
3. What is the advantage of crystal-controlled clock generator circuits over *RC*-controlled circuits?

5-24 TROUBLESHOOTING FLIP-FLOP CIRCUITS

Flip-flop ICs are susceptible to the same kinds of internal and external faults that occur in combinational logic circuits. All of the troubleshooting ideas that were discussed in Chapter 4 can readily be applied to circuits that contain FFs as well as logic gates.

Because of their memory characteristic and their clocked operation, FF circuits are subject to several types of faults and associated symptoms that do not occur in combinational circuits. In particular, FF circuits are susceptible to timing problems that are generally not a concern in combinational circuits. The most common types of FF circuit faults are described.

Open Inputs

Unconnected or floating inputs of any logic circuit are particularly susceptible to picking up spurious voltage fluctuations called *noise*. If the noise is large enough in amplitude and long enough in duration, the logic circuit's output may change states in response to the noise. In a logic gate, the output will return to its original state when the noise signal subsides. In a FF, however, the output will remain in its new state because of its memory characteristic. Thus, the effect of noise pickup at any open input is usually more critical for a FF or latch than it is for a logic gate.

The most susceptible FF inputs are those that can trigger the FF to a different state—such as the *CLK*, PRESET, and CLEAR. Whenever you see a FF output that is changing states erratically, you should consider the possibility of an open connection at one of these inputs.

EXAMPLE 5-19

Figure 5-57 shows a three-bit shift register made up of TTL flip-flops. Initially, all of the FFs are in the LOW state before clock pulses are applied.

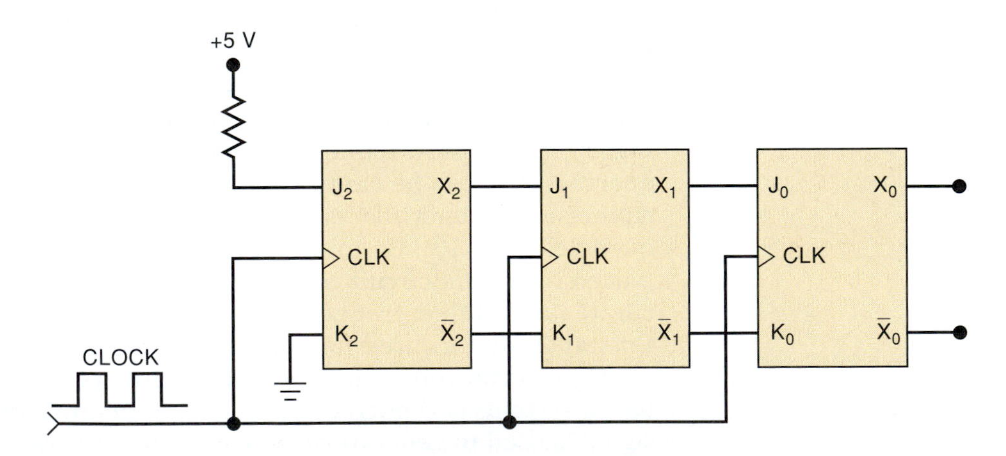

FIGURE 5-57 Example 5-19.

Clock pulse number	"Expected"			"Actual"		
	X_2	X_1	X_0	X_2	X_1	X_0
0	0	0	0	0	0	0
1	1	0	0	1	0	0
2	1	1	0	1	1	0
3	1	1	1	1	1	1
4	1	1	1	1	1	0
5	1	1	1	1	1	1
6	1	1	1	1	1	0
7	1	1	1	1	1	1
8	1	1	1	1	1	0

As clock pulses are applied, each PGT will cause the information to shift from each FF to the one on its right. The diagram shows the "expected" sequence of FF states after each clock pulse. Since $J_2 = 1$ and $K_2 = 0$, flip-flop X_2 will go HIGH on clock pulse 1 and will stay there for all subsequent pulses. This HIGH will shift into X_1, and then X_0 on clock pulses 2 and 3, respectively. Thus, after the third pulse, all FFs will be HIGH and should remain there as pulses are continually applied.

Now let's suppose that the "actual" response of the FF states is as shown in the diagram. Here the FFs change as expected for the first three clock pulses. From then on, flip-flop X_0, instead of staying HIGH, alternates between HIGH and LOW. What possible circuit fault can produce this operation?

Solution

On the second pulse, X_1 goes HIGH. This should make $J_0 = 1, K_0 = 0$ so that all subsequent clock pulses should set $X_0 = 1$. Instead, we see X_0 changing states (toggling) on all pulses after the second one. This toggle operation would occur if J_0 and K_0 were both HIGH. The most probable fault is a break in the connection between $\overline{X}_1$ and K_0. Recall that a TTL device responds to an open input as if it were a logic HIGH, so that an open at K_0 is the same as a HIGH.

Shorted Outputs

The following example illustrates how a fault in a FF circuit can cause a misleading symptom that may result in a longer time to isolate the fault.

EXAMPLE 5-20

Consider the circuit in Figure 5-58 and examine the logic probe indications shown in the accompanying table. There is a LOW at the D input of the FF when pulses are applied to its CLK input, but the Q output fails to go to the LOW state. The technician testing this circuit considers each of the following possible circuit faults:

1. Z2-5 is internally shorted to V_{CC}.
2. Z1-4 is internally shorted to V_{CC}.

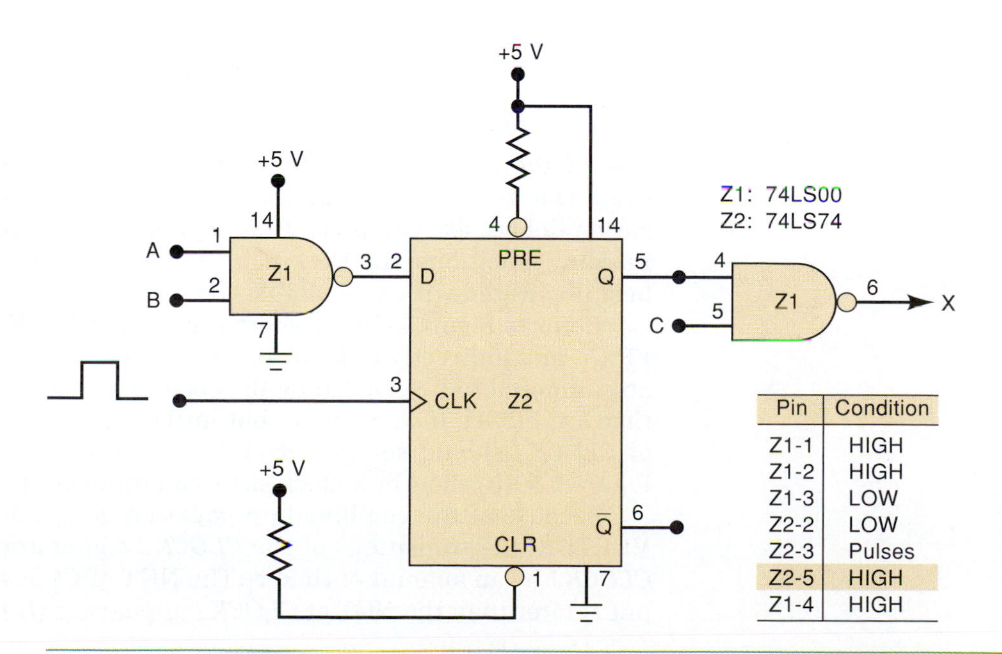

Pin	Condition
Z1-1	HIGH
Z1-2	HIGH
Z1-3	LOW
Z2-2	LOW
Z2-3	Pulses
Z2-5	HIGH
Z1-4	HIGH

FIGURE 5-58 Example 5-20.

3. Z2-5 or Z1-4 is externally shorted to V_{CC}.
4. Z2-4 is internally or externally shorted to GROUND. This would keep $\overline{PRE}$ activated and would override the *CLK* input.
5. There is an internal failure in Z2 that prevents *Q* from responding properly to its inputs.

The technician, after making the necessary ohmmeter checks, rules out the first four possibilities. He also checks Z2's V_{CC} and GROUND pins and finds that they are at the proper voltages. He is reluctant to unsolder Z2 from the circuit until he is certain that it is faulty, and so he decides to look at the clock signal. He uses an oscilloscope to check its amplitude, frequency, pulse width, and transition times. He finds that they are all within the specifications for the 74LS74. Finally, he concludes that Z2 is faulty.

He removes the 74LS74 chip and replaces it with another one. To his dismay, the circuit with the new chip behaves in exactly the same way. After scratching his head, he decides to change the NAND gate chip, although he doesn't know why. As expected, he sees no change in the circuit operation.

Becoming more puzzled, he recalls that his electronics lab instructor emphasized the value of performing a thorough visual check on the circuit board, and so he begins to examine it carefully. While he is doing that, he detects a solder bridge between pins 6 and 7 of Z2. He removes it and tests the circuit, and it functions correctly. Explain how this fault produced the operation observed.

Solution

The solder bridge was shorting the $\overline{Q}$ output to GROUND. This means that $\overline{Q}$ is permanently stuck LOW. Recall that in all latches and FFs, the $\overline{Q}$ and Q outputs are internally cross-coupled so that the level on one will affect the other. For example, take another look at the internal circuitry for a J-K flip-flop in Figure 5-25. Note that a constant LOW at $\overline{Q}$ would keep a LOW at one input of NAND gate 3 so that Q would have to stay HIGH, regardless of the conditions at *J*, *K*, and *CLK*.

The technician learned a valuable lesson about troubleshooting FF circuits. He learned that both outputs should be checked for faults, even those that are not connected to other devices.

Clock Skew

One of the most common timing problems in sequential circuits is **clock skew**. One type of clock skew occurs when a clock signal, because of propagation delays, arrives at the *CLK* inputs of different FFs at different times. In many situations, the skew can cause a FF to go to a wrong state. This is best illustrated with an example.

Refer to Figure 5-59(a), where the signal *CLOCK1* is connected directly to FF Q_1, and indirectly to Q_2 through a NAND gate and INVERTER. Both FFs are supposed to be clocked by the occurrence of a NGT of *CLOCK1* provided that *X* is HIGH. If we assume that initially $Q_1 = Q_2 = 0$ and $X = 1$, the NGT of *CLOCK1* should set $Q_1 = 1$ and have no effect on Q_2. The waveforms in Figure 5-59(b) show how clock skew can produce incorrect triggering of Q_2.

Because of the combined propagation delays of the NAND gate and INVERTER, the transitions of the *CLOCK2* signal are delayed with respect to *CLOCK1* by an amount of time t_1. The NGT of *CLOCK2* arrives at Q_2's *CLK* input t_1 later than the NGT of *CLOCK1* appears at Q_1's *CLK* input. This t_1 is the

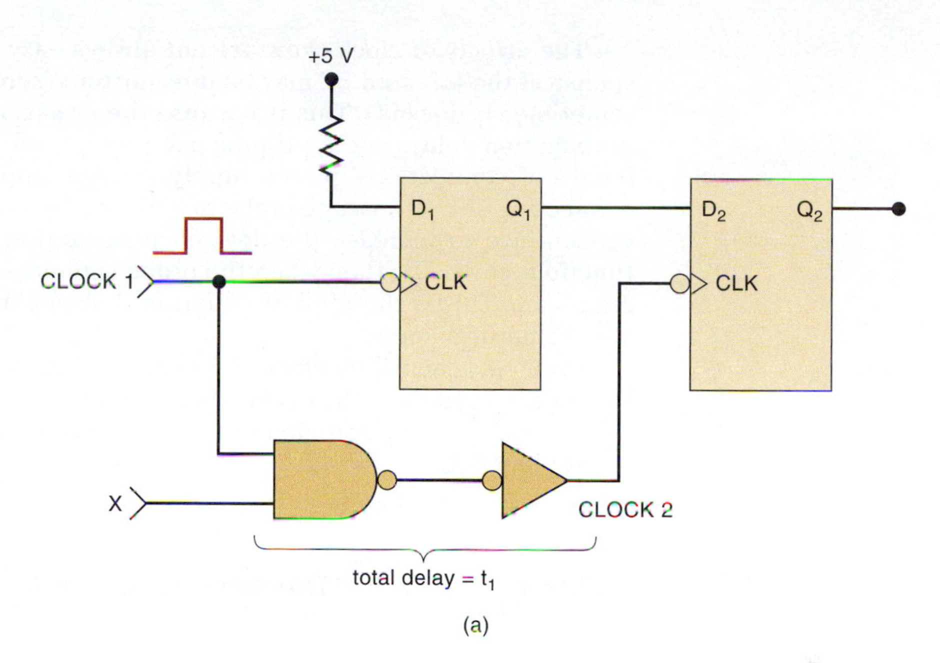

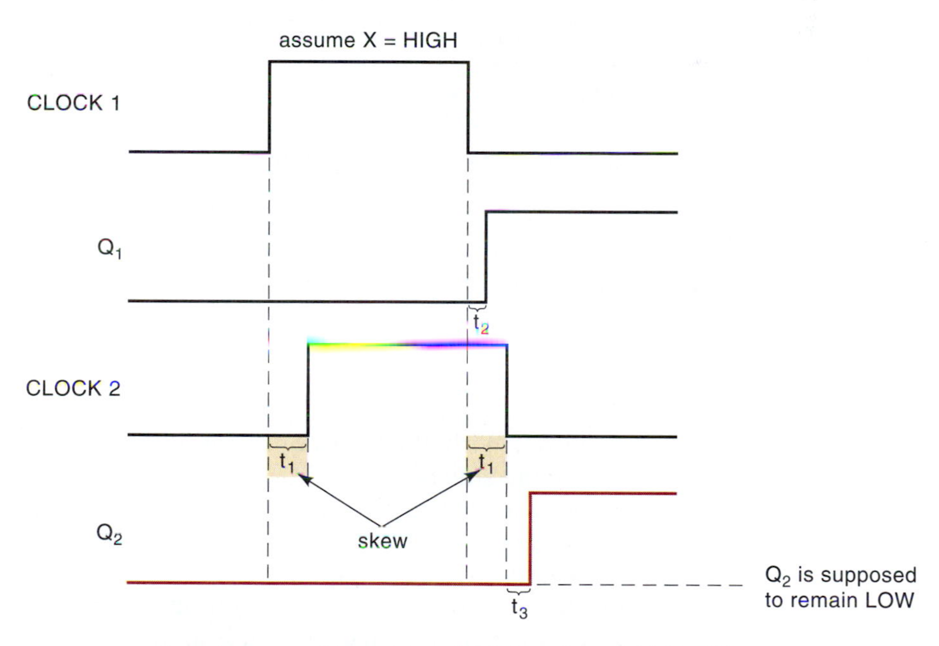

FIGURE 5-59 Clock skew occurs when two flip-flops that are supposed to be clocked simultaneously are clocked at slightly different times due to a delay in the arrival of the clock signal at the second flip-flop. (a) Extra gating circuits that can cause clock skew; (b) timing showing the later arrival of CLOCK 2.

clock skew. The NGT of *CLOCK1* will cause Q_1 to go HIGH after a time t_2 that is equal to Q_1's t_{PLH} propagation delay. If t_2 is less than the skew t_1, Q_1 will be HIGH when the NGT of *CLOCK2* occurs, and this may incorrectly set $Q_2 = 1$ if its setup time requirement, t_S, is met.

For example, assume that the clock skew is 40 ns and the t_{PLH} of Q_1 is 25 ns. Thus, Q_1 will go HIGH 15 ns before the NGT of *CLOCK2*. If Q_2's setup time requirement is smaller than 15 ns, Q_2 will respond to the HIGH at its D input when the NGT of *CLOCK2* occurs, and Q_2 will go HIGH. This, of course, is not the expected response of Q_2. It is supposed to remain LOW.

The effects of clock skew are not always easy to detect because the response of the affected FF may be intermittent (sometimes it works correctly, sometimes it doesn't). This is because the situation is dependent on circuit propagation delays and FF timing parameters, which vary with temperature, length of connections, power supply voltage, and loading. Sometimes just connecting an oscilloscope probe to a FF or gate output will add enough load capacitance to increase the device's propagation delay so that the circuit functions correctly; then when the probe is removed, the incorrect operation reappears. This is the kind of situation that explains why some technicians are prematurely gray.

Problems caused by clock skew can be eliminated by equalizing the delays in the various paths of the clock signal so that the active transition arrives at each FF at approximately the same time. This situation is examined in Problem 5-52.

REVIEW QUESTION

1. What is clock skew? How can it cause a problem?

5-25 SEQUENTIAL CIRCUITS USING HDL*

In Chapters 3 and 4, we used HDL to program simple combinational logic circuits. In this chapter, we have studied logic circuits that latch and clocked flip-flop circuits that sequence through various states in response to a clock edge. These latching and sequential circuits can also be implemented using PLDs and described using HDL.

Section 5-1 of this chapter described a NAND gate latch. You will recall that the unique characteristic of this circuit is the fact that its outputs are cross coupled back to its gates' inputs. This causes the circuit to respond differently depending on which state its output happens to be in. Describing circuits that have outputs that *feed back* to the input with Boolean equations or HDL involves using the output variables in the conditional portion of the description. With Boolean equations it means including output terms in the right-hand side of the equation. Using IF/THEN constructs it means including output variables in the IF clause. Most PLDs have the ability to feed back the output signal to the input circuitry in order to accommodate latching action.

When writing equations that use feedback, some languages, such as VHDL, require a special designation for the output port. In these cases the port bit is not only an output; it is an output with feedback. The difference is shown in Figure 5-60.

FIGURE 5-60 Three input/output modes.

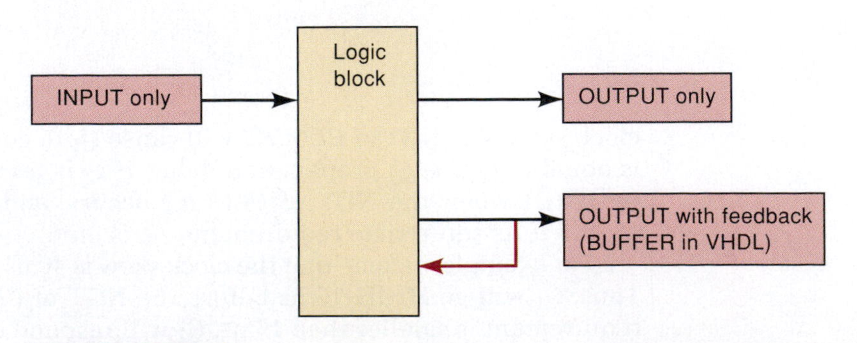

*As stated in Chapter 3, this section and all sections covering PLDs and HDLs may be skipped if desired.

Rather than describing the operation of a latch using Boolean equations, let's try to think of a behavioral description of how the latch should operate. The situations we need to address are when SBAR is activated, when RBAR is activated, and when neither is activated. Recall that the invalid state occurs when both inputs are activated simultaneously. If we can describe a circuit that always recognizes one of the inputs as the winner when both are active, we can avoid the undesirable results of having an invalid input condition. To describe such a circuit, let's ask ourselves under what conditions the latch should be set ($Q = 1$). Certainly, the latch should be set if the SET input is active, but what about after SET goes back to its inactive level? How does the latch know to stay in the SET state? The description needs to use the *condition of the output now* to determine the *future* condition of the output. The following statement describes the conditions that should make the output HIGH on an SR latch:

IF *SET* is active, THEN Q should be HIGH.

What conditions should make the output LOW?

IF *RESET* is active, THEN Q should be LOW.

What if neither input is activated? Then the output should remain the same and we can express this as $Q = Q$. This expression provides the feedback of the output state to be combined with input conditions for the purpose of deciding what happens next to the output.

What if both inputs are activated (i.e., the invalid input combination)? The structure of the IF/ELSE decision shown graphically in Figure 5-61 makes sure that the latch never tries to respond to both inputs. If the SET is active, regardless of what is on RESET, the output will be forced HIGH. The invalid input will always default to a set condition this way. The ELSIF clause is considered only when SET is not active. The use of the feedback term ($Q = Q$) affects the operation (holding action) only when neither input is active.

When you design sequential circuits that feed the output value back to the inputs, it is possible to create an unstable system. A change in the output

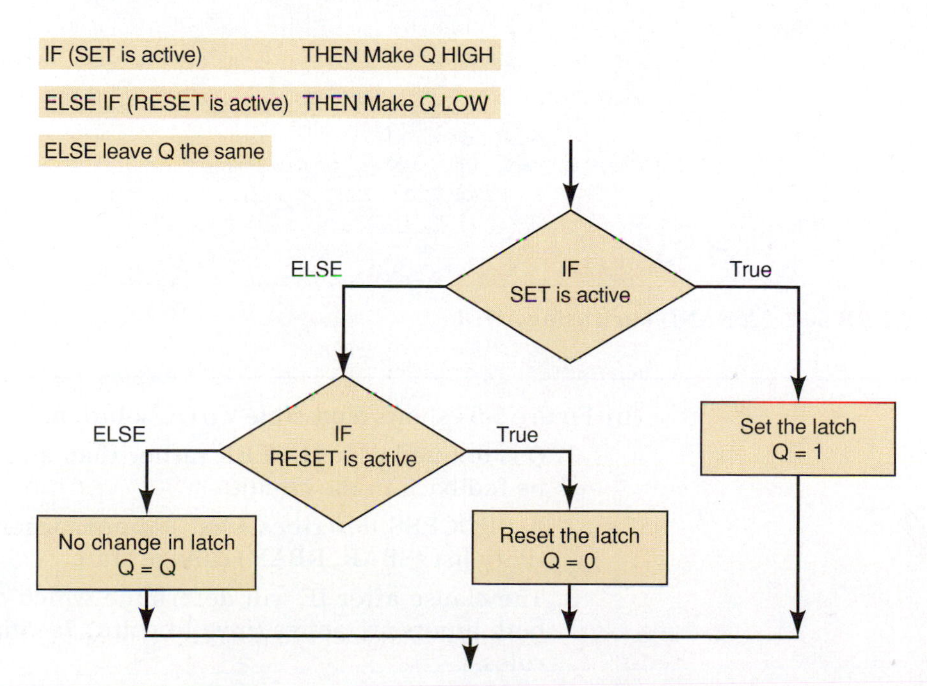

FIGURE 5-61 The logic of a behavioral description of an SR latch.

state might be fed back to the inputs, which changes the output state again, which feeds back to the inputs, which changes the output back again. This oscillation is obviously undesirable and so it is very important to make sure that no combination of inputs and outputs can make this happen. Careful analysis, simulation, and testing should be used to ensure that your circuit is stable under all conditions. For this circuit it is necessary to enable multilevel synthesis for the MAX 7000 series of components before compiling in order to avoid oscillation (at least in the simulator) when changing from the "invalid" input to the "no change" input.

EXAMPLE 5-21

Describe an active-LOW input SR latch with inputs named SBAR, RBAR, and one output named Q. It should follow the function table of a NAND latch (see Figure 5-6) and the invalid input combination should produce $Q = 1$.

(a) Use AHDL.
(b) Use VHDL.

AHDL

Solution

(a) Figure 5-62 shows a possible AHDL solution. Important items to note are:
1. Q is defined as an OUTPUT, even though it is *fed back* in the circuit. AHDL allows outputs to be fed back into the circuit.
2. The clause after IF will determine which output state occurs when both inputs are active (invalid state). In this code the SET command rules.
3. To evaluate equality, the double equal sign is used. In other words, SBAR == 0 evaluates TRUE when SBAR is active (LOW).

```
SUBDESIGN fig5_62
(
      sbar, rbar                  :INPUT;
      q                           :OUTPUT;
)
BEGIN
      IF    sbar == 0   THEN  q = VCC;    -- set or illegal command
      ELSIF rbar == 0   THEN  q = GND;    -- reset
      ELSE                    q = q;      -- hold
      END IF;
END;
```

FIGURE 5-62 A NAND latch using AHDL.

VHDL

(b) Figure 5-63 shows a possible VHDL solution. Important items to note are:
1. Q is defined as a BUFFER rather than an OUTPUT. This allows it to be fed back in the circuit.
2. A PROCESS describes what happens when the values in the sensitivity list (SBAR, RBAR) change state.
3. The clause after IF will determine which output state occurs when both inputs are active (invalid state). In this code the SET command rules.

```
-- must compile with Multi-Level Synthesis for 7000 enabled
ENTITY fig5_63 IS
PORT (   sbar, rbar   :IN BIT;
         q            :BUFFER BIT);
END fig5_63;
ARCHITECTURE behavior OF fig5_63 IS
BEGIN
   PROCESS (sbar, rbar)
      BEGIN
         IF    sbar = '0'  THEN  q <= '1';    -- set or illegal command
         ELSIF rbar = '0'  THEN  q <= '0';    -- reset
         ELSE                    q <= q;      -- hold
         END IF;
      END PROCESS;
END behavior;
```

FIGURE 5-63 A NAND latch using VHDL.

The *D* Latch

The transparent *D* latch can also be easily implemented with HDLs. Altera's software has a library primitive called LATCH that is available. The AHDL module below illustrates using this LATCH primitive. All that is needed is to connect the primitive's enable (.ena) and data (.d) ports to the appropriate module signals. The VHDL module also shown below is a behavioral description of the *D* latch function. You can also use the LATCH primitive as a component in VHDL.

AHDL D latch

```
SUBDESIGN dlatch_ahdl
(enable, din                    :INPUT;
   q                            :OUTPUT;)

VARIABLE
   q                            :LATCH;
BEGIN
   q.ena = enable;
   q.d = din;
END;
```

VHDL D latch

```
ENTITY  dlatch_vhdl  IS
PORT  (enable, din          :IN BIT;
         q                  :OUT BIT);
END  dlatch_vhdl;

ARCHITECTURE v OF dlatch_vhdl IS
BEGIN
        PROCESS (enable, din)
        BEGIN
           IF  enable = '1'  THEN
               q <= din;
           END IF;
        END PROCESS;
END v;
```

REVIEW QUESTIONS

1. What is the distinguishing hardware characteristic of latching logic circuits?
2. What is the major characteristic of sequential circuits?

5-26 EDGE-TRIGGERED DEVICES

Earlier in this chapter, we introduced edge-triggered devices whose outputs respond to the inputs when the clock input sees an "edge." An edge simply means a transition from HIGH to LOW, or vice versa, and is often referred to as an **event**. If we are writing statements in the code that are concurrent, how can outputs change only when a clock input detects an edge event? The answer to this question differs substantially, depending on the HDL you use. In this section, we want to concentrate on creating clocked logic circuits in their simplest form using HDL. We will use JK flip-flops to correlate with many of the examples found earlier in this chapter.

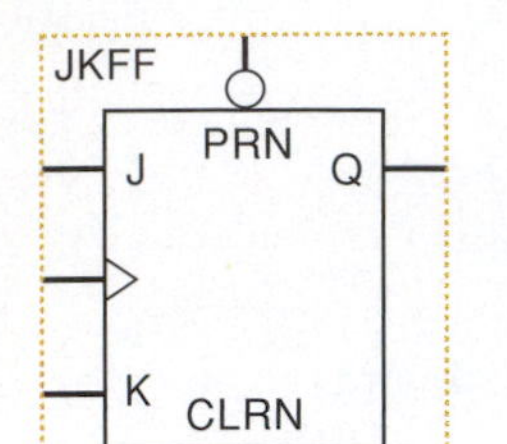

FIGURE 5-64 JK flip-flop logic primitive.

The JK flip-flop is a standard building block of clocked (sequential) logic circuits known as a **logic primitive**. In its most common form, it has five inputs and one output, as shown in Figure 5-64. The input/output names can be standardized to allow us to refer to the connections of this primitive or fundamental circuit. The actual operation of the primitive circuit is defined in a library of components that is available to the HDL compiler as it generates a circuit from our description. AHDL uses logic primitives to describe flip-flop operation. VHDL offers something similar, but it also allows the designer to describe the clocked logic circuit's operation explicitly in the code.

AHDL

AHDL FLIP-FLOPS

A flip-flop can be used in AHDL by declaring a register (even one flip-flop is called a register). Several different types of register primitives are available for use in AHDL, including JKFF, DFF, SRFF, and latch. Each different type of register primitive has its own official names (according to Altera software) for the ports of these primitives. These can be found by using the HELP menu in the ALTERA software and looking under Primitives. Table 5-3 lists some of these names. Registers that use these primitives are declared in the VARIABLE section of the code. The register is given an instance name, just as we have named intermediate variables or buried nodes in previous examples. Instead of declaring it as a node, however, it is declared by the type of the register primitive. For example, a JK flip-flop can be declared as:

```
VARIABLE
     ff1 :JKFF;
```

TABLE 5-3 Altera primitive port identifiers.

Standard Part Function	Primitive Port Name
Clock input	clk
Asynchronous preset (active-LOW)	prn
Asynchronous clear (active-LOW)	clrn
J, K, S, R, D inputs	j, k, s, r, d
Level triggered ENABLE input	ena
Q output	q

The instance name is ff1 (which you can make up) and the register primitive type is JKFF (which Altera requires you to use). Once you have declared a register, it is connected to the other logic in the design using its standard

```
1   %     JK flip-flop circuit     %
2   SUBDESIGN fig5_65
3   (
4         jin, kin, clkin, preset, clear     :INPUT;
5         qout                               :OUTPUT;
6   )
7   VARIABLE
8   ff1          :JKFF;       -- define this flip-flop as a JKFF type
9   BEGIN
10        ff1.prn = preset; -- these are optional and default to vcc
11        ff1.clrn = clear;
12        ff1.j = jin;      -- connect primitive to the input signal
13        ff1.k = kin;
14        ff1.clk = clkin;
15        qout = ff1.q;     -- connect the output pin to the primitive
16  END;
```

FIGURE 5-65 Single JK flip-flop using AHDL.

port names. The ports (or pins) on the flip-flop are referred to using the instance name, with a dot extension that designates the particular input or output. An example for a JK flip-flop in AHDL is shown in Figure 5-65. Notice that we have made up our own input/output names for this SUBDESIGN in order to distinguish them from the primitive port names. The single flip-flop is declared on line 8, as previously described. The J input or port for this device is then labeled *ff1.j*, the K input is *ff1.k*, the clock input is *ff1.clk*, and so on. Each of the given port assignment statements will make the needed wiring connections for this design block. The *prn* and *clrn* ports are both active-LOW, asynchronous controls such as those commonly found on a standard flip-flop. In fact, these asynchronous controls on an FF primitive can be used to implement an SR latch more efficiently than the code in Figure 5-62. The *prn* and *clrn* controls are optional in AHDL and will default to a disabled condition (at a logic 1) if they are omitted from the logic section. In other words, if lines 10 and 11 were deleted, the *prn* and *clrn* ports of ff1 would automatically be tied to V_{CC}.

VHDL LIBRARY COMPONENTS

The Altera software comes with some extensive libraries of components and primitives that can be used by a designer. The graphic description of a JKFF component in the Altera library is shown in Figure 5-66(a). After placing the component on the worksheet, each of its ports is connected to inputs and outputs of the module. This same concept can be implemented in VHDL using a library component. The inputs and outputs of these library components can be found by looking under the HELP/Primitives menu. Figure 5-66(b) shows the VHDL **COMPONENT** declaration for a JK flip-flop primitive. The key things to notice are the name of the component (JKFF) and the names of the ports. They are the same names as those used in the graphic symbol of Figure 5-66(a). Also, notice that the *type* of each input and output variable is STD_LOGIC. This is one of the IEEE standard data types defined in the library and used by many components in the library.

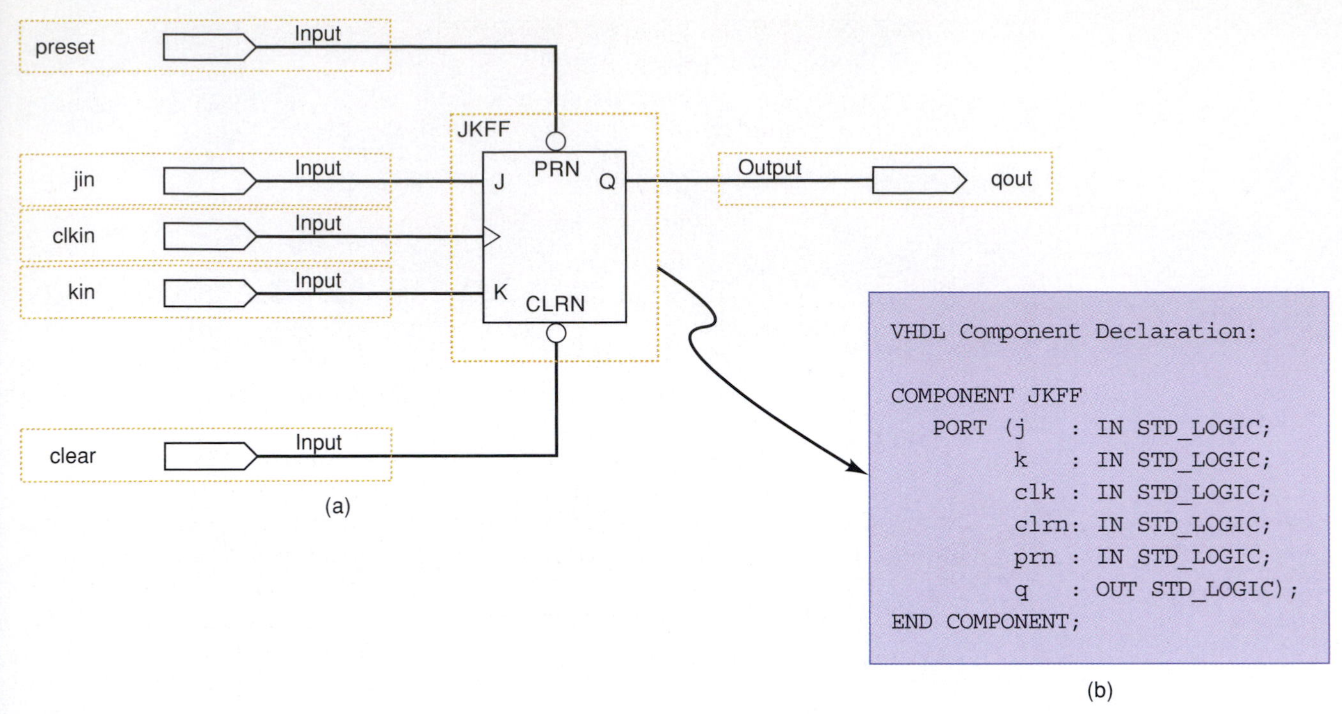

FIGURE 5-66 (a) Graphic representation using a component. (b) VHDL component declaration.

Figure 5-67 uses a JKFF component from the library in VHDL to create a circuit equivalent to the graphic design of Figure 5-66(a). The first two lines tell the compiler to use the IEEE library to find the definitions of the std_logic data types. The next two lines tell the compiler that it should look in the Altera library for any standard library components that are used later on in the code. The module inputs and outputs are declared as they were in previous examples, except that the type is now STD_LOGIC rather than BIT. This is because the module port types must match the component port types. Within the architecture section, a name (ff1) is given to this instance of the

```
LIBRARY ieee;
USE   ieee.std_logic_1164.all;          --defines std_logic types
LIBRARY  altera;
USE      altera.maxplus2.all;           -- provides standard components

ENTITY fig5_67 IS
PORT( clkin, jin, kin, preset, clear       :IN std_logic;
      qout                                 :OUT std_logic);
END fig5_67;

ARCHITECTURE a OF fig5_67 IS
BEGIN
   ff1:  JKFF PORT MAP (   clk    => clkin,
                           j      => jin,
                           k      => kin,
                           prn    => preset,
                           clrn   => clear,
                           q      => qout);
end a;
```

FIGURE 5-67 A JK flip-flop using VHDL.

component JKFF. The keywords **PORT MAP** are followed by a list of all the connections that must be made to the component ports. Notice that the component ports (e.g., clk) are listed on the left of the symbol =>and the objects they are connected to (e.g., clkin) are listed on the right.

VHDL FLIP-FLOPS

Now that we have seen how to use standard components that are available in the library, let's look next at how to create our own component that can be used over and over again. For the sake of comparison we will describe the VHDL code for a JK flip-flop that is identical to the library component JKFF.

VHDL was created as a very flexible language and it allows us to define the operation of clocked devices explicitly in the code, without relying on logic primitives. The key to edge-triggered sequential circuits in VHDL is the PROCESS. As you recall, this keyword is followed by a sensitivity list in parentheses. Whenever a variable in the sensitivity list changes state, the code in the process block determines how the circuit should respond. This is very much like a flip-flop that does nothing until the clock input changes state, at which time it evaluates its inputs and updates its outputs. If the flip-flop needs to respond to inputs other than the clock (e.g., preset and clear), they can be added to the sensitivity list. The code in Figure 5-68 demonstrates a JK flip-flop written in VHDL.

On line 9 of the figure, a signal is declared with a name of *qstate.* Signals can be thought of as wires that connect two points in the circuit description, but they also have characteristics of implied "memory." This means that once a value is assigned to the signal, it will stay at that value until a different value is assigned in the code. In VHDL, a VARIABLE is often used to

```
1   -- JK Flip-Flop circuit
2   ENTITY jk IS
3   PORT(
4         clk, j, k, prn, clrn :IN BIT;
5         q                    :OUT BIT);
6   END jk ;
7
8   ARCHITECTURE a OF jk IS
9   SIGNAL qstate  :BIT;
10  BEGIN
11     PROCESS(clk, prn, clrn)    -- respond to any of these signals
12     BEGIN
13        IF prn = '0' THEN qstate <= '1'; -- asynch preset
14        ELSIF clrn = '0' THEN qstate <= '0';-- asynch clear
15        ELSIF clk = '1' AND clk'EVENT THEN  -- on PGT clock edge
16           IF j = '1' AND k = '1' THEN qstate <= NOT qstate;
17           ELSIF j = '1' AND k = '0' THEN qstate <= '1';
18           ELSIF j = '0' AND k = '1' THEN qstate <= '0';
19           END IF;
20        END IF;
21     END PROCESS;
22     q <= qstate;         -- update output pin
23  END a;
```

FIGURE 5-68 Single JK flip-flop using VHDL.

implement this feature of "memory," but variables must be declared and used within the same description block. In this example, if *qstate* were declared as a VARIABLE, it would need to be declared within the PROCESS (after line 11) and must be assigned to *q* before the end of the PROCESS (line 21). Our example uses a SIGNAL that can be declared and used throughout the architecture description.

Notice that the PROCESS sensitivity list contains the asynchronous preset and clear signals. The flip-flop must respond to these inputs as soon as they are asserted (LOW), and these inputs should override the *J*, *K*, and clock inputs. To accomplish this, we can use the sequential nature of the IF/ELSE constructs. First, the PROCESS will describe what happens only when one of the three signals—*clk*, *prn*, or *clrn*—changes state. The highest priority input in this example is *prn* because it is evaluated first in line 13. If it is asserted, *qstate* will be set HIGH and the other inputs will not even be evaluated because they are in the else branch of the decision. If *prn* is HIGH, *clrn* will be evaluated in line 14 to see if it is LOW. If it is, the flip-flop will be cleared and nothing else will be evaluated in the PROCESS. Line 15 will be evaluated only if both *prn* and *clrn* are HIGH. The term clk' **EVENT** in line 15 evaluates as TRUE only if there has been a transition on *clk*. Because *clk* = '1' must be TRUE also, this condition responds only to a rising edge transition on the clock. The next three conditions of lines 16, 17, and 18 are evaluated only following a rising edge on *clk* and serve to update the flip-flop's state. In other words, they are **nested** within the ELSIF statement of line 15. Only the JK input commands for toggle, set, and reset are evaluated by the IF/ELSIF on lines 16–18. Of course, with a JK there is a fourth command, hold. The "missing" ELSE condition will be interpreted by VHDL as an implied memory device that will then hold the PRESENT state if none of the given JK conditions is TRUE. Note that each IF/ELSIF structure has its own END IF statement. Line 19 ends the decision structure that decides to set, clear, or toggle. Line 20 ends the IF/ELSIF structure that decides among the preset, clear, and clock edge responses. As soon as the PROCESS ends, the flip-flop's state is transferred to the output port *q*.

Regardless of whether you develop your description in AHDL or VHDL, the circuit's proper operation can be verified using a simulator. The most important and challenging part of verification using a simulator is creating a set of hypothetical input conditions that will prove that the circuit does everything it is intended to do. There are many ways to do this, and it is up to the designer to decide which way is best. The simulation used to verify the operation of the JKFF primitive is shown in Figure 5-69. The *preset* input is initially activated and then, at t1, the *clear* input is activated. These tests ensure that *preset* and *clear* are operating asynchronously. The *jin* input is HIGH at t2 and *kin* is HIGH at t3. In between these points, the inputs on *jin* and *kin*

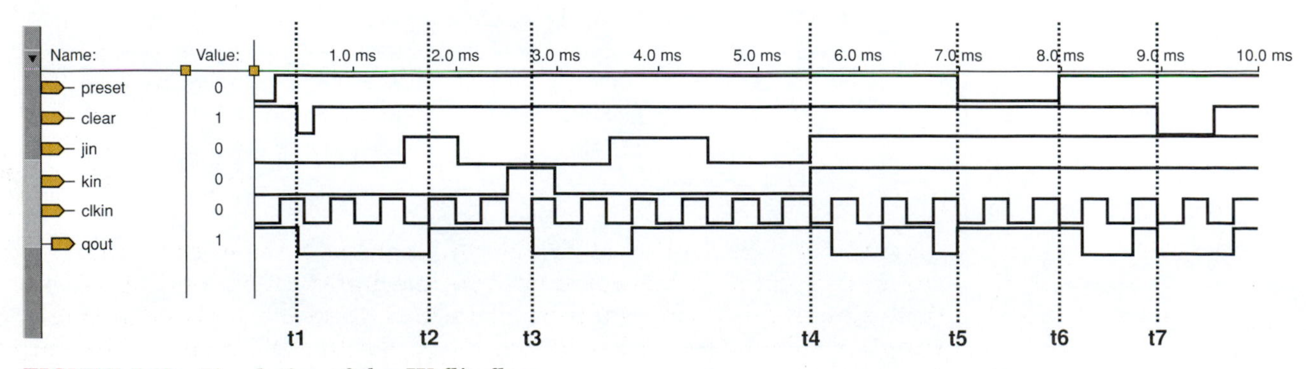

FIGURE 5-69 Simulation of the JK flip-flop.

are both LOW. This portion of the simulation tests the synchronous modes of set, hold, and reset. Starting at t4, the toggle command is tested with *jin* = *kin* = 1. Notice at t5, *preset* is asserted (LOW) to test whether *preset* overrides the toggle command. After t6, the output starts toggling again, and at t7, the *clear* input is shown overriding the synchronous inputs. Testing of all modes of operation and the interaction of various controls is very important when you are simulating.

REVIEW QUESTIONS

1. What is a logic primitive?
2. What does the designer need to know in order to use a logic primitive?
3. In the Altera system, where can you find information on primitives and library functions?
4. What is the key VHDL element that allows the explicit description of clocked logic circuits?
5. Which library defines the std_logic data types?
6. Which library defines the logic primitives and common components?

5-27 HDL CIRCUITS WITH MULTIPLE COMPONENTS

We began this chapter by studying latches. Latches were used to make flip-flops and flip-flops were used to make many circuits, including binary counters. A graphic description (logic diagram) of a simple binary up counter is shown in Figure 5-70. This circuit is functionally the same as Figure 5-47, which was drawn with the LSB on the right to make it easier to visualize the numeric value of the binary count. The circuit has been redrawn here to show the signal flow in the more conventional format, with inputs on the left and outputs on the right. Notice that these logic symbols are negative edge-triggered. These flip-flops also do not have asynchronous inputs prn or clrn. Our goal is to describe this counter circuit using HDL by interconnecting three instances of the same JK flip-flop component.

FIGURE 5-70 A three-bit binary counter.

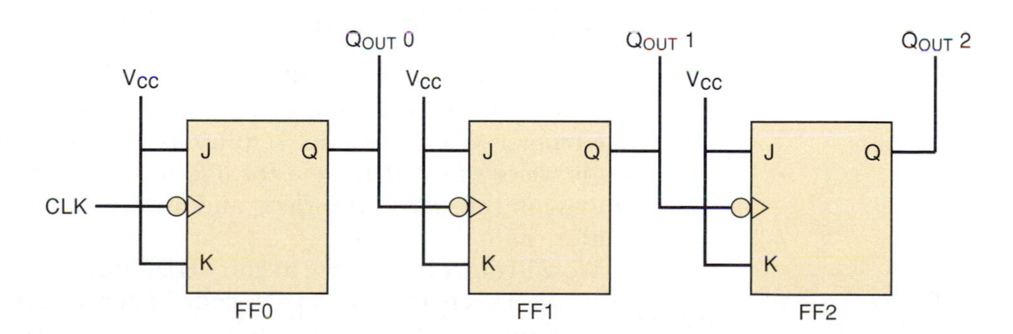

AHDL RIPPLE-UP COUNTER

A text-based description of this circuit requires three of the same type of flip-flop, just like the graphic description. Refer to Figure 5-71. On line 8 of the figure, bit array notation is used to declare a register of three JK flip-flops. The name of this register is *q*, just like the name of the output port. AHDL can interpret this to mean that the output of each flip-flop

AHDL

```
1   % MOD 8 ripple up counter. %
2   SUBDESIGN fig5_71
3   (
4       clock                    :INPUT;
5       q[2..0]                  :OUTPUT;
6   )
7   VARIABLE
8       q[2..0] :JKFF;           -- defines three JK FFs
9   BEGIN
10                               -- note: prn, clrn default to vcc!
11      q[2..0].j = VCC;         -- toggle mode J=K=1 for all FFs
12      q[2..0].k = VCC;
13      q[0].clk = !clock;
14      q[1].clk = !q[0].q;
15      q[2].clk = !q[1].q;      -- connect clocks in ripple form
16  END;
```

FIGURE 5-71 MOD-8 ripple counter in AHDL.

should be connected to the output port. Each bit of the array *q* has all the attributes of a JKFF primitive. AHDL is very flexible in its use of indexed sets like this. As an example of the use of this set notation, notice how all the J and K inputs for all the flip-flops are tied to VCC in lines 11 and 12. If the flip-flops had been named A, B, and C rather than using a bit array, then individual assignments would be necessary for each J and K input, making the code much longer. Next, the key interconnections are made between the flip-flops to make this a ripple-up counter. The clock signal is inverted and assigned to FF0 clock input (line 13), the Q output of FF0 is inverted and assigned to FF1 clock input (line 14), and so on, forming a ripple counter.

VHDL

VHDL RIPPLE-UP COUNTER

We described in Figure 5-68 the VHDL code for a positive-edge-triggered JKFF with preset and clear controls. The counter in Figure 5-70 is negative-edge-triggered and does not require asynchronous preset or clear. Our goal now is write the VHDL code for one of these flip-flops, represent three instances of the same flip-flop, and interconnect the ports to create the counter.

We will start by looking at the VHDL description in Figure 5-72, starting at line 18. This module of VHDL code is describing the operation of a single JK flip-flop component. The name of the component is neg_jk (line 18) and it has inputs *clk*, *j*, and *k* (line 19) and output *q* (line 20). A signal named *qstate* is used to hold the state of the flip-flop and connect it to the *q* output. On line 25, the PROCESS has only *clk* in its sensitivity list, so it only responds to changes in the *clk* (PGTs and NGTs). The statement that makes this flip-flop negative-edge-triggered is on line 27. IF (*clk*'EVENT AND *clk* = '0') is true, then a *clk* edge has just occurred and *clk* is now LOW, meaning it must have been an NGT of *clk*. The IF/ELSE decisions that follow implement the four states of a JK flip-flop.

```
1   ENTITY fig5_72 IS
2   PORT (    clock           :IN BIT;
3             qout            :BUFFER BIT_VECTOR (2 DOWNTO 0));
4   END fig5_72;
5   ARCHITECTURE counter OF fig5_72 IS
6      SIGNAL high            :BIT;
7      COMPONENT neg_jk
8      PORT (    clk, j, k    :IN BIT;
9                q            :OUT BIT);
10     END COMPONENT;
11  BEGIN
12     high <= '1';        -- connect to Vcc
13  ff0:  neg_jk    PORT MAP (j => high, k => high, clk => clock,   q => qout(0));
14  ff1:  neg_jk    PORT MAP (j => high, k => high, clk => qout(0),q => qout(1));
15  ff2:  neg_jk    PORT MAP (j => high, k => high, clk => qout(1),q => qout(2));
16  END counter;
17
18  ENTITY neg_jk IS
19  PORT (    clk, j, k       :IN BIT;
20        q                   :OUT BIT);
21  END neg_jk;
22  ARCHITECTURE simple of neg_jk IS
23     SIGNAL qstate          :BIT;
24  BEGIN
25     PROCESS (clk)
26     BEGIN
27        IF (clk'EVENT AND clk = '0') THEN
28           IF j = '1' AND k = '1'  THEN qstate <= NOT qstate; -- toggle
29           ELSIF j ='1' AND k = '0'    THEN qstate <= '1';      -- set
30           ELSIF j = '0' AND k = '1'  THEN qstate <= '0';      -- reset
31           END IF;
32        END IF;
33     END PROCESS;
34     q <= qstate                  -- connect flip-flop state to output
35  END simple;;
```

FIGURE 5-72 MOD-8 ripple counter in VHDL.

Now that we know how one flip-flop named neg_jk works, let's see how we can use it three times in a circuit and hook all the ports together. Line 1 defines the ENTITY that will make up the three-bit counter. Lines 2–3 contain the definitions of the inputs and outputs. Notice that the outputs are in the form of a three-bit array (bit vector). On line 6 the SIGNAL *high* can be thought of as a wire used to connect points in the circuit to V_{CC}. Line 7 is very important because this is where we declare that we plan to use a component in our design whose name is neg_jk. In this example, the actual code is written at the bottom of the page, but it could be in a separate file or even in a library. This declaration tells the compiler all the important facts about the component and its port names.

The final part of the description is the concurrent section of lines 12–15. First, the signal *high* is connected to V_{CC} on line 12. The next three lines are instantiations of the flip-flop components. The three instances are named ff0, ff1, and ff2. Each instance is followed by a PORT MAP which lists each port of the component and describes what it is connected to in the module.

Connecting components together using HDL is not difficult, but it is very tedious. As you can see, the file for even a very simple circuit can be quite long. This method of describing circuits is referred to as the **structural level of abstraction**. It requires the designer to account for each pin of each component and define signals for each wire that is to interconnect the components. People who are accustomed to using logic diagrams to describe circuits generally find it easy to understand the structural level, but not as easy to read at a glance as the equivalent logic circuit diagram. In fact, it is safe to say that if the structural level of description was all that was available, most people would prefer using graphic descriptions (schematics) rather than HDL. The real advantage of HDL is found in the use of higher levels of abstraction and the ability to tailor components to fit the needs of the project exactly. We will explore the use of these methods, as well as graphical tools to connect modules, in the following chapters.

REVIEW QUESTIONS

1. Can the same component be used more than once in the same circuit?
2. In AHDL, where are multiple instances of a component declared?
3. How do you distinguish between multiple instances of a component?
4. In AHDL, what operator is used to "connect" signals?
5. In VHDL, what serves as "wires" that connect components?
6. In VHDL, what keyword identifies the section of code where connections are specified for instances of components?

SUMMARY

1. A flip-flop is a logic circuit with a memory characteristic such that its Q and $\overline{Q}$ outputs will go to a new state in response to an input pulse and will remain in that new state after the input pulse is terminated.
2. A NAND latch and a NOR latch are simple FFs that respond to logic levels on their SET and RESET inputs.
3. Clearing (resetting) a FF means that its output ends up in the $Q = 0/\overline{Q} = 1$ state. Setting a FF means that it ends up in the $Q = 1/\overline{Q} = 0$ state.
4. Clocked FFs have a clock input (*CLK*, *CP*, *CK*) that is edge-triggered, meaning that it triggers the FF on a positive-going transition (PGT) or a negative-going transition (NGT).
5. Edge-triggered (clocked) FFs can be triggered to a new state by the active edge of the clock input according to the state of the FF's synchronous control inputs (S, R or J, K or D).
6. Most clocked FFs also have asynchronous inputs that can set or clear the FF independently of the clock input.
7. The D latch is a modified NAND latch that operates like a D flip-flop except that it is not edge-triggered.
8. Some of the principal uses of FFs include data storage and transfer, data shifting, counting, and frequency division. They are used in sequential circuits that follow a predetermined sequence of states.

9. A one-shot (OS) is a logic circuit that can be triggered from its normal resting state ($Q = 0$) to its triggered state ($Q = 1$), where it remains for a time interval proportional to an *RC* time constant.
10. Circuits that have a Schmitt-trigger type of input will respond reliably to slow-changing signals and will produce outputs with clean, sharp edges.
11. A variety of circuits can be used to generate clock signals at a desired frequency, including Schmitt-trigger oscillators, a 555 timer, and a crystal-controlled oscillator.
12. A complete summary of the various types of FFs can be found inside the back cover.
13. Programmable logic devices can be programmed to operate as latching circuits and sequential circuits.
14. Fundamental building blocks called logic primitives are available in the Altera library to help implement larger systems.
15. Clocked flip-flops are available as logic primitives.
16. VHDL code can be written to describe clocked logic explicitly without using logic primitives.
17. VHDL allows HDL files to be used as components in larger systems. Prefabricated components are available in the Altera library.
18. HDL can be used to describe interconnected components in a manner much like a graphic schematic capture tool.

IMPORTANT TERMS

flip-flop
SET (states/inputs)
CLEAR (states/inputs)
RESET (states/inputs)
NAND gate latch
contact bounce
NOR gate latch
pulses
clock
positive-going transition (PGT)
negative-going transition (NGT)
clocked flip-flop
period frequency
edge-triggered
control inputs
synchronous control inputs
setup time, t_S
hold time, t_H
clocked S-R flip-flop
trigger
pulse-steering circuit
edge-detector circuit
clocked J-K flip-flop
toggle mode
clocked D flip-flop
parallel data transfer
D latch
asynchronous inputs
override inputs
common-control block
propagation delay
sequential circuits
registers
data transfer
synchronous transfer
asynchronous (jam) transfer
jam transfer
serial data transfer
shift register
frequency division
binary counter
state table
state transition diagram
MOD number
Schmitt-trigger circuit
one-shot (OS)
quasi-stable state
nonretriggerable OS
retriggerable OS
astable or free-running multivibrator
555 timer
crystal-controlled clock generator
clock skew
EVENT
logic primitive
nested
COMPONENT
PORT MAP
structural level of abstraction

PROBLEMS

SECTIONS 5-1 TO 5-3

B 5-1.* Assuming that $Q = 0$ initially, apply the x and y waveforms of Figure 5-73 to the SET and RESET inputs of a NAND latch, and determine the Q and $\overline{Q}$ waveforms.

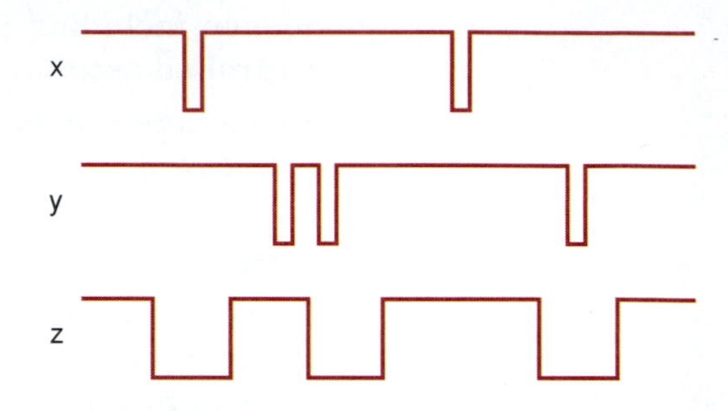

FIGURE 5-73 Problems 5-1 to 5-3.

B 5-2. Invert the x and y waveforms of Figure 5-73, apply them to the SET and RESET inputs of a NOR latch, and determine the Q and $\overline{Q}$ waveforms. Assume that $Q = 0$ initially.

5-3.* The waveforms of Figure 5-73 are connected to the circuit of Figure 5-74. Assume that $Q = 0$ initially, and determine the Q waveform.

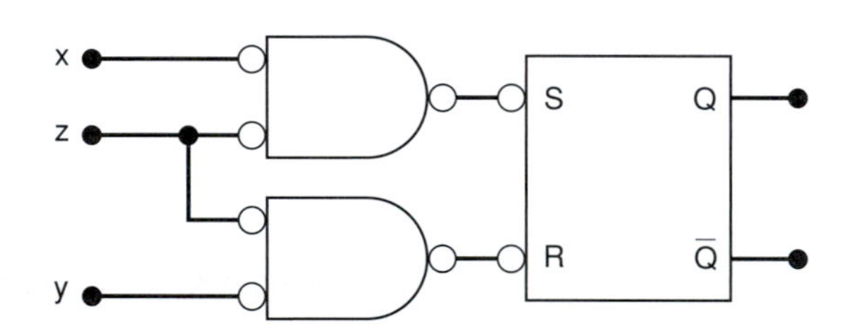

FIGURE 5-74 Problem 5-3.

D 5-4. Modify the circuit of Figure 5-9 to use a NOR gate latch.

D 5-5. Modify the circuit of Figure 5-12 to use a NAND gate latch.

T 5-6.* Refer to the circuit of Figure 5-13. A technician tests the circuit operation by observing the outputs with a storage oscilloscope while the switch is moved from A to B. When the switch is moved from A to B, the scope display of X_B appears as shown in Figure 5-75. What circuit fault could produce this result? (*Hint:* What is the function of the NAND latch?)

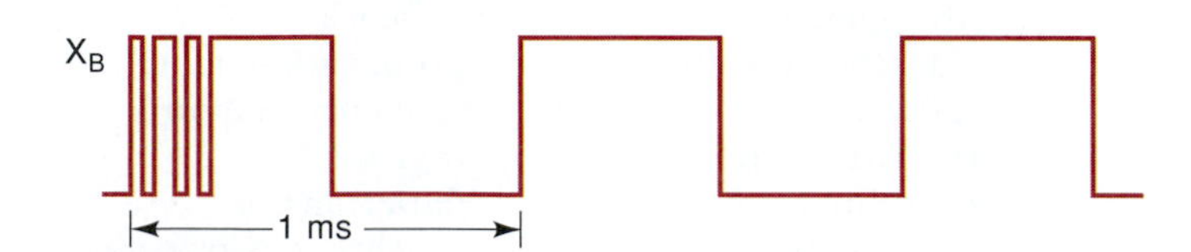

FIGURE 5-75 Problem 5-6.

SECTIONS 5-4 THROUGH 5-6

B 5-7. A certain clocked FF has minimum $t_S = 20$ ns and $t_H = 5$ ns. How long must the control inputs be stable prior to the active clock transition?

---

*Answers to problems marked with an asterisk can be found in the back of the text.

B 5-8. Apply the S, R, and CLK waveforms of Figure 5-19 to the FF of Figure 5-20, and determine the Q waveform.

B 5-9.* Apply the waveforms of Figure 5-76 to the FF of Figure 5-19 and determine the waveform at Q. Repeat for the FF of Figure 5-20. Assume $Q = 0$ initially.

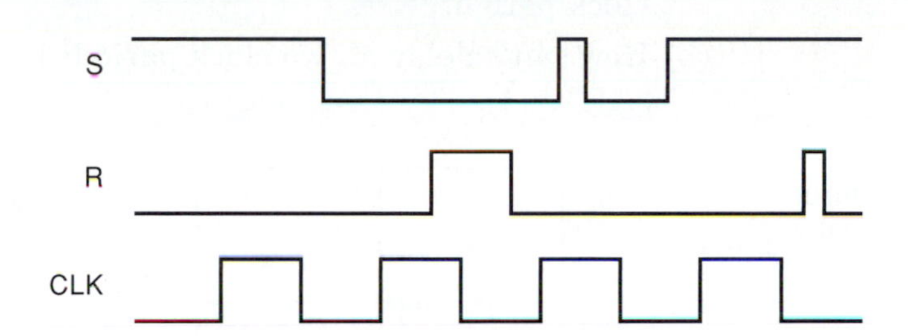

FIGURE 5-76 Problem 5-9.

5-10. Draw the following digital pulse waveforms. Label t_r, t_f, and t_w, leading edge, and trailing edge.

(a) A negative TTL pulse with t_r = 20 ns, t_f = 5 ns, and t_W = 50 ns.

(b) A positive TTL pulse with t_r = 5 ns, t_f = 1 ns, t_W = 25 ns.

(c) A positive pulse with t_w = 1 ms whose leading edge occurs every 5 ms. Give the frequency of this waveform.

SECTION 5-7

B 5-11.* Apply the J, K, and CLK waveforms of Figure 5-23 to the FF of Figure 5-24. Assume that $Q = 1$ initially, and determine the Q waveform.

D 5-12. (a)* Show how a J-K flip-flop can operate as a *toggle* FF (changes states on each clock pulse). Then apply a 10-kHz clock signal to its CLK input and determine the waveform at Q.

(b) Connect Q from this FF to the CLK input of a second J-K FF that also has $J = K = 1$. Determine the frequency of the signal at this FF's output.

B 5-13. The waveforms shown in Figure 5-77 are to be applied to two different FFs:

(a) positive-edge-triggered J-K

(b) negative-edge-triggered J-K

Draw the Q waveform response for each of these FFs, assuming that $Q = 0$ initially. Assume that each FF has $t_H = 0$.

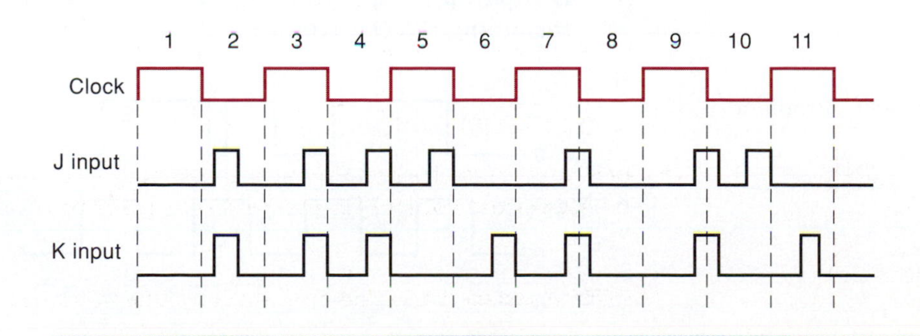

FIGURE 5-77 Problem 5-13.

SECTION 5-8

N 5-14. A D FF is sometimes used to *delay* a binary waveform so that the binary information appears at the output a certain amount of time after it appears at the *D* input.

(a)*Determine the *Q* waveform in Figure 5-78, and compare it with the input waveform. Note that it is delayed from the input by one clock period.

(b) How can a delay of two clock periods be obtained?

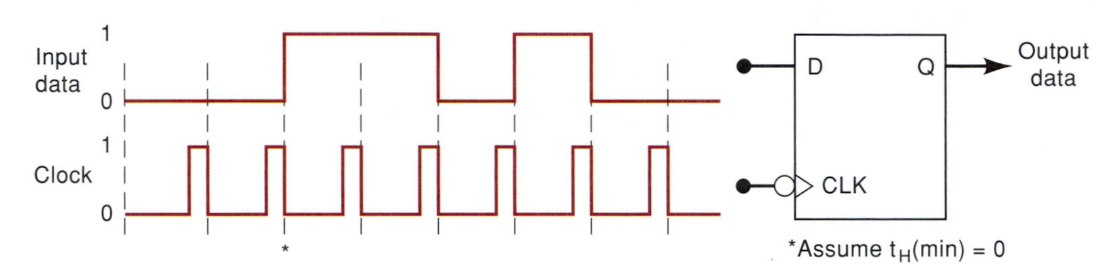

FIGURE 5-78 Problem 5-14.

B 5-15. (a) Apply the *S* and *CLK* waveforms of Figure 5-76 to the *D* and *CLK* inputs of a D FF that triggers on PGTs. Then determine the waveform at *Q*.

(b) Repeat using the *C* waveform of Figure 5-76 for the *D* input.

B 5-16.*An edge-triggered D flip-flop can be made to operate in the toggle mode by connecting it as shown in Figure 5-79. Assume that $Q = 0$ initially, and determine the *Q* waveform.

FIGURE 5-79 D flip-flop connected to toggle (Problem 5-16).

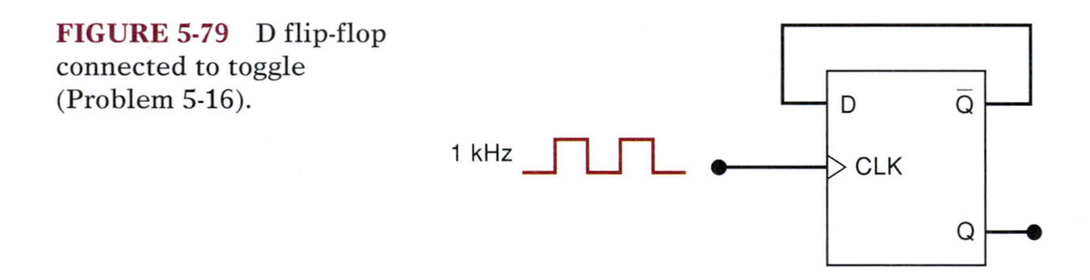

SECTION 5-9

B 5-17. (a) Apply the *S* and *CLK* waveforms of Figure 5-76 to the *D* and *EN* inputs of a D latch, respectively, and determine the waveform at *Q*.

(b) Repeat using the *C* waveform applied to *D*.

5-18. Compare the operation of the *D* latch with a negative-edge-triggered D flip-flop by applying the waveforms of Figure 5-80 to each and determining the *Q* waveforms.

FIGURE 5-80 Problem 5-18.

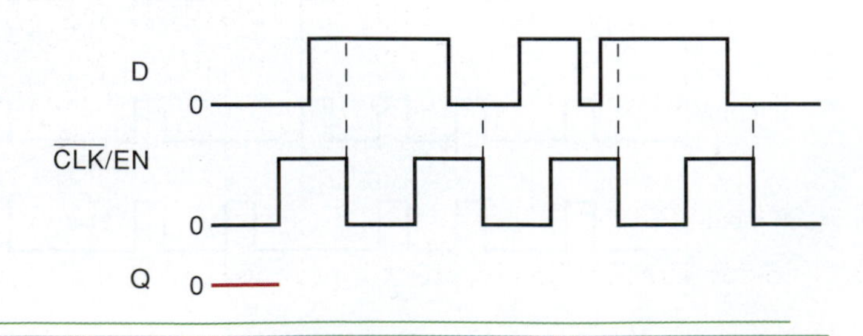

5-19. In Problem 5-16, we saw how an edge-triggered D flip-flop can be operated in the toggle mode. Explain why this same idea will not work for a *D* latch.

SECTION 5-10

B 5-20. Determine the *Q* waveform for the FF in Figure 5-81. Assume that $Q = 0$ initially, and remember that the asynchronous inputs override all other inputs.

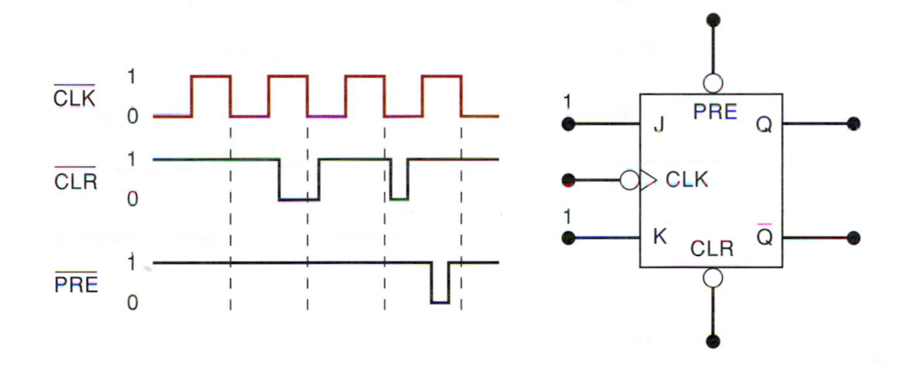

FIGURE 5-81 Problem 5-20.

B 5-21.* Apply the $\overline{CLK}$, $\overline{PRE}$, and $\overline{CLR}$ waveforms of Figure 5-32 to a positive-edge-triggered D flip-flop with active-LOW asynchronous inputs. Assume that *D* is kept HIGH and *Q* is initially LOW. Determine the *Q* waveform.

B 5-22. Apply the waveforms of Figure 5-81 to a D flip-flop that triggers on NGTs and has active-LOW asynchronous inputs. Assume that *D* is kept LOW and that *Q* is initially HIGH. Draw the resulting *Q* waveform.

SECTION 5-12

B 5-23. Use Table 5-2 in Section 5-12 to determine the following.

(a)* How long can it take for the *Q* output of a 74C74 to switch from 0 to 1 in response to an active *CLK* transition?

(b)* Which FF in Table 5-2 requires its control inputs to remain stable for the longest time *after* the active *CLK* transition? *Before* the transition?

(c) What is the narrowest pulse that can be applied to the $\overline{PRE}$ of a 7474 FF?

B 5-24. Use Table 5-2 to determine the following:

(a) How long does it take to asynchronously clear a 74LS112?

(b) How long does it take to asynchronously set a 74HC112?

(c) What is the shortest acceptable interval between active clock transitions for a 7474?

(d) The D input of a 74HC112 goes HIGH 15 ns before the active clock edge. Will the data be stored reliably in the flip-flop?

(e) How long does it take (after the clock edge) to synchronously store a 1 in a cleared 7474 D flip-flop?

SECTIONS 5-15 AND 5-16

D 5-25.* Modify the circuit of Figure 5-40 to use a J-K flip-flop.

D 5-26. In the circuit of Figure 5-82, inputs A, B, and C are all initially LOW. Output Y is supposed to go HIGH only when A, B, and C go HIGH in a certain sequence.

(a) Determine the sequence that will make Y go HIGH.

(b) Explain why the START pulse is needed.

(c) Modify this circuit to use D FFs.

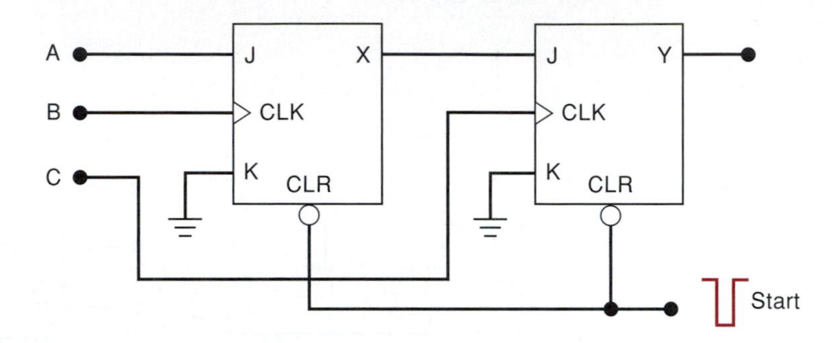

FIGURE 5-82 Problem 5-26.

SECTIONS 5-17 AND 5-18

D 5-27.* (a) Draw a circuit diagram for the synchronous parallel transfer of data from one three-bit register to another using J-K flip-flops.

(b) Repeat for asynchronous parallel transfer.

N, D 5-28. A *recirculating* shift register is a shift register that keeps the binary information circulating through the register as clock pulses are applied. The shift register of Figure 5-45 can be made into a circulating register by connecting X_0 to the DATA IN line. No external inputs are used. Assume that this circulating register starts out with 1011 stored in it (i.e., $X_3 = 1$, $X_2 = 0$, $X_1 = 1$, and $X_0 = 1$). List the sequence of states that the register FFs go through as eight shift pulses are applied.

D 5-29.* Refer to Figure 5-46, where a three-bit number stored in register X is serially shifted into register Y. How can the circuit be modified so that, at the end of the transfer operation, the original number stored in X is present in both registers? (*Hint:* See Problem 5-28.)

SECTION 5-19

B 5-30. Refer to the counter circuit of Figure 5-47 and answer the following:

(a)* If the counter starts at 000, what will be the count after 13 clock pulses? After 99 pulses? After 256 pulses?

(b) If the counter starts at 100, what will be the count after 13 pulses? After 99 pulses? After 256 pulses?

(c) Connect a fourth J-K FF (X_3) to this counter and draw the state transition diagram for this 4-bit counter. If the input clock frequency is 80 MHz, what will the waveform at X_3 look like?

B 5-31. Refer to the binary counter of Figure 5-47. Change it by connecting $\overline{X_0}$ to the *CLK* of flip-flop X_1, and $\overline{X_1}$ to the *CLK* of flip-flop X_2. Start with all FFs in the 1 state, and draw the various FF output waveforms (X_0, X_1, X_2) for 16 input pulses. Then list the sequence of FF states as was done in Figure 5-48. This counter is called a *down counter*. Why?

B 5-32. Draw the state transition diagram for this down counter, and compare it with the diagram of Figure 5-49. How are they different?

B 5-33.* (a) How many FFs are required to build a binary counter that counts from 0 to 1023?

(b) Determine the frequency at the output of the last FF of this counter for an input clock frequency of 2 MHz.

(c) What is the counter's MOD number?

(d) If the counter is initially at zero, what count will it hold after 2060 pulses?

B 5-34. A binary counter is being pulsed by a 256-kHz clock signal. The output frequency from the last FF is 2 kHz.

(a) Determine the MOD number.

(b) Determine the counting range.

B 5-35. A photodetector circuit is being used to generate a pulse each time a customer walks into a certain establishment. The pulses are fed to an eight-bit counter. The counter is used to count these pulses as a means for determining how many customers have entered the store. After closing the store, the proprietor checks the counter and finds that it shows a count of $00001001_2 = 9_{10}$. He knows that this is incorrect because there were many more than nine people in his store. Assuming that the counter circuit is working properly, what could be the reason for the discrepancy?

SECTION 5-20

D 5-36.* Modify the circuit of Figure 5-50 so that only the presence of address code 10110110 will allow data to be transferred to register *X*.

T 5-37. Suppose that the circuit of Figure 5-50 is malfunctioning so that data are being transferred to *X* for either of the address codes 11111110 or 11111111. What are some circuit faults that could be causing this?

N, D 5-38. Many microcontrollers share the same pins to output the lower address and transfer data. In order to hold the address constant while the data are transferred. The address information is stored in a latch which is enabled by the control signal ALE (address latch enable) as shown in Figure 5-83. Connect this latch to the microcontroller such that it takes what is on the lower address and data lines while ALE is HIGH and holds it on the lower address only lines when ALE is LOW.

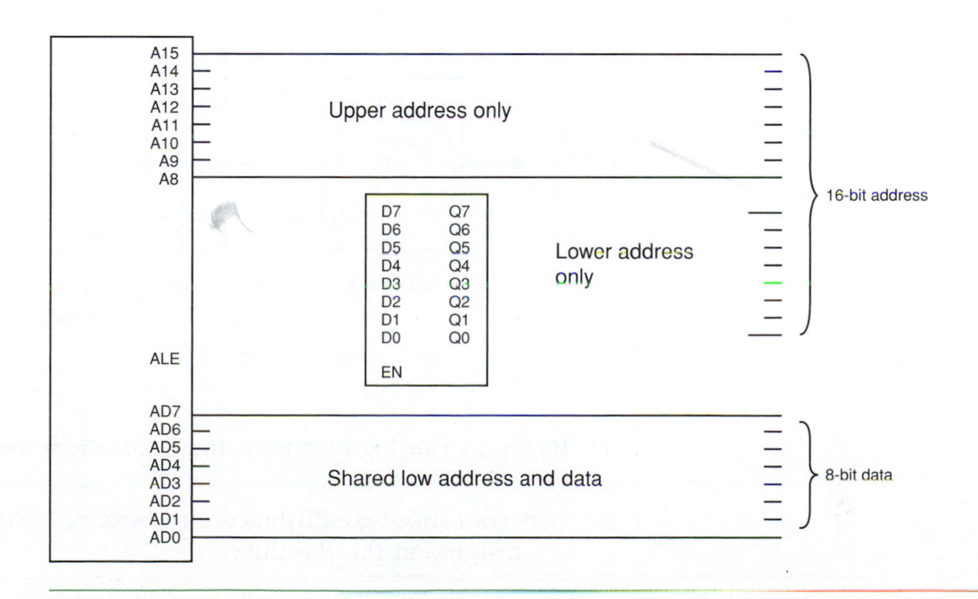

FIGURE 5-83 Problem 5-38.

D 5-39. Modify the circuit of Figure 5-50 so that the MPU has eight data output lines connected to transfer eight bits of data to an eight-bit register made up of two 74HC175 ICs [Figure 5-34(b)]. Show all circuit connections.

SECTION 5-22

B 5-40. Refer to the waveforms in Figure 5-53(a). Change the OS pulse duration to 0.5 ms and determine the Q output for both types of OS. Then repeat using a OS pulse duration of 1.5 ms.

N 5-41.*Figure 5-84 shows three nonretriggerable one-shots connected in a timing chain that produces three sequential output pulses. Note the "1" in front of the pulse on each OS symbol to indicate nonretriggerable operation. Draw a timing diagram showing the relationship between the input pulse and the three OS outputs. Assume an input pulse duration of 10 ms.

FIGURE 5-84 Problem 5-41.

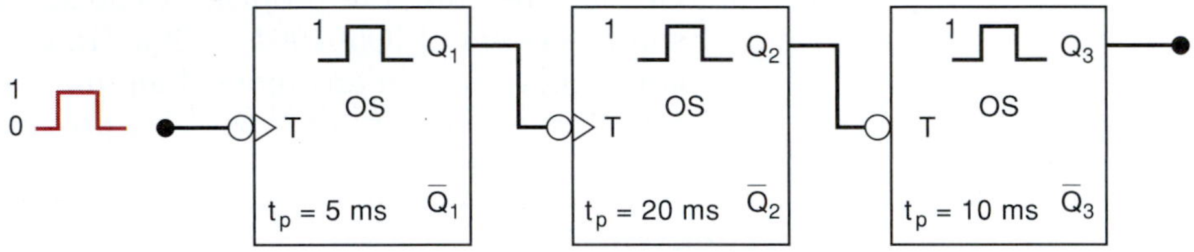

5-42. A *retriggerable* OS can be used as a pulse-frequency detector that detects when the frequency of a pulse input is below a predetermined value. A simple example of this application is shown in Figure 5-85. The operation begins by momentarily closing switch SW1.

(a) Describe how the circuit responds to input frequencies above 1 kHz.

(b) Describe how the circuit responds to input frequencies below 1 kHz.

(c) How would you modify the circuit to detect when the input frequency drops below 50 kHz?

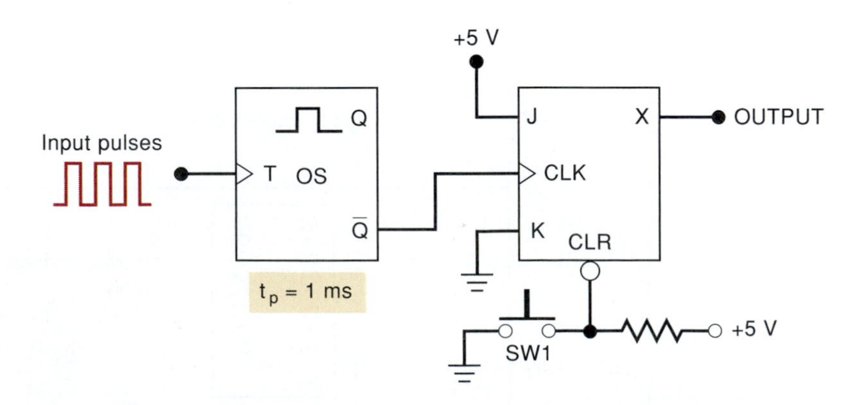

FIGURE 5-85 Problem 5-42.

5-43. Refer to the logic symbol for a 74121 nonretriggerable one-shot in Figure 5-54(a).

(a)*What input conditions are necessary for the OS to be triggered by a signal at the B input?

(b) What input conditions are necessary for the OS to be triggered by a signal at the A_1 input?

C, D 5-44. The output pulse width from a 74121 OS is given by the approximate formula

$$t_p \approx 0.7\, R_T C_T$$

where R_T is the resistance connected between the R_{EXT}/C_{EXT} pin and V_{CC}, and C_T is the capacitance connected between the C_{EXT} pin and the R_{EXT}/C_{EXT} pin. The value for R_T can be varied between 2 and 40 kΩ, and C_T can be as large as 1000 μF.

(a) Show how a 74121 can be connected to produce a negative-going pulse with a 5-ms duration whenever either of two logic signals (*E* or *F*) makes a NGT. Both *E* and *F* are normally in the HIGH state.

(b) Modify the circuit so that a control input signal, *G*, can disable the OS output pulse, regardless of what occurs at *E* or *F*.

SECTION 5-23

B, D 5-45.* Show how to use a 74LS14 Schmitt-trigger INVERTER to produce an approximate square wave with a frequency of 10 kHz.

B, D 5-46. Design a 555 free-running oscillator to produce an approximate square wave at 40 kHz. *C* should be kept at 500 pF or greater.

D 5-47. A 555 oscillator can be combined with a J-K flip-flop to produce a perfect (50 percent duty cycle) square wave. Modify the circuit of Problem 5-46 to include a J-K flip-flop. The final output is still to be a 40-kHz square wave.

5-48. Design a 555 timer circuit that will produce a 10 percent duty-cycle 5-kHz waveform. Choose a capacitor greater than 500 pF and resistors less than 100 kΩ. Draw the circuit diagram with pin numbers labeled.

C, N 5-49. The circuit in Figure 5-86 can be used to generate two nonoverlapping clock signals at the same frequency. These clock signals were used in early microprocessor systems that required four different clock transitions to synchronize their operations.

(a) Draw the CP1 and CP2 timing waveforms if *CLOCK* is a 1-MHz square wave. Assume that t_{PLH} and t_{PHL} are 20 ns for the FF and 10 ns for the AND gates.

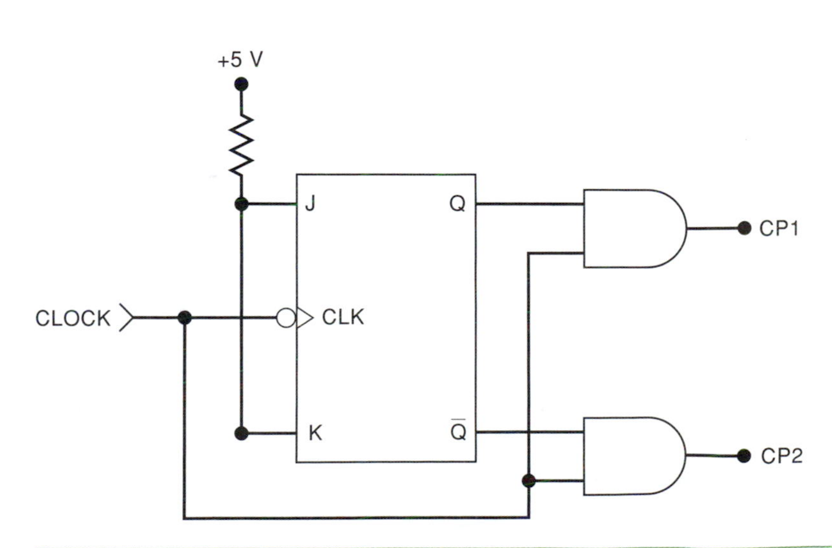

FIGURE 5-86 Problem 5-49.

(b) This circuit would have a problem if the FF were changed to one that responds to a PGT at *CLK*. Draw the CP1 and CP2 waveforms for that situation. Pay particular attention to conditions that can produce glitches.

SECTION 5-24

T 5-50. Refer to the counter circuit in Figure 5-47. Assume that all asynchronous inputs are connected to V_{CC}. When tested, the circuit waveforms appear as shown in Figure 5-87. Consider the following list of possible faults. For each one, indicate "yes" or "no" as to whether it could cause the observed results. Explain each response.

(a)* *CLR* input of X_2 is open.

(b)* X_1 output's transition times are too long, possibly due to loading.

(c) X_2 output is shorted to ground.

(d) X_2's hold time requirement is not being met.

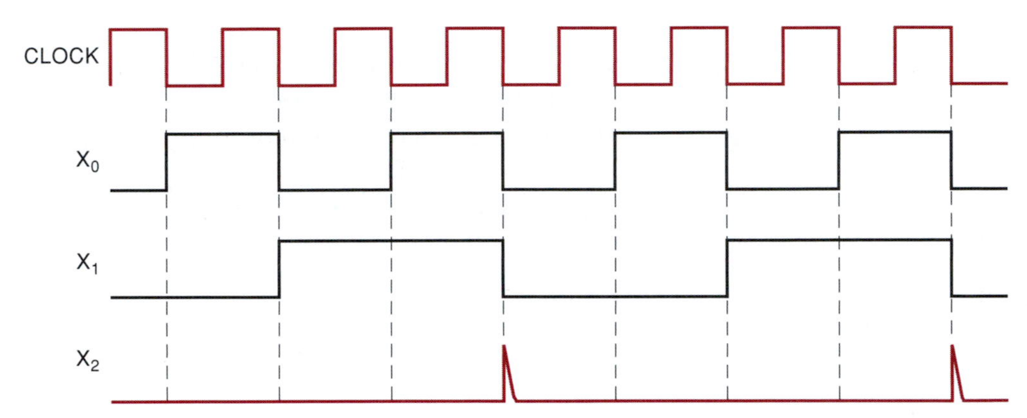

FIGURE 5-87 Problem 5-50.

C, T 5-51. Consider the situation of Figure 5-59 for each of the following sets of timing values. For each, indicate whether or not flip-flop Q_2 will respond correctly.

(a)*Each FF: $t_{PLH} = 12$ ns; $t_{PHL} = 8$ ns; $t_S = 5$ ns; $t_H = 0$ ns
NAND gate: $t_{PLH} = 8$ ns; $t_{PHL} = 6$ ns
INVERTER: $t_{PLH} = 7$ ns; $t_{PHL} = 5$ ns

(b) Each FF: $t_{PLH} = 10$ ns; $t_{PHL} = 8$ ns; $t_S = 5$ ns; $t_H = 0$ ns
NAND gate: $t_{PLH} = 12$ ns; $t_{PHL} = 10$ ns
INVERTER: $t_{PLH} = 8$ ns; $t_{PHL} = 6$ ns

D 5-52. Show and explain how the clock skew problem in Figure 5-59 can be eliminated by the appropriate insertion of two INVERTERs.

T 5-53. Refer to the circuit of Figure 5-88. Assume that the ICs are of the TTL logic family. The *Q* waveform was obtained when the circuit was tested with the input signals shown and with the switch in the "up" position; it is not correct. Consider the following list of faults, and for each indicate "yes" or "no" as to whether it could be the actual fault. Explain each response.

(a)*Point *X* is always LOW due to a faulty switch.

(b)*Z1 pin 1 is internally shorted to V_{CC}.

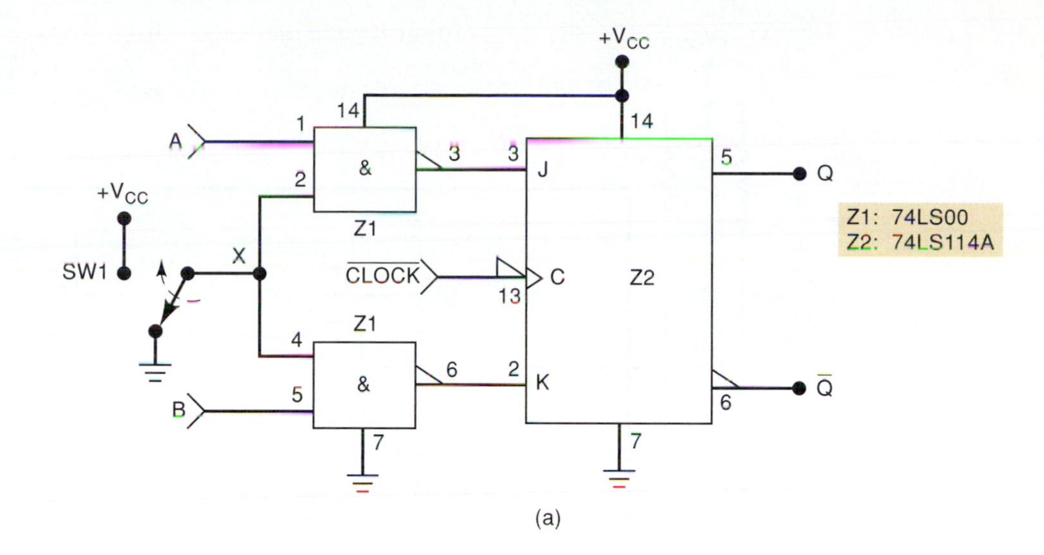

FIGURE 5-88 Problem 5-53.

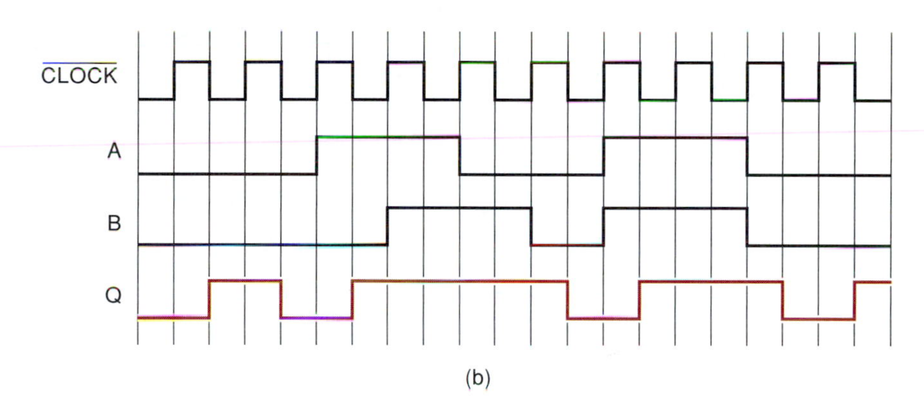

(c) The connection from Z1-3 to Z2-3 is broken.

(d) There is a solder bridge between pins 6 and 7 of Z1.

C 5-54. The circuit of Figure 5-89 functions as a sequential combination lock. To operate the lock, proceed as follows:

1. Momentarily activate the RESET switch.
2. Set the switches SWA, SWB, and SWC to the first part of the combination. Then momentarily toggle the ENTER switch back and forth.
3. Set the switches to the second part of the combination, and toggle ENTER again. This should produce a HIGH at Q_2 to open the lock.

If the incorrect combination is entered in either step, the operator must start the sequence over. Analyze the circuit and determine the correct sequence of combinations that will open the lock.

C, T 5-55.* When the combination lock of Figure 5-89 is tested, it is found that entering the correct combination does not open the lock. A logic probe check shows that entering the correct first combination sets Q_1 HIGH, but entering the correct second combination produces only a momentary pulse at Q_2. Consider each of the following faults and indicate which one(s) could produce the observed operation. Explain each choice.

(a) Switch bounce at SWA, SWB, or SWC.

(b) *CLR* input of Q_2 is open.

(c) Connection from NAND gate 4 output to NAND gate 3 input is open.

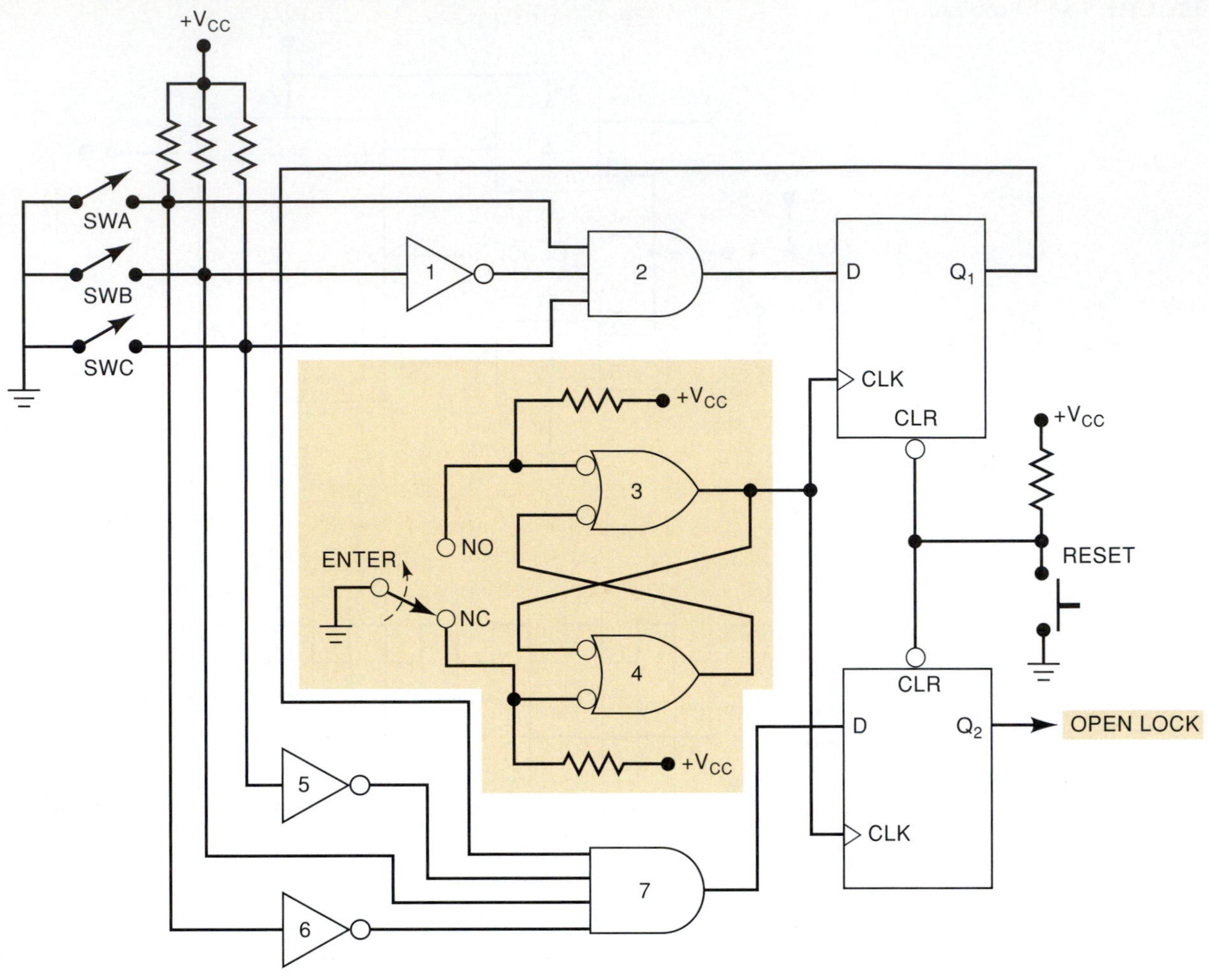

FIGURE 5-89 Problems 5-54 and 5-55.

DRILL QUESTIONS

B 5-56. For each statement indicate what type of FF is being described.

(a)*Has a SET and a CLEAR input but does not have a *CLK* input

(b)*Will toggle on each *CLK* pulse when its control inputs are both HIGH

(c)*Has an ENABLE input instead of a *CLK* input

(d)*Is used to transfer data easily from one FF register to another

(e) Has only one control input

(f) Has two outputs that are complements of each other

(g) Can change states only on the active transition of *CLK*

(h) Is used in binary counters

B 5-57. Define the following terms.

(a) Asynchronous inputs

(b) Edge-triggered

(c) Shift register

(d) Frequency division

(e) Asynchronous (jam) transfer
(f) State transition diagram
(g) Parallel data transfer
(h) Serial data transfer
(i) Retriggerable one-shot
(j) Schmitt-trigger inputs

SECTION 5-25

B 5-58. Simulate the HDL design for a NAND latch given in Figure 5-62 (AHDL) or Figure 5-63 (VHDL). What does this SR latch do if an "invalid" input command is applied? Since we know that any SR latch can have an unusual output result when an invalid input command is applied, you should simulate that input condition as well as the latch's normal set, reset, and hold commands. Some latch designs can have a tendency for the output to oscillate when an invalid command is followed by a hold command, so be sure to check for that in your simulation.

B, H 5-59.* Write an HDL design file for an active-HIGH input SR latch.

B, H 5-60. Modify the latch description given in Figure 5-62 (AHDL) or Figure 5-63 (VHDL) to make the SR reset if an invalid input is applied. Simulate the design.

B, H 5-61.* Add inverted outputs to the HDL NAND latch designs given in Figure 5-62 or Figure 5-63. Verify correct operation by simulation.

B 5-62. Simulate the AHDL or VHDL design for a D latch given in Section 5-25.

D, H 5-63. Create a four-bit transparent latch with one *enable* input based on the AHDL or VHDL design for a single D latch device given in Section 5-25. Simulate the four-bit latch.

D, H, N 5-64. A toggle (T) flip-flop has a single control input (T). When T = 0, the flip-flop is in the no change state, similar to a JKFF with J = K = 0. When T = 1, the flip-flop is in the toggle mode, similar to a JKFF with J = K = 1. Write the design file in

(a) AHDL
(b) VHDL

H 5-65. (a) Write an AHDL design file for the shift register shown in Figure 5-45.

(b) Write a VHDL design file for the shift register shown in Figure 5-45.

H 5-66. (a)*Write an AHDL design file for the shift register shown in Figure 5-46.

(b)*Write a VHDL design file for the shift register shown in Figure 5-46.

H 5-67. (a) Write an AHDL design file for the FF circuit shown in Figure 5-59.

(b) Write a VHDL design file for the FF circuit shown in Figure 5-59.

5-68. Simulate the operation of either Problem 5-74 or 5-75. (The simulations should be identical and match the results in Figure 5-58.)

H 5-69. (a) Write an AHDL design file to implement the entire circuit of Figure 5-89.

(b) Write a VHDL design file to implement the entire circuit of Figure 5-89.

ANSWERS TO SECTION REVIEW QUESTIONS

SECTION 5-1

1. HIGH; LOW 2. $Q = 0, \overline{Q} = 1$ 3. True 4. Apply a momentary LOW to $\overline{\text{SET}}$ input.

SECTION 5-2

1. LOW, HIGH 2. $Q = 1$ and $\overline{Q} = 0$ 3. Make CLEAR = 1 4. $\overline{\text{SET}}$ and $\overline{\text{RESET}}$ would both be normally in their active-LOW state.

SECTION 5-4

1. Synchronous control inputs and clock input 2. The FF output can change only when the appropriate clock transition occurs. 3. False 4. Setup time is the required interval immediately prior to the active edge of the *CLK* signal during which the control inputs must be held stable. Hold time is the required interval immediately following the active edge of *CLK* during which the control inputs must be held stable.

SECTION 5-5

1. HIGH; LOW; HIGH 2. Because *CLK** is HIGH only for a few nanoseconds

SECTION 5-6

1. True 2. No 3. $J = 1, K = 0$

SECTION 5-7

1. Q will go LOW at point a and remain LOW. 2. False. The D input can change without affecting Q because Q can change only on the active *CLK* edge. 3. Yes, by converting to D FFs (Figure 5-25).

SECTION 5-8

1. In a D latch, the Q output can change while *EN* is HIGH. In a D flip-flop, the output can change only on the active edge of *CLK*. 2. False 3. True

SECTION 5-9

1. Asynchronous inputs work independently of the *CLK* input. 2. Yes, because $\overline{PRE}$ is active-LOW 3. $J = K = 1$, $\overline{PRE} = \overline{CLR} = 1$, and a PGT at *CLK*

SECTION 5-10

1. The triangle inside the rectangle indicates edge-triggered operation; the right triangle outside the rectangle indicates triggering on a NGT. 2. It is used to indicate the function of those inputs that are common to more than one circuit on the chip.

SECTION 5-11

1. t_{PLH} and t_{PHL} 2. False; the waveform must also satisfy $t_W(L)$ and $t_W(H)$ requirements.

SECTION 5-17

1. False 2. D flip-flop 3. Six 4. True

SECTION 5-18

1. True 2. Fewer interconnections between registers 3. $X_2X_1X_0 = 111$; $Y_2Y_1Y_0 = 101$ 4. Parallel

SECTION 5-19

1. 10 kHz 2. Eight 3. 256 4. 2 kHz 5. $00001000_2 = 8_{10}$

SECTION 5-21

1. The output may contain oscillations. 2. It will produce clean, fast output signals even for slow-changing input signals.

SECTION 5-22

1. $Q = 0, \overline{Q} = 1$ 2. True 3. External *R* and *C* values 4. For a retriggerable OS, each new trigger pulse begins a new t_p interval, regardless of the state of the *Q* output.

SECTION 5-23

1. 24 kHz 2. 109.3 kHz; 66.7 percent 3. Frequency stability

SECTION 5-24

1. Clock skew is the arrival of a clock signal at the *CLK* inputs of different FFs at different times. It can cause a FF to go to an incorrect state.

SECTION 5-25

1. Feedback: The outputs are combined with the inputs to determine the next state of the outputs. 2. It progresses through a predetermined sequence of states in response to an input clock signal.

SECTION 5-26

1. A standard building block from a library of components that performs some fundamental logic function. 2. The names of each input and output and the primitive name that is recognized by the development system. 3. Under the HELP menu. 4. The PROCESS allows sequential IF constructs and the EVENT attribute detects transitions. 5. ieee.std_logic_1164. 6. altera.maxplus2

SECTION 5-27

1. Yes 2. In the VARIABLE section. 3. Each is assigned a variable name. 4. = 5. SIGNALs 6. PORT MAP

CHAPTER 6

DIGITAL ARITHMETIC: OPERATIONS AND CIRCUITS

OUTLINE

OBJECTIVES

Upon completion of this chapter, you will be able to:

- Perform binary addition, subtraction, multiplication, and division on two binary numbers.
- Add and subtract hexadecimal numbers.
- Know the difference between binary addition and OR addition.
- Compare the advantages and disadvantages among three different systems of representing signed binary numbers.
- Manipulate signed binary numbers using the 2's-complement system.
- Understand the BCD addition process.
- Describe the basic operation of an arithmetic/logic unit.
- Employ full adders in the design of parallel binary adders.
- Cite the advantages of parallel adders with the look-ahead carry feature.
- Explain the operation of a parallel adder/subtractor circuit.
- Use an ALU integrated circuit to perform various logic and arithmetic operations on input data.
- Analyze troubleshooting case studies of adder/subtractor circuits.
- Use HDL forms of standard TTL parts from libraries to implement more complicated circuits.
- Use the Boolean equation form of description to perform operations on entire sets of bits.
- Apply software engineering techniques to expand the capacity of a hardware description.

INTRODUCTION

Digital computers and calculators perform the various arithmetic operations on numbers that are represented in binary form. The subject of digital arithmetic can be a very complex one if we want to understand all the various methods of computation and the theory behind them. Fortunately, this level of knowledge is not required by most technicians, at least not until they become experienced computer programmers. Our approach in this chapter will be to concentrate on those basic principles that are necessary for understanding how digital machines (i.e., computers) perform the basic arithmetic operations.

First, we will see how the various arithmetic operations are performed on binary numbers using "pencil and paper," and then we will study the actual logic circuits that perform these operations in a digital system. Finally, we will learn how to describe these simple circuits using HDL techniques. Several methods of expanding the capacity of these circuits will also be covered. The focus will be on the fundamentals of HDL, using arithmetic circuits as an example. The powerful capability of HDL combined with PLD hardware will provide the basis for further study, design, and experimentation with much more sophisticated arithmetic circuits in more advanced courses.

6-1 BINARY ADDITION

The addition of two binary numbers is performed in exactly the same manner as the addition of decimal numbers. In fact, binary addition is simpler because there are fewer cases to learn. Let us first review decimal addition:

```
        ↓‾‾‾‾‾↓
  3 7 6    LSD
 +4 6 1
 ------
  8 3 7
```

The least-significant-digit (LSD) position is operated on first, producing a sum of 7. The digits in the second position are then added to produce a sum of 13, which produces a **carry** of 1 into the third position. This produces a sum of 8 in the third position.

The same general steps are followed in binary addition. However, only four cases can occur in adding the two binary digits (bits) in any position. They are:

$$0 + 0 = 0$$
$$1 + 0 = 1$$
$$1 + 1 = 10 = 0 + \text{carry of 1 into next position}$$
$$1 + 1 + 1 = 11 = 1 + \text{carry of 1 into next position}$$

The last case occurs when the two bits in a certain position are 1 and there is a carry from the previous position. Here are several examples of the addition of two binary numbers (decimal equivalents are in parentheses):

```
   011 (3)        1001 (9)        11.011 (3.375)
 + 110 (6)     + 1111 (15)     + 10.110 (2.750)
 ---------     -----------     ----------------
  1001 (9)     11000 (24)      110.001 (6.125)
```

It is not necessary to consider the addition of more than two binary numbers at a time because in all digital systems the circuitry that actually performs the addition can handle only two numbers at a time. When more than two numbers are to be added, the first two are added together and then their sum is added to the third number, and so on. This is not a serious drawback because modern digital computers can typically perform an addition operation in several nanoseconds.

Addition is the most important arithmetic operation in digital systems. As we shall see, the operations of subtraction, multiplication, and division as they are performed in most modern digital computers and calculators actually use only addition as their basic operation.

REVIEW QUESTION

1. Add the following pairs of binary numbers.
   (a) 10110 + 00111
   (b) 011.101 + 010.010
   (c) 10001111 + 00000001

6-2 REPRESENTING SIGNED NUMBERS

In digital computers, the binary numbers are represented by a set of binary storage devices (e.g., flip-flops). Each device represents one bit. For example, a six-bit FF register can store binary numbers ranging from 000000 to 111111 (0 to 63 in decimal). This represents the *magnitude* of the number. Because most digital computers and calculators handle negative as well as positive numbers, some means is required for representing the *sign* of the number (+ or −). This is usually done by adding to the number another bit called the **sign bit**. In general, the common convention is that a 0 in the sign bit represents a positive number and a 1 in the sign bit represents a negative number. This is illustrated in Figure 6-1. Register A contains the bits 0110100. The 0 in the leftmost bit (A_6) is the sign bit that represents +. The other six bits are the magnitude of the number 110100_2, which is equal to 52 in decimal. Thus, the number stored in the A register is +52. Similarly, the number stored in the B register is −52 because the sign bit is 1, representing −.

The sign bit is used to indicate the positive or negative nature of the stored binary number. The numbers in Figure 6-1 consist of a sign bit and six magnitude bits. The magnitude bits are the true binary equivalent of the decimal value being represented. This is called the **sign-magnitude system** for representing signed binary numbers.

Although the sign-magnitude system is straightforward, calculators and computers do not normally use it because the circuit implementation is more complex than in other systems. The most commonly used system for representing signed binary numbers is the **2's-complement system**. Before we see how this is done, we must first see how to form the 1's complement and 2's complement of a binary number.

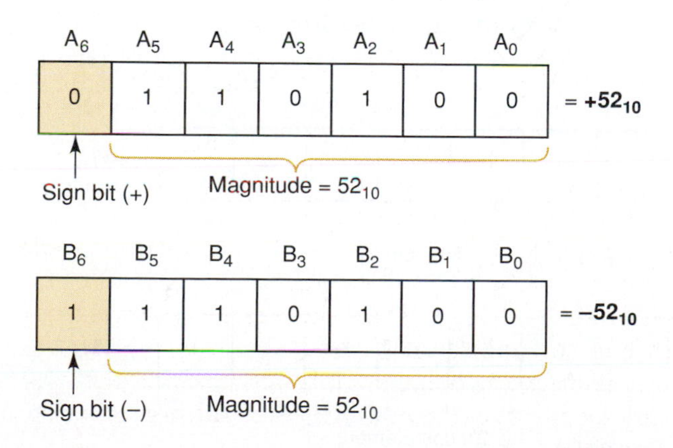

FIGURE 6-1 Representation of signed numbers in sign-magnitude form.

1's-Complement Form

The 1's complement of a binary number is obtained by changing each 0 to a 1 and each 1 to a 0. In other words, change each bit in the number to its complement. The process is shown below.

```
1 0 1 1 0 1  original binary number
↓ ↓ ↓ ↓ ↓ ↓
0 1 0 0 1 0  complement each bit to form 1's complement
```

Thus, we say that the 1's complement of 101101 is 010010.

2's Complement Form

The 2's complement of a binary number is formed by taking the 1's complement of the number and adding 1 to the least-significant-bit position. The process is illustrated below for $101101_2 = 45_{10}$.

```
  1 0 1 1 0 1   binary equivalent of 45
  0 1 0 0 1 0   complement each bit to form 1's complement
+           1   add 1 to form 2's complement
  -----------
  0 1 0 0 1 1   2's complement of original binary number
```

Thus, we say that 010011 is the 2's complement representation of 101101.

Here's another example of converting a binary number to its 2's-complement representation:

```
  1 0 1 1 0 0   original binary number
  0 1 0 0 1 1   1's complement
+           1   add 1
  -----------
  0 1 0 1 0 0   2's complement of original number
```

Representing Signed Numbers Using 2's Complement

The 2's-complement system for representing signed numbers works like this:

- If the number is positive, the magnitude is represented in its true binary form, and a sign bit of 0 is placed in front of the MSB. This is shown in Figure 6-2 for $+45_{10}$.
- If the number is negative, the magnitude is represented in its 2's-complement form, and a sign bit of 1 is placed in front of the MSB. This is shown in Figure 6-2 for -45_{10}.

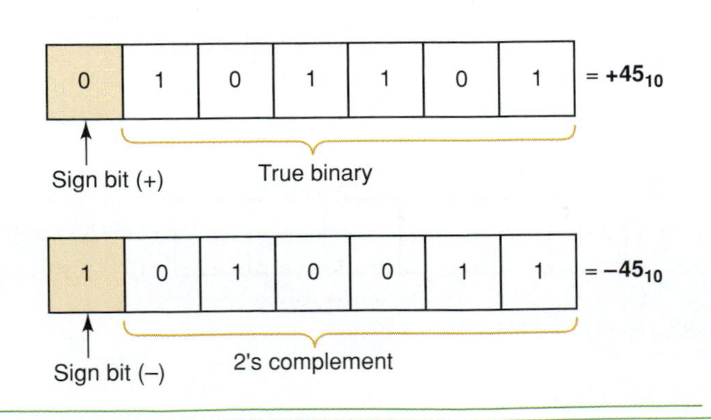

FIGURE 6-2 Representation of signed numbers in the 2's-complement system.

The 2's-complement system is used to represent signed numbers because, as we shall see, it allows us to perform the operation of subtraction by actually performing addition. This is significant because it means that a digital computer can use the same circuitry both to add and to subtract, thereby realizing a saving in hardware.

EXAMPLE 6-1

Represent each of the following signed decimal numbers as a signed binary number in the 2's-complement system. Use a total of five bits, including the sign bit.

(a) +13 (b) −9 (c) +3 (d) −2 (e) −8

Solution

(a) The number is positive, so the magnitude (13) will be represented in its true-magnitude form, that is, $13 = 1101_2$. Attaching the sign bit of 0, we have

$$+13 = 01101$$

sign bit (leftmost bit)

(b) The number is negative, so the magnitude (9) must be represented in 2's-complement form:

$$9_{10} = 1001_2$$

0110 (1's complement)
+ 1 (add 1 to LSB)
0111 (2's complement)

When we attach the sign bit of 1, the complete signed number becomes

$$-9 = 10111$$

sign bit (leftmost bit)

The procedure we have just followed required two steps. First, we determined the 2's complement of the magnitude, and then we attached the sign bit. This can be accomplished in one step if we include the sign bit in the 2's-complement process. For example, to find the representation for −9, we start with the representation for +9, *including the sign bit,* and we take the 2's complement of it in order to obtain the representation for −9.

+9 = 01001
10110 (1's complement of each bit including sign bit)
+ 1 (add 1 to LSB)
−9 = 10111 (2's-complement representation of −9)

The result is, of course, the same as before.

(c) The decimal value 3 can be represented in binary using only two bits. However, the problem statement requires a four-bit magnitude preceded by a sign bit. Thus, we have

$$+3_{10} = 00011$$

In many situations the number of bits is fixed by the size of the registers that will be holding the binary numbers, so that 0s may have to be added in order to fill the required number of bit positions.

(d) Start by writing +2 using five bits:

+2 =	00010	
	11101	(1's complement)
+	1	(add 1)
−2 =	11110	(2's-complement representation of −2)

(e) Start with +8:

+8 =	01000	
	10111	(complement each bit)
+	1	(add 1)
−8 =	11000	(2's-complement representation of −8)

Sign Extension

Example 6-1 required that we use a total of five bits to represent the signed numbers. The size of a register (*number of flip-flops*) determines the number of binary digits that are stored for each number. Most digital systems today store numbers in registers sized in even multiples of four bits. In other words, the storage registers will be made up of 4, 8, 12, 16, 32, or 64 bits. In a system that stores eight-bit numbers, seven bits represent the magnitude and the MSB represents the sign. If we need to store a positive five-bit number in an eight-bit register, it makes sense to simply add leading zeros. The MSB (sign bit) is still 0, indicating a positive value.

0000 1001

appended leading 0s — binary value for 9

What happens when we try to store five-bit negative numbers in an eight-bit register? In the previous section we found that the five-bit, 2's-complement binary representation for −9 is 10111.

1 0111

If we appended leading 0s, this would no longer be a negative number in eight-bit format. The proper way to extend a negative number is to append leading 1's. Thus, the value stored for negative 9 is

Negation

Negation is the operation of converting a positive number to its negative equivalent or a negative number to its positive equivalent. When signed binary numbers are represented in the 2's-complement system, negation is performed simply by performing the 2's-complement operation. To illustrate, let's start with +9 in eight-bit binary form. Its signed representation is 00001001. If we take its 2's complement we get 11110111, which represents the signed value −9. Likewise, we can start with the representation of −9, which is 11110111, and take its 2's complement to get 00001001, which represents +9. These steps are diagrammed below.

Start with	00001001	+9
2's complement (negate)	11110111	−9
negate again	00001001	+9

Thus, we negate a signed binary number by 2's-complementing it.

This negation changes the number to its equivalent of opposite sign. We used negation in steps (d) and (e) of Example 6-1 to convert positive numbers to their negative equivalents.

EXAMPLE 6-2

Each of the following numbers is a five-bit signed binary number in the 2's-complement system. Determine the decimal value in each case:

(a) 01100 (b) 11010 (c) 10001

Solution

(a) The sign bit is 0, so the number is *positive* and the other four bits represent the true magnitude of the number. That is, $1100_2 = 12_{10}$. Thus, the decimal number is +12.

(b) The sign bit of 11010 is a 1, so we know that the number is negative, but we can't tell what the magnitude is. We can find the magnitude by negating (2's-complementing) the number to convert it to its positive equivalent.

11010	(original negative number)
00101	(1's complement)
+ 1	(add 1)
00110	(+6)

Because the result of the negation is 00110 = +6, the original number 11010 must be equivalent to −6.

(c) Follow the same procedure as in (b):

10001	(original negative number)
01110	(1's complement)
+ 1	(add 1)
01111	(+15)

Thus, 10001 = −15.

Special Case in 2's-Complement Representation

Whenever a signed number has a 1 in the sign bit and all 0s for the magnitude bits, its decimal equivalent is -2^N, where N is the number of bits in the *magnitude.* For example,

$$1000 = -2^3 = -8$$
$$10000 = -2^4 = -16$$
$$100000 = -2^5 = -32$$

and so on. Notice that in this special case, taking the 2's complement of these numbers produces the value we started with because we are at the negative limit of the range of numbers that can be represented by this many bits. If we extend the sign of these special numbers, the normal negation procedure works fine. For example, extending the number 1000 (−8) to 11000 (five-bit negative 8) and taking its 2's complement we get 01000 (8), which is the magnitude of the negative number.

Thus, we can state that the complete range of values that can be represented in the 2's-complement system having N magnitude bits is

$$-2^N \text{ to } +(2^N - 1)$$

There are a total of 2^{N+1} different values, *including* zero.

For example, Table 6-1 lists all signed numbers that can be represented in four bits using the 2's-complement system (note there are three magnitude bits, so $N = 3$). Note that the sequence starts at $-2^N = -2^3 = -8_{10} = 1000_2$ and proceeds upward to $+(2^N - 1) = +2^3 - 1 = +7_{10} = 0111_2$ by adding 0001 at each step as in an up counter.

TABLE 6-1

Decimal Value	Signed Binary Using 2's Complement
$+7 = 2^3 - 1$	0111
+6	0110
+5	0101
+4	0100
+3	0011
+2	0010
+1	0001
0	0000
−1	1111
−2	1110
−3	1101
−4	1100
−5	1011
−6	1010
−7	1001
$-8 = -2^3$	1000

EXAMPLE 6-3

What is the range of *unsigned* decimal values that can be represented in a byte?

Solution

Recall that a byte is eight bits. We are interested in unsigned numbers here, so there is no sign bit, and all of the eight bits are used for the magnitude. Therefore, the values will range from

$$00000000_2 = 0_{10}$$

to

$$11111111_2 = 255_{10}$$

This is a total of 256 different values, which we could have predicted because $2^8 = 256$.

EXAMPLE 6-4

What is the range of *signed* decimal values that can be represented in a byte?

Solution

Because the MSB is to be used as the sign bit, there are seven bits for the magnitude. The largest negative value is

$$10000000_2 = -2^7 = -128_{10}$$

The largest positive value is

$$01111111_2 = +2^7 - 1 = +127_{10}$$

Thus, the range is −128 to +127; this is a total of 256 different values, including zero. Alternatively, because there are seven magnitude bits ($N = 7$), then there are $2^{N+1} = 2^8 = 256$ different values.

EXAMPLE 6-5

A certain computer is storing the following two signed numbers in its memory using the 2's-complement system:

$$00011111_2 = +31_{10}$$
$$11110100_2 = -12_{10}$$

While executing a program, the computer is instructed to convert each number to its opposite sign; that is, change the +31 to −31 and change the −12 to +12. How will it do this?

Solution

This is the negation operation whereby a signed number can have its polarity changed simply by performing the 2's-complement operation on the *complete* number, including the sign bit. The computer circuitry will take the signed number from memory, find its 2's complement, and put the result back in memory.

REVIEW QUESTIONS

1. Represent each of the following values as an eight-bit signed number in the 2's-complement system.
   (a) +13 (b) −7 (c) −128
2. Each of the following is a signed binary number in the 2's-complement system. Determine the decimal equivalent for each.
   (a) 100011 (b) 1000000 (c) 01111110
3. What range of signed decimal values can be represented in 12 bits (including the sign bit)?
4. How many bits are required to represent decimal values ranging from −50 to +50?
5. What is the largest negative decimal value that can be represented by a two-byte number?
6. Perform the 2's-complement operation on each of the following.
   (a) 10000 (b) 10000000 (c) 1000
7. Define the negation operation.

6-3 ADDITION IN THE 2's-COMPLEMENT SYSTEM

We will now investigate how the operations of addition and subtraction are performed in digital machines that use the 2's-complement representation for negative numbers. In the various cases to be considered, it is important to note that the sign bit of each number is operated on in the same manner as the magnitude bits.

Case I: Two Positive Numbers. The addition of two positive numbers is straightforward. Consider the addition of +9 and +4:

```
+9 → 0 1001      (augend)
+4 → 0 0100      (addend)
     ------
     0 1101      (sum = +13)
     ↑
     └ sign bits
```

Note that the sign bits of the **augend** and the **addend** are both 0 and the sign bit of the sum is 0, indicating that the sum is positive. Also note that the augend and the addend are made to have the same number of bits. This must *always* be done in the 2's-complement system.

Case II: Positive Number and Smaller Negative Number. Consider the addition of +9 and −4. Remember that the −4 will be in its 2's-complement form. Thus, +4 (00100) must be converted to −4 (11100).

```
       ┌ sign bits
       ↓
+9 →   0 1001      (augend)
−4 →   1 1100      (addend)
     --------
     1 0 0101
     ↑
     └ This carry is disregarded; the result is 00101 (sum = +5).
```

In this case, the sign bit of the addend is 1. Note that the sign bits also participate in the addition process. In fact, a carry is generated in the last position of addition. *This carry is always disregarded,* so that the final sum is 00101, which is equivalent to +5.

Case III: Positive Number and Larger Negative Number. Consider the addition of −9 and +4:

```
−9 →   10111
+4 →   00100
       -----
       11011    (sum = −5)
       ↑
       └── negative sign bit
```

The sum here has a sign bit of 1, indicating a negative number. Because the sum is negative, it is in 2's-complement form, so that the last four bits, 1011, actually represent the 2's complement of the sum. To find the true magnitude of the sum, we must negate (2's-complement) 11011; the result is 00101 = +5. Thus, 11011 represents −5.

Case IV: Two Negative Numbers

```
−9 →    10111
−4 →    11100
        -----
     1  10011
     ↑  ↑
     |  └── sign bit
     └───── This carry is disregarded; the result is 10011 (sum = −13).
```

This final result is again negative and in 2's-complement form with a sign bit of 1. Negating (2's-complementing) this result produces 01101 = +13.

Case V: Equal and Opposite Numbers

```
−9 →    10111
+9 →    01001
        -----
0    1  00000
     ↑
     └────── Disregard; the result is 00000 (sum = +0).
```

The result is obviously +0, as expected.

REVIEW QUESTIONS

Assume the 2's-complement system for both questions.

1. *True or false:* Whenever the sum of two signed binary numbers has a sign bit of 1, the magnitude of the sum is in 2's-complement form.
2. Add the following pairs of signed numbers. Express the sum as a signed binary number and as a decimal number.
   (a) 100111 + 111011 (b) 100111 + 011001

6-4 SUBTRACTION IN THE 2's-COMPLEMENT SYSTEM

The subtraction operation using the 2's-complement system actually involves the operation of addition and is really no different from the various cases for addition considered in Section 6-3. When subtracting one binary number (the **subtrahend**) from another binary number (the **minuend**), use the following procedure:

1. *Negate the subtrahend.* This will change the subtrahend to its equivalent value of opposite sign.
2. *Add this to the minuend.* The result of this addition will represent the *difference* between the subtrahend and the minuend.

Once again, as in all 2's-complement arithmetic operations, it is necessary that both numbers have the same number of bits in their representations.

Let us consider the case where +4 is to be subtracted from +9.

```
   minuend (+9) →   01001
subtrahend (+4) →   00100
```

Negate the subtrahend to produce 11100, which represents −4. Now add this to the minuend.

```
  01001  (+9)
+ 11100  (−4)
---------
1 00101  (+5)
↑
└Disregard, so the result is 00101 = +5.
```

When the subtrahend is changed to its 2's complement, it actually becomes −4, so that we are *adding* −4 and +9, which is the same as subtracting +4 from +9. This is the same as case II of Section 6-3. Any subtraction operation, then, actually becomes one of addition when the 2's-complement system is used. This feature of the 2's-complement system has made it the most widely used of the methods available because it allows addition and subtraction to be performed by the same circuitry.

Here's another example showing +9 subtracted from −4:

```
  11100  (−4)
− 01001  (+9)
```

Negate the subtrahend (+9) to produce 10111 (−9) and add this to the minuend (−4).

```
  11100  (−4)
+ 10111  (−9)
---------
1 10011  (−13)
↑
└Disregard
```

The reader should verify the results of using the above procedure for the following subtractions: (a) +9 − (−4); (b) −9 − (+4); (c) −9 − (−4); (d) +4 − (−4). Remember that when the result has a sign bit of 1, it is negative and in 2's-complement form.

Arithmetic Overflow

In each of the previous addition and subtraction examples, the numbers that were added consisted of a sign bit and four magnitude bits. The answers also consisted of a sign bit and four magnitude bits. Any carry into the sixth bit position was disregarded. In all of the cases considered, the magnitude of the answer was small enough to fit into four bits. Let's look at the addition of +9 and +8.

$$
\begin{array}{rc|l}
+9 \rightarrow & 0 & 1001 \\
+8 \rightarrow & 0 & 1000 \\
\hline
 & 1 & \underline{0001}
\end{array}
$$

incorrect sign ↑ ↑ incorrect magnitude

The answer has a negative sign bit, which is obviously incorrect because we are adding two positive numbers. The answer should be +17, but the magnitude 17 requires more than four bits and therefore *overflows* into the sign-bit position. This **overflow** condition can occur only when two positive or two negative numbers are being added, and it always produces an incorrect result. Overflow can be detected by checking to see that the sign bit of the result is the same as the sign bits of the numbers being added.

Subtraction in the 2's-complement system is performed by negating the minuend and *adding* it to the subtrahend, so overflow can occur only when the minuend and subtrahend have different signs. For example, if we are subtracting −8 from +9, the −8 is negated to become +8 and is added to +9, just as shown above, and overflow produces an erroneous negative result because the magnitude is too large.

A computer will have a special circuit to detect any overflow condition when two numbers are added or subtracted. This detection circuit will signal the computer's control unit that overflow has occurred and the result is incorrect. We will examine such a circuit in an end-of-chapter problem.

Number Circles and Binary Arithmetic

The concept of signed arithmetic and overflow can be illustrated by taking the numbers from Table 6-1 and "bending" them into a number circle as shown in Figure 6-3. Notice that there are two ways to look at this circle. It can be thought of as a circle of unsigned numbers (as shown in the outer ring) with minimum value 0 and maximum 15, or as signed 2's-complement numbers (as shown in the inner ring) with maximum value 7 and minimum −8. To add using a number circle, simply start at the value of the augend and

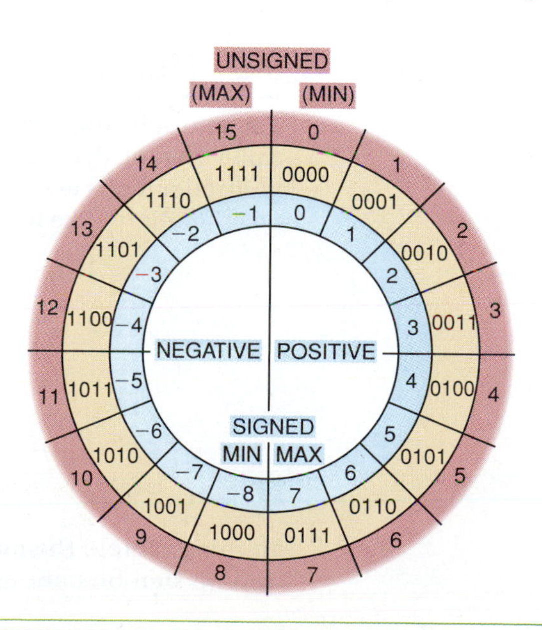

FIGURE 6-3 A four-bit number circle.

advance around the number circle clockwise by the number of spaces in the addend. For example, to add 2 + 3, start at 2 (0010) and then advance clockwise three more spaces to arrive at 5 (0101). Overflow occurs when the sum is too big to fit into four-bit signed format, meaning we have exceeded the maximum value of 7. On the number circle this is indicated when adding two positive values causes us to cross the line between 0111 (max positive) and 1000 (max negative).

The number circle can also illustrate how 2's-complement subtraction really works. For example, let's perform the subtraction of 5 from 3. Of course, we know the answer is −2, but let's run the problem through the number circle. First we start at the number 3 (0011) on the number circle. The most apparent way to subtract is to move *counterclockwise* around the circle five spaces, which lands us on the number 1110 (−2). The less obvious operation that illustrates 2's-complement arithmetic is to add −5 to the number 3. Negative five (the 2's complement of 0101) is 1011 which, interpreted as an unsigned binary number, represents the value 11 (eleven) in decimal. Start at the number 3 (0011) and move clockwise around the circle 11 spaces and you will once again find yourself arriving at the number 1110 (−2), which is the correct result.

Any subtraction operation between four-bit numbers of opposite sign that produces a result greater than 7 or less than −8 is an overflow of the four-bit format and results in an incorrect answer. For example, 3 minus −6 should produce the answer 9, but moving clockwise six spaces from 3 lands us on the signed number −7: an overflow condition has occurred, giving us an incorrect answer.

REVIEW QUESTIONS

1. Perform the subtraction on the following pairs of signed numbers using the 2's-complement system. Express the results as signed binary numbers and as decimal values.

   (a) 01001 − 11010 (b) 10010 − 10011

2. How can arithmetic overflow be detected when signed numbers are being added? Subtracted?

6-5 MULTIPLICATION OF BINARY NUMBERS

The multiplication of binary numbers is done in the same manner as the multiplication of decimal numbers. The process is actually simpler because the multiplier digits are either 0 or 1 and so we are always multiplying by 0 or 1 and no other digits. The following example illustrates for unsigned binary numbers:

```
      1001    ← multiplicand = 9₁₀
      1011    ← multiplier = 11₁₀
   -------
      1001  ⎫
     1001   ⎪
    0000    ⎬ partial products
   1001     ⎭
   -------
   1100011 }  final product = 99₁₀
```

In this example the multiplicand and the multiplier are in true binary form and no sign bits are used. The steps followed in the process are exactly the same as in decimal multiplication. First, the LSB of the multiplier is examined; in our example, it is a 1. This 1 multiplies the multiplicand to produce 1001, which is written down as the first partial product. Next, the second bit of the multiplier is examined. It is a 1, and so 1001 is written for the second partial product. Note that this second partial product is *shifted* one place to the left relative to the first one. The third bit of the multiplier is 0, and 0000 is written as the third partial product; again, it is shifted one place to the left relative to the previous partial product. The fourth multiplier bit is 1, and so the last partial product is 1001 shifted again one position to the left. The four partial products are then summed to produce the final product.

Most digital machines can add only two binary numbers at a time. For this reason, the partial products formed during multiplication cannot all be added together at the same time. Instead, they are added together two at a time; that is, the first is added to the second, their sum is added to the third, and so on. This process is now illustrated for the example above:

```
Add {     1001   ← first partial product
         1001    ← second partial product shifted left
        ______
Add {    11011   ← sum of first two partial products
        0000     ← third partial product shifted left
        ______
Add {   011011   ← sum of first three partial products
       1001      ← fourth partial product shifted left
       _______
       1100011   ← sum of four partial products, which
                   equals final total product
```

Multiplication in the 2's-Complement System

In computers that use the 2's-complement representation, multiplication is carried on in the manner described above, provided that both the multiplicand and the multiplier are put in true binary form. If the two numbers to be multiplied are positive, they are already in true binary form and are multiplied as they are. The resulting product is, of course, positive and is given a sign bit of 0. When the two numbers are negative, they will be in 2's-complement form. The 2's complement of each is taken to convert it to a positive number, and then the two numbers are multiplied. The product is kept as a positive number and is given a sign bit of 0.

When one of the numbers is positive and the other is negative, the negative number is first converted to a positive magnitude by taking its 2's complement. The product will be in true-magnitude form. However, the product must be negative because the original numbers are of opposite sign. Thus, the product is then changed to 2's-complement form and is given a sign bit of 1.

REVIEW QUESTION

1. Multiply the unsigned numbers 0111 and 1110.

6-6 BINARY DIVISION

The process for dividing one binary number (the *dividend*) by another (the *divisor*) is the same as that followed for decimal numbers, that which we usually refer to as "long division." The actual process is simpler in binary because when we are checking to see how many times the divisor "goes into" the dividend, there are only two possibilities, 0 or 1. To illustrate, consider the following simple division examples:

$$
\begin{array}{r}
0011 \\
11\overline{)1001} \\
\underline{011} \\
0011 \\
\underline{11} \\
0
\end{array}
\qquad\qquad
\begin{array}{r}
0010.1 \\
100\overline{)1010.0} \\
\underline{100} \\
100 \\
\underline{100} \\
0
\end{array}
$$

In the first example, we have 1001_2 divided by 11_2, which is equivalent to $9 \div 3$ in decimal. The resulting quotient is $0011_2 = 3_{10}$. In the second example, 1010_2 is divided by 100_2, or $10 \div 4$ in decimal. The result is $0010.1_2 = 2.5_{10}$.

In most modern digital machines, the subtractions that are part of the division operation are usually carried out using 2's-complement subtraction, that is, taking the 2's complement of the subtrahend and then adding.

The division of signed numbers is handled in the same way as multiplication. Negative numbers are made positive by complementing, and the division is then carried out. If the dividend and the divisor are of opposite sign, the resulting quotient is changed to a negative number by taking its 2's-complement and is given a sign bit of 1. If the dividend and the divisor are of the same sign, the quotient is left as a positive number and is given a sign bit of 0.

6-7 BCD ADDITION

In Chapter 2, we stated that many computers and calculators use the BCD code to represent decimal numbers. Recall that this code takes *each* decimal digit and represents it by a four-bit code ranging from 0000 to 1001. The addition of decimal numbers that are in BCD form can be best understood by considering the two cases that can occur when two decimal digits are added.

Sum Equals 9 or Less

Consider adding 5 and 4 using BCD to represent each digit:

$$
\begin{array}{r}
5 \\
+\underline{4} \\
9
\end{array}
\qquad
\begin{array}{r}
0101 \\
+\ \underline{0100} \\
1001
\end{array}
\quad
\begin{array}{l}
\leftarrow \text{BCD for } 5 \\
\leftarrow \text{BCD for } 4 \\
\leftarrow \text{BCD for } 9
\end{array}
$$

The addition is carried out as in normal binary addition, and the sum is 1001, which is the BCD code for 9. As another example, take 45 added to 33:

$$
\begin{array}{r}
45 \\
\underline{+33} \\
78
\end{array}
\qquad
\begin{array}{r}
0100\ 0101 \\
+\ \underline{0011\ 0011} \\
0111\ 1000
\end{array}
\quad
\begin{array}{l}
\leftarrow \text{BCD for } 45 \\
\leftarrow \text{BCD for } 33 \\
\leftarrow \text{BCD for } 78
\end{array}
$$

In this example, the four-bit codes for 5 and 3 are added in binary to produce 1000, which is BCD for 8. Similarly, adding the second-decimal-digit positions produces 0111, which is BCD for 7. The total is 01111000, which is the BCD code for 78.

In the examples above, none of the sums of the pairs of decimal digits exceeded 9; therefore, *no decimal carries were produced.* For these cases, the BCD addition process is straightforward and is actually the same as binary addition.

Sum Greater than 9

Consider the addition of 6 and 7 in BCD:

```
  6      0110  ← BCD for 6
 +7    + 0111  ← BCD for 7
+13      1101  ← invalid code group for BCD
```

The sum 1101 does not exist in the BCD code; it is one of the six forbidden or invalid four-bit code groups. This has occurred because the sum of the two digits exceeds 9. Whenever this occurs, the sum must be corrected by the addition of six (0110) to take into account the skipping of the six invalid code groups:

```
         0110  ← BCD for 6
       + 0111  ← BCD for 7
         1101  ← invalid sum
         0110  ← add 6 for correction
 0001    0011  ← BCD for 13
   1       3
```

As shown above, 0110 is added to the invalid sum and produces the correct BCD result. Note that with the addition of 0110, a carry is produced in the second decimal position. This addition must be performed whenever the sum of the two decimal digits is greater than 9.

As another example, take 47 plus 35 in BCD:

```
 47     0100     0111  ← BCD for 47
+35   + 0011     0101  ← BCD for 35
 82     0111     1100  ← invalid sum in first digit
           1←┐   0110  ← add 6 to correct
        1000 └─0010    ← correct BCD sum
          8        2
```

The addition of the four-bit codes for the 7 and 5 digits results in an invalid sum and is corrected by adding 0110. Note that this generates a carry of 1, which is carried over to be added to the BCD sum of the second-position digits.

Consider the addition of 59 and 38 in BCD:

```
               ┌──┐
               ↓  │
               1  │
 59      0101     │ 1001  ← BCD for 59
+38    + 0011     │ 1000  ← BCD for 38
 97      1001     └─0001  ← perform addition
                    0110  ← add 6 to correct
         1001       0111    BCD for 97
           9          7
```

Here, the addition of the least significant digits (LSDs) produces a sum of 17 = 10001. This generates a carry into the next digit position to be added to the codes for 5 and 3. Since 17 > 9, a correction factor of 6 must be added to the LSD sum. Addition of this correction does not generate a carry; the carry was already generated in the original addition.

To summarize the BCD addition procedure:

1. Using ordinary binary addition, add the BCD code groups for each digit position.
2. For those positions where the sum is 9 or less, no correction is needed. The sum is in proper BCD form.
3. When the sum of two digits is greater than 9, a correction of 0110 should be added to that sum to get the proper BCD result. This case always produces a carry into the next digit position, either from the original addition (step 1) or from the correction addition.

The procedure for BCD addition is clearly more complicated than straight binary addition. This is also true of the other BCD arithmetic operations. Readers should perform the addition of 275 + 641. Then check the correct procedure below.

```
 275      0010   0111   0101   ← BCD for 275
+641    + 0110   0100   0001   ← BCD for 641
 916      1000   1011   0110   ← perform addition
        +        0110          ← add 6 to correct second digit
          1001   0001   0110   ← BCD for 916
```

BCD Subtraction

The process of subtracting BCD numbers is more difficult than addition. It involves a complement-then-add procedure similar to the 2's-complement method. We do not cover it in this book.

REVIEW QUESTIONS

1. How can you tell when a correction is needed in BCD addition?
2. Represent 135_{10} and 265_{10} in BCD and then perform BCD addition. Check your work by converting the result back to decimal.

6-8 HEXADECIMAL ARITHMETIC

Hex numbers are used extensively in machine-language computer programming and in conjunction with computer memories (i.e., addresses). When working in these areas, you will encounter situations where hex numbers must be added or subtracted.

Hex Addition

Addition of hexadecimal numbers is done in much the same way as decimal addition, as long as you remember that the largest hex digit is F instead of 9. The following procedure is suggested:

1. Add the two hex digits in decimal, mentally inserting the decimal equivalent for those digits larger than 9.

2. If the sum is 15 or less, it can be directly expressed as a hex digit.
3. If the sum is greater than or equal to 16, subtract 16 and carry a 1 to the next digit position.

The following examples will illustrate the procedure.

EXAMPLE 6-6

Add the hex numbers 58 and 24.

Solution

```
 58
+24
 7C
```

Adding the LSDs (8 and 4) produces 12, which is C in hex. There is no carry into the next digit position. Adding 5 and 2 produces 7.

EXAMPLE 6-7

Add the hex numbers 58 and 4B.

Solution

```
 58
+4B
 A3
```

Start by adding 8 and B, mentally substituting decimal 11 for B. This produces a sum of 19. Because 19 is greater than 16, subtract 16 to get 3; write down the 3 and carry a 1 into the next position. This carry is added to the 5 and 4 to produce a sum of 10_{10}, which is then converted to hexadecimal A.

EXAMPLE 6-8

Add 3AF to 23C.

Solution

```
 3AF
+23C
 5EB
```

The sum of F and C is considered as $15 + 12 = 27_{10}$. Because this is greater than 16, subtract 16 to get 11_{10}, which is hexadecimal B, and carry a 1 into the second position. Add this carry to A and 3 to obtain E. There is no carry into the MSD position.

Hex Subtraction

Remember that hex numbers are just an efficient way to represent binary numbers. Thus, we can subtract hex numbers using the same method we used for binary numbers. The 2's complement of the hex subtrahend will be taken and then *added* to the minuend, and any carry out of the MSD position will be disregarded.

How do we find the 2's complement of a hex number? One way is to convert it to binary, take the 2's complement of the binary equivalent, and then convert it back to hex. This process is illustrated below.

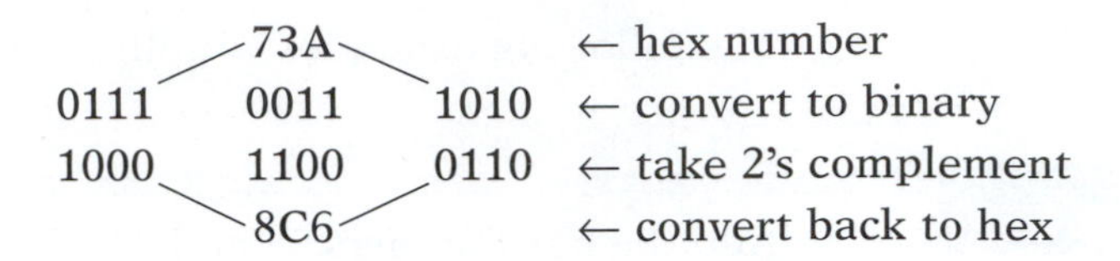

There is a quicker procedure: subtract *each* hex digit from F; then add 1. Let's try this for the same hex number from the example above.

```
 F    F     F  }
-7   -3    -A  } ← subtract each digit from F
 8    C     5  }
           +1    ← add 1
-----------------
 8    C     6    ← hex equivalent of 2's complement
```

Try either of the procedures above on the hex number E63. The correct result for the 2's complement is 19D.

CALCULATOR HINT

On a hex calculator, you can subtract the hex digits from a string of F's and then add one as we just demonstrated, or you can add one to the string of all F's and then subtract. For example, adding 1 to FFF_{16} yields 1000_{16}. On the hex calculator enter:

1000 − 73A =   The answer is 8C6

EXAMPLE 6-9

Subtract $3A5_{16}$ from 592_{16}.

Solution

First, convert the subtrahend (3A5) to its 2's-complement form by using either method presented above. The result is C5B. Then add this to the minuend (592):

```
   592
 + C5B
  ----
  11ED
  ↑
  └── Disregard carry.
```

Ignore the carry out of the MSD addition; the result is 1ED. We can prove that this is correct by adding 1ED to 3A5 and checking to see that it equals 592_{16}.

Hex Representation of Signed Numbers

The data stored in a microcomputer's internal working memory or on a hard disk or CD ROM are typically stored in bytes (groups of eight bits). The data

TABLE 6-2

Hex Address	Stored Binary Data	Hex Value	Decimal Value
4000	00111010	3A	+58
4001	11100101	E5	−29
4002	01010111	57	+87
4003	10000000	80	−128

byte stored in a particular memory location is often expressed in hexadecimal because it is more efficient and less error-prone than expressing it in binary. When the data consist of *signed* numbers, it is helpful to be able to recognize whether a hex value represents a positive or a negative number. For example, Table 6-2 lists the data stored in a small segment of memory starting at address 4000.

Each memory location stores a single byte (eight bits), which is the binary equivalent of a signed decimal number. The table also shows the hex equivalent of each byte. For a negative data value, the sign bit (MSB) of the binary number will be a 1; this will always make the MSD of the hex number 8 or greater. When the data value is positive, the sign bit will be a 0, and the MSD of the hex number will be 7 or less. The same holds true no matter how many digits are in the hex number. *When the MSD is 8 or greater, the number being represented is negative; when the MSD is 7 or less, the number is positive.*

REVIEW QUESTIONS

1. Add 67F + 2A4.
2. Subtract 67F − 2A4.
3. Which of the following hex numbers represent positive values: 2F, 77EC, C000, 6D, FFFF?

6-9 ARITHMETIC CIRCUITS

One essential function of most computers and calculators is the performance of arithmetic operations. These operations are all performed in the arithmetic/logic unit of a computer, where logic gates and flip-flops are combined so that they can add, subtract, multiply, and divide binary numbers. These circuits perform arithmetic operations at speeds that are not humanly possible. Typically, an addition operation will take less than 100 ns.

We will now study some of the basic arithmetic circuits that are used to perform the arithmetic operations discussed earlier. In some cases, we will go through the actual design process, even though the circuits may be commercially available in integrated-circuit form, to provide more practice in the use of the techniques learned in Chapter 4.

Arithmetic/Logic Unit

All arithmetic operations take place in the **arithmetic/logic unit (ALU)** of a computer. Figure 6-4 is a block diagram showing the major elements included in a typical ALU. The main purpose of the ALU is to accept binary

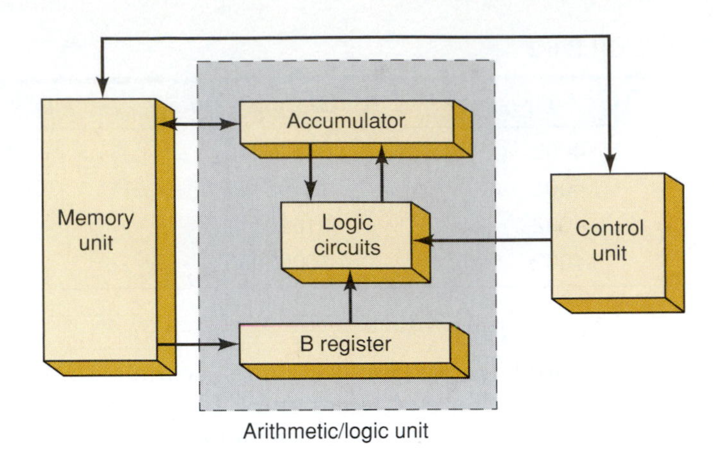

FIGURE 6-4 Functional parts of an ALU.

data that are stored in the memory and to execute arithmetic and logic operations on these data according to instructions from the control unit.

The arithmetic/logic unit contains at least two flip-flop registers: the *B register* and the **accumulator register**. It also contains combinational logic, which performs the arithmetic and logic operations on the binary numbers that are stored in the *B* register and the accumulator. A typical sequence of operations may occur as follows:

1. The control unit receives an instruction (from the memory unit) specifying that a number stored in a particular memory location (address) is to be added to the number presently stored in the accumulator register.
2. The number to be added is transferred from memory to the *B* register.
3. The number in the *B* register and the number in the accumulator register are added together in the logic circuits (upon command from the control unit). The resulting sum is then sent to the accumulator to be stored.
4. The new number in the accumulator can remain there so that another number can be added to it or, if the particular arithmetic process is finished, it can be transferred to memory for storage.

These steps should make it apparent how the accumulator register derives its name. This register "accumulates" the sums that occur when performing successive additions between new numbers acquired from memory and the previously accumulated sum. In fact, for any arithmetic problem containing several steps, the accumulator usually contains the results of the intermediate steps as they are completed as well as the final result when the problem is finished.

6-10 PARALLEL BINARY ADDER

Computers and calculators perform the addition operation on two binary numbers at a time, where each binary number can have several binary digits. Figure 6-5 illustrates the addition of two five-bit numbers. The **augend** is stored in the accumulator register; that is, the accumulator contains five FFs, storing the values 10101 in successive FFs. Similarly, the **addend**, the number that is to be added to the augend, is stored in the *B* register (in this case, 00111).

The addition process starts by adding the least significant bits (LSBs) of the augend and addend. Thus, $1 + 1 = 10$, which means that the *sum* for that position is 0, with a *carry* of 1.

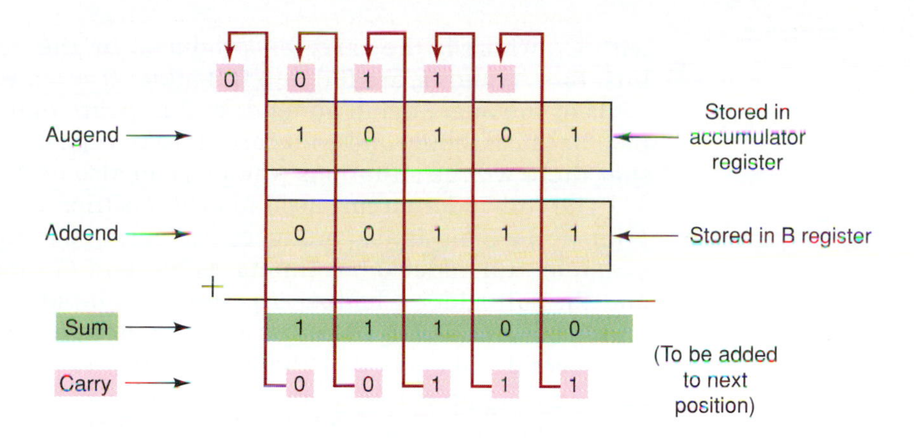

FIGURE 6-5 Typical binary addition process.

This carry must be added to the next position along with the augend and addend bits in that position. Thus, in the second position, $1 + 0 + 1 = 10$, which is again a sum of 0 and a carry of 1. This carry is added to the next position together with the augend and addend bits in that position, and so on, for the remaining positions, as shown in Figure 6-5.

At each step in this addition process, we are performing the addition of three bits: the augend bit, the addend bit, and a carry bit from the previous position. The result of the addition of these three bits produces two bits: a *sum* bit, and a *carry* bit that is to be added to the next position. It should be clear that the same process is followed for each bit position. Thus, if we can design a logic circuit that can duplicate this process, then all we have to do is to use the identical circuit for each of the bit positions. This is illustrated in Figure 6-6.

In this diagram, variables A_4, A_3, A_2, A_1, and A_0 represent the bits of the augend that are stored in the accumulator (which is also called the *A* register). Variables B_4, B_3, B_2, B_1, and B_0 represent the bits of the addend stored in the *B* register. Variables C_4, C_3, C_2, C_1, and C_0 represent the carry bits into the corresponding positions. Variables S_4, S_3, S_2, S_1, S_0 are the sum output bits for each position. Corresponding bits of the augend and addend are fed to a logic circuit called a **full adder (FA)**, along with a carry bit from the previous position. For example, bits A_1 and B_1 are fed into full adder 1 along

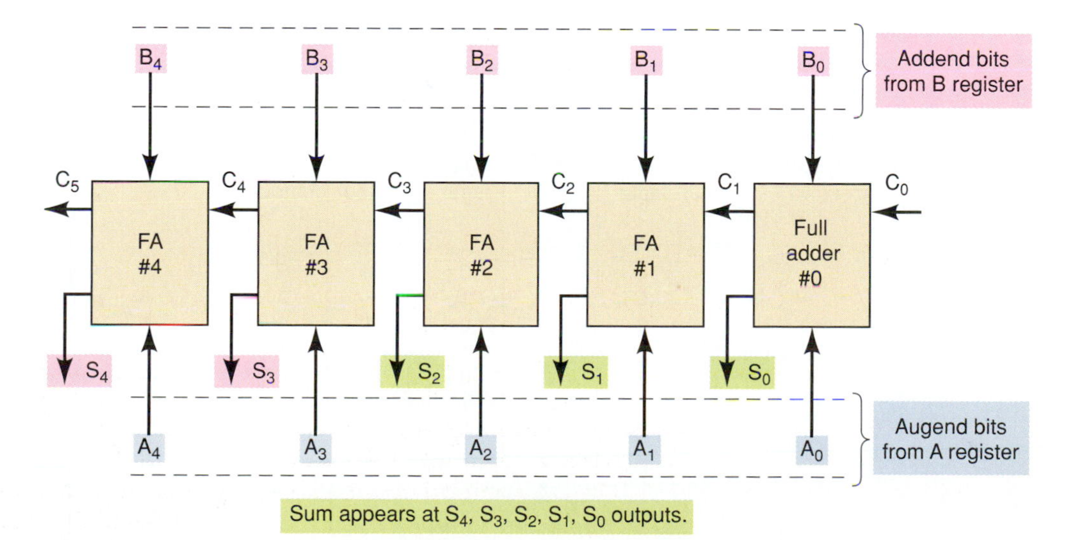

FIGURE 6-6 Block diagram of a five-bit parallel adder circuit using full adders.

with C_1, which is the carry bit produced by the addition of the A_0 and B_0 bits. Bits A_0 and B_0 are fed into full adder 0 along with C_0. A_0 and B_0 are the LSBs of the augend and addend, so it appears that C_0 would always have to be 0 because there can be no carry into that position. We shall see, however, that there will be situations when C_0 can also be 1.

The full-adder circuit used in each position has three inputs: an A bit, a B bit, and a C bit. It also produces two outputs: a sum bit and a carry bit. For example, full adder 0 has inputs A_0, B_0, and C_0, and it produces outputs S_0 and C_1. Full adder 1 had A_1, B_1, and C_1 as inputs and S_1 and C_2 as outputs, and so on. This arrangement is repeated for as many positions as there are in the augend and addend. Although this illustration is for five-bit numbers, in modern computers the numbers usually range from 8 to 64 bits.

The arrangement in Figure 6-6 is called a **parallel adder** because all of the bits of the augend and addend are present and are fed into the adder circuits *simultaneously*. This means that the additions in each position are taking place at the same time. This is different from how we add on paper, taking each position one at a time starting with the LSB. Clearly, parallel addition is extremely fast. More will be said about this later.

REVIEW QUESTIONS

1. How many inputs does a full adder have? How many outputs?
2. Assume the following input levels in Figure 6-6: $A_4A_3A_2A_1A_0 = 01001$; $B_4B_3B_2B_1B_0 = 00111$; $C_0 = 0$.
   (a) What are the logic levels at the outputs of FA #2?
   (b) What is the logic level at the C_5 output?

6-11 DESIGN OF A FULL ADDER

Now that we know the function of the full adder, we can design a logic circuit that will perform this function. First, we must construct a truth table showing the various input and output values for all possible cases. Figure 6-7 shows the truth table having three inputs, A, B, and C_{IN}, and two outputs, S and C_{OUT}. There are eight possible cases for the three inputs, and for each

FIGURE 6-7 Truth table for a full-adder circuit.

Augend bit input	Addend bit input	Carry bit input	Sum bit output	Carry bit output
A	B	C_{IN}	S	C_{OUT}
0	0	0	0	0
0	0	1	1	0
0	1	0	1	0
0	1	1	0	1
1	0	0	1	0
1	0	1	0	1
1	1	0	0	1
1	1	1	1	1

case the desired output values are listed. For example, consider the case $A = 1$, $B = 0$, and $C_{IN} = 1$. The full adder (FA) must add these bits to produce a sum (S) of 0 and a carry (C_{OUT}) of 1. The reader should check the other cases to be sure they are understood.

Because there are two outputs, we will design the circuitry for each output individually, starting with the S output. The truth table shows that there are four cases where S is to be a 1. Using the sum-of-products method, we can write the expression for S as

$$S = \overline{A}\,\overline{B}C_{IN} + \overline{A}B\overline{C}_{IN} + A\overline{B}\,\overline{C}_{IN} + ABC_{IN} \qquad \textbf{(6-1)}$$

We can now try to simplify this expression by factoring. Unfortunately, none of the terms in the expression has two variables in common with any of the other terms. However, $\overline{A}$ can be factored from the first two terms, and A can be factored from the last two terms:

$$S = \overline{A}(\overline{B}C_{IN} + B\overline{C}_{IN}) + A(\overline{B}\,\overline{C}_{IN} + BC_{IN})$$

The first term in parentheses should be recognized as the exclusive-OR combination of B and C_{IN}, which can be written as $B \oplus C_{IN}$. The second term in parentheses should be recognized as the exclusive-NOR of B and C_{IN}, which can be written as $\overline{B \oplus C_{IN}}$. Thus, the expression for S becomes

$$S = \overline{A}(B \oplus C_{IN}) + A(\overline{B \oplus C_{IN}})$$

If we let $X = B \oplus C_{IN}$, this can be written as

$$S = \overline{A} \cdot X + A \cdot \overline{X} = A \oplus X$$

which is simply the exclusive-OR of A and X. Replacing the expression for X, we have

$$S = A \oplus [B \oplus C_{IN}] \qquad \textbf{(6-2)}$$

Consider now the output C_{OUT} in the truth table of Figure 6-7. We can write the sum-of-products expression for C_{OUT} as follows:

$$C_{OUT} = \overline{A}BC_{IN} + A\overline{B}C_{IN} + AB\overline{C}_{IN} + ABC_{IN}$$

This expression can be simplified by factoring. We will employ the trick introduced in Chapter 4, whereby we will use the ABC_{IN} term *three* times because it has common factors with each of the other terms. Hence,

$$\begin{aligned} C_{OUT} &= BC_{IN}(\overline{A} + A) + AC_{IN}(\overline{B} + B) + AB(\overline{C}_{IN} + C_{IN}) \qquad \textbf{(6-3)} \\ &= BC_{IN} + AC_{IN} + AB \end{aligned}$$

This expression cannot be simplified further.

Expressions (6-2) and (6-3) can be implemented as shown in Figure 6-8. Several other implementations can be used to produce the same expressions for S and C_{OUT}, none of which has any particular advantage over those shown. The complete circuit with inputs A, B, and C_{IN} and outputs S and C_{OUT} represents the full adder. Each of the FAs in Figure 6-6 contains this same circuitry (or its equivalent).

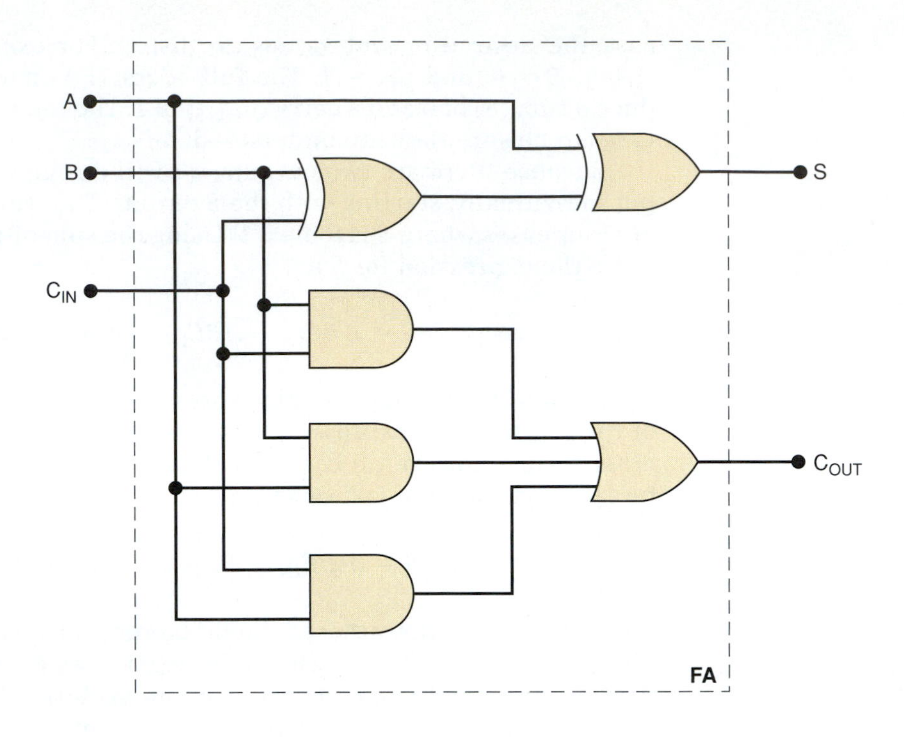

FIGURE 6-8 Complete circuitry for a full adder.

K-Map Simplification

We simplified the expressions for S and C_{OUT} using algebraic methods. The K-map method can also be used. Figure 6-9(a) shows the K map for the S output. This map has no adjacent 1s, and so there are no pairs or quads to loop. Thus, the expression for S cannot be simplified using the K map. This points out a limitation of the K-map method compared with the algebraic method. We were able to simplify the expression for S through factoring and the use of XOR and XNOR operations.

The K map for the C_{OUT} output is shown in Figure 6-9(b). The three pairs that are looped will produce the same expression obtained from the algebraic method.

Half Adder

The FA operates on three inputs to produce a sum and carry output. In some cases, a circuit is needed that will add only two input bits, to produce

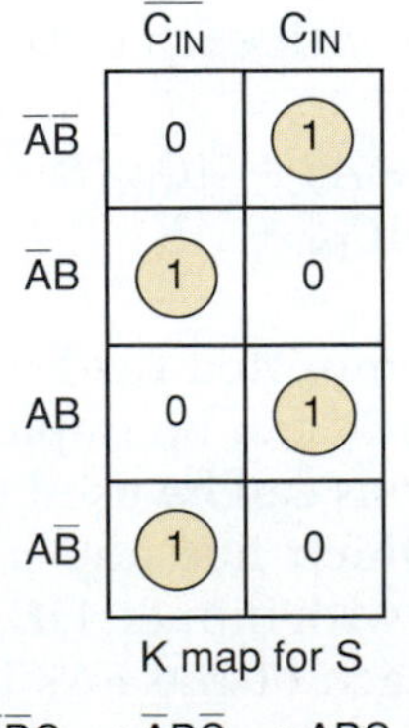

$S = \overline{A}\overline{B}C_{IN} + \overline{A}B\overline{C}_{IN} + ABC_{IN} + A\overline{B}\overline{C}_{IN}$

(a)

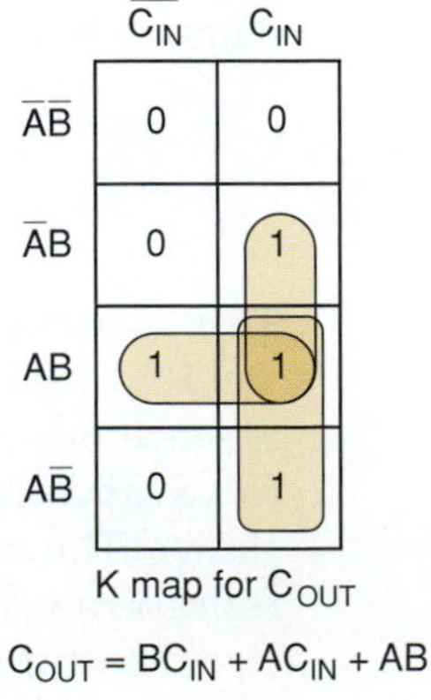

$C_{OUT} = BC_{IN} + AC_{IN} + AB$

(b)

FIGURE 6-9 K mappings for the full-adder outputs.

a sum and carry output. An example would be the addition of the LSB position of two binary numbers where there is no carry input to be added. A special logic circuit can be designed to take *two* input bits, A and B, and to produce sum (S) and carry (C_{OUT}) outputs. This circuit is called a **half adder (HA)**. Its operation is similar to that of an FA except that it operates on only two bits. We shall leave the design of the HA as an exercise at the end of the chapter.

6-12 COMPLETE PARALLEL ADDER WITH REGISTERS

In a computer, the numbers that are to be added are stored in FF registers. Figure 6-10 shows the diagram of a four-bit parallel adder, including the storage registers. The augend bits A_3 through A_0 are stored in the accumulator (A register); the addend bits B_3 through B_0 are stored in the B register. Each of these registers is made up of D flip-flops for easy transfer of data.

The contents of the A register (i.e., the binary number stored in A_3 through A_0) is added to the contents of the B register by the four FAs, and the sum is produced at outputs S_3 through S_0. C_4 is the carry out of the fourth FA, and it can be used as the carry input to a fifth FA, or as an *overflow* bit to indicate that the sum exceeds 1111.

Note that the sum outputs are connected to the D inputs of the A register. This will allow the sum to be parallel-transferred into the A register on the positive-going transition (PGT) of the TRANSFER pulse. In this way, the sum can be stored in the A register.

Also note that the D inputs of the B register are coming from the computer's memory, so that binary numbers from memory will be parallel-transferred into the B register on the PGT of the LOAD pulse. In most computers, there is also provision for parallel-transferring binary numbers from memory into the accumulator (A register). For simplicity, the circuitry necessary for performing this transfer is not shown in this diagram; it will be addressed in an end-of-chapter exercise.

Finally, note that the A register outputs are available for transfer to other locations such as another computer register or the computer's memory. This will make the adder circuit available for a new set of numbers.

Register Notation

Before we go through the complete process of how this circuit adds two binary numbers, it will be helpful to introduce some notation that makes it easy to describe the contents of a register and data transfer operations.

Whenever we want to give the levels that are present at each FF in a register or at each output of a group of outputs, we will use brackets, as illustrated below:

$$[A] = 1011$$

This is the same as saying that $A_3 = 1$, $A_2 = 0$, $A_1 = 1$, $A_0 = 1$. In other words, think of $[A]$ as representing "the contents of register A."

Whenever we want to indicate the transfer of data to or from a register, we will use an arrow, as illustrated below:

$$[B] \rightarrow [A]$$

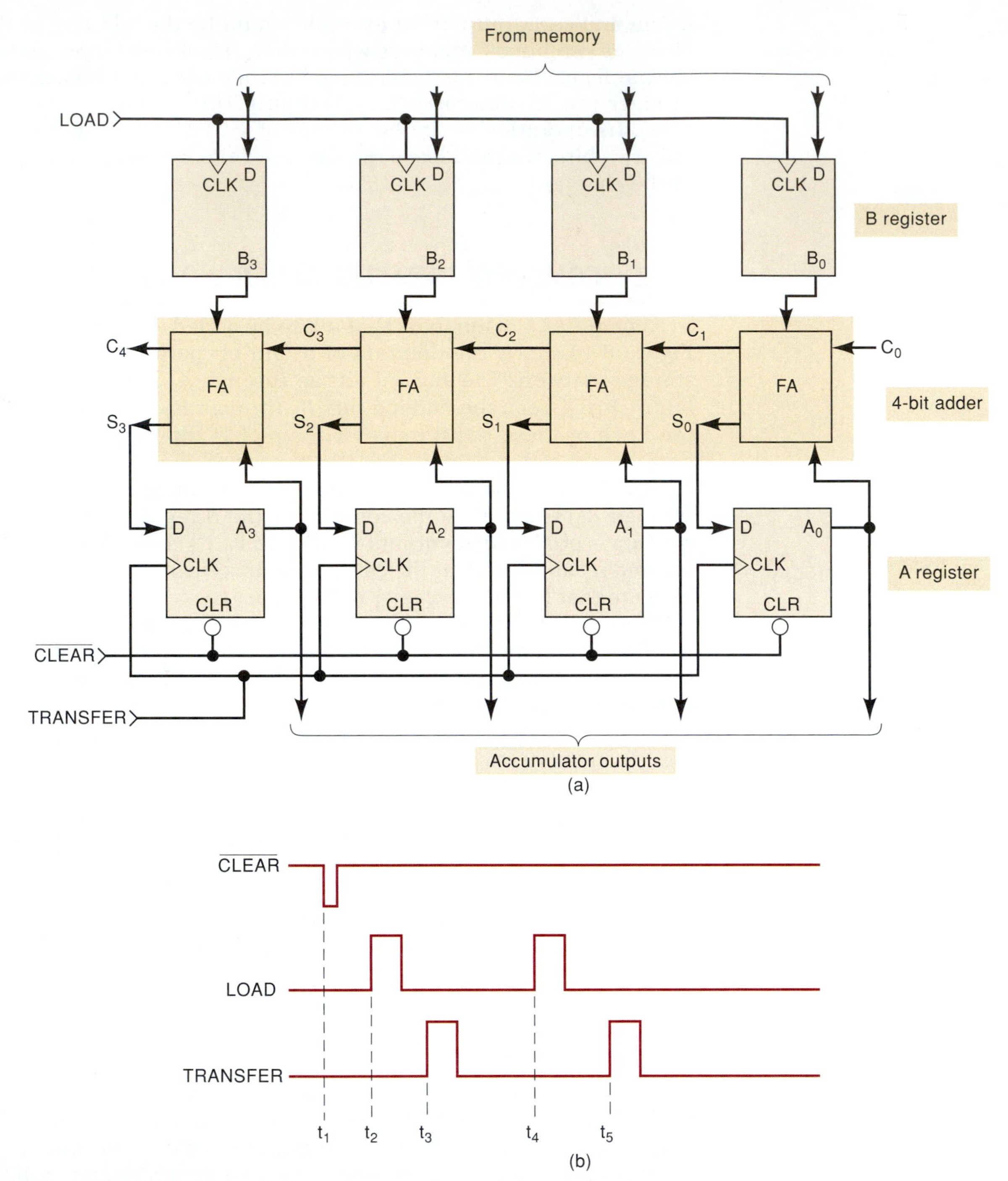

FIGURE 6-10 (a) Complete four-bit parallel adder with registers; (b) signals used to add binary numbers from memory and store their sum in the accumulator.

This means that the contents of the *B* register have been transferred to the *A* register. The old contents of the *A* register will be lost as a result of this operation, and the *B* register will be unchanged. This type of notation is very common, especially in data books describing microprocessor and microcontroller operations. In many ways, it is very similar to the notation used to refer to bit-array data objects using hardware description languages.

Sequence of Operations

We will now describe the process by which the circuit of Figure 6-10 will add the binary numbers 1001 and 0101. Assume that $C_0 = 0$; that is, there is no carry into the LSB position.

1. $[A] = 0000$. A $\overline{\text{CLEAR}}$ pulse is applied to the asynchronous inputs ($\overline{CLR}$) of each FF in register A. This occurs at time t_1.
2. $[M] \rightarrow [B]$. This first binary number is transferred from memory (M) to the B register. In this case, the binary number 1001 is loaded into register B on the PGT of the LOAD pulse at t_2.
3. $[S]^* \rightarrow [A]$. With $[B] = 1001$ and $[A] = 0000$, the full adders produce a sum of 1001; that is, $[S] = 1001$. These sum outputs are transferred into the A register on the PGT of the TRANSFER pulse at t_3. This makes $[A] = 1001$.
4. $[M] \rightarrow [B]$. The second binary number, 0101, is transferred from memory into the B register on the PGT of the second LOAD pulse at t_4. This makes $[B] = 0101$.
5. $[S] \rightarrow [A]$. With $[B] = 0101$ and $[A] = 1001$, the FAs produce $[S] = 1110$. These sum outputs are transferred into the A register when the second TRANSFER pulse occurs at t_5. Thus, $[A] = 1110$.
6. At this point, the sum of the two binary numbers is present in the accumulator. In most computers, the contents of the accumulator, $[A]$, will usually be transferred to the computer's memory so that the adder circuit can be used for a new set of numbers. The circuitry that performs this $[A] \rightarrow [M]$ transfer is not shown in Figure 6-10.

REVIEW QUESTIONS

1. Suppose that four different four-bit numbers are to be taken from memory and added by the circuit of Figure 6-10. How many CLEAR pulses will be needed? How many TRANSFER pulses? How many LOAD pulses?
2. Determine the contents of the A register after the following sequence of operations: $[A] = 0000$, $[0110] \rightarrow [B]$, $[S] \rightarrow [A]$, $[1110] \rightarrow [B]$, $[S] \rightarrow [A]$.

6-13 CARRY PROPAGATION

The parallel adder of Figure 6-10 performs additions at a relatively high speed because it adds the bits from each position simultaneously. However, its speed is limited by an effect called **carry propagation** or **carry ripple**, which can best be explained by considering the following addition:

$$\begin{array}{r} 0111 \\ +\ \underline{0001} \\ 1000 \end{array}$$

Addition of the LSB position produces a carry into the second position. This carry, when added to the bits of the second position, produces a carry into

*Even though S is not a register, we will use $[S]$ to represent the group of S outputs.

the third position. The latter carry, when added to the bits of the third position, produces a carry into the last position. The key point to notice in this example is that the sum bit generated in the *last* position (MSB) depended on the carry that was generated by the addition in the *first* position (LSB).

Looking at this from the viewpoint of the circuit of Figure 6-10, S_3 out of the last full adder depends on C_1 out of the first full adder. But the C_1 signal must pass through three FAs before it produces S_3. What this means is that the S_3 output will not reach its correct value until C_1 has propagated through the intermediate FAs. This represents a time delay that depends on the propagation delay produced in each FA. For example, if each FA has a propagation delay of 40 ns, then S_3 will not reach its correct level until 120 ns after C_1 is generated. This means that the add command pulse cannot be applied until 160 ns after the augend and addend numbers are present in the FF registers (the extra 40 ns is due to the delay of the LSB full adder, which generates C_1).

Obviously, the situation becomes much worse if we extend the adder circuitry to add a greater number of bits. If the adder were handling 32-bit numbers, the carry propagation delay could be 1280 ns = 1.28 μs. The add pulse could not be applied until at least 1.28 μs after the numbers were present in the registers.

This magnitude of delay is prohibitive for high-speed computers. Fortunately, logic designers have come up with several ingenious schemes for reducing this delay. One of the schemes, called **look-ahead carry**, utilizes logic gates to look at the lower-order bits of the augend and addend to see if a higher-order carry is to be generated. For example, it is possible to build a logic circuit with B_2, B_1, B_0, A_2, A_1, and A_0 as inputs and C_3 as an output. This logic circuit would have a shorter delay than is obtained by the carry propagation through the FAs. This scheme requires a large amount of extra circuitry but is necessary to produce high-speed adders. The extra circuitry is not a significant consideration with the present use of integrated circuits. Many high-speed adders available in integrated-circuit form utilize the look-ahead carry or a similar technique for reducing overall propagation delays.

6-14 INTEGRATED-CIRCUIT PARALLEL ADDER

Several parallel adders are available as ICs. The most common is a four-bit parallel adder IC that contains four interconnected FAs and the look-ahead carry circuitry needed for high-speed operation. The 7483A, 74LS83A, 74LS283, and 74HC283 are all four-bit parallel-adder chips.

Figure 6-11(a) shows the functional symbol for the 74HC283 four-bit parallel adder (and its equivalents). The inputs to this IC are two four-bit numbers, $A_3A_2A_1A_0$ and $B_3B_2B_1B_0$, and the carry, C_0, into the LSB position. The outputs are the sum bits and the carry, C_4, out of the MSB position. The sum bits are labeled $\Sigma_3\Sigma_2\Sigma_1\Sigma_0$, where Σ is the Greek capital letter *sigma*. The Σ label is just a common alternative to the S label for a sum bit.

Cascading Parallel Adders

Two or more IC adders can be connected together (cascaded) to accomplish the addition of larger binary numbers. Figure 6-11(b) shows two 74HC283 adders connected to add two 8-bit numbers $A_7\,A_6\,A_5\,A_4\,A_3\,A_2\,A_1\,A_0$ and $B_7\,B_6\,B_5\,B_4\,B_3\,B_2\,B_1\,B_0$. The adder on the right adds the lower-order bits of the numbers. The adder on the left adds the higher-order bits *plus* the C_4 carry out of the lower-order adder. The eight sum outputs are the resultant sum of the two

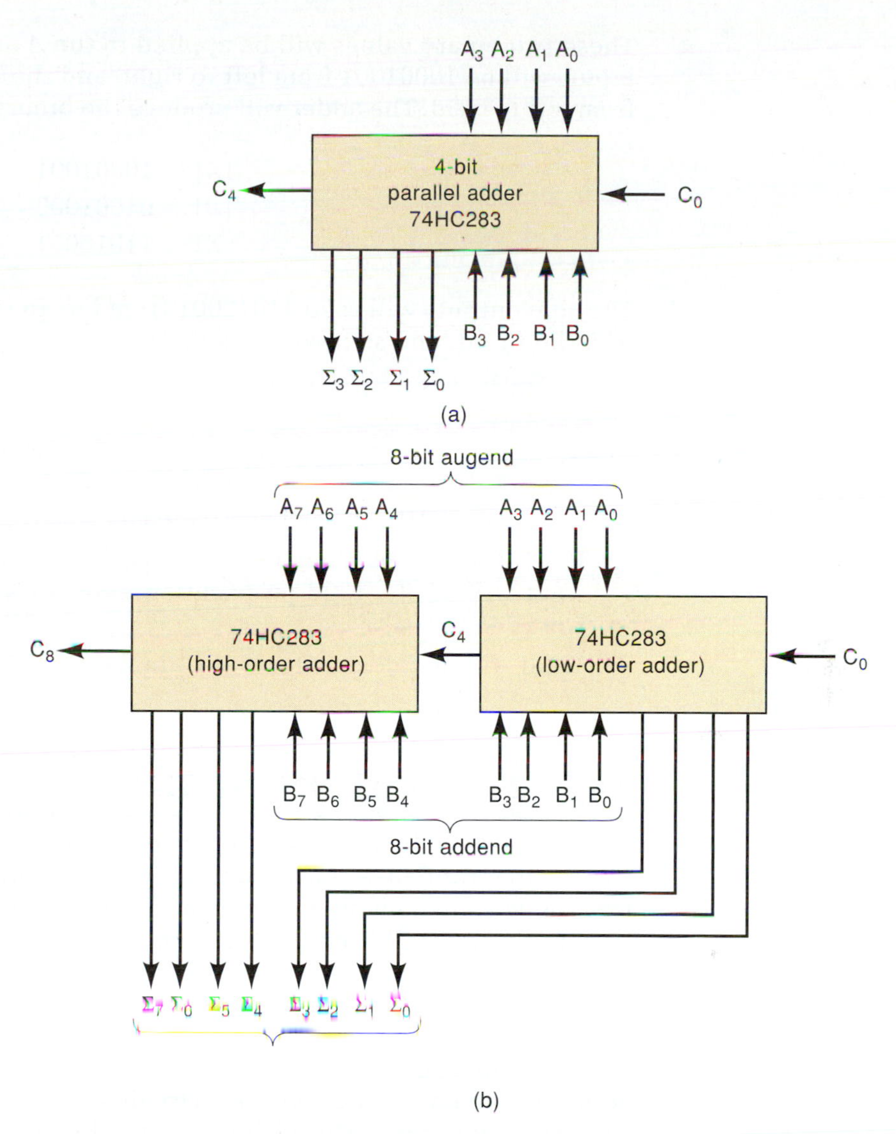

FIGURE 6-11 (a) Block symbol for the 74HC283 four-bit parallel adder; (b) cascading two 74HC283s.

8-bit numbers. C_8 is the carry out of the MSB position. It can be used as the carry input to a third adder stage if larger binary numbers are to be added.

The look-ahead carry feature of the 74HC283 speeds up the operation of this two-stage adder because the logic level at C_4, the carry out of the lower-order stage, is generated more rapidly than it would be if there were no look-ahead carry circuitry on the 74HC283 chip. This allows the higher-order stage to produce its sum outputs more quickly.

EXAMPLE 6-10

Determine the logic levels at the inputs and outputs of the eight-bit adder in Figure 6-11(b) when 72_{10} is added to 137_{10}.

Solution

First, convert each number to an eight-bit binary number:

$$137 = 10001001$$
$$72 = 01001000$$

These two binary values will be applied to the A and B inputs; that is, the A inputs will be 10001001 from left to right, and the B inputs will be 01001000 from left to right. The adder will produce the binary sum of the two numbers:

$$\begin{aligned} [A] &= 10001001 \\ [B] &= \underline{01001000} \\ [\Sigma] &= 11010001 \end{aligned}$$

The sum outputs will read 11010001 from left to right. There is no overflow into the C_8 bit, and so it will be a 0.

REVIEW QUESTIONS

1. How many 74HC283 chips are needed to add two 20-bit numbers?
2. If a 74HC283 has a maximum propagation delay of 30 ns from C_0 to C_4, what will be the total propagation delay of a 32-bit adder constructed from 74HC283s?
3. What will be the logic level at C_4 in Example 6-10?

6-15 2's-COMPLEMENT SYSTEM

Most modern computers use the 2's-complement system to represent negative numbers and to perform subtraction. The operations of addition and subtraction of signed numbers can be performed using only the addition operation if we use the 2's-complement form to represent negative numbers.

Addition

Positive and negative numbers, including the sign bits, can be added together in the basic parallel-adder circuit when the negative numbers are in 2's-complement form. This is illustrated in Figure 6-12 for the addition of -3 and $+6$. The -3 is represented in its 2's-complement form as 1101, where the first 1 is the sign bit; the $+6$ is represented as 0110, with the first zero as the sign bit. These numbers are stored in their corresponding registers. The four-bit parallel adder produces sum outputs of 0011, which represents $+3$. The C_4 output is 1, but remember that it is disregarded in the 2's-complement method.

Subtraction

When the 2's-complement system is used, the number to be subtracted (the subtrahend) is changed to its 2's complement and then *added* to the minuend (the number the subtrahend is being subtracted from). For example, we can assume that the minuend is already stored in the accumulator (A register). The subtrahend is then placed in the B register (in a computer it would be transferred here from memory) and is changed to its 2's-complement form before it is added to the number in the A register. The sum outputs of the adder circuit now represent the *difference* between the minuend and the subtrahend.

The parallel-adder circuit that we have been discussing can be adapted to perform the subtraction described above if we provide a means for taking

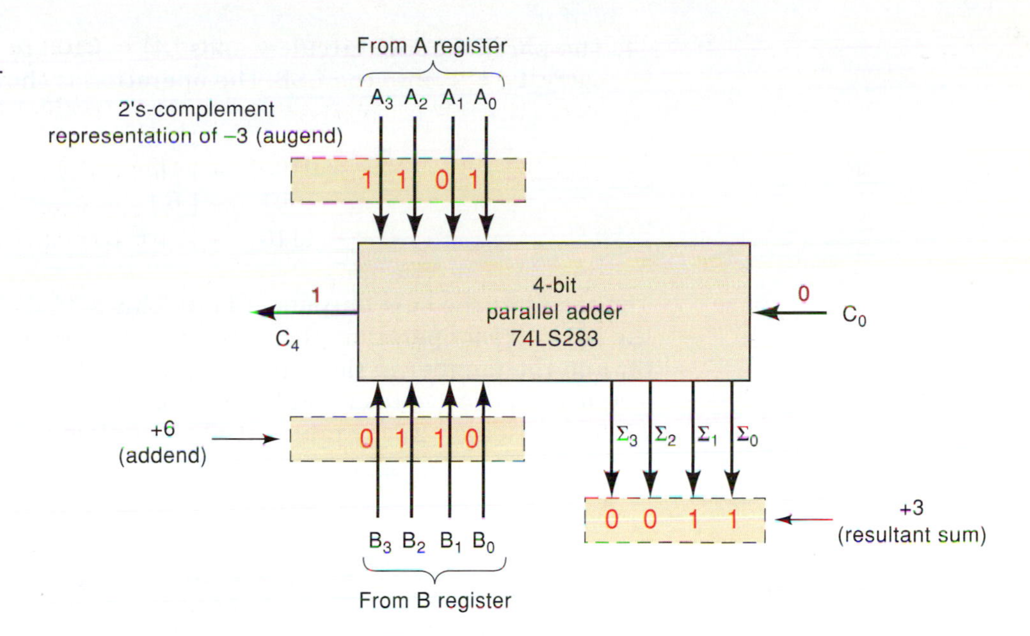

FIGURE 6-12 Parallel adder used to add and subtract numbers in 2's-complement system.

the 2's complement of the B register number. The 2's complement of a binary number is obtained by complementing (inverting) each bit and then adding 1 to the LSB. Figure 6-13 shows how this can be accomplished. The *inverted* outputs of the B register are used rather than the normal outputs; that is, $\overline{B}_0$, $\overline{B}_1$, $\overline{B}_2$, and $\overline{B}_3$ are fed to the adder inputs (remember, B_3 is the sign bit). This takes care of complementing each bit of the B number. Also, C_0 is made a logical 1, so that it adds an extra 1 into the LSB of the adder; this accomplishes the same effect as adding 1 to the LSB of the B register for forming the 2's complement.

The outputs Σ_3 to Σ_0 represent the results of the subtraction operation. Of course, Σ_3 is the sign bit of the result and indicates whether the result is + or −. The carry output C_4 is again disregarded.

To help clarify this operation, study the following steps for subtracting +6 from +4:

1. +4 is stored in the A register as 0100.
2. +6 is stored in the B register as 0110.
3. The inverted outputs of the B-register FFs (1001) are fed to the adder.

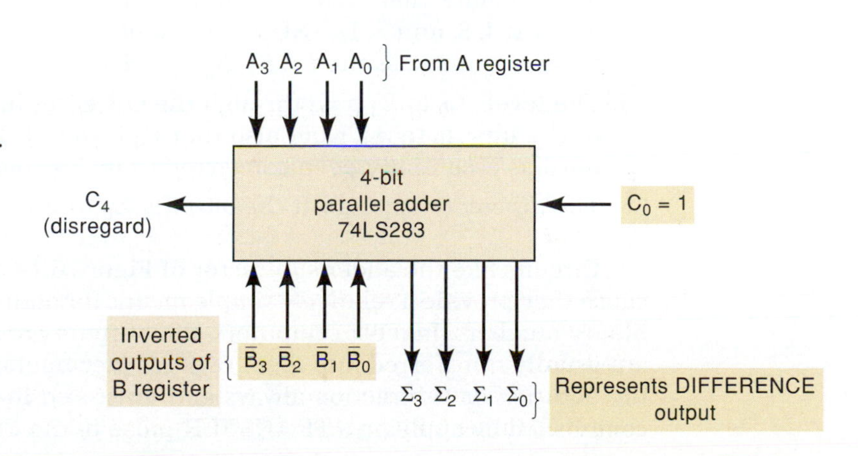

FIGURE 6-13 Parallel adder used to perform subtraction ($A - B$) using the 2's-complement system. The bits of the subtrahend (B) are inverted, and $C_0 = 1$ to produce the 2's complement.

4. The parallel-adder circuitry adds $[A] = 0100$ to $[\overline{B}] = 1001$ along with a carry, $C_0 = 1$, into the LSB. The operation is shown below.

$$
\begin{array}{rl}
1 & \leftarrow C_0 \\
0100 & \leftarrow [A] \\
+\ \underline{1001} & \leftarrow [\overline{B}] \\
1110 & \leftarrow [\Sigma] = [A] - [B]
\end{array}
$$

The result at the sum outputs is 1110. This actually represents the result of the *subtraction* operation, the *difference* between the number in the A register and the number in the B register, that is, $[A] - [B]$. Because the sign bit = 1, it is a negative result and is in 2's-complement form. We can verify that 1110 represents -2_{10} by taking its 2's complement and obtaining $+2_{10}$:

$$
\begin{array}{rl}
 & 1110 \\
 & 0001 \\
+ & \underline{1} \\
 & 0010 = +2_{10}
\end{array}
$$

Combined Addition and Subtraction

It should now be clear that the basic parallel-adder circuit can be used to perform addition or subtraction depending on whether the B number is left unchanged or is converted to its 2's complement. A complete circuit that can perform *both* addition and subtraction in the 2's-complement system is shown in Figure 6-14.

This **adder/subtractor** circuit is controlled by the two control signals ADD and SUB. When the ADD level is HIGH, the circuit performs addition of the numbers stored in the A and B registers. When the SUB level is HIGH, the circuit subtracts the B-register number from the A-register number. The operation is described as follows:

1. Assume that ADD = 1 and SUB = 0. The SUB = 0 *disables* (inhibits) AND gates 2, 4, 6, and 8, holding their outputs at 0. The ADD = 1 *enables* AND gates 1, 3, 5, and 7, allowing their outputs to pass the B_0, B_1, B_2, and B_3 levels, respectively.
2. The levels B_0 to B_3 pass through the OR gates into the four-bit parallel adder to be added to the bits A_0 to A_3. The *sum* appears at the outputs Σ_0 to Σ_3.
3. Note that SUB = 0 causes a carry $C_0 = 0$ into the adder.
4. Now assume that ADD = 0 and SUB = 1. The ADD = 0 inhibits AND gates 1, 3, 5, and 7. The SUB = 1 enables AND gates 2, 4, 6, and 8, so that their outputs pass the $\overline{B}_0$, $\overline{B}_1$, $\overline{B}_2$, and $\overline{B}_3$ levels, respectively.
5. The levels $\overline{B}_0$ to $\overline{B}_3$ pass through the OR gates into the adder to be added to the bits A_0 to A_3. Note also that C_0 is now 1. Thus, the B-register number has essentially been converted to its 2's complement.
6. The *difference* appears at the *outputs* Σ_0 to Σ_3.

Circuits like the adder/subtractor of Figure 6-14 are used in computers because they provide a relatively simple means for adding and subtracting signed binary numbers. In most computers, the outputs present at the Σ output lines are usually transferred into the A register (accumulator), so that the results of the addition or subtraction always end up stored in the A register. This is accomplished by applying a TRANSFER pulse to the CLK inputs of register A.

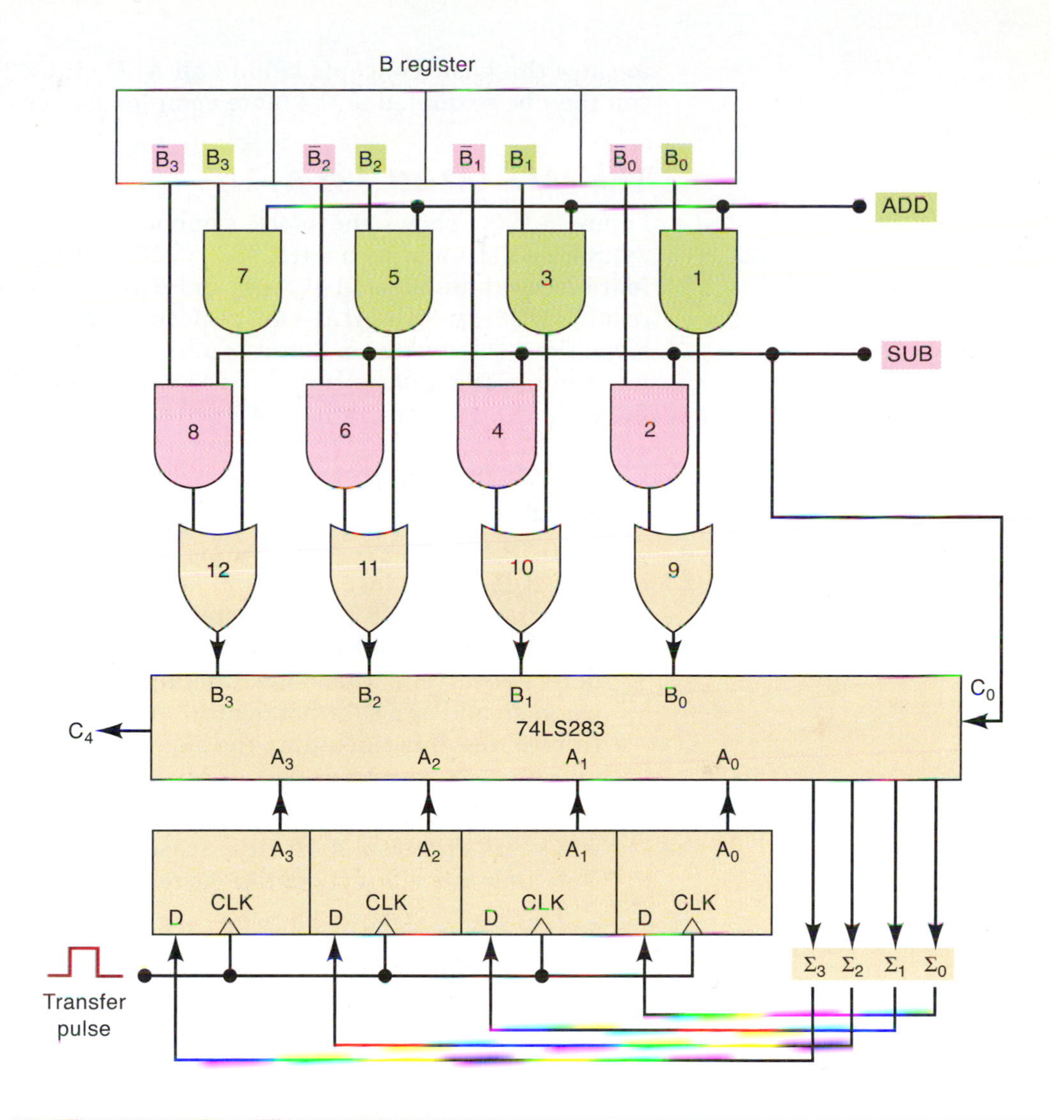

FIGURE 6-14 Parallel adder/subtractor using the 2's-complement system.

REVIEW QUESTIONS

1. Why does C_0 have to be a 1 in order to use the adder circuit in Figure 6-13 as a subtractor?
2. Assume that $[A] = 0011$ and $[B] = 0010$ in Figure 6-14. If ADD = 1 and SUB = 0, determine the logic levels at the OR gate outputs.
3. Repeat question 2 for ADD = 0, SUB = 1.
4. *True or false:* When the adder/subtractor circuit is used for subtraction, the 2's complement of the subtrahend appears at the input of the adder.

6-16 ALU INTEGRATED CIRCUITS

Several integrated circuits are called arithmetic/logic units (ALUs), even though they do not have the full capabilities of a computer's arithmetic/logic unit. These ALU chips are capable of performing several different arithmetic and logic operations on binary data inputs. The specific operation that an ALU IC is to perform is determined by a specific binary code applied to its function-select inputs. Some of the ALU ICs are fairly complex, and it would require a great amount of time and space to explain and illustrate their operation. In this section, we will use a relatively simple, yet useful, ALU chip to show the basic concepts behind all ALU chips. The ideas presented here can then be extended to the more complex devices.

The 74LS382/HC382 ALU

Figure 6-15(a) shows the block symbol for an ALU that is available as a 74LS382 (TTL) and as a 74HC382 (CMOS). This 20-pin IC operates on two four-bit input numbers, $A_3A_2A_1A_0$ and $B_3B_2B_1B_0$, to produce a four-bit output result $F_3F_2F_1F_0$. This ALU can perform *eight* different operations. At any given time, the operation that it is performing depends on the input code applied to the function-select inputs $S_2S_1S_0$. The table in Figure 6-15(b) shows the eight available operations. We will now describe each of these operations.

CLEAR OPERATION With $S_2S_1S_0 = 000$, the ALU will *clear* all of the bits of the F output so that $F_3F_2F_1F_0 = 0000$.

ADD OPERATION With $S_2S_1S_0 = 011$, the ALU will add $A_3A_2A_1A_0$ to $B_3B_2B_1B_0$ to produce their sum at $F_3F_2F_1F_0$. For this operation, C_N is the carry into the LSB position, and it must be made a 0. C_{N+4} is the carry output from the MSB position. *OVR* is the overflow indicator output; it detects overflow when signed numbers are being used. *OVR* will be a 1 when an add or a subtract operation produces a result that is too large to fit into four bits (including the sign bit).

SUBTRACT OPERATIONS With $S_2S_1S_0 = 001$, the ALU will subtract the A input number from the B input number. With $S_2S_1S_0 = 010$, the ALU will subtract B from A. In either case, the difference appears at $F_3F_2F_1F_0$. Note that the subtract operations require that the C_N input be a 1.

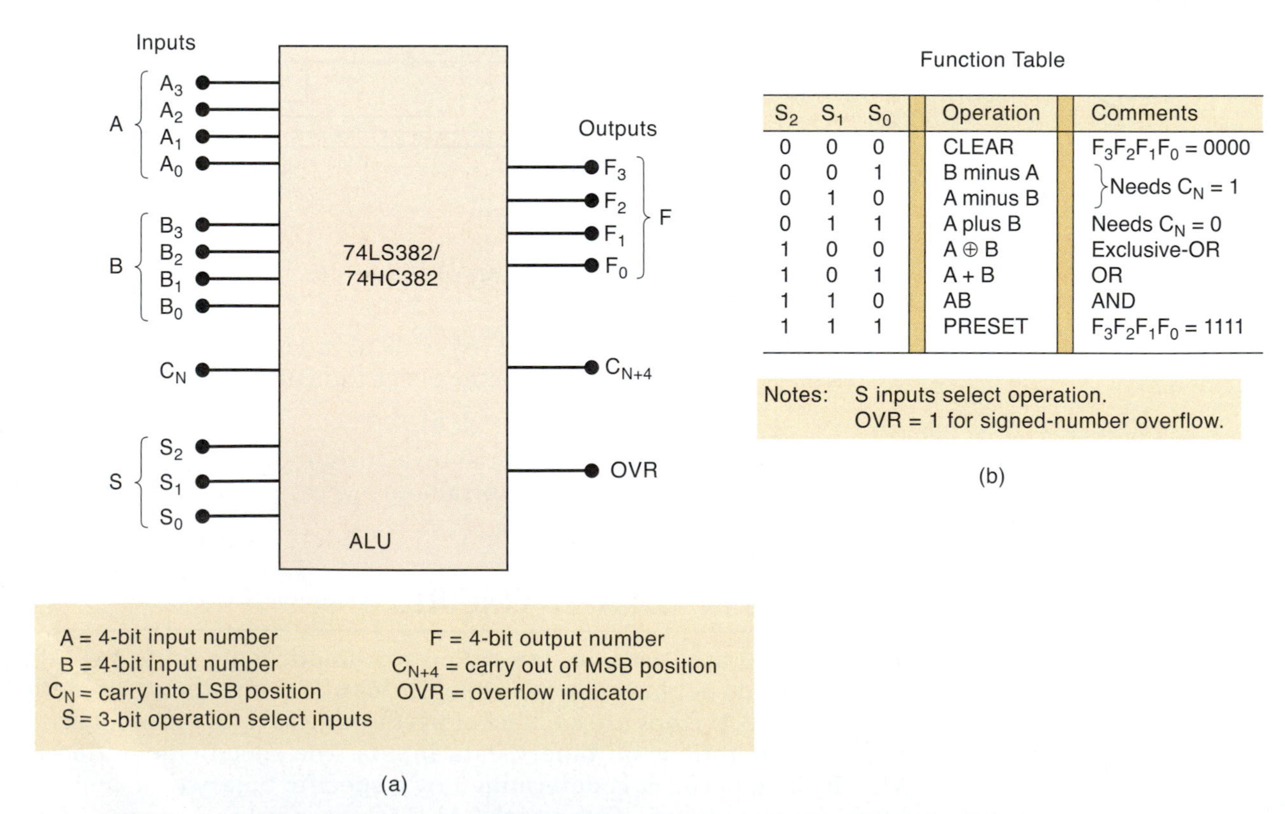

Function Table

S_2	S_1	S_0	Operation	Comments
0	0	0	CLEAR	$F_3F_2F_1F_0 = 0000$
0	0	1	B minus A	Needs $C_N = 1$
0	1	0	A minus B	Needs $C_N = 1$
0	1	1	A plus B	Needs $C_N = 0$
1	0	0	$A \oplus B$	Exclusive-OR
1	0	1	$A + B$	OR
1	1	0	AB	AND
1	1	1	PRESET	$F_3F_2F_1F_0 = 1111$

Notes: S inputs select operation.
OVR = 1 for signed-number overflow.

(b)

FIGURE 6-15 (a) Block symbol for 74LS382/HC382 ALU chip; (b) function table showing how select inputs (S) determine what operation is to be performed on A and B inputs.

XOR OPERATION With $S_2S_1S_0 = 100$, the ALU will perform a bit-by-bit XOR operation on the A and B inputs. This is illustrated below for $A_3A_2A_1A_0 = 0110$ and $B_3B_2B_1B_0 = 1100$.

$$\begin{aligned} A_3 \oplus B_3 &= 0 \oplus 1 = 1 = F_3 \\ A_2 \oplus B_2 &= 1 \oplus 1 = 0 = F_2 \\ A_1 \oplus B_1 &= 1 \oplus 0 = 1 = F_1 \\ A_0 \oplus B_0 &= 0 \oplus 0 = 0 = F_0 \end{aligned}$$

The result is $F_3F_2F_1F_0 = 1010$.

OR OPERATION With $S_2S_1S_0 = 101$, the ALU will perform a bit-by-bit OR operation on the A and B inputs. For example, with $A_3A_2A_1A_0 = 0110$ and $B_3B_2B_1B_0 = 1100$, the ALU will generate a result of $F_3F_2F_1F_0 = 1110$.

AND OPERATION With $S_2S_1S_0 = 110$, the ALU will perform a bit-by-bit AND operation on the A and B inputs. For example, with $A_3A_2A_1A_0 = 0110$ and $B_3B_2B_1B_0 = 1100$, the ALU will generate a result of $F_3F_2F_1F_0 = 0100$.

PRESET OPERATIONS With $S_2S_1S_0 = 111$, the ALU will *set* all of the bits of the output so that $F_3F_2F_1F_0 = 1111$.

EXAMPLE 6-11

(a) Determine the 74HC382 outputs for the following inputs: $S_2S_1S_0 = 010$, $A_3A_2A_1A_0 = 0100$, $B_3B_2B_1B_0 = 0001$, and $C_N = 1$.

(b) Change the select code to 011 and repeat.

Solution

(a) From the function table in Figure 6-15(b), 010 selects the $(A - B)$ operation. The ALU will perform the 2's-complement subtraction by complementing B and adding it to A and C_N. Note that $C_N = 1$ is needed to complete the 2's complement of B effectively.

$$\begin{array}{rl} 1 & \leftarrow C_N \\ 0100 & \leftarrow A \\ +\ \underline{1110} & \leftarrow \overline{B} \\ 10011 & \end{array}$$

C_{N+4} ← (MSB carry) ; $F_3\,F_2\,F_1\,F_0$ ← (remaining four bits)

As always in 2's-complement subtraction, the CARRY OUT of the MSB is discarded. The correct result of the $(A - B)$ operation appears at the F outputs.

The OVR output is determined by considering the input numbers to be signed numbers. Thus, we have $A_3A_2A_1A_0 = 0100 = +4_{10}$ and $B_3B_2B_1B_0 = 0001 = +1_{10}$. The result of the subtract operation is $F_3F_2F_1F_0 = 0011 = +3_{10}$, which is correct. Therefore, no overflow has occurred, and $OVR = 0$. If the result had been negative, it would have been in 2's-complement form.

(b) A select code of 011 will produce the sum of the A and B inputs. However, because $C_N = 1$, there will be a carry of 1 added into the LSB position. This will produce a result of $F_3F_2F_1F_0 = 0110$, which is 1 greater than $(A + B)$. The C_{N+4} and OVR outputs will both be 0. For the correct sum to appear at F, the C_N input must be at 0.

Expanding the ALU

A single 74LS382 or 74HC382 operates on four-bit numbers. Two or more of these chips can be connected together to operate on larger numbers. Figure 6-16 shows how two four-bit ALUs can be combined to add two eight-bit numbers, $B_7B_6B_5B_4B_3B_2B_1B_0$ and $A_7A_6A_5A_4A_3A_2A_1A_0$, to produce the output sum $\Sigma_7\Sigma_6\Sigma_5\Sigma_4\Sigma_3\Sigma_2\Sigma_1\Sigma_0$. Study the circuit diagram and note the following points:

1. Chip Z1 operates on the four lower-order bits of the two input numbers. Chip Z2 operates on the four higher-order bits.
2. The sum appears at the F outputs of Z1 and Z2. The lower-order bits appear at Z1, and the higher-order bits appear at Z2.
3. The C_N input of Z1 is the carry into the LSB position. For addition, it is made a 0.
4. The carry output $[C_{N+4}]$ of Z1 is connected to the carry input $[C_N]$ of Z2.
5. The OVR output of Z2 is the overflow indicator when signed eight-bit numbers are being used.
6. The corresponding select inputs of the two chips are connected together so that Z1 and Z2 are always performing the same operation. For addition, the select inputs are shown as 011.

EXAMPLE 6-12

How would the arrangement of Figure 6-16 have to be changed in order to perform the subtraction $(B - A)$?

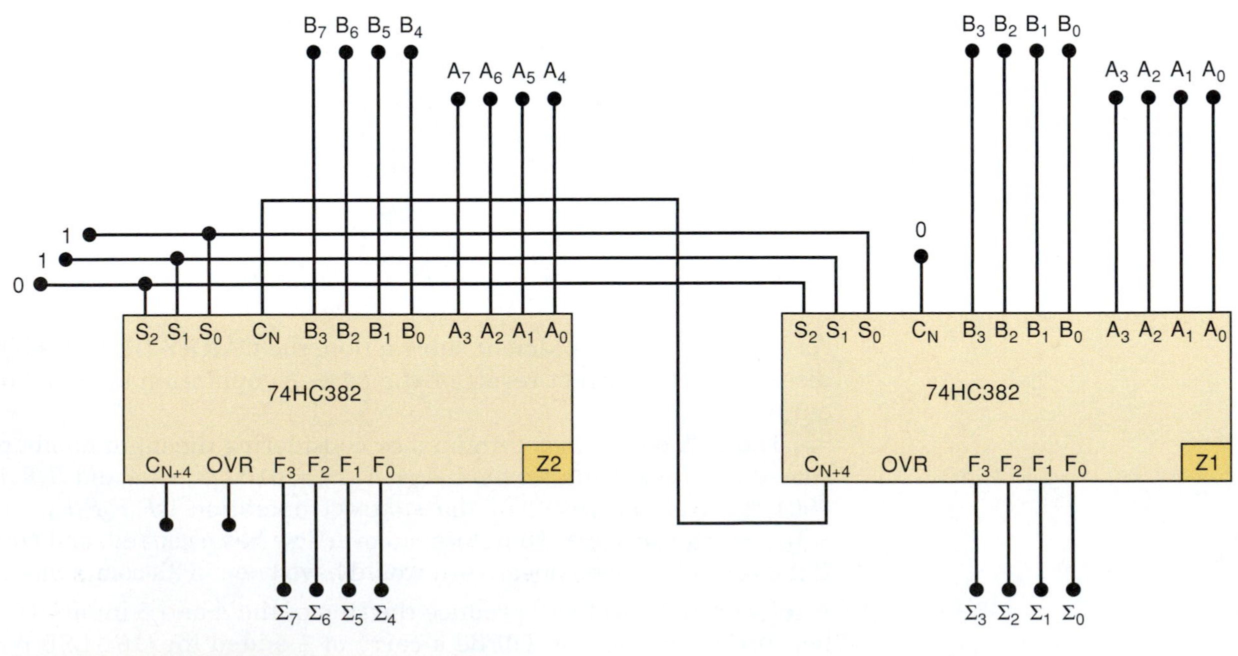

FIGURE 6-16 Two 74HC382 ALU chips connected as an eight-bit adder.

Solution

The select input code [see the table in Figure 6-15(b)] must be changed to 001, and the C_N input of Z1 must be made a 1.

Other ALUs

The 74LS181/HC181 is another four-bit ALU. It has four select inputs that can select any of 16 different operations. It also has a mode input bit that can switch between logic operations and arithmetic operations (add and subtract). This ALU has an $A = B$ output that is used to compare the magnitudes of the A and B inputs. When the two input numbers are exactly equal, the $A = B$ output will be a 1; otherwise, it is a 0.

The 74LS881/HC881 is similar to the 181 chip, but it has the capability of performing some additional logic operations.

REVIEW QUESTIONS

1. Apply the following inputs to the ALU of Figure 6-15, and determine the outputs: $S_2S_1S_0$ = 001, $A_3A_2A_1A_0$ = 1110, $B_3B_2B_1B_0$ = 1001, C_N = 1.
2. Change the select code to 011 and C_N to 0, and repeat review question 1.
3. Change the select code to 110, and repeat review question 1.
4. Apply the following inputs to the circuit of Figure 6-16, and determine the outputs: B = 01010011, A = 00011000.
5. Change the select code to 111, and repeat review question 4.
6. How many 74HC382s are needed to add two 32-bit numbers?

6-17 TROUBLESHOOTING CASE STUDY

A technician is testing the adder/subtractor redrawn in Figure 6-17 and records the following test results for the various operating modes:

Mode 1: ADD = 0, SUB = 0. The sum outputs are always equal to the number in the A register *plus one*. For example, when $[A]$ = 0110, the sum is $[\Sigma]$ = 0111. This is incorrect because the OR outputs and C_0 should all be 0 in this mode to produce $[\Sigma] = [A]$.

Mode 2: Add = 1, SUB = 0. The sum is always 1 more than it should be. For example, with $[A]$ = 0010 and $[B]$ = 0100, the sum output is 0111 instead of 0110.

Mode 3: Add = 0, SUB = 1. The Σ outputs are always equal to $[A] - [B]$, as expected.

When she examines these test results, the technician sees that the sum outputs exceed the expected results by 1 for the first two modes of operation. At first, she suspects a possible fault in one of the LSB inputs to the adder, but she dismisses this because such a fault would also affect the subtraction operation, which is working correctly. Eventually, she realizes that there is another fault that could add an extra 1 to the results for the first two modes without causing an error in the subtraction mode.

Recall that C_0 is made a 1 in the subtraction mode as part of the 2's-complement operation on $[B]$. For the other modes, C_0 is to be a 0. The technician

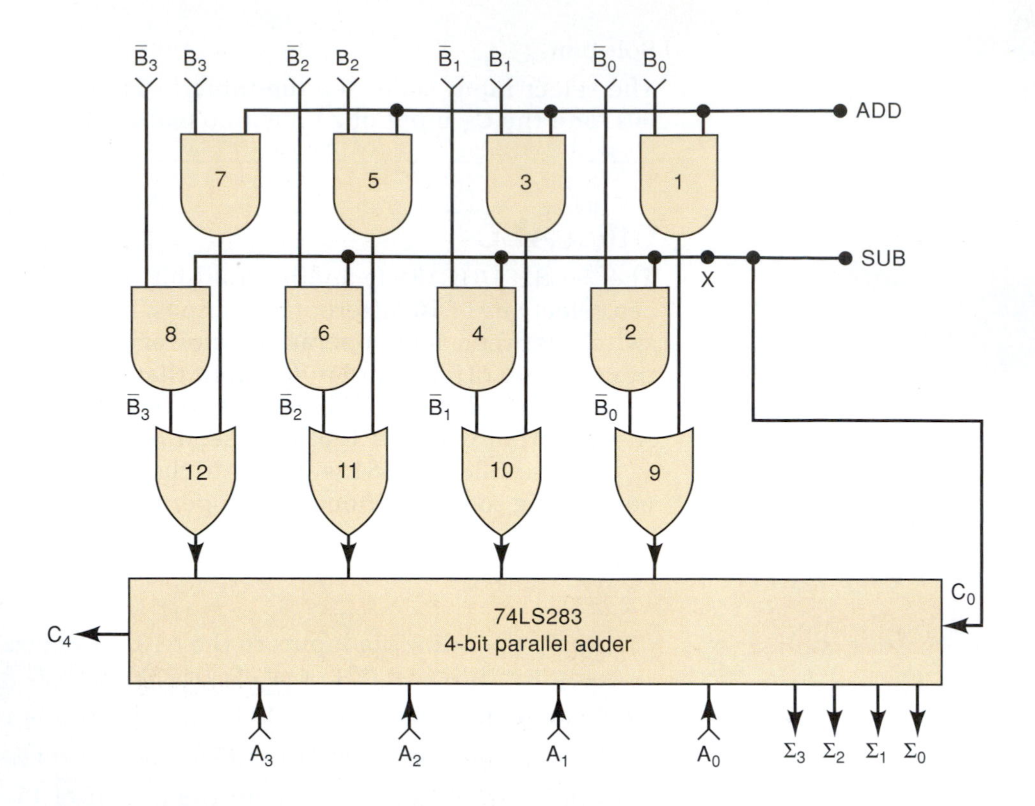

FIGURE 6-17 Parallel adder/subtractor circuit.

checks the connection between the *SUB* signal and the C_0 input to the adder and finds that it is open due to a bad solder connection. This open connection explains the observed results because the TTL adder responds as if C_0 were a constant logic 1, causing an extra 1 to be added to the result in modes 1 and 2. The open connection would have no effect on mode 3 because C_0 is supposed to be a 1 anyway.

EXAMPLE 6-13

Consider again the adder/subtractor circuit. Suppose that there is a break in the connection path between the SUB input and the AND gates at point *X* in Figure 6-17. Describe the effects of this open on the circuit operation for each mode.

Solution

First, realize that this fault will produce a logic 1 at the affected input of AND gates 2, 4, 6, and 8, which will permanently enable each of these gates to pass its $\bar{B}$ input to the following OR gate as shown.

Mode 1: ADD = 0, SUB = 0. The fault will cause the circuit to perform subtraction—almost. The 1's complement of [*B*] will reach the OR gate outputs and be applied to the adder along with [*A*]. With $C_0 = 0$, the 2's complement of [*B*] will not be complete; it will be short by 1. Thus, the adder will produce [*A*] − [*B*] − 1. To illustrate, let's try [*A*] = +6 = 0110 and [*B*] = +3 = 0011. The adder will add as follows:

$$\begin{aligned} \text{1's complement of } [B] &= 1100 \\ [A] &= 0110 \\ \text{result} &= \cancel{1}0010 \end{aligned}$$

↑ Disregard carry.

The result is 0010 = +2 instead of 0011 = +3, as it would be for normal subtraction.

Mode 2: ADD = 1, SUB = 0. With ADD = 1, AND gates 1, 3, 5, and 7 will pass the B inputs to the following OR gate. Thus, each OR gate will have a $\overline{B}$ and a B at its inputs, thereby producing a 1 output. For example, the inputs to OR gate 9 will be $\overline{B}_0$ coming from AND gate 2 (because of the fault), and B_0 coming from AND gate 1 (because ADD = 1). Thus, OR gate 9 will produce an output of $\overline{B}_0 + B_0$, which will always be a logic 1.

The adder will add the 1111 from the OR gates to the [A] to produce a sum that is 1 less than [A]. Why? Because $1111_2 = -1_{10}$.

Mode 3: ADD = 0, SUB = 1. This mode will work correctly because SUB = 1 is supposed to enable AND gates 2, 4, 6, and 8 anyway.

6-18 USING TTL LIBRARY FUNCTIONS WITH ALTERA

The adder and ALU ICs that we have looked at in this chapter are just a few of the many MSI chips that have served as the building blocks of digital systems for decades. Whenever a technology serves such a long and useful lifetime, it has a lasting impact on the field and the people who use it. TTL integrated circuits certainly fall into this category and continue in various forms today. Experienced engineers and technicians (we want to avoid the word *old*) are familiar with the standard parts. Existing designs can be remanufactured and upgraded using the same basic circuits if they can be implemented in a VLSI PLD. Data sheets for these devices are readily available, and studying these old TTL parts is still an excellent way to learn the fundamentals of any digital system.

For all of these reasons, the Altera development system offers what they refer to as old-style macrofunctions. A **macrofunction** is a self-contained description of a logic circuit with all its inputs, outputs, and operational characteristics defined. In other words, they have gone to the trouble of writing the code necessary to get a PLD to emulate the operation of many conventional TTL MSI devices. All the designer needs to know is how to hook it into the rest of the system. In this section, we will expand on the concepts of logic primitives and libraries presented in Chapter 5 to see how we can use standard MSI parts in our designs.

The 74382 arithmetic logic unit (ALU) is a fairly sophisticated IC. The task of describing its operation using HDL code is challenging but certainly within our reach. Refer again to the examples of this IC and its operation, which were covered in Section 6-16. Specifically, look at Figure 6-16, which shows how to cascade two four-bit ALU chips to make an eight-bit ALU that could serve as the heart of a microcontroller's central processing unit (CPU). Figure 6-18 shows the graphic method of describing the eight-bit circuit using Altera's graphic description file and macrofunction blocks from its library of components. The 74382 symbols are simply chosen from the list in the macrofunction library and placed on the screen. Wiring these chips together is simple and intuitive.

It is possible to connect standard library MSI parts using only HDL. Just as we demonstrated connecting flip-flop primitives together to make more complex circuits, we can connect ICs such as the 74382 to other parts. The names of all the input and output ports on these standard parts are defined

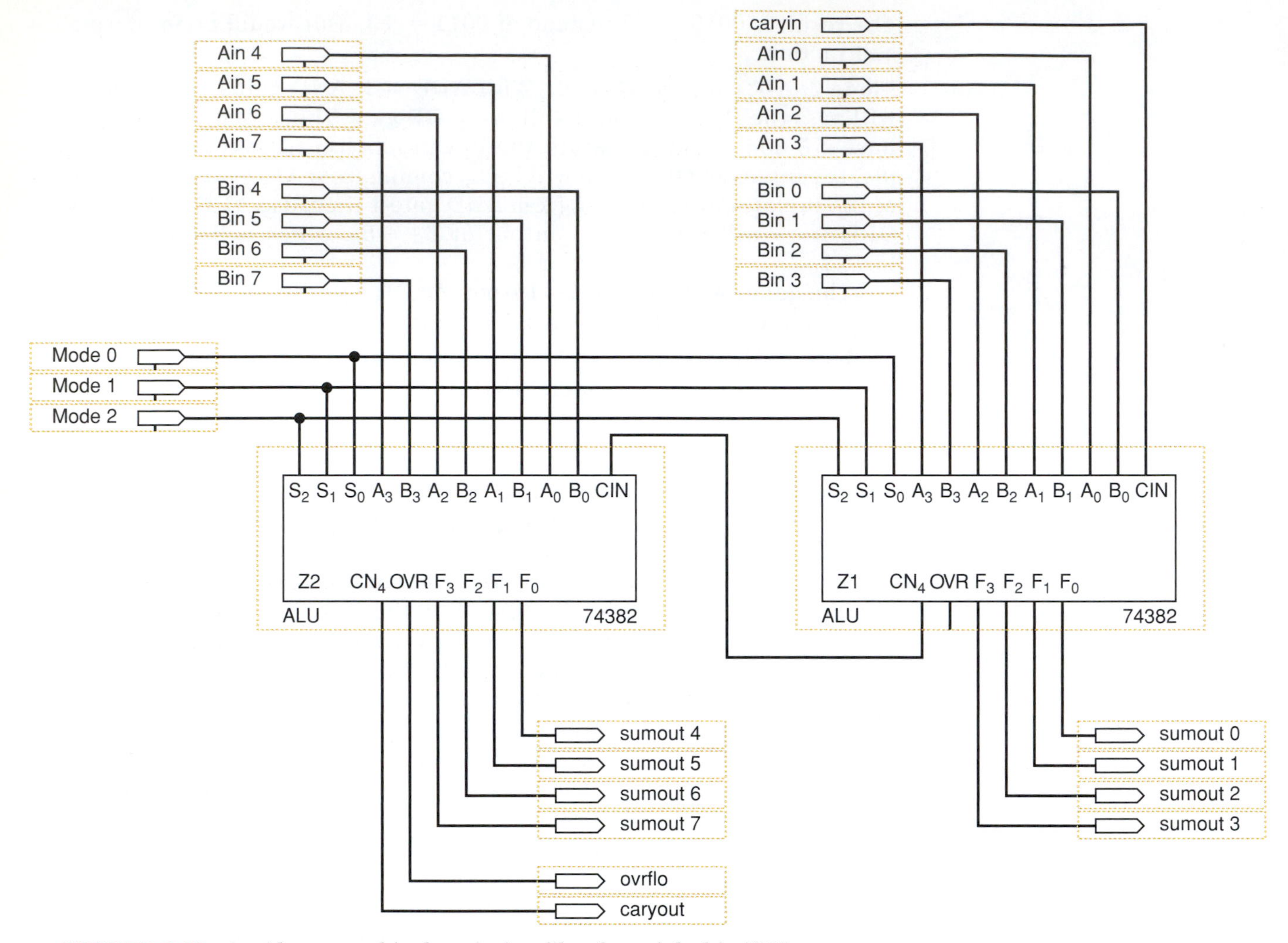

FIGURE 6-18 An Altera graphic description file of an eight-bit ALU.

in a **function prototype**, which can be found in the help menu. The function prototype given for a 74382 is

AHDL Function Prototype (port name and order also apply to Verilog HDL):

```
FUNCTION 74382 (s[2..0], a[3..0], b[3..0], cin)
RETURNS (ovr, cn4, f[3..0]);
```

REVIEW QUESTIONS

1. Where can you find information about using a 74283 full adder in your HDL design?
2. What is a macrofunction?

6-19 LOGICAL OPERATIONS ON BIT ARRAYS

In the previous section, we examined the use of macrofunctions to build systems from standard parts. Now we need to practice writing HDL code rather than using a macrofunction to make an adder similar to Figure 6-6. In this section, we will expand our understanding of the HDL techniques in two main areas: specifying groups of bits in an array and using logical operations to combine arrays of bits using Boolean expressions.

In Section 6-12, we discussed register notation, which makes it easy to describe the contents of registers and signals consisting of multiple bits. The HDLs use arrays of bits in a similar notation to describe signals, as we discussed in Chapter 4. For example, in AHDL, the four-bit signal named *d* is defined as:

```
VARIABLE d[3..0] :NODE.
```

In VHDL, the same data format is expressed as:

```
SIGNAL d :BIT_VECTOR (3 DOWNTO 0).
```

Each bit in these data types is designated by an element number. In this example of a bit array named *d*, the bits can be referred to as d3, d2, d1, d0. Bits can also be grouped into **sets**. For example, if we want to refer to the three most significant bits of *d* as a set, we can use the expression d[3..1] in AHDL and the expression d (3 DOWNTO 1) in VHDL. Once a value is assigned to the array and the desired set of bits is identified, logical operations can be performed on the entire set of bits. As long as the sets are the same size (the same number of bits), two sets can be combined in a logical expression, just like you would combine single variables in a Boolean equation. Each of the corresponding pairs of bits in the two sets is combined as stated in the logic equation. This allows one equation to describe the logical operation performed on each bit in a set.

EXAMPLE 6-14

Assume D_3, D_2, D_1, D_0 has the value 1011 and G_3, G_2, G_1, G_0 has the value 1100. Let's define Dnum = $[D_3, D_2, D_1, D_0]$ and Gnum = $[G_3, G_2, G_1, G_0]$. Let's also define Y = $[Y_3, Y_2, Y_1, Y_0]$ where Y is related to Dnum and Gnum as follows:

Y = Dnum • Gnum;

What is the value of X after this operation?

Solution

D_3, D_2, D_1, D_0	1 0 1 1	
↕ ↕ ↕ ↕	↕ ↕ ↕ ↕	AND each bit position together
G_3, G_2, G_1, G_0	1 1 0 0	
Y_3, Y_2, Y_1, Y_0	1 0 0 0	

Thus, Y is a set of four bits with value 1000.

EXAMPLE 6-15

For the register values described in Example 6-14, declare each *d*, *g*, and *x*. Then write an expression using your favorite HDL that performs the ANDing operation on all bits.

Solution

```
SUBDESIGN bitwise_and
(    d[3..0], g[3..0]          :INPUT;
     y[3..0]                   :OUTPUT;)
BEGIN
     y[] = d[] & g[];
END;
```

```
ENTITY bitwise_and IS
PORT(d, g            :IN BIT_VECTOR (3 DOWNTO 0);
     y               :OUT BIT_VECTOR (3 DOWNTO 0));
END bitwise_and;
ARCHITECTURE a OF bitwise_and IS
BEGIN
     y <= d AND g;
END a;
```

6-20 HDL ADDERS

In this section, we will see how to create a parallel adder circuit that can be used to add bit arrays using the logic equation for a single-bit full adder. Figure 6-19 shows the basic block diagram with signals labeled to create a four-bit adder. Notice that each bit of the augend [A], addend [B], carry [C], and sum [S] are bit array variables and have index numbers associated with them. The equation for the sum using register notation is:

$$[S] = [A] \oplus [B] \oplus [C];$$

Notice that the carry signals between stages are not inputs or outputs of this overall circuit, but rather intermediate variables. A strategy must be developed for labeling the carry bits so that they can be used in an array. We have chosen to let each carry bit serve as an input to its corresponding adder stage, as shown in Figure 6-19. For example, C0 is an input to the bit 0 stage, C1 is the carry input to the bit 1 stage, and so on. The bits of the carry array can be thought of as the "wires" that connect the adders. This array must have a carry in for each stage and also a carry out for the most significant state. In this example, there will be five bits, labeled C4 through C0, in the carry array. The set of data that represents the carry outputs would be C4 through C1, and the set of data that represents the carry inputs would be C3 through C0.

$$\text{Cin} = \text{C[3..0]}$$
$$\text{Cout} = \text{C[4..1]}$$

The HDLs allow us to specify which sets of bits out of the entire array we want to use in an equation. To ensure that all variables being combined in a logic equation contain the same number of bits, we can start with the general equation for the carry out of a one-bit adder as follows:

$$\text{Cout} = \text{AB} + \text{ACin} + \text{BCin}$$

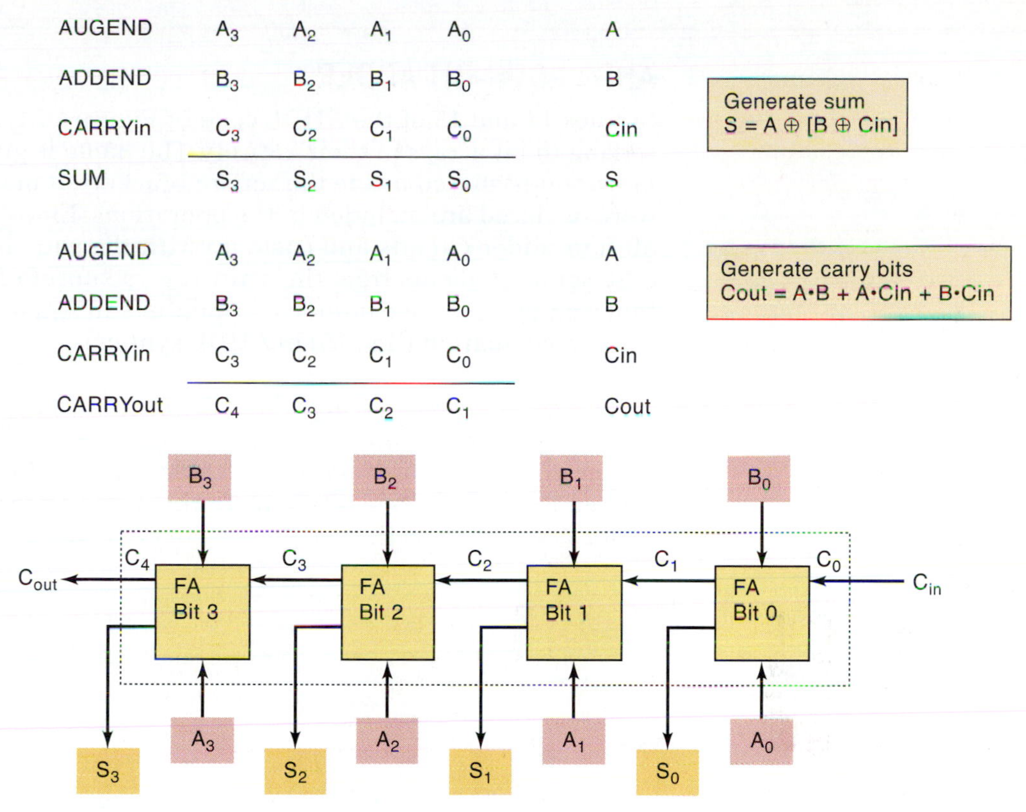

FIGURE 6-19 Four-bit parallel adder.

Substituting our definition of Cin and Cout (above), we have the following equation for the four-bit adder carry out:

$$C[4..1] = A[3..0] \,\&\, B[3..0] + A[3..0] \,\&\, C[3..0] + B[3..0] \,\&\, C[3..0]$$

The graphic symbol for this device is shown in Figure 6-20. Notice that it does not show the carry array. It is a variable or signal *inside* the block. Altera allows any SUBDESIGN in AHDL or ENTITY in VHDL, even those that you create, to be represented by a graphic block diagram symbol such as this. This is all part of the hierarchical design scheme of the Altera development system (described in Chapter 4).

To summarize this information, Figure 6-19 shows the "insides" of the block diagram in Figure 6-20 and summarizes the operation described by the two equations. Now let's look at text-based files that can be used to generate a block symbol like the one in Figure 6-20 using AHDL and VHDL.

FIGURE 6-20 Block symbol generated by Altera MAX+PLUS.

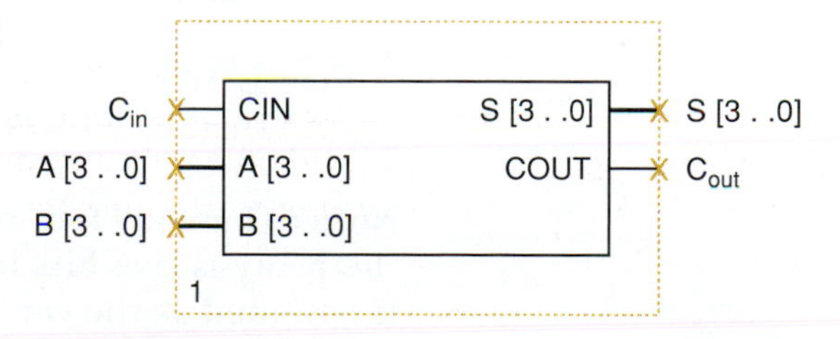

AHDL

AHDL FOUR-BIT ADDER

In lines 14 and 15 of the AHDL code of Figure 6-21, notice the syntax for referring to bit arrays in their entirety. The name is given, followed by [ ]. If no bits are designated inside the square brackets, it means that all the bits that were declared are included in the operations. Lines 14 and 15 describe fully all four adder circuits and come up with the sum. In order to choose a specific set of elements from the array (i.e., a subset of the array), the name is followed by the range of element numbers in square brackets. For example, the carry equation (line 15) in AHDL syntax is:

c[4..1] = a[] & b[] # a[] & c[3..0] # b[] & c[3..0];

Notice that only four bits of the carry array *c[ ]* are being assigned, even though the array is five bits long. In this way, the carry out of each single-bit adder is assigned as the carry in of the next stage.

```
1   SUBDESIGN fig6_21
2   (
3         cin          :INPUT;      -- carry in
4         a[3..0]      :INPUT;      -- augend
5         b[3..0]      :INPUT;      -- addend
6         s[3..0]      :OUTPUT;     -- sum
7         cout         :OUTPUT;     -- carry OUT
8   )
9   VARIABLE
10        c[4..0]      :NODE;       -- carry array is 5 bits long!
11
12  BEGIN
13        c[0] = cin;
14        s[] = a[] $ b[] $ c[3..0];     -- generate sum
15        c[4..1] = (a[] & b[]) # (a[] & c[3..0]) # (b[] & c[3..0]);
16        cout = c[4];                   -- carry out
17  END;
```

FIGURE 6-21 AHDL adder.

VHDL

VHDL FOUR-BIT ADDER

In the VHDL code of Figure 6-22, notice the syntax for referring to bit arrays in their entirety. The name is simply used with no bit designations. Lines 15 and 16 describe fully all four adder circuits that will come up with the sum. In order to choose a specific set of elements from the array (i.e., a subset of the array), the name is followed by the range of element numbers in parentheses. The carry equation (line 16) in VHDL syntax is:

c(4 DOWNTO 1) <= (a AND b) OR (a AND c(3 DOWNTO 0)) OR (b AND c(3 DOWNTO 0));

Notice that only four bits of the carry array *c* are being assigned, even though the array is five bits long. In this way, the carry out of each single-bit adder is assigned as the carry in of the next stage.

```
1   ENTITY fig6_22 IS
2   PORT(
3      cin   :IN BIT;
4      a     :IN BIT_VECTOR(3 DOWNTO 0);
5      b     :IN BIT_VECTOR(3 DOWNTO 0);
6      s     :OUT BIT_VECTOR(3 DOWNTO 0);
7      cout  :OUT BIT);
8   END fig6_22;
9
10  ARCHITECTURE a OF fig6_22 IS
11  SIGNAL c :BIT_VECTOR (4 DOWNTO 0);   -- carries require 5 bit array
12
13  BEGIN
14     c(0) <= cin;              -- Read the carry in to bit array
15     s <= a XOR b  XOR c(3 DOWNTO 0); -- Generate the sum bits
16     c(4 DOWNTO 1) <=          (a AND b)
17                     OR        (a AND c(3 DOWNTO 0))
18                     OR        (b AND c(3 DOWNTO 0));
19     cout <= c(4);             -- output the carry of the MSB.
20  END a;
```

FIGURE 6-22 VHDL adder.

REVIEW QUESTIONS

1. If $[A] = 1001$ and $[B] = 0011$, what is the value of (a) $[A] \cdot [B]$? (b) $[A] + [B]$?(Note that · means AND; + means OR.)
2. If $A[7..0] = 1010\,1100$, what is the value of (a) $A[7..4]$? (b) $A[5..2]$?
3. In AHDL, the following object is declared: toggles[7..0] :INPUT. Give an expression for the least significant four bits using AHDL syntax.
4. In VHDL, the following object is declared: toggles :IN BIT_VECTOR (7 DOWNTO 0). Give an expression for the least significant four bits using VHDL syntax.
5. What would be the result of ORing the two registers of Example 6-14?
6. Write an HDL statement that would OR the two objects *d* and *g* together. Use your favorite HDL.
7. Write an HDL statement that would XOR the two most significant bits of *d* with the two least significant bits of *g* and put the result in the middle two bits of *x*.

6-21 EXPANDING THE BIT CAPACITY OF A CIRCUIT

One way we have learned to expand the capacity of a circuit is to cascade stages, like we did with the 74382 ALU chip in the previous section. This can be done using the Altera graphic design file approach (like Figure 6-18) or the structural text-based HDL approach. With either of these methods, we need to specify all the inputs, outputs, and interconnections between blocks. In the case of this adder circuit, it would be much easier to start with the HDL file for a four-bit adder and simply increase the size of each operand variable in the equation. For example, if we wanted an eight-bit adder, we simply need to expand *a*, *b*, and *s* to eight bits. The code would remain almost identical to the four-bit adder shown above. This is just a glimpse of some of the efficiency improvements that HDL offers. The way this code is written, however, the indices of each signal, and each *bit array* specification in the equation, would also have to be redefined. In other words, the designer would need to examine the code carefully and change all the 3s to 7s, all the 4s to 8s, and so on.

An important principle in software engineering is symbolic representation of the **constants** that are used throughout the code. Constants are simply fixed numbers represented by a name (symbol). If we can define a symbol (i.e., make up a name) at the top of the source code that is assigned the value for the total number of bits and then use this symbol (name) throughout the code, it is much easier to modify the circuit. Only one line of code needs to be changed to expand the capacity of the circuit. The examples that follow add this feature to the code and also upgrade the code to implement the adder/subtractor circuit like the one in Figure 6-14. It should be noted here that expanding the capacity of an adder circuit such as this one will also reduce the speed of the circuit because of carry propagation (described in Section 6-13). In order to keep these examples simple, we have not added any logic to generate a look-ahead carry.

AHDL

AHDL ADDER/SUBTRACTOR

In AHDL, using constants is very simple, as shown on lines 1 and 2 of Figure 6-23. The keyword CONSTANT is followed by the symbolic name and the value it is to be assigned. Notice that we can allow the compiler to do some

```
1   CONSTANT number_of_bits = 8;                      -- set total number of bits
2   CONSTANT n = number_of_bits - 1;                  -- n is highest bit index
3
4   SUBDESIGN fig6_23
5   (
6      add          :INPUT;      -- add control
7      sub          :INPUT;      -- subtract control and LSB Carry in
8      a[n..0]      :INPUT;      -- Augend bits
9      bin[n..0]    :INPUT;      -- Addend bits
10     s[n..0]      :OUTPUT;     -- Sum bits
11     caryout      :OUTPUT;     -- MSB carry OUT
12  )
13  VARIABLE
14     c[n+1..0]   :NODE;       -- intermediate carry vector
15     b[n..0]      :NODE;      -- intermediate operand vector
16  BEGIN
17     b[] = bin[] & add # NOT bin[] & sub;
18     c[0] = sub;                           --Read the carry in to group variable
19     s[] = a[] $ b[] $ c[n..0];            --Generate the sums
20     c[n+1..1] = (a[] & b[]) # (a[] & c[n..0]) # (b[] & c[n..0]);
21     caryout = c[n+1];                     -- output the carry of the MSB.
22  END;
```

FIGURE 6-23 An *n*-bit adder/subtractor description in AHDL.

simple math calculations to establish a value for one constant based on another. We can also use this feature as we refer to the constant in the code, as shown on lines 14, 20, and 21. For example, we can refer to *c[7]* as *c[n]* and *c[8]* as *c[n+1]*. The size of this adder/subtractor can be expanded by simply changing the value of *number_of_bits* to the desired number of bits and then recompiling.

As mentioned, this code has been upgraded from the previous example to make it an adder/subtractor like the one in Figure 6-14. The *add* and *sub* inputs have been included on lines 6 and 7, and a new intermediate variable named *b[ ]* has been included on line 15. The first concurrent statement on line 17 describes all the SOP logic that drives the *b* inputs to the adder in Figure 6-14. First, it describes a logical AND operation between every bit of *bin[ ]* and the logic level on *add*. This result is ORed (bit for bit) with the result of ANDing the complement of every bit of *bin[ ]* with *sub*. In other words, it creates the following Boolean function for each bit: $b = bin \cdot add + \overline{bin} \cdot sub$. The signal *b[ ]* is then used in the adder equations instead of *bin[ ]*, as was used in the adder examples. Notice on line 18 that *sub* is also used to connect the carry array LSB (carry into bit 0) with the value on *sub*, which needs to be 0 when adding and 1 when subtracting.

VHDL ADDER/SUBTRACTOR

In VHDL, using constants is a little bit more involved. Constants must be included in a **PACKAGE**, as shown in Figure 6-24, lines 1–4. Packages are also used to contain component definitions and other information that must be available to all entities in the design file. Notice on line 6 that the keyword USE tells the compiler to use the definitions in this package throughout this design file. Inside the package, the keyword CONSTANT is followed by the symbolic name, its type, and the value it is to be assigned using the := operator. Notice on line 3 that we can allow the compiler to do some simple math calculations to establish a value of one constant based on another. We can also use this feature as we refer to the constant in the code, as shown on lines 34 and 37. For example, we can refer to *c(7)* as *c(n)* and *c(8)* as *c(n+1)*. The size of this adder/subtractor can be expanded by simply changing the value of *number_of_bits* to the desired number of bits and then recompiling.

As mentioned, this code has been upgraded from the previous example to make it an adder/subtractor like the one in Figure 6-14. The *add* and *sub* inputs have been included on lines 10 and 11, and a new signal named *b* has been included on line 20, *bnot* has been included on line 21, and *mode* has been included on line 22. The first concurrent statement on line 24 serves to create the 1's complement of *bin*. The SOP circuits in Figure 6-14 that drive the *b* inputs to the adder select the *bin* inputs if *add* = 1 or the 1's complement (*bnot*) if *sub* = 1. This is an excellent application of the VHDL selected signal assignment, as shown on lines 27–30. When *add* is 1, *bin* is channeled to *b*. When *sub* is 1, *bnot* is channeled to *b*. The signal *b* is then used in the adder equations instead of *bin*, as was used in the previous adder examples. Notice on line 32 that *sub* is also used to connect the carry array LSB (carry into bit 0) with the value on *sub*, which needs to be 0 when adding and 1 when subtracting.

VHDL

```
1   PACKAGE const IS
2     CONSTANT number_of_bits :INTEGER:=8;  -- set total number of bits
3     CONSTANT n :INTEGER:= number_of_bits - 1;  -- MSB index number
4   END const;
5
6   USE work.const.all;
7
8   ENTITY fig6_24 IS
9   PORT(
10     add      :IN BIT; -- add control
11     sub      :IN BIT; -- subtract control and LSB carry in
12     a        :IN BIT_VECTOR(n DOWNTO 0);
13     bin      :IN BIT_VECTOR(n DOWNTO 0);
14     s        :OUT BIT_VECTOR(n DOWNTO 0);
15     carryout :OUT BIT);
16  END fig6_24;
17
18  ARCHITECTURE a OF fig6_24 IS
19  SIGNAL c :BIT_VECTOR (n+1 DOWNTO 0);  -- define intermediate carries
20  SIGNAL b :BIT_VECTOR (n DOWNTO 0);      -- define intermediate operand
21  SIGNAL bnot :BIT_VECTOR (n DOWNTO 0);
22  SIGNAL mode :BIT_VECTOR (1 DOWNTO 0);
23  BEGIN
24     bnot <= NOT bin;
25     mode <= add & sub;
26
27     WITH mode SELECT
28        b <= bin       WHEN "10",     -- add
29             bnot      WHEN "01",     -- sub
30             "0000"    WHEN OTHERS;
31
32     c(0) <= sub;                        -- read the carry_in to bit array
33     s <= a XOR b XOR c(n DOWNTO 0); -- generate the sum bits
34     c(n+1 DOWNTO 1) <= (a AND b) OR
35                        (a AND c(n DOWNTO 0)) OR
36                        (b AND c(n DOWNTO 0)); --generate carries
37     carryout <= c(n+1);             -- output the carry of the MSB.
38
39  END a;
```

FIGURE 6-24 An *n*-bit adder/subtractor description in VHDL.

VHDL GENERATE Statement

Another way to make circuits that handle more bits is the VHDL **GENERATE** statement. It is a very concise way of telling the compiler to replicate several components that are cascaded together. As we have shown, there are many other ways to accomplish the same thing and if the abstract nature of this method seems difficult, use another method. The GENERATE statement is offered here for the sake of completeness. The adder circuits we have been discussing are cascaded chains of single-bit full adder modules. The VHDL code for a single-bit full adder module is shown in Figure 6-25. Multiple instances of this module need to be connected to each other to form

FIGURE 6-25 Single-bit full adder in VHDL.

```
1   ENTITY add1 IS
2   PORT(
3         cin    :IN BIT;
4         a      :IN BIT;
5         b      :IN BIT;
6         s      :OUT BIT;
7         cout   :OUT BIT);
8   END add1;
9
10  ARCHITECTURE a OF add1 IS
11  BEGIN
12
13        s      <= a XOR b XOR cin;
14        cout   <= (a AND b) OR (a AND cin) OR (b AND cin);
15  END a;
```

an n-bit adder circuit. Of course, you can do it using the same component techniques that we have discussed previously, but it would result in very lengthy code.

To make the code more concise and easier to modify, a strategy is needed for the way we label the inputs and outputs for each module. As we mentioned previously, the bit-0 adder has inputs that have an index of 0 (e.g., *a0*, *b0*, *c0*, *s0*). The carry out of bit 0 is labeled *c1*, and it becomes the carry input for the bit 1 adder module. Each time we instantiate another component for the next bit of the multibit adder, the index number of all connections goes up by 1 (*a1*, *b1*, *c1*, *s1*). The GENERATE statement allows us to repeat an instantiation of a component n times, increasing the index number by 1 for each instantiation up to n. In line 27 of Figure 6-26, the GENERATE keyword is used in an **iterative loop (FOR loop)**, which means that a set of descriptive actions (PORT MAP) will be repeated a certain number of times. The variable i represents an index number that starts at 0 (for the first iteration) and ends at n (the last iteration). The advantages of this method are code compactness and the ease with which the number of bits can be expanded. The code in Figure 6-26 shows how to use a single-bit adder (Figure 6-25) as a component to generate an eight-bit adder circuit. Remember that the file for the single-bit adder (add1.vhd in Figure 6-25) must be saved in the same folder as the design file that uses it to generate multiple instances of the adder (fig6_26.vhd).

The single-bit adder component is defined on lines 17–23 of Figure 6-26. For the first iteration of lines 29 and 30, the value of i is 0, creating an adder stage for bit 0. The second iteration, i, has been increased to 1 to form adder stage 1. This continues until i is equal to n, generating each stage of the $(n + 1)$-bit adder. Notice that VHDL allows us the option of placing labels at the beginning of a line of code to help describe its purpose. For example, on line 27, the label *repeat* is used and on line 29, the label *casc* is used. Labels are optional but they must always end with a colon.

Libraries of Parameterized Modules

Using HDL techniques clearly makes it easy to alter the bit capacity of a generic circuit. In this chapter, we can see that it is easy to change from a

```
1   PACKAGE const IS
2      CONSTANT number_of_bits :INTEGER:=8;    -- Specify number of bits
3      CONSTANT n :INTEGER:=number_of_bits - 1;      --n is MSB bit number
4   END const;
5   USE work.const.all;
6   ENTITY fig6_26 IS
7   PORT(
8      caryin    :IN bit;
9      ain       :IN BIT_VECTOR (n DOWNTO 0);
10     bin       :IN BIT_VECTOR (n DOWNTO 0);
11     sout      :OUT BIT_VECTOR (n DOWNTO 0);
12    carryout  :OUT bit);
13  END fig6_26;
14
15  ARCHITECTURE a OF fig6_26 IS
16
17  COMPONENT add1             -- declare single bit full adder
18     PORT   (
19              cin    :IN BIT;
20              a,b    :IN BIT;
21              s      :OUT BIT;
22              cout   :OUT BIT);
23  END COMPONENT;
24  SIGNAL c :BIT_VECTOR (n+1 DOWNTO 0);    -- declare bit array for carries
25  BEGIN
26     c(0)   <= caryin;                    -- put LSB in array (carry in)
27     repeat:FOR i IN 0 TO n GENERATE      -- instantiate n+1 adders
28          -- cascade them
29     casc:add1 PORT MAP (cin=> c(i), a=> ain(i), b=> bin(i),
30                          s=> sout(i), cout=> c(i+1));
31     END GENERATE;
32     carryout    <= c(n+1);              -- out the carry from nth bit stage
33  END a;
```

FIGURE 6-26 Use of the VHDL GENERATE statement.

four-bit adder to an eight-, 12-, or 16-bit adder. When Altera was creating its library of useful functions, they also took advantage of these techniques and created what they refer to as **megafunctions**, which include a **library of parameterized modules (LPMs)**. These functions do not attempt to imitate a particular standard IC like the old-style macrofunctions; instead, they offer a generic solution for the various types of logic circuits that are useful in digital systems. Examples of these generic circuits that we have covered so far are logic gates (AND, OR, XOR), latches, counters, shift registers, and adders. The term *parameterized* means that when you instantiate a function from the library, you also specify some parameters that define certain attributes (bit capacity, for example) for the circuit you are describing. The various LPMs that are available can be found through the HELP menu under megafunctions/LPM. This documentation describes the parameters that the user can specify as well as the port features of the device.

REVIEW QUESTIONS

1. What keyword is used to assign a symbolic name to a fixed number?
2. In AHDL, where are constants defined? Where are they defined in VHDL?
3. Why are constants useful?
4. If the constant max_val has a value of 127, how will a compiler interpret the expression max_val −5?
5. What is the GENERATE statement used for in VHDL?

SUMMARY

1. To represent signed numbers in binary, a sign bit is attached as the MSB. A + sign is a 0, and a − sign is a 1.
2. The 2's complement of a binary number is obtained by complementing each bit and then adding 1 to the result.
3. In the 2's-complement method of representing signed binary numbers, positive numbers are represented by a sign bit of 0 followed by the magnitude in its true binary form. Negative numbers are represented by a sign bit of 1 followed by the magnitude in 2's-complement form.
4. A signed binary number is negated (changed to a number of equal value but opposite sign) by taking the 2's complement of the number, including the sign bit.
5. Subtraction can be performed on signed binary numbers by negating (2's complementing) the subtrahend and adding it to the minuend.
6. In BCD addition, a special correction step is needed whenever the sum of a digit position exceeds 9 (1001).
7. When signed binary numbers are represented in hexadecimal, the MSD of the hex number will be 8 or greater when the number is negative; it will be 7 or less when the number is positive.
8. The arithmetic/logic unit (ALU) of a computer contains the circuitry needed to perform arithmetic and logic operations on binary numbers stored in memory.
9. The accumulator is a register in the ALU. It holds one of the numbers being operated upon, and it also is where the result of the operation is stored in the ALU.
10. A full adder performs the addition on two bits plus a carry input. A parallel binary adder is made up of cascaded full adders.
11. The problem of excessive delays caused by carry propagation can be reduced by a look-ahead carry logic circuit.
12. IC adders such as the 74LS83/HC83 and the 74LS283/HC283 can be used to construct high-speed parallel adders and subtractors.
13. A BCD adder circuit requires special correction circuitry.
14. Integrated-circuit ALUs are available that can be commanded to perform a wide range of arithmetic and logic operations on two input numbers.
15. Prefabricated functions are available in the Altera libraries.
16. These library parts and the HDL circuits you create can be interconnected using either graphic or structural HDL techniques.

17. Logical operations can be performed on all the bits in a set using Boolean equations.
18. Practicing good software engineering techniques, specifically the use of symbols to represent constants, allows for easy code modification and expansion of the bit capacity of circuits such as full adders.
19. Libraries of parameterized modules (LPMs) offer a flexible, easily modified or expanded solution for many types of digital circuits.

IMPORTANT TERMS

carry
sign bit
sign-magnitude system
2's-complement system
negation
augend
addend
subtrahend
minuend
overflow
arithmetic/logic unit (ALU)
accumulator register
full adder (FA)
parallel adder
half adder (HA)
carry propagation (carry ripple)
look-ahead carry
adder/subtractor
macrofunction
function prototype
sets
constants
PACKAGE
GENERATE
iterative loop
FOR loop
library of parameterized functions (LPMs)
megafunctions

PROBLEMS

SECTION 6-1

B 6-1. Add the following in binary. Check your results by doing the addition in decimal.

(a)* 1010 + 1011
(b)* 1111 + 0011
(c)* 1011.1101 + 11.1
(d) 0.1011 + 0.1111
(e) 10011011 + 10011101
(f) 1010.01 + 10.111

SECTION 6-2

B 6-2. Represent each of the following signed decimal numbers in the 2's-complement system. Use a total of eight bits, including the sign bit.

(a)* +32
(b)* −14
(c)* +63
(d)* −104
(e)* +127
(f)* −127
(g)* +89
(h)* −55
(i) −1
(j) −128
(k) +169
(l) 0
(m) +84
(n) +3
(o) −3
(p) −190

B 6-3. Each of the following numbers represents a signed decimal number in the 2's-complement system. Determine the decimal value in each case. (*Hint:* Use negation to convert negative numbers to positive.)

(a)* 01101
(b)* 11101
(c)* 01111011
(d)* 10011001
(e)* 01111111
(f) 10000000
(g) 11111111
(h) 10000001
(i) 01100011
(j) 11011001

*Answers to problems marked with an asterisk can be found in the back of the text.

6-4. (a) What range of signed decimal values can be represented using 12 bits, including the sign bit?

(b) How many bits would be required to represent decimal numbers from −32,768 to +32,767?

6-5.* List, in order, all of the signed numbers that can be represented in five bits using the 2's-complement system.

6-6. Represent each of the following decimal values as an eight-bit signed binary value. Then negate each one.

(a)* +73 (b)* −12 (c) +15 (d) −1 (e) −128 (f) +127

6-7. (a)* What is the range of unsigned decimal values that can be represented in 10 bits? What is the range of signed decimal values using the same number of bits?

(b) Repeat both problems using eight bits.

SECTIONS 6-3 AND 6-4

6-8. The reason why the sign-magnitude method for representing signed numbers is not used in most computers can readily be illustrated by performing the following.

(a) Represent +12 in eight bits using the sign-magnitude form.

(b) Represent −12 in eight bits using the sign-magnitude form.

(c) Add the two binary numbers and note that the sum does not look anything like zero.

6-9. Perform the following operations in the 2's-complement system. Use eight bits (including the sign bit) for each number. Check your results by converting the binary result back to decimal.

(a)* Add +9 to +6.
(b)* Add +14 to −17.
(c)* Add +19 to −24.
(d)* Add −48 to −80.
(e)* Subtract +16 from +17.
(f) Subtract +21 from −13.
(g) Subtract +47 from +47.
(h) Subtract −36 from −15.
(i) Add +17 to −17.
(j) Subtract −17 from −17.

6-10. Repeat Problem 6-9 for the following cases, and show that overflow occurs in each case.

(a) Add +37 to +95.
(b) Subtract +37 from −95.
(c) Add −37 to −95.
(d) Subtract −37 from +95.

SECTIONS 6-5 AND 6-6

B 6-11. Multiply the following pairs of binary numbers, and check your results by doing the multiplication in decimal.

(a)* 111 × 101
(b)* 1011 × 1011
(c) 101.101 × 110.010
(d) .1101 × .1011

B 6-12. Perform the following divisions. Check your results by doing the division in decimal.

(a)* 1100 ÷ 100
(b)* 111111 ÷ 1001
(c) 10111 ÷ 100
(d) 10110.1101 ÷ 1.1

SECTIONS 6-7 AND 6-8

B 6-13. Add the following decimal numbers after converting each to its BCD code.

(a)* 74 + 23
(b)* 58 + 37
(c)* 147 + 380
(d) 385 + 118
(e) 998 + 003
(f) 623 + 599

B 6-14. Find the sum of each of the following pairs of hex numbers.

(a)* 3E91 + 2F93
(b)* 91B + 6F2
(c)* ABC + DEF
(d) 2FFE + 0002
(e) FFF + 0FF
(f) D191 + AAAB

B 6-15. Perform the following subtractions on the pairs of hex numbers.

(a)* 3E91 − 2F93
(b)* 91B − 6F2
(c)* 0300 − 005A
(d) 0200 − 0003
(e) F000 − EFFF
(f) 2F00 − 4000

6-16. The owner's manual for a small microcomputer states that the computer has usable memory locations at the following hex addresses: 0200 through 03FF, and 4000 through 7FD0. What is the total number of available memory locations?

6-17. (a)* A certain eight-bit memory location holds the hex data 77. If this represents an *unsigned* number, what is its decimal value?

(b)* If this represents a *signed* number, what is its decimal value?

(c) Repeat (a) and (b) if the data value is E5.

SECTION 6-11

6-18. Convert the FA circuit of Figure 6-8 to all NAND gates.

6-19.* Write the function table for a half adder (inputs A and B; outputs SUM and CARRY). From the function table, design a logic circuit that will act as a half adder.

6-20. A full adder can be implemented in many different ways. Figure 6-27 shows how one may be constructed from two half adders. Construct a function table for this arrangement, and verify that it operates as a FA.

FIGURE 6-27 Problem 6-20.

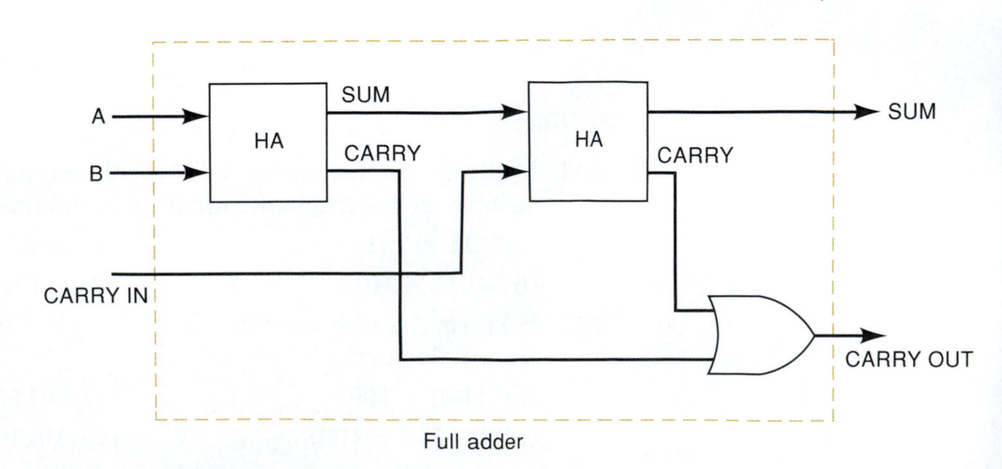

SECTION 6-12

6-21.* Refer to Figure 6-10. Determine the contents of the A register after the following sequence of operations: $[A] = 0000$, $[0100] \rightarrow [B]$, $[S] \rightarrow [A]$, $[1011] \rightarrow [B]$, $[S] \rightarrow [A]$.

6-22. Refer to Figure 6-10. Assume that each FF has $t_{PLH} = t_{PHL} = 30$ ns and a setup time of 10 ns, and that each FA has a propagation delay of 40 ns. What is the minimum time allowed between the PGT of the LOAD pulse and the PGT of the TRANSFER pulse for proper operation?

D 6-23. In the adder and subtractor circuits discussed in this chapter, we gave no consideration to the possibility of *overflow*. Overflow occurs when the two numbers being added or subtracted produce a result that contains more bits than the capacity of the accumulator. For example, using four-bit registers, including a sign bit, numbers ranging from +7 to −8 (in 2's complement) can be stored. Therefore, if the result of an addition or subtraction exceeds +7 or −8, we would say that an overflow has occurred. When an overflow occurs, the results are useless because they cannot be stored correctly in the accumulator register. To illustrate, add +5 (0101) and +4 (0100), which results in 1001. This 1001 would be interpreted incorrectly as a negative number because there is a 1 in the sign-bit position.

In computers and calculators, there are usually circuits that are used to detect an overflow condition. There are several ways to do this. One method that can be used for the adder that operates in the 2's-complement system works as follows:

1. Examine the sign bits of the two numbers being added.
2. Examine the sign bit of the result.
3. Overflow occurs whenever the numbers being added are *both positive* and the sign bit of the result is 1 *or* when the numbers are *both negative* and the sign bit of the result is 0.

This method can be verified by trying several examples. Readers should try the following cases for their own clarification: (1) 5 + 4; (2) −4 + (−6); (3) 3 + 2. Cases 1 and 2 will produce an overflow, and case 3 will not. Thus, by examining the sign bits, one can design a logic circuit that will produce a 1 output whenever the overflow condition occurs. Design this overflow circuit for the adder of Figure 6-10.

C, D 6-24. Add the necessary logic circuitry to Figure 6-10 to accommodate the transfer of data from memory into the A register. The data values from memory are to enter the A register through its D inputs on the PGT of the *first* TRANSFER pulse; the data from the sum outputs of the FAs will be loaded into A on the PGT of the *second* TRANSFER. In other words, a LOAD pulse followed by two TRANSFER pulses is required to perform the complete sequence of loading the B register from memory, loading the A register from memory, and then transferring their sum into the A register. (*Hint:* Use a flip-flop X to control which source of data gets loaded into the D inputs of the accumulator.)

SECTION 6-13

C, D 6-25.* Design a look-ahead carry circuit for the adder of Figure 6-10 that generates the carry C_3 to be fed to the FA of the MSB position based on the values of A_0, B_0, C_0, A_1, B_1, A_2, and B_2. In other words, derive an expression for C_3 in terms of A_0, B_0, C_0, A_1, B_1, A_2, and B_2. (*Hint:* Begin by writing the expression for C_1 in terms of A_0, B_0, and C_0. Then write the expression for C_2 in terms of A_1, B_1, and C_1. Substitute the expression for C_1 into the expression for C_2. Then write the expression for C_3 in terms of A_2, B_2, and C_2. Substitute the expression for C_2 into the expression for C_3. Simplify the final expression for C_3 and put it in sum-of-products form. Implement the circuit.)

SECTION 6-14

6-26. Show the logic levels at each input and output of Figure 6-11(a) when 354_8 is added to 103_8.

SECTION 6-15

6-27. For the circuit of Figure 6-14, determine the sum outputs for the following cases.

(a)* A register = 0101 (+5), B register = 1110 (−2); SUB = 1, ADD = 0

(b) A register = 1100 (−4), B register = 1110 (−2); SUB = 0, ADD = 1

(c) Repeat (b) with ADD = SUB = 0.

6-28. For the circuit of Figure 6-14 determine the sum outputs for the following cases.

(a) A register = 1101 (−3), B register = 0011 (+3), SUB = 1, ADD = 0.

(b) A register = 1100 (−4), B register = 0010 (+2), SUB = 0, ADD = 1.

(c) A register = 1011 (−5), B register = 0100 (+4), SUB = 1, ADD = 0.

6-29. For each of the calculations of Problem 6-27, determine if overflow has occurred.

6-30. For each of the calculations of Problem 6-28, determine if overflow has occurred.

D 6-31. Show how the gates of Figure 6-14 can be implemented using three 74HC00 chips.

D 6-32.* Modify the circuit of Figure 6-14 so that a single control input, X, is used in place of ADD and SUB. The circuit is to function as an adder when $X = 0$, and as a subtractor when $X = 1$. Then simplify each set of gates. (*Hint:* Note that now each set of gates is functioning as a controlled inverter.)

SECTION 6-16

B 6-33. Determine the F, C_{N+4}, and OVR outputs for each of the following sets of inputs applied to a 74LS382.

(a)* $[S] = 011$, $[A] = 0110$, $[B] = 0011$, $C_N = 0$

(b) $[S] = 001$, $[A] = 0110$, $[B] = 0011$, $C_N = 1$

(c) $[S] = 010$, $[A] = 0110$, $[B] = 0011$, $C_N = 1$

D 6-34. Show how the 74HC382 can be used to produce $[F] = [\overline{A}]$. (*Hint:* Recall that special property of an XOR gate.)

6-35. Determine the Σ outputs in Figure 6-16 for the following sets of inputs.

(a)* $[S] = 110$, $[A] = 10101100$, $[B] = 00001111$

(b) $[S] = 100$, $[A] = 11101110$, $[B] = 00110010$

C, D 6-36. Add the necessary logic to Figure 6-16 to produce a single HIGH output whenever the binary number at A is exactly the same as the binary number at B. Apply the appropriate select input code (three codes can be used).

SECTION 6-17

T 6-37. Consider the circuit of Figure 6-10. Assume that the A_2 output is stuck LOW. Follow the sequence of operations for adding two numbers, and determine the results that will appear in the A register after the second TRANSFER pulse for each of the following cases. Note that the numbers are given in decimal, and the first number is the one loaded into B by the first LOAD pulse.

(a)*2 + 3

(b)*3 + 7

T (c) 7 + 3

(d) 8 + 3

(e) 9 + 3

T 6-38. A technician breadboards the adder/subtractor of Figure 6-14. During testing, she finds that whenever an addition is performed, the result is 1 more than expected, and when a subtraction is performed, the result is 1 less than expected. What is the likely error that the technician made in connecting this circuit?

6-39.*Describe the symptoms that would occur at the following points in the circuit of Figure 6-14 if the ADD and SUB lines were shorted together.

(a) B[3..0] inputs of the 74LS283 IC

(b) C_0 input of the 74LS283 IC

(c) SUM (Σ) [3..0] outputs

(d) C_4

SECTION 6-19

Problems 6-40 through 6-45 deal with the same two arrays, a and b, which we will assume have been defined in an HDL source file and have the following values: [a] = [10010111], [b] = [00101100]. Output array [z] is also an eight-bit array. Answer Problems 6-40 through 6-45 based on this information. (Assume undefined bits in z are 0.)

B, H 6-40. Declare these data objects using your favorite HDL syntax.

B, H 6-41. Give the value of z for each expression (identical AHDL and VHDL expressions are given):

(a)*z[] = a[] & b[]; z <= a AND b;

(b)*z[] = a[] # b[]; z <= a OR b;

(c) z[] = a[] $!b[]; z <= a XOR NOT b;

(d) z[7..4] = a[3..0] & b[3..0]; z(7 DOWNTO 4) <= a(3 DOWNTO 0) AND b(3 DOWNTO 0);

(e) z[7..1] = a[6..0]; z[0] = GND; z(7 DOWNTO 1) <= a(6 DOWNTO 0); z(0) <= '0';

6-42. What is the value of each of the following:

(a) a[3..0] a(3 DOWNTO 0)

(b) b[0] b(0)

(c) a[7] b(7)

B, H 6-43. What is the value of each of the following?

(a)* a[5] a(5)

(b)* b[2] b(2)

(c)* b[7..1] b(7 DOWNTO 1)

H 6-44.* Write one or more statements in HDL that will shift all the bits in [a] one position to the right. The LSB should move to the MSB position. The rotated data should end up in z[].

6-45. Write one or more HDL statements that will take the upper nibble of b and place it in the lower nibble of z. The upper nibble of z should be zero.

D, H 6-46. Refer to Problem 6-23. Modify the code of Figure 6-21 or Figure 6-22 to add an overflow output.

D, H, N 6-47.* Another way to detect 2's-complement overflow is to XOR the carry into the MSB with the carry out of the MSB of an adder/subtractor. Use the same numbers given in Problem 6-23 to verify this. Modify Figure 6-21 or Figure 6-22 to detect overflow using this method.

D, H 6-48.* Modify Figure 6-21 or Figure 6-22 to implement Figure 6-10.

SECTION 6-20

B, H 6-49. Modify Figure 6-21 or Figure 6-22 to make it a 12-bit adder without using constants.

B, H 6-50. Modify Figure 6-21 or Figure 6-22 to make it a versatile *n*-bit adder module with a constant defining the number of bits.

D, C, H 6-51. Write an HDL file to create the equivalent of a 74382 ALU without using a built-in macrofunction.

DRILL QUESTION

6-52. Define each of the following terms.

(a) Full adder

(b) 2's complement

(c) Arithmetic/logic unit

(d) Sign bit

(e) Overflow

(f) Accumulator

(g) Parallel adder

(h) Look-ahead carry

(i) Negation

(j) *B* register

MICROCOMPUTER APPLICATIONS

C, D 6-53.* In a typical microprocessor ALU, the results of every arithmetic operation are usually (but not always) transferred to the accumulator register, as in Figures 6-10, 6-14, and 6-15. In most microprocessor ALUs, the result of each arithmetic operation is also used to control the states of several special flip-flops called *flags*. These flags are used by the microprocessor when it is making decisions during the execution of certain types of instructions. The three most common flags are:

S (sign flag). This FF is always in the same state as the sign of the last result from the ALU.

Z (zero flag). This flag is set to 1 whenever the result from an ALU operation is exactly 0. Otherwise, it is cleared to 0.

C (carry flag). This FF is always in the same state as the carry from the MSB of the ALU.

Using the adder/subtractor of Figure 6-14 as the ALU, design the logic circuit that will implement these flags. The sum outputs and C_4 output are to be used to control what state each flag will go to upon the occurrence of the TRANSFER pulse. For example, if the sum is exactly 0 (i.e., 0000), the Z flag should be set by the PGT of TRANSFER; otherwise, it should be cleared.

6-54.*In working with microcomputers, it is often necessary to move binary numbers from an eight-bit register to a 16-bit register. Consider the numbers 01001001 and 10101110, which represent +73 and −82, respectively, in the 2's-complement system. Determine the 16-bit representations for these decimal numbers.

6-55. Compare the eight- and 16-bit representations for +73 from Problem 6-53. Then compare the two representations for −82. There is a general rule that can be used to convert easily from eight-bit to 16-bit representations. Can you see what it is? It has something to do with the sign bit of the eight-bit number.

ANSWERS TO SECTION REVIEW QUESTIONS

SECTION 6-1

1. (a) 11101 (b) 101.111 (c) 10010000

SECTION 6-2

1. (a) 00001101 (b) 11111001 (c) 10000000 2. (a) −29 (b) −64 (c) +126 3. −2048 to +2047 4. Seven 5. −32768 6. (a) 10000 (b) 10000000 (c) 1000 7. Refer to text.

SECTION 6-3

1. True 2. (a) $100010_2 = -30_{10}$ (b) $000000_2 = 0_{10}$

SECTION 6-4

1. (a) $01111_2 = +15_{10}$ (b) $11111_2 = -1_{10}$ 2. By comparing the sign bit of the sum with the sign bits of the numbers being added

SECTION 6-5

1. 1100010

SECTION 6-7

1. The sum of at least one decimal digit position is greater than 1001 (9).
2. The correction factor is added to both the units and the tens digit positions.

SECTION 6-8

1. 923 2. 3DB 3. 2F, 77EC, 6D

SECTION 6-10

1. Three; two 2. (a) $S_2 = 0, C_3 = 1$ (b) $C_5 = 0$

SECTION 6-12

1. One; four; four 2. 0100

SECTION 6-14

1. Five chips 2. 240 ns 3. 1

SECTION 6-15

1. To add the 1 needed to complete the 2's-complement representation of the number in the B register 2. 0010 3. 1101 4. False; the 1's complement appears there.

SECTION 6-16

1. $F = 1011$; $OVR = 0$; $C_{N+4} = 0$ 2. $F = 0111$; $OVR = 1$; $C_{N+4} = 1$ 3. $F = 1000$ 4. $\Sigma = 01101011$; $C_{N+4} = OVR = 0$ 5. $\Sigma = 11111111$ 6. Eight

SECTION 6-18

1. See the MAX+PLUS HELP menu under old-style macrofunctions/adders.
2. An HDL description of a standard IC that can be used from the library.

SECTION 6-19

1. (a) 0001 (b) 1011 2. (a) 1010 (b) 1011 3. toggles[3..0] 4. toggles(3 DOWNTO 0) 5. [X] = [1,1,1,1] 6. AHDL: xx[] = d[] # g[]; VHDL: x <= d OR g; 7. AHDL: xx[2..1] = d[3..2] $ g[1..0]; VHDL: x(2 DOWNTO 1) <= d(3 DOWNTO 2) XOR g(1 DOWNTO 0);

SECTION 6-21

1. CONSTANT. 2. In AHDL, near the top of the source file. In VHDL, in a PACKAGE near the top of the source file. 3. They allow for global changes of the value of a symbol used throughout the code. 4. max_val −5 represents the number 122. 5. GENERATE is used with an iterative FOR statement to instantiate duplicate code modules that can be connected together or cascaded.

CHAPTER 7

COUNTERS AND REGISTERS

OUTLINE

OBJECTIVES

Upon completion of this chapter, you will be able to:

- Understand the operation and characteristics of synchronous and asynchronous counters.
- Construct counters with MOD numbers less than 2^N.
- Construct both up and down counters.
- Connect multistage counters.
- Analyze and evaluate various types of counters.
- Design arbitrary-sequence synchronous counters.
- Understand several types of schemes used to decode different types of counters.
- Describe counter circuits using different levels of abstraction in HDL.
- Compare the major differences between ring and Johnson counters.
- Recognize and understand the operation of various types of IC registers.
- Describe shift registers and shift register counters using HDL.
- Apply existing troubleshooting techniques used for combinational logic systems to troubleshoot sequential logic systems.

INTRODUCTION

In Chapter 5, we saw how flip-flops could be connected to function as counters and registers. At that time we studied only the basic counter and register circuits. Digital systems employ many variations of these basic circuits, mostly in integrated-circuit form. In this chapter, we will look at how FFs and logic gates can be combined to produce different types of counters and registers.

Because there are a great number of topics in this chapter, it has been divided into two parts. In **PART 1**, we will cover the principles of counter operation, the various counter circuit arrangements, and representative IC counters. **PART 2** will present several types of IC registers, shift register counters, and troubleshooting. Each part includes a section containing HDL descriptions of counters and registers.

As you progress through this chapter, you will find that you are constantly drawing on your understanding of the material we have covered in the preceding chapters. It is a good idea to go back and review previously learned material whenever you need to.

PART 1

7-1 ASYNCHRONOUS (RIPPLE) COUNTERS

Figure 7-1 shows a four-bit binary counter circuit such as the one discussed in Chapter 5. Recall the following points concerning its operation:

1. The clock pulses are applied only to the *CLK* input of flip-flop *A*. Thus, flip-flop *A* will toggle (change to its opposite state) each time the clock pulses make a negative (HIGH-to-LOW) transition. Note that $J = K = 1$ for all FFs.
2. The normal output of flip-flop *A* acts as the *CLK* input for flip-flop *B*, and so flip-flop *B* will toggle each time the *A* output goes from 1 to 0. Similarly, flip-flop *C* will toggle when *B* goes from 1 to 0, and flip-flop *D* will toggle when *C* goes from 1 to 0.
3. FF outputs *D*, *C*, *B*, and *A* represent a four-bit binary number, with *D* as the MSB. Let's assume that all FFs have been cleared to the 0 state (CLEAR inputs are not shown). The waveforms in Figure 7-1 show that a binary counting sequence from 0000 to 1111 is followed as clock pulses are continuously applied.
4. After the NGT of the fifteenth clock pulse has occurred, the counter FFs are in the 1111 condition. On the sixteenth NGT, flip-flop *A* goes from 1 to 0, which causes flip-flop *B* to go from 1 to 0, and so on, until the

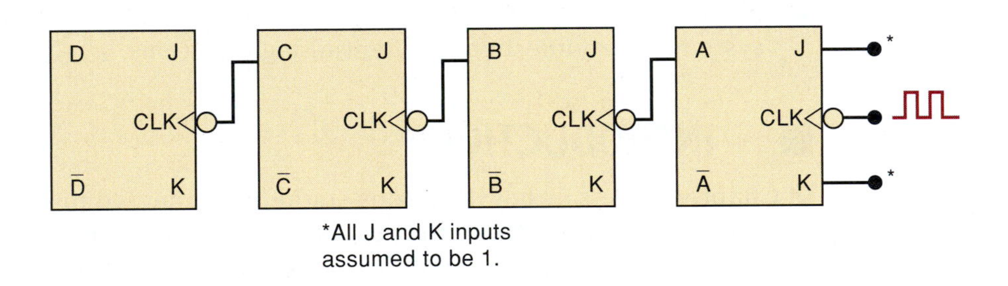

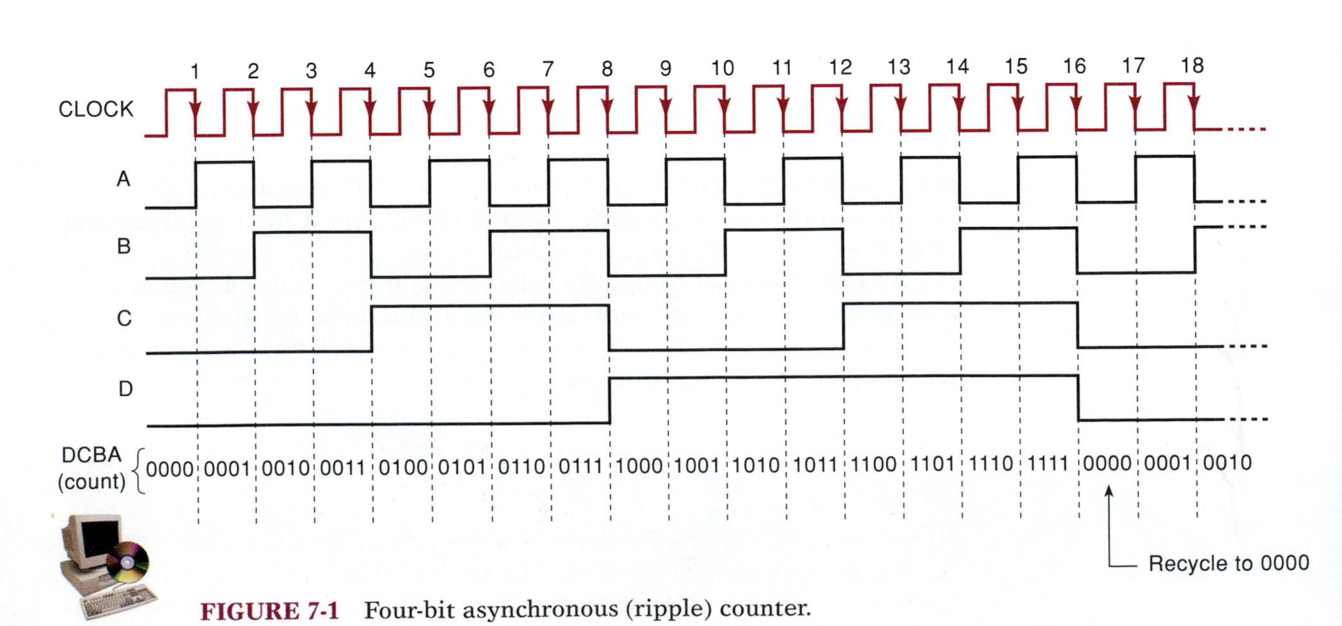

FIGURE 7-1 Four-bit asynchronous (ripple) counter.

counter is in the 0000 state. In other words, the counter has gone through one complete cycle (0000 through 1111) and has *recycled* back to 0000. From this point, it will begin a new counting cycle as subsequent clock pulses are applied.

In this counter, each FF output drives the *CLK* input of the next FF. This type of counter arrangement is called an **asynchronous counter** because the FFs do not change states in exact synchronism with the applied clock pulses; only flip-flop *A* responds to the clock pulses. FF *B* must wait for FF *A* to change states before it can toggle; FF *C* must wait for FF *B*, and so on. Thus, there is a delay between the responses of successive FFs. This delay is typically 5–20 ns per FF. In some cases, as we shall see, this delay can be troublesome. This type of counter is also often referred to as a **ripple counter** because of the way the FFs respond one after another in a kind of rippling effect. We will use the terms *asynchronous counter* and *ripple counter* interchangeably.

Signal Flow

It is conventional in circuit schematics to draw the circuits (wherever possible) so that the signal flow is from left to right, with inputs on the left and outputs on the right. In this chapter, we will often break with this convention, especially in diagrams showing counters. For example, in Figure 7-1, the *CLK* inputs of each FF are on the right, the outputs are on the left, and the input clock signal is shown coming in from the right. We will use this arrangement because it makes the counter operation easier to understand and follow (because the order of the FFs is the same as the order of the bits in the binary number that the counter represents). In other words, FF *A* (which is the LSB) is the rightmost FF, and FF *D* (which is the MSB) is the leftmost FF. If we adhered to the conventional left-to-right signal flow, we would have to put FF *A* on the left and FF *D* on the right, which is opposite to their positions in the binary number that the counter represents. In some of the counter diagrams later in the chapter, we will employ the conventional left-to-right signal flow so that you will get used to seeing it.

EXAMPLE 7-1

The counter in Figure 7-1 starts off in the 0000 state, and then clock pulses are applied. Some time later the clock pulses are removed, and the counter FFs read 0011. How many clock pulses have occurred?

Solution

The apparent answer seems to be 3 because 0011 is the binary equivalent of 3. With the information given, however there is no way to tell whether or not the counter has recycled. This means that there could have been 19 clock pulses; the first 16 pulses bring the counter back to 0000, and the last 3 bring it to 0011. There could have been 35 pulses (two complete cycles and then three more), or 51 pulses, and so on.

MOD Number

The counter in Figure 7-1 has 16 distinctly different states (0000 through 1111). Thus, it is a *MOD-16 ripple counter.* Recall that the **MOD number** is generally equal to the number of states that the counter goes through in each complete cycle before it recycles back to its starting state. The MOD number can be increased simply by adding more FFs to the counter. That is,

$$\text{MOD number} = 2^N \tag{7-1}$$

where N is the number of FFs connected in the arrangement of Figure 7-1.

EXAMPLE 7-2

A counter is needed that will count the number of items passing on a conveyor belt. A photocell and light source combination is used to generate a single pulse each time an item crosses its path. The counter must be able to count as many as one thousand items. How many FFs are required?

Solution

It is a simple matter to determine what value of N is needed so that $2^N \geq 1000$. Since $2^9 = 512$, 9 FFs will not be enough. $2^{10} = 1024$, so 10 FFs would produce a counter that could count as high as $1111111111_2 = 1023_{10}$. Therefore, we should use 10 FFs. We could use more than 10, but it would be a waste of FFs because any FF past the tenth one will not be needed.

Frequency Division

In Chapter 5, we saw that in the basic counter each FF provides an output waveform that is exactly *half* the frequency of the waveform at its *CLK* input. To illustrate, suppose that the clock signal in Figure 7-1 is 16 kHz. Figure 7-2 shows the FF output waveforms. The waveform at output A is an 8-kHz *square wave,* at output B it is 4 kHz, at output C it is 2 kHz, and at output D it is 1 kHz. Notice that the output of flip-flop D has a frequency equal to the original clock frequency divided by 16. In general,

> **In any counter, the signal at the output of the last FF (i.e., the MSB) will have a frequency equal to the input clock frequency divided by the MOD number of the counter.**

For example, in a MOD-16 counter, the output from the last FF will have a frequency of 1/16 of the input clock frequency. Thus, it can also be called a *divide-by-16 counter.* Likewise, a MOD-8 counter has an output frequency of $\frac{1}{8}$ the input frequency; it is a *divide-by-8 counter.*

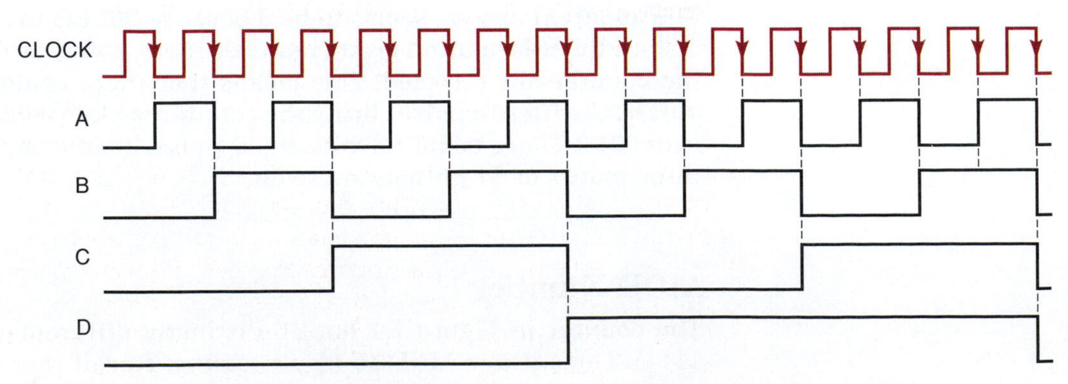

FIGURE 7-2 Counter waveforms showing frequency division by 2 for each FF.

EXAMPLE 7-3

The first step involved in building a digital clock is to take the 60-Hz signal and feed it into a Schmitt-trigger, pulse-shaping circuit* to produce a square wave, as illustrated in Figure 7-3. The 60-Hz square wave is then put into a MOD-60 counter, which is used to divide the 60-Hz frequency by exactly 60 to produce a 1-Hz waveform. This 1-Hz waveform is fed to a series of counters, which then count seconds, minutes, hours, and so on. How many FFs are required for the MOD-60 counter?

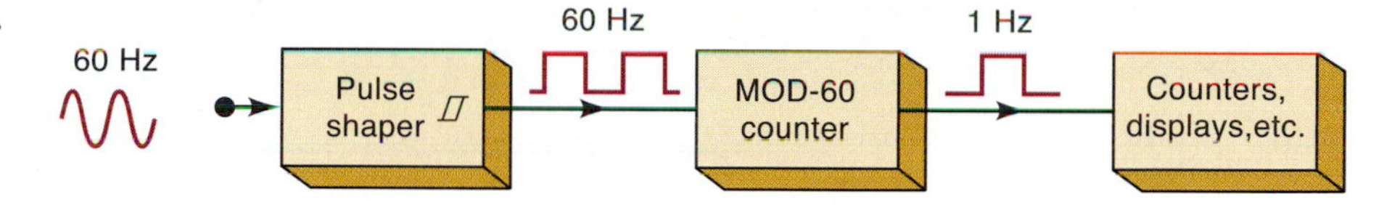

FIGURE 7-3 Example 7-3.

Solution

There is no integer power of 2 that will equal 60. The closest is $2^6 = 64$. Thus, a counter using six FFs would act as a MOD-64 counter. Obviously, this will not satisfy the requirement. It seems that there is no solution using a counter of the type shown in Figure 7-1. This is partly true; in Section 7-4, we will see how to modify basic binary counters so that almost *any* MOD number can be obtained and we will not be limited to values of 2^N.

REVIEW QUESTIONS

1. *True or false:* In an asynchronous counter, all FFs change states at the same time.
2. Assume that the counter in Figure 7-1 is holding the count 0101. What will be the count after 27 clock pulses?
3. What would be the MOD number of the counter if three more FFs were added?

7-2 PROPAGATION DELAY IN RIPPLE COUNTERS

Ripple counters are the simplest type of binary counters because they require the fewest components to produce a given counting operation. They do, however, have one major drawback, which is caused by their basic principle of operation: each FF is triggered by the transition at the output of the preceding FF. Because of the inherent propagation delay time (t_{pd}) of each FF, this means that the second FF will not respond until a time t_{pd} after the first FF receives an active clock transition; the third FF will not respond until a time equal to $2 \times t_{pd}$ after that clock transition; and so on. In other words, the propagation delays of the FFs accumulate so that the Nth FF cannot change states until a time equal to $N \times t_{pd}$ after the clock transition occurs. This is illustrated in Figure 7-4, where the waveforms for a three-bit ripple counter are shown.

The first set of waveforms in Figure 7-4(a) shows a situation where an input pulse occurs every 1000 ns (the clock period T = 1000 ns) and it is assumed that each FF has a propagation delay of 50 ns (t_{pd} = 50 ns). Notice

*See Section 5-21.

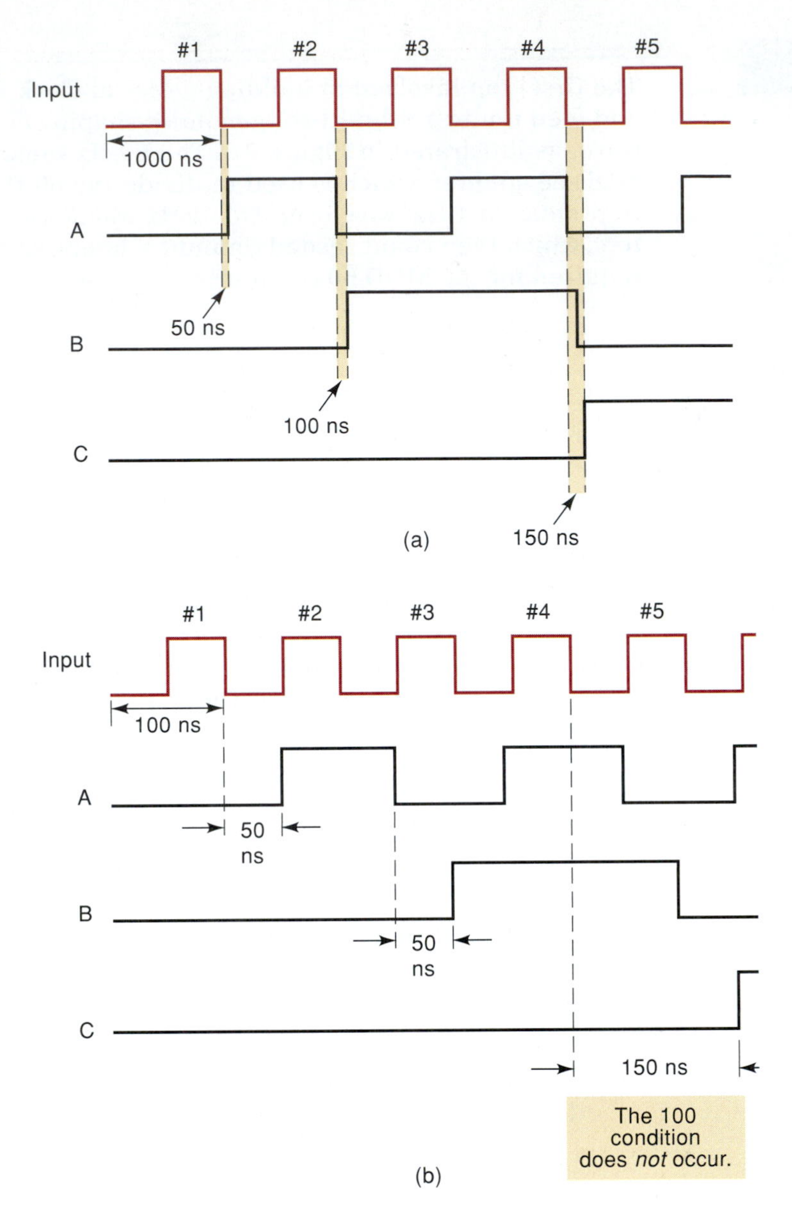

FIGURE 7-4 Waveforms of a three-bit ripple counter illustrating the effects of FF propagation delays for different input pulse frequencies.

that the A flip-flop output toggles 50 ns after the NGT of each input pulse. Similarly, B toggles 50 ns after A goes from 1 to 0, and C toggles 50 ns after B goes from 1 to 0. As a result, when the fourth input NGT occurs, the C output goes HIGH after a delay of 150 ns. In this situation, the counter does operate properly in the sense that the FFs do eventually get to their correct states, representing the binary count. However, the situation worsens if the input pulses are applied at a much higher frequency.

The waveforms in Figure 7-4(b) show what happens if the input pulses occur once every 100 ns. Again, each FF output responds 50 ns after the 1-to-0 transition at its *CLK* input (note the change in the relative time scale). Of particular interest is the situation after the falling edge of the *fourth* input pulse, where the C output does not go HIGH until 150 ns later, which is the same time that the A output goes HIGH in response to the *fifth* input pulse. In other words, the condition $C = 1, B = A = 0$ (count of 100) never appears because the input frequency is too high. This could cause a serious problem if this condition were supposed to be used to control some other operation in a digital system. Problems such as this can be avoided if the period between input pulses is made longer than the total propagation delay of the counter. That is, for proper counter operation we need

$$T_{\text{clock}} \geq N \times t_{\text{pd}} \qquad \textbf{(7-2)}$$

where N = the number of FFs. Stated in terms of input-clock frequency, the maximum frequency that can be used is given by

$$f_{\text{max}} = \frac{1}{N \times t_{\text{pd}}} \qquad \textbf{(7-3)}$$

For example, suppose that a four-bit ripple counter is constructed using the 74LS112 J-K flip-flop. Table 5-2 shows that the 74LS112 has $t_{PLH} = 16$ ns and $t_{PHL} = 24$ ns as the propagation delays from *CLK* to *Q*. To calculate f_{max}, we will assume the "worst case"; that is, we will use $t_{\text{pd}} = t_{PHL} = 24$ ns, so that

$$f_{\text{max}} = \frac{1}{4 \times 24 \text{ ns}} = 10.4 \text{ MHz}$$

Clearly, as the number of FFs in the counter increases, the total propagation delay increases and f_{max} decreases. For example, a ripple counter that uses six 74LS112 FFs will have

$$f_{\text{max}} = \frac{1}{6 \times 24 \text{ ns}} = 6.9 \text{ MHz}$$

Thus, asynchronous counters are not useful at very high frequencies, especially for counters with large numbers of bits. Another problem caused by propagation delays in asynchronous counters occurs when we try to electronically detect (*decode*) the counter's output states. If you look closely at Figure 7-4(a), for a short period of time (50 ns in our example) right after state 011, you see that state 010 occurs before 100. This is obviously not the correct binary counting sequence, and while the human eye is much too slow to see this temporary state, our digital circuits will be fast enough to detect it. These erroneous count patterns can generate what are called **glitches** in the signals that are produced by digital systems using asynchronous counters. In spite of their simplicity, these problems limit the usefulness of asynchronous counters in digital applications.

REVIEW QUESTIONS

1. Explain why a ripple counter's maximum frequency limitation decreases as more FFs are added to the counter.
2. A certain J-K flip-flop has $t_{\text{pd}} = 12$ ns. What is the largest MOD counter that can be constructed from these FFs and still operate up to 10 MHz?

7-3 SYNCHRONOUS (PARALLEL) COUNTERS

The problems encountered with ripple counters are caused by the accumulated FF propagation delays; stated another way, the FFs do not all change states simultaneously in synchronism with the input pulses. These limitations can be overcome with the use of **synchronous** or **parallel counters** in which all of the FFs are triggered simultaneously (in parallel) by the clock input pulses.

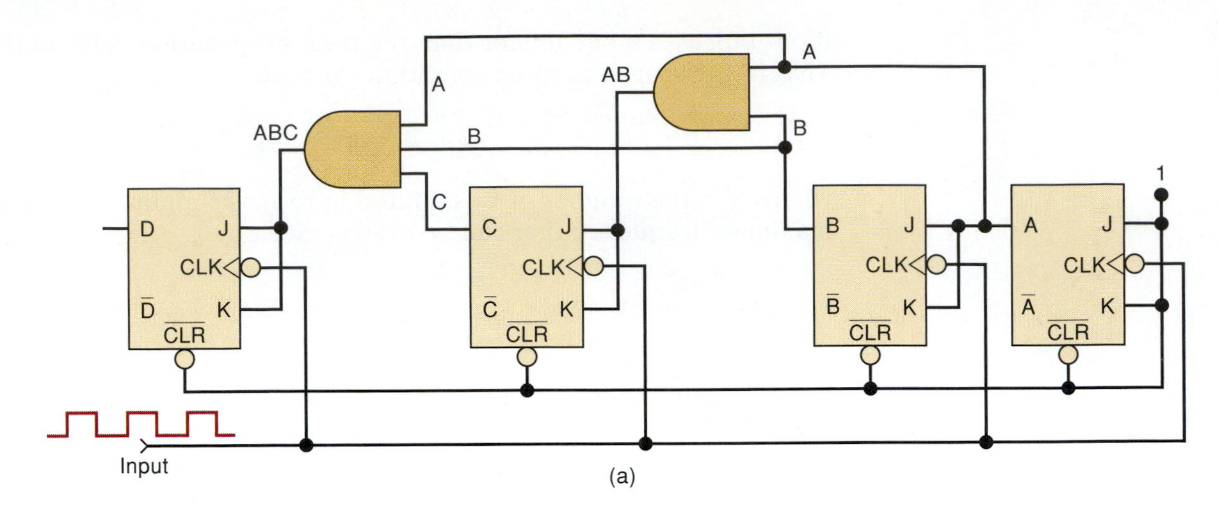

(a)

Count	D	C	B	A
0	0	0	0	0
1	0	0	0	1
2	0	0	1	0
3	0	0	1	1
4	0	1	0	0
5	0	1	0	1
6	0	1	1	0
7	0	1	1	1
8	1	0	0	0
9	1	0	0	1
10	1	0	1	0
11	1	0	1	1
12	1	1	0	0
13	1	1	0	1
14	1	1	1	0
15	1	1	1	1
0	0	0	0	0
.	.	.	.	.
.	.	.	.	.
.	.	etc.	.	.

(b)

FIGURE 7-5 Synchronous MOD-16 counter. Each FF is clocked by the NGT of the clock input signal so that all FF transitions occur at the same time.

Because the input pulses are applied to all the FFs, some means must be used to control when an FF is to toggle and when it is to remain unaffected by a clock pulse. This is accomplished by using the *J* and *K* inputs and is illustrated in Figure 7-5 for a four-bit, MOD-16 synchronous counter.

If we compare the circuit arrangement for this synchronous counter with its asynchronous counterpart in Figure 7-1, we can see the following notable differences:

- The *CLK* inputs of all of the FFs are connected together so that the input clock signal is applied to each FF simultaneously.
- Only flip-flop *A*, the LSB, has its *J* and *K* inputs permanently at the HIGH level. The *J*, *K* inputs of the other FFs are driven by some combination of FF outputs.
- The synchronous counter requires more circuitry than does the asynchronous counter.

Circuit Operation

For this circuit to count properly, on a given NGT of the clock, only those FFs that are supposed to toggle on that NGT should have $J = K = 1$ when that NGT occurs. Let's look at the counting sequence in Figure 7-5(b) to see what this means for each FF.

The counting sequence shows that the A flip-flop must change states at each NGT. For this reason, its J and K inputs are permanently HIGH so that it will toggle on each NGT of the clock input.

The counting sequence shows that flip-flop B must change states on each NGT that occurs while $A = 1$. For example, when the count is 0001, the next NGT must toggle B to the 1 state; when the count is 0011, the next NGT must toggle B to the 0 state; and so on. This operation is accomplished by connecting output A to the J and K inputs of flip-flop B so that $J = K = 1$ only when $A = 1$.

The counting sequence shows that flip-flop C must change states on each NGT that occurs while $A = B = 1$. For example, when the count is 0011, the next NGT must toggle C to the 1 state; when the count is 0111, the next NGT must toggle C to the 0 state; and so on. By connecting the logic signal AB to FF C's J and K inputs, this FF will toggle only when $A = B = 1$.

In a like manner, we can see that flip-flop D must toggle on each NGT that occurs while $A = B = C = 1$. When the count is 0111, the next NGT must toggle D to the 1 state; when the count is 1111, the next NGT must toggle D to the 0 state. By connecting the logic signal ABC to FF D's J and K inputs, this FF will toggle only when $A = B = C = 1$.

The basic principle for constructing a synchronous counter can therefore be stated as follows:

> **Each FF should have its J and K inputs connected so that they are HIGH only when the outputs of all lower-order FFs are in the HIGH state.**

Advantage of Synchronous Counters over Asynchronous

In a parallel counter, all of the FFs will change states simultaneously; that is, they are all synchronized to the NGTs of the input clock pulses. Thus, unlike the asynchronous counters, the propagation delays of the FFs do not add together to produce the overall delay. Instead, the total response time of a synchronous counter like the one in Figure 7-5 is the time it takes *one* FF to toggle plus the time for the new logic levels to propagate through a *single* AND gate to reach the J, K inputs. That is, for a synchronous counter,

$$\text{total delay} = \text{FF } t_{pd} + \text{AND gate } t_{pd}$$

This total delay is the same no matter how many FFs are in the counter, and it will generally be much lower than with an asynchronous counter with the same number of FFs. Thus, a synchronous counter can operate at a much higher input frequency. Of course, the circuitry of the synchronous counter is more complex than that of the asynchronous counter.

Actual ICs

There are many synchronous IC counters in both the TTL and the CMOS logic families. Some of the most commonly used devices are:

- 74ALS160/162, 74HC160/162: synchronous decade counters
- 74ALS161/163, 74HC161/163: synchronous MOD-16 counters

EXAMPLE 7-4

(a) Determine f_{max} for the counter of Figure 7-5(a) if t_{pd} for each FF is 50 ns and t_{pd} for each AND gate is 20 ns. Compare this value with f_{max} for a MOD-16 ripple counter.

(b) What must be done to convert this counter to MOD-32?

(c) Determine f_{max} for the MOD-32 parallel counter.

Solution

(a) The total delay that must be allowed between input clock pulses is equal to FF t_{pd} + AND gate t_{pd}. Thus, $T_{clock} \geq 50 + 20 = 70$ ns, and so the parallel counter has

$$f_{max} = \frac{1}{70 \text{ ns}} = 14.3 \text{ MHz (parallel counter)}$$

A MOD-16 ripple counter uses four FFs with $t_{pd} = 50$ ns. Thus, f_{max} for the ripple counter is

$$f_{max} = \frac{1}{4 \times 50 \text{ ns}} = 5 \text{ MHz (ripple counter)}$$

(b) A fifth FF must be added because $2^5 = 32$. The *CLK* input of this FF is also tied to the input pulses. Its *J* and *K* inputs are fed by the output of a four-input AND gate whose inputs are *A*, *B*, *C*, and *D*.

(c) f_{max} is still determined as in (a) regardless of the number of FFs in the parallel counter. Thus, f_{max} is still 14.3 MHz.

REVIEW QUESTIONS

1. What is the advantage of a synchronous counter over an asynchronous counter? What is the disadvantage?
2. How many logic devices are required for a MOD-64 parallel counter?
3. What logic signal drives the *J*, *K* inputs of the MSB flip-flop for the counter of question 2?

7-4 COUNTERS WITH MOD NUMBERS $< 2^N$

The basic synchronous counter of Figure 7-5 is limited to MOD numbers that are equal to 2^N, where *N* is the number of FFs. This value is actually the maximum MOD number that can be obtained using *N* flip-flops. The basic counter can be modified to produce MOD numbers less than 2^N by allowing the counter to *skip states* that are normally part of the counting sequence. One of the most common methods for doing this is illustrated in Figure 7-6, where a three-bit counter is shown. Disregarding the NAND gate for a moment, we can see that the counter is a MOD-8 binary counter that will count in sequence from 000 to 111. However, the presence of the NAND gate will alter this sequence as follows:

1. The NAND output is connected to the asynchronous CLEAR inputs of each FF. As long as the NAND output is HIGH, it will have no effect on the counter. When it goes LOW, however, it will clear all of the FFs so that the counter immediately goes to the 000 state.

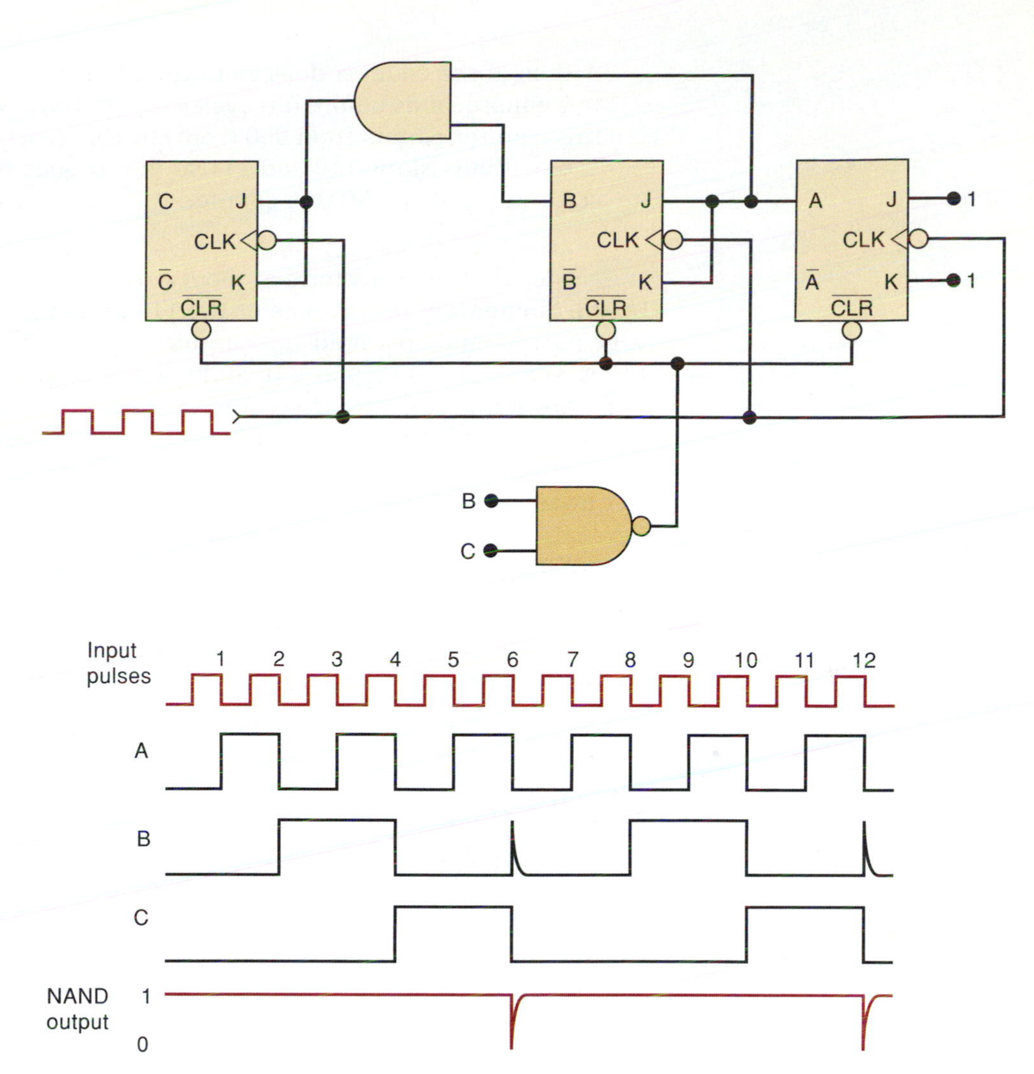

FIGURE 7-6 MOD-6 counter produced by clearing a MOD-8 counter when a count of six (110) occurs.

2. The inputs to the NAND gate are the outputs of the B and C flip-flops, and so the NAND output will go LOW whenever $B = C = 1$. This condition will occur when the counter goes from the 101 state to the 110 state on the NGT of input pulse 6. The LOW at the NAND output will immediately (generally within a few nanoseconds) clear the counter to the 000 state. Once the FFs have been cleared, the NAND output goes back HIGH because the $B = C = 1$ condition no longer exists.
3. The counting sequence is, therefore,

CBA
000 ←
001
010
011
100
101
110 → (temporary state needed to clear counter)

Although the counter does go to the 110 state, it remains there for only a few nanoseconds before it recycles to 000. Thus, we can essentially say that this counter counts from 000 (zero) to 101 (five) and then recycles to 000. It essentially skips 110 and 111 so that it goes through only six different states; thus, it is a MOD-6 counter.

Notice that the waveform at the B output contains a *spike* or *glitch* caused by the momentary occurrence of the 110 state before clearing. This glitch is very narrow and so would not produce any visible indication on indicator LEDs or numerical displays. It could, however, cause a problem if the B output were being used to drive other circuitry outside the counter. It should also be noted that the C output has a frequency equal to one-sixth of the input frequency; in other words, this MOD-6 counter has divided the input frequency by *six*. The waveform at C is *not* a symmetrical square wave (50 percent duty cycle) because it is HIGH for only two clock cycles, whereas it is LOW for four cycles.

State Transition Diagram

Figure 7-7(a) is the state transition diagram for the MOD-6 counter of Figure 7-6, showing how FFs C, B, and A change states as pulses are applied to the CLK input of flip-flop A. Recall that each circle represents one of the possible counter states and that the arrows indicate how one state changes to another in response to an input clock pulse.

If we assume a starting count of 000, the diagram shows that the states of the counter change normally up until the count of 101. When the next clock pulse occurs, the counter temporarily goes to the 110 count before going to the stable 000 count. The dotted lines indicate the temporary nature of the 110 state. As stated earlier, the duration of this temporary state is so short that for most purposes we can consider that the counter goes directly from 101 to 000 (solid arrow).

Note that there is no arrow into the 111 state because the counter can never advance to that state. However, the 111 state can occur on power-up when the FFs come up in random states. If that happens, the 111 condition will produce a LOW at the NAND gate output and immediately clear the counter to 000. Thus, the 111 state is also a temporary condition that ends up at 000.

Displaying Counter States

Sometimes during normal operation, and very often during testing, it is necessary to have a visible display of how a counter is changing states in response to the input pulses. We will take a detailed look at several ways of doing this later in the text. For now, Figure 7-7(b) shows one of the simplest methods using individual indicator LEDs for each FF output. Each FF output is connected to an INVERTER whose output provides the current path for the LED. For example, when output A is HIGH, the INVERTER output goes LOW and the LED turns ON. An LED that is turned on indicates $A = 1$. When output A is LOW, the INVERTER output is HIGH and the LED turns OFF. When the LED is turned off, it indicates $A = 0$.

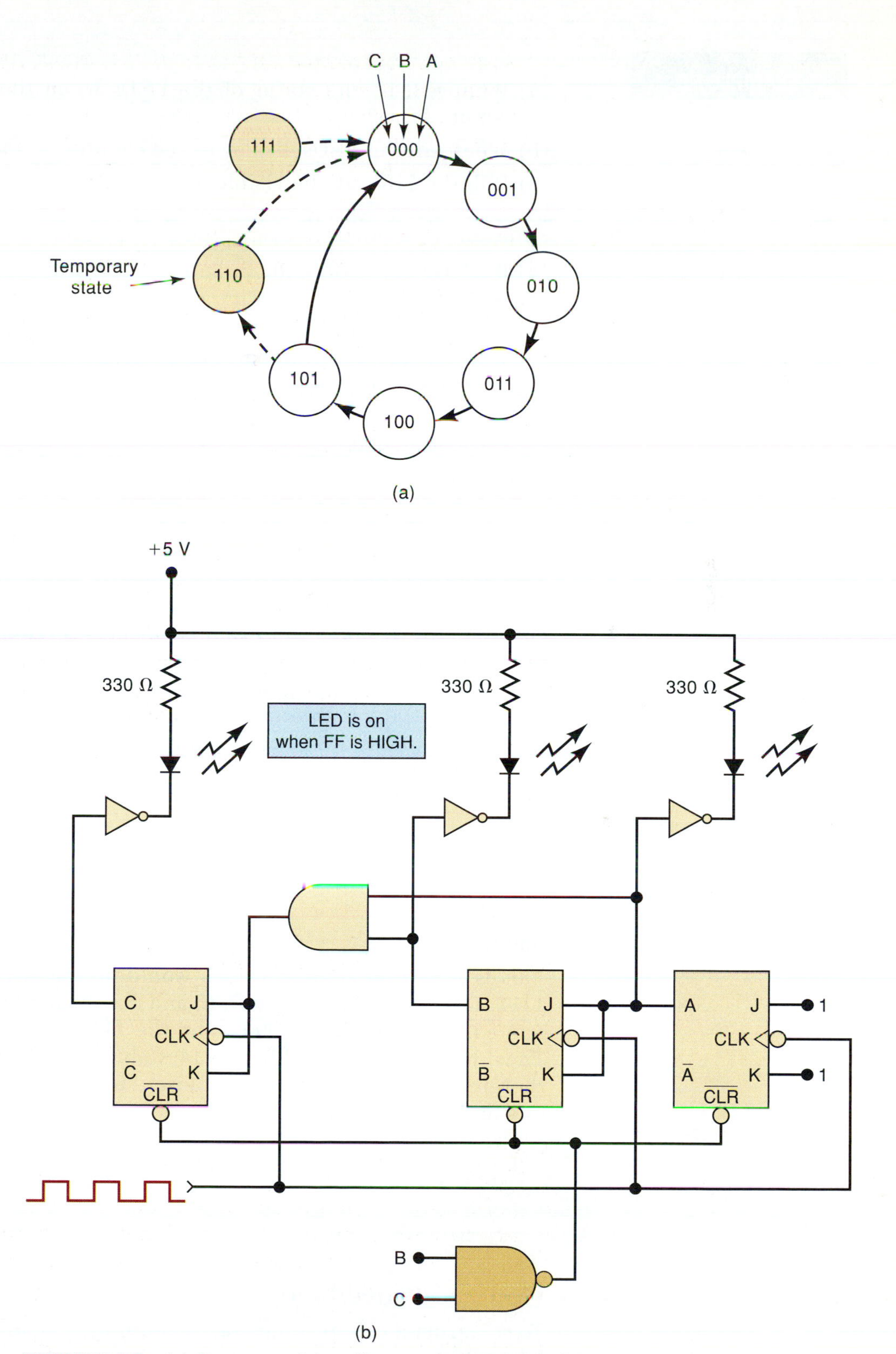

FIGURE 7-7 (a) State transition diagram for the MOD-6 counter of Figure 7-6. (b) LEDs are often used to display the states of a counter.

EXAMPLE 7-5

(a) What will be the status of the LEDs when the counter is holding the count of five?
(b) What will the LEDs display as the counter is clocked by a 1-kHz input?
(c) Will the 110 state be visible on the LEDs?

Solution

(a) Because $5_{10} = 101_2$, the 2^0 and 2^2 LEDs will be ON, and the 2^1 LED will be OFF.
(b) At 1 kHz, the LEDs will be switching ON and OFF so rapidly that they will appear to the human eye to be ON all the time at about half the normal brightness.
(c) No; the 110 state will persist for only a few nanoseconds as the counter recycles to 000.

Changing the MOD Number

The counter of Figures 7-6 and 7-7 is a MOD-6 counter because of the choice of inputs to the NAND gate. Any desired MOD number can be obtained by changing these inputs. For example, using a three-input NAND gate with inputs A, B, and C, the counter would function normally until the 111 condition was reached, at which point it would immediately reset to the 000 state. Ignoring the very temporary excursion into the 111 state, the counter would go from 000 through 110 and then recycle back to 000, resulting in a MOD-7 counter (seven states).

EXAMPLE 7-6

Determine the MOD number of the counter in Figure 7-8(a). Also determine the frequency at the D output.

Solution

This is a four-bit counter, which would normally count from 0000 through 1111. The NAND inputs are D, C, and B, which means that the counter will immediately recycle to 0000 when the 1110 (decimal 14) count is reached. Thus, the counter actually has 14 stable states 0000 through 1101 and is therefore a *MOD-14* counter. Because the input frequency is 30 kHz, the frequency at output D will be

$$\frac{30\text{ kHz}}{14} = 2.14\text{ kHz}$$

General Procedure

To construct a counter that starts counting from all 0s and has a MOD number of X:

1. Find the smallest number of FFs such that $2^N \geq X$, and connect them as a counter. If $2^N = X$, do not do steps 2 and 3.
2. Connect a NAND gate to the asynchronous CLEAR inputs of all the FFs.
3. Determine which FFs will be in the HIGH state at a count $= X$; then connect the normal outputs of these FFs to the NAND gate inputs.

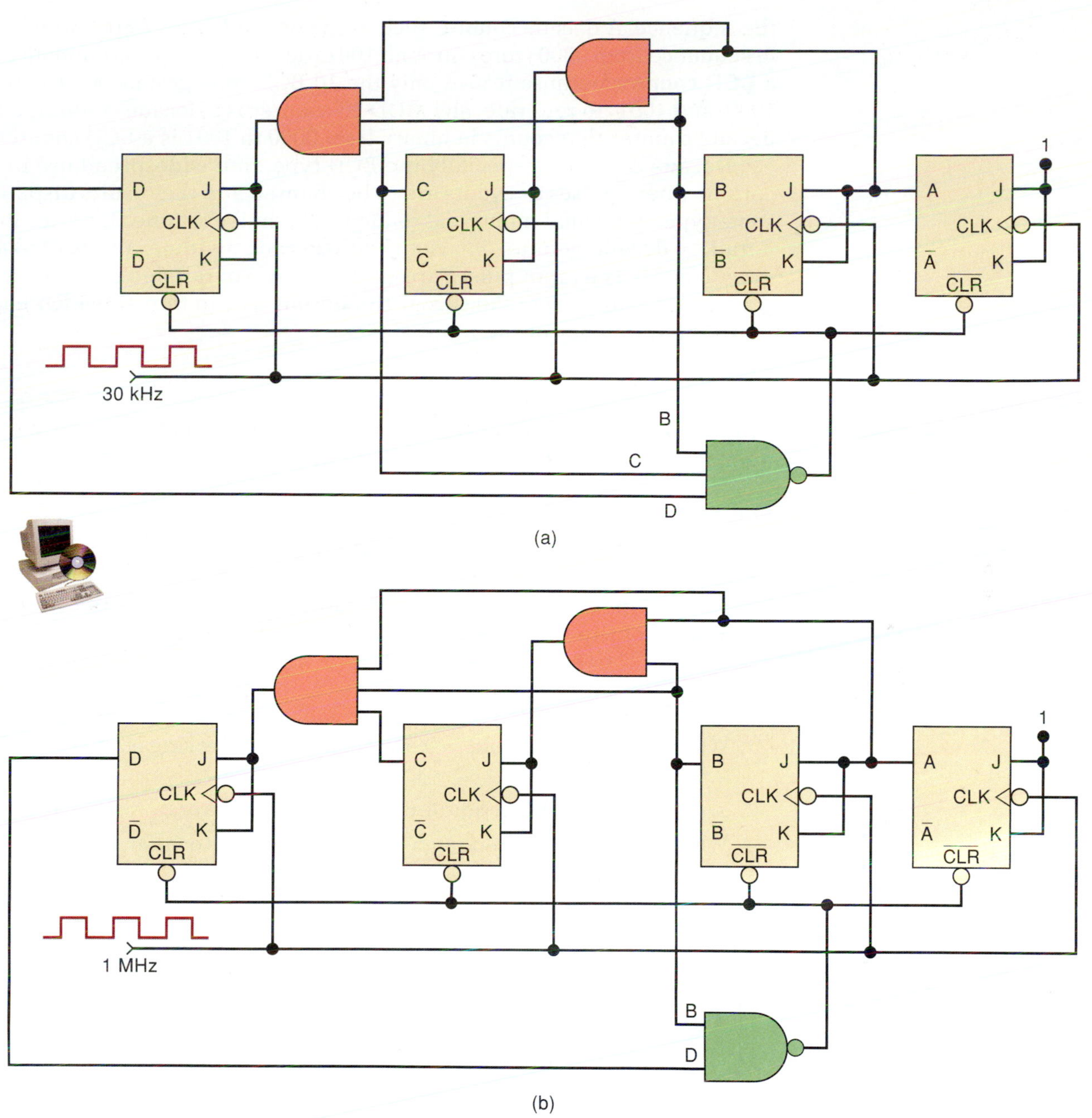

FIGURE 7-8 (a) MOD-14 ripple counter; (b) MOD-10 (decade) ripple counter.

EXAMPLE 7-7

Construct a MOD-10 counter that will count from 0000 (zero) through 1001 (decimal 9).

Solution

$2^3 = 8$ and $2^4 = 16$; thus, four FFs are required. Because the counter is to have stable operation up to the count of 1001, it must be reset to zero when the count of 1010 is reached. Therefore, FF outputs *D* and *B* must be connected as the NAND gate inputs. Figure 7-8(b) shows the arrangement.

Decade Counters/BCD Counters

The MOD-10 counter of Example 7-7 is also referred to as a **decade counter**. In fact, a decade counter is any counter that has 10 distinct states, no matter what the sequence. A decade counter such as the one in Figure 7-8(b), which counts in sequence from 0000 (zero) through 1001 (decimal 9), is also commonly called a **BCD counter** because it uses only the 10 BCD code groups 0000, 0001, . . . , 1000, and 1001. To reiterate, any MOD-10 counter is a decade counter; and any decade counter that counts in binary from 0000 to 1001 is a BCD counter.

Decade counters, especially the BCD type, find widespread use in applications where pulses or events are to be counted and the results displayed on some type of decimal numerical readout. We shall examine this later in more detail. A decade counter is also often used for dividing a pulse frequency *exactly* by 10. The input pulses are applied to the paralleled clock inputs, and the output pulses are taken from the output of flip-flop *D*, which has one-tenth the frequency of the input signal.

EXAMPLE 7-8

In Example 7-3, a MOD-60 counter was needed to divide the 60-Hz line frequency down to 1 Hz. Construct an appropriate MOD-60 counter.

Solution

$2^5 = 32$ and $2^6 = 64$, and so we need six FFs, as shown in Figure 7-9. The counter is to be cleared when it reaches the count of 60 (111100). Thus, the outputs of flip-flops Q_5, Q_4, Q_3, and Q_2 must be connected to the NAND gate. The output of flip-flop Q_5 will have a frequency of 1 Hz.

REVIEW QUESTIONS

1. What FF outputs should be connected to the clearing NAND gate to form a MOD-13 counter?
2. *True or false:* All BCD counters are decade counters.
3. What is the output frequency of a decade counter that is clocked from a 50-kHz signal?

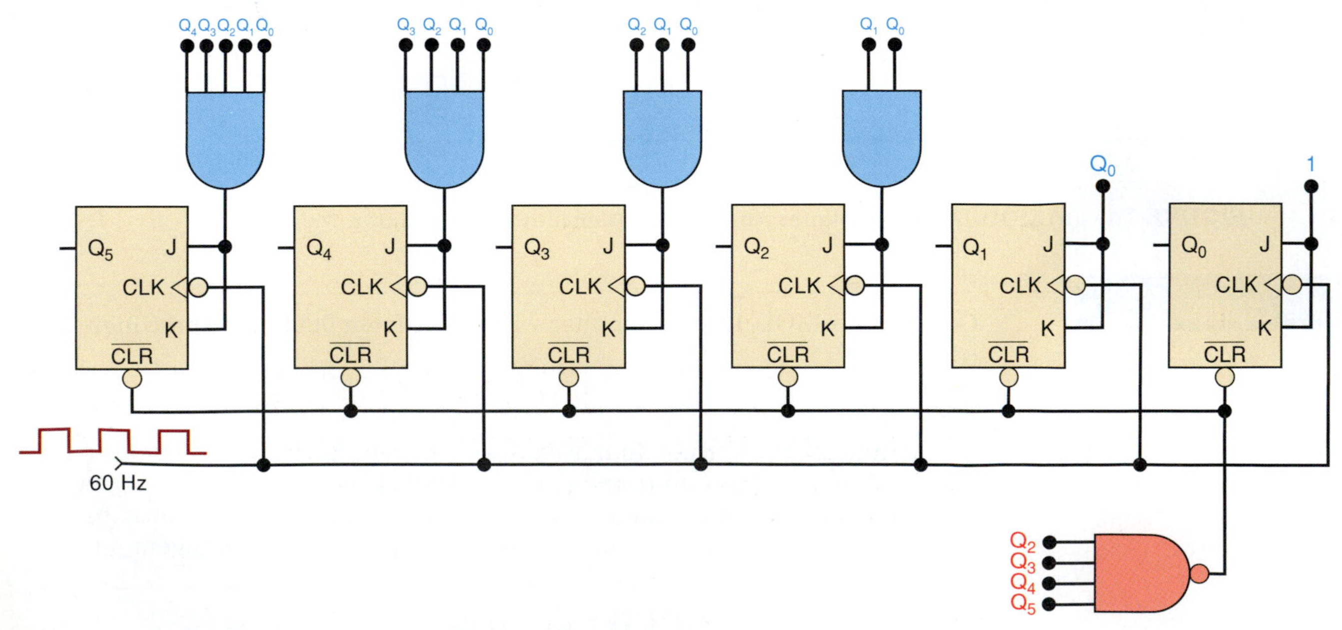

FIGURE 7-9 MOD-60 counter.

7-5 SYNCHRONOUS DOWN AND UP/DOWN COUNTERS

In Section 7-3, we saw that using the output of lower-order FFs to control the toggling of each FF creates a synchronous **up counter**. A synchronous **down counter** is constructed in a similar manner except that we use the inverted FF outputs to control the higher-order *J*, *K* inputs. Comparing the synchronous, MOD-16, down counter in Figure 7-10 with the up counter in Figure 7-5 shows that we need only to substitute the corresponding inverted FF output in place of the *A*, *B*, and *C* outputs. For a down count sequence, the LSB FF (A) still needs to toggle with each NGT of the clock input signal. Flip-flop B must change states on the next NGT of the clock when $A = 0$ ($\overline{A} = 1$). Flip-flop C changes states when $A = B = 0$ ($\overline{A} \cdot \overline{B} = 1$), and flip-flop *D* changes states when $A = B = C = 0$ ($\overline{A} \cdot \overline{B} \cdot \overline{C} = 1$). This circuit configuration will produce the count sequence: 15, 14, 13, 12, . . . , 3, 2, 1, 0, 15, 14, and so on, as shown in the timing diagram.

Figure 7-11(a) shows how to form a parallel **up/down counter**. The control input Up/$\overline{\text{Down}}$ controls whether the normal FF outputs or the inverted FF outputs are fed to the *J* and *K* inputs of the successive FFs. When Up/$\overline{\text{Down}}$ is held HIGH, AND gates 1 and 2 are enabled while AND gates 3 and 4 are disabled (note the inverter). This allows the *A* and *B* outputs through gates 1 and 2 to the *J* and *K* inputs of FFs *B* and *C*. When Up/$\overline{\text{Down}}$ is held LOW, AND gates 1 and 2 are disabled while AND gates 3 and 4 are enabled. This allows the inverted *A* and *B* outputs through gates 3 and 4 into the *J* and *K* inputs of FFs *B* and *C*. The waveforms in Figure 7-11(b) illustrate the operation. Notice that for the first five clock pulses, Up/$\overline{\text{Down}}$ = 1 and the counter counts up; for the last five pulses, Up/$\overline{\text{Down}}$ = 0, and the counter counts down.

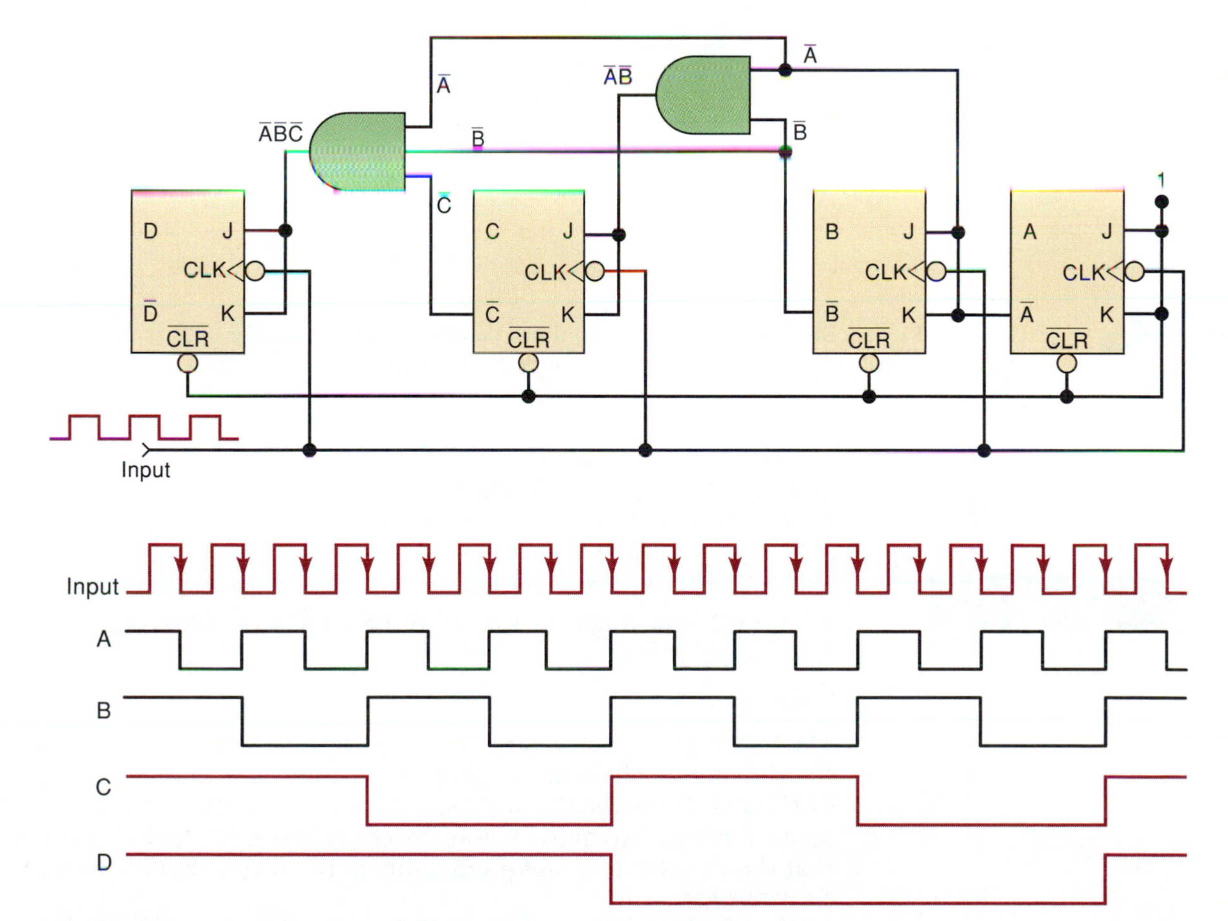

FIGURE 7-10 Synchronous, MOD-16, down counter and output waveforms.

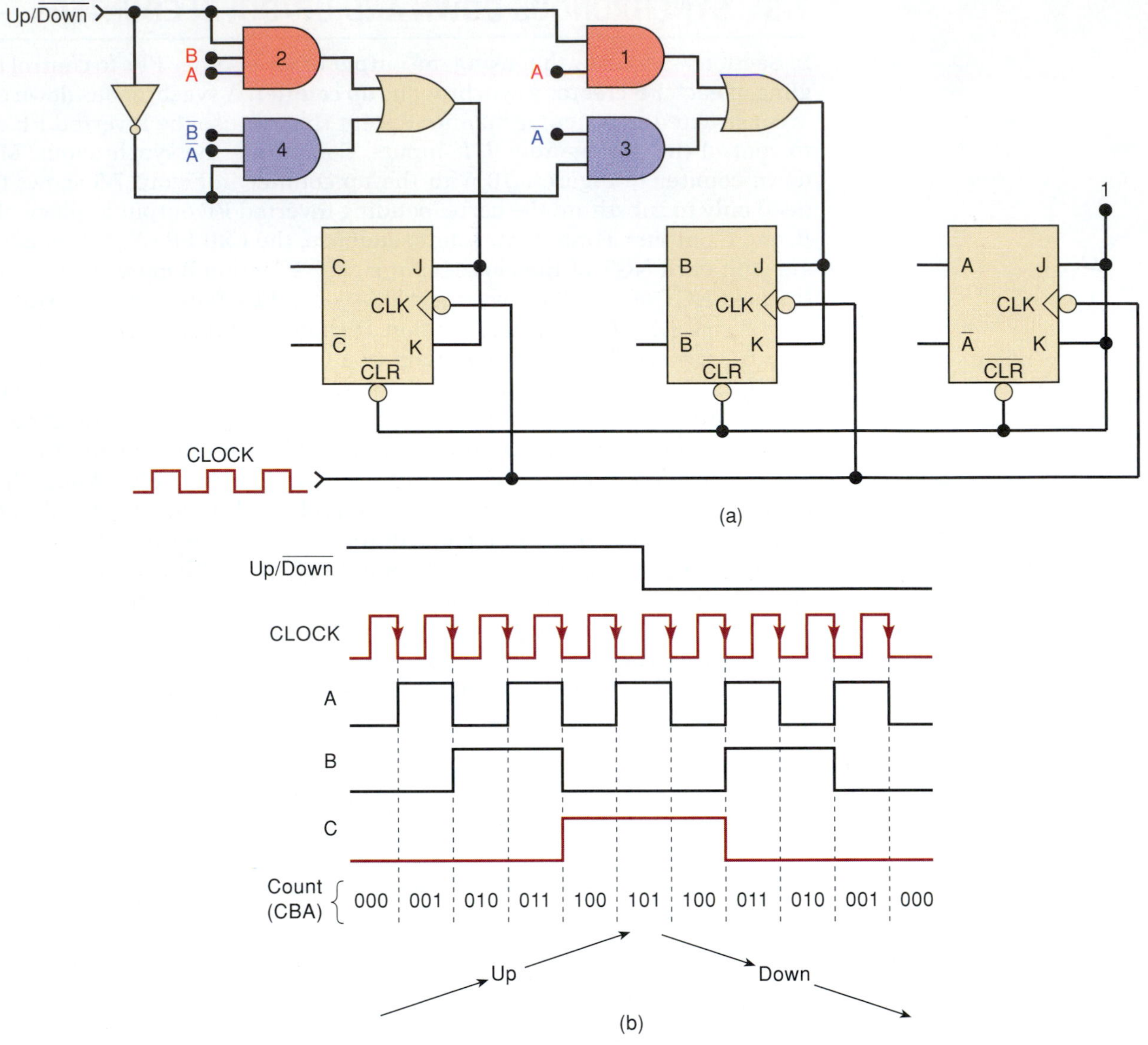

FIGURE 7-11 (a) MOD-8 synchronous up/down counter. (b) The counter counts up when the control input Up/$\overline{\text{Down}}$ = 1; it counts down when the control input Up/$\overline{\text{Down}}$ = 0.

The nomenclature used for the control signal (Up/$\overline{\text{Down}}$) was chosen to make it clear how it affects the counter. The count-up operation is active-HIGH; the count-down operation is active-LOW.

EXAMPLE 7-9

What problems might be caused if the Up/$\overline{\text{Down}}$ signal changes levels on the NGT of the clock?

Solution

The FFs might operate unpredictably because some of them would have their *J* and *K* inputs changing at about the same time that a NGT occurs at their *CLK* input. However, the effects of the change in the control signal must propagate through two gates before reaching the *J*, *K* inputs, so it is more likely that the FFs will respond predictably to the levels that are at *J*, *K* prior to the NGT of *CLK*.

REVIEW QUESTIONS

1. What is the difference between the counting sequence of an up counter and a down counter?
2. What circuit changes will convert a synchronous, binary up counter into a binary down counter?

7-6 PRESETTABLE COUNTERS

Many synchronous (parallel) counters that are available as ICs are designed to be **presettable**; in other words, they can be preset to any desired starting count either asynchronously (independent of the clock signal) or synchronously (on the active transition of the clock signal). This presetting operation is also referred to as **parallel loading** the counter.

Figure 7-12 shows the logic circuit for a three-bit presettable parallel up counter. The *J*, *K*, and *CLK* inputs are wired for operation as a parallel up counter. The asynchronous PRESET and CLEAR inputs are wired to perform asynchronous presetting. The counter is loaded with any desired count at any time by doing the following:

1. Apply the desired count to the parallel data inputs, P_2, P_1, and P_0.
2. Apply a LOW pulse to the PARALLEL LOAD input, $\overline{PL}$.

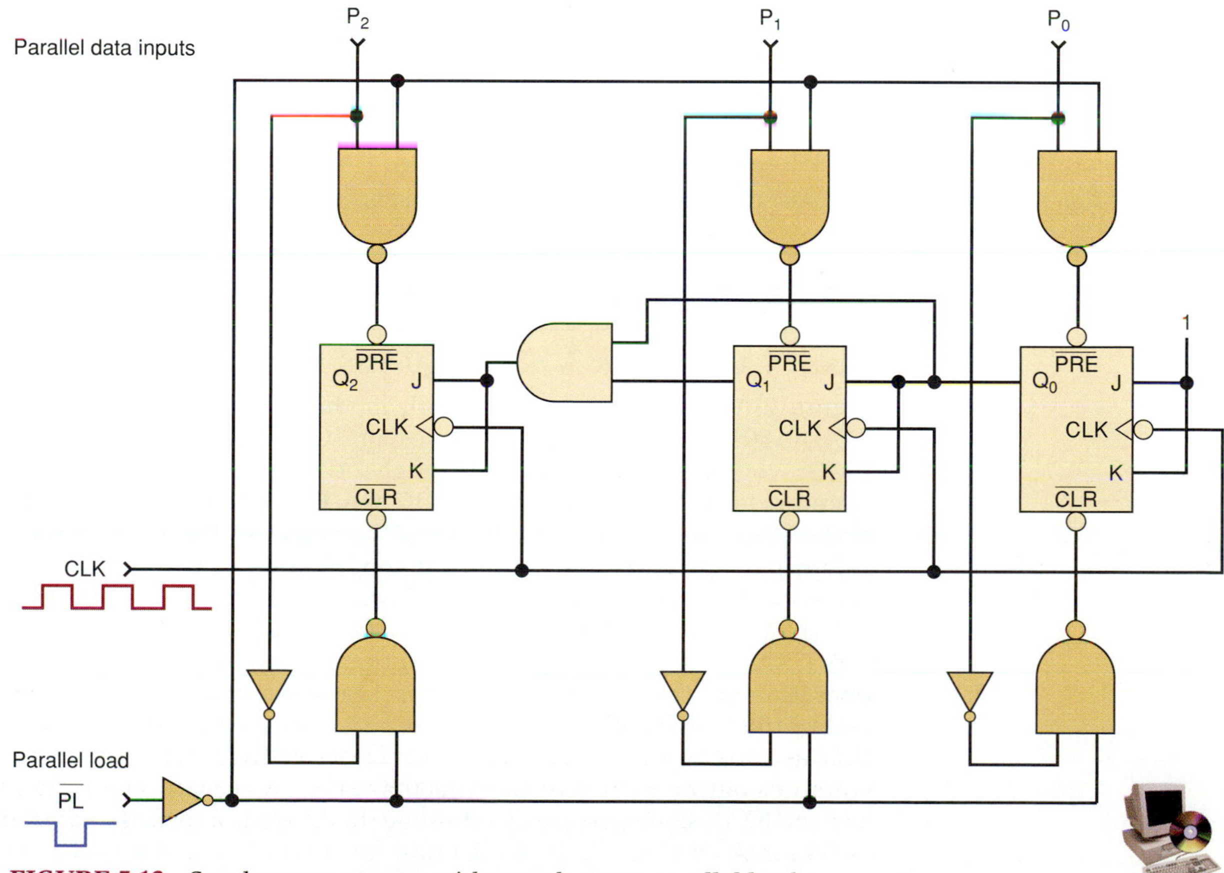

FIGURE 7-12 Synchronous counter with asynchronous parallel load.

This procedure will perform an asynchronous transfer of the P_2, P_1, and P_0 levels into flip-flops Q_2, Q_1, and Q_0, respectively (Section 5-17). This *jam transfer* occurs independently of the *J*, *K*, and *CLK* inputs. The effect of the *CLK* input will be disabled as long as $\overline{PL}$ is in its active-LOW state because each FF will have one of its asynchronous inputs activated while $\overline{PL} = 0$. Once $\overline{PL}$ returns HIGH, the FFs can respond to their *CLK* inputs and can resume the counting-up operation starting from the count that was loaded into the counter.

For example, let's say that $P_2 = 1$, $P_1 = 0$, and $P_0 = 1$. While $\overline{PL}$ is HIGH, these parallel data inputs have no effect. If clock pulses are present, the counter will perform the normal count-up operation. Now let's say that $\overline{PL}$ is pulsed LOW when the counter is at the 010 count (i.e., $Q_2 = 0$, $Q_1 = 1$, and $Q_0 = 0$). This LOW at $\overline{PL}$ will produce LOWs at the $\overline{CLR}$ input of Q_1 and at the $\overline{PRE}$ inputs of Q_2 and Q_0 so that the counter will go to the 101 count *regardless of what is occurring at the CLK input.* The count will hold at 101 until $\overline{PL}$ is deactivated (returned HIGH); at that time the counter will resume counting up at each clock pulse from the count of 101.

This asynchronous presetting is used by several IC counters, such as the TTL 74ALS190, 74ALS191, 74ALS192, and 74ALS193 and the CMOS equivalents, 74HC190, 74HC191, 74HC192, and 74HC193.

Synchronous Presetting

Many IC parallel counters use *synchronous presetting* whereby the counter is preset on the active transition of the same clock signal that is used for counting. The logic level on the parallel load control input determines if the counter is preset with the applied input data at the next active clock transition.

Examples of IC counters that use synchronous presetting include the TTL 74ALS160, 74ALS161, 74ALS162, and 74ALS163 and their CMOS equivalents, 74HC160, 74HC161, 74HC162, and 74HC163.

REVIEW QUESTIONS

1. What is meant when we say that a counter is presettable?
2. Describe the difference between asynchronous and synchronous presetting.

7-7 IC SYNCHRONOUS COUNTERS

The 74ALS160-163/74HC160-163 Series

Figure 7-13 shows the logic symbol, modulus, and function table for the 74ALS160 through 74ALS163 series of IC counters (and the equivalent CMOS counterparts, 74HC160 through 74HC163). These recycling, four-bit counters have outputs labeled QD, QC, QB, QA, where QA is the LSB and QD is the MSB. They are clocked by a PGT applied to CLK. Each of the four different part numbers has a different combination of two feature variations. As seen in Figure 7-13(b), two of the counters are MOD-10 counters (74ALS160 and 74ALS162), while the other two are MOD-16 binary counters (74ALS161 and 74ALS163). The other variation for these parts is in the operation of the clear function [as highlighted in Figure 7-13(c)]. The 74ALS160 and 74ALS161 each has an asynchronous clear input. This means that as soon as $\overline{\text{CLR}}$ goes LOW ($\overline{\text{CLR}}$ is active-LOW for all four parts), the counter's output will be reset to 0000. On the other hand, the 74ALS162 and 74ALS163 IC counters are synchronously cleared. For these counters to be synchronously cleared, the $\overline{\text{CLR}}$ input must be LOW and a PGT must be applied to the clock input. The clear input has priority over all other functions

FIGURE 7-13 74ALS160-74ALS163 series synchronous counters: (a) logic symbol; (b) modules; (c) function table.

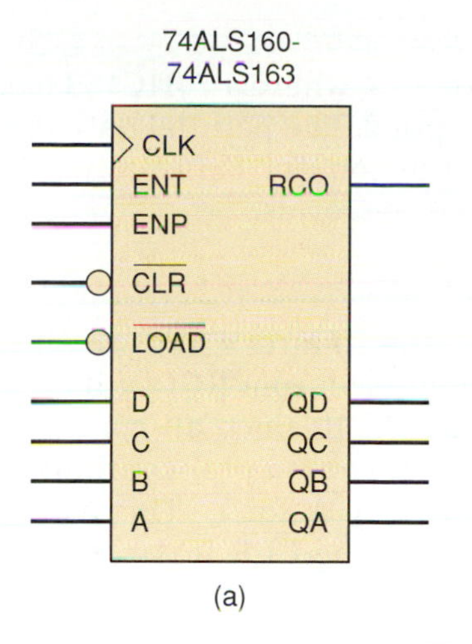

(a)

Part Number	Modulus
74ALS160	10
74ALS161	16
74ALS162	10
74ALS163	16

(b)

74ALS160-74ALS163 Function Table

$\overline{CLR}$	$\overline{LOAD}$	ENP	ENT	CLK	Function	Part Numbers
L	X	X	X	X	Asynch. Clear	74ALS160 & 74ALS161
L	X	X	X	↑	Synchr. Clear	74ALS162 & 74ALS163
H	L	X	X	↑	Synchr. Load	All
H	H	H	H	↑	Count up	All
H	H	L	X	X	No change	All
H	H	X	L	X	No change	All

(c)

for this series of IC counters. Clear will override all other control inputs, as indicated by the Xs in the Figure 7-13(c) function table.

The second priority function available in this series of IC counters is the parallel loading of data into the counter's flip-flops. To preset a data value, make the clear input inactive (HIGH), apply the desired four-bit value to the data input pins D, C, B, A (A is LSB and D is MSB), apply a LOW to the $\overline{LOAD}$ input control, and then clock the chip with a PGT. The load function is therefore synchronous and has priority over counting, so it does not matter what logic levels are applied to ENT or ENP. To count from the preset state it will be necessary to disable the load (with a HIGH) and enable the count function. If the load function is inactive, it does not matter what is applied to the data input pins.

To enable counting, the lowest-priority function, both $\overline{CLR}$ and $\overline{LOAD}$ control inputs must be inactive. Additionally, there are two active-HIGH count enable controls, ENT and ENP. ENT and ENP are essentially ANDed together to control the count function. If either or both of the **count enable** controls is inactive (LOW), the counter will hold the current state. Therefore, to increment the count with each PGT on CLK, all four of the control inputs must be HIGH. When counting, the decade counters (74ALS160 and 74ALS162) will automatically recycle to 0000 after state 1001 (9) and the binary counters (74ALS161 and 74ALS163) will automatically recycle after 1111 (15).

This series of IC counter chips has one more output pin, RCO. The function of this active-HIGH output is to detect (*decode*) the last or terminal state of the counter. The terminal state for a decade counter is 1001 (9), while the terminal state for a MOD-16 counter is 1111 (15). ENT, the primary count enable input, also controls the operation of RCO. ENT must be HIGH for the counter to indicate with the RCO output that it has reached its terminal state. You will see that this feature is very useful in connecting two or more counter chips together in a multistage arrangement to create larger counters.

EXAMPLE 7-10

Refer to Figure 7-14, where a 74HC163 has the input signals given in the timing diagram applied. The parallel data inputs are permanently connected as 1100. Assume the counter is initially in the 0000 state, and determine the counter output waveforms.

Solution

Initially (at t_0), the counter's FFs are all LOW. Since this is not the terminal state for the counter, output RCO will be LOW also. The first PGT on the CLK input occurs at t_1 and, since all control inputs are HIGH, the counter will increment to 0001. The counter continues to count up with each PGT until t_2. The $\overline{CLR}$ input is LOW for t_2. This will synchronously reset the counter to 0000 at t_2. After t_2, the $\overline{CLR}$ input goes inactive (HIGH) so the counter will

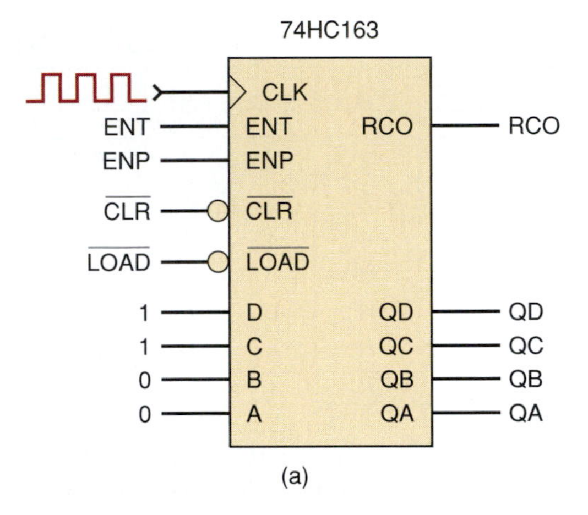

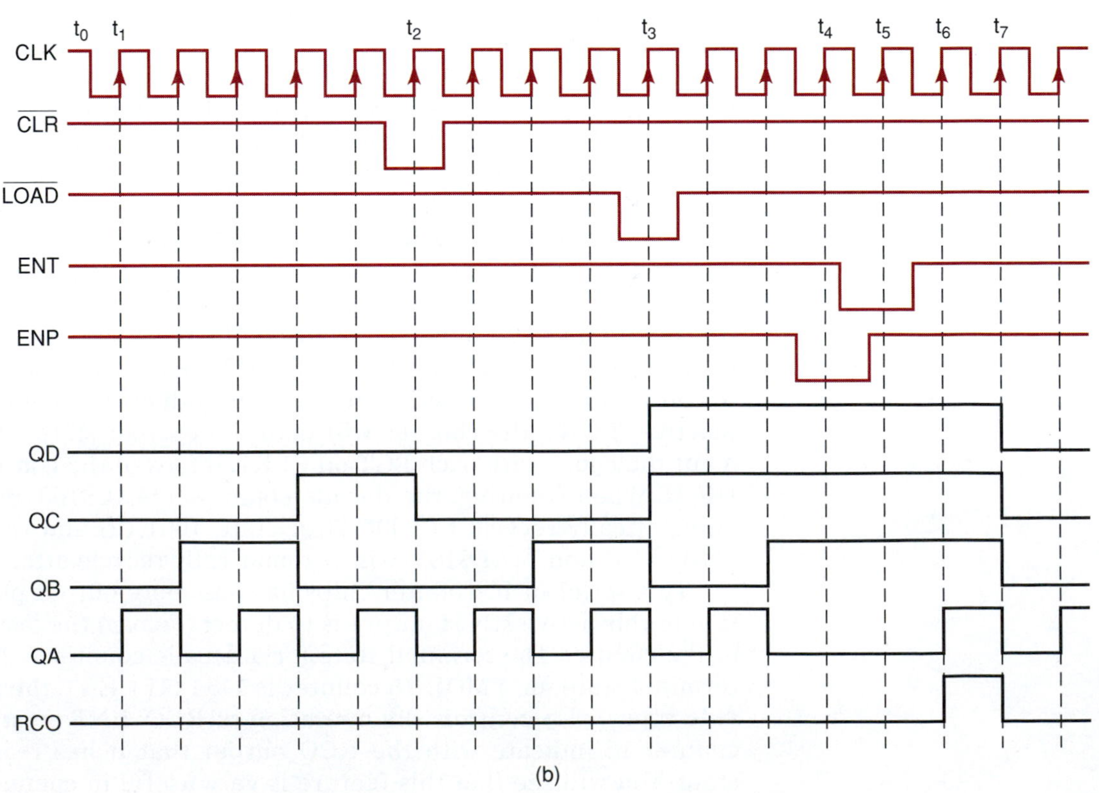

FIGURE 7-14 Example 7-10.

start counting up again from 0000 with each subsequent PGT. The $\overline{\text{LOAD}}$ input is LOW for t_3. This will synchronously load the applied data value 1100 (12) into the counter at t_3. After t_3, the $\overline{\text{LOAD}}$ input goes inactive (HIGH), so the counter will continue counting up from 1100 with each subsequent PGT until t_4. The counter output does not change at t_4 or t_5, since either ENP or ENT (the count enable inputs) is LOW. This holds the count at 1110 (14). At t_6, the counter is enabled again and counts up to 1111 (15), its terminal state. As a result, the RCO output now goes HIGH. At t_7, another PGT on CLK will make the counter recycle to 0000 and RCO returns to a LOW output.

EXAMPLE 7-11

Refer to Figure 7-15, where a 74HC160 has the input signals given in the timing diagram applied. The parallel data inputs are permanently connected as

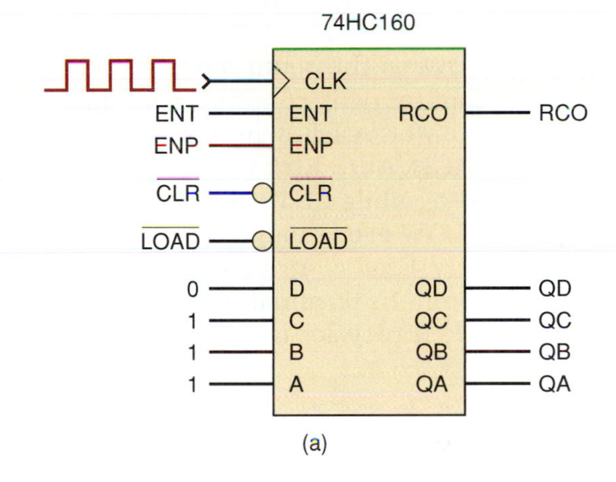

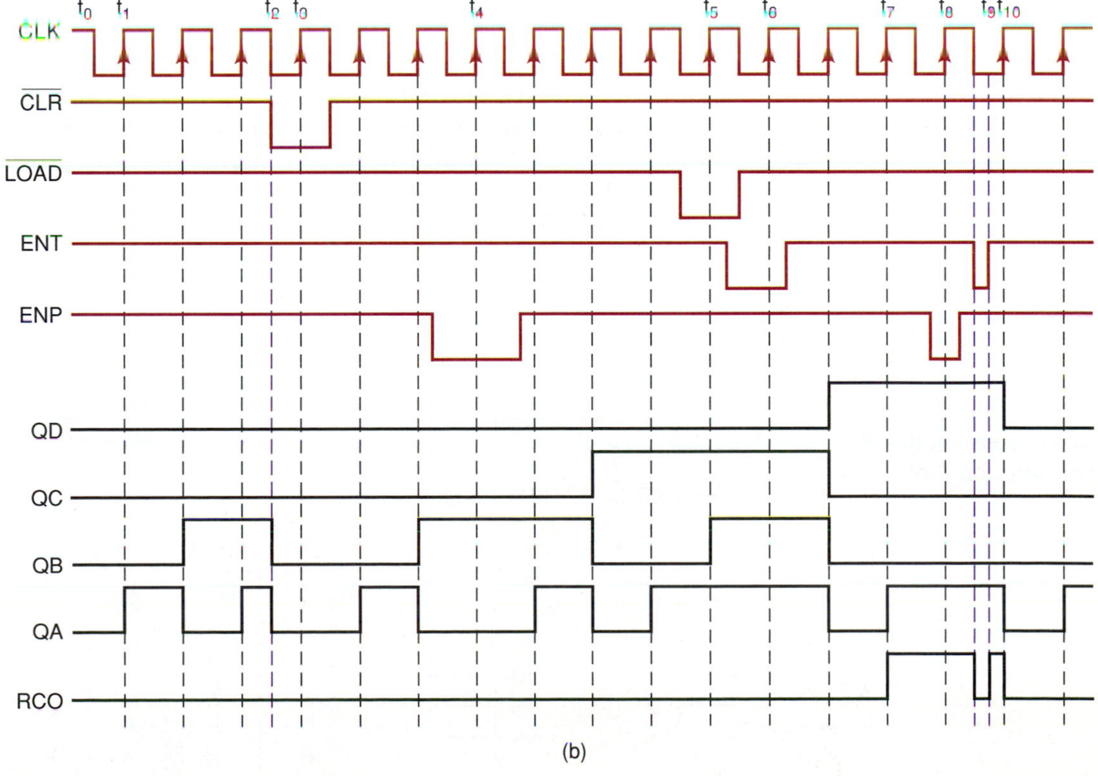

FIGURE 7-15 Example 7-11.

0111. Assume the counter is initially in the 0000 state, and determine the counter output waveforms.

Solution

Initially (at t_0) the counter's FFs are all LOW. Since this is not the terminal state for the BCD counter, output RCO will be LOW also. The first PGT on the CLK input occurs at t_1 and, since all control inputs are HIGH, the counter will increment to 0001. The counter continues to count up with each PGT until t_2. The asynchronous $\overline{\text{CLR}}$ input goes LOW at t_2 and will immediately reset the counter to 0000 at that point. At t_3, the $\overline{\text{CLR}}$ input is still active (LOW), so the PGT of the CLK input will be ignored and the counter will stay at 0000. Later the $\overline{\text{CLR}}$ input goes inactive again and the counter will count up to 0001 and then to 0010. At t_4, the count enable ENP is LOW, so the count holds at 0010. For subsequent PGTs of the CLK input, the counter is enabled and counts up until t_5. The $\overline{\text{LOAD}}$ input is LOW for t_5. This will synchronously load the applied data value 0111 (7) into the counter at t_5. At t_6, the count enable ENT is LOW, so the count holds at 0111. For the two subsequent PGTs after t_6, the counter will continue counting up since it is re-enabled. At t_7, the BCD counter reaches its terminal state 1001 (9) and the RCO output now goes HIGH. At t_8, ENP is LOW and the counter stops counting (remaining at 1001). At t_9, while ENT is LOW, the RCO output will be disabled so that it returns to a LOW even though the counter is still at its terminal state (1001). Recall that only ENT controls the RCO output. When ENT returns HIGH during the counter's terminal state, RCO goes HIGH again. At t_{10} the counter is enabled, and it recycles to 0000 and then counts to 0001 on the last PGT.

The 74ALS190-191/74HC190-191 Series

Figure 7-16 shows the logic symbol, modulus, and function table for the 74ALS190 and 74ALS191 series of IC counters (and the equivalent CMOS counterparts, 74HC190 and 74HC191). These recycling, four-bit counters have outputs labeled QD, QC, QB, QA, where QA is the LSB and QD is the MSB. They are clocked by a PGT applied to CLK. The only difference between the two part numbers is the counter's modulus. The 74ALS190 is a MOD-10 counter and the 74ALS191 is a MOD-16 binary counter. Both chips are up/down counters and have an asynchronous, active-LOW load input. This

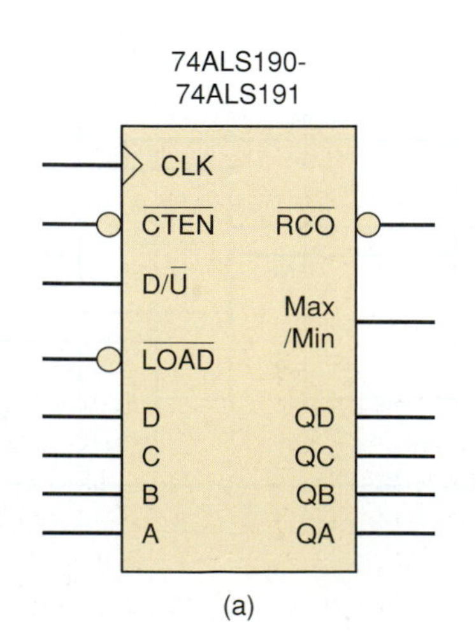

Part Number	Modulus
74ALS190	10
74ALS191	16

(b)

74ALS190-74ALS191 Function Table

$\overline{\text{LOAD}}$	$\overline{\text{CTEN}}$	D/$\overline{\text{U}}$	CLK	Function
L	X	X	X	Asynch. Load
H	L	L	↑	Count up
H	L	H	↑	Count down
H	H	X	X	No change

(c)

FIGURE 7-16 74ALS190-74ALS191 series synchronous counters: (a) logic symbol; (b) modulus; (c) function table.

means that as soon as $\overline{\text{LOAD}}$ goes LOW, the counter will be preset to the parallel data on the D, C, B, A (A is LSB and D is MSB) input pins. If the load function is inactive, it does not matter what is applied to the data input pins. The load input has priority over the counting function.

To count, the $\overline{\text{LOAD}}$ control input must be inactive (HIGH) and the count enable control $\overline{\text{CTEN}}$ must be LOW. The count direction is controlled by the D/$\overline{\text{U}}$ control input. If D/$\overline{\text{U}}$ is LOW, the count is incremented with each PGT on CLK, while a HIGH on D/$\overline{\text{U}}$ will decrement the count. Both counters automatically recycle in either count direction. The decade counter recycles to 0000 after state 1001 (9) when counting up or to 1001 after state 0000 when counting down. The binary counter will recycle to 0000 after 1111 (15) when counting up or to 1111 after state 0000 when counting down.

These counter chips have two more output pins, MAX/MIN and RCO. MAX/MIN is an active-HIGH output that detects (decodes) the terminal state of the counter. Since they are up/down counters, the terminal state depends on the direction of the count. The terminal state (MIN) for either counter when counting down is 0000 (0). However when counting up, the terminal state (MAX) for a decade counter is 1001 (9), while the terminal state for a MOD-16 counter is 1111 (15). Note that MAX/MIN detects only one state in the count sequence—it just depends on whether it is counting up or down. The active-LOW $\overline{\text{RCO}}$ output also detects the appropriate terminal state for the counter, but it is a bit more complicated. First, it is only enabled when $\overline{\text{CTEN}}$ is LOW. Additionally, $\overline{\text{RCO}}$ will only be LOW while the CLK input is also LOW. So essentially $\overline{\text{RCO}}$ will mimic the CLK waveform only during the terminal state while the counter is enabled.

EXAMPLE 7-12

Refer to Figure 7-17, where a 74HC190 has the input signals given in the timing diagram applied. The parallel data inputs are permanently connected as 0111. Assume the counter is initially in the 0000 state, and determine the counter output waveforms.

Solution

Initially (at t_0), the counter's FFs are all LOW. Since the counter is enabled ($\overline{\text{CTEN}} = 0$) and the count direction control D/$\overline{\text{U}}$ = 0, the BCD counter will start counting up on the first PGT applied to CLK at t_1 and continues to count up with each PGT until t_2, where the count has reached 0101. The asynchronous $\overline{\text{LOAD}}$ input goes LOW at t_2 and will immediately load 0111 into the counter at that point. At t_3, the $\overline{\text{LOAD}}$ input is still active (LOW), so the PGT of the CLK input will be ignored and the counter will stay at 0111. Later the $\overline{\text{LOAD}}$ input goes HIGH again and the counter will count up to 1000 at the next PGT. At t_4, the counter increments to 1001, which is the terminal state for a BCD up counter and the MAX/MIN output goes HIGH. During t_5, the counter is at its terminal state and the CLK input is LOW, so $\overline{\text{RCO}}$ goes LOW. For subsequent PGTs of the CLK input, the counter recycles to 0000 and continues to count up until t_6. Just prior to t_6, the D/$\overline{\text{U}}$ control changes to a HIGH. This will make the counter count down at t_6 and again at t_7, where it will be at state 0000, which now is the terminal state since we are counting down, and MAX/MIN will output a HIGH. During t_8, when the CLK input goes LOW, the $\overline{\text{RCO}}$ output again will be LOW. At t_9, the counter is disabled with $\overline{\text{CTEN}}$ = 1 and the counter holds at 1001. For the subsequent CLK pulses, the counter continues to count down.

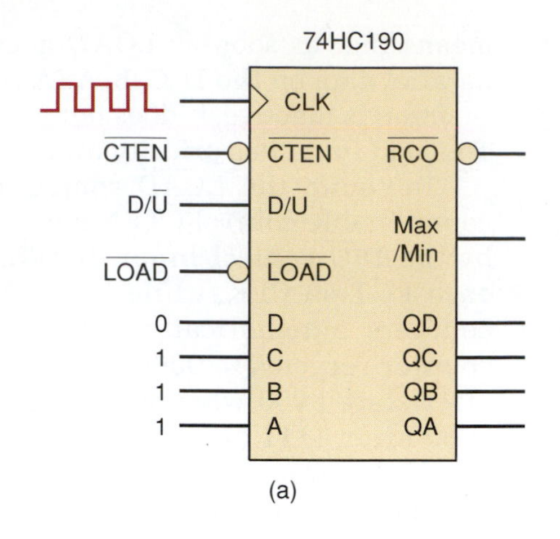

(a)

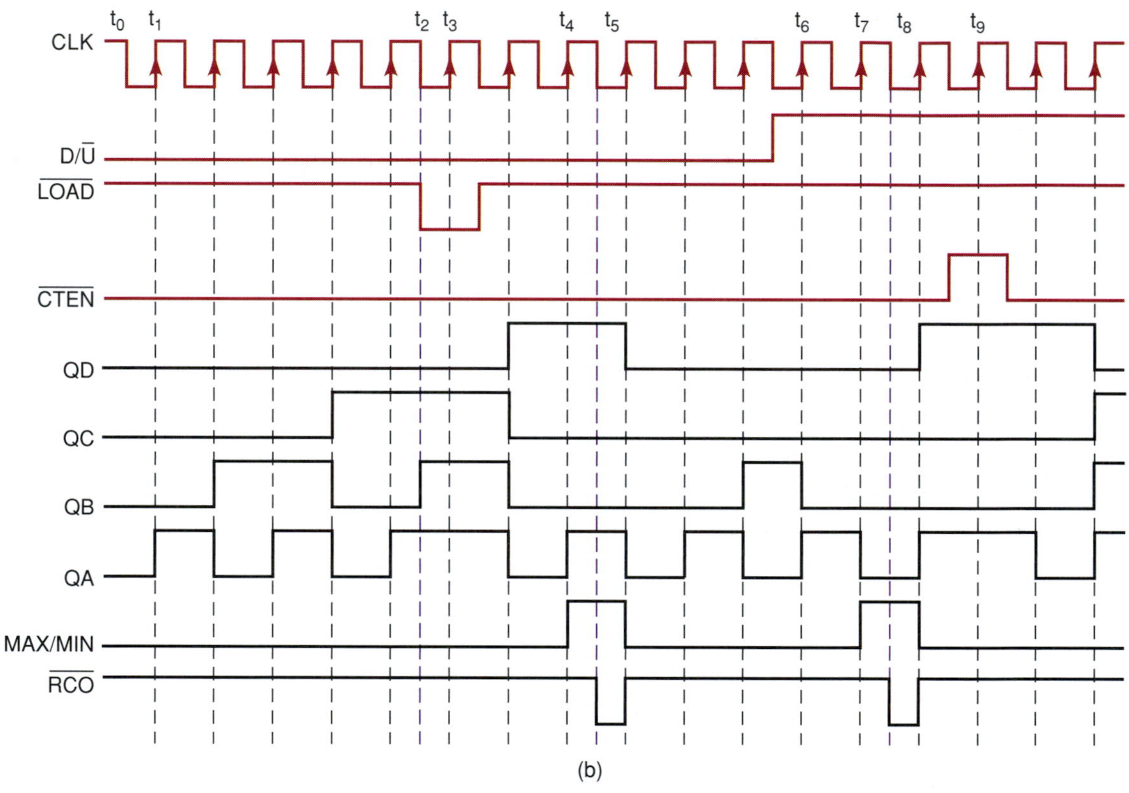

(b)

FIGURE 7-17 Example 7-12.

EXAMPLE 7-13

Compare the operation of two counters, one with synchronous load and the other with asynchronous load. Refer to Figure 7-18(a), in which a 74ALS163 and a 74ALS191 have been wired in a similar fashion to count up in binary. Both chips are driven by the same clock signal and have their QD and QC outputs NANDed together to control the respective $\overline{\text{LOAD}}$ input control. Assume that both counters are initially in the 0000 state.

(a) Determine the output waveform for each counter.

(b) What is the recycling count sequence and modulus for each counter?

(c) Why do they have different count sequences?

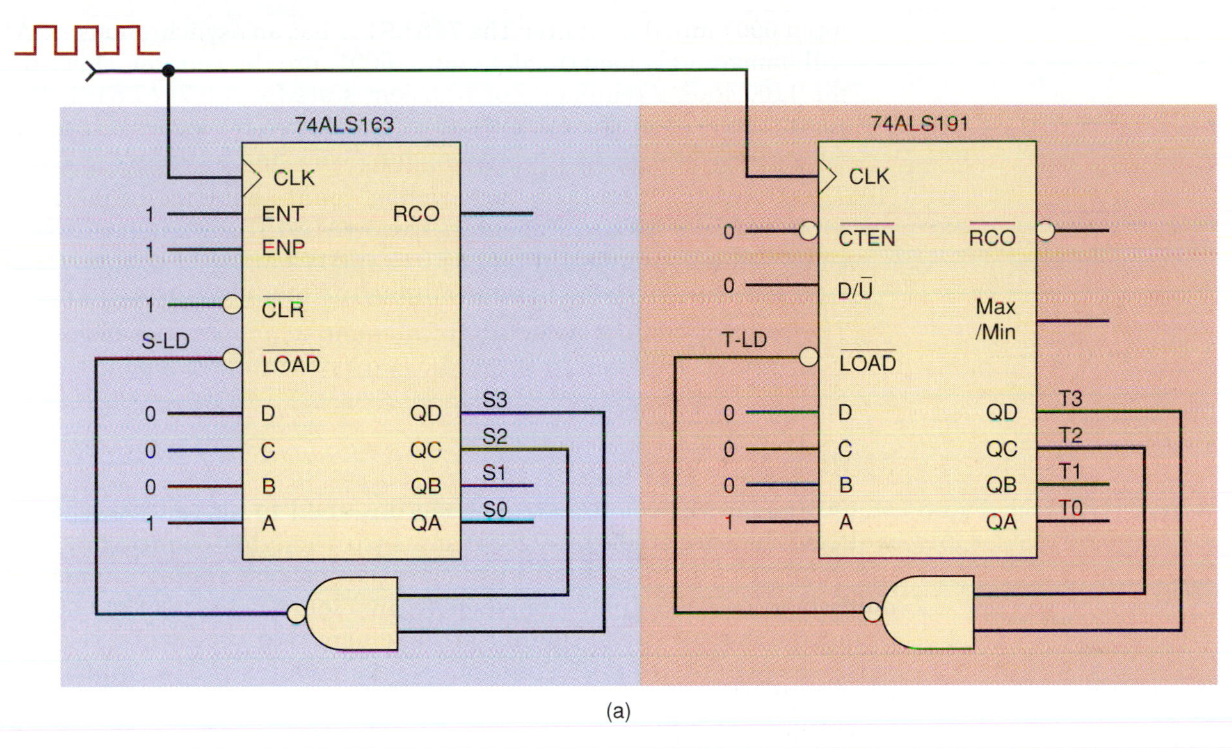

(a)

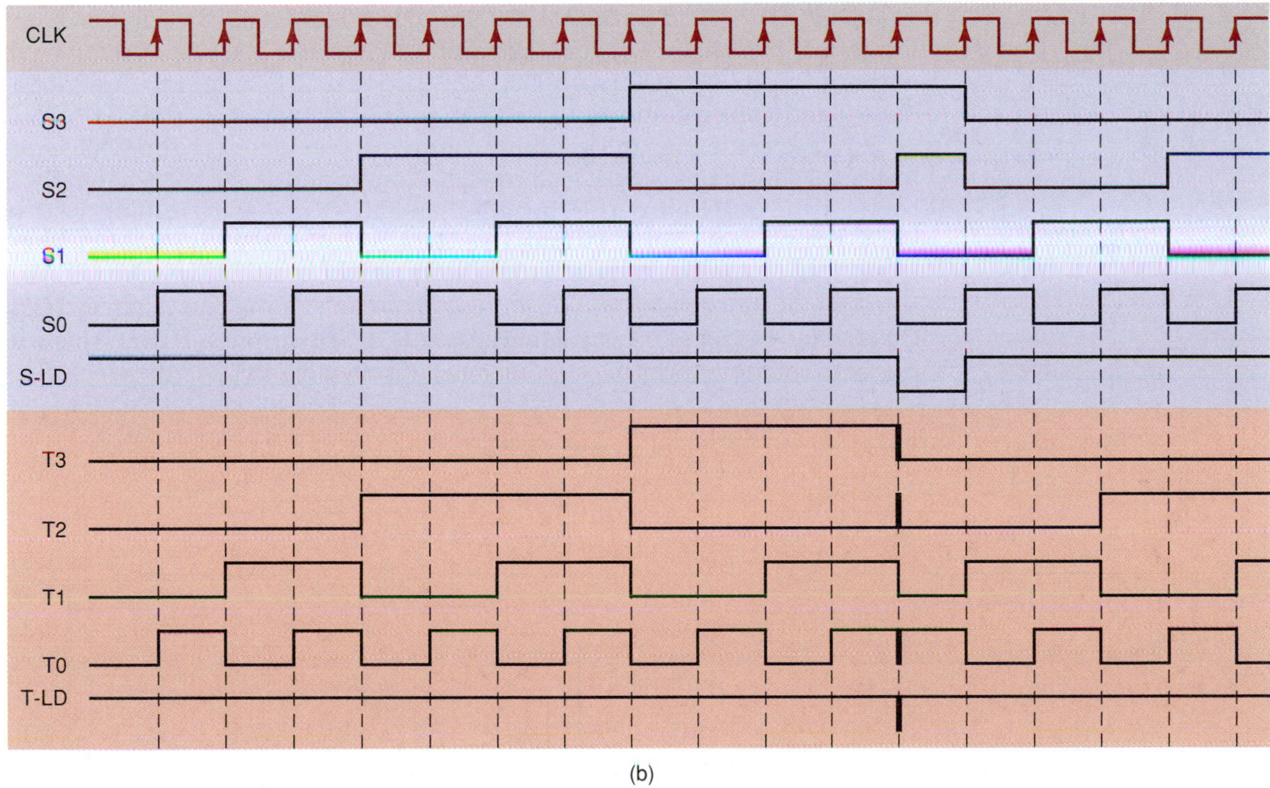

(b)

FIGURE 7-18 Example 7-13.

Solution

(a) Starting at state 0000, each counter will count up until it reaches state 1100 (12) as shown in Figure 7-18(b). The output of each NAND gate will apply a LOW to the respective $\overline{\text{LOAD}}$ input at that time. The 74ALS163 has a synchronous $\overline{\text{LOAD}}$ and will wait until the next PGT on CLK to load the data input 0001 into the counter. The 74ALS191 has an asynchronous $\overline{LOAD}$ and will immediately load the data input 0001 into the counter. This will make the 1100 state a temporary or transient state for the 74ALS191. The transient state will produce some spikes or glitches for some of the counter's outputs because of their rapid switching back and forth.

(b) The 74ALS163 circuit has a recycling count sequence of 0001 through 1100 and is a MOD-12 counter. The 74ALS191 circuit has a recycling count sequence of 0001 through 1011 and is a MOD-11 counter. Transient states are not included in determining the modulus for a counter.

(c) The counter circuits have different count sequences because one has a synchronous load and the other has an asynchronous load.

Multistage Arrangement

Many standard IC counters have been designed to make it easy to connect multiple chips together to create circuits with a higher counting range. All of the counter chips presented in this section can be simply connected in a **multistage** or **cascading** arrangement. In Figure 7-19, two 74ALS163s are connected in a two-stage counter arrangement that produces a recycling, binary sequence from 0 to 255 for a maximum modulus of 256. Applying a LOW to the $\overline{CLR}$ input will synchronously clear both counter stages, and applying a LOW to $\overline{LD}$ will synchronously preset the eight-bit counter to the binary value on inputs D7, D6, D5, D4, D3, D2, D1, D0 (D0 = LSB). The block on the left (stage 1) is the low-order stage and provides the least-significant counter outputs Q3, Q2, Q1, Q0 (with Q0 = LSB). Stage 2 on the right provides the most-significant counter outputs Q7, Q6, Q5, Q4 (with Q7 = MSB).

EN, the enable for the eight-bit counter, is connected to the ENT input on stage 1. Note that we must use the ENT input and not ENP, since only ENT controls the RCO output. Using ENT and RCO makes cascading very easy. Both counter blocks are clocked together synchronously, but the block on the right (stage 2) is disabled until the least-significant output nibble has reached its terminal state, which will be indicated by the TC1 output. When Q3, Q2, Q1, Q0 reaches 1111 and if EN is HIGH, then TC1 will output a HIGH. This will allow both counter stages to count up one with the next PGT on the clock. Stage 1

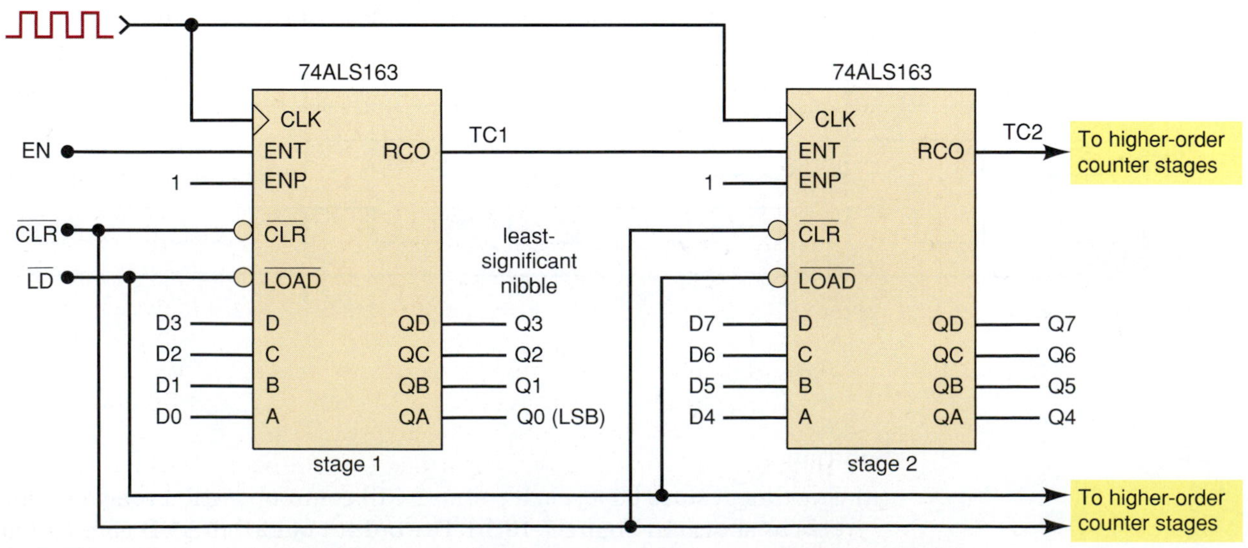

FIGURE 7-19 Two 74ALS163s connected in a two-stage arrangement to extend the maximum counting range.

will recycle back to 0000 and stage 2 will increment from its previous output state. TC1 will return to a LOW, since stage 1 is no longer at its terminal state. With subsequent clock pulses, stage 1 will continue to count up if EN=1 until it again reaches 1111 and the process repeats. When the eight-bit counter reaches 11111111, it will recycle back to 00000000 on the next clock pulse.

Additional 74ALS163 counter chips can be cascaded in the same fashion. TC2 would be connected to the ENT control on the next chip, and so on. TC2 will be HIGH when Q7, Q6, Q5, Q4 is equal to 1111 and TC1 is HIGH, which in turn means that Q3, Q2, Q1, Q0 is also equal to 1111 and EN is HIGH. This cascading technique works for all chips (TTL or CMOS families) in this series, even for the BCD counters. The 74ALS190-191 (or 74HC190-191) series also can be cascaded similarly using the active-LOW $\overline{CTEN}$ and $\overline{RCO}$ pins. A multistage counter using 74ALS190-191 chips connected in this fashion can count up or down.

REVIEW QUESTIONS

1. Describe the function of the inputs $\overline{LOAD}$ and D, C, B, A.
2. Describe the function of the $\overline{CLR}$ input.
3. *True or false:* The 74HC161 cannot be preset while $\overline{CLR}$ is active.
4. What logic levels must be present on the control inputs in order for the 74ALS162 to count pulses that appear on the CLK?
5. What logic levels must be present on the control inputs in order for the 74HC190 to count down with pulses that appear on the CLK?
6. What would be the maximum counting range for a four-stage counter made up of 74HC163 ICs? What is the maximum counting range for 74ALS190 ICs?

7-8 DECODING A COUNTER

Digital counters are often used in applications where the count represented by the states of the FFs must somehow be determined or displayed. One of the simplest means for displaying the contents of a counter involves just connecting the output of each FF to a small indicator LED [see Figure 7-7(b)]. In this way the states of the FFs are visibly represented by the LEDs (on = 1, off = 0), and the count can be mentally determined by **decoding** the binary states of the LEDs. For instance, suppose that this method is used for a BCD counter and the states of the LEDs are off–on–on–off, respectively. This would represent 0110, which we would mentally decode as decimal 6. Other combinations of LED states would represent the other possible counts.

The indicator LED method becomes inconvenient as the size (number of bits) of the counter increases because it is much harder to decode the displayed results mentally. For this reason, it is preferable to develop a means for *electronically* decoding the contents of a counter and displaying the results in a form that is immediately recognizable and requires no mental operations.

An even more important reason for electronic decoding of a counter occurs because of the many applications in which counters are used to control the timing or sequencing of operations *automatically* without human intervention. For example, a certain system operation might have to be initiated when a counter reaches the 101100 state (count of 44_{10}). A logic circuit can be used to decode for or detect when this particular count is present and then initiate the operation. Many operations may have to be controlled in this manner in a digital system. Clearly, human intervention in this process would be undesirable except in extremely slow systems.

Active-HIGH Decoding

A MOD-*X* counter has *X* different states; each state is a particular pattern of 0s and 1s stored in the counter FFs. A decoding network is a logic circuit that generates *X* different outputs, each of which detects (decodes) the presence of one particular state of the counter. The decoder outputs can be designed to produce either a HIGH or a LOW level when the detection occurs. An active-HIGH decoder produces HIGH outputs to indicate detection. Figure 7-20 shows the complete active-HIGH decoding logic for a MOD-8 counter. The decoder

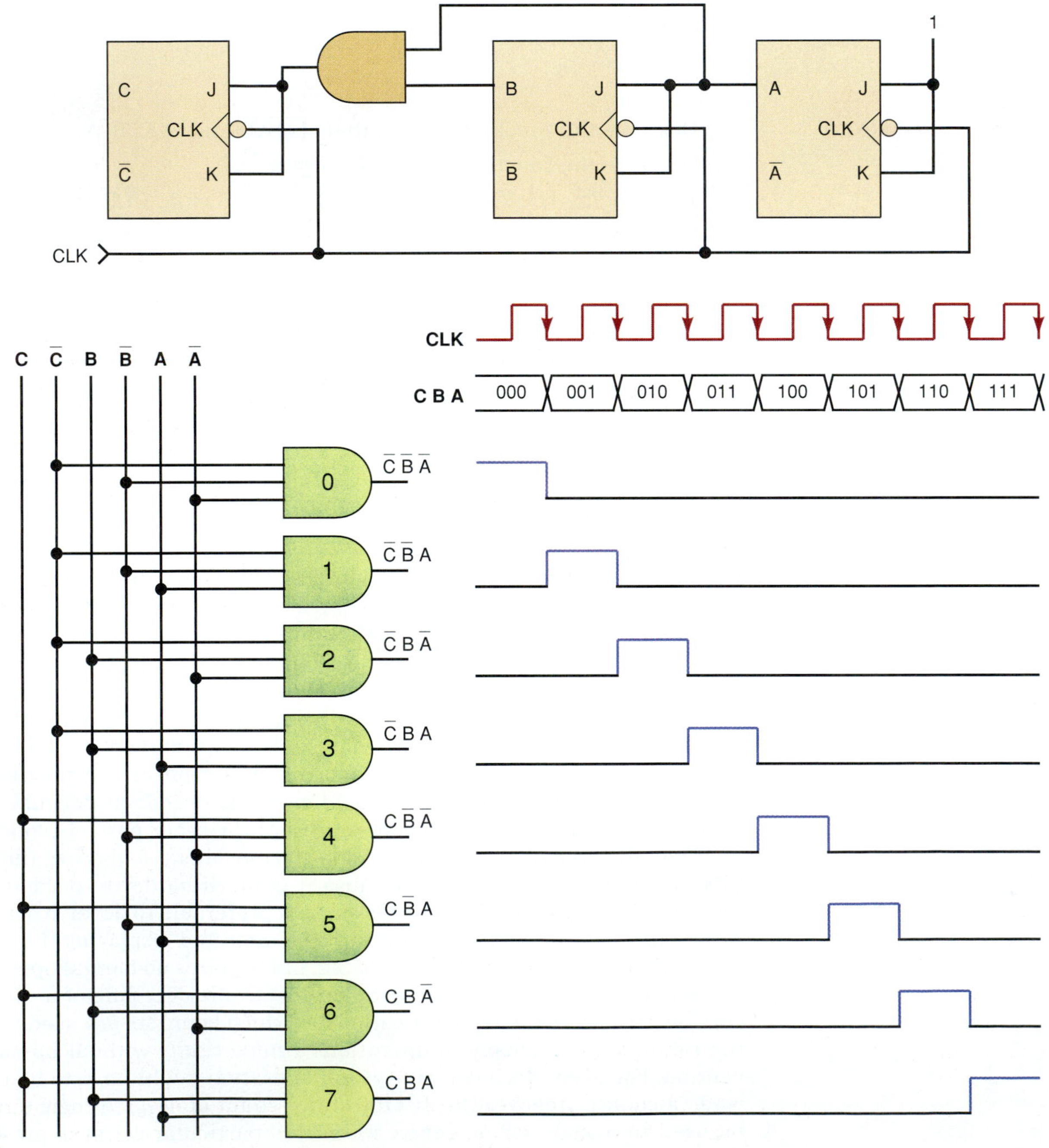

FIGURE 7-20 Using AND gates to decode a MOD-8 counter.

consists of eight three-input AND gates. Each AND gate produces a HIGH output for one particular state of the counter.

For example, AND gate 0 has at its inputs the FF outputs $\overline{C}$, $\overline{B}$, and $\overline{A}$. Thus, its output will be LOW at all times *except* when $A = B = C = 0$, that is, on the count of 000 (zero). Similarly, AND gate 5 has as its inputs the FF outputs C, $\overline{B}$, and A, so that its output will go HIGH only when $C = 1$, $B = 0$, and $A = 1$, that is, on the count of 101 (decimal 5). The rest of the AND gates perform in the same manner for the other possible counts. At any one time, only one AND gate output is HIGH: the one that is decoding for the particular count present in the counter. The waveforms in Figure 7-20 show this clearly.

The eight AND outputs can be used to control eight separate indicator LEDs, which represent the decimal numbers 0 through 7. Only one LED will be on at a given time, indicating the proper count.

The AND gate decoder can be extended to counters with any number of states. The following example illustrates.

EXAMPLE 7-14

How many AND gates are required to decode completely all of the states of a MOD-32 binary counter? What are the inputs to the gate that decodes for the count of 21?

Solution

A MOD-32 counter has 32 possible states. One AND gate is needed to decode for each state; therefore, the decoder requires 32 AND gates. Because $32 = 2^5$, the counter contains five FFs. Thus, each gate will have five inputs, one from each FF. Decoding for the count of 21 (that is, 10101_2) requires AND gate inputs of E, $\overline{D}$, C, $\overline{B}$, and A, where E is the MSB flip-flop.

Active-LOW Decoding

If NAND gates are used in place of AND gates, the decoder outputs produce a normally HIGH signal, which goes LOW only when the number being decoded occurs. Both types of decoders are used, depending on the type of circuits being driven by the decoder outputs.

EXAMPLE 7-15

Figure 7-21 shows a common situation in which a counter is used to generate a control waveform, which could be used to control devices such as a motor, solenoid valve, or heater. The MOD-16 counter cycles and recycles through its counting sequence. Each time it goes to the count of 8 (1000), the upper NAND gate will produce a LOW output, which sets flip-flop X to the 1 state. Flip-flop X stays HIGH until the counter reaches the count of 14 (1110), at which time the lower NAND gate decodes it and produces a LOW output to clear X to the 0 state. Thus, the X output is HIGH between the counts of 8 and 14 for each cycle of the counter.

BCD Counter Decoding

A BCD counter has 10 states that can be decoded using the techniques described previously. BCD decoders provide 10 outputs corresponding to the decimal digits 0 through 9 and represented by the states of the counter

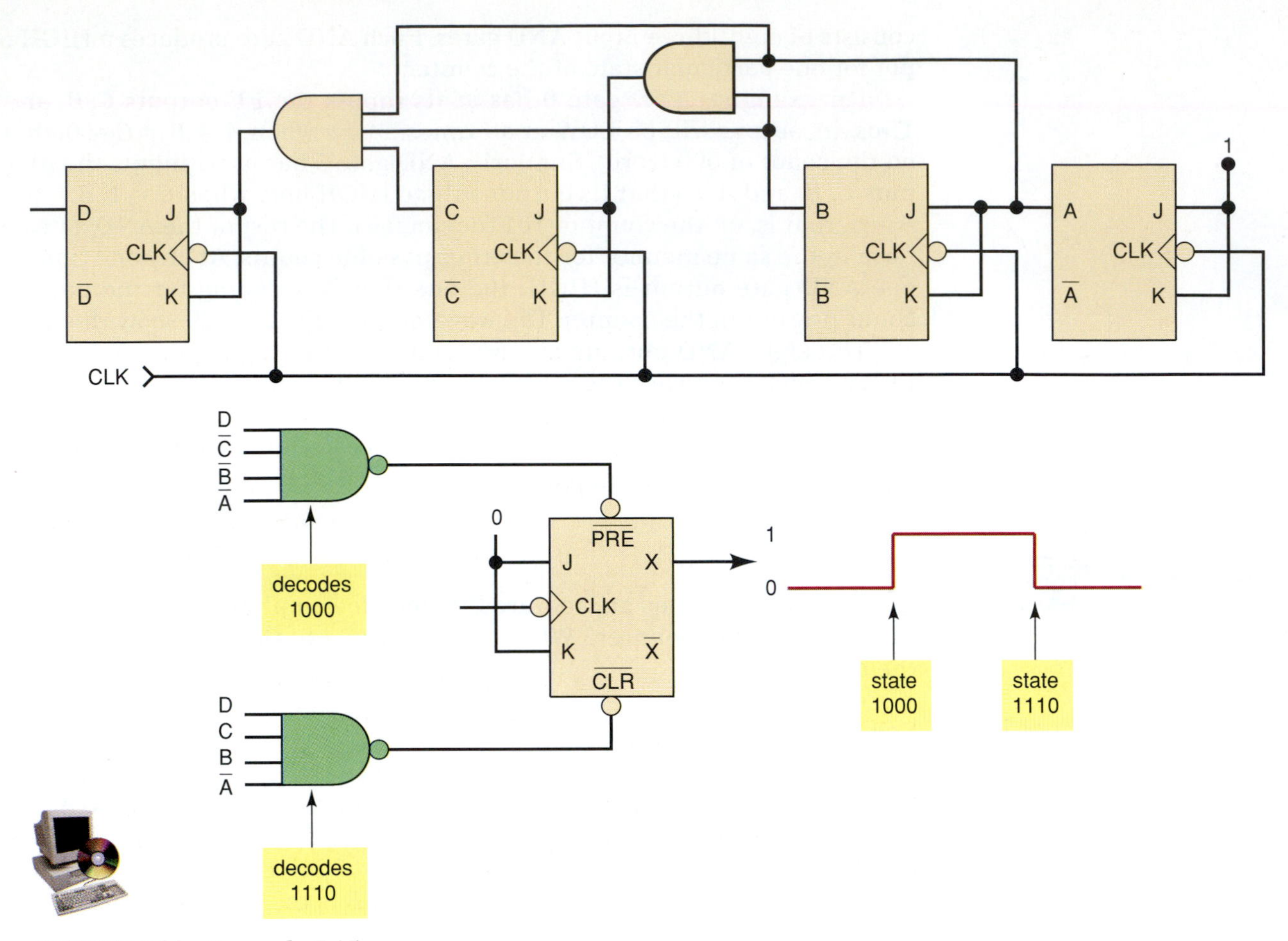

FIGURE 7-21 Example 7-15.

FFs. These 10 outputs can be used to control 10 individual indicator LEDs for a visual display. More often, instead of using 10 separate LEDs, a single display device is used to display the decimal numbers 0 through 9. One class of decimal displays contains seven small segments made of a material (usually LEDs or liquid-crystal displays) that either emits light or reflects ambient light. The BCD decoder outputs control which segments are illuminated in order to produce a pattern representing one of the decimal digits.

We will go into more detail concerning these types of decoders and displays in Chapter 9. However, because BCD counters and their associated decoders and displays are very commonplace, we will use the decoder/display unit (see Figure 7-22) to represent the complete circuitry used to display visually the contents of a BCD counter as a decimal digit.

FIGURE 7-22 BCD counters usually have their count displayed on a single display device.

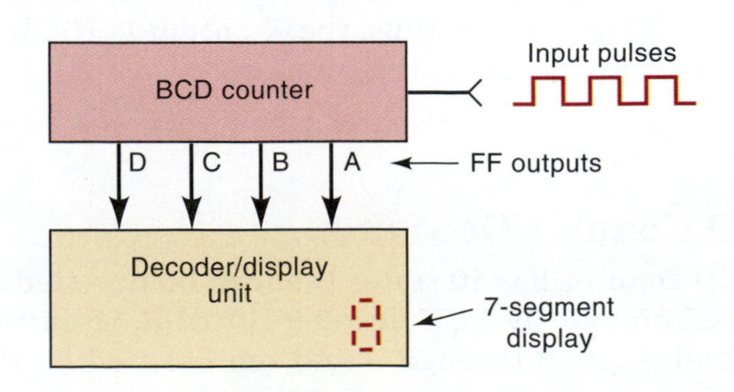

REVIEW QUESTIONS

1. How many gates are needed to decode a six-bit counter fully?
2. Describe the decoding gate needed to produce a LOW output when a MOD-64 counter is at the count of 23.

7-9 ANALYZING SYNCHRONOUS COUNTERS

Synchronous counter circuits can be custom-designed to generate any desired count sequence. We can use just the synchronous inputs that are applied to the individual flip-flops to produce the counter's sequence. By not using asynchronous FF controls, such as the clears, to change the counter's sequence, we will never have to deal with transient states and possible glitches in output waveforms. The process of designing completely synchronous counters will be investigated in the next section. First, let's see how to analyze a counter design of this type by predicting the FF control inputs for each state of the counter. A **PRESENT state/NEXT state table** is a very useful tool in this analysis process. The first step is to write the logic expression for each FF control input. Next assume a PRESENT state for the counter and apply that combination of bits to the control logic expressions. The outputs from the control expressions will allow us to predict the commands to each FF and the resulting NEXT state for the counter after clocking. Repeat the analysis process until the entire count sequence is determined.

Figure 7-23 is a synchronous counter that has slightly different J and K inputs than we saw in Section 7-3 for a regular binary up counter. These minor changes to the control circuitry will cause the counter to produce a different count sequence. The control input expressions for this counter are:

$$J_C = A \cdot B$$
$$K_C = C$$
$$J_B = K_B = A$$
$$J_A = K_A = \overline{C}$$

Let us assume that the PRESENT state for the counter is CBA = 000. Applying this combination to the control expressions above will yield $J_C\,K_C = 0\ 0$, $J_B\,K_B = 0\ 0$, and $J_A\,K_A = 1\ 1$. These control inputs will tell FFs C and B to hold and FF A to toggle on the next NGT on CLK. Our predicted NEXT state is 001 for CBA. This information has been entered in the first line of the PRESENT state/NEXT state table shown in Table 7-1. Next we can use the state 001

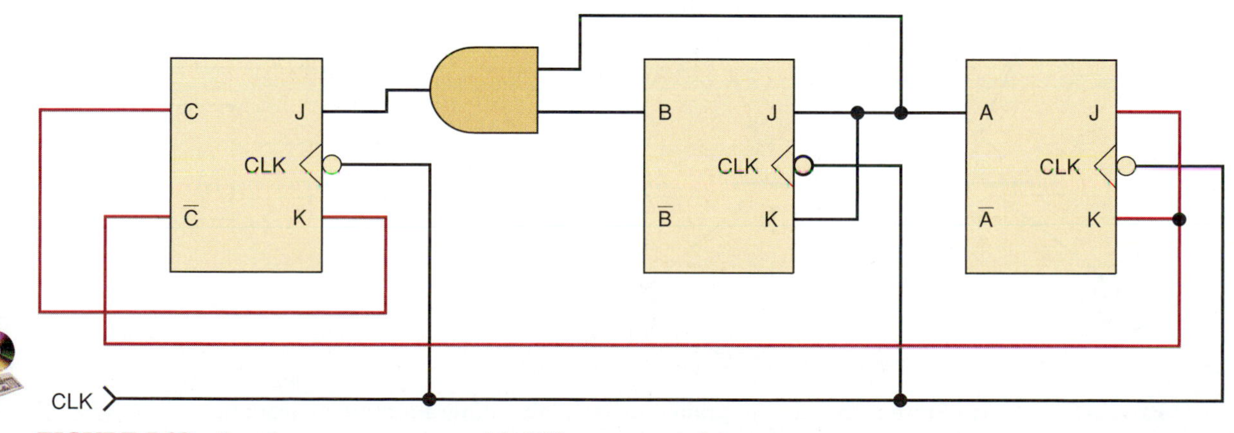

FIGURE 7-23 Synchronous counter with different control inputs.

TABLE 7-1

PRESENT State			Control Inputs						NEXT State		
C	B	A	J_C	K_C	J_B	K_B	J_A	K_A	C	B	A
0	0	0	0	0	0	0	1	1	0	0	1
0	0	1	0	0	1	1	1	1	0	1	0
0	1	0	0	0	0	0	1	1	0	1	1
0	1	1	1	0	1	1	1	1	1	0	0
1	0	0	0	1	0	0	0	0	0	0	0
1	0	1	0	1	1	1	0	0	0	1	1
1	1	0	0	1	0	0	0	0	0	1	0
1	1	1	1	1	1	1	0	0	0	0	1

as our PRESENT state. Analyzing the control expressions with this new combination will now yield $J_C\,K_C = 0\;0$, $J_B\,K_B = 1\;1$, and $J_A\,K_A = 1\;1$ giving us a hold command for FF C and toggle commands for FFs B and A. This will produce a NEXT state of 010 for CBA, which we have listed on the second line of Table 7-1. Continuing with this process will result in a recycling count sequence of 000, 001, 010, 011, 100, 000. This would be a MOD-5 count sequence. We can also predict the NEXT states for the remaining three possible state combinations in the same way. By doing so, we can determine if the counter design is *self-correcting.* A **self-correcting counter** is one in which normally unused states will all somehow return to the normal count sequence. If any of these unused states cannot return to the normal sequence, the counter is said to be not self-correcting. Our NEXT-state predictions for all possible states have been entered into Table 7-1. The highlighted information indicates that this counter design happens to be self-correcting. The complete state transition diagram and timing diagram for this counter is shown in Figure 7-24.

We can likewise analyze the operation of counter circuits that use D flip-flops to store the present state of the counter. The control circuitry for a D-type will typically be more complex than for an equivalent JK-type counter that produces the same count sequence, but we will also have half the number of

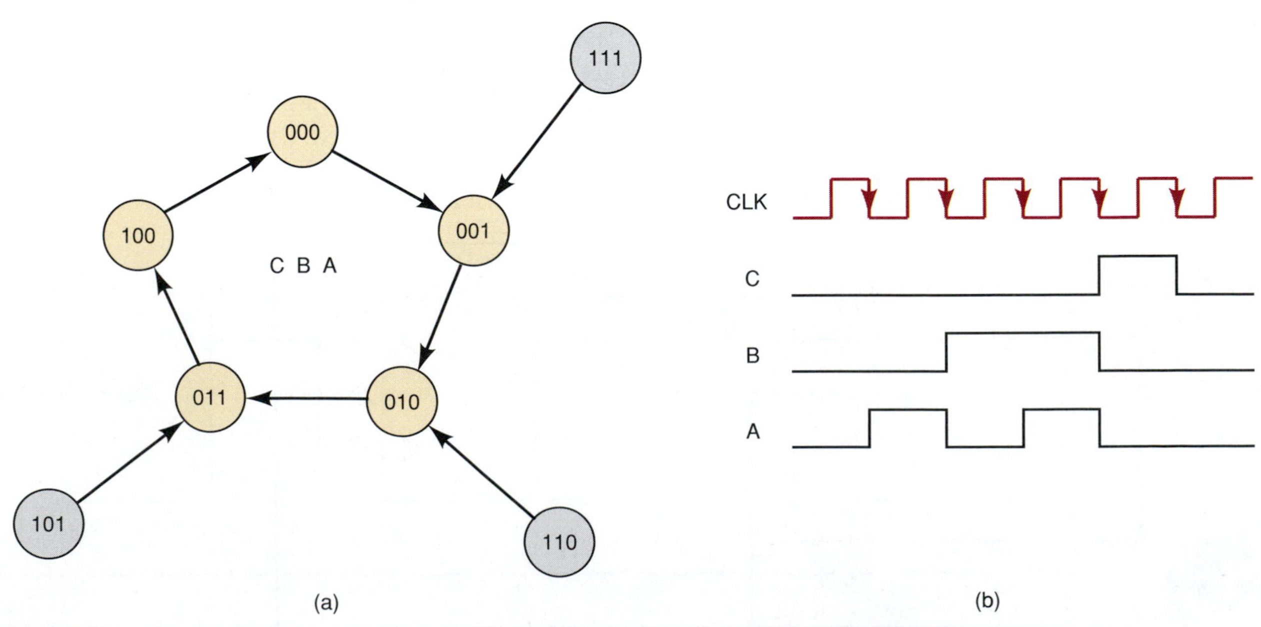

FIGURE 7-24 (a) State transition diagram and (b) timing diagram for synchronous counter in Figure 7-23.

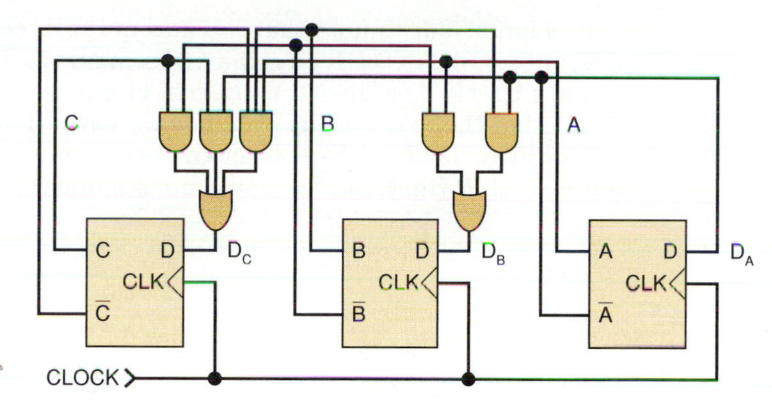

FIGURE 7-25 Synchronous counter using *D* flip-flops.

synchronous inputs to control. Most PLDs utilize *D* flip-flops for their memory elements, so the analysis of this type of counter circuit will give us some insight into how counters are actually programmed inside a PLD.

A synchronous counter designed with *D* flip-flops is shown in Figure 7-25. The first step is to write the logic expressions for the *D* inputs:

$$D_C = C\overline{B} + C\overline{A} + \overline{C}BA$$
$$D_B = \overline{B}A + B\overline{A}$$
$$D_A = \overline{A}$$

Then we will determine the PRESENT state/NEXT state table for the counter circuit by assuming a state and applying that set of bit values to the input expressions given above. If we pick CBA = 000 for the initial counter state, we will find that $D_C = 0$, $D_B = 0$, and $D_A = 1$. With a PGT on CLOCK, the flip-flops will "load" in the value 001, which becomes the counter's NEXT state. Using 001 as a PRESENT state will produce inputs of $D_C = 0$, $D_B = 1$, and $D_A = 0$ so that 010 will be the NEXT state, and so on. The completed PRESENT state/NEXT state table, shown in Table 7-2, indicates that this circuit is a recycling MOD-8 binary counter. By applying a little Boolean algebra to the input expressions, we can see that there is actually a fairly simple circuit pattern in creating binary counters from D flip-flops:

$$D_C = C\overline{B} + C\overline{A} + \overline{C}BA = C(\overline{B} + \overline{A}) + \overline{C}BA$$
$$= C\overline{BA} + \overline{C}(BA) = C \oplus (AB)$$
$$D_B = \overline{B}A + B\overline{A} = B \oplus A$$
$$D_A = \overline{A}$$

TABLE 7-2

PRESENT State			Control Inputs			NEXT State		
C	*B*	*A*	D_C	D_B	D_A	*C*	*B*	*A*
0	0	0	0	0	1	0	0	1
0	0	1	0	1	0	0	1	0
0	1	0	0	1	1	0	1	1
0	1	1	1	0	0	1	0	0
1	0	0	1	0	1	1	0	1
1	0	1	1	1	0	1	1	0
1	1	0	1	1	1	1	1	1
1	1	1	0	0	0	0	0	0

It is important to note that the gating resources for most PLDs actually consist of sets of AND-OR circuit arrangements and the SOP logic expression more accurately describes the internal circuit implementation. However, we can see that the expressions have been greatly simplified by using the XOR function. This leads us to predict correctly that to create a MOD-16 binary counter with D flip-flops, we would need a fourth FF with:

$$D_D = D \oplus (A\,B\,C)$$

REVIEW QUESTIONS

1. Why is it desirable to avoid having asynchronous controls on counters?
2. What tool is useful in the analysis of synchronous counters?
3. What determines the count sequence for a counter circuit?
4. What counter characteristic is described by saying that it is self-correcting?

7-10 SYNCHRONOUS COUNTER DESIGN*

Many different counter arrangements are available as ICs—asynchronous, synchronous, and combined asynchronous/synchronous. Most of these count in a normal binary or BCD count sequence, although their counting sequences can be somewhat altered using the clearing or loading methods we demonstrated for the 74ALS160-163 and 74ALS190-191 series of ICs. There are situations, however, where a custom counter is required that follows a sequence that is not a regular binary count pattern, for example, 000, 010, 101, 001, 110, 000, . . .

Several methods exist for designing counters that follow arbitrary sequences. We will present the details for one common method that uses J-K flip-flops in a synchronous counter configuration. The same method can be used in designs with D flip-flops. The technique is one of several design procedures that are part of an area of digital circuit design called **sequential circuit design**, which is normally part of an advanced course.

Basic Idea

In synchronous counters, all of the FFs are clocked at the same time. Before each clock pulse, the J and K input of each FF in the counter must be at the correct level to ensure that the FF goes to the correct state. For example, consider the situation where state 101 for counter CBA is to be followed by state 011. When the next clock pulse occurs, the J and K inputs of the FFs must be at the correct levels that will cause flip-flop C to change from 1 to 0, flip-flop B from 0 to 1, and flip-flop A from 1 to 1 (i.e., no change).

The process of designing a synchronous counter thus becomes one of designing the logic circuits that *decode* the various states of the counter to supply the proper logic levels to each J and K input at the correct time. The inputs to these decoder circuits will come from the outputs of one or more of the FFs. To illustrate, for the synchronous counter of Figure 7-5, the AND gate that feeds the J and K inputs of flip-flop C decodes the states of flip-flops A and B. Likewise, the AND gate that feeds the J and K inputs of flip-flop D decodes the states of A, B, and C.

*This topic may be omitted without affecting the continuity of the remainder of the book.

J-K Excitation Table

Before we begin the process of designing the decoder circuits for each J and K input, we must first review the operation of the J-K flip-flop using a different approach, one called an *excitation table* (Table 7-3). The leftmost column of this table lists each possible FF output transition. The second and third columns list the FF's PRESENT state, symbolized as Q_n, and the NEXT state, symbolized as Q_{n+1}, for each transition. The last two columns list the J and K levels required to produce each transition. Let's examine each case.

0 → 0 TRANSITION The FF PRESENT state is at 0 and is to remain at 0 when a clock pulse is applied. From our understanding of how a J-K flip-flop works, this can happen when either $J = K = 0$ (no-change condition) or $J = 0$ and $K = 1$ (clear condition). Thus, J must be at 0, but K can be at either level. The table indicates this with a "0" under J and an "x" under K. Recall that "x" means the don't-care condition.

0 → 1 TRANSITION The PRESENT state is 0 and is to change to a 1, which can happen when either $J = 1$ and $K = 0$ (set condition) or $J = K = 1$ (toggle condition). Thus, J must be a 1, but K can be at either level for this transition to occur.

1 → 0 TRANSITION The PRESENT state is 1 and is to change to a 0, which can happen when either $J = 0$ and $K = 1$ or $J = K = 1$. Thus, K must be a 1, but J can be at either level.

1 → 1 TRANSITION The PRESENT state is a 1 and is to remain a 1, which can happen when either $J = K = 0$ or $J = 1$ and $K = 0$. Thus, K must be a 0 while J can be at either level.

The use of this **J-K excitation table** (Table 7-3) is a principal part of the synchronous counter design procedure.

TABLE 7-3 J-K flip-flop excitation table.

Transition at FF Output	PRESENT State Q_n	NEXT State Q_{n+1}	J	K
0 → 0	0	0	0	x
0 → 1	0	1	1	x
1 → 0	1	0	x	1
1 → 1	1	1	x	0

Design Procedure

We will now go through a complete synchronous counter design procedure. Although we will do it for a specific counting sequence, the same steps can be followed for any desired sequence.

TABLE 7-4

C	B	A
0	0	0
0	0	1
0	1	0
0	1	1
1	0	0
0	0	0
0	0	1
	etc.	

Step 1. Determine the desired number of bits (FFs) and the desired counting sequence.

For our example, we will design a three-bit counter that goes through the sequence shown in Table 7-4. Notice that this sequence does not include the 101, 110, and 111 states. We will refer to these states as *undesired states.*

Step 2. Draw the state transition diagram showing *all* possible states, including those that are not part of the desired counting sequence.

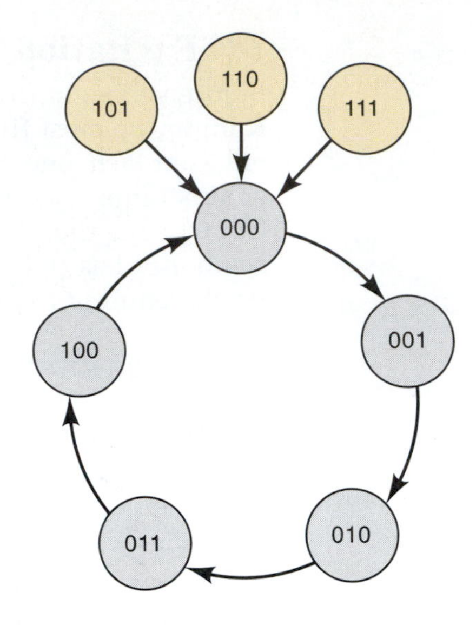

FIGURE 7-26 State transition diagram for the synchronous counter design example.

For our example, the state transition diagram appears as shown in Figure 7-26. The 000 through 100 states are connected in the expected sequence. We have also included a defined NEXT state for each of the undesired states. This was done in case the counter accidentally gets into one of these states upon power-up or due to noise. The circuit designer can choose to have each of these undesired states go to any state upon the application of the next clock pulse. Alternatively, the designer may choose not to define the counter's action for the undesired states at all. In other words, we may not care about the NEXT state for any undesired state. Using the latter "don't care" design approach will generally produce a simpler design but can be a potential problem in the application where this counter is to be used. For our design example, we will choose to have all undesired states go to the 000 state. This will make our design self-correcting but slightly different from the example MOD-5 counter that was analyzed in Section 7-9.

Step 3. Use the state transition diagram to set up a table that lists *all* PRESENT states and their NEXT states.

For our example, the information is shown in Table 7-5. The left-hand portion of the table lists *every* possible state, even those that are not part of the sequence. We label these as the PRESENT states. The right-hand portion lists the NEXT state for each PRESENT state. These are obtained from the state transition

TABLE 7-5

	PRESENT State			NEXT State		
	C	*B*	*A*	*C*	*B*	*A*
Line 1	0	0	0	0	0	1
2	0	0	1	0	1	0
3	0	1	0	0	1	1
4	0	1	1	1	0	0
5	1	0	0	0	0	0
6	1	0	1	0	0	0
7	1	1	0	0	0	0
8	1	1	1	0	0	0

diagram in Figure 7-26. For instance, line 1 shows that the PRESENT state of 000 has the NEXT state of 001, and line 5 shows that the PRESENT state of 100 has the NEXT state of 000. Lines 6, 7, and 8 show that the undesired PRESENT states 101, 110, and 111 all have the NEXT state of 000.

Step 4. Add a column to this table for each J and K input. For each PRESENT state, indicate the levels required at each J and K input in order to produce the transition to the NEXT state.

Our design example uses three FFs—C, B, and A—and each one has a J and a K input. Therefore, we must add six new columns as shown in Table 7-6. This completed table is called the **circuit excitation table.** The six new columns are the J and K inputs of each FF. The entries under each J and K are obtained from Table 7-3, the J-K flip-flop excitation table that we developed earlier. We will demonstrate this for several of the cases, and you can verify the rest.

Let's look at line 1 in Table 7-6. The PRESENT state of 000 is to go to the NEXT state of 001 on the occurrence of a clock pulse. For this state transition, the C flip-flop goes from 0 to 0. From the J-K excitation table, we see that J_C must be at 0 and K_C at "x" for this transition to occur. The B flip-flop also goes from 0 to 0, and so $J_B = 0$ and $K_B = x$. The A flip-flop goes from 0 to 1. Also from Table 7-3, we see that $J_A = 1$ and $K_A = x$ for this transition.

In line 4 in Table 7-6, the PRESENT state of 011 has a NEXT state of 100. For this state transition, flip-flop C goes from 0 to 1, which requires $J_C = 1$ and $K_C = x$. Flip-flops B and A are both going from 1 to 0. The J-K excitation table indicates that these two FFs need $J = x$ and $K = 1$ for this to occur.

The required J and K levels for all other lines in Table 7-6 can be determined in the same manner.

Step 5. Design the logic circuits needed to generate the levels required at each J and K input.

Table 7-6, the circuit excitation table, lists six J, K inputs—J_C, K_C, J_B, K_B, J_A, and K_A. We must consider each of these as an output from its own logic circuit with inputs from flip-flops C, B, and A. Then we must design the circuit for each one. Let's design the circuit for J_A.

To do this, we need to look at the PRESENT states of C, B, and A and the desired levels at J_A for each case. This information has been extracted from Table 7-6 and presented in Figure 7-27(a). This truth table shows the desired

TABLE 7-6
Circuit excitation table.

	PRESENT State			NEXT State								
	C	B	A	C	B	A	J_C	K_C	J_B	K_B	J_A	K_A
Line 1	0	0	0	0	0	1	0	x	0	x	1	x
2	0	0	1	0	1	0	0	x	1	x	x	1
3	0	1	0	0	1	1	0	x	x	0	1	x
4	0	1	1	1	0	0	1	x	x	1	x	1
5	1	0	0	0	0	0	x	1	0	x	0	x
6	1	0	1	0	0	0	x	1	0	x	x	1
7	1	1	0	0	0	0	x	1	x	1	0	x
8	1	1	1	0	0	0	x	1	x	1	x	1

FIGURE 7-27 (a) Portion of circuit excitation table showing J_A for each PRESENT state; (b) K map used to obtain the simplified expression for J_A.

PRESENT			
C	B	A	J_A
0	0	0	1
0	0	1	x
0	1	0	1
0	1	1	x
1	0	0	0
1	0	1	x
1	1	0	0
1	1	1	x

(a)

	$\overline{A}$	A
$\overline{C}\,\overline{B}$	1	X
$\overline{C}B$	1	X
CB	0	X
$C\overline{B}$	0	X

$J_A = \overline{C}$

(b)

levels at J_A for each PRESENT state. Of course, for some of the cases, J_A is a don't-care. To develop the logic circuit for J_A, we must first determine its expression in terms of C, B, and A. We will do this by transferring the truth-table information to a three-variable Karnaugh map and performing the K-map simplification, as in Figure 7-27(b).

There are only two 1s in this K map, and they can be looped to obtain the term $\overline{A}\,\overline{C}$, but if we use the don't-care conditions at $A\,\overline{B}\,\overline{C}$ and $AB\overline{C}$ as 1s, we can loop a quad to obtain the simpler term $\overline{C}$. Thus, the final expression is

$$J_A = \overline{C}$$

Now let's consider K_A. We can follow the same steps as we did for J_A. However, a look at the entries under K_A in the circuit excitation table shows only 1s and don't-cares. If we change all the don't-cares to 1s, then K_A is always a 1. Thus, the final expression is

$$K_A = 1$$

In a similar manner, we can derive the expressions for J_C, K_C, J_B, and K_B. The K maps for these expressions are given in Figure 7-28. You might want to confirm their correctness by checking them against the circuit excitation table.

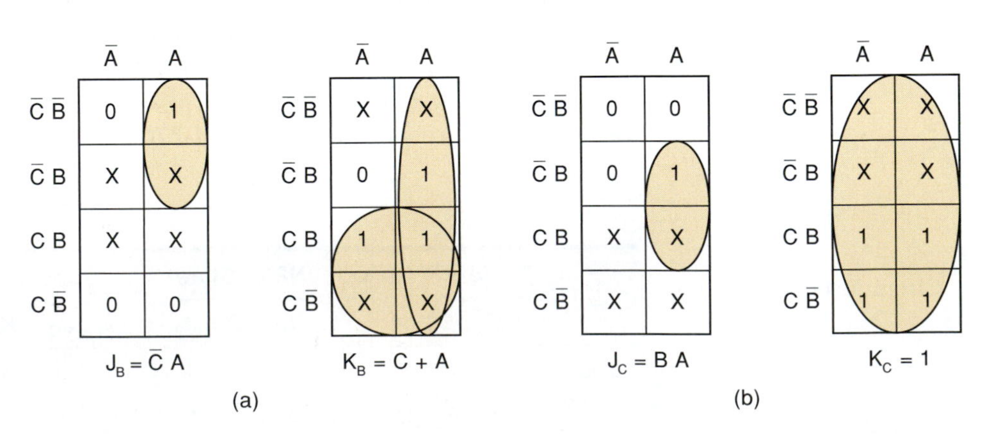

FIGURE 7-28 (a) K maps for the J_B and K_B logic circuits; (b) K maps for the J_C and K_C logic circuits.

Step 6. Implement the final expressions.

The logic circuits for each J and K input are implemented from the expressions obtained from the K map. The complete synchronous counter design is implemented in Figure 7-29. Note that all FFs are clocked in parallel. You might want to verify that the logic for the J and K inputs agrees with Figures 7-27 and 7-28.

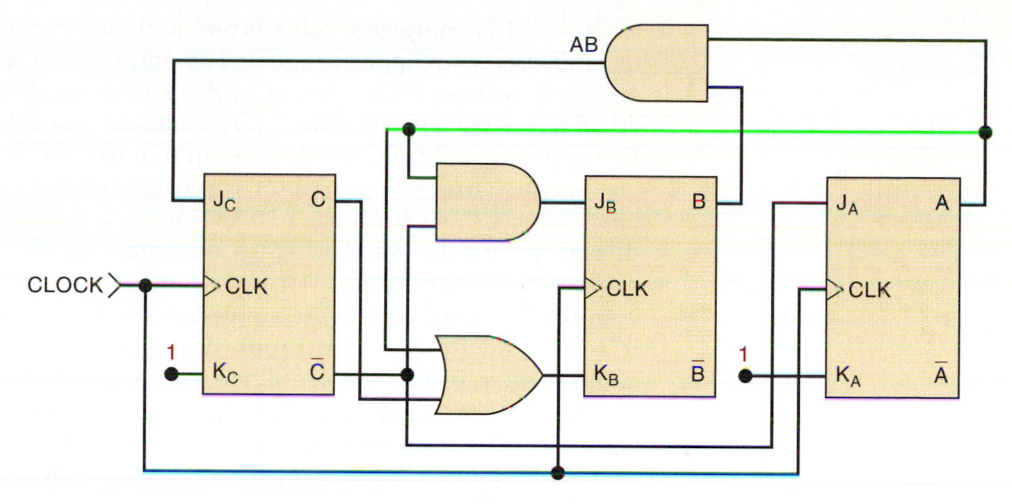

FIGURE 7-29 Final implementation of the synchronous counter design example.

Stepper Motor Control

We will now apply this design procedure to a practical situation—driving a *stepper motor.* A stepper motor is a motor that rotates in steps, typically 15° per step, rather than in a continuous motion. Magnetic coils or windings within the motor must be energized and deenergized in a specific sequence in order to produce this stepping action. Digital signals are normally used to control the current in each of the motor's coils. Stepper motors are used extensively in situations where precise position control is needed, such as in positioning of read/write heads on magnetic disks, in controlling print heads in printers, and in robots.

Figure 7-30(a) is a diagram of a typical stepper motor with four coils. For the motor to rotate properly, coils 1 and 2 must always be in opposite states; that is,

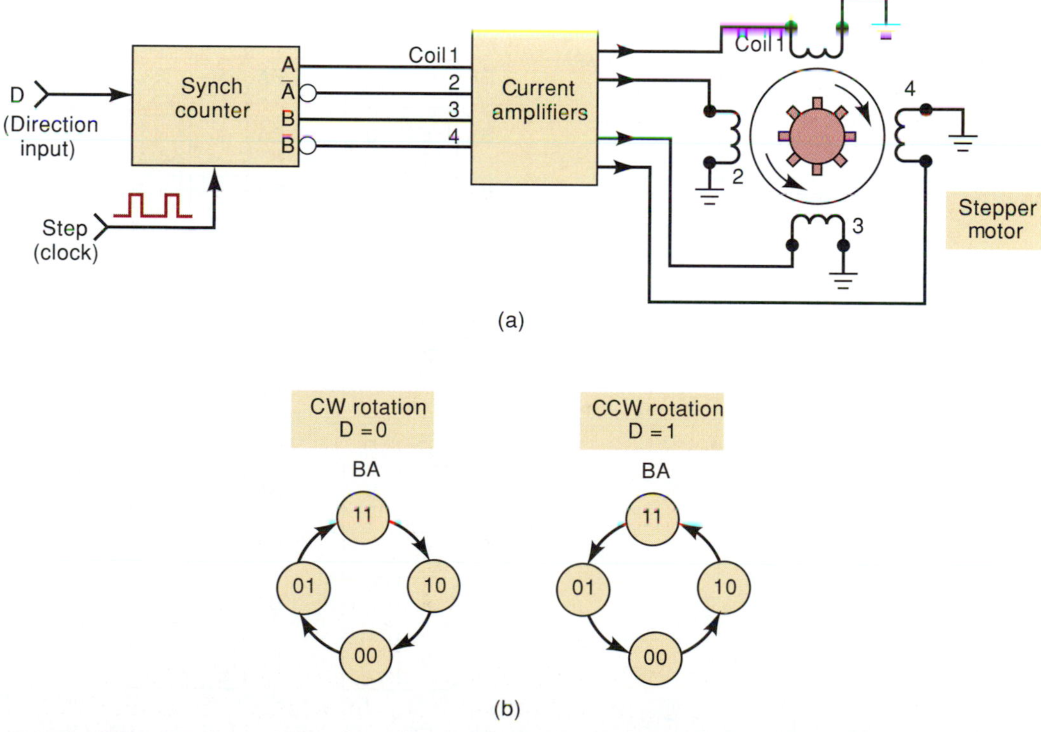

FIGURE 7-30 (a) A synchronous counter supplies the appropriate sequential outputs to drive a stepper motor; (b) state transition diagrams for both states of Direction input, *D*.

when coil 1 is energized, coil 2 is not, and vice versa. Likewise, coil 3 and coil 4 must always be in opposite states. The outputs of a two-bit synchronous counter are used to control the current in the four coils; A and $\overline{A}$ control coils 1 and 2, and B and $\overline{B}$ control coils 3 and 4. The current amplifiers are needed because the FF outputs cannot supply the amount of current that the coils require.

Because this stepper motor can rotate either clockwise (CW) or counter-clockwise (CCW), we have a Direction input, D, which is used to control the direction of rotation. The state diagrams in Figure 7-30(b) show the two cases. For CW rotation to occur, we must have $D = 0$, and the state of the counter, BA, must follow the sequence 11, 10, 00, 01, 11, 10, . . . , and so on, as it is clocked by the Step input signal. For CCW rotation, $D = 1$, and the counter must follow the sequence 11, 01, 00, 10, 11, 01, . . . , and so on.

We are now ready to follow the six steps of the synchronous counter design procedure. Steps 1 and 2 have already been done, so we can proceed with steps 3 and 4. Table 7-7 shows each possible PRESENT state of D, B, and

TABLE 7-7

PRESENT State			NEXT State		Control Inputs			
D	B	A	B	A	J_B	K_B	J_A	K_A
0	0	0	0	1	0	x	1	x
0	0	1	1	1	1	x	x	0
0	1	0	0	0	x	1	0	x
0	1	1	1	0	x	0	x	1
1	0	0	1	0	1	x	0	x
1	0	1	0	0	0	x	x	1
1	1	0	1	1	x	0	1	x
1	1	1	0	1	x	1	x	0

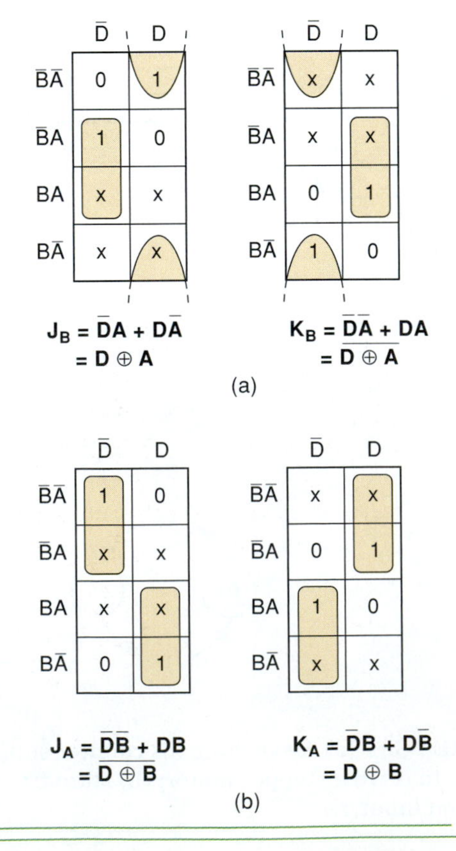

FIGURE 7-31 (a) K maps for J_B and K_B; (b) K maps for J_A and K_A.

A and the desired NEXT state, along with the levels at each *J* and *K* input needed to achieve the transitions. Note that in all cases, the Direction input, *D*, does not change in going from the PRESENT to the NEXT state because it is an independent input that is held HIGH or LOW as the counter goes through its sequence.

Step 5 of the design process is presented in Figure 7-31, where the information in Table 7-7 has been transferred to the K maps showing how each *J* and *K* signal is related to the PRESENT states of *D*, *B*, and *A*. Using the appropriate looping, the simplified logic expressions for each *J* and *K* signal are obtained.

The final step is shown in Figure 7-32, where the two-bit synchronous counter is implemented using the *J*, *K* expressions obtained from the K maps.

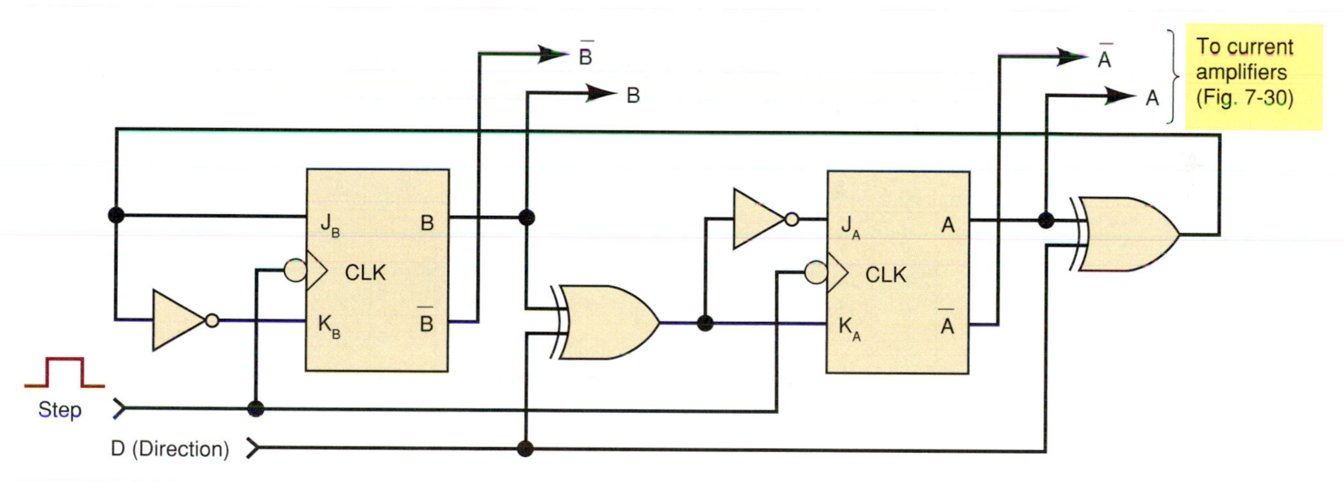

FIGURE 7-32 Synchronous counter implemented from the *J*, *K* equations.

Synchronous Counter Design with D FF

We have provided a detailed procedure for designing synchronous counters using J-K flip-flops. Historically, J-K flip-flops have been used to implement counters because the logic circuits needed for the J and K inputs are usually simpler than the logic circuits needed to control an equivalent synchronous counter using D flip-flops. When designing counters that will be implemented in PLDs, where abundant gates are generally available, it makes sense to use D flip-flops instead of J-Ks. Let us now look at synchronous counter design using D FFs.

Designing counter circuits using D flip-flops is even easier than using J-K flip-flops. We will demonstrate by designing a D FF circuit that produces the same count sequence as is given in Figure 7-26. The first three steps for synchronous D counter design are identical to the J-K technique. Step 4 for D FF design is trivial since the necessary D inputs are the same as the desired NEXT state as seen in Table 7-8. Step 5 is to generate the logic expressions

TABLE 7-8

PRESENT State			NEXT State			Control Inputs		
C	B	A	C	B	A	D_C	D_B	D_A
0	0	0	0	0	1	0	0	1
0	0	1	0	1	0	0	1	0
0	1	0	0	1	1	0	1	1
0	1	1	1	0	0	1	0	0
1	0	0	0	0	0	0	0	0
1	0	1	0	0	0	0	0	0
1	1	0	0	0	0	0	0	0
1	1	1	0	0	0	0	0	0

from the PRESENT state/NEXT state table for the D inputs. The K maps and simplified expressions are given in Figure 7-33. Finally, for step 6, the counter can be implemented with the circuit shown in Figure 7-34.

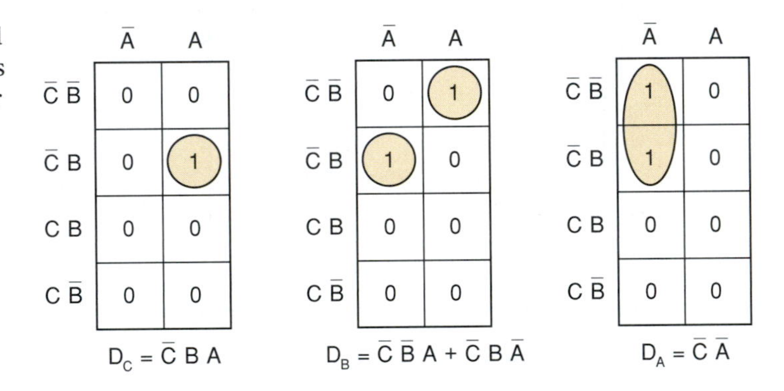

$D_C = \overline{C}\,B\,A$
$D_B = \overline{C}\,\overline{B}\,A + \overline{C}\,B\,\overline{A}$
$D_A = \overline{C}\,\overline{A}$

FIGURE 7-33 K maps and simplified logic expressions for MOD-5 flip-flop counter design.

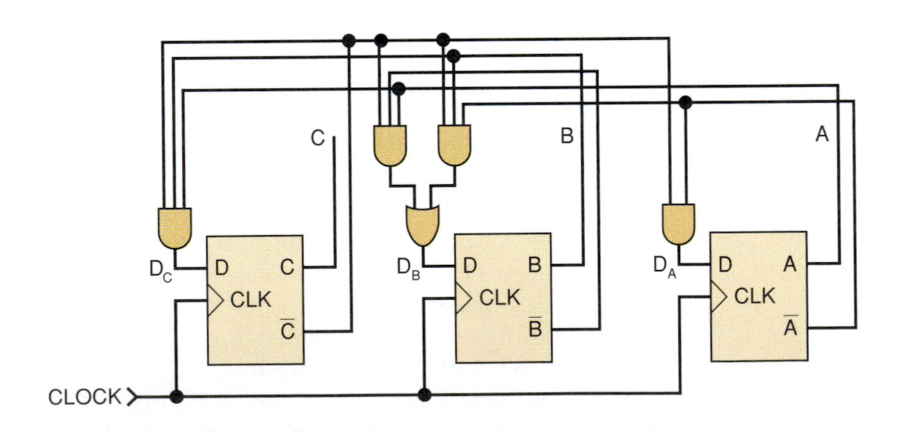

FIGURE 7-34 Circuit implementation of MOD-5 D flip-flop counter design.

REVIEW QUESTIONS

1. List the six steps in the procedure for designing a synchronous counter.
2. What information is contained in a PRESENT state-NEXT state table?
3. What information is contained in the circuit excitation table?
4. *True or false:* The synchronous counter design procedure can be used for the following sequence: 0010, 0011, 0100, 0111, 1010, 1110, 1111, and repeat.

7-11 BASIC COUNTERS USING HDL

In Chapter 5, we studied flip-flops and the methods used with HDLs to represent flip-flop circuits. The last section in Chapter 5 illustrated how to connect FF components very much like you would wire integrated circuits to one another. By connecting the Q output of one FF to the clock input of the next FF, we found that a counter circuit can be created. Using an HDL to describe component connections is referred to as the structural level of abstraction. It is obvious that constructing a complicated circuit using the structural methods would be very tedious and also very difficult to read and interpret. In this section, we will broaden our use of HDL to describe circuits using methods that are considered higher levels of abstraction. This term sounds intimidating, but it only means that there are much more concise and sensible ways to describe what we want a counter to do without worrying about all the details of how to wire flip-flop circuits to do it.

It is still vital that we understand the fundamental principles of flip-flop operation compared with combinational logic gates. As you recall, flip-flops have the following unique characteristics. The output is normally updated according to the condition of the synchronous control inputs when the *active edge* of the clock occurs, which means there is a logic state on the Q output before the clock edge (PRESENT state) and potentially a different state on the Q output after the clock edge (NEXT state). A flip-flop "remembers," or holds its state between clocks, regardless of changes in the synchronous control inputs (e.g., J and K).

Counter circuits using HDL rely on this basic understanding of a circuit going through a sequence of states in response to the event of a clock edge. Ripple counters provide an easy circuit to analyze and understand. They are also much less complicated to build using flip-flops and logic gates than their synchronous counterparts. The problem with ripple counters is the combination of time delay and spurious temporary states that occur when the counter changes state. When we advance to the next level of abstraction and plan to use PLDs to implement our design, we are no longer focusing on wiring issues but rather on describing the circuit's operation concisely. Consequently, the methods we use to describe counter circuits using HDL primarily use synchronous techniques, where all flip-flops update simultaneously in response to the same clock event. All the bits in a count sequence go from their PRESENT state to their prescribed NEXT state simultaneously, thereby preventing any intermediate, spurious states.

State Transition Description Methods

The next method of describing circuits that we need to examine uses tables. This method is not concerned with connecting ports of components but rather with assigning values to objects like ports, signals, and variables. In other words, it describes how the output data relates to the input data throughout the circuit. We have already used this method in several of the introductory circuits in Chapters 3 and 4, in the form of truth tables. With sequential counter circuits, the equivalent of the truth table is the PRESENT state/NEXT state table, as we saw in the last section. We can use the HDL essentially to describe the PRESENT state/NEXT state table and thus avoid the tedious details of generating the Boolean equations, as we did in Section 7-10 to design with standard logic devices.

STATE DESCRIPTIONS IN AHDL

As an example of a simple counter circuit, we will implement the MOD-5 counter of Figure 7-26 in AHDL. The inputs and outputs are defined in the SUBDESIGN section of Figure 7-35, as always. In the VARIABLE section on line 7, we have declared (or instantiated) a three-bit array of DFF primitives that are given the instance name *count[ ]*. This array will be treated basically as a three-bit register in the design and we will essentially define what value should be stored for each NEXT state. Because this is a synchronous counter, we need to tie all the DFF *clk* inputs to the SUBDESIGN's *clock* input. This is accomplished in AHDL by the following statement in the logic section:

```
count[].clk = clock;
```

The flip-flop primitives provided in AHDL have standard inputs and outputs that are referred to as "ports." These ports are labeled by a standard port name that is attached to the instance name of the flip-flops. As seen in Table 5-3, the clock port name is *.clk*, a D input is named *.d*, and the FF's output has the name *.q*. To implement the PRESENT state/NEXT state table, a CASE construct is used. For each of the possible values of the register *count[ ]*, we determine the value that should be placed on the *D* inputs of the flip-flops, which will determine the NEXT state of the counter. The statement on line 21 assigns the value on *count[ ]* to the output pins. Without this line, the counter would be "buried" in the SUBDESIGN and would not be visible to the outside world.

An alternative design solution is given in Figure 7-36. There are two modifications from Figure 7-35. The first is seen on line 7, where the array name for the D flip-flops is now the same as the output port for the SUBDESIGN.

```
1   SUBDESIGN fig7_35
2   (
3         clock           :INPUT;
4         q[2..0]         :OUTPUT;
5   )
6   VARIABLE
7         count[2..0]   :DFF;          --create a 3-bit register
8   BEGIN
9         count[].clk = clock;        --connect all clocks in parallel
10
11              CASE count[] IS
12  --                 Present            Next
13  ------------------------------------------------------------
14                      WHEN  0      =>     count[].d = 1;
15                      WHEN  1      =>     count[].d = 2;
16                      WHEN  2      =>     count[].d = 3;
17                      WHEN  3      =>     count[].d = 4;
18                      WHEN  4      =>     count[].d = 0;
19                      WHEN OTHERS =>     count[].d = 0;
20               END CASE;
21         q[] = count[];              -- assign register to output pins
22  END;
```

FIGURE 7-35 AHDL MOD-5 counter.

FIGURE 7-36 Another version of the MOD-5 counter described in Figure 7-26.

```
1   SUBDESIGN fig7_36
2   (
3       clock       :INPUT;
4       q[2..0]     :OUTPUT;
5   )
6   VARIABLE
7       q[2..0]     :DFF;     -- create a 3-bit register
8   BEGIN
9       q[].clk = clock;      -- connect all clocks in parallel
10      TABLE
11          q[].q =>     q[].d;
12          0      =>    1;
13          1      =>    2;
14          2      =>    3;
15          3      =>    4;
16          4      =>    0;
17          5      =>    0;
18          6      =>    0;
19          7      =>    0;
20      END TABLE;
21  END;
```

This will automatically connect the flip-flop outputs to the SUBDESIGN outputs and eliminate the need to include an assignment statement like line 21 in the first solution. The second modification is the use of an AHDL TABLE instead of the CASE statement used in Figure 7-35. In line 11, the *.q* port on the *q[ ]* DFF array represents the PRESENT state side of the table, while the *.d* port for *q[ ]* represents the NEXT state that will be entered into the array's set of D inputs when a PGT is applied to *clock*.

STATE DESCRIPTIONS IN VHDL

As an example of a simple counter circuit, we will implement the MOD-5 counter of Figure 7-26 in VHDL. Our purpose in this example is to demonstrate a counter using a control structure similar to a PRESENT state/NEXT state table. Two key tasks must be accomplished in VHDL: detecting the desired clock edge, and assigning the proper NEXT state to the counter. Recall from our study of flip-flops that a PROCESS can be used to respond to a transition of an input signal. Also, we have learned that a CASE construct can evaluate an expression and, for any valid input value, assign a corresponding value to another signal. The code in Figure 7-37 uses a PROCESS and a CASE construct to implement this counter. The inputs and outputs are defined in the ENTITY declaration, as in the past.

When VHDL is used to describe a counter, we must find a way to "store" the state of the counter between clock pulses (i.e., the action of a flip-flop). This is done in one of two ways: using SIGNALs, or using VARIABLEs. We have used SIGNALs extensively in previous examples that operated concurrently. A SIGNAL in VHDL holds the last value that was assigned to it, very much like a flip-flop. Consequently, we can use a SIGNAL as the data object representing the counter value. This SIGNAL can then be used to

```
1    ENTITY fig7_37 IS
2    PORT   (
3              clock   :IN BIT;
4              q       :OUT BIT_VECTOR(2 DOWNTO 0)
5          );
6    END fig7_37 ;
7
8    ARCHITECTURE a OF fig7_37    IS
9    BEGIN
10      PROCESS (clock)                              -- respond to clk input
11      VARIABLE count: BIT_VECTOR(2 DOWNTO 0);      -- create a 3-bit register
12      BEGIN
13         IF (clock = '1' AND clock'EVENT) THEN     -- rising edge trigger
14            CASE count IS
15   --            Present              Next
16   -------------------------------------------------------------------------
17               WHEN "000"    =>    count := "001";
18               WHEN "001"    =>    count := "010";
19               WHEN "010"    =>    count := "011";
20               WHEN "011"    =>    count := "100";
21               WHEN "100"    =>    count := "000";
22               WHEN OTHERS   =>    count := "000";
23            END CASE;
24         END IF;
25         q <= count;            -- assign register to output pins
26      END PROCESS;
27   END a;
```

FIGURE 7-37 VHDL MOD-5 counter.

connect the counter value to any other elements in the architecture description.

In this design, we have chosen to use a **VARIABLE** instead of a SIGNAL as the data object that stores the counter value. VARIABLEs are not exactly like SIGNALs because they are not used to connect various parts of the design. Instead, they are used as a local place to "store" a value. Variables are considered to be local data objects because they are recognized only within the PROCESS in which they are declared. On line 11 of Figure 7-37, the variable named *count* is declared within the PROCESS before BEGIN. Its type is the same as the output port *q*. The keyword PROCESS on line 10 is followed by the sensitivity list containing the input signal *clock*. Whenever *clock* changes state, the PROCESS is invoked, and the statements within the PROCESS will be evaluated to produce a result. A 'EVENT (read as "tick-event") attribute will evaluate as TRUE if the signal preceding it has just changed states. Line 13 states that if *clock* has just changed states and right now it is '1', then we know it was a rising edge. To implement the PRESENT state/NEXT state table, a CASE construct is used. For each of the possible values of the variable *count,* we determine the NEXT state of the counter. Notice that the = operator is used to assign a value to a variable. Line 25 assigns the value stored in *count* to the output pins. Because *count* is a local variable, this assignment must be done before END PROCESS on line 26.

Behavioral Description

The **behavioral level of abstraction** is a way to describe a circuit by describing its behavior in terms very similar to the way you might describe its operation in English. Think about the way a counter circuit's operation might be described by someone who knows nothing about flip-flops or logic gates. Perhaps that person's description would sound something like, "When the counter input changes from LOW to HIGH, the number on the output counts up by 1." This level of description deals more with cause-and-effect relationships than with the path of data flow or wiring details. However, we cannot really use just any description in English to describe the circuit's behavior. The proper syntax must be used within the constraints of the HDL.

AHDL

In AHDL, the first important step in this description method is to declare the counter output pins properly. They should be declared as a bit array, with indices decreasing left to right and with 0 as the least significant index in the array, as opposed to individual bits named a, b, c, d, and so on. In this way, the numeric value associated with the bit array's name is interpreted as a binary number upon which certain arithmetic operations can be performed. For example, the bit array *count* shown in Figure 7-38 might contain the bits 1001, as shown. The AHDL compiler interprets this bit pattern as having the value of 9 in decimal.

In order to create our MOD-5 counter in AHDL, we will need a three-bit register that will store the current counter state. This three-bit array, named *count*, is declared using D flip-flops on line 7 in Figure 7-39. Recall from Figure 7-36 that we could name the DFF array the same as the output port *q[2..0]* and thereby eliminate line 15, but we would also need to change *count[ ]* to *q[ ]* everywhere in the logic section. In other words, the statement on line 7 can be changed to

```
q[2..0] :DFF;.
```

If this were done, all references to *count* thereafter would be changed to *q*. This can make the code shorter, but it does not demonstrate universal HDL concepts as clearly. In AHDL, all the clocks can be specified as being tied together and connected to a common clock source using the statement on line 10, *count[ ].clk = clock.* In this example, *count[ ].clk* refers to the clock input of each flip-flop in the array called *count.*

The behavioral description of this counter is very simple. The current state of the counter is evaluated (*count[ ].q*) on line 11, and if it is less than the highest desired count value, it uses the description *count[ ].d = count.q* + 1 (line 12). This means that the current state of the *D* inputs must be equal to a value one count greater than the current state of the *Q* outputs. When the current state of the counter has reached the highest desired state (or higher), the *IF* statement test will be false, resulting in a NEXT-state input

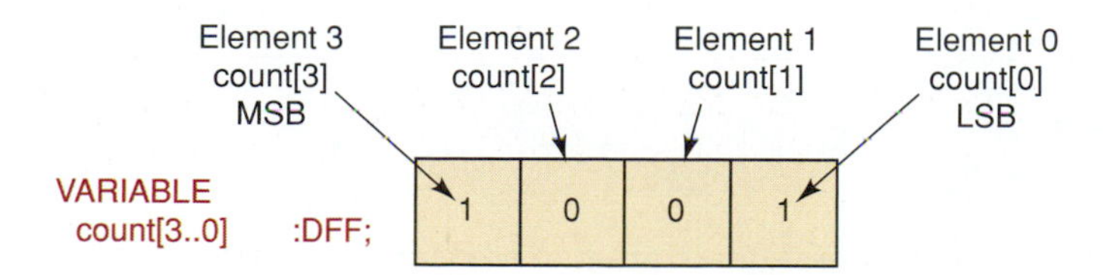

FIGURE 7-38 The elements of a D register storing the number 9.

```
1   SUBDESIGN fig7_39
2   (
3      clock     :INPUT;
4      q[2..0]   :OUTPUT;   -- declare 3-bit array of output bits
5   )
6   VARIABLE
7      count[2..0]   :DFF; -- declare a register of D flip flops.
8
9   BEGIN
10     count[].clk = clock;  -- connect all clocks to synchronous source
11     IF count[].q < 4 THEN    -- note; count[] is the same as count[].q
12        count[].d = count[].q + 1; -- increment current value by one
13     ELSE count[].d = 0;         -- recycle to zero: force unused states to 0
14     END IF;
15     q[] = count[];              -- transfer register contents to outputs
16  END;
```

FIGURE 7-39 Behavioral description of a counter in AHDL.

value of zero (line 13), which recycles the counter. The last statement on line 15 simply connects the counter value to the output pins of the device.

VHDL

VHDL

In VHDL, the first important step in this description method is to declare properly the counter output port, as shown in Figure 7-40. The data type of

```
1   ENTITY fig7_40 IS
2   PORT( clock    :IN BIT;
3         q  :OUT INTEGER RANGE 0 TO 7  );
4   END  fig7_40;
5
6   ARCHITECTURE a OF fig7_40 IS
7   BEGIN
8      PROCESS (clock)
9      VARIABLE count: INTEGER RANGE 0 to 7;  -- define a numeric VARIABLE
10         BEGIN
11           IF (clock = '1' AND clock'EVENT) THEN   -- rising edge?
12              IF count < 4 THEN              -- less than max?
13                 count := count + 1;         -- increment value
14              ELSE                           -- must be at max or bigger
15                 count := 0;                 -- recycle to zero
16              END IF;
17           END IF;
18     q <= count;                 -- transfer register contents to outputs
19     END PROCESS;
20  END a;
```

FIGURE 7-40 Behavioral description of a counter in VHDL.

the output port (line 3) must match the type of the counter variable (line 9), and it must be a type that allows arithmetic operations. Recall that VHDL treats BIT_VECTORS as just a string of bits, not as a binary numeric quantity. In order to recognize the signal as a numeric quantity, the data object must be typed as an INTEGER. The compiler looks at the RANGE 0 TO 7 clause on line 3 and knows that the counter needs three bits. A similar declaration is needed for the register variable on line 9 that will actually be counting up. This is called *count*. The first statement after BEGIN in the PROCESS responds to the rising edge of the clock as in the previous examples. It then uses behavioral description methods to define the counter's response to the clock edge. If the counter has not reached its maximum (line 12), then it should be incremented (line 13). Otherwise (line 14), it should recycle the counter to zero (line 15). The last statement on line 18 simply connects the counter value to the output pins of the device.

Simulation of Basic Counters

Simulation of any of our MOD-5 counter designs is pretty straightforward. The counters have only one input bit (*clock*) and three output bits (*q2 q1 q0*) to display in the simulation. The clock frequency has not been specified, so we can use any frequency that we wish for a functional simulation—although we probably should avoid a high-frequency clock unless we want to investigate the effects of propagation delays. About the only decision that we must make is to determine how many clock pulses to apply. Since the counter is a MOD-5 counter, we should apply at least five clock pulses to verify that the HDL design has the correct count sequence and that it recycles. The simulation will start with the initial state 000 because the Altera PLDs have a built-in power-on reset feature. We will not be able to test for any of the unused states because the HDL designs did not provide for a way to preset the counter to any of the unused states. Our simulation results for the HDL design of a MOD-5 counter are shown in Figure 7-41.

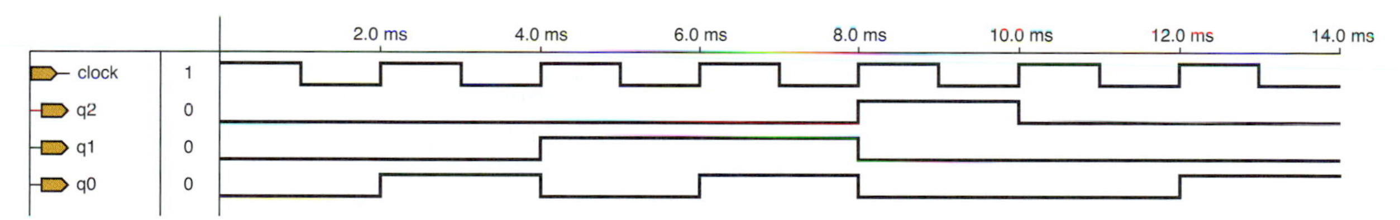

FIGURE 7-41 Simulation results for HDL design of MOD-5 counter.

REVIEW QUESTIONS

1. What type of table is used to describe a counter's operation?
2. When designing a counter with D flip-flops, what is applied to the D inputs in order to drive it to the NEXT state on the next active clock edge?
3. How would you write the HDL description to trigger a storage device (flip-flop) on a falling edge instead of a rising edge of the clock?
4. Which method describes the circuit's operation using cause-and-effect relationships?

7-12 FULL-FEATURED COUNTERS IN HDL

The examples we have chosen so far have been very basic counters. All they do is count up to four and then roll over to zero. The standard IC counters that we have examined have many other features that make them very useful for numerous digital applications. For example, consider the 74161 and the 74191 IC counters that were discussed in Section 7-7. These devices have combinations of various features including count enable, up/down counting, parallel loading (preset to any count), and clearing. In addition, these counters have been designed to easily cascade synchronously to create larger counters. In this section, we will explore the techniques that allow us to include these features in an HDL counter. We are going to create a counter that will combine more features than are found in either the 74161 or the 74191. We will use this example to demonstrate the methods of designing a counter with capabilities that specifically suit our needs. When we use HDLs to create digital designs, we are not limited to features that happen to be included with a certain IC.

Let's review the specifications for our more complex counter example. The recycling, MOD-16 binary counter is to change states on the rising edge of the clock input when the counter is enabled with a HIGH level. A direction control input will make the counter count up when it is LOW or count down when it is HIGH. The counter will have an active-HIGH, asynchronous clear to reset the counter immediately when the control input is activated. The counter can be synchronously loaded with a number on the data input pins when the load control is HIGH. The priority of the input control functions, from highest to lowest, will be clearing, loading, and counting. And finally, the counter will also include an active-HIGH output that will detect the terminal state of the counter when the count function is enabled. Remember, the terminal state will be dependent on the count direction. As we will see, the correct operation of these features is determined by the way we write the HDL code, so we will have to pay very close attention to the details.

AHDL

AHDL FULL-FEATURED COUNTER

The code in Figure 7-42 implements all of the features we have discussed. This is a four-bit counter, but it can easily be expanded in size. Read through the inputs and outputs on lines 3 and 4 to make sure you understand what each one is supposed to do. If you do not, reread the previous paragraphs of this section. Line 7 defines a four-bit register of D flip-flops that will serve as the counter. It should be noted again here that this register could have been named the same as the output variable (*q*). The code is written with different names to distinguish between ports (inputs and outputs) of the circuit and the devices that are operating within the circuit. The clock input is connected to all the *clk* inputs of all the D flip-flops on line 10. All the active-LOW clear inputs (*clrn*) to the DFF primitive are connected to the complement of the *clear* input signal on line 11. This clears the flip-flops immediately when the *clear* input goes HIGH because the *prn* and *clrn* inputs to the DFF primitive are not dependent on the clock (i.e., they are asynchronous).

In order to make the load function synchronously, the *D* inputs to the flip-flops must be controlled so that the input data (*din*) is present on the *D* inputs when the load line is HIGH. This way, when the next active clock edge comes along, the data will be loaded into the counter. This action must happen regardless of whether the counter is enabled or not. Consequently, the first conditional decision (IF) on line 12 evaluates the load input. Recall

```
1   SUBDESIGN fig7_42
2   (
3      clock, clear, load, cntenabl, down, din[3..0]       :INPUT;
4      q[3..0], term_ct :OUTPUT;  -- declare 4-bit array of output bits
5   )
6   VARIABLE
7      count[3..0]     :DFF;        -- declare a register of D flip flops
8
9   BEGIN
10     count[].clk = clock;        -- connect all clocks to synch source
11     count[].clrn= !clear;       -- connect for asynch active HIGH clear
12     IF load THEN count[].d = din[]; -- synchronous load
13        ELSIF !cntenabl THEN count[].d = count[].q; -- hold count
14        ELSIF !down THEN count[].d = count[].q + 1; -- increment
15        ELSE count[].d = count[].q - 1;              -- decrement
16     END IF;
17     IF ((count[].q == 0) & down # (count[].q == 15) & !down)& cntenabl
18     THEN      term_ct = VCC;     -- synchronous cascade output signal
19     ELSE term_ct = GND;
20     END IF;
21     q[] = count[];               -- transfer register contents to outputs
22  END;
```

FIGURE 7-42 Full-featured counter in AHDL.

from Chapter 4 that the IF/ELSE decision structure gives precedence to the first condition that is found to be true because, once it finds a condition that is true, it does not go on to evaluate the conditions in subsequent ELSE clauses. In this case, it means that if the load line is activated, it does not matter whether the count is enabled, or it is trying to count up or down. It will do a parallel load on the next clock edge.

Assuming that the load line is not active, the ELSIF clause on line 13 is evaluated to see if the count is disabled. In AHDL, it is very important to realize that the Q output must be fed back to the D input so that, on the next clock edge, the register will hold its previous value. Forgetting to insert this clause results in the D inputs defaulting to zero, thus resetting the counter. If the counter is enabled, the ELSIF clause on line 14 is evaluated and either increments *count* (line 14) or decrements *count* (line 15). To summarize these decisions, first decide if it is time to load, next decide if the count should hold or change, then decide whether to count up or down.

The next function described is the detecting (or decoding) of the terminal count. Lines 17–20 decide whether the terminal count has been reached while counting up or down. The double equals (==) operator is the symbol that tests for equality between the expressions on each side of the operator. Which counter state is the terminal state depends on the counting direction. This is determined by ANDing the appropriate terminal state detection of 0 or 15 with the correct expression, *down* or !*down*. *Term_ct* will output a HIGH if the correct state has been reached, otherwise it will be LOW. Line 21 will connect the output for *count* to the output pins for the SUBDESIGN.

One of the key concepts of using HDLs is that it is generally very easy to expand the size of a logic module. Let us look at the necessary changes to this AHDL design to increase the binary counter modulus to 256. Since $2^8 = 256$, we will need to increase the number of bits to eight. Only four modifications to Figure 7-42 will be required to make this change in counter modulus:

Line #	*Modification*
3	din [~~3~~ 7 . . 0]
4	q [~~3~~ 7 . . 0]
7	count [~~3~~ 7 . . 0]
17	(count [ ]. q == ~~15~~ 255)

VHDL FULL-FEATURED COUNTER

The code in Figure 7-43 implements all the features we have discussed. This is a four-bit counter, but it can easily be expanded in size. Read through the inputs and outputs on lines 2–5 to make sure you understand what each one is supposed to do. If you do not, reread the previous paragraphs of this section. The PROCESS statement on line 10 is the key to all clocked circuits described in VHDL, but it also plays an important role in determining whether the circuit responds synchronously or asynchronously to its inputs. We want

```
1   ENTITY fig7_43 IS
2   PORT( clock, clear, load, cntenabl, down   :IN BIT;
3         din         :IN INTEGER RANGE 0 TO 15;
4         q           :OUT INTEGER RANGE 0 TO 15;
5         term_ct     :OUT BIT);
6   END fig7_43;
7
8   ARCHITECTURE a OF fig7_43 IS
9      BEGIN
10        PROCESS ( clock, clear, down)
11        VARIABLE count :INTEGER RANGE 0 to 15;    -- define a numeric signal
12           BEGIN
13              IF clear = '1' THEN count := 0;     -- asynch clear
14              ELSIF (clock = '1' AND clock'EVENT)  THEN  -- rising edge?
15                 IF load = '1' THEN count := din;     -- parallel load
16                 ELSIF cntenabl = '1' THEN            -- enabled?
17                    IF down = '0' THEN count := count + 1;  -- increment
18                    ELSE           count := count - 1;  -- decrement
19                    END IF;
20                 END IF;
21              END IF;
22              IF (((count = 0) AND (down = '1')) OR
23                 ((count = 15) AND (down = '0'))) AND cntenabl = '1'
24                 THEN  term_ct <= '1';
25              ELSE     term_ct <= '0';
26              END IF;
27              q <= count;    -- transfer register contents to outputs
28           END PROCESS;
29       END a;
```

FIGURE 7-43 Full-featured counter in VHDL.

this circuit to respond immediately to transitions on the *clock, clear,* and *down* inputs. With these signals in the sensitivity list, we assure that the code inside the PROCESS will be evaluated as soon as any of these inputs change states. The variable *count* is defined on line 11 as an INTEGER so it can be incremented and decremented easily. Variables are declared within the PROCESS and can be used within the PROCESS only.

The *clear* input is given precedence by evaluating it with the first IF statement on line 13. Recall from Chapter 4 that the IF/ELSE decision structure gives precedence to the first condition that is found to be true because it does not go on to evaluate the conditions in subsequent ELSE clauses. In this case, if the *clear* is active, the other conditions will not matter. The output will be zero. In order to make the *load* function operate synchronously, it must be evaluated after detecting the clock edge. The clock edge is detected on line 14, and the circuit checks immediately to see if *load* is active. If *load* is active, the *count* is loaded from *din,* regardless of whether or not the counter is enabled. Consequently, the conditional decision (IF) on line 15 evaluates the *load* input; only if it is inactive does it evaluate line 16 to see if the counter is enabled. If the counter is enabled, the *count* will be incremented or decremented (lines 17 and 18, respectively).

The next issue is detecting the terminal count. Lines 22–25 decide whether the maximum or minimum terminal count has been reached and drive the output to the appropriate level. The decision-making structure here is very important because we want to evaluate this situation, regardless of whether the decision-making process was invoked by *clock, clear,* or *down.* Notice that this decision is not another ELSE branch of the previous IF decisions but is evaluated for each signal in the sensitivity list *after* the clearing or counting has occurred. After all these decisions are made, *count* should have the right value in the register, and line 27 effectively connects the register to the output pins.

One of the key concepts of using HDLs is that it is generally very easy to expand the size of a logic module. Let us look at the necessary changes to this VHDL design to increase the binary counter modulus to 256. Only four modifications to Figure 7-43 will be required to make this change in counter modulus:

Line #	***Modification***
3	RANGE 0 TO ~~15~~ 255
4	RANGE 0 TO ~~15~~ 255
11	RANGE 0 TO ~~15~~ 255
23	(count = ~~15~~ 255)

Simulation of Full-Featured Counter

Simulation of our full-featured counter design will require some planning to generate appropriate input waveforms. While it may not be necessary to exhaustively simulate every conceivable input combination, we do need to test enough of the possible input conditions to be convinced that it works properly. This is exactly what we should also do to test our prototype design on the bench. The counter has five different input signals (*clock, clear, load, cntenabl,* and *din*) and two different output signals (*q* and *term_ct*) to display in our simulation. One of the input signals and one of the output signals actually is four bits wide. We will pick a convenient clock frequency since none has been specified for our functional simulation of the counter. We will need to provide enough clock pulses to allow us to look at several operational conditions. The simulation should test the functions of enabling and disabling the counter, counting up and counting down, clearing the counter, loading a value into the counter and counting from that value, and terminal count state detection.

There are some general simulation issues that we should consider in creating our input waveforms. Since the target PLDs have power-on reset, the simulation will start with the initial output state at 0000. Therefore, it would be better to wait until the count has reached another state before applying a clear input so that we can see a change in the output. Likewise, loading in the same value as the counter's NEXT state does not really convince us that *load* is working correctly. Changing input control signals at the same time as the clocking edge occurs may create some setup time problems and produce questionable results. Asynchronous controls should be applied at a time other than the proper clocking edge to show clearly that the resultant circuit action is immediate and not dependent on the clock. In general, we should apply common sense in creating our input waveforms and consider what we are trying to verify with the simulation. Simulation will be valuable in the design process only if we apply appropriate input conditions and evaluate the results critically.

Some simulation results for the full-featured counter are shown in Figure 7-44. The four-bit input *din* and the four-bit output *q* are displayed in hexadecimal. The counter is initially enabled (*cntenabl* = 1) to count up (*down* = 0), and we see the output is incrementing 0, 1, 2, 3, 4, 5. At t_1, the counter synchronously (i.e., on the PGT of *clock*) responds to the HIGH applied to the *load* input. The counter is preset to the parallel data input (*din*) value of 8. This also shows that loading has priority over counting, since they are both active at the same time. After t_1, load is LOW again and the counter continues to count up from 8. A LOW input to *cntenabl* makes the counter hold at state 9 for an extra clock cycle. The count is continued when *cntenabl* goes HIGH again until t_2, when the counter is asynchronously cleared. Notice the shortened time for the output state A due to the immediate clearing of the counter. We would have to zoom in to actually see that state A is displayed. We can also see that the clear function has the highest priority when all three controls, *clear, load,* and *cntenabl,* are simultaneously high. The count-up sequence continues and recycles to 0 after state F to verify that the counter is a MOD-16 binary counter. At t_3, the counter reaches its terminal state F when counting up, and *term_ct* outputs a HIGH. At t_4, the counter starts counting down because *down* has been switched to a HIGH. Again, *term_ct* outputs a HIGH since the counter is now at state 0, which is the terminal state when counting down. Notice that, by the action of *term_ct,* the terminal state for the counter depends on its direction of counting, which is

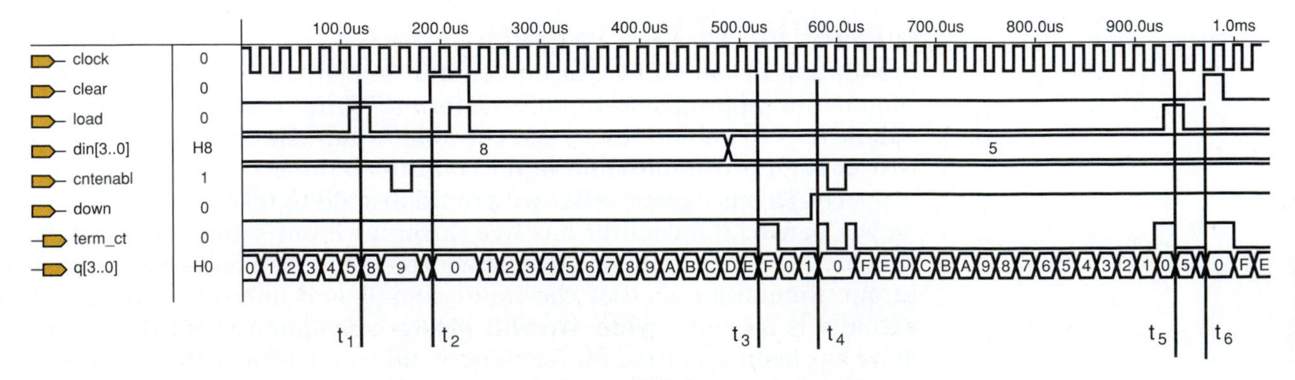

FIGURE 7-44 Simulation results for HDL design of full-featured counter.

controlled by the input *down*. The count holds at state 0 for an extra clock period when *cntenabl* goes LOW. The output *term_ct* is also disabled while *cntenabl* = 0. The down count sequence continues correctly when *cntenabl* again goes HIGH. At t_5, the counter synchronously loads the parallel data value 5. At t_6, the counter is asynchronously cleared. Again the priority of loading or clearing over a down count is verified at t_5 and t_6. Did we verify that our design operates correctly in comparison to the specifications? We did a pretty good job, but there are a couple of test conditions that could also be added for completeness. Will the counter clear or load when the *cntenabl* is LOW? It appears that we neglected to verify those scenarios. As you can see, complex designs may require a lot of thought to verify their operation adequately by simulation or bench testing. Can you think of any other tests that we should make?

REVIEW QUESTIONS

1. What is the difference between asynchronous clear and synchronous load?
2. How do you create an asynchronous clear function in an HDL?
3. How do you create functions priority in an HDL description of a counter?

7-13 WIRING HDL MODULES TOGETHER

In the previous two sections we have looked at how to implement common counter features using an HDL. We should also investigate how we can connect these counter circuits to other digital modules to create larger systems. Designing large digital systems becomes much easier if the system is subdivided into smaller, more manageable modules that are then interconnected. This is the essence of the concept of **hierarchical design**, and we will readily see its benefits with example projects in Chapter 10. Let us now look at the basic techniques for wiring modules together.

AHDL

DECODING THE AHDL MOD-5 COUNTER

We looked briefly at the idea of decoding a counter in Section 7-8. You should recall that a decoding circuit detects a counter's state by the unique bit pattern for that state. Let's see how to connect a decoder circuit to the MOD-5 counter design in Figure 7-35 (or Figure 7-36). We will rename the counter SUBDESIGN mod5 to be a bit more descriptive in the block diagram for the overall circuit that we will draw later. Since the counter does not produce all eight possible states for a three-bit counter, our decoder design shown in Figure 7-45 will only decode the states that are used, 000 through 100. The three input bits (c = MSB) declared on line 3 will be connected later to the MOD-5 counter's outputs. The five outputs for the decoder are named *state0* through *state4* on line 4. A CASE statement (lines 7–14) describes the behavior of the decoder by checking the $c\ b\ a$ input combination to determine which one of the decoder outputs should be HIGH. When the $c\ b\ a$ input is 000, only the *state0* output will be HIGH or, when $c\ b\ a$ is 001, only the *state1* output will be HIGH, and so on. Any input value greater than 100, which is covered by OTHERS and actually should not occur in this application, will produce LOWs on all outputs.

```
1   SUBDESIGN decode5
2   (
3       c, b, a                 : INPUT;
4       state[0..4]             : OUTPUT;
5   )
6   BEGIN
7       CASE (c,b,a) IS             -- decode binary value
8           WHEN B"000"     =>   state[] = B"10000";
9           WHEN B"001"     =>   state[] = B"01000";
10          WHEN B"010"     =>   state[] = B"00100";
11          WHEN B"011"     =>   state[] = B"00010";
12          WHEN B"100"     =>   state[] = B"00001";
13          WHEN OTHERS     =>   state[] = B"00000";
14      END CASE;
15  END;
```

FIGURE 7-45 AHDL MOD-5 counter decoder module.

We will instruct the Altera software to create symbols for our two design files, mod5 and decode5. This will allow us to draw a block diagram (see Figure 7-46) for our complete circuit that consists of these two modules, input and output ports, and the wiring between them. Each symbol is labeled with its respective SUBDESIGN name mod5 or decode5. Notice that some of the wiring is drawn with heavier-weight lines. This is to represent a bus, which is a collection of signal lines. The lighter-weight lines are individual signals. The symbols created by Altera will automatically have ports drawn to indicate whether they represent individual signals or buses. This will be determined by the signal declarations in the SUBDESIGN section. Ports with group names will be drawn as buses. Since the counter output port is a bus but the decoder input ports are individual signals, it will be necessary to split the bus into individual signal lines to wire the two modules together. Whenever a bus is split, you must label both the group signal name of the bus and the individual signals that are being used. Our block diagram has a bus labeled *q[2..0]* and the corresponding individual signals *q2, q1,* and *q0.* The simulation results for this counter and decoder circuit are shown in Figure 7-47.

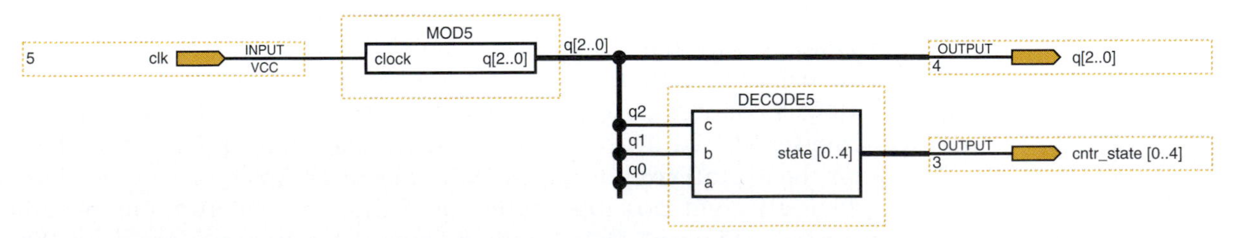

FIGURE 7-46 Block diagram design for the MOD-5 counter and decoder circuit.

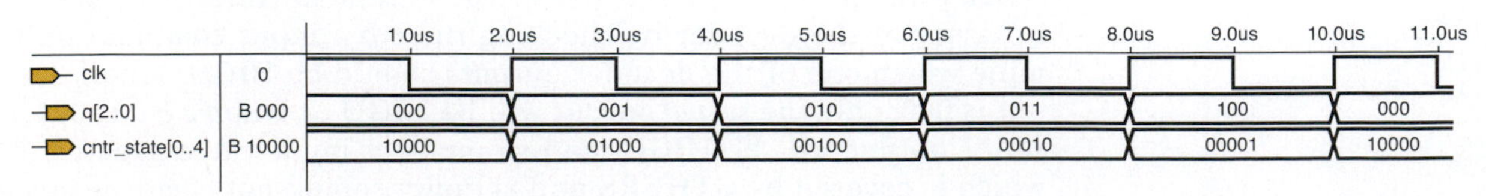

FIGURE 7-47 Simulation of MOD-5 counter and decoder circuit.

DECODING THE VHDL MOD-5 COUNTER

We looked briefly at the idea of decoding a counter in Section 7-8. You should recall that a decoding circuit detects a counter's state by the unique bit pattern for that state. Let's see how to connect a decoder circuit to the MOD-5 counter design in Figure 7-37. We will rename the counter ENTITY mod5 to make it easier to identify the module in our overall circuit. Since the counter does not produce all eight possible states for a three-bit counter, our decoder design shown in Figure 7-48 will only decode the states that are used, 000 through 100. The three input bits (c = MSB) declared on line 3 will be connected later to the MOD-5 counter's outputs. The five outputs for the decoder are named *state*, a bit vector, on line 4. An internal bit vector signal named *input* is declared on line 9. Then line 11 combines the three input port bits (c b a) together as a bit vector called *input*, which then can be evaluated by the CASE statement on lines 14–21. If any of the input bits changes logic level, the PROCESS will be invoked to determine the resultant output. The CASE statement describes the behavior of the decoder by checking the *input* combination (representing c b a) to determine which one of the decoder outputs should be HIGH. When the *input* is 000, only the *state(0)* output will be HIGH; when *input* is 001, only the *state(1)* output will be HIGH; and so on. Any *input* value greater than 100, which is covered by OTHERS and actually should not occur in this application, will produce LOWs on all outputs.

Since we are using the Altera PLD Development software, we can connect the two modules graphically. To do this, you will need to instruct the software to create symbols for our two design files, mod5 and decode5. This will allow us to draw a block diagram (see Figure 7-46) for our complete circuit that consists of these two modules, input and output ports, and the wiring between them. Notice that some of the wiring is drawn with heavier-weight lines. This is to represent a bus, which is a collection of signal lines. The lighter-weight lines are individual signals. The symbols created by Altera will automatically have ports drawn to indicate whether they represent individual signals or

```
1   ENTITY  decode5  IS
2   PORT (
3           c, b, a    : IN BIT;
4           state      : OUT BIT_VECTOR (0 TO 4)
5   );
6   END decode5;
7
8   ARCHITECTURE a OF decode5 IS
9   SIGNAL    input    : BIT_VECTOR (2 DOWNTO 0);
10  BEGIN
11      input <= (c & b & a);    -- combine inputs into bit vector
12      PROCESS (c, b, a)
13      BEGIN
14          CASE input IS
15              WHEN "000" =>       state <= "10000";
16              WHEN "001" =>       state <= "01000";
17              WHEN "010" =>       state <= "00100";
18              WHEN "011" =>       state <= "00010";
19              WHEN "100" =>       state <= "00001";
20              WHEN OTHERS =>      state <= "00000";
21          END CASE;
22      END PROCESS;
23  END a;
```

FIGURE 7-48 VHDL MOD-5 counter decoder module.

buses. This will be determined by the data type declarations for each port of the ENTITY. BIT_VECTOR ports will be drawn as buses and BIT type ports will be drawn as individual signal lines. Since the counter output port is a bus but the decoder input ports are individual signals, it will be necessary to split the bus into individual signal lines to wire the two modules together. Whenever a bus is split, you must label both the group signal name of the bus and the individual signals that are being used. Our block diagram has a bus labeled *q[2..0]* and the corresponding individual signals *q2*, *q1*, and *q0*. The simulation results for this counter and decoder circuit are shown in Figure 7-47.

The standard VHDL technique (and an alternative with Altera's software) to connect design modules is to use VHDL to describe the connections between the modules in a text file. The desired modules are instantiated in a higher-level design file using COMPONENTs in which the module's PORTs are declared. The wiring connections for each instance where the module is utilized are listed in a PORT MAP. A VHDL file that connects the mod5 and decode5 modules together is shown in Figure 7-49. Even though *q* is an output port for our top-level design file, it is typed as a BUFFER on line 4 due to the fact that it is necessary to "read" the bit vector array for an input to the *decode5* COMPONENT in its PORT MAP (line 25). VHDL does not permit output ports to be used as inputs. The BUFFER data type declaration provides a port that can be used for both input and output. The mod5 module is declared on lines 10–15 and the decode5 module is declared on lines 16–21. The mod5 and decode5 ENTITY/ARCHITECTURE descriptions may be included within the top-level design file, or instead they may be saved in the same folder as the top-level file as was done here. The PORT MAP for each instance of the modules is listed on lines 23 and 24–25. The word to the left of the colon is a unique label for each instance and the module name is on the right, then the keywords PORT MAP, and finally, in parentheses, are the named associations between the design signals and ports. The => operator indicates which module ports (on the left side) are connected to which

```
1   ENTITY mod5decoded1 IS
2   PORT (
3        clk              :IN BIT;
4        q                :BUFFER BIT_VECTOR (2 DOWNTO 0);
5        cntr_state       :OUT BIT_VECTOR (0 TO 4)
6        );
7   END mod5decoded1;
8
9   ARCHITECTURE toplevel OF mod5decoded1 IS
10  COMPONENT mod5
11       PORT (
12            clock       :IN BIT;
13            q           :OUT BIT_VECTOR (2 DOWNTO 0)
14            );
15  END COMPONENT;
16  COMPONENT decode5
17  PORT (
18            c, b, a     :IN BIT;
19            state       :OUT BIT_VECTOR (0 TO 4)
20            );
21  END COMPONENT;
22  BEGIN
23  counter:  mod5        PORT MAP (clock => clk, q => q);
24  decoder:  decode5     PORT MAP
25       (c => q(2), b => q(1), a => q(0), state => cntr_state);
26  END toplevel;
```

FIGURE 7-49 Higher-level VHDL file to connect mod5 and decode5 together.

higher-level system signals (on the right side). This circuit produces the simulation results shown in Figure 7-47.

MOD-100 BCD Counter

We wish to design a recycling, MOD-100 BCD counter that has a synchronous clear. Creating a MOD-10 BCD counter module and synchronously cascading two of these modules together in a higher-level design file is the easiest way to do this. The clock inputs to the two MOD-10 modules will both be connected to the system clock to achieve synchronous cascading of the two counter modules. Remember, there are significant benefits to using synchronous counter design rather than asynchronous clocking techniques. Also, if we did not employ synchronous clocking, the synchronous clear would not work properly. Even though the design specifications did not require a count enable or terminal count detection for the MOD-100 counter, it will be necessary to include these features in our design. In order to synchronously cascade two counters, the enable and decoding features will be needed. The count enable input causes the counter to ignore clock edges unless it is enabled. The terminal count output indicates that the counting sequence has reached its limit and will roll over on the next clock. To synchronously cascade counter stages together, the terminal count output is connected to the next higher-order stage's enable input. By using the count enable to also control the decoding of the terminal count, our MOD-10 module can be used to create even larger BCD counters.

CASCADING AHDL BCD COUNTERS

Our MOD-10 BCD counter SUBDESIGN is shown in Figure 7-50. The terminal state for a BCD counter is 9. Lines 10–13 will detect this terminal state only when the counter is enabled with a HIGH. ANDing the *enable* control

```
1   SUBDESIGN  mod10
2   (
3        clock, enable, clear              :INPUT;
4        counter[3..0], tc                 :OUTPUT;
5   )
6   VARIABLE
7        counter[3..0]                     :DFF;
8   BEGIN
9        counter[].clk  = clock;
10       IF counter[].q == 9 & enable == VCC  THEN
11                 tc = VCC;                  -- detect terminal count
12       ELSE      tc = GND;
13       END IF;
14       IF  clear  THEN
15            counter[].d = B"0000";          -- synchronous clear
16       ELSIF  enable  THEN                  -- clear has priority
17            IF counter[].q == 9  THEN       -- check for last state
18                 counter[].d = B"0000";
19            ELSE
20                 counter[].d = counter[].q + 1;       -- increment
21            END IF;
22       ELSE                             -- hold count when disabled
23            counter[].d = counter[].q;
24       END IF;
25  END;
```

FIGURE 7-50 MOD-10 BCD counter in AHDL.

AHDL

in the decoding function will allow more than two counter modules to be cascaded synchronously if necessary and makes our mod10 design more versatile. The *clear* function will operate synchronously in AHDL by including it in the IF statement as shown on lines 14–15. If *clear* is inactive, we next check to see if the counter is enabled (line 16). If *enable* is HIGH, the counter checks, using a nested IF on lines 17–21, to see if the last state 9 has been reached. After state 9, the counter synchronously recycles to 0. Otherwise, the count will be incremented. If the counter is disabled, lines 22–23 will hold the current count value by feeding the current output back to the counter's input. This holding action will be necessary in the cascaded MOD-100 counter for the 10s digit to hold its current state while the 1s digit progresses through its count sequence. An appropriate design strategy would be for us to simulate this module to determine if it functions correctly before we use it in a more complex circuit application. From the simulation results for mod10, given in Figure 7-51, we see that the count sequence is correct, the *clear* is synchronous and has priority, and *enable* controls both the count function and the decoding output tc.

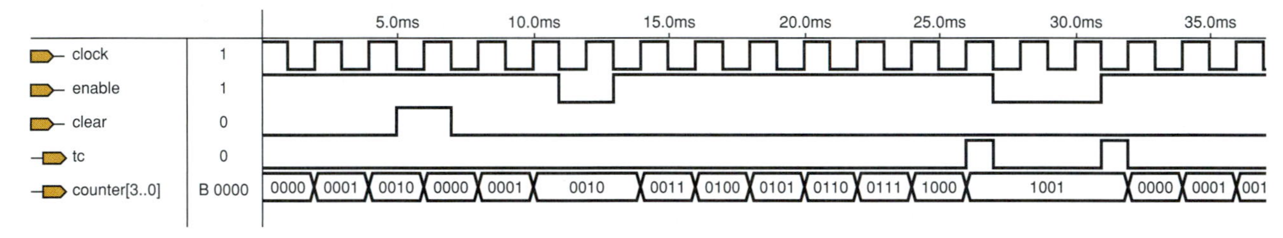

FIGURE 7-51 MOD-10 simulation results.

After creating a default symbol for our mod10 counter module, we can now draw the block diagram for the MOD-100 BCD counter application. The input ports, output ports, and wiring have also been added to create the design in Figure 7-52. Notice that the counter outputs representing the 1s and 10s digits are drawn as buses. The mod10 modules are clocked synchronously. They are cascaded by using the terminal count output from the 1s digit to control the enable input on the 10s digit. The *en* input port controls the enabling/disabling of the entire MOD-100 counter circuit. The BCD counter design can be easily expanded with an additional mod10 stage by connecting the *tc* output to the next *enable* input for each digit needed. A sample of simulation results can be seen in Figure 7-53. The simulation shows that the MOD-100 counter has a correct BCD count sequence and can be synchronously cleared.

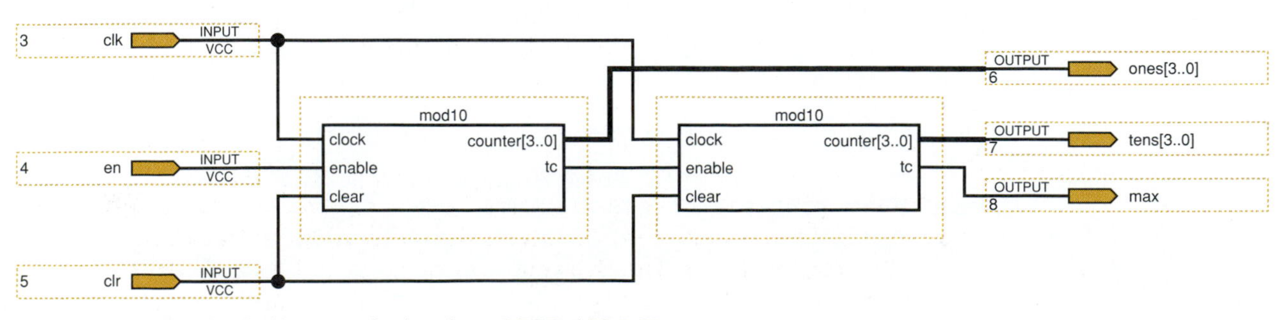

FIGURE 7-52 Block diagram design for a MOD-100 BCD counter.

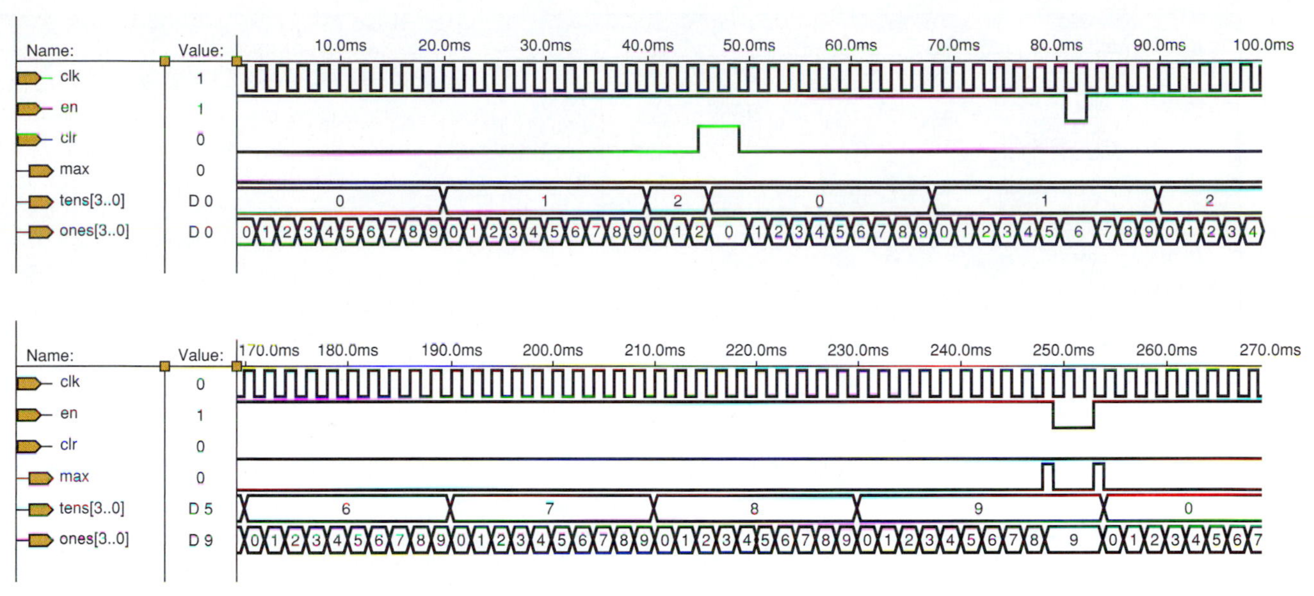

FIGURE 7-53 Simulation results for MOD-100 BCD counter design.

CASCADING VHDL BCD COUNTERS

The ENTITY and ARCHITECTURE for our MOD-10 BCD counter is shown in lines 26–51 of Figure 7-54. The terminal state for a BCD counter is 9. Lines 38–40 will detect this terminal state only when the counter is enabled with a HIGH. ANDing the *enable* control in the decoding function will allow more than two counter modules to be cascaded synchronously if necessary and makes our mod10 design more versatile. The *clear* function will be synchronous in VHDL by placing it in the nested IF statement (line 42) after the clock edge has been detected in line 41. If *clear* is inactive, we next check to see if the counter is enabled (line 43). If *enable* is HIGH, the counter checks, using another nested IF on lines 44–46, to see if the last state 9 has been reached. After state 9, the counter synchronously recycles to 0. Otherwise, the count will be incremented. If the counter is disabled, VHDL will automatically hold the current count value. This holding action will be necessary in the cascaded MOD-100 counter for the 10s digit to hold its current state while the 1s digit progresses through its count sequence. An appropriate design strategy would be for us to simulate this module as a separate ENTITY to determine if it functions correctly before we use it in a more complex circuit application. Simulation results for the mod10 ENTITY, given in Figure 7-51, show that the count sequence is correct, the *clear* is synchronous and has priority, and *enable* controls both the count function and the decoding output.

We have two choices for implementing the MOD-100 counter. One technique is to represent the design graphically in a block diagram as seen in Figure 7-52. The mod10 counter modules, input ports, output ports, and wiring have also been added to create the MOD-100 counter. Notice that the counter outputs representing the 1s and 10s digits are drawn as buses. The mod10 modules are clocked synchronously. They are cascaded by using the terminal count output from the 1s digit to control the enable input on the 10s digit. The *en* input port controls the enabling/disabling of the entire MOD-100 counter circuit. The BCD counter design can be easily expanded with an additional mod10 stage by connecting the *tc* output to the next *enable* input for each digit needed. A sample of simulation results can be seen in Figure 7-53.

```
1   ENTITY mod100 IS
2   PORT (
3         clk, en, clr                  :IN BIT;
4         ones                          :OUT INTEGER RANGE 0 TO 15;
5         tens                          :OUT INTEGER RANGE 0 TO 15;
6         max                           :OUT BIT
7   );
8   END mod100;
9   ARCHITECTURE toplevel OF mod100 IS
10  COMPONENT mod10
11        PORT (
12              clock, enable, clear    :IN BIT;
13              q                       :OUT INTEGER RANGE 0 TO 15;
14              tc                      :OUT BIT
15              );
16  END COMPONENT;
17  SIGNAL rco                          :BIT;
18  BEGIN
19  digit1:  mod10  PORT MAP (clock => clk, enable => en,
20                         clear => clr, q => ones, tc => rco);
21  digit2:  mod10  PORT MAP (clock => clk, enable => rco,
22                         clear => clr, q => tens, tc => max);
23  END toplevel;
24
25
26  ENTITY mod10 IS
27  PORT (
28        clock, enable, clear          :IN BIT;
29        q                             :OUT INTEGER RANGE 0 TO 15;
30        tc                            :OUT BIT
31  );
32  END mod10;
33  ARCHITECTURE lowerblk OF mod10 IS
34  BEGIN
35        PROCESS (clock, enable)
36              VARIABLE  counter       :INTEGER RANGE 0 TO 15;
37        BEGIN
38              IF ((counter = 9) AND (enable = '1'))  THEN  tc <= '1';
39              ELSE tc <= '0';
40              END IF;
41              IF (clock'EVENT AND clock = '1')  THEN
42                    IF (clear = '1')  THEN  counter := 0;
43                    ELSIF  (enable = '1')  THEN
44                          IF (counter = 9)  THEN  counter := 0;
45                          ELSE     counter := counter + 1;
46                          END IF;
47                    END IF;
48              END IF;
49              q <= counter;
50        END PROCESS;
51  END lowerblk;
```

FIGURE 7-54 MOD-100 BCD counter in VHDL.

The simulation shows that the MOD-100 counter has a correct BCD count sequence and can be synchronously cleared.

The second technique for creating the MOD-100 counter is to make the necessary connections between design modules by describing the circuit structure with VHDL. The listing for this system design file is given in Figure 7-54. The ENTITY/ARCHITECTURE description for the mod10 sub-block is contained within the overall mod100 design file (but could be in a separate file within this project's folder). The mod100 design file would be the top level for the hierarchical design of this system. It contains lower-level sub-blocks, which are actually two copies of the lower-level mod10 counter. The mod10 COMPONENT is declared in this higher-level design file (lines 10–16). The wiring connections for each instance where the module is utilized are listed in a PORT MAP. Since we need two instances of mod10, there is a PORT MAP for each instance (lines 19–20 and 21–22). Each instance must have a unique label (*digit1* or *digit2*) to distinguish them from each other. The PORT MAPs contain named associations between the lower-level module ports, given on the left, and the higher-level signals to which they are connected, given on the right. This circuit produces the same simulation results shown in Figure 7-53.

REVIEW QUESTIONS

1. Describe how to connect HDL modules together to create a digital system.
2. What is a bus and how is it represented in a graphical block diagram design file in Altera?
3. What counter features must be included to synchronously cascade counter modules together?

7-14 STATE MACHINES

The term **state machine** refers to a circuit that sequences through a set of predetermined states controlled by a clock and other input signals. So the counter circuits we have been studying so far in Chapter 7 are state machines. Generally, we use the term *counter* for sequential circuits that have a regular numeric count sequence. They may count up or count down, they may have a full 2^N modules or they may have a $< 2^N$ modulus, or they may recycle or stop automatically at some predetermined state. A counter, as its name implies, is used to count things. The things that are counted are actually called clock pulses, but the pulses may represent many kinds of events. The pulses may be the cycles of a signal for frequency division or they may be seconds, minutes, and hours of a day for a digital clock. They may indicate that an item has moved down the conveyer in a factory or that a car has passed a particular spot on the highway.

The term *state machine* is more often used to describe other kinds of sequential circuits. They may have an irregular counting pattern like our stepper motor control circuit in Section 7-10. The objective for that design was to drive a stepper motor so that it would rotate in precise angular steps. The control circuit had to produce the required specific sequence of states for that movement, rather than count numerically. There are also many applications where we do not care about the specific binary value for each state because we will use appropriate decoding logic to identify specific states of interest and to generate desired output signals. The general distinction between the two terms is that a *counter* is commonly used to count events, while a *state machine* is commonly used to control events. The correct descriptive term depends on how we wish to use the sequential circuit.

The block diagram shown in Figure 7-55 may represent a state machine or a counter. In Section 7-10 we found out that the classic sequential circuit design process was to figure out how many flip-flops would be needed and then determine the necessary combinational circuit to produce the desired sequence. The output produced by a counter or a state machine may come

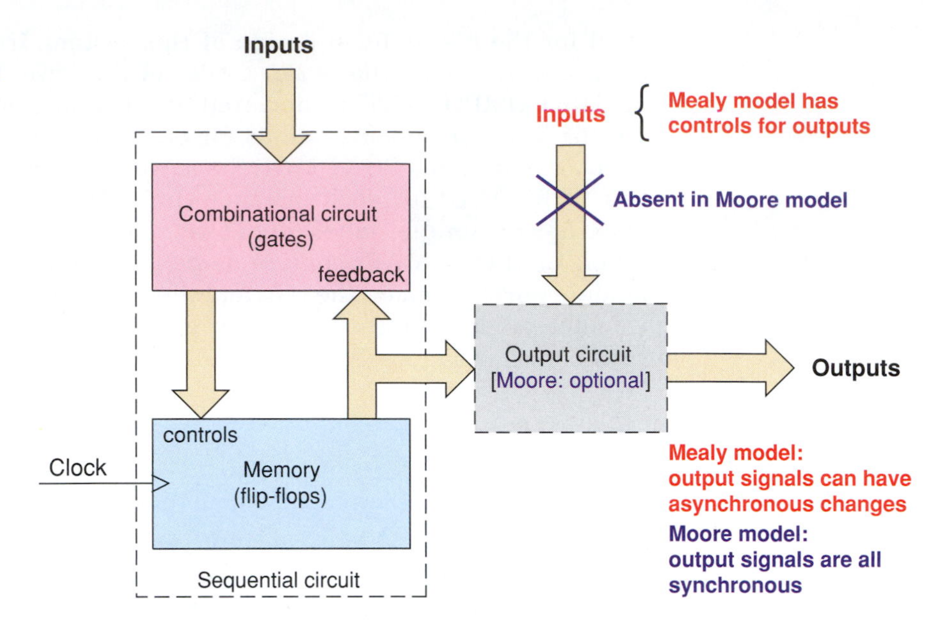

FIGURE 7-55 Block diagram for counters and state machines.

directly from the flip-flop outputs or there may be some gating circuitry needed, as indicated in the block diagram. The two variations are described as either a **Mealy model** for a sequential circuit or a **Moore model**. In the Mealy model the output signals are also controlled by additional input signals, while the Moore model does not have any external controls for the generated output signals. The Moore output is a function only of the current flip-flop state. An example of a Moore-type design would be the decoded MOD-5 circuit in Section 7-13. On the other hand, the BCD counter design in the same section would be a Mealy-type design because of the external input (*enable*) that controls the terminal state decoding output (*tc*). One significant consequence of this subtle design variation is that Moore-type circuit outputs will be completely synchronous to the circuit's clock, while outputs produced by a Mealy-type circuit can change asynchronously. The enable input is not synchronized to the system clock in our MOD-10 design.

HDLs, of course, can make state machines easy and intuitive to describe. As an oversimplified example that everyone can relate to, the following hardware description deals with four states through which a typical washing machine might progress. Although a real washing machine is more complex than this example, it will serve to demonstrate the techniques. This washing machine is idle until the start button is pressed, then it fills with water until the tub is full, then it runs the agitator until a timer expires, and finally it spins the tub until the water is spun out, at which time it goes back to idle. The point of this example focuses on the use of a set of named states for which no binary values are defined. The name of the counter variable is *wash*, which can be in any of the named states: *idle, fill, agitate,* or *spin.*

AHDL

SIMPLE AHDL STATE MACHINE

The AHDL code in Figure 7-56 shows the syntax for declaring a counter with named states on lines 6 and 7. The name of this counter is *cycle*. The keyword **MACHINE** is used in AHDL to define *cycle* as a state machine. The number of bits needed for this counter to produce the named states will be determined by the compiler. Notice that in line 7 the states are named, but the binary value for each state is also left for the compiler to determine. The

FIGURE 7-56 State machine example using AHDL.

```
1   SUBDESIGN fig7_56
2   (  clock, start, full, timesup, dry      :INPUT;
3      water_valve, ag_mode, sp_mode         :OUTPUT;
4   )
5   VARIABLE
6   cycle:  MACHINE
7             WITH STATES (idle, fill, agitate, spin);
8   BEGIN
9   cycle.clk = clock;
10
11     CASE cycle IS
12        WHEN idle =>IF start THEN cycle = fill;
13                    ELSE      cycle = idle;
14                    END IF;
15        WHEN fill =>IF full THEN cycle = agitate;
16                    ELSE      cycle = fill;
17                    END IF;
18        WHEN agitate=> IF timesup THEN cycle = spin;
19                    ELSE      cycle = agitate;
20                    END IF;
21        WHEN spin => IF dry THEN cycle = idle;
22                    ELSE      cycle = spin;
23                    END IF;
24        WHEN OTHERS => cycle = idle;
25     END CASE;
26
27     TABLE
28        cycle    => water_valve,    ag_mode, sp_mode;
29        idle     => GND,            GND,     GND;
30        fill     => VCC,            GND,     GND;
31        agitate  => GND,            VCC,     GND;
32        spin     => GND,            GND,     VCC;
33     END TABLE;
34  END;
```

designer does not need to worry about this level of detail. The CASE structure on lines 11–25 and the decoding logic that drives the outputs (lines 27–33) refer to the states by name. This makes the description easy to read and allows the compiler more freedom to minimize the circuitry. If the design requires the state machine also to be connected to an output port, then line 6 can be changed to:

```
cycle: MACHINE OF BITS (st [1..0])
```

and the output port *st[1..0]* can be added to the SUBDESIGN section. A second state machine option that is available is the ability for the designer to define a binary value for each state. This can be accomplished in this example by changing line 7 to:

```
WITH STATES (idle = B"00", fill = B"01", agitate = B"11", spin =
B"10");
```

SIMPLE VHDL STATE MACHINE

The VHDL code in Figure 7-57 shows the syntax for declaring a counter with named states. On line 6, a data object is declared named *state_machine.* Notice the keyword TYPE. This is called an **enumerated type** in VHDL, in which the designer lists by symbolic names all possible values that a signal, variable, or port that is declared to be of that type is allowed to have. Notice also that on line 6, the states are named, but the binary value for each state is left for the compiler to determine. The designer does not need to worry about this level of detail. The CASE structure on lines 12–29 and the decoding logic that drives the outputs (lines 31–36) refer to the states by name. This makes the description easy to read and allows the compiler more freedom to minimize the circuitry.

Using the simulator to verify our HDL designs produces the results given in Figure 7-58. The Altera simulator allows us to also simulate intermediate nodes in our design modules. The "buried" state machine named *cycle* has been included in the simulation in order to confirm that it operates correctly. Note that the results for *cycle* are given twice, since it will be displayed differently

```
 1  ENTITY  fig7_57  IS
 2  PORT (   clock, start, full, timesup, dry          :IN BIT;
 3        water_valve, ag_mode, sp_mode                :OUT BIT);
 4  END fig7_57;
 5  ARCHITECTURE  vhdl  OF  fig7_57  IS
 6  TYPE  state_machine  IS  (idle, fill, agitate, spin);
 7  BEGIN
 8     PROCESS (clock)
 9     VARIABLE  cycle              :state_machine;
10     BEGIN
11     IF (clock'EVENT  AND  clock = '1')  THEN
12        CASE  cycle  IS
13           WHEN idle =>
14              IF start = '1' THEN      cycle := fill;
15              ELSE                     cycle := idle;
16              END IF;
17           WHEN fill =>
18              IF full = '1' THEN       cycle := agitate;
19              ELSE                     cycle := fill;
20              END IF;
21           WHEN agitate =>
22              IF timesup = '1' THEN    cycle := spin;
23              ELSE                     cycle := agitate;
24              END IF;
25           WHEN spin =>
26              IF dry = '1' THEN        cycle := idle;
27              ELSE                     cycle := spin;
28              END IF;
29        END CASE;
30     END IF;
31     CASE  cycle  IS
32        WHEN idle    =>  water_valve <= '0';  ag_mode <= '0';  sp_mode <= '0';
33        WHEN fill    =>  water_valve <= '1';  ag_mode <= '0';  sp_mode <= '0';
34        WHEN agitate =>  water_valve <= '0';  ag_mode <= '1';  sp_mode <= '0';
35        WHEN spin    =>  water_valve <= '0';  ag_mode <= '0';  sp_mode <= '1';
36     END CASE;
37     END PROCESS;
38  END vhdl;
```

FIGURE 7-57 State machine example using VHDL.

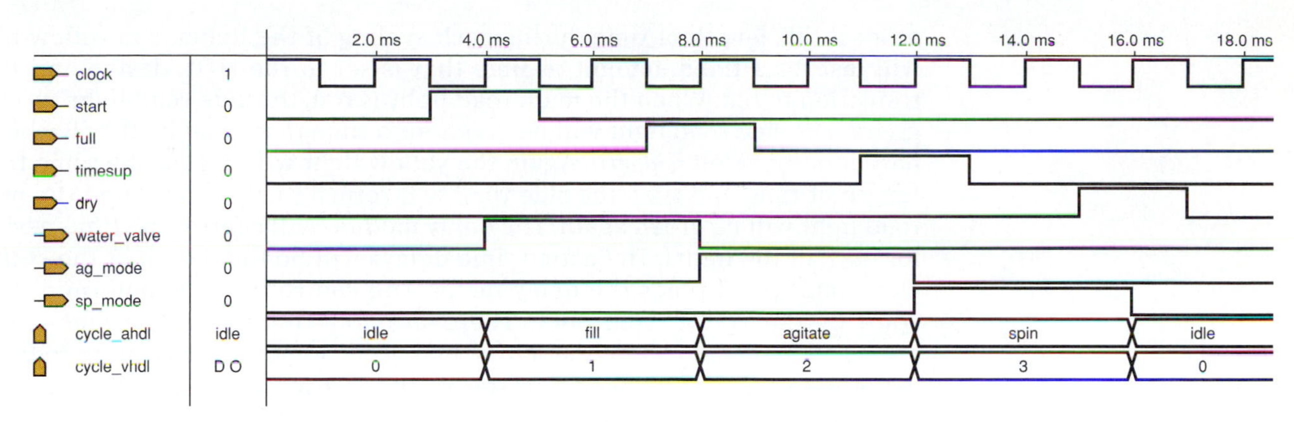

FIGURE 7-58 Simulation of washing machine HDL design example for a state machine.

for the two HDLs. The simulator cannot actually show the simulations for both AHDL and VHDL together. The second buried node information has been merely copied and pasted for a composite figure here. In AHDL the machine state names are displayed, while in VHDL the compiler-assigned values for the enumerated state names are displayed instead.

Traffic Light Controller State Machine

Let us investigate a state machine design that is a little more complicated, a traffic light controller. The block diagram is shown in Figure 7-59. Our simple controller is designed to control the flow of traffic at the intersection of a main road with a less busy side road. Traffic will flow uninterrupted on the main road with a green light, until a car is sensed on the side road (indicated by the input labeled *car*). After a time delay that is set by the five-bit binary input labeled *tmaingrn*, the main road light will change to yellow. The *tmaingrn* time delay ensures that the main road will receive a green light for

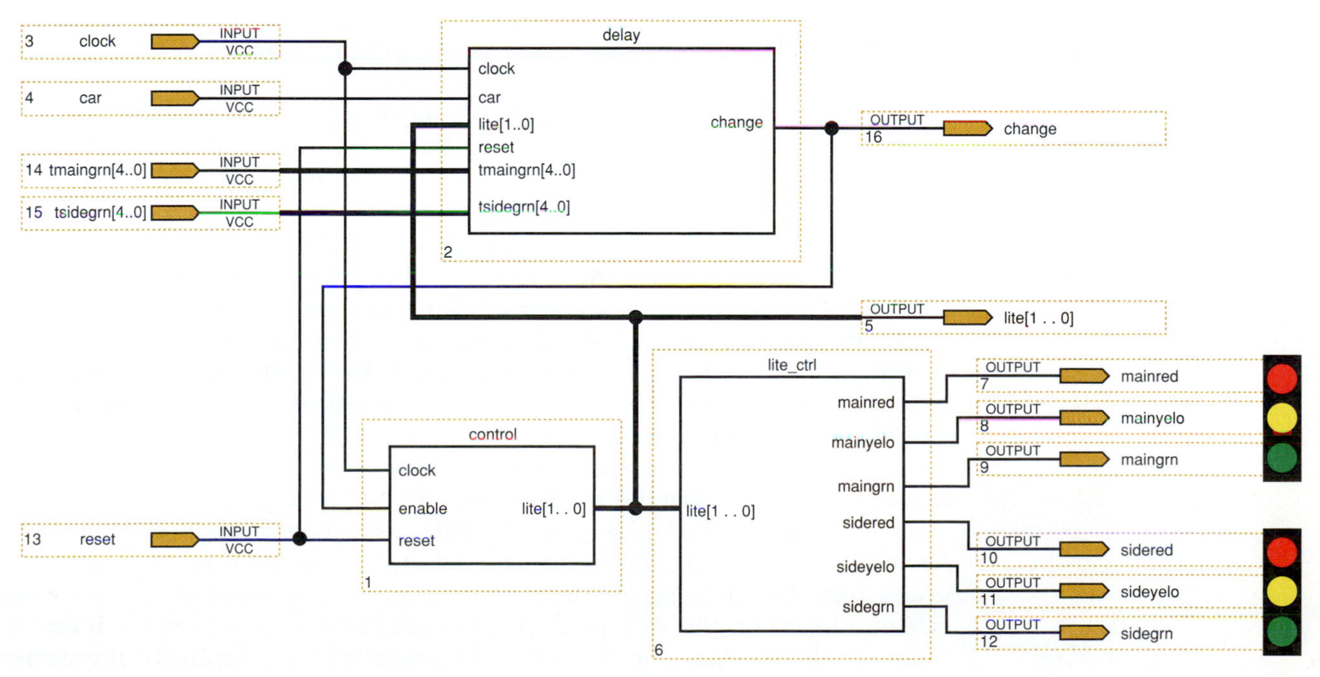

FIGURE 7-59 Traffic light controller.

at least this length of time during each cycling of the lights. The yellow light will last for a fixed amount of time that is set in the HDL design and then transition to red. When the main road light is red, the side road light turns to green. The side road light will be green for a time that is set by the five-bit binary input labeled *tsidegrn*. Again the yellow light will last for the same fixed length of time and then the side road will return to a red light and the main road light will be green again. The delay module will control the time periods for each of the lights. The actual time delays will be the period of the system clock multiplied times the delay factor. The control module determines the state of the traffic controller. There are four light combinations—main-green/side-red, main-yellow/side-red, main-red/side-green, and main-red/side-yellow—so control will need four states. The traffic light states are translated into the proper on–off patterns for each of the six pairs of lights by the lite_ctrl module. The outputs labeled *change* and *lite* are provided for diagnostic purposes. *Reset* is used to initialize each of the two sequential circuits.

AHDL

AHDL TRAFFIC LIGHT CONTROLLER

The three design modules for our AHDL traffic light controller are listed together in Figure 7-60. They are actually three separate design files that are interconnected with the block diagram design shown in Figure 7-59. The delay module (lines 1–23) is basically a buried down counter (line 20) named *mach,* which waits at zero when the main road has a green light (*lite* = 0) until it is triggered by the car sensor (line 13) to load the delay factor *tmaingrn*−1 on line 14. Since the counter decrements all the way to zero, one is subtracted from each delay factor to make the delay counter's modulus equal to the value of the delay factor. For example, if we wish to have a delay factor of 25, the counter must count from 24 down to 0. The actual length of time represented by the delay factors depends on the clock frequency. With a 1-Hz clock frequency, the period would be 1 s, and the delay factors would then be in seconds. Line 22 defines an output signal called *change* that detects when *mach* is equal to one. *Change* will be HIGH to indicate that the test condition is true, which in turn will enable the state machine in the control module to move to its next state (*lite* = 1) when clocked to indicate a yellow light on the main road. As the delay counter *mach* counts down and reaches zero, CASE determines that *lite* has a new value and the fixed time delay factor of 5 for a yellow light is loaded (actually loading one less than 5, as previously discussed) into *mach* (line 16) on the next clock. The count down continues from this new delay time, with *change* again enabling the control module to move to its next state (*lite* = 2) when *mach* is equal to 1, resulting in a green light for the side road. When *mach* again reaches zero, the time delay (*tsidegrn*−1) for a green light on the side road will be loaded into the down counter (line 17). When *change* again goes active, *lite* will advance to state 3 for a yellow light on the side road. *Mach* will recycle to the value 4 (5–1) on line 18 for the fixed time delay for a yellow light. When *change* goes active this time, the control module will return to the *lite* = 0 state (green light on main). When *mach* decrements to its terminal state (zero) this time, lines 13–15 will determine by the status of the *car* sensor input whether to wait for another car or to load in the delay factor for a green light on main (*tmaingrn*−1) to start the cycle over again. The main road will receive a green light for at least this length of time, even if there is a continuous stream of cars on the side road. It is obvious that we could make improvements to this design, but that, of course, would also complicate the design further.

```
1   SUBDESIGN  delay
2   (   clock, car, lite[1..0], reset         :INPUT;
3       tmaingrn[4..0], tsidegrn[4..0]        :INPUT;
4       change                                :OUTPUT;  )
5   VARIABLE
6       mach[4..0]                            :DFF;
7   BEGIN
8       mach[].clk = clock;                          -- with 1 Hz clock, times in seconds
9       mach[].clrn = reset;
10      IF  mach[] == 0  THEN
11          CASE  lite[]  IS                         -- check state of light controller
12              WHEN 0 =>
13                  IF  !car  THEN  mach[].d = 0;    -- wait for car on side road
14                  ELSE  mach[].d = tmaingrn[] - 1; -- set time for main's green
15                  END IF;
16              WHEN 1 => mach[].d = 5 - 1;          -- set time for main's yellow
17              WHEN 2 => mach[].d = tsidegrn[] - 1; -- set time for side's green
18              WHEN 3 => mach[].d = 5 - 1;          -- set time for side's yellow
19          END CASE;
20      ELSE  mach[].d = mach[].q - 1;               -- decrement timer counter
21      END IF;
22      change = mach[] == 1;                        -- change lights on control module
23  END;
24  ---------------------------------------------------------------------------------
25  SUBDESIGN  control
26  (   clock, enable, reset    :INPUT;
27      lite[1..0]              :OUTPUT;  )
28  VARIABLE
29      light:    MACHINE OF BITS (lite[1..0])       -- need 4 states for light combinations
30                WITH STATES (mgrn = B"00", myel = B"01", sgrn = B"10", syel = B"11");
31  BEGIN
32      light.clk = clock;
33      light.reset = !reset;            -- MACHINEs have asynchronous, active-high reset
34      CASE  light  IS          -- wait for enable to change light states
35          WHEN mgrn     =>     IF enable THEN light = myel;  ELSE light = mgrn;  END IF;
36          WHEN myel     =>     IF enable THEN light = sgrn;  ELSE light = myel;  END IF;
37          WHEN sgrn     =>     IF enable THEN light = syel;  ELSE light = sgrn;  END IF;
38          WHEN syel     =>     IF enable THEN light = mgrn;  ELSE light = syel;  END IF;
39      END CASE;
40  END;
41  ---------------------------------------------------------------------------------
42  SUBDESIGN lite_ctrl
43  (   lite[1..0]                      :INPUT;
44      mainred, mainyelo, maingrn      :OUTPUT;
45      sidered, sideyelo, sidegrn      :OUTPUT;  )
46  BEGIN
47      CASE lite[]       IS         -- determine which lights to turn on
48          WHEN B"00"    =>     maingrn = VCC; mainyelo = GND; mainred = GND;
49                               sidegrn = GND; sideyelo = GND; sidered = VCC;
50          WHEN B"01"    =>     maingrn = GND; mainyelo = VCC; mainred = GND;
51                               sidegrn = GND; sideyelo = GND; sidered = VCC;
52          WHEN B"10"    =>     maingrn = GND; mainyelo = GND; mainred = VCC;
53                               sidegrn = VCC; sideyelo = GND; sidered = GND;
54          WHEN B"11"    =>     maingrn = GND; mainyelo = GND; mainred = VCC;
55                               sidegrn = GND; sideyelo = VCC; sidered = GND;
56      END CASE;
57  END;
```

FIGURE 7-60 AHDL design files for traffic light controller.

The control module (lines 25–40) contains a state machine named *light* that will sequence through the four states for the traffic light combinations. The bits for the state machine are named and connected as an output port for this module (lines 27 and 29). The four states for *light* are named *mgrn, myel, sgrn,* and *syel* on line 30. Each state represents which road, main or side, is to receive a green or yellow light. The other road will have a red light. The values for each state of the control module have also been specified on line 30 so that we can identify them as inputs to the other two modules, delay and lite_ctrl. The *enable* input is connected to the *change* output signal produced by the delay module. When enabled, the *light* state machine will advance to the next state when clocked as described by the CASE and nested IF statements on lines 34–39. Otherwise, *light* will hold at the current state.

The lite_ctrl module (lines 42–57) inputs *lite[1..0],* which represents the state of the *light* state machine from the control module, and will output the signals that will turn on the proper combinations of green, yellow, and red lights for the main and side roads. Each output from the lite_ctrl module will actually be connected to lamp driver circuits to control the higher voltages and currents necessary for real lamps in a traffic light. The CASE statement on lines 47–55 determines which main road/side road light combination to turn on for each state of *light.* The function of the lite_ctrl module is very much like a decoder. It essentially decodes each state combination of *lite* to turn on a green or yellow light for one road and a red light for the other road. A unique output combination is produced for each input state.

VHDL

VHDL TRAFFIC LIGHT CONTROLLER

The VHDL design for the traffic light controller is listed in Figure 7-61. The top level of the design is described structurally on lines 1–34. There are three COMPONENT modules to declare (lines 10–24). The PORT MAPs giving the wiring interconnects between each module and the top level design are listed on lines 26–33.

The delay module (lines 36–66) is basically a buried down counter (line 59) created with the integer variable *mach* that waits at zero when the main road has a green light (*lite* = "00") until it is triggered by the car sensor (line 52) to load the delay factor *tmaingrn*−1 on line 53. Since the counter decrements all the way to zero, one is subtracted from each delay factor to make the delay counter's modulus equal to the value of the delay factor. For example, if we wish to have a delay factor of 25, the counter must count from 24 down to 0. The actual length of time represented by the delay factors depends on the clock frequency. With a 1-Hz clock frequency, the period would be 1 s, and the delay factors would then be in seconds. Lines 62–64 define an output signal called *change* that detects when *mach* is equal to one. *Change* will be HIGH to indicate that the test condition is true, which in turn will enable the state machine in the control module to move to its next state (*lite* = "01") when clocked to indicate a yellow light on the main road. When *mach* reaches zero now, CASE determines that *lite* has a new value and the fixed time delay factor of 5 for a yellow light is loaded (actually loading one less, as previously discussed) into *mach* (line 55) on the next clock. The count down continues from this new delay time, with *change* again enabling the control module to move to its next state (*lite* = "10"), resulting in a green light for the side road. When *mach* again reaches zero, the time delay (*tsidegrn*−1) for a green light on the side road will be loaded into the down

```
1   ENTITY  traffic  IS
2   PORT  (   clock, car, reset              :IN BIT;
3             tmaingrn, tsidegrn             :IN INTEGER RANGE 0 TO 31;
4             lite                           :BUFFER INTEGER RANGE 0 TO 3;
5             change                         :BUFFER BIT;
6             mainred, mainyelo, maingrn     :OUT BIT;
7             sidered, sideyelo, sidegrn     :OUT BIT);
8   END traffic;
9   ARCHITECTURE  toplevel OF  traffic  IS
10  COMPONENT delay
11      PORT ( clock, car, reset             :IN BIT;
12             lite                          :IN INTEGER RANGE 0 TO 3;
13             tmaingrn, tsidegrn            :IN INTEGER RANGE 0 TO 31;
14             change                        :OUT BIT);
15  END COMPONENT;
16  COMPONENT control
17      PORT ( clock, enable, reset          :IN BIT;
18             lite                          :OUT INTEGER RANGE 0 TO 3);
19  END COMPONENT;
20  COMPONENT lite_ctrl
21      PORT ( lite                          :IN INTEGER RANGE 0 TO 3;
22             mainred, mainyelo, maingrn    :OUT BIT;
23             sidered, sideyelo, sidegrn    :OUT BIT);
24  END COMPONENT;
25  BEGIN
26  module1:  delay     PORT MAP (clock => clock, car => car, reset => reset,
27                      lite => lite, tmaingrn => tmaingrn, tsidegrn => tsidegrn,
28                      change => change);
29  module2:  control   PORT MAP (clock => clock, enable => change, reset => reset,
30                      lite => lite);
31  module3:  lite_ctrl PORT MAP (lite => lite, mainred => mainred, mainyelo => mainyelo,
32                      maingrn => maingrn, sidered => sidered, sideyelo => sideyelo,
33                      sidegrn => sidegrn);
34  END toplevel;
35  ---------------------------------------------------------------------------------
36  ENTITY  delay  IS
37  PORT  (   clock, car, reset        :IN BIT;
38            lite                     :IN BIT_VECTOR (1 DOWNTO 0);
39            tmaingrn, tsidegrn       :IN INTEGER RANGE 0 TO 31;
40            change                   :OUT BIT);
41  END delay;
42  ARCHITECTURE  time  OF delay  IS
43  BEGIN
44      PROCESS (clock, reset)
45      VARIABLE  mach                 :INTEGER RANGE 0 TO 31;
46      BEGIN
47      IF  reset = '0'  THEN  mach := 0;
48      ELSIF (clock = '1' AND clock'EVENT)  THEN   -- with 1 Hz clock, times in seconds
49          IF  mach = 0  THEN
50              CASE  lite  IS
51                  WHEN "00"
52                      IF car = '0'  THEN  mach := 0;       -- wait for car on side road
53                      ELSE            mach := tmaingrn - 1;  -- set time for main's green
54                      END IF;
55                  WHEN "01"    =>    mach := 5 - 1;        -- set time for main's yellow
56                  WHEN "10"    =>    mach := tsidegrn - 1;  -- set time for side's green
57                  WHEN "11"    =>    mach := 5 - 1;        -- set time for side's yellow
58              END CASE;
59          ELSE  mach := mach - 1;                          -- decrement timer counter
60          END IF;
61      END IF;
```

FIGURE 7-61 VHDL design for traffic light controller.

```
 62     IF  mach = 1  THEN  change <= '1';                -- change lights on control
 63     ELSE  change <= '0';
 64     END IF;
 65     END PROCESS;
 66  END time;
 67  -------------------------------------------------------------------------------
 68  ENTITY  control  IS
 69  PORT  (   clock, enable, reset    :IN BIT;
 70            lite                    :OUT BIT_VECTOR (1 DOWNTO 0));
 71  END control;
 72  ARCHITECTURE  a  OF  control  IS
 73  TYPE  enumerated  IS (mgrn, myel, sgrn, syel); -- need 4 states for light combinations
 74  BEGIN
 75     PROCESS (clock, reset)
 76     VARIABLE  lights :enumerated;
 77     BEGIN
 78        IF  reset = '0'  THEN  lights := mgrn;
 79        ELSIF (clock = '1' AND clock'EVENT)  THEN
 80           IF  enable = '1'  THEN        -- wait for enable to change light states
 81              CASE  lights  IS
 82                 WHEN  mgrn      =>     lights := myel;
 83                 WHEN  myel      =>     lights := sgrn;
 84                 WHEN  sgrn      =>     lights := syel;
 85                 WHEN  syel      =>     lights := mgrn;
 86              END CASE;
 87           END IF;
 88        END IF;
 89        CASE  lights  IS                   -- patterns for light states
 90           WHEN  mgrn=>     lite <= "00";
 91           WHEN  myel=>     lite <= "01";
 92           WHEN  sgrn=>     lite <= "10";
 93           WHEN  syel=>     lite <= "11";
 94        END CASE;
 95     END PROCESS;
 96  END a;
 97  -------------------------------------------------------------------------------
 98  ENTITY  lite_ctrl  IS
 99  PORT  (   lite                          :IN BIT_VECTOR (1 DOWNTO 0);
100            mainred, mainyelo, maingrn    :OUT BIT;
101            sidered, sideyelo, sidegrn    :OUT BIT);
102  END lite_ctrl;
103  ARCHITECTURE  patterns  OF  lite_ctrl  IS
104  BEGIN
105     PROCESS (lite)
106     BEGIN
107     CASE  lite  IS   -- control state determines which lights to turn on/off
108        WHEN "00" => maingrn <= '1';     mainyelo <= '0';     mainred <= '0';
109                     sidegrn <= '0';     sideyelo <= '0';     sidered <= '1';
110        WHEN "01" => maingrn <= '0';     mainyelo <= '1';     mainred <= '0';
111                     sidegrn <= '0';     sideyelo <= '0';     sidered <= '1';
112        WHEN "10" => maingrn <= '0';     mainyelo <= '0';     mainred <= '1';
113                     sidegrn <= '1';     sideyelo <= '0';     sidered <= '0';
114        WHEN "11" => maingrn <= '0';     mainyelo <= '0';     mainred <= '1';
115                     sidegrn <= '0';     sideyelo <= '1';     sidered <= '0';
116     END CASE;
117     END PROCESS;
118  END patterns;
```

FIGURE 7-61 *Continued*

counter (line 56). When *change* again goes active, *lite* will advance to "11" for a yellow light on the side road. *Mach* will recycle to the value 4 (5 – 1) on line 57 for the fixed time delay for a yellow light. When *change* goes active this time, the control module will return to *lite* = "00" (green light on main). When *mach* decrements to its terminal state (zero) this time, lines 52–54 will determine by the status of the *car* sensor input whether to wait for another car or to load in the delay factor for a green light on main (*tmaingrn*−1) to start the cycle over again. The main road will receive a green light for at least this length of time, even if there is a continuous stream of cars on the side road. It is obvious that we could make improvements to this design, but that, of course, would also complicate the design further.

The control module (lines 68–96) contains a state machine named *lights* that will sequence through four enumerated states for the traffic light combinations. The four enumerated states for *lights* are *mgrn, myel, sgrn,* and *syel* (lines 73 and 76). Each state represents which road, main or side, is to receive a green or yellow light. The other road will have a red light. The *enable* input is connected to the *change* output signal produced by the delay module. When enabled, the *lights* state machine will advance to the next state when clocked, as described by the nested IF and CASE statements on lines 79–88. Otherwise, *lights* will hold at the current state. The bit patterns for output port *lite* have been specified for each state of *lights* with the CASE statement on lines 89–94 so that we can identify them as inputs to the other two modules, delay and lite_ctrl.

The lite_ctrl module (lines 98–118) inputs *lite,* which represents the state of the *lights* state machine from the control module, and will output the signals that will turn on the proper combinations of green, yellow, and red lights for the main and side roads. Each output from the lite_ctrl module will actually be connected to lamp driver circuits to control the higher voltages and currents necessary for real lamps in a traffic light. The CASE statement on lines 107–116, invoked by the PROCESS when the *lite* input changes, determines which main road/side road light combination to turn on for each state of *lights.* The function of the lite_ctrl module is very much like a decoder. It essentially decodes each state combination of *lite* to turn on a green or yellow light for one road and a red light for the other road. A unique output combination is produced for each input state.

By this time, you may be wondering why there are so many ways to describe logic circuits. If one way is easier than the others, why not just study that one? The answer, of course, is that each level of abstraction offers advantages over the others in certain cases. The structural method provides the most complete control over interconnections. The use of Boolean equations, truth tables, and PRESENT state/NEXT state tables allows us to describe the way data flows through the circuit using HDL. Finally, the behavioral method allows a more abstract description of the circuit's operation in terms of cause and effect. In practice, each source file may have portions that can be categorized under each level of abstraction. Choosing the right level when writing code is not an issue of right and wrong as much as it is an issue of style and preference.

There are also several ways to approach any task from a standpoint of choosing control structures. Should we use selected signal assignments or Boolean equations, IF/ELSE or CASE, sequential processes or concurrent statements, macrofunctions or megafunctions? Or should we write our own code? The answers to these questions ultimately define your personal strategy in solving the problem. Your preferences and the advantages you find in using one method over another will be established with practice and experience.

REVIEW QUESTIONS

1. What is the fundamental difference between a counter and a state machine?
2. What is the difference between describing a counter and describing a state machine in an HDL?
3. If the actual binary states for a state machine are not defined in the HDL code, how are they assigned?
4. What is the advantage of using state machine description?

PART 1 SUMMARY

1. In asynchronous (ripple) counters, the clock signal is applied to the LSB FF, and all other FFs are clocked by the output of the preceding FF.
2. A counter's MOD number is the number of stable states in its counting cycle; it is also the maximum frequency-division ratio.
3. The normal (maximum) MOD number of a counter is 2^N. One way to modify a counter's MOD number is to add circuitry that will cause it to recycle before it reaches its normal last count.
4. Counters can be cascaded (chained together) to produce greater counting ranges and frequency-division ratios.
5. In a synchronous (parallel) counter, all of the FFs are simultaneously clocked from the input clock signal.
6. The maximum clock frequency for an asynchronous counter, f_{max}, decreases as the number of bits increases. For a synchronous counter, f_{max} remains the same, regardless of the number of bits.
7. A decade counter is any MOD-10 counter. A BCD counter is a decade counter that sequences through the 10 BCD codes (0–9).
8. A presettable counter can be loaded with any desired starting count.
9. An up/down counter can be commanded to count up or count down.
10. Logic gates can be used to decode (detect) any or all states of a counter.
11. The count sequence for a synchronous counter can be easily determined by using a PRESENT state/NEXT state table that lists all possible states, the flip-flop input control information, and the resulting NEXT states.
12. Synchronous counters with arbitrary counting sequences can be implemented by following a standard design procedure.
13. Counters can be described in many different ways using HDL, including structural wiring descriptions, PRESENT state/NEXT state tables, and behavioral descriptions.
14. All the features available on the various standard IC counter chips, such as asynchronous or synchronous loading or clearing, count enabling, and terminal count decoding, can be described using HDL. HDL counters can be easily modified for higher MOD numbers or changes in the active levels for controls.
15. Digital systems can be subdivided into smaller modules or blocks that can be interconnected as a hierarchical design.
16. State machines can be represented in HDL using descriptive names for each state rather than specifying a numeric sequence of states.

PART 1 IMPORTANT TERMS

asynchronous (ripple) counter
MOD number
glitches
synchronous (parallel) counters
decade counter
BCD counter
up counter
down counter
up/down counters
presettable counters
parallel load
count enable
multistage counters
cascading
decoding
PRESENT state/NEXT state table
self-correcting counter
sequential circuit design
J-K excitation table
circuit excitation table
VARIABLE
behavioral level of abstraction
hierarchical design
state machine
mealy model
Moore model
MACHINE
enumerated type

PART 2

7-15 INTEGRATED-CIRCUIT REGISTERS

The various types of registers can be classified according to the manner in which data can be entered into the register for storage and the manner in which data are outputted from the register. The various classifications are listed below.

1. Parallel in/parallel out (PIPO)
2. Serial in/serial out (SISO)
3. Parallel in/serial out (PISO)
4. Serial in/parallel out (SIPO)

Each of these types and several variations are available in IC form so that a logic designer can usually find exactly what is required for a given application. In the following sections, we will examine a representative IC from each of the above categories.

7-16 PARALLEL IN/PARALLEL OUT—THE 74ALS174/74HC174

A group of flip-flops that can store multiple bits simultaneously and in which all bits of the stored binary value are directly available is referred to as a **parallel in/parallel out** register. Figure 7-62(a) shows the logic diagram for the 74ALS174 (also the 74HC174), a six-bit register that has parallel inputs D_5 through D_0 and parallel outputs Q_5 through Q_0. Parallel data are loaded into the register on the PGT of the clock input *CP*. A master reset input $\overline{MR}$ can be used to reset asynchronously all of the register FFs to 0. The logic symbol for the 74ALS174 is shown in Figure 7-62(b). This symbol is used in circuit diagrams to represent the circuitry of Figure 7-62(a).

The 74ALS174 is normally used for synchronous parallel data transfer whereby the logic levels present at the *D* inputs are transferred to the corresponding *Q* outputs when a PGT occurs at the clock *CP*. This IC, however, can be wired for serial data transfer, as the following examples will show.

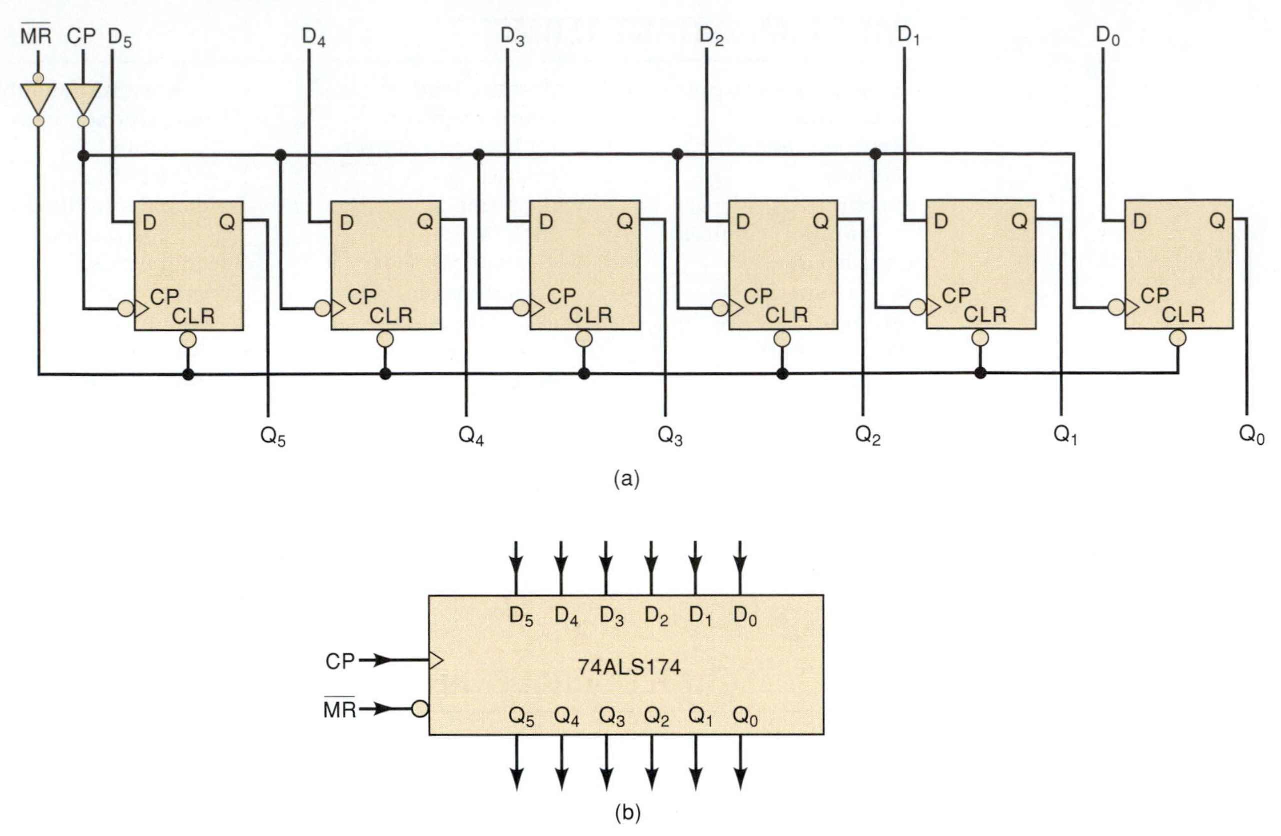

FIGURE 7-62 (a) Circuit diagram of the 74ALS174; (b) logic symbol.

EXAMPLE 7-16

Show how to connect the 74ALS174 so that it operates as a serial shift register with data shifting on each PGT of *CP* as follows: Serial input $\rightarrow Q_5 \rightarrow Q_4 \rightarrow Q_3 \rightarrow Q_2 \rightarrow Q_1 \rightarrow Q_0$. In other words, serial data will enter at D_5 and will output at Q_0.

Solution

Looking at Figure 7-62(a), we can see that to connect the six FFs as a serial shift register, we have to connect the *Q* output of one to the *D* input of the next so that data is transferred in the required manner. Figure 7-63 shows how this is accomplished. Note that data shifts left to right, with input data applied at D_5 and output data appearing at Q_0.

EXAMPLE 7-17

How would you connect two 74ALS174s to operate as a 12-bit shift register?

Solution

Connect a second 74ALS174 IC as a shift register, and connect Q_0 from the first IC to D_5 of the second IC. Connect the *CP* inputs of both ICs so that they will be clocked from the same signal. Also connect the MR inputs together if using the asynchronous reset.

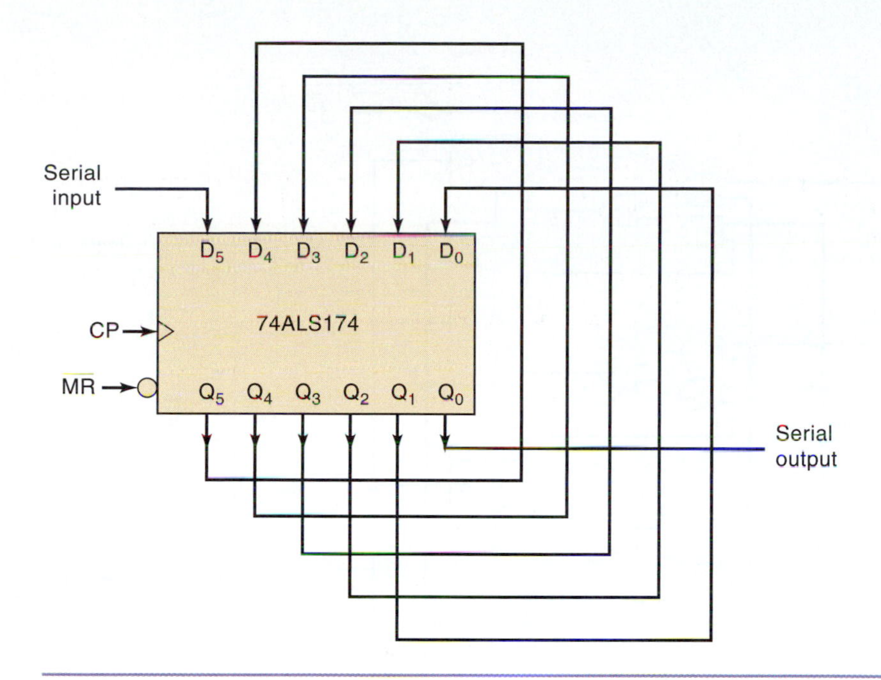

FIGURE 7-63 Example 7-16: The 74ALS174 wired as a shift register.

7-17 SERIAL IN/SERIAL OUT—THE 74ALS166/74HC166

A **serial in/serial out** shift register will have data loaded into it one bit at a time. The data will move one bit at a time with each clock pulse through the set of flip-flops toward the other end of the register. With continued clocking, the data will then exit the register one bit at a time in the same order as it was originally loaded. The 74HC166 (and also the 74ALS166) can be used as a serial in/serial out register. The logic diagram and schematic symbol for the 74HC166 is shown in Figure 7-64. It is an eight-bit shift register of which only FF QH is accessible. The serial data is input on *SER* and will be stored in FF QA. The serial output is obtained at the other end of the shift register on Q_H. As can be seen from the function table for this shift register in Figure 7-64(c), parallel data can also be synchronously loaded into it. If SH/$\overline{\text{LD}}$ = 1, the register function will be serial shifting, while a LOW will instead parallel load data via the *A* through *H* inputs. The synchronous serial shifting and parallel loading functions can be inhibited (disabled) by applying a HIGH to the *CLK INH* control input. The register also has an active-LOW, asynchronous clear input (*CLR*).

EXAMPLE 7-18

A shift register is often used as a way to delay a digital signal by an integral number of clock cycles. The digital signal is applied to the shift register's serial input and is shifted through the shift register by successive clock pulses until it reaches the end of the shift register, where it appears as the output signal. This method for delaying the effect of a digital signal is common in the digital communications field. For instance, the digital signal might be the digitized version of an audio signal that is to be delayed before it is transmitted. The input waveforms given in Figure 7-65 are applied to a 74HC166. Determine the resultant output waveform.

Solution

QH starts at a LOW, since all flip-flops are initially cleared by the LOW applied to the asynchronous *CLR* input at the beginning of the timing diagram.

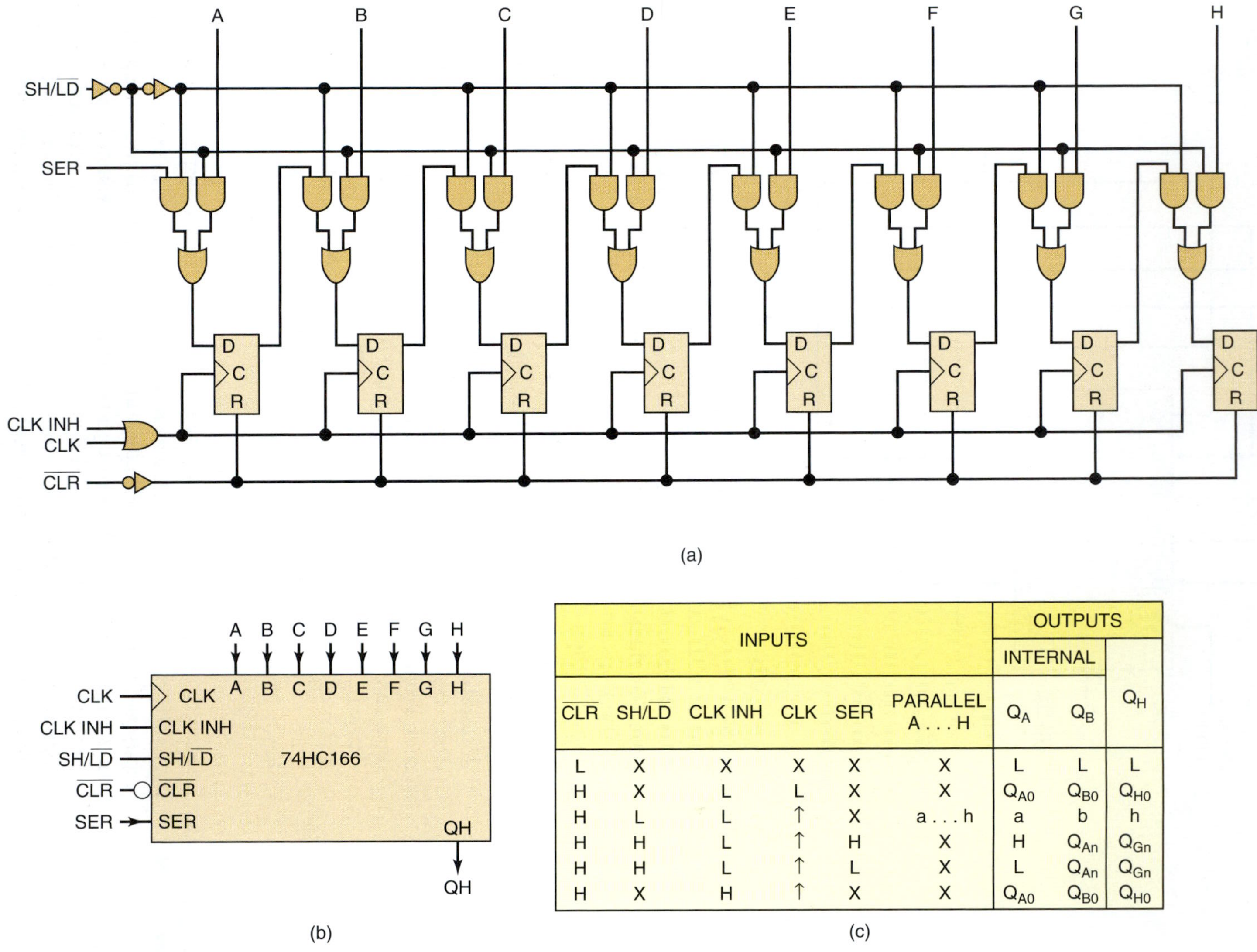

INPUTS						OUTPUTS		
						INTERNAL		
$\overline{CLR}$	SH/$\overline{LD}$	CLK INH	CLK	SER	PARALLEL A . . . H	Q_A	Q_B	Q_H
L	X	X	X	X	X	L	L	L
H	X	L	L	X	X	Q_{A0}	Q_{B0}	Q_{H0}
H	L	L	↑	X	a . . . h	a	b	h
H	H	L	↑	H	X	H	Q_{An}	Q_{Gn}
H	H	L	↑	L	X	L	Q_{An}	Q_{Gn}
H	X	H	↑	X	X	Q_{A0}	Q_{B0}	Q_{H0}

(c)

FIGURE 7-64 (a) Circuit diagram of the 74HC166; (b) logic symbol; (c) function table.

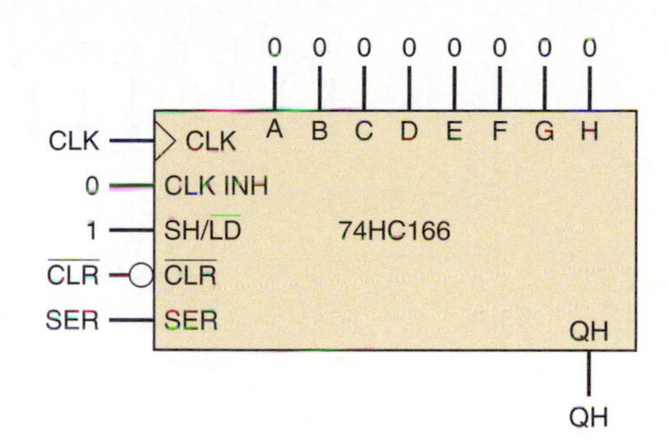

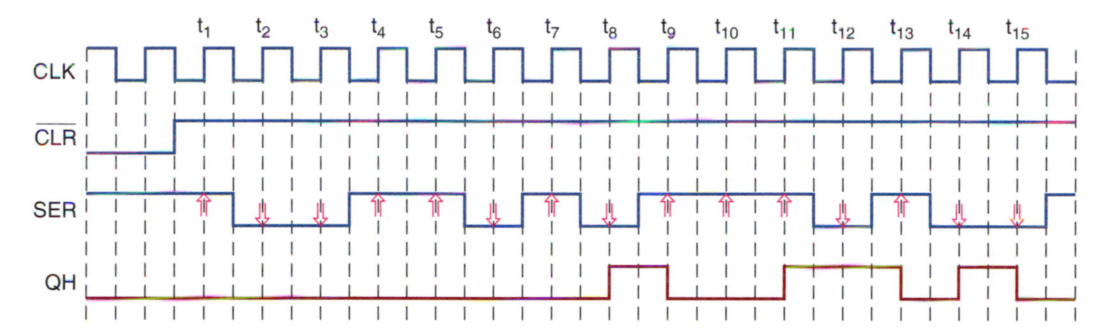

FIGURE 7-65 Example 7-18.

At t_1, the shift register will input the current bit applied to *SER*. This will be stored in QA. At t_2, the first bit will move to QB and a second bit on *SER* will be stored in QA. At t_3, the first bit will now move to QC and a third bit on *SER* will be stored in QA. The first data input bit will finally show up at the output QH at t_8. Each successive input bit on *SER* will follow at QH delayed by eight clock cycles.

7-18 PARALLEL IN/SERIAL OUT—THE 74ALS165/74HC165

The logic symbol for the 74HC165 is shown in Figure 7-66(a). This IC is an eight-bit **parallel in/serial out** register. It actually has serial data entry via D_S and asynchronous parallel data entry via P_0 through P_7. The register contains eight FFs—Q_0 through Q_7—internally connected as a shift register, but the only accessible FF outputs are Q_7 and $\overline{Q}_7$. *CP* is the clock input used for the shifting operation. The clock inhibit input, $\overline{CP\ INH}$ is used to inhibit the effect of the *CP* input. The shift/load input, $SH/\overline{LD}$, controls which operation is taking place—shifting or parallel loading. The function table in Figure 7-66(b) shows how the various input combinations determine what operation, if any, is being performed. Parallel loading is asynchronous and serial shifting is synchronous. Note that the serial shifting function will always be synchronous, since the clock is required to ensure that the input data moves only one bit at a time with each appropriate clocking edge.

EXAMPLE 7-19

Examine the 74HC165 function table and determine (a) the conditions necessary to load the register with parallel data; (b) the conditions necessary for the shifting operation.

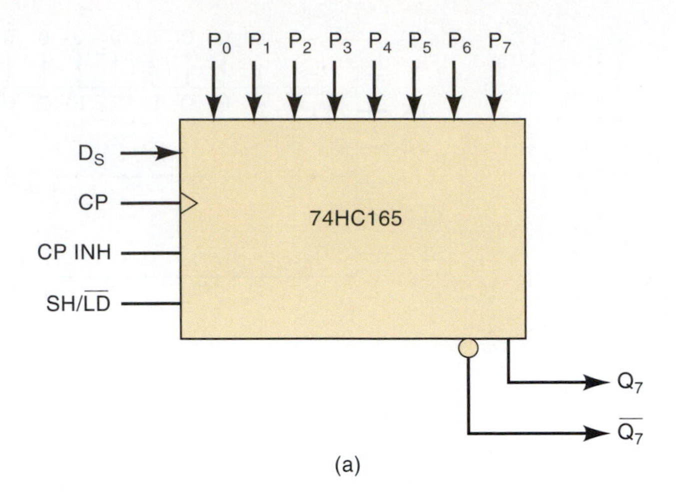

(a)

Function Table

Inputs			Operation
SH/$\overline{LD}$	CP	CP INH	
L	X	X	Parallel load
H	H	X	No change
H	X	H	No change
H	↑ (PGT)	L	Shifting
H	L	↑ (PGT)	Shifting

H = high level
L = low level
X = immaterial
↑ = PGT

(b)

FIGURE 7-66 (a) Logic symbol for the 74HC165 parallel in/serial out register; (b) function table.

Solution

(a) The first entry in the table shows that the $SH/\overline{LD}$ input has to be LOW for the parallel load operation. When this input is LOW, the data present at the P inputs are *asynchronously* loaded into the register FFs, independent of the *CP* and the *CP INH* inputs. Of course, only the outputs from the last FF are externally available.

(b) The shifting operation cannot take place unless the $SH/\overline{LD}$ input is HIGH and a PGT occurs at *CP* while *CP INH* is LOW [see the fourth table entry in Figure 7-66(b)]. A HIGH at *CP INH* will inhibit the effect of any clock pulses. Note that the roles of the *CP* and *CP INH* inputs can be reversed, as indicated by the last table entry, because these two signals are ORed together inside the IC.

EXAMPLE 7-20

Determine the output signal at Q_7 if we connect a 74HC165 with $D_S = 0$ and CP INH $= 0$ and then apply the input waveforms given in Figure 7-67. P_0–P_7 represent the parallel data on $P_0\ P_1\ P_2\ P_3\ P_4\ P_5\ P_6\ P_7$.

Solution

We have drawn the timing diagram for all eight FFs so that we could track their contents over time even though only Q_7 will be accessible. The parallel load is asynchronous and will occur as soon as SH/LD goes LOW. After SH/LD returns to a HIGH, the data stored in the register will move one FF to the right (toward Q_7) with each PGT on CP.

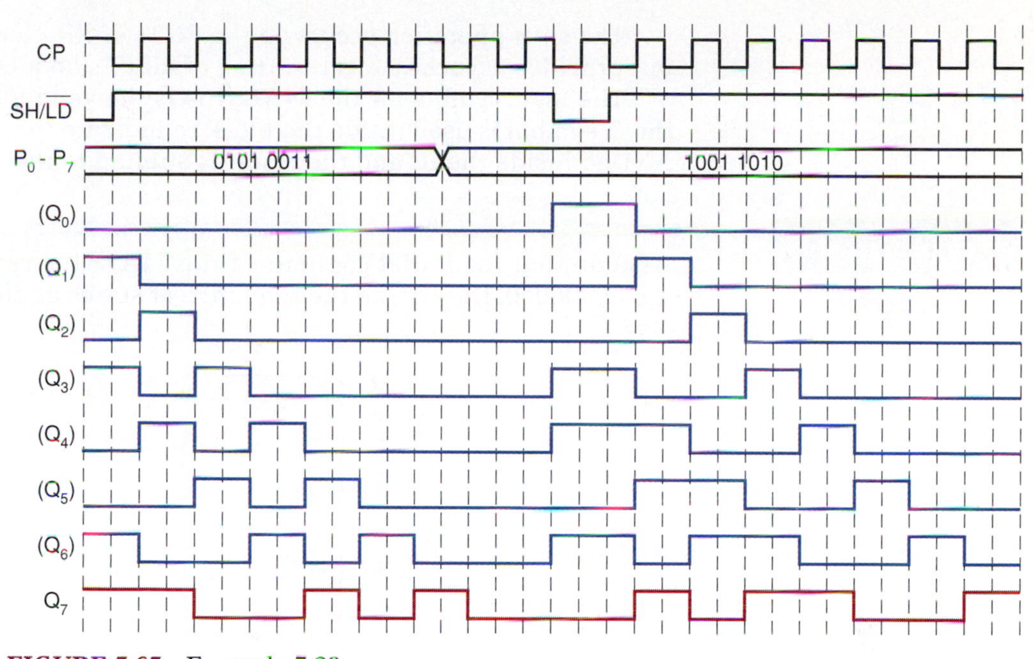

FIGURE 7-67 Example 7-20.

7-19 SERIAL IN/PARALLEL OUT—THE 74ALS164/74HC164

The logic diagram for the 74ALS164 is shown in Figure 7-68(a). It is an eight-bit **serial in/parallel out** shift register with each FF output externally accessible. Instead of a single serial input, an AND gate combines inputs A and B to produce the serial input to flip-flop Q_0.

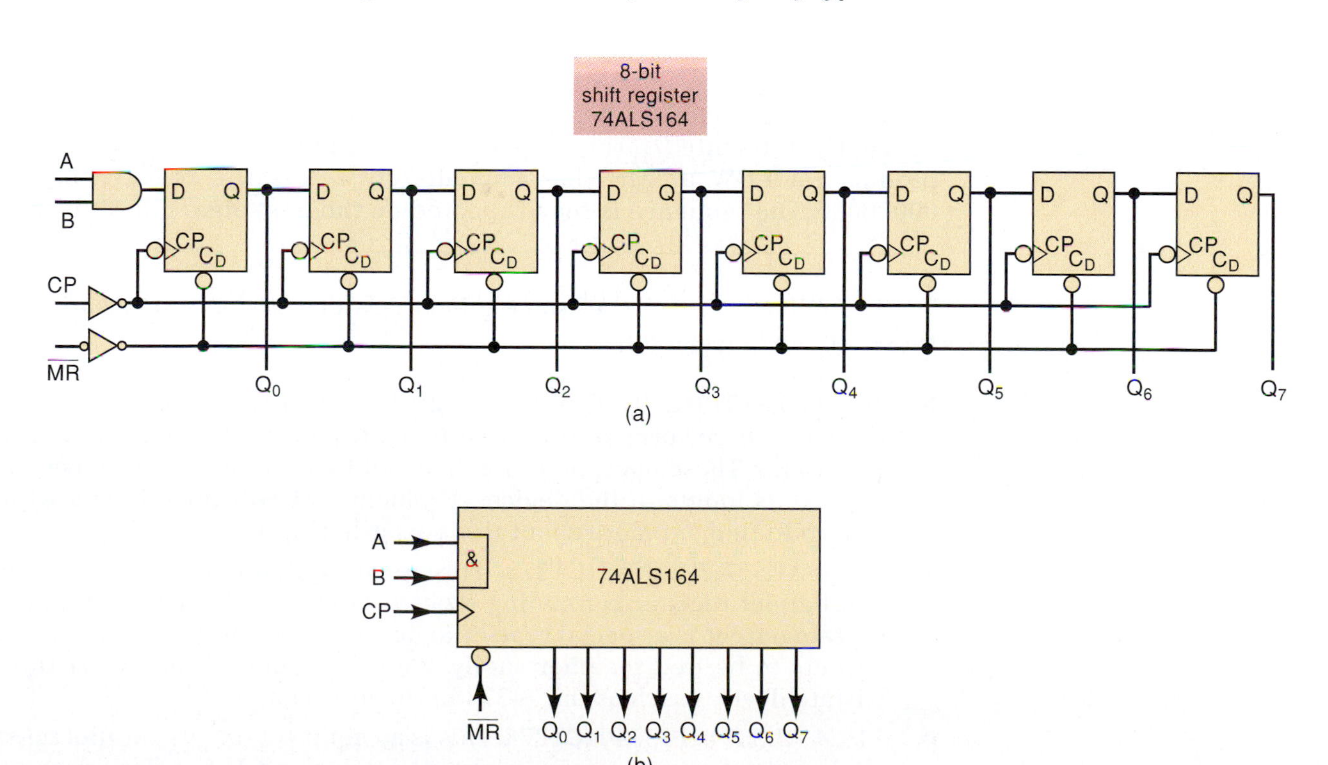

FIGURE 7-68 (a) Logic diagram for the 74ALS164; (b) logic symbol.

The shift operation occurs on the PGTs of the clock input *CP*. The $\overline{MR}$ input provides asynchronous resetting of all FFs on a LOW level.

The logic symbol for the 74ALS164 is shown in Figure 7-68(b). Note that the & symbol is used inside the block to indicate that the *A* and *B* inputs are ANDed inside the IC and the result is applied to the *D* input of Q_0.

EXAMPLE 7-21

Assume that the initial contents of the 74ALS164 register in Figure 7-69(a) are 00000000. Determine the sequence of states as clock pulses are applied.

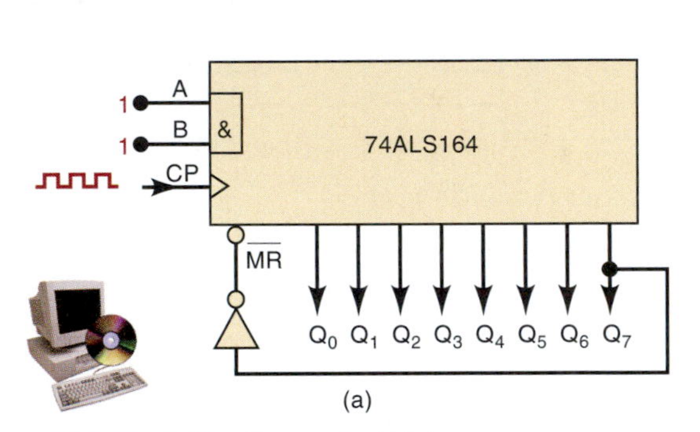

Input pulse number	Q_0	Q_1	Q_2	Q_3	Q_4	Q_5	Q_6	Q_7
0	0	0	0	0	0	0	0	0
1	1	0	0	0	0	0	0	0
2	1	1	0	0	0	0	0	0
3	1	1	1	0	0	0	0	0
4	1	1	1	1	0	0	0	0
5	1	1	1	1	1	0	0	0
6	1	1	1	1	1	1	0	0
7	1	1	1	1	1	1	1	0
8	1	1	1	1	1	1	1	1

Recycles

Temporary state

(b)

FIGURE 7-69 Example 7-21.

Solution

The correct sequence is given in Figure 7-69(b). With $A = B = 1$, the serial input is 1, so that 1s will shift into the register on each PGT of *CP*. Because Q_7 is initially at 0, the $\overline{MR}$ input is inactive.

On the eighth pulse, the register tries to go to the 11111111 state as the 1 from Q_6 shifts into Q_7. This state occurs only momentarily because $Q_7 = 1$ produces a LOW at $\overline{MR}$ that immediately resets the register back to 00000000. The sequence is then repeated on the next eight clock pulses.

The following is a list of some other register ICs that are variations on those already presented:

- 74194/ALS194/HC194. This is a four-bit *bidirectional universal shift-register* IC that can perform shift-left, shift-right, parallel in, and parallel out operations. These operations are selected by a two-bit mode-select code applied as inputs to the device. (Problem 7-71 will provide you with a chance to find out more about this versatile chip.)
- 74373/ALS373/HC373/HCT373. This is an eight-bit (octal) parallel in/-parallel out register containing eight *D* latches with *tristate* outputs. A tristate output is a special type of logic circuit output that allows device outputs to be tied together safely. We will cover the characteristics of tristate devices such as the 74373 in the next chapter.
- 74374/ALS374/HC374/HCT374. This is an eight-bit (octal) parallel in/parallel out register containing eight edge-triggered D flip-flops with tristate outputs.

The IC registers that have been presented here are representative of the various types that are commercially available. Although there are many variations on these basic registers, most of them should now be relatively easy to understand from the manufacturers' data sheets.

We will present several register applications in the end-of-chapter problems and in the material covered in subsequent chapters.

REVIEW QUESTIONS

1. What kind of register can have a complete binary number loaded into it in one operation, and then have it shifted out one bit at a time?
2. *True or false:* A serial in/parallel out register can have all of its bits displayed at one time.
3. What type of register can have data entered into it only one bit at a time, but has all data bits available as outputs?
4. In what type of register do we store data one bit at a time and have access to only one output bit at a time?
5. How does the parallel data entry differ for the 74165 and the 74174?
6. How does the *CP INH* input of the 74ALS165 work?

7-20 SHIFT-REGISTER COUNTERS

In Section 5-18, we saw how to connect FFs in a shift-register arrangement to transfer data left to right, or vice versa, one bit at a time (serially). Shift-register counters use *feedback,* which means that the output of the last FF in the register is connected back to the first FF in some way.

Ring Counter

The simplest shift-register counter is essentially a **circulating shift register** connected so that the last FF shifts its value into the first FF. This arrangement is shown in Figure 7-70 using D-type FFs (J-K flip-flops can also be used). The FFs are connected so that information shifts from left to right and back around from Q_0 to Q_3. In most instances, only a single 1 is in the register, and it is made to circulate around the register as long as clock pulses are applied. For this reason, it is called a **ring counter**.

The waveforms, sequence table, and state diagram in Figure 7-70 show the various states of the FFs as pulses are applied, assuming a starting state of $Q_3 = 1$ and $Q_2 = Q_1 = Q_0 = 0$. After the first pulse, the 1 has shifted from Q_3 to Q_2 so that the counter is in the 0100 state. The second pulse produces the 0010 state, and the third pulse produces the 0001 state. On the *fourth* clock pulse, the 1 from Q_0 is transferred to Q_3, resulting in the 1000 state, which is, of course, the initial state. Subsequent pulses cause the sequence to repeat.

This counter functions as a MOD-4 counter because it has *four* distinct states before the sequence repeats. Although this circuit does not progress through the normal binary counting sequence, it is still a counter because each count corresponds to a unique set of FF states. Note that each FF output waveform has a frequency equal to one-fourth of the clock frequency because this is a MOD-4 ring counter.

Ring counters can be constructed for any desired MOD number; a MOD-N ring counter uses N flip-flops connected in the arrangement of Figure 7-70.

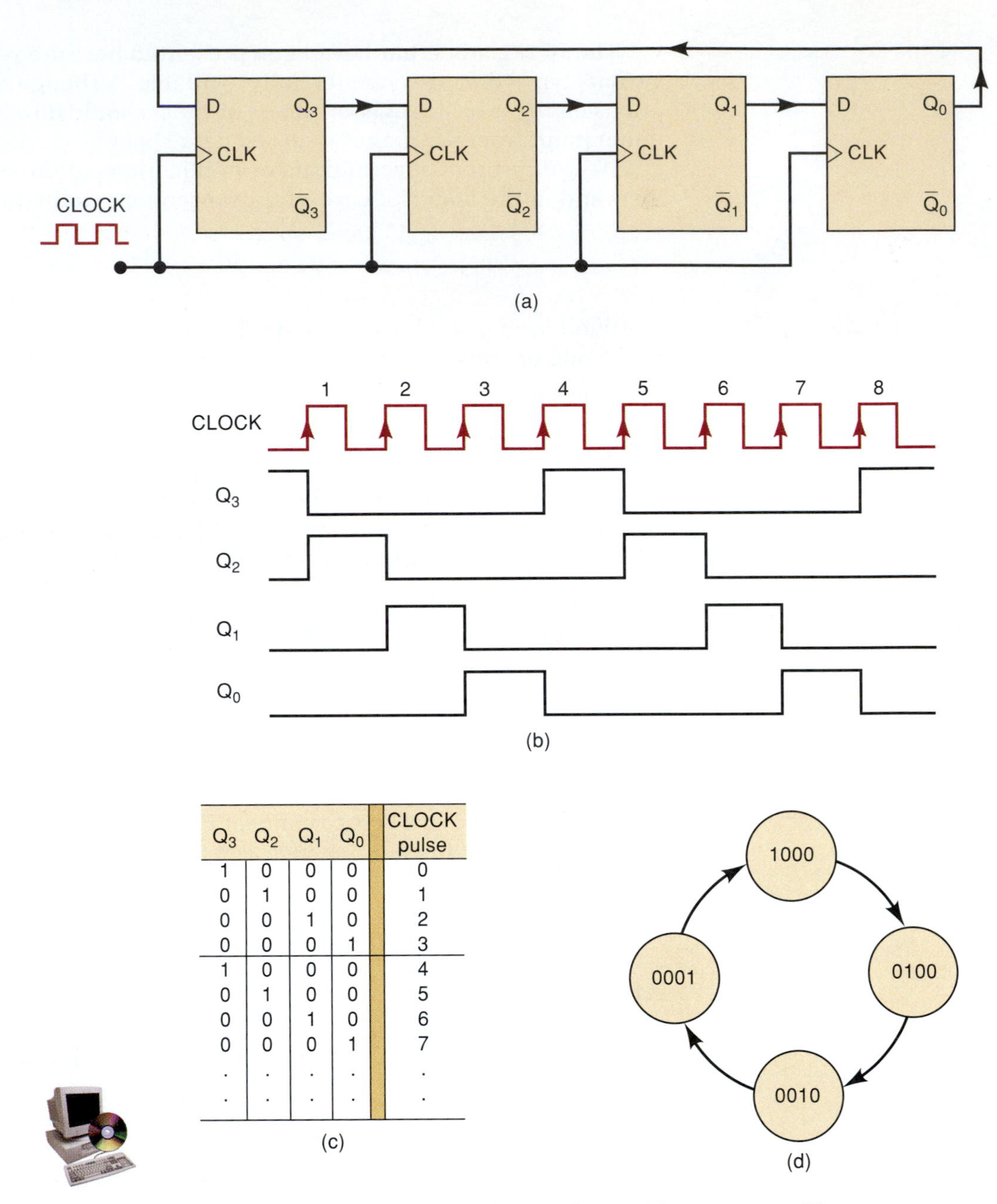

Q_3	Q_2	Q_1	Q_0	CLOCK pulse
1	0	0	0	0
0	1	0	0	1
0	0	1	0	2
0	0	0	1	3
1	0	0	0	4
0	1	0	0	5
0	0	1	0	6
0	0	0	1	7
.	.	.	.	.
.	.	.	.	.

FIGURE 7-70 (a) Four-bit ring counter; (b) waveforms; (c) sequence table; (d) state diagram.

In general, a ring counter requires more FFs than a binary counter for the same MOD number; for example, a MOD-8 ring counter requires eight FFs, while a MOD-8 binary counter requires only three.

Despite the fact that it is less efficient in the use of FFs, a ring counter is still useful because it can be decoded without the use of decoding gates. The decoding signal for each state is obtained at the output of its corresponding FF. Compare the FF waveforms of the ring counter with the decoding waveforms in Figure 7-20. In some cases, a ring counter might be a better choice than a binary counter with its associated decoding gates. This is especially true in applications where the counter is being used to control the sequencing of operations in a system.

Starting a Ring Counter

To operate properly, a ring counter must start off with only one FF in the 1 state and all the others in the 0 state. Because the starting states of the FFs will be unpredictable on power-up, the counter must be preset to the required starting state before clock pulses are applied. One way to do this is to apply a momentary pulse to the asynchronous $\overline{PRE}$ input of one of the FFs (e.g., Q_3 in Figure 7-70) and to the $\overline{CLR}$ input of all other FFs. Another method is shown in Figure 7-71. On power-up, the capacitor will charge up relatively slowly toward $+V_{CC}$. The output of Schmitt-trigger INVERTER 1 will stay HIGH, and the output of INVERTER 2 will remain LOW until the capacitor voltage exceeds the positive-going threshold voltage (V_{T+}) of the INVERTER 1 input (about 1.7 V). This will hold the $\overline{PRE}$ input of Q_3 and the $\overline{CLR}$ inputs of Q_2, Q_1, and Q_0 in the LOW state long enough during power-up to ensure that the counter starts at 1000.

Johnson Counter

The basic ring counter can be modified slightly to produce another type of shift-register counter, which will have somewhat different properties. The **Johnson** or **twisted-ring counter** is constructed exactly like a normal ring counter except that the *inverted* output of the last FF is connected to the input of the first FF. A three-bit Johnson counter is shown in Figure 7-72. Note that the $\overline{Q}_0$ output is connected back to the D input of Q_2, which means that the *inverse* of the level stored in Q_0 will be transferred to Q_2 on the clock pulse.

The Johnson-counter operation is easy to analyze if we realize that on each positive clock-pulse transition, the level at Q_2 shifts into Q_1, the level at Q_1 shifts into Q_0, and the *inverse* of the level at Q_0 shifts into Q_2. Using these ideas and assuming that all FFs are initially 0, the waveforms, sequence table, and state diagram of Figure 7-72 can be generated.

Examination of the waveforms and sequence table reveals the following important points:

1. This counter has six distinct states—000, 100, 110, 111, 011, and 001—before it repeats the sequence. Thus, it is a MOD-6 Johnson counter. Note that it does not count in a normal binary sequence.
2. The waveform of each FF is a square wave (50 percent duty cycle) at one-sixth the frequency of the clock. In addition, the FF waveforms are shifted by one clock period with respect to each other.

The MOD number of a Johnson counter will always be equal to *twice* the number of FFs. For example, if we connect five FFs in the arrangement of

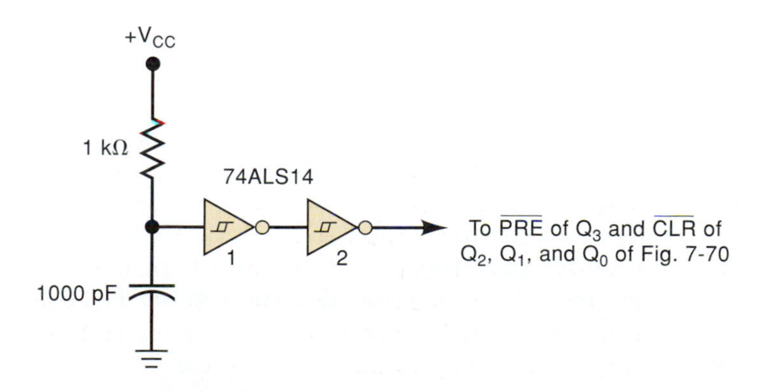

FIGURE 7-71 Circuit for ensuring that the ring counter of Figure 7-70 starts in the 1000 state on power-up.

FIGURE 7-72 (a) MOD-6 Johnson counter; (b) waveform; (c) sequence table; (d) state diagram.

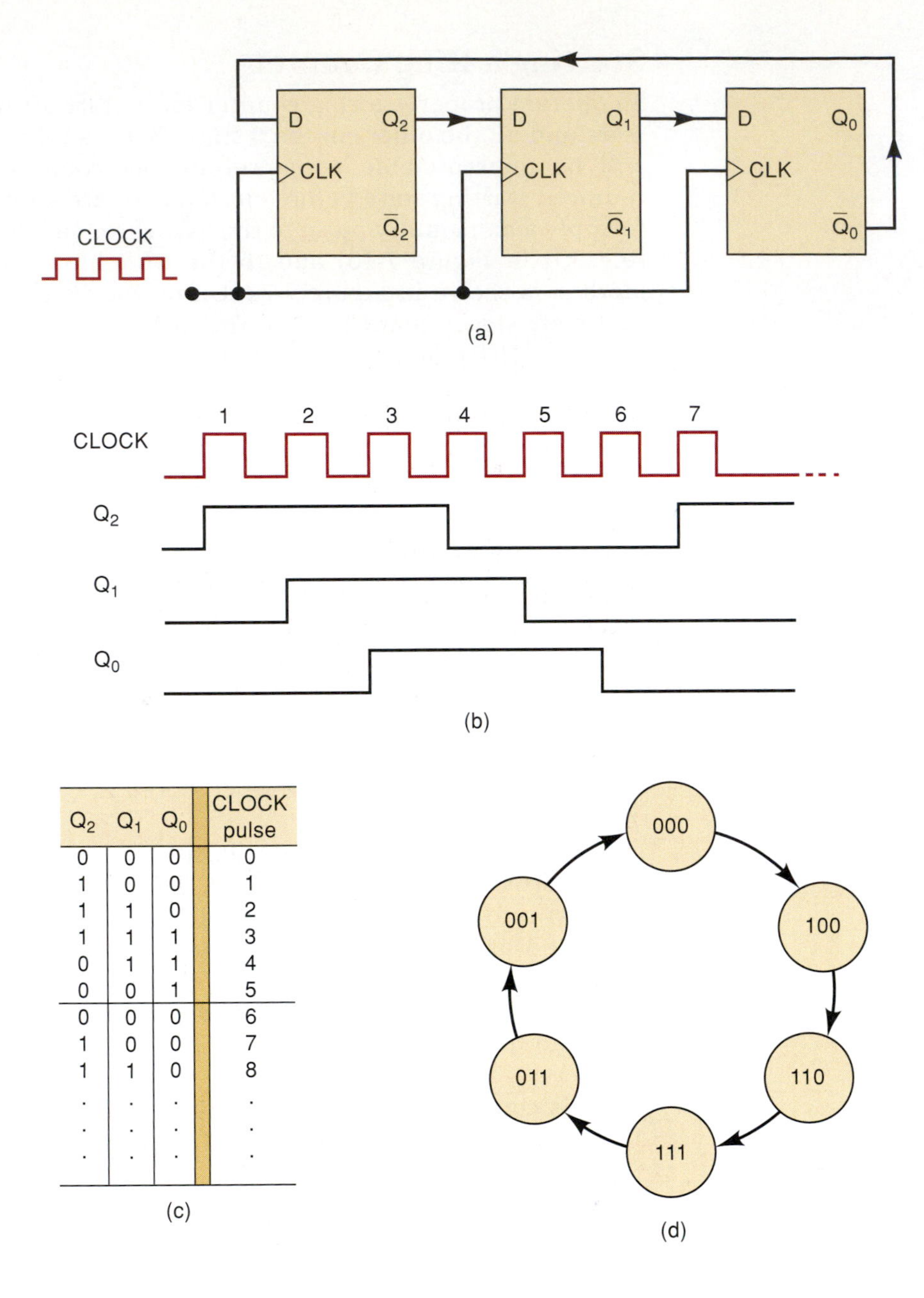

Q_2	Q_1	Q_0	CLOCK pulse
0	0	0	0
1	0	0	1
1	1	0	2
1	1	1	3
0	1	1	4
0	0	1	5
0	0	0	6
1	0	0	7
1	1	0	8
.	.	.	.
.	.	.	.
.	.	.	.

Figure 7-72, the result is a MOD-10 Johnson counter, where each FF output waveform is a square wave at one-tenth the clock frequency. Thus, it is possible to construct a MOD-N counter (where N is an even number) by connecting $N/2$ flip-flops in a Johnson-counter arrangement.

Decoding a Johnson Counter

For a given MOD number, a Johnson counter requires only half the number of FFs that a ring counter requires. However, a Johnson counter requires decoding gates, whereas a ring counter does not. As in the binary counter, the Johnson counter uses one logic gate to decode for each count, but each gate requires only two inputs, regardless of the number of FFs in the counter. Figure 7-73 shows the decoding gates for the six states of the Johnson counter of Figure 7-72.

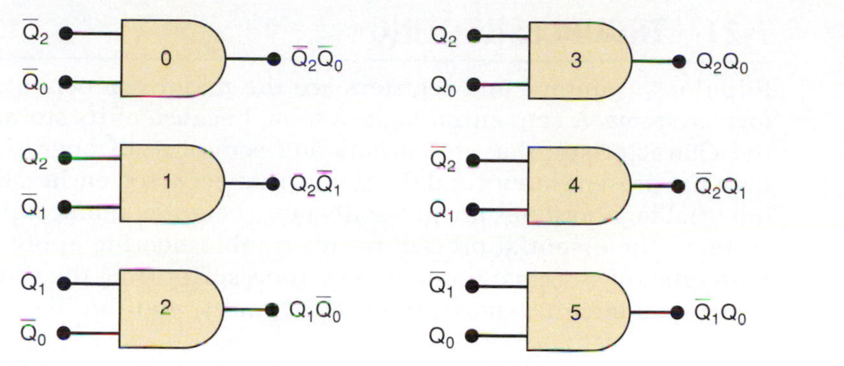

Q_2	Q_1	Q_0	Active gate
0	0	0	0
1	0	0	1
1	1	0	2
1	1	1	3
0	1	1	4
0	0	1	5

FIGURE 7-73 Decoding logic for a MOD-6 Johnson counter.

Notice that each decoding gate has only two inputs, even though there are three FFs in the counter, because for each count, two of the three FFs are in a unique combination of states. For example, the combination $Q_2 = Q_0 = 0$ occurs only once in the counting sequence, at the count of 0. Thus, AND gate 0, with inputs $\overline{Q}_2$ and $\overline{Q}_0$, can be used to decode for this count. This same characteristic is shared by all of the other states in the sequence, as the reader can verify. In fact, for *any size* Johnson counter, the decoding gates will have only two inputs.

Johnson counters represent a middle ground between ring counters and binary counters. A Johnson counter requires fewer FFs than a ring counter but generally more than a binary counter; it has more decoding circuitry than a ring counter but less than a binary counter. Thus, it sometimes represents a logical choice for certain applications.

IC Shift-Register Counters

Very few ring counters or Johnson counters are available as ICs because it is relatively simple to take a shift-register IC and to wire it as either a ring counter or a Johnson counter. Some of the CMOS Johnson-counter ICs (74HC4017, 74HC4022) include the complete decoding circuitry on the same chip as the counter.

REVIEW QUESTIONS

1. Which shift-register counter requires the most FFs for a given MOD number?
2. Which shift-register counter requires the most decoding circuitry?
3. How can a ring counter be converted to a Johnson counter?
4. *True or false:*
   (a) The outputs of a ring counter are always square waves.
   (b) The decoding circuitry for a Johnson counter is simpler than for a binary counter.
   (c) Ring and Johnson counters are synchronous counters.
5. How many FFs are needed in a MOD-16 ring counter? How many are needed in a MOD-16 Johnson counter?

7-21 TROUBLESHOOTING

Flip-flops, counters, and registers are the major components in **sequential logic systems.** A sequential logic system, because of its storage devices, has the characteristic that its outputs and sequence of operations depend on both the present inputs and the inputs that occurred earlier. Even though sequential logic systems are generally more complex than combinational logic systems, the essential procedures for troubleshooting apply equally well to both types of systems. Sequential systems suffer from the same types of failures (open circuits, shorts, internal IC faults, and the like) as do combinational systems.

Many of the same steps used to isolate faults in a combinational system can be applied to sequential systems. One of the most effective troubleshooting techniques begins with the troubleshooter observing the system operation and, by analytical reasoning, determining the possible causes of the system malfunction. Then he or she uses available test instruments to isolate the exact fault. The following examples will show the kinds of analytical reasoning that should be the initial step in troubleshooting sequential systems. After studying these examples, you should be ready to tackle the troubleshooting problems at the end of the chapter.

EXAMPLE 7-22

Figure 7-74(a) shows a 74ALS161 wired as a MOD-12 counter, but it produces the count sequence given in Figure 7-74(b). Determine the cause of the incorrect circuit behavior.

Solution

Outputs QB and QA seem to be operating correctly but QC and QD stay LOW. Our first choice for the fault is that QC is shorted to ground, but an ohmmeter check does not confirm this. The 74ALS161 might have an internal fault that prevents it from counting above 0011. We try removing the

FIGURE 7-74 Example 7-22.

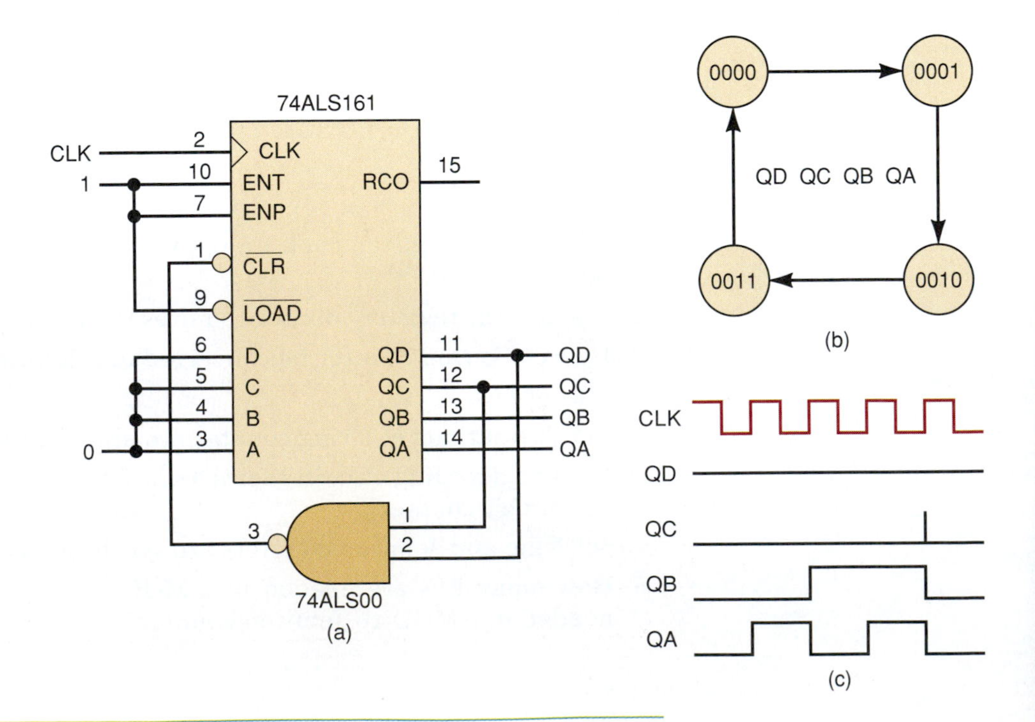

7400 NAND chip from its socket and shorting the CLR pin to a HIGH. The counter now counts a regular MOD-16 sequence, so at least the counter's outputs seem to be ok. Next we decide to look at the CLR pin with the NAND reconnected. Using a logic probe with its "pulse capture" turned on shows us that the CLR pin is receiving pulses. Connecting a scope to the outputs, we see that the counter produces the waveforms shown in Figure 7-74(c). A glitch is observed on QC when the counter should be going to state 0100. That indicates that 0100 is a transient state when the transient state should actually be 1100. The QD connection to the NAND gate is now suspected, so we use the logic probe to check pin 2. There is no logic signal at all indicated on pin 2, which now leads us to the conclusion that the fault is an open between the QD output and pin 2 on the NAND. The NAND input is floating HIGH, causing the circuit to detect state 0100 instead of 1100 as it should be doing.

EXAMPLE 7-23

A technician receives a "trouble ticket" for a circuit board that says the variable frequency divider operates "sometimes." Sounds like a dreaded intermittent fault problem—often the hardest problems to find! His first thought is to send it back with the note "Use only when operating correctly!" but he decides to investigate further since he feels up to a good challenge today. The schematic for the circuit block is shown in Figure 7-75. The desired

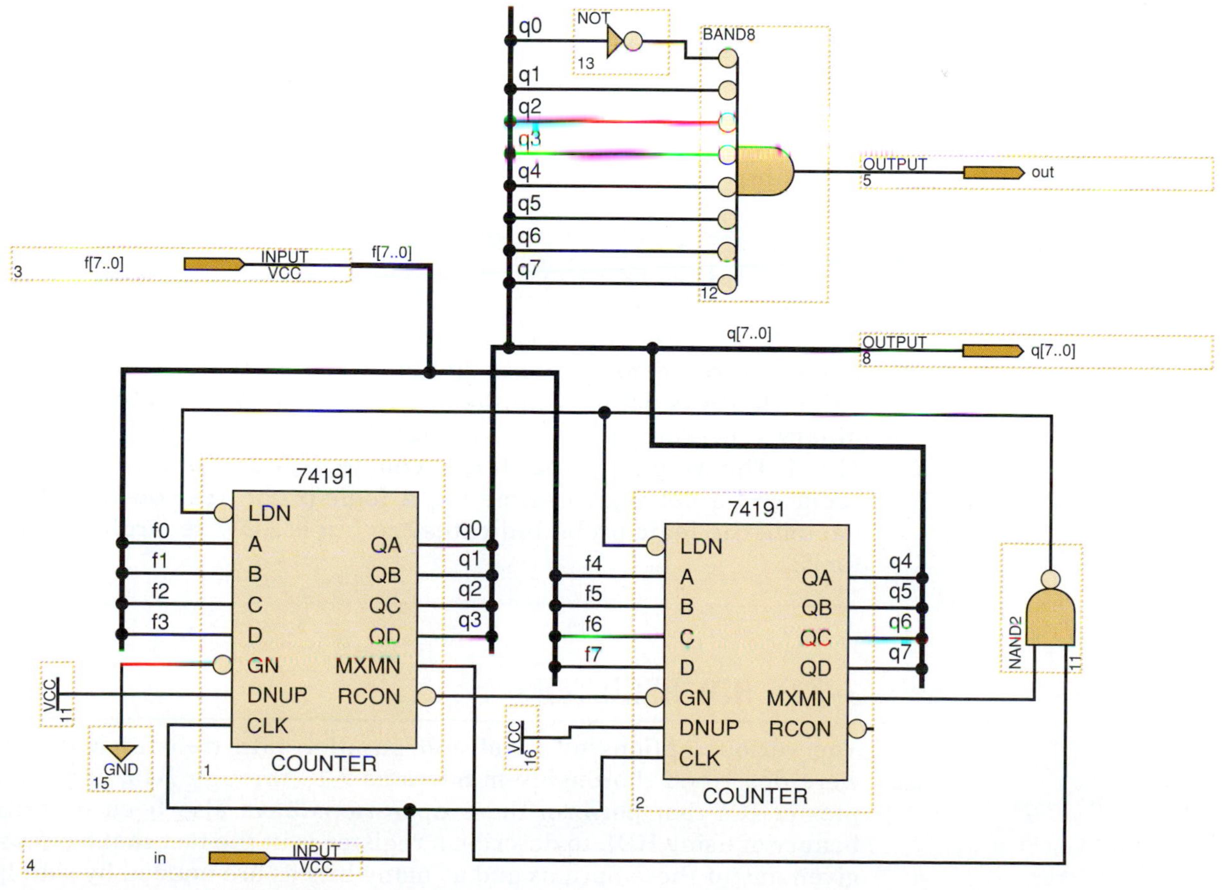

FIGURE 7-75 Example 7-23.

divide-by factor is applied to input *f[7..0]* in binary. The eight-bit counter counts down from this number until it reaches zero and then asynchronously loads in *f[ ]* again, making zero a transient state. The resulting modulus will be equal to the value on *f[ ]*. The output frequency signal is obtained by decoding state 00000001, making the frequency of *out* equal to the frequency of *in* divided by the binary value *f[ ]*. In the application, the frequency of *in* is 100 kHz. Change *f[ ]* and a new frequency will be output.

Solution

The technician decides that he needs to obtain some test results to look at. He picks some easy divide-by factors to apply to *f* and records the results listed in Table 7-9.

TABLE 7-9

f[ ] (decimal)	f[ ] (binary)	Measured f_{out}	OK?
255	11111111	398.4 Hz	
240	11110000	416.7 Hz	✓
200	11001000	500.0 Hz	✓
100	01100100	1041.7 Hz	
50	00110010	2000.0 Hz	✓
25	00011001	4000.0 Hz	✓
15	00001111	9090.9 Hz	

He observes that the circuit produces correct results for some test cases but incorrect results for others. The problem does not seem to be intermittent after all. Instead, it appears to be dependent on the value for *f*. The technician decides to calculate the relationship between input and output frequencies for the three tests that failed and obtains the following:

```
100 kHz/398.4 Hz = 251
100 kHz/1041.7 Hz = 96
100 kHz/9090.9 Hz = 11
```

Each failure seems to be a divide-by factor that is four less than the value that was actually applied to the input. After looking again at the binary representation for *f*, he notes that every failure occurred when *f2* = 1. The weight for that bit, of course, is four. Eureka! That bit doesn't seem to be getting in—time for a logic-probe test on the *f2* pin. Sure enough, the logic probe indicates the pin is LOW regardless of the value for *f2*.

7-22 HDL REGISTERS

The various options of serial and parallel data transfer within registers were described thoroughly in Sections 7-15 through 7-19, and some examples of ICs that perform these operations have also been described. The beauty of using HDL to describe a register is in the fact that a circuit can be given any of these options and as many bits as are needed by simply changing a few words.

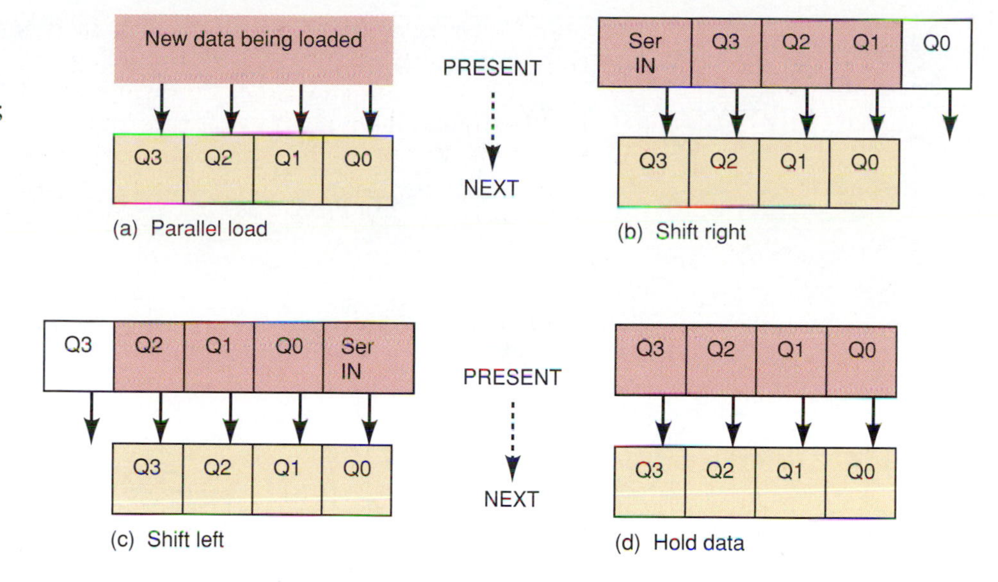

FIGURE 7-76 Data transfers made in shift registers: (a) parallel load; (b) shift right; (c) shift left; (d) hold data.

HDL techniques use bit arrays to describe a register's data and to transfer that data in a parallel or serial format. To understand how data are shifted in HDL, consider the diagrams in Figure 7-76, which shows four flip-flops performing transfer operations of parallel load, shift right, shift left, and hold data. For all of these diagrams, the bits are transferred synchronously, which means that they all move simultaneously on a single clock edge. In Figure 7-76(a), the data that is to be parallel loaded into the register is presented to the *D* inputs, and on the next clock pulse, it will be transferred to the *q* outputs. Shifting data right means that each bit is transferred to the bit location to its immediate right, while a new bit is transferred in on the left end and the last bit on the right end is lost. This situation is depicted in Figure 7-76(b). Notice that the data set that we want in the NEXT state is made up of the new serial input and three of the four bits in the PRESENT state array. This data simply needs to move over and overwrite the four data bits of the register. The same operation occurs in Figure 7-76(c), but it is moving data to the left. The key to shifting the contents of the register to the right or left is to group the appropriate three PRESENT state data bits in correct order with the serial input bit so that these four bits can be loaded in parallel into the register. **Concatenation** (grouping together in a specific sequence) of the desired set of data bits can be used to describe the necessary data movement for serial shifting in either direction. The last possibility is called the hold data mode and is shown in Figure 7-76(d). It may seem unnecessary because registers (flip-flops) hold data by their very nature. We must consider, however, what must be done to a register in order to hold its value as it is clocked. The *Q* outputs must be tied back to the *D* inputs for each flip-flop so that the old data is reloaded on each clock. Let's look at some example HDL shift register circuits.

AHDL SISO REGISTER

A four-bit serial in/serial out (SISO) register in AHDL is listed in Figure 7-77. An array of four D flip-flops is instantiated in line 7 and the serial output is obtained from the last FF *q0* (line 10). If the *shift* control is HIGH, *serial_in* will be shifted into the register and the other bits will move to the right (lines 11-15). Concatenating *serial_in* and FF output bits *q3*, *q2*, and *q1*

```
1   SUBDESIGN  fig7_77
2   (
3         clk, shift, serial_in           :INPUT;
4         serial_out                      :OUTPUT;
5   )
6   VARIABLE
7         q[3..0]                         :DFF;
8   BEGIN
9         q[].clk = clk;
10        serial_out = q0.q;                            -- output last register bit
11        IF (shift == VCC)  THEN
12              q[3..0].d = (serial_in, q[3..1].q);   -- concatenates for shift
13        ELSE
14              q[3..0].d = (q[3..0].q);               -- hold data
15        END IF;
16  END;
```

FIGURE 7-77 Serial in/serial out register using AHDL.

together in that order creates the proper shift-right data input bit pattern (line 12). If the *shift* control is LOW, the register will hold the current data (line 14). The simulation results are shown in Figure 7-78.

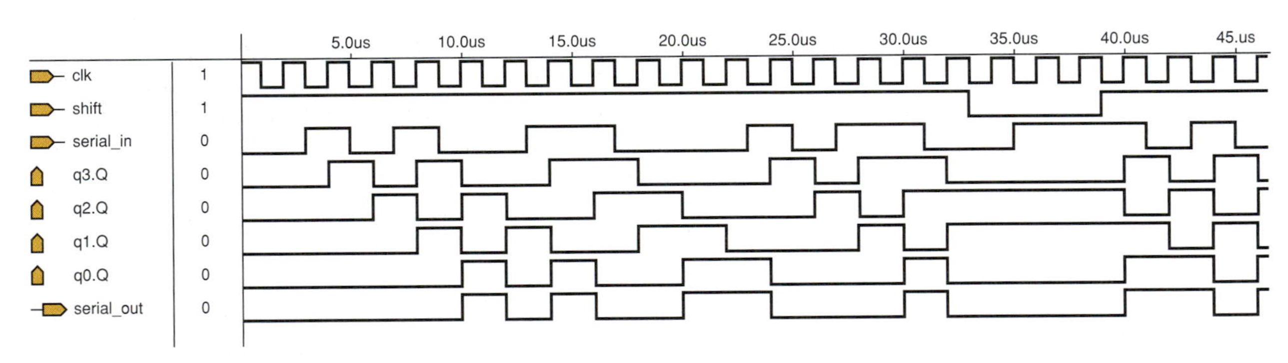

FIGURE 7-78 SISO register simulation.

VHDL

VHDL SISO REGISTER

A four-bit serial in/serial out (SISO) register in VHDL is listed in Figure 7-79. A register is created with the declaration of the variable *q* on line 8 and the serial output is obtained from the register's last bit or *q(0)* (line 10). If the *shift* control is HIGH, *serial_in* will be shifted into the register and the other bits will move to the right (lines 12–14). Concatenating *serial_in* and register bits *q(3)*, *q(2)*, and *q(1)* together in that order creates the proper shift-right data input bit pattern (line 13). If the shift control is LOW, VHDL will assume that the variable stays the same and will therefore hold the current data. Simulation results are shown in Figure 7-78.

```
1   ENTITY  fig7_79  IS
2   PORT (       clk, shift, serial_in          :IN BIT;
3                serial_out                     :OUT BIT    );
4   END fig 7-79;
5   ARCHITECTURE  vhdl  OF  fig 7-79  IS
6   BEGIN
7   PROCESS (clk)
8         VARIABLE  q                     :BIT_VECTOR (3 DOWNTO 0);
9         BEGIN
10        serial_out <= q(0);                          -- output last register bit
11        IF (clk'EVENT AND clk = '1')   THEN
12              IF (shift = '1')   THEN
13                    q := (serial_in & q(3 DOWNTO 1));  -- concatenate for shift
14              END IF;                                -- otherwise, hold data
15        END IF;
16  END PROCESS;
17  END vhdl;
```

FIGURE 7-79 Serial in/serial out register using VHDL.

AHDL PISO REGISTER

A four-bit parallel in/serial out (PISO) register in AHDL is listed in Figure 7-80. The register named *q* is created on line 8 using four D FFs, and the serial output from *q0* is described on line 11. The register has separate parallel *load* and serial *shift* controls. The register's functions are defined in lines 12–15. If *load* is HIGH, the external input *data[3..0]* will be synchronously loaded. *Load* has priority and must be LOW to serial-shift the register's contents on each PGT of *clk* when *shift* is HIGH. The pattern for shifting data right is created by concatenation on line 13. Note that a constant LOW will be the serial data input for a shift operation. If neither *load* nor *shift* is HIGH, the register will hold the current data value (line 14). Simulation results are shown in Figure 7-81.

```
1   SUBDESIGN  fig7_80
2   (
3         clk, shift, load        :INPUT;
4         data[3..0]              :INPUT;
5         serial_out              :OUTPUT;
6   )
7   VARIABLE
8         q[3..0]                 :DFF;
9   BEGIN
10        q[].clk = clk;
11        serial_out = q0.q;                          -- output last register bit
12        IF (load == VCC)  THEN  q[3..0].d = data[3..0];        -- parallel load
13        ELSIF (shift == VCC)  THEN  q[3..0].d = (GND, q[3..1].q);  -- shift
14        ELSE  q[3..0].d = q[3..0].q;                               -- hold
15        END IF;
16  END;
```

FIGURE 7-80 Parallel in/serial out register using AHDL.

AHDL

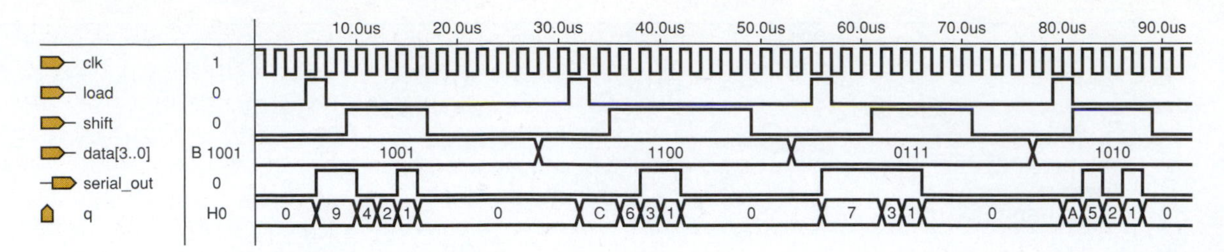

FIGURE 7-81 PISO register simulation.

VHDL

VHDL PISO REGISTER

A four-bit parallel in/serial out (PISO) register in VHDL is listed in Figure 7-82. The register is created with the variable declaration for *q* on line 11, and the serial output from *q(0)* is described on line 13. The register has separate parallel *load* and serial *shift* controls. The register's functions are defined in lines 14–18. If *load* is HIGH, the external input *data* will be synchronously loaded. *Load* has priority and must be LOW to serial-shift the register's contents on each PGT of *clk* when *shift* is HIGH. The pattern for shifting data right is created by concatenation on line 16. Note that a constant LOW will be the serial data input for a shift operation. If neither *load* nor *shift* is HIGH, the register will hold the current data value by VHDL's implied operation. Simulation results are shown in Figure 7-81.

```
1   ENTITY  fig7_82  IS
2   PORT (
3         clk, shift, load         :IN BIT;
4         data                     :IN BIT_VECTOR (3 DOWNTO 0);
5         serial_out               :OUT BIT
6   );
7   END fig 7-82;
8   ARCHITECTURE vhdl OF fig 7-82 IS
9   BEGIN
10  PROCESS (clk)
11        VARIABLE  q        :BIT_VECTOR (3 DOWNTO 0);
12        BEGIN
13        serial_out <= q(0);                       -- output last register bit
14        IF (clk'EVENT AND clk = '1')  THEN
15              IF (load = '1')  THEN  q := data;   -- parallel load
16              ELSIF (shift = '1')  THEN  q := ('0' & q(3 DOWNTO 1));  -- shift
17              END IF;                             -- otherwise, hold
18        END IF;
19  END PROCESS;
20  END vhdl;
```

FIGURE 7-82 Parallel in/serial out register using VHDL.

EXAMPLE 7-24

Suppose we want to design a universal four-bit shift register, using HDL, that has four synchronous modes of operation: Hold Data, Shift Left, Shift Right, and Parallel Load. Two input bits will select the operation that is to be performed on each rising edge of the clock. To implement a shift register, we can use structural code to describe a string of flip-flops. Making the shift register versatile by allowing it to shift right or left or to parallel load would make this file quite long and thus hard to read and understand using structural methods. A much better approach is to use the more abstract and intuitive methods available in HDL to describe the circuit concisely. To do this, we must develop a strategy that will create the shifting action. The concept is very similar to the one presented in Example 7-16, where a D flip-flop register chip (74174) was wired to form a shift register. Rather than thinking of the shift register as a serial string of flip-flops, we consider it as a parallel register whose contents are being transferred in parallel to a set of bits that is offset by one bit position. Figure 7-76 demonstrates the concept of each transfer needed in this design.

Solution

A very reasonable first step is to define a two-bit input named *mode* with which we can specify mode 0, 1, 2, or 3. The next challenge is deciding how to choose among the four operations using HDL. Several methods can work here. The CASE structure was chosen because it allows us to choose a different set of HDL statements for each and every possible mode value. There is no priority associated with checking for the existing mode settings or overlapping ranges of mode numbers, so we do not need the advantages of the IF/ELSE construct. The HDL solutions are given in Figures 7-83 and 7-84. The same inputs and outputs are defined in each approach: a clock, four bits of parallel load data, a single bit for the serial input to the register, two bits for the mode selection, and four output bits.

```
1   SUBDESIGN fig7_83
2   (
3      clock        :INPUT;
4      din[3..0]    :INPUT;  -- parallel data in
5      ser_in       :INPUT;  -- serial data in from Left or Right
6      mode [1..0]  :INPUT;  -- MODE Select: 0=hold, 1=right, 2=left, 3=load
7      q[3..0]      :OUTPUT;
8   )
9   VARIABLE
10     ff[3..0] :DFF;        -- define register set
11  BEGIN
12     ff[].clk = clock;     -- synchronous clock
13     CASE mode[] IS
14        WHEN 0  => ff[].d      = ff[].q;        -- hold shift
15        WHEN 1  => ff[2..0].d = ff[3..1].q);   -- shift right
16                   ff[3].d    = ser_in;        -- new data from left
17        WHEN 2  => ff[3..1].d = ff[2..0].q;    -- shift left
18                   ff[0].d    = ser_in;        -- new data bit from right
19        WHEN 3  => ff[].d      = din[];         -- parallel load
20     END CASE;
21     q[] = ff[];               -- update outputs
22  END;
```

FIGURE 7-83 AHDL universal shift register.

```
1   ENTITY  fig7_84  IS
2   PORT (
3      clock            :IN BIT;
4      din              :IN BIT_VECTOR (3 DOWNTO 0);   -- parallel data in
5      ser_in           :IN BIT;                       -- serial data in L or R
6      mode             :IN INTEGER RANGE 0 TO 3;      -- 0=hold 1=rt 2=lt 3=load
7      q                :OUT BIT_VECTOR (3 DOWNTO 0));
8   END fig7_84;
9   ARCHITECTURE a  OF fig7_84  IS
10  BEGIN
11     PROCESS (clock)                                -- respond to clock
12     VARIABLE  ff   :BIT_VECTOR (3 DOWNTO 0);
13     BEGIN
14        IF (clock'EVENT AND clock = '1')  THEN
15           CASE mode  IS
16              WHEN 0  => ff := ff;                                 -- hold data
17              WHEN 1  => ff(2 DOWNTO 0)  := ff(3 DOWNTO 1);  -- shift right
18                         ff(3) := ser_in;
19              WHEN  2 => ff(3 DOWNTO 1)  := ff(2 DOWNTO 0);  -- shift left
20                         ff(0) := ser_in;
21              WHEN  3 => ff := din;                                -- parallel load
22           END CASE;
23        END IF;
24     q <= ff;                                       -- update outputs
25     END PROCESS;
26  END a;
```

FIGURE 7-84 VHDL universal shift register.

AHDL

AHDL SOLUTION

The AHDL solution of Figure 7-83 uses a register of D flip-flops declared by the name *ff* on line 10, representing the current state of the register. Because the flip-flops all need to be clocked at the same time (synchronously), all the clock inputs are assigned to *clock* on line 12. The CASE construct selects a different transfer configuration for each value of the *mode* inputs. Mode 0 (hold data) uses a direct parallel transfer from the current state to the same bit positions on the *D* inputs to produce the identical NEXT state. Mode 1 (shift right), which is described on lines 15 and 16, transfers bits 3, 2, and 1 to bit positions 2, 1, and 0, respectively, and loads bit 3 from the serial input. Mode 2 (shift left) performs a similar operation in the opposite direction (see lines 17 and 18). Mode 3 (parallel load) transfers the value on the parallel data inputs to become the NEXT state of the register. The code creates the circuitry that chooses one of these logical operations on the actual register, and the proper data is transferred to the output pins on the next clock. This code can be shortened by combining lines 15 and 16 into a single statement that concatenates the ser_in with the three data bits and groups them as a set of four bits. The statement that can replace lines 15 and 16 is:

```
WHEN 1 => ff[].d = (ser_in, ff[3..1].q);
```

Lines 17 and 18 can also be replaced by:

```
WHEN 2 => ff[].d = (ff[2..0].q,ser_in);
```

VHDL

VHDL SOLUTION

The VHDL solution of Figure 7-84 defines an internal variable by the name *ff* on line 12, representing the current state of the register. Because all the transfer operations need to take place in response to a rising clock edge, a PROCESS is used, with *clock* specified in the sensitivity list. The CASE construct selects a different transfer configuration for each value of the *mode* inputs. Mode 0 (hold data) uses a direct parallel transfer from the current state to the same bit positions to produce the identical NEXT state. Mode 1 (shift right) transfers bits 3, 2, and 1 to bit positions 2, 1, and 0, respectively (line 17), and loads bit 3 from the serial input (line 18). Mode 2 (shift left) performs a similar operation in the opposite direction. Mode 3 (parallel load) transfers the value on the parallel data inputs to the NEXT state of the register. After choosing one of these operations on the actual register, the data is transferred to the output pins on line 24. This code can be shortened by combining lines 17 and 18 into a single statement that concatenates the ser_in with the three data bits and groups them as a set of four bits. The statement that can replace lines 17 and 18 is:

```
WHEN 1 => ff := ser_in & ff(3 DOWNTO 1);
```

Lines 19 and 20 can also be replaced by:

```
WHEN 2 => ff := ff(2 DOWNTO 0) & ser_in;
```

REVIEW QUESTIONS

1. Write a HDL expression that can implement a shift left of an eight-bit array *reg[7..0]* with serial input *dat*.
2. Why is it necessary to reload the current data during the hold data mode on a shift register?

7-23 HDL RING COUNTERS

In Section 7-20 we used a shift register to make a counter that circulates a single active logic level through all of its flip-flops. This was referred to as a ring counter. One characteristic of ring counters is that the modulus is equal to the number of flip-flops in the register and thus there are always many unused and invalid states. We have already discussed ways of describing counters using the CASE construct to specify PRESENT state and NEXT state transitions. In those examples, we took care of invalid states by including them under "others." This method also works for ring counters. In this section, however, we look at a more intuitive way to describe shift counters.

These methods use the same techniques described in Section 7-22 in order to make the register shift one position on each clock. The main feature of this code is the method of completing the "ring" by driving the *ser_in* line of the shift register. With a little planning, we should also be able to ensure that the counter eventually reaches the desired sequence, no matter what state it is in initially. For this example, we re-create the operation of the ring counter whose state diagram is shown in Figure 7-70(d). In order to make this counter self-start without using asynchronous inputs, we control the *ser_in* line of the shift register using an IF/ELSE construct. Any time we detect that the upper three bits are all LOW, we assume the lowest order bit is HIGH, and on the next clock, we want to shift in a HIGH to *ser_in*. For all other states (valid and invalid), we shift in a LOW. Regardless of the state to which the counter is initialized, it eventually fills with zeros; at which time, our logic shifts in a HIGH to start the ring sequence.

AHDL

AHDL RING COUNTER

The AHDL code shown in Figure 7-85 should look familiar by now. Lines 11 and 12 control the serial input using the strategy we just described. Notice the use of the double equals (==) operator on line 11. This operator evaluates whether the expressions on each side are equal or not. Remember, the single equals (=) operator assigns (i.e., connects) one object to another. Line 14 implements the shift right action that we described in the previous section. Simulation results are shown in Figure 7-86.

FIGURE 7-85 AHDL four-bit ring counter.

```
1   SUBDESIGN fig7_85
2   (
3      clk          :INPUT;
4      q[3..0]      :OUTPUT;
5   )
6   VARIABLE
7      ff[3..0]     :DFF;
8      ser_in       :NODE;
9   BEGIN
10    ff[].clk = clk;
11    IF ff[3..1].q == B"000" THEN ser_in = VCC;   -- self start
12    ELSE ser_in = GND;
13    END IF;
14    ff[3..0].d = (ser_in, ff[3..1].q);          -- shift right
15    q[] = ff[];
16  END;
```

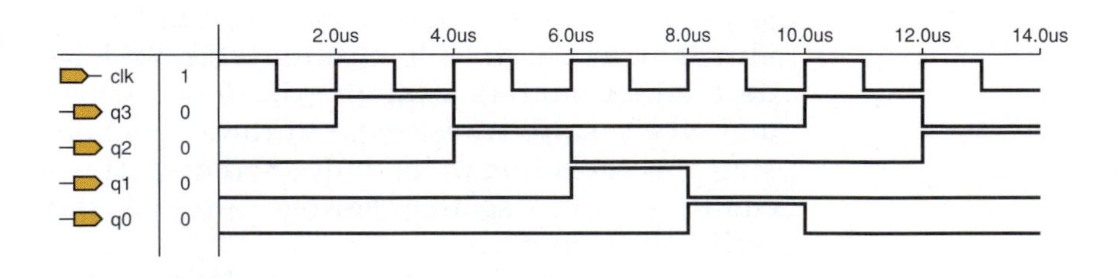

FIGURE 7-86 Simulation of HDL ring counter.

VHDL

VHDL RING COUNTER

The VHDL code shown in Figure 7-87 should look familiar by now. Lines 12 and 13 control the serial input using the strategy we just described. Line 16 implements the shift right action that we described in the previous section. Simulation results are shown in Figure 7-86.

```
1   ENTITY  fig7_87  IS
2   PORT (       clk          :IN BIT;
3                q            :OUT BIT_VECTOR (3 DOWNTO 0));
4   END fig7_87;
5
6   ARCHITECTURE vhdl OF fig7_87 IS
7   SIGNAL  ser_in            :BIT;
8   BEGIN
9   PROCESS (clk)
10     VARIABLE  ff           :BIT_VECTOR (3 DOWNTO 0);
11     BEGIN
12        IF (ff(3 DOWNTO 1) = "000")  THEN   ser_in <= '1';       -- self-start
13        ELSE  ser_in <= '0';
14        END IF;
15        IF (clk'EVENT AND clk = '1')  THEN
16           ff(3 DOWNTO 0) := (ser_in & ff(3 DOWNTO 1));          -- shift right
17        END IF;
18     q <= ff;
19  END PROCESS;
20  END vhdl;
```

FIGURE 7-87 VHDL four-bit ring counter.

REVIEW QUESTIONS

1. What does it mean for a ring counter to self-start?
2. Which lines of Figure 7-85 ensure that the ring counter self-starts?
3. Which lines of Figure 7-87 ensure that the ring counter self-starts?

7-24 HDL ONE-SHOTS

Another important circuit that we have studied is the one-shot. We can apply the concept of a counter to implement a **digital one-shot** using HDL. Recall from Chapter 5 that one-shots were devices that produce a pulse of a predefined width every time the trigger input is activated. A *nonretriggerable* one-shot ignores the trigger input as long as the pulse output is still active. A *retriggerable* one-shot starts a pulse in response to a trigger and restarts the internal pulse timer every time a subsequent trigger edge occurs before the pulse is complete. The first example we investigate is a nonretriggerable, HIGH-level-triggered digital one-shot. The one-shots that we studied in Chapter 5 used a resistor and capacitor as the internal pulse timing mechanism. In order to create a one-shot using HDL techniques, we use a four-bit counter to determine the width of the pulse. The inputs are a clock signal, trigger, clear, and pulse width value. The only output is the pulse out, Q. The idea is quite simple. Whenever a trigger is detected, make the pulse go HIGH and load a down-counter with a number from the pulse width inputs. The larger this number, the longer it will take to count down to zero. The advantage of this one-shot is that the pulse width can be adjusted easily by changing the value loaded into the counter. As you read the sections below, consider the following question: "What makes this circuit nonretriggerable and what makes it level-triggered?"

AHDL

SIMPLE AHDL ONE-SHOTS

A nonretriggerable, level-sensitive, one-shot description in AHDL is shown in Figure 7-88. A register of four flip-flops is created on line 8, and it serves as the counter that counts down during the pulse. The *clock* is connected in parallel to all the flip-flops on line 10. The reset function is implemented by connecting the *reset* control line directly to the asynchronous clear input of each flip-flop on line 11. After these assignments, the first condition that is tested is the trigger. If it is activated (HIGH) at any time while the count value is 0 (i.e., the previous pulse is done), then the delay value is loaded into the counter. On line 14, it tests to see if the pulse is done by checking to see if the counter is down to zero. If it is, then the counter should not roll over but rather stay at zero. If the count is not at zero, then it must be counting, so line 15 sets up the flip-flops to decrement on the next clock. Finally, line 17 generates the output pulse. This Boolean expression can be thought of as follows: "Make the pulse (Q) HIGH when the *count* is anything other than zero."

FIGURE 7-88 AHDL nonretriggerable one-shot.

```
1   SUBDESIGN fig7_88
2   (
3      clock, trigger, reset    : INPUT;
4      delay[3..0]              : INPUT;
5      q                        : OUTPUT;
6   )
7   VARIABLE
8      count[3..0]     : DFF;
9   BEGIN
10     count[].clk = clock;
11     count[].clrn = reset;
12     IF trigger & count[].q == b"0000" THEN
13           count[].d = delay[];
14     ELSIF count[].q == B"0000" THEN count[].d = B"0000";
15     ELSE count[].d = count[].q - 1;
16     END IF;
17     q = count[].q != B"0000";   -- make output pulse
18  END;
```

VHDL

SIMPLE VHDL ONE-SHOTS

A nonretriggerable, level-sensitive, one-shot description in VHDL is shown in Figure 7-89. The inputs and outputs are shown on lines 3–5, as previously described. In the architecture description, a PROCESS is used (line 11) to respond to either of two inputs: the clock, or the reset. Within this PROCESS, a variable is used to represent the value on the counter. The input that should have overriding precedence is the *reset* signal. This is tested first (line 14) and if it is active, the *count* is cleared immediately. If the *reset* is not active, line 15 is evaluated and looks for a rising edge on the *clock*. Line 16 checks for the trigger. If it is activated at any time while the count value is 0 (i.e., the previous pulse is done), then the width value is loaded into the counter. On line 18, it tests to see if the pulse is done by checking to see if the counter is down to zero. If it is, then the counter should not roll over but rather stay at zero. If the

```
1   ENTITY fig7_89 IS
2   PORT (
3         clock, trigger, reset     :IN BIT;
4         delay                     :IN INTEGER RANGE 0 TO 15;
5         q                         :OUT BIT
6         );
7   END fig 7_89;
8
9   ARCHITECTURE vhdl OF fig7_89 IS
10  BEGIN
11     PROCESS (clock, reset)
12     VARIABLE count        : INTEGER RANGE 0 TO 15;
13     BEGIN
14        IF reset = '0' THEN count := 0;
15        ELSIF (clock'EVENT AND clock = '1' ) THEN
16           IF trigger = '1' AND count = 0 THEN
17              count := delay;                      -- load counter
18           ELSIF count = 0 THEN count := 0;
19           ELSE count := count - 1;
20           END IF;
21        END IF;
22        IF count /= 0 THEN q <= '1';
23        ELSE q <= '0';
24        END IF;
25     END PROCESS;
26  END vhdl;
```

FIGURE 7-89 VHDL nonretriggerable one-shot.

count is not at zero, then it must be counting, so line 19 sets up the flip-flops to decrement on the next clock. Finally, lines 22 and 23 generate the output pulse. This Boolean expression can be thought of as follows: "Make the pulse (q) HIGH when the count is anything other than zero."

Now that we have reviewed the code that describes this one-shot, let's evaluate its performance. Converting a traditionally analog circuit to digital usually offers some advantages and some disadvantages. On a standard one-shot chip, the output pulse starts immediately after the trigger. For the digital one-shot described here, the output pulse starts on the next clock edge and lasts as long as the counter is greater than zero. This situation is shown in Figure 7-90 within the first ms of the simulation. Notice that the trigger goes high almost 0.5 ms before the *q* out responds. If another trigger event occurs while it is counting down (like the one just before 3 ms), it is ignored. This is the nonretriggerable characteristic.

Another point to make for this digital one-shot is that the trigger pulse must be long enough to be seen as a HIGH on the rising clock edge. At about the 4.5-ms mark, a pulse occurs on the trigger input but goes LOW before the rising edge of the clock. This circuit does *not* respond to this input event. At just past 5 ms, the trigger goes HIGH and stays there. The pulse lasts exactly 6 ms, but because the trigger input remains HIGH, it responds with another output pulse one clock later. The reason for this situation is that this circuit is level-triggered rather than edge-triggered, like most of the conventional one-shot ICs.

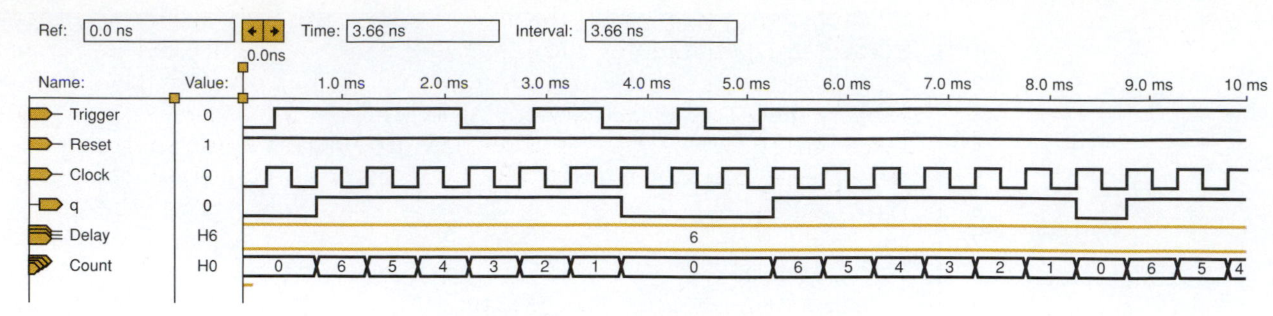

FIGURE 7-90 Simulation of the nonretriggerable one-shots.

Retriggerable, Edge-Triggered One-Shots in HDL

Many applications of one-shots require the circuit to respond to an edge rather than a level. How can HDL code be used to make the circuit respond once to each positive transition on its trigger input? The technique described here is called *edge-trapping* and has been a very useful tool in programming microcontrollers for years. As we will see, it is equally useful for describing edge-triggering for a digital circuit using HDL. This section illustrates an example of a retriggerable one-shot while also demonstrating edge-trapping, which can be useful in many other situations.

The general operation of this retriggerable one-shot requires that it responds to a rising edge of the trigger input. As soon as the edge is detected, it should start timing the pulse. In the digital one-shot, this means that it loads the counter as soon as possible after the trigger edge and starts counting down toward zero. If another trigger event (rising edge) occurs before the pulse is terminated, the counter is immediately reloaded, and the pulse timing starts again from the beginning, thus sustaining the pulse. Activating the clear at any point should force the counter to zero and terminate the pulse. The minimum output pulse width is simply the number applied to the width input multiplied by the clock period.

The strategy behind edge-trapping for a one-shot is demonstrated in Figure 7-91. On each active clock edge are two important pieces of information that are needed. The first is the state of the *trigger* input *now* and the second is the state of the *trigger* input when the last active clock edge occurred. Start with point *a* on the diagram of Figure 7-91 and determine these two values, then move to point *b*, and so on. By completing this task, you should have concluded that, at point *c*, a unique result has been obtained. The *trigger* is HIGH now but it was LOW on the last active clock edge. This is the point where we have detected the *trigger* edge event.

FIGURE 7-91 Detecting edges.

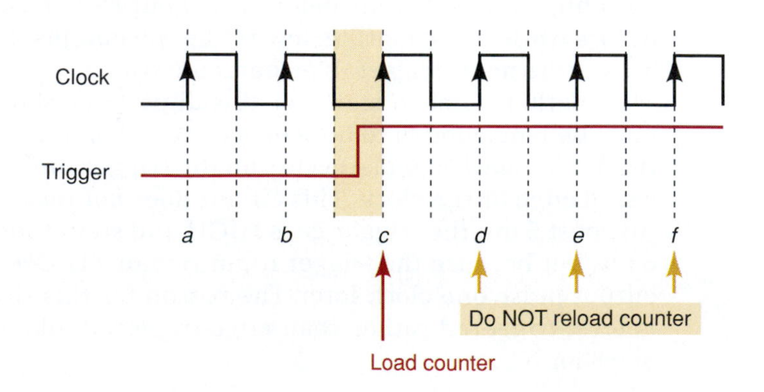

In order to know what the *trigger* was on the last active clock edge, the system must remember the last value that *the trigger had at that point.* This is done by storing the value of the trigger bit in a flip-flop. Recall that we discussed a similar concept in Chapter 5 when we talked about using a flip-flop to detect a sequence. The code for a one-shot is written so that the counter is loaded only after a rising edge is detected on the *trigger* input.

AHDL RETRIGGERABLE, EDGE-TRIGGERED ONE-SHOT

The first five lines of Figure 7-92 are identical to the previous nonretriggerable example. In AHDL, the only way to remember a value obtained in the past is to store the value on a flip-flop. This section uses a flip-flop named *trig_was* (line 9) to store the value that was on the trigger on the last active clock edge. This flip-flop is simply connected so that the trigger is on its *D* input (line 14) and the clock is connected to its *clk* input (line 13). The *Q* output of *trig_was* remembers the value of the *trigger* right up to the next clock edge. At this point, we use line 16 to evaluate if a triggering edge has occurred. If *trigger* is HIGH (now), but *trigger* was LOW (last clock), it is time to load the counter (line 17). Line 18 ensures that, once the *count* reaches zero, it will remain at zero until a new trigger comes along. If the decisions allow line 19 to be evaluated, it means that there is a value loaded into the counter and it is not zero, so it needs to be decremented. Finally, the output pulse is made HIGH any time a value other than 0000 is still on the counter, like we saw previously.

FIGURE 7-92 AHDL retriggerable one-shot with edge trigger.

```
1   SUBDESIGN fig7_92
2   (
3      clock, trigger, reset  : INPUT;
4      delay[3..0]            : INPUT;
5      q                      : OUTPUT;
6   )
7   VARIABLE
8                count[3..0]  : DFF;
9                trig_was     : DFF;
10  BEGIN
11     count[].clk = clock;
12     count[].clrn = reset;
13     trig_was.clk = clock;
14     trig_was.d = trigger;
15
16     IF trigger & !trig_was.q THEN
17           count[].d = delay[];
18     ELSIF count[].q == B"0000" THEN count[].d = B"0000";
19     ELSE count[].d = count[].q - 1;
20     END IF;
21     q = count[].q != B"0000";
22  END;
```

VHDL RETRIGGERABLE, EDGE-TRIGGERED ONE-SHOT

The ENTITY description in Figure 7-93 is exactly like the previous nonretriggerable example. In fact, the only differences between this example and the one shown in Figure 7-89 have to do with the logic of the decision process. When we want to remember a value in VHDL, it must be stored in a VARIABLE. Recall that we can think of a PROCESS as a description of what happens each time a signal in the sensitivity list changes state. A VARIABLE retains the last value assigned to it between the times the process is invoked. In this sense, it acts like a flip-flop. For the one-shot, we need to store a value that tells us what the trigger was on the last active clock edge. Line 11 declares a variable bit to serve this purpose. The first decision (line 13) is the overriding decision that checks and responds to the *reset* input. Notice that this is an asynchronous control because it is evaluated before the clock edge is detected on line 14. Line 14 determines that a rising clock edge has occurred, and then the main logic of this process is evaluated between lines 15 and 20.

When a clock edge occurs, one of three conditions exists:

1. A trigger edge has occurred and we must load the counter.
2. The counter is zero and we need to keep it at zero.
3. The counter is not zero and we need to count down by one.

```
1   ENTITY fig7_93 IS
2   PORT (   clock, trigger, reset     : IN BIT;
3         delay                        : IN INTEGER RANGE 0 TO 15;
4         q                            : OUT BIT);
5   END fig7_93;
6
7   ARCHITECTURE vhdl OF fig7_93 IS
8   BEGIN
9      PROCESS (clock, reset)
10     VARIABLE count        : INTEGER RANGE 0 TO 15;
11     VARIABLE trig_was     : BIT;
12     BEGIN
13        IF reset = '0' THEN count := 0;
14        ELSIF (clock'EVENT AND clock = '1' ) THEN
15           IF trigger = '1' AND trig_was = '0' THEN
16              count := delay;                -- load counter
17              trig_was := '1';               -- "remember" edge detected
18           ELSIF count = 0 THEN count := 0; -- hold @ 0
19           ELSE count := count - 1;          -- decrement
20           END IF;
21           IF trigger = '0' THEN trig_was := '0';
22           END IF;
23        END IF;
24        IF count /= 0  THEN q <= '1';
25        ELSE q <= '0';
26        END IF;
27     END PROCESS;
28  END vhdl;
```

FIGURE 7-93 VHDL retriggerable one-shot with edge trigger.

Recall that it is very important to consider the order in which questions are asked and assignments are made in VHDL PROCESS statements because the *sequence* affects the operation of the circuit we are describing. The code that updates the *trig_was* variable must occur after the evaluation of its previous condition. For this reason, the conditions necessary to detect a rising edge on *trigger* are evaluated on line 15. If an edge occurred, then the counter is loaded (line 16) and the variable is updated (line 17) to remember this for the next time. If a trigger edge has not occurred, the code either holds at zero (line 18) or counts down (line 19). Line 21 makes sure that, as soon as the trigger input goes LOW, the variable *trig_was* remembers this by resetting. Finally, lines 24–25 are used to create the output pulse during the time the counter is not zero.

The two improvements that were made in this one-shot over the last example are the edge-triggering and the retriggerable feature. Figure 7-94 evaluates the new performance features. Notice in the first ms of the timing diagram that a trigger edge is detected, but the response is not immediate. The output pulse goes high on the next clock edge. This is a drawback to the digital one-shot. The retriggerable feature is demonstrated at about the 2-ms mark. Notice that *trigger* goes high and on the next clock edge, the *count* starts again at 5, sustaining the output pulse. Also notice that even after the *q* output pulse is complete and the *trigger* is still HIGH, the one-shot does not fire another pulse because it is not level-triggered but rather rising edge-triggered. At the 6-ms mark, a short trigger pulse occurs but is ignored because it does not stay HIGH until the next clock. On the other hand, an even shorter trigger pulse occurring just after the 7-ms mark does fire the one-shot because it is present during the rising clock edge. The resulting output pulse lasts exactly five clock cycles because no other triggers occur during this period.

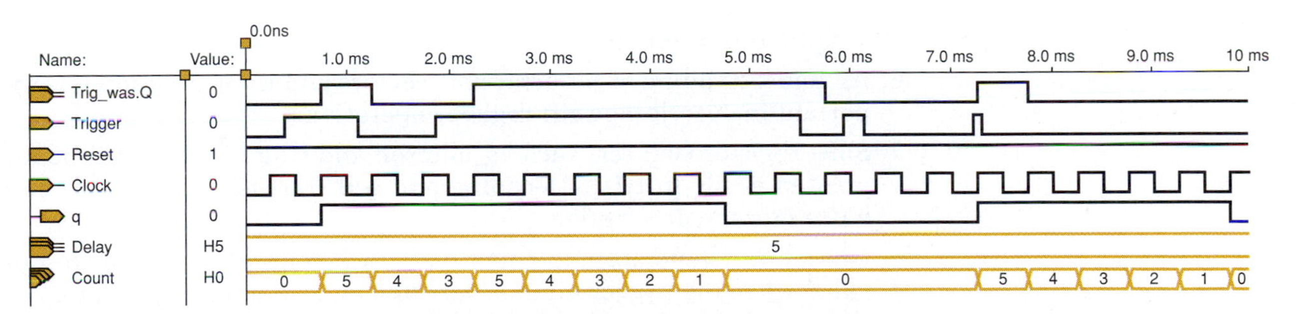

FIGURE 7-94 Simulation of the edge-triggered retriggerable one-shot.

To minimize the effects of delayed response to trigger edges and the possibility of missing trigger edges that are too short, this circuit can be improved quite simply. The clock frequency and the number of bits used to load the delay value can both be increased to provide the same range of pulse widths (with more precise control) while reducing the minimum trigger pulse width. In order to cure this problem completely, the one-shot must respond asynchronously to the trigger input. This is possible in both AHDL and VHDL, but it will always result in a pulse that fluctuates in width by up to one clock period.

REVIEW QUESTIONS

1. Which control input signal holds the highest priority for each of the one-shot descriptions?
2. Name two factors that determine how long a pulse from a digital one-shot will last.
3. For the one-shots shown in this section, are the counters loaded synchronously or asynchronously?
4. What is the advantage of loading a counter synchronously?
5. What is the advantage of loading the counter asynchronously?
6. What two pieces of information are necessary to detect an edge?

PART 2 SUMMARY

1. Numerous IC registers are available and can be classified according to whether their inputs are parallel (all bits entered simultaneously), serial (one bit at a time), or both. Likewise, registers can have outputs that are parallel (all bits available simultaneously) or serial (one bit available at a time).
2. A sequential logic system uses FFs, counters, and registers, along with logic gates. Its outputs and sequence of operations depend on present and past inputs.
3. Troubleshooting a sequential logic system begins with observation of the system operation, followed by analytical reasoning to determine the possible causes of any malfunction, and finally test measurements to isolate the actual fault.
4. A ring counter is actually an *N*-bit shift register that recirculates a single 1 continuously, thereby acting as a MOD-*N* counter. A Johnson counter is a modified ring counter that operates as MOD-2*N* counter.
5. Shift registers can be implemented with HDL by writing custom descriptions of their operation.
6. An understanding of bit arrays/bit vectors and their notation is very important in describing shift register operations.
7. Shift register counters such as Johnson and ring counters can be implemented easily in HDL. Decoding and self-starting features are easy to write into the description.
8. Digital one-shots are implemented with a counter loaded with a delay value when the trigger input is detected and counts down to zero. During the countdown time, the output pulse is held HIGH.
9. With strategic placement of the hardware description statements, HDL one-shots can be made edge- or level-triggered and retriggerable or nonretriggerable. They produce an output pulse that responds synchronously or asynchronously to the trigger.

PART 2 IMPORTANT TERMS

parallel in/parallel out
serial in/serial out
parallel in/serial out
serial in/parallel out
circulating shift register
ring counter
Johnson counter (twisted ring counter)
sequential logic system
concatenation
digital one-shot

PROBLEMS

PART 1

SECTION 7-1

B 7-1.* Add another flip-flop, E, to the counter of Figure 7-1. The clock signal is an 8-MHz square wave.

(a) What will be the frequency at the E output? What will be the duty cycle of this signal?

(b) Repeat (a) if the clock signal has a 20 percent duty cycle.

(c) What will be the frequency at the C output?

(d) What is the MOD number of this counter?

B 7-2. Draw a binary counter that will convert a 64-kHz pulse signal into a 1-kHz square wave.

B 7-3.* Assume that a five-bit binary counter starts in the 00000 state. What will be the count after 144 input pulses?

B 7-4. A 10-bit ripple counter has a 256-kHz clock signal applied.

(a) What is the MOD number of this counter?

(b) What will be the frequency at the MSB output?

(c) What will be the duty cycle of the MSB signal?

(d) Assume that the counter starts at zero. What will be the count in hexadecimal after 1000 input pulses?

SECTION 7-2

7-5.* A four-bit ripple counter is driven by a 20-MHz clock signal. Draw the waveforms at the output of each FF if each FF has t_{pd} = 20 ns. Determine which counter states, if any, will not occur because of the propagation delays.

7-6. (a) What is the maximum clock frequency that can be used with the counter of Problem 7-5?

(b) What would f_{max} be if the counter were expanded to six bits?

SECTIONS 7-3 AND 7-4

B 7-7.* (a) Draw the circuit diagram for a MOD-32 synchronous counter.

(b) Determine f_{max} for this counter if each FF has t_{pd} = 20 ns and each gate has t_{pd} = 10 ns.

B 7-8. (a) Draw the circuit diagram for a MOD-64 synchronous counter.

(b) Determine f_{max} for this counter if each FF has t_{pd} = 20 ns and each gate has t_{pd} = 10 ns.

B 7-9.* Draw the waveforms for all the FFs in the decade counter of Figure 7-8(b) in response to a 1-kHz clock frequency. Show any glitches that might appear on any of the FF outputs. Determine the frequency at the D output.

B 7-10. Repeat Problem 7-9 for the counter of Figure 7-8(a).

7-11.* Change the inputs to the NAND gate of Figure 7-9 so that the counter divides the input frequency by 50.

D 7-12. Draw a synchronous counter that will output a 10-kHz signal when a 1-MHz clock is applied.

*Answers to problems marked with an asterisk can be found in the back of the text.

SECTIONS 7-5 AND 7-6

B 7-13.* Draw a synchronous, MOD-32, down counter.

B 7-14. Draw a synchronous, MOD-16, up/down counter. The count direction is controlled by *dir* (*dir* = 0 to count up).

C, T 7-15.* Determine the count sequence of the up/down counter in Figure 7-11 if the INVERTER output were stuck HIGH. Assume the counter starts at 000.

7-16. Complete the timing diagram in Figure 7-95 for the presettable counter in Figure 7-12. Note that the initial condition for the counter is given in the timing diagram.

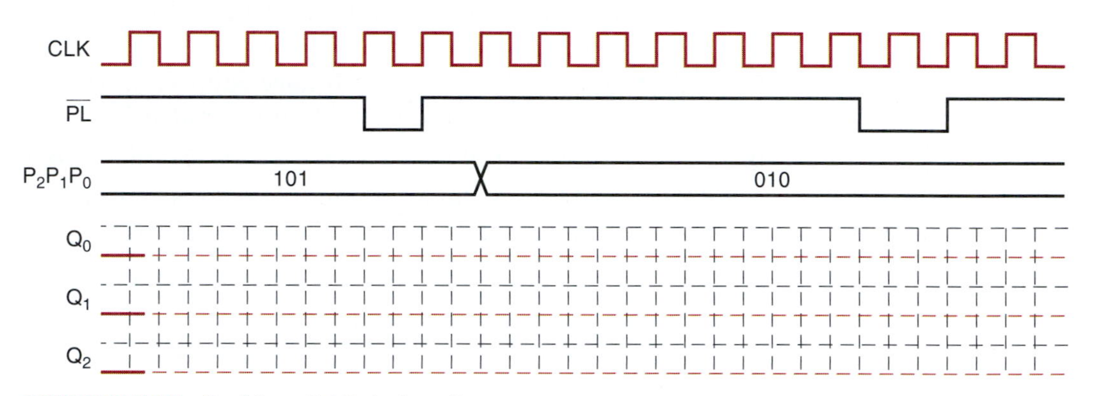

FIGURE 7-95 Problem 7-16 timing diagram.

SECTION 7-7

7-17.* Complete the timing diagram in Figure 7-96 for a 74ALS161 with the indicated input waveforms applied. Assume the initial state is 0000.

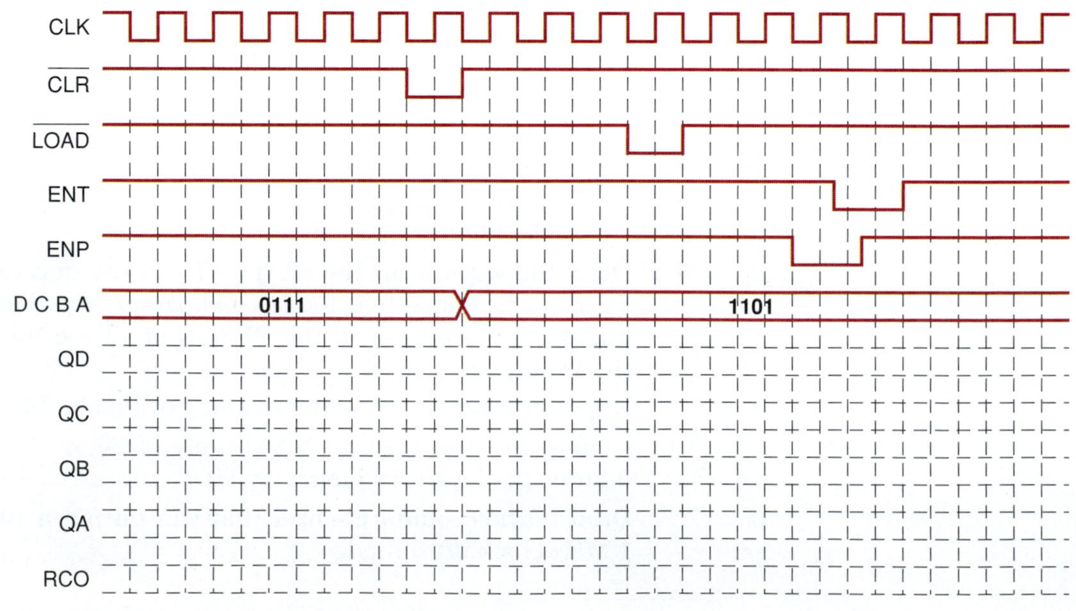

FIGURE 7-96 Problem 7-17 timing diagram.

7-18. Complete the timing diagram in Figure 7-97 for a 74ALS162 with the indicated input waveforms applied. Assume the initial state is 0000.

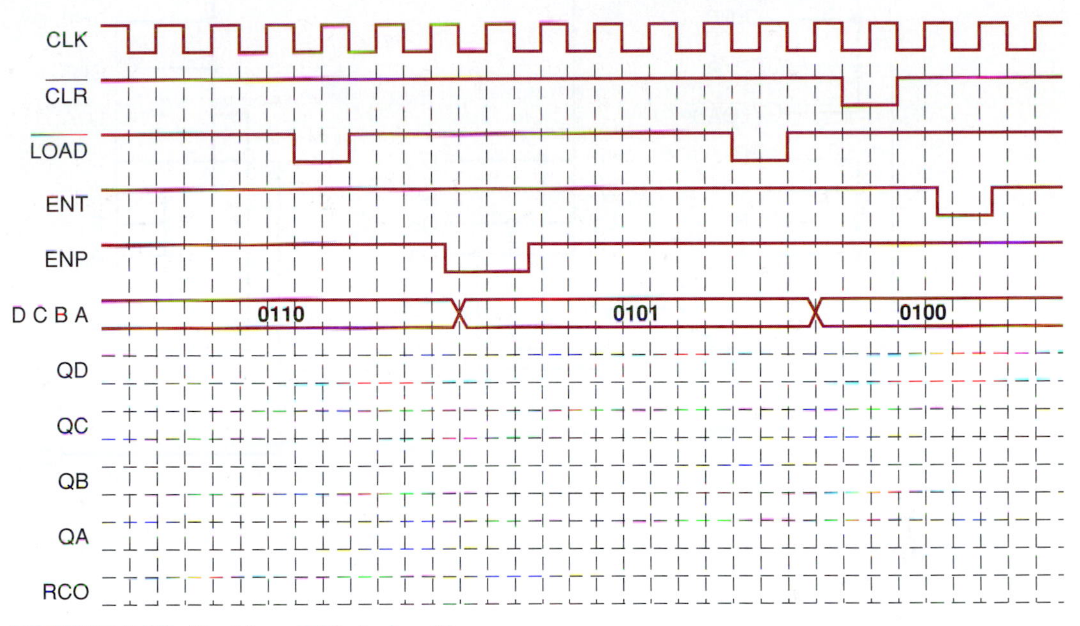

FIGURE 7-97 Problem 7-18 timing diagram.

7-19.*Complete the timing diagram in Figure 7-98 for a 74ALS190 with the indicated input waveforms applied. The *DCBA* input is 0101.

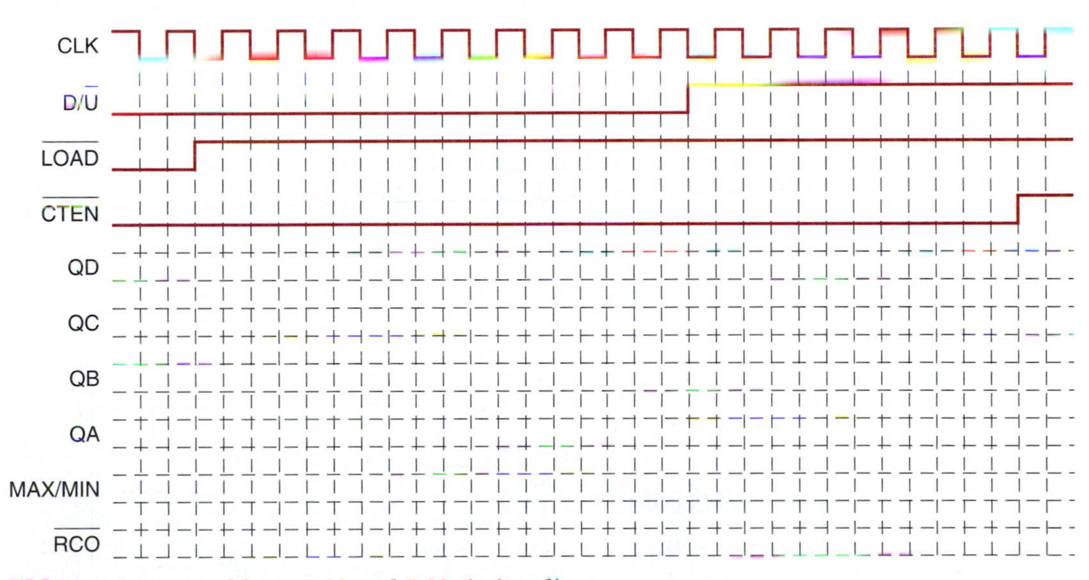

FIGURE 7-98 Problems 7-19 and 7-20 timing diagram.

7-20. Repeat Problem 7-19 for a 74ALS191 and a *DCBA* input of 1100.

B 7-21.*Refer to the IC counter circuit in Figure 7-99(a):

(a) Draw the state transition diagram for the counter's *QD QC QB QA* outputs.

(b) Determine the counter's modulus.

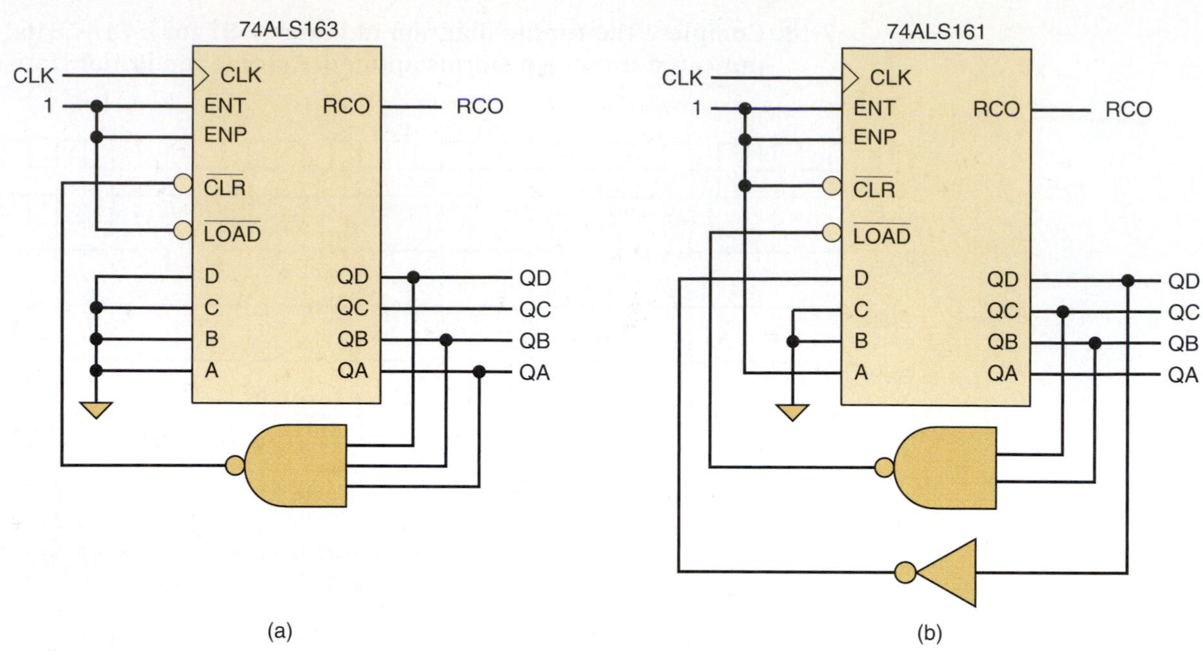

FIGURE 7-99 Problems 7-21 and 7-22.

(c) What is the relationship of the output frequency of the MSB to the input *CLK* frequency?

(d) What is the duty cycle of the MSB output waveform?

B 7-22. Repeat Problem 7-21 for the IC counter circuit in Figure 7-99(b).

B 7-23.* Refer to the IC counter circuit in Figure 7-100(a).

(a) Draw the timing diagram for outputs *QD QC QB QA*.

(b) What is the counter's modulus?

(c) What is the count sequence? Does it count up or down?

(d) Can we produce the same modulus with a 74HC190? Can we produce the same count sequence with a 74HC190?

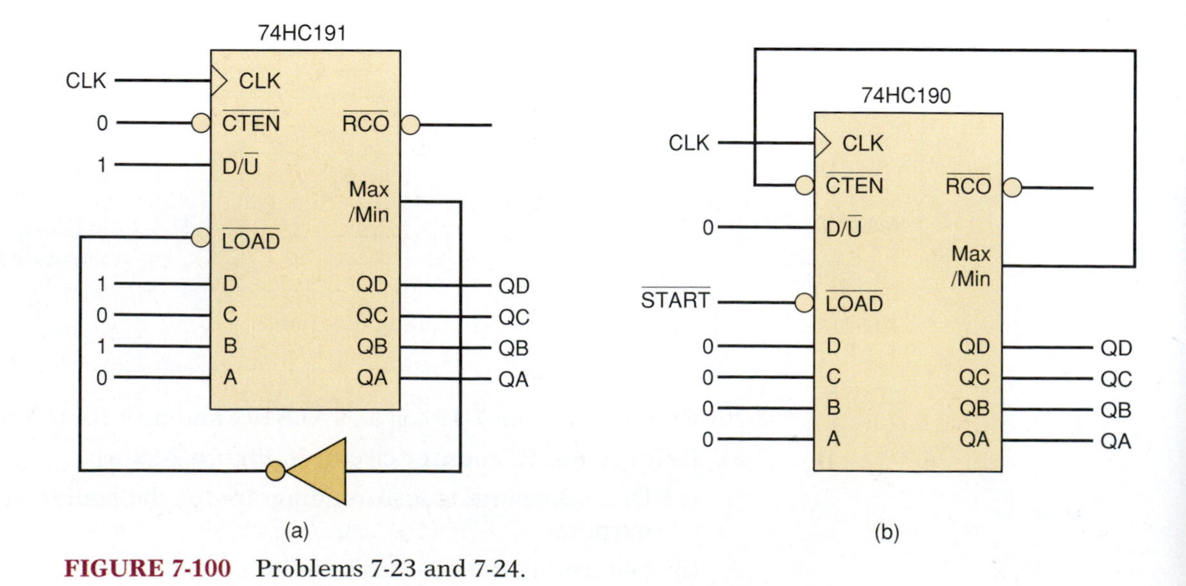

FIGURE 7-100 Problems 7-23 and 7-24.

7-24. Refer to the IC counter circuit in Figure 7-100(b):

(a) Describe the counter's output on *QD QC QB QA* if $\overline{START}$ is LOW.

(b) Describe the counter's output on *QD QC QB QA* if $\overline{START}$ is momentarily pulsed LOW and then returns to a HIGH.

(c) What is the counter's modulus? Is this a recycling counter?

D 7-25.*Draw a schematic to create a recycling, MOD-6 counter that uses:

(a) the clear control on a 74ALS160

(b) the clear control on a 74ALS162

D 7-26. Draw a schematic to create a recycling, MOD-6 counter that produces the count sequence:

(a) 1, 2, 3, 4, 5, 6, and repeats with a 74ALS162

(b) 5, 4, 3, 2, 1, 0, and repeats with a 74ALS190

(c) 6, 5, 4, 3, 2, 1, and repeats with a 74ALS190

D 7-27.*Design a MOD-100, binary counter using either two 74HC161 or two 74HC163 chips and any necessary gates. The IC counter chips are to be synchronously cascaded together to produce the binary count sequence for 0 to 99. The MOD-100 is to have two control inputs, an active-LOW count enable ($\overline{EN}$) and an active-LOW, asynchronous clear ($\overline{CLR}$). Label the counter outputs *Q0*, *Q1*, *Q2*, etc., with *Q0* = LSB. Which output is the MSB?

D 7-28. Design a MOD-100, BCD counter using either two 74HC160 or two 74HC162 chips and any necessary gates. The IC counter chips are to be synchronously cascaded together to produce the BCD count sequence for 0 to 99. The MOD-100 is to have two control inputs, an active-HIGH count enable (*EN*) and an active-HIGH, synchronous load (*LD*). Label the counter outputs *Q0*, *Q1*, *Q2*, etc., with *Q0* = LSB. Which set of outputs represents the 10s digit?

B 7-29.*With a 6-MHz clock input to a 74ALS163 that has all four control inputs HIGH, determine the output frequency and duty cycle for each of the *five* outputs (including RCO).

B 7-30. With a 6-MHz clock input to a 74ALS162 that has all four control inputs HIGH, determine the output frequency and duty cycle for each of the following outputs: QA, QC, QD, RCO. What is unusual about the waveform pattern that would be produced by the QB output? This pattern characteristic results in an undefined duty cycle.

B 7-31.*The frequency of f_{in} is 6 MHz in Figure 7-101. The two IC counter chips have been cascaded asynchronously so that the output frequency produced by counter U1 is the input frequency for counter U2. Determine the output frequency for f_{out1} and f_{out2}.

B 7-32. The frequency of f_{in} is 1.5 MHz in Figure 7-102. The two IC counter chips have been cascaded asynchronously so that the output frequency produced by counter U1 is the input frequency for counter U2. Determine the output frequency for f_{out1} and f_{out2}.

D 7-33.*Design a frequency divider circuit that will produce the following three output signal frequencies: 1.5 MHz, 150 kHz, and 100 kHz. Use 74HC162 and 74HC163 counter chips and any necessary gates. The input frequency is 12 MHz.

D 7-34. Design a frequency divider circuit that will produce the following three output signal frequencies: 1 MHz, 800 kHz, and 100 kHz. Use 74HC160 and 74HC161 counter chips and any necessary gates. The input frequency is 12 MHz.

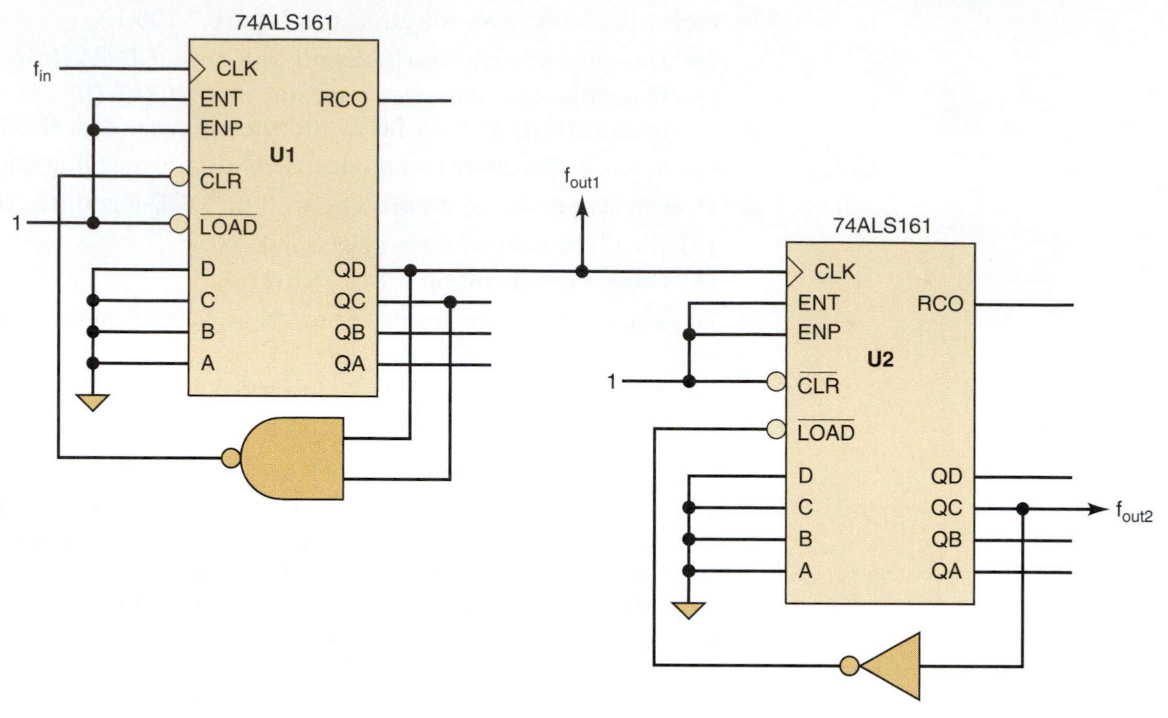

FIGURE 7-101 Problem 7-31.

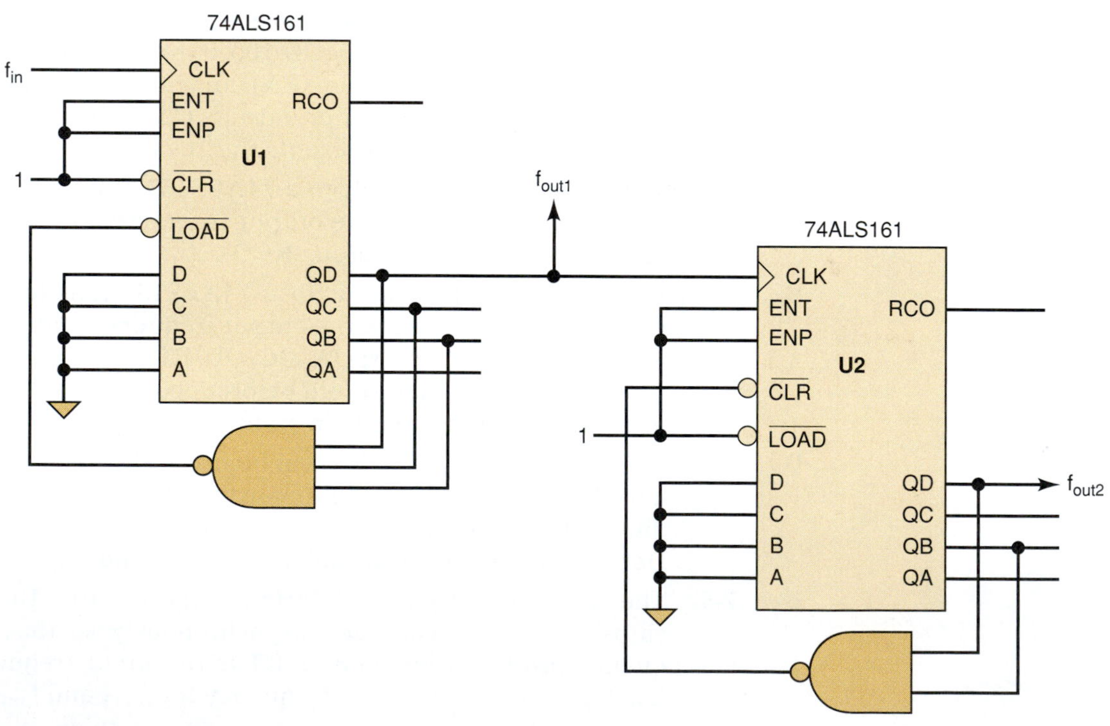

FIGURE 7-102 Problem 7-32.

SECTION 7-8

B 7-35.*Draw the gates necessary to decode all of the states of a MOD-16 counter using active-LOW outputs.

B 7-36. Draw the AND gates necessary to decode the 10 states of the BCD counter of Figure 7-8(b).

SECTION 7-9

C 7-37.* Analyze the synchronous counter in Figure 7-103(a). Draw its timing diagram and determine the counter's modulus.

C 7-38. Repeat Problem 7-37 for Figure 7-103(b).

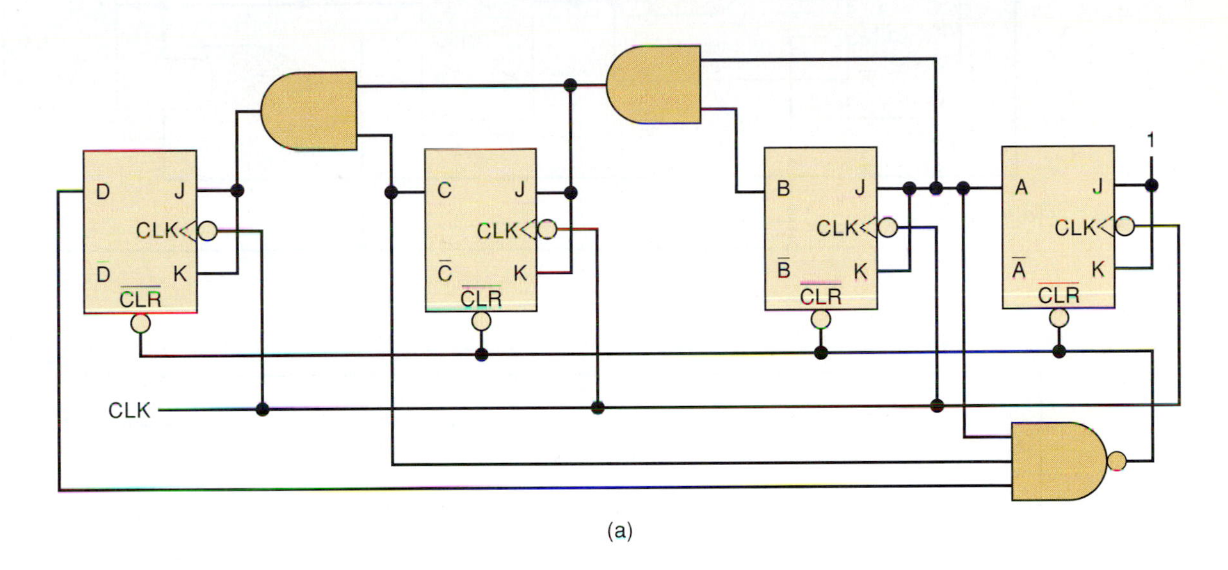

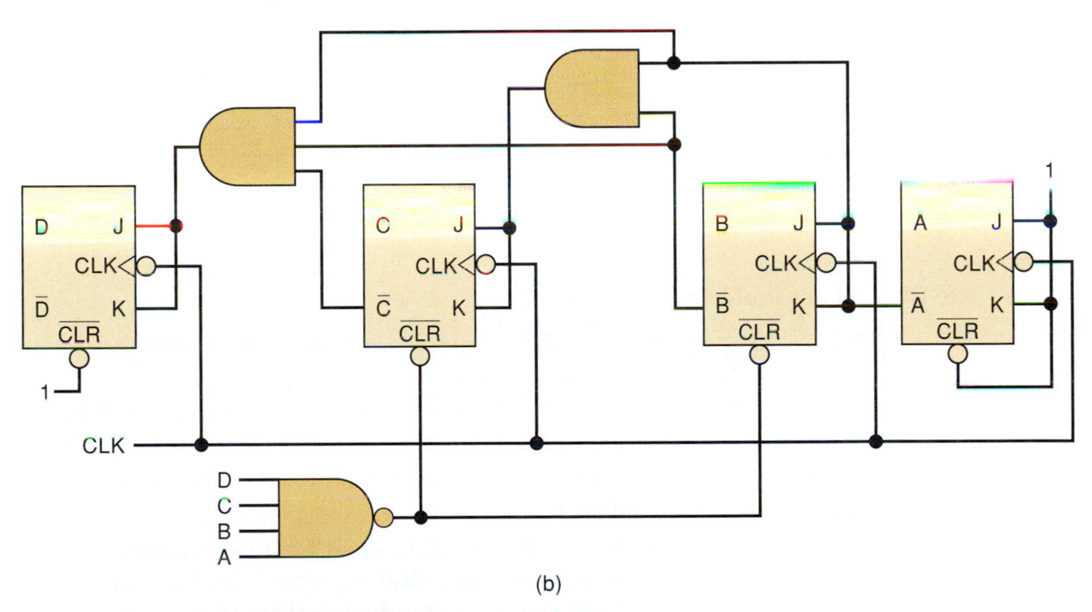

FIGURE 7-103 Problems 7-37 and 7-38.

C 7-39.* Analyze the synchronous counter in Figure 7-104(a). Draw its timing diagram and determine the counter's modulus.

C 7-40. Repeat Problem 7-39 for Figure 7-104(b).

C 7-41.* Analyze the synchronous counter in Figure 7-105(a). F is a control input. Draw its state transition diagram and determine the counter's modulus.

C 7-42. Analyze the synchronous counter in Figure 7-105(b). Draw its complete state transition diagram and determine the counter's modulus. Is the counter self-correcting?

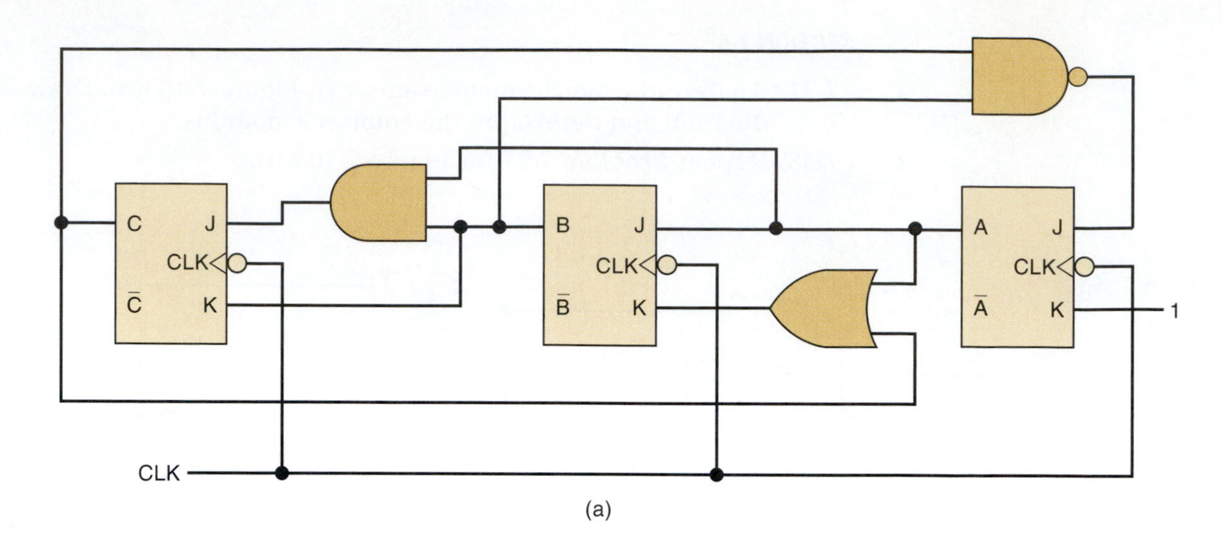

(a)

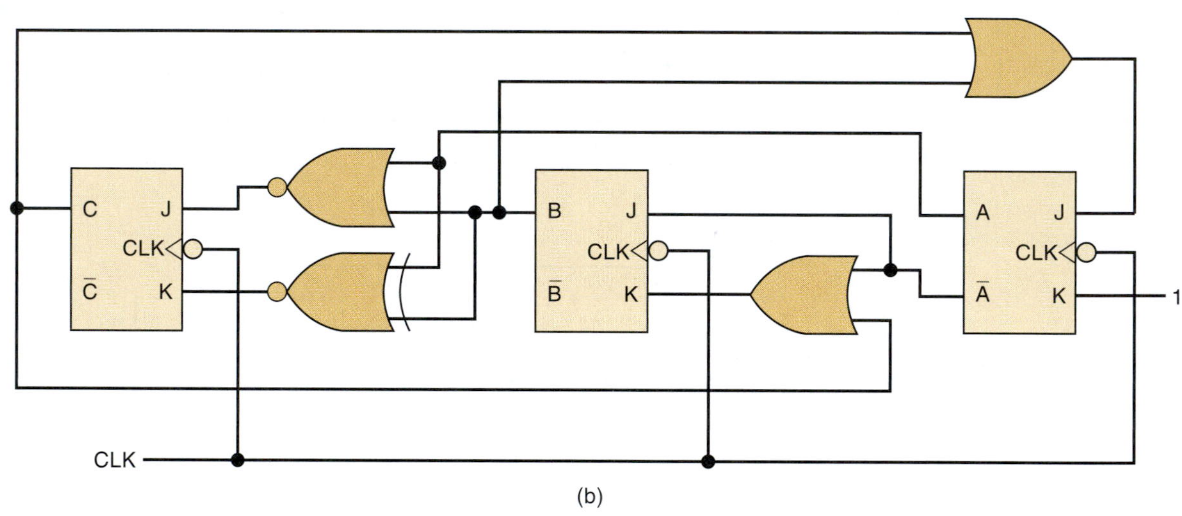

(b)

FIGURE 7-104 Problems 7-39 and 7-40.

SECTION 7-10

D 7-43.* (a) Design a synchronous counter using J-K FFs that has the following sequence: 000, 010, 101, 110, and repeat. The undesired (unused) states 001, 011, 100, and 111 must always go to 000 on the next clock pulse.

(b) Redesign the counter of part (a) without any requirement on the unused states; that is, their NEXT states can be don't cares. Compare with the design from (a).

D 7-44. Design a synchronous, recycling, MOD-5 down counter that produces the sequence 100, 011, 010, 001, 000, and repeat. Use J-K flip-flops.

(a) Force the unused states to 000 on the next clock pulse.

(b) Use don't-care NEXT states for the unused states. Is this design self-correcting?

D 7-45.* Design a synchronous, recycling, BCD down counter with J-K FFs using don't-care NEXT states.

D 7-46. Design a synchronous, recycling, MOD-7 up/down counter with J-K FFs. Use the states 000 through 110 in the counter. Control the count direction with input D ($D = 0$ to count up and $D = 1$ to count down).

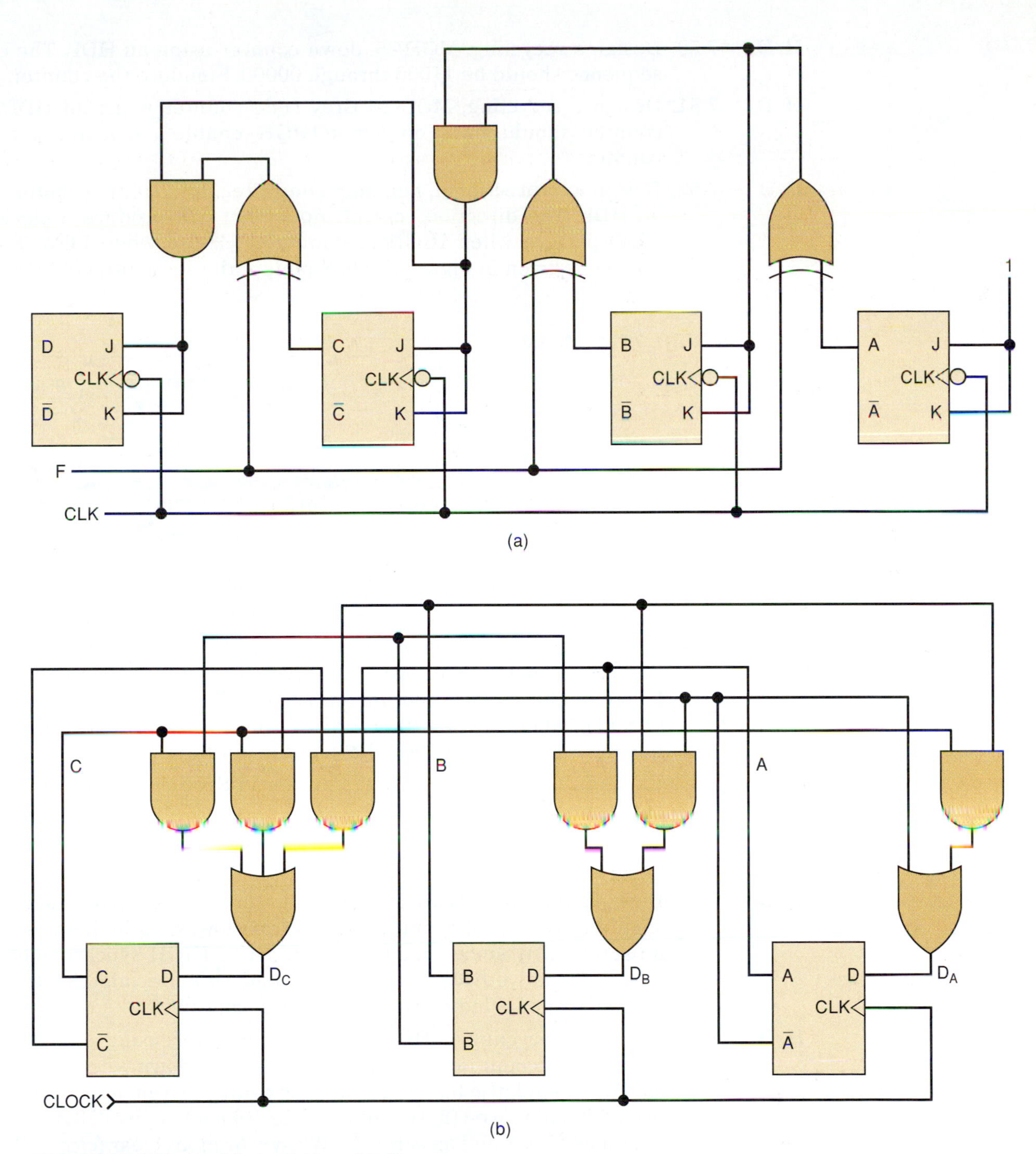

FIGURE 7-105 Problems 7-41 and 7-42.

D 7-47.* Design a synchronous, recycling, MOD-8, binary down counter with D flip-flops.

D 7-48. Design a synchronous, recycling, MOD-12 counter with D FFs. Use the states 0000 through 1011 in the counter.

SECTIONS 7-11 AND 7-12

H, D 7-49.* Design a recycling, MOD-13, up counter using an HDL. The count sequence should be 0000 through 1100. Simulate the counter.

H, D 7-50. Design a recycling, MOD-25, down counter using an HDL. The count sequence should be 11000 through 00000. Simulate the counter.

H, D 7-51.* Design a recycling, MOD-16 Gray code counter using an HDL. The counter should have an active-HIGH enable (*cnt*). Simulate the counter.

H, D 7-52. Design a bidirectional, half-step controller for a stepper motor using an HDL. The direction control input (*dir*) will produce a clockwise (CW) pattern when HIGH or counterclockwise when LOW. The sequence is given in Figure 7-106. Simulate the sequential circuit.

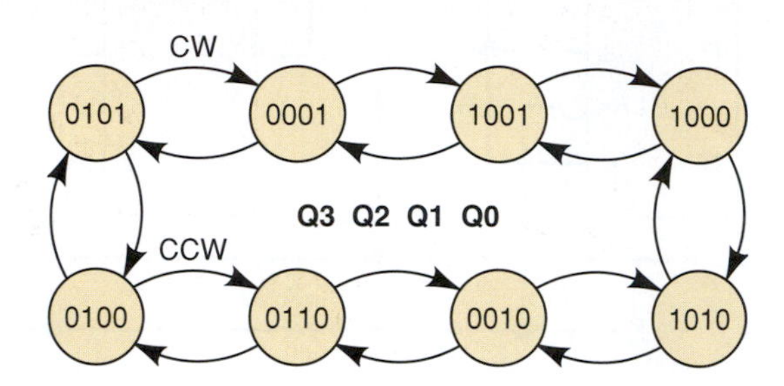

FIGURE 7-106 Problem 7-52.

H, D 7-53.* Design a frequency divider circuit to output a 100-kHz signal using an HDL. The input frequency is 5 MHz. Simulate the counter.

H, D 7-54. Design a frequency divider circuit that will output either of two specified frequency signals using an HDL. The output frequency is selected by the control input *fselect*. The divider will output a frequency of 5 kHz when *fselect* = 0 or 12 kHz when *fselect* = 1. The input frequency is 60 kHz. Simulate the counter.

H, B 7-55.* Expand the full-featured HDL counter in Section 7-12 to a MOD-256 counter. Simulate the counter.

H, B 7-56. Expand the full-featured HDL counter in Section 7-12 to a MOD-1024 counter. Simulate the counter.

H, B 7-57.* Design a recycling, MOD-16, down counter using an HDL. The counter should have the following controls (from lowest to highest priority): an active-LOW count enable ($\overline{en}$), an active-HIGH synchronous clear (*clr*), and active-LOW synchronous load ($\overline{ld}$). Decode the terminal count when enabled by $\overline{en}$. Simulate the counter.

H, D 7-58. Design a recycling, MOD-10, up/down counter using an HDL. The counter will count up when *up* = 1 and counts down when *up* = 0. The counter should also have the following controls (from lowest to highest priority): an active-HIGH count enable (*enable*), active-HIGH synchronous load (*load*), and an active-LOW asynchronous clear ($\overline{clear}$). Decode the terminal count when enabled by *enable*. Simulate the counter.

SECTION 7-13

H 7-59.* Create a MOD-1000 BCD counter by cascading together three of the HDL BCD counter modules (described in Section 7-13). Simulate the counter.

H 7-60. Create a MOD-256 binary counter by cascading together two of the full-featured, MOD-16, HDL counter modules (described in Section 7-12). Simulate the counter.

H, D 7-61.* Design a synchronous, MOD-50 BCD counter by cascading the HDL designs for a MOD-10 and a MOD-5 counter together. The MOD-50 counter should have an active-HIGH count enable (*enable*) and an active-LOW, synchronous clear ($\overline{clrn}$). Be sure to include the terminal count detection for the 1s digit to cascade with the 10s digit. Simulate the counter.

H, D 7-62. Design a synchronous, MOD-100, BCD down counter by cascading two MOD-10 HDL down counter modules together. The MOD-100 counter should have a synchronous parallel load (*load*). Simulate the counter.

SECTION 7-14

H 7-63.* Modify the HDL description in Figure 7-56 or Figure 7-57 to add a rinse sequence after the clothes are washed. The new state machine sequence should be *idle* → *wash_fill* → *wash_agitate* → *wash_spin* → *rinse_fill* → *rinse_agitate* → *rinse _spin* → *idle.* Use hot water to wash, and cold water to rinse (add output bits to control two water valves). Simulate the modified HDL design.

H 7-64. Simulate the HDL traffic light controller design presented in Section 7-14.

PART 2

SECTIONS 7-15 THROUGH 7-19

B 7-65.* A set of 74ALS174 registers is connected as shown in Figure 7-107. What type of data transfer is performed with each register? Determine the output of each register when the $\overline{MR}$ is pulsed momentarily LOW and after each of the indicated clock pulses (CP#) in Table 7-10. How many clock pulses must be applied before data that are input on *I5–I0* are available at *Z5–Z0*?

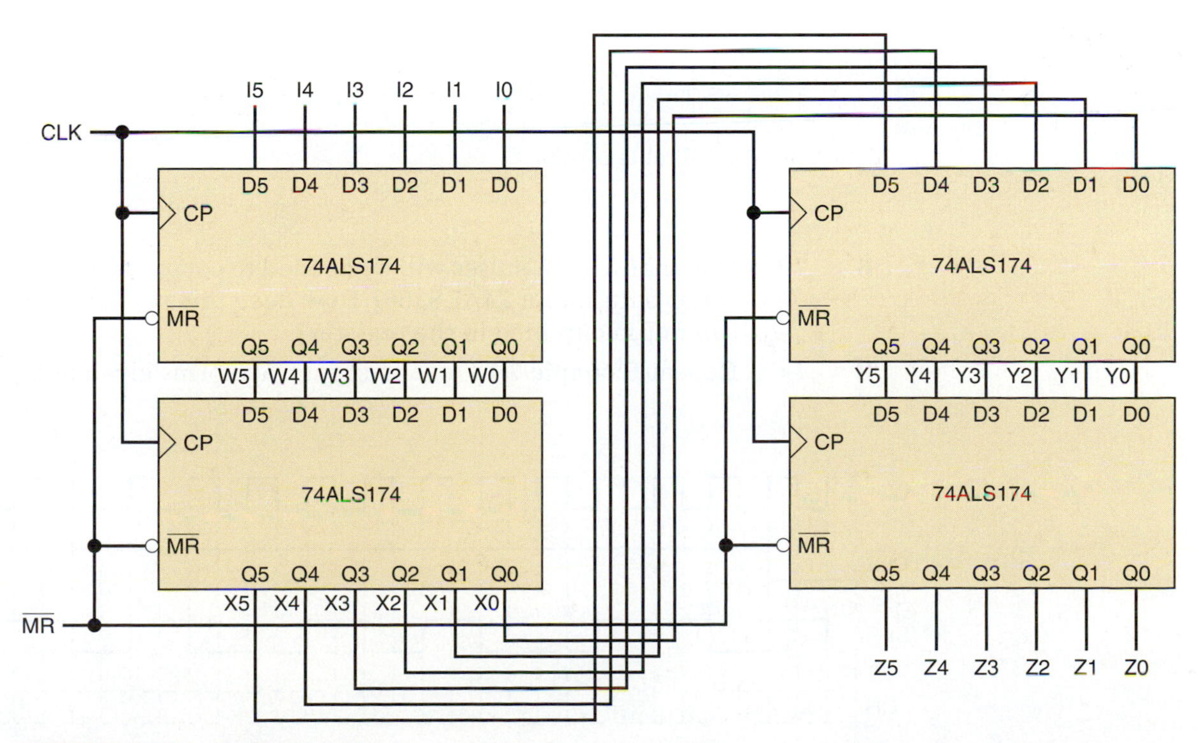

FIGURE 7-107 Problem 7-65.

TABLE 7-10

↑ CLK	$\overline{MR}$	I5–I10	W5–W0	X5–X0	Y5–Y0	Z5–Z0
X	0	101010				
CP1	1	101010				
CP2	1	010101				
CP3	1	000111				
CP4	1	111000				
CP5	1	011011				
CP6	1	001101				
CP7	1	000000				
CP8	1	000000				

B 7-66. Complete the timing diagram in Figure 7-108 for a 74HC174. How does the timing diagram show that the master reset is asynchronous?

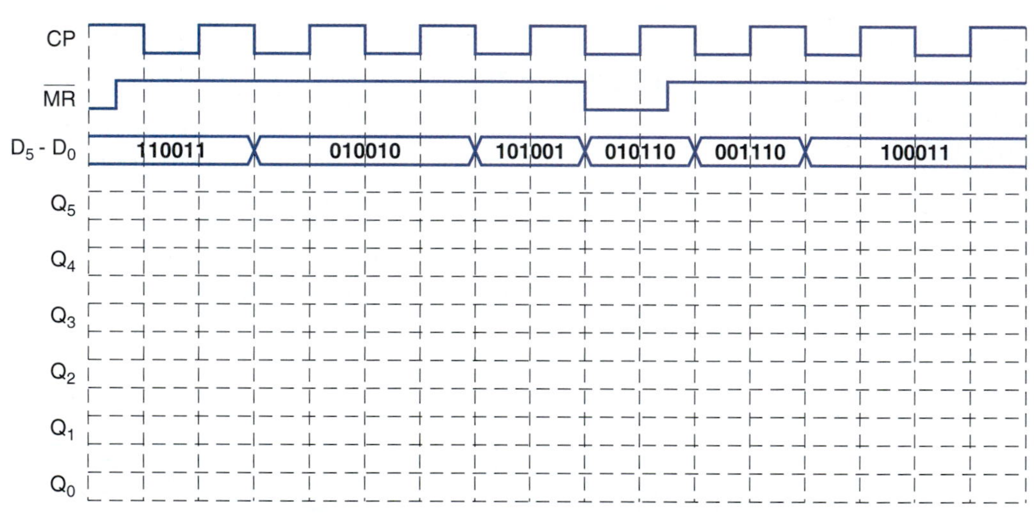

FIGURE 7-108 Problem 7-66.

B 7-67.* How many clock pulses will be needed to completely load eight bits of serial data into a 74ALS166? How does this relate to the number of flip-flops contained in the register?

B 7-68. Repeat Example 7-18 for the input waveforms given in Figure 7-109.

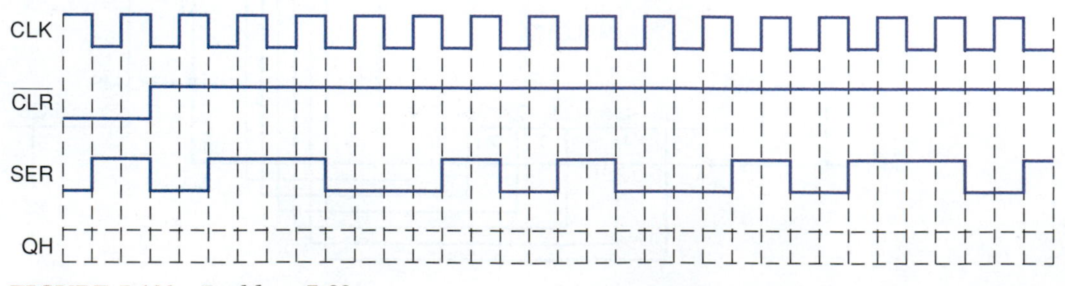

FIGURE 7-109 Problem 7-68.

7-69.* Repeat Example 7-20 with $D_S = 1$ and the input waveforms given in Figure 7-110.

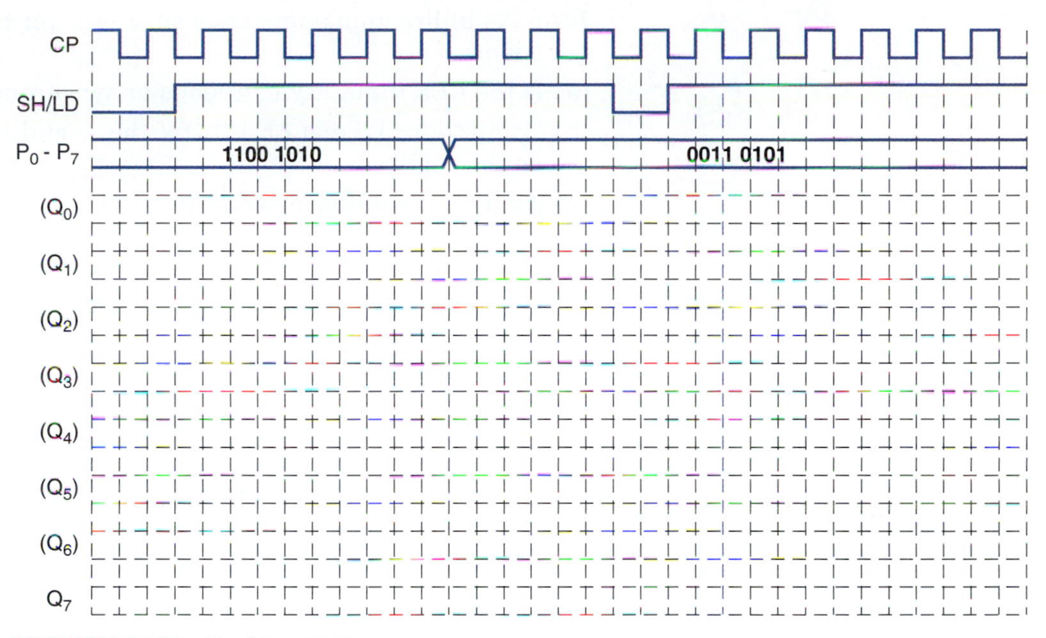

FIGURE 7-110 Problem 7-69.

7-70. Apply the input waveforms given in Figure 7-111 to a 74ALS166 and determine the output produced.

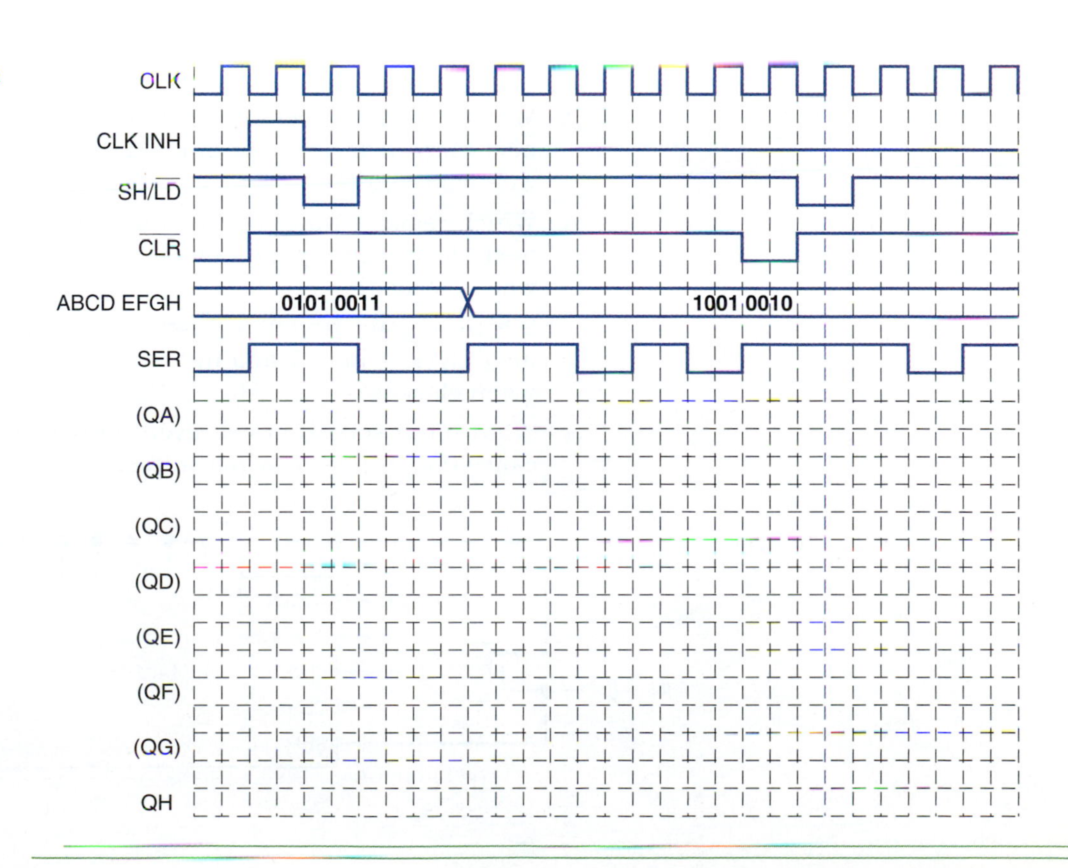

FIGURE 7-111 Problem 7-70.

B 7-71.* While examining the schematic for a piece of equipment, a technician or an engineer will often come across an IC that is unfamiliar. In such cases, it is often necessary to consult the manufacturer's data sheets for specifications on the device. Research the data sheet for the 74AS194 bidirectional universal shift register to answer the following questions:

(a) Is the $\overline{CLR}$ input asynchronous or synchronous?

(b) *True or false:* When *CLK* is LOW, the S_0 and S_1 inputs have no effect on the register.

(c) Assume the following conditions:

$$Q_A\ Q_B\ Q_C\ Q_D = 1\ 0\ 1\ 1$$
$$A\ B\ C\ D = 0\ 1\ 1\ 0$$
$$\overline{CLR} = 1$$
$$SR\ SER = 0$$
$$SL\ SER = 1$$

If $S_0 = 0$ and $S_1 = 1$, determine the register outputs after one *CLK* pulse. After two *CLK* pulses. After three. After four.

(d) Use the same conditions except $S_0 = 1$ and $S_1 = 0$ and repeat part (c).

(e) Repeat part (c) with $S_0 = 1$ and $S_1 = 1$.

(f) Repeat part (c) with $S_0 = 0$ and $S_1 = 0$.

(g) Use the same conditions as in part (c), except assume that Q_A is connected to *SL SER*. What will be the register outputs after four *CLK* pulses?

C 7-72. Refer to Figure 7-112 to answer the following questions:

(a) Which register function (load or shift) will be performed on the next clock if in = 1 and out = 0? What data value will be input when clocked?

(b) Which register function (load or shift) will be performed on the next clock if in = 0 and out = 1? What data value will be input when clocked?

(c) Which register function (load or shift) will be performed on the next clock if in = 0 and out = 0? What data value will be input when clocked?

(d) Which register function (load or shift) will be performed on the next clock if in = 1 and out = 1? What data value will be input when clocked?

(e) What input condition will eventually (after several clock pulses) cause the output to switch states?

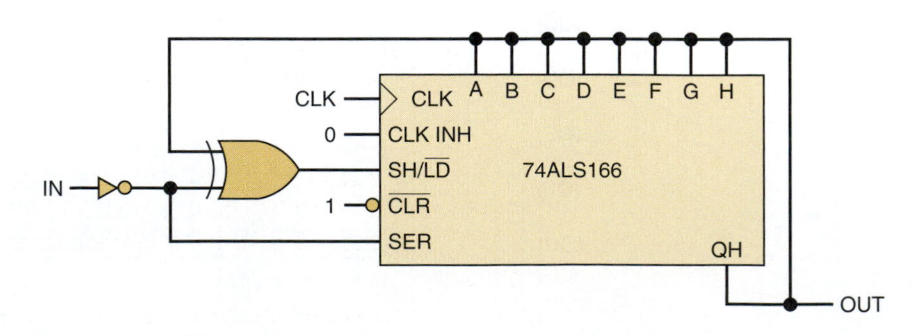

FIGURE 7-112 Problem 7-72.

(f) To change the output logic level requires the new input condition to last for at least how many clock pulses?

(g) If the input signal changes levels and then goes back to its original logic level before the number of clock pulses specified in part (f), what happens to the output signal.

(h) Explain why this circuit can be used to debounce switches.

SECTION 7-20

B 7-73.* Draw the diagram for a MOD-5 ring counter using J-K flip-flops. Make sure that the counter will start the proper count sequence when it is turned on.

7-74. Add one more J-K flip-flop to convert the MOD-5 ring counter in Problem 7-73 into a MOD-10 counter. Determine the sequence of states for this counter. This is an example of a decade counter that is not a BCD counter. Draw the decoding circuit for this counter.

B 7-75.* Draw the diagram for a MOD-10 Johnson counter using a 74HC164. Make sure that the counter will start the proper count sequence when it is turned on. Determine the count sequence for this counter and draw the decoding circuit needed to decode each of the 10 states. This is another example of a decade counter that is not a BCD counter.

7-76. The clock input to the Johnson counter in Problem 7-75 is 10 Hz. What is the frequency and duty cycle for each of the counter outputs?

SECTION 7-21

T 7-77.* The MOD-10 counter in Figure 7-8(b) produces the count sequence 0000, 0001, 0010, 0011, 0100, 0101, 0110, 0111, and repeats. Identify some possible fault conditions that might produce this result.

T 7-78. The MOD-10 counter in Figure 7-8(b) produces the count sequence 0000, 0101, 0010, 0111, 1000, 1101, 1010, 1111, and repeats. Identify some possible fault conditions that might produce this result.

SECTIONS 7-22 AND 7-23

H 7-79.* Create an eight-bit SISO shift register using an HDL. The serial input is called *ser* and the serial output is called *qout*. An active-LOW enable ($\overline{en}$) controls the shift register. Simulate the design.

H 7-80. Create an eight-bit PIPO shift register using an HDL. The data in is *d[7..0]* and the outputs are *q[7..0]*. An active-HIGH enable (*ld*) controls the shift register. Simulate the design.

H 7-81.* Create an eight-bit PISO shift register using an HDL. The data in is *d[7..0]* and the output is *q0*. The shift register function is controlled by *sh_ld* (*sh_ld* = 0 to synchronously parallel load and *sh_ld* = 1 to serial shift). The register also should have an active-LOW asynchronous clear ($\overline{clrn}$). Simulate the design.

H 7-82. Create an eight-bit SIPO shift register using an HDL. The data in is *ser_in* and the outputs are *q[7..0]*. The shift register function is enabled by an active-HIGH control named *shift*. The shift register also has a higher priority active-HIGH synchronous clear (*clear*). Simulate the design.

H 7-83.* Simulate the universal shift register design from Example 7-24.

H 7-84. Create an eight-bit universal shift register by cascading two of the modules in Example 7-24. Simulate the design.

H, D 7-85.* Design a MOD-10, self-starting Johnson counter with an active-HIGH, asynchronous reset (*reset*) using an HDL. Simulate the design.

H, D 7-86. Sometimes a digital application may need a ring counter that recirculates a single zero instead of a single one. The ring counter would then have an active-LOW output instead of an active-HIGH. Design a MOD-8, self-starting ring counter with an active-LOW output using an HDL. The ring counter should also have an active-HIGH *hold* control to disable the counting. Simulate the design.

SECTION 7-24

H 7-87.* Use Altera's simulator to test the nonretriggerable, level-sensitive, one-shot design example in either Figure 7-88 (AHDL) or 7-89 (VHDL). Use a 1-kHz clock and create a 10-ms output pulse for the simulation. Verify that:

(a) The correct pulse width is created when triggered.

(b) The output can be terminated early with the reset input.

(c) The one-shot design is nonretriggerable and cannot be triggered again until it has timed out.

(d) The trigger signal must last long enough for the clock to catch it.

(e) The pulse width can be changed to a different value.

H 7-88. Modify the nonretriggerable, level-sensitive, one-shot design example from either Figure 7-88 (AHDL) or Figure 7-89 (VHDL) so that the one-shot is retriggerable but still level-sensitive. Simulate the design.

DRILL QUESTION

B 7-89.* For each of the following statements, indicate the type(s) of counter being described.

(a) Each FF is clocked at the same time.

(b) Each FF divides the frequency at its CLK input by 2.

(c) The counting sequence is 111, 110, 101, 100, 011, 010, 001, 000.

(d) The counter has 10 distinct states.

(e) The total switching delay is the sum of the individual FFs' delays.

(f) This counter requires no decoding logic.

(g) The MOD number is always twice the number of FFs.

(h) This counter divides the input frequency by its MOD number.

(i) This counter can begin its counting sequence at any desired starting state.

(j) This counter can count in any direction.

(k) This counter can suffer from decoding glitches due to its propagation delays.

(l) This counter only counts from 0 to 9.

(m) This counter can be designed to count through arbitrary sequences by determining the logic circuit needed at each flip-flop's synchronous control inputs.

ANSWERS TO SECTION REVIEW QUESTIONS

PART 1

SECTION 7-1

1. False 2. 0000 3. 128

SECTION 7-2

1. Each FF adds its propagation delay to the total counter delay in response to a clock pulse. 2. MOD-256

SECTION 7-3

1. Can operate at higher clock frequencies and has more complex circuitry
2. Six FFs and four AND gates 3. ABCDE

SECTION 7-4

1. D, C, and A 2. True, because a BCD counter has 10 distinct states 3. 5 kHz

SECTION 7-5

1. In an up counter, the count is increased by 1 with each clock pulse; in a down counter, the count is decreased by 1 with each pulse. 2. Change connections to respective inverted outputs instead of *Q*s.

SECTION 7-6

1. It can be preset to any desired starting count. 2. Asynchronous presetting is independent of the clock input, while synchronous presetting occurs on the active edge of the clock signal.

SECTION 7-7

1. $\overline{LOAD}$ is the control that enables the parallel loading of the data inputs D C B A (A = LSB). 2. $\overline{CLR}$ is the control that enables the resetting of the counter to 0000. 3. True 4. All control inputs ($\overline{CLR}$, $\overline{LOAD}$, ENT, and ENP) on the 74162 must be HIGH. 5. $\overline{LOAD} = 1$, $\overline{CTEN} = 0$, and $D/\overline{U} = 1$ to count down.
6. 74HC163: 0 to 65,535; 74ALS190: 0 to 9999 or 9999 to 0.

SECTION 7-8

1. Sixty-four 2. A six-input NAND gate with inputs A, B, C, $\overline{D}$, E, and $\overline{F}$.

SECTION 7-9

1. We will not have to deal with transient states and possible glitches in output waveforms. 2. PRESENT state/NEXT state table 3. The gates control the count sequence. 4. Unused states all lead back to the count sequence of the counter.

SECTION 7-10

1. See text. 2. It associates every possible PRESENT state with its desired NEXT state. 3. It shows the necessary levels at each flip-flop's synchronous input to produce the counter's state transitions. 4. True

SECTION 7-11

1. PRESENT state/NEXT state tables 2. The desired NEXT state 3. AHDL:
```
ff[ ].clk = !clock
```
VHDL:
```
IF (clock = '0' AND clock' EVENT) THEN
```
4. Behavioral description

SECTION 7-12

1. Asynchronous clear causes the counter to clear immediately. Synchronous load occurs on the next active clock edge. 2. AHDL: Use .clrn port on FFs; VHDL: Define clear function before checking for clock edge 3. By the order of evaluation in an IF statement.

SECTION 7-13

1. Both HDLs can use a block diagram to connect modules; VHDL can also use a text file that describes the connections between components. 2. A bus is a collection of signal lines; it is represented graphically by a heavy-weight line
3. Count enable and terminal count decoding

SECTION 7-14

1. A counter is commonly used to count events, while a state machine is commonly used to control events. 2. A state machine can be described using symbols to describe its states rather than actual binary states. 3. The compiler assigns the optimal values to minimize the circuitry. 4. The description is much easier to write and understand.

PART 2

SECTION 7-19

1. Parallel in/serial out 2. True 3. Serial in/parallel out 4. Serial in/serial out
5. The 74165 uses asynchronous parallel data transfer; the 74174 uses synchronous parallel data transfer. 6. A HIGH prevents shifting on CPs.

SECTION 7-20

1. Ring counter 2. Johnson counter 3. The inverted output of the last FF is connected to the input of the first FF. 4. (a) False (b) True (c) True
5. Sixteen; eight

SECTION 7-22

1. AHDL:
```
reg [ ] .d = (reg [6..0], dat)
```
VHDL:
```
reg := reg (6 DOWNTO 0) & dat
```
2. Because the register may continue to receive clock edges during hold

SECTION 7-23

1. It can start in any state, but it will eventually reach the desired ring sequence.
2. Lines 11 and 12 3. Lines 12 and 13

SECTION 7-24

1. The reset input 2. The clock frequency and the delay value loaded into the counter 3. Synchronously 4. The output pulse width is very consistent.
5. The output pulse responds to the trigger edge immediately. 6. The state of the trigger on the current clock edge and its state on the previous edge.

CHAPTER 8

INTEGRATED-CIRCUIT LOGIC FAMILIES

OUTLINE

OBJECTIVES

Upon completion of this chapter, you will be able to:

- Read and understand digital IC terminology as specified in manufacturers' data sheets.
- Compare the characteristics of standard TTL and the various TTL series.
- Determine the fan-out for a particular logic device.
- Use logic devices with open-collector outputs.
- Analyze circuits containing tristate devices.
- Compare the characteristics of the various CMOS series.
- Analyze circuits that use a CMOS bilateral switch to allow a digital system to control analog signals.
- Describe the major characteristics of and differences among TTL, ECL, MOS, and CMOS logic families.
- Cite and implement the various considerations that are required when interfacing digital circuits from different logic families.
- Use voltage comparators to allow a digital system to be controlled by analog signals.
- Use a logic pulser and a logic probe as digital circuit troubleshooting tools.

INTRODUCTION

As we described in Chapter 4, digital IC technology has advanced rapidly from small-scale integration (SSI), with fewer than 12 gates per chip; through medium-scale integration (MSI), with 12 to 99 equivalent gates per chip; on to large-scale and very large scale integration (LSI and VLSI, respectively), which can have tens of thousands of gates per chip; and, most recently, to ultra-large-scale integration (ULSI), with over 100,000 gates per chip, and giga-scale integration (GSI), with 1 million or more gates.

Most of the reasons that modern digital systems use integrated circuits are obvious. ICs pack a lot more circuitry in a small package, so that the overall size of almost any digital system is reduced. The cost is dramatically reduced because of the economies of mass-producing large volumes of similar devices. Some of the other advantages are not so apparent.

ICs have made digital systems more reliable by reducing the number of external interconnections from one device to another. Before we had ICs, every circuit connection was from one discrete component (transistor, diode, resistor, etc.) to another. Now most of the connections are internal to the ICs, where they are protected from poor soldering, breaks or shorts in connecting paths on a circuit board, and other physical problems. ICs have also drastically reduced the amount of electrical power needed to perform a given function because their miniature circuitry typically requires less power than their discrete counterparts. In addition to the savings in power-supply costs, this reduction in power has also meant that a system does not require as much cooling.

There are some things that ICs cannot do. They cannot handle very large currents or voltages because the heat generated in such small spaces would cause temperatures to rise beyond acceptable limits. In addition, ICs cannot easily implement certain electrical devices such as inductors, transformers, and large capacitors. For these reasons, ICs are principally used to perform low-power circuit operations that are commonly called *information processing*. The operations that require high power levels or devices that cannot be integrated are still handled by discrete components.

With the widespread use of ICs comes the necessity to know and understand the electrical characteristics of the most common IC logic families. Remember that the various logic families differ in the major components that they use in their circuitry. TTL and ECL use *bipolar* transistors as their major circuit element; PMOS, NMOS, and CMOS use unipolar *MOS*FET transistors as their principal component. In this chapter, we will present the important characteristics of each of these IC families and their subfamilies. The most important point is understanding the nature of the input circuitry and output circuitry for each logic family. Once these are understood, you will be much better prepared to do analysis, troubleshooting, and some design of digital circuits that contain any combination of IC families. We will study the inner workings of devices in each family with the simplest circuitry that conveys the critical characteristics of all members of the family.

8-1 DIGITAL IC TERMINOLOGY

Although there are many digital IC manufacturers, much of the nomenclature and terminology is fairly standardized. The most useful terms are defined and discussed below.

Current and Voltage Parameters (See Figure 8-1)

- **V_{IH}(min)—High-Level Input Voltage.** The minimum voltage level required for a logical 1 at an *input*. Any voltage below this level will not be accepted as a HIGH by the logic circuit.
- **V_{IL}(max)—Low-Level Input Voltage.** The maximum voltage level required for a logic 0 at an *input*. Any voltage above this level will not be accepted as a LOW by the logic circuit.
- **V_{OH}(min)—High-Level Output Voltage.** The minimum voltage level at a logic circuit *output* in the logical 1 state under defined load conditions.
- **V_{OL}(max)—Low-Level Output Voltage.** The maximum voltage level at a logic circuit *output* in the logical 0 state under defined load conditions.
- **I_{IH}—High-Level Input Current.** The current that flows into an input when a specified high-level voltage is applied to that input.
- **I_{IL}—Low-Level Input Current.** The current that flows into an input when a specified low-level voltage is applied to that input.
- **I_{OH}—High-Level Output Current.** The current that flows from an output in the logical 1 state under specified load conditions.

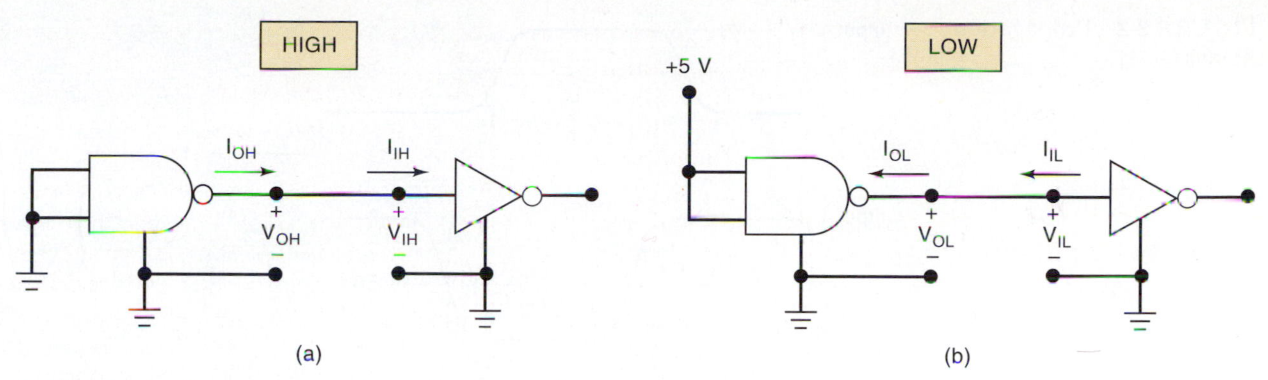

FIGURE 8-1 Currents and voltages in the two logic states.

- **I_{OL}—Low-Level Output Current.** The current that flows from an output in the logical 0 state under specified load conditions.

Note: The actual current directions may be opposite to those shown in Figure 8-1, depending on the logic family. All descriptions of current flow in this text refer to conventional current flow (from higher potential to lower potential). In keeping with the conventions of most data books, current flowing into a node or device is considered positive, and current flowing out of a node or device is considered negative.

Fan-Out

In general, a logic-circuit output is required to drive several logic inputs. Sometimes all ICs in the digital system are from the same logic family, but many systems have a mix of various logic families. The **fan-out** (also called *loading factor*) is defined as the *maximum* number of logic inputs that an output can drive reliably. For example, a logic gate that is specified to have a fan-out of 10 can drive 10 logic inputs. If this number is exceeded, the output logic-level voltages cannot be guaranteed. Obviously, fan-out depends on the nature of the input devices that are connected to an output. Unless a different logic family is specified as the load device, fan-out is assumed to refer to load devices of the same family as the driving output.

Propagation Delays

A logic signal always experiences a delay in going through a circuit. The two propagation delay times are defined as follows:

- t_{PLH}. Delay time in going from logical 0 to logical 1 state (LOW to HIGH)
- t_{PHL}. Delay time in going from logical 1 to logical 0 state (HIGH to LOW)

Figure 8-2 illustrates these propagation delays for an INVERTER. Note that t_{PHL} is the delay in the output's response as it goes from HIGH to LOW. It is measured between the 50 percent points on the input and output transitions. The t_{PLH} value is the delay in the output's response as it goes from LOW to HIGH.

In some logic circuits, t_{PHL} and t_{PLH} are not the same value, and both will vary depending on capacitive loading conditions. The values of propagation times are used as a measure of the relative speed of logic circuits. For example, a logic circuit with values of 10 ns is a faster logic circuit than one with values of 20 ns under specified load conditions.

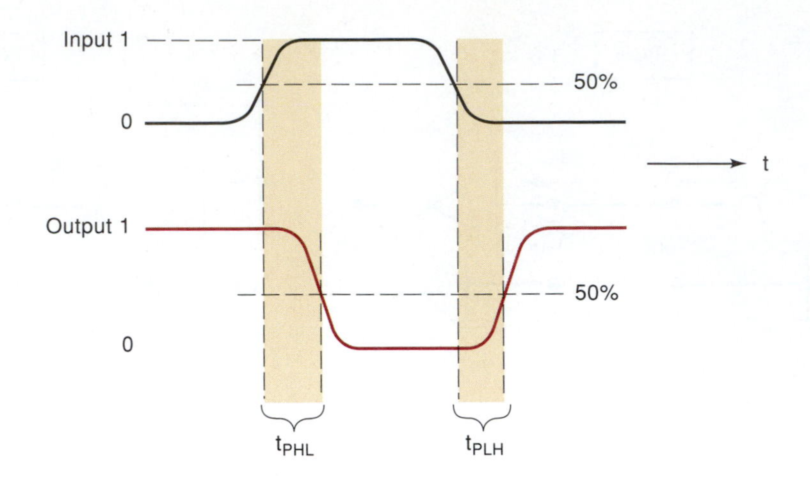

FIGURE 8-2 Propagation delays.

Power Requirements

Every IC requires a certain amount of electrical power to operate. This power is supplied by one or more power-supply voltages connected to the power pin(s) on the chip labeled V_{CC} (for TTL) or V_{DD} (for MOS devices).

The amount of power that an IC requires is determined by the current, I_{CC} (or I_{DD}), that it draws from the V_{CC} (or V_{DD}) supply, and the actual power is the product $I_{CC} \times V_{CC}$. For many ICs, the current drawn from the supply varies depending on the logic states of the circuits on the chip. For example, Figure 8-3(a) shows a NAND chip where *all* of the gate *outputs* are HIGH. The current drain on the V_{CC} supply for this case is called I_{CCH}. Likewise, Figure 8-3(b) shows the current when *all* of the gate *outputs* are LOW. This current is called I_{CCL}. The values are always measured with the outputs open circuit (no load) because the size of the load will also have an effect on I_{CCH}.

In some logic circuits, I_{CCH} and I_{CCL} will be different values. For these devices, the average current is computed based on the assumption that gate outputs are LOW half the time and HIGH half the time.

$$I_{CC}(\text{avg}) = \frac{I_{CCH} + I_{CCL}}{2}$$

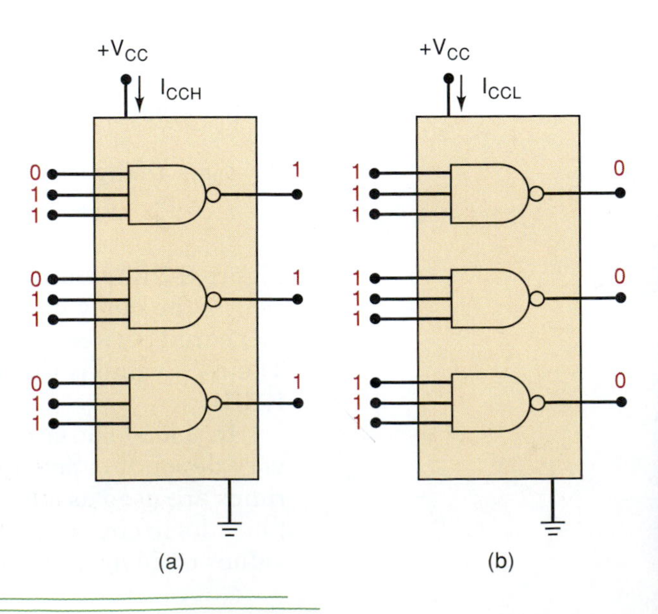

FIGURE 8-3 I_{CCH} and I_{CCL}.

This equation can be rewritten to calculate average power dissipated:

$$P_D(\text{avg}) = I_{CC}(\text{avg}) \times V_{CC}$$

Noise Immunity

Stray electric and magnetic fields can induce voltages on the connecting wires between logic circuits. These unwanted, spurious signals are called *noise* and can sometimes cause the voltage at the input to a logic circuit to drop below V_{IH}(min) or rise above V_{IL}(max), which could produce unpredictable operation. The **noise immunity** of a logic circuit refers to the circuit's ability to tolerate noise without causing spurious changes in the output voltage. A quantitative measure of noise immunity is called **noise margin** and is illustrated in Figure 8-4.

Figure 8-4(a) is a diagram showing the range of voltages that can occur at a logic-circuit output. Any voltages greater than V_{OH}(min) are considered a logic 1, and any voltages lower than V_{OL}(max) are considered a logic 0. Voltages in the indeterminate range should not appear at a logic circuit output under normal conditions. Figure 8-4(b) shows the voltage requirements at a logic circuit input. The logic circuit responds to any input greater than V_{IH}(min) as a logic 1, and it responds to voltages lower than V_{IL}(max) as a logic 0. Voltages in the indeterminate range produce an unpredictable response and should not be used.

The *high-state noise margin* V_{NH} is defined as

$$V_{NH} = V_{OH}(\text{min}) - V_{IH}(\text{min}) \quad \textbf{(8-1)}$$

and is illustrated in Figure 8-4. V_{NH} is the difference between the lowest possible HIGH output and the minimum input voltage required for a HIGH. When a HIGH logic output is driving a logic-circuit input, any negative noise spikes greater than V_{NH} appearing on the signal line can cause the voltage to drop into the indeterminate range, where unpredictable operation can occur.

The *low-state noise margin* V_{NL} is defined as

$$V_{NL} = V_{IL}(\text{max}) - V_{OL}(\text{max}) \quad \textbf{(8-2)}$$

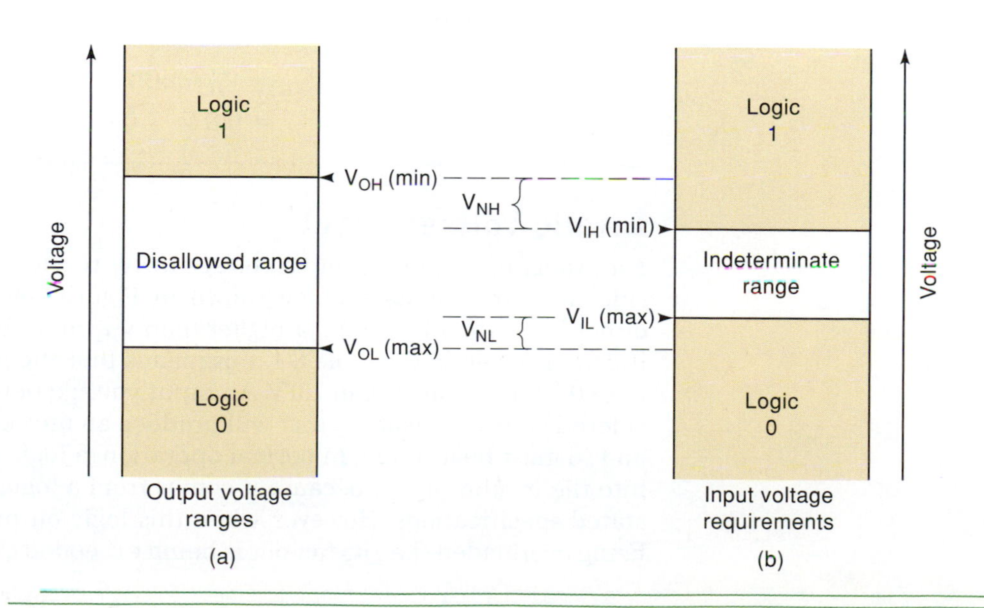

FIGURE 8-4 dc noise margins.

and it is the difference between the largest possible LOW output and the maximum input voltage required for a LOW. When a LOW logic output is driving a logic input, any positive noise spikes greater than V_{NL} can cause the voltage to rise into the indeterminate range.

EXAMPLE 8-1

The input/output voltage specifications for the standard TTL family are listed in Table 8-1. Use these values to determine the following.

(a) The maximum-amplitude noise spike that can be tolerated when a HIGH output is driving an input.

(b) The maximum-amplitude noise spike that can be tolerated when a LOW output is driving an input.

TABLE 8-1

Parameter	Min (V)	Typical (V)	Max (V)
V_{OH}	2.4	3.4	
V_{OL}		0.2	0.4
V_{IH}	2.0*		
V_{IL}			0.8*

*Normally only the minimum V_{IH} and maximum V_{IL} values are given.

Solution

(a) When an output is HIGH, it may be as low as $V_{OH}(\text{min}) = 2.4$ V. The minimum voltage that an input responds to as a HIGH is $V_{IH}(\text{min}) = 2.0$ V. A negative noise spike can drive the actual voltage below 2.0 V if its amplitude is greater than

$$\begin{aligned} V_{NH} &= V_{OH}(\text{min}) - V_{IH}(\text{min}) \\ &= 2.4\text{ V} - 2.0\text{ V} = 0.4\text{ V} \end{aligned}$$

(b) When an output is LOW, it may be as high as $V_{OL}(\text{max}) = 0.4$ V. The maximum voltage that an input responds to as a LOW is $V_{IL}(\text{max}) = 0.8$ V. A positive noise spike can drive the actual voltage above the 0.8-V level if its amplitude is greater than

$$\begin{aligned} V_{NL} &= V_{IL}(\text{max}) - V_{OL}(\text{max}) \\ &= 0.8\text{ V} - 0.4\text{ V} = 0.4\text{ V} \end{aligned}$$

Invalid Voltage Levels

For proper operation the input voltage levels to a logic circuit must be kept outside the indeterminate range shown in Figure 8-4(b); that is, they must be either lower than $V_{IL}(\text{max})$ or higher than $V_{IH}(\text{min})$. For the standard TTL specifications given in Example 8-1, this means that the input voltage must be less than 0.8 V or greater than 2.0 V. An input voltage between 0.8 and 2.0 V is considered an *invalid* voltage that will produce an unpredictable output response, and so must be avoided. In normal operation, a logic input voltage will not fall into the invalid region because it comes from a logic output that is within the stated specifications. However, when this logic output is malfunctioning or is being overloaded (i.e., its fan-out is being exceeded), then its voltage may be in

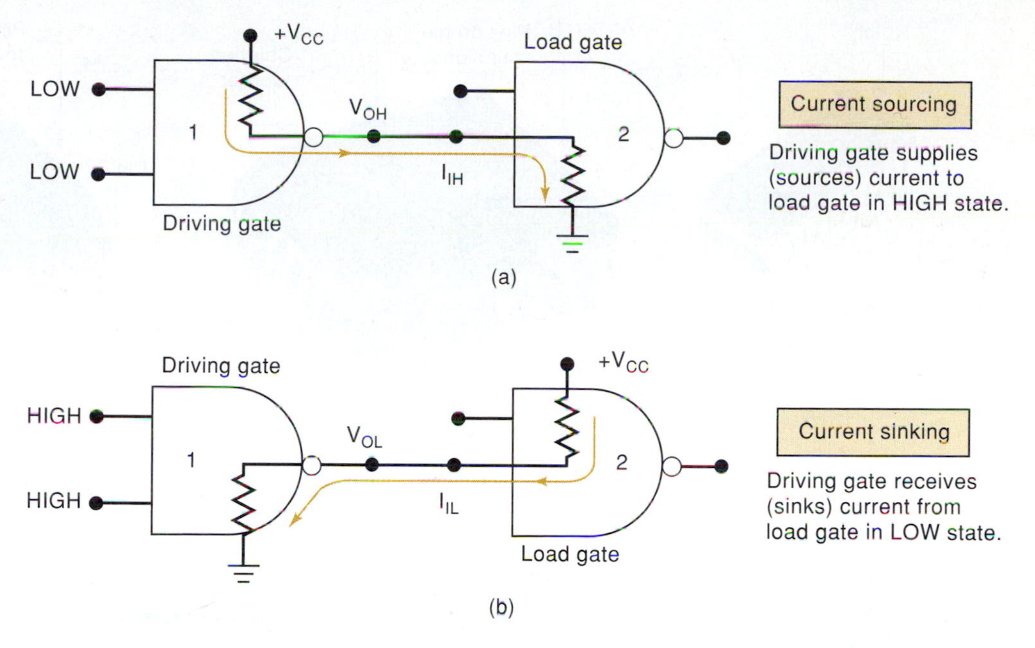

FIGURE 8-5 Comparison of current-sourcing and current-sinking actions.

the invalid region. Invalid voltage levels in a digital circuit can also be caused by power-supply voltages that are outside the acceptable range. It is important to know the valid voltage ranges for the logic family being used so that invalid conditions can be recognized when testing or troubleshooting.

Current-Sourcing and Current-Sinking Action

Logic families can be described according to how current flows between the output of one logic circuit and the input of another. Figure 8-5(a) illustrates **current-sourcing** action. When the output of gate 1 is in the HIGH state, it supplies a current I_{IH} to the input of gate 2, which acts essentially as a resistance to ground. Thus, the output of gate 1 is acting as a *source* of current for the gate 2 input. We can think of it as being like a faucet that acts as a *source* of water.

Current-sinking action is illustrated in Figure 8-5(b). Here the input circuitry of gate 2 is represented as a resistance tied to $+V_{CC}$, the positive terminal of a power supply. When the gate 1 output goes to its LOW state, current will flow in the direction shown from the input circuit of gate 2 back through the output resistance of gate 1 to ground. In other words, in the LOW state, the circuit output that drives the input of gate 2 must be able to *sink* a current, I_{IL}, coming from that input. We can think of this as acting like a *sink* into which water is flowing.

The distinction between current sourcing and current sinking is an important one, which will become more apparent as we examine the various logic families.

IC Packages

Developments and advancements in integrated circuits continue at a rapid pace. The same is true of IC packaging. There are various types of packages, which differ in physical size, the environmental and power-consumption conditions under which the device can be operated reliably, and the way in which the IC package is mounted to the circuit board. Figure 8-6 shows five representative IC packages.

The package in Figure 8-6(a) is the **DIP** (dual-in-line package), which has been around for a long time. Its pins (or leads) run down the two long sides of the rectangular package. The device shown is a 24-pin DIP. Note the presence

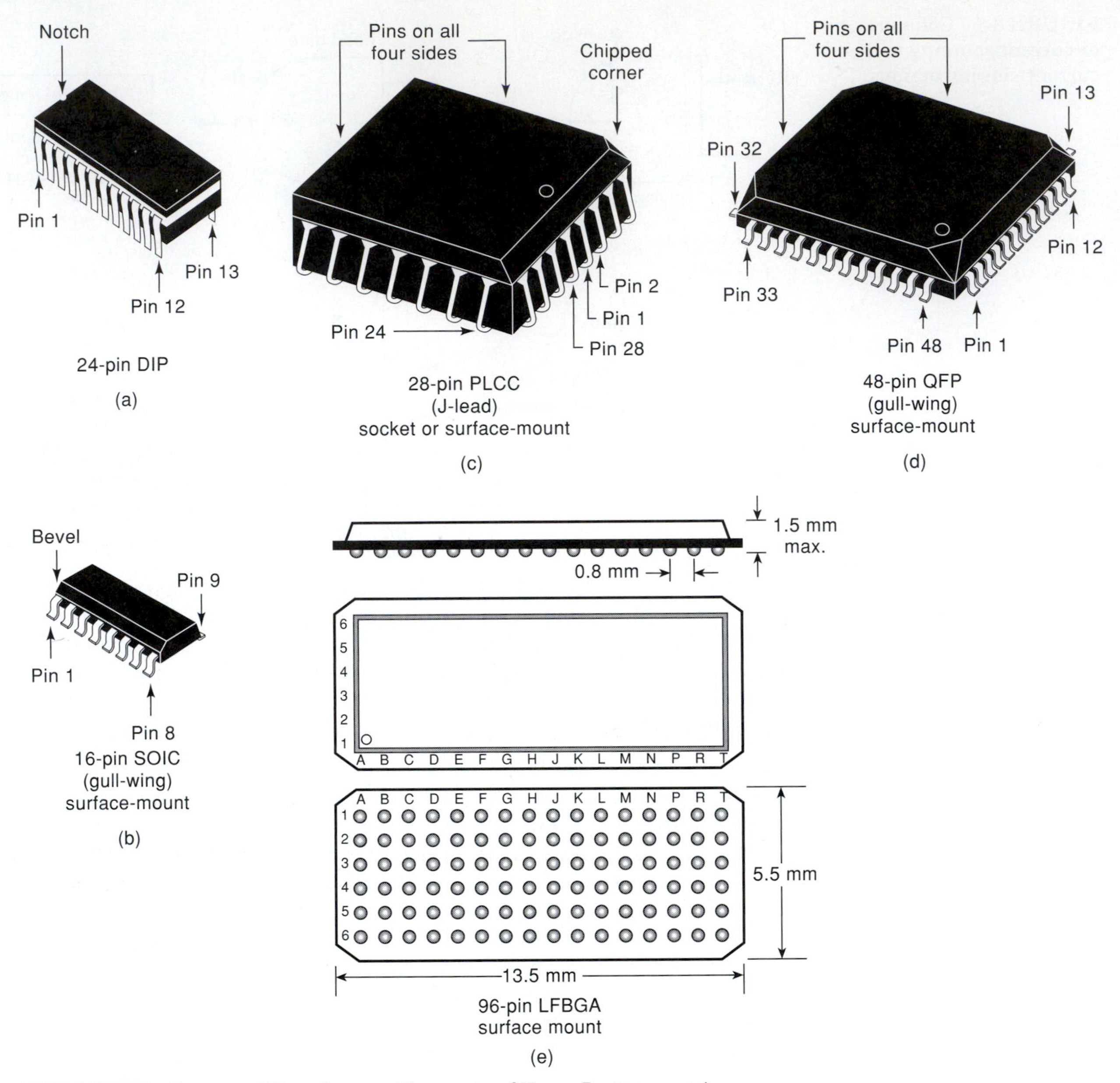

FIGURE 8-6 Common IC packages. (Courtesy of Texas Instruments)

of the notch on one end, which is used to locate pin 1. Some DIPs use a small dot on the top surface of the package to locate pin 1. The leads extend straight out of the DIP package so that the IC can be plugged into an IC socket or inserted into holes drilled through a printed circuit board. The spacing between pins **(lead pitch)** is typically 100 mils (a mil is a thousandth of an inch). DIP packages are still the most popular package for prototyping, breadboarding, and educational experimentation.

Nearly all new circuit boards that are produced using automated manufacturing equipment have moved away from using DIP packages whose leads are inserted through holes in the board. New manufacturing methods use **surface-mount technology**, which places an IC onto conductive pads on the surface of the board. They are held in place by a solder paste, and the entire board is heated to create a soldered connection. The precision of the placement machine allows for very tight lead spacing. The leads on these surface-mount packages are bent out from the plastic case, providing adequate surface area

TABLE 8-2 IC packages.

Abbreviation	Package Name	Height	Lead Pitch
DIP	Dual-in-line package	200 mils (5.1 mm)	100 mils (2.54 mm)
SOIC	Small outline integrated circuit	2.65 mm	50 mils (1.27 mm)
SSOP	Shrink small outline package	2.0 mm	0.65 mm
TSSOP	Thin shrink small outline package	1.1 mm	0.65 mm
TVSOP	Thin very small outline package	1.2 mm	0.4 mm
PLCC	Plastic leaded chip carrier	4.5 mm	1.27 mm
QFP	Quad flat pack	4.5 mm	0.635 mm
TQFP	Thin quad flat pack	1.6 mm	0.5 mm
LFBGA	Low-profile fine-pitch ball grid array	1.5 mm	0.8 mm

for the solder joint. The shape of these leads has resulted in the nickname of "gull-wing" package. Many different packages are available for surface-mount devices. Some of the most common packages used for logic ICs are shown in Figure 8-6. Table 8-2 gives the definition of each abbreviation along with its dimensions.

The need for more and more connections to a complex IC has resulted in another very popular package that has pins on all four sides of the chip. The PLCC has J-shaped leads that curl under the IC, as shown in Figure 8-6(c). These devices can be surface-mounted to a circuit board but can also be placed in a special PLCC socket. This is commonly used for components that are likely to need to be replaced for repair or upgrade, such as programmable logic devices or central processing units in computers. The QFP and TQFP packages have pins on all four sides in a gull-wing surface-mount package, as shown in Figure 8-6(d). The ball grid array (BGA) shown in Figure 8-6(e) is a surface-mount package that offers even more density. The pin grid array (PGA) is a similar package that is used when components must be in a socket to allow easy removal. The PGA has a long pin instead of a contact ball (BGA) at each position in the grid.

The proliferation of small, handheld consumer equipment such as digital video cameras, cellular phones, computers (PDAs), portable audio systems, and other devices has created a need for logic circuits in very small packages. Logic gates are now available in individual surface-mount packages containing one, two, or three gates (1G, 2G, 3G, respectively). These devices may have as few as five or six pins (power, ground, two to three inputs, and an output) and take up less space than an individual letter on this page.

REVIEW QUESTIONS

1. Define each of the following: V_{OH}, V_{IL}, I_{OL}, I_{IH}, t_{PLH}, t_{PHL}, I_{CCL}, I_{CCH}.
2. *True or false:* If a logic circuit has a fan-out of 5, the circuit has five outputs.
3. *True or false:* The HIGH-stage noise margin is the difference between V_{IH}(min) and V_{CC}.
4. Describe the difference between current sinking and current sourcing.
5. Which IC package can be plugged into sockets?
6. Which package has leads bent under the IC?
7. How do surface-mount packages differ from DIPs?
8. Will a standard TTL device work with an input level of 1.7 V?

8-2 THE TTL LOGIC FAMILY

At this writing, many small- to medium-scale ICs (SSI and MSI) can still be obtained in the standard **TTL** technology series that has been available for over 30 years. This original series of devices and their descendants in the TTL family have had a tremendous influence on the characteristics of all logic devices today. TTL devices are still used as "glue" logic that connects the more complex devices in digital systems. They are also used as interface circuits to devices that require high current drive. Even though the bipolar TTL family as a whole is on the decline, we will begin our discussion of logic ICs with the devices that shaped digital technology.

The basic TTL logic circuit is the NAND gate, shown in Figure 8-7(a). Even though the standard TTL family is nearly obsolete, we can learn a great deal about the more current family members by studying the original circuitry in its simplest form. The characteristics of TTL inputs come from the multiple-emitter (diode junction) configuration of transistor Q_1. Forward biasing either (or both) of these diode junctions will turn on Q_1. Only when all junctions are reverse biased will the transistor be off. This *multiple-emitter* input transistor can have up to eight emitters for an eight-input NAND gate.

Also note that on the output side of the circuit, transistors Q_3 and Q_4 are in a **totem-pole** arrangement. The totem pole is made up of two transistor switches, Q_3 and Q_4. The job of Q_3 is to connect V_{CC} to the output, making a logic HIGH. The job of Q_4 is to connect the output to ground, making a logic LOW. As we will see shortly, in normal operation, either Q_3 or Q_4 will be conducting, depending on the logic state of the output.

Circuit Operation—LOW State

Although this circuit looks extremely complex, we can simplify its analysis somewhat by using the diode equivalent of the multiple-emitter transistor Q_1, as shown in Figure 8-7(b). Diodes D_2 and D_3 represent the two E–B junctions of Q_1, and D_4 is the collector-base (C–B) junction. In the following analysis, we will use this representation for Q_1.

FIGURE 8-7 (a) Basic TTL NAND gate; (b) diode equivalent for Q_1.

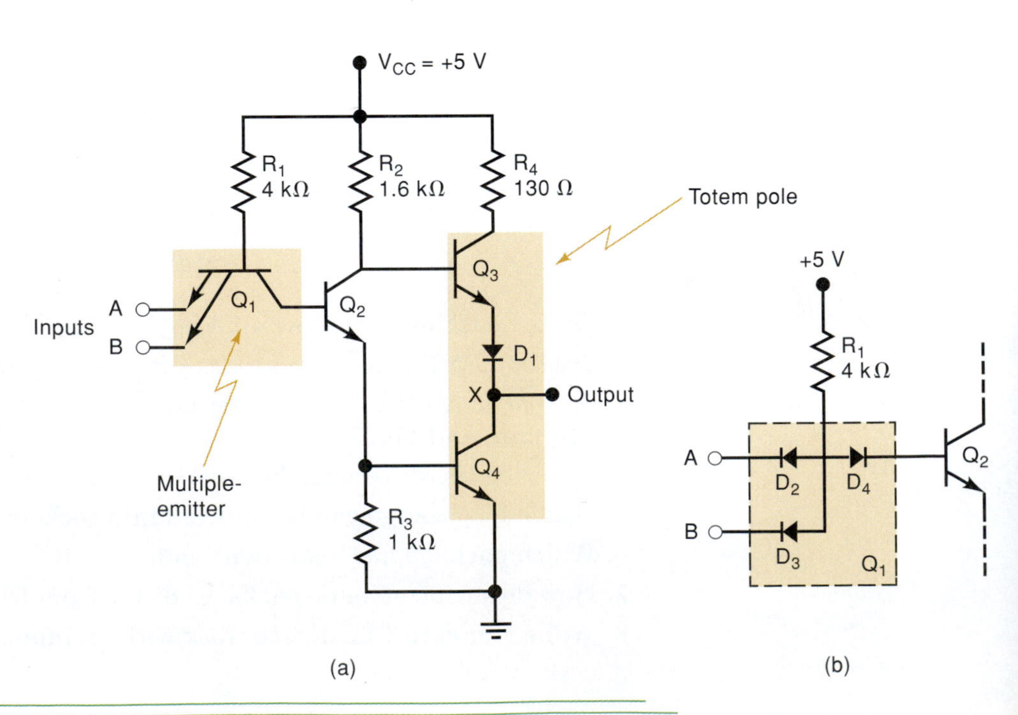

First, let's consider the case where the output is LOW. Figure 8-8(a) shows this situation with inputs A and B both at +5 V. The +5 V at the cathodes of D_2 and D_3 will turn these diodes off, and they will conduct almost no current. The +5 V supply will push current through R_1 and D_4 into the base of Q_2, which turns on. Current from Q_2's emitter will flow into the base of Q_4 and turn Q_4 on. At the same time, the flow of Q_2 collector current produces a voltage drop across R_2 that reduces Q_2's collector voltage to a low value that is insufficient to turn Q_3 on.

The voltage at Q_2's collector is shown as approximately 0.8 V. This is because Q_2's emitter is at 0.7 V relative to ground due to Q_4's E–B forward voltage,

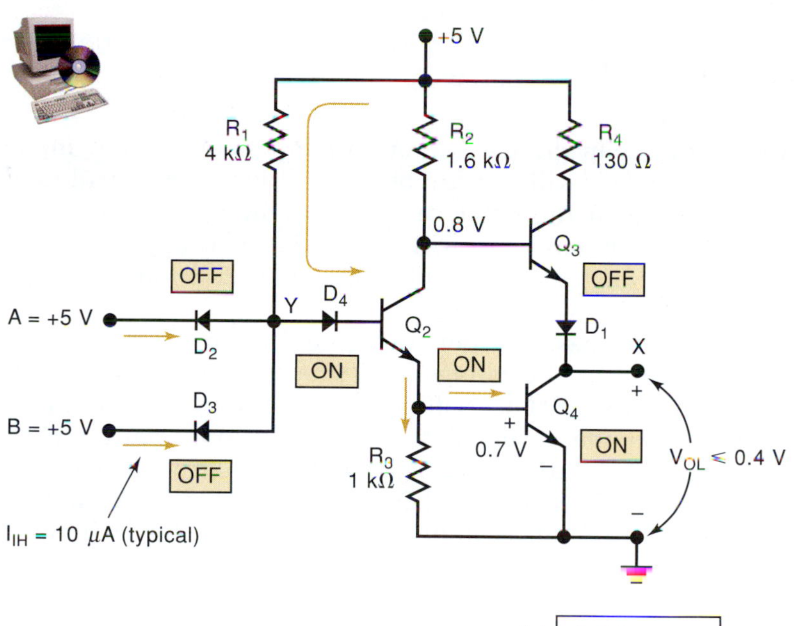

Input conditions	Output conditions
A and B are both HIGH (≥ 2 V)	Q_3 OFF
Input currents are very low I_{IH} = 10 μA	Q_4 ON so that V_X is LOW (≤ 0.4 V)

(a) LOW output

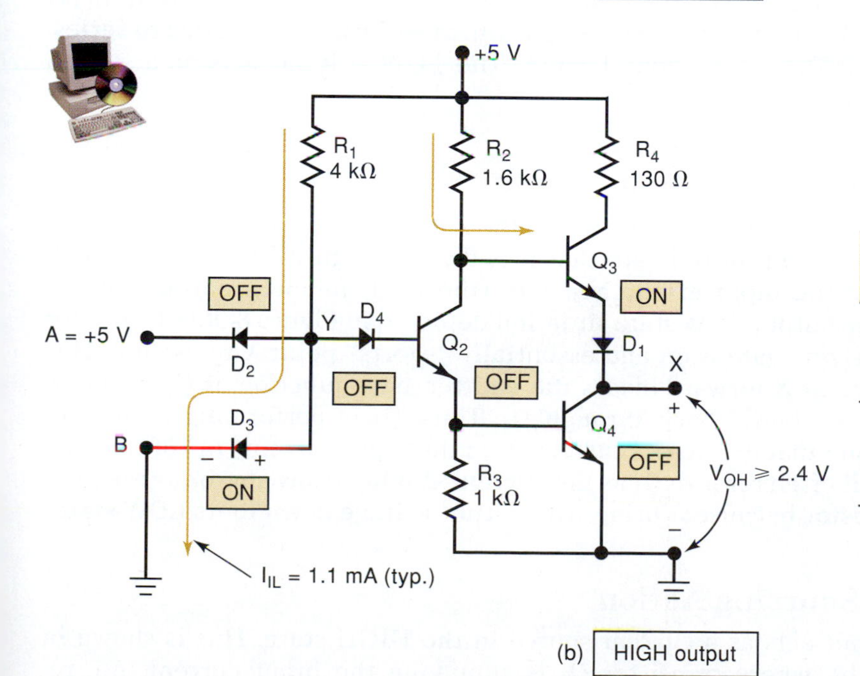

Input conditions	Output conditions
A or B or both are LOW (≤ 0.8 V)	Q_4 OFF
Current flows back to ground through LOW input terminal. I_{IL} = 1.1 mA	Q_3 acts as emitter-follower and V_{OH} ≥ 2.4 V, typically 3.6 V

(b) HIGH output

FIGURE 8-8 TTL NAND gate in its two output states.

and Q_2's collector is at 0.1 V relative to its emitter due to V_{CE}(sat). This 0.8 V at Q_3's base is not enough to forward-bias both Q_3's E–B junction and diode D_1. In fact, D_1 is needed to keep Q_3 off in this situation.

With Q_4 on, the output terminal, X, will be at a very low voltage because Q_4's ON-state resistance will be low (1 to 25 Ω). Actually, the output voltage, V_{OL}, will depend on how much collector current Q_4 conducts. With Q_3 off, there is no current coming from the +5 V terminal through R_4. As we shall see, Q_4's collector current will come from the TTL inputs that terminal X is connected to.

It is important to note that the HIGH inputs at A and B will have to supply only a very small diode leakage current. Typically, this current I_{IH} is only around 10 μA at room temperature.

Circuit Operation—HIGH State

Figure 8-8(b) shows the situation where the circuit output is HIGH. This situation can be produced by connecting either or both inputs LOW. Here, input B is connected to ground. This will forward-bias D_3 so that current will flow from the +5 V source terminal, through R_1 and D_3, and through terminal B to ground. The forward voltage across D_3 will hold point Y at approximately 0.7 V. This voltage is not enough to forward-bias D_4 and the E–B junction of Q_2 sufficiently for conduction.

With Q_2 off, there is no base current for Q_4, and it turns off. Because there is no Q_2 collector current, the voltage at Q_3's base will be large enough to forward-bias Q_3 and D_1, so that Q_3 will conduct. Actually, Q_3 acts as an emitter follower because output terminal X is essentially at its emitter. With no load connected from point X to ground, V_{OH} will be around 3.4 to 3.8 V because two 0.7-V diode drops (E–B of Q_3, and D_1) subtract from the 5 V applied to Q_3's base. This voltage will decrease under load because the load will draw emitter current from Q_3, which draws base current through R_2, thereby increasing the voltage drop across R_2.

It's important to note that there is a substantial current flowing back through input terminal B to ground when B is held LOW. This current, I_{IL}, is determined by the value of resistor R_1, which will vary from series to series. For standard TTL, it is about 1.1 mA. The LOW B input acts as a *sink* to ground for this current.

Current-Sinking Action

A TTL output acts as a current sink in the LOW state because it *receives* current from the input of the gate that it is driving. Figure 8-9 shows one TTL gate driving the input of another gate (the load) for both output voltage states. In the output LOW state situation depicted in Figure 8-9(a), transistor Q_4 of the driving gate is on and essentially "shorts" point X to ground. This LOW voltage at X forward-biases the emitter–base junction of Q_1, and current flows, as shown, back through Q_4. Thus, Q_4 is performing a current-sinking action that derives its current from the input current (I_{IL}) of the load gate. We will often refer to Q_4 as the **current-sinking transistor** or as the **pull-down transistor** because it brings the output voltage down to its LOW state.

Current-Sourcing Action

A TTL output acts as a current source in the HIGH state. This is shown in Figure 8-9(b), where transistor Q_3 is supplying the input current, I_{IH}, required by the Q_1 transistor of the load gate. As stated above, this current is a

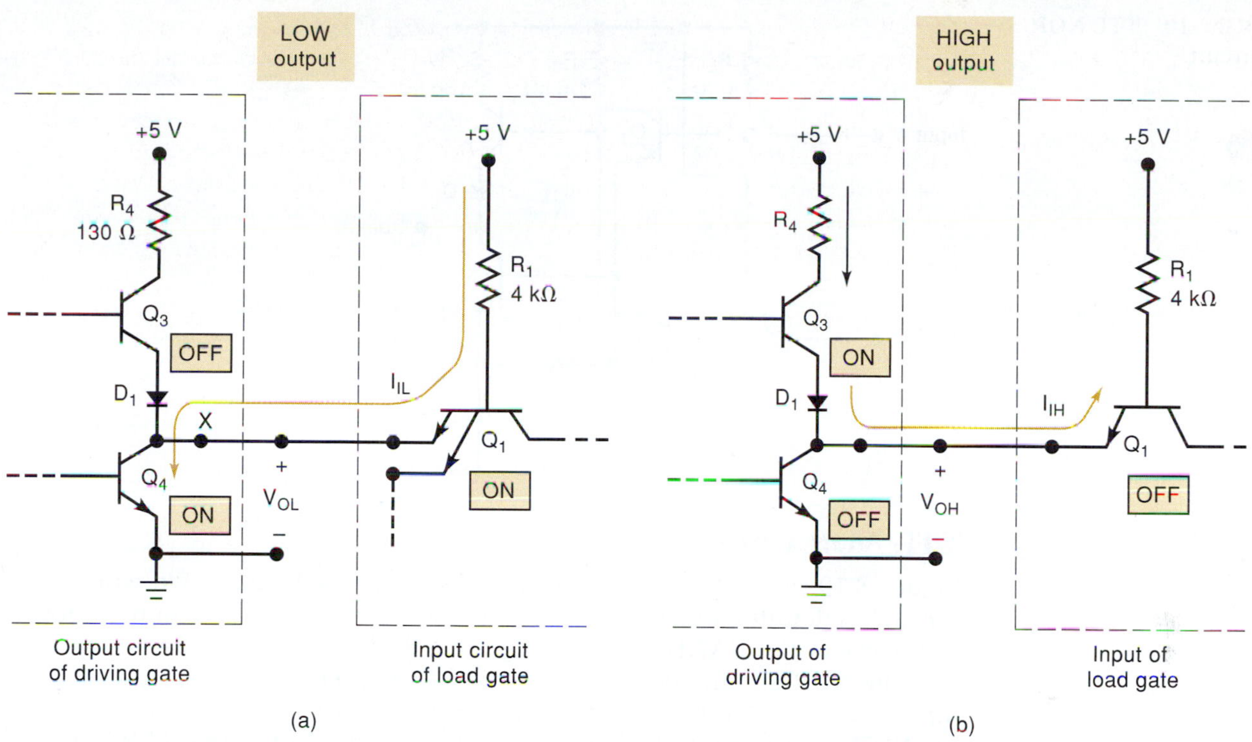

FIGURE 8-9 (a) When the TTL output is in the LOW state, Q_4 acts as a current sink, deriving its current from the load. (b) In the output HIGH state, Q_3 acts as a current source, providing current to the load gate.

small reverse-bias leakage current (typically 10 μA). We will often refer to Q_3 as the **current-sourcing transistor** or **pull-up transistor**. In some of the more modern TTL series, the pull-up circuit is made up of two transistors, rather than a transistor and diode.

Totem-Pole Output Circuit

Several points should be mentioned concerning the totem-pole arrangement of the TTL output circuit, as shown in Figure 8-9, because it is not readily apparent why it is used. The same logic can be accomplished by eliminating Q_3 and D_1 and connecting the bottom of R_4 to the collector of Q_4. But this arrangement would mean that Q_4 would conduct a fairly heavy current in its saturation state (5 V/130 Ω $\approx$ 40 mA). With Q_3 in the circuit, there will be no current through R_4 in the output LOW state. This is important because it keeps the circuit power dissipation down.

Another advantage of this arrangement occurs in the output HIGH state. Here Q_3 is acting as an emitter follower with its associated low output impedance (typically 10 Ω). This low output impedance provides a short time constant for charging up any capacitive load on the output. This action (commonly called *active pull-up*) provides very fast rise-time waveforms at TTL outputs.

A disadvantage of the totem-pole output arrangement occurs during the transition from LOW to HIGH. Unfortunately, Q_4 turns off more slowly than Q_3 turns on, and so there is a period of a few nanoseconds during which both transistors are conducting and a relatively large current (30 to 40 mA) will be drawn from the 5-V supply. This can present a problem that will be examined later.

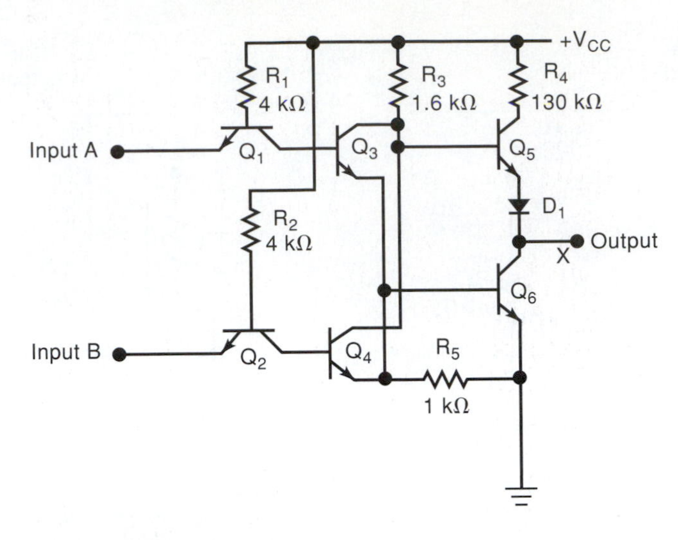

FIGURE 8-10 TTL NOR gate circuit.

TTL NOR Gate

Figure 8-10 shows the internal circuit for a TTL NOR gate. We will not go through a detailed analysis of this circuit, but it is important to note how it compares to the NAND circuit of Figure 8-8. On the input side, we can see that the NOR circuit *does not use a multiple-emitter* transistor; instead, each input is applied to the emitter of a separate transistor. On the output side, the NOR circuit uses the same totem-pole arrangement as the NAND circuit.

Summary

All TTL circuits have a similar structure. NAND and AND gates use multiple-emitter transistor or multiple diode junction inputs; NOR and OR gates use separate input transistors. In either case, the input will be the cathode (N-region) of a P–N junction, so that a HIGH input voltage will turn off the junction and only a small leakage current (I_{IH}) will flow. Conversely, a LOW input voltage turns on the junction, and a relatively large current (I_{IL}) will flow back through the signal source. Most, but not all, TTL circuits will have some type of totem-pole output configuration. There are some exceptions that will be discussed later.

REVIEW QUESTIONS

1. *True or false:* A TTL output acts as a current sink in the LOW state.
2. In which TTL input state does the largest amount of input current flow?
3. State the advantages and disadvantages of a totem-pole output.
4. Which TTL transistor is the pull-up transistor in the NAND circuit?
5. Which TTL transistor is the pull-down transistor in the NOR circuit?
6. How does the TTL NOR circuit differ from the NAND circuit?

8-3 TTL DATA SHEETS

In 1964, Texas Instruments Corporation introduced the first line of standard TTL ICs. The 54/74 series, as it is called, has been one of the most widely used IC logic families. We will simply refer to it as the 74 series because the major difference between the 54 and 74 versions is that devices in the 54 series can operate over a wider range of temperatures and power-supply voltages. Many semiconductor manufacturers still produce TTL ICs. Fortunately, they all use the same numbering system, so that the basic IC number is the same from one manufacturer to another. Each manufacturer, however, usually attaches its own special prefix to the IC number. For example, Texas Instruments uses the prefix SN, National Semiconductor uses DM, and Signetics uses S. Thus, depending on the manufacturer, you may see a quad NOR gate chip labeled as a DM7402, SN7402, S7402, or some other similar designation. The important part is the number 7402, which is the same for all manufacturers.

As we learned in Chapter 4, there are several series in the TTL family of logic devices (74, 74LS, 74S, etc.). The original standard series and its immediate descendants (74, 74LS, 74S) are no longer recommended by the manufacturers for use in new designs. In spite of this, enough demand in the market keeps them in production. An understanding of the characteristics that define the capabilities and limitations of any logic device is vital. This section will define those characteristics using the advanced low-power Schottky (ALS) series and help you understand a typical data sheet. Later we introduce the other TTL series and compare their characteristics.

We can find all of the information we need on any IC by consulting the manufacturer's published data sheets for that particular IC family. These data sheets can be obtained from data books, CD ROMs, or the IC manufacturer's Internet web site. Figure 8-11 is the manufacturer's data sheet for the 74ALS00 NAND gate IC showing the recommended operating conditions, electrical characteristics, and switching characteristics. Most of the quantities discussed in the following paragraphs in this section can be found on this data sheet. As we discuss each quantity, you should refer to this data sheet to see where the information came from.

Supply Voltage and Temperature Range

Both the 74ALS series and the 54ALS series use a nominal supply voltage (V_{CC}) of 5 V, but can tolerate a supply variation of 4.5 to 5.5 V. The 74ALS series is designed to operate properly in ambient temperatures ranging from 0 to 70°C, while the 54ALS series can handle −55 to +125°C. Because of its greater tolerance of voltage and temperature variations, the 54ALS series is more expensive. It is employed only in applications where reliable operation must be maintained over an extreme range of conditions. Examples are military and space applications.

Voltage Levels

The input and output logic voltage levels for the 74ALS series can be found on the data sheet of Figure 8-11. Table 8-3 presents them in summary form. The minimum and maximum values shown are for worst-case conditions of power supply, temperature, and loading conditions. Inspection of the table reveals a guaranteed maximum logical 0 output $V_{OL} = 0.5$ V, which is 300 mV less than the logical 0 voltage needed at the input $V_{IL} = 0.8$ V. This means that the guaranteed LOW-state dc noise margin is 300 mV. That is,

$$V_{NL} = V_{IL}(\text{max}) - V_{OL}(\text{max}) = 0.8\text{ V} - 0.5\text{ V} = 0.3\text{ V} = 300\text{ mV}$$

Similarly, the logical 1 output V_{OH} is a guaranteed minimum of 2.5 V, which is 500 mV greater than the logical 1 voltage needed at the input, $V_{IH} = 2.0$ V. Thus, the HIGH-state dc noise margin is 500 mV.

$$V_{NH} = V_{OH}(\text{min}) - V_{IH}(\text{min}) = 2.5\text{ V} - 2.0\text{ V} = 0.5\text{ V} = 500\text{ mV}$$

Thus, the *guaranteed worst-case* dc noise margin for the 74ALS series is 300 mV.

recommended operating conditions

		SN54ALS00A			SN74ALS00A			UNIT
		MIN	NOM	MAX	MIN	NOM	MAX	
V_{CC}	Supply voltage	4.5	5	5.5	4.5	5	5.5	V
V_{IH}	High-level input voltage	2			2			V
V_{IL}	Low-level input voltage			0.8‡ 0.7§			0.8	V
I_{OH}	High-level output current			−0.4			−0.4	mA
I_{OL}	Low-level output current			4			8	mA
T_A	Operating free-air temperature	−55		125	0		70	°C

‡ Applies over temperature range −55°C to 70°C
§ Applies over temperature range 70°C to 125°C

electrical characteristics over recommended operating free-air temperature range unless otherwise noted

PARAMETER	TEST CONDITIONS		SN54ALS00A			SN74ALS00A			UNIT
			MIN	TYP†	MAX	MIN	TYP†	MAX	
V_{IK}	V_{CC} = 4.5 V,	I_I = −18 mA			−1.2			−1.5	V
V_{OH}	V_{CC} = 4.5 V to 5.5 V,	I_{OH} = −0.4 mA	V_{CC} −2			V_{CC} −2			V
V_{OL}	V_{CC} = 4.5 V	I_{OL} = 4 mA		0.25	0.4		0.25	0.4	V
		I_{OL} = 8 mA					0.35	0.5	
I_I	V_{CC} = 5.5 V,	V_I = 7 V			0.1			0.1	mA
I_{IH}	V_{CC} = 5.5 V,	V_I = 2.7 V			20			20	μA
I_{IL}	V_{CC} = 5.5 V,	V_I = 0.4 V			−0.1			−0.1	mA
I_O‡	V_{CC} = 5.5 V,	V_O = 2.25 V	−20		−112	−30		−112	mA
I_{CCH}	V_{CC} = 5.5 V,	V_I = 0		0.5	0.85		0.5	0.85	mA
I_{CCL}	V_{CC} = 5.5 V,	V_I = 4.5 V		1.5	3		1.5	3	mA

† All typical values are at V_{CC} = 5 V, T_A = 25°C.
‡ The output conditions have been chosen to produce a current that closely approximates one half of the true short-circuit output current, I_{OS}.

switching characteristics (see Figure 1)

PARAMETER	FROM (INPUT)	TO (OUTPUT)	V_{CC} = 4.5 V to 5.5 V, C_L = 50 pF, R_L = 500 Ω, T_A = MIN to MAX§				UNIT
			SN54ALS00A		SN74ALS00A		
			MIN	MAX	MIN	MAX	
t_{PLH}	A or B	Y	3	15	3	11	ns
t_{PHL}			2	9	2	8	

§ For conditions shown as MIN or MAX, use the appropriate value specified under recommended operating conditions.

FIGURE 8-11 Data sheet for the 74ALS00 NAND gate IC. (Courtesy of Texas Instruments)

TABLE 8-3 74ALS series voltage levels.

	Minimum	Typical	Maximum
V_{OL}	—	0.35	0.5
V_{OH}	2.5	3.4	—
V_{IL}	—	—	0.8
V_{IH}	2.0	—	—

Maximum Voltage Ratings

The voltage values in Table 8-3 *do not include* the absolute maximum ratings beyond which the useful life of the IC may be impaired. The absolute maximum operating conditions are generally given at the top of a data sheet (not shown in Figure 8-11). The voltages applied to any input of this series IC must never exceed +7.0 V. A voltage greater than +7.0 V applied to an input emitter can cause reverse breakdown of the E–B junction of Q_1.

There is also a limit on the maximum *negative* voltage that can be applied to a TTL input. This limit, −0.5 V, is caused by the fact that most TTL circuits employ protective shunt diodes on each input. These diodes were purposely left out of our earlier analysis because they do not enter into the normal circuit operation. They are connected from each input to ground to limit the negative input voltage excursions that often occur when logic signals have excessive ringing. With these diodes, we should not apply more than −0.5 V to an input because the protective diodes would begin to conduct and draw substantial current, probably causing the diode to short out, resulting in a permanently faulty input.

Power Dissipation

An ALS TTL NAND gate draws an average power of 2.4 mW. This is a result of I_{CCH} = 0.85 mA and I_{CCL} = 3 mA, which produces I_{CC}(avg) = 1.93 mA and P_D(avg) = 1.93 mA × 5 V = 9.65 mW. This 9.65 mW is the total power required by all four gates on the chip. Thus, one NAND gate requires an average power of 2.4 mW.

Propagation Delays

The data sheet gives minimum and maximum propagation delays. Assuming the typical value is midway between gives a t_{PLH} = 7 ns and t_{PHL} = 5 ns. The typical *average* propagation delay t_{pd}(avg) = 6 ns.

EXAMPLE 8-2

Refer to the data sheet for the 74ALS00 quad two-input NAND IC in Figure 8-11. Determine the *maximum* average power dissipation and the *maximum* average propagation delay of a *single* gate.

Solution

Look under the electrical characteristics for the *maximum* I_{CCH} and I_{CCL} values. The values are 0.85 mA and 3 mA, respectively. The average I_{CC} is therefore 1.9 mA. The average power is obtained by multiplying by V_{CC}. The data sheet indicates that these I_{CC} values were obtained when V_{CC} was at its maximum value (5.5 V for the 74ALS series). Thus, we have

$$P_D(\text{avg}) = 1.9\text{ mA} \times 5.5\text{ V} = 10.45\text{ mW}$$

as the power drawn by the *complete* IC. We can determine the power drain of one NAND gate by dividing this by 4:

$$P_D(\text{avg}) = 2.6\text{ mW per gate}$$

Because this average power drain was calculated using the maximum current and voltage values, it is the maximum average power that a 74ALS00 NAND gate will draw under worst-case conditions. Designers often use worst-case values to ensure that their circuits will work under all conditions.

The maximum propagation delays for a 74ALS00 NAND gate are listed as

$$t_{PLH} = 11\text{ ns} \qquad t_{PHL} = 8\text{ ns}$$

so that the maximum average propagation delay is

$$t_{pd}(\text{avg}) = \frac{11 + 8}{2} = 9.5\text{ ns}$$

Again, this is a worst-case maximum possible average propagation delay.

8-4 TTL SERIES CHARACTERISTICS

The standard 74 series of TTL has evolved into several other series. All of them offer a wide variety of gates and flip-flops in the small-scale integration (SSI) line, and counters, registers, multiplexers, decoders/encoders, and other logic functions in their medium scale integration (MSI) line. The following TTL series—often called "subfamilies"—provide a wide range of speed and power capabilities.

Standard TTL, 74 Series

The original standard 74 series of TTL logic was described in Section 8-2. These devices are still readily available, but in most cases they are no longer a reasonable choice for new designs because other devices are now available that perform much better at a lower cost.

Schottky TTL, 74S Series

The 7400 series operates using saturated switching in which many of the transistors, when conducting, will be in the saturated condition. This operation causes a storage-time delay, t_S, when the transistors switch from ON to OFF, and it limits the circuit's switching speed.

The 74S series reduces this storage-time delay by not allowing the transistor to go as deeply into saturation. It accomplishes this by using a Schottky barrier diode (SBD) connected between the base and the collector of each transistor, as shown in Figure 8-12(a). The SBD has a forward voltage of only 0.25 V. Thus, when the C–B junction becomes forward-biased at the onset of saturation, the SBD will conduct and divert some of the input current away from the base. This reduces the excess base current and decreases the storage-time delay at turn-off.

As shown in Figure 8-12(a), the transistor/SBD combination is given a special symbol. This symbol is used for all of the transistors in the circuit diagram for the 74S00 NAND gate shown in Figure 8-12(b). This 74S00 NAND gate has an average propagation delay of only 3 ns, which is six times as fast as the 7400. Note the presence of shunt diodes D_1 and D_2 to limit negative input voltages.

Circuits in the 74S series also use smaller resistor values to help improve switching times. This increases the circuit average power dissipation to about 20 mW, about two times greater than the 74 series. The 74S circuits also use a Darlington pair (Q_3 and Q_4) to provide a shorter output rise time when switching from ON to OFF.

Low-Power Schottky TTL, 74LS Series (LS-TTL)

The 74LS series is a lower-powered, slower-speed version of the 74S series. It uses the Schottky-clamped transistor, but with larger resistor values than the 74S series. The larger resistor values reduce the circuit power requirement, but at the expense of an increase in switching times. A NAND gate in the

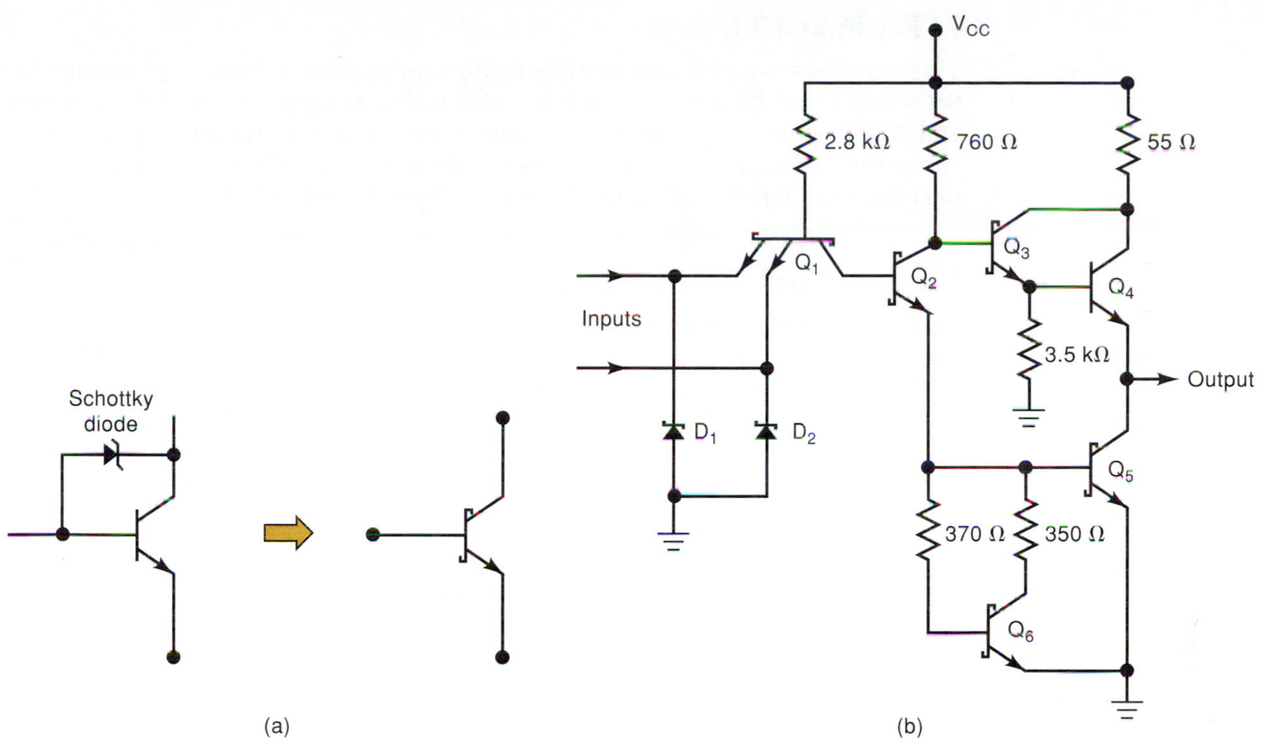

FIGURE 8-12 (a) Schottky-clamped transistor; (b) basic NAND gate in S-TTL series.

74LS series will typically have an average propagation delay of 9.5 ns and an average power dissipation of 2 mW.

Advanced Schottky TTL, 74AS Series (AS-TTL)

Innovations in integrated-circuit design led to the development of two improved TTL series: advanced Schottky (74AS) and advanced low-power Schottky (74ALS). The 74AS series provides a considerable improvement in speed over the 74S series at a much lower power requirement. The comparison is shown in Table 8-4 for a NAND gate in each series. This comparison clearly shows the advantage of the 74AS series. It is the fastest TTL series, and its power dissipation is significantly lower than that of the 74S series. The 74AS has other improvements, including lower input current requirements (I_{IL}, I_{IH}), that result in a greater fan-out than in the 74S series.

Advanced Low-Power Schottky TTL, 74ALS Series

This series offers an improvement over the 74LS series in both speed and power dissipation, as the numbers in Table 8-5 illustrate. The 74ALS series has the lowest gate power dissipation of all the TTL series.

TABLE 8-4

	74S	74AS
Propagation delay	3 ns	1.7 ns
Power dissipation	20 mW	8 mW

TABLE 8-5

	74LS	74ALS
Propagation delay	9.5 ns	4 ns
Power dissipation	2 mW	1.2 mW

74F—Fast TTL

This series uses a new integrated-circuit fabrication technique to reduce interdevice capacitances and thus achieve reduced propagation delays. A typical NAND gate has an average propagation delay of 3 ns and a power consumption of 6 mW. ICs in this series are designated with the letter F in their part number. For instance, the 74F04 is a hex-inverter chip.

Comparison of TTL Series Characteristics

Table 8-6 gives the typical values for some of the more important characteristics of each of the TTL series. All of the performance ratings, except for the maximum clock rate, are for a NAND gate in each series. The maximum clock rate is specified as the maximum frequency that can be used to toggle a J-K flip-flop. This gives a useful measure of the frequency range over which each IC series can be operated.

TABLE 8-6 Typical TTL series characteristics.

	74	74S	74LS	74AS	74ALS	74F
Performance ratings						
Propagation delay (ns)	9	3	9.5	1.7	4	3
Power dissipation (mW)	10	20	2	8	1.2	6
Max. clock rate (MHz)	35	125	45	200	70	100
Fan-out (same series)	10	20	20	40	20	33
Voltage parameters						
V_{OH}(min)	2.4	2.7	2.7	2.5	2.5	2.5
V_{OL}(max)	0.4	0.5	0.5	0.5	0.5	0.5
V_{IH}(min)	2.0	2.0	2.0	2.0	2.0	2.0
V_{IL}(max)	0.8	0.8	0.8	0.8	0.8	0.8

EXAMPLE 8-3

Use Table 8-6 to calculate the dc noise margins for a typical 74LS IC. How does this compare with the standard TTL noise margins?

Solution

74LS

$$V_{NH} = V_{OH}(\text{min}) - V_{IH}(\text{min}) = 2.7\text{ V} - 2.0\text{ V} = 0.7\text{ V}$$

$$V_{NL} = V_{IL}(\text{max}) - V_{OL}(\text{max}) = 0.8\text{ V} - 0.5\text{ V} = 0.3\text{ V}$$

74

$$V_{NH} = 2.4\text{ V} - 2.0\text{ V} = 0.4\text{ V}$$

$$V_{NL} = 0.8\text{ V} - 0.4\text{ V} = 0.4\text{ V}$$

EXAMPLE 8-4

Which TTL series can drive the most device inputs of the same series?

Solution

The 74AS series has the highest fan-out (40), which means that a 74AS00 NAND gate can drive 40 inputs of other 74AS devices. If we want to determine the number of inputs of a *different* TTL series that an output can drive, we will need to know the input and output currents of the two series. This will be dealt with in the next section.

REVIEW QUESTIONS

1. (a) Which TTL series is the best at high frequencies?
   (b) Which TTL series has the largest HIGH-state noise margin?
   (c) Which series has essentially become obsolete in new designs?
   (d) Which series uses a special diode to reduce switching time?
   (e) Which series would be best for a battery-powered circuit operating at 10 MHz?
2. Assuming the same cost for each, why should you choose to use a 74ALS193 counter over a 74LS193 or a 74AS193 in a circuit operating from a 40-MHz clock?
3. Identify the pull-up and pull-down transistors for the 74S circuit in Figure 8-12.

8-5 TTL LOADING AND FAN-OUT

It is important to understand what determines the fan-out or load drive capability of an IC output. Figure 8-13(a) shows a standard TTL output in the LOW state connected to drive several standard TTL inputs. Transistor Q_4 is on and is acting as a current sink for an amount of current I_{OL} that is the sum of the I_{IL} currents from each input. In its ON state, Q_4's collector–emitter resistance is very small, but it is not zero, and so the current I_{OL} will produce a voltage drop V_{OL}. This voltage must not exceed the $V_{OL}(\text{max})$ limit of the IC, which limits the maximum value of I_{OL} and thus the number of loads that can be driven.

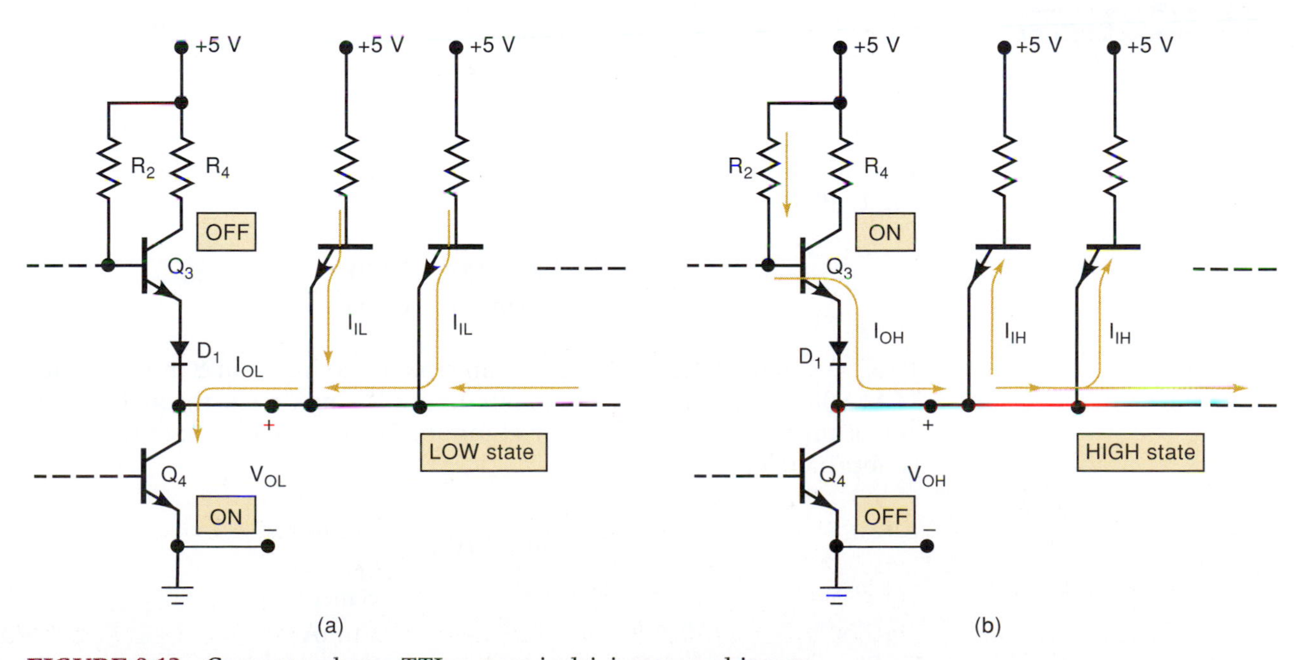

FIGURE 8-13 Currents when a TTL output is driving several inputs.

To illustrate, suppose that the ICs are in the 74 series and each I_{IL} is 1.6 mA. From Table 8-6, we see that the 74 series has $V_{OL}(\text{max}) = 0.4$ V and $V_{IL}(\text{max}) = 0.8$ V. Let's suppose further that Q_4 can sink up to 16 mA before its output voltage reaches $V_{OL}(\text{max}) = 0.4$ V. This means that it can sink the current from up to 16 mA/1.6 mA = 10 loads. If it is connected to more than 10 loads, its I_{OL} will increase and cause V_{OL} to increase above 0.4 V. This is usually undesirable because it reduces the noise margin at the IC inputs [remember, $V_{NL} = V_{IL}(\text{max}) - V_{OL}(\text{max})$]. In fact, if V_{OL} rises above $V_{IL}(\text{max}) = 0.8$ V, it will be in the indeterminate range.

A similar situation occurs in the HIGH state depicted in Figure 8-13(b). Here, Q_3 is acting as an emitter follower that is sourcing (supplying) a total current I_{OH} that is the sum of the I_{IH} currents of the different TTL inputs. If too many loads are being driven, this current I_{OH} will become large enough to cause the voltage drops across R_2, Q_3's emitter–base junction, and D_1 to bring V_{OH} below $V_{OH}(\text{min})$. This too is undesirable because it reduces the HIGH-state noise margin and could even cause V_{OH} to go into the indeterminate range.

What this all means is that a TTL output has a limit, $I_{OL}(\text{max})$, on how much current it can sink in the LOW state. It also has a limit, $I_{OH}(\text{max})$, on how much current it can source in the HIGH state. These output current limits must not be exceeded if the output voltage levels are to be maintained within their specified ranges.

Determining the Fan-Out

To determine how many different inputs an IC output can drive, you need to know the current drive capability of the output [i.e., $I_{OL}(\text{max})$ and $I_{OH}(\text{max})$] and the current requirements of each input (i.e., I_{IL} and I_{IH}). This information is always presented in some form on the manufacturer's IC data sheet. The following examples will illustrate one type of situation.

EXAMPLE 8-5

How many 74ALS00 NAND gate inputs can be driven by a 74ALS00 NAND gate output?

Solution

We will consider the LOW state first as depicted in Figure 8-14. Refer to the 74ALS00 data sheet in Figure 8-11 and find

$$I_{OL}(\text{max}) = 8 \text{ mA}$$
$$I_{IL}(\text{max}) = 0.1 \text{ mA}$$

This says that a 74ALS00 output can sink a maximum of 8 mA and that each 74ALS00 input will source a maximum of 0.1 mA back through the driving gate's output. Thus, the number of inputs that can be driven in the LOW state is obtained as

$$\text{fan-out (LOW)} = \frac{I_{OL}(\text{max})}{I_{IL}(\text{max})} = \frac{8 \text{ mA}}{0.1 \text{ mA}} = 80$$

FIGURE 8-14 Example 8-5.

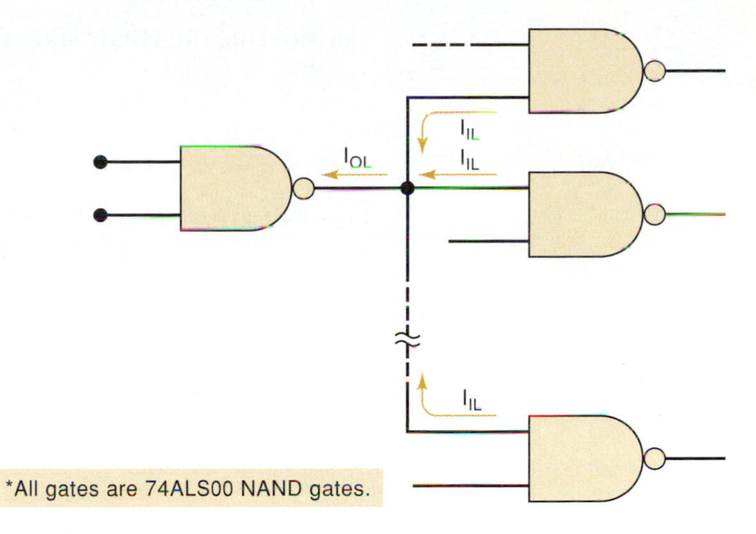

(*Note:* The entry for I_{IL} is actually −0.1 mA. The negative sign is used to indicate that this current flows *out of* the input terminal; we can ignore this sign for our purposes here.) The HIGH state is analyzed in the same manner. Refer to the data sheet to find values for I_{OH} and I_{IH}, ignoring any negative signs.

$$I_{OH}(\text{max}) = 0.4\ \text{mA} = 400\ \mu\text{A}$$
$$I_{IH}(\text{max}) = 20\ \mu\text{A}$$

Thus, the number of inputs that can be driven in the HIGH state is

$$\text{fan-out (HIGH)} = \frac{I_{OH}(\text{max})}{I_{IH}(\text{max})} = \frac{400\ \mu\text{A}}{20\ \mu\text{A}} = 20$$

If fan-out (LOW) and fan-out (HIGH) are not the same, as will sometimes occur, the fan-out is chosen as the smaller of the two. Thus, the 74ALS00 NAND gate can drive up to 20 other 74ALS00 NAND gates.

EXAMPLE 8-6

Refer to the data sheet on the TI CD ROM (or Table 8-7) and determine how many 74AS20 NAND gates can be driven by the output of another 74AS20.

Solution

The 74AS20 data sheet gives the following values:

$$I_{OH}(\text{max}) = 2\ \text{mA}$$
$$I_{OL}(\text{max}) = 20\ \text{mA}$$
$$I_{IH}(\text{max}) = 20\ \mu\text{A}$$
$$I_{IL}(\text{max}) = 0.5\ \text{mA}$$

Considering the HIGH state first, we have

$$\text{fan-out (HIGH)} = \frac{2\text{ mA}}{20\ \mu\text{A}} = 100$$

For the LOW state, we have

$$\text{fan-out (LOW)} = \frac{20\text{ mA}}{0.5\text{ mA}} = 40$$

In this case, the overall fan-out is chosen to be 40 because it is the lower of the two values. Thus, one 74AS20 can drive 40 other 74AS20 inputs.

In older equipment, you will notice that most of the logic ICs were often chosen from the same logic family. In today's digital systems, there is much more likely to be a combination of various logic families. Consequently, loading and fan-out calculations are not as straightforward as they once were. A good method for determining the loading of any digital output is as follows:

Step 1. Add the I_{IH} for all inputs connected to an output. This sum must be less than the output's I_{OH} specification.

Step 2. Add the I_{IL} for all inputs connected to an output. This sum must be less than the output's I_{OL} specification.

Table 8-7 shows the limiting specifications for input and output currents in simple logic gates of the various TTL families. Notice that some of the current values are given as negative numbers. This convention is used to show the direction of current flow. Positive values indicate current flowing into the specified node, whether it is an input or an output. Negative values indicate current flowing out of the specified node. Consequently, all I_{OH} values are negative as current flows out of the output (sourcing current), and all I_{OL} values are positive as load current flows into the output pin on its way to ground (sinking current). Likewise, I_{IH} is positive, while I_{IL} is negative. When calculating loading and fan-out as described above, you should ignore these signs.

TABLE 8-7 Current ratings of TTL series logic gates.*

	Outputs		Inputs	
TTL Series	I_{OH}	I_{OL}	I_{IH}	I_{IL}
74	−0.4 mA	16 mA	40 μA	−1.6 mA
74S	−1 mA	20 mA	50 μA	−2 mA
74LS	−0.4 mA	8 mA	20 μA	−0.4 mA
74AS	−2 mA	20 mA	20 μA	−0.5 mA
74ALS	−0.4 mA	8 mA	20 μA	−0.1 mA
74F	−1 mA	20 mA	20 μA	−0.6 mA

*Some devices may have different input or output current ratings. Always consult the data sheet.

EXAMPLE 8-7

A 74ALS00 NAND gate output is driving three 74S gate inputs and one 7406 input. Determine if there is a loading problem.

Solution

1. Add all of the I_{IH} values:

$$3 \cdot (I_{IH} \text{ for } 74S) + 1 \cdot (I_{IH} \text{ for } 74)$$
$$\text{Total} = 3 \cdot (50\ \mu A) + 1 \cdot (40\ \mu A) = 190\ \mu A$$

The I_{OH} for the 74ALS output is 400 μA (max), which is greater than the sum of the loads (190 μA). This poses no problem when the output is HIGH.

2. Add all of the I_{IL} values:

$$3 \cdot (I_{IL} \text{ for } 74S) + 1 \cdot (I_{IL} \text{ for } 74)$$
$$\text{Total} = 3 \cdot (2\text{ mA}) + 1 \cdot (1.6\text{ mA}) = 7.6\text{ mA}$$

The I_{OH} for the 74ALS output is 8 mA (max), which is greater than the sum of the loads (7.6 mA). This poses no problem when the output is LOW.

EXAMPLE 8-8

The 74ALS00 NAND gate output in Example 8-7 needs to be used to drive some 74ALS inputs in addition to the load inputs described in Example 8-7. How many additional 74ALS inputs could the output drive without being overloaded?

Solution

From the calculations of Example 8-7, only in the LOW state are we close to being overloaded. A 74ALS input has an I_{IL} of 0.1 mA. The maximum sink current (I_{OL}) is 8 mA, and the load current is 7.6 mA (as calculated in Example 8-7). The additional current that the output can sink is found by

$$\text{Additional current} = I_{OLmax} - \text{sum of loads } (I_{IL})$$
$$= 8\text{ mA} - 7.6\text{ mA} = 0.4\text{ mA}$$

This output can drive up to four more 74ALS inputs that have an I_{IL} of 0.1 mA.

EXAMPLE 8-9

The output of a 74AS04 inverter is providing the CLEAR signal to a parallel register made up of 74AS74 D flip-flops. What is the maximum number of FF *CLR* inputs that this gate can drive?

Solution

The input specifications for flip-flop inputs are not always the same as those for a logic gate input in the same family. Refer to the 74AS74 data sheet on the TI CD ROM. The clock and *D* inputs are similar to the gate inputs in Table 8-7. However, the *PRE* and *CLR* inputs have specifications of $I_{IH} = 40\ \mu A$ and $I_{IL} = 1.8$ mA. The 74AS04 has specifications of $I_{OH} = 2$ mA and $I_{OL} = 20$ mA.

$$\text{Maximum number of inputs (HIGH)} = 2\text{ mA}/40\ \mu A = 50$$
$$\text{Maximum number of inputs (LOW)} = 20\text{ mA}/1.8\text{ mA} = 11.11$$

We must limit the fan-out to 11 *CLR* inputs.

REVIEW QUESTIONS

1. What factors determine the $I_{OL}(max)$ rating of a device?
2. How many 7407 inputs can a 74AS chip drive?
3. What can happen if a TTL output is connected to more gate inputs than it is rated to handle?
4. How many 74S112 $\overline{CP}$ inputs can be driven by a 74LS04 output? By a 74F00 output?

8-6 OTHER TTL CHARACTERISTICS

Several other characteristics of TTL logic must be understood if one is to use TTL intelligently in a digital-system application.

Unconnected Inputs (Floating)

Any input to a TTL circuit that is left disconnected (open) acts exactly like a logical 1 applied to that input because in either case the emitter–base junction or diode at the input will not be forward-biased. This means that on *any* TTL IC, *all* of the inputs are 1s if they are not connected to some logic signal or to ground. When an input is left unconnected, it is said to be **floating**.

Unused Inputs

Frequently, not all of the inputs on a TTL IC are being used in a particular application. A common example is when not all the inputs to a logic gate are needed for the required logic function. For example, suppose that we needed the logic operation $\overline{AB}$ and we were using a chip that had a three-input NAND gate. The possible ways of accomplishing this are shown in Figure 8-15.

In Figure 8-15(a), the unused input is left disconnected, which means that it acts as a logical 1. The NAND gate output is therefore $x = \overline{A \cdot B \cdot 1} = \overline{A \cdot B}$, which is the desired result. Although the logic is correct, it is highly undesirable to leave an input disconnected because it will act like an antenna, which is liable to pick up stray radiated signals that could cause the gate to operate improperly. A better technique is shown in Figure 8-15(b). Here, the unused input is connected to +5 V through a 1-kΩ resistor, so that the logic level is a 1. The 1-kΩ resistor is simply for current protection of the emitter–base junctions of the gate inputs in case of spikes on the power-supply line. This same technique can be used for AND gates because a 1 on an unused input will not affect the output. As many as 30 unused inputs can share the same 1-kΩ resistor tied to V_{CC}.

A third possibility is shown in Figure 8-15(c), where the unused input is tied to a used input. This is satisfactory provided that the circuit driving

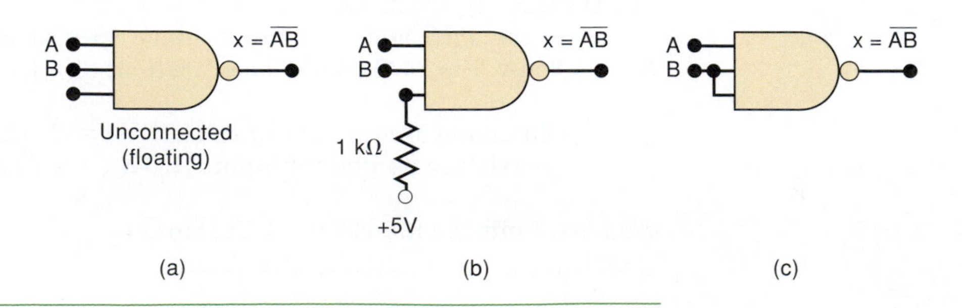

FIGURE 8-15 Three ways to handle unused logic inputs.

input *B* is not going to have its fan-out exceeded. This technique can be used for *any* type of gate. For OR gates and NOR gates, the unused inputs cannot be left disconnected or tied to +5 V because this would produce a constant-output logic level (1 for OR, 0 for NOR) regardless of the other inputs. Instead, for these gates, the unused inputs must either be connected to ground (0 V) for a logic 0 or be tied to a used input, as in Figure 8-15(c).

Tied-Together Inputs

When two (or more) TTL inputs on the same gate are connected together to form a common input, as in Figure 8-15(c), the common input will generally represent a load that is the sum of the load current rating of each individual input. The only exception is for NAND and AND gates. For these gates, the LOW-state input load *will be the same as a single input* no matter how many inputs are tied together.

To illustrate, assume that each input of the three-input NAND gate in Figure 8-15(c) is rated at 0.5 mA for I_{IL} and 20 μA for I_{IH}. The common input *B* will therefore represent an input load of 40 μA in the HIGH state but only 0.5 mA in the LOW state. The same would be true if this were an AND gate. If it were an OR or a NOR gate, the common *B* input would present an input load 40 μA in the HIGH state and 1 mA in the LOW state.

The reason for this characteristic can be found by looking back at the circuit diagram of the TTL NAND gate in Figure 8-8(b). The current I_{IL} is limited by the resistance R_1. Even if inputs *A* and *B* were tied together and grounded, this current would not change; it would merely divide and flow through the parallel paths provided by diodes D_2 and D_3. The situation is different for OR and NOR gates because they do not use multiple-emitter transistors but rather have a separate input transistor for each input, as we saw in Figure 8-10.

EXAMPLE 8-10

Determine the load that the *X* output is driving in Figure 8-16. Assume that each gate is a 74LS series device with I_{IH} = 20 μA and I_{IL} = 0.4 mA.

Solution

The loading on the output of gate 1 is equivalent to six 74LS input loads in the HIGH state but only five 74LS input loads in the LOW state because the NAND gate represents only a single input load in the LOW state.

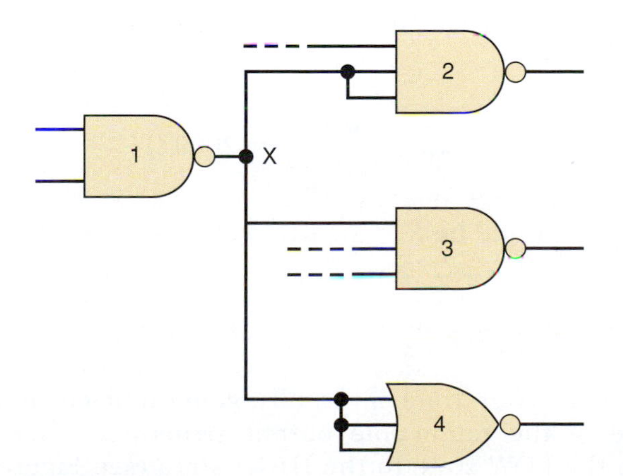

Loading on gate 1 output			
HIGH		LOW	
Load Current	Gate	Load Current	Gate
40 μA	2	0.4 mA	2
20 μA	3	0.4 mA	3
60 μA	4	1.2 mA	4
120 μA	Total	2.0 mA	Total

FIGURE 8-16 Example 8-10.

Biasing TTL Inputs Low

Occasionally, the situation arises where a TTL input must be held normally LOW and then caused to go HIGH by the actuation of a mechanical switch. This situation is illustrated in Figure 8-17 for the input to a one-shot. This OS triggers on a positive transition that occurs when the switch is momentarily closed. The resistor R serves to keep the T input LOW while the switch is open. Care must be taken to keep the value of R low enough so that the voltage developed across it by the current I_{IL} that flows out of the OS input to ground will not exceed V_{IL}(max). Thus, the largest value of R is given by

$$I_{IL} \times R_{max} = V_{IL}(max)$$

$$R_{max} = \frac{V_{IL}(max)}{I_{IL}} \qquad (8\text{-}3)$$

R must be kept below this value to ensure that the OS input will be at an acceptable LOW level while the switch is open. The minimum value of R is determined by the current drain on the 5-V supply when the switch is closed. In practice, this current drain should be minimized by keeping R just slightly below R_{max}.

FIGURE 8-17

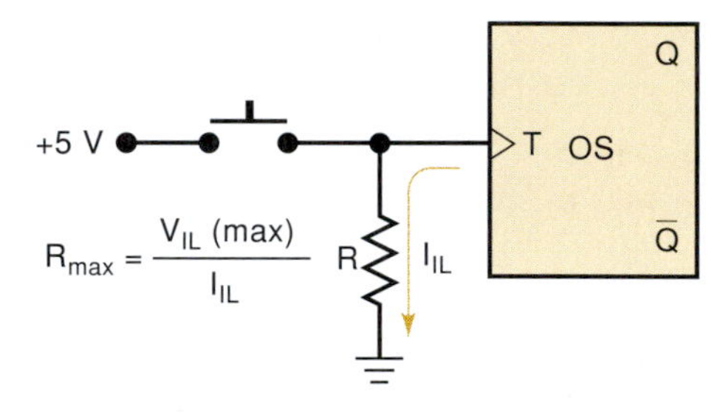

EXAMPLE 8-10

Determine an acceptable value for R if the OS is a 74LS TTL IC with an I_{IL} input rating of 0.4 mA.

Solution

The value of I_{IL} will be a maximum of 0.4 mA. This maximum value should be used to calculate R_{max}. From Table 8-6, $V_{IL}(max) = 0.8$ V for the 74LS series. Thus, we have

$$R_{max} = \frac{0.8 \text{ V}}{0.4 \text{ mA}} = 2000\ \Omega$$

A good choice here would be $R = 1.8$ kΩ, a standard resistor value.

Current Transients

TTL logic circuits suffer from internally generated current transients or spikes because of the totem-pole output structure. When the output is switching from the LOW state to the HIGH state (see Figure 8-18), the two output transistors are changing states: Q_3 OFF to ON, and Q_4 ON to OFF. Because Q_4 is changing from the saturated condition, it will take longer than

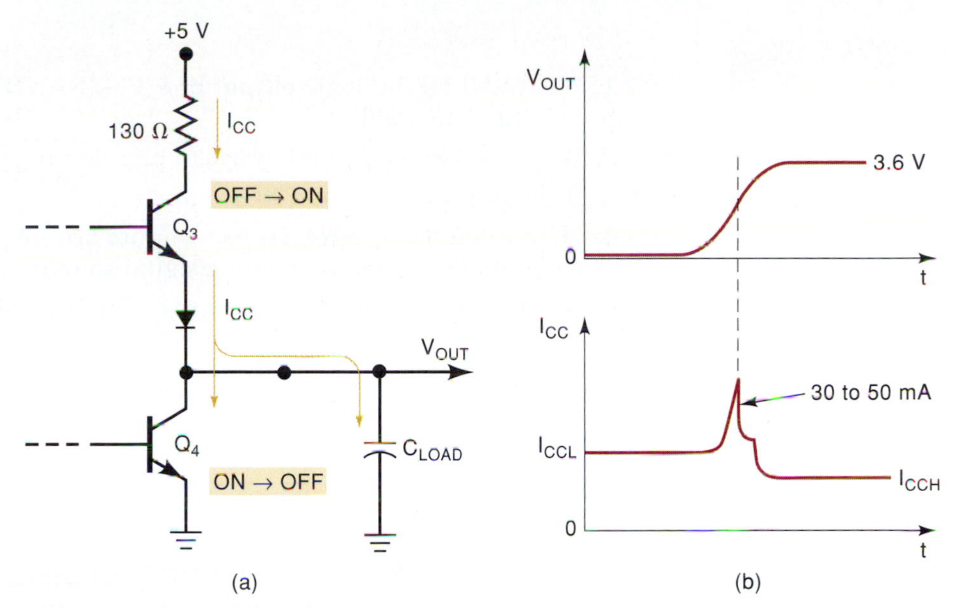

FIGURE 8-18 A large current spike is drawn from V_{CC} when a totem-pole output switches from LOW to HIGH.

Q_3 to switch states. Thus, there is a short interval of time (about 2 ns) during the switching transition when both transistors are conducting and a relatively large surge of current (30 to 50 mA) is drawn from the +5 V supply. The duration of this current transient is extended by the effects of any load capacitance on the circuit output. This capacitance consists of stray wiring capacitance and the input capacitance of any load circuits and must be charged up to the HIGH-state output voltage. This overall effect can be summarized as follows:

Whenever a totem-pole TTL output goes from LOW to HIGH, a high-amplitude current spike is drawn from the V_{CC} supply.

In a complex digital circuit or system, there may be many TTL outputs switching states at the same time, each one drawing a narrow spike of current from the power supply. The accumulative effect of all of these current spikes will be to produce a voltage spike on the common V_{CC} line, mostly due to the distributed inductance on the supply line [remember: $V = L(di/dt)$ for inductance, and di/dt is very large for a 2-ns current spike]. This voltage spike can cause serious malfunctions during switching transitions unless some type of filtering is used. The most common technique uses small radio-frequency capacitors connected from V_{CC} to GROUND essentially to "short out" these high-frequency spikes. This is called **power-supply decoupling**.

It is standard practice to connect a 0.01-μF or 0.1-μF low-inductance, ceramic disk capacitor between V_{CC} and ground near each TTL IC on a circuit board. The capacitor leads are kept very short to minimize series inductance.

In addition, it is standard practice to connect a single large capacitor (2 to 20 μF) between V_{CC} and ground on each board to filter out relatively low-frequency variations in V_{CC} caused by the large changes in I_{CC} levels as outputs switch states.

REVIEW QUESTIONS

1. What will be the logic output of a TTL NAND gate that has all of its inputs unconnected?
2. What are two acceptable ways to handle unused inputs to an AND gate?
3. Repeat question 2 for a NOR gate.
4. *True or false:* When NAND gate inputs are tied together, they are always treated as a single load on the signal source.
5. What is power-supply decoupling? Why is it used?

8-7 MOS TECHNOLOGY

MOS (metal-oxide-semiconductor) technology derives its name from the basic MOS structure of a metal electrode over an oxide insulator over a semiconductor substrate. The transistors of MOS technology are field-effect transistors called **MOSFETs**. This means that the electric *field* on the *metal* electrode side of the *oxide* insulator has an *effect* on the resistance of the substrate. Most of the MOS digital ICs are constructed entirely of MOSFETs and no other components.

The chief advantages of the MOSFET are that it is relatively simple and inexpensive to fabricate, it is small, and it consumes very little power. The fabrication of MOS ICs is approximately one-third as complex as the fabrication of bipolar ICs (TTL, ECL, etc.). In addition, MOS devices occupy much less space on a chip than do bipolar transistors. More important, MOS digital ICs normally do not use the IC resistor elements that take up so much of the chip area of bipolar ICs.

All of this means that MOS ICs can accommodate a much larger number of circuit elements on a single chip than bipolar ICs. This advantage is illustrated by the fact that MOS ICs have dominated bipolar ICs in the area of large-scale integration (LSI, VLSI). The high packing density of MOS ICs makes them especially well suited for complex ICs such as microprocessor and memory chips. Improvements in MOS IC technology have led to devices that are faster than 74, 74LS, and 74ALS TTL with comparable current drive characteristics. Consequently, MOS devices (specifically CMOS) have also become dominant in the SSI and MSI market. The 74AS TTL family is still as fast as the best CMOS devices, but at the price of much greater power dissipation.

The principal disadvantage of MOS devices is their susceptibility to static-electricity damage. Although this can be minimized by proper handling procedures, TTL is still more durable for laboratory experimentation. Consequently, you are likely to see TTL devices used in education as long as they are available.

The MOSFET

There are presently two general types of MOSFETs: *depletion* and *enhancement*. MOS digital ICs use enhancement MOSFETs exclusively, and so only this type will be considered in the following discussion. Furthermore, we will concern ourselves only with the operation of these MOSFETs as on/off switches.

Figure 8-19 shows the schematic symbols for the N-channel and P-channel enhancement MOSFETs, where the direction of the arrow indicates either

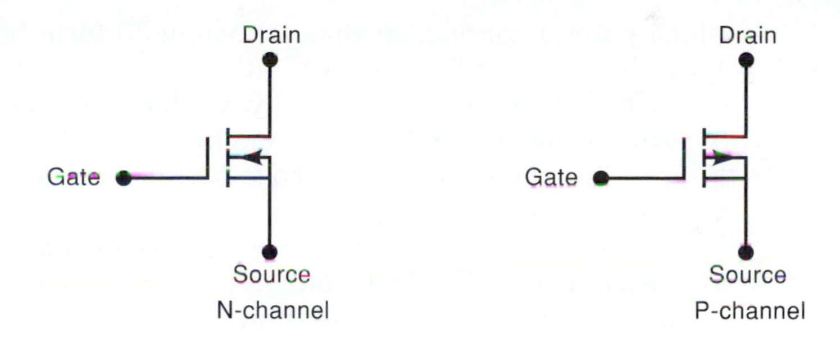

FIGURE 8-19 Schematic symbols for enhancement MOSFETs.

P- or N-channel. The symbols show a broken line between the *source* and the *drain* to indicate that there is *normally* no conducting channel between these electrodes. The symbol also shows a separation between the *gate* and the other terminals to indicate the very high resistance (typically around 10^{12} Ω) of the oxide layer between the gate and the channel, which is formed in the substrate.

Basic MOSFET Switch

Figure 8-20 shows the switching operation of an N-channel MOSFET, the basic element in a family of devices known as **N-MOS**. For the N-channel device, the drain is always biased positive relative to the source. The gate-to-source voltage V_{GS} is the input voltage, which is used to control the resistance between drain and source (i.e., the channel resistance) and therefore determines whether the device is on or off.

When $V_{GS} = 0$ V, there is no conductive channel between source and drain, and the device is off, as shown in Figure 8-20(b). Typically the channel resistance in this OFF state is 10^{10} Ω, which for most purposes is an *open circuit*. The MOSFET will remain off as long as V_{GS} is zero or negative. As V_{GS} is made positive (gate positive relative to source), a threshold voltage (V_T) is reached, at

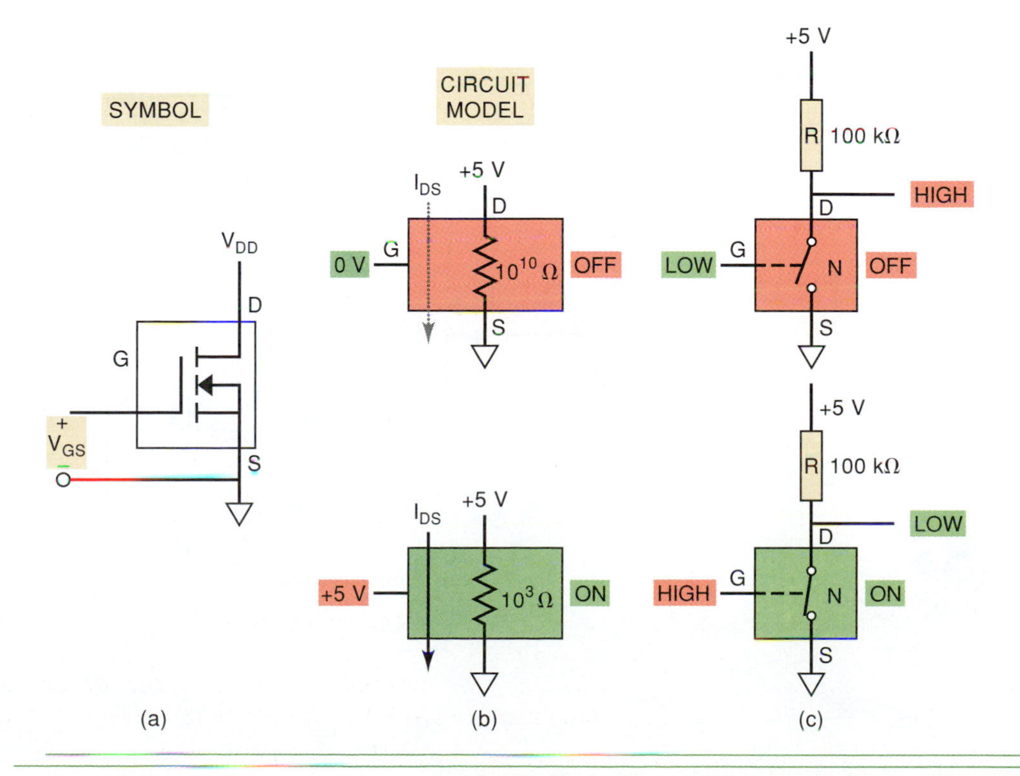

FIGURE 8-20 N-channel MOSFET used as a switch: (a) symbol; (b) circuit model; (c) N-MOS inverter operation.

which point a conductive channel begins to form between source and drain. Typically $V_T = +1.5$ V for an N-MOSFET, and so any $V_{GS} \geq 1.5$ V will cause the MOSFET to conduct. Generally, a value of V_{GS} much larger than V_T is used to turn on the MOSFET more completely. As shown in Figure 8-20(b), when $V_{GS} = +5$ V, the channel resistance between source and drain has dropped to a value of $R_{ON} = 1000\ \Omega$.

In essence, then, the N-MOS will switch from a very high resistance to a low resistance as the gate voltage switches from a LOW voltage to a HIGH voltage. It is helpful simply to think of the MOSFET as a switch that is either opened or closed between source and drain. Figure 8-20(c) shows how an inverter can be formed using one N-MOS transistor as a switch. The first N-MOS logic devices were built using this approach. The drawback to this circuit, as with TTL, is that when the transistor is ON, there will always be current flowing from the supply to ground, producing heat.

The P-channel MOSFET, or P-MOS, shown in Figure 8-21(a) operates in exactly the same manner as the N-channel except that it uses voltages of opposite polarity. For P-MOSFETs, the drain is connected to the lower side of the circuit so that it is biased with a more negative voltage relative to the source. To turn the P-MOSFET ON, a voltage *lower* than the source by V_T must be applied to the gate, meaning the voltage at the gate, relative to the source, must be negative.

Figure 8-21(b) shows that when the gate is at 5 V with respect to ground (the same voltage as applied to the source), the transistor is OFF and has a very high resistance from drain to source. When the gate is at 0 V (relative to ground), then the gate-to-source voltage $V_{GS} = -5$ V and it turns the transistor ON, lowering its resistance from drain to source. The circuit of Figure 8-20(c) shows the switching action of an inverter using P-MOS logic.

Table 8-8 summarizes the P- and N-channel switching characteristics.

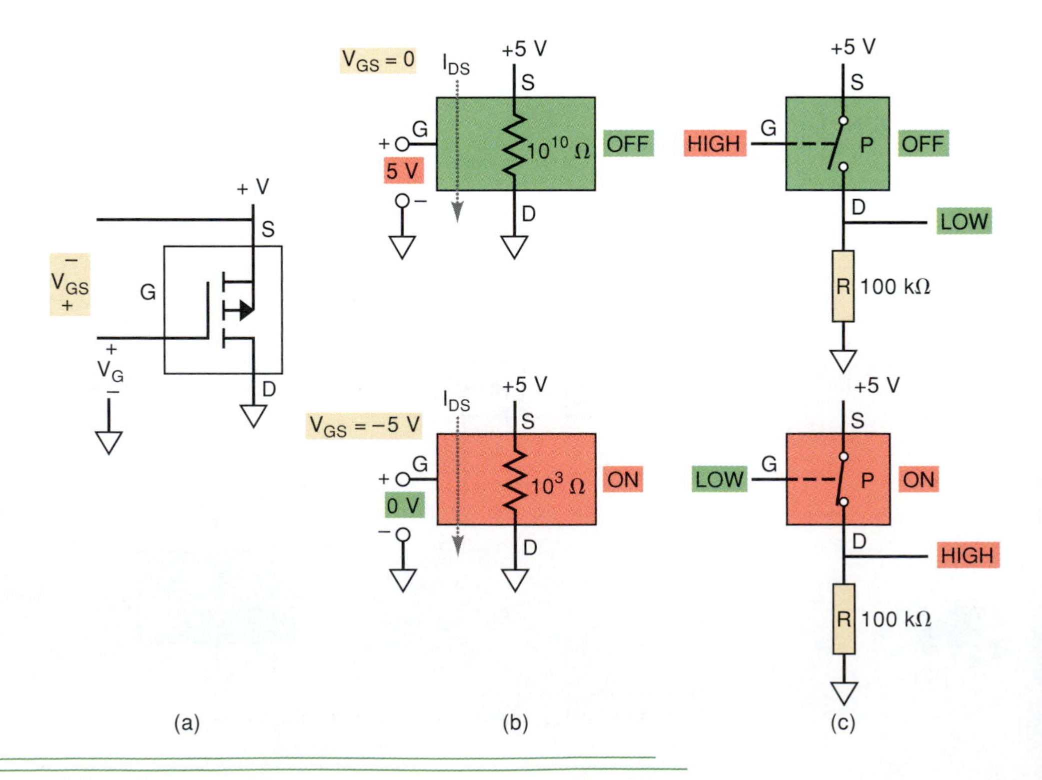

FIGURE 8-21 P-channel MOSFET used as a switch: (a) symbol; (b) circuit model for OFF and ON; (c) P-MOS inverter circuit.

TABLE 8-8

	Drain-to-Source Bias	Gate-to-Source Voltage (V_{GS}) Needed for Conduction	R_{ON} (Ω)	R_{OFF} (Ω)
P-channel	Negative	Typically more negative than −1.5 V	1000 (typical)	10^{10}
N-channel	Positive	Typically more positive than +1.5 V	1000 (typical)	10^{10}

8-8 COMPLEMENTARY MOS LOGIC

P-MOS and N-MOS logic circuits use fewer components and are much simpler to manufacture than TTL circuits. As a result, they began to dominate the LSI and VLSI markets in the 1970s and 1980s. During this era, a new technology began to emerge that used both P-MOS transistors (as high-side switches) and N-MOS transistors (as low-side switches) in the same logic circuit. This is referred to as complementary MOS, or **CMOS**, technology. CMOS logic circuits are not quite as simple and easy to manufacture as P-MOS or N-MOS, but they are faster, use much less power, and are the dominant technology in the market today.

CMOS Inverter

The circuitry for the basic CMOS INVERTER is shown in Figure 8-22. For this diagram and those that follow, the standard symbols for the MOSFETs have been replaced by blocks labeled P and N to denote a P-MOS and an N-MOS, respectively. This is done simply for convenience in analyzing the circuits. The CMOS INVERTER has two MOSFETs in series so that the P-channel device has its source connected to $+V_{DD}$ (a positive voltage), and the N-channel device has its source connected to ground.* The gates of the two devices are connected together as a common input. The drains of the two devices are connected together as the common output.

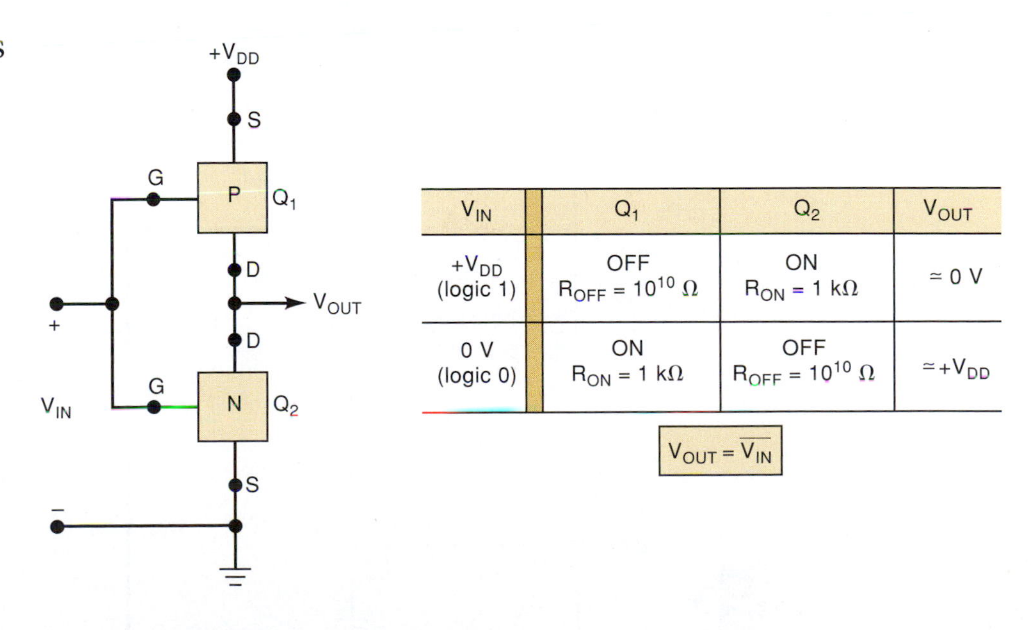

V_{IN}	Q_1	Q_2	V_{OUT}
$+V_{DD}$ (logic 1)	OFF $R_{OFF} = 10^{10}$ Ω	ON $R_{ON} = 1$ kΩ	≈ 0 V
0 V (logic 0)	ON $R_{ON} = 1$ kΩ	OFF $R_{OFF} = 10^{10}$ Ω	≈ $+V_{DD}$

$V_{OUT} = \overline{V_{IN}}$

FIGURE 8-22 Basic CMOS INVERTER.

*Most manufacturers label this terminal V_{SS}.

The logic levels for CMOS are essentially $+V_{DD}$ for logical 1 and 0 V for logical 0. Consider, first, the case where $V_{IN} = +V_{DD}$. In this situation, the gate of Q_1 (P-channel) is at 0 V relative to the source of Q_1. Thus, Q_1 will be in the OFF state with $R_{OFF} \approx 10^{10}\ \Omega$. The gate of Q_2 (N-channel) will be at $+V_{DD}$ relative to its source. Thus, Q_2 will be on with typically $R_{ON} = 1\ \mathrm{k}\Omega$. The voltage divider between Q_1's R_{OFF} and Q_2's R_{ON} will produce $V_{OUT} \approx 0$ V.

Next, consider the case where $V_{IN} = 0$ V. Q_1 now has its gate at a negative potential relative to its source, while Q_2 has $V_{GS} = 0$ V. Thus, Q_1 will be on with $R_{ON} = 1\ \mathrm{k}\Omega$, and Q_2 will be off with $R_{OFF} = 10^{10}\ \Omega$, producing a V_{OUT} of approximately $+V_{DD}$. These two operating states are summarized in the table on Figure 8-22, showing that the circuit does act as a logic INVERTER.

CMOS NAND Gate

Other logic functions can be constructed by modifying the basic INVERTER. Figure 8-23 shows a NAND gate formed by adding a parallel P-channel MOSFET and a series N-channel MOSFET to the basic INVERTER. To analyze this circuit, it helps to realize that a 0-V input turns on its corresponding P-MOS and turns off its corresponding N-MOS, and vice versa, for a $+V_{DD}$ input. Thus, you can see that the only time a LOW output will occur is when inputs A and B are both HIGH ($+V_{DD}$) to turn on both N-MOSFETs, thereby providing a low resistance from the output terminal to ground. For all other input conditions, at least one P-MOS will be on while at least one N-MOS will be off. This produces a HIGH output.

CMOS NOR Gate

A CMOS NOR gate is formed by adding a series P-MOS and a parallel N-MOS to the basic INVERTER, as shown in Figure 8-24. Once again, this circuit can be analyzed by realizing that a LOW at any input turns on its corresponding

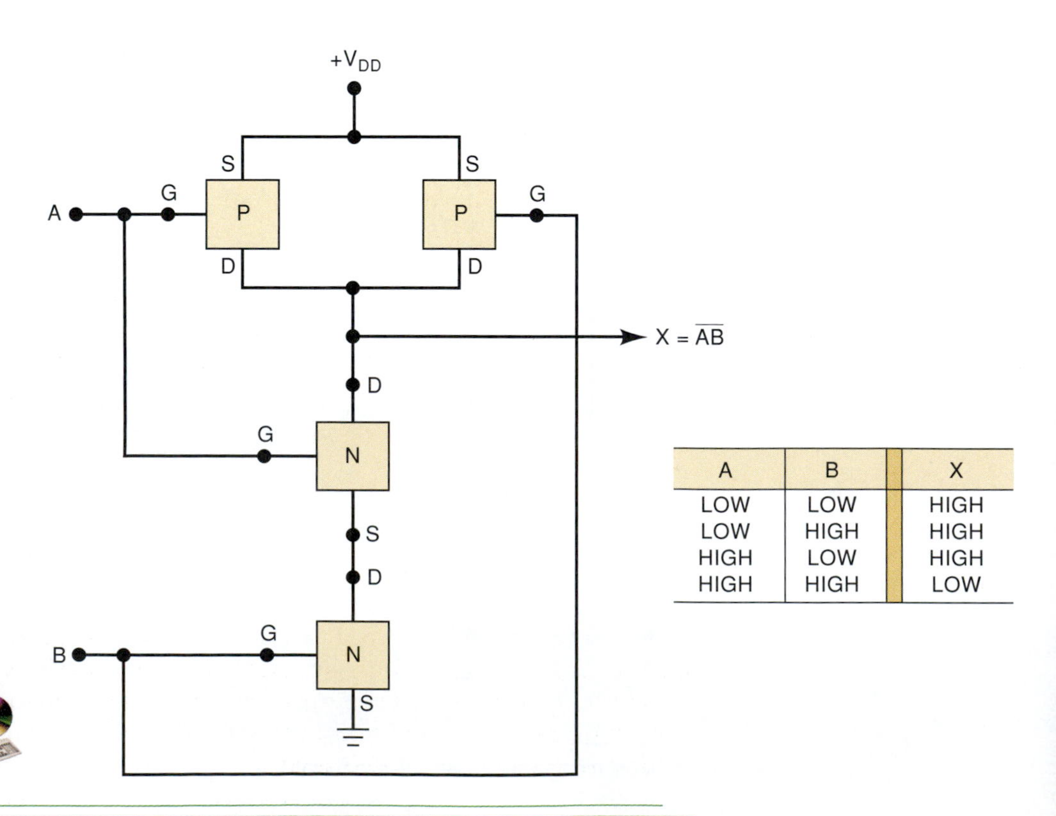

A	B	X
LOW	LOW	HIGH
LOW	HIGH	HIGH
HIGH	LOW	HIGH
HIGH	HIGH	LOW

FIGURE 8-23 CMOS NAND gate.

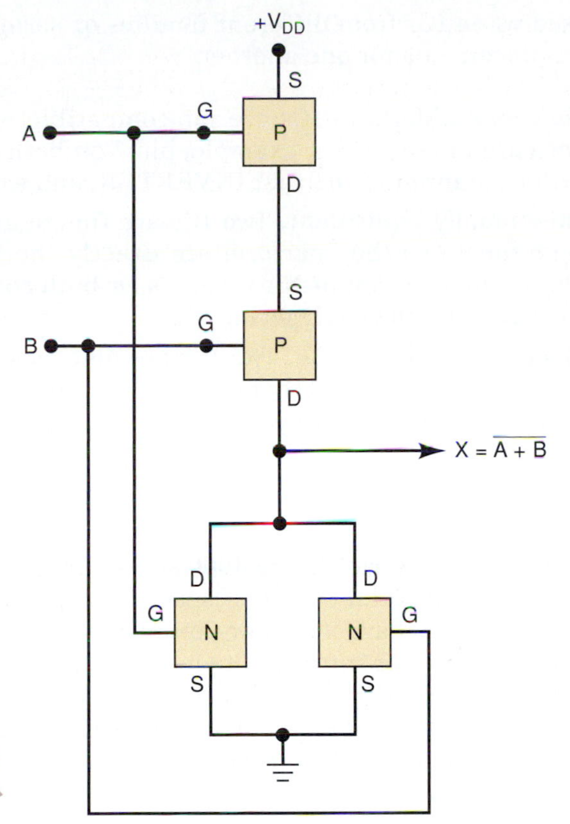

FIGURE 8-24 CMOS NOR gate.

A	B	X
LOW	LOW	HIGH
LOW	HIGH	LOW
HIGH	LOW	LOW
HIGH	HIGH	LOW

P-MOS and turns off its corresponding N-MOS, and vice versa, for a HIGH input. It is left to the reader to verify that this circuit operates as a NOR gate.

CMOS AND and OR gates can be formed by combining NANDs and NORs with INVERTERs.

CMOS SET-RESET FF

Two CMOS NOR gates or NAND gates can be cross-coupled to form a simple SET-RESET latch. Additional gating circuitry is used to convert the basic SET-RESET latch to clocked D and J-K flip-flops.

REVIEW QUESTIONS

1. How does CMOS internal circuitry differ from N-MOS?
2. How many P-channel MOSFETs are in a CMOS INVERTER?
3. How many MOSFETs are in a three-input CMOS NAND gate?

8-9 CMOS SERIES CHARACTERISTICS

CMOS ICs provide not only all of the same logic functions that are available in TTL but also several special-purpose functions not provided by TTL. Several different CMOS series have been developed over time because manufacturers have sought to improve performance characteristics. Before we look at the various CMOS series, it will be helpful to define a few terms that are used when ICs from different families or series are to be used together or as replacements for one another.

- **Pin-compatible.** Two ICs are pin-compatible when their pin configurations are the same. For example, pin 7 on both ICs is GROUND, pin 1 on both is an input to the first INVERTER, and so on.
- **Functionally equivalent.** Two ICs are functionally equivalent when the logic functions they perform are exactly the same. For example, both contain four two-input NAND gates, or both contain six D flip-flops with positive-edge clock triggering.
- **Electrically compatible.** Two ICs are electrically compatible when they can be connected directly to each other without taking any special measures to ensure proper operation.

4000/14000 Series

The oldest CMOS series is the 4000 series first introduced by RCA, and its functionally equivalent 14000 series from Motorola. Devices in the 4000/14000 series have very low power dissipation and can operate over a wide range of power-supply voltages (3 to 15 V). They are very slow compared to TTL and other CMOS series and have very low output current capabilities. They are not pin-compatible or electrically compatible with any TTL series. The 4000/14000 series devices are rarely used in new designs except when a special-purpose IC is available that is not available in other series.

74HC/HCT (High-Speed CMOS)

The 74HC series has a 10-fold increase in switching speed, comparable to that of the 74LS devices, and a much higher output current capability than the first 7400 CMOS series ICs. 74HC/HCT ICs are pin-compatible with and functionally equivalent to TTL ICs with the same device number. 74HCT devices are electrically compatible with TTL, but 74HC devices are not. This means, for example, that a 74HCT04 hex-INVERTER chip can replace a 74LS04 chip, and vice versa. It also means that a 74HCT IC can be connected directly to any TTL IC.

74AC/ACT (Advanced CMOS)

This series is often referred to as ACL for advanced CMOS logic. The series is functionally equivalent to the various TTL series but is *not* pin-compatible with TTL because the pin placements on 74AC or 74ACT chips have been chosen to improve noise immunity so that the device inputs are less sensitive to signal changes occurring on other IC pins. 74AC devices are not electrically compatible with TTL; 74ACT devices can be connected directly to TTL. ACL offers advantages over the HC series in noise immunity, propagation delay, and maximum clock speed.

Device numbering for this series differs slightly from TTL, 74C, and 74HC/HCT numbering. It uses a five-digit device number beginning with the digits 11. The following examples illustrate:

74AC11 004 ≡ 74HC 04
74ACT11 293 ≡ 74HCT 293

74AHC/AHCT (Advanced High-Speed CMOS)

This series of CMOS devices offers a natural migration path from the HC series to faster, lower-power, low-drive applications. The devices in this series are three times faster and can be used as direct replacements for HC series devices. They offer similar noise immunity to HC without the overshoot/undershoot problems often associated with higher drive characteristics required for comparable speed.

BiCMOS 5-V Logic

Several IC manufacturers have developed logic series that combine the best features of bipolar and CMOS logic—called BiCMOS logic. The low-power characteristics of CMOS and the high-speed characteristics of bipolar circuits are integrated to produce an extremely low-power, high-speed logic family. BiCMOS ICs are not available in most SSI and MSI functions, but are limited to functions that are used in microprocessor and bus interfacing applications such as latches, buffers, drivers, and transceivers. The 74BCT (BiCMOS bus-interface technology) series offers 75 percent reduction in power consumption over the 74F family while maintaining similar speed and drive characteristics. Parts in this series are pin-compatible with industry standard TTL parts and operate on standard 5-V logic levels. The 74ABT (advanced BiCMOS technology) series is the second generation of BiCMOS bus-interface devices. Details of bus interface logic will be presented Section 8-13.

Power-Supply Voltage

The 4000/14000 series and 74C series devices operate with V_{DD} values ranging from 3 to 15 V, which makes them very versatile. They can be used in low-voltage battery-operated circuits, in standard 5-V circuits, and in circuits where a higher supply voltage is used to attain the noise margins required for operation in a high-noise environment. The 74HC/HCT, 74AC/ACT, and 74AHC/AHCT series operate over a much narrower range of supply voltages, typically between 2 and 6 V.

Logic series that are designed to operate at lower voltages (e.g., 2.5 or 3.3 V) are also available. Whenever devices that use different power supply voltages are interconnected in the same digital system, special measures must be taken. The low-voltage devices and the special interfacing techniques will be covered in Section 8-10.

Logic Voltage Levels

The input and output voltage levels will be different for the different CMOS series. Table 8-9 lists these voltage values for the various CMOS series as well as those for the TTL series. The values listed in the table assume that all devices are operating from a supply voltage of 5 V and that all device outputs are driving inputs of the same logic family.

Examination of this table discloses some important points. First, note that V_{OL} for the CMOS devices is very close to 0 V, and V_{OH} is very close to 5 V. The reason why is that the CMOS outputs do not have to source or sink any significant amount of current when they are driving CMOS inputs with their extremely high input resistance ($10^{12}\ \Omega$). Also note that, except for 74HCT and 74ACT, the required input voltage levels are greater for CMOS than for TTL. Recall that 74HCT and 74ACT are designed to be electrically

TABLE 8-9 Input/output voltage levels (in volts) with $V_{DD} = V_{CC} = +5\,\text{V}$.

	CMOS							TTL			
Parameter	**4000B**	**74HC**	**74HCT**	**74AC**	**74ACT**	**74AHC**	**74AHCT**	**74**	**74LS**	**74AS**	**74ALS**
V_{IH}(min)	3.5	3.5	2.0	3.5	2.0	3.85	2.0	2.0	2.0	2.0	2.0
V_{IL}(max)	1.5	1.0	0.8	1.5	0.8	1.65	0.8	0.8	0.8	0.8	0.8
V_{OH}(min)	4.95	4.9	4.9	4.9	4.9	4.4	3.15	2.4	2.7	2.7	2.5
V_{OL}(max)	0.05	0.1	0.1	0.1	0.1	0.44	0.1	0.4	0.5	0.5	0.5
V_{NH}	1.45	1.4	2.9	1.4	2.9	0.55	1.15	0.4	0.7	0.7	0.7
V_{NL}	1.45	0.9	0.7	1.4	0.7	1.21	0.7	0.4	0.3	0.3	0.4

compatible with TTL, so they must be able to accept the same input voltage levels as TTL.

Noise Margins

The noise margins for each series are also given in Table 8-9. They are calculated using

$$V_{NH} = V_{OH}(\text{min}) - V_{IH}(\text{min})$$
$$V_{NL} = V_{IL}(\text{max}) - V_{OL}(\text{max})$$

Note that, in general, the CMOS devices have greater noise margins than TTL. The difference would be even greater if the CMOS devices were operated at a supply voltage greater than 5 V.

Power Dissipation

When a CMOS logic circuit is in a static state (not changing), its power dissipation is extremely low. We can see the reason by examining each of the circuits shown in Figures 8-22 to 8-24. Note that, regardless of the state of the output, there is always a very high resistance between the V_{DD} terminal and ground because there is always an off MOSFET in the current path. This results in a typical CMOS dc power dissipation of only 2.5 nW per gate when $V_{DD} = 5\,\text{V}$; even at $V_{DD} = 10\,\text{V}$, this power increases to only 10 nW. With these values for P_D, it is easy to see why CMOS is ideally suited for applications using battery power or battery backup power.

P_D Increases with Frequency

The power dissipation of a CMOS IC will be very low as long as it is in a dc condition. Unfortunately, P_D will increase in proportion to the frequency at which the circuits are switching states. For example, a CMOS NAND gate that has $P_D = 10\,\text{nW}$ under dc conditions will have $P_D = 0.1\,\text{mW}$ at a frequency of 100 kpps, and 1 mW at 1 MHz. The reason for this dependence on frequency is illustrated in Figure 8-25.

Each time a CMOS output switches from LOW to HIGH, a transient charging current must be supplied to the load capacitance. This capacitance consists of the combined input capacitances of any loads being driven and the device's own output capacitance. These narrow spikes of current are

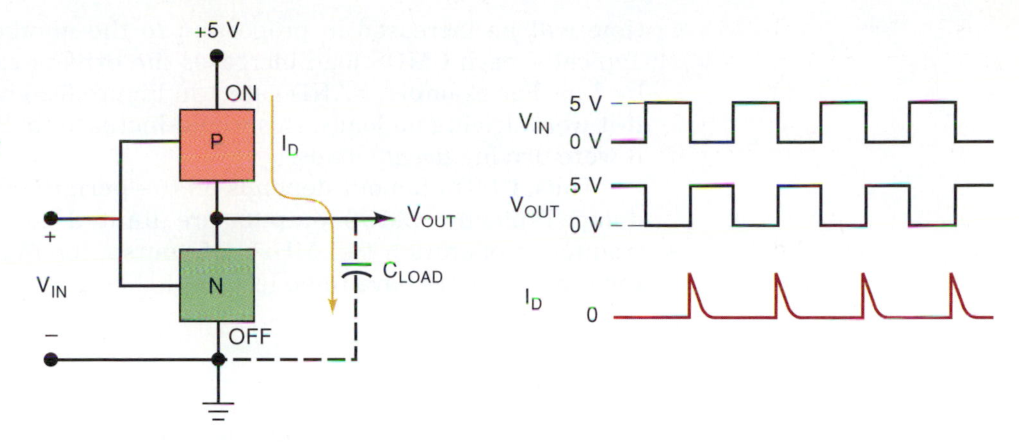

FIGURE 8-25 Current spikes are drawn from the V_{DD} supply each time the output switches from LOW to HIGH. This is due mainly to the charging current of the load capacitance.

supplied by V_{DD} and can have a typical amplitude of 5 mA and a duration of 20 to 30 ns. Clearly, as the switching frequency increases, there will be more of these current spikes occurring per second, and the average current drawn from V_{DD} will increase. Even with very low capacitive loads, there is a brief point in the transition from LOW to HIGH or HIGH to LOW when the two output transistors are partially turned on. This effectively lowers the resistance from the supply to ground, causing a current spike as well.

Thus, at higher frequencies, CMOS begins to lose some of its advantage over other logic families. As a general rule, a CMOS gate will have the same average P_D as a 74LS gate at frequencies near 2 to 3 MHz. Above these frequencies, TTL power also increases with frequency because of the current required to reverse the charge on the load capacitance. For MSI chips, the situation is somewhat more complex than stated here, and a logic designer must do a detailed analysis to determine whether or not CMOS has a power-dissipation advantage at a particular frequency of operation.

Fan-Out

Like N-MOS and P-MOS, CMOS inputs have an extremely large resistance ($10^{12}\ \Omega$) that draws essentially no current from the signal source. Each CMOS input, however, typically presents a 5-pF load to ground. This input capacitance limits the number of CMOS inputs that one CMOS output can drive (see Figure 8-26). The CMOS output must charge and discharge the parallel combination of all of the input capacitances, so that the output switching

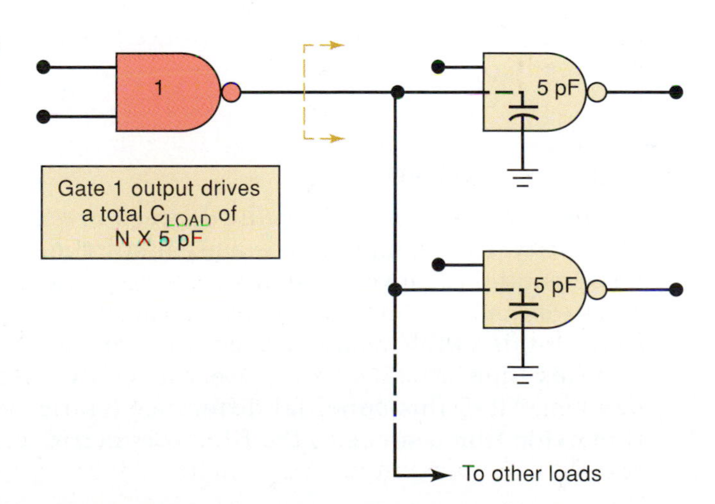

FIGURE 8-26 Each CMOS input adds to the total load capacitance seen by the driving gate's output.

time will be increased in proportion to the number of loads being driven. Typically, each CMOS load increases the driving circuit's propagation delay by 3 ns. For example, NAND gate 1 in Figure 8-26 might have a t_{PLH} of 25 ns if it were driving no loads; this would increase to 25 ns + 20(3 ns) = 85 ns if it were driving *twenty* loads.

Thus, CMOS fan-out depends on the permissible maximum propagation delay. Typically, CMOS outputs are limited to a fan-out of 50 for low-frequency operation ($\leq$1 MHz). Of course, for higher-frequency operation, the fan-out would have to be less.

Switching Speed

Although CMOS, like N-MOS and P-MOS, must drive relatively large load capacitances, its switching speed is somewhat faster because of its low output resistance in each state. An N-MOS output must charge the load capacitance through a relatively large (100-kΩ) resistance. In the CMOS circuit, the output resistance in the HIGH state is the R_{ON} of the P-MOSFET, which is typically 1 kΩ or less. This allows more rapid charging of load capacitance.

A 4000 series NAND gate will typically have an average t_{pd} of 50 ns at V_{DD} = 5 V, and 25 ns at V_{DD} = 10 V. The reason for the improvement in t_{pd} as V_{DD} is increased is that the R_{ON} on the MOSFETs decreases significantly at higher supply voltages. Thus, it appears that V_{DD} should be made as large as possible for operation at higher frequencies. However, the larger V_{DD} will result in increased power dissipation.

A typical NAND gate in the 74HC or 74HCT series has an average t_{pd} of around 8 ns when operated at V_{DD} = 5 V. A 74AC/ACT NAND gate has an average t_{pd} of around 4.7 ns. A 74AHC NAND gate has an average t_{pd} of around 4.3 ns.

Unused Inputs

CMOS inputs should never be left disconnected. All CMOS inputs must be tied either to a fixed voltage level (0 V or V_{DD}) or to another input.

This rule applies even to the inputs of extra unused logic gates on a chip. An unconnected CMOS input is susceptible to noise and static charges that could easily bias both the P-channel and the N-channel MOSFETs in the conductive state, resulting in increased power dissipation and possible overheating.

Static Sensitivity

All electronic devices, to varying degrees, are sensitive to damage by static electricity. The human body is a great storehouse of electrostatic charges. For example, when you walk across a carpet, a static charge of over 30,000 V can be built up on your body. If you then touch an electronic device, some of this sizable charge can be transferred to the device. The MOS logic families (and all MOSFETs) are especially susceptible to static-charge damage. All of this potential difference (static charge) applied across the thin oxide film overcomes the film's dielectric insulation capability. When it breaks down, the resulting flow of current (discharge) is like a lightning strike, blowing a hole in the oxide layer and permanently damaging the device.

Electrostatic discharge (ESD) is responsible for billions of dollars of damage to electronic equipment annually, and equipment manufacturers have devoted considerable attention to developing special handling procedures for all electronic devices and circuits. Even though most modern ICs have on-chip resistor–diode networks to protect inputs and outputs from the effects of ESD, the following precautions are used by most engineering labs, production facilities, and field service departments:

1. Connect the chassis of all test instruments, soldering-iron tips, and your workbench (if metal) to earth ground (i.e., the round prong in the 120-VAC plug). This prevents the buildup of static charge on these devices that could be transferred to any circuit board or IC that they come in contact with.
2. Connect yourself to earth ground with a special wrist strap. This will allow potentially dangerous charges from your body to be discharged to ground. The wrist strap contains a 1-MΩ resistor that limits current to a nonlethal value should you accidentally touch a “live” voltage while working with the equipment.
3. Keep ICs (especially MOS) in conductive foam or aluminum foil. This will keep all IC pins shorted together so that no dangerous voltages can be developed between any two pins.
4. Avoid touching IC pins, and insert the IC into the circuit immediately after removing it from the protective carrier.
5. Place shorting straps across the edge connectors of PC boards when the boards are being carried or transported. Avoid touching the edge connectors. Store PC boards in conductive plastic or metallic envelopes.
6. Do not leave any unused IC inputs unconnected because open inputs tend to pick up stray static charges.

Latch-Up

Because of the unavoidable existence of *parasitic* (unwanted) PNP and NPN transistors embedded in the substrate of CMOS ICs, a condition known as **latch-up** can occur under certain circumstances. If these parasitic transistors on a CMOS chip are triggered into conduction, they will latch-up (stay ON permanently), and a large current may flow and destroy the IC. Most modern CMOS ICs are designed with protection circuitry that helps prevent latch-up, but it can still occur when the device's maximum voltage ratings are exceeded. Latch-up can be triggered by high-voltage spikes or ringing at the device inputs and outputs. Clamping diodes can be connected externally to protect against such transients, especially when the ICs are used in industrial environments where high-voltage and/or high-current load switching takes place (motor controllers, relays, etc.). A well-regulated power supply will minimize spikes on the V_{DD} line; if the supply also has current limiting, it will limit current should latch-up occur. Modern CMOS fabrication techniques have greatly reduced ICs' susceptibility to latch-up.

REVIEW QUESTIONS

1. Which CMOS series is pin-compatible with TTL?
2. Which CMOS series is electrically compatible with TTL?
3. Which CMOS series is functionally equivalent to TTL?
4. What logic family combines the best features of CMOS and bipolar logic?
5. What factors determine CMOS fan-out?
6. What precautions should be taken when handling CMOS ICs?
7. Which IC family (CMOS, TTL) is best suited for battery-powered applications?
8. *True or false:*
   (a) CMOS power drain increases with operating frequency.
   (b) Unused CMOS inputs can be left unconnected.
   (c) TTL is better suited than CMOS for operation in high-noise environments.
   (d) CMOS switching speed increases with operating frequency.
   (e) CMOS switching speed increases with supply voltage.
   (f) The latch-up condition is an advantage of CMOS over TTL.

8-10 LOW-VOLTAGE TECHNOLOGY

IC manufacturers are continually looking for ways to put semiconductor devices (diodes, resistors, transistors, etc.) closer together on a chip, that is, to increase the chip density. This higher density has at least two major benefits. First, it allows more circuits to be packed onto the chip; second, with the circuits closer together, the time for signals to propagate from one circuit to another will decrease, thereby improving overall circuit operating speed. There are also drawbacks to higher chip density. When circuits are placed closer together, the insulating material that isolates one circuit from another is narrower. This decreases the amount of voltage that the device can withstand before dielectric breakdown occurs. Increasing the chip density increases the overall chip power dissipation, which can raise the chip temperature above the maximum level allowed for reliable operation.

These drawbacks can be neutralized by operating the chip at lower voltage levels, thereby reducing power dissipation. Several series of logic on the market operate on 3.3 V. The newer series are optimized to run on 2.5 V. This low-voltage technology may very well signal the beginning of a gradual transition in the digital equipment field that will eventually find all digital ICs operating from a new low-voltage standard.

Low-voltage devices are currently designed for applications ranging from electronic games to engineering workstations. The newer CPUs are 2.5-V devices, and 3.3-V dynamic RAM chips are used in memory modules for personal computers.

Several low-voltage logic series are currently available. It is not possible to cover all of the families and series from all manufacturers, so we will describe those currently offered by Texas Instruments.

CMOS Family

- The *74LVC (Low-Voltage CMOS)* series contains the widest assortment of the familiar SSI gates and MSI functions of the 5-V families, along with many bus-interface devices such as buffers, latches, drivers, and so on. This series can handle 5-V logic levels on its inputs, so it can convert from 5-V systems to 3-V systems. As long as the current drive is kept low enough to keep the output voltage within acceptable limits, the 74LVC can also drive 5-V TTL inputs. The V_{IH} input requirements of 5-V CMOS parts such as the 74HC/AHC do not allow LVC devices to drive them.

- The *74ALVC (Advanced Low-Voltage CMOS)* series currently offers the highest performance. The devices in this series are intended primarily for bus-interface applications that use 3.3-V logic only.
- The *74LV (Low-Voltage)* series offers CMOS technology and many of the common SSI gates and MSI logic functions, along with some popular octal buffers, latches, and flip-flops. It is intended to operate only with other 3.3-V devices.
- The *74AVC (Advanced Very-Low-Voltage CMOS)* series has been introduced with tomorrow's systems in mind. It is optimized for 2.5-V systems, but it can operate on supplies as low as 1.2 V or as high as 3.3 V. This broad range of supply voltage makes it useful in mixed-voltage systems. It has propagation delays of less than 2 ns, which rivals 74AS bipolar devices. It has many of the bus interface features of the BiCMOS series that will make it useful in future generations of low-voltage workstations, PCs, networks, and telecommunications equipment.
- The *74AUC (Advanced Ultra-Low-Voltage CMOS)* series is optimized to operate at 1.8-V logic levels.
- The *74AUP* (Advanced Ultra-low Power) series is the industries lowest-power logic series and is used in battery-operated portable applications.
- The *74CBT (Cross Bar Technology)* series offers high-speed bus-interface circuits that can switch quickly when enabled and not load the bus when they are disabled.
- The *74CBTLV (Cross Bar Technology Low Voltage)* is the 3.3 V complement to the 74CBT series.
- The *74GTLP* (Gunning Transceiver Logic Plus) series is made for high-speed parallel backplane applications. This series will be covered in a later section.
- The *74SSTV (Stub Series Terminated Logic)* is useful in the high-speed advanced-memory systems of today's computers.
- The *TS Switch* (TI Signal Switch) series is made for mixed-signal applications and offers some analog and digital switching and multiplexing solutions.
- The *74TVC (Translation Voltage Clamp)* series is used to protect the inputs and outputs of sensitive devices from voltage overshoot on the bus lines.

BiCMOS Family

- The *74LVT (Low-Voltage BiCMOS Technology)* contains BiCMOS parts that are intended for 8- and 16-bit bus-interface applications. As with the LVC series, the inputs can handle 5-V logic levels and serve as a 5-V to 3-V translator. Because the output levels [V_{OH}(min) and V_{OL}(max)] are equivalent to TTL levels, they are fully electrically compatible with TTL. Table 8-10 compares the various features.
- The *74ALVT (Advanced Low-Voltage BiCMOS Technology)* series is an improvement over the LVT series. It offers 3.3-V or 2.5-V operation at 3 ns and is pin-compatible with existing ABT and LVT series. It is also intended for bus-interface applications.

TABLE 8-10 Low-voltage series characteristics.

	LV	ALVC	AVC	ALVT	ALB
V_{CC} (recommended)	2.7–3.6	2.3–3.6	1.65–3.6	2.3–2.7	3–3.6
T_{PD} (ns)	18	3	1.9	3.5	2
V_{IH} (V)	2 to V_{CC} + 0.5	2.0 to 4.6	1.2 to 4.6	2 to 7	2.2 to 4.6
V_{IL} (V)	0.8	0.8	0.7	0.8	0.6
I_{OH} (mA)	6	12	8	32	25
I_{OL} (mA)	6	12	8	32	25

- The *74ALB (Advanced Low-Voltage BiCMOS)* series is designed for 3.3-V bus-interface applications. It provides 25 mA output drive and propagation delays of only 2.2 ns.
- The *74VME* (*VERSA Module Eurocard*) series is designed to operate with the standard VME bus technology.

Digital technicians and engineers can no longer assume that every IC in a digital circuit, system, or piece of equipment is operating at 5 V, and they must be prepared to deal with the necessary interfacing considerations in mixed-voltage systems. The interfacing skills you learn in this chapter will allow you to accomplish this, regardless of what develops as low-voltage systems become more common.

The continued development of low-voltage technology promises to bring about a complete revolution from the original 5-V system, to mixed-voltage systems, and finally to pure 3.3-V, 2.5-V, or even lower-voltage digital systems. To put all of this in perspective, Figure 8-27 shows Texas Instruments' perception of the life cycle of the various logic families.

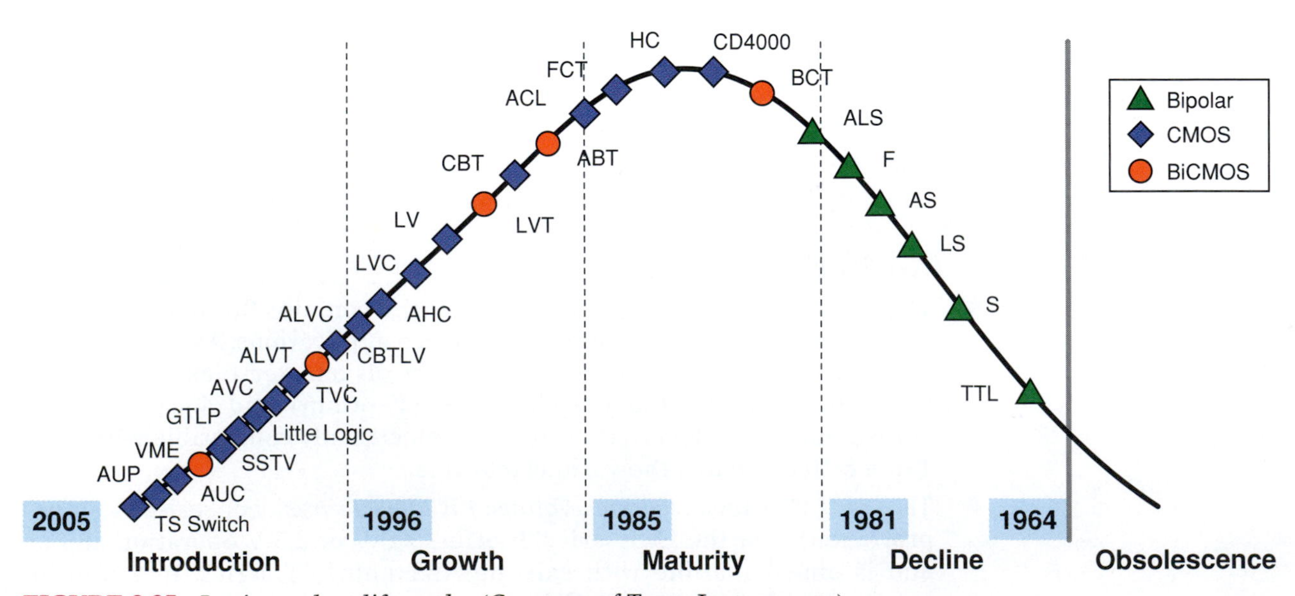

FIGURE 8-27 Logic product life cycle. (Courtesy of Texas Instruments)

REVIEW QUESTIONS

1. What are the two advantages of higher-density ICs?
2. What are the drawbacks?
3. What is the minimum HIGH voltage at a 74LVT input?
4. Which low-voltage series can work only with other low-voltage series ICs?
5. Which low-voltage series is fully electrically compatible with TTL?

8-11 OPEN-COLLECTOR/OPEN-DRAIN OUTPUTS

Several digital devices must sometimes share the use of a single wire in order to transmit a signal to some destination device, very much like several neighbors sharing the same street. This means that several devices must have their outputs connected to the same wire, which essentially connects them all to each other. For all of the logic devices we have considered so far, this presents a problem. Each output has two states, HIGH and LOW. When one output is HIGH while the other is LOW and when they are connected together, we have a HIGH/LOW conflict. Which one will win? Just like arm wrestling, the stronger of the two wins. In this case, the transistor circuit whose output transistor has the lowest "ON" resistance will pull the output voltage in its direction.

Figure 8-28 shows a generic block diagram of two logic devices with their outputs connected to a common wire. If the two logic devices were CMOS, then the ON resistance of the pull-up circuit that outputs the HIGH would be approximately the same as the ON resistance of the pull-down circuit that outputs the LOW. The voltage on the common wire will be about half the supply voltage. This voltage is in the indeterminate range for most CMOS series and is unacceptable for driving a CMOS input. Furthermore, the current through the two conducting MOSFETs will be much greater than normal, especially at higher values of V_{DD}, and it can damage the ICs.

Conventional CMOS outputs should never be connected together.

If the two devices were TTL totem-pole outputs, as shown in Figure 8-29, a similar situation would occur but with different results because of the

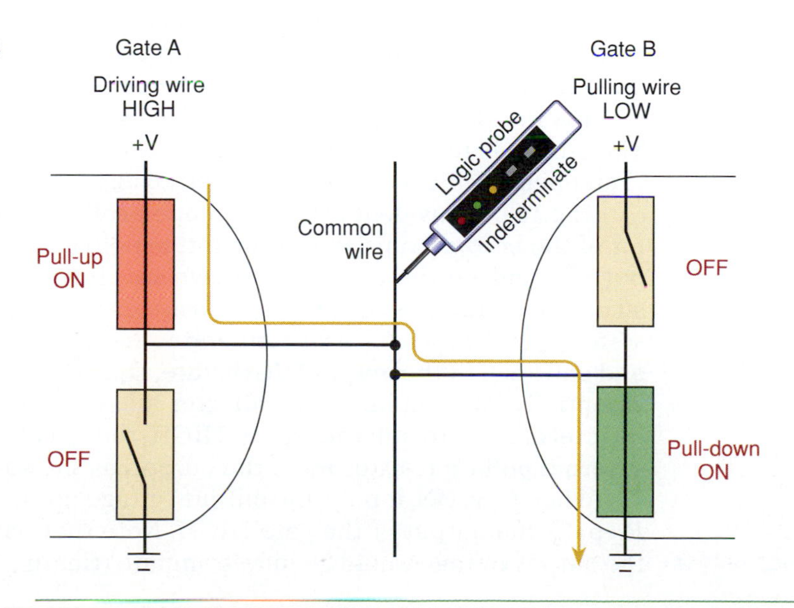

FIGURE 8-28 Two outputs contending for control of a wire.

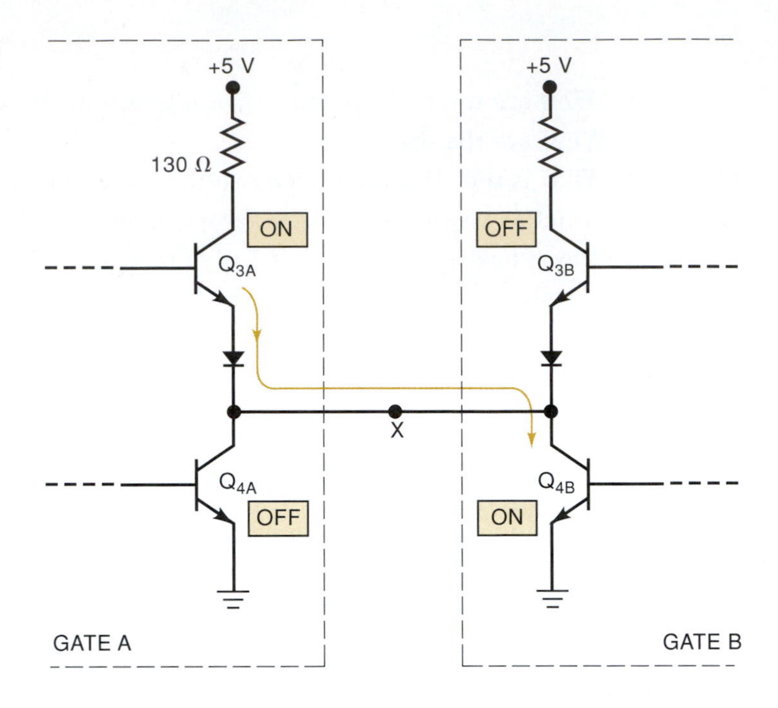

FIGURE 8-29 Totem-pole outputs tied together can produce harmful current through Q_4.

difference in output circuitry. Suppose that gate A output is in the HIGH state (Q_{3A} ON, Q_{4A} OFF) and the gate B output is in the LOW state (Q_{3B} OFF, Q_{4B} ON). In this situation, Q_{4B} is a very low resistance load on Q_{3A} and will draw a current that is far greater than it is rated to handle. This current might not damage Q_{3A} or Q_{4B} immediately, but over a period of time it can cause overheating and deterioration in performance and eventual device failure.

Another problem caused by this relatively high current flowing through Q_{4B} is that it will produce a larger voltage drop across the transistor collector emitter, making V_{OL} of between 0.5 and 1 V. This is greater than the allowable V_{OL} (max). For these reasons:

TTL totem-pole outputs should never be tied together.

Open-Collector/Open-Drain Outputs

One solution to the problem of sharing a common wire among gates is to remove the active pull-up transistor from each gate's output circuit. In this way, none of the gates will ever try to assert a logic HIGH. TTL outputs that have been modified in this way are called **open-collector outputs**. CMOS output circuits that have been modified in this way are called open-drain outputs. The output is taken at the drain of the N-channel pull-down MOSFET, which is an open circuit (i.e., not connected to any other circuitry).

The TTL equivalent is called an open-collector output because the collector of the bottom transistor in the totem pole is connected directly to the output pin and nowhere else, as shown in Figure 8-30(a). The open-collector structure eliminates the pull-up transistors Q_3, D_1, and R_4. In the output LOW state, Q_4 is ON (has base current and is essentially a short between collector and emitter); in the output HIGH state, Q_4 is OFF (has no base current and is essentially an open between collector and emitter). Because this circuit has no internal way to pull the output HIGH, the circuit designer must connect an external pull-up resistor R_P to the output, as shown in Figure 8-30(b).

When Q_4 is ON, it pulls the output voltage down to a LOW. When Q_4 is OFF, R_P pulls the output of the gate HIGH. Note that without the pull-up resistor, the output voltage would be indeterminate (floating). The value of the resistor

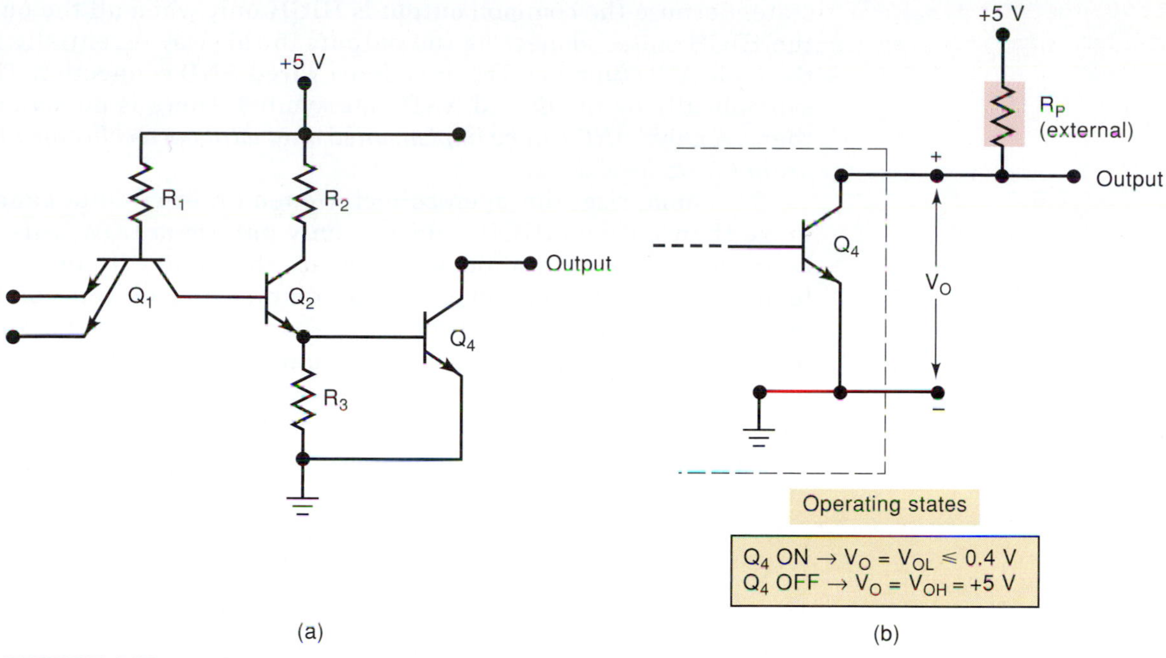

FIGURE 8-30 (a) Open-collector TTL circuit; (b) with external pull-up resistor.

R_P is usually chosen to be 10 kΩ. This value is small enough so that, in the HIGH state, the voltage dropped across it due to load current will not lower the output voltage below the minimum V_{OH}. It is large enough so that, in the LOW state, it will limit the current through Q_4 to a value below I_{OL}(max).

When several open-collector or open-drain gates share a common connection, as shown in Figure 8-31, the common wire is HIGH by default due to the pull-up resistor. When any one (or more) of the gate outputs pulls it LOW, the 5 V are dropped across R_P and the common connection is in the LOW

FIGURE 8-31 Wired-AND operation using open-collector gates.

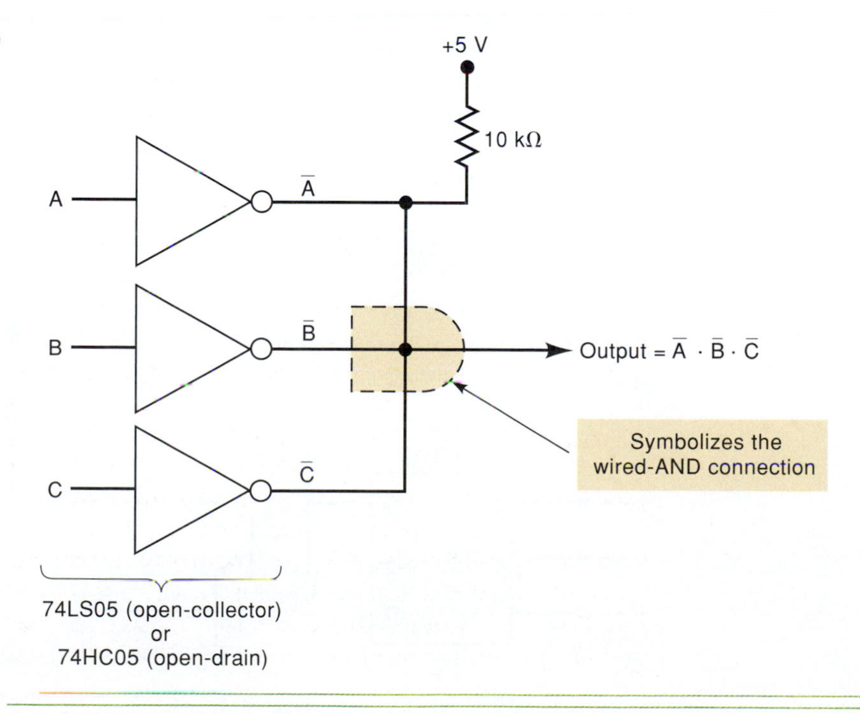

state. Because the common output is HIGH only when all the outputs are in the HIGH state, connecting the outputs in this way essentially implements the logic AND function. This is called a **wired-AND** connection. This is shown symbolically by the dotted AND gate symbol. There is no actual AND gate there. *A wired-AND can be implemented only with open-collector TTL and open-drain CMOS logic devices.*

To summarize, the open-collector/open-drain circuits cannot actively make their outputs HIGH; they can only pull them LOW. This feature can be used to allow several devices to share the same wire for transmitting a logic level to another device or to combine the outputs of the devices effectively in a logic AND function. As we mentioned before, the purpose of the active pull-up transistor in the output circuit of conventional gates is to charge up the load capacitance rapidly and allow for fast switching. Open-collector and open-drain devices have a much slower switching speed from LOW to HIGH and consequently are not used in high-speed applications.

Open-Collector/Open-Drain Buffer/Drivers

The applications of open-collector/drain outputs that we have described were more prevalent in the early days of logic circuits than they are today. A more common use of these circuits now is as a **buffer/driver**. A buffer or a driver is a logic circuit that is designed to have a greater output current and/or voltage capability than an ordinary logic circuit. They allow a weaker output circuit to drive a heavy load. Open-collector/drain circuits offer some unique flexibility as buffer/drivers.

Due to their high I_{OL} and V_{OH} specifications, the 7406 and 7407 are the only standard TTL devices that are still being recommended for new designs. The 7406 is an open-collector buffer/driver IC that contains six INVERTERs with open-collector outputs that can sink up to 40 mA in the LOW state. In addition, the 7406 can handle output voltages up to 30 V in the HIGH state. This means that the output can be connected to a load that operates on a voltage greater than 5 V. This is illustrated in Figure 8-32, where a 7406 is used as a buffer between a 74LS112 flip-flop and an incandescent indicator lamp that is rated at 24 V, 25 mA. The 7406 controls the lamp's ON/OFF status to indicate the state of FF output Q. Note that the lamp is powered from +24 V, and it acts as the pull-up resistor for the open-collector output.

When $Q = 1$, the 7406 output goes LOW, its output transistor sinks the 25 mA of lamp current supplied by the 24-V source, and the lamp is on. When

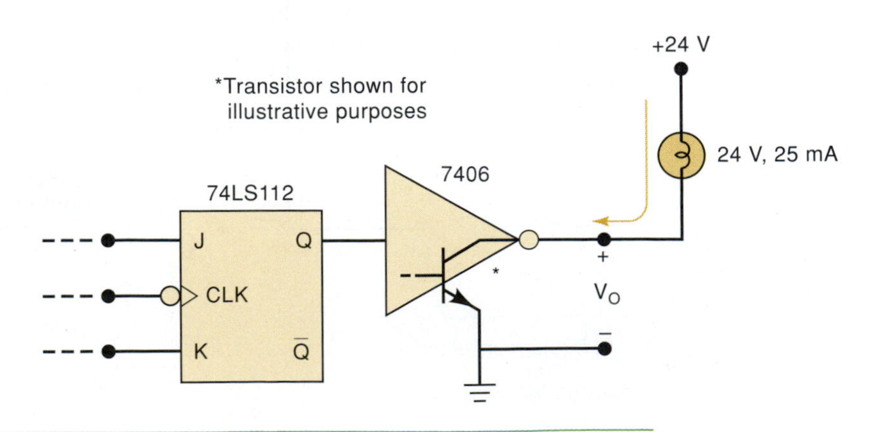

FIGURE 8-32 An open-collector buffer/driver drives a high-current, high-voltage load.

$Q = 0$, the 7406 output transistor turns off; there is no path for current, and the lamp turns off. In this state, the full 24 V will appear across the OFF output transistor so that $V_{OH} = 24$ V, which is lower than the 7406 maximum V_{OH} rating.

Open-collector outputs are often used to drive indicator LEDs, as shown in Figure 8-33(a). The resistor is used to limit the current to a safe value. When the INVERTER output is LOW, its output transistor provides a low-resistance path to ground for the LED current, so that the LED is on. When the INVERTER output is HIGH, its output transistor is off, and there is no path for LED current; in this state, the LED is off.

The 7407 is an open-collector, noninverting buffer with the same voltage and current ratings as a 7406.

The 74HC05 is an open-drain hex inverter with 25 mA current sink capability. Figure 8-33(b) shows a way to interface a 74AHC74 D-FF to a control relay. A control relay is an electromagnetic switch. The contacts close magnetically when the rated current flows through the coil. The 74HC05 can handle the relay's relatively high voltage and current so that the 74AHC74 output can turn the relay on and off.

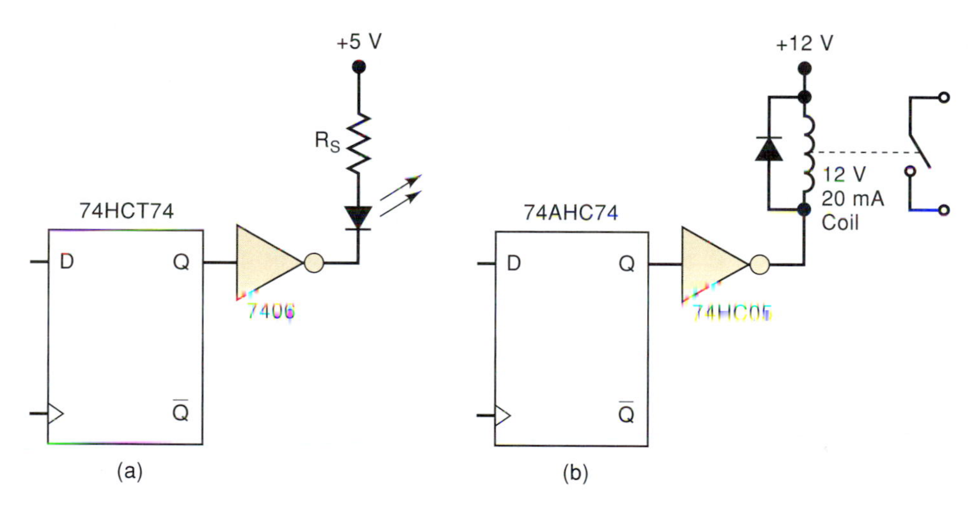

FIGURE 8-33 (a) An open-collector output can be used to drive an LED indicator; (b) an open-drain CMOS output.

IEEE/ANSI Symbol for Open-Collector/Drain Outputs

The new IEEE/ANSI symbology uses a distinctive notation to identify open-collector/drain outputs. Figure 8-34 shows the standard IEEE/ANSI designation for an open-collector/drain output. It is an underlined diamond. Although we will not normally use the complete IEEE/ANSI symbology in this book, we will use this underlined diamond to indicate open-collector and open-drain outputs.

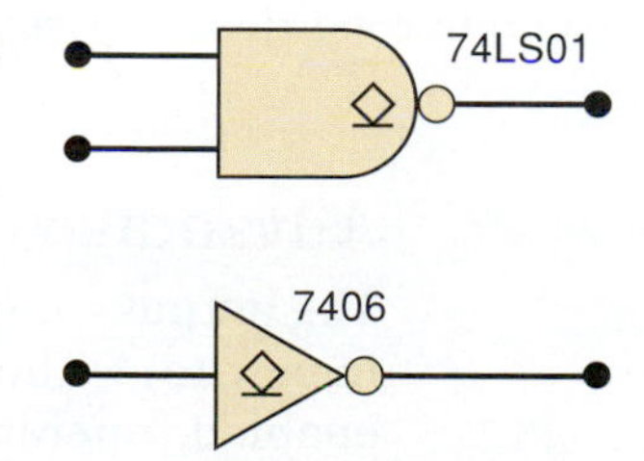

FIGURE 8-34 IEEE/ANSI notation for open-collector and open-drain outputs.

REVIEW QUESTIONS

1. When does a HIGH/LOW conflict occur?
2. Why shouldn't totem-pole outputs be tied together?
3. How do open-collector outputs differ from totem-pole outputs?
4. Why do open-collector outputs need a pull-up resistor?
5. What is the logic expression for the wired-AND connection of six 7406 outputs?
6. Why are open-collector outputs generally slower than totem-pole outputs?
7. What is the IEEE/ANSI symbol for open-collector outputs?

8-12 TRISTATE (THREE-STATE) LOGIC OUTPUTS

The **tristate** configuration is a third type of output circuitry used in TTL and CMOS families. It takes advantage of the high-speed operation of the pull-up/pull-down output arrangement, while allowing outputs to be connected together to share a common wire. It is called tristate because it allows three possible output states: HIGH, LOW, and high-impedance (Hi-Z). The Hi-Z state is a condition in which both the pull-up and the pull-down transistors are turned OFF so that the output terminal is a high impedance to both ground and the power supply +V. Figure 8-35 illustrates these three states for a simple inverter circuit.

Devices with tristate outputs have an *enable* input. It is often labeled *E* for enable or *OE* for output enable. When $OE = 1$, as shown in Figures 8-35(a) and (b), the circuit operates as a normal INVERTER because the HIGH logic level at *OE* enables the output. The output will be either HIGH or LOW, depending on the input level. When $OE = 0$, as shown in Figure 8-35(c), the circuit's output is *disabled*. It goes into its Hi-Z state with both transistors in the nonconducting state. In this state, the output terminal is essentially an open circuit (not connected to anything).

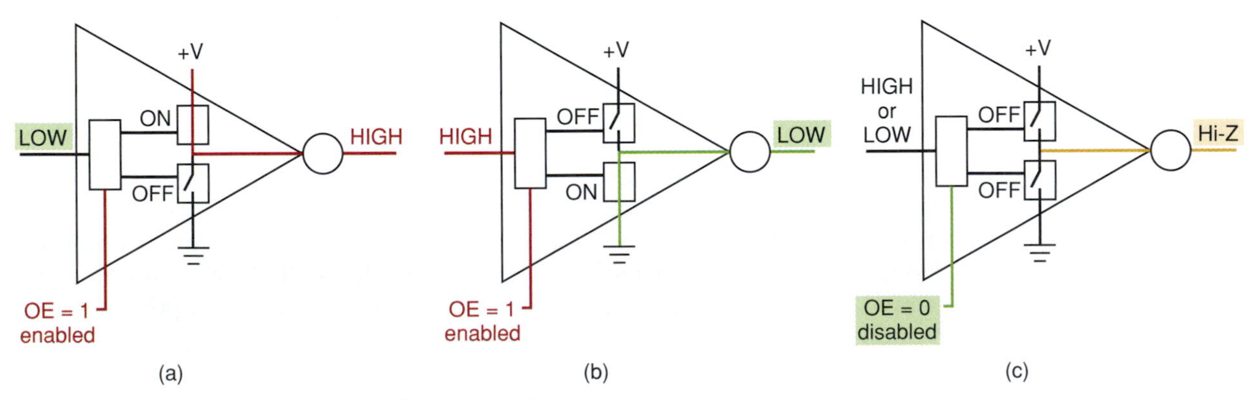

FIGURE 8-35 Three output conditions of tristate.

Advantage of Tristate

The outputs of tristate ICs can be connected together (share the use of a common wire) without sacrificing switching speed because a tristate output, when enabled, operates as a totem pole for TTL or an active pull-up/pull-down

CMOS output with its associated low-impedance, high-speed characteristics. It is important to realize, however, that when tristate outputs are connected together, only one of them should be enabled at one time. Otherwise, two active outputs could fight for control of the common wire, as we discussed earlier, causing damaging currents to flow and producing invalid logic levels.

In our discussion of open-collector/open-drain and tristate circuits, we have referred to cases when the outputs of several devices must share a single wire to transmit information to another device. The shared wire is referred to as a bus wire. An entire bus is made up of several wires that are used to carry digital information between two or more devices that share the use of the bus.

Tristate Buffers

A *tristate buffer* is a circuit used to control the passage of a logic signal from input to output. Some tristate buffers also invert the signal as it goes through. The circuits in Figure 8-35 can be called *inverting tristate buffers.*

Two commonly used tristate buffer ICs are the 74LS125 and the 74LS126. Both contain four *noninverting* tristate buffers like those shown in Figure 8-36. The 74LS125 and 74LS126 differ only in the active state of their ENABLE inputs. The 74LS125 allows the input signal A to reach the output when $\overline{E} = 0$, while the 74LS126 passes the input when $E = 1$.

Tristate buffers have many applications in circuits where several signals are connected to common lines (buses). We will examine some of these applications in Chapter 9, but we can get the basic idea from Figure 8-37(a). Here, we have three logic signals A, B, and C connected to a common bus line through 74AHC126 tristate buffers. This arrangement permits us to transmit any one of these signals over the bus line to other circuits by enabling the appropriate buffer.

For example, consider the situation in Figure 8-37(b), where $E_B = 1$ and $E_A = E_C = 0$. This disables the upper and lower buffers so that their outputs are in the Hi-Z state and are essentially disconnected from the bus. This is symbolized by the X's on the diagram. The middle buffer is enabled so that its input, B, is passed through to its output and onto the bus, from which it is routed to other circuits connected to the bus. When tristate outputs are connected together as in Figure 8-37, it is important to remember that no more than one output should be enabled at one time. Otherwise, two or more active totem-pole outputs would be connected, which could

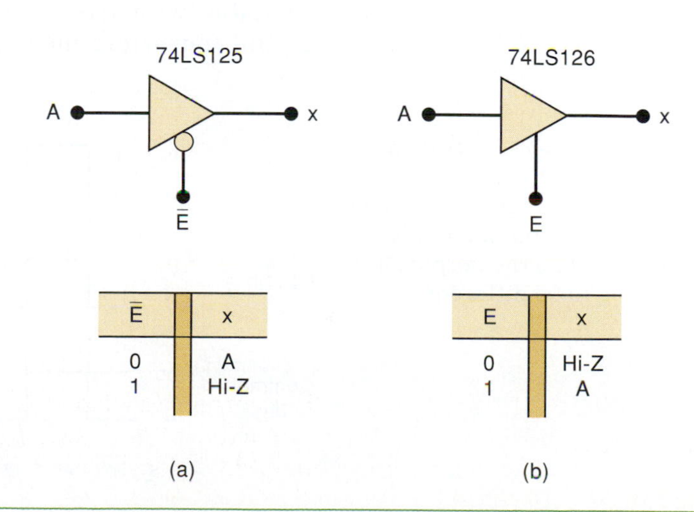

$\overline{E}$	x
0	A
1	Hi-Z

E	x
0	Hi-Z
1	A

FIGURE 8-36 Tristate noninverting buffers.

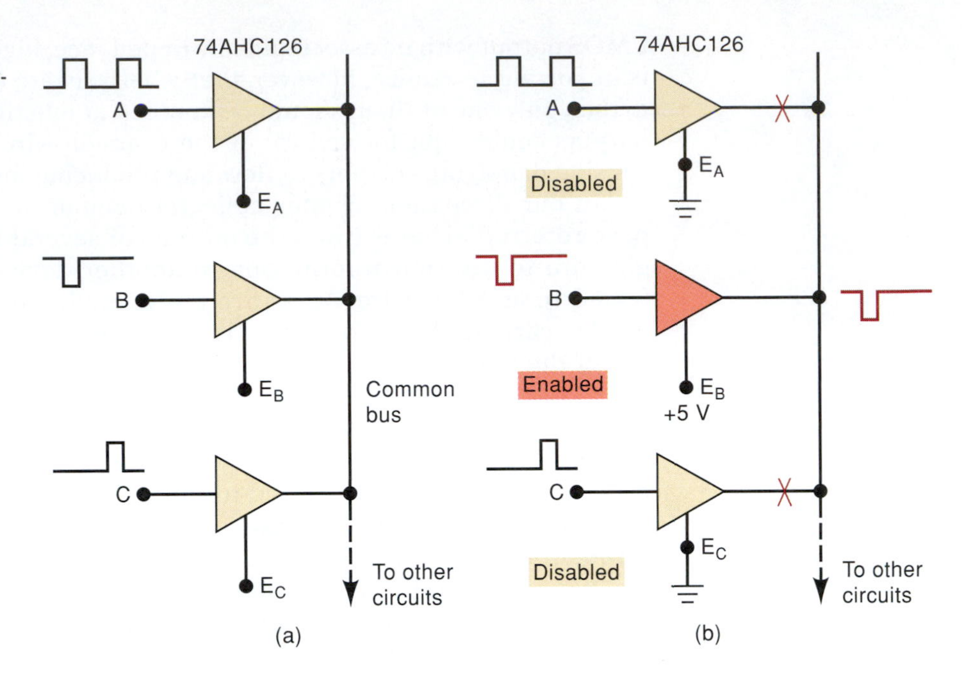

FIGURE 8-37 (a) Tristate buffers used to connect several signals to a common bus; (b) conditions for transmitting B to the bus.

produce damaging currents. Even if damage did not occur, this situation would produce a signal on the bus that is a combination of more than one signal. This is commonly referred to as **bus contention**. Figure 8-38 shows the effect of enabling outputs A and B simultaneously. In Figure 8-37, when inputs A and B are in opposite states, they contend for control of the bus. The resulting voltage on the bus is an invalid logic state. In tristate bus systems, the designer must make sure that the enable signals do not allow bus contention to occur.

Tristate ICs

In addition to tristate buffers, many ICs are designed with tristate outputs. For example, the 74LS374 is an octal D-type FF register IC with tristate outputs. This means that it is an eight-bit register made up of D-type FFs whose outputs are connected to tristate buffers. This type of register can be connected to common bus lines along with the outputs from other, similar devices to allow efficient transfer of data over the bus. We examine this *tristate data bus* arrangement in Chapter 9. Other types of logic devices that are available with tristate outputs include decoders, multiplexers, analog-to-digital converters, memory chips, and microprocessors.

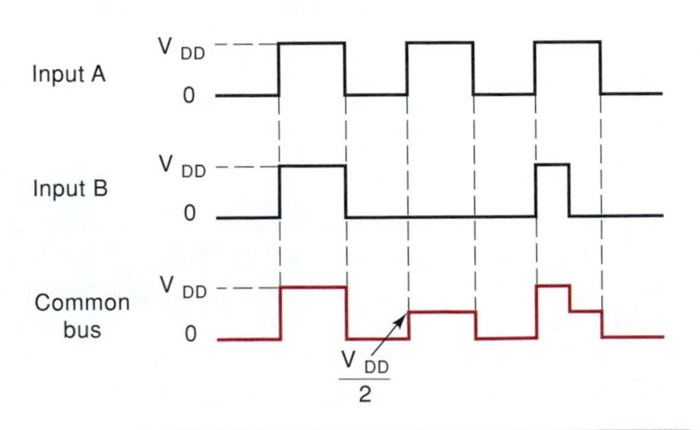

FIGURE 8-38 If two enabled CMOS outputs are connected together, the bus will be at approximately $V_{DD}/2$ when the outputs are trying to be different.

IEEE/ANSI Symbol for Tristate Outputs

The traditional logic symbology has no special notation for tristate outputs. Figure 8-39 shows the notation used in the IEEE/ANSI symbology to indicate a tristate output. It is a triangle that points downward. Although it is not part of the traditional symbology, we will use this triangle to designate tristate outputs throughout the remainder of the book.

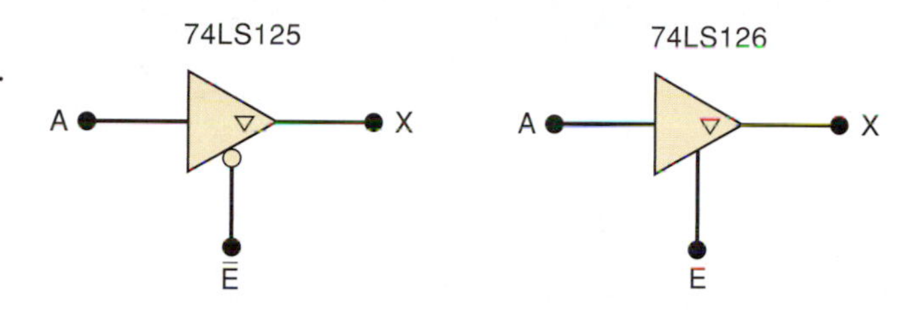

FIGURE 8-39 IEEE/ANSI notation for tristate outputs.

REVIEW QUESTIONS

1. What are the three possible states of a tristate output?
2. What is the state of a tristate output when it is disabled?
3. What is bus contention?
4. What conditions are necessary to transmit signal *C* onto the bus in Figure 8-37?
5. What is the IEEE/ANSI designation for tristate outputs?

8-13 HIGH-SPEED BUS INTERFACE LOGIC

Many digital systems use a shared bus to transfer digital signals and data between the various components of the system. As you can see from our discussion of CMOS technology development, systems are getting faster and faster. Many of the newer high-speed logic series are designed specifically to interface to a tristate bus system. The components in these series are primarily tristate buffers, bidirectional transceivers, latches, and high-current line drivers.

A significant distance often physically separates the components in these systems. If this distance is more than about 4 inches, the bus wires between them need to be viewed as a transmission line. Although transmission line theory could fill up a whole book and is beyond the scope of this text, the general idea is simple enough. Wires have inductance, capacitance, and resistance, which means that for changing signals (ac), they have a characteristic impedance that can affect a signal placed on one end and distort it by the time it reaches the other end. At the high speeds we are discussing, the travel time down the wire and the effects of reflected waves (like echoes) and ringing become real concerns. There are several ways to combat the problems associated with transmission lines. In order to prevent reflected pulse waves, the end of the bus must be terminated with a resistance that is equal to the line impedance (about 50 Ω), as shown in Figure 8-40(a). This method is not feasible because too much current is required to maintain logic level voltages across such a low resistance. Another technique uses a capacitor to block the dc current when the line is not changing, but effectively appears as just a resistor to the rising or falling pulse. This method is shown in Figure 8-40(b).

Using a voltage divider, as in Figure 8-40(c), with resistances larger than the line, impedance helps reduce reflections, but with hundreds of individual bus lines, it obviously makes a heavy load on the system power supply. The

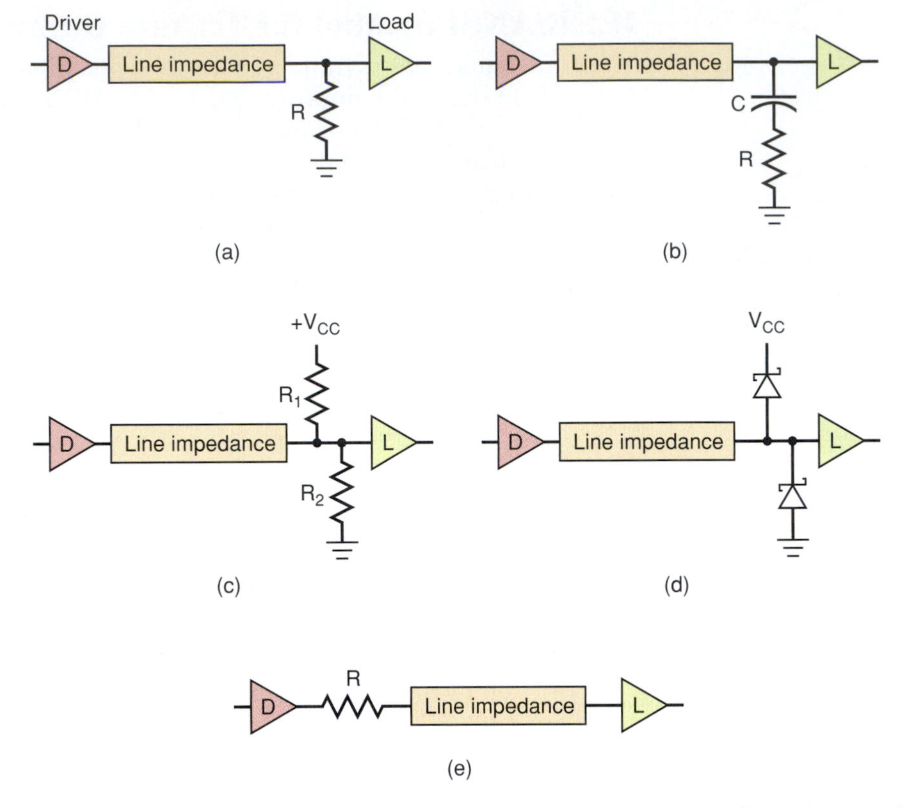

FIGURE 8-40 Bus termination techniques.

diode termination shown in Figure 8-40(d) simply clips off or clamps the overshoot/undershoot of the ringing caused by the reactive LC nature of the line. Series termination at the source, as shown in Figure 8-40(e), slows down the switching speed, which reduces the frequency limits of the bus but substantially improves the reliability of the bus signals.

As you can see, none of these methods are ideal. IC manufacturers are designing new series of logic circuits that overcome many of these problems. Texas Instruments' bus interface logic series offers new output circuits that dynamically lower the output impedance during signal transition to provide fast transition times, then raises the impedance during the steady state (like a series termination) to damp any ringing and reduce reflections on the bus line. The GTLP (Gunning Transceiver Logic Plus) series of bus interface devices is specially designed to drive the relatively long buses that connect modules of a large digital system. The backplane refers to the interconnections between modules in the back of an industry standard, 19-inch rack mounting system.

Another major player in the high-speed bus-interface arena is known as **low-voltage differential signaling (LVDS)**. It uses two wires for each signal, and differential signaling means it responds to the difference between the two wires. Unwanted noise signals are usually present on both lines, and have no effect on the difference between the two. To represent the two logic states, LVDS uses a low voltage swing but switches polarity to clearly distinguish a 1 from a 0.

REVIEW QUESTIONS

1. How close together do components need to be to ignore "transmission line" effects?
2. What three characteristics of real wires add up to distort signals that move through them?
3. What is the purpose of bus terminations?

8-14 THE ECL DIGITAL IC FAMILY

The TTL family uses transistors operating in the saturated mode. As a result, their switching speed is limited by the storage delay time associated with a transistor that is driven into saturation. Another *bipolar* logic family has been developed that prevents transistor saturation, thereby increasing overall switching speed. This logic family is called **emitter-coupled logic (ECL)**, and it operates on the principle of current switching whereby a fixed bias current less than I_C(sat) is switched from one transistor's collector to another. Because of this current-mode operation, this logic form is also referred to as *current-mode logic* (CML).

Basic ECL Circuit

The basic circuit for emitter-coupled logic is essentially the differential amplifier configuration of Figure 8-41(a). The V_{EE} supply produces an essentially fixed current I_E, which remains around 3 mA during normal operation. This current is allowed to flow through either Q_1 or Q_2, depending on the voltage level at V_{IN}. In other words, this current switches between Q_1's collector and Q_2's collector as V_{IN} switches between its two logic levels of -1.7 V (logical 0 for ECL) and -0.8 V (logical 1 for ECL). The table in Figure 8-41(a) shows the resulting output voltages for these two conditions at V_{IN}. Two important points should be noted: (1) V_{C1} and V_{C2} are the *complements* of each other, and (2) the output voltage levels are not the same as the input logic levels.

The second point noted above is easily taken care of by connecting V_{C1} and V_{C2} to emitter-follower stages (Q_3 and Q_4), as shown in Figure 8-41(b). The emitter followers perform two functions: (1) they subtract approximately 0.8 V from V_{C1} and V_{C2} to shift the output levels to the correct ECL logic levels; and (2) they provide a very low output impedance (typically 7 Ω), which provides for large fan-out and fast charging of load capacitance. This circuit produces two complementary outputs: V_{OUT1}, which is equal to $\overline{V_{IN}}$, and V_{OUT2}, which is equal to V_{IN}.

ECL OR/NOR Gate

The basic ECL circuit of Figure 8-41(b) can be used as an INVERTER if the output is taken at V_{OUT1}. This basic circuit can be expanded to more than one input by paralleling transistor Q_1 with other transistors for the other inputs, as in Figure 8-42(a). Here, either Q_1 or Q_3 can cause the current to be switched out of Q_2, resulting in the two outputs V_{OUT1} and V_{OUT2} being the logical NOR and OR operations, respectively. This OR/NOR gate is symbolized in Figure 8-42(b) and is the fundamental ECL gate.

ECL Characteristics

The latest ECL series by Motorola is called ECLin PS. This stands for ECL in pico seconds. This logic series boasts a maximum gate propagation delay of 500 ps (that's half a nanosecond!) and FF toggle rates of 1.4 GHz. Some devices in this series have gate delays of only 100 ps at an average power of 5 mW. The following are the most important characteristics of the ECLin PS series of Motorola's MECL family of logic circuits:

1. The transistors never saturate, and so switching speed is very high. Typical propagation delay time is 360 ps, which makes ECL faster than any TTL or CMOS family members.

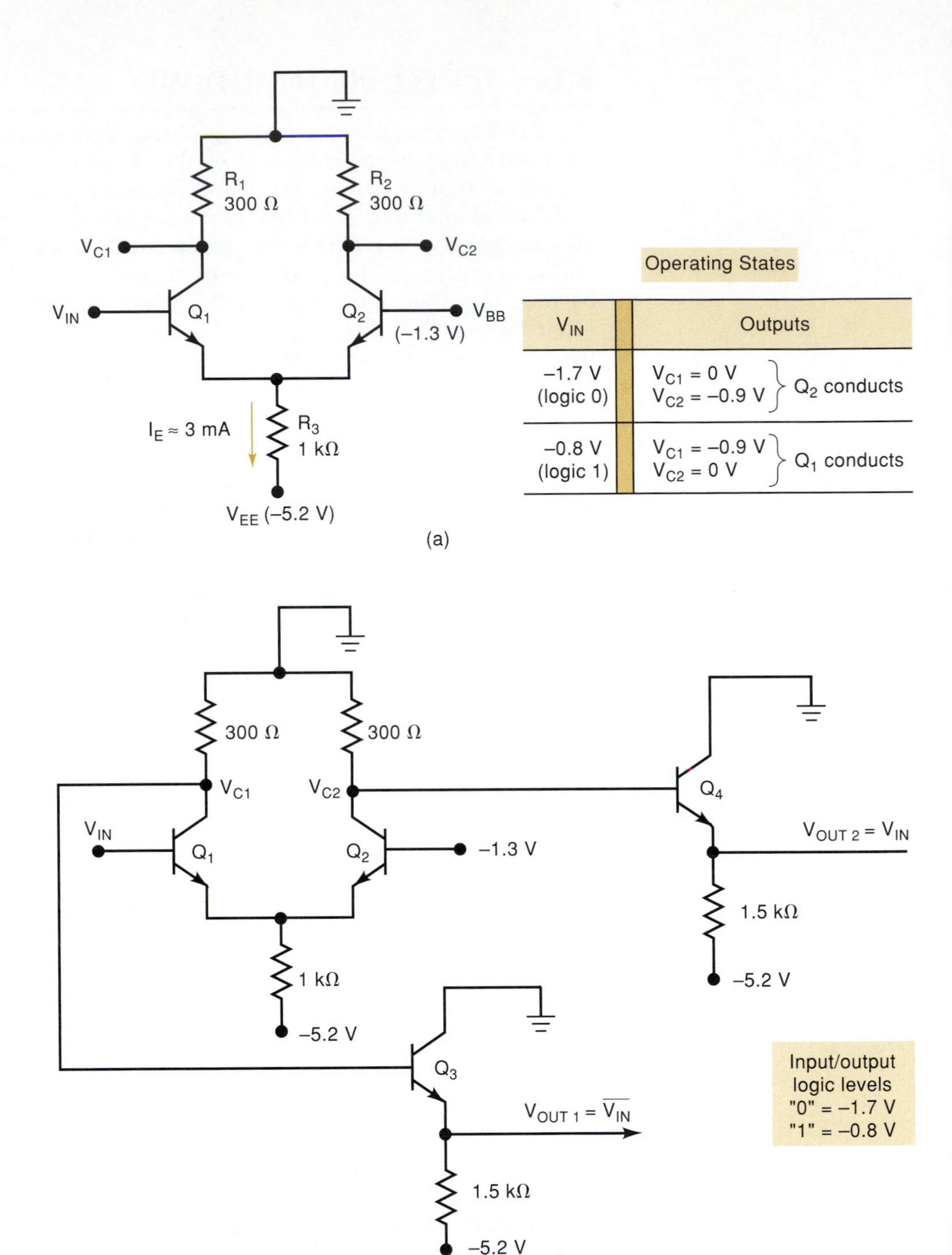

FIGURE 8-41 (a) Basic ECL circuit; (b) with addition of emitter followers.

2. The logic levels are nominally −0.8 V and −1.7 V for the logical 1 and 0, respectively. ECLin PS is fully voltage-compatible with former series of ECL.
3. Worst-case ECL noise margins are approximately 150 mV. These low noise margins make ECL somewhat unreliable for use in heavy industrial environments.
4. An ECL logic block usually produces an output and its complement. This eliminates the need for inverters. High current complementary drive

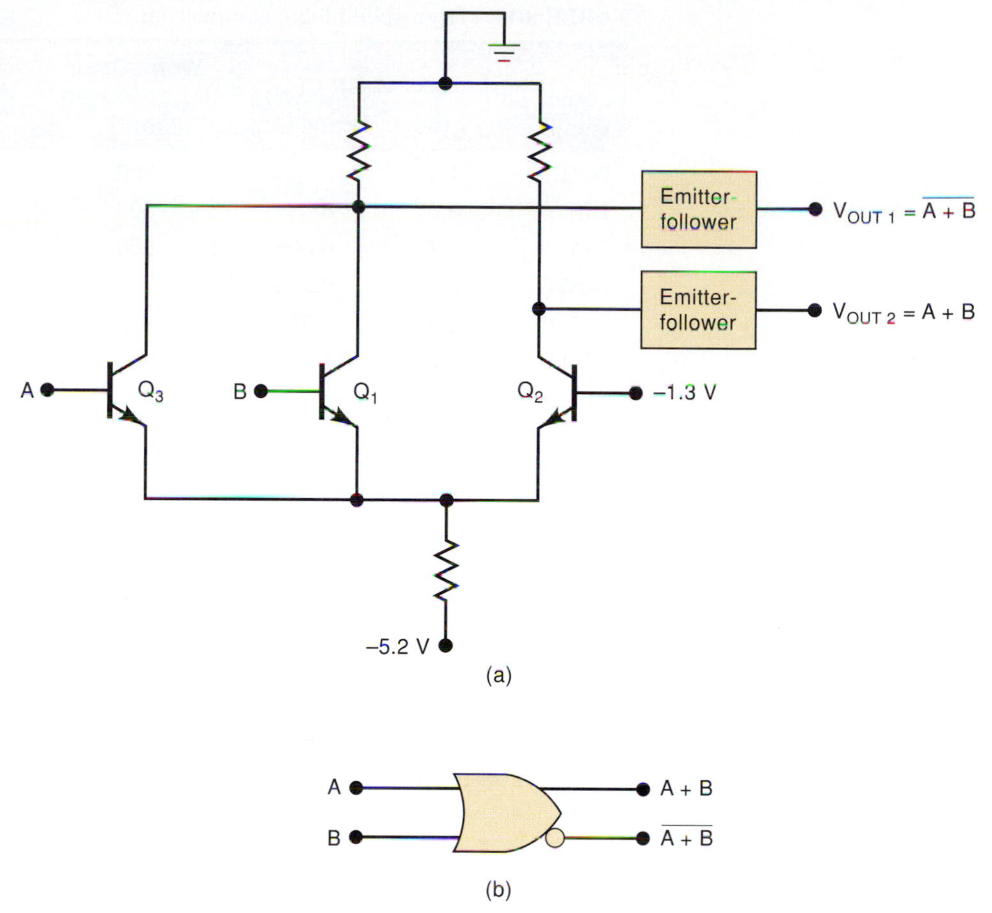

FIGURE 8-42 (a) ECL NOR/OR circuit; (b) logic symbol.

also makes ECL an excellent line driver for twisted-pair cable (such as telephone wires).

5. Fan-out is typically around 25, owing to the low-impedance emitter–follower outputs.
6. Typical power dissipation is 25 mW, somewhat higher than the 74AS series.
7. The total current flow in an ECL circuit remains relatively constant, regardless of its logic state. This helps to maintain an unvarying current drain on the power supply even during switching transitions. Thus, no noise spikes will be generated internally, like those produced by TTL and CMOS switching.

Table 8-11 shows how ECL compares with the important TTL logic families. The ECL family of ICs does not include a wide range of general-purpose logic devices, as do the TTL and CMOS families. ECL does include complex, special-purpose ICs used in applications such as high-speed data transmission, high-speed memories, and high-speed arithmetic units. The relatively low noise margins and high power drain of ECL are disadvantages compared with TTL and CMOS. Another drawback is its negative power-supply voltage and logic levels, which are not compatible with those of the other logic families. This makes it difficult to use ECL devices in conjunction with TTL and/or CMOS ICs; special level-shifting circuits must be connected between ECL devices and the TTL (or CMOS) devices on both input and output.

TABLE 8-11 High-speed logic comparison.

Logic Family	t_{pd} (ns)	P_D (mW) <100 kHz	Worst-Case Noise Margin (mV)	Maximum Clock Rate (MHz)
74AS	1.7	8	300	200
74F	3.8	6	300	100
74AHC	3.7	0.006	550	130
74AVC	2	0.006	250	*
74ALVT	2.4	0.33	400	*
74ALB	2.2	1	400	*
ECL	0.3	25	150	1400

*Flip-flops not available in this series.

REVIEW QUESTIONS

1. *True or false:*
   (a) ECL obtains high-speed operation by preventing transistor saturation.
   (b) ECL circuits usually have complementary outputs.
   (c) The noise margins for ECL circuits are larger than TTL noise margins.
   (d) ECL circuits do not generate noise spikes during state transitions.
   (e) ECL devices require less power than standard TTL.
   (f) ECL can easily be used with TTL.

8-15 CMOS TRANSMISSION GATE (BILATERAL SWITCH)

A special CMOS circuit that has no TTL or ECL counterpart is the **transmission gate** or **bilateral switch**, which acts essentially as a single-pole, single-throw switch controlled by an input logic level. This transmission gate passes signals in both directions and is useful for digital and analog applications.

Figure 8-43(a) is the basic arrangement for the bilateral switch. It consists of a P-MOSFET and an N-MOSFET in parallel so that both polarities of input voltage can be switched. The CONTROL input and its inverse are used to turn the switch on (closed) and off (open). When the CONTROL is HIGH, both MOSFETs are turned on and the switch is closed. When CONTROL is LOW, both MOSFETs are turned off and the switch is open. Ideally, this circuit operates like an electromechanical relay. In practice, however, it is not a perfect short circuit when the switch is closed; the switch resistance R_{ON} is typically 200 Ω. In the open state, the switch resistance is very large, typically 10^{12} Ω, which for most purposes is an open circuit. The symbol in Figure 8-43(b) is used to represent the bilateral switch.

This circuit is called a *bilateral* switch because the input and output terminals can be interchanged. The signals applied to the switch input can be either digital or analog signals, provided that they stay within the limits of 0 to V_{DD} volts.

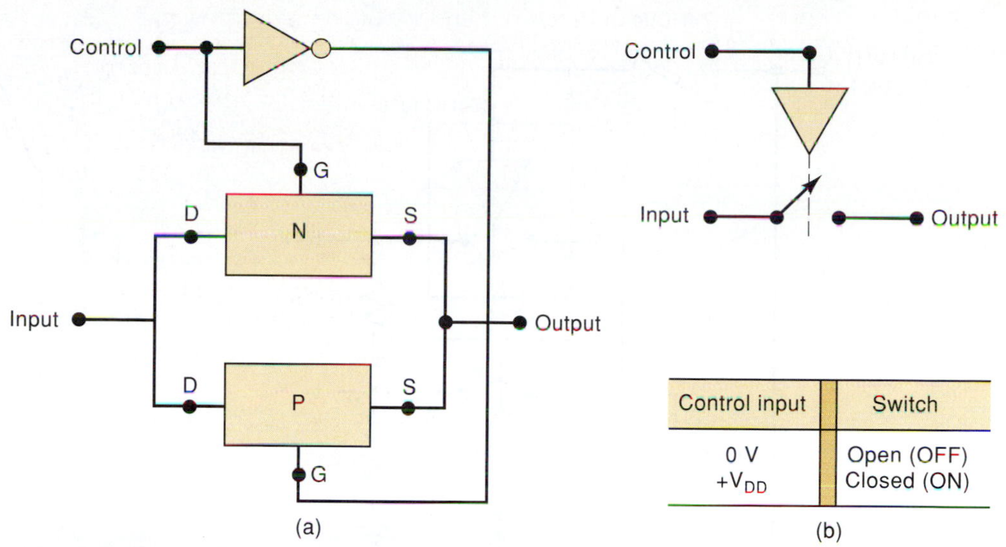

Control input	Switch
0 V $+V_{DD}$	Open (OFF) Closed (ON)

FIGURE 8-43 CMOS bilateral switch (transmission gate).

Figure 8-44(a) shows the traditional logic diagram for a 4016 quad bilateral switch IC, which is also available in the 74HC series as a 74HC4016. The IC contains four bilateral switches that operate as described above. Each switch is independently controlled by its own control input. For example, the ON/OFF status of the top switch is controlled by input $CONT_A$. Because the switches are bidirectional, either switch terminal can serve as input or output, as the labeling indicates.

FIGURE 8-44 The 4016/74HC4016 quad bilateral switch.

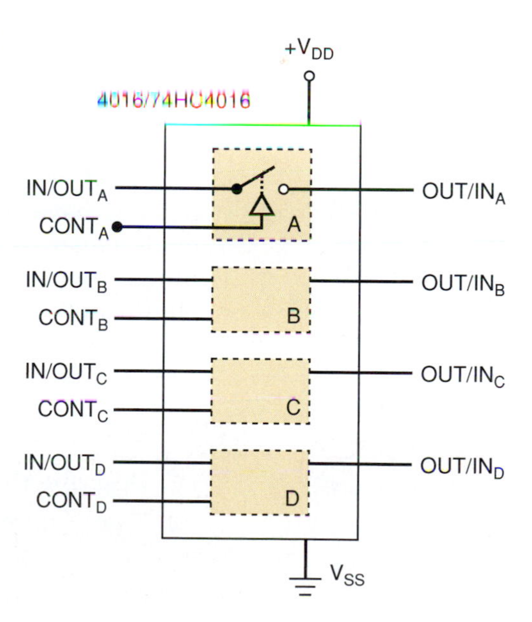

EXAMPLE 8-12

Describe the operation of the circuit in Figure 8-45.

Solution

Here, two of the bilateral switches are connected so that a common analog input signal can be switched to either output X or output Y, depending on the logic state of the OUTPUT SELECT input. When OUTPUT SELECT is LOW, the upper switch is closed and the lower one is open so that V_{IN} is connected

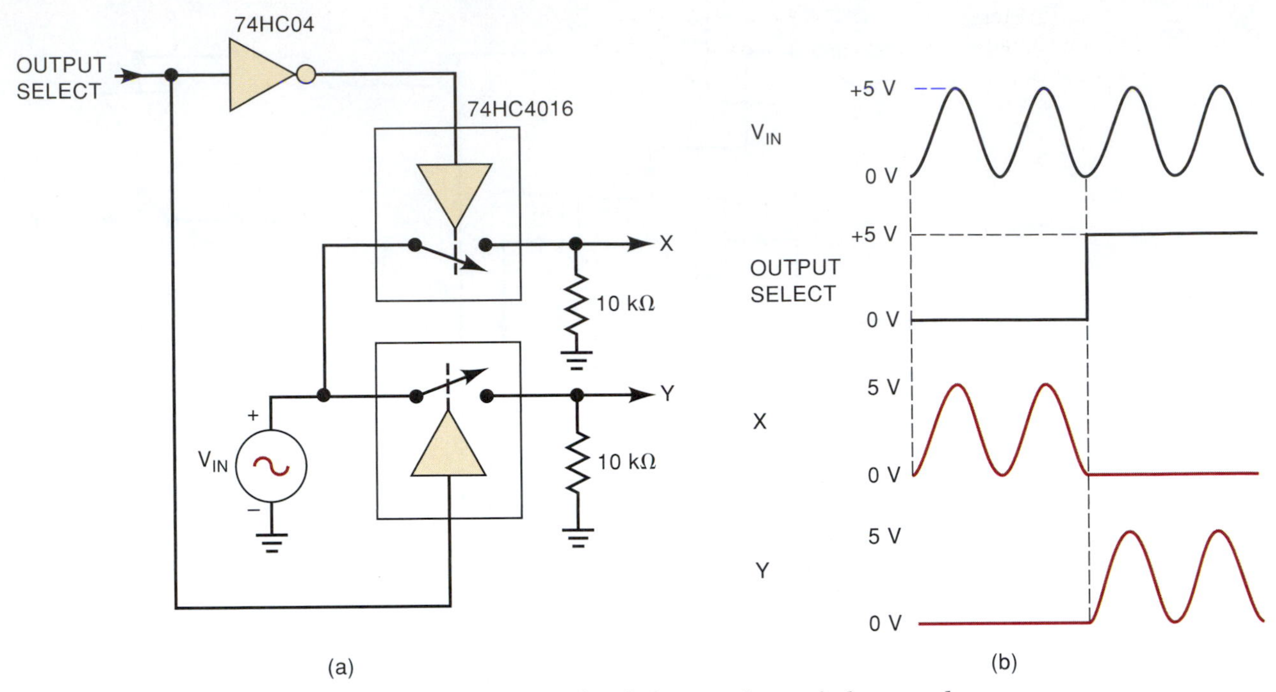

FIGURE 8-45 Example 8-12: 74HC4016 bilateral switches used to switch an analog signal to two different outputs.

to output X. When OUTPUT SELECT is HIGH, the upper switch is open and the lower one is closed so that V_{IN} is connected to output Y. Figure 8-45(b) shows some typical waveforms. Note that for proper operation, V_{IN} must be within the range 0 V to $+V_{DD}$.

The 4016/74HC4016 bilateral switch can switch only input voltages that lie between 0 V and V_{DD}, and so it could not be used for signals that were both positive and negative relative to ground. The 4316 and 74HC4316 ICs are quad bilateral switches that can switch *bipolar* analog signals. These devices have a second power-supply terminal called V_{EE}, which can be made negative with respect to ground. This permits input signals that can range from V_{EE} to V_{DD}. For example, with $V_{EE} = -5$ V and $V_{DD} = +5$ V, the analog input signal can be anywhere from −5 V to +5 V.

REVIEW QUESTIONS

1. Describe the operation of a CMOS bilateral switch.
2. *True or false:* There is no TTL bilateral switch.

8-16 IC INTERFACING

Interfacing means connecting the output(s) of one circuit or system to the input(s) of another circuit or system that has different electrical characteristics. Often a direct connection cannot be made because of the difference in the electrical characteristics of the *driver* circuit that is providing the output signal and the *load* circuit that is receiving the signal. An interface circuit is connected between the driver and the load as shown in Figure 8-46. Its function is to take the driver output signal and condition it so that it is compatible with

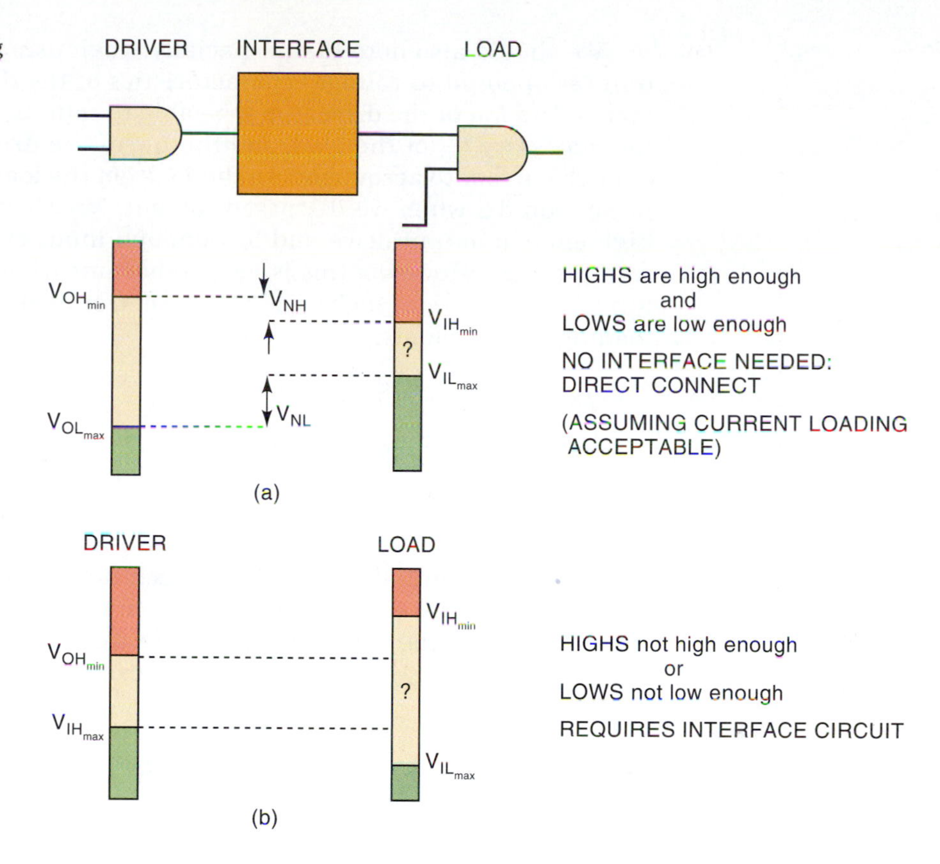

FIGURE 8-46 Interfacing logic ICs: (a) no interface needed; (b) requires interface.

the requirements of the load. In digital systems this is pretty simple, because each device is either on or off. The interface must ensure that when the driver outputs a HIGH, the load receives a signal it perceives to be HIGH; *and*, when the driver outputs a LOW, the load receives a signal it perceives to be LOW.

The simplest and most desirable interface circuit between a driver and a load is a direct connection. Of course, devices that are in the same series are designed to interface directly with each other. Today, however, many systems involve mixed families, mixed voltages, and mixed series. In these systems the challenge is to make sure that the driver is able to consistently activate the load in both the LOW and HIGH states.

For any case such as shown in Figure 8-46(a), where V_{OH} of the driver is *enough greater* than the V_{OH}(min) of the load and V_{OL} of the driver is *enough less* than the V_{IL}(max) of the load, there is no need for an interface circuit other than a direct connection. "How much *greater?*" and "How much *less?*" are questions related to how much noise is expected in the system. Recall that the noise margins (V_{NH} and V_{NL}) are measures of this difference between output and input characteristics. (Refer back to Figure 8-4.) The minimum acceptable noise margin for any system is a judgment call that must be made by the system designer. Whenever the V_{NH} or V_{NL} is determined to be too small (or even negative), then an interface circuit is necessary in order to ensure that the driver and load can work together. This situation is depicted in Figure 8-46(b) To summarize this:

Driver	***Load***
V_{OH}(min)	$> V_{IH}(\text{min}) + V_{NH}$
$V_{OL}(\text{max}) + V_{NL}$	$< V_{IL}$(max)

We should also note that, especially when using older families, the current (as opposed to voltage) characteristics of the driver and load must also match. The I_{OH} of the driver must be able to source enough current to supply the necessary I_{IH} of the load, and the I_{OL} of the driver must be able to sink enough current to accommodate the I_{IL} from the load. This topic was covered in Section 8-5 when we discussed fan-out. Most modern logic devices have high enough output drive and low enough input current to make loading a rare problem. However, this is very important when interfacing to external input/output devices such as motors, lights, or heaters. To summarize current loading requirements:

Driver	***Load***
I_{OH}(max)	$> I_{IH}$(total)
I_{OL}(max)	$> I_{IL}$(total)

Table 8-12 lists some nominal values for a number of different families and series of digital devices. Within each family there will be exceptions to these listed values, and so in practice it is important that you look up the data sheet values for the specific ICs you are working with. For the sake of convenience we will use these values in the examples that follow.

Interfacing 5-V TTL and CMOS

When interfacing different types of ICs, we must check that the driving device can meet the current and voltage requirements of the load device. Examination of Table 8-12 indicates that the input current values for CMOS are extremely low compared with the output current capabilities of any TTL series. Thus, TTL has no problem meeting the CMOS input current requirements.

There is a problem, however, when we compare the TTL output voltages with the CMOS input voltage requirements. Table 8-9 shows that V_{OH}(min) of every TTL series is too low when compared with the V_{IH}(min) requirement of the 4000B, 74HC, and 74AC series. For these situations, something must be done to raise the TTL output voltage to an acceptable level for CMOS.

The most common solution to this interface problem is shown in Figure 8-47, where the TTL output is connected to +5 V with a pull-up resistor. The presence of the pull-up resistor causes the TTL output to rise to approximately 5 V in the HIGH state, thereby providing an adequate CMOS input voltage level. This pull-up resistor is not required if the CMOS device is a 74HCT or a 74ACT because these series are designed to accept TTL outputs directly, as Table 8-9 shows.

TABLE 8-12 Input/output currents for standard devices with a supply voltage of 5 V.

	CMOS				TTL				
Parameter	4000B	74HC/HCT	74AC/ACT	74AHC/AHCT	74	74LS	74AS	74ALS	74F
I_{IH}(max)	1 μA	1 μA	1 μA	1 μA	40 A	20 μA	20 μA	20 μA	20 μA
I_{IL}(max)	1 μA	1 μA	1 μA	1 μA	1.6 mA	0.4 mA	0.5 mA	100 μA	0.6 mA
I_{OH}(max)	0.4 mA	4 mA	24 mA	8 mA	0.4 mA	0.4 mA	2 mA	400 mA	1.0 mA
I_{OL}(max)	0.4 mA	4 mA	24 mA	8 mA	16 mA	8 mA	20 mA	8 mA	20 mA

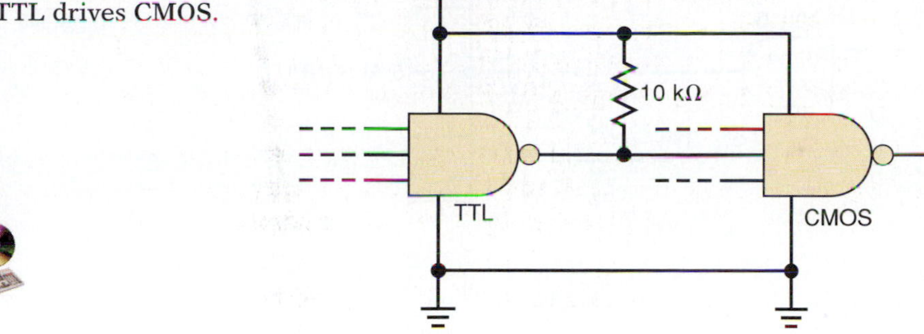

FIGURE 8-47 External pull-up resistor is used when TTL drives CMOS.

CMOS Driving TTL

Before we consider the problem of interfacing CMOS outputs to TTL inputs, it will be helpful to review the CMOS output characteristics for the two logic states. Figure 8-48(a) shows the equivalent output circuit in the HIGH state. The R_{ON} of the P-MOSFET connects the output terminal to V_{DD} (remember, the N-MOSFET is off). Thus, the CMOS output circuit acts like a V_{DD} source with a source resistance of R_{ON}. The value of R_{ON} typically ranges from 100 to 1000 Ω.

Figure 8-48(b) shows the equivalent output circuit in the LOW state. The R_{ON} of the N-MOSFET connects the output terminal to ground (remember, the P-MOSFET is off). Thus, the CMOS output acts as a low resistance to ground; that is, it acts as a current sink.

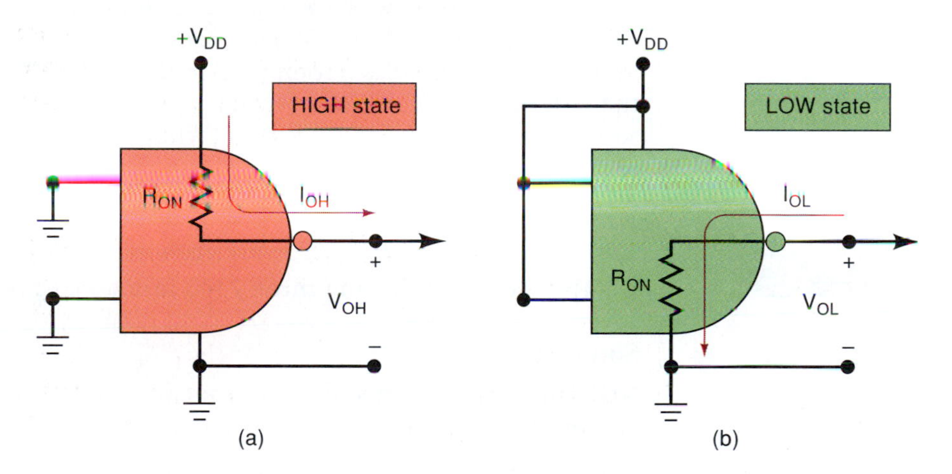

FIGURE 8-48 Equivalent CMOS output circuits for both logic states.

CMOS Driving TTL in the HIGH State

Table 8-9 shows that CMOS outputs can easily supply enough voltage (V_{OH}) to satisfy the TTL input requirement in the HIGH state (V_{IH}). Table 8-12 shows that CMOS outputs can supply more than enough current (I_{OH}) to meet the TTL input current requirements (I_{IH}). Thus, no special consideration is needed for the HIGH state.

CMOS Driving TTL in the LOW State

Table 8-12 shows that TTL inputs have a relatively high input current in the LOW state, ranging from 100 μA to 2 mA. The 74HC and 74HCT series can sink up to 4 mA, and so they would have no trouble driving a *single* TTL load of any series. The 4000B series, however, is much more limited. Its low I_{OL} capability is not sufficient to drive even one input of the 74 or 74AS series. The 74AHC series has output drive comparable to that of the 74LS series.

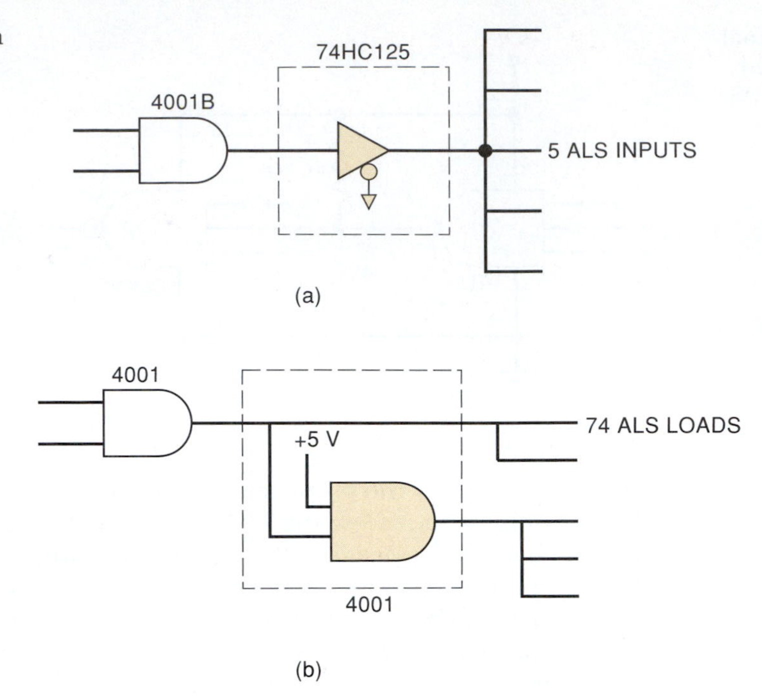

FIGURE 8-49 (a) Using an HC series interface IC. (b) Using a similar gate to share the load.

For the situation in which a driver cannot supply enough current to the load, the interface solution is to select a buffer that has input specifications that are compatible with the driver and enough output drive current to supply the load. Figure 8-49(a) shows an example of this situation. The 4001B's maximum output current is not enough to drive five ALS inputs. It is able to drive the 74HC125 input, which in turn can drive the other inputs. Another possible solution is shown in Figure 8-49(b), where the load is divided up among multiple 4001 series parts such that no output needs to drive more than three loads.

EXAMPLE 8-13

A 74HC output is driving three 7406 inputs. Is this a good design?

Solution

NO! The 74HC00 can sink 4 mA, but the 7406 input I_{IL} is 1.6 mA. Total load current when LOW is 1.6 mA × 3 = 4.8 mA. . . . Too much load current.

EXAMPLE 8-14

A 4001B output is driving three 74LS inputs. Is this a well-designed circuit?

Solution

NO! The 4001B can sink 0.4 mA, but each 74LS input accounts for 0.4 mA × 3 = 1.2 mA. . . . Too much load current.

REVIEW QUESTIONS

1. What must be done to interface a standard TTL output to a 74AC or a 74HC input? Assume V_{DD} = +5 V.
2. What is usually the problem with CMOS driving TTL?

8-17 MIXED-VOLTAGE INTERFACING

As we discussed in Section 8-10, many new logic devices operate on less than 5 V. In many situations, these devices need to communicate with each other. In this section we will look specifically at how to interface logic devices that operate on different voltage standards.

Low-Voltage Outputs Driving High-Voltage Loads

In some situations, the V_{OH} of the driver is just slightly lower than the load requires to recognize it as a HIGH. This situation was discussed when interfacing TTL outputs to 5-V CMOS inputs. The only interface component needed was a 10-kΩ pull-up resistor, which will cause the TTL output voltage to be boosted above the 3.3-V level when the output is HIGH.

When there is a need for a more substantial shift in voltage because the driver and load operate on different power supply voltages, a **voltage-level translator** interface circuit is required. An example of this is a low-voltage (1.8-V) CMOS device driving a 5-V CMOS input. The driver can put out a maximum of only 1.8 V as a HIGH, and the load gate requires 3.33V for a HIGH. We need an interface that can accept 1.8-V logic levels and translate them to 5-V CMOS levels. The simplest way to accomplish this is with a buffer that has an open drain, such as the 74LVC07 shown in Figure 8-50(a). Notice that the pull-up resistor is connected to the 5-V supply, while the power supply for the interface buffer is 1.8 V. Another solution is to utilize a dual-supply-level translator circuit such as the 74AVC1T45, as shown in Figure 8-50(b). This device uses two different power supply voltages, one for the inputs and the other for the outputs, and translates between the two levels.

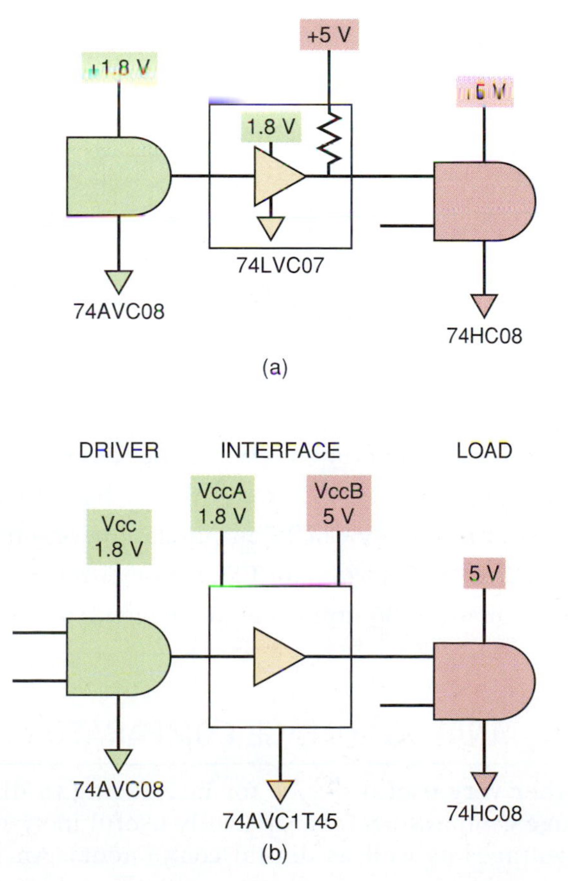

FIGURE 8-50 (a) Using an open drain with pullup to high voltage. (b) Using a voltage-level translator.

High-Voltage Outputs Driving Low-Voltage Loads

When logic circuits that operate on a higher voltage supply must drive other logic circuits that operate on a lower voltage supply, the output voltage of the driver often exceeds the safe limits that the load gate can handle. In these situations, a dual-supply-voltage-level translator can be used just as it was in Figure 8-50(b). Another common solution to this problem is to interface them using a buffer from a series that can withstand the higher input voltage. Figure 8-51 demonstrates this with a 5-V CMOS part driving a 1.8-V AUC series input. The highest voltage the AUC input (load gate) can handle is 3.6 V. However, a 74LVC07A can handle up to 5.5 V on its input without damaging it, even though it is operating on 1.8 V. Figure 8-51 shows how we can use the higher voltage tolerance of the 74LVC07A to translate a 5-V logic level down to a 1.8-V logic level.

At this point you may be wondering, "Why in the world would anyone choose to use such an assortment of incompatible parts?" The answer lies in considering larger systems and trying to balance performance and cost. In a computer system, for example, you may have a 2.5-V CPU, a 3.3-V memory module, and a 5-V hard drive controller all working on the same mother board. The low-voltage components may be necessary to obtain the desired performance, but the 5-V hard drive may be the least expensive or the only type available. The driver and load devices may not be standard logic gates at all, but may be a VLSI component in our system. Using the data sheet for those devices, we must look up the output characteristics and interface them using the techniques we have shown. As logic standards continue to evolve, it is important that we can make systems work using any of the diverse components available to us.

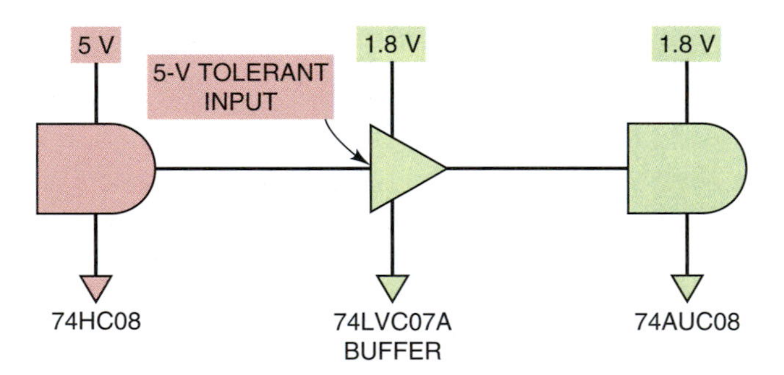

FIGURE 8-51 (a) A low-voltage series with 5-V tolerant inputs as an interface.

REVIEW QUESTIONS

1. What is the function of an *interface* circuit?
2. *True or false:* All CMOS outputs can drive TTL in the HIGH state.
3. *True or false:* Any CMOS output can drive any single TTL input.
4. Which CMOS series can TTL drive without a pull-up resistor?
5. How many 7400 inputs can be driven from a 74HCT00 output?

8-18 ANALOG VOLTAGE COMPARATORS

Another very useful device for interfacing to digital systems is the **analog voltage comparator**. It is especially useful in systems that contain both analog voltages as well as digital components. An analog voltage comparator compares two voltages. If the voltage on the (+) input is greater than the voltage on the (−) input, the output is HIGH. If the voltage on the (−) input is greater than the voltage on the (+) input, the output is LOW. The inputs to a comparator can be thought of as analog inputs, but the output is digital because it will always be either HIGH or LOW. For this reason, the comparator is often referred to as a one-bit analog-to-digital (A/D) converter. We will examine A/D converters in detail in Chapter 10.

An LM339 is an analog linear IC that contains four voltage comparators. The output of each comparator is an open-collector transistor just like an open-collector TTL output. V_{CC} can range from 2 to 36 V but is usually set slightly greater than the analog input voltages that are being compared. A pull-up resistor must be connected from the output to the same supply that the digital circuits use (normally 5 V).

EXAMPLE 8-15

Suppose that an incubator must have an emergency alarm to warn if the temperature exceeds a dangerous level. The temperature-measuring device is an LM34 that puts out a voltage directly proportional to the temperature. The output voltage goes up 10 mV per degree F. The digital system alarm must sound when the temperature exceeds 100°F. Design a circuit to interface the temperature sensor to the digital circuit.

Solution

We need to compare the voltage from the sensor with a fixed threshold voltage. First, we must calculate the proper threshold voltage. We want the comparator output to go HIGH when the temperature exceeds 100°F. The voltage out of the LM34 at 100°F will be

$$100°\text{F} \cdot 10\ \text{mV/°F} = 1.0\ \text{V}$$

This means that we must set the (−) input pin of the comparator to 1.0 V and connect the LM34 to the (+) input. In order to create a 1.0-V reference voltage, we can use a voltage-divider circuit and choose a bias current of 100 μA. The LM339 input current will be relatively negligible because it will draw less than 1 μA. This means that $R_1 + R_2$ must total 10 kΩ. In this example, we can operate everything from a +5 V power supply. Figure 8-52 shows the

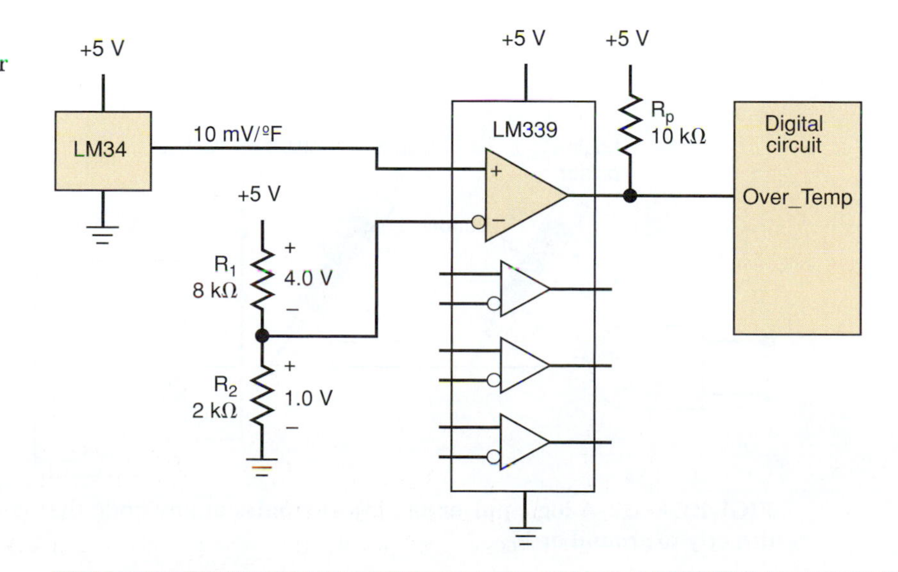

FIGURE 8-52 A temperature-limit detector using an LM339 analog voltage comparator.

complete circuit. The calculations are as follows:

$$V_{R2} = V_{CC} \cdot \frac{R_2}{R_1 + R_2}$$
$$R_2 = V_{R2} \cdot (R_1 + R_2)/V_{CC}$$
$$= 1.0\,\text{V}(10\,\text{k}\Omega)/5\,\text{V} = 2\,\text{k}\Omega$$
$$R_1 = 10\,\text{k}\Omega - R_2 = 10\,\text{k}\Omega - 2\,\text{k}\Omega = 8\,\text{k}\Omega$$

REVIEW QUESTIONS

1. What causes the output of a comparator to go to the HIGH logic state?
2. What causes the output of a comparator to go to the LOW logic state?
3. Is an LM339 output more similar to a TTL totem-pole or an open-collector output?

8-19 TROUBLESHOOTING

A **logic pulser** is a testing and troubleshooting tool that generates a short-duration pulse when manually actuated, usually by pressing a push button. The logic pulser shown in Figure 8-53 has a needle-shaped tip that is touched to the circuit node that is to be pulsed. The logic pulser is designed so that it senses the existing voltage level at the node and produces a voltage pulse in the opposite direction. In other words, if the node is LOW, the logic pulser produces a narrow positive-going pulse; if the node is HIGH, it produces a narrow negative-going pulse.

The logic pulser is used to change the logic level at a circuit node momentarily, even though the output of another device may be connected to that same node. In Figure 8-53, the logic pulser is contacting node X, which is also connected to the output of the NAND gate. The logic pulser has a very low output impedance (typically 2 Ω or less), so that it can overcome the NAND gate's output and can change the voltage at the node. The logic pulser, however, cannot produce a voltage pulse at a node that is shorted directly to ground or V_{CC} (e.g., through a solder bridge).

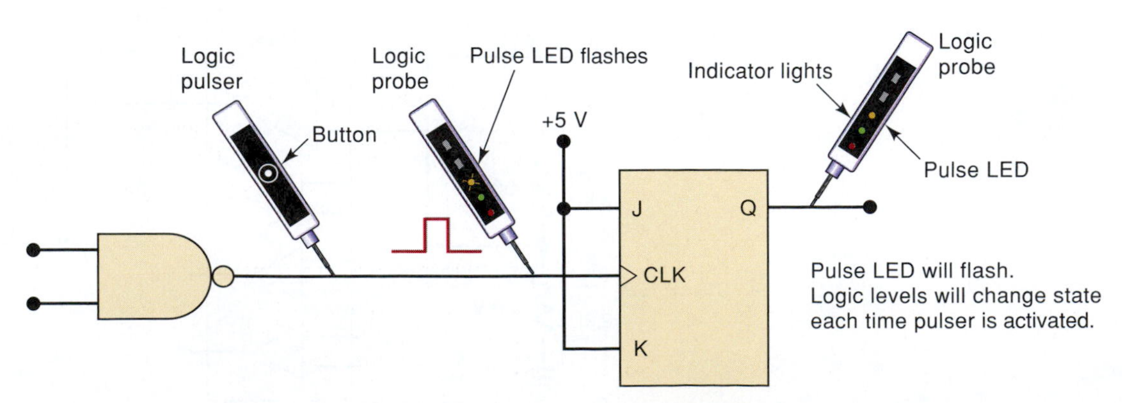

FIGURE 8-53 A logic pulser can inject a pulse at any node that is not shorted directly to ground or V_{CC}.

Using a Logic Pulser and Probe to Test a Circuit

A logic pulser can be used to inject a pulse or a series of pulses manually into a circuit in order to test the circuit's response. A logic probe is almost always used to monitor the circuit's response to the logic pulser. In Figure 8-53, the J-K flip-flop's toggle operation is being tested by applying pulses from the logic pulser and monitoring Q with the logic probe. This logic pulser/logic probe combination is very useful for checking the operation of a logic device while it is wired into a circuit. Note that the logic pulser is applied to the circuit node *without* disconnecting the output of the NAND gate that is driving that node. When the probe is placed on the same node as the pulser, the logic level indicators appear to remain unchanged (LOW in this example), but the yellow pulse indicator flashes once each time the pulser button is pressed. When the probe is placed on the Q output, the pulse LED flashes once (indicating a transition), and the logic level indicators change state each time the pulser button is pushed.

Finding Shorted Nodes

The logic pulser and logic probe can be used to check for nodes that are shorted directly to ground or V_{CC}, as shown in Figure 8-54. When you touch a logic pulser and a logic probe to the same node and press the logic pulser button, the logic probe should indicate the occurrence of a pulse at the node. If the probe indicates a constant LOW and the pulse LED does not flash, the node is shorted to ground, as shown in Figure 8-54(a). If the probe indicates a constant HIGH and the pulse LED does not flash, the node is shorted to V_{CC} as shown in Figure 8-54(b).

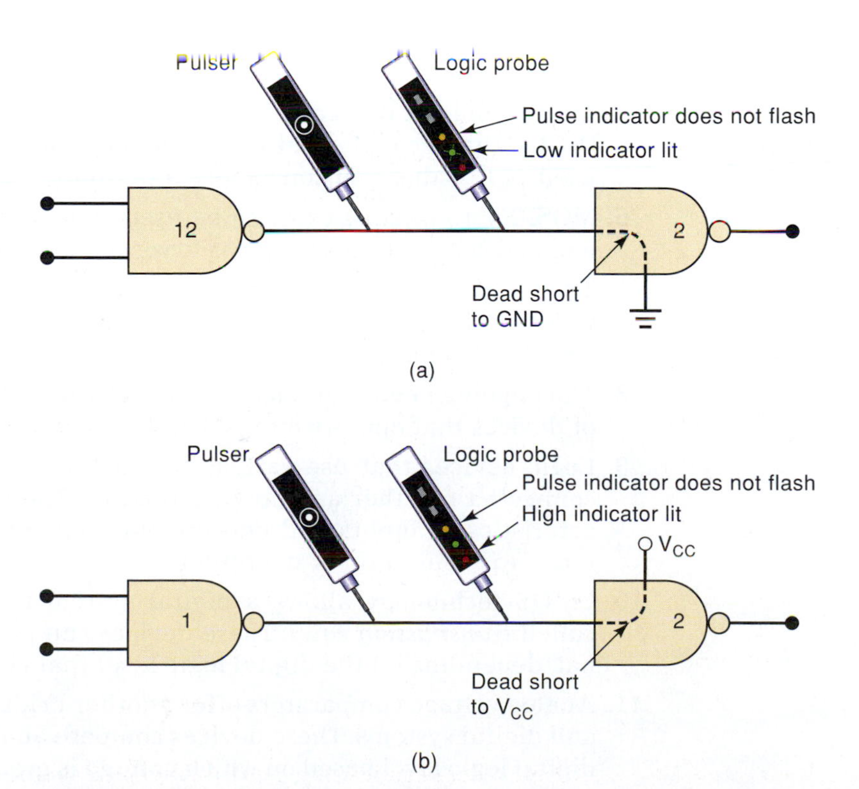

FIGURE 8-54 A logic pulser and a logic probe can be used to trace shorted nodes.

REVIEW QUESTIONS

1. What is the function of a logic pulser?
2. *True or false:* A logic pulser will produce a voltage pulse at any node.
3. *True or false:* A logic pulser can force a node LOW or HIGH for extended periods of time.
4. How does a logic probe respond to the logic pulser?

SUMMARY

1. All digital logic devices are similar in nature but very much different regarding the details of their characteristics. An understanding of the terms used to describe these characteristics is important and allows us to compare and contrast the performance of devices. By understanding the capabilities and limitations of each type of device, we can intelligently combine devices to take advantage of each device's strengths in building reliable digital systems.
2. The TTL family of logic devices has been in use for over 30 years. The circuitry uses bipolar transistors. This family offers many SSI logic gates, and MSI devices. Numerous series of similarly numbered devices have been developed because advances in technology have offered improved characteristics.
3. When you are connecting devices together, it is vital to know how many inputs a given output can drive without compromising reliability. This is referred to as *fan-out.*
4. Open-collector and open-drain outputs can be wired together to implement a wired-AND function. Tristate outputs can be wired together to allow numerous devices to share a common data path known as a *bus.* In such a case, only one device is allowed to assert a logic level on the bus (i.e., drive the bus) at any one time.
5. The *fastest* logic devices are from a family that uses emitter-coupled logic (ECL). This technology also uses bipolar transistors but is not as widely used as TTL due to inconvenient input/output characteristics.
6. MOSFET transistors can also be used to implement logic functions. The main advantage of MOS logic is lower power and greater packing density.
7. The use of complementary MOSFETs has produced CMOS logic families. CMOS technology has captured the market due to its very low power and competitive speed.
8. The ongoing need to reduce power and size has led to several new series of devices that operate on 3.3 V and 2.5 V.
9. Logic devices that use various technologies cannot always be directly connected together and operate reliably. The voltage and current characteristics of inputs and outputs must be considered and precautions taken to ensure proper operation.
10. CMOS technology allows a digital system to control analog switches called *transmission gates.* These devices can pass or block an analog signal, depending on the digital logic level that controls it.
11. Analog voltage comparators offer another bridge between analog signals and digital systems. These devices compare analog voltages and output a digital logic level based on which voltage is greater. They allow an analog system to control a digital system.

IMPORTANT TERMS

fan-out
noise immunity
noise margin
current sourcing
current sinking
DIP
lead pitch
surface mount technology
TTL
totem pole
current-sinking transistor (pull-down transistor)
current-sourcing transistor (pull-up transistor)
floating inputs
power-supply decoupling
MOSFETs
N-MOS
P-MOS
CMOS
electrostatic discharge (ESD)
latch-up
open-collector output
wired-AND
buffer/driver
tristate
bus contention
low-voltage differential signaling (LVDS)
emitter-coupled logic (ECL)
transmission gate (bilateral switch)
interfacing
voltage-level translator
analog voltage comparator
logic pulser

PROBLEMS

SECTIONS 8-1 TO 8-3

B 8-1.* Two different logic circuits have the characteristics shown in Table 8-13.

(a) Which circuit has the best LOW-state dc noise immunity? The best HIGH-state dc noise immunity?

(b) Which circuit can operate at higher frequencies?

(c) Which circuit draws the most supply current?

TABLE 8-13

	Circuit A	Circuit B
V_{supply} (V)	6	5
V_{IH}(min) (V)	1.6	1.8
V_{IL}(max) (V)	0.9	0.7
V_{OH}(min) (V)	2.2	2.5
V_{OL}(max) (V)	0.4	0.3
t_{PLH} (ns)	10	18
t_{PHL} (ns)	8	14
P_D (mW)	16	10

B 8-2. Look up the IC data sheets, and use *maximum* values to determine P_D(avg) and t_{pd}(avg) for one gate on each of the following TTL ICs. (See Example 8-2 in Section 8-3.)

(a)*7432

(b)*74S32

(c) 74LS20

(d) 74ALS20

(e) 74AS20

*Answers to problems marked with an asterisk can be found in the back of the text.

8-3. A certain logic family has the following voltage parameters:

$$V_{IH}(\text{min}) = 3.5\text{ V} \qquad V_{IL}(\text{max}) = 1.0\text{ V}$$
$$V_{OH}(\text{min}) = 4.9\text{ V} \qquad V_{OL}(\text{max}) = 0.1\text{ V}$$

(a)* What is the largest positive-going noise spike that can be tolerated?

(b) What is the largest negative-going noise spike that can be tolerated?

DRILL QUESTION

B 8-4.* For each statement, indicate the term or parameter being described.

(a) Current at an input when a logic 1 is applied to that input

(b) Current drawn from the V_{CC} source when all outputs are LOW

(c) Time required for an output to switch from the 1 to the 0 state

(d) The size of the voltage spike that can be tolerated on a HIGH input without causing indeterminate operation

(e) An IC package that does not require holes to be drilled in the printed circuit board

(f) When a LOW output receives current from the input of the circuit it is driving

(g) Number of different inputs that an output can safely drive

(h) Arrangement of output transistors in a standard TTL circuit

(i) Another term that describes pull-down transistor Q_4

(j) Range of V_{CC} values allowed for TTL

(k) $V_{OH}(\text{min})$ and $V_{IH}(\text{min})$ for the 74ALS series

(l) $V_{IL}(\text{max})$ and $V_{OL}(\text{max})$ for the 74ALS series

(m) When a HIGH output supplies current to a load

SECTION 8-4

8-5.* (a) From Table 8-6, determine the noise margins when a 74LS device is driving a 74ALS input.

(b) Repeat part (a) for a 74ALS driving a 74LS.

(c) What will be the overall noise margin of a logic circuit that uses 74LS and 74ALS circuits in combination?

(d) A certain logic circuit has $V_{IL}(\text{max}) = 450$ mV. Which TTL series can be used with this circuit?

SECTIONS 8-5 AND 8-6

B 8-6. **DRILL QUESTION**

(a) Define *fan-out.*

(b)* In which type of gates do tied-together inputs always count as a single input load in the LOW state?

(c)* Define "floating" inputs.

(d) What causes current spikes in TTL? What undesirable effect can they produce? What can be done to reduce this effect?

(e) When a TTL output drives a TTL input, where does I_{OL} come from? Where does I_{OH} go?

8-7. Use Table 8-12 to find the fan-out for interfacing the first logic family to drive the second.

(a)*74AS to 74AS

(b)*74F to 74F

(c) 74AHC to 74AS

(d) 74HC to 74ALS

B 8-8. Refer to the data sheet for the 74LS112 J-K flip-flop.

(a)*Determine the HIGH and LOW load current at the J and K inputs.

(b) Determine the HIGH and LOW load current at the clock and clear inputs.

(c) How many 74LS112 clock inputs can the output of one 74LS112 drive?

B 8-9.*Figure 8-55(a) shows a 74LS112 J-K flip-flop whose output is required to drive a total of eight standard TTL inputs. Because this exceeds the fan-out of the 74LS112, a buffer of some type is needed. Figure 8-55(b) shows one possibility using one of the NAND gates from the 74LS37 quad NAND buffer, which has a much higher fan-out than the 74LS112. Note that $\overline{Q}$ is used because the NAND is acting as an INVERTER. Refer to the data sheet for the 74LS37.

(a) Determine its maximum fan-out to standard TTL.

(b) Determine its maximum sink current in the LOW state.

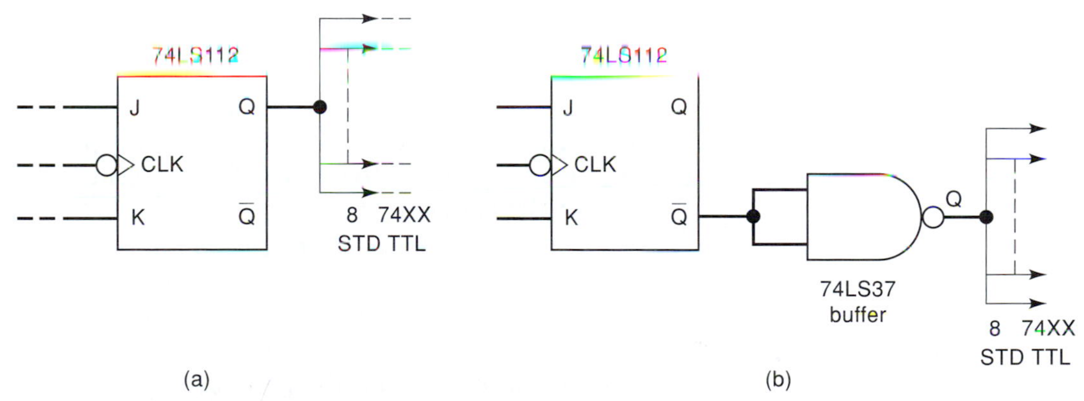

FIGURE 8-55 Problems 8-9 and 8-10.

D 8-10. Buffer gates are generally more expensive than ordinary gates, and sometimes there are unused ordinary gates available that can be used to solve a loading problem such as that in Figure 8-55(a). Show how 74LS00 NAND gates can be used to solve this problem.

B 8-11.*Refer to the logic diagram of Figure 8-56, where the 74LS86 exclusive-OR output is driving several 74LS20 inputs. Determine whether the fan-out of the 74LS86 is being exceeded, and explain. Repeat using all 74AS devices. Use Table 8-7.

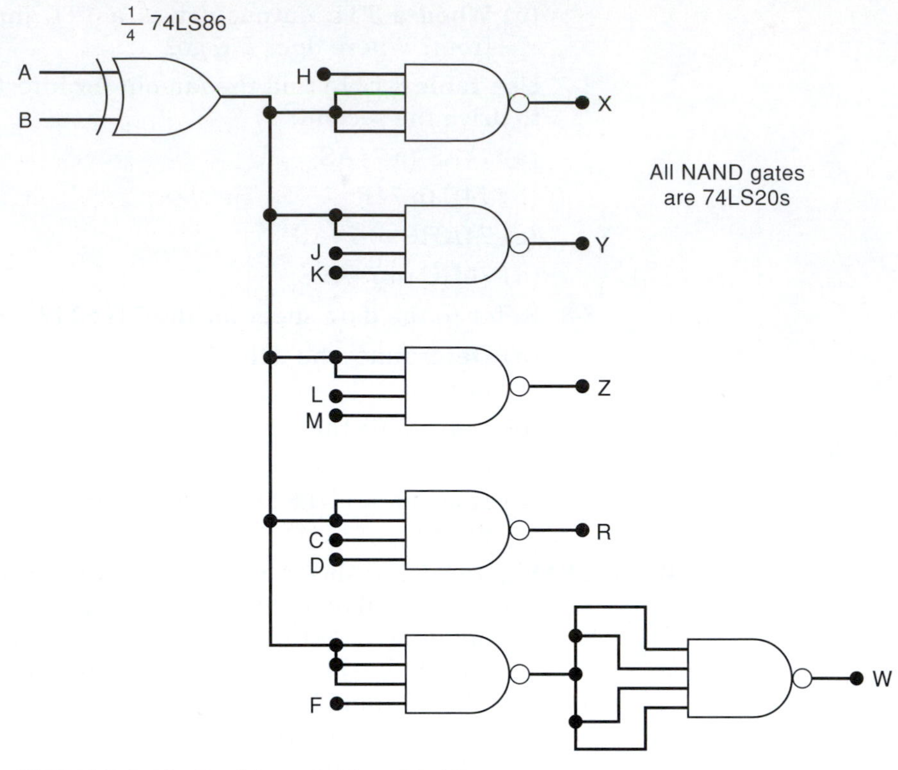

FIGURE 8-56 Problems 8-11 and 8-13.

B 8-12. How long does it take for the output of a typical 74LS04 to change states in response to a positive-going transition at its input?

C 8-13.* For the circuit of Figure 8-56, determine the longest time it will take for a change in the *A* input to be felt at output *W*. Use all worst-case conditions and maximum values of gate propagation delays. (*Hint:* Remember that NAND gates are inverting gates.) Repeat using all 74ALS devices.

8-14.* (a) Figure 8-57 shows a 74LS193 counter with its active-HIGH master reset input activated by a push-button switch. Resistor *R* is used to hold *MR* LOW while the switch is open. What is the maximum value that can be used for *R*?

(b) Repeat part (a) for the 74ALS193.

FIGURE 8-57 Problem 8-14.

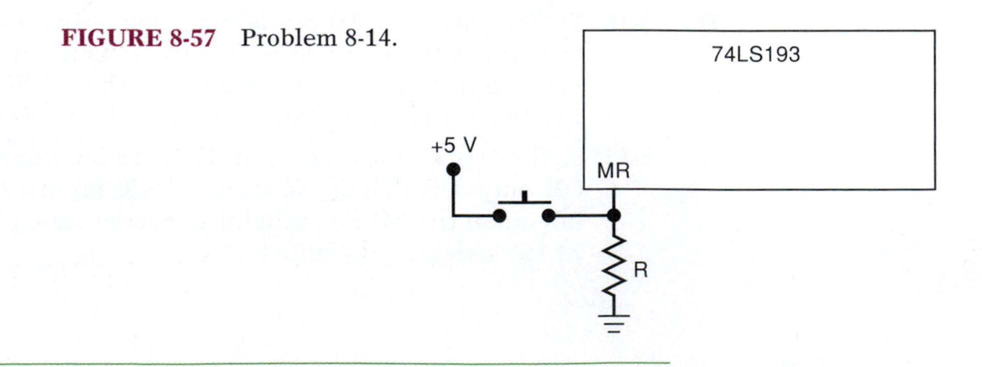

C, T 8-15. Figure 8-58(a) shows a circuit used to convert a 60-Hz sine wave to a 60-pps signal that can reliably trigger FFs and counters. This type of circuit might be used in a digital clock.

(a) Explain the circuit operation.

(b)*A technician is testing this circuit and observes that the 74LS14 output stays LOW. He checks the waveform at the INVERTER input, and it appears as shown in Figure 8-58(b). Thinking that the INVERTER is faulty, he replaces the chip and observes the same results. What do you think is causing the problem, and how can it be fixed? (*Hint:* Examine the v_x waveform carefully.)

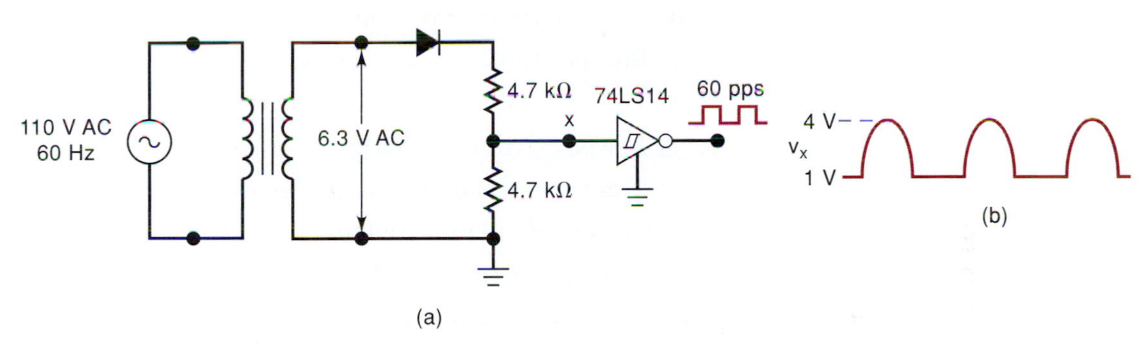

FIGURE 8-58 Problem 8-15.

T 8-16. For each waveform in Figure 8-59, determine *why* it will *not* reliably trigger a 74LS112 flip-flop at its *CLK* input.

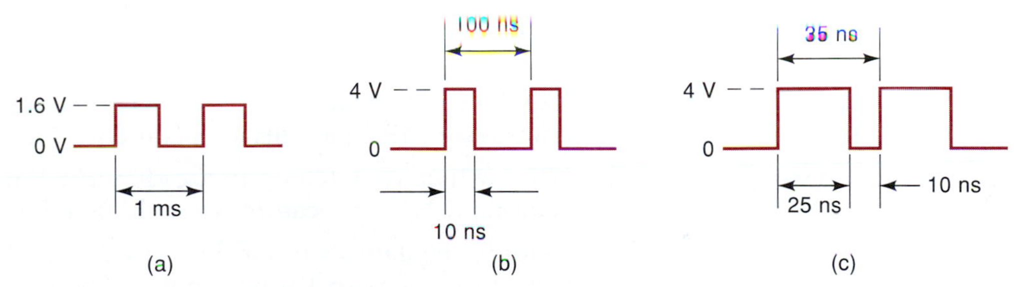

FIGURE 8-59 Problem 8-16.

T 8-17. A technician breadboards a logic circuit for testing. As she tests the circuit's operation, she finds that many of the FFs and counters are triggering erratically. Like any good technician, she checks the V_{CC} line with a dc meter and reads 4.97 V, which is acceptable for TTL. She then checks all circuit wiring and replaces each IC one by one, but the problem persists. Finally she decides to observe V_{CC} on the scope and sees the waveform shown in Figure 8-60. What is the probable cause of the noise on V_{CC}? What did the technician forget to include when she breadboarded the circuit?

FIGURE 8-60 Problem 8-17. V_{CC} 4.97 V 1.3 V

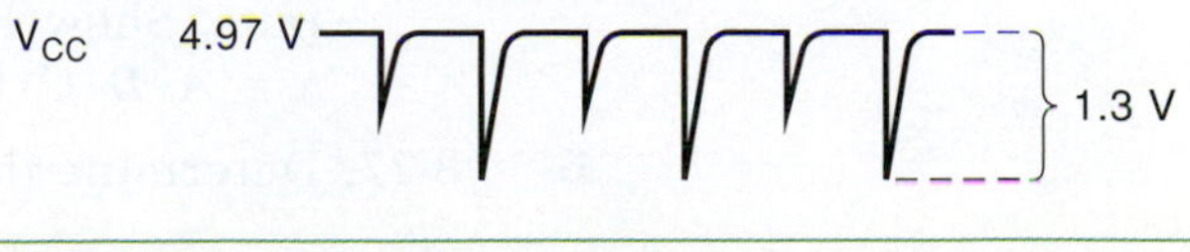

SECTIONS 8-7 TO 8-10

B 8-18. Which type of MOSFET is turned on by placing

(a) 5 V on the gate and 0 V on the source?

(b) 0 V on the gate and 5 V on the source?

B 8-19.* Which of the following are advantages that CMOS generally has over TTL?

(a) Greater packing density

(b) Higher speed

(c) Greater fan-out

(d) Lower output impedance

(e) Simpler fabrication process

(f) More suited for LSI

(g) Lower P_D (below 1 MHz)

(h) Transistors as only circuit element

(i) Lower input capacitance

(j) Less susceptible to ESD

8-20. Which of the following operating conditions will probably result in the lowest average P_D for a CMOS logic system? Explain.

(a) $V_{DD} = 5$ V, switching frequency $f_{max} = 1$ MHz

(b) $V_{DD} = 5$ V, $f_{max} = 10$ kHz

(c) $V_{DD} = 10$ V, $f_{max} = 10$ kHz

C 8-21.* The output of each INVERTER on a 74LS04 IC is driving two 74HCT08 inputs. The input to each INVERTER is LOW over 99 percent of the time. What is the maximum power that the 74LS04 chip is dissipating?

8-22. Use the values from Table 8-9 to calculate the HIGH-state noise margin when a 74HC gate drives a standard 74LS input.

8-23. What will cause latch-up in a CMOS IC? What might happen in this condition? What precautions should be taken to prevent latch-up?

8-24. Refer to the data sheet for the 74HC20 NAND gate IC. Use maximum values to calculate P_D(avg) and t_{pd}(avg). Compare with the values calculated in Problem 8-2 for TTL.

SECTIONS 8-11 AND 8-12

B 8-25. **DRILL QUESTION**

(a) Define wired-AND.

(b) What is a pull-up resistor? Why is it used?

(c) What types of TTL outputs can safely be tied together?

(d) What is bus contention?

D 8-26. The 74LS09 TTL IC is a quad two-input AND with open-collector outputs. Show how 74LS09s can be used to implement the operation $x = A \cdot B \cdot C \cdot D \cdot E \cdot F \cdot G \cdot H \cdot I \cdot J \cdot K \cdot M$.

B 8-27.* Determine the logic expression for output X in Figure 8-61.

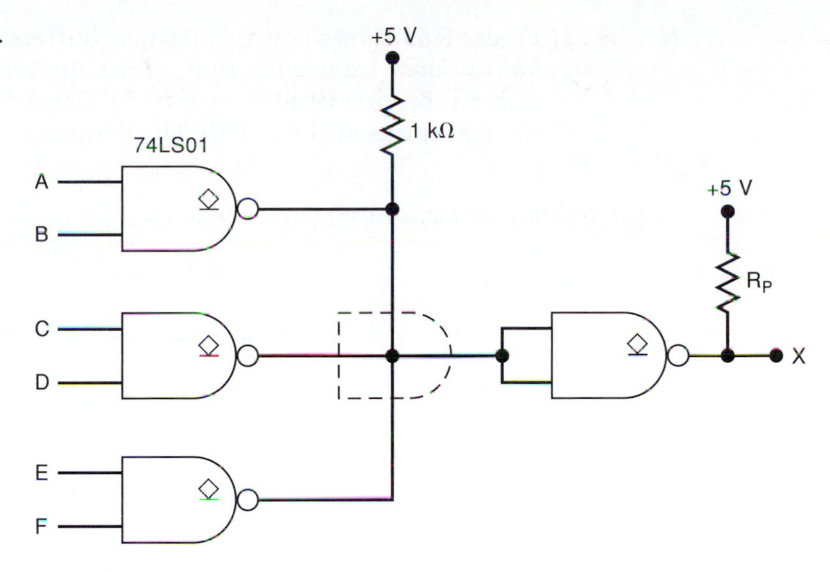

FIGURE 8-61 Problem 8-27.

C 8-28. Which of the following would be most likely to destroy a TTL totem-pole output while it is trying to switch from HIGH to LOW?

(a) Tying the output to +5 V

(b) Tying the output to ground

(c) Applying an input of 7 V

(d) Tying the output to another TTL totem-pole output

D 8-29.* Figure 8-62(a) shows a 7406 open-collector inverting buffer used to control the ON/OFF status of an LED to indicate the state of FF output Q. The LED's nominal specification is $V_F = 2.4$ V at $I_F = 20$ mA, and $I_F(\max) = 30$ mA.

(a) What voltage will appear at the 7406 output when $Q = 0$?

(b) Choose an appropriate value for the series resistor for proper operation.

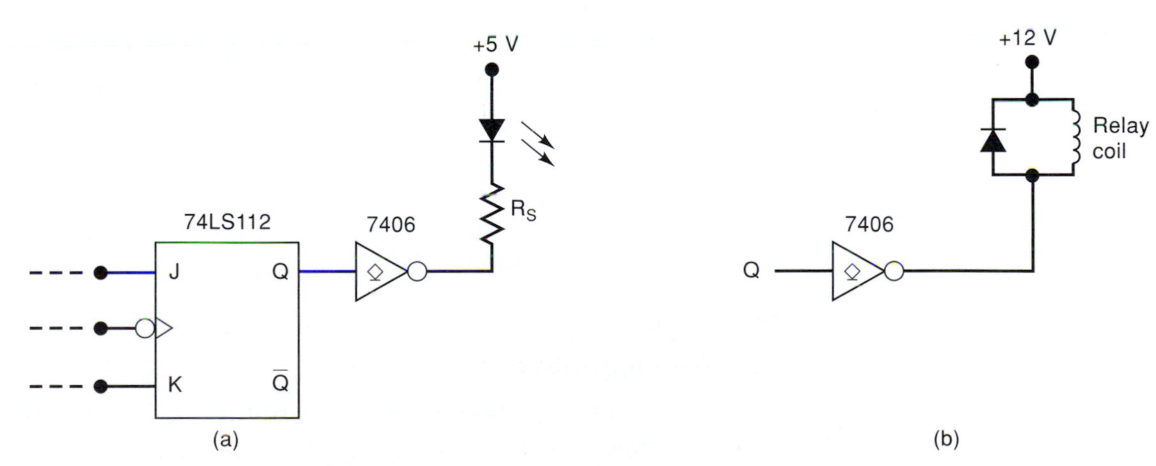

FIGURE 8-62 Problems 8-29 and 8-30.

8-30. In Figure 8-62(b), the 7406 output is used to switch current to a relay.

(a)* What voltage will be at the 7406 output when $Q = 0$?

(b)* What is the largest current relay that can be used?

(c) How can we modify this circuit to use a 7407?

N 8-31. Figure 8-63 shows how two tristate buffers can be used to construct a *bidirectional transceiver* that allows digital data to be transmitted in either direction (*A* to *B*, or *B* to *A*). Describe the circuit operation for the two states of the DIRECTION input.

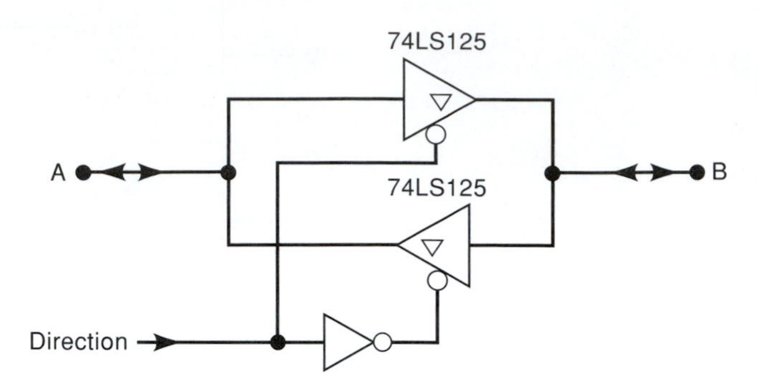

FIGURE 8-63 Problem 8-31.

8-32. The circuit of Figure 8-64 is used to provide the enable inputs for the circuit of Figure 8-37.

(a) Determine which of the data inputs (*A*, *B*, or *C*) will appear on the bus for each combination of inputs *x* and *y*.

(b) Explain why the circuit will not work if the NOR is changed to an XNOR.

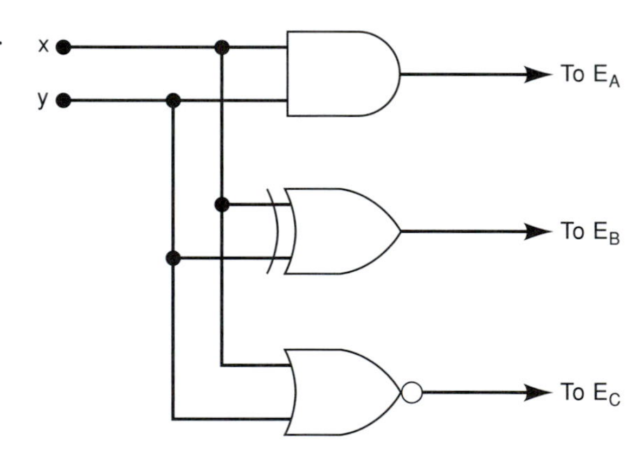

FIGURE 8-64 Problem 8-32.

8-33.* What type of counter circuit from Chapter 7 could control the enables in Figure 8-37 so that only one buffer is on at any time, and the buffers are enabled sequentially?

SECTION 8-14

B 8-34. **DRILL QUESTIONS**

(a) Which logic family must be used if maximum speed is of utmost importance?

(b) Which logic family uses the most power?

(c) Which TTL series is the fastest?

(d) Which pure CMOS series is the fastest?

(e) Which family has the best speed–power product?

8-35. Name two radical differences between ECL outputs and either TTL or CMOS outputs.

SECTION 8-15

8-36.* Determine the approximate values of V_{OUT} for both states of the CONTROL input in Figure 8-65.

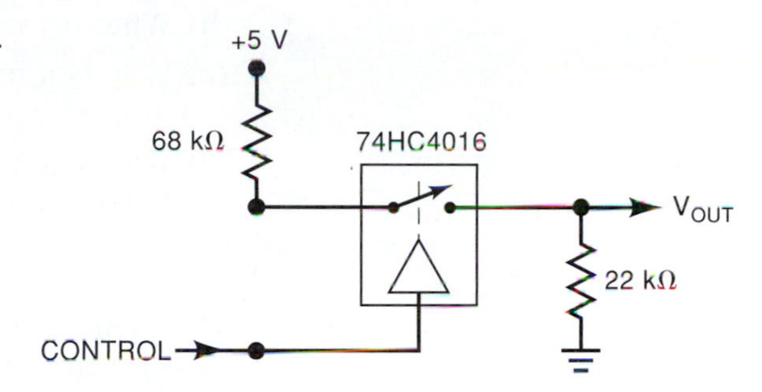

FIGURE 8-65 Problem 8-36.

8-37.* Determine the waveform at output X in Figure 8-66 for the given input waveforms. Assume that $R_{ON} \approx 200\ \Omega$ for the bilateral switch.

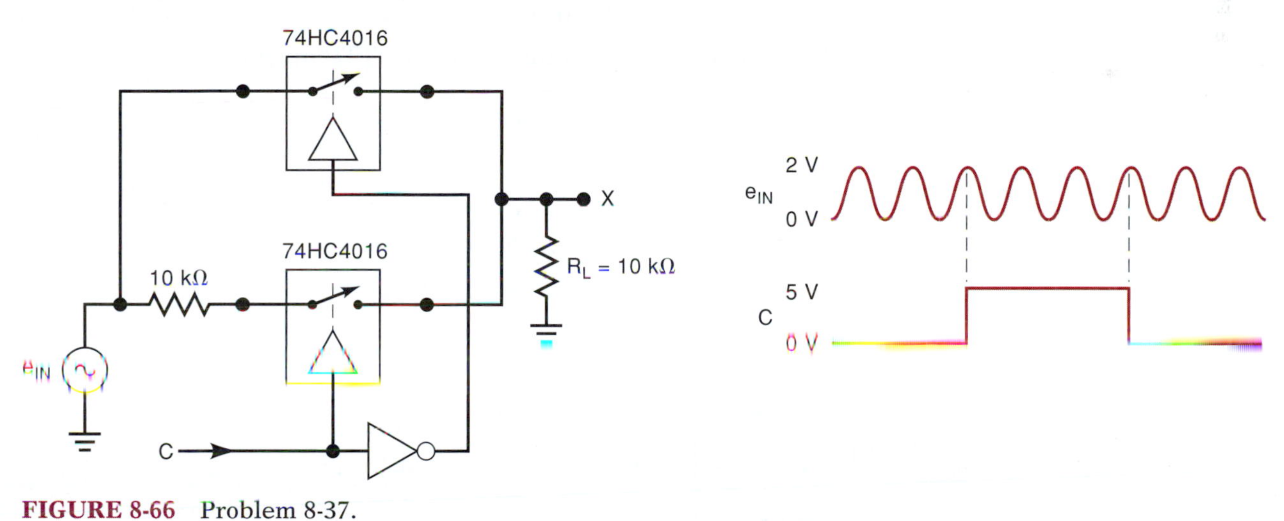

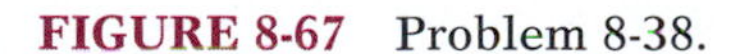

FIGURE 8-66 Problem 8-37.

N, D, C 8-38.* Determine the gain of the op-amp circuit of Figure 8-67 for the two states of the GAIN SELECT input. This circuit shows the basic principle of digitally controlled signal amplification.

FIGURE 8-67 Problem 8-38.

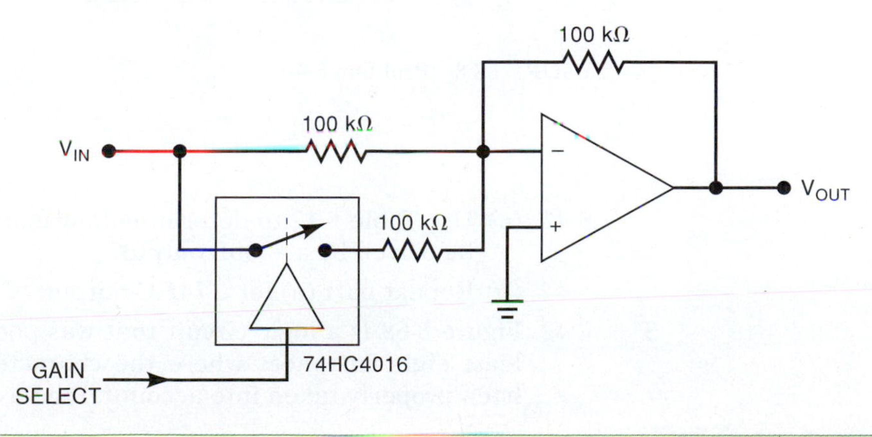

SECTION 8-16

B 8-39. **DRILL QUESTION**

(a) Which CMOS series can have its inputs driven directly from a TTL output?

(b) What is the function of a level translator? When is it used?

(c) Why is a buffer required between some CMOS outputs and TTL inputs?

(d) *True or false:* Most CMOS outputs have trouble supplying the TTL HIGH-state input current.

T 8-40. Refer to Figure 8-68(a), where a 74LS TTL output, Q, is driving a CMOS INVERTER operating at V_{DD} = 10 V. The waveforms at Q and X appear as shown in Figure 8-68(b). Which of the following is a possible reason why X stays HIGH?

(a) The 10-V supply is faulty.

(b) The pull-up resistor is too large.

(c) The 74LS112 output breaks down at well below 10 V and maintains a 5.5-V level in the HIGH state. This is in the indeterminate range for the CMOS input.

(d) The CMOS input is loading down the TTL output.

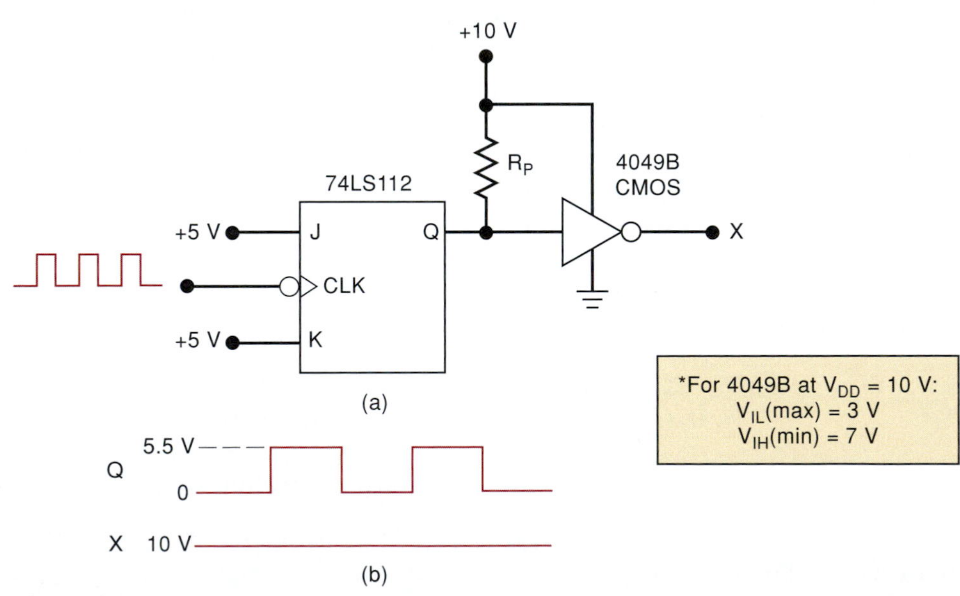

FIGURE 8-68 Problem 8-40.

8-41. (a)* Use Table 8-12 to determine how many 74AS inputs can typically be driven by a 4000B output.

(b) Repeat part (a) for a 74HC output.

T 8-42. Figure 8-69 is a logic circuit that was poorly designed. It contains at least eight instances where the characteristics of the ICs have not been properly taken into account. Find as many of these as you can.

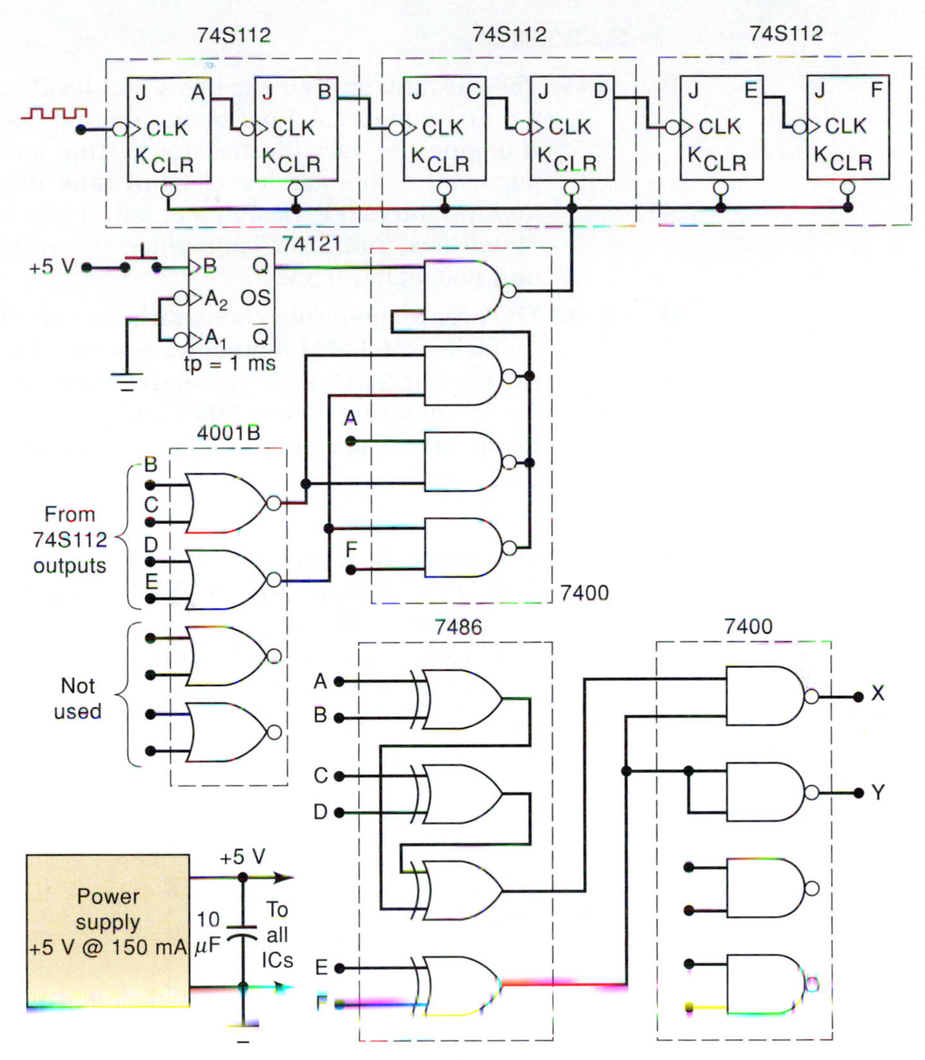

FIGURE 8-69 Problems 8-42 and 8-43.

T 8-43. Repeat Problem 8-42 with the following changes in the circuit:

- Each TTL IC is replaced with its 74LS equivalent.
- The 4001B is replaced with a 74HCT02.

8-44.* Use Table 8-12 to explain why the circuit of Figure 8-70 will not work as it is. How can the problem be corrected?

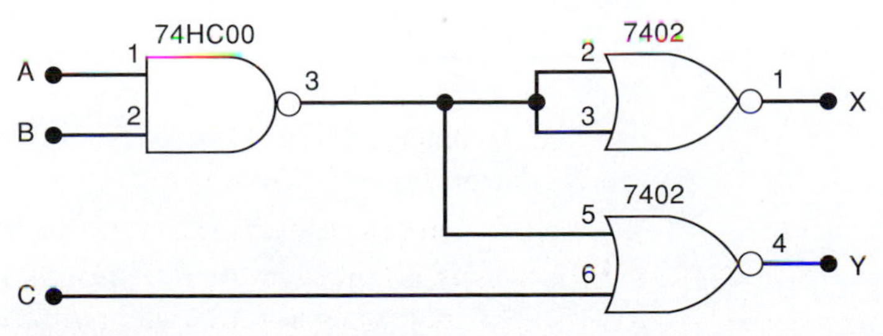

FIGURE 8-70 Problem 8-44.

SECTION 8-18

D 8-45. The gas tank on your car has a fuel-level sending unit that works like a potentiometer. A float moves up and down with the gasoline level, changing the variable resistor setting and producing a voltage proportional to the gas level. A full tank produces 12 V, and an empty tank produces 0 V. Design a circuit using an LM339 that turns on the "Fuel Low" indicator lamp when the voltage level from the sending unit gets below 0.5 V.

D 8-46.* The over-temperature comparator circuit in Figure 8-52 is modified by replacing the LM34 temperature sensor with an LM35 that outputs 10 mV per degree Celsius. The alarm must still be activated (HIGH) when the temperature is over 100°F, which is equal to approximately 38°C. Recalculate the values of R_1 and R_2 to complete the modification.

SECTION 8-19

T 8-47. The circuit in Figure 8-71 uses a 74HC05 IC that contains six open-drain INVERTERs. The INVERTERs are connected in a wired-AND arrangement. The output of the NAND gate is always HIGH, regardless of the inputs A–H. Describe a procedure that uses a logic probe and pulser to isolate this fault.

FIGURE 8-71 Problem 8-47.

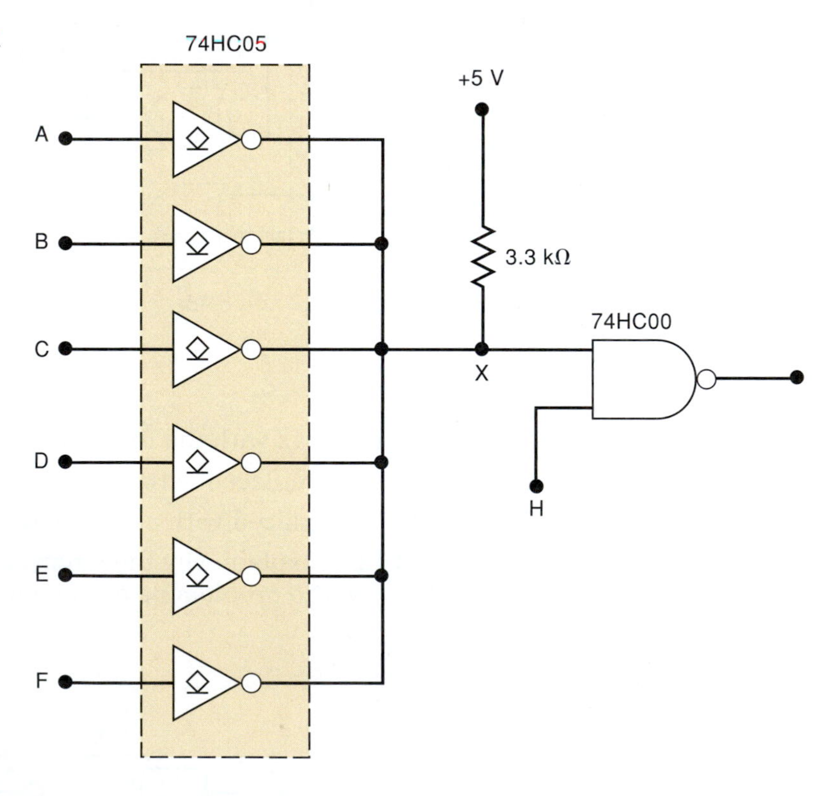

T 8-48. The circuit of Figure 8-53 has a solder bridge to ground somewhere between the output of the NAND gate and the input of the FF. Describe a procedure for a test that could be performed to indicate that the fault is on the circuit board and probably not in either the NAND or the FF ICs.

T 8-49.*In Figure 8-46, a logic probe indicates that the lower end of the pull-up resistor is stuck in the LOW state. Which of the following is the possible fault?

(a) The TTL gate's current-sourcing transistor is open.

(b) The TTL gate's current-sinking transistor has a collector–emitter short.

(c) There is a break in the connection from R_P to the CMOS gate.

MICROCOMPUTER APPLICATION

C, N 8-50.*In Chapter 5, we saw how a microprocessor (MPU), under software control, transfers data to an external register. The circuit diagram is repeated in Figure 8-72. Once the data are in the register, they are stored there and used for whatever purpose they are needed. Sometimes, each individual bit in the register has a unique function. For example, in automobile computers, each bit could represent the status of a different physical variable being monitored by the MPU. One bit might indicate when the engine temperature is too high. Another bit might signal when oil pressure is too low. In other applications, the bits in the register are used to produce an analog output that can be used to drive devices requiring analog inputs that have many different voltage levels.

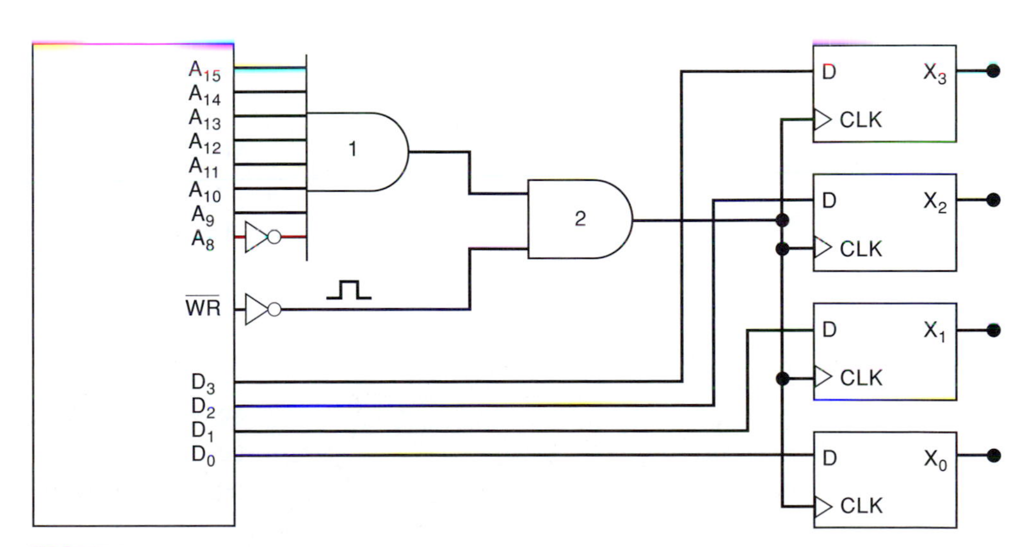

FIGURE 8-72 Problem 8-50.

Figure 8-73 shows how we can use the MPU to generate an analog voltage by taking the register data from Figure 8-72 and using them to control the inputs to a summing amplifier. Assume that the MPU is executing a program that is transferring a new set of data to the register every 10 μs according to Table 8-14. Sketch the resulting waveform at V_{OUT}.

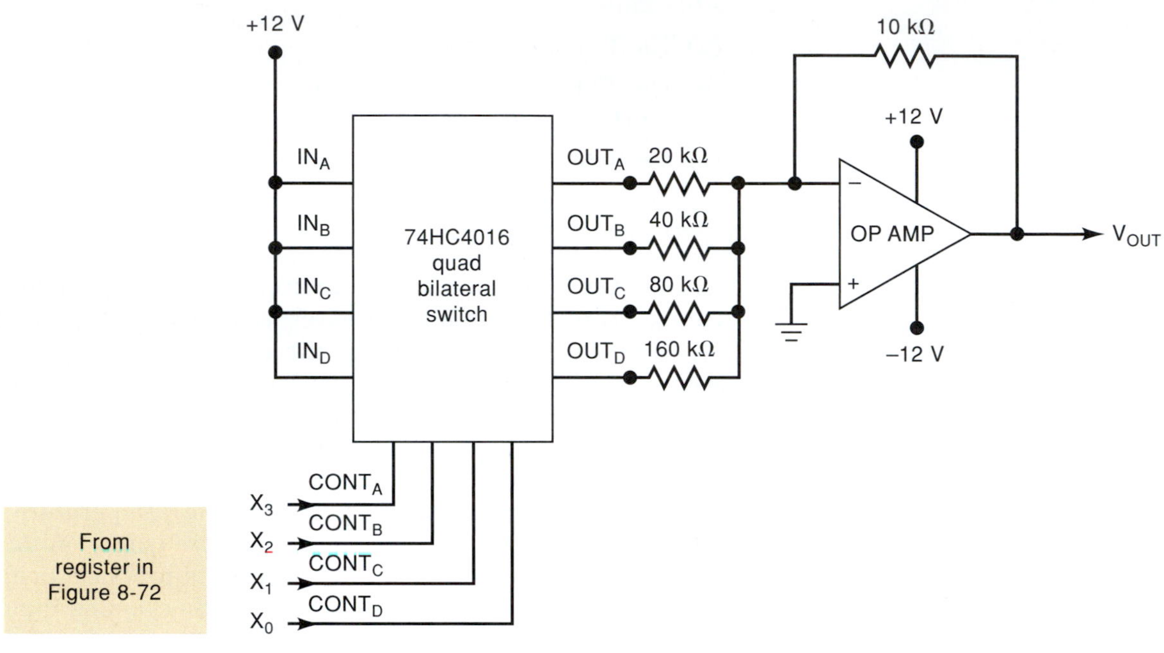

FIGURE 8-73 Problem 8-50.

TABLE 8-14 Problem 8-50.

Time (μs)	MPU Data
0	0000
10	0010
20	0100
30	0111
40	1010
50	1110
60	1111
70	1111
80	1110
90	1100
100	1000

ANSWERS TO SECTION REVIEW QUESTIONS

SECTION 8-1

1. See text. 2. False 3. False; V_{NH} is the difference between V_{OH}(min) and V_{IH}(min). 4. Current sinking: an output actually receives (sinks) current from the input of the circuit it is driving. Current sourcing: an output supplies (sources) current to the circuit it is driving. 5. DIP 6. J-lead 7. Its leads are bent. 8. No

SECTION 8-2

1. True 2. LOW 3. Fast switching times, low power dissipation; large current spike during switching from LOW to HIGH 4. Q_3 5. Q_6 6. No multiple-emitter transistor

SECTION 8-4

1. (a) 74AS (b) 74S, 74LS (c) standard 74 (d) 74S, 74LS, 74AS, 74ALS (e) 74ALS 2. All three can operate at 40 MHz, but the 74ALS193 will use less power. 3. Q_4, Q_5, respectively

SECTION 8-5

1. Q_4's ON-state resistance and V_{OL}(max) 2. 12 3. Its output voltages may not remain in the allowed logic 0 and 1 ranges. 4. Two; five

SECTION 8-6

1. LOW 2. Connect to $+V_{CC}$ through a 1-kΩ resistor; connect to another used input 3. Connect to ground; connect to another used input 4. False; only in the LOW state 5. Connecting small RF capacitors between V_{CC} and ground near each TTL IC to filter out voltage spikes caused by rapid current changes during output transitions from LOW to HIGH

SECTION 8-8

1. CMOS uses both P- and N-channel MOSFETs. 2. One 3. Six

SECTION 8-9

1. 74C, HC, HCT, AHC, AHCT 2. 74ACT, HCT, AHCT 3. 74C, HC/HCT, AC/ACT, AHC/AHCT 4. BiCMOS 5. Maximum permissible propagation delay; input capacitance of each load 6. See text. 7. CMOS 8. (a) True (b) False (c) False (d) False (e) True (f) False

SECTION 8-10

1. More circuits on chip; higher operating speed 2. Can't handle higher voltages: increased power dissipation can overheat the chip. 3. Same as standard TTL: 2.0 V 4. 74ALVC, 74LV 5. 74LVT

SECTION 8-11

1. When two or more circuit outputs are connected together 2. Damaging current can flow; V_{OL} exceeds V_{OL}(max). 3. Current-sinking transistor Q_4's collector is unconnected (there is no Q_3). 4. To produce a V_{OH} level 5. $\overline{A}\,\overline{B}\,\overline{C}\,\overline{D}\,\overline{E}\,\overline{F}$ 6. No active pull-up transistor 7. See Figure 8-34.

SECTION 8-12

1. HIGH, LOW, Hi-Z 2. Hi-Z 3. When two or more tristate outputs tied to a common bus are enabled at the same time 4. $E_A = E_B = 0$, $E_C = 1$ 5. See Figure 8-39.

SECTION 8-13

1. Less than 4 inches 2. Resistance, capacitance, inductance 3. To reduce reflections and reactive ringing on the line.

SECTION 8-14

1. (a) True (b) True (c) False (d) True (e) False (f) False

SECTION 8-15

1. The logical level at the control input controls the open/closed status of a bidirectional switch that can pass analog signals in either direction. 2. True

SECTION 8-16

1. A pull-up resistor must be connected to +5 V at the TTL output. 2. CMOS I_{OH} or I_{OL} may be too low.

SECTION 8-17

1. It takes the output from a driver circuit and conditions it so that it is compatible with the input requirements of the load. 2. True 3. False; for example, the 4000B series cannot sink I_{IL} of a 74 or a 74AS device. 4. 74HCT and ACT 5. Two

SECTION 8-18

1. $V^{(+)} > V^{(-)}$ 2. $V^{(-)} > V^{(+)}$ 3. Open-collector

SECTION 8-19

1. It injects a voltage pulse of selected polarity at a node that is not shorted to V_{CC} or ground. 2. False 3. False 4. The pulse LED flashes once each time the pulser is activated.

CHAPTER 9

MSI LOGIC CIRCUITS

OUTLINE

OBJECTIVES

Upon completion of this chapter, you will be able to:

- Analyze and use decoders and encoders in various types of circuit applications.
- Compare the advantages and disadvantages of LEDs and LCDs.
- Utilize the observation/analysis technique for troubleshooting digital circuits.
- Understand the operation of multiplexers and demultiplexers by analyzing several circuit applications.
- Compare two binary numbers by using the magnitude comparator circuit.
- Understand the function and operation of code converters.
- Cite the precautions that must be considered when connecting digital circuits using the data bus concept.
- Use HDL to implement the equivalent of MSI logic circuits.

INTRODUCTION

Digital systems obtain binary-coded data and information that are continuously being operated on in some manner. Some of the operations include: (1) *decoding and encoding,* (2) *multiplexing,* (3) *demultiplexing,* (4) *comparison,* (5) *code conversion,* and (6) *data busing.* All of these operations and others have been facilitated by the availability of numerous ICs in the MSI (medium-scale-integration) category.

In this chapter, we will study many of the common types of MSI devices. For each type, we will start with a brief discussion of its basic operating principle and then introduce specific ICs. We then show how they can be used alone or in combination with other ICs in various applications.

9-1 DECODERS

A **decoder** is a logic circuit that accepts a set of inputs that represents a binary number and activates only the output that corresponds to that input number. In other words, a decoder circuit looks at its inputs, determines which binary number is present there, and activates the one output that corresponds to that number; all other outputs remain inactive. The diagram for a general decoder is shown in Figure 9-1 with N inputs and M outputs. Because each of the N inputs can be 0 or 1, there are 2^N possible input combinations or codes. For each of these input combinations, only one of the M outputs will be active (HIGH); all the other outputs are LOW. Many decoders are designed to produce active-LOW outputs, where only the selected output is LOW while all others are HIGH. This situation is indicated by the presence of small circles on the output lines in the decoder diagram.

Some decoders do not utilize all of the 2^N possible input codes but only certain ones. For example, a BCD-to-decimal decoder has a four-bit input code and *ten* output lines that correspond to the *ten* BCD code groups 0000 through 1001. Decoders of this type are often designed so that if any of the unused codes are applied to the input, *none* of the outputs will be activated.

In Chapter 7, we saw how decoders are used in conjunction with counters to detect the various states of the counter. In that application, the FFs in the counter provided the binary code inputs for the decoder. The same basic decoder circuitry is used no matter where the inputs come from. Figure 9-2 shows the circuitry for a decoder with three inputs and $2^3 = 8$ outputs. It uses all AND gates, and so the outputs are active-HIGH. Note that for a given input code, the only output that is active (HIGH) is the one corresponding to the decimal equivalent of the binary input code (e.g., output O_6 goes HIGH only when $CBA = 110_2 = 6_{10}$).

This decoder can be referred to in several ways. It can be called a *3-line-to-8-line decoder* because it has three input lines and eight output lines. It can also be called a *binary-to-octal decoder* or *converter* because it takes a three-bit binary input code and activates one of the eight (octal) outputs corresponding to that code. It is also referred to as a *1-of-8 decoder* because only 1 of the 8 outputs is activated at one time.

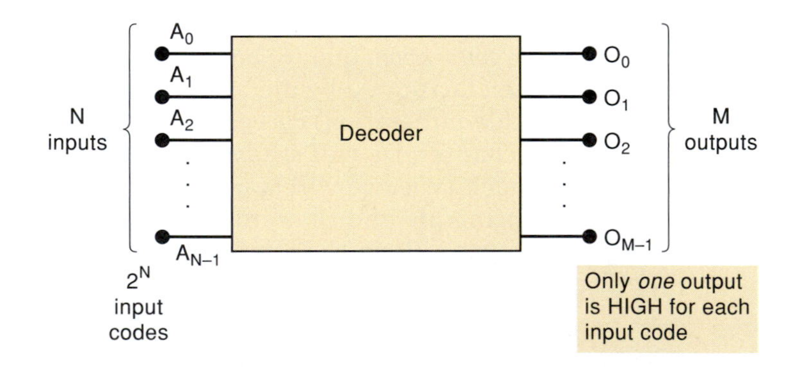

FIGURE 9-1 General decoder diagram.

ENABLE Inputs

Some decoders have one or more ENABLE inputs that are used to control the operation of the decoder. For example, refer to the decoder in Figure 9-2 and visualize having a common ENABLE line connected to a fourth input of each gate. With this ENABLE line held HIGH, the decoder will function normally, and the A, B, C input code will determine which output is HIGH. With ENABLE held LOW, however, *all* of the outputs will be forced to the LOW state regardless of the levels at the A, B, C inputs. Thus, the decoder is enabled only if ENABLE is HIGH.

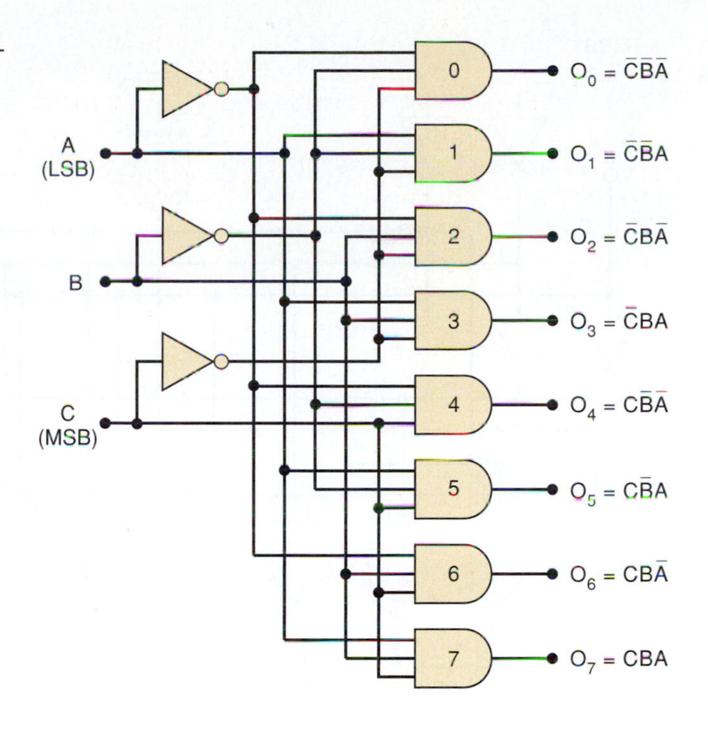

FIGURE 9-2 Three-line-to-8-line (or 1-of-8) decoder.

C	B	A	O_7	O_6	O_5	O_4	O_3	O_2	O_1	O_0
0	0	0	0	0	0	0	0	0	0	1
0	0	1	0	0	0	0	0	0	1	0
0	1	0	0	0	0	0	0	1	0	0
0	1	1	0	0	0	0	1	0	0	0
1	0	0	0	0	0	1	0	0	0	0
1	0	1	0	0	1	0	0	0	0	0
1	1	0	0	1	0	0	0	0	0	0
1	1	1	1	0	0	0	0	0	0	0

Figure 9-3(a) shows the logic diagram for the 74ALS138 decoder. By examining this diagram carefully, we can determine exactly how this decoder functions. First, notice that it has NAND gate outputs, so its outputs are active-LOW. Another indication is the labeling of the outputs as $\overline{O}_7, \overline{O}_6, \overline{O}_5$, and so on; the overbar indicates active-LOW outputs.

The input code is applied at A_2, A_1, and A_0, where A_2 is the MSB. With three inputs and eight outputs, this is a 3-to-8 decoder or, equivalently, a 1-of-8 decoder.

Inputs $\overline{E}_1, \overline{E}_2$, and E_3 are separate enable inputs that are combined in the AND gate. In order to enable the output NAND gates to respond to the input code at $A_2A_1A_0$, this AND gate output must be HIGH. This will occur only when $\overline{E}_1 = \overline{E}_2 = 0$ and $E_3 = 1$. In other words, $\overline{E}_1$ and $\overline{E}_2$ are active-LOW, E_3 is active-HIGH, and all three must be in their active states to activate the decoder outputs. If one or more of the enable inputs is in its inactive state, the AND output will be LOW, which will force all NAND outputs to their inactive HIGH state regardless of the input code. This operation is summarized in the truth table in Figure 9-3(b). Recall that x represents the don't-care condition.

The logic symbol for the 74ALS138 is shown in Figure 9-3(c). Note how the active-LOW outputs are represented and how the enable inputs are represented. Even though the enable AND gate is shown as external to the decoder block, it is part of the IC's internal circuitry. The 74HC138 is the high-speed CMOS version of this decoder.

FIGURE 9-3 (a) Logic diagram for the 74ALS138 decoder; (b) truth table; (c) logic symbol.

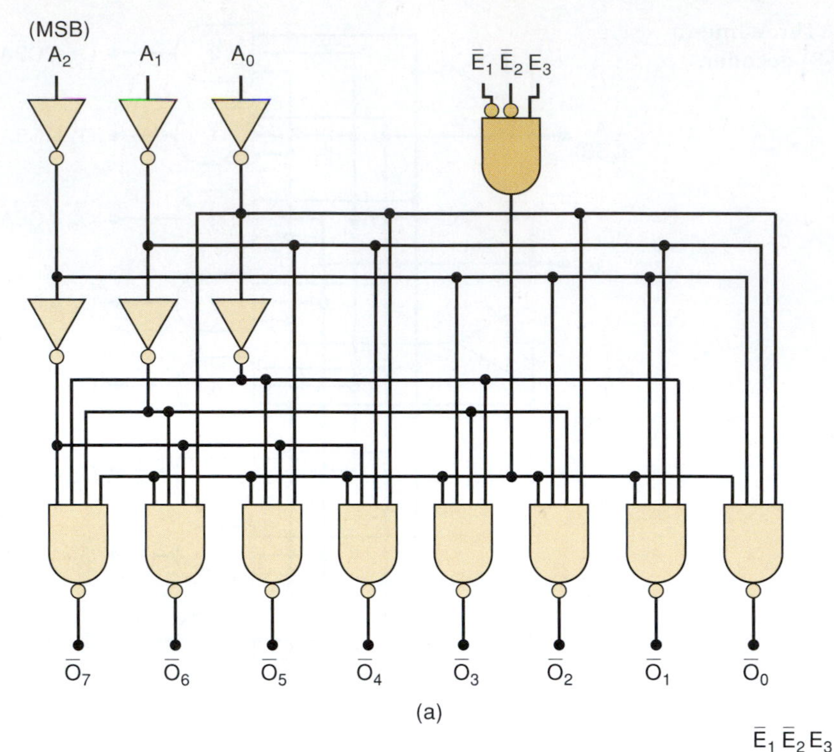

$\overline{E}_1$	$\overline{E}_2$	E_3	Outputs
0	0	1	Respond to input code $A_2A_1A_0$
1	X	X	Disabled – all HIGH
X	1	X	Disabled – all HIGH
X	X	0	Disabled – all HIGH

(b)

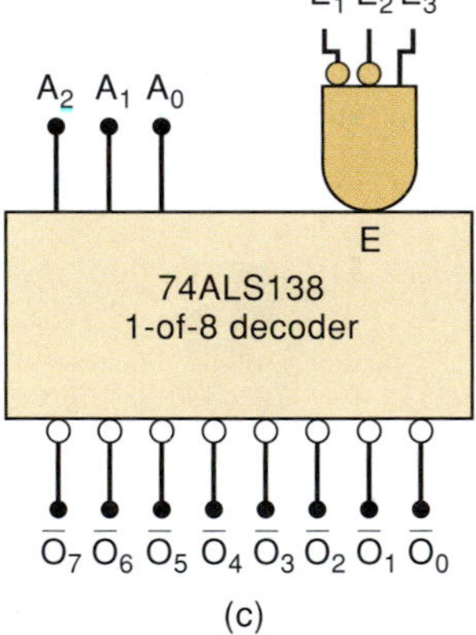

EXAMPLE 9-1

Indicate the states of the 74ALS138 outputs for each of the following sets of inputs.

(a) $E_3 = \overline{E}_2 = 1, \overline{E}_1 = 0, A_2 = A_1 = 1, A_0 = 0$
(b) $E_3 = 1, \overline{E}_2 = \overline{E}_1 = 0, A_2 = 0, A_1 = A_0 = 1$

Solution

(a) With $\overline{E}_2 = 1$, the decoder is disabled and all of its outputs will be in their inactive HIGH state. This can be determined from the truth table or by following the input levels through the circuit logic.

(b) All of the enable inputs are activated, so the decoding portion is enabled. It will decode the input code $011_2 = 3_{10}$ to activate output $\overline{O}_3$. Thus, $\overline{O}_3$ will be LOW and all other outputs will be HIGH.

EXAMPLE 9-2

Figure 9-4 shows how four 74ALS138s and an INVERTER can be arranged to function as a 1-of-32 decoder. The decoders are labeled Z_1 to Z_4 for easy reference, and the eight outputs from each one are combined into 32 outputs. Z_1's outputs are $\overline{O}_0$ to $\overline{O}_7$; Z_2's outputs $\overline{O}_0$ to $\overline{O}_7$ are renamed $\overline{O}_8$ to $\overline{O}_{15}$, respectively; Z_3's outputs are renamed $\overline{O}_{16}$ to $\overline{O}_{23}$; and Z_4's are renamed $\overline{O}_{24}$ to $\overline{O}_{31}$. A five-bit input code $A_4A_3A_2A_1A_0$ will activate only one of these 32 outputs for each of the 32 possible input codes.

(a) Which output will be activated for $A_4A_3A_2A_1A_0 = 01101$?

(b) What range of input codes will activate the Z_4 chip?

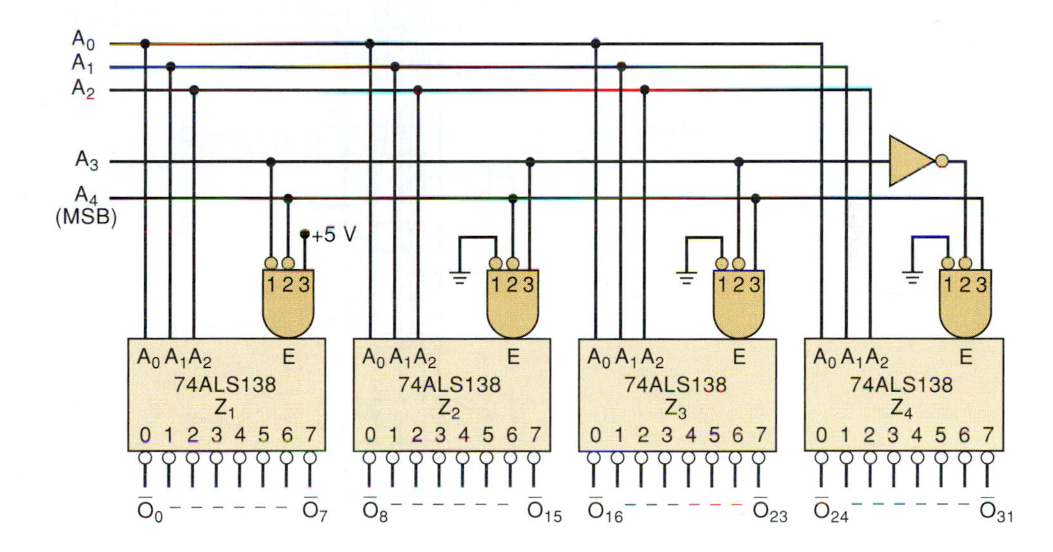

FIGURE 9-4 Four 74ALS138s forming a 1-of-32 decoder.

Solution

(a) The five-bit code has two distinct portions. The A_4 and A_3 bits determine which one of the decoder chips Z_1 to Z_4 will be enabled, while $A_2A_1A_0$ determine which output of the enabled chip will be activated. With $A_4A_3 = 01$, only Z_2 has all of its enable inputs activated. Thus, Z_2 responds to the $A_2A_1A_0 = 101$ code and activates its $\overline{O}_5$ output, which has been renamed $\overline{O}_{13}$. Thus, the input code 01101, which is the binary equivalent of decimal 13, will cause output $\overline{O}_{13}$ to go LOW, while all others stay HIGH.

(b) To enable Z_4, both A_4 and A_3 must be HIGH. Thus, all input codes ranging from 11000 (24_{10}) to 11111 (31_{10}) will activate Z_4. This corresponds to outputs $\overline{O}_{24}$ to $\overline{O}_{31}$.

BCD-to-Decimal Decoders

Figure 9-5(a) shows the logic diagram for a 7442 **BCD-to-decimal decoder**. It is also available as a 74LS42 and a 74HC42. Each output goes LOW only when its corresponding BCD input is applied. For example, $\overline{O}_5$ will go LOW only when inputs $DCBA = 0101$; $\overline{O}_8$ will go LOW only when $DCBA = 1000$. For input combinations that are invalid for BCD, none of the outputs will be activated. This decoder can also be referred to as a *4-to-10 decoder* or a *1-of-10 decoder*. The logic symbol and the truth table for the 7442 are also shown in

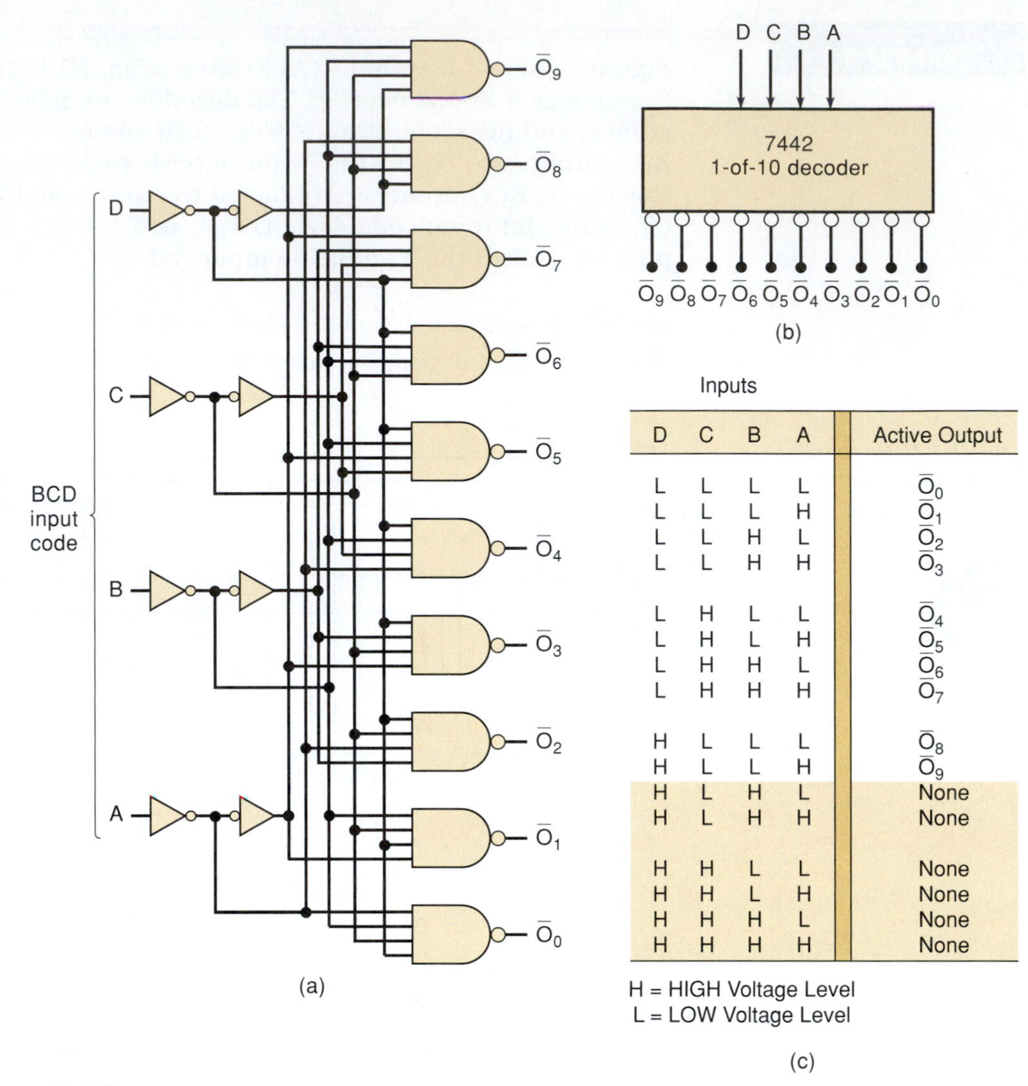

D	C	B	A	Active Output
L	L	L	L	$\overline{O}_0$
L	L	L	H	$\overline{O}_1$
L	L	H	L	$\overline{O}_2$
L	L	H	H	$\overline{O}_3$
L	H	L	L	$\overline{O}_4$
L	H	L	H	$\overline{O}_5$
L	H	H	L	$\overline{O}_6$
L	H	H	H	$\overline{O}_7$
H	L	L	L	$\overline{O}_8$
H	L	L	H	$\overline{O}_9$
H	L	H	L	None
H	L	H	H	None
H	H	L	L	None
H	H	L	H	None
H	H	H	L	None
H	H	H	H	None

FIGURE 9-5 (a) Logic diagram for the 7442 BCD-to-decimal decoder; (b) logic symbol; (c) truth table.

the figure. Note that this decoder does not have an enable input. In Problem 9-7, we will see how the 7442 can be used as a 3-to-8 decoder, with the *D* input used as an enable input.

BCD-to-Decimal Decoder/Driver

The TTL 7445 is a BCD-to-decimal decoder/**driver**. The term *driver* is added to its description because this IC has open-collector outputs that can operate at higher current and voltage limits than a normal TTL output. The 7445's outputs can sink up to 80 mA in the LOW state, and they can be pulled up to 30 V in the HIGH state. This makes them suitable for directly driving loads such as indicator LEDs or lamps, relays, or dc motors.

Decoder Applications

Decoders are used whenever an output or a group of outputs is to be activated only on the occurrence of a specific combination of input levels. These input levels are often provided by the outputs of a counter or a register.

When the decoder inputs come from a counter that is being continually pulsed, the decoder outputs will be activated sequentially, and they can be used as timing or sequencing signals to turn devices on or off at specific times. An example of this operation is shown in Figure 9-6 using the 74ALS163 counter and the 7445 decoder/driver described above.

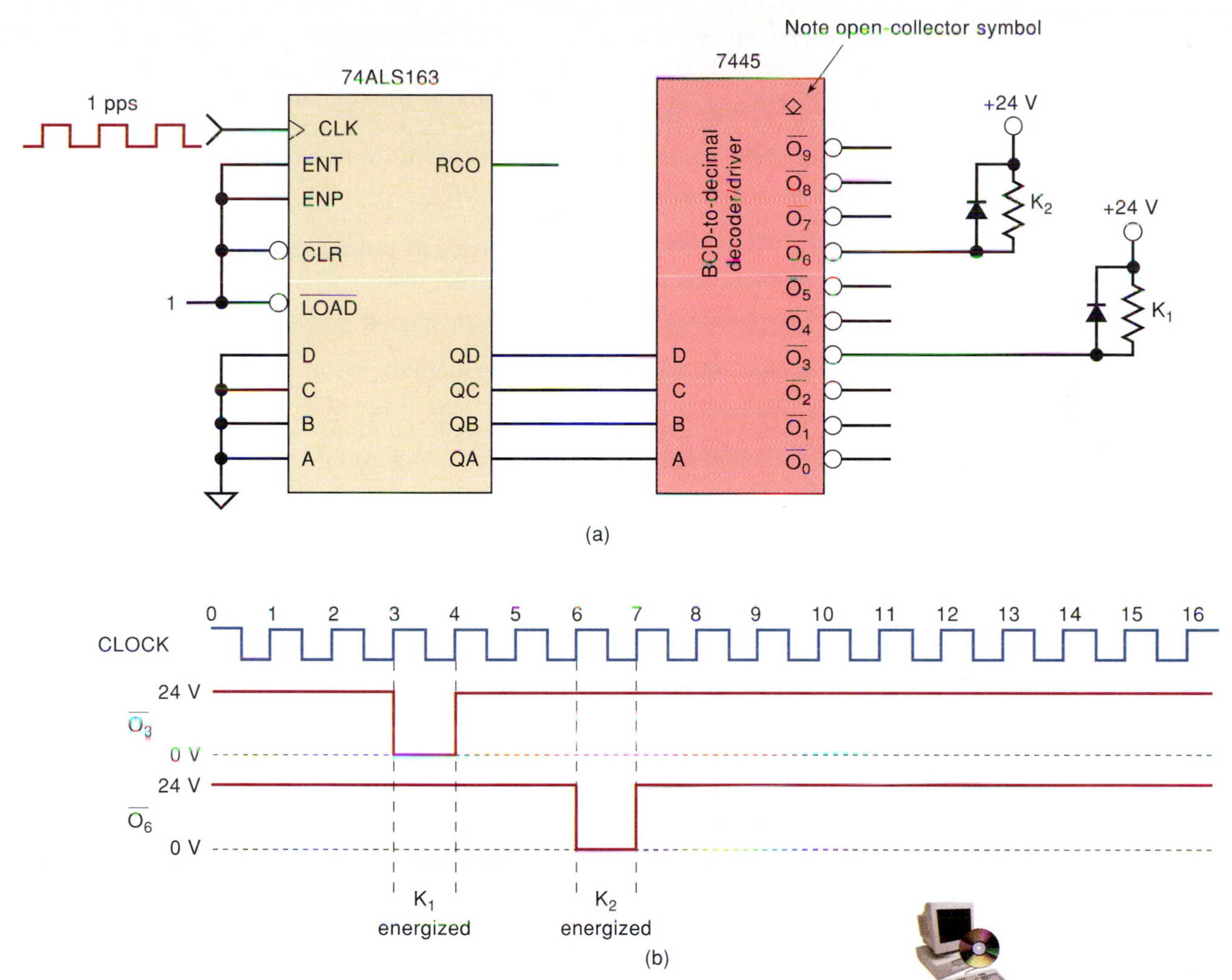

FIGURE 9-6 Example 9-3: counter/decoder combination used to provide timing and sequencing operations.

EXAMPLE 9-3

Describe the operation of the circuit in Figure 9-6(a).

Solution

The counter is being pulsed by a 1-pps signal so that it will sequence through the binary counts at the rate of 1 count/s. The counter FF outputs are connected as the inputs to the decoder. The 7445 open-collector outputs $\overline{O}_3$ and $\overline{O}_6$ are used to switch relays K_1 and K_2 on and off. For instance, when $\overline{O}_3$ is in its inactive HIGH state, its output transistor will be off (nonconducting) so that no current can flow through relay K_1 and it will be deenergized. When $\overline{O}_3$ is in its active-LOW state, its output transistor is on and acts as a current sink for current through K_1 so that K_1 is energized. Note that the relays operate from +24 V. Also note the presence of the diodes across the relay coils; these protect the decoder's output transistors from the large "inductive kick" voltage that would be produced when coil current is stopped abruptly.

The timing diagram in Figure 9-6(b) shows the sequence of events. If we assume that the counter is in the 0000 state at time 0, then both outputs $\overline{O}_3$ and $\overline{O}_6$ are initially in the inactive HIGH state, where their output transistors are off and both relays are deenergized. As clock pulses are applied, the counter will be incremented once per second. On the NGT of the third pulse (time 3), the counter will go to the 0011 (3) state. This will activate decoder output $\overline{O}_3$ and thereby energize K_1. On the NGT of the fourth pulse, the counter goes to the 0100 (4) state. This will deactivate $\overline{O}_3$ and deenergize relay K_1.

Similarly, at time 6, the counter will go to the 0110 (6) state; this will make $\overline{O}_6 = 0$ and energize K_2. At time 7, the counter goes to 0111 (7) and deactivates O_6 to deenergize K_2.

The counter will continue counting as pulses are applied. After 16 pulses, the sequence just described will start over.

Decoders are widely used in the memory system of a computer where they respond to the address code generated by the central processor to activate a particular memory location. Each memory IC contains many registers that can store binary numbers (data). Each register needs to have its own unique address to distinguish it from all the other registers. A decoder is built into the memory IC's circuitry and allows a particular storage register to be activated when a unique combination of inputs (i.e., its address) is applied. In a system, there are usually several memory ICs combined to make up the entire storage capacity. A decoder is used to select a memory chip in response to a range of addresses by decoding the most significant bits of the system address and enabling (selecting) a particular chip. We will examine this application in Problem 9-63, and we will study it in much more depth when we read about memories in Chapter 12.

In more complicated memory systems, the memory chips are arranged in multiple banks that must be selected individually or simultaneously, depending on whether the microprocessor wants one or more bytes at a time. This means that under certain circumstances, more than one output of the decoder must be activated. For systems such as this, a programmable logic device is often used to implement the decoder because a simple 1-of-8 decoder alone is not sufficient. Programmable logic devices can be used easily for custom decoding applications.

REVIEW QUESTIONS

1. Can more than one decoder output be activated at one time?
2. What is the function of a decoder's enable input(s)?
3. How does the 7445 differ from the 7442?
4. The 74154 is a 4-to-16 decoder with two active-LOW enable inputs. How many pins (including power and ground) does this IC have?

9-2 BCD-TO-7-SEGMENT DECODER/DRIVERS

Most digital equipment has some means for displaying information in a form that can be understood readily by the user or operator. This information is often numerical data but can also be alphanumeric (numbers and letters). One

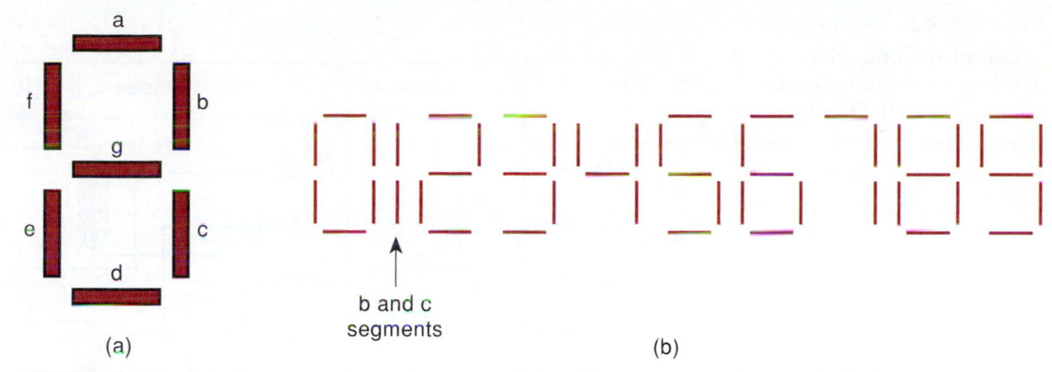

FIGURE 9-7 (a) 7-segment arrangement; (b) active segments for each digit.

of the simplest and most popular methods for displaying numerical digits uses a 7-segment configuration [Figure 9-7(a)] to form the decimal characters 0 through 9 and sometimes the hex characters A through F. One common arrangement uses light-emitting diodes (LEDs) for each segment. By controlling the current through each LED, some segments will be light and others will be dark so that the desired character pattern will be generated. Figure 9-7(b) shows the segment patterns that are used to display the various digits. For example, to display a "6," the segments *a*, *c*, *d*, *e*, *f*, and *g* are made bright while segment *b* is dark.

A **BCD-to-7-segment decoder/driver** is used to take a four-bit BCD input and provide the outputs that will pass current through the appropriate segments to display the decimal digit. The logic for this decoder is more complicated than the logic of decoders that we have looked at previously because each output is activated for more than one combination of inputs. For example, the *e* segment must be activated for any of the digits 0, 2, 6, and 8, which means whenever any of the codes 0000, 0010, 0110, or 1000 occurs.

Figure 9-8(a) shows a BCD-to-7-segment decoder/driver (TTL 7446 or 7447) being used to drive a 7-segment LED readout. Each segment consists of an LED (light-emitting diode). Diodes are solid-state devices that allow current to flow through them in one direction, but block the flow in the other direction. Whenever the anode of an LED is more positive than the cathode by approximately 2 V, the LED will light up. The anodes of the LEDs are all tied to V_{CC} (+5 V). The cathodes of the LEDs are connected through current-limiting resistors to the appropriate outputs of the decoder/driver. The decoder/driver has active-LOW outputs that are open-collector driver transistors and can sink a fairly large current because LED readouts may require 10 to 40 mA per segment, depending on their type and size.

To illustrate the operation of this circuit, let us suppose that the BCD input is $D = 0, C = 1, B = 0, A = 1$, which is BCD for 5. With these inputs, the decoder/driver outputs $\overline{a}, \overline{f}, \overline{g}, \overline{c}$, and $\overline{d}$ will be driven LOW (connected to ground), allowing current to flow through the *a*, *f*, *g*, *c*, and *d* LED segments and thereby displaying the numeral 5. The $\overline{b}$ and $\overline{e}$ outputs will be HIGH (open), so that LED segments *b* and *e* cannot conduct.

The 7446/47 decoder/drivers are designed to activate specific segments even for non-BCD input codes (greater than 1001). Figure 9-8(b) shows the activated segment patterns for all possible input codes from 0000 to 1111. Note that an input code of 1111 (15) will blank out all the segments.

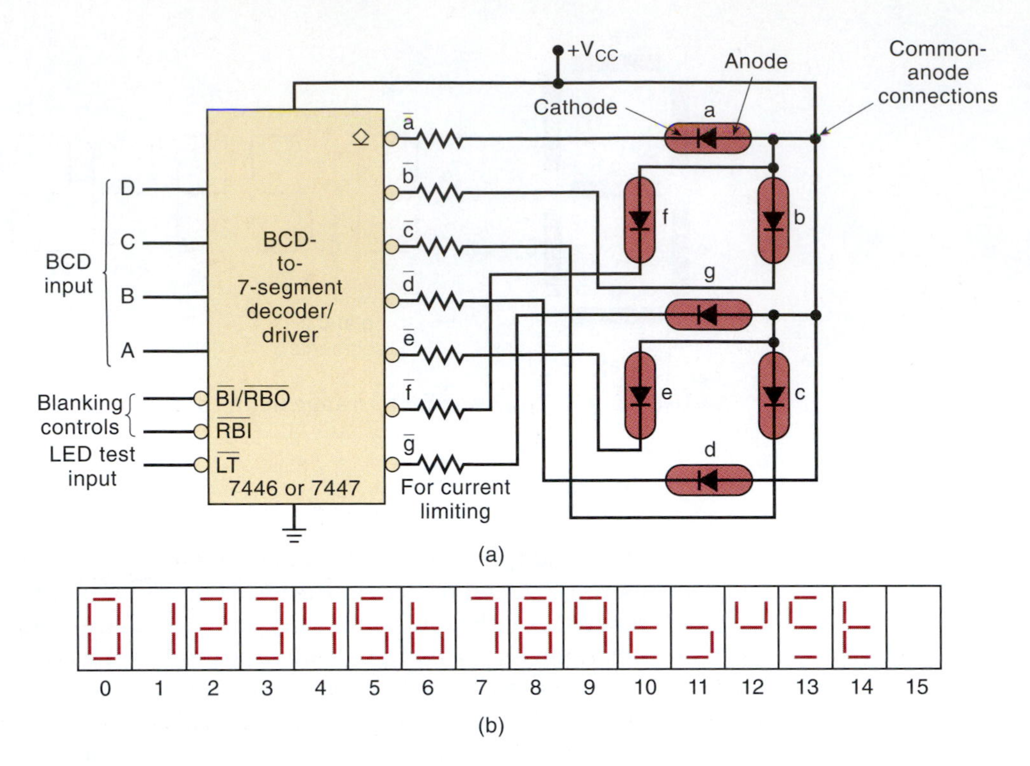

FIGURE 9-8 (a) BCD-to-7-segment decoder/driver driving a common-anode 7-segment LED display; (b) segment patterns for all possible input codes.

Seven-segment decoder/drivers such as the 7446/47 are exceptions to the rule that decoder circuits activate only one output for each combination of inputs. Rather, they activate a unique pattern of outputs for each combination of inputs.

Common-Anode Versus Common-Cathode LED Displays

The LED display used in Figure 9-8 is a **common-anode** type because the anodes of all of the segments are tied together to V_{CC}. Another type of 7-segment LED display uses a **common-cathode** arrangement where the cathodes of all of the segments are tied together and connected to ground. This type of display must be driven by a BCD-to-7-segment decoder/driver with active-HIGH outputs that apply a HIGH voltage to the anodes of those segments that are to be activated. Because each segment requires 10 to 20 mA of current to light it, TTL and CMOS devices are normally not used to drive the common-cathode display directly. Recall from Chapter 8 that TTL and CMOS outputs are not able to source large amounts of current. A transistor interface circuit is often used between decoder chips and the common-cathode display.

EXAMPLE 9-4

Each segment of a typical 7-segment LED display is rated to operate at 10 mA at 2.7 V for normal brightness. Calculate the value of the current-limiting resistor needed to produce approximately 10 mA per segment.

Solution

Referring to Figure 9-8(a), we can see that the series resistor must have a voltage drop equal to the difference between $V_{CC} = 5$ V and the segment voltage of 2.7 V. This 2.3 V across the resistor must produce a current of about 10 mA. Thus, we have

$$R_S = \frac{2.3\ \text{V}}{10\ \text{mA}} = 230\ \Omega$$

A standard resistor value close to this can be used. A 220-Ω resistor would be a good choice.

REVIEW QUESTIONS

1. Which LED segments will be on for a decoder/driver input of 1001?
2. *True or false:* More than one output of a BCD-to-7-segment decoder/driver can be active at one time.

9-3 LIQUID-CRYSTAL DISPLAYS

An LED display generates or emits light energy as current is passed through the individual segments. A liquid-crystal display **(LCD)** controls the reflection of available light. The available light may simply be ambient (surrounding) light such as sunlight or normal room lighting; *reflective* LCDs use ambient light. Or the available light might be provided by a small light source that is part of the display unit; *backlit* LCDs use this method. In any case, LCDs have gained wide acceptance because of their very low power consumption compared to LEDs, especially in battery-operated equipment such as calculators, digital watches, and portable electronic measuring instruments. LEDs have the advantage of a much brighter display that, unlike reflective LCDs, is easily visible in dark or poorly lit areas.

Basically, LCDs operate from a low voltage (typically 3 to 15 V rms), low-frequency (25 to 60 Hz) ac signal and draw very little current. They are often arranged as 7-segment displays for numerical readouts as shown in Figure 9-9(a). The ac voltage needed to turn on a segment is applied between the segment and the **backplane,** which is common to all segments. The segment and the backplane form a capacitor that draws very little current as long as the ac frequency is kept low. It is generally not lower than 25 Hz because this would produce visible flicker.

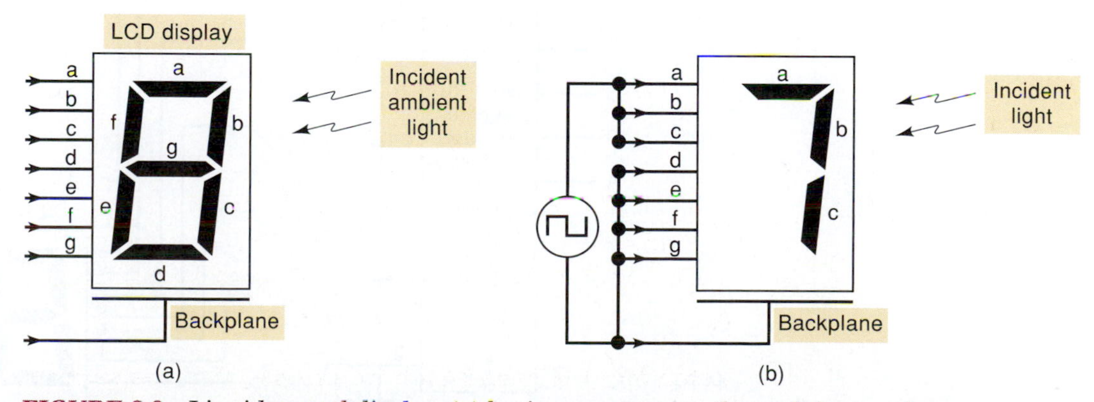

FIGURE 9-9 Liquid-crystal display: (a) basic arrangement; (b) applying a voltage between the segment and the backplane turns ON the segment. Zero voltage turns the segment OFF.

An admittedly simplified explanation of how an LCD operates goes something like this. When there is no difference in voltage between a segment and the backplane, the segment is said to be *nonactivated* (OFF). Segments *d*, *e*, *f*, and *g* in Figure 9-9(b) are OFF and will reflect incident light so that they appear invisible against their background. When an appropriate ac voltage is applied between a segment and the backplane, the segment is activated (ON). Segments *a*, *b*, and *c* in Figure 9-9(b) are ON and will not reflect the incident light, and thus they appear dark against their background.

Driving an LCD

An LCD segment will turn ON when an ac voltage is applied between the segment and the backplane, and will turn OFF when there is no voltage between the two. Rather than generating an ac signal, it is common practice to produce the required ac voltage by applying out-of-phase square waves to the segment and the backplane. This is illustrated in Figure 9-10(a) for one segment. A 40-Hz square wave is applied to the backplane and also to the input of a CMOS 74HC86 XOR. The other input to the XOR is a CONTROL input that will control whether the segment is ON or OFF.

When the CONTROL input is LOW, the XOR output will be exactly the same as the 40-Hz square wave, so that the signals applied to the segment and

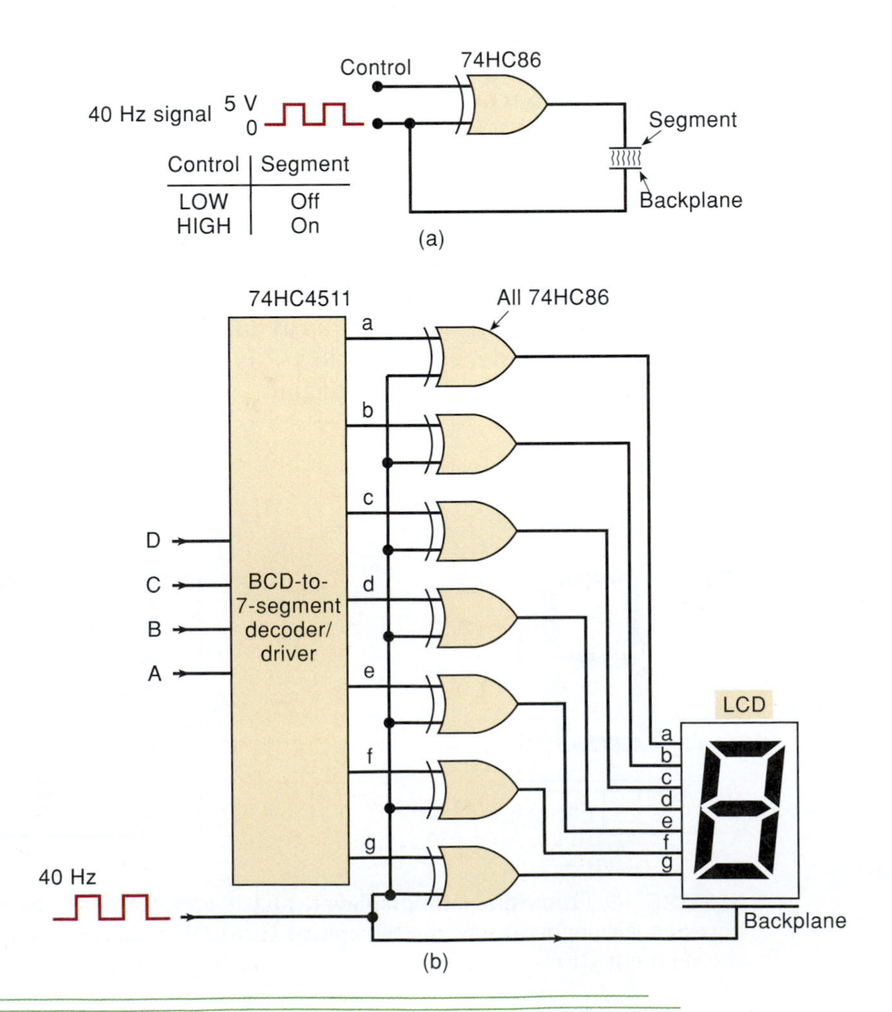

FIGURE 9-10 (a) Method for driving an LCD segment; (b) driving a 7-segment display.

the backplane are equal. Because there is no difference in voltage, the segment will be OFF. When the CONTROL input is HIGH, the XOR output will be the INVERSE of the 40-Hz square wave, so that the signal applied to the segment is out of phase with the signal applied to the backplane. As a result, the segment voltage will alternately be at +5 V and at −5 V relative to the backplane. This ac voltage will turn ON the segment.

This same idea can be extended to a complete 7-segment LCD display, as shown in Figure 9-10(b). Here, the CMOS 74HC4511 BCD-to-7-segment decoder/driver supplies the CONTROL signals to each of seven XOR for the seven segments. The 74HC4511 has active-HIGH outputs because a HIGH is required to turn on a segment. The decoder/driver and XOR gates of Figure 9-10(b) are available on a single chip. The CMOS 74HC4543 is one such device. It takes the BCD input code and provides the outputs to drive the LCD segments directly.

In general, CMOS devices are used to drive LCDs for two reasons: (1) they require much less power than TTL and are more suited to the battery-operated applications where LCDs are used; (2) the TTL LOW-state voltage is not exactly 0 V and can be as much as 0.4 V. This will produce a dc component of voltage between the segment and the backplane that considerably shortens the life of an LCD.

Types of LCDs

Liquid crystals are available as multidigit 7-segment decimal numeric displays. They come in many sizes and with many special characters such as colons (:) for clock displays, + and − indicators for digital voltmeters, decimal points for calculators, and battery-low indicators because many LCD devices are battery-powered. These displays must be driven by a decoder/driver chip such as the 74HC4543.

A more complicated but readily available LCD display is the alphanumeric LCD module. These modules are available from many companies in numerous formats such as 1-line-by-16-characters up to 4-lines-by-40-characters. The interface to these modules has been standardized so that an LCD module from any manufacturer will use the same signals and data format. The module includes some VLSI chips that make this device simple to use. Eight data lines are used to send the ASCII code for whatever you wish to display. These data lines also carry special control codes to the LCD command register. Three other inputs (Register Select, Read/Write, and Enable) are used to control the location, direction, and timing of the data transfer. As characters are sent to the module, it stores them in its own memory and types them across the display screen.

Other LCD modules allow the user to create a graphical display by controlling individual dots on the screen called **pixels**. Larger LCD panels can be scanned at a high rate, producing high-quality video motion pictures. In these displays, the control lines are arranged in a grid of rows and columns. At the intersection of each row and column is a pixel that acts like a "window" or "shutter" that can be electronically opened and closed to control the amount of light that is transmitted through the cell. The voltage from a row to a column determines the brightness of each pixel. In a laptop computer, a binary number for each pixel is stored in the "video" memory. These numbers are converted to voltages that are applied to the display.

Each pixel on a color display is actually made up of three subpixels. These subpixels control the light that passes through a red, green, or blue filter to produce the color of each pixel. On a 640-by-480 LCD screen there would be 640 × 3 connections for columns and 480 connections for rows, for a total of 2400 connections to the LCD. Obviously, the driver circuitry for such a device is a very complicated VLSI circuit.

The advances in technology for LCD displays have increased the speed at which the pixels can be turned on and off. The older screens are called Twisted Nematic (TN) or Super Twisted Nematic (STN). These devices are referred to as passive LCDs. Instead of using a uniform backplane like the 7-segment LCD displays, they have conducting parallel lines manufactured onto two pieces of glass. The two glass sheets are used to sandwich the liquid crystal material with the conducting lines at 90°, forming a grid of rows and columns, as shown in Figure 9-11. The intersection of each row and column forms a pixel. The actual switching of the current on and off is done in the driver IC that is connected to the rows and columns of the display. Passive matrix displays are rather slow at turning off. This limits the rate at which objects can move on the screen without leaving a shadow trail behind them.

The newer displays are called active matrix TFT LCDs. The active matrix means that an active element on the display is used to switch the pixels on and off. The active component is a thin film transistor (TFT) that is manufactured directly onto one piece of glass. The other piece of glass has a uniform coating to form a backplane. The control lines for these transistors run in rows and columns between the pixels. The technology that allows these transistors to be manufactured in a matrix on a thin film the size of a laptop computer screen has made these displays possible. They provide a much faster-response, higher-resolution display. The use of polysilicon technology allows the driver circuits to be integrated into the display unit, reducing connection problems and requiring very little perimeter space around the LCD.

Other display technologies are being refined, including vacuum fluorescent, gas discharge plasma, and electroluminescence. The optical physics for each of these displays varies, but the means of controlling all of them is the same. A digital system must activate a row and a column of a matrix in order to control the amount of light at the pixel located at the row/column intersection.

FIGURE 9-11 A passive matrix LCD panel.

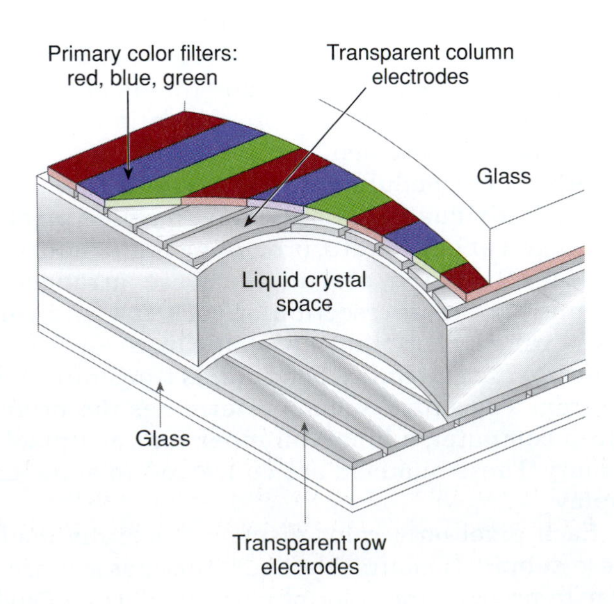

REVIEW QUESTIONS

1. Indicate which of the following statements refer to LCD displays and which refer to LED displays.
   (a) Emit light
   (b) Reflect ambient light
   (c) Are best for low-power applications
   (d) Require an ac voltage
   (e) Use a 7-segment arrangement to produce digits
   (f) Require current-limiting resistors
2. What form of data is sent to each of the following?
   (a) A 7-segment LCD display with a decoder/driver
   (b) An alphanumeric LCD module
   (c) An LCD computer display

9-4 ENCODERS

Most decoders accept an input code and produce a HIGH (or a LOW) at *one and only one* output line. In other words, we can say that a decoder identifies, recognizes, or detects a particular code. The opposite of this decoding process is called **encoding** and is performed by a logic circuit called an **encoder**. An encoder has a number of input lines, only one of which is activated at a given time, and produces an N-bit output code, depending on which input is activated. Figure 9-12 is the general diagram for an encoder with M inputs and N outputs. Here, the inputs are active-HIGH, which means that they are normally LOW.

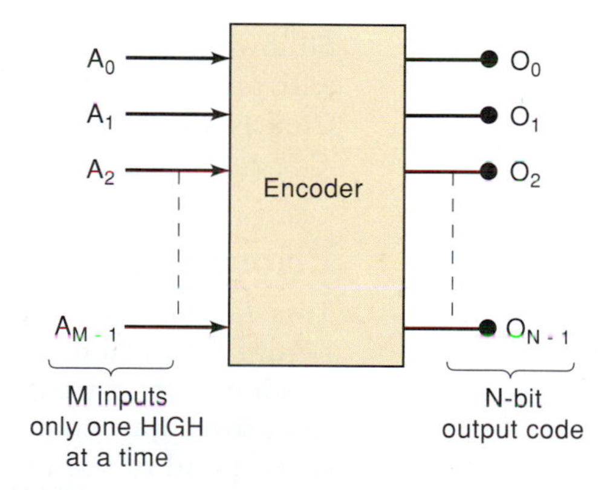

FIGURE 9-12 General encoder diagram.

We saw that a *binary-to-octal decoder (3-line-to-8-line decoder)* accepts a three-bit input code and activates one of eight output lines corresponding to that code. An *octal-to-binary encoder (8-line-to-3-line encoder)* performs the opposite function: it accepts eight input lines and produces a three-bit output code corresponding to the activated input. Figure 9-13 shows the logic circuit and the truth table for an octal-to-binary encoder with active-LOW inputs.

By following through the logic, you can verify that a LOW at any single input will produce the output binary code corresponding to that input. For instance, a LOW at $\overline{A}_3$ (while all other inputs are HIGH) will produce

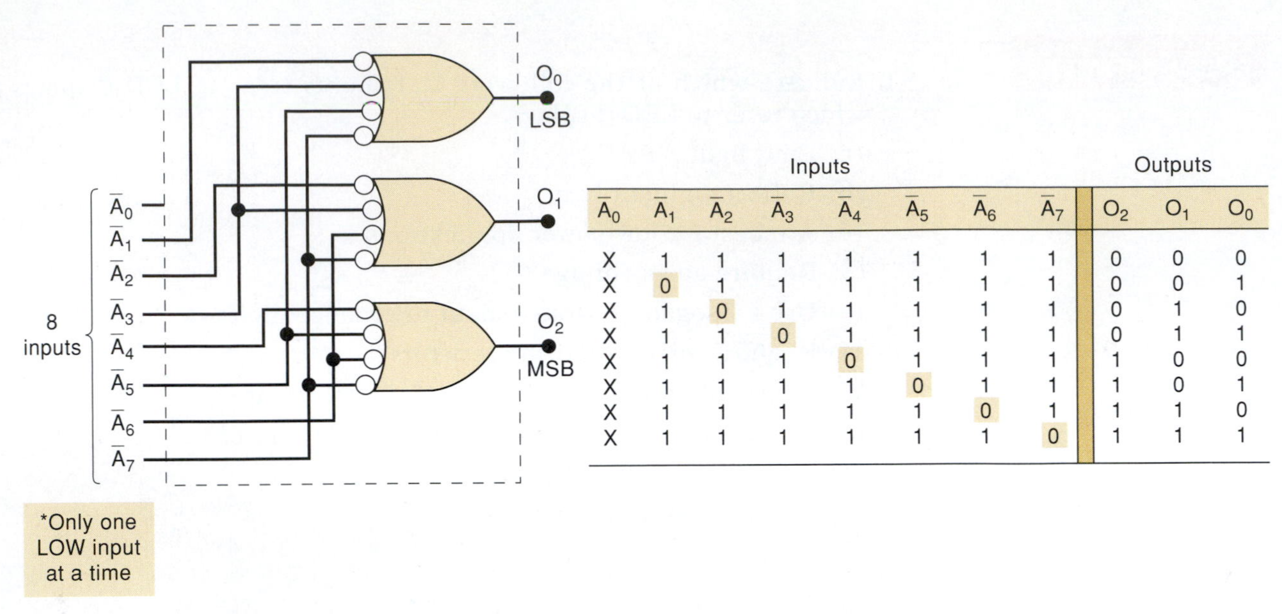

Inputs								Outputs		
$\overline{A}_0$	$\overline{A}_1$	$\overline{A}_2$	$\overline{A}_3$	$\overline{A}_4$	$\overline{A}_5$	$\overline{A}_6$	$\overline{A}_7$	O_2	O_1	O_0
X	1	1	1	1	1	1	1	0	0	0
X	0	1	1	1	1	1	1	0	0	1
X	1	0	1	1	1	1	1	0	1	0
X	1	1	0	1	1	1	1	0	1	1
X	1	1	1	0	1	1	1	1	0	0
X	1	1	1	1	0	1	1	1	0	1
X	1	1	1	1	1	0	1	1	1	0
X	1	1	1	1	1	1	0	1	1	1

FIGURE 9-13 Logic circuit for an octal-to-binary (8-line-to-3-line) encoder. For proper operation, only one input should be active at one time.

$O_2 = 0, O_1 = 1$, and $O_0 = 1$, which is the binary code for 3. Notice that $\overline{A}_0$ is not connected to the logic gates because the encoder outputs will normally be at 000 when none of the inputs $\overline{A}_1$ to $\overline{A}_9$ is LOW.

EXAMPLE 9-5

Determine the outputs of the encoder in Figure 9-13 when $\overline{A}_3$ and $\overline{A}_5$ are simultaneously LOW.

Solution

Following through the logic gates, we see that the LOWs at these two inputs will produce HIGHs at each output, in other words, the binary code 111. Clearly, this is not the code for either activated input.

Priority Encoders

This last example identifies a drawback of the simple encoder circuit of Figure 9-13 when more than one input is activated at one time. A modified version of this circuit, called a **priority encoder**, includes the necessary logic to ensure that when two or more inputs are activated, the output code will correspond to the highest-numbered input. For example, when both $\overline{A}_3$ and $\overline{A}_5$ are LOW, the output code will be 101 (5). Similarly, when $\overline{A}_6$, $\overline{A}_2$, and $\overline{A}_0$ are all LOW, the output code is 110 (6). The 74148, 74LS148, and 74HC148 are all octal-to-binary priority encoders.

74147 Decimal-to-BCD Priority Encoder

Figure 9-14 shows the logic symbol and the truth table for the 74147 (74LS147, 74HC147), which functions as a decimal-to-BCD priority encoder. It has nine active-LOW inputs representing the decimal digits 1 through 9, and it produces the *inverted* BCD code corresponding to the highest-numbered activated input.

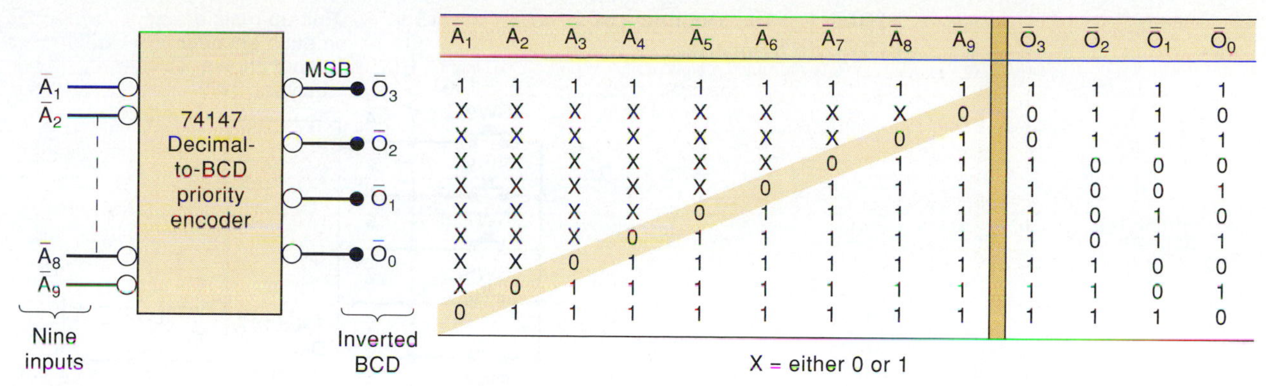

$\overline{A}_1$	$\overline{A}_2$	$\overline{A}_3$	$\overline{A}_4$	$\overline{A}_5$	$\overline{A}_6$	$\overline{A}_7$	$\overline{A}_8$	$\overline{A}_9$	$\overline{O}_3$	$\overline{O}_2$	$\overline{O}_1$	$\overline{O}_0$
1	1	1	1	1	1	1	1	1	1	1	1	1
X	X	X	X	X	X	X	X	0	0	1	1	0
X	X	X	X	X	X	X	0	1	0	1	1	1
X	X	X	X	X	X	0	1	1	1	0	0	0
X	X	X	X	X	0	1	1	1	1	0	0	1
X	X	X	X	0	1	1	1	1	1	0	1	0
X	X	X	0	1	1	1	1	1	1	0	1	1
X	X	0	1	1	1	1	1	1	1	1	0	0
X	0	1	1	1	1	1	1	1	1	1	0	1
0	1	1	1	1	1	1	1	1	1	1	1	0

X = either 0 or 1

FIGURE 9-14 74147 decimal-to-BCD priority encoder.

Let's examine the truth table to see how this IC works. The first line in the table shows all inputs in their inactive HIGH state. For this condition, the outputs are 1111, which is the inverse of 0000, the BCD code for 0. The second line in the table indicates that a LOW at $\overline{A}_9$, regardless of the states of the other inputs, will produce an output code of 0110, which is the inverse of 1001, the BCD code for 9. The third line shows that a LOW at $\overline{A}_8$, provided that $\overline{A}_9$ is HIGH, will produce an output code of 0111, the inverse of 1000, the BCD code for 8. In a similar manner, the remaining lines in the table show that a LOW at any input, provided that all higher-numbered inputs are HIGH, will produce the inverse of the BCD code for that input.

The 74147 outputs will normally be HIGH when none of the inputs are activated. This corresponds to the decimal 0 input condition. There is no $\overline{A}_0$ input because the encoder assumes the decimal 0 input state when all other inputs are HIGH. The 74147 inverted BCD outputs can be converted to normal BCD by putting each one through an INVERTER.

EXAMPLE 9-6

Determine the states of the outputs in Figure 9-14 when $\overline{A}_5$, $\overline{A}_7$, and $\overline{A}_3$ are LOW and all other inputs are HIGH.

Solution

The truth table shows that when $\overline{A}_7$ is LOW, the levels at $\overline{A}_5$ and $\overline{A}_3$ do not matter. Thus, the outputs will each be 1000, the inverse of 0111 (7).

Switch Encoder

Figure 9-15 shows how a 74147 can be used as a *switch encoder.* The 10 switches might be the keyboard switches on a calculator representing digits 0 through 9. The switches are of the normally open type, so that the encoder inputs are all normally HIGH and the BCD output is 0000 (note the INVERTERs). When a digit key is depressed, the circuit will produce the BCD code for that digit. Because the 74LS147 is a priority encoder, simultaneous key depressions will produce the BCD code for the higher-numbered key.

The switch encoder of Figure 9-15 can be used whenever BCD data must be entered manually into a digital system. A prime example would be in an electronic calculator, where the operator depresses several keyboard switches in succession to enter a decimal number. In a simple, basic calculator, the BCD code for each decimal digit is entered into a four-bit storage register. In other

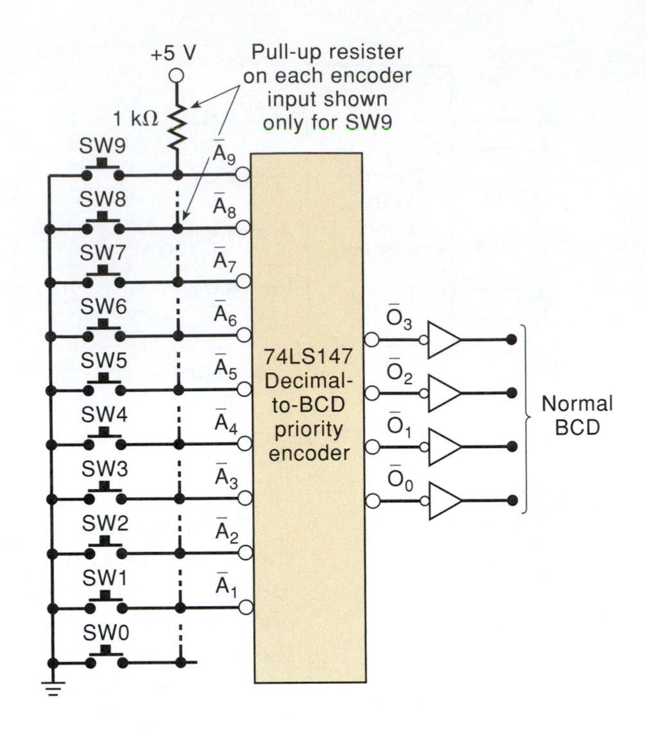

FIGURE 9-15 Decimal-to-BCD switch encoder.

words, when the first key is depressed, the BCD code for that digit is sent to a four-bit FF register; when the second switch is depressed, the BCD code for that digit is sent to *another* four-bit FF register, and so on. Thus, a calculator that can handle eight digits will have eight four-bit registers to store the BCD codes for these digits. Each four-bit register drives a decoder/driver and a numerical display so that the eight-digit number can be displayed.

The operation described above can be accomplished with the circuit in Figure 9-16. This circuit will take three decimal digits entered from the keyboard in sequence, encode them in BCD, and store the BCD in three FF output registers. The 12 D-type flip-flops Q_0 to Q_{11} are used to receive and store the BCD codes for the digits. Q_8 to Q_{11} store the BCD code for the most significant digit (MSD), which is the first one entered on the keyboard. Q_4 to Q_7 store the second entered digit, and Q_0 to Q_3 store the third entered digit. Flip-flops X, Y, and Z form a ring counter (Chapter 7) that controls the transfer of data from the encoder outputs to the appropriate output register. The OR gate produces a HIGH output any time one of the keys is depressed. This output may be affected by switch contact bounce, which would produce several pulses before settling down to the HIGH state. The OS is used to neutralize the switch bounce by triggering on the first positive transition from the OR gate and remaining HIGH for 20 ms, well past the time duration of the switch bounce. The OS output clocks the ring counter.

The circuit operation is described as follows for the case where the decimal number 309 is being entered:

1. The CLEAR key is depressed. This clears all storage flip-flops Q_0 to Q_{11} to 0. It also clears flip-flops X and Y and presets flip-flop Z to 1, so that the ring counter begins in the 001 state.
2. The CLEAR key is released and the "3" key is depressed. The encoder outputs 1100 are inverted to produce 0011, the BCD code for 3. These binary values are sent to the D inputs of the three four-bit output registers.
3. The OR output goes HIGH (because two of its inputs are HIGH) and triggers the OS output $Q = 1$ for 20 ms. After 20 ms, Q returns LOW and clocks

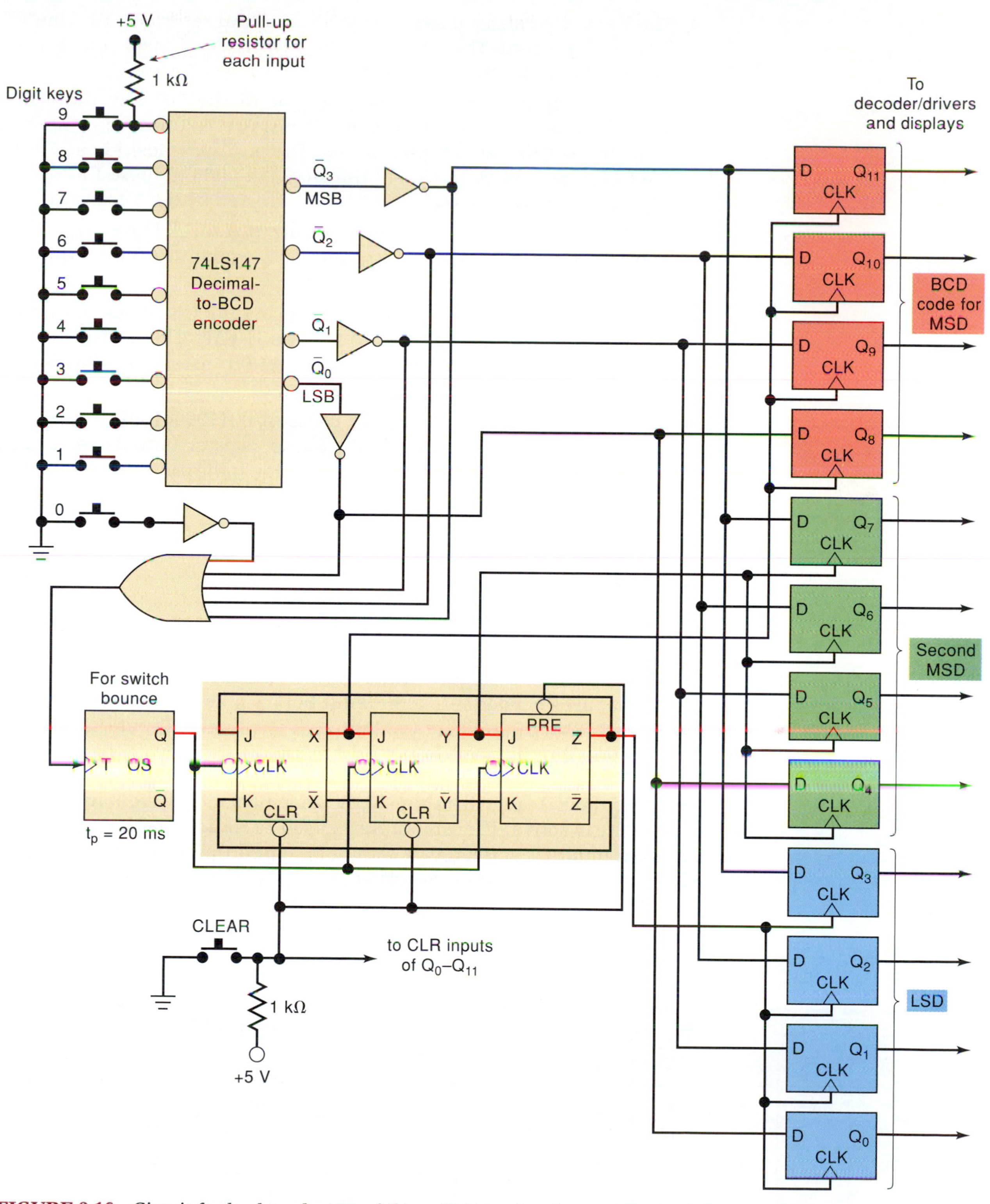

FIGURE 9-16 Circuit for keyboard entry of three-digit number into storage registers.

the ring counter to the 100 state (X goes HIGH). The positive transition at X is fed to the CLK inputs of flip-flops Q_8 to Q_{11}, so that the encoder outputs are transferred to these FFs. That is, $Q_{11} = 0, Q_{10} = 0, Q_9 = 1$, and $Q_8 = 1$. Note that flip-flops Q_0 to Q_7 are not affected because their CLK inputs have not received a positive transition.

4. The "3" key is released and the OR gate output returns LOW. The "0" key is then depressed. This produces the BCD code of 0000, which is fed to the inputs of the three registers.
5. The OR output goes HIGH in response to the "0" key (note the INVERTER) and triggers the OS for 20 ms. After 20 ms, the ring counter shifts to the 010 state (Y goes HIGH). The positive transition at Y is fed to the *CLK* inputs of Q_4 to Q_7 and transfers the 0000 to these FFs. Note that flip-flops Q_0 to Q_3 and Q_8 to Q_{11} are not affected by the Y transition.
6. The "0" key is released and the OR output returns LOW. The "9" key is depressed, producing BCD outputs 1001, which are fed to the storage registers.
7. The OR output goes HIGH again, triggering the OS, which in turn clocks the ring counter to the 001 state (Z goes HIGH). The positive transition at Z is fed to the *CLK* inputs of Q_0 to Q_3 and transfers the 1001 into these FFs. The other storage FFs are unaffected.
8. At this point, the storage register contains 001100001001, beginning with Q_{11}. This is the BCD code of 309. These register outputs feed decoder/drivers that drive appropriate displays for indicating the decimal digits 309.
9. The storage FF outputs are also fed to other circuits in the system. In a calculator, for example, these outputs would be sent to the arithmetic section to be processed.

Several problems at the end of the chapter will deal with some other aspects of this circuit, including troubleshooting exercises.

The 74ALS148 is slightly more sophisticated than the '147. It has eight inputs that are encoded into a three-bit binary number. This IC also provides three control pins as indicated in Table 9-1. The Enable Input ($\overline{EI}$) and Enable Output ($\overline{EO}$) can be used to cascade two IC's producing a hexadecimal-to-binary encoder. The $\overline{EI}$ pin must be LOW in order for any output pin to go LOW, and the $\overline{EO}$ pin will only go LOW when none of the eight inputs is active and the $\overline{EI}$ is active. The $\overline{GS}$ output is used to indicate when at least one of the eight inputs is activated. It should be noted that the outputs A_2 through A_0 are inverted, just as in the 74147.

TABLE 9-1 74ALS148 function table.

	INPUTS								OUTPUTS				
$\overline{EI}$	$\overline{0}$	$\overline{1}$	$\overline{2}$	$\overline{3}$	$\overline{4}$	$\overline{5}$	$\overline{6}$	$\overline{8}$	$\overline{A}_2$	$\overline{A}_1$	$\overline{A}_0$	$\overline{GS}$	$\overline{EO}$
H	x	x	x	x	x	x	x	x	H	H	H	H	H
L	H	H	H	H	H	H	H	H	H	H	H	H	L
L	x	x	x	x	x	x	x	L	L	L	L	L	H
L	x	x	x	x	x	x	L	H	L	L	H	L	H
L	x	x	x	x	x	L	H	H	L	H	L	L	H
L	x	x	x	x	L	H	H	H	L	H	H	L	H
L	x	x	x	L	H	H	H	H	H	L	L	L	H
L	x	x	L	H	H	H	H	H	H	L	H	L	H
L	x	L	H	H	H	H	H	H	H	H	L	L	H
L	L	H	H	H	H	H	H	H	H	H	H	L	H

REVIEW QUESTIONS

1. How does an encoder differ from a decoder?
2. How does a priority encoder differ from an ordinary encoder?
3. What will the outputs be in Figure 9-15 when SW6, SW5, and SW2 are all closed?
4. Describe the functions of each of the following parts of the keyboard entry circuit of Figure 9-16.
   (a) OR gate
   (b) 74147 encoder
   (c) One-shot
   (d) Flip-flops X, Y, Z
   (e) Flip-flops Q_0 to Q_{11}
5. What is the purpose of each control input and output on a 74148 encoder?

9-5 TROUBLESHOOTING

As circuits and systems become more complex, the number of possible causes of failure obviously increases. Whereas the procedure for fault isolation and correction remains essentially the same, the application of the **observation/analysis** process is more important for complex circuits because it helps the troubleshooter narrow the location of the fault to a small area of the circuit. This reduces to a reasonable amount the testing steps and resulting data that must be analyzed. By understanding the circuit operation, observing the symptoms of the failure, and reasoning through the operation, the troubleshooter can often predict the possible faults before ever picking up a logic probe or an oscilloscope. This observation/analysis process is one that inexperienced troubleshooters are hesitant to apply, probably because of the great variety and capabilities of modern test equipment available to them. It is easy to become overly reliant on these tools while not adequately utilizing the human brain's reasoning and analytical skills.

The following examples illustrate how the observation/analysis process can be applied. Many of the end-of-chapter troubleshooting problems will provide you with the opportunity to develop your skill at applying this process.

Another vital strategy in troubleshooting is known as **divide-and-conquer**. It is used to identify the location of the problem after observation/analysis has generated several possibilities. A less efficient method would be to investigate each possible cause, one by one. The divide-and-conquer method finds a point in the circuit that can be tested, thereby dividing the total possible number of causes in half. In simple systems, this may seem unnecessary, but as complexity increases, the total number of possible causes also increases. If there are eight possible causes, then a test should be performed that eliminates four of them. The next test should eliminate two more, and the third test should identify the problem.

EXAMPLE 9-7

A technician tests the circuit of Figure 9-4 by using a set of switches to apply the input code at A_4 through A_0. She runs through each possible input code and checks the corresponding decoder output to see if it is activated. She observes that all of the odd-numbered outputs respond correctly, but all of the even-numbered outputs fail to respond when their code is applied. What are the most probable faults?

Solution

In a situation where so many outputs are failing, it is unreasonable to expect that each of these outputs has a fault. It is much more likely that some faulty input condition is causing the output failures. What do all of the even-numbered outputs have in common? The input codes for several of them are listed in Table 9-2.

TABLE 9-2

Output	Input Code
$\overline{O}_0$	00000
$\overline{O}_4$	00100
$\overline{O}_{14}$	01110
$\overline{O}_{18}$	10010

Clearly, each even-numbered output requires an input code with an $A_0 = 0$ in order to be activated. Thus, the most probable faults would be those that prevent A_0 from going LOW. These include:

1. A faulty switch connected to the A_0 input
2. A break in the path between the switch and the A_0 line
3. An external short from the A_0 line to V_{CC}
4. An internal short to V_{CC} at the A_0 inputs of any one of the decoder chips

Through observation and analysis, the technician has identified several possible causes. Potential causes 1 and 2 are in the switches generating the address. Causes 3 and 4 are in the decoder circuit itself. The circuit can be divided by opening the connection between the least significant switch and the A_0 input, as shown in Figure 9-17. A logic probe can be used to see if the switch can generate a LOW as well as a HIGH. Regardless of the outcome, two of the four possible causes have been eliminated.

Thus, the fault is narrowed to a specific area of the circuit. The exact fault can be traced with the testing and measurement techniques that we are already familiar with.

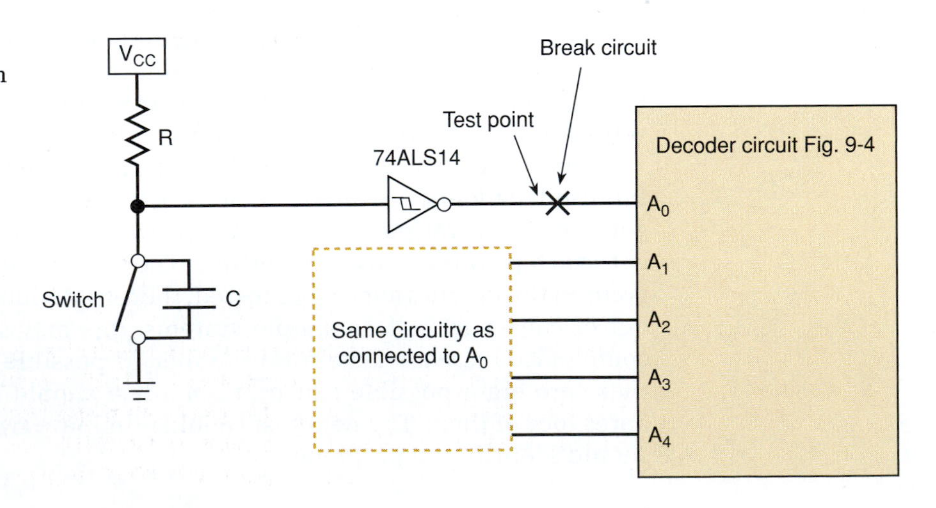

FIGURE 9-17 Troubleshooting circuitry in Example 9-7.

EXAMPLE 9-8

A technician wires the outputs from a BCD counter to the inputs of the decoder/driver of Figure 9-8. He applies pulses to the counter at a very slow rate and observes the LED display, which is shown below, as the counter counts up from 0000 to 1001. Examine this observed sequence carefully and try to predict the most probable fault.

COUNT	0	1	2	3	4	5	6	7	8	9
Observed display										
Expected display										

Solution

Comparing the observed display with the expected display for each count, we see several important points:

- For those counts where the observed display is incorrect, the observed display is not one of the segment patterns that correspond to counts greater than 1001.
- This rules out a faulty counter or faulty wiring from the counter to the decoder/driver.
- The correct segment patterns (0, 1, 3, 6, 7, and 8) have the common property that segments *e* and *f* are either both on or both off.
- The incorrect segment patterns have the common property that segments *e* and *f* are in opposite states, and if we interchange the states of these two segments, the correct pattern is obtained.

Giving some thought to these points should lead us to conclude that the technician has probably "crossed" the connections to the *e* and *f* segments.

9-6 MULTIPLEXERS (DATA SELECTORS)

A modern home stereo system may have a switch that selects music from one of four sources: a cassette tape, a compact disc (CD), a radio tuner, or an auxilliary input such as audio from a VCR or DVD. The switch selects one of the electronic signals from one of these four sources and sends it to the power amplifier and speakers. In simple terms, this is what a **multiplexer (MUX)** does: it selects one of several input signals and passes it on to the output.

A *digital multiplexer* or *data selector* is a logic circuit that accepts several digital data inputs and selects one of them at any given time to pass on to the output. The routing of the desired data input to the output is controlled by SELECT inputs (often referred to as ADDRESS inputs). Figure 9-18 shows the functional diagram of a general digital multiplexer. The inputs and outputs are drawn as wide arrows rather than lines; this indicates that they may actually be more than one signal line.

The multiplexer acts like a digitally controlled multiposition switch where the digital code applied to the SELECT inputs controls which data inputs will be switched to the output. For example, output Z will equal data input I_0 for some particular SELECT input code, Z will equal I_1 for another particular SELECT input code, and so on. Stated another way, a multiplexer selects 1 out of N input data sources and transmits the selected data to a single output channel. This is called **multiplexing**.

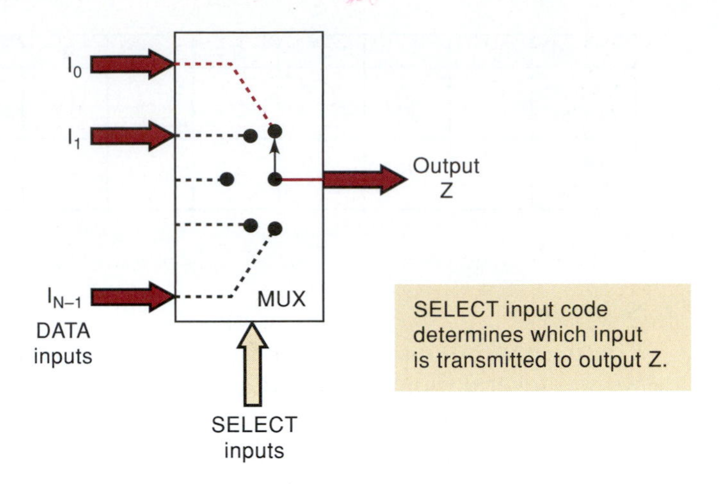

FIGURE 9-18 Functional diagram of a digital multiplexer (MUX).

Basic Two-Input Multiplexer

Figure 9-19 shows the logic circuitry for a two-input multiplexer with data inputs I_0 and I_1 and SELECT input S. The logic level applied to the S input determines which AND gate is enabled so that its data input passes through the OR gate to output Z. Looking at it another way, the Boolean expression for the output is

$$Z = I_0\overline{S} + I_1 S$$

With $S = 0$, this expression becomes

$$Z = I_0 \cdot 1 + I_1 \cdot 0 \quad \text{[gate 2 enabled]}$$
$$= I_0$$

which indicates that Z will be identical to input signal I_0, which in turn can be a fixed logic level or a time-varying logic signal. With $S = 1$, the expression becomes

$$Z = I_0 \cdot 0 + I_1 \cdot 1 = I_1 \quad \text{[gate 1 enabled]}$$

showing that output Z will be identical to input signal I_1.

An example of where a two-input MUX could be used is in a digital system that uses two different MASTER CLOCK signals: a high-speed clock (say, 10 MHz) in one mode and a slow-speed clock (say, 4.77 MHz) for the

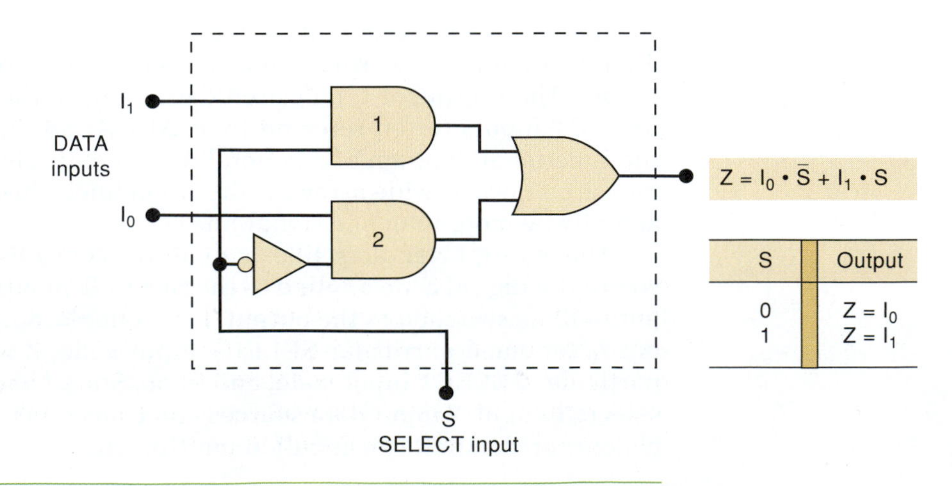

S	Output
0	$Z = I_0$
1	$Z = I_1$

FIGURE 9-19 Two-input multiplexer.

other. Using the circuit of Figure 9-19, the 10-MHz clock would be tied to I_0, and the 4.77-MHz clock would be tied to I_1. A signal from the system's control logic section would drive the SELECT input to control which clock signal appears at output Z for routing to the other parts of the circuit.

Four-Input Multiplexer

The same basic idea can be used to form the four-input multiplexer shown in Figure 9-20(a). Here, four inputs are selectively transmitted to the output according to the four possible combinations of the S_1S_0 select inputs. Each data input is gated with a different combination of select input levels. I_0 is gated with $\overline{S}_1\overline{S}_0$ so that I_0 will pass through its AND gate to output Z only when $S_1 = 0$ and $S_0 = 0$. The table in the figure gives the outputs for the other three input-select codes.

Another circuit that performs exactly the same function is shown in Figure 9-20(b). This approach uses tristate buffers to select one of the signals. The decoder ensures that only one buffer can be enabled at any time. S_1 and S_0 are used to specify which of the input signals is allowed to pass through its buffer and arrive at the output.

Two-, four-, eight-, and 16-input multiplexers are readily available in the TTL and CMOS logic families. These basic ICs can be combined for multiplexing a larger number of inputs.

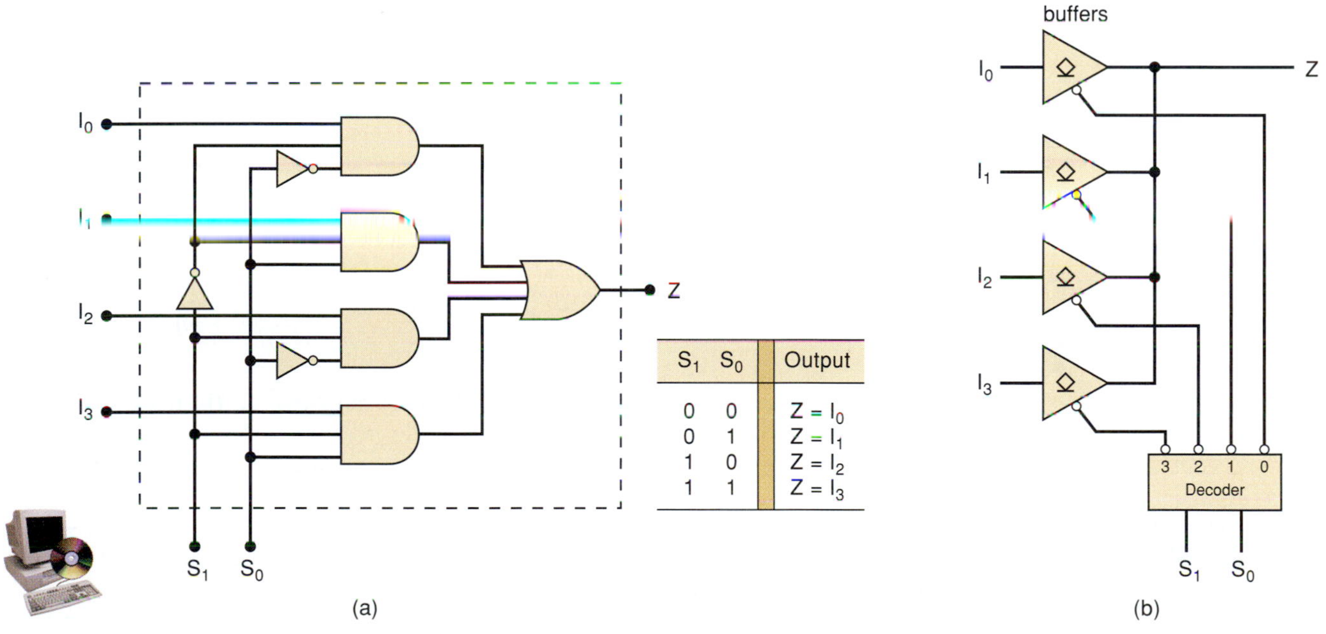

S_1	S_0	Output
0	0	$Z = I_0$
0	1	$Z = I_1$
1	0	$Z = I_2$
1	1	$Z = I_3$

FIGURE 9-20 Four-input multiplexer: (a) using sum of products logic; (b) using tristate buffers.

Eight-Input Multiplexer

Figure 9-21(a) shows the logic diagram for the 74ALS151 (74HC151) eight-input multiplexer. This multiplexer has an enable input $\overline{E}$ and provides both the normal and the inverted outputs. When $\overline{E} = 0$, the select inputs $S_2S_1S_0$ will select one data input (from I_0 through I_7) for passage to output Z. When $\overline{E} = 1$, the multiplexer is disabled so that $Z = 0$ regardless of the select input code. This operation is summarized in Figure 9-21(b), and the 74151 logic symbol is shown in Figure 9-21(c).

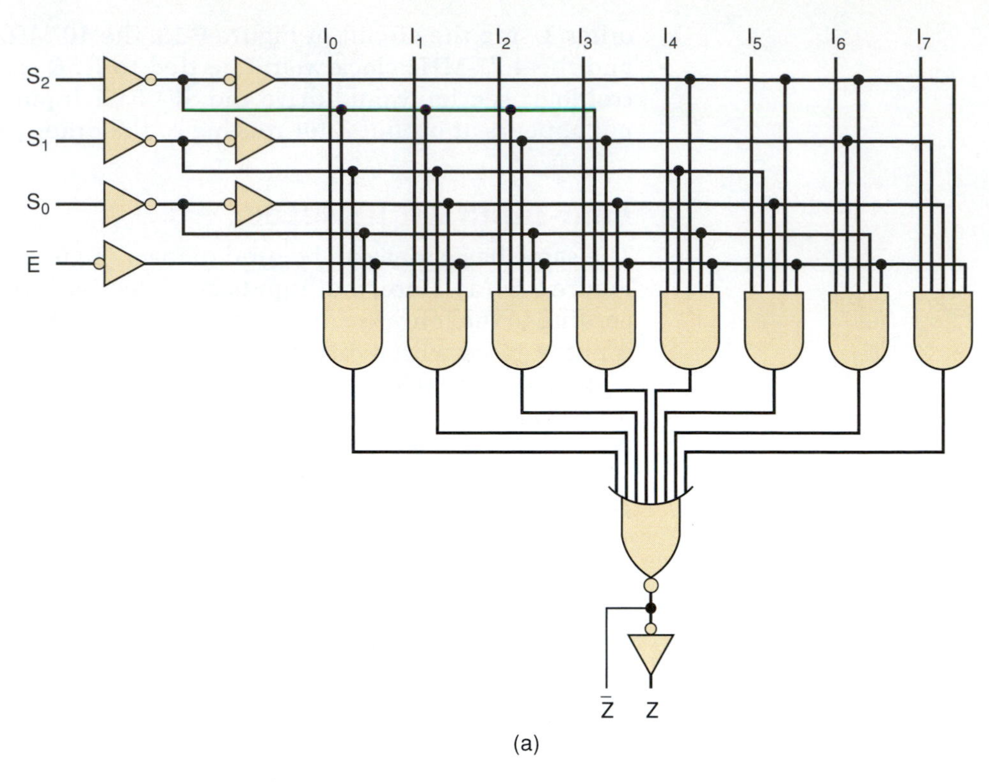

Inputs				Outputs	
$\overline{E}$	S_2	S_1	S_0	$\overline{Z}$	Z
H	X	X	X	H	L
L	L	L	L	$\overline{I_0}$	I_0
L	L	L	H	$\overline{I_1}$	I_1
L	L	H	L	$\overline{I_2}$	I_2
L	L	H	H	$\overline{I_3}$	I_3
L	H	L	L	$\overline{I_4}$	I_4
L	H	L	H	$\overline{I_5}$	I_5
L	H	H	L	$\overline{I_6}$	I_6
L	H	H	H	$\overline{I_7}$	I_7

(b)

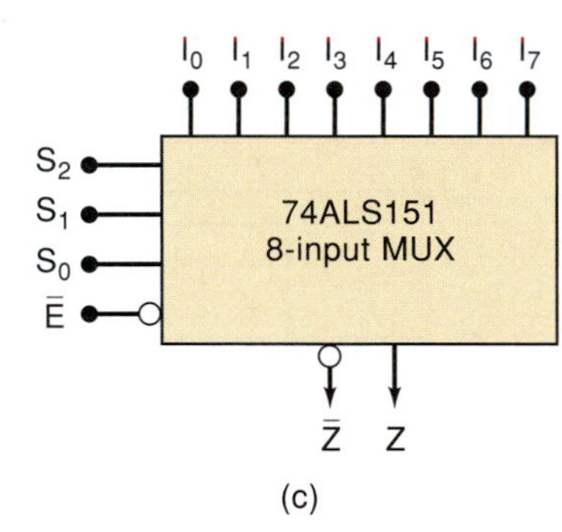

FIGURE 9-21 (a) Logic diagram for the 74ALS151 multiplexer; (b) truth table; (c) logic symbol.

EXAMPLE 9-9

The circuit in Figure 9-22 uses two 74HC151s, an INVERTER, and an OR gate. Describe this circuit's operation.

Solution

This circuit has a total of 16 data inputs, eight applied to each multiplexer. The two multiplexer outputs are combined in the OR gate to produce a single output X. The circuit functions as a 16-input multiplexer. The four select inputs $S_3S_2S_1S_0$ will select one of the 16 inputs to pass through to X.

The S_3 input determines which multiplexer is enabled. When $S_3 = 0$, the top multiplexer is enabled, and the $S_2S_1S_0$ inputs determine which of its data inputs will appear at its output and pass through the OR gate to X. When $S_3 = 1$, the bottom multiplexer is enabled, and the $S_2S_1S_0$ inputs select one of its data inputs for passage to output X.

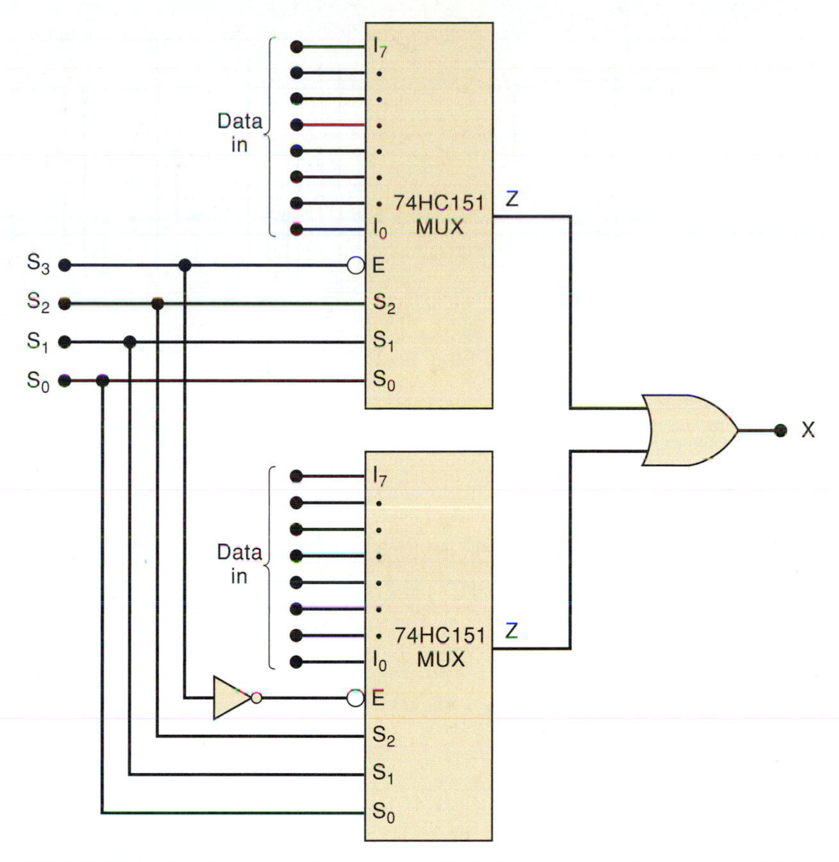

FIGURE 9-22 Example 9-9: two 74HC151s combined to form a 16-input multiplexer.

Quad Two-Input MUX (74ALS157/HC157)

The 74ALS157 is a very useful multiplexer IC that contains four two-input multiplexers like the one in Figure 9-19. The logic diagram for the 74ALS157 is shown in Figure 9-23(a). Note the manner in which the data inputs and outputs are labeled.

EXAMPLE 9-10

Determine the input conditions required for each Z output to take on the logic level of its corresponding I_0 input. Repeat for I_1.

Solution

First of all, the enable input must be active; that is, $\overline{E} = 0$. In order for Z_a to equal I_{0a}, the select input must be LOW. These same conditions will produce $Z_b = I_{0b}$, $Z_c = I_{0c}$, and $Z_d = I_{0d}$.

With $\overline{E} = 0$ and $S = 1$, the Z outputs will follow the set of I_1 inputs; that is, $Z_a = I_{1a}$, $Z_b = I_{1b}$, $Z_c = I_{1c}$, and $Z_d = I_{1d}$.

All of the outputs will be disabled (LOW) when $\overline{E} = 1$.

It is helpful to think of this multiplexer as being a simple two-input multiplexer, but one in which each input is four lines and the output is four lines. The four output lines switch back and forth between the two sets of four input lines under the control of the select input. This operation is represented by the 74ALS157's logic symbol in Figure 9-23(b).

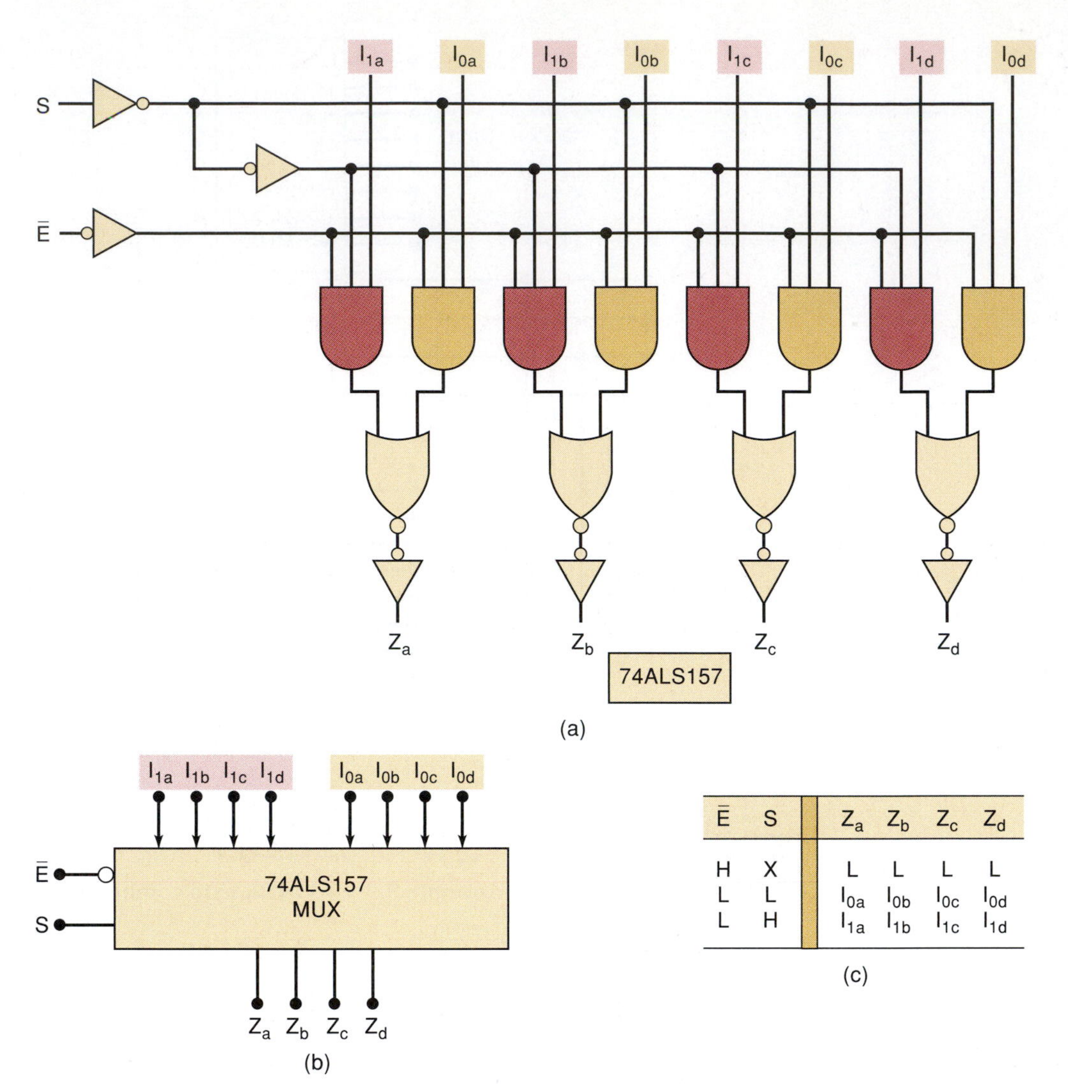

FIGURE 9-23 (a) Logic diagram for the 74ALS157 multiplexer; (b) logic symbol; (c) truth table.

REVIEW QUESTIONS

1. What is the function of a multiplexer's select inputs?
2. A certain multiplexer can switch one of 32 data inputs to its output. How many different inputs does this MUX have?

9-7 MULTIPLEXER APPLICATIONS

Multiplexer circuits find numerous and varied applications in digital systems of all types. These applications include data selection, data routing, operation sequencing, parallel-to-serial conversion, waveform generation, and logic-function generation. We shall look at some of these applications here and several more in the problems at the end of the chapter.

Data Routing

Multiplexers can route data from one of several sources to one destination. One typical application uses 74ALS157 multiplexers to select and display the

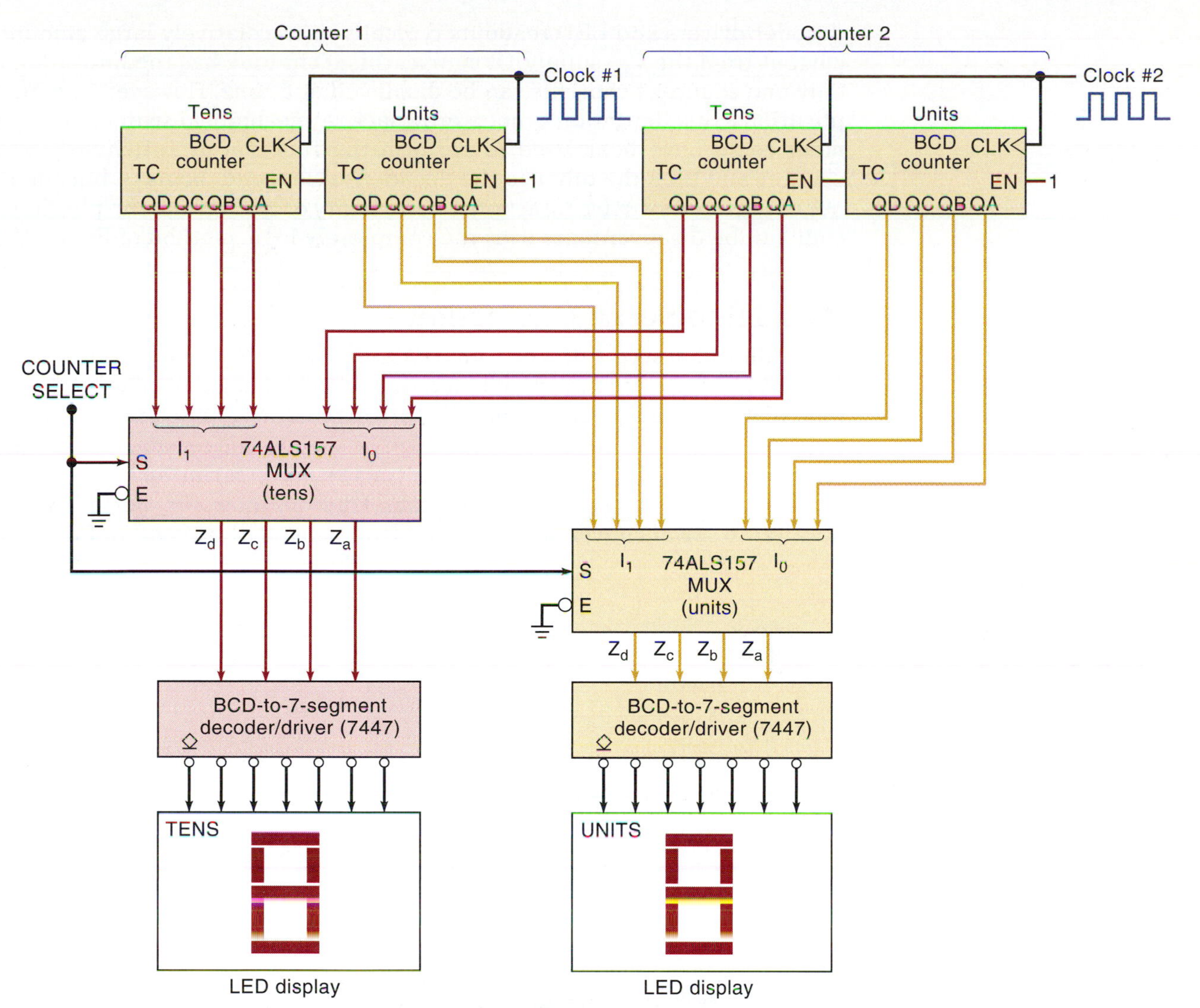

FIGURE 9-24 System for displaying two multidigit BCD counters one at a time.

contents of either of two BCD counters using a *single* set of decoder/drivers and LED displays. The circuit arrangement is shown in Figure 9-24.

Each counter consists of two cascaded BCD stages, and each one is driven by its own clock signal. When the COUNTER SELECT line is HIGH, the outputs of counter 1 will be allowed to pass through the multiplexers to the decoder/drivers to be displayed on the LED readouts. When COUNTER SELECT = 0, the outputs of counter 2 will pass through the multiplexers to the displays. In this way, the decimal contents of one counter or the other will be displayed under the control of the COUNTER SELECT input. A common situation where this might be used is in a digital watch. The digital watch circuitry contains many counters and registers that keep track of seconds, minutes, hours, days, months, alarm settings, and so on. A multiplexing scheme such as this one allows different data to be displayed on the limited number of decimal readouts.

The purpose of the multiplexing technique, as it is used here, is to *time-share* the decoder/drivers and display circuits between the two counters rather than have a separate set of decoder/drivers and displays for each counter. This results in a significant saving in the number of wiring connections, especially when more BCD stages are added to each counter. Even more important, it represents a significant decrease in power consumption because decoder/drivers and LED readouts typically draw relatively large amounts of current from the V_{CC} supply. Of course, this technique has the limitation that only one counter's contents can be displayed at a time. However, in many applications, this limitation is not a drawback. A mechanical switching arrangement could have been used to perform the function of switching first one counter and then the other to the decoder/drivers and displays, but the number of required switch contacts, the complexity of wiring, and the physical size could all be disadvantages over the completely logic method of Figure 9-24.

Parallel-to-Serial Conversion

Many digital systems process binary data in parallel form (all bits simultaneously) because it is faster. When data are to be transmitted over relatively long distances, however, the parallel arrangement is undesirable because it requires a large number of transmission lines. For this reason, binary data or information in parallel form is often converted to serial form before being transmitted to a remote destination. One method for performing this **parallel-to-serial conversion** uses a multiplexer, as illustrated in Figure 9-25.

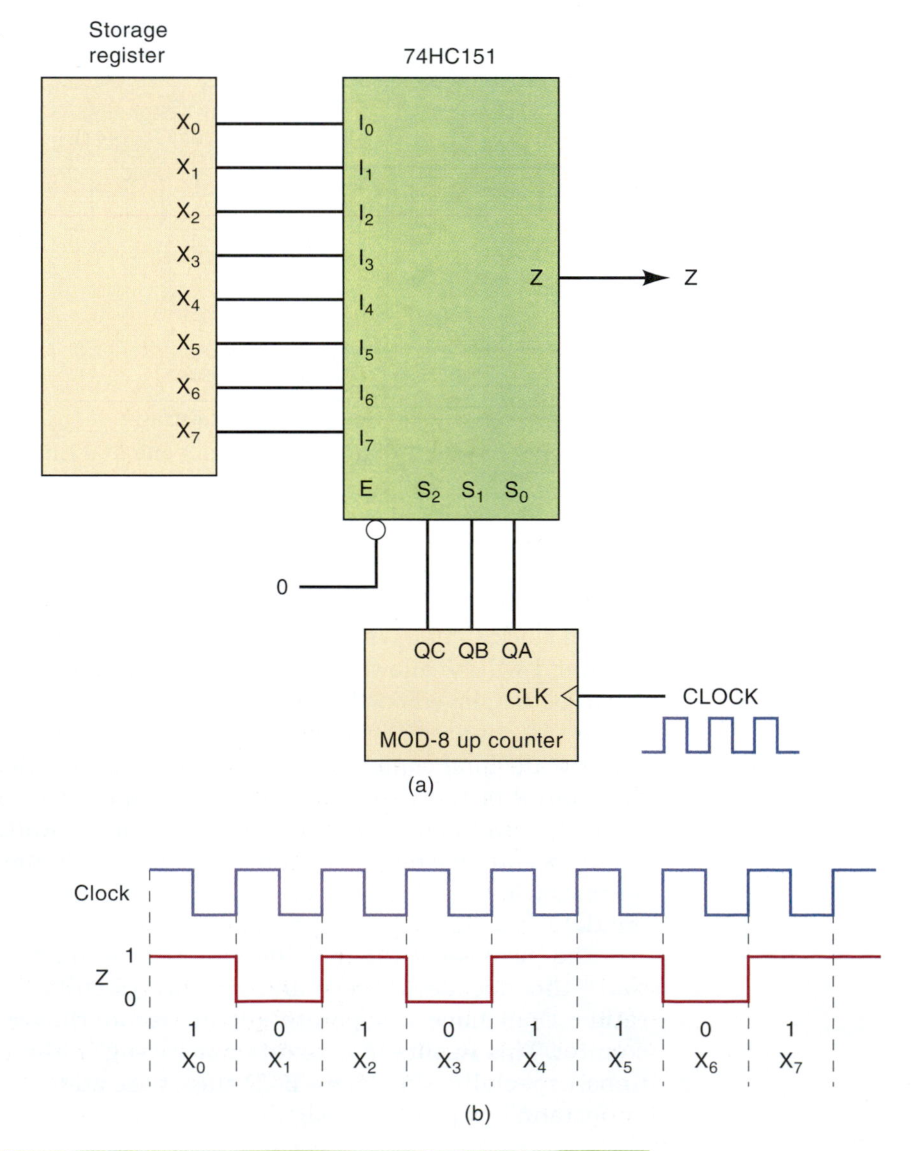

FIGURE 9-25 (a) Parallel-to-serial converter; (b) waveforms for $X_7X_6X_5X_4X_3X_2X_1X_0$ = 10110101.

The data are present in parallel form at the outputs of the X register and are fed to the eight-input multiplexer. A three-bit (MOD-8) counter is used to provide the select code bits $S_2S_1S_0$ so that they cycle through from 000 to 111 as clock pulses are applied. In this way, the output of the multiplexer will be X_0 during the first clock period, X_1 during the second clock period, and so on. The output Z is a waveform that is a serial representation of the parallel input data. The waveforms in the figure are for the case where $X_7X_6X_5X_4X_3X_2X_1X_0 = 10110101$. This conversion process takes a total of eight clock cycles. Note that X_0 (the LSB) is transmitted first and the X_7 (MSB) is transmitted last.

Operation Sequencing

The circuit of Figure 9-26 uses an eight-input multiplexer as part of a control sequencer that steps through seven steps, each of which actuates some portion of the physical process being controlled. This could be, for example, a process that mixes two liquid ingredients and then cooks the mixture. The circuit also uses a 3-line-to-8-line decoder and a MOD-8 binary counter. The operation is described as follows.

1. Initially the counter is reset to the 000 state. The counter outputs are fed to the select inputs of the multiplexer and to the inputs of the decoder. Thus, the decoder output $\overline{O}_0 = 0$ and the others are all 1, so that all the ACTUATOR inputs of the process are LOW. The SENSOR outputs of the process all start out LOW. The multiplexer output $\overline{Z} = \overline{I}_0 = 1$ because the S inputs are 000.
2. The START pulse initiates the sequencing operation by setting flip-flop Q_0 HIGH, bringing the counter to the 001 state. This causes decoder output $\overline{O}_1$ to go LOW, thereby activating actuator 1, which is the first step in the process (opening fill valve 1).
3. Some time later, SENSOR output 1 goes HIGH, indicating the completion of the first step (the float switch indicates that the tank is full). This HIGH is now present at the I_1 input of the multiplexer. It is inverted and reaches the $\overline{Z}$ output because the select code from the counter is 001.
4. The LOW transition at $\overline{Z}$ is fed to the *CLK* of flip-flop Q_0. This negative transition advances the counter to the 010 state.
5. Decoder output $\overline{O}_2$ now goes LOW, activating actuator 2, which is the second step in the process (opening fill valve 2). $\overline{Z}$ now equals $\overline{I}_2$ (the select code is 010). Because SENSOR output 2 is still LOW, $\overline{Z}$ will go HIGH.
6. When the second process step is complete, SENSOR output 2 goes HIGH, producing a LOW at $\overline{Z}$ and advancing the counter to 011.
7. This same action is repeated for each of the other steps. When the seventh step is completed, SENSOR output 7 goes HIGH, causing the counter to go from 111 to 000, where it will remain until another START pulse reinitiates the sequence.

Logic Function Generation

Multiplexers can be used to implement logic functions directly from a truth table without the need for simplification. When a multiplexer is used for this purpose, the select inputs are used as the logic variables, and each data input is connected permanently HIGH or LOW as necessary to satisfy the truth table.

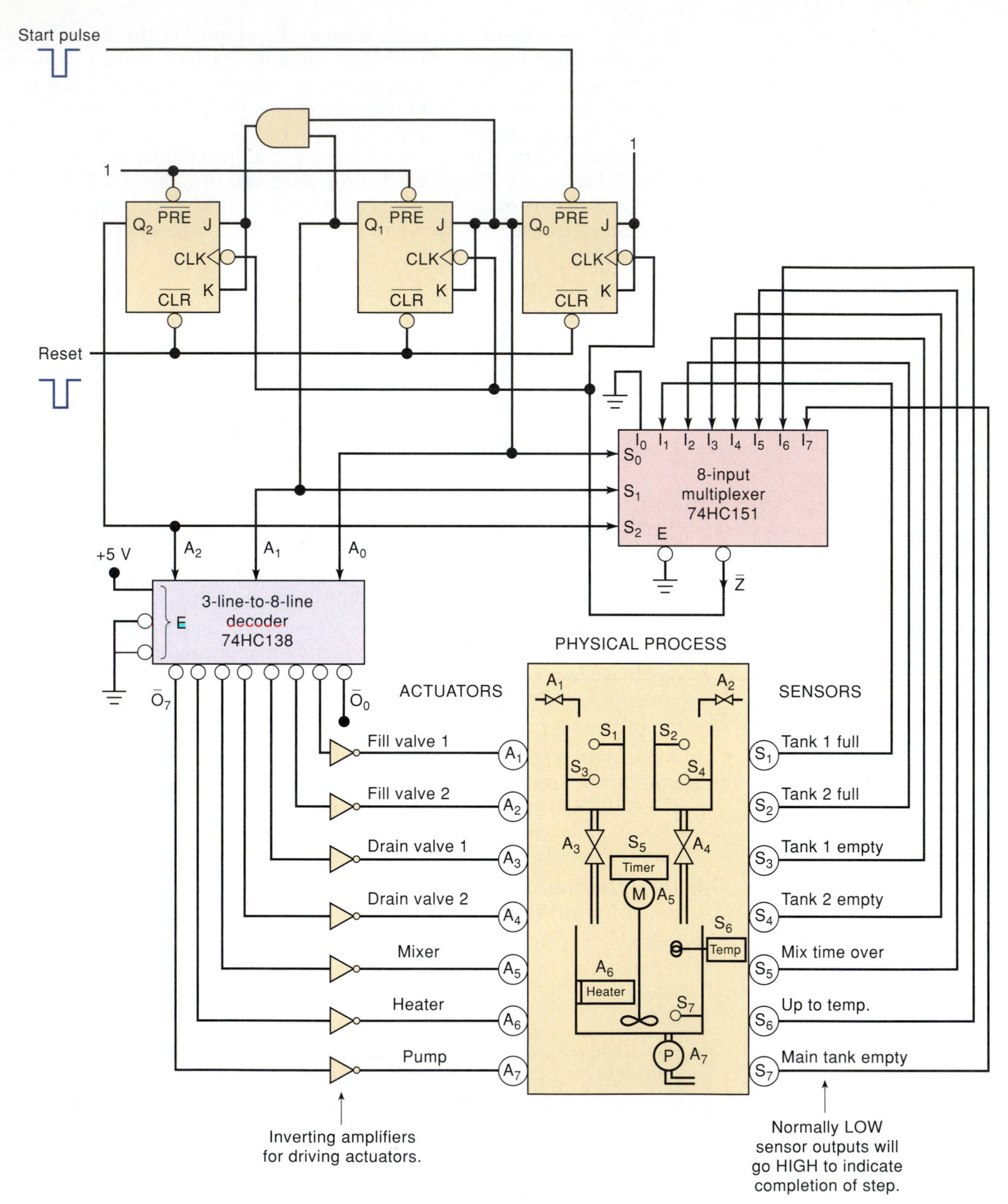

FIGURE 9-26 Seven-step control sequencer.

Figure 9-27 illustrates how an eight-input multiplexer can be used to implement the logic circuit that satisfies the given truth table. The input variables A, B, C are connected to S_0, S_1, S_2, respectively, so that the levels on these inputs determine which data input appears at output Z. According to the truth table, Z is supposed to be LOW when $CBA = 000$. Thus, multiplexer input I_0 should be

FIGURE 9-27 Multiplexer used to implement a logic function described by the truth table.

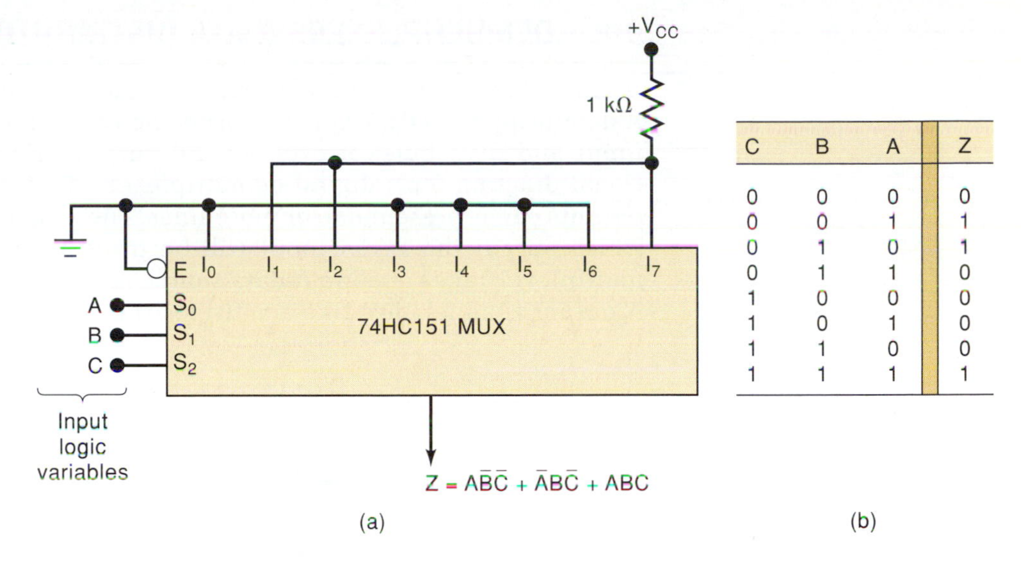

C	B	A	Z
0	0	0	0
0	0	1	1
0	1	0	1
0	1	1	0
1	0	0	0
1	0	1	0
1	1	0	0
1	1	1	1

connected LOW. Likewise, Z is supposed to be LOW for CBA = 011, 100, 101, and 110, so that inputs I_3, I_4, I_5, and I_6 should also be connected LOW. The other sets of CBA conditions must produce $Z = 1$, and so multiplexer inputs I_1, I_2, and I_7 are connected permanently HIGH.

It is easy to see that any three-variable truth table can be implemented with this eight-input multiplexer. This method of implementation is often more efficient than using separate logic gates. For example, if we can write the sum-of-products expression for the truth table in Figure 9-27, we have

$$Z = A\overline{B}\,\overline{C} + \overline{A}B\overline{C} + ABC$$

This *cannot* be simplified either algebraically or by K mapping, and so its gate implementation would require three INVERTERs and four NAND gates, for a total of three ICs.

There is an even more efficient method for using multiplexers to implement logic functions. This method will allow the logic designer to use a multiplexer with three select inputs (e.g., a 74HC151) to implement a *four-variable* logic function. We will present this method in Problem 9-37.

The most important concept to be gained from using a MUX to implement a sum-of-products expression is the fact that the logic function can be very easily changed by simply changing the 1s and 0s on the MUX inputs. In other words, a MUX can very easily be used as a programmable logic device (PLD). Many PLDs use this strategy in hardware blocks that are generally referred to as look-up tables (LUTs). We will discuss look-up tables in more detail in Chapters 12 and 13.

REVIEW QUESTIONS

1. What are some of the major applications of multiplexers?
2. *True or false:* When a multiplexer is used to implement a logic function, the logic variables are applied to the multiplexer's data inputs.
3. What type of circuit provides the select inputs when a MUX is used as a parallel-to-serial converter?

9-8 DEMULTIPLEXERS (DATA DISTRIBUTORS)

A multiplexer takes several inputs and transmits *one* of them to the output. *A* **demultiplexer (DEMUX)** performs the reverse operation: it takes a single input and distributes it over several outputs. Figure 9-28 shows the functional diagram for a digital demultiplexer. The large arrows for inputs and outputs can represent one or more lines. The select input code determines to which output the DATA input will be transmitted. In other words, the demultiplexer takes one input data source and selectively distributes it to 1 of N output channels just like a multiposition switch.

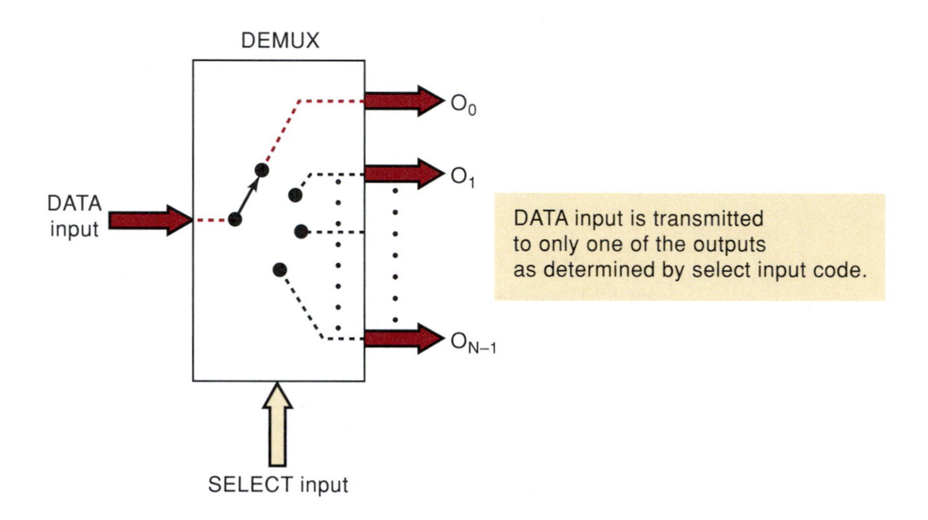
DEMUX
O_0
O_1
O_{N-1}
DATA input
DATA input is transmitted to only one of the outputs as determined by select input code.
SELECT input

FIGURE 9-28 General demultiplexer.

1-Line-to-8-Line Demultiplexer

Figure 9-29 shows the logic diagram for a demultiplexer that distributes one input line to eight output lines. The single data input line I is connected to all eight AND gates, but only one of these gates will be enabled by the SELECT input lines. For example, with $S_2S_1S_0 = 000$, only AND gate 0 will be enabled, and data input I will appear at output O_0. Other SELECT codes cause input I to reach the other outputs. The truth table summarizes the operation.

The demultiplexer circuit of Figure 9-29 is very similar to the 3-line-to-8-line decoder circuit in Figure 9-2 except that a fourth input (I) has been added to each gate. It was pointed out earlier that many IC decoders have an ENABLE input, which is an extra input added to the decoder gates. This type of decoder chip can therefore be used as a demultiplexer, with the binary code inputs (e.g., A, B, C in Figure 9-2) serving as the SELECT inputs and the ENABLE input serving as the data input I. For this reason, IC manufacturers often call this type of device a *decoder/demultiplexer,* and it can be used for either function.

We saw earlier how the 74ALS138 is used as a 1-of-8 decoder. Figure 9-30 shows how it can be used as a demultiplexer. The enable input $\overline{E}_1$ is used as the data input I, while the other two enable inputs are held in their active states. The $A_2A_1A_0$ inputs are used as the select code. To illustrate the operation, let's assume that the select inputs are 000. With this input code, the only output that can be activated is $\overline{O}_0$, while all other outputs are HIGH. $\overline{O}_0$ will go LOW only if $\overline{E}_1$ goes LOW and will be HIGH if $\overline{E}_1$ goes HIGH. In other words, $\overline{O}_0$ *will follow the signal on* $\overline{E}_1$ (i.e., the data input, I) while all other outputs stay HIGH. In a similar manner, a different select code applied to $A_2A_1A_0$ will cause the corresponding output to follow the data input, I.

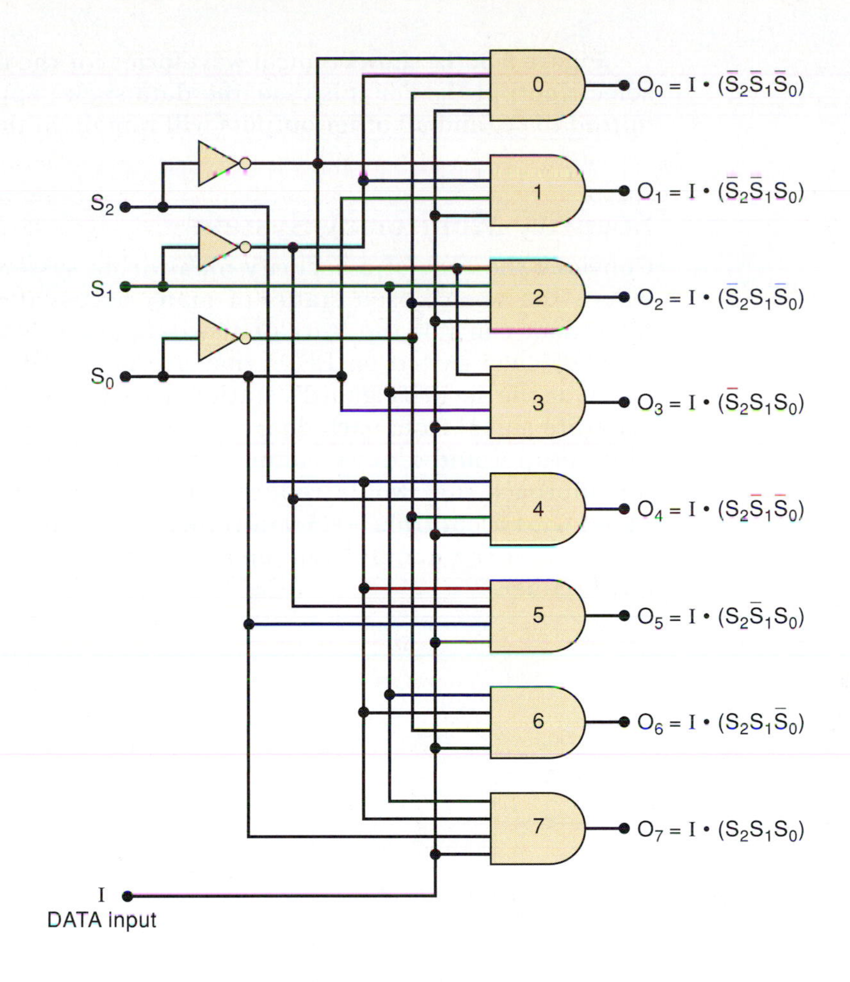

FIGURE 9-29 A 1-line-to-8-line demultiplexer.

SELECT code			OUTPUTS							
S_2	S_1	S_0	O_7	O_6	O_5	O_4	O_3	O_2	O_1	O_0
0	0	0	0	0	0	0	0	0	0	I
0	0	1	0	0	0	0	0	0	I	0
0	1	0	0	0	0	0	0	I	0	0
0	1	1	0	0	0	0	I	0	0	0
1	0	0	0	0	0	I	0	0	0	0
1	0	1	0	0	I	0	0	0	0	0
1	1	0	0	I	0	0	0	0	0	0
1	1	1	I	0	0	0	0	0	0	0

Note: I is the data input

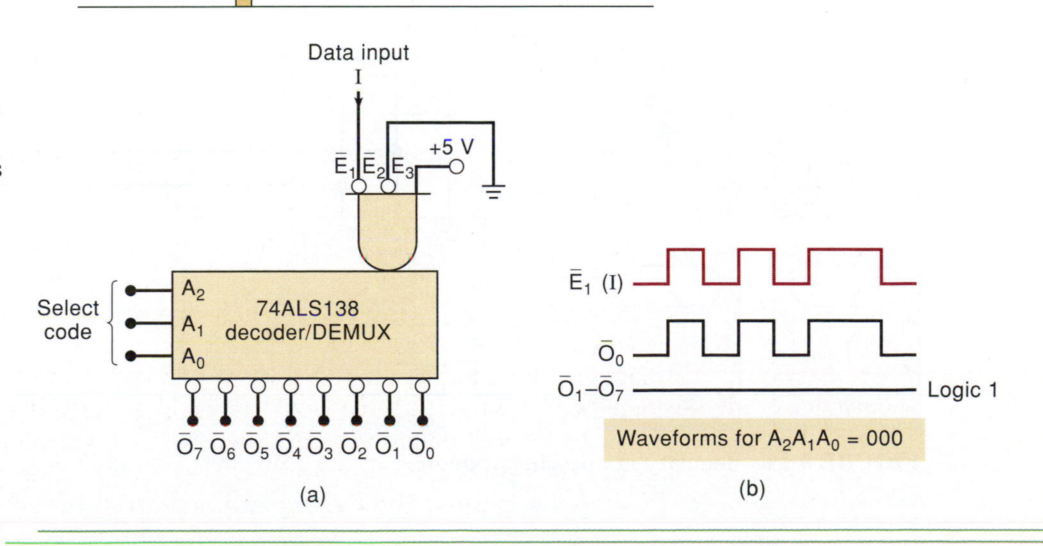

FIGURE 9-30 (a) The 74ALS138 decoder can function as a demultiplexer with $\overline{E}_1$ used as the data input; (b) typical waveforms for a select code of $A_2A_1A_0 = 000$ show that $\overline{O}_0$ is identical to the data input I on $\overline{E}_1$.

Figure 9-30(b) shows typical waveforms for the case where $A_2A_1A_0 = 000$ selects output $\overline{O}_0$. For this case, the data signal applied to $\overline{E}_1$ will be transmitted to $\overline{O}_0$, and all other outputs will remain in their inactive HIGH state.

Security Monitoring System

Consider the case of a security monitoring system in an industrial plant where the open/closed status of many access doors is to be monitored. Each door controls the state of a switch, and it is necessary to display the state of each switch on LEDs that are mounted on a remote monitoring panel at the security guard's station. One way to do this would be to run a separate signal from each door switch to an LED on the monitoring panel. This setup would require running many wires over a long distance. A better approach that would reduce the amount of wiring to the monitoring panel uses a multiplexer/demultiplexer combination. Figure 9-31 shows a system that can handle eight doors, but the basic idea can be expanded to any number.

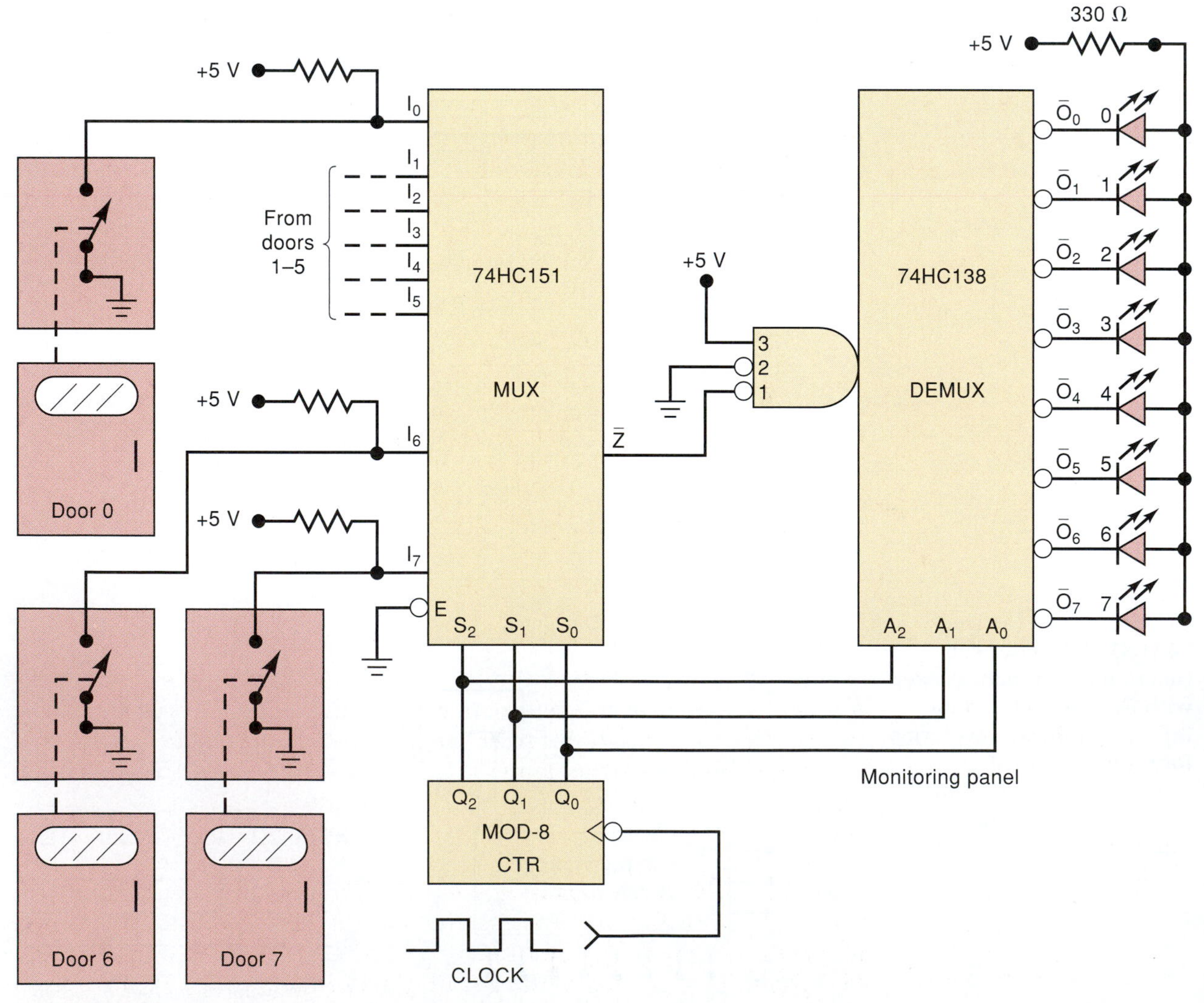

FIGURE 9-31 Security monitoring system.

EXAMPLE 9-11

Examine Figure 9-31 carefully and describe the complete operation.

Solution

The eight door switches are the data inputs to the MUX; they produce a HIGH when a door is open and a LOW when it is closed. The MOD-8 counter provides the select inputs to the MUX and also to the DEMUX on the remote monitoring panel. Each DEMUX output is connected to an indicator LED that will be on when the output is LOW. Clock pulses applied to the counter will cause the select inputs to sequence through all of the possible states 000 through 111. At each number of the counter, the switch status for the door of the same number will be inverted by the MUX and passed to output $\overline{Z}$. From there, it is transmitted to the DEMUX input, which passes it through to the output corresponding to the same number.

For example, let's say that the counter is at the count of 110 (6). While the counter is in this state, let's say that door 6 is closed. The LOW at I_6 will pass through the MUX and be inverted to produce a HIGH at $\overline{Z}$. This HIGH will be passed through the DEMUX to output $\overline{O}_6$ so that LED 6 will be off, indicating that door 6 is closed. Now let's say that door 6 is open. A LOW will appear at $\overline{Z}$ and $\overline{O}_6$ so that LED 6 will be on to signal that door 6 is open. Of course, all other LEDs will be off during this time because $\overline{O}_6$ is the only active output.

As the counter is clocked through its eight states 000 through 111, the LEDs will sequentially indicate the status of the eight doors. If all the doors are closed, none of the LEDs will be on even when the corresponding DEMUX output is selected. If a door is open, its LED will turn on only during the time interval that the counter is at the appropriate count; it will be off at all other counts. Thus, the LED will be flashing on and off if its door is open. The flashing rate can be adjusted by changing the frequency of the clock.

Note that there are only four signal lines going from the "door-sensing" circuitry to the remote monitoring panel: the $\overline{Z}$ output and the three select lines. This is a saving of four lines when compared with the alternative of having one line per door. The MUX/DEMUX combination is used to transmit the status of each door to its LED one at a time (serially) instead of all at once (parallel).

Synchronous Data Transmission System

Figures 9-32 and 9-33 show the logic diagrams for a synchronous data transmission system that is used to transmit four, four-bit words serially from a transmitter to a remote receiver. To operate this system, four data words are parallel-loaded into the input registers of the transmitter block and the transmit signal is activated. The 16 data bits are then sent over a single data line, one bit at a time, reassembled by the receiver, and stored in output registers. Let's look at the transmitter details in Figure 9-32 first. The *clock* input is a high-frequency, constantly running, periodic clock signal that synchronizes all activities in the system. The four-bit data words are stored individually (synchronously) in the PISO registers when enabled by the appropriate *ld_x* input. For simplicity, the parallel data inputs to the PISO registers are not shown in the diagram. These input registers are designed to shift the data to the right and also recirculate the LSB (rightmost bit) to the MSB

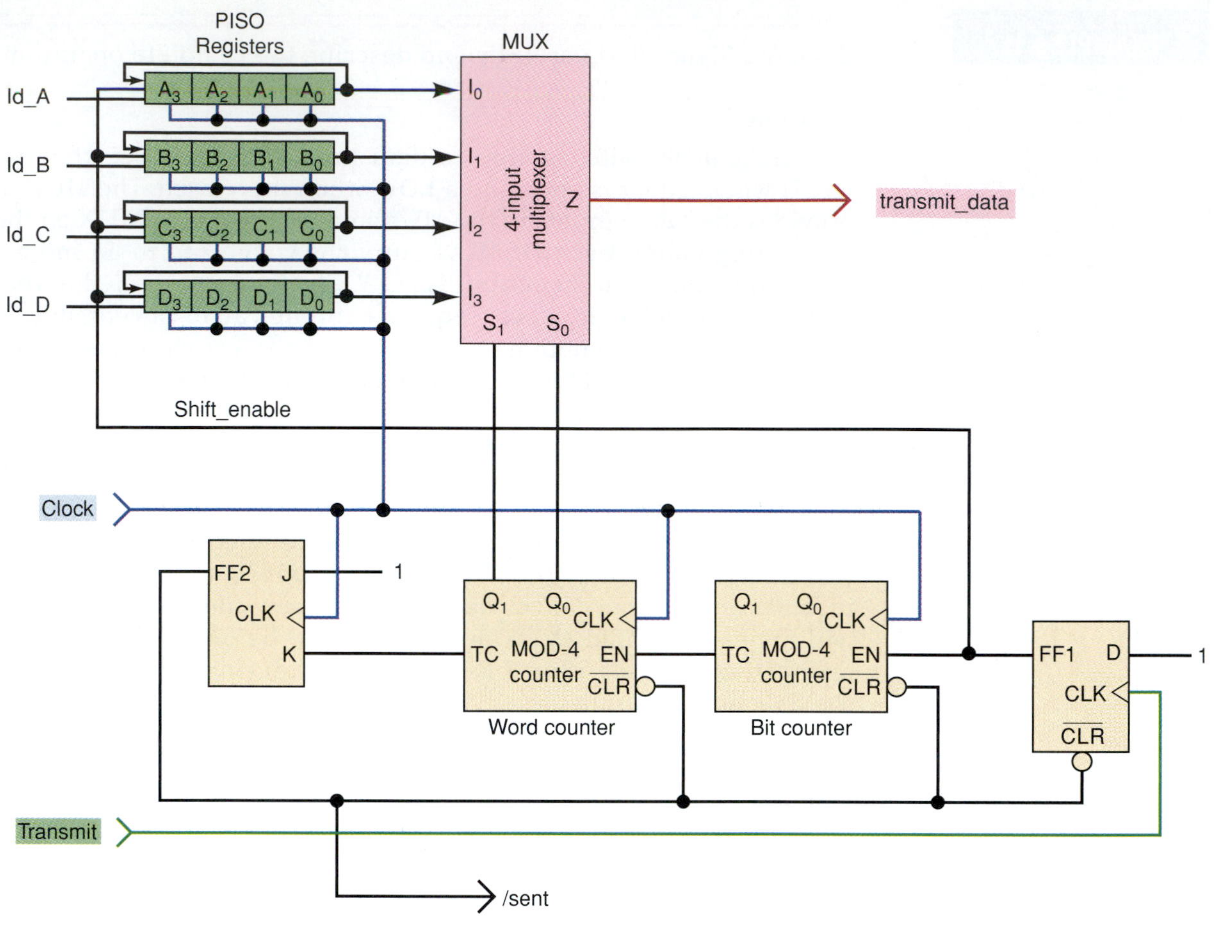

FIGURE 9-32 Transmitter block in synchronous data transmission system.

(leftmost bit). With this arrangement the bits are all shifted to the serial output and also end up back in their proper locations after four clock pulses.

TRANSMITTER OPERATION Initially, let's assume that all the flip-flops and the two MOD-4 counters in Figure 9-32 are all cleared. On the next PGT of clock, *FF2* is SET, removing the asynchronous clear command from the counters and *FF1*. When the *transmit* signal goes HIGH, *FF1* is SET, putting all the shift registers in the shift mode. The MUX selects input 0 (register *A*) because the MOD-4 Word counter is at 0. At this point the LSB of register *A* is on the *transmit_data* line. The next three clock pulses (counted by the Bit counter) shift the other bits of register *A* to the serial output. As a result, the *transmit_data* line outputs each of the register *A* bits, one at a time from the least to the most significant. On the fourth PGT, the Bit counter rolls over to zero, the Word counter increments to 1, all of the shift registers have recirculated their data back to the original position, and the MUX now selects the LSB data from register *B* to output on the *transmit_data* line. The next three clocks shift out the contents of register *B*, followed by registers *C* and *D*. On the 16th PGT, *FF2* toggles to a zero state, resetting all the counters and disabling any further counting by also clearing *FF1*. The next PGT sets *FF2* again, and the system is waiting for new data to be loaded and the next *transmit* signal.

RECEIVER OPERATION The receiver circuit shown in Figure 9-33 is very similar in operation to the transmitter. Notice that all flip-flops, counters,

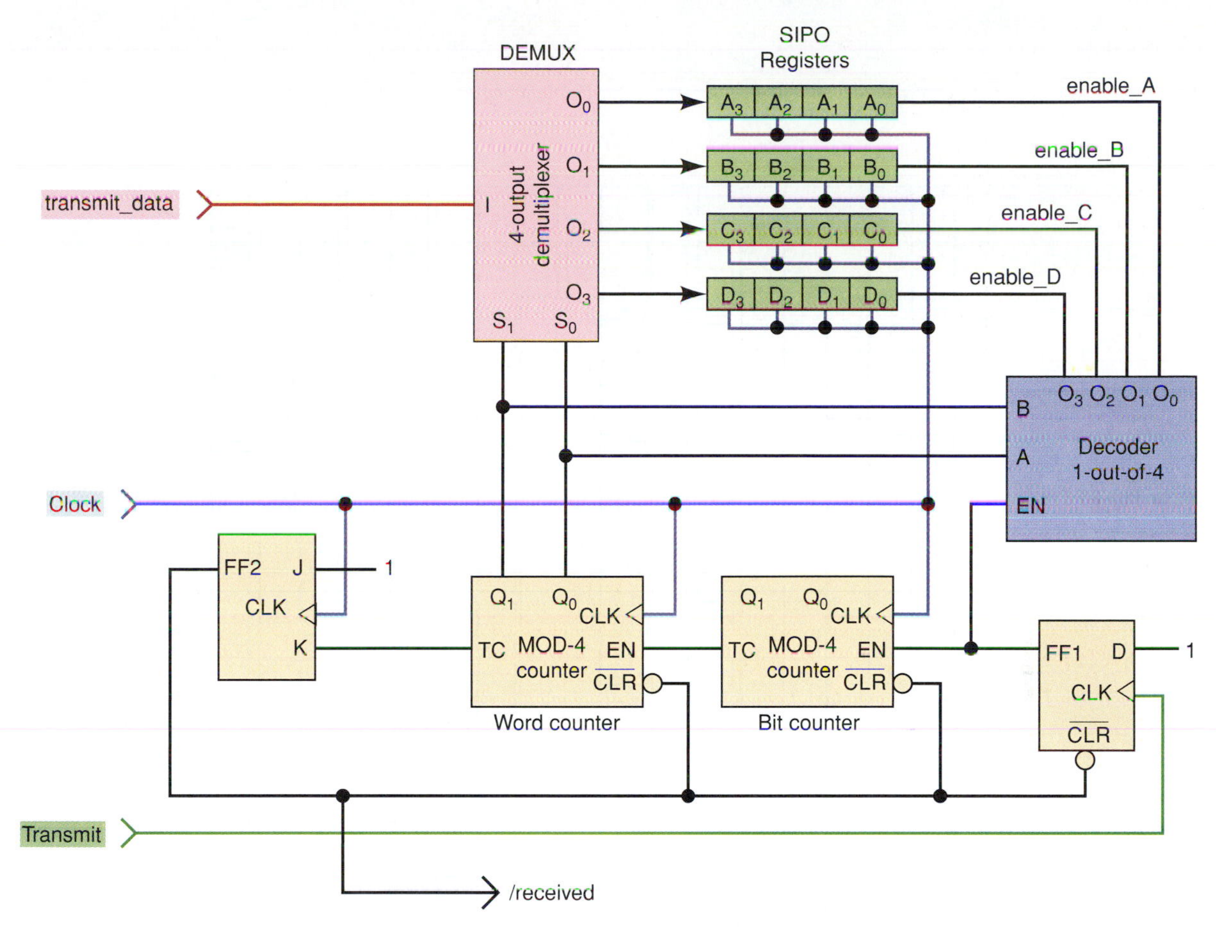

FIGURE 9-33 Receiver block in synchronous data transmission system.

and registers use the same clock as the transmitter. The receiver uses a DEMUX to distribute the serial data to the appropriate SIPO register and a decoder to enable one register at a time. Let's begin analyzing this circuit with all counters and flip-flops at zero. The next *clock* sets *FF2*, removing the asynchronous clear command from the counters and *FF1*. When the *transmit* line goes HIGH, *FF1* is SET, enabling the Bit counter, Word counter, and also the decoder. With the Word counter at zero, the decoder enables register *A* and the DEMUX connects the serial data line (which currently contains the LSB of transmit register *A*) to the serial data input of receive register *A*. The next PGT shifts the least significant data bit into register *A* and advances the Bit counter. The next three PGTs shift the next three data bits into register *A*, the Bit counter rolls over to zero, the Word counter increments to 1, and the decoder and DEMUX switch to register *B*. After the 16th PGT, all four registers contain the proper data, *FF2* has toggled to a zero state, *FF1* is cleared and disables the decoder, which disables all the SIPO registers. On the next PGT, *FF2* is set and the system is waiting for the next transmission of data.

SYSTEM TIMING The timing diagram in Figure 9-34 shows the parallel data that is loaded into the transmitter, the serial data stream, and the distribution and storage of the four data values in the receiver registers. At times t_{1-4}, the binary data values (shown as hex 3, 5, 6, and D) are loaded into transmit registers *A*, *B*, *C*, and *D*, respectively. The system is idle until the *transmit* line goes HIGH at t_5. At this point the LSB from register *A* (A_0) is already on the *transmit_data* line. Also notice that at t_5–t_8, the data on output

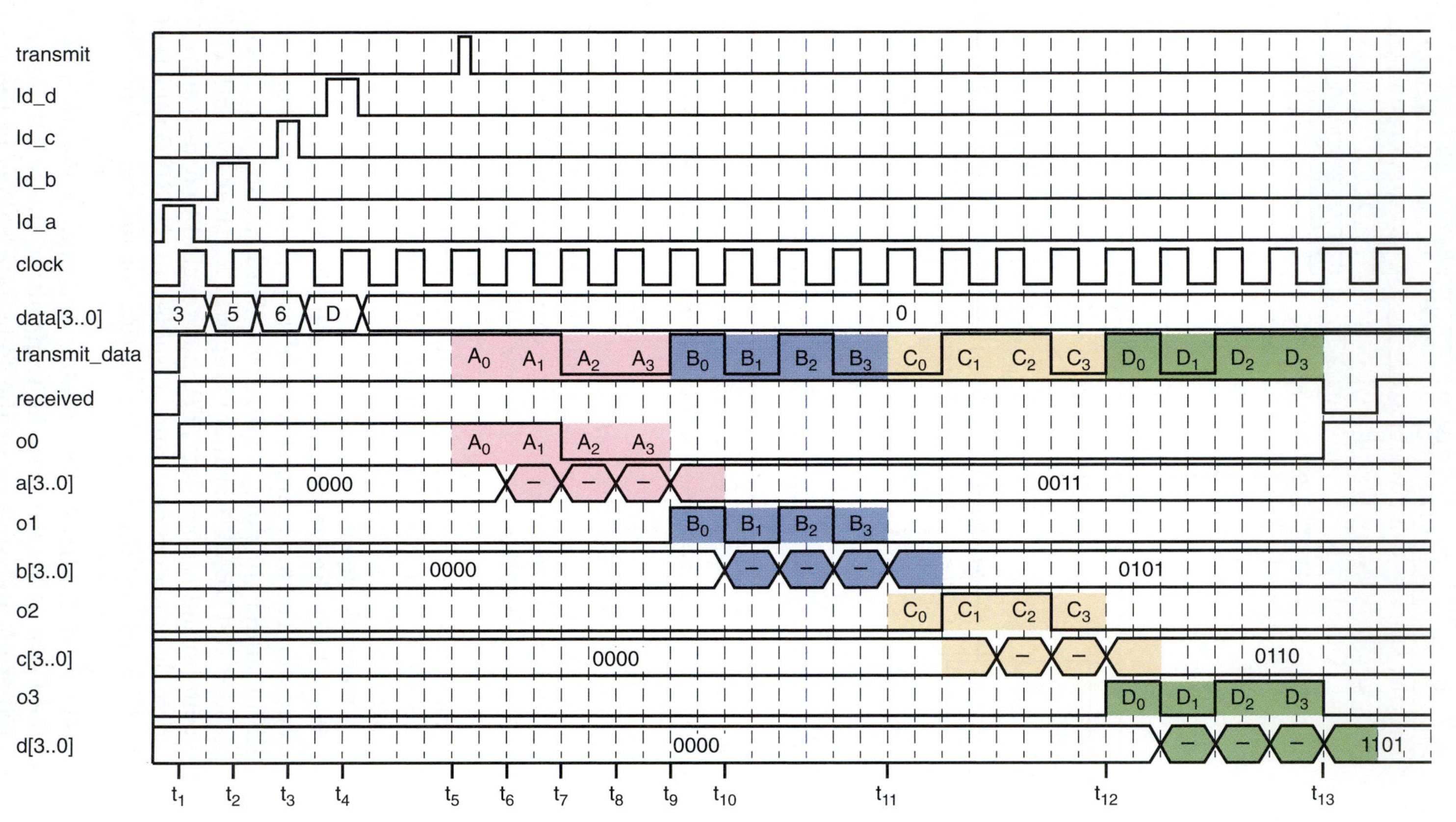

FIGURE 9-34 Timing diagram for one complete transmission cycle.

O_0 of the DEMUX is identical to the *transmit_data* line. This shows that the DEMUX has distributed the *transmit_data* to shift register A. At t_6, the PGT of the clock shifts A_0 into the MSB of receive register A, all transmit data registers (not shown in the timing) are shifted, and data bit A_1 appears on the *transmit_data* line. At times t_7, t_8, and t_9, the other three bits are shifted into register A such that after t_9, receive register A contains the data bits that were stored in transmit register A. The diagram shows that the DEMUX has switched to distribute data to register B because the DEMUX output O_1 is now identical to *transmit_data* from t_9 through t_{11}. Starting at t_{10}, the data are shifted into receive register B, which at t_{11} contains the value that was originally stored in transmit register B. Register C and Register D are sent and stored from t_{11} to t_{12} and from t_{12} to t_{13}, respectively.

REVIEW QUESTIONS

1. Explain the difference between a DEMUX and a MUX.
2. *True or false:* The circuit for a DEMUX is basically the same as for a decoder.
3. For the system of Figure 9-31, what will the security guard see on the monitoring panel when all of the doors are open?

9-9 MORE TROUBLESHOOTING

Here are three more examples to illustrate the observation/reasoning process that is such an important initial step when troubleshooting. For each case, try to determine the circuit fault before looking at the solution.

EXAMPLE 9-12

Consider the circuit of Figure 9-24. A test performed on this circuit yields the result shown in Table 9-3. What is the probable circuit fault?

TABLE 9-3

		Actual Count	Displayed Count
Case 1	Counter 1	25	25
	Counter 2	37	35
Case 2	Counter 1	49	49
	Counter 2	72	79
Case 3	Counter 1	96	96
	Counter 2	14	16

Solution

In each of the test cases, the display of counter 1 matches the counter's actual count. This indicates that the I_1 inputs, all MUX outputs, and both displays are probably working correctly. On the other hand, each test case shows that counter 2's *tens* digit is displayed correctly but its *units* digit is displayed incorrectly. This could mean that there is a fault somewhere between the output of the units section of counter 2 and the I_0 inputs of the units MUX. We should compare the bit patterns of the actual and displayed values of the units for counter 2 (Table 9-4). The idea is to look for things such as a bit that does not change (stuck LOW or HIGH) or two bits that are reversed (crossed connections). The data in Table 9-4 reveal no obvious pattern.

TABLE 9-4

	Actual Units	Displayed Units
Case 1	0111 (7)	0101 (5)
Case 2	0010 (2)	1001 (9)
Case 3	0100 (4)	0110 (6)

If we take another look at the recorded test results, we see that the displayed units digit of counter 2 is always the same as the units digit of counter 1. This symptom is probably the result of a constant logic HIGH at the select input of the units MUX because that would continually pass the units digit of counter 1 to the units MUX output. This constant HIGH at the select input is most likely caused by an open path somewhere between the select input of the tens MUX and the select input of the units MUX. It could not be caused by a short to V_{CC} because that would also keep the select input of the tens MUX at a constant HIGH, and we know that the tens MUX is working.

EXAMPLE 9-13

The security monitoring system of Figure 9-31 is tested and the results are recorded in Table 9-5. What are the possible faults that could produce these results?

TABLE 9-5

Condition	LEDs
All doors closed	All LEDs off
Door 0 open	LED 4 flashing
Door 1 open	LED 5 flashing
Door 2 open	LED 6 flashing
Door 3 open	LED 7 flashing
Door 4 open	LED 4 flashing
Door 5 open	LED 5 flashing
Door 6 open	LED 6 flashing
Door 7 open	LED 7 flashing

Solution

Again, the data should be reviewed to see if there is some pattern that could help to narrow down the search for the fault to a small area of the circuit. The data in Table 9-5 reveal that the correct LEDs flash for open doors 4 through 7. They also show that for open doors 0 through 3, the number of the flashing LED is *four* more than the number of the door, and LEDs 0 through 3 are always off. This is most probably caused by a constant logic HIGH at A_2, the MSB of the select input of the DEMUX, because this would always make the select code 4 or greater, and it would add 4 to the select codes 0 through 3.

Thus, we have two possibilities: A_2 is somehow shorted to V_{CC}, or there is an open connection at A_2. A little thought will eliminate the first choice as a possibility because this would also mean that S_2 of the MUX would also be stuck HIGH. If that were so, then the status of doors 0 through 3 would not get through the MUX and into the DEMUX. We know that this is not true because the data show that when any of these doors is open, it affects one of the DEMUX outputs.

EXAMPLE 9-14

An extremely important principle of troubleshooting, called *divide-and-conquer*, was introduced in Section 9-5. It is really not about military strategy, but rather describes the most efficient way to eliminate from consideration all the parts of the circuit that are working correctly. Assume that data have been loaded into the four transmit registers of Figure 9-32 and the transmit pulse has occurred, but after the next 16 clock pulses, no new data have appeared in the receive registers shown in Figure 9-33. How can we most efficiently find the problem?

Solution

In a synchronous digital system that is simply not functioning, it is reasonable first to check to see if the power supply and clock are working, just as you might check for a pulse if you found a person lying on the ground. However, assuming the clock is oscillating, there is a much more efficient way to isolate the problem than randomly picking points in the circuit and determining if the correct signal is present. We want to perform a test on this circuit such that, if we obtain the desired results, we know that half of the circuit is working correctly and we can eliminate that half from consideration. In this circuit the best place to look is at the *transmit_data* line. A logic probe should be placed on the *transmit_data* line and the *transmit* signal should be activated. If a burst of pulses is observed on the logic probe, it means that the transmit section is functioning. We may not know if the data are correct, but remember, the receiver is not getting incorrect data but rather no data at all. However, if no burst of pulses is observed, there is certainly a problem in the transmit section.

A troubleshooting tree diagram as shown in Figure 9-35 is helpful in isolating problems in a system. Let's assume there were no pulses on *transmit_data*. Now we need to perform a test on the transmitter to prove that half of the transmitter is working properly. In this case the circuit does not divide exactly in half easily. A good choice might be to examine the output of the word counter. A logic probe should be placed on the select inputs of the MUX and the *transmit* signal activated. If brief pulses occur immediately after transmit, then the entire control section (made up of two counters and two flip-flops) is probably functioning properly and we can look elsewhere. The next place to look is at the outputs of the PISO registers (or data inputs of the MUX). If data pulses are present on each line after *transmit* is activated, the problem must be in the MUX. If not, we can further break down the PISO section. Each test that is performed should eliminate the largest possible amount of the remaining circuitry until all that is left is a small block containing the fault.

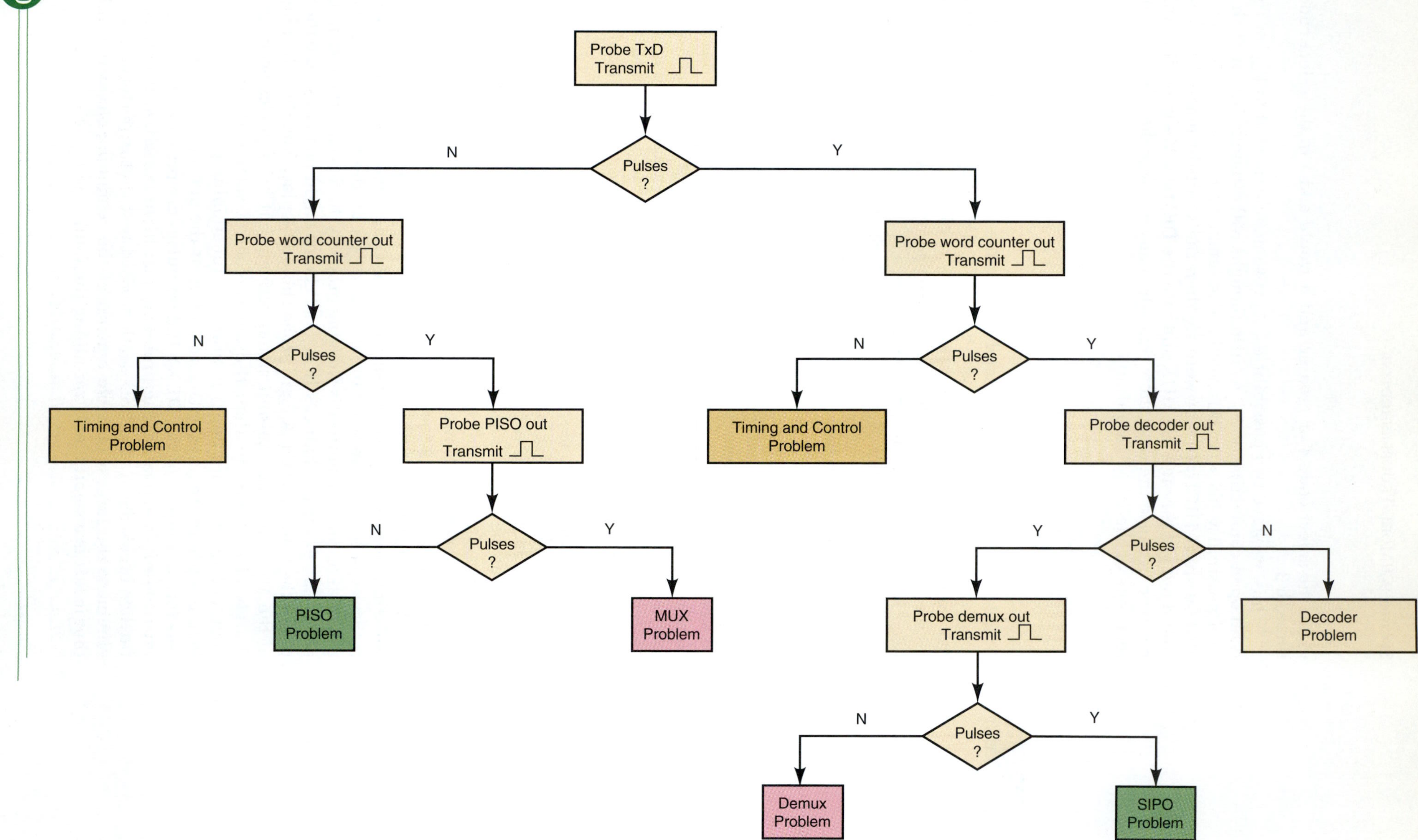

FIGURE 9-35 Example 9-14: A troubleshooting tree diagram.

9-10 MAGNITUDE COMPARATOR

Another useful member of the MSI category of ICs is the **magnitude comparator**. It is a combinational logic circuit that compares two input binary quantities and generates outputs to indicate which one has the greater magnitude. Figure 9-36 shows the logic symbol and the truth table for the 74HC85 four-bit magnitude comparator, which is also available as the 74LS85.

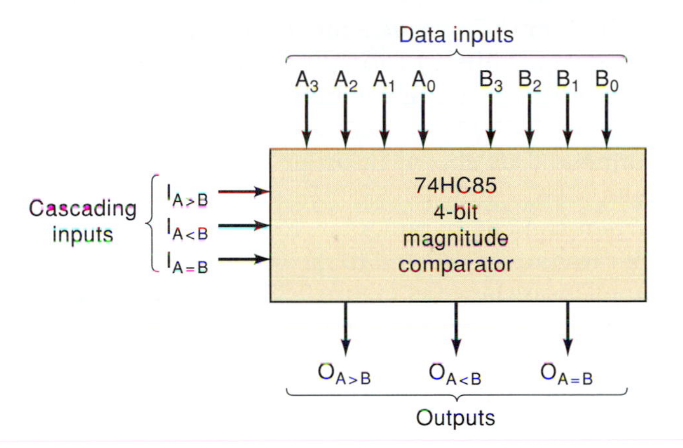

TRUTH TABLE

COMPARING INPUTS				CASCADING INPUTS			OUTPUTS		
A_3, B_3	A_2, B_2	A_1, B_1	A_0, B_0	$I_{A>B}$	$I_{A<B}$	$I_{A=B}$	$O_{A>B}$	$O_{A<B}$	$O_{A=B}$
$A_3>B_3$	X	X	X	X	X	X	H	L	L
$A_3<B_3$	X	X	X	X	X	X	L	H	L
$A_3=B_3$	$A_2>B_2$	X	X	X	X	X	H	L	L
$A_3=B_3$	$A_2<B_2$	X	X	X	X	X	L	H	L
$A_3=B_3$	$A_2=B_2$	$A_1>B_1$	X	X	X	X	H	L	L
$A_3=B_3$	$A_2=B_2$	$A_1<B_1$	X	X	X	X	L	H	L
$A_3=B_3$	$A_2=B_2$	$A_1=B_1$	$A_0>B_0$	X	X	X	H	L	L
$A_3=B_3$	$A_2=B_2$	$A_1=B_1$	$A_0<B_0$	X	X	X	L	H	L
$A_3=B_3$	$A_2=B_2$	$A_1=B_1$	$A_0=B_0$	H	L	L	H	L	L
$A_3=B_3$	$A_2=B_2$	$A_1=B_1$	$A_0=B_0$	L	H	L	L	H	L
$A_3=B_3$	$A_2=B_2$	$A_1=B_1$	$A_0=B_0$	X	X	H	L	L	H
$A_3=B_3$	$A_2=B_2$	$A_1=B_1$	$A_0=B_0$	L	L	L	H	H	L
$A_3=B_3$	$A_2=B_2$	$A_1=B_1$	$A_0=B_0$	H	H	L	L	L	L

H = HIGH Voltage Level
L = LOW Voltage Level
X = Immaterial

FIGURE 9-36 Logic symbol and truth table for a 74HC85 (7485, 74LS85) four-bit magnitude comparator.

Data Inputs

The 74HC85 compares two *unsigned* four-bit binary numbers. One of them is $A_3A_2A_1A_0$, which is called word A; the other is $B_3B_2B_1B_0$, which is called word B. The term *word* is used in the digital computer field to designate a group of bits that represents some specific type of information. Here, word A and word B represent numerical quantities.

Outputs

The 74HC85 has three active-HIGH outputs. Output $O_{A>B}$ will be HIGH when the magnitude of word A is greater than the magnitude of word B. Output $O_{A<B}$ will be HIGH when the magnitude of word A is less than the magnitude of word B. Output $O_{A=B}$ will be HIGH when word A and word B are identical.

Cascading Inputs

Cascading inputs provide a means for expanding the comparison operation to more than four bits by cascading two or more four-bit comparators. Note that the cascading inputs are labeled the same as the outputs. When a four-bit comparison is being made, as in Figure 9-37(a), the cascading inputs should be connected as shown in order for the comparator to produce the correct outputs.

When two comparators are to be cascaded, the outputs of the lower-order comparator are connected to the corresponding inputs of the higher-order comparator. This is shown in Figure 9-37(b), where the comparator on the left is comparing the lower-order four bits of the two eight-bit words: $A_7A_6A_5A_4A_3A_2A_1A_0$ and $B_7B_6B_5B_4B_3B_2B_1B_0$. Its outputs are fed to the cascade inputs of the comparator on the right, which is comparing the high-order bits. The outputs of the high-order comparator are the final outputs that indicate the result of the eight-bit comparison.

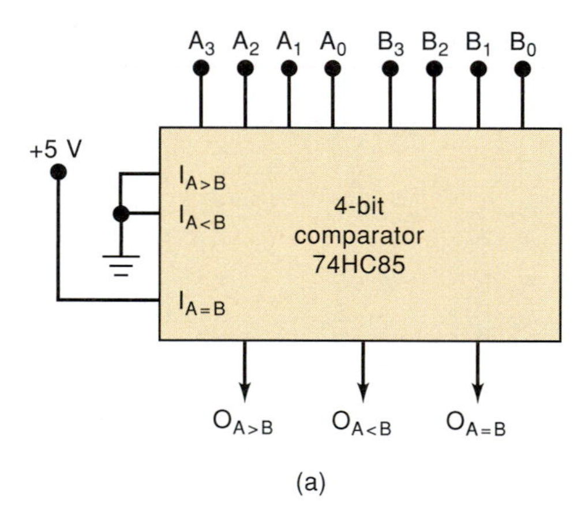

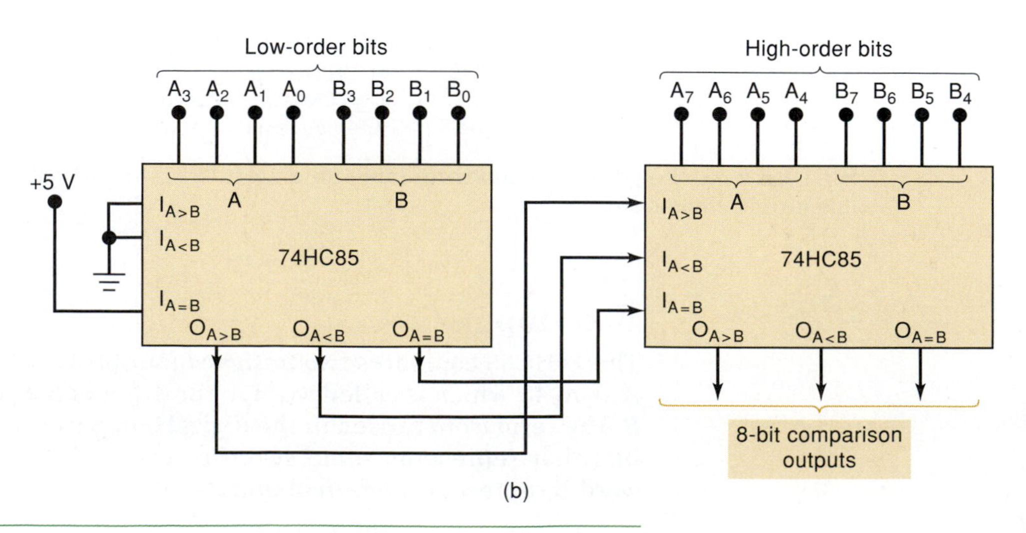

FIGURE 9-37 (a) 74HC85 wired as a four-bit comparator; (b) two 74HC85s cascaded to perform an eight-bit comparison.

EXAMPLE 9-15

Describe the operation of the eight-bit comparison arrangement in Figure 9-37(b) for the following cases:

(a) $A_7A_6A_5A_4A_3A_2A_1A_0 = 10101111; B_7B_6B_5B_4B_3B_2B_1B_0 = 10110001$
(b) $A_7A_6A_5A_4A_3A_2A_1A_0 = 10101111; B_7B_6B_5B_4B_3B_2B_1B_0 = 10101001$

Solution

(a) The high-order comparator compares its inputs $A_7A_6A_5A_4 = 1010$ and $B_7B_6B_5B_4 = 1011$ and produces $O_{A<B} = 1$ regardless of what levels are applied to its cascade inputs from the low-order comparator. In other words, once the high-order comparator senses a difference in the high-order bits of the two eight-bit words, it knows which eight-bit word is greater without having to look at the results of the low-order comparison.

(b) The high-order comparator sees $A_7A_6A_5A_4 = B_7B_6B_5B_4 = 1010$, so it must look at its cascade inputs to see the result of the low-order comparison. The low-order comparator has $A_3A_2A_1A_0 = 1111$ and $B_3B_2B_1B_0 = 1001$, which produces a 1 at its $O_{A>B}$ output and the $I_{A>B}$ input of the high-order comparator. The high-order comparator senses this 1, and because its data inputs are equal, it produces a HIGH at its $O_{A>B}$ to indicate the result of the eight-bit comparison.

Applications

Magnitude comparators are also useful in control applications where a binary number representing the physical variable being controlled (e.g., position, speed, or temperature) is compared with a reference value. The comparator outputs are used to actuate circuitry to drive the physical variable toward the reference value. The following example will illustrate one application. We will examine another comparator application in Problem 9-52.

EXAMPLE 9-16

Consider a digital thermostat in which the measured room temperature is converted to a digital number and applied to the A inputs of a comparator. The desired room temperature, entered from a keypad, is stored in a register that is connected to the B inputs. If $A < B$, the furnace should be activated to heat the room. The furnace should continue to heat while $A = B$ and shut off when $A > B$. As the room cools off, the furnace should stay off while $A = B$ and turn on again when $A < B$. What digital circuit can be used to interface a magnitude comparator to a furnace to perform the thermostat control application described above?

Solution

Using the $O_{A<B}$ output to drive the furnace directly would cause it to turn off as soon as the values became equal. This can cause severe on/off cycling of the furnace when the actual temperature is very close to the boundary between $A < B$ and $A = B$. By using a NOR gate SET-CLEAR latch circuit (refer to Chapter 5) as shown in Figure 9-38, the system will operate as described. Notice that $O_{A<B}$ is connected to the SET input and $O_{A>B}$ is connected to the CLEAR input of the latch. When the temperature is hotter than desired, it clears the latch, shutting off the furnace. When the temperature is cooler than desired, it sets the latch, turning the furnace on.

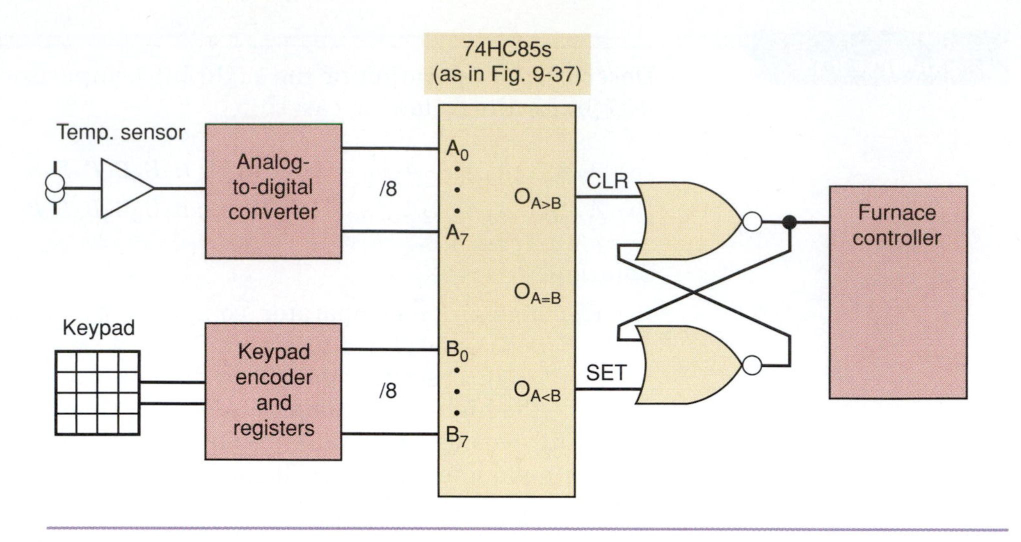

FIGURE 9-38 Magnitude comparator used in a digital thermostat.

REVIEW QUESTIONS

1. What is the purpose of the cascading inputs of the 74HC85?
2. What are the outputs of a 74HC85 with the following inputs: $A_3A_2A_1A_0 = B_3B_2B_1B_0 = 1001$, $I_{A>B} = I_{A<B} = 0$, and $I_{A=B} = 1$?

9-11 CODE CONVERTERS

A code converter is a logic circuit that changes data presented in one type of binary code to another type of binary code. The BCD-to-7-segment decoder-driver that we presented earlier is a code converter because it changes a BCD input code to the 7-segment code needed by the LED display. A partial list of some of the more common code conversions is given in Table 9-6.

TABLE 9-6

BCD to 7-segment
BCD to binary
Binary to BCD
Binary to Gray code
Gray code to binary
ASCII to EBCDIC*
EBCDIC to ASCII

*EBCDIC is an alphanumeric code developed by IBM and is similar to ASCII.

As an example of a code converter circuit, let's consider a BCD-to-binary converter. Before we get started on the circuit implementation, we should review the BCD representation.

Two-digit decimal values ranging from 00 to 99 can be represented in BCD by two four-bit code groups. For example, 57_{10} is represented as

$$\overbrace{0101}^{5} \quad \overbrace{0111}^{7} \quad \text{(BCD)}$$

The straight binary representation for decimal 57 is

$$57_{10} = 111001_2$$

The largest two-digit decimal value of 99 has the following representations:

$$99_{10} = 10011001\text{ (BCD)} = 1100011_2$$

Note that the binary representation requires only seven bits.

Basic Idea

The diagram of Figure 9-39 shows the basic idea for a two-digit BCD-to-binary converter. The inputs to the converter are the two four-bit code groups

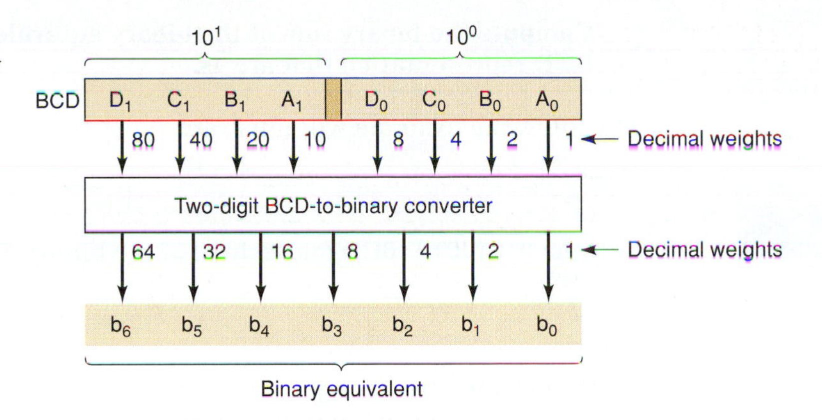

FIGURE 9-39 Basic idea of a two-digit BCD-to-binary converter.

$D_0C_0B_0A_0$, representing the 10^0 or units digit, and $D_1C_1B_1A_1$, representing the 10^1 or tens digit of the decimal value. The outputs from the converter are $b_6b_5b_4b_3b_2b_1b_0$, the seven bits of the binary equivalent of the same decimal value. Note the difference in the weights of the BCD bits and those of the binary bits.

A typical use of a BCD-to-binary converter would be where BCD data from an instrument such as a DMM (digital multimeter) are being transferred to a computer for storage or processing. The data must be converted to binary so that they can be operated on in binary by the computer ALU, which may not have the capability of performing arithmetic operations on BCD data. The BCD-to-binary conversion can be accomplished with either hardware or software. The hardware method (which we will look at momentarily) is generally faster but requires extra circuitry. The software method uses no extra circuitry, but it takes more time because the software does the conversion step by step. The method chosen in a particular application depends on whether or not conversion time is an important consideration.

Conversion Process

The bits in a BCD representation have decimal weights that are 8, 4, 2, 1 within each code group but that differ by a factor of 10 from one code group (decimal digit) to the next. Figure 9-39 shows the bit weights for the two-digit BCD representation.

The decimal weight of each bit in the BCD representation can be converted to its binary equivalent. The results are given in Table 9-7. Using these weights, we can perform the BCD-to-binary conversion by simply doing the following:

TABLE 9-7 Binary equivalents of decimal weights of each BCD bit.

BCD Bit	Decimal Weight	Binary Equivalent						
		b_6	b_5	b_4	b_3	b_2	b_1	b_0
A_0	1	0	0	0	0	0	0	1
B_0	2	0	0	0	0	0	1	0
C_0	4	0	0	0	0	1	0	0
D_0	8	0	0	0	1	0	0	0
A_1	10	0	0	0	1	0	1	0
B_1	20	0	0	1	0	1	0	0
C_1	40	0	1	0	1	0	0	0
D_1	80	1	0	1	0	0	0	0

Compute the binary sum of the binary equivalents of all bits in the BCD representation that are 1s.

The following example will illustrate.

EXAMPLE 9-17

Convert 01010010 (BCD for decimal 52) to binary. Repeat for 10010101 (decimal 95).

Solution

Write down the binary equivalents for all the 1s in the BCD representation. Then add them all together in binary.

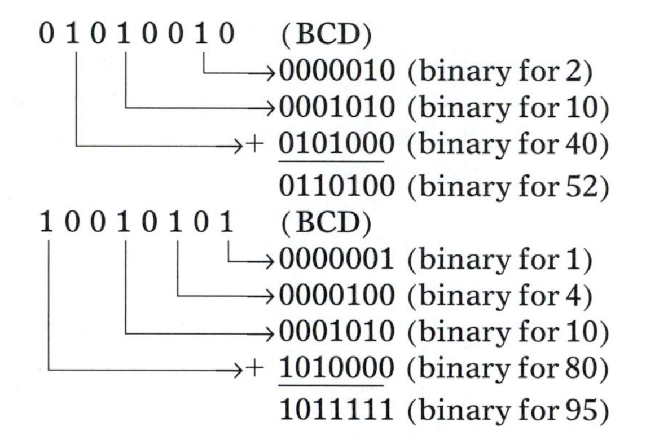

Circuit Implementation

Clearly, one way to implement the logic circuit that performs this conversion process is to use binary adder circuits. Figure 9-40 shows how two 74HC83

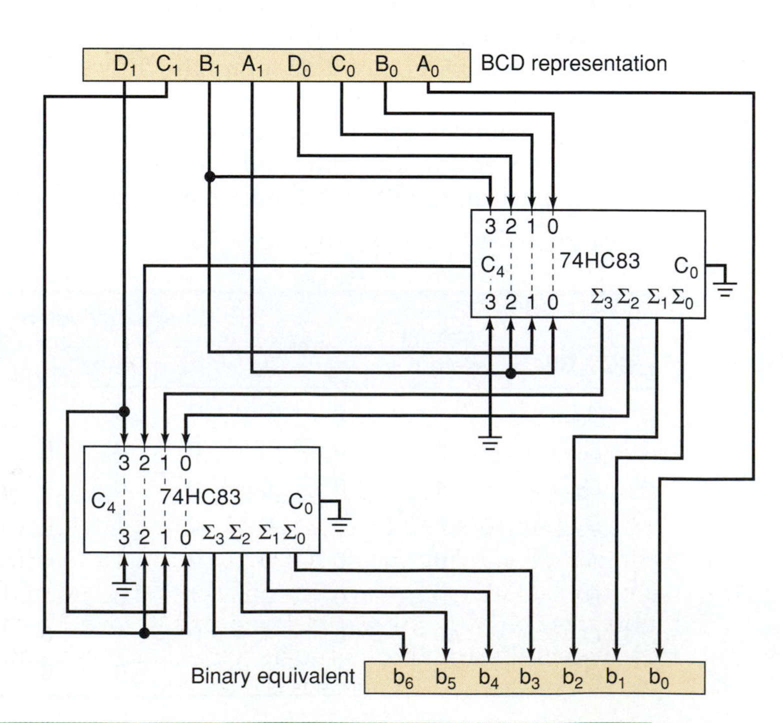

FIGURE 9-40 BCD-to-binary converter implemented with 74HC83 four-bit parallel adders.

four-bit parallel adders can be wired to perform the conversion. This is one of several possible adder arrangements that will work. You may want to review the operation of this IC in Section 6-14.

The two adder ICs perform the addition of the BCD bits in the proper combinations according to Table 9-7. For instance, Table 9-7 shows that A_0 is the only BCD bit that contributes to the LSB, b_0, of the binary equivalent. Because there is no carry into this bit position, A_0 is connected directly as output b_0. The table also shows that only BCD bits B_0 and A_1 contribute to bit b_1 of the binary output. These two bits are combined in the upper adder to produce output b_1. Likewise, only BCD bits D_0, A_1, and C_1 contribute to bit b_3. The upper adder combines D_0 and A_1 to generate Σ_2, which is connected to the lower adder, where C_1 is added to it to produce b_3.

EXAMPLE 9-18

The BCD representation for decimal 56 is applied to the converter of Figure 9-40. Determine the Σ outputs from each adder and the final binary output.

Solution

Write down the bits of the BCD representation 01010110 on the circuit diagram. Because $A_0 = 0$, the b_0 bit of the output is 0.

The top inputs to the upper adder are 0011. The bottom inputs are 0101. This adder adds these to produce

$$\begin{array}{r} 0011 \\ +\underline{0101} \\ 1000 \end{array} = \Sigma_3\Sigma_2\Sigma_1\Sigma_0 \text{ outputs of the upper adder}$$

The Σ_1 and Σ_0 bits become binary outputs b_2 and b_1, respectively. The Σ_3 and Σ_2 bits are fed to the lower adder. The top inputs to the lower adder are therefore 0010. The bottom inputs are 0101. This adder adds these to produce

$$\begin{array}{r} 0010 \\ +\underline{0101} \\ 0111 \end{array} = \Sigma_3\Sigma_2\Sigma_1\Sigma_0 \text{ outputs of the lower adder}$$

These bits become $b_6b_5b_4b_3$, respectively.

Thus, we have $b_6b_5b_4b_3b_2b_1b_0 = 0111000$ as the correct binary equivalent for decimal 56.

Other Code Converter Implementations

Whereas all types of code converters can be made by combining logic gates, adder circuits, or other combinational logic, the circuitry can become quite complex, requiring many ICs. It is often more efficient to use a read-only memory (ROM) or programmable logic device (PLD) to function as a code converter. As we will see in Chapters 12 and 13, these devices contain the equivalent of hundreds of logic gates, and they can be programmed to provide a wide range of logic functions.

REVIEW QUESTIONS

1. What is a code converter?
2. How many binary outputs would a three-digit BCD-to-binary converter have?

9-12 DATA BUSING

In most modern computers, the transfer of data takes place over a common set of connecting lines called a **data bus**. In these bus-organized computers, many different devices can have their outputs and inputs tied to the common data bus lines. Because of this, the devices that are tied to the data bus will often have tristate outputs, or they will be tied to the data bus through tristate buffers.

Some of the devices that are commonly connected to a data bus are (1) microprocessors; (2) semiconductor memory chips, covered in Chapter 12; and (3) digital-to-analog converters (DACs) and analog-to-digital converters (ADCs), described in Chapter 11.

Figure 9-41 illustrates a typical situation in which a microprocessor (the CPU chip in a microcomputer) is connected to several devices over an eight-line

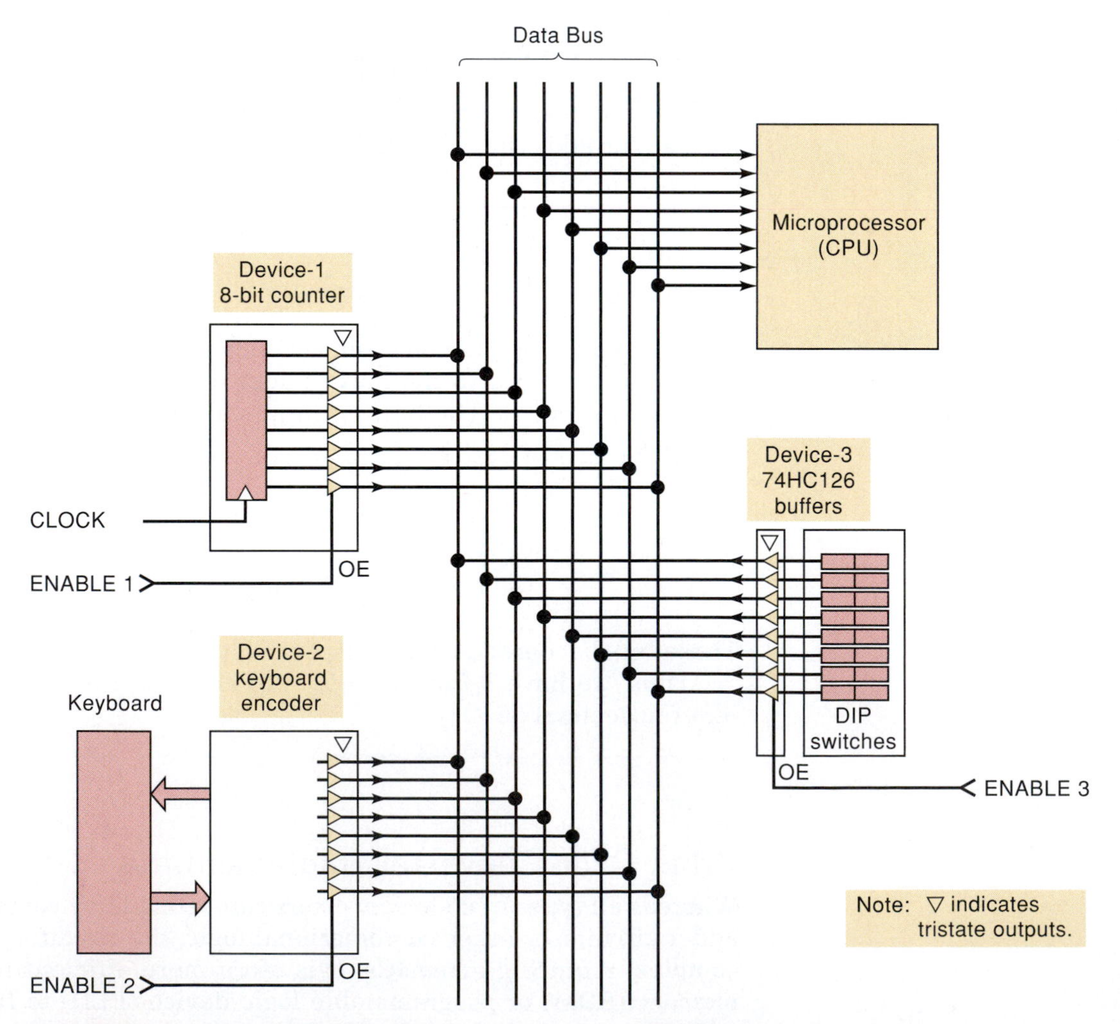

FIGURE 9-41 Three different devices can transmit eight-bit data over an eight-line data bus to a microprocessor; only one device at a time is enabled so that bus contention is avoided.

data bus. The data bus is simply a collection of conducting paths over which digital data are transmitted from one device to another. Each device provides an eight-bit output that is sent to the inputs of the microprocessor over the eight-line data bus. Clearly, because the outputs of each of the three devices are connected to the same microprocessor inputs over the data bus conducting paths, we must be aware of bus contention problems (Section 8-12), where two or more signals tied to the same bus line are active and are essentially fighting each other. Bus contention is avoided if the devices have tristate outputs or are connected to the bus through tristate buffers (Section 8-12). The output enable inputs (*OE*) to each device (or its buffer) are used to ensure that no more than one device's outputs are active at a given time.

EXAMPLE 9-19

(a) For Figure 9-41, describe the conditions necessary to transmit data from device 3 to the microprocessor.

(b) What will the status of the data bus be when none of the devices is enabled?

Solution

(a) ENABLE 3 must be activated; ENABLE 1 and ENABLE 2 must be in their inactive state. This will put the outputs of device 1 and device 2 in the Hi-Z state and essentially disconnect them from the bus. The outputs of device 3 will be activated so that their logic levels will appear on the data bus lines and be transmitted to the inputs of the microprocessor. We can visualize this by covering up device 1 and device 2 as if they are not even part of the circuit; then we are left with device 3 alone connected to the microprocessor over the data bus.

(b) If none of the device enable inputs are activated, all of the device outputs are in the Hi-Z state. This disconnects all device outputs from the bus. Thus, there is no definite logic level on any of the data bus lines; they are in the indeterminate state. This condition is known as a **floating bus**, and each data bus line is said to be in a *floating* (indeterminate) state. An oscilloscope display of a floating bus line would be unpredictable. A logic probe would indicate an indeterminate logic level.

REVIEW QUESTIONS

1. What is meant by the term *data bus*?
2. What is *bus contention*, and what must be done to prevent it?
3. What is a *floating bus*?

9-13 THE 74ALS173/HC173 TRISTATE REGISTER

The devices connected to a data bus will contain registers (usually flip-flops) that hold the device data. The outputs of these registers are usually connected to tristate buffers that allow them to be tied to a data bus. We will demonstrate the details of data bus operation by using an IC register that includes the tristate buffers on the same chip: the TTL 74ALS173 (also available in CMOS 74HC173 versions). Its logic diagram and truth table are shown in Figure 9-42.

Inputs					FF Outputs
MR	CP	$\overline{IE}_1$	$\overline{IE}_2$	D_n	Q
H	X	X	X	X	L
L	L	X	X	X	Q_0
L	↑	H	X	X	Q_0
L	↑	X	H	X	Q_0
L	↑	L	L	L	L
L	↑	L	L	H	H

When either $\overline{OE}_1$ or $\overline{OE}_2$ is HIGH, the output is in the OFF state (high impedance); however, this does not affect the contents or sequential operating of the register.

H = HIGH voltage level
L = LOW voltage level
X = immaterial
Q_0 = output prior to PGT

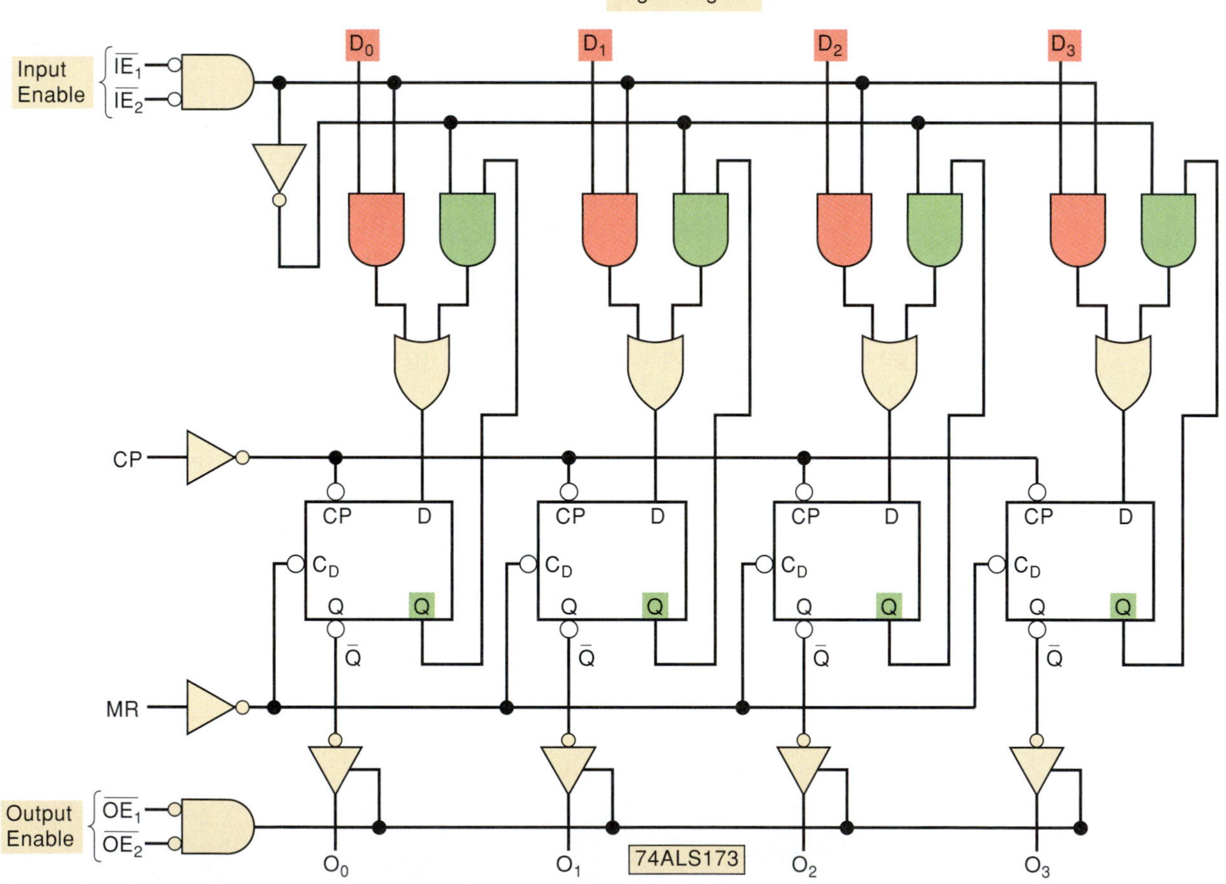

FIGURE 9-42 Truth table and logic diagram for the 74ALS173 tristate register.

The 74ALS173 is a four-bit register with parallel in/parallel out capability. Note that the FF outputs are connected to tristate buffers that provide outputs O_0 through O_3. Also note that the data inputs D_0 through D_3 are connected to the D inputs of the register FFs through logic circuitry. This logic allows two modes of operation: (1) *load*, where the data at inputs D_0 to D_3 are transferred into the FFs on the PGT of the clock pulse at CP; and (2) *hold*, where the data in the register do not change when the PGT of CP occurs.

EXAMPLE 9-20

(a) What input conditions will produce the load operation?

(b) What input conditions will produce the hold operation?

(c) What input conditions will allow the internal register outputs to appear at O_0 to O_3?

Solution

(a) The last two entries in the truth table show that each Q output takes on the value present at its D input when a PGT occurs at CP provided that MR is LOW and *both* input-enable inputs, $\overline{IE}_1$ and $\overline{IE}_2$, are LOW.

(b) The third and fourth lines of the truth table state that when either $\overline{IE}$ input is HIGH, the D inputs have no effect, and the Q outputs will retain their current values when the PGT occurs.

(c) The output buffers are enabled when *both* output-enable inputs, $\overline{OE}_1$ and $\overline{OE}_2$, are LOW. This will pass the register outputs through to the external outputs O_0 to O_3. If either output-enable input is HIGH, the buffers will be disabled, and the outputs will be in the Hi-Z state.

Note that the $\overline{OE}$ inputs have no effect on the data load operation. They are used only to control whether or not the register outputs are passed to the external outputs.

The logic symbol for the 74ALS173/HC173 is given in Figure 9-43. We have included the IEEE/ANSI "&" notation to indicate the AND relationship of the two pairs of enable inputs.

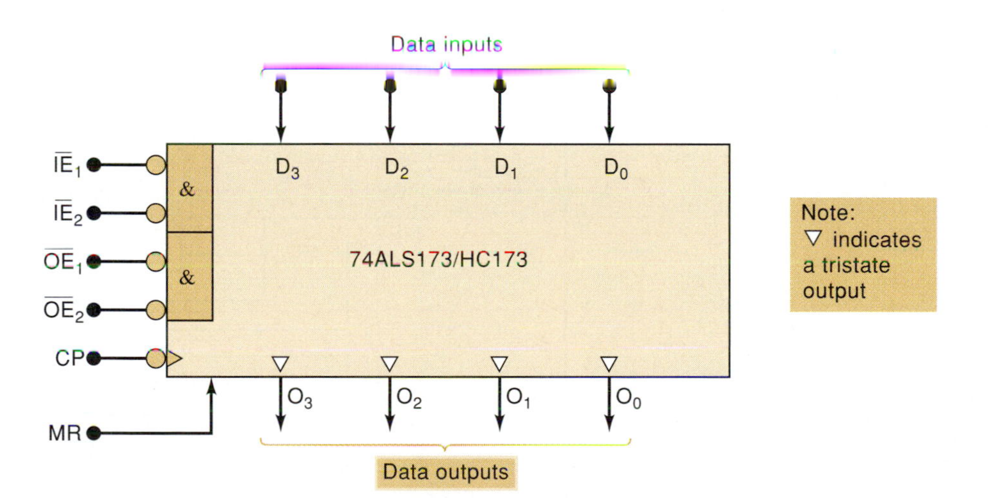

FIGURE 9-43 Logic symbol for the 74ALS173/HC173 IC.

REVIEW QUESTIONS

1. Assume that both IE inputs are LOW and that $D_0D_1D_2D_3$ = 1011. What logic levels are present at the FF D inputs?
2. *True or false:* The register cannot be loaded when the master reset input (MR) is held HIGH.
3. What will the output levels be when MR = HIGH and both OE inputs are held low?

9-14 DATA BUS OPERATION

The data bus is very important in computer systems, and its significance will not be appreciated until our later studies of memories and microprocessors. For now, we will illustrate the data bus operation for register-to-register data transfer. Figure 9-44 shows a bus-organized system for three 74HC173 tristate registers. Note that each register has its pair of $\overline{OE}$ inputs tied together as one $\overline{OE}$ input, and likewise for the $\overline{IE}$ inputs. Also note that the registers will be referred to as registers A, B, and C from top to bottom. This is indicated by the subscripts on each input and output.

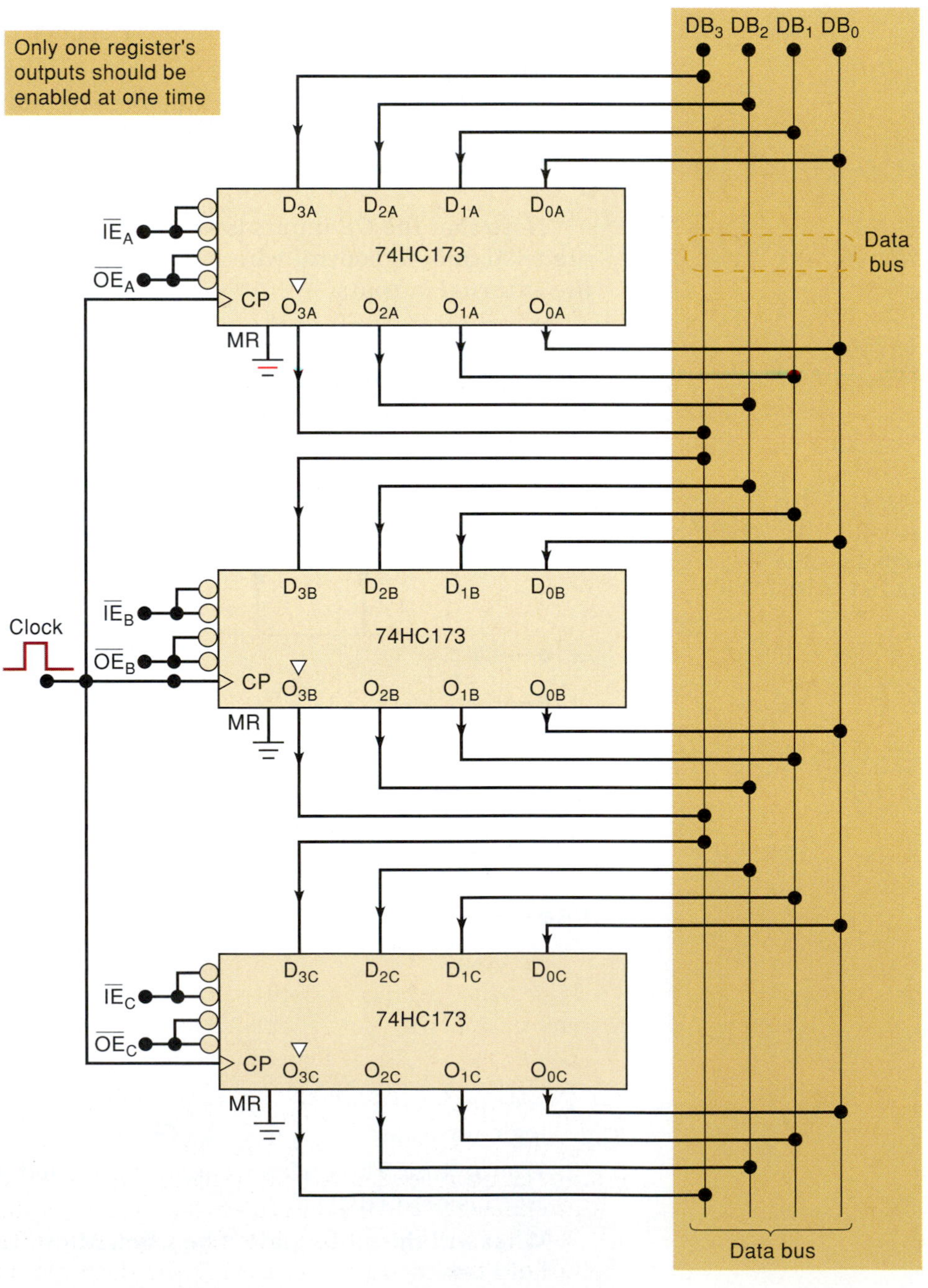

FIGURE 9-44 Tristate registers connected to a data bus.

In this arrangement, the data bus consists of four lines labeled DB_0 to DB_3. Corresponding outputs of each register are connected to the same data bus line (e.g., O_{3A}, O_{3B}, and O_{3C} are connected to DB_3). Because the three registers have their outputs connected together, it is imperative that only one register have its outputs enabled and that the other two register outputs remain in the Hi-Z state. Otherwise, there will be bus contention (two or more sets of outputs fighting each other), producing uncertain levels on the bus and possible damage to the register output buffers.

Corresponding register inputs are also tied to the same bus line (e.g., D_{3A}, D_{3B}, and D_{3C} are tied to DB_3). Thus, the levels on the bus will always be ready to be transferred to one or more of the registers depending on the $\overline{IE}$ inputs.

Data Transfer Operation

The contents of any one of the three registers can be parallel-transferred over the data bus to one of the other registers through the proper application of logic levels to the register enable inputs. In a typical system, the control unit of a computer (i.e., the CPU) will generate the signals that select which register will put its data on the data bus and which one will take the data from the data bus. The following example will illustrate this.

EXAMPLE 9-21

Describe the input signal requirements for transferring $[A] \rightarrow [C]$.

Solution

First of all, only register *A* should have its outputs enabled. That is, we need

$$\overline{OE}_A = 0 \qquad \overline{OE}_B = \overline{OE}_C = 1$$

This will place the contents of register *A* onto the data bus lines.

Next, only register *C* should have its inputs enabled. For this, we want

$$\overline{IE}_C = 0 \qquad \overline{IE}_A = \overline{IE}_B = 1$$

This will allow only register *C* to accept data from the data bus when the PGT of the clock signal occurs.

Finally, a clock pulse is required to transfer the data from the bus into the register *C* flip-flops.

Bus Signals

The timing diagram in Figure 9-45 shows the various signals involved in the transfer of the data 1011 from register *A* to register *C*. The $\overline{IE}$ and $\overline{OE}$ lines that are not shown are assumed to be in their inactive HIGH state. Prior to time t_1, the $\overline{IE}_C$ and $\overline{OE}_A$ lines are also HIGH, so that all of the register outputs are disabled, and none of the registers will be placing their data on the bus lines. In other words, the data bus lines are in the Hi-Z or "floating" state as represented by the hatched lines on the timing diagram. The Hi-Z state does not correspond to any particular voltage level.

At t_1 the $\overline{IE}_C$ and $\overline{OE}_A$ inputs are activated. The outputs of register *A* are enabled, and they start changing the data bus lines DB_3 through DB_0 from the Hi-Z state to the logic levels 1011. After allowing time for the logic levels

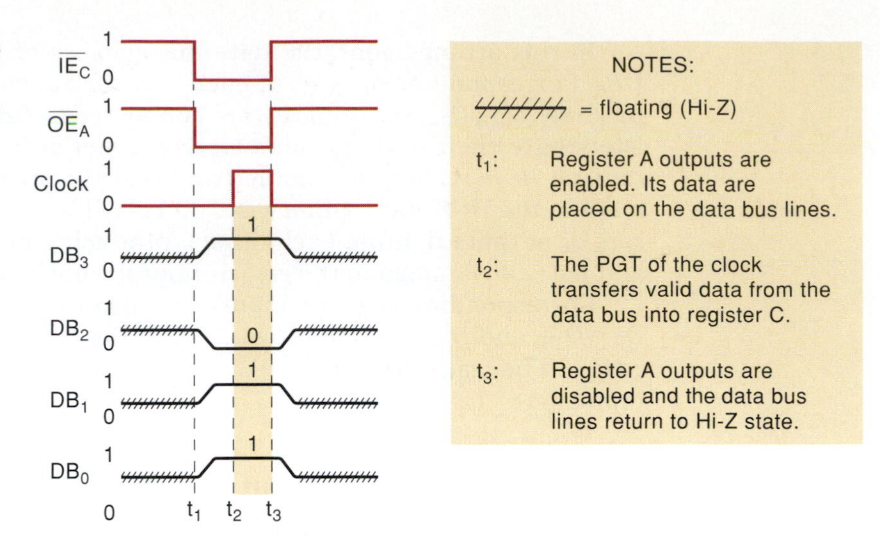

$\overline{IE}_C$
$\overline{OE}_A$
Clock
DB_3
DB_2
DB_1
DB_0
1
0
t_1 t_2 t_3
NOTES:
= floating (Hi-Z)
t_1: Register A outputs are enabled. Its data are placed on the data bus lines.
t_2: The PGT of the clock transfers valid data from the data bus into register C.
t_3: Register A outputs are disabled and the data bus lines return to Hi-Z state.

FIGURE 9-45 Signal activity during the transfer of the data 1011 from register *A* to register *C*.

to stabilize on the bus, the PGT of the clock is applied at t_2. This PGT will transfer these logic levels into register *C* because $\overline{IE}_C$ is active. If the PGT occurs before the data bus has valid logic levels, unpredictable data will be transferred into *C*.

At t_3, the $\overline{IE}_C$ and $\overline{OE}_A$ lines return to the inactive state. As a result, register *A*'s outputs go to the Hi-Z state. This removes the register *A* output data from the bus lines, and the bus lines return to the Hi-Z state.

Note that the data bus lines show valid logic levels only during the time interval when register *A*'s outputs are enabled. At all other times, the data bus lines are floating, and there is no way to predict easily what they would look like if displayed on an oscilloscope. A logic probe would give an "indeterminate" indication if it were monitoring a floating bus line. Also note the relatively slow rate at which the signals on the data bus lines are changing. Although this effect has been somewhat exaggerated in the diagram, it is a characteristic common to bus systems and is caused by the capacitive load on each line. This load consists of a combination of parasitic capacitance and the capacitances contributed by each input and output connected to the line.

Simplified Bus Timing Diagram

The timing diagram in Figure 9-45 shows the signals on each of the four data bus lines. This same kind of signal activity occurs in digital systems that use the more common data buses of 8, 16, or 32 lines. For these larger buses, the timing diagrams like Figure 9-45 would get excessively large and cumbersome. There is a simplified method for showing the signal activity that occurs on a set of bus lines that uses only a single timing waveform to represent the complete set of bus lines. This is illustrated in Figure 9-46 for the same data transfer situation depicted in Figure 9-45. Notice how the data bus activity is represented. Especially note how the valid data 1011 are indicated on the diagram during the t_2–t_3 interval. We will generally use this simplified bus timing diagram from now on.

Expanding the Bus

The data transfer operation of the four-line data bus of Figure 9-44 is typical of the operation of larger data buses found in most computers and other dig-

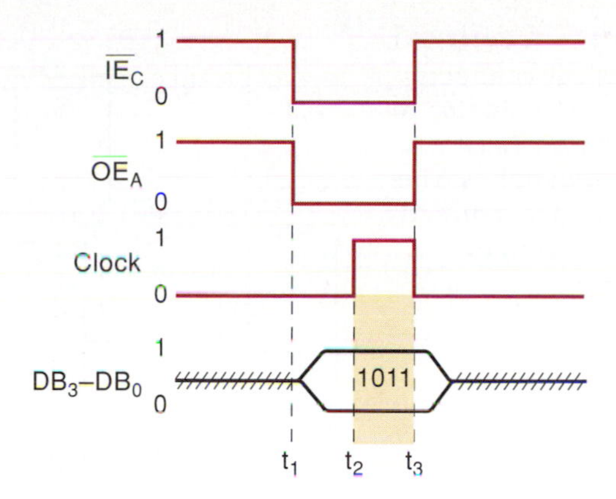

FIGURE 9-46 Simplified way to show signal activity on data bus lines.

ital systems, usually the 8-, 16-, or 32-line data buses. These larger buses generally have many more than three devices tied to the bus, but the basic data transfer operation is the same: *one device has its outputs enabled so that its data are placed on the data bus; another device has its inputs enabled so that it can take these data off the bus and latch them into its internal circuitry on the appropriate clock edge.*

The number of lines on the data bus will depend on the size of the data **word** (unit of data) that is to be transferred over the bus. A computer that has an 8-bit word size will have an eight-line data bus, a computer that has a 16-bit word size will have a 16-line data bus, and so on. The number of devices connected to a data bus varies from one computer to another and depends on factors such as how much memory the computer has and the number of input and output devices that must communicate with the CPU over the data bus.

All device outputs must be tied to the bus through tristate buffers. Some devices, such as the 74173 register, have these buffers on the same chip. Other devices will need to be connected to the bus through an IC called a **bus driver**. A bus driver IC has tristate outputs with a very low output impedance that can rapidly charge and discharge the bus capacitance. This bus capacitance represents the cumulative effect of all of the parasitic capacitances of the different inputs and outputs tied to the bus, and it can cause deterioration of the bus signal transition times if they are not driven from a low-impedance signal source. Figure 9-47 shows a 74HC541 octal bus driver IC connecting the outputs of an eight-bit analog-to-digital converter (ADC) to a data bus. The ADC has tristate outputs but lacks the drive capability to charge the bus capacitance (shown as capacitors to ground in the drawing). Notice that data bit 0 is driving the bus directly, without the assistance of the bus driver. If the transition time is slow enough, the voltage may never reach a HIGH logic level in the allotted enable time. The bus driver's two enable inputs are tied together so that a LOW on the common enable line will allow the ADC's outputs through the buffers and onto the data bus, from which they can be transferred to another device.

Simplified Bus Representation

Usually, many devices are connected to the same data bus. On a circuit schematic, this can produce a confusing array of lines and connections. For this reason, a more simplified representation of data bus connections is often used on block diagrams and in some circuit schematics. One type of simplified representation is shown in Figure 9-48 for an eight-line data bus.

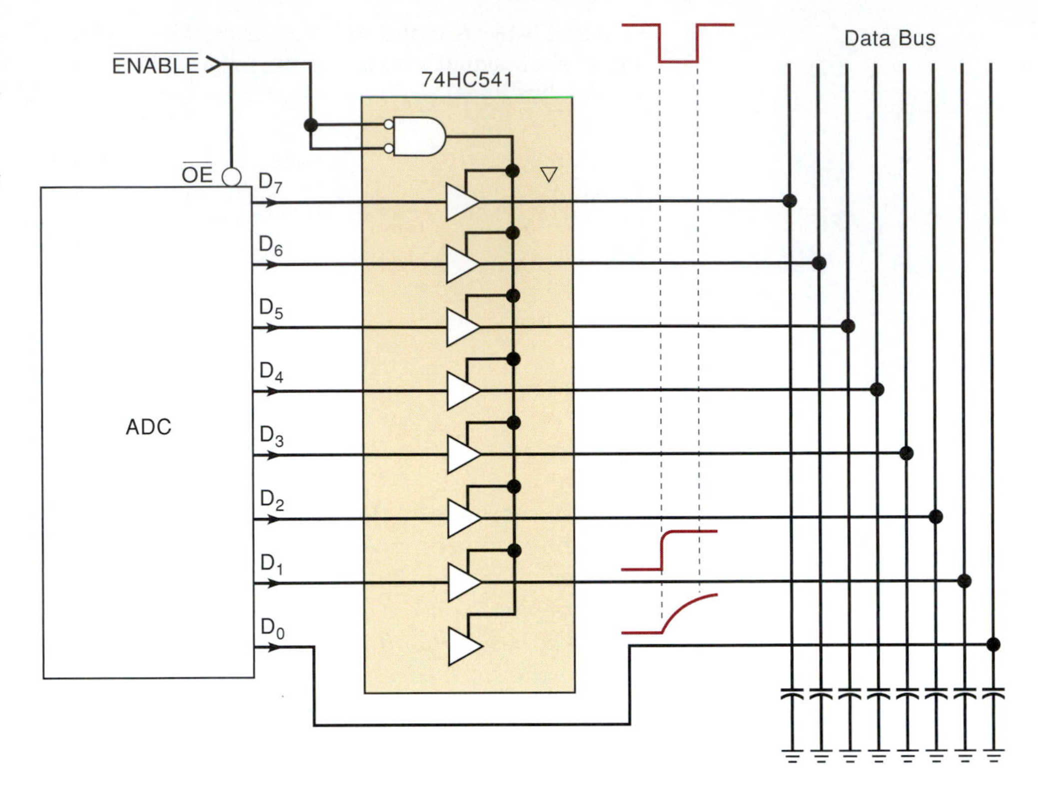

FIGURE 9-47 A 74HC541 octal bus driver connects the outputs of an analog-to-digital converter (ADC) to an eight-line data bus. The D_0 output connects directly to the bus showing the capacitive effects.

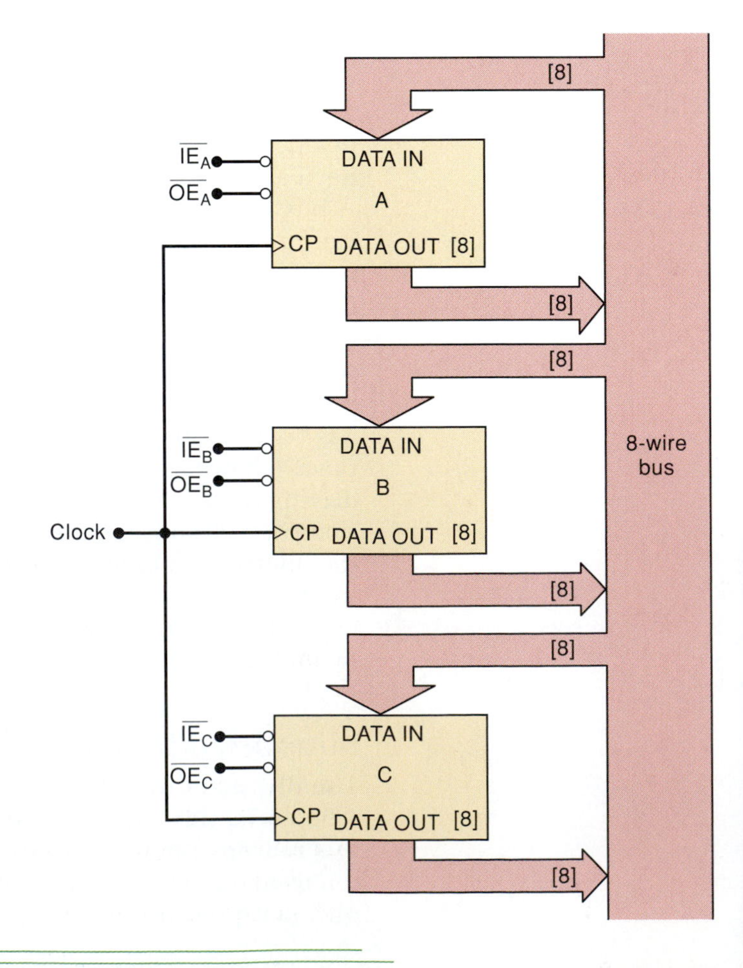

FIGURE 9-48 Simplified representation of bus arrangement.

The connections to and from the data bus are represented by wide arrows. The numbers in brackets indicate the number of bits that each register contains, as well as the number of lines connecting the register inputs and outputs to the bus.

Another common method for representing buses on a schematic is presented in Figure 9-49 for an eight-line data bus. It shows the eight individual output lines from a 74HC541 bus driver labeled D_7–D_0 bundled (not connected) together and shown as a single line. These bundled data output lines are then connected to the data bus, which is also shown as one line (i.e., the eight data bus lines are bundled together). The "/8" notation indicates the number of lines represented by each bundle. This bundle method is used to represent the connections from the data bus to the eight microprocessor data inputs. When the bundle method is used, it is very important to label both ends of every wire that is in the bundle because the connection cannot be traced visually on the diagram.

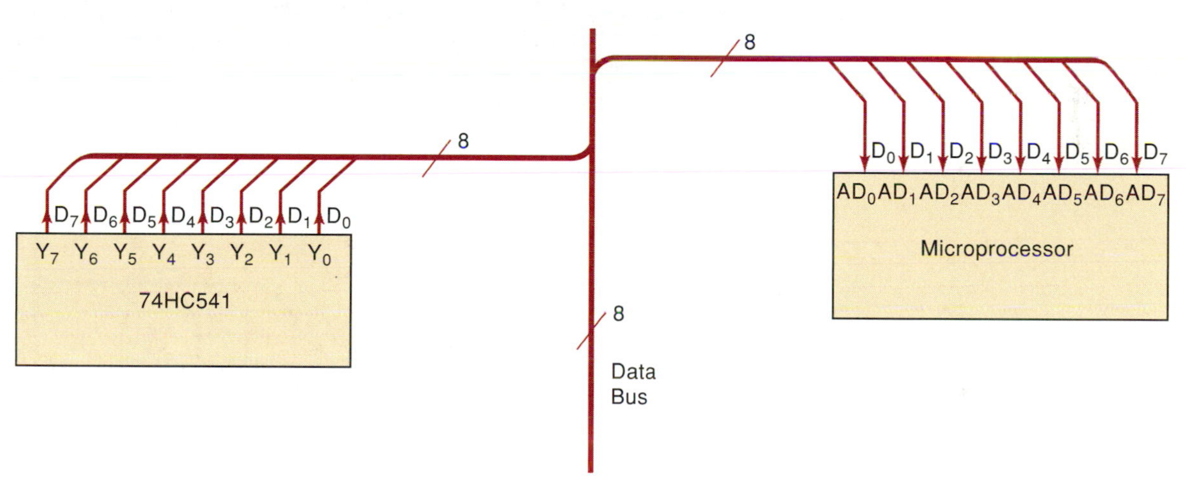

FIGURE 9-49 Bundle method for simplified representation of data bus connections. The "/8" denotes an eight-line data bus.

Bidirectional Busing

Each register in Figure 9-44 has both its inputs and its outputs connected to the data bus, so that corresponding inputs and outputs are shorted together. For example, each register has output O_2 connected to input D_2 because of their common connection to DB_2. This, of course, would not be true if external bus drivers were connected between the register outputs and the data bus.

Because inputs and outputs are often connected together in bus systems, IC manufacturers have developed ICs that connect inputs and outputs together *internal* to the chip in order to reduce the number of IC pins and the number of connections to the bus. Figure 9-50 illustrates this for a four-bit register. The separate data input lines (D_0 to D_3) and output lines (O_0 to O_3) have been replaced by input/output lines (I/O_0 to I/O_3).

Each I/O line will function as either an input or an output depending on the states of the enable inputs. Thus, they are called **bidirectional data lines**. The 74ALS299 is an eight-bit register with common I/O lines. Many memory ICs and microprocessors have bidirectional transfer of data.

We will return to the important topic of data busing in our comprehensive coverage of memory systems in Chapter 12.

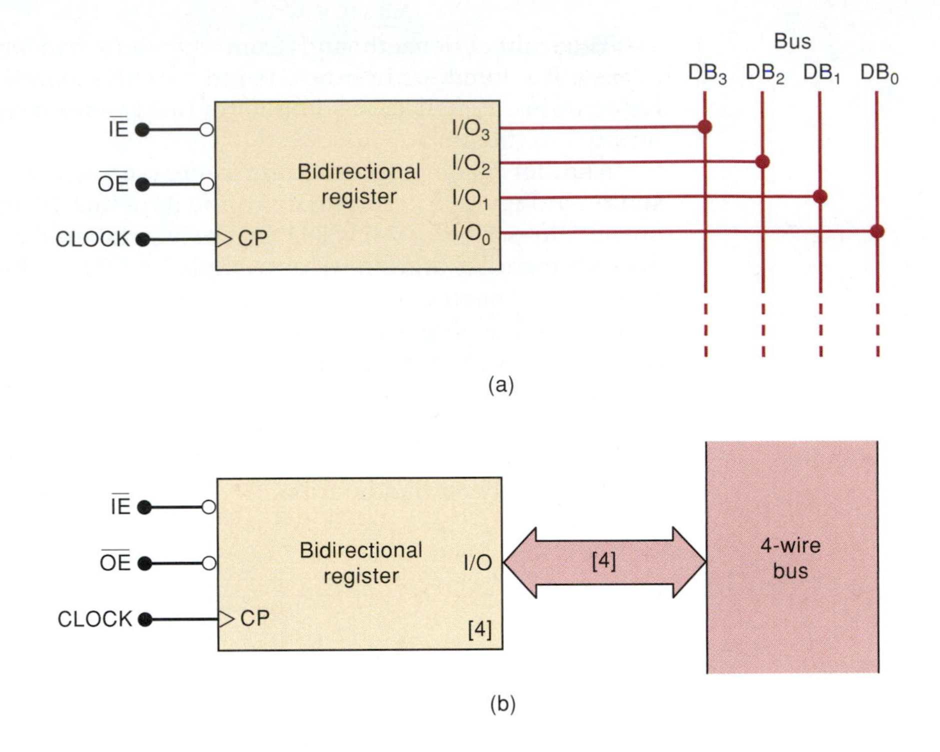

FIGURE 9-50 Bidirectional register connected to data bus.

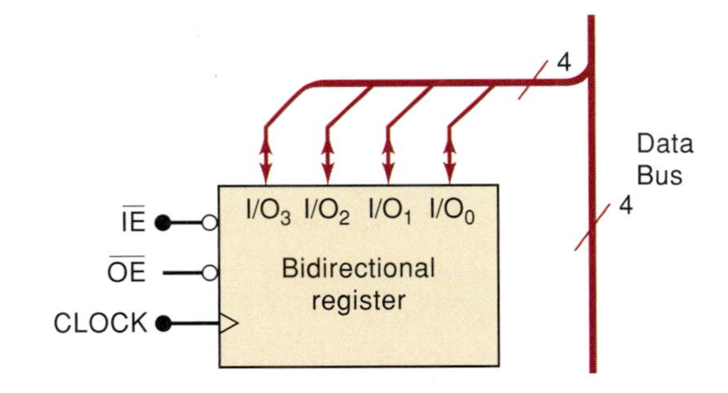

FIGURE 9-51

REVIEW QUESTIONS

1. What will happen if $\overline{OE}_A = \overline{OE}_B =$ LOW in Figure 9-44?
2. What logic level is on a data bus line when all devices tied to the bus are disabled?
3. What is the function of a bus driver?
4. What are the reasons for having registers with common I/O lines?
5. Redraw Figure 9-50(a) using the bundled line representation. (The answer is shown in Figure 9-51.)

9-15 DECODERS USING HDL

Section 9-1 introduced the decoder as a device that can recognize a binary number on its input and activate a corresponding output. Specifically, the 74138 1-of-8 decoder was presented. It uses three binary inputs to activate one of the eight outputs when the chip is enabled. In order to study HDL methods for implementing the types of digital devices that are covered in this chapter, we will focus primarily on conventional MSI parts, which have been discussed earlier. Not only is the operation of these devices already described in this book, but further reference material is readily available in logic data books. In all of these cases, it is vital that you understand what the device is supposed to do before trying to dissect the HDL code that describes it.

In actual practice, we are not recommending, for example, that new code be written to perform the task of a 74138. After all, there is a macrofunction already available that works exactly like this standard part. Using these devices as examples and showing the HDL techniques used to create them opens the door for embellishment of these devices so that a circuit that will uniquely fit the application at hand can be described. In some instances, we will add our own embellishments to a circuit that has been described; in other instances, we will describe a simpler version of a part in order to focus on the core principle in HDL and avoid other confusing features.

The methods used to define the inputs and outputs should take into consideration the purpose of these signals. In the case of a 1-of-8 decoder such as the 74138 described in Figure 9-3, there are three enable inputs ($\overline{E}_1, \overline{E}_2$, and E_3) that should be described as individual inputs to the device. On the other hand, the binary inputs that are to be decoded (A_2, A_1, A_0) should be described as three-bit numbers. The outputs can be described as eight individual bits. They can also be described as an array of eight bits, with output 0 represented by element 0 in the array, and so on, to output 7 represented by element 7. Depending on the way the code is written, one strategy may be easier to write than the other. Generally, using individual names can make the purpose of each I/O bit clearer, and using bit arrays makes it easier to write the code.

When an application such as a decoder calls for a unique response from the circuit corresponding to each combination of its input variables, the two methods that best serve this purpose are the CASE construct and the truth TABLE. The interesting aspect of this decoder is that the output response should happen only when *all* the enables are activated. If any of the enables are not in their active state, it should cause all the outputs to go HIGH. Each of the examples that follow will demonstrate ways to decode the input number only when *all* of the enables are activated.

AHDL DECODERS

The first illustration of an AHDL decoder, shown in Figure 9-52, is intended to demonstrate the use of a CASE construct that is evaluated only under the condition that all enables are active. The outputs must all revert back to HIGH as soon as any enable is deactivated. This example also illustrates a way to accomplish this without explicitly assigning a value to each output for each case, and it uses individually named output bits.

Line 3 defines the three-bit binary number that will be decoded. Line 4 defines the three enable inputs, and line 5 specifically names each output. The unique property of this solution is the use of the **DEFAULTS** keyword in AHDL (lines 10 to 13) to establish a value for variables that are not specified elsewhere in the code. This maneuver allows each case to force one bit LOW without specifically stating that the others must go HIGH.

The next illustration, in Figure 9-53, is intended to demonstrate the same decoder using the truth table approach. Notice that the outputs are defined as bit arrays but are still numbered *y[7]* down to *y[0]*. The unique aspect of this code is the use of the don't-care values in the truth table. Line 11 is used to concatenate the six input bits into a single variable (bit array) named *inputs[ ]*. Notice that in lines 14, 15, and 16 of the table, only one bit value is specified as 1 or 0. The others are all in the don't-care state (X). Line 14 says, "As long as *e3*

FIGURE 9-52 AHDL equivalent to the 74138 decoder.

```
1   SUBDESIGN fig9_52
2   (
3       a[2..0]                       :INPUT;      -- binary inputs
4       e3, e2bar, e1bar              :INPUT;      -- enable inputs
5       y7,y6,y5,y4,y3,y2,y1,y0       :OUTPUT;     -- decoded outputs
6   )
7   VARIABLE
8       enable                        :NODE;
9   BEGIN
10      DEFAULTS
11         y7=VCC;y6=VCC;y5=VCC;y4=VCC;
12         y3=VCC;y2=VCC;y1=VCC;y0=VCC;     -- defaults all HIGH out
13      END DEFAULTS;
14      enable = e3 & !e2bar & !e1bar;     -- all enables activated
15      IF enable    THEN
16         CASE a[] IS
17            WHEN 0    =>    y0 = GND;
18            WHEN 1    =>    y1 = GND;
19            WHEN 2    =>    y2 = GND;
20            WHEN 3    =>    y3 = GND;
21            WHEN 4    =>    y4 = GND;
22            WHEN 5    =>    y5 = GND;
23            WHEN 6    =>    y6 = GND;
24            WHEN 7    =>    y7 = GND;
25         END CASE;
26      END IF;
27  END;
```

FIGURE 9-53 AHDL decoder using a TABLE.

```
1   SUBDESIGN fig9_53
2   (
3       a[2..0]              :INPUT;      -- decoder inputs
4       e3, e2bar, e1bar     :INPUT;      -- enable inputs
5       y[7..0]              :OUTPUT;     -- decoded outputs
6   )
7   VARIABLE
8      inputs[5..0]          :NODE;       -- all 6 inputs combined
9
10  BEGIN
11     inputs[] = (e3, e2bar, e1bar, a[]); -- concatenate the inputs
12     TABLE
13        inputs[]        =>     y[];
14        B"0XXXXX"       =>     B"11111111";     -- e1 not enabled
15        B"X1XXXX"       =>     B"11111111";     -- e2bar disabled
16        B"XX1XXX"       =>     B"11111111";     -- e3bar disabled
17        B"100000"       =>     B"11111110";     -- Y0 active
18        B"100001"       =>     B"11111101";     -- Y1 active
19        B"100010"       =>     B"11111011";     -- Y2 active
20        B"100011"       =>     B"11110111";     -- Y3 active
21        B"100100"       =>     B"11101111";     -- Y4 active
22        B"100101"       =>     B"11011111";     -- Y5 active
23        B"100110"       =>     B"10111111";     -- Y6 active
24        B"100111"       =>     B"01111111";     -- Y7 active
25     END TABLE;
26  END;
```

is *not* enabled, it does not matter what the other inputs are doing; the outputs will be HIGH." Lines 15 and 16 do the same thing, making sure that if *e2bar* or *e1bar* is HIGH (disabled), the outputs will be HIGH. Lines 17 through 24 state that as long as the first three bits (enables) are "100," the proper decoder output will be activated to correspond with the lower three bits of *inputs[ ]*.

VHDL DECODERS

The VHDL solution presented in Figure 9-54 essentially uses a truth table approach. The key strategy in this solution involves the concatenation of the three enable bits (*e3, e2bar, e1bar*) with the binary input *a* on line 11. The VHDL selected signal assignment is used to assign a value to a signal when a specific combination of inputs is present. Line 12 (WITH inputs SELECT) indicates that we are using the value of the intermediate signal *inputs* to determine which value is assigned to *y*. Each of the *y* outputs is listed on lines 13–20. Notice that only combinations that begin with 100 follow the WHEN clause on lines 13–20. This combination of *e3, e2bar,* and *e1bar* is necessary to make each of the enables active. Line 21 assigns a disabled state to each output when any combination other than 100 is present on the enable inputs.

```
1   ENTITY fig9_54 IS
2   PORT(
3        a                    :IN BIT_VECTOR (2 DOWNTO 0);
4        e3, e2bar, e1bar     :IN BIT;
5        y                    :OUT BIT_VECTOR (7 DOWNTO 0)
6        );
7   END fig9_54 ;
8   ARCHITECTURE truth OF fig9_54 IS
9   SIGNAL inputs:  BIT_VECTOR (5 DOWNTO 0); --combine enables w/ binary in
10     BEGIN
11        inputs <= e3 & e2bar & e1bar & a;
12        WITH inputs SELECT
13           y  <= "11111110"  WHEN "100000",   --Y0 active
14                 "11111101"  WHEN "100001",   --Y1 active
15                 "11111011"  WHEN "100010",   --Y2 active
16                 "11110111"  WHEN "100011",   --Y3 active
17                 "11101111"  WHEN "100100",   --Y4 active
18                 "11011111"  WHEN "100101",   --Y5 active
19                 "10111111"  WHEN "100110",   --Y6 active
20                 "01111111"  WHEN "100111",   --Y7 active
21                 "11111111"  WHEN OTHERS;     --disabled
22      END truth;
```

FIGURE 9-54 VHDL equivalent to the 74138 decoder.

REVIEW QUESTIONS

1. What is the purpose of the three inputs *e3, e2bar,* and *e1bar*?
2. Name two AHDL methods to describe a decoder's operation.
3. Name two VHDL methods to describe a decoder's operation.

9-16 THE HDL 7-SEGMENT DECODER/DRIVER

Section 9-2 described a BCD-to-7-segment decoder/driver. The standard part number for the circuit described is a 7447. In this section, we look into the HDL code necessary to produce a device that meets the same criteria as the 7447. Recall that the $\overline{BI}$ (blanking input) is the overriding control that turns all segments off regardless of other input levels. The $\overline{LT}$ (lamp test) input is used to test all the segments on the display by lighting them up. The $\overline{RBO}$ (ripple blanking output) is designed to go LOW when $\overline{RBI}$ (ripple blanking input) is LOW and the BCD input value is 0. Typically, in multiple-digit display applications, each $\overline{RBO}$ pin is connected to the $\overline{RBI}$ pin of the next digit to the right. This setup creates the feature of blanking all leading zeros in a display value without blanking zeros in the middle of a number. For example, the number 2002 would display as 2002, but the number 0002 would *not* display as 0002, but rather _ _ _ 2. One feature of the 7447 that would be difficult to replicate in HDL is the combination input/output pin named $\overline{BI}/\overline{RBO}$. Rather than complicate the code, we have decided to create a separate input ($\overline{BI}$) and an output ($\overline{RBO}$) on two different pins. This discussion also makes no attempt to replicate the non-BCD display characters of a 7447 but simply blanks all segments for values greater than 9.

Several decisions must be made when designing a circuit such as this one. The first involves the type of display we intend to use. If it is a common cathode, then a logic 1 lights the LED segment. If it is a common anode, then a logic 0 is required to turn on a segment. Next, we must decide on the type of inputs, outputs, and intermediate variables. We have decided that the outputs for each individual segment should be assigned a bit name (*a*–*g*) rather than using a bit array. This arrangement will make it clearer when connecting the display to the IC. These individual bits can be grouped as a set of bits and assigned binary values, as we have done in AHDL, or an intermediate variable bit array can be used to make it convenient when assigning all seven bit levels in a single statement, as we have done in VHDL. The BCD inputs are treated as a four-bit number, and the blanking controls are individual bits. The other issue that greatly affects the bit patterns in the HDL code is the arbitrary decision of the order of the segment names *a*–*g*. In this discussion, we have assigned segment *a* to the leftmost bit in the binary bit pattern, with the bits moving alphabetically left to right.

Some of the controls must have precedence over other controls. For example, the $\overline{LT}$ (lamp test) should override any regular digit display, and the $\overline{BI}$ (blanking input) should override even the lamp test input. In these illustrations, the IF/ELSE control structure is used to establish precedence. The first condition that is evaluated as true will determine the resulting output, regardless of the other input levels. Subsequent ELSE statements will have no effect, which is why the code tests first for $\overline{BI}$, then $\overline{LT}$, then $\overline{RBI}$, and finally determines the correct segment pattern.

AHDL DECODER/DRIVER

The AHDL code for this circuit is shown in Figure 9-55. AHDL allows output bits to be grouped in a set by separating the bits with commas and enclosing them in parentheses. A group of binary states can be assigned directly to these bit sets, as shown on lines 9, 11, 13, and 15. This convention avoids the need for an intermediate variable and is much shorter than eight separate assignment statements. The TABLE feature of AHDL is useful in this application to correlate an input BCD value to a 7-segment bit pattern.

```
1   SUBDESIGN fig9_55
2   (
3      bcd[3..0]           :INPUT;         -- 4-bit number
4      lt, bi, rbi         :INPUT;         -- 3 independent controls
5      a,b,c,d,e,f,g,rbo   :OUTPUT;        -- individual outputs
6   )
7   BEGIN
8       IF !bi THEN
9           (a,b,c,d,e,f,g,rbo) = (1,1,1,1,1,1,1,0);     % blank all %
10      ELSIF     !lt THEN
11          (a,b,c,d,e,f,g,rbo) = (0,0,0,0,0,0,0,1);     % test segments %
12      ELSIF !rbi & bcd[] == 0  THEN
13          (a,b,c,d,e,f,g,rbo) = (1,1,1,1,1,1,1,0);     % blank leading 0's %
14      ELSIF bcd[] > 9 THEN
15          (a,b,c,d,e,f,g,rbo) = (1,1,1,1,1,1,1,1);     % blank non BCD input %
16      ELSE
17          TABLE                     % display 7 segment Common Anode pattern %
18          bcd[]      =>     a,b,c,d,e,f,g,rbo;
19          0          =>     0,0,0,0,0,0,1,1;
20          1          =>     1,0,0,1,1,1,1,1;
21          2          =>     0,0,1,0,0,1,0,1;
22          3          =>     0,0,0,0,1,1,0,1;
23          4          =>     1,0,0,1,1,0,0,1;
24          5          =>     0,1,0,0,1,0,0,1;
25          6          =>     1,1,0,0,0,0,0,1;
26          7          =>     0,0,0,1,1,1,1,1;
27          8          =>     0,0,0,0,0,0,0,1;
28          9          =>     0,0,0,1,1,0,0,1;
29          END TABLE;
30      END IF;
31  END;
```

FIGURE 9-55 AHDL 7-segment BCD display decoder.

VHDL DECODER/DRIVER

The VHDL code for this circuit is shown in Figure 9-56. This illustration demonstrates the use of a VARIABLE as opposed to a SIGNAL. A VARIABLE can be thought of as a piece of scrap paper used to write down some numbers that will be needed later. A SIGNAL, on the other hand, is usually thought of as a wire connecting two points in the circuit. In line 12, the keyword VARIABLE is used to declare *segments* as a bit vector with seven bits. Take special note of the order of the indices for this variable. They are declared as 0 TO 6. In VHDL, this means that element 0 appears on the left end of the binary bit pattern and element 6 appears on the right end. This is exactly opposite of the way most examples in this text have presented variables, but it is important to realize the significance of the declaration statement in VHDL. For this illustration, segment *a* is bit 0 (on the left), segment *b* is bit 1 (moving to the right), and so on.

Notice that on line 3, the BCD input is declared as an INTEGER. This allows us to refer to it by its numeric value in decimal rather than being limited to bit pattern references. A PROCESS is employed here in order to allow us to use the IF/ELSE constructs to establish the precedence of one input over the other. Notice that the sensitivity list contains all the inputs. The code within the

```
 1  ENTITY fig9_56 IS
 2  PORT  (
 3          bcd                  :IN INTEGER RANGE 0 TO 15;
 4          lt, bi, rbi          :IN BIT;
 5          a,b,c,d,e,f,g,rbo :OUT BIT
 6       );
 7  END fig9_56 ;
 8
 9  ARCHITECTURE vhdl OF fig9_56 IS
10  BEGIN
11  PROCESS  (bcd, lt, bi, rbi)
12  VARIABLE  segments        :BIT_VECTOR (0 TO 6);
13      BEGIN
14         IF  bi = '0' THEN
15            segments := "1111111";  rbo <= '0'; -- blank all
16         ELSIF lt = '0' THEN
17            segments := "0000000";  rbo <= '1'; -- test segments
18         ELSIF (rbi = '0' AND bcd = 0) THEN
19            segments := "1111111";  rbo <= '0'; -- blank leading 0's
20         ELSE
21            rbo <= '1';
22            CASE bcd IS     -- display 7 segment Common Anode pattern
23               WHEN 0       => segments := "0000001";
24               WHEN 1       => segments := "1001111";
25               WHEN 2       => segments := "0010010";
26               WHEN 3       => segments := "0000110";
27               WHEN 4       => segments := "1001100";
28               WHEN 5       => segments := "0100100";
29               WHEN 6       => segments := "1100000";
30               WHEN 7       => segments := "0001111";
31               WHEN 8       => segments := "0000000";
32               WHEN 9       => segments := "0001100";
33               WHEN OTHERS => segments := "1111111";
34            END CASE;
35         END IF;
36      a <= segments(0); --assign bits of array to output pins
37      b <= segments(1);
38      c <= segments(2);
39      d <= segments(3);
40      e <= segments(4);
41      f <= segments(5);
42      g <= segments(6);
43         END PROCESS;
44   END vhdl;
```

FIGURE 9-56 VHDL 7-segment BCD display decoder.

PROCESS describes the behavioral operation of the circuit that is necessary whenever any of the inputs in the sensitivity list changes state. Another very important point in this illustration is the assignment operator for variables. Notice in line 15, for example, the statement *segments := "1111111"*. The variable assignment operator := is used for variables in place of the <= operator that was used for signal assignments. In lines 36–42, the individual bits that were established in the IF/ELSE decisions are assigned to the proper output bits.

REVIEW QUESTIONS

1. What feature of a 7447 is very difficult to duplicate in PLD hardware and HDL code?
2. Are these illustrations intended to drive common-anode or common-cathode 7-segment displays?
3. How are certain inputs (e.g., lamp test) given precedence over other inputs (e.g., RBI) in the HDL code in this section?

9-17 ENCODERS USING HDL

In Section 9-4, we discussed encoders and priority encoders. Similarities exist, of course, between decoders and encoders. Decoders take a binary number and activate one output that corresponds to that number. An encoder works in the other direction by monitoring one of its several inputs; when one of the inputs is activated, it produces a binary number corresponding to that input. If more than one of its inputs is activated at the same time, a priority encoder ignores the input of lower significance and produces the binary value that corresponds to the most significant input. In other words, it gives more significant inputs priority over less significant inputs. This section focuses on the methods that can be used in HDL to describe circuits that have this characteristic of priority for some inputs over others.

Another very important concept, which was presented in Chapter 8, was the tristate output circuit. Devices with tristate outputs can produce a logic HIGH or a logic LOW, just like a normal circuit, when their output is enabled. However, these devices can have their outputs disabled, which puts them in a "disconnected" or a high-impedance state. This is very important for devices connected to common buses, as described in Section 9-12. The next logical question is, "How do we describe tristate outputs using HDL?" This section incorporates tristate outputs in the encoder design to address this issue. In order to keep the discussion focused on the essentials, we create a circuit that emulates the 74147 priority encoder, with one added feature of having active-HIGH tristate outputs. Other features like cascading inputs and outputs (such as those found on a 74148) are left for you to try later. A symbol for the circuit we are describing is shown in Figure 9-57. Because the inputs are all labeled in a manner very similar to bit array notation, it makes sense to use a bit array to describe the encoder inputs. The tristate enable must be a single bit, and the encoded outputs can be described as an integer numeric value.

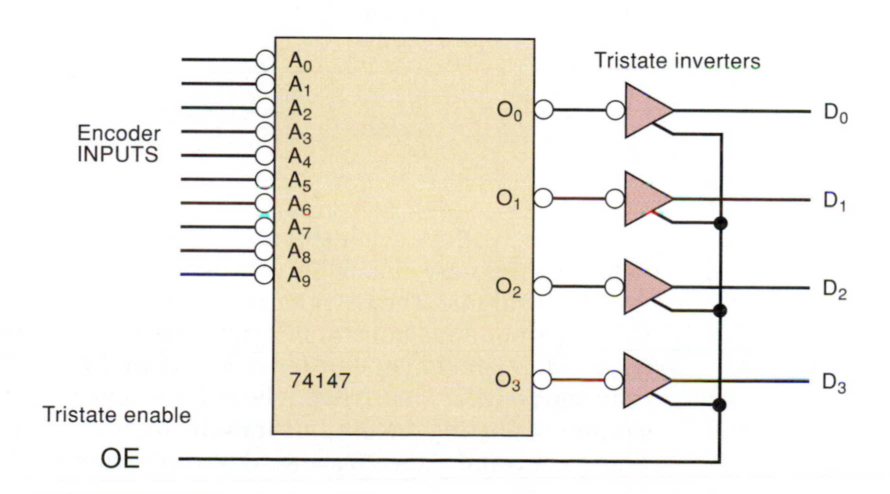

FIGURE 9-57 Graphic description of an encoder with tristate outputs.

AHDL ENCODER

The most important point to be made from Figure 9-58 is the method of establishing priority, but also note the I/O assignments. The AHDL input/output descriptions do not provide a separate type for integers but allow a bit array to be referred to as an integer. Consequently, line 4 describes the outputs as a bit array. In this illustration, a TABLE is used that is very similar to the tables often found in data books describing this circuit's operation. The key to this table is the use of the don't-care state (X) on inputs. The priority is described by the way we position these don't-care states in the truth table. Reading line 15, for instance, we see that as soon as we encounter an active input (LOW on input *a[4]*), the lower order input bits do not matter. The output has been determined to be 4. The tristate outputs are made possible by using the built-in primitive function :TRI on line 6. This line assigns the attributes of a tristate buffer to the variable that has been named *buffer*. Recall that this is the same way a flip-flop is described in AHDL. The ports of a tristate buffer are quite straightforward. They represent the input (*in*), the output (*out*), and the tristate output enable (*oe*).

FIGURE 9-58 AHDL priority encoder with tristate outputs.

```
1   SUBDESIGN fig9_58
2   (
3      a[9..0], oe            :INPUT;
4      d[3..0]                :OUTPUT;
5   )
6   VARIABLE buffer[3..0]     :TRI;
7   BEGIN
8      TABLE
9         a[]            => buffer[].in;
10        B"1111111111"  => B"1111";     -- no input active
11        B"1111111110"  => B"0000";     -- 0
12        B"111111110X"  => B"0001";     -- 1
13        B"11111110XX"  => B"0010";     -- 2
14        B"1111110XXX"  => B"0011";     -- 3
15        B"111110XXXX"  => B"0100";     -- 4
16        B"11110XXXXX"  => B"0101";     -- 5
17        B"1110XXXXXX"  => B"0110";     -- 6
18        B"110XXXXXXX"  => B"0111";     -- 7
19        B"10XXXXXXXX"  => B"1000";     -- 8
20        B"0XXXXXXXXX"  => B"1001";     -- 9
21     END TABLE;
22     buffer[].oe = oe;      -- hook up enable line
23     d[] = buffer[].out;    -- hook up outputs
24  END;
```

The next illustration (Figure 9-59) uses the IF/ELSE construct to establish priority, very much like the method demonstrated in the 7-segment decoder example. The first IF condition that evaluates TRUE will THEN cause the corresponding value to be applied to the tristate buffer inputs. The priority is established by the order in which we list the IF conditions. Notice that they start with input 9, the highest-order input. This illustration adds another feature of putting the outputs into the high-impedance state when no input is being activated. Line 20 shows that the output enables will be activated only

```
1   SUBDESIGN fig9_59
2   (
3      sw[9..0], oe     :INPUT;
4      d[3..0]          :OUTPUT;
5   )
6   VARIABLE
7      buffers[3..0]    :TRI;
8   BEGIN
9      IF    !sw[9]    THEN  buffers[].in = 9;
10     ELSIF !sw[8]    THEN  buffers[].in = 8;
11     ELSIF !sw[7]    THEN  buffers[].in = 7;
12     ELSIF !sw[6]    THEN  buffers[].in = 6;
13     ELSIF !sw[5]    THEN  buffers[].in = 5;
14     ELSIF !sw[4]    THEN  buffers[].in = 4;
15     ELSIF !sw[3]    THEN  buffers[].in = 3;
16     ELSIF !sw[2]    THEN  buffers[].in = 2;
17     ELSIF !sw[1]    THEN  buffers[].in = 1;
18     ELSE                  buffers[].in = 0;
19     END IF;
20     buffers[].oe = oe & sw[]!=b"1111111111";   -- enable on any input
21     d[] = buffers[].out;                        -- connect to outputs
22  END;
```

FIGURE 9-59 AHDL priority encoder using IF/ELSE.

when the *oe* pin is activated and one of the inputs is activated. Another item of interest in this illustration is the use of bit array notation to describe individual inputs. For example, line 9 states that IF switch input 9 is activated (LOW), THEN the inputs to the tristate buffer will be assigned the value 9 (in binary, of course).

VHDL ENCODER

Two very important VHDL techniques are demonstrated in this description of a priority encoder. The first is the use of tristate outputs in VHDL, and the second is a new method of describing priority. Figure 9-60 shows the input/output definitions for this encoder circuit. Notice on line 6 that the input switches are defined as bit vectors with indices from 9 to 0. Also note that the *d* output is defined as an IEEE standard bit array (std_logic_vector type). This definition is necessary to allow the use of high-impedance states (tristate) on the outputs and also explains the need for the LIBRARY and USE statements on lines 1 and 2. As we mentioned, a very important point of this illustration is the method of describing precedence for the inputs. This code uses the **conditional signal assignment statement** starting on line 14 and continuing through line 24. On line 14, it assigns the value listed to the right of <= to the variable *d* on the left, assuming the condition following WHEN is true. If this clause is not true, the clauses following ELSE are evaluated one at a time until one that is true is found. The value preceding WHEN will then be assigned to *d*. A very important attribute of the conditional signal assignment statement is the sequential evaluation. The precedence of these statements is established by the order in which they are listed. Notice that in this illustration, the first condition being tested (line 14) is the enabling of the tristate outputs. Recall from Chapter 8 that the three states of a tristate out-

FIGURE 9-60 VHDL priority encoder using conditional signal assignment.

```
1   LIBRARY    ieee;
2   USE ieee.std_logic_1164.ALL;
3
4   ENTITY fig9_60 IS
5   PORT(
6        sw     :IN BIT_VECTOR (9 DOWNTO 0);   -- standard logic not needed
7        oe     :IN BIT;                       -- standard logic not needed
8        d      :OUT STD_LOGIC_VECTOR (3 DOWNTO 0)   -- std logic for hi-Z
9        );
10  END fig9_60;
11
12  ARCHITECTURE a OF fig9_60 IS
13     BEGIN
14       d  <= "ZZZZ"  WHEN  ((oe = '0') OR (sw = "1111111111")) ELSE
15             "1001"  WHEN  sw(9) = '0' ELSE
16             "1000"  WHEN  sw(8) = '0' ELSE
17             "0111"  WHEN  sw(7) = '0' ELSE
18             "0110"  WHEN  sw(6) = '0' ELSE
19             "0101"  WHEN  sw(5) = '0' ELSE
20             "0100"  WHEN  sw(4) = '0' ELSE
21             "0011"  WHEN  sw(3) = '0' ELSE
22             "0010"  WHEN  sw(2) = '0' ELSE
23             "0001"  WHEN  sw(1) = '0' ELSE
24             "0000"  WHEN  sw(0) = '0';
25  END a;
```

put are HIGH, LOW, and high impedance, which is referred to as high Z. When the value "ZZZZ" is assigned to the output, each output is in the high-impedance state. If the outputs are to be disabled (high Z), then none of the other encoding matters. Line 15 tests the highest priority input, which is bit 9 of the *sw* input array. If it is active (LOW), then a value of 9 is output regardless of whether other inputs are being activated at the same time.

REVIEW QUESTIONS

1. Name two methods in AHDL for giving priority to some inputs over others.
2. Name two methods in VHDL for giving priority to some inputs over others.
3. In AHDL, how are tristate outputs implemented?
4. In VHDL, how are tristate outputs implemented?

9-18 HDL MULTIPLEXERS AND DEMULTIPLEXERS

A multiplexer is a device that acts like a selector switch for digital signals. The select inputs are used to specify the input channel that is to be "connected" to the output pins. A demultiplexer works in the opposite direction by taking a digital signal as an input and distributing it to one of its outputs. Figure 9-61 shows a multiplexer/demultiplexer system with four data input channels. Each input is a four-bit number. These devices are not exactly like any of the multiplexers or demultiplexers described earlier in this chapter, but they operate in the same way. The system in this illustration allows the four digital signals to share a common "pipeline" in order to get data from

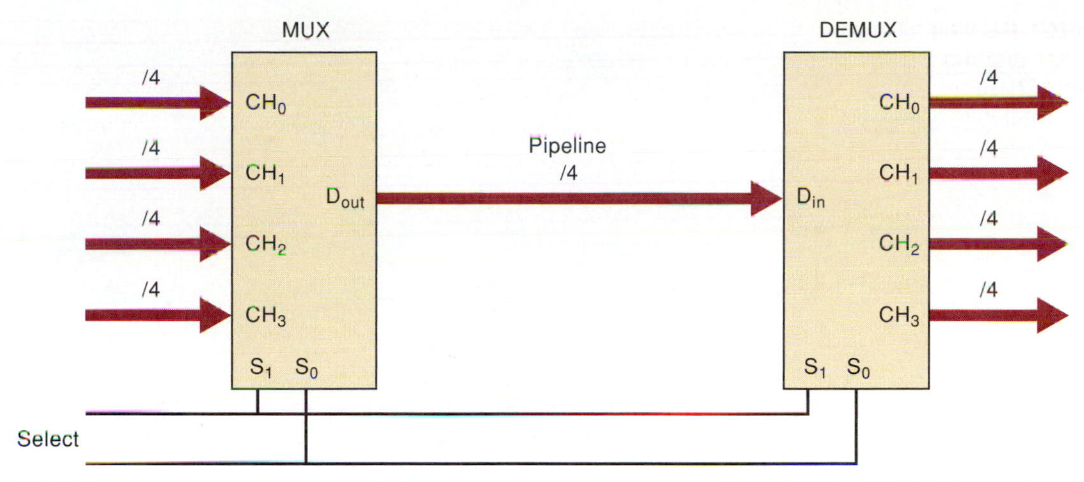

FIGURE 9-61 Four channels of data sharing a common data path.

one point to the other. The select inputs are used to decide which signal is going through the pipeline at any time.

In this section, we examine some code that implements both the multiplexer and the demultiplexer. The key HDL issue in both the MUX and DEMUX is assigning signals under certain conditions. For the demux, another issue is assigning a state to whichever outputs are not currently selected to distribute data. In other words, when an output is not being used for data (not selected), do we want it to have all bits HIGH, all bits LOW, or the tristate disabled? In the following descriptions, we have chosen to make them all HIGH when not selected, but with the structure shown, it would be a simple matter to change to one of the other possibilities.

AHDL MUX AND DEMUX

We will implement the multiplexer first. Figure 9-62 describes a multiplexer with four inputs of four bits each. Each input channel is named in a way that identifies its channel number. In this figure, each input is described as a four-bit array. The select input (*s[ ]*) requires two bits to specify the four channel numbers (0–3). A CASE construct is used here to assign an input channel conditionally to the output pins. Line 9, for example, states that in the case when the select inputs (*s[ ]*) are set to 0 (that is, binary 00), the

FIGURE 9-62 Four-bit × four-channel MUX in AHDL.

```
1   SUBDESIGN fig9_62
2   (
3     ch0[3..0], ch1[3..0], ch2[3..0], ch3[3..0]:INPUT;
4     s[1..0]                                   :INPUT; -- select inputs
5     dout[3..0]                                :OUTPUT;
6   )
7   BEGIN
8      CASE s[] IS
9            WHEN 0 =>     dout[] = ch0[];
10           WHEN 1 =>     dout[] = ch1[];
11           WHEN 2 =>     dout[] = ch2[];
12           WHEN 3 =>     dout[] = ch3[];
13      END CASE;
14  END;
```

AHDL

FIGURE 9-63 Four-bit × four-channel DEMUX in AHDL.

```
1   SUBDESIGN fig9_63
2   (
3      ch0[3..0], ch1[3..0], ch2[3..0], ch3[3..0]   :OUTPUT;
4      s[1..0]                                        :INPUT;
5      din[3..0]                                      :INPUT;
6   )
7   BEGIN
8      DEFAULTS
9         ch0[] = B"1111";
10        ch1[] = B"1111";
11        ch2[] = B"1111";
12        ch3[] = B"1111";
13     END DEFAULTS;
14
15     CASE S[] IS
16           WHEN 0 =>     ch0[] = din[];
17           WHEN 1 =>     ch1[] = din[];
18           WHEN 2 =>     ch2[] = din[];
19           WHEN 3 =>     ch3[] = din[];
20      END CASE;
21  END;
```

circuit should connect the channel 0 input to the data output. Notice that when assigning connections, the destination (output) of the signal is on the left of the = sign and the source (input) is on the right.

The demultiplexer code works in a similar way but has only one input channel and four output channels. It must also ensure that the outputs are all HIGH when they are not selected. In Figure 9-63, the inputs and outputs are declared as usual on lines 3–5. The default condition for each channel is specified after the keyword DEFAULTS, which tells the compiler to generate a circuit that will have HIGHs on the outputs unless specifically assigned a value elsewhere in the code. If this default section is not specified, the output values would default automatically to all LOW. Notice on lines 16–19 that the input signal is assigned conditionally to one of the output channels. Consequently, the output channel is on the left of the = sign and the input signal is on the right.

VHDL

VHDL MUX AND DEMUX

Figure 9-64 shows the code that creates a four-channel MUX with four bits per channel. The inputs are declared as bit arrays on line 3. They could have been declared just as easily as integers ranging from 0 to 15. Whichever way the inputs are declared, the outputs must be of the same type. Notice on line 4 that the select input (*s*) is declared as a decimal integer from 0 to 3 (equivalent to binary 00 to 11). This allows us to refer to it by its decimal channel number in the code, making it easier to understand. Lines 11–15 use the selected signal assignment statement to "connect" the appropriate input to the output, depending on the value on the select inputs. For example, line 15 states that channel 3 should be selected to connect to the data outputs when the select inputs are set to 3.

The demultiplexer code works in a similar way but has only one input channel and four output channels. In Figure 9-65, the inputs and outputs are declared as usual on lines 3–5. Notice that in line 3, the select input(s) is typed as an integer, just like the MUX code in Figure 9-64. The operation of a DEMUX is described most easily using several conditional signal assignment

FIGURE 9-64 Four-bit × four-channel MUX in VHDL.

```
1   ENTITY fig9_64 IS
2   PORT     (
3               ch0, ch1, ch2, ch3   :IN BIT_VECTOR (3 DOWNTO 0);
4               s                    :IN INTEGER RANGE 0 TO 3;
5               dout                 :OUT BIT_VECTOR (3 DOWNTO 0)
6             );
7   END fig9_64;
8
9   ARCHITECTURE selecter OF fig9_64 IS
10      BEGIN
11         WITH s SELECT
12         dout  <=  ch0 WHEN 0,  -- switch channel 0 to output
13                   ch1 WHEN 1,  -- switch channel 1 to output
14                   ch2 WHEN 2,  -- switch channel 2 to output
15                   ch3 WHEN 3;  -- switch channel 3 to output
16      END selecter;
```

FIGURE 9-65 Four-bit × four-channel DEMUX in VHDL.

```
1   ENTITY fig9_65 IS
2   PORT    (
3             s                    :IN INTEGER RANGE 0 TO 3;
4             din                  :IN BIT_VECTOR (3 DOWNTO 0);
5             ch0, ch1, ch2, ch3   :OUT BIT_VECTOR(3 DOWNTO 0)
6           );
7   END fig9_65;
8
9   ARCHITECTURE selecter OF fig9_65 IS
10      BEGIN
11         ch0 <= din WHEN s = 0  ELSE "1111";
12         ch1 <= din WHEN s = 1  ELSE "1111";
13         ch2 <= din WHEN s = 2  ELSE "1111";
14         ch3 <= din WHEN s = 3  ELSE "1111";
15      END selecter;
```

statements, as shown in lines 11–14. We decided earlier that the code for this DEMUX must ensure that the outputs are all HIGH when they are not selected. This is accomplished with the ELSE clause of each conditional signal assignment. If the ELSE clause is not used, the output values would default automatically to all LOW. For example, line 13 states that channel 2 will be connected to the data inputs whenever the select inputs are set to 2. If *s* is set to any other value, then channel 2 will be forced to have all bits HIGH.

REVIEW QUESTIONS

1. For the four-bit by four-channel MUX, name the data inputs, the data outputs, and the control inputs that choose one channel of the four.
2. For the four-bit by four-channel DEMUX, name the data inputs, the data outputs, and the control inputs that choose one channel of the four.
3. In the AHDL example, how are the logic states determined for the channels that are not selected?
4. In the VHDL example, how are the logic states determined for the channels that are not selected?

9-19 HDL MAGNITUDE COMPARATORS

In Section 9-10, we studied a 7485 magnitude comparator chip. As the name implies, this device compares the magnitude of two binary numbers and indicates the relationship between the two (greater than, less than, equal to). Control inputs are provided for the purpose of cascading these chips. These chips are interconnected so that the chip comparing the lower-order bits has its outputs connected to the control inputs of the next higher-order chip, as shown in Figure 9-37. When the highest-order stage detects that its data inputs have equal magnitude, it will look to the next lower stage and use these control inputs to make the final decision. This gives us a chance to look at one of the defining differences between using traditional logic ICs and using HDL to design a circuit. If we need to compare bigger values using HDL we can simply adjust the size of the comparator input ports to be whatever we need, rather than trying to cascade several four-bit comparators. Consequently, there is no need for cascading input controls in the HDL version.

There are many possible ways to describe the operation of a comparator. However, it is best to use an IF/ELSE construct because each IF clause evaluates a relationship between two values, as opposed to looking for the single value of a variable, like a CASE. The two inputs being compared should definitely be declared as numerical values. The three comparator outputs should be declared as individual bits in order to label each bit's purpose clearly.

AHDL

AHDL COMPARATOR

The AHDL code in Figure 9-66 follows the algorithm we have described using IF/ELSE constructs. Notice in line 3 that the data values are declared as four-bit numbers. Also note in lines 8, 10, and 11 that several statements can be used to specify the circuit's operation when the IF clause is true. Each statement is used to set the level on one of the outputs. These three statements are considered concurrent, and the order in which they are listed makes no difference. For example, in line 8, when *A* is greater than *B*, the *agtb* output will go HIGH at the same time the other two outputs (*altb, aeqb*) go LOW.

```
1    SUBDESIGN fig9_66
2    (
3       a[3..0], b[3..0]      :INPUT;
4       agtb, altb, aeqb      :OUTPUT;
5    )
6    BEGIN
7       IF    a[] > b[] THEN
8                agtb = VCC;    altb = GND;   aeqb = GND;
9       ELSIF a[] < b[] THEN
10               agtb = GND;    altb = VCC;   aeqb = GND;
11      ELSE     agtb = GND ;   altb = GND  ; aeqb = VCC;
12      END IF;
13   END;
```

FIGURE 9-66 Magnitude comparator in AHDL.

VHDL COMPARATOR

The VHDL code in Figure 9-67 follows the algorithm we have described using IF/ELSE constructs. Notice in line 2 that the data values are declared as four-bit integers. Remember, in VHDL, the IF/ELSE constructs can be used only inside a PROCESS. In this case, we want to evaluate the PROCESS whenever any of the inputs change state. Consequently, each input is listed in the sensitivity list within the parentheses. Also note in lines 10, 11, and 12 that several statements can be used to specify the circuit's operation when the IF clause is true. Each statement is used to set the level on one of the outputs. These three statements are considered concurrent, and the order in which they are listed makes no difference. For example, on line 11, when A is greater than B, the *agtb* output will go HIGH at the same time the other two outputs (*altb, aeqb*) go LOW.

```
1   ENTITY fig9_67 IS
2   PORT (    a, b                 : IN INTEGER RANGE 0 TO 15;
3             agtb, altb, aeqb     : OUT BIT);
4   END fig9_67;
5
6   ARCHITECTURE vhdl OF fig9_67 IS
7   BEGIN
8      PROCESS (a, b)
9      BEGIN
10        IF    a < b THEN     altb <= '1';  agtb <= '0';  aeqb <= '0';
11        ELSIF a > b THEN     altb <= '0';  agtb <= '1';  aeqb <= '0';
12        ELSE                 altb <= '0';  agtb <= '0';  aeqb <= '1';
13        END IF;
14     END PROCESS;
15  END vhdl;
```

FIGURE 9-67 Magnitude comparator in VHDL.

REVIEW QUESTIONS

1. What type of data objects must be declared for data inputs to a comparator?
2. What is the key control structure used to describe a comparator?
3. What are the key operators used?

9-20 HDL CODE CONVERTERS

Section 9-11 demonstrated some methods using adder circuits in an interesting but not at all intuitive way to create a BCD-to-binary conversion. In Chapter 6, we discussed adder circuits, and the circuit of Figure 9-40 can certainly be implemented using HDL and 7483 macrofunctions or adder descriptions that we know how to write. However, this is an excellent opportunity to point out the tremendous advantage that HDL can offer because it allows a circuit to be described in a way that makes the most sense. In the case of BCD-to-binary conversion, the sensible method of conversion is to use the concepts that we all learned in the third grade about the decimal number system. You were once taught that the number 275 was actually:

$$
\begin{array}{rrcrcr}
 & 2 & \times & 100 & = & 200 \\
+ & 7 & \times & 10 & = & 70 \\
+ & 5 & \times & 1 & = & \underline{5} \\
 & & & & & 275
\end{array}
$$

Now we have studied the BCD number system and realize that 275 is represented in BCD as 0010 0111 0101. Each digit is simply represented in binary. If we could multiply these binary digits by the decimal weight (represented in binary) and add them, we would have a binary answer that is equivalent to the BCD quantity. For example, let's try using the BCD representation for 275:

BCD	*Decimal Weight (in binary)*	*Partial Product (in binary)*
0010 ×	1100100 =	11001000
+ 0111 ×	1010 =	01000110
+ 0101 ×	1 =	0101
		100010011 = 275_{10}

The solution presented here for our eight-bit (two BCD digits) HDL code converter will use the following strategy:

Take the most significant BCD digit (the tens place) and multiply it by 10. Add this product to the least significant BCD digit (the ones place).

The answer will be a binary number representing the BCD quantity. It is important to realize that the HDL compiler does not necessarily try to implement an actual multiplier circuit in its solution. It will create the most efficient circuit that will do the job, which allows the designer to describe its behavior in the most sensible way.

AHDL

AHDL BCD-TO-BINARY CODE CONVERTER

The key to this strategy is in being able to multiply by 10. AHDL does not offer a multiplication operator, so in order to use this overall strategy, we need some math tricks. We will use the shifting of bits to perform multiplication and then employ the distributive property from algebra to multiply by 10. In the same way that we can shift a decimal number left by one digit, thus multiplying it by 10, we can likewise shift a binary number one place to the left and multiply it by 2. Shifting two places multiplies a binary number by 4, and shifting three places multiplies by 8. The distributive property tells us that:

$$\text{num} \times 10 = \text{num} \times (8 + 2) = (\text{num} \times 8) + (\text{num} \times 2)$$

If we can take the BCD tens digit and shift it left three bit positions (i.e., multiply it by 8), then take the same number and shift it left one place (i.e., multiply it by 2), and then add them together, the result will be the same as multiplying the BCD digit by 10. This value is then added to the BCD ones digit to produce the binary equivalent of the two-digit BCD input.

The next challenge is to shift the BCD digit left using AHDL. Because AHDL allows us to make up sets of variables, we can shift the bits by appending zeros to the right end of the array. For example, if we have the number 5 in BCD (0101) and we want to shift it three places, we can concatenate the number 0101 with the number 000 in a set, as follows:

(B"0101", B"000") = B"0101000"

The AHDL code in Figure 9-68 begins by declaring inputs for the BCD ones and tens digits. The binary output must be able to represent 99_{10}, which requires seven bits. We also need a variable to hold the product of the BCD digit multiplied by 10. Line 5 declares this variable as a seven-bit number. Line 8 performs the shifting of the *tens[ ]* array three times and adds it to the *tens[ ]* array shifted one place to the left. Notice that this latter set must have seven bits in order to be added to the first set, thus the need to concatenate B"00" on the left end. Finally, in line 10, the result from line 8 is added to the BCD ones digit with leading zero extensions (to make seven bits) to form the binary output.

```
1   SUBDESIGN fig9_68
2   (  ones[3..0], tens[3..0]     :INPUT;
3      binary[6..0]               :OUTPUT;  )
4
5   VARIABLE times10[6..0]        :NODE;    % variable for tens digit times 10%
6
7   BEGIN
8      times10[] = (tens[],B"000") + (B"00",tens[],B"0");
9            % shift left 3X (times 8) + shift left 1X (times 2) %
10     binary[] = times10[] + (B"000",ones[]);
11           % tens digit times 10 + ones digit %
12  END;
```

FIGURE 9-68 BCD-to-binary code converter in AHDL.

VHDL BCD-TO-BINARY CODE CONVERTER

The VHDL solution in Figure 9-69 is very simple due to the powerful math operations available in the language. The inputs and outputs must be declared as integers because we intend to perform arithmetic operations on them. Notice that the range is specified based on the largest valid BCD number using two digits. In line 9, the tens digit is multiplied by ten, and in line 10, the ones digit is added to form the binary equivalent of the BCD input.

FIGURE 9-69 BCD-to-binary code converter in VHDL.

```
1   ENTITY fig9_69 IS
2   PORT (    ones, tens    :IN INTEGER RANGE 0 TO 9;
3             binary        :OUT INTEGER RANGE 0 TO 99);
4   END fig9_69;
5
6   ARCHITECTURE vhdl OF fig9_69 IS
7   SIGNAL times10         :INTEGER RANGE 0 TO 90;
8   BEGIN
9      times10 <= tens * 10;
10     binary <= times10 + ones;
11  END vhdl;
```

AHDL

VHDL

REVIEW QUESTIONS

1. For a two-digit BCD (eight-bit) number, what is the decimal weight of the most significant digit?
2. In AHDL, how is multiplication by 10 accomplished?
3. In VHDL, how is multiplication by 10 accomplished?

SUMMARY

1. A decoder is a device whose output is activated only when a unique binary combination (code) is presented on its inputs. Many MSI decoders have several outputs, each one corresponding to only one of the many possible input combinations.
2. Digital systems often need to display decimal numbers. This is done using 7-segment displays that are driven by special chips that decode the binary number and translate it into segment patterns that represent decimal numbers to people. The segment elements can be light-emitting diodes, liquid crystals, or glowing electrodes surrounded by neon gas.
3. Graphical LCDs use a matrix of picture elements called pixels to create an image on a large screen. Each pixel is controlled by activating the row and column that have that pixel in common. The brightness level of each pixel is stored as a binary number in the video memory. A fairly complex digital circuit must scan through the video memory and all the row/column combinations, controlling the amount of light that can pass through each pixel.
4. An encoder is a device that generates a unique binary code in response to the activation of each individual input.
5. Troubleshooting a digital system involves *observation/analysis* to identify the possible causes, and a process of elimination called *divide-and-conquer* to isolate and identify the cause.
6. Multiplexers act like digitally controlled switches that select and connect one logic input at a time to the output pin. By taking turns, many different data signals can share the same data path using multiplexers. Demultiplexers are used at the other end of the data path to separate the signals that are sharing a data path and distribute them to their respective destinations.
7. Magnitude comparators serve as an indicator of the relationship between two binary numbers, with outputs that show $>$, $<$, and $=$.
8. It is often necessary to translate between and among various methods of representing quantities with binary numbers. Code converters are devices that take in one form of binary representation and convert it to another form.
9. In digital systems, many devices must often share the same data path. This data path is often called a *data bus*. Even though many devices can be "riding" on the bus, there can be only one bus "driver" at any one time. Thus, devices must take turns applying logic signals to the data bus.
10. In order to take turns, the devices must have *tristate outputs* that can be disabled when another device is driving the bus. In the disabled state, the device's output is essentially electrically disconnected from the bus by going into a state that offers a high-impedance path to both ground and the positive power supply. Devices designed to interface to a bus have outputs that can be HIGH, LOW, or disabled (high impedance).

11. PLDs offer an alternative to the use of MSI circuits to implement digital systems. Boolean equations can be used to describe the operation of these circuits, but HDLs also offer high-level language constructs.
12. HDL macrofunctions are available for many MSI standard parts described in this chapter.
13. Custom code can be written in HDL to describe each of the common logic functions presented in this chapter.
14. Priority and precedence can be established in AHDL using don't-care entries in truth tables and using IF/ELSE decisions. Priority and precedence can be established in VHDL using conditional signal assignments or using a PROCESS containing IF/ELSE or CASE decisions.
15. Tristate outputs can be created in HDL. AHDL uses :TRI primitives that drive the outputs. VHDL assigns Z (high impedance) as a valid state for STD_LOGIC outputs.
16. The DEFAULTS statement in AHDL can be used to define the proper level for outputs that are not explicitly defined in the code.
17. The ELSE clause in the conditional signal assignment statement of VHDL can be used to define the default state of an output.

IMPORTANT TERMS

decoder
BCD-to-decimal decoder
driver
BCD-to-7-segment decoder/driver
common anode
common cathode
LCD
backplane
pixel
encoding
encoder
priority encoder
observation/analysis
divide-and-conquer
multiplexer (MUX)
multiplexing
parallel to serial conversion
demultiplexer (DEMUX)
magnitude comparator
data bus
floating bus
word
bus driver
bidirectional data lines
DEFAULTS
conditional signal assignment statement

PROBLEMS

SECTION 9-1

B 9-1. Refer to Figure 9-3. Determine the levels at each decoder output for the following sets of input conditions.

(a)* All inputs LOW

(b)* All inputs LOW except E_3 = HIGH

(c) All inputs HIGH except $\overline{E}_1 = \overline{E}_2$ = LOW

(d) All inputs HIGH

B 9-2.* What is the number of inputs and outputs of a decoder that accepts 64 different input combinations?

*Answers to problems marked with an asterisk can be found in the back of the text.

B 9-3. For a 74ALS138, what input conditions will produce the following outputs:

(a)*LOW at $\overline{O}_6$

(b)*LOW at $\overline{O}_3$

(c) LOW at $\overline{O}_5$

(d) LOW at $\overline{O}_0$ and $\overline{O}_7$, simultaneously

D 9-4. Show how to use 74LS138s to form a 1-of-16 decoder.

9-5.*Figure 9-70 shows how a decoder can be used in the generation of control signals. Assume that a RESET pulse has occurred at time t_0, and determine the CONTROL waveform for 32 clock pulses.

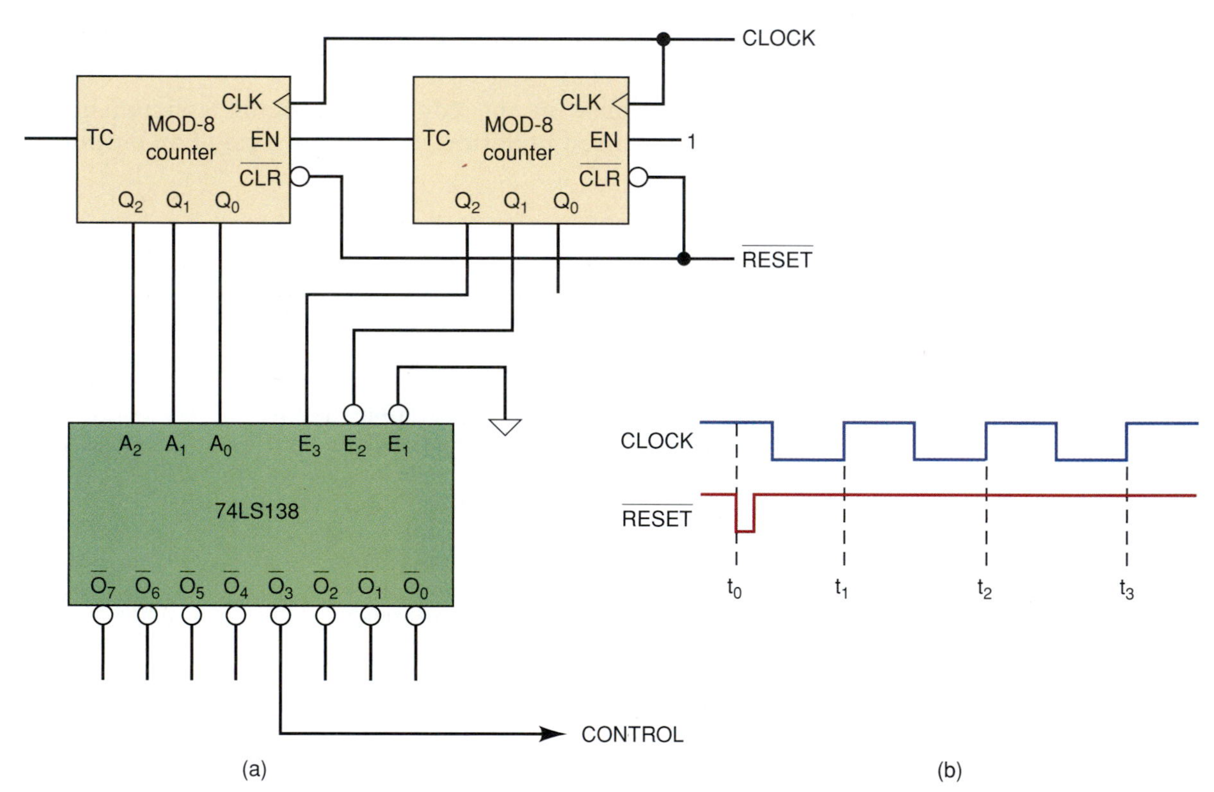

FIGURE 9-70 Problems 9-5 and 9-6.

D 9-6. Modify the circuit of Figure 9-70 to generate a CONTROL waveform that goes LOW from t_{20} to t_{24}. (*Hint:* The modification does not require additional logic.)

9-7.*The 7442 decoder of Figure 9-5 does not have an ENABLE input. However, we can operate it as a 1-of-8 decoder by not using outputs $\overline{O}_8$ and $\overline{O}_9$ and by using the *D* input as an ENABLE. This is illustrated in Figure 9-71. Describe how this arrangement works as an enabled 1-of-8 decoder, and state how the level on *D* either enables or disables the outputs.

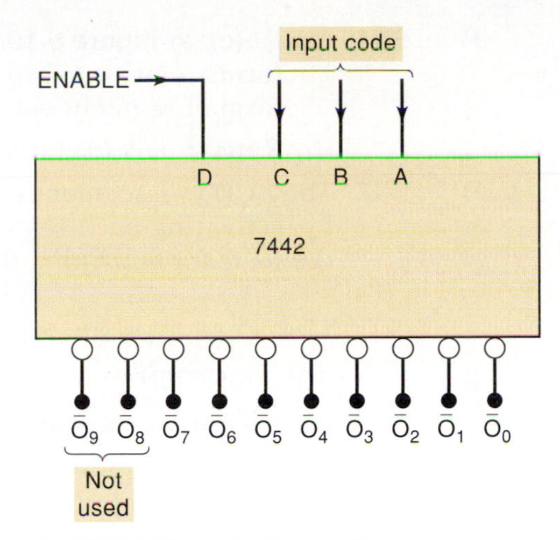

FIGURE 9-71 Problem 9-7.

9-8. Consider the waveforms in Figure 9-72. Apply these signals to the 74LS138 as follows:

$$A \rightarrow A_0 \qquad B \rightarrow A_1 \qquad C \rightarrow A_2 \qquad D \rightarrow E_3$$

Assume that $\overline{E}_1$ and $\overline{E}_2$ are tied LOW, and draw the waveforms for outputs $\overline{O}_0$, $\overline{O}_3$, $\overline{O}_6$, and $\overline{O}_7$.

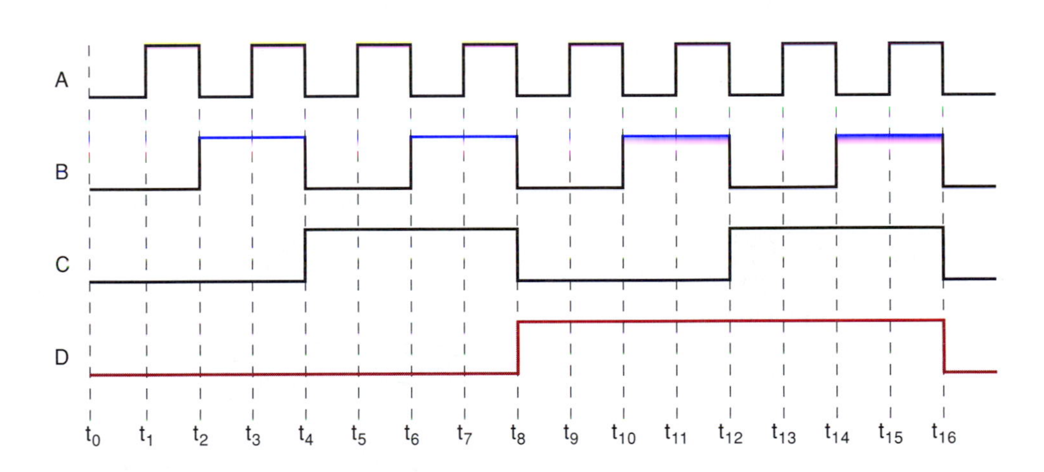

FIGURE 9-72 Problems 9-8, 9-15, and 9-41.

D 9-9. Modify the circuit of Figure 9-6 so that relay K_1 stays energized from PGT 3 to 5 and K_2 stays energized from PGT 6 to 9. (*Hint:* This modification requires no additional circuitry.)

SECTIONS 9-2 AND 9-3

B, D 9-10.* Show how to connect BCD-to-7-segment decoder/drivers and LED 7-segment displays to the counter circuit of Figure 7-22. Assume that each segment is to operate at approximately 10 mA at 2.5 V.

B 9-11. (a) Refer to Figure 9-10 and draw the segment and backplane waveforms relative to ground for CONTROL = 0. Then draw the waveform of segment voltage relative to backplane voltage.

(b) Repeat part (a) for CONTROL = 1.

C, D 9-12.* The BCD-to-7-segment decoder/driver of Figure 9-8 contains the logic for activating each segment for the appropriate BCD inputs. Design the logic for activating the *g* segment.

SECTION 9-4

B 9-13.* **DRILL QUESTION**

For each item, indicate whether it is referring to a decoder or an encoder.

(a) Has more inputs than outputs.

(b) Is used to convert key actuations to a binary code.

(c) Only one output can be activated at one time.

(d) Can be used to interface a BCD input to an LED display.

(e) Often has driver-type outputs to handle large *I* and *V*.

9-14. Determine the output levels for the 74147 encoder when $\overline{A}_8 = \overline{A}_4 = 0$ and all other inputs are HIGH.

9-15. Apply the signals of Figure 9-72 to the inputs of a 74147 as follows:

$$A \rightarrow \overline{A}_7 \qquad B \rightarrow \overline{A}_4 \qquad C \rightarrow \overline{A}_2 \qquad D \rightarrow \overline{A}_1$$

Draw the waveforms for the encoder's outputs.

C, D 9-16. Figure 9-73 shows the block diagram of a logic circuit used to control the number of copies made by a copy machine. The machine operator selects the number of desired copies by closing one of the selector switches S_1 to S_9. This number is encoded in BCD by the encoder and is sent to a comparator circuit. The operator then hits a momentary-contact START switch, which clears the counter and initiates a HIGH

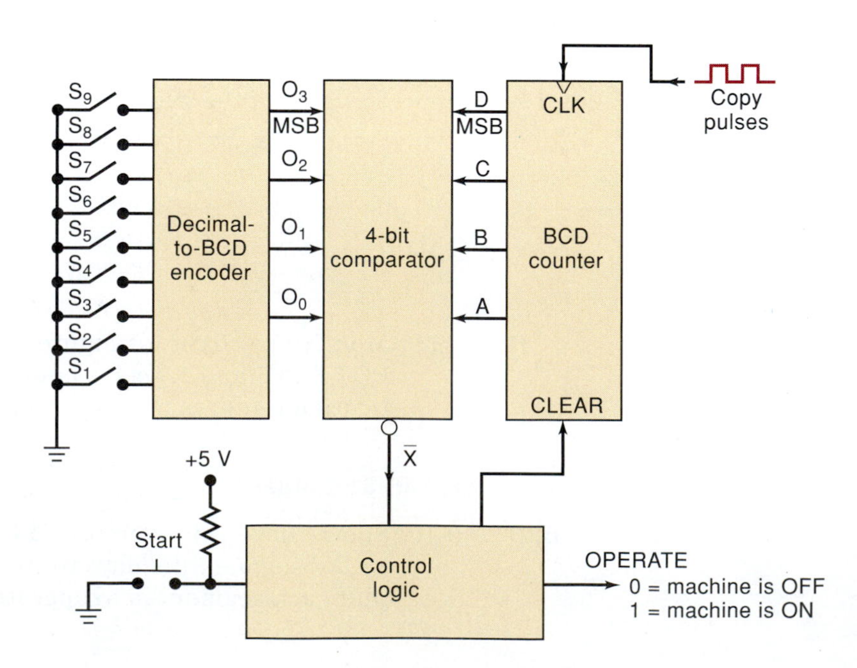

FIGURE 9-73 Problems 9-16 and 9-52.

OPERATE output that is sent to the machine to signal it to make copies. As the machine makes each copy, a copy pulse is generated and fed to the BCD counter. The counter outputs are continually compared with the switch encoder outputs by the comparator. When the two BCD numbers match, indicating that the desired number of copies has been made, the comparator output $\overline{X}$ goes LOW; this causes the OPERATE level to return LOW and stop the machine so that no more copies are made. Activating the START switch will cause this process to be repeated. Design the complete logic circuitry for the comparator and control sections of this system.

C, D 9-17.* The keyboard circuit of Figure 9-16 is designed to accept a three-digit decimal number. What would happen if *four* digit keys were activated (e.g., 3095)? Design the necessary logic to be added to this circuit so that after three digits have been entered, any additional digits will be ignored until the CLEAR key is depressed. In other words, if 3095 is entered on the keyboard, the output registers will display 309 and will ignore the 5 and any subsequent digits until the circuit is cleared.

SECTION 9-5

T 9-18.* A technician breadboards the keyboard entry circuit of Figure 9-16 and tests its operation by trying to enter a series of three-digit numbers. He finds that sometimes the digit 0 is entered instead of the digit he pressed. He also observes that it happens with all of the keys more or less randomly, although it is worse for some keys than others. He replaces all of the ICs, and the malfunction persists. Which of the following circuit faults would explain his observations? Explain each choice.

(a) The technician forgot to ground the unused inputs of the OR gate.

(b) He has mistakenly used $\overline{Q}$ instead of Q from the one-shot.

(c) The switch bounce from the digit keys lasts longer than 20 ms.

(d) The Y and Z outputs are shorted together.

T 9-19. Repeat Problem 9-18 with the following symptom: the registers and displays stay at 0 no matter how many times a key is pressed.

T 9-20.* While testing the circuit of Figure 9-16, a technician finds that every odd-numbered key results in the correct digit being entered, but every even-numbered key results in the wrong digit being entered as follows: the 0 key causes a 1 to be entered, the 2 key causes a 3 to be entered, the 4 key causes a 5 to be entered, and so on. Consider each of the following faults as possible causes of the malfunction. For each one, explain why it can or cannot be the actual cause.

(a) There is an open connection from the output of the LSB inverter to the D inputs of the FFs.

(b) The D input of flip-flop Q_8 is internally shorted to V_{CC}.

(c) A solder bridge is shorting $\overline{O_0}$ to ground.

T 9-21.* A technician tests the circuit of Figure 9-4 as described in Example 9-7, and she obtains the following results: all of the outputs work except $\overline{O_{16}}$ to $\overline{O_{19}}$ and $\overline{O_{24}}$ to $\overline{O_{27}}$, which are permanently HIGH. What is the most probable circuit fault?

T 9-22. A technician tests the circuit of Figure 9-4 as described in Example 9-7 and finds that the correct output is activated for each possible input

TABLE 9-8

Input Code					
A_4	A_3	A_2	A_1	A_0	**Activated Outputs**
1	0	0	0	0	$\overline{O}_{16}$ and $\overline{O}_{24}$
1	0	0	0	1	$\overline{O}_{17}$ and $\overline{O}_{25}$
1	0	0	1	0	$\overline{O}_{18}$ and $\overline{O}_{26}$
1	0	0	1	1	$\overline{O}_{19}$ and $\overline{O}_{27}$
1	0	1	0	0	$\overline{O}_{20}$ and $\overline{O}_{28}$
1	0	1	0	1	$\overline{O}_{21}$ and $\overline{O}_{29}$
1	0	1	1	0	$\overline{O}_{22}$ and $\overline{O}_{30}$
1	0	1	1	1	$\overline{O}_{23}$ and $\overline{O}_{31}$

code except those listed in Table 9-8. Examine this table and determine the probable cause of the malfunction.

T 9-23.*Suppose that a 22-Ω resistor was mistakenly used for the *g* segment in Figure 9-8. How would this affect the display? What possible problems could occur?

T 9-24. Repeat Example 9-8 with the observed sequence shown below:

COUNT	0	1	2	3	4	5	6	7	8	9
Observed display	0	1	2	3	8	9	c	ɔ	4	5

T 9-25.*Repeat Example 9-8 with the observed sequence shown below:

COUNT	0	1	2	3	4	5	6	7	8	9
Observed display	0	7	2	3	9	9	8	7	8	9

T 9-26.*To test the circuit of Figure 9-11, a technician connects a BCD counter to the 74HC4511 inputs and pulses the counter at a very slow rate. She notices that the *f* segment works erratically, and no particular pattern is evident. What are some of the possible causes of the malfunction? (*Hint:* Remember, the ICs are CMOS.)

SECTIONS 9-6 AND 9-7

B 9-27. The timing diagram in Figure 9-74 is applied to Figure 9-19. Draw the output waveform Z.

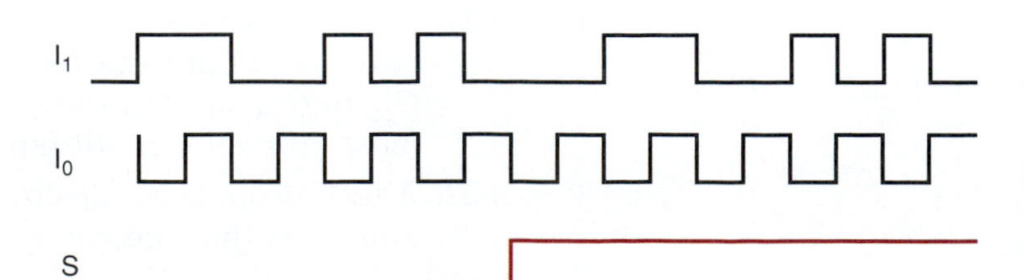

FIGURE 9-74 Problem 9-27.

9-28. Figure 7-68 shows an eight-bit shift register that could be used to delay a signal by 1 to 8 clock periods. Show how to wire a 74151 to this shift register in order to select the desired Q output and indicate the logic level necessary on the select inputs to provide a delay of $6 \times T_{clk}$.

9-29.*The circuit in Figure 9-75 uses three two-input multiplexers (Figure 9-19). Determine the function performed by this circuit.

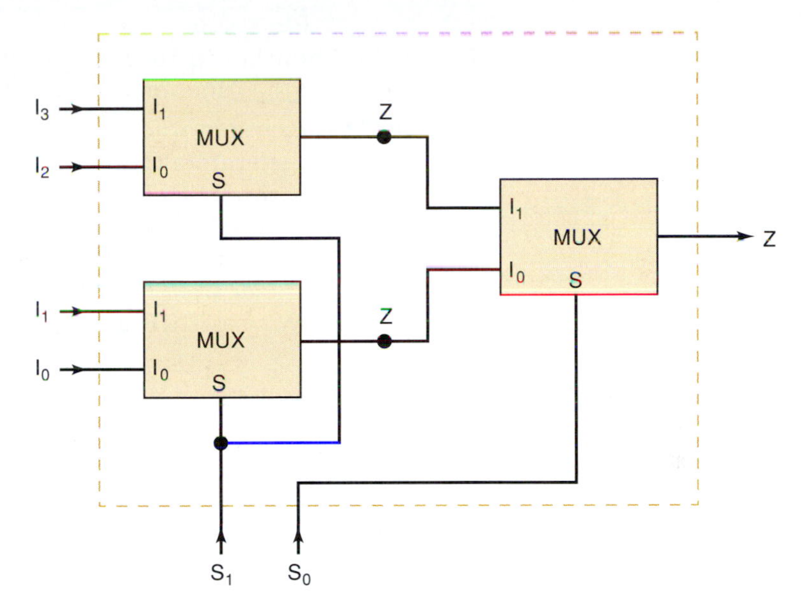

FIGURE 9-75 Problem 9-29.

D 9-30. Use the idea from Problem 9-29 to arrange several 74151 1-of-8 multiplexers to form a 1-of-64 multiplexer.

C, D 9-31.*Show how two 74157s and a 74151 can be arranged to form a 1-of-16 multiplexer with no other required logic. Label the inputs I_0 to I_{15} to show how they correspond to the select code.

D 9-32. (a) Expand the circuit of Figure 9-24 to display the contents of two three-stage BCD counters.

(b)*Count the number of connections in this circuit, and compare it with the number required if a separate decoder/driver and display were used for each stage of each counter.

9-33.*Figure 9-76 shows how a multiplexer can be used to generate logic waveforms with any desirable pattern. The pattern is programmed

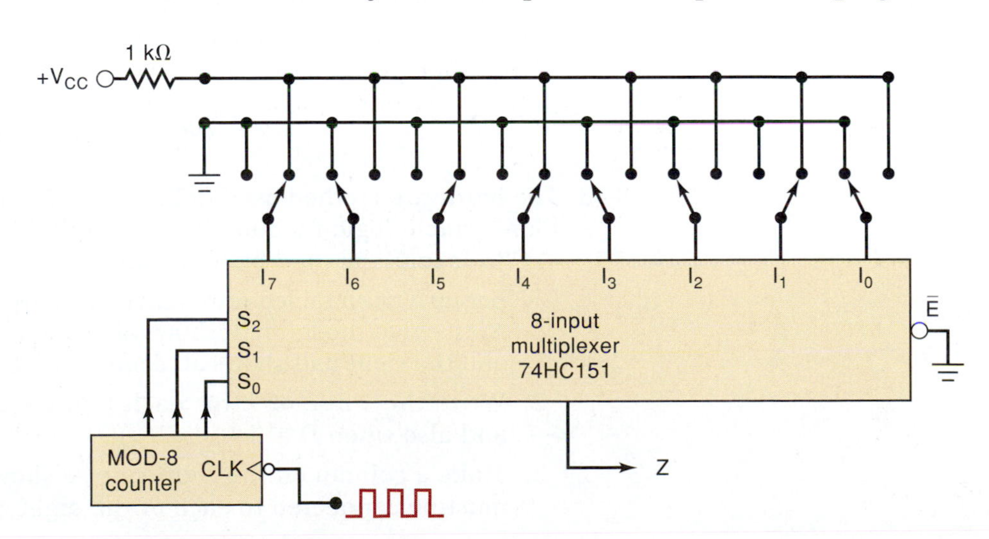

FIGURE 9-76 Problems 9-33 and 9-34.

using eight SPDT switches, and the waveform is repetitively produced by pulsing the MOD-8 counter. Draw the waveform at Z for the given switch positions.

9-34. Change the MOD-8 counter in Figure 9-76 to a MOD-16 counter, and connect the MSB to the multiplexer $\overline{E}$ input. Draw the Z waveform.

D 9-35.*Show how a 74151 can be used to generate the logic function $Z = AB + BC + AC$.

D 9-36. Show how a 16-input multiplexer such as the 74150 is used to generate the function $Z = \overline{A}\,\overline{B}\,\overline{C}D + BCD + A\overline{B}\,\overline{D} + AB\overline{C}D$.

N 9-37.*The circuit of Figure 9-77 shows how an eight-input MUX can be used to generate a four-variable logic function, even though the MUX has only three SELECT inputs. Three of the logic variables A, B, and C are connected to the SELECT inputs. The fourth variable D and its inverse $\overline{D}$ are connected to selected data inputs of the MUX as required by the desired logic function. The other MUX data inputs are tied to a LOW or a HIGH as required by the function.

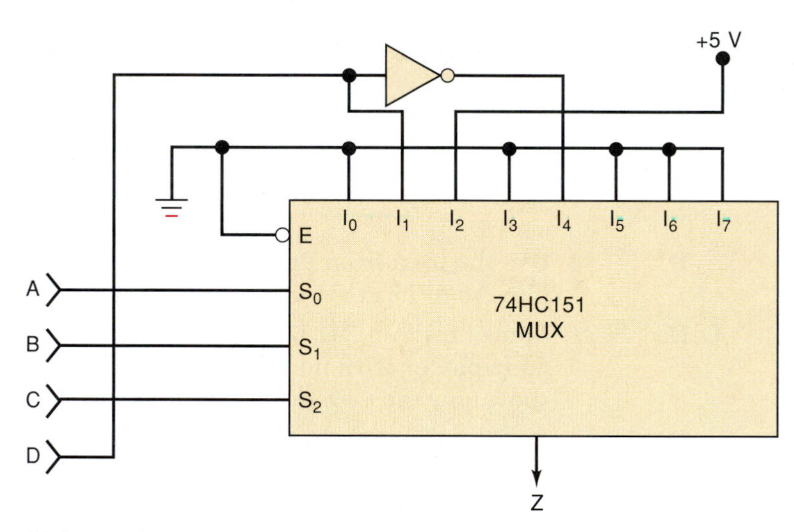

FIGURE 9-77 Problems 9-37 and 9-38.

(a) Set up a truth table showing the output Z for the 16 possible combinations of input variables.

(b) Write the sum-of-products expression for Z and simplify it to verify that

$$Z = \overline{C}B\overline{A} + D\overline{C}\,\overline{B}A + \overline{D}C\overline{B}\,\overline{A}$$

9-38. The hardware method used in Figure 9-77 can be used to generate any four-variable logic function. For example, $Z = \overline{D}\,\overline{B}\,\overline{C}A + \overline{C}BA + DC\overline{B}A + CB\overline{A}$ is implemented by following these steps:

1. Set up a truth table in two halves, side by side as shown in Table 9-9. Notice that the left half shows all combinations of CBA when $D = 0$, and the right half shows all combinations of CBA when $D = 1$.
2. Write the value of Z for each four-bit combination when $D = 0$ and also when $D = 1$.
3. Make a column on the right side as shown, which describes what must be connected to each of the eight MUX inputs I_n.

TABLE 9-9

	D = 0		D = 1	
DCBA	Z	DCBA	Z	I_n
0000	0	1000	0	$I_0 = 0$
0001	1	1001	0	$I_1 = \overline{D}$
0010	0	1010	0	$I_2 = 0$
0011	1	1011	1	$I_3 = 1$
0100	0	1100	0	$I_4 = 0$
0101	0	1101	1	$I_5 = D$
0110	1	1110	1	$I_6 = 1$
0111	0	1111	0	$I_7 = 0$

4. For each line of this table, compare the value for Z when $D = 0$ with the value for Z when $D = 1$. Enter the appropriate information for I_n as follows:

   When $Z = 0$ regardless of whether $D = 0$ or 1, THEN $I_n = 0$ (GND).

   When $Z = 1$ regardless of whether $D = 0$ or 1, THEN $I_n = 1$ (VCC).

   When $Z = 0$ when $D = 0$ AND $Z = 1$ when $D = 1$, THEN $I_n = D$.

   When $Z = 1$ when $D = 0$ AND $Z = 0$ when $D = 1$, THEN $I_n = \overline{D}$.

   (a) Verify the design of Figure 9-77 using this method.

   (b) Use this method to implement a function that will produce a HIGH only when the four input variables are at the same level or when the B and C variables are at different levels.

SECTION 9-8

B 9-39.* **DRILL QUESTION**

For each item, indicate whether it is referring to a decoder, an encoder, a MUX, or a DEMUX.

(a) Has more inputs than outputs.

(b) Uses SELECT inputs.

(c) Can be used in parallel-to-serial conversion.

(d) Produces a binary code at its output.

(e) Only one of its outputs can be active at one time.

(f) Can be used to route an input signal to one of several possible outputs.

(g) Can be used to generate arbitrary logic functions.

9-40. Show how the 7442 decoder can be used as 1-to-8 demultiplexer. (*Hint:* See Problem 9-7.)

9-41.* Apply the waveforms of Figure 9-72 to the inputs of the 74LS138 DEMUX of Figure 9-30(a) as follows:

$D \rightarrow A_2$ $\quad C \rightarrow A_1$ $\quad B \rightarrow A_0$ $\quad A \rightarrow \overline{E}_1$

Draw the waveforms at the DEMUX outputs.

9-42. Consider the system of Figure 9-31. Assume that the clock frequency is 10 pps. Describe what the monitoring panel indications will be for each of the following cases.

(a) All doors closed

(b) All doors open

(c) Doors 2 and 6 open

C, D 9-43.* Modify the system of Figure 9-31 to handle 16 doors. Use a 74150 16-input MUX and two 74LS138 DEMUXes. How many lines are going to the remote monitoring panel?

9-44. Draw the waveforms at transmit_data, and DEMUX outputs O_0, O_1, O_2, and O_3 in Figure 9-33 for the following register data loaded into the transmit registers in Figure 9-32: $[A] = 0011$, $[B] = 0110$, $[C] = 1001$, $[D] = 0111$.

9-45. Figure 9-78 shows an 8 × 8 graphic LCD display grid controlled by a 74HC138 configured as a decoder, and a 74HC138 configured as a demultiplexer. Draw 48 cycles of the clock and the data input necessary to activate the pixels shown on the display.

FIGURE 9-78 Problem 9-45.

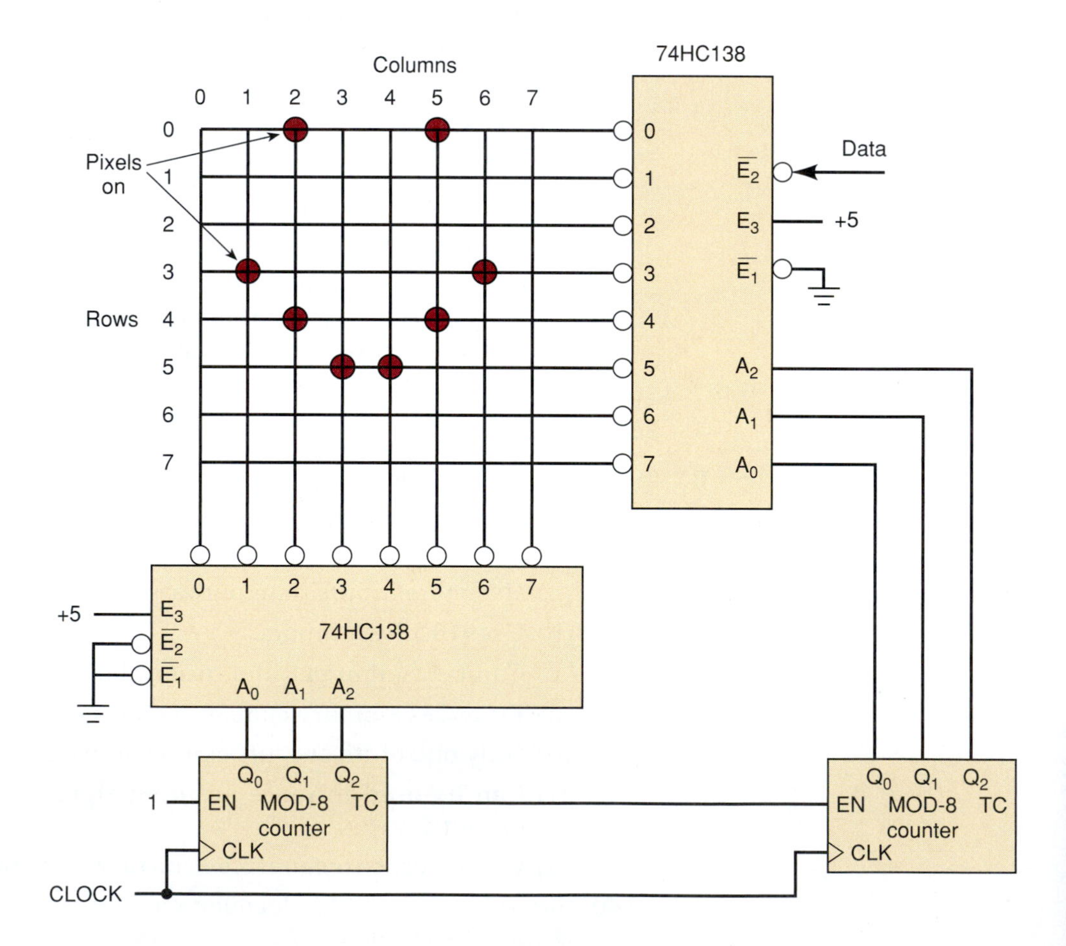

SECTION 9-9

T 9-46. Consider the control sequencer of Figure 9-26. Describe how each of the following faults will affect the operation.

(a)* The I_3 input of the MUX is shorted to ground.

(b) The connections from sensors 3 and 4 to the MUX are reversed.

TABLE 9-10

		Actual Count	Displayed Count
Case 1	Counter 1	33	33
	Counter 2	47	47
Case 2	Counter 1	82	02
	Counter 2	64	64
Case 3	Counter 1	63	63
	Counter 2	95	15

T 9-47.* Consider the circuit of Figure 9-24. A test of the circuit yields the results shown in Table 9-10. What are the possible causes of the malfunction?

T 9-48.* A test of the security monitoring system of Figure 9-31 produces the results recorded in Table 9-11. What are the possible faults that could cause this operation?

TABLE 9-11

Condition	LEDs
All doors closed	All LEDs off
Door 0 open	LED 0 flashing
Door 1 open	LED 2 flashing
Door 2 open	LED 1 flashing
Door 3 open	LED 3 flashing
Door 4 open	LED 4 flashing
Door 5 open	LED 6 flashing
Door 6 open	LED 5 flashing
Door 7 open	LED 7 flashing

T 9-49.* A test of the security monitoring system of Figure 9-31 produces the results recorded in Table 9-12. What are the possible faults that could cause this operation? How can this be verified or eliminated as a fault?

TABLE 9-12

Condition	LEDs
All doors closed	All LEDs off
Door 0 open	LED 0 flashing
Door 1 open	LED 1 flashing
Door 2 open	LED 2 flashing
Door 3 open	LED 3 flashing
Door 4 open	LED 4 flashing
Door 5 open	LED 5 flashing
Door 6 open	No LED flashing
Door 7 open	No LED flashing
Doors 6 and 7 open	LEDs 6 and 7 flashing

T 9-50.* The synchronous data transmission system of Figure 9-32 and Figure 9-33 is malfunctioning. An oscilloscope is used to monitor the MUX and

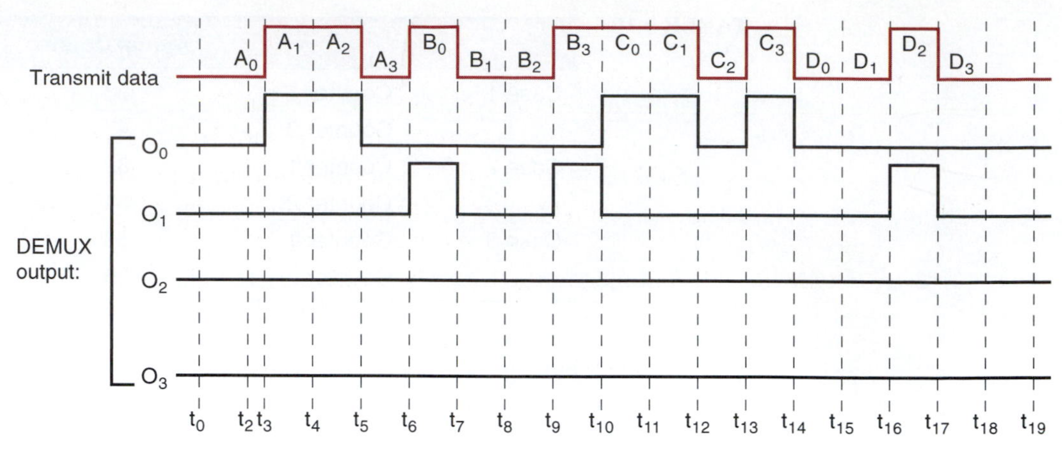

FIGURE 9-79 Problem 9-50.

DEMUX outputs during the transmission cycle, with the results shown in Figure 9-79. What are the possible causes of the malfunction?

T 9-51. The synchronous data transmission system of Figures 9-32 and 9-33 is not working properly and the troubleshooting tree diagram of Figure 9-35 has been used to isolate the problem to the timing and control section of the receiver. Draw a troubleshooting tree diagram that will isolate the problem further to one of the four blocks in that section (*FF*1, Bit counter, Word counter, or *FF*2). Assume that all wires are connected as shown, with no wiring errors.

SECTION 9-10

C, D 9-52. Redesign the circuit of Problem 9-16 using a 74HC85 magnitude comparator. Add a "copy overflow" feature that will activate an ALARM output if the OPERATE output fails to stop the machine when the requested number of copies is done.

D 9-53.* Show how to connect 74HC85s to compare two 10-bit numbers.

SECTION 9-11

9-54. Assume a BCD input of 69 to the code converter of Figure 9-40. Determine the levels at each Σ output and at the final binary output.

T 9-55.* A technician tests the code converter of Figure 9-40 and observes the following results:

BCD Input	*Binary Output*
52	0110011
95	1100000
27	0011011

What is the probable circuit fault?

SECTIONS 9-12 TO 9-14

B 9-56. **DRILL QUESTION**

True or false:

(a) A device connected to a data bus should have tristate outputs.

(b) Bus contention occurs when more than one device takes data from the bus.

(c) Larger units of data can be transferred over an eight-line data bus than over a four-line data bus.

(d) A bus driver IC generally has a high output impedance.

(e) Bidirectional registers and buffers have common I/O lines.

9-57.* For the bus arrangement of Figure 9-44, describe the input signal requirements for simultaneously transferring the contents of register *C* to both of the other registers.

9-58. Assume that the registers in Figure 9-44 are initially [*A*] = 1011, [*B*] = 1000, and [*C*] = 0111. The signals in Figure 9-80 are applied to the register inputs.

(a) Determine the contents of each register at times t_1, t_2, t_3, and t_4.

(b) Describe what would happen if $\overline{IE_A}$ were LOW when the third clock pulse occurred.

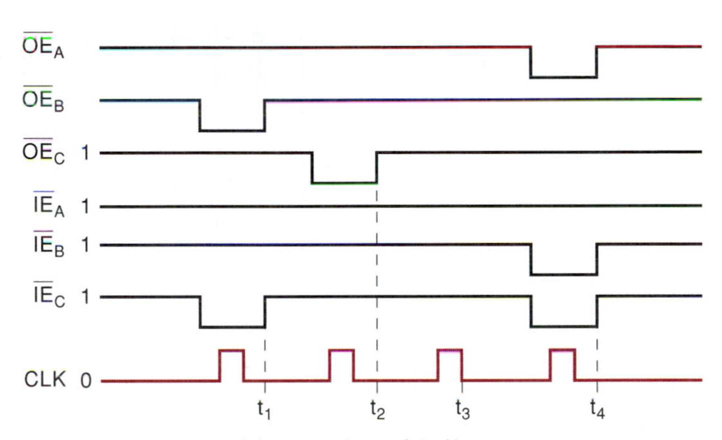

FIGURE 9-80 Problems 9-58 and 9-59.

9-59. Assume the same initial conditions of Problem 9-58, and sketch the signal on DB_3 for the waveforms of Figure 9-80.

9-60. Figure 9-81 shows two more devices that are to be added to the data bus of Figure 9-44. One is a set of buffered switches that can be used to enter data manually into any of the bus registers. The other device is an output register that is used to latch any data that are on the bus during a data transfer operation and display them on a set of LEDs.

(a) Assume that all registers contain 0000. Outline the sequence of operations needed to load the registers with the following data from the switches: [*A*] = 1011, [*B*] = 0001, [*C*] = 1110.

(b) What will the state of the LEDs be at the end of this sequence?

C 9-61. Now that the circuitry of Figure 9-81 has been added to Figure 9-44, a total of five devices are connected to the data bus. The circuit in Figure 9-82(a) will now be used to generate the enable signals needed to perform the different data transfers over the data bus. It uses a 74HC139 chip that contains two identical independent 1-of-4 decoders with an active-LOW enable. The top decoder is used to select the device that will put data on the data bus (output select), and the bottom decoder is used to select the device that is to take the data from the data bus (input select). Assume that the decoder outputs are connected to the corresponding enable inputs of the devices tied to the data bus. Also assume that all registers initially

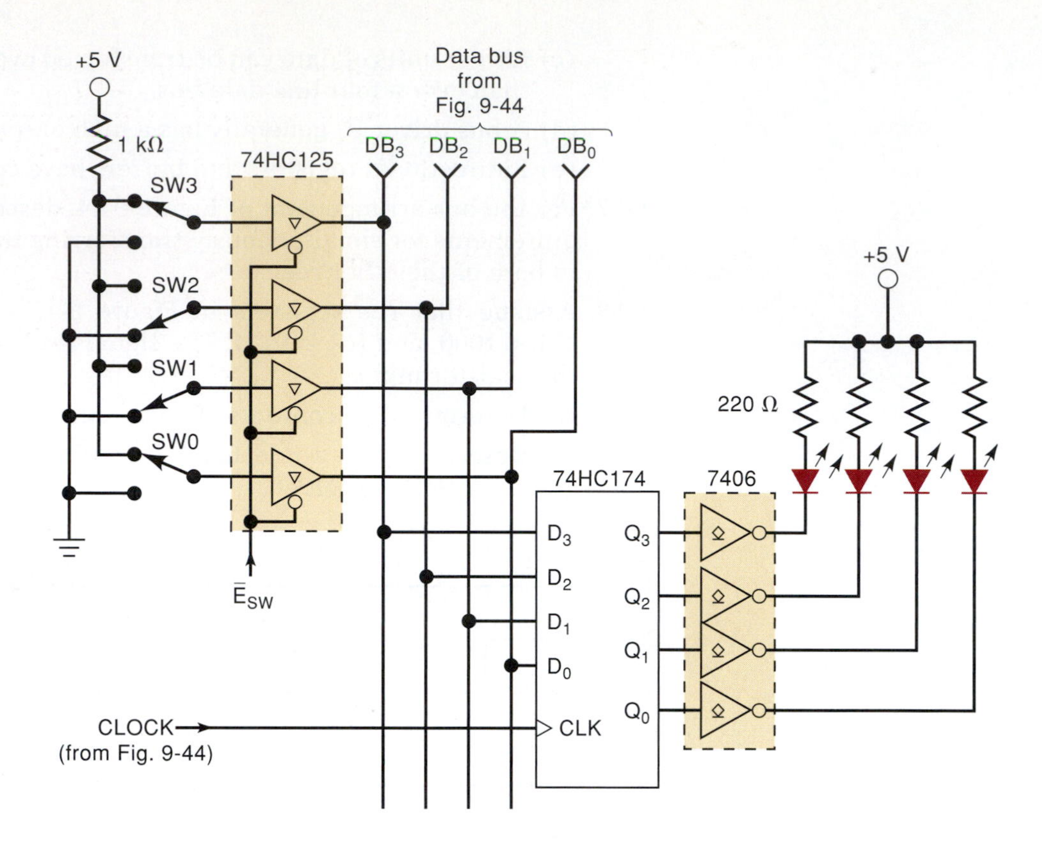

FIGURE 9-81 Problems 9-60, 9-61, and 9-62.

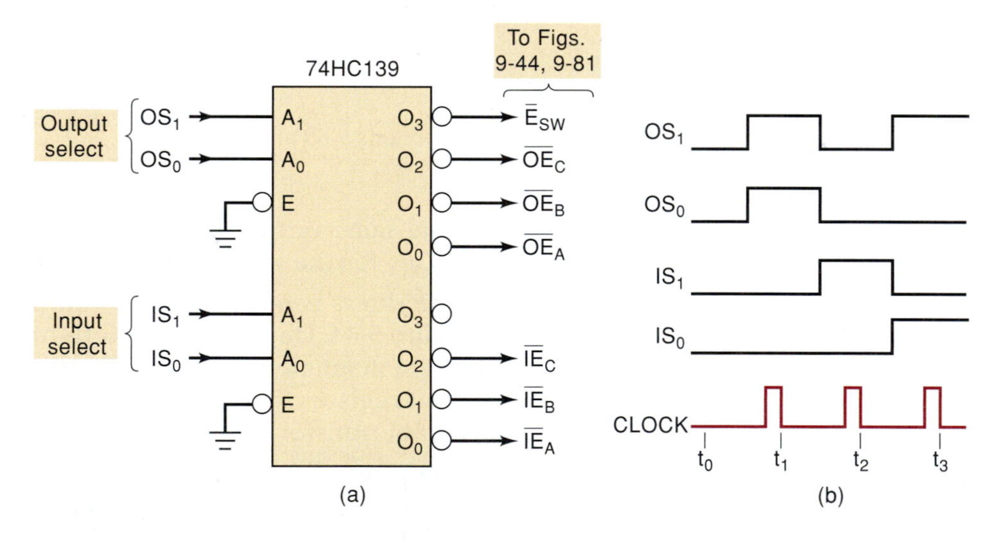

FIGURE 9-82 Problem 9-61.

contain 0000 at time t_0, and the switches are in the positions shown in Figure 9-81.

(a)* Determine the contents of each register at times t_1, t_2, and t_3 in response to the waveforms in Figure 9-82(b).

(b) Can bus contention ever occur with this circuit? Explain.

9-62. Show how a 74HC541 (Figure 9-47) can be used in the circuit of Figure 9-81.

MICROCOMPUTER APPLICATIONS

C, N 9-63.* Figure 9-83 shows the basic circuitry to interface a microprocessor (MPU) to a memory module. The memory module will contain one or

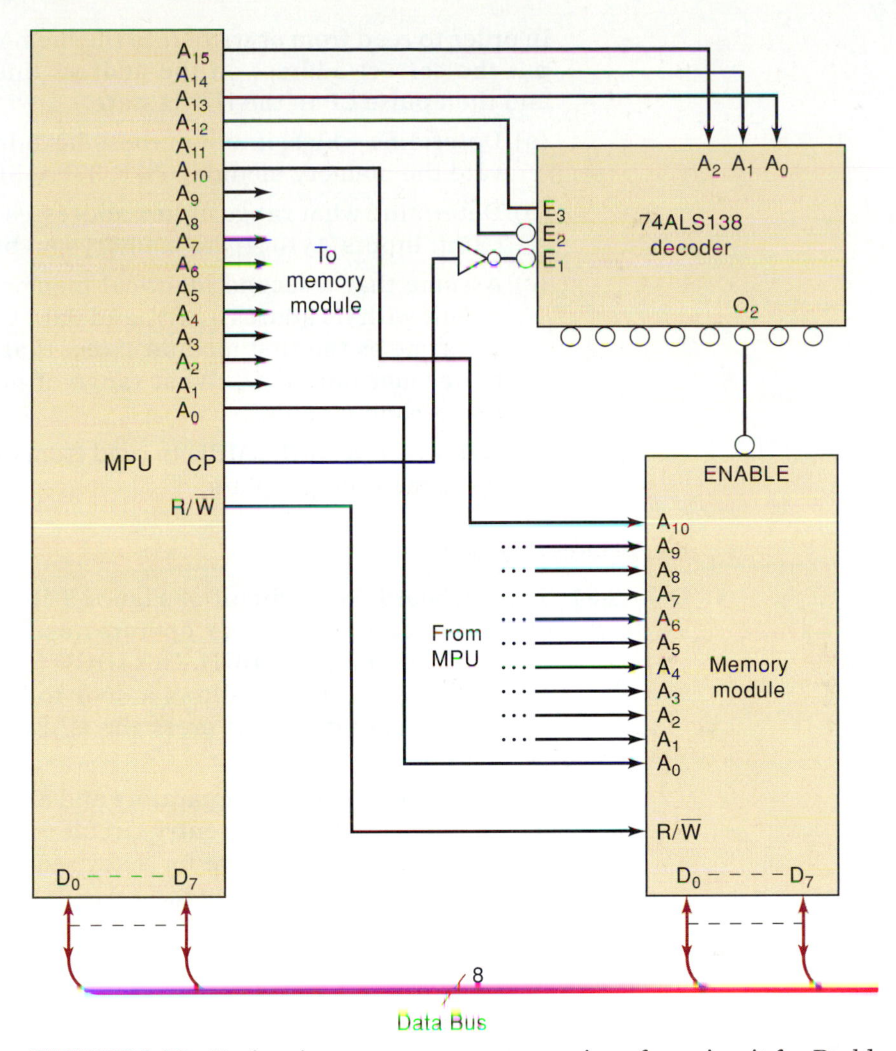

FIGURE 9-83 Basic microprocessor-to-memory interface circuit for Problem 9-63.

more memory ICs (Chapter 12) that can either receive data from the MPU (a WRITE operation) or send data to the MPU (a READ operation). The data are transferred over the eight-line data bus. The MPU's data lines and the memory's I/O data lines are connected to this common bus. For now we will be concerned with how the MPU controls the selection of the memory module for a READ or WRITE operation. The steps involved are as follows:

1. The MPU places the memory address on its address output lines A_{15} to A_0.
2. The MPU generates the $R/\overline{W}$ signal to inform the memory module which operation is to be performed: $R/\overline{W} = 1$ for READ, $R/\overline{W} = 0$ for WRITE.
3. The upper five bits of the MPU address lines are decoded by the 74ALS138, which controls the ENABLE input of the memory module. This ENABLE input must be active in order for the memory module to do a READ or WRITE operation.
4. The other 11 address bits are connected to the memory module, which uses them to select the specific *internal* memory location being accessed by the MPU, provided that ENABLE is active.

In order to read from or write into the memory module, the MPU must put the correct address on the address lines to enable the memory, and then pulse *CP* to the HIGH state.

(a) Determine which, if any, of these hexadecimal addresses will activate the memory module: 607F, 57FA, 5F00.

(b) Determine what range of hex addresses will activate the memory. (*Hint:* Inputs A_0 to A_{10} to memory can be any combination.)

(c) Assume that a second identical memory module is added to the circuit with its address, $R/\overline{W}$, and data I/O lines connected exactly the same as the first module *except* that its ENABLE input is tied to decoder output $\overline{O}_4$. What range of hex addresses will activate this second module?

(d) Is it possible for the MPU to read from or write to both modules at the same time? Explain.

DESIGN PROBLEM

C, D 9-64. The keyboard entry circuit of Figure 9-16 is to be used as part of an electronic digital lock that operates as follows: when activated, an UNLOCK output goes HIGH. This HIGH is used to energize a solenoid that retracts a bolt and allows a door to be opened. To activate UNLOCK, the operator must press the CLEAR key and then enter the correct three-key sequence.

(a) Show how 74HC85 comparators and any other needed logic can be added to the keyboard entry circuit to produce the digital lock operation described above for a key sequence of CLEAR-3-5-8.

(b) Modify the circuit to activate an ALARM output if the operator enters something other than the correct three-key sequence.

SECTIONS 9-15 TO 9-20

H, D, N 9-65.* Write the HDL code for a BCD-to-decimal decoder (the equivalent of a 7442).

H, D, N 9-66. Write the HDL code for a HEX decoder/driver for a 7-segment display. The first 10 characters should appear as shown in Figure 9-7. The last six characters should appear as shown in Figure 9-84.

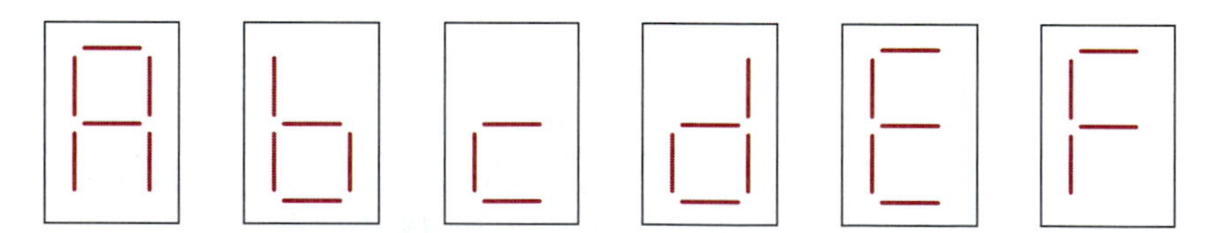

FIGURE 9-84 HEX characters for Problem 9-66.

B, H, N 9-67. Write a low-priority ENCODER description that will always encode the lowest number if two inputs are activated simultaneously.

H, N 9-68. Rewrite the code of the four-bit comparator of Figures 9-66 or 9-67 to make an eight-bit comparator without using macrofunctions.

H, N 9-69. Use HDL to describe a four-bit binary number to a two-digit BCD code converter.

H, N 9-70. Use HDL to describe a three-digit BCD code to eight-bit binary number converter. (Maximum BCD input is 255.)

ANSWERS TO SECTION REVIEW QUESTIONS

SECTION 9-1

1. No 2. The enable input controls whether or not the decoder logic responds to the input binary code. 3. The 7445 has open-collector outputs that can handle up to 30 V and 80 mA. 4. 24 pins: 2 enables, 4 inputs, 16 outputs, V_{CC}, and ground

SECTION 9-2

1. *a*, *b*, *c*, *f*, *g* 2. True

SECTION 9-3

1. LEDs: (a), (e), (f). LCDs: (b), (c), (d), (e) 2. (a) four-bit BCD, (b) seven- or eight-bit ASCII, (c) binary value for pixel intensity

SECTION 9-4

1. An encoder produces an output code corresponding to the activated input. A decoder activates one output corresponding to an applied input code. 2. In a priority encoder, the output code corresponds to the *highest*-numbered input that is activated. 3. Normal BCD = 0110 4. (a) produces a PGT when a key is pressed, (b) converts key actuation to its BCD code, (c) generates bounce-free pulse to trigger the ring counter, (d) form a ring counter that sequentially clocks output registers, (e) store BCD codes generated by key actuations 5. E_1 and E_0 are used for cascading and GS indicates an active input.

SECTION 9-6

1. The binary number at the select inputs determines which data input will pass through to the output. 2. Thirty-two data inputs and five select inputs

SECTION 9-7

1. Parallel-to-serial conversion, data routing, logic-function generation, operations sequencing 2. False; they are applied to the select inputs. 3. Counter

SECTION 9-8

1. A MUX selects one of many input signals to be passed to its output; a DEMUX selects one of many outputs to receive the input signal. 2. True, provided that the decoder has an ENABLE input 3. The LEDs will go on and off in sequence.

SECTION 9-10

1. To provide a means for expanding the compare operations to numbers with more than four bits. 2. $O_{A=B} = 1$; other outputs are 0.

SECTION 9-11

1. A code converter takes input data represented in one type of binary code and converts it to another type of binary code. 2. Three digits can represent decimal values up to 999. To represent 999 in straight binary requires 10 bits.

SECTION 9-12

1. A set of connecting lines to which the inputs and outputs of many different devices can be connected 2. Bus contention occurs when the outputs of more than one device connected to a bus are enabled at the same time. It is prevented by controlling the device enable inputs so that this cannot happen. 3. A condition in which all devices connected to a bus are in the Hi-Z state

SECTION 9-13

1. 1011 2. True 3. 0000

SECTION 9-14

1. Bus contention 2. Floating, Hi-Z 3. Provides tristate low-impedance outputs 4. Reduces the number of IC pins and the number of connections to the data bus 5. See Figure 9-51.

SECTION 9-15

1. They are enable inputs. All must be active for the decoder to work. 2. The CASE and the TABLE 3. The selected signal assignment statement and the CASE

SECTION 9-16

1. The combination input/output pin BI/RBO 2. Common anode. The outputs connect to the cathodes and go LOW to light the segments. 3. The IF/ELSE control structure is evaluated sequentially and gives precedence in the order in which decisions are listed.

SECTION 9-17

1. The don't-care entry in a truth table and the IF/ELSE control structure 2. The IF/ELSE control structure and the conditional signal assignment statement 3. By use of the :TRI primitive and assigning a value to OE 4. By use of the IEEE STD_LOGIC data type that has a possible value of *Z*

SECTION 9-18

1. Inputs: *ch0, ch1, ch2, ch3*; output: *dout*; control inputs (Select): s 2. Input *din*; outputs: *ch0, ch1, ch2, ch3*; control inputs (Select): *s* 3. DEFAULTS 4. ELSE

SECTION 9-19

1. numerical data objects (e.g., INTEGER in VHDL) 2. IF/ ELSE 3. Relational operators (<, >)

SECTION 9-20

1. 10 2. By multiplying by 8 + 2. Shifting the BCD digit three places left multiplies by 8, and shifting the same BCD digit one place left multiplies by 2. Adding these results produced the BCD digit multiplied by 10. 3. VHDL simply uses the * operator to multiply.

CHAPTER 10

DIGITAL SYSTEM PROJECTS USING HDL

OUTLINE

OBJECTIVES

Upon completion of this chapter, you will be able to:

- Analyze the operation of systems made of several components that have been covered earlier in this textbook.
- Describe an entire project with one HDL file.
- Describe the process of hierarchical project management.
- Understand how to break a project into manageable pieces.
- Use MAX+PLUS II or Quartus II software tools to implement a hierarchical modular project.
- Plan ways to test the operation of the circuits you build.

INTRODUCTION

Throughout the first nine chapters of this book, we have explained the fundamental building blocks of digital systems. Now that we have taken out each block and looked it over, we do not want to put them all away and forget them; it is time to build something with the blocks. Some of the examples we have used to demonstrate the operation of individual circuits are really digital systems in their own right, and we have studied how they work. In this chapter, we want to focus more on the building process.

Surveys of graduates show us that most of the professionals in the electrical and computer engineering and technology field have the responsibility of project management. Experience with students has also shown us that the most efficient way to manage a project is not intuitively obvious to everyone, which explains why so many of us end up attending the school of hard knocks (learning through trial and error). This chapter is intended to give you a strategic plan for managing projects while learning about digital systems and the modern tools used to develop them. The principles here are not limited to digital or even electronic projects in general. They could apply to building a house or building your own business. They will definitely improve your success rate and reduce the frustration factor.

Hardware description languages were really created for the purpose of managing large digital systems: for documentation, simulation testing, and the synthesis of working circuits. Likewise, the Altera software tools are specifically designed to work with managing projects that go far beyond the scope of this text. We will describe some of the features of the Altera software packages as we go through the steps of developing these small projects. This concept of modular project development, which was introduced in Chapter 4, will be demonstrated here through a series of examples.

10-1 SMALL-PROJECT MANAGEMENT

The first projects described here are relatively small systems that consist of a small number of building blocks. These projects can be developed in separate modules, but this approach would only add to the complexity. They are small enough that it makes sense to implement the entire project in a single HDL design file. This does not mean, however, that a structured process should *not* be followed to complete the project. In fact, most of the same steps that should be employed in a large modular project are also applicable in these examples. The steps that should be followed are (1) overall definition, (2) strategic planning to break the project into small pieces, (3) synthesis and testing of each piece, and (4) system integration and testing.

Definition

The first step in any project is the thorough definition of its scope. In this step, the following issues should be determined:

- How many bits of data are needed?
- How many devices are controlled by the outputs?
- What are the names of each input and output?
- Are the inputs and outputs active-HIGH or active-LOW?
- What are the speed requirements?
- Do I understand fully how this device should operate?
- What will define successful completion of this project?

From this step should come a complete and thorough description of the overall project's operation, a definition of its inputs and outputs, and complete numeric specifications that define its capabilities and limitations.

Strategic Planning

The second step involves developing a strategy for dividing this overall project into manageable pieces. The requirements of the pieces are:

- A way to test each piece must be developed.
- Each piece must fit together to make up the whole system.
- We must know the nature of all the signals that connect the pieces.
- The exact operation of each block must be thoroughly defined and understood.
- We must have a clear vision of how to make each block work.

This last requirement might seem obvious, but it is amazing how many projects are planned around one central block that involves a not-yet-discovered technical miracle or violates silly little laws like conservation of energy. In this stage, each subsystem (section block) becomes somewhat of a project in and of itself, with the possibility of additional subsystems defined within its boundaries. This is the concept of hierarchical design.

Synthesis and Testing

Each subsystem should be built starting at the simplest level. In the case of a digital system designed using HDL, it means writing pieces of code. It also means developing a plan for testing that code to make sure it meets all the criteria. This is often accomplished through some sort of simulation. When a circuit is simulated on a computer, the designer must create all the different scenarios that will be experienced by the actual circuit and must also know what the proper response to those inputs should be. This testing often takes a great deal of thought and is not an area that should be overlooked. The worst mistake you can make is to conclude that a fundamental block works perfectly, only to find later those few situations where it fails. This predicament often forces you to rethink many of the other blocks, thus nullifying much of your work.

System Integration and Testing

The last step is to put the blocks together and test them as a unit. Blocks are added and tested at each stage until the entire project is working. This area is often trivialized but rarely goes smoothly. Even if you took care of all the details you thought about, there are always the "gotchas" that nobody thought about.

Some aspects of project planning and management go beyond the scope of this text. One is the selection of a hardware platform that will best fit your application. In Chapter 13, we will explore the broad field of digital systems and look specifically at the capabilities and limitations of PLDs in various categories. Another very critical dimension in project management is time. Your boss will give you only a certain amount of time to complete your project, and you must plan your work (and effort) to meet this deadline. We will not be able to cover time management in this text, but as a general rule you will find that most facets of the project will actually take two to three times longer than you think they will when you begin.

REVIEW QUESTIONS

1. Name the steps of project management.
2. At what stage should you decide how to measure success?

10-2 STEPPER MOTOR DRIVER PROJECT

The purpose of this section is to demonstrate a typical application of counters combined with decoding circuits. A digital system often contains a counter that cycles through a specified sequence and whose output states are decoded by a combinational logic circuit, which in turn controls the operation of the system. Many applications also have external inputs that are used to put the system into various modes of operation. This section discusses all these features to control a stepper motor.

In a real project, the first step of definition often involves some research on the part of the project manager. In this section (or project), it is vital that we understand what a stepper motor is and how it works before we try to create a circuit that is supposed to control it. In Section 7-10, we showed you how to design a simple synchronous counter that could be used to drive a stepper motor. The sequence demonstrated in that section is called the *full-step sequence.* As you recall, it involved two flip-flops and their Q and $\overline{Q}$ outputs driving the four coils of the motor. The full-step sequence always has two coils of the stepper motor energized in any state of the sequence and typically causes 15° of shaft rotation per step. Other sequences, however, will also cause a stepper to rotate. If you look at the full-step sequence, you will notice that each state transition involves turning off one coil and simultaneously turning on another coil. For example, look at the first state (1010) in the full-step sequence of Table 10-1. When it switches to the second state in the sequence, *coil 1* is turned off and *coil 0* is turned on. The *half-step sequence* is created by inserting a state with only one coil energized between full steps, as shown in the middle column of Table 10-1. In this sequence, one coil is de-energized before the other is energized. The first state is 1010 and the second state is 1000, meaning that *coil 1* is turned off for one state before *coil 0* is turned on. This intermediate state causes the stepper shaft to rotate half as far (7.5°) as it would in the full-step sequence (15°). The half-step sequence is used when smaller steps are desirable and more steps per revolution are acceptable. As it turns out, the stepper motor will rotate in a manner similar to the full-step sequence (15° per step) if you apply only the sequence of intermediate states with one coil energized at a time. This sequence, called the *wave-drive sequence,* has less torque but operates more smoothly than the full-step sequence at moderate speeds. The wave-drive sequence is shown in the right-hand column of Table 10-1.

TABLE 10-1 Stepper motor coil drive sequences.

Full-Step	Half-Step	Wave-Drive
1010	1010	
	1000	1000
1001	1001	
	0001	0001
0101	0101	
	0100	0100
0110	0110	
	0010	0010

Problem Statement

A microprocessor laboratory needs a universal interface to drive a stepper motor. In order to experiment with microcontrollers driving stepper motors, it would be useful to have a single universal interface IC wired to the stepper motor. This circuit needs to accept any of the typical forms of stepper drive signals from a microcontroller and activate the windings of the motor to make it move in the desired manner. The interface needs to operate in one of four modes: decoded full-step, decoded half-step, decoded wave-drive, or nondecoded direct drive. The mode is selected by controlling the logic levels on the *M1, M0* input pins. In the first three modes, the interface receives just two control bits—a step pulse and a direction control bit—from the microcontroller. Each time it sees a *rising* edge on the step input, the circuit must cause the motor to move one increment of motion clockwise or counterclockwise, depending on the level present on the direction bit. Depending on the mode that the IC is in, the outputs will respond to each step pulse by changing state according to the sequences shown in Table 10-1. The fourth mode of operation of this circuit must allow the microcontroller to control each winding of the motor directly. In this mode, the circuit accepts four control bits from the microcontroller and passes these logic levels directly to its outputs, which are used to energize the stepper coils. The four modes are summarized in Table 10-2.

TABLE 10-2

Mode	M1 M0	Input Signals	Output
0	0 0	Step, direction	Full-step count sequence
1	0 1	Step, direction	Wave-drive count sequence
2	1 0	Step, direction	Half-step count sequence
3	1 1	Four control inputs	Direct drive from control inputs

In modes 0, 1, and 2, the outputs count through the corresponding count sequence on each rising edge of the step input. The direction input determines whether the sequence moves forward or backward through the states in Table 10-1, thus moving the motor clockwise or counterclockwise. From this description, we can make some decisions about the project.

Inputs

step: rising edge trigger

direction: 0 = backward through table, 1 = forward through table

cin0, cin1, cin2, cin3, m1, m0: active-HIGH control inputs

Outputs

cout0, cout1, cout2, cout3: active-HIGH control outputs

Strategic Planning

This project has two key requirements. It requires a sequential counter circuit that will control the outputs in three of the modes. In the last mode, the output does not follow a counter but rather follows the control inputs. While there are many ways to divide this project and still fulfill these requirements, we will choose to have a simple up/down binary counter that responds to the step and direction inputs. A separate combinational logic circuit will translate (decode) the binary count into the appropriate output state, depending on the mode input setting. This circuit will also ignore the counter inputs and pass the control inputs directly to the outputs when the mode is set to 3. The circuit diagram is shown in Figure 10-1.

Breaking this problem into manageable pieces is also fairly straightforward. The first step is to build an up/down counter. This counter should be tested on a simulator using only the direction and step inputs. Next, try to make each decoded sequence work individually with the counter. Then try to get the mode inputs to select one of the decoder sequences and add the direct-drive option (which is fairly trivial). When the circuit can follow the states shown in Table 10-1 in either direction, for each mode sequence, and pass the four *cin* signals directly to *cout* in mode 3, we will be successful.

Synthesis and Testing

The code in Figures 10-2 and 10-3 shows the first stage of development: designing and testing an up/down counter. We will use an intermediate integer variable for the counter value and test it by outputting the count directly to *q*. To test this part of the design, we simply need to make sure it can count up and down through the eight states. Figure 10-4 shows the simulation results. We only need to provide the clock pulses and make up a direction control signal, and the simulator demonstrates the counter's response.

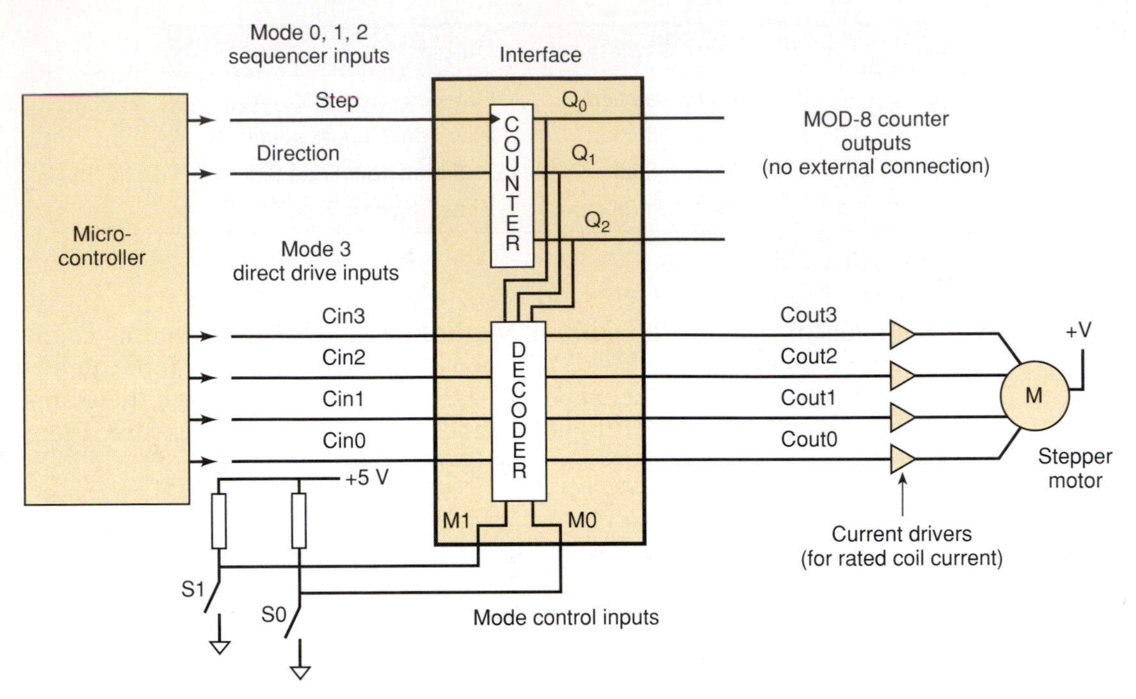

FIGURE 10-1 A universal stepper motor interface circuit.

```
SUBDESIGN  fig10_2
(
   step, dir       :INPUT;
   q[2..0]         :OUTPUT;
)
VARIABLE
count[2..0]       : DFF;

BEGIN
   count[].clk = step;
   IF dir THEN count[].d = count[].q + 1;
   ELSE        count[].d = count[].q - 1;
   END IF;
   q[] = count[].q;
END;
```

FIGURE 10-2 AHDL MOD-8.

```
ENTITY fig10_3 IS
PORT( step, dir     :IN BIT;
      q             :OUT INTEGER RANGE 0 TO 7);
END fig10_3;

ARCHITECTURE vhdl OF fig10_3 IS
BEGIN
   PROCESS (step)
   VARIABLE count     :INTEGER RANGE 0 TO 7;
   BEGIN
      IF (step'EVENT AND step = '1') THEN
         IF dir = '1' THEN count := count + 1;
         ELSE              count := count - 1;
         END IF;
      END IF;
      q <= count;
   END PROCESS;
END vhdl;
```

FIGURE 10-3 VHDL MOD-8.

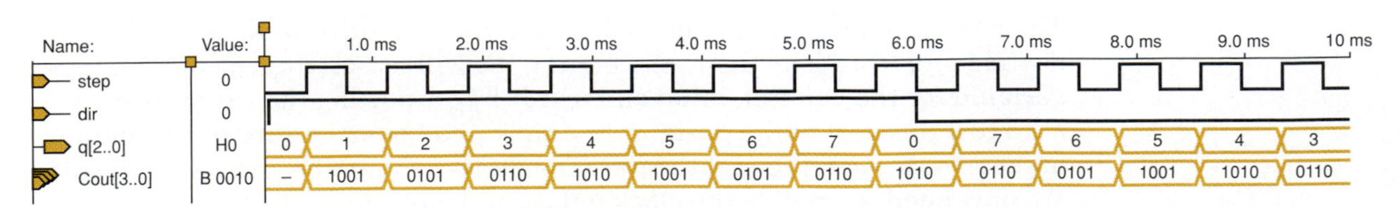

FIGURE 10-4 Simulation testing of a basic MOD-8.

The next step is to add one of the decoded outputs and test it, which will require adding the four-bit *cout* output specification. The *q* output bits of the MOD-8 counter are kept for the sake of continuity. Figure 10-5 shows the AHDL code for this stage of testing, and Figure 10-6 shows the VHDL code for the same stage of testing. Notice that a CASE construct is used to decode the counter and drive the outputs. In the VHDL code, the *cout* outputs have been declared as bit_vector type because we now want to assign binary bit patterns to them. Figure 10-7 shows the simulated test of its operation with enough clock cycles included to test an entire counter cycle up and down.

The other count sequences are simply variations of the code we just tested. It is probably not necessary to test each one independently, so now is a good time to bring in the mode selector inputs (*m*) and direct-drive coil control inputs (*cin*). Notice that the new inputs have been defined in Figures 10-8 (AHDL) and 10-9 (VHDL). Because the mode control has four possible states and we want to do something different for each state, another CASE construct works best. In other words, we have chosen to use a CASE structure to select the mode and a CASE structure within each mode to select the

```
SUBDESIGN  fig10_5
(
   step, dir        :INPUT;
   q[2..0]          :OUTPUT;
   cout[3..0]       :OUTPUT;
)
VARIABLE
   count[2..0]      : DFF;

BEGIN
   count[].clk = step;
   IF dir THEN count[].d = count[].q + 1;
   ELSE        count[].d = count[].q - 1;
   END IF;
   q[] = count[].q;
   CASE count[] IS
      WHEN  B"000"   => cout[] = B"1010";
      WHEN  B"001"   => cout[] = B"1001";
      WHEN  B"010"   => cout[] = B"0101";
      WHEN  B"011"   => cout[] = B"0110";
      WHEN  B"100"   => cout[] = B"1010";
      WHEN  B"101"   => cout[] = B"1001";
      WHEN  B"110"   => cout[] = B"0101";
      WHEN  B"111"   => cout[] = B"0110";
   END CASE;
END;
```

FIGURE 10-5 AHDL full-step sequence decoder.

```
ENTITY fig10_6 IS
PORT (   step, dir :IN BIT;
         q          :OUT INTEGER RANGE 0 TO 7;
         cout       :OUT BIT_VECTOR (3 downto 0));
END fig10_6;

ARCHITECTURE vhdl OF fig10_6 IS
BEGIN
   PROCESS (step)
   VARIABLE count    :INTEGER RANGE 0 TO 7;
   BEGIN
      IF (step'EVENT AND step = '1') THEN
         IF dir = '1' THEN count := count + 1;
         ELSE              count := count - 1;
         END IF;
         q <= count;
      END IF;
         CASE count IS
            WHEN 0    => cout <= B"1010";
            WHEN 1    => cout <= B"1001";
            WHEN 2    => cout <= B"0101";
            WHEN 3    => cout <= B"0110";
            WHEN 4    => cout <= B"1010";
            WHEN 5    => cout <= B"1001";
            WHEN 6    => cout <= B"0101";
            WHEN 7    => cout <= B"0110";
         END CASE;
      END PROCESS;
END vhdl;
```

FIGURE 10-6 VHDL full-step sequence decoder.

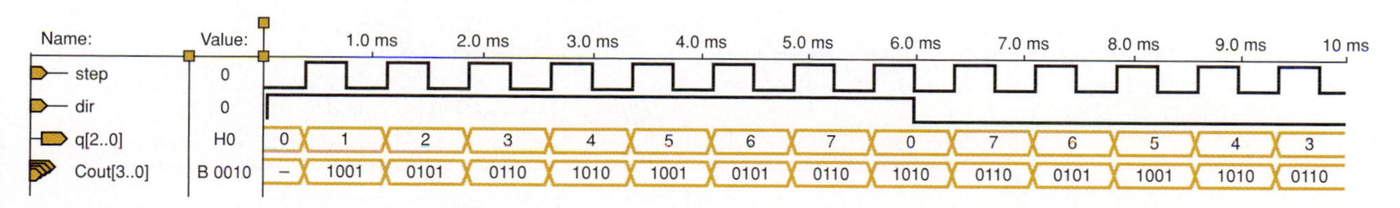

FIGURE 10-7 Simulation testing of decoded sequence.

FIGURE 10-8 AHDL stepper driver.

```
SUBDESIGN  fig10_8
(
   step, dir                 :INPUT;
   m[1..0], cin[3..0]        :INPUT;
   cout[3..0], q[2..0]       :OUTPUT;
)
VARIABLE
   count[2..0]      : DFF;
BEGIN
   count[].clk = step;
   IF dir THEN count[].d = count[].q + 1;
   ELSE        count[].d = count[].q - 1;
   END IF;
   q[] = count[].q;
   CASE m[] IS
   WHEN 0 =>
         CASE count[] IS           -- FULL STEP
         WHEN  B"000"   => cout[] = B"1010";
         WHEN  B"001"   => cout[] = B"1001";
         WHEN  B"010"   => cout[] = B"0101";
         WHEN  B"011"   => cout[] = B"0110";
         WHEN  B"100"   => cout[] = B"1010";
         WHEN  B"101"   => cout[] = B"1001";
         WHEN  B"110"   => cout[] = B"0101";
         WHEN  B"111"   => cout[] = B"0110";
         END CASE;
   WHEN 1 =>
         CASE count[] IS           -- WAVE DRIVE
         WHEN  B"000"   => cout[] = B"1000";
         WHEN  B"001"   => cout[] = B"0001";
         WHEN  B"010"   => cout[] = B"0100";
         WHEN  B"011"   => cout[] = B"0010";
         WHEN  B"100"   => cout[] = B"1000";
         WHEN  B"101"   => cout[] = B"0001";
         WHEN  B"110"   => cout[] = B"0100";
         WHEN  B"111"   => cout[] = B"0010";
         END CASE;
   WHEN 2 =>
         CASE count[] IS           -- HALF STEP
         WHEN  B"000"   => cout[] = B"1010";
         WHEN  B"001"   => cout[] = B"1000";
         WHEN  B"010"   => cout[] = B"1001";
         WHEN  B"011"   => cout[] = B"0001";
         WHEN  B"100"   => cout[] = B"0101";
         WHEN  B"101"   => cout[] = B"0100";
         WHEN  B"110"   => cout[] = B"0110";
         WHEN  B"111"   => cout[] = B"0010";
         END CASE;
   WHEN 3 =>   cout[] = cin[];   -- Direct Drive
   END CASE;
END;
```

FIGURE 10-9 VHDL stepper driver.

```
ENTITY fig10_9 IS
PORT (   step, dir     :IN BIT;
         m             :IN BIT_VECTOR (1 DOWNTO 0);
         cin           :IN BIT_VECTOR (3 DOWNTO 0);
         q             :OUT INTEGER RANGE 0 TO 7;
         cout          :OUT BIT_VECTOR (3 DOWNTO 0));
END fig10_9;

ARCHITECTURE vhdl OF fig10_9 IS
BEGIN
   PROCESS (step)
   VARIABLE count     :INTEGER RANGE 0 TO 7;
   BEGIN
      IF (step'EVENT AND step = '1') THEN
         IF dir = '1' THEN count := count + 1;
         ELSE              count := count - 1;
         END IF;
      END IF;
      q <= count;
   CASE m IS
      WHEN "00" =>                 -- FULL STEP
         CASE count IS
            WHEN 0   => cout <= "1010";
            WHEN 1   => cout <= "1001";
            WHEN 2   => cout <= "0101";
            WHEN 3   => cout <= "0110";
            WHEN 4   => cout <= "1010";
            WHEN 5   => cout <= "1001";
            WHEN 6   => cout <= "0101";
            WHEN 7   => cout <= "0110";
         END CASE;
      WHEN "01" =>                 -- WAVE DRIVE
         CASE count IS
            WHEN 0   => cout <= "1000";
            WHEN 1   => cout <= "0001";
            WHEN 2   => cout <= "0100";
            WHEN 3   => cout <= "0010";
            WHEN 4   => cout <= "1000";
            WHEN 5   => cout <= "0001";
            WHEN 6   => cout <= "0100";
            WHEN 7   => cout <= "0010";
         END CASE;
      WHEN "10" =>                 -- HALF STEP
         CASE count IS
            WHEN 0   => cout <= "1010";
            WHEN 1   => cout <= "1000";
            WHEN 2   => cout <= "1001";
            WHEN 3   => cout <= "0001";
            WHEN 4   => cout <= "0101";
            WHEN 5   => cout <= "0100";
            WHEN 6   => cout <= "0110";
            WHEN 7   => cout <= "0010";
         END CASE;
      WHEN "11" =>   cout <= cin;--Direct Drive
   END CASE;
   END PROCESS;
END vhdl;;
```

proper output. Using one construct inside another is known as **nesting**. The use of indentation is very important to show the structure and logic of the code, especially when nesting is used.

The simulations of Figure 10-10 verify that the circuit seems to be working properly. Figure 10-10(a) shows each state decoding in mode 0 (full-step) and completing the cycle in both directions. Notice that after the mode (*m*) changes to 01_2, the output (*cout*) is decoded as the wave-drive sequence. Figure 10-10(b) shows the wave-drive (mode 1) sequence in both directions and then changes the mode to 10_2, resulting in the half-step sequence being decoded from the MOD-8 counter. Finally, Figure 10-10(c) shows the half-step mode cycling up and starting back down. It then switches to mode 3 (direct-drive) at 7.5 ms, showing that the data on *cin* is transferred asynchronously to the outputs. Notice that the values chosen for *cin* ensure that each bit can go HIGH and LOW.

Final integration and testing should involve more than just simulation. A real stepper motor and current driver should be connected to the circuit and tested. In this case, the step rate that the simulation used would probably be faster than the actual stepper motor could handle and would need to be slowed down for a real hardware functional test.

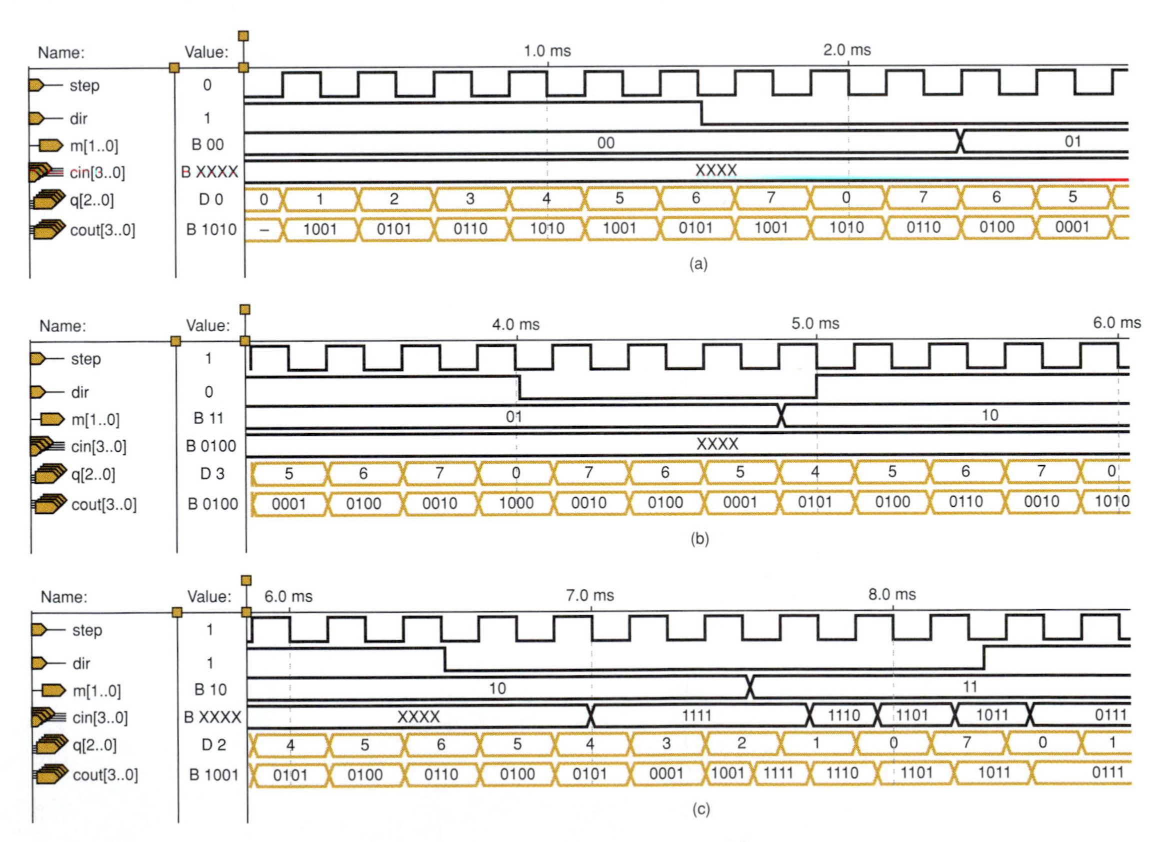

FIGURE 10-10 Simulation testing of the complete stepper driver.

REVIEW QUESTIONS

1. What are the four modes of operation for this stepper motor driver?
2. What are the inputs for the direct-drive mode?
3. What are the inputs for the wave-drive mode?
4. How many states are in the half-step sequence?

10-3 KEYPAD ENCODER PROJECT

Another important skill that we are trying to reinforce is circuit analysis. That may sound like something out of an analog textbook, but we really need to be able to analyze and understand how existing digital circuits operate. In this section, we present a circuit and analyze how it operates. Then we use the skills we have acquired to redesign this circuit and write the code for it in HDL.

Problem Analysis

To reinforce the encoding concepts of Chapter 9, we present a very useful digital circuit that encodes a hexadecimal (16-key) keypad into a four-bit binary output. Encoders such as this generally have a strobe output that indicates when someone presses and releases a key. Because keypads are often interfaced to a microcomputer's bus system, the encoded outputs should have tristate enables. Figure 10-11 shows the block diagram of the keypad encoder.

The priority encoder method shown in Chapter 9, Figure 9-15, is effective for small keypads. However, large keyboards such as those found on personal computers must use a different technique. In these keyboards, each key is not an independent switch to V_{CC} or ground. Instead, each key switch is used

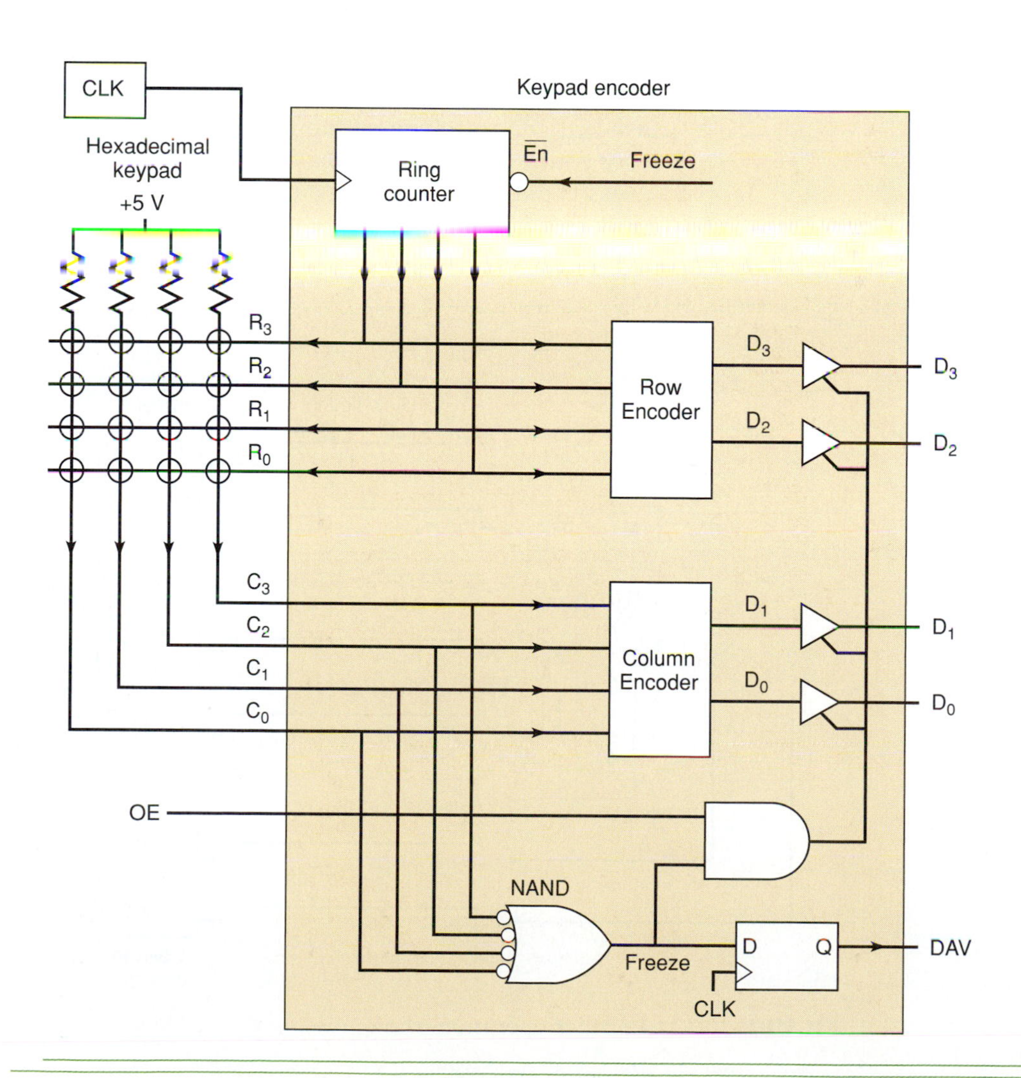

FIGURE 10-11 Keypad encoder block diagram.

to connect a row to a column in the keyboard matrix. When keys are not pressed, there are no connections between the rows and columns. The trick of knowing which key is pressed is accomplished by activating (pulling LOW) one row at a time and then checking to see if any of the columns have gone LOW. If one of the columns has a LOW on it, then the key being pressed is at the intersection of the activated row and the column that is currently LOW. If no columns are LOW, we know that no keys in the activated row are being pressed and we can check the next row by pulling it LOW. Sequentially activating rows is called *scanning* the keyboard. The advantage of this method is the reduction in connections to the keypad. In this case, 16 keys can be encoded using eight inputs/outputs.

Each key represents a unique combination of a row number and a column number. By strategically numbering the rows and columns, we can combine the binary row and column numbers to create the binary value of the hexadecimal keys as shown in Figure 10-12. In this figure, row 1 (01_2) is pulled LOW and the data on the column encoder is 10_2 so the button at row 1, column 2 is evidently pressed. The NAND gate in Figure 10-11 is used to determine if any column is LOW, indicating that a key is pressed in the currently active row. The output of this gate is named *FREEZE* because when a key is pressed, we want to freeze the ring counter and quit scanning until the key is released. As the encoders go through their propagation delay and the tristate buffers become enabled, the data outputs are in a transient state. On the next rising edge of the clock, the D flip-flop will transfer a HIGH from *FREEZE* to the *DAV* output, indicating that a key is being pressed and the valid data is available.

A shift register counter (ring counter), as we studied in Chapter 7, is used to generate the sequential scan of the four rows. The count sequence uses four states, each state having a different bit pulled LOW. When a key press is detected, the ring counter must hold in its current state (freeze) until the key is released. Figure 10-13 shows the state transition diagram. Each state of this counter must be encoded to generate a two-bit binary row num-

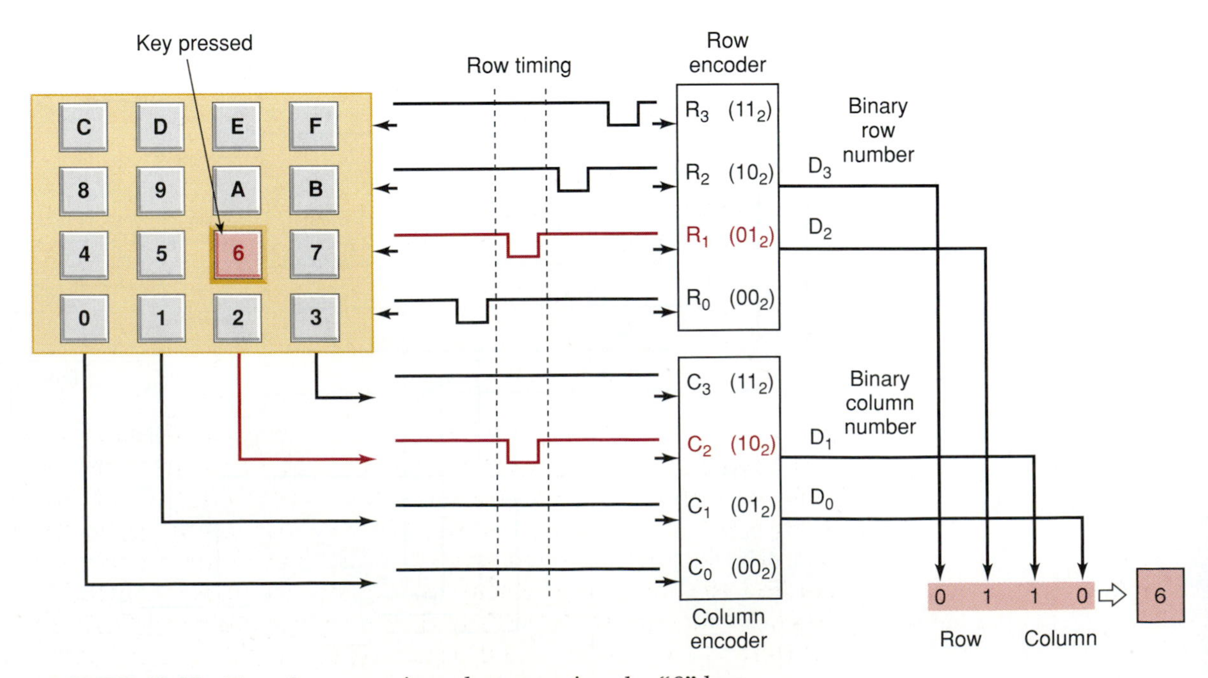

FIGURE 10-12 Encoder operation when pressing the "6" key.

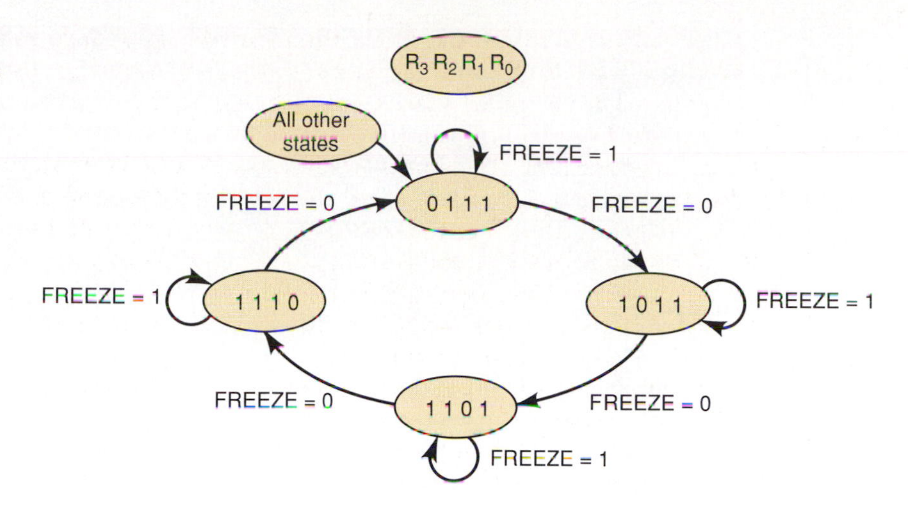

FIGURE 10-13 Row drive ring counter state diagram.

ber. Each column value must also be encoded to generate a two-bit binary column number. The system will require the following inputs and outputs.

4	Row drive outputs	R_0–R_3
4	Column read inputs	C_0–C_3
4	Encoded data outputs	D_0–D_3
1	Data available strobe output	*DAV*
1	Tristate enable input	*OE*
1	Clock input	*CLK*

Strategic Planning

This circuit is already structured so that we can easily write pieces of HDL code to emulate each section of the system. The major blocks are as follows:

A ring counter with active-LOW outputs.

Two encoders for the row and column numbers.

Key-press detection and tristate enable circuits.

Because these circuits have been explored in previous chapters, we will not show the development and testing of each block here. The solutions that follow jump directly to the integration and testing phase of the project.

AHDL SOLUTION

The inputs and outputs (see Figure 10-14) are defined on lines 3–8 and follow the description obtained from analyzing the schematic. The VARIABLE section defines several features of this encoder circuit. The freeze bit detects when a key is pressed. The data node is used to combine the row and column encoder data. The *ts* bit array (line 13) represents a tristate buffer, as we studied in Chapter 9. Recall that each bit of this buffer has an input, (*ts[ ].IN*), an output (*ts.OUT*), and an output enable (*ts[ ].OE*). The *data_avail* bit (line 14) represents a D flip-flop with inputs *data_avail.CLK, data_avail.D,* and output *data_avail.Q*.

AHDL

```
1   SUBDESIGN fig10_14
2   (
3      clk            :INPUT;
4      col[3..0]      :INPUT;
5      oe             :INPUT;    --tristate output enable
6      row[3..0]      :OUTPUT;
7      d[3..0]        :OUTPUT;
8      dav            :OUTPUT;   --data available
9   )
10  VARIABLE
11  freeze            :NODE;
12  data[3..0]        :NODE;
13  ts[3..0]          :TRI;
14  data_avail        :DFF;
15  ring: MACHINE OF BITS (row[3..0])
16  WITH STATES (s1 = B"1110", s2 = B"1101", s3 = B"1011", s4 = B"0111",
17                       % s = ring states %
18               f1 = B"0001", f2 = B"0010", f3 = B"0011", f4 = B"0100",
19               f5 = B"0101", f6 = B"0110", f7 = B"1000", f8 = B"1001",
20               f9 = B"1010", fa = B"1100", fb = B"1111", fc = B"0000");
21                     % f = unused states --> self-correcting design %
22  BEGIN
23     ring.CLK = clk;
24     ring.ENA = !freeze;
25     data_avail.CLK = clk;
26     data_avail.D = freeze;
27     dav = data_avail.Q;
28     ts[].OE = oe & freeze;
29     ts[].IN = data[];
30     d[] = ts[].OUT;
31
32     CASE ring IS
33        WHEN s1 =>   ring = s2;    data[3..2] = B"00";
34        WHEN s2 =>   ring = s3;    data[3..2] = B"01";
35        WHEN s3 =>   ring = s4;    data[3..2] = B"10";
36        WHEN s4 =>   ring = s1;    data[3..2] = B"11";
37        WHEN OTHERS => ring = s1;
38     END CASE;
39
40     CASE col[] IS
41        WHEN B"1110" =>   data[1..0] = B"00";     freeze = VCC;
42        WHEN B"1101" =>   data[1..0] = B"01";     freeze = VCC;
43        WHEN B"1011" =>   data[1..0] = B"10";     freeze = VCC;
44        WHEN B"0111" =>   data[1..0] = B"11";     freeze = VCC;
45        WHEN OTHERS  =>   data[1..0] = B"00";     freeze = GND;
46     END CASE;
47  END;
```

FIGURE 10-14 AHDL scanning keypad encoder.

Lines 15–20 demonstrate a powerful feature of AHDL that allows us to define a state machine, with each state made up of the bit pattern we need. On line 15, the name *ring* was given to this state machine because it acts like a ring counter. The bits that make up this ring counter machine are the four row bits that were defined on line 6. These states are labeled *s1–s4* and have their bit patterns assigned to them so that one bit of the four is LOW for each state, like an active-LOW ring counter. The other twelve states are specified by an arbitrary label that starts with *f* to indicate they are not valid states. Lines 23 through 30 essentially connect all the components as shown in the circuit drawing of Figure 10-11. Both the ring count sequence and the encoding of the row value are described on lines 32–38. For each PRESENT state value of *ring*, the NEXT state is defined as well as the proper output of the row encoder (*data[3..2]*). Line 37 ensures that this counter will self-start by sending it to *s1* from any state other than *s1–s4*. The encoding of the column value is described on lines 40–46. Notice that the generation of the *freeze* signal in this design does not follow the diagram of Figure 10-11 exactly. In this design, rather than NANDing the columns, the CASE structure activates *freeze* only when one (and only one) column is LOW. Thus, if multiple keys in the same row were pressed, the encoder would not recognize any as a valid key press and would not activate *dav*.

VHDL SOLUTION

Compare the VHDL description in Figure 10-15 with the circuit drawing of Figure 10-11. The inputs and outputs are defined on lines 5–9 and follow the description obtained from analyzing the schematic. Two SIGNALs are defined on lines 13 and 14 for this design. The *freeze* bit detects when a key is pressed. The *data* signal is used to combine the row and column encoder data to make a four-bit value representing the key that was pressed. The ring counter is implemented using a PROCESS that responds to the *clk* input. Line 26 ensures that this counter will self-start by sending it to state "1110" from any state other than those in the *ring* sequence. Notice that on line 20, the status of *freeze* is checked before a CASE is used to assign a NEXT state to *ring*. This is the way the count enable is implemented in this design. On line 29, the data available output (*dav*) is updated synchronously with the value of *freeze*. It is synchronous because it is within the IF structure (lines 19–30) that detects the active clock edge. The remaining statements (lines 31–52) do not depend on the active clock edge but describe what the circuit will do on either edge of the clock.

The encoding of the row value is described on lines 33–39. For each PRESENT state value of *ring*, the output of the row encoder *data*(3 DOWNTO 2) is defined. The encoding of the column value is described on lines 41–47. Notice that the generation of the *freeze* signal in this design does not follow the diagram of Figure 10-11 exactly. In this design, rather than NANDing the columns, the CASE structure activates *freeze* only when one (and only one) column is LOW. Thus, if multiple keys in the same row were pressed, the encoder would not recognize any as a valid key press and would not activate *dav*.

FIGURE 10-15 VHDL scanning keypad encoder.

```
1   LIBRARY ieee;
2   USE ieee.std_logic_1164.all;
3
4   ENTITY fig10_15 IS
5   PORT (   clk          :IN STD_LOGIC;
6            col          :IN STD_LOGIC_VECTOR (3 DOWNTO 0);
7            row          :OUT STD_LOGIC_VECTOR (3 DOWNTO 0);
8            d            :OUT STD_LOGIC_VECTOR (3 DOWNTO 0);
9            dav          :OUT STD_LOGIC                       );
10  END fig10_15;
11
12  ARCHITECTURE vhdl OF fig10_15 IS
13  SIGNAL freeze         :STD_LOGIC;
14  SIGNAL data           :STD_LOGIC_VECTOR (3 DOWNTO 0);
15  BEGIN
16     PROCESS (clk)
17     VARIABLE ring      :STD_LOGIC_VECTOR (3 DOWNTO 0);
18     BEGIN
19        IF (clk'EVENT AND clk = '1')  THEN
20           IF freeze = '0'   THEN
21              CASE ring IS
22                 WHEN "1110" => ring := "1101";
23                 WHEN "1101" => ring := "1011";
24                 WHEN "1011" => ring := "0111";
25                 WHEN "0111" => ring := "1110";
26                 WHEN OTHERS => ring := "1110";
27              END CASE;
28           END IF;
29           dav <= freeze;
30        END IF;
31        row <= ring;
32
33        CASE ring IS
34           WHEN "1110" => data(3 DOWNTO 2) <= "00";
35           WHEN "1101" => data(3 DOWNTO 2) <= "01";
36           WHEN "1011" => data(3 DOWNTO 2) <= "10";
37           WHEN "0111" => data(3 DOWNTO 2) <= "11";
38           WHEN OTHERS => data(3 DOWNTO 2) <= "00";
39        END CASE;
40
41        CASE col IS
42           WHEN "1110" => data(1 DOWNTO 0) <= "00";      freeze <= '1';
43           WHEN "1101" => data(1 DOWNTO 0) <= "01";      freeze <= '1';
44           WHEN "1011" => data(1 DOWNTO 0) <= "10";      freeze <= '1';
45           WHEN "0111" => data(1 DOWNTO 0) <= "11";      freeze <= '1';
46           WHEN OTHERS => data(1 DOWNTO 0) <= "00";      freeze <= '0';
47        END CASE;
48
49        IF freeze = '1'   THEN  d <= data;
50        ELSE                    d <= "ZZZZ";
51        END IF;
52     END PROCESS;
53  END vhdl;
```

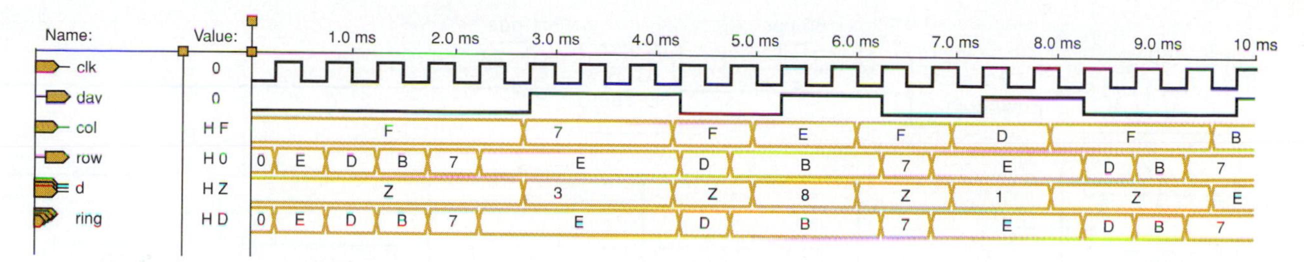

FIGURE 10-16 Simulation of the scanning keypad encoder.

The simulation of the project is shown in Figure 10-16. The column values (*col*) are entered by the designer as a test input that simulates the value being read from the columns of the keypad as the rows are being scanned. As long as all columns are HIGH (i.e., the hex value F is on *col*), the *ring* counter is enabled and counting, *dav* is LOW, and the *d* outputs are in the Hi-Z state. Just before the 3.0-ms mark, a 7 is simulated as a *col* input, which means that one of the columns went LOW. This simulates a key being detected in the most significant column (C3) of the keypad matrix. Notice that as a result of the column going LOW, on the next active (rising) clock edge, the *dav* line goes HIGH and the ring counter does not change state. It is disabled from going to its NEXT state as long as the key is pressed. At this point, the *row* value is E hex (1110_2), which means that the least significant row (R0) is being pulled LOW by the ring counter. The row encoder translates this into the binary row number (00). The key located at the intersection of the least significant row (00_2) and the most significant column (11_2) is the 3 key (see Figure 10-12). At this point, the *d* outputs hold the encoded key value of 3 (0011_2). Just after the 4-ms mark, the simulation imitates the release of the key by changing the column value back to F hex, which causes the *d* output to go into its Hi-Z state. On the next rising clock edge, the *dav* line goes LOW and the ring counter resumes its count sequence.

REVIEW QUESTIONS

1. How many rows on the scanned keyboard are activated at any point in time?
2. If two keys in the same column are pressed simultaneously, which key will be encoded?
3. What is the purpose of the D flip-flop on the DAV pin?
4. Will the time between the key being pressed and DAV going HIGH always be the same?
5. When are the data output pins in the Hi-Z state?

10-4 DIGITAL CLOCK PROJECT

One of the most common applications of counters is the digital clock—a time clock that displays the time of day in hours, minutes, and sometimes seconds. In order to construct an accurate digital clock, a closely controlled basic clock frequency is required. For battery-operated digital clocks or watches, the basic frequency is normally obtained from a quartz-crystal oscillator. Digital clocks operated from the ac power line can use the *60-Hz* power frequency as the basic clock frequency. In either case, the basic frequency must

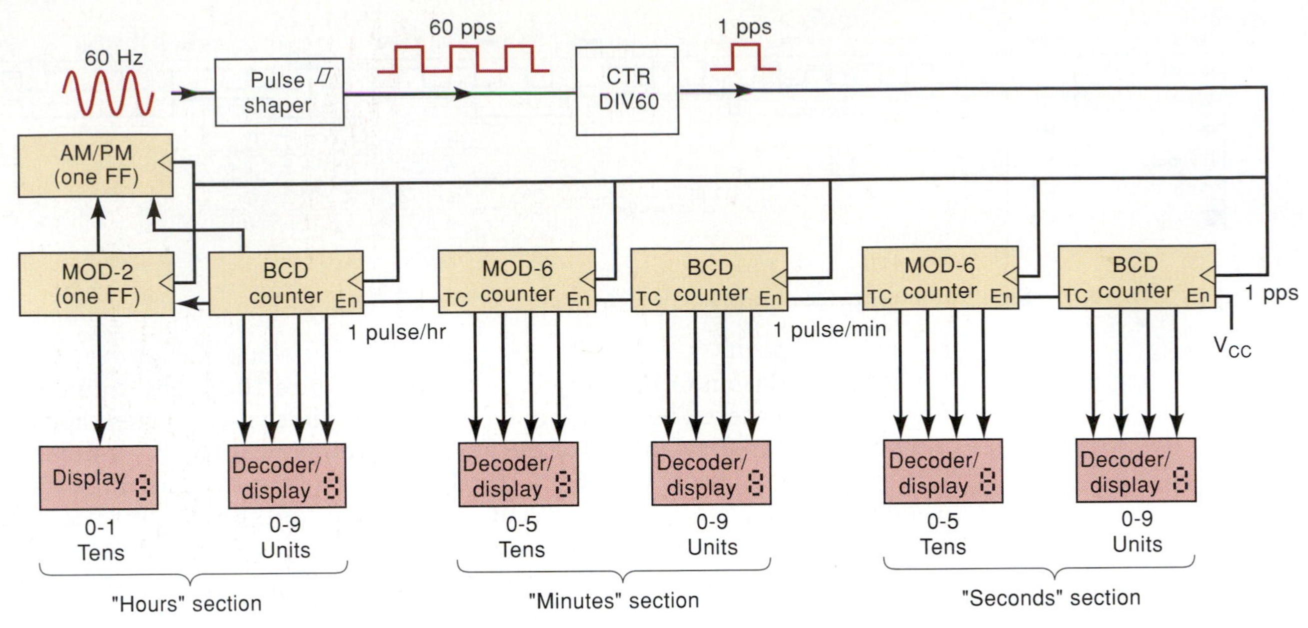

FIGURE 10-17 Block diagram for a digital clock.

be divided to a frequency of 1 Hz or 1 pulse per second (pps). Figure 10-17 shows the basic block diagram for a digital clock operating from 60 Hz.

The *60-Hz* signal is sent through a Schmitt-trigger circuit to produce square pulses at the rate of 60 pps. This 60-pps waveform is fed into a MOD-60 counter that is used to divide the 60 pps down to 1 pps. The 1-pps signal is used as a synchronous clock for all of the counter stages, which are synchronously cascaded. The first stage is the SECONDS section, which is used to count and display seconds from 0 through 9. The *BCD* counter advances one count per second. When this stage reaches 9 seconds, the BCD counter activates its terminal count output (*tc*), and on the next active clock edge, it recycles to 0. The BCD terminal count enables the MOD-6 counter and causes it to advance by one count at the same time that the BCD counter recycles. This process continues for 59 seconds, at which point the MOD-6 counter is at the 101 (5) count and the BCD counter is at 1001 (9) so that the display reads 59 s and *tc* of the MOD-6 is HIGH. The next pulse recycles the BCD counter and the MOD-6 counter to zero (remember, the MOD-6 counts from 0 through 5).

The *tc* output of the MOD-6 counter in the SECONDS section has a frequency of 1 pulse per minute (i.e, the MOD-6 recycles every 60 s). This signal is fed to the MINUTES section, which counts and displays minutes from 0 through 59. The MINUTES section is identical to the SECONDS section and operates in exactly the same manner.

The *tc* output of the MOD-6 counter in the MINUTES section has a frequency of 1 pulse per hour (i.e., the MOD-6 recycles every 60 min). This signal is fed to the HOURS section, which counts and displays hours from 1 through 12. This HOURS section is different from the SECONDS and MINUTES sections because it never goes to the 0 state. The circuitry in this section is sufficiently unusual to warrant a closer investigation.

Figure 10-18 shows the detailed circuitry contained in the HOURS section. It includes a BCD counter to count units of hours and a single FF (MOD-2) to count tens of hours. The BCD counter is a 74160, which has two active-HIGH inputs, ENT and ENP, that are ANDed together internally to enable the

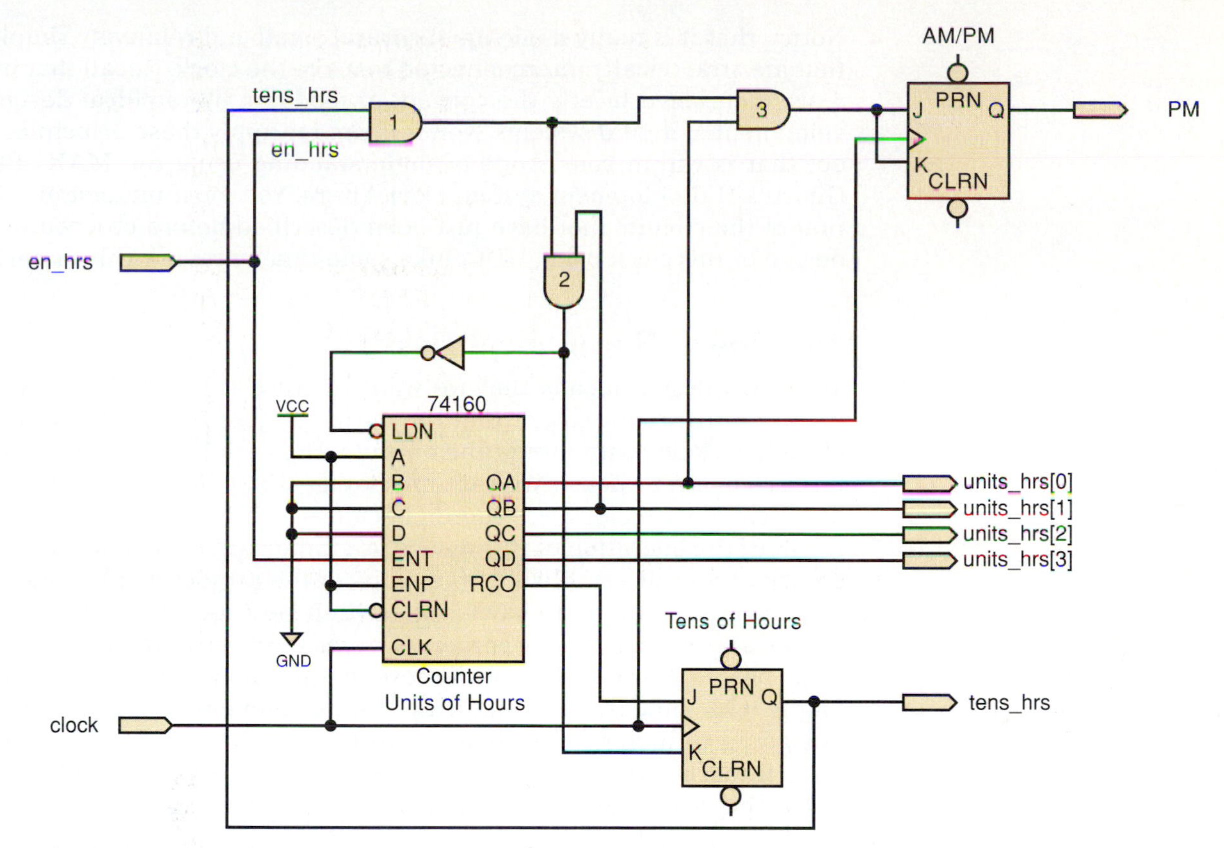

FIGURE 10-18 Detailed circuitry for the HOURS section.

count. The ENT input also enables the active-HIGH ripple carry out (RCO) that detects the BCD terminal count of 9. The ENT input and RCO output can therefore be used for synchronous counter cascading. The ENP input is tied HIGH so that the counter will increment whenever ENT is HIGH.

The hours counter is enabled by the minutes and seconds stages for only one clock pulse every hour. When this condition occurs, ENT is HIGH, which means that the minutes:seconds stages are at 59:59. For example, at 9:59:59, the tens of hours flip-flop holds a 0, the 74160 holds 1001_2 (9), and the RCO output is HIGH, putting the tens of hours flip-flop in the SET mode. The two display digits for the hours show 09. On the next rising clock edge, the BCD counter advances to its natural NEXT state of 0000_2, RCO goes LOW, and the tens of hours flip-flop advances to 1 so that the hours display digits now show 10.

When it is 11:59:59, AND gate 1 detects that the tens of hours is 1 and the enable input is active (previous stages are at 59:59). AND gate 3 combines the conditions of AND gate 1 and the condition that the BCD counter is in the state 0001_2. The output of AND gate 3 will be HIGH only at 11:59:59 in the hours count sequence. On the next clock pulse, the AM/PM flip-flop toggles, indicating noon (HIGH) or midnight (LOW). At the same time, the BCD counter advances to 2 and the minutes:seconds stages roll over to 00:00, resulting in a BCD display of 12:00:00. At 12:59:59, AND gate 1 detects that the tens digit is 1 and it is time to advance the hours. AND gate 2 detects that the BCD counter is at 2. The output of AND gate 2 prepares to do two tasks on the next clock edge: reset the tens of hours flip-flop, and load the 74160 counter with the value 0001_2. After the next clock pulse, it is 01:00:00 o'clock.

The operation of counter circuits should make sense now, and you should have a good grasp on how you can connect MSI chips to make this digital clock.

Notice that it is really made up of several small and relatively simple circuits that are strategically interconnected to make the clock. Recall that in Chapter 4, we mentioned briefly the concept of modular, hierarchical design and development of digital systems. Now it is time to apply these principles to a project that is within your scope of understanding using the MAX+PLUS II or Quartus II development system from Altera. You must understand the operation of the circuits that have just been described before proceeding with the design of this clock using HDL. Take some time to review this material.

Top-Down Hierarchical Design

Top-down design means that we want to start at the highest level of complexity in the hierarchy, or that the entire project is considered to exist in a closed, dark box with inputs and outputs. The details regarding what is in the box are not yet known. We can only say at this point how we want it to behave. The digital clock was chosen because everyone is familiar with the end result of the operation of this device. An important aspect of this stage of the design process is establishing the scope of the project. For example, this digital clock is not going to have a way to set the time, set an alarm time, shut off the alarm, snooze, or incorporate other features that you may find on the clock beside your bed. To add all these features now would only clutter the example with unnecessary complexity for our immediate purpose. We are also not going to include the signal conditioning that transforms a 60-Hz sine wave into a 60-pulse-per-second digital waveform, or the decoder/display circuits. The project we are tackling has the following specifications:

Inputs: 60 pps CMOS compatible waveform (accuracy dependent on line frequency)

Outputs: BCD Hours: 1 bit TENS 4 bits UNITS
BCD Minutes: 3 bits TENS 4 bits UNITS
BCD Seconds: 3 bits TENS 4 bits UNITS
PM indicator

Minutes and Seconds sequence: BCD MOD 60
00–59 (decimal representation of BCD)
Hours sequence BCD MOD 12
01–12 (decimal representation of BCD)
Overall range of display
01:00:00–12:59:59

AM/PM indicator toggles at 12:00:00

A **hierarchy** is a group of objects arranged in rank order of magnitude, importance, or complexity. A block diagram of the overall project (highest level of the hierarchy) is shown in Figure 10-19. Notice that there are four bits for each of the BCD units outputs and only three bits for each of the

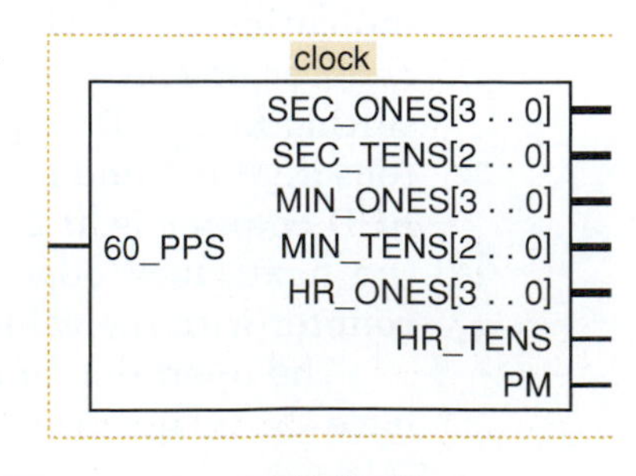

FIGURE 10-19 The top level block of the hierarchy.

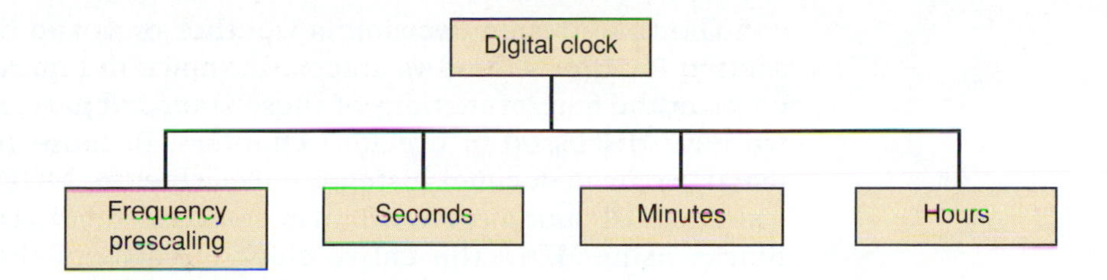

FIGURE 10-20 The section level of the hierarchy.

minute and second BCD tens outputs. Because the most significant BCD digit for the tens place is 5 (101_2), only three bits are needed. Notice also that the tens place for the hours (HR_TENS) is only one bit. It will never have a value other than 0 or 1.

The next phase is to break this problem into more manageable sections. First, we need to take the 60-pps input and transform it into a 1-pulse-per-second timing signal. A circuit that divides a reference frequency to a rate required by the system is called a **prescaler**. Next, it makes sense to have individual sections for a seconds counter, minutes counter, and hours counter. So far, the hierarchy diagram looks like Figure 10-20, which shows the project broken into four subsections.

The entire purpose of the frequency prescaler section is to divide the 60-pps input to a frequency of one pulse every second. This requires a MOD-60 counter, and the sequence of the count does not really matter. In this example, the minutes and seconds sections both require MOD-60 counters that count from 00–59 in BCD. Looking for similarities like this is very important in the design process. In this case, we can use the exact same circuit design to implement the frequency prescaler, the minutes counter, and the seconds counters.

A MOD-60 BCD counter can be made quite easily from a MOD-10 (decade) counter cascaded to a MOD-6 BCD counter, as we saw in the diagram of Figure 10-17. This means that inside each of these MOD-60 blocks, we would find a diagram similar to Figure 10-21. The hierarchy of the project now appears as shown in Figure 10-22.

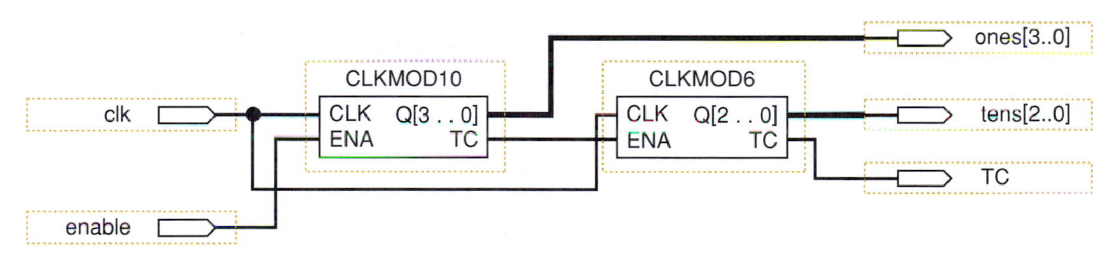

FIGURE 10-21 The blocks inside the MOD-60 section.

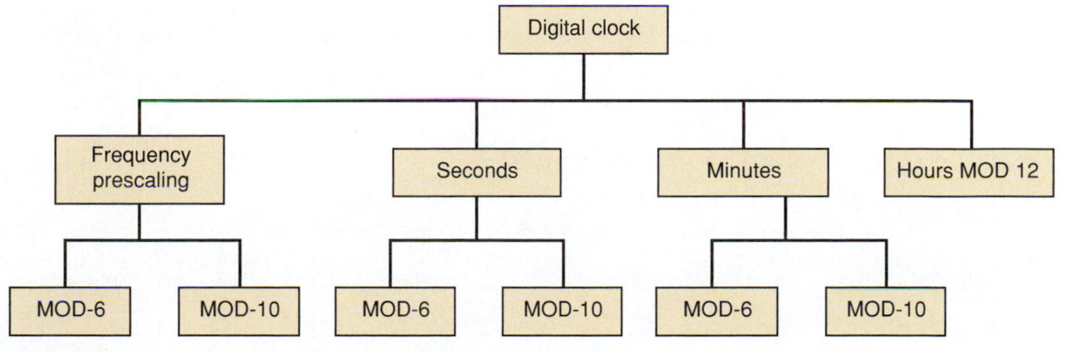

FIGURE 10-22 The complete hierarchy of the clock project.

The final design decision is whether or not to break down the MOD-12 section for Hours into two stages, as shown in Figure 10-18. One option is to connect the macrofunctions of these standard parts from the HDL library, as we have discussed in previous chapters. Because this circuit is rather unusual, we have decided instead to describe the MOD-12 hours counter using a single HDL module. We will also describe the MOD-6 and MOD-10 building blocks using HDL. The entire clock circuit can then be built using these three basic circuit descriptions. Of course, even these blocks can be broken down into smaller flip-flop blocks and designed using the schematic entry, but it will be much easier using HDL at this level.

Building the Blocks from the Bottom Up

Each of the basic blocks are presented here in both AHDL and VHDL. We present the MOD-6 as a simple modification of the MOD-5 synchronous counter descriptions presented earlier in Chapter 7 (see Figures 7-39 and 7-40). Then we modify this code further to create the MOD-10 counter and finally design the MOD-12 Hours counter from the ground up. We construct the entire clock from these three basic blocks.

AHDL

MOD-6 COUNTER AHDL

The only additional features that this design needs that are not covered in Figure 7-39 are the count *enable* input and terminal count (*tc*) output shown in Figure 10-23. Notice that the extra input (*enable*, line 3) and output (*tc*, line 4) are included in the I/O definition. A new line (line 11) in the architecture description tests *enable* before deciding how to update the value of

```
1   SUBDESIGN fig10_23
2   (
3      clock, enable      :INPUT;      -- synch clock and enable.
4      q[2..0], tc        :OUTPUT;     -- 3-bit counter
5   )
6   VARIABLE
7      count[2..0]    :DFF;    -- declare a register of D flip-flops.
8
9   BEGIN
10     count[].clk = clock;       -- connect all clocks to synchronous source
11     IF enable THEN
12        IF count[].q < 5 THEN
13           count[].d = count[].q + 1;  -- increment current value by one
14        ELSE count[].d = 0;            -- recycle,force unused states to 0
15        END IF;
16     ELSE count[].d = count[].q;       -- not enabled: hold at this count
17     END IF;
18     tc = enable & count[].q == 5;     -- detect maximum count if enabled
19     q[] = count[].q;                  -- connect register to outputs
20  END;
```

FIGURE 10-23 The MOD-6 design in AHDL.

count (lines 12–15). If *enable* is LOW, the same value is held on *count* at every clock edge by the ELSE branch (line 16). Remember always to match an IF with an END IF, as we did on lines 15 and 17. Terminal count (*tc*, line 18) will be HIGH when it is *true* that *count* = = 5 AND *enable* is active. Notice the use of double equal signs (= =) to evaluate equality in AHDL.

MOD-6 COUNTER VHDL

The only additional features that this design needs that are not covered in Figure 7-40 are the count *enable* input and terminal count (*tc*) output shown in Figure 10-24. Notice that the extra input (*enable*, line 2) and output (*tc*, line 4) are included in the I/O definition. A new line (line 15) in the architecture description tests *enable* before deciding how to update the value of *count* (lines 16–20). In the case that *enable* is LOW, the current value is held in the variable *count* and does not count up. Remember always to match an IF with an END IF, as we did on lines 20–22. The terminal count indicator (*tc*, lines 23 and 24) will be HIGH when it is *true* that *count* = 5 AND *enable* is active.

```
1   ENTITY fig10_24 IS
2   PORT( clock, enable       :IN BIT ;
3         q                   :OUT INTEGER RANGE 0 TO 5;
4         tc                  :OUT BIT
5       );
6   END fig10_24;
7
8   ARCHITECTURE a OF fig10_24 IS
9   BEGIN
10     PROCESS (clock)                       -- respond to clock
11     VARIABLE count    :INTEGER RANGE 0 TO 5;
12
13     BEGIN
14        IF (clock = '1' AND clock'event) THEN
15           IF enable = '1' THEN            -- synchronous cascade input
16              IF count < 5 THEN            -- < max (terminal) count?
17                 count := count + 1;
18              ELSE
19                 count := 0;
20              END IF;
21           END IF;
22        END IF;
23        IF (count = 5) AND (enable = '1') THEN     -- synch cascade output
24              tc <= '1';                           -- indicate terminal ct
25        ELSE  tc <= '0';
26        END IF;
27        q <= count;                                -- update outputs
28     END PROCESS;
29  END a;
```

FIGURE 10-24 The MOD-6 design in VHDL.

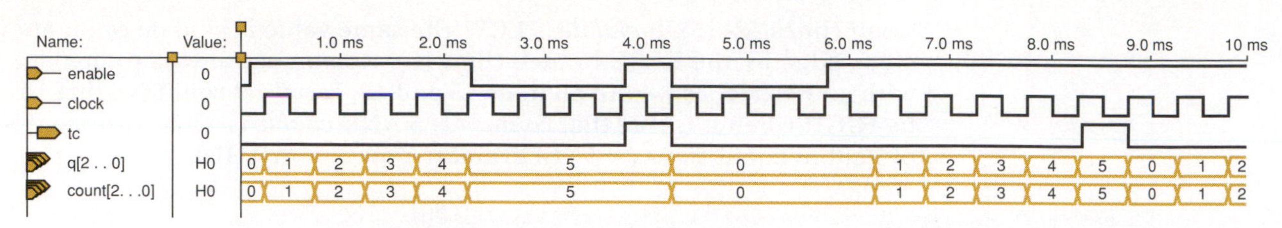

FIGURE 10-25 Simulation of the MOD-6 counter.

The simulation testing of the MOD-6 counter in Figure 10-25 verifies that it counts 0–5 and that it responds to the count enable input by ignoring the clock pulses and freezing the count whenever *enable* is LOW. It also generates the *tc* output when it is enabled at its maximum count of 5.

AHDL

MOD-10 COUNTER AHDL

The MOD-10 counter varies only slightly from the MOD-6 counter that was described in Figure 10-23. The only changes that are necessary involve changing the number of bits in the output port and the register (in the VARIABLE section) along with the maximum value that the counter should reach before rolling over. Figure 10-26 presents the MOD-10 design.

```
1   SUBDESIGN fig10_26
2   (
3      clock, enable        :INPUT;       -- synch clock and enable.
4      q[3..0], tc          :OUTPUT;      -- 4-bit Decade counter
5   )
6   VARIABLE
7      count[3..0] :DFF;            -- declare a register of D flip flops.
8
9   BEGIN
10     count[].clk = clock;         -- connect all clocks to synchronous source
11     IF enable THEN
12        IF count[].q < 9 THEN
13           count[].d = count[].q + 1;    -- increment current value by one
14        ELSE count[].d = 0;              -- recycle,force unused states to 0
15        END IF;
16     ELSE count[].d = count[].q;         -- not enabled: hold at this count
17     END IF;
18     tc = enable & count[].q == 9;       -- detect maximum count
19     q[] = count[].q;                    -- connect register to outputs
20  END;
```

FIGURE 10-26 The MOD-10 design in AHDL.

VHDL

MOD-10 COUNTER VHDL

The MOD-10 counter varies only slightly from the MOD-6 counter that was described in Figure 10-24. The only changes that are necessary involve changing the number of bits in the output port and the variable *count* (using INTEGER RANGE) along with the maximum value that the counter should reach before rolling over. Figure 10-27 presents the MOD-10 design.

```
1   ENTITY fig10_27 IS
2   PORT( clock, enable     :IN BIT ;
3         q                 :OUT INTEGER RANGE 0 TO 9;
4         tc                :OUT BIT
5      );
6   END fig10_27;
7
8   ARCHITECTURE a OF fig10_27 IS
9   BEGIN
10     PROCESS (clock)                          -- respond to clock
11     VARIABLE count   :INTEGER RANGE 0 TO 9;
12
13     BEGIN
14        IF (clock = '1' AND clock'event) THEN
15           IF enable = '1' THEN              -- synchronous cascade input
16              IF count < 9 THEN              -- decade counter
17                 count := count + 1;
18              ELSE
19                 count := 0;
20              END IF;
21           END IF;
22        END IF;
23        IF (count = 9) AND (enable = '1') THEN     -- synch cascade output
24              tc <= '1';
25        ELSE  tc <= '0';
26        END IF;
27        q <= count;                                -- update outputs
28     END PROCESS;
29  END a;
```

FIGURE 10-27 The MOD-10 design in VHDL.

MOD 12 Design

We have already decided that the hours counter is to be implemented as a single design file using HDL. It must be a MOD-12 BCD counter that follows the hours sequence of a clock (1–12) and provides the AM/PM indicator. Recall from the initial design step that the BCD outputs need to be a four-bit array for the low-order digit and a single bit for the high-order digit. To design this counter circuit, consider how it needs to operate. Its sequence is:

01 02 03 04 05 06 07 08 09 10 11 12 01 . . .

By observing this sequence, we can conclude that there are four critical areas that define the operations needed to produce the proper NEXT state:

1. When the value is 01 through 08, increment the low digit and keep the high digit the same.
2. When the value is 09, reset the low digit to 0 and force the high digit to 1.
3. When the value is 10 or 11, increment the low digit and keep the high digit the same.
4. When the value is 12, reset the low digit to 1 and the high digit to 0.

Because these conditions need to evaluate a range of values, it is most appropriate to use an IF/ELSIF construct rather than a CASE construct. There is also a need to identify when it is time to toggle the AM/PM indicator. This time occurs when the hour state is 11 and the enable is HIGH, which means that the lower-order counters are at their maximum (59:59).

MOD-12 COUNTER IN AHDL

The AHDL counter needs a bank of four D flip-flops for the low-order BCD digit and only a single D flip-flop for the high-order BCD digit because its value will always be 0 or 1. A flip-flop is also needed to keep track of A.M. and P.M. These primitives are declared on lines 7–9 of Figure 10-28. Also note that in this design, the same names are used for the output ports. This is a convenient feature of AHDL. When the enable input *(ena)* is active, the circuit evaluates the IF/ELSE statements of lines 16–28 and performs the proper

```
1   SUBDESIGN fig10_28
2   (
3      clk, ena            :INPUT;
4      low[3..0], hi, pm :OUTPUT;
5   )
6   VARIABLE
7      low[3..0]      :DFF;
8      hi             :DFF;
9      am_pm          :JKFF;
10     time           :NODE;
11  BEGIN
12     low[].clk = clk;          -- synchronous clocking
13     hi.clk = clk;
14     am_pm.clk = clk;
15     IF ena THEN               -- use enable to count
16         IF low[].q < 9 & hi.q == 0   THEN
17           low[].d = low[].q + 1; --inc lo digit
18           hi.d = hi.q;    -- hold hi digit
19        ELSIF low[].q == 9 THEN
20           low[].d = 0;
21           hi.d = VCC;
22        ELSIF hi.q == 1 & low[].q < 2 THEN
23                 low[].d = low[].q + 1;
24                 hi.d = hi.q;
25        ELSIF hi.q == 1 & low[].q == 2 THEN
26                 low[].d = 1;
27                 hi.d = GND;
28        END IF;
29     ELSE
30        low[].d = low[].q;
31        hi.d = hi.q;
32     END IF;
33     time = hi.q == 1 & low[3..0].q == 1 & ena;    -- detect 11:59:59
34     am_pm.j = time;       -- toggle am/pm at noon and midnight
35     am_pm.k = time;
36     pm = am_pm.q;
37  END;
```

FIGURE 10-28 The MOD-12 hours counter in AHDL.

operation on the high and low nibble of the BCD number. Whenever the enable input is LOW, the value remains the same, as shown on lines 30 and 31. Line 33 detects when the count reaches 11 while the counter is enabled. This signal is applied to the J and K inputs of the am_pm flip-flop to cause it to toggle at 11:59:59.

MOD-12 COUNTER IN VHDL

The VHDL counter of Figure 10-29 needs a four-bit output for the low-order BCD digit and a single output bit for the high-order BCD digit because its

```
1   ENTITY fig10_29 IS
2   PORT( clk, ena        :IN BIT ;
3         low             :OUT INTEGER RANGE 0 TO 9;
4         hi              :OUT INTEGER RANGE 0 TO 1;
5         pm              :OUT BIT               );
6   END fig10_29;
7
8   ARCHITECTURE a OF fig10_29 IS
9   BEGIN
10     PROCESS (clk)                              -- respond to clock
11     VARIABLE am_pm :BIT;
12     VARIABLE ones  :INTEGER RANGE 0 TO 9;  -- 4-bit units signal
13     VARIABLE tens  :INTEGER RANGE 0 TO 1;  -- 1-bit tens signal
14     BEGIN
15        IF (clk = '1' AND clk'EVENT) THEN
16           IF ena = '1' THEN                    -- synchronous cascade input
17              IF (ones = 1) AND (tens = 1) THEN   -- at 11:59:59
18                 am_pm := NOT am_pm;              -- toggle am/pm
19              END IF;
20              IF (ones < 9) AND  (tens = 0) THEN  -- states 00-08
21                 ones := ones + 1;              -- increment units
22              ELSIF ones = 9 THEN               -- state 09...set to 10:00
23                 ones := 0;                     -- units reset to zero
24                 tens := 1;                     -- tens bump up to 1
25              ELSIF (tens = 1) AND (ones < 2) THEN-- states 10, 11
26                 ones := ones + 1;              -- increment units
27              ELSIF (tens = 1) AND (ones = 2) THEN -- state 12
28                 ones := 1;                     -- set to 01:00
29                 tens := 0;
30              END IF;
31     -----------------------------------------------------------
32     -- This space is the alternate location for updating am/pm
33     -----------------------------------------------------------
34           END IF;
35        END IF;
36        pm <= am_pm;
37        low <= ones;                                  -- update outputs
38        hi <= tens;
39     END PROCESS;
40  END a;
```

FIGURE 10-29 The MOD-12 hours counter in VHDL.

value will always be 0 or 1. These outputs (lines 3 and 4) and also the variables that will produce the outputs (lines 12 and 13) are declared as integers because this makes "counting" possible by simply adding 1 to the variable value. On each active edge of the clock, when the enable input is active, the circuit needs to decide what to do with the BCD units-of-hours counter, the single bit tens-of-hours flip-flop, and also the AM/PM flip-flop.

This example is an excellent opportunity to point out some of the advanced features of VHDL that allow the designer to describe precisely the operation of the final hardware circuit. In previous chapters, we discussed the issue of statements within a PROCESS being evaluated sequentially. Recall that statements outside the PROCESS are considered concurrent, and the order in which they are written in the design file has no effect on the operation of the final circuit. In this example, we must evaluate the current state to decide whether to toggle the AM/PM indicator and also advance the counter to the NEXT state. The issues involved include the following:

1. How do we "remember" the current count value in VHDL?
2. Do we evaluate the current count to see if it is 11 (to determine if we need to toggle the AM/PM flip-flop) and then increment to 12, or do we increment the counter's state from 11 to 12 and then evaluate the count to see if it is 12 (to know we need to toggle the AM/PM flip flop)?

Regarding the first issue, there are two ways to remember the current state of a counter in VHDL. Both SIGNALs and VARIABLEs hold their value until they are updated. Generally, SIGNALs are used to connect nodes in the circuit like wires, and VARIABLEs are used like a register to store data between updates. Consequently, VARIABLEs are generally used to implement counters. The major differences are that VARIABLEs are local to the PROCESS in which they are declared and SIGNALs are global. Also, VARIABLEs are considered to be updated immediately within a sequence of statements in a PROCESS, whereas SIGNALs referred to in a PROCESS are updated when the PROCESS suspends. In this example, we have chosen to use VARIABLEs, which are local to the PROCESS that describes what should happen when the active clock edge occurs.

For the second issue, either of these strategies will work, but how do we describe them using VHDL? If we want the circuit to toggle A.M. and P.M. by detecting 11 prior to the counter updating (like a synchronous cascade), then the test must occur in the code before the VARIABLEs are updated. This test is demonstrated in the design file of Figure 10-29 on lines 17–19. On the other hand, if we want the circuit to toggle A.M. and P.M. by detecting when the hour 12 has arrived after the clock edge (more like a ripple cascade), then the VARIABLEs must be updated prior to testing for the value 12. To modify the design in Figure 10-29 to accomplish this task, the IF construct of lines 17–19 can be moved to the blank area of lines 31–33 and edited as shown in bold below:

```
31   IF (ones = 2) AND (tens = 1) THEN -- at 12:00:00
32      am_pm := NOT am_pm;            -- toggle am/pm
33      END IF;
```

The order of the statements and the value that is decoded makes all the difference in how the circuit operates. On lines 36–38, the *am_pm* VARIABLE is connected to the *pm* port, the units BCD digit is applied to the lower four bits of the output (*low*), and the tens digit (a single-bit variable) is applied to the most significant digit (*hi*) of the output port. Because all these VARIABLEs are local, these statements must occur prior to END PROCESS on line 39.

After the design is compiled, it must be simulated to verify its operation, especially at the critical areas. Figure 10-30 shows an example of a simulation to test this counter. On the left side of the timing diagram, the counter is disabled and is holding the hour 11 because the *hi* digit is at 1 and the *low[ ]* digit is at 1. On the rising edge of the clock, after the enable goes HIGH, the hour goes from 11 to 12 and causes the PM indicator to go HIGH, which means it is noon. The next active edge causes the count to roll over from 12 to 01. On the right half of the timing, the same sequence is simulated, showing that there would actually be many clock pulses between the times the hour increments. On the clock cycle before it must increment, the enable is driven HIGH by the terminal count of the previous stage.

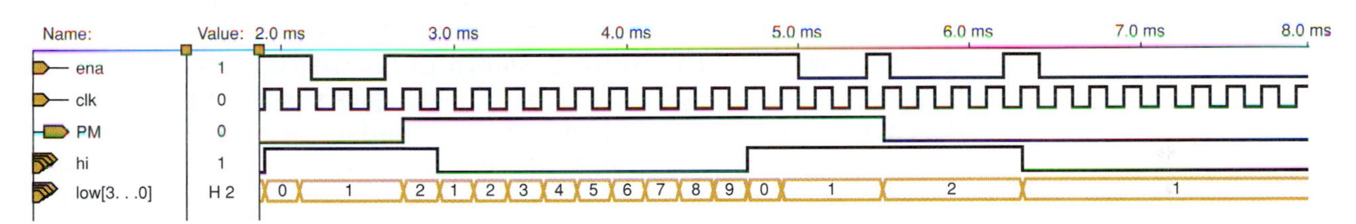

FIGURE 10-30 Simulation of the MOD-12 hours counter.

Combining Blocks Graphically

The building blocks of the project have been defined, created, and individually simulated to verify that they work correctly. Now it is time to combine the blocks to make sections and to combine the sections to make the final product. Altera's software offers several ways to accomplish the integration of all the pieces of a project. In Chapter 4, we mentioned that all different types of design files (AHDL, VHDL, VERILOG, Schematic) can be combined graphically. This technique is made possible by a feature that allows us to create a "symbol" to represent a particular design file. For example, the MOD-6 counter design file that was written in the VHDL design file fig10_24 can be represented in the software as the circuit block, as shown in Figure 10-31(a). The MAX+PLUS II or Quartus II software creates this symbol at the click of a button. From that point, it will recognize the symbol as operating according to the design specified in the HDL code. The symbol of Figure 10-31(b) was created from the AHDL file for the MOD-10 counter of Figure 10-26, and the symbol of Figure 10-31(c) was created from the VHDL file for the MOD-12 counter of Figure 10-29. (The reason these blocks are named by figure number is simply to make it easier to locate the design files on the enclosed CD. In a design environment [rather than in a textbook], they should be named according to their purpose, with names like MOD6, MOD10, and CLOCK_HOURS.)

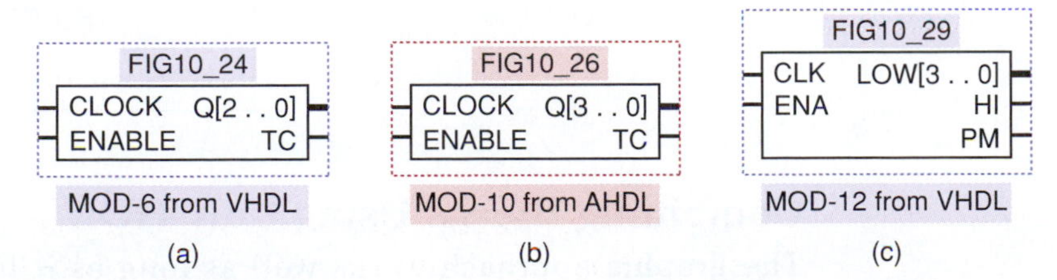

FIGURE 10-31 Graphic block symbols generated from HDL design files: (a) MOD-6 from VHDL; (b) MOD-10 from AHDL; (c) MOD-12 from VHDL.

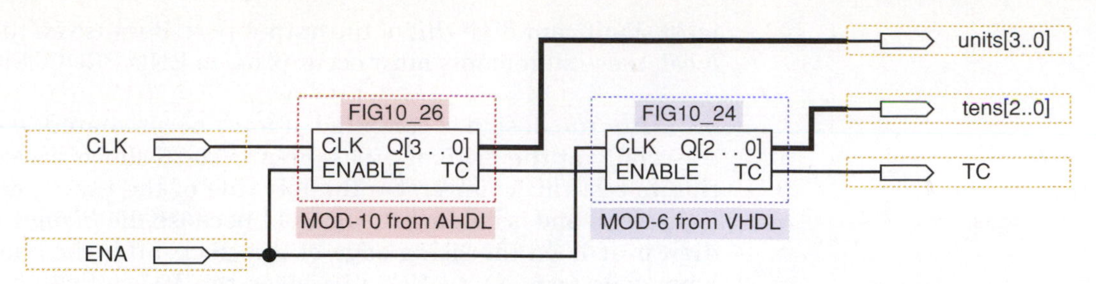

FIGURE 10-32 Graphically combining HDL blocks to make a MOD-60.

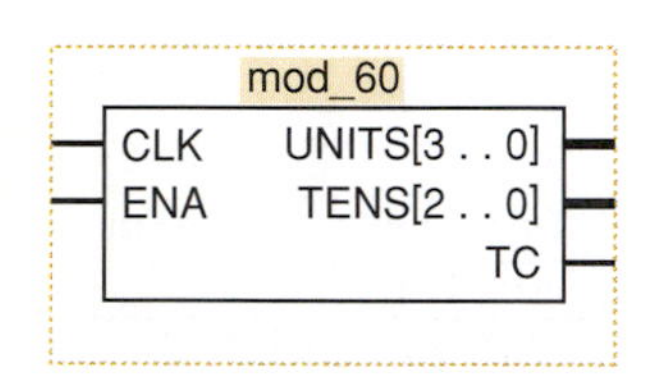

FIGURE 10-33 The MOD-60 counter.

Following the design hierarchy that we established, the next step is to combine the MOD-6 and MOD-10 counters to make a MOD-60 block. MAX+PLUS II software uses graphic design files (.gdf) to integrate the block symbols by drawing lines that connect the input ports, symbols, and output ports. Quartus II software provides the same feature but uses block design files (.bdf). The result is shown in Figure 10-32, which represents a GDF file in MAX+PLUS II or a BDF file in Quartus II. This graphic or block design file can be compiled and used to simulate the operation of the MOD-60 counter. When the design has been verified as working properly, the MAX+PLUS II or Quartus II system allows us to take this circuit and create a block symbol for it, as shown in Figure 10-33.

The MOD-60 symbol can be used repeatedly along with the MOD-12 symbol to create the system-level block symbol diagram shown in Figure 10-34. Even this system-level diagram can be represented by a block symbol for the entire project, as shown in Figure 10-35.

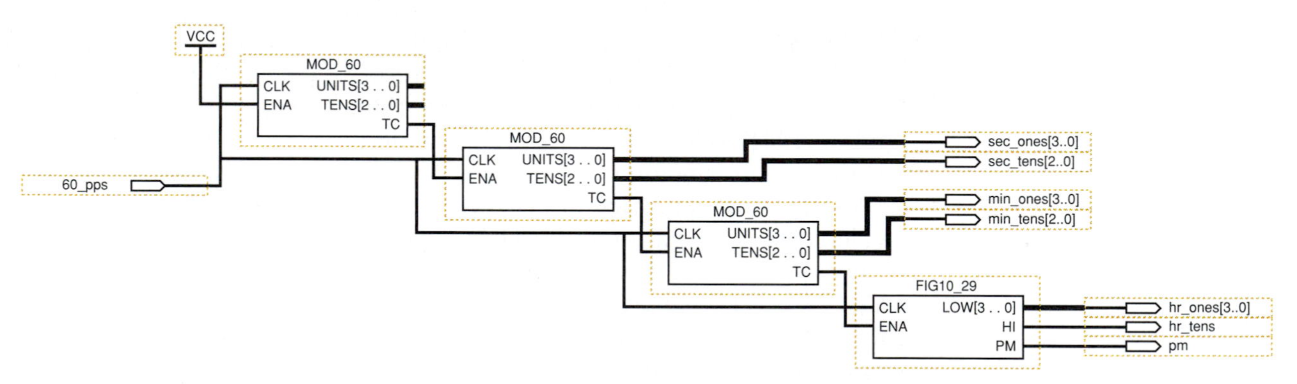

FIGURE 10-34 The complete clock project connected using block symbols.

FIGURE 10-35 The entire clock represented by one symbol.

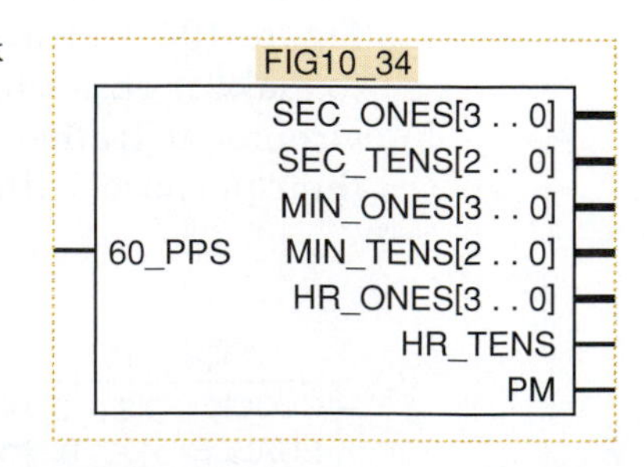

Combining Blocks Using Only HDL

The graphic approach works well as long as it is available and adequate for the purpose at hand. As we mentioned previously, HDL was developed to provide a convenient way to document complex systems and to store the information in a more timeless and software-independent manner. It is reasonable to assume that with AHDL, the option of graphic integration of subdesigns will always be available with the tools from Altera; however, this assumption is not reasonable for users of VHDL. Many VHDL development systems do not offer any equivalent to the graphic block integration of Altera, which is why it is important to address the same concept of modular, hierarchical development and project integration using only text-based language tools. Our coverage of AHDL integration will not be as in-depth as our coverage of VHDL because the graphic method is generally preferred.

AHDL MODULE INTEGRATION

Let's go back to the two AHDL files for the MOD-6 and MOD-10 counters. How do we combine these files into a MOD-60 counter using only text-based AHDL? The method is really very similar to that of graphic integration. Instead of creating a "symbol" representation of the MOD-6 and MOD-10 files, a new kind of file called an "INCLUDE" file is created. It contains all the important information about the AHDL file it represents. To describe a MOD-60 counter, a new TDF file, shown in Figure 10-36, is opened. The building block files are "included" at the top, as shown on lines 1 and 2. Next, the names that were used for the building blocks are used like library components or primitives to define the nature of a variable. On line 10, the variable *mod10* is now used to represent the MOD-10 counter in the other module (fig10_26). *MOD10* now has all the attributes (inputs, outputs, functional operation) described in fig10_26.tdf. Likewise, on line 11, the variable *mod6* is given the attributes of the MOD-6 counter of fig10_23.tdf. Lines 13–19 accomplish the exact same task as drawing lines on the GDF or BDF file to connect the components to one another and to the input/output ports.

```
1   INCLUDE "fig10_26.inc";       -- mod-10 counter module
2   INCLUDE "fig10_23.inc";       -- mod-6 counter module
3
4   SUBDESIGN fig10_36
5   (
6      clk, ena                         :INPUT;
7      ones[3..0], tens[2..0], tc       :OUTPUT;
8   )
9   VARIABLE
10     mod10           :fig10_26;       -- mod-10 for units
11     mod6            :fig10_23;       -- mod-6 for tens
12  BEGIN
13     mod10.clock = clk;               -- synchronous clocking
14     mod6.clock = clk;
15     mod10.enable = ena;
16     mod6.enable = mod10.tc;          -- cascade
17     ones[3..0] = mod10.q[3..0];      -- 1s
18     tens[2..0] = mod6.q[2..0];       -- 10s
19     tc = mod6.tc;                    -- Make terminal count at 59
20  END;
```

FIGURE 10-36 The MOD-60 made from MOD-10 and MOD-6 in AHDL.

This file (FIG10_36.TDF) can be translated into an "include" file (fig10_36.inc) by the compiler and then used in another tdf file that describes the interconnection of major sections to make up the system. Each level of the hierarchy refers back to the constituent modules of the lower levels.

VHDL

VHDL MODULE INTEGRATION

Let's go back to the two VHDL files for the MOD-6 and MOD-10 counters, which were shown in Figures 10-24 and 10-27, respectively. How do we combine these files into a MOD-60 counter using only text-based VHDL? The method is really very similar to that of graphic integration. Instead of creating a "symbol" representation of the MOD-6 and MOD-10 files, these design files are described as a COMPONENT, like we studied in Chapter 5. It contains all the important information about the VHDL file it represents. To describe a MOD-60 counter, a new VHDL file, shown in Figure 10-37, is opened. The building block files are described as "components," as shown on lines

```
1   ENTITY fig10_37 IS
2   PORT( clk, ena         :IN BIT ;
3         tens             :OUT INTEGER RANGE 0 TO 5;
4         ones             :OUT INTEGER RANGE 0 TO 9;
5         tc               :OUT BIT              );
6   END fig10_37;
7
8   ARCHITECTURE a OF fig10_37 IS
9   SIGNAL cascade_wire    :BIT;
10  COMPONENT fig10_24                                 -- MOD-6 module
11  PORT( clock, enable    :IN BIT ;
12        q                :OUT INTEGER RANGE 0 TO 5;
13        tc               :OUT BIT);
14  END COMPONENT;
15  COMPONENT fig10_27                                 -- MOD-10 module
16  PORT( clock, enable    :IN BIT ;
17        q                :OUT INTEGER RANGE 0 TO 9;
18        tc               :OUT BIT);
19  END COMPONENT;
20
21  BEGIN
22     mod10:fig10_27
23        PORT MAP(     clock => clk,
24                      enable => ena,
25                      q  => ones,
26                      tc => cascade_wire);
27
28     mod6:fig10_24
29        PORT MAP(     clock => clk,
30                      enable => cascade_wire,
31                      q  => tens,
32                      tc => tc);
33  END a;
```

FIGURE 10-37 The MOD-60 made from MOD-10 and MOD-6 in VHDL.

```
1   ENTITY fig10_38 IS
2   PORT( pps_60                          :IN BIT ;
3         hour_tens                       :OUT INTEGER RANGE 0 TO 1;
4         hour_ones, min_ones, sec_ones   :OUT INTEGER RANGE 0 TO 9;
5         min_tens, sec_tens              :OUT INTEGER RANGE 0 to 5;
6         pm                              :OUT BIT                 );
7   END fig10_38;
8
9   ARCHITECTURE a OF fig10_38 IS
10  SIGNAL cascade_wire1, cascade_wire2, cascade_wire3     :BIT;
11  SIGNAL enabled                                         :BIT;
12  COMPONENT fig10_37   -- MOD-60
13  PORT( clk, ena        :IN BIT ;
14        tens            :OUT INTEGER RANGE 0 TO 5;
15        ones            :OUT INTEGER RANGE 0 TO 9;
16        tc              :OUT BIT          );
17  END COMPONENT;
18  COMPONENT fig10_29   -- MOD-12
19  PORT( clk, ena        :IN BIT ;
20        low             :OUT INTEGER RANGE 0 TO 9;
21        hi              :OUT INTEGER RANGE 0 TO 1;
22        pm              :OUT BIT          );
23  END COMPONENT;
24  BEGIN
25     enabled <= '1';
26
27     prescale:fig10_37          -- MOD-60 prescaler
28        PORT MAP(   clk    => pps_60,
29                    ena    => enabled,
30                    tc     => cascade_wire1);
31
32     second:fig10_37            -- MOD-60 seconds counter
33        PORT MAP(   clk    => pps_60,
34                    ena    => cascade_wire1,
35                    ones   => sec_ones,
36                    tens   => sec_tens,
37                    tc     => cascade_wire2);
38
39     minute:fig10_37            -- MOD-60 minutes counter
40        PORT MAP(   clk    => pps_60,
41                    ena    => cascade_wire2,
42                    ones   => min_ones,
43                    tens   => min_tens,
44                    tc     => cascade_wire3);
45
46     hour:fig10_29              -- MOD12 Hours Counter
47        PORT MAP(   clk    => pps_60,
48                    ena    => cascade_wire3,
49                    low    => hour_ones,
50                    hi     => hour_tens,
51                    pm     => pm);
52  END a;
```

FIGURE 10-38 The complete clock in VHDL.

10–14 and lines 15–19 in the architecture description. Next, the names that were used for the building blocks (components) are used along with the PORT MAP keywords to describe the interconnection of these components. The information in the PORT MAP sections describes the exact same operations as drawing wires on a schematic diagram in a GDF file or BDF file.

Finally, the VHDL file that represents the block at the top of the hierarchy is created using components from Figure 10-37 (MOD-60) and Figure 10-29 (MOD-12). This file is shown in Figure 10-38. Notice that the general form is as follows:

Define I/O: lines 1–7

Define signals: lines 10–11

Define components: lines 12–23

Instantiate components and connect them together: lines 27–52

REVIEW QUESTIONS

1. What is being defined at the top level of a hierarchical design?
2. Where does the design process start?
3. Where does the building process start?
4. At which stage(s) should simulation testing be done?

10-5 FREQUENCY COUNTER PROJECT

The project in this section demonstrates the use of counters and other standard logic functions to implement a system called a frequency counter, which is similar to the piece of test equipment that you have probably used in the laboratory. The theory of operation will be described in terms of conventional MSI logic devices and then related to the building blocks that can be developed using HDL. As with most projects, this example consists of several circuits that we have studied in earlier chapters. They are combined here to form a digital system with a unique purpose. First, let us define a frequency counter.

A **frequency counter** is a circuit that can measure and display the frequency of a signal. As you know, the frequency of a periodic waveform is simply the number of cycles per second. Shaping each cycle of the unknown frequency into a digital pulse allows us to use a digital circuit to count the cycles. The general idea behind measuring frequency involves enabling a counter to count the number of cycles (pulses) of the incoming waveform during a precisely specified period of time called the **sampling interval**. The length of the sampling interval determines the range of frequencies that can be measured. A longer interval provides improved precision for low frequencies but will overflow the counter at high frequencies. A shorter sample interval provides a less precise measurement of low frequencies but can measure a much higher maximum frequency without exceeding the upper limit of the counter.

EXAMPLE 10-1

Assume that a frequency counter uses a four-digit BCD counter. Determine the maximum frequency that can be measured using each of the following sample intervals:

(a) 1 second (b) 0.1 second (c) 0.01 second

Solution

(a) With a sampling interval of 1 second, the four-digit counter can count up to 9999 pulses. The frequency is 9999 pulses per second or 9.999 kHz.

(b) The counter can count up to 9999 pulses within the sampling interval of 0.1 second. This translates into a frequency of 99,990 pulses per second or 99.99 kHz.

(c) The counter can count up to 9999 pulses within the sampling interval of 0.01 second. This translates into a frequency of 999,900 pulses per second or 999.9 kHz.

EXAMPLE 10-2

If a frequency of 3792 pps is applied to the input of the frequency counter, what will the counter read under each of the following sample intervals?

(a) 1 second (b) 0.1 second (c) 10 ms

Solution

(a) During a sampling interval of 1 second, the counter will count 3792 cycles. The frequency will read 3.792 kpps.

(b) During a sampling interval of 0.1 second, the number of pulses that will be counted is 379 or 380 cycles, depending on where the sample interval begins. The frequency will read 03.79 kpps or 03.80 kpps.

(c) During a sampling interval of 0.01 second, the number of pulses that will be counted is 37 or 38 cycles, depending on where the sample interval begins. The frequency will read 003.7 kpps or 003.8 kpps.

One of the most straightforward methods for constructing a frequency counter is shown as a block diagram in Figure 10-39. The major blocks are the counter, the display register, the decoder/display, and the timing and control unit. The counter block contains several cascaded BCD counters that are used to count the number of pulses produced by the unknown signal applied to the clock input. The counter block has count enable and clear controls. The time period for counting (sample interval) is controlled by an enable signal that is produced by the timing and control block. The length

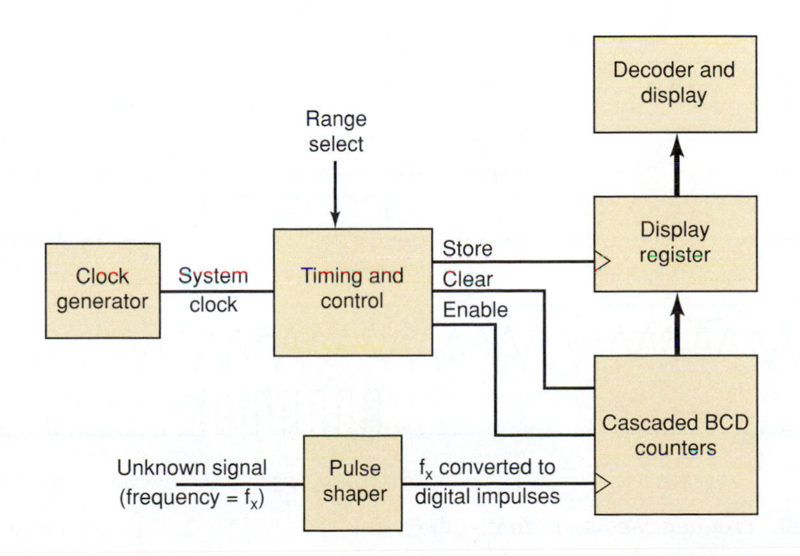

FIGURE 10-39 Basic frequency counter block diagram.

of time for the BCD counters to be enabled can be selected with the range select input to the timing and control block. This allows the user to select the desired frequency range to be measured and effectively determines the location of the decimal point in the digital readout. The pulse width of the enable signal (sample interval) is critical for taking an accurate frequency measurement. The counter must be cleared before it is enabled for a new frequency measurement of the unknown signal. After a new count has been taken, the counter is disabled, and the most recent frequency measurement is stored in the display register. The output of the display register is input to the decoder and display block, where the BCD values are converted into decimal for the display readout. Using a separate display register allows the frequency counter to take a new measurement in the background so that the user does not watch the counter while it is totaling the number of pulses for a new reading. The display is instead updated periodically with the last frequency reading.

The accuracy of this frequency counter depends almost entirely on the accuracy of the system clock frequency, which is used to create the proper pulse width for the counter enable signal. A crystal-controlled clock generator is used in Figure 10-39 to produce an accurate system clock for the timing and control block.

A pulse shaper block is needed to ensure that the unknown signal whose frequency is to be measured will be compatible with the clock input for the counter block. A Schmitt-trigger circuit may be used to convert "nonsquare" waveforms (sine, triangle, etc.) as long as the unknown input signal is of satisfactory amplitude. If the unknown signal might have a larger or smaller amplitude than is compatible with a given Schmitt trigger, then additional analog signal conditioning circuitry, such as an automatic gain control, will be required for the pulse shaper block.

The timing diagram for the control of the frequency counter is shown in Figure 10-40. The control clock is derived from the system clock signal by frequency dividers contained in the control and timing block. The period of the control clock signal is used to create the desired enable pulse width. A recycling control counter inside the control and timing block is clocked by the control clock signal. It has selected states decoded to produce the repeating

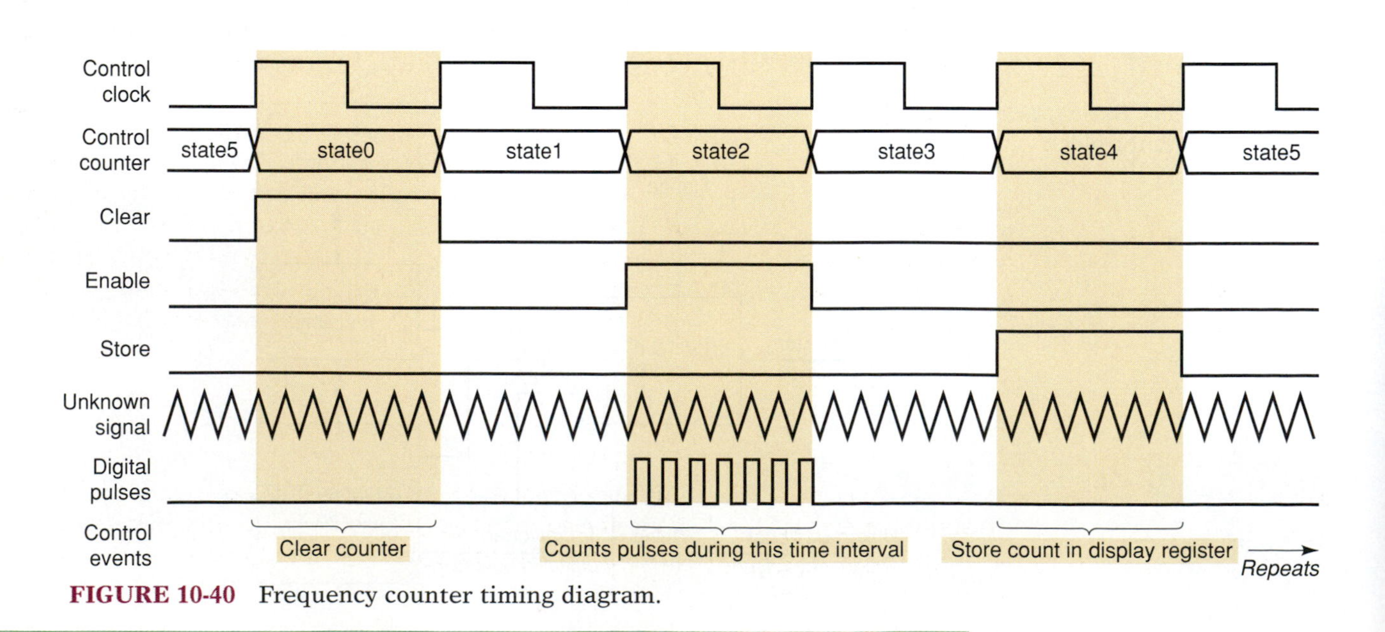

FIGURE 10-40 Frequency counter timing diagram.

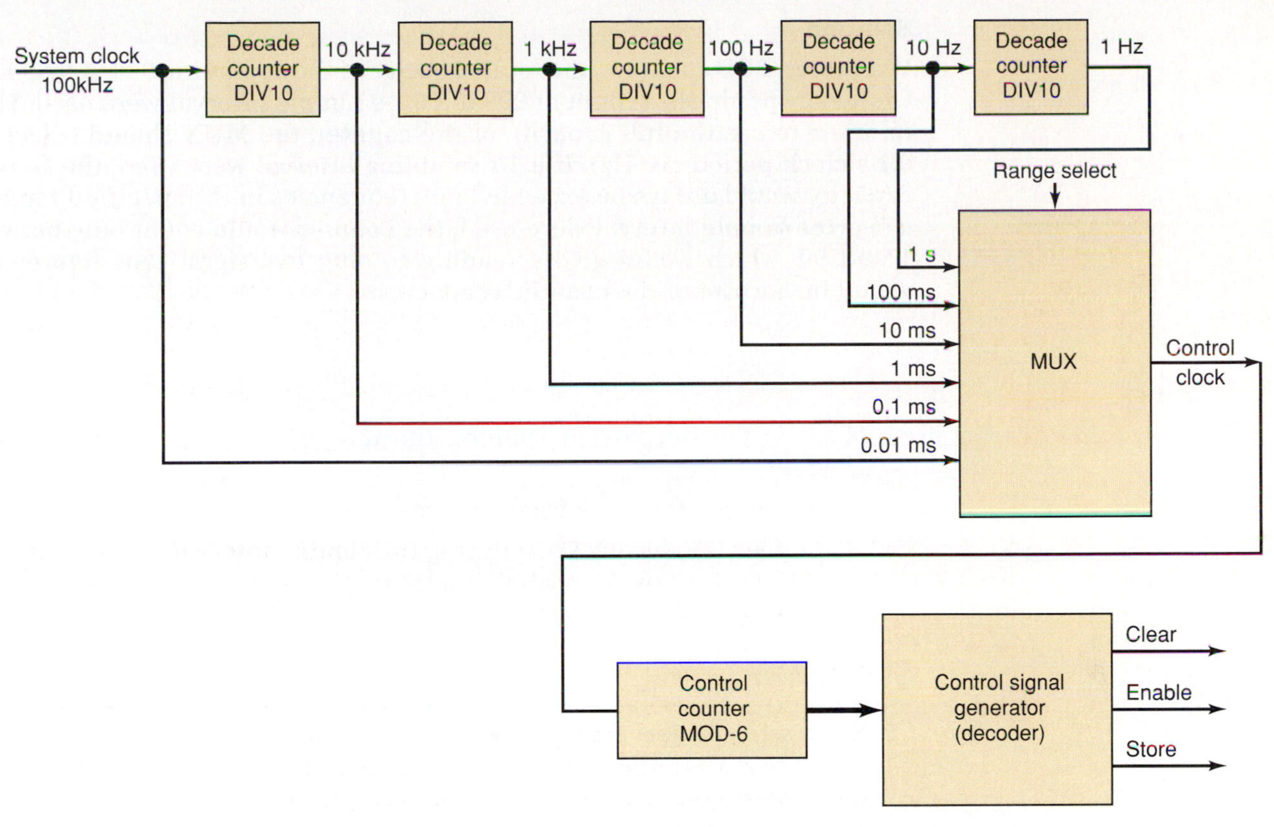

FIGURE 10-41 Timing and control block for frequency counter.

control signal sequence (clear, enable, and store). The counter (cascaded BCD stages) is first cleared. Then the counter is enabled for the proper sample interval to count the digital pulses, which have the same frequency as the unknown signal. After disabling the counter, the new count is stored in the display register.

The counter, display register, and decoder/display sections are straightforward and are not described any further here. The timing and control block provides the "brains" for our frequency counter and deserves a little more discussion to explain its operation. Figure 10-41 shows the sub-blocks within the timing and control block. For our example design, we will assume that the clock generator produces a 100-kHz system clock signal. The system clock frequency is divided by a set of five decade counters (MOD-10). This gives the user six different frequencies that can be selected by the multiplexer for the control clock frequency using the range select control. Because the period of the control clock is the same as the pulse width of the counter enable, this setup allows the frequency counter to have six different frequency measurement ranges. The control counter is a MOD-6 counter that has three selected states decoded by the control signal generator to produce the clear, enable, and store control signals.

EXAMPLE 10-3

Assume that the BCD counter in Figure 10-39 consists of three cascaded BCD stages and their associated displays. If the unknown frequency is between 1 kpps and 9.99 kpps, which range (sample interval) should be selected using the MUX of Figure 10-41?

Solution

With three BCD counters, the total capacity of the counter is 999. A 9.99-kpps frequency produces a count of 999 if a 0.1-s sample interval were used. Thus, in order to use the full capacity of the counter, the MUX should select the 0.1-s clock period (10 Hz). If a 1-s sampling interval were used, the counter capacity would always be exceeded for frequencies in the specified range. If a shorter sample interval were used, the counter would count only between 1 and 99, which would give a reading to only two significant figures and would be a waste of the counter's capacity.

REVIEW QUESTIONS

1. What is the purpose of running the unknown signal through a pulse shaper?
2. What are the units of a frequency measurement?
3. What does the display show during the sample interval?

SUMMARY

1. Successful project management can be accomplished by the following steps: overall project definition; breaking the project into small, strategic pieces; synthesis and testing of each piece; and system integration.
2. Small projects like the stepper motor driver can be completed in a single design file, even though these projects are developed modularly.
3. Projects that consist of several simple building blocks, like the keypad encoder, can produce very useful systems.
4. Larger projects like the digital clock can often take advantage of standard common modules that can be used repeatedly in the overall design.
5. Projects should be built and tested in modules starting at the lowest levels of hierarchy.
6. Preexisting modules can easily be combined with new custom modules using both graphical and text-based description methods.
7. Modules can be combined and represented as a single block in the next higher level of the hierarchy using the Altera design tools.

IMPORTANT TERMS

nesting
hierarchy
prescaler
frequency counter
sampling interval

PROBLEMS

SECTION 10-1

B 10-1. The security monitoring system of Section 9-8 in Chapter 9 can be developed as a project.
(a) Write a project definition with specifications for this system.
(b) Define three major blocks of this project.
(c) Identify the signals that interconnect the blocks.

(d)*At what frequency must the oscillator run for a 2.5-Hz flash rate?

(e)*Why is it reasonable to use only one current limiting resistor for all eight LEDs?

SECTION 10-2

Problems 10-2 through 10-7 refer to stepper motors described in Section 10-2.

B 10-2.*How many full steps must occur for a complete revolution?

B 10-3.*How many degrees of rotation result from one compete cycle through the full-step sequence in Table 10-1?

B 10-4. How many degrees of rotation result from one complete cycle through the half-step sequence in Table 10-1?

B 10-5. The *cout* lines of Figure 10-1 started at 1010 and have just progressed through the following sequence: 1010, 1001, 0101, 0110.

(a)*How many degrees has the shaft rotated?

(b) What sequence will reverse the rotation and return the shaft to its original position?

B 10-6. Describe a method to test the stepper driver in:

(a) Full-step mode

(b) Half-step mode

(c) Wave-drive mode

(d) Direct-drive mode

D, H 10-7. Rewrite the stepper driver design file of Figure 10-8 or 10-9 without using a CASE statement. Use your favorite HDL.

D, H 10-8. Modify the stepper design file of Figure 10-8 or 10-9 to add an enable input that puts the outputs in the Hi-Z state (tristate) when enable = 0.

SECTION 10-3

B 10-9. Write the state table for the ring counter shown in Figure 10-11 and described in Figure 10-13.

B 10-10.*With no keys pressed, what is the value on c[3..0]?

B 10-11. Assume that the ring counter is in state 0111 when someone presses the 7 key. Will the ring counter advance to the NEXT state?

B 10-12. Assume the 9 key is pressed and held until DAV = 1.

(a)*What is the value on the ring counter?

(b) What is the value encoded by the row encoder?

(c) What is the value encoded by the column encoder?

(d) What binary number is on the D[3..0] lines?

B 10-13.*In Problem 10-12, will the data be valid on the falling edge of DAV?

B, D 10-14. If you wanted to latch data from the keypad into a 74174 register, which signal from the keypad would you connect to the clock of the register? Draw the circuit.

T 10-15.*The keypad is connected to a 74373 octal transparent latch as shown in Figure 10-42. The output is correct as long as a key is held. However, it is unable to latch data between key presses. Why will this circuit *not* work correctly?

---

*Answers to problems marked with an asterisk can be found in the back of the text.

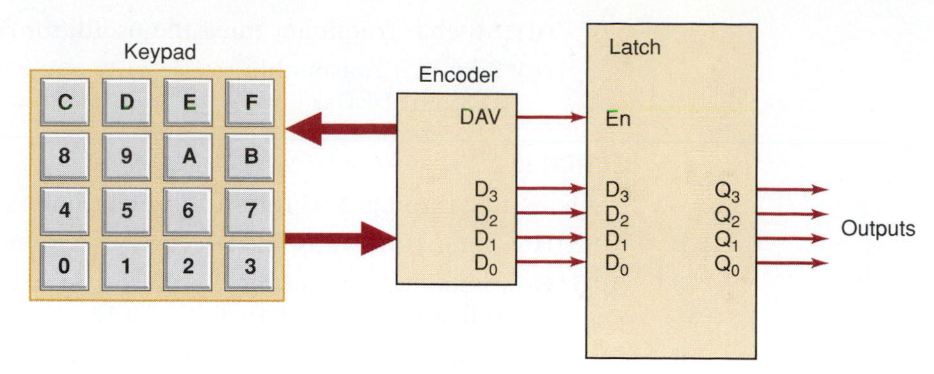

FIGURE 10-42 Problem 10-15.

SECTION 10-4

B 10-16. Assume a 1-Hz clock is applied to the seconds stage of the clock in Figure 10–17. The MOD-10 *units of seconds* counter's terminal count (*tc*) output is shown in Figure 10-43. Draw a similar diagram showing the number of clock cycles between the *tc* output pulses of each of the following:

(a)*Tens of seconds counter

(b) Units of minutes counter

(c) Tens of minutes counter

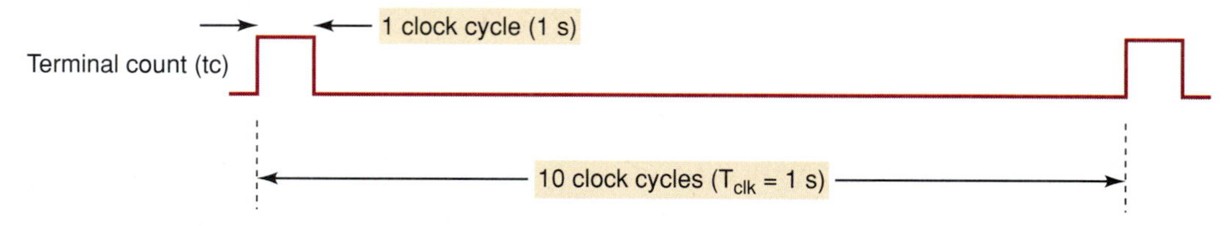

FIGURE 10-43 Problem 10-16.

B 10-17.*How many cycles of the 60-Hz power line will occur in a 24-hour period? What problem do you think will result if we attempt to simulate the operation of the entire clock circuit?

D 10-18.*Many digital clocks are set by simply making them count faster while a push button is held down. Modify the design to add this feature.

D, H 10-19. Modify the hours stage of Figure 10-18 to keep military time (00–23 hours).

SECTION 10-5

B 10-20. Draw the hierarchy diagram for the frequency counter project.

D, H 10-21. Write the HDL code for the MOD-6 control counter and control signal generator in Figure 10-41.

D, H 10-22.*Write the HDL code for the MUX of Figure 10-41.

D 10-23. Use graphic design techniques and the BCD counter described in Figure 10-31, the MUX, and the control signal generator design to create the entire timing and control block for the frequency counter project.

D, H 10-24. Write the HDL code for the timing and control section of the frequency counter.

ANSWERS TO SECTION REVIEW QUESTIONS

SECTION 10-1

1. Definition, strategic planning, synthesis and testing, system integration and testing 2. The definition stage

SECTION 10-2

1. Full-step, half-step, wave-drive, and direct-drive 2. cin_0-cin_3 [mode selector switches set to (1,1)] 3. Step, direction [mode selector switches set to (0,1)] 4. Eight states

SECTION 10-3

1. Only one 2. The first one scanned after being pressed (usually the first one pressed) 3. To make DAV go HIGH after the data stabilizes 4. No, it goes HIGH on the next clock after the key is pressed. 5. Whenever OE is LOW or when no keys are pressed

SECTION 10-4

1. The overall operating specifications and the system inputs and outputs. 2. At the top of the hierarchy 3. At the bottom, building the simplest blocks first 4. At each stage of modular implementation

SECTION 10-5

1. To change the shape of the analog signal into a digital signal of the same frequency 2. Cycles per second (Hz) or pulses per second (pps) 3. The display shows the frequency measured during the previous sample interval.

CHAPTER 11

INTERFACING WITH THE ANALOG WORLD

OUTLINE

11-1 Review of Digital Versus Analog
11-2 Digital-to-Analog Conversion
11-3 D/A-Converter Circuitry
11-4 DAC Specifications
11-5 An Integrated-Circuit DAC
11-6 DAC Applications
11-7 Troubleshooting DACs
11-8 Analog-to-Digital Conversion
11-9 Digital-Ramp ADC
11-10 Data Acquisition
11-11 Successive-Approximation ADC
11-12 Flash ADCs
11-13 Other A/D Conversion Methods
11-14 Sample-and-Hold Circuits
11-15 Multiplexing
11-16 Digital Storage Oscilloscope
11-17 Digital Signal Processing (DSP)

■ OBJECTIVES

Upon completion of this chapter, you will be able to:

- Understand the theory of operation and the circuit limitations of several types of digital-to-analog converters (DACs).
- Read and understand the various DAC manufacturer specifications.
- Use different test procedures to troubleshoot DAC circuits.
- Compare the advantages and disadvantages among the digital-ramp analog-to-digital converter (ADC), successive-approximation ADC, and flash ADC.
- Analyze the process by which a computer, in conjunction with an ADC, digitizes an analog signal and then reconstructs that analog signal from the digital data.
- Describe the basic operation of a digital voltmeter.
- Understand the need for using sample-and-hold circuits in conjunction with ADCs.
- Describe the operation of an analog multiplexing system.
- Understand the features and basic operation of a digital storage oscilloscope.
- Understand the basic concepts of digital signal processing.

11-1 REVIEW OF DIGITAL VERSUS ANALOG

A **digital quantity** has a value that is specified as one of two possibilities, such as 0 or 1, LOW or HIGH, true or false, and so on. In practice, a digital quantity such as a voltage may actually have a value that is anywhere within specified ranges, and we define values within a given range to have the same digital value. For example, for TTL logic, we know that

$$0 \text{ V to } 0.8 \text{ V} = \text{logic } 0$$
$$2 \text{ V to } 5 \text{ V} = \text{logic } 1$$

Any voltage falling in the range from 0 to 0.8 V is given the digital value 0, and any voltage in the range 2 to 5 V is assigned the digital value 1. The exact voltage values are not significant because the digital circuits respond in the same way to all voltage values within a given range.

By contrast, an **analog quantity** can take on any value over a continuous range of values and, most important, its exact value is significant. For example, the output of an analog temperature-to-voltage converter might be measured as 2.76 V, which may represent a specific temperature of 27.6°C. If the voltage were measured as something different, such as 2.34 V or 3.78 V, this would represent a completely different temperature. In other words, each possible value of an analog quantity has a different meaning. Another example of this is the output voltage from an audio amplifier into a speaker. This voltage is an analog quantity because each of its possible values produces a different response in the speaker.

Most physical variables are analog in nature and can take on any value within a continuous range of values. Examples include temperature, pressure, light intensity, audio signals, position, rotational speed, and flow rate. Digital systems perform all of their internal operations using digital circuitry and digital operations. Any information that must be input to a digital system must first be put into digital form. Similarly, the outputs from a digital system are always in digital form. When a digital system such as a computer is to be used to monitor and/or control a physical process, we must deal with the difference between the digital nature of the computer and the analog nature of the process variables. Figure 11-1 illustrates the situation. This diagram shows the five elements that are involved when a computer is monitoring and controlling a physical variable that is assumed to be analog:

1. **Transducer.** The physical variable is normally a nonelectrical quantity. A **transducer** is a device that converts the physical variable to an electrical variable. Some common transducers include thermistors, photocells, photodiodes, flow meters, pressure transducers, and tachometers. The electrical output of the transducer is an analog current or voltage that is proportional to the physical variable that it is monitoring. For example, the physical variable could be the temperature of water in a large tank that is being filled from cold and hot water pipes. Let's say that the water temperature varies from 80 to 150°F and that a thermistor and its associated circuitry convert this water temperature to a voltage ranging from 800 to 1500 mV. Note that the transducer's output is directly proportional to temperature such that each 1°F produces a 10-mV output. This proportionality factor was chosen for convenience.
2. **Analog-to-digital converter (ADC).** The transducer's electrical analog output serves as the analog input to the **analog-to-digital converter (ADC).** The ADC converts this analog input to a digital output. This digital output consists of a number of bits that represent the value of the analog input. For example, the ADC might convert the transducer's 800- to 1500-mV analog values to binary values ranging from 01010000 (80) to 10010110 (150). Note that the binary output from the ADC is proportional to the analog input voltage so that each unit of the digital output represents 10 mV.

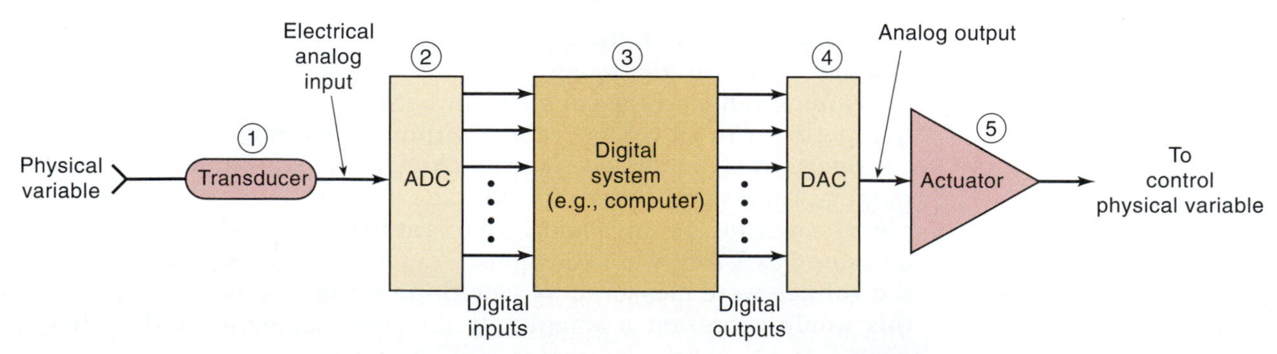

FIGURE 11-1 Analog-to-digital converter (ADC) and digital-to-analog converter (DAC) are used to interface a computer to the analog world so that the computer can monitor and control a physical variable.

3. **Computer.** The digital representation of the process variable is transmitted from the ADC to the digital computer, which stores the digital value and processes it according to a program of instructions that it is executing. The program might perform calculations or other operations on this digital representation of temperature to come up with a digital output that will eventually be used to control the temperature.
4. **Digital-to-analog converter (DAC).** This digital output from the computer is connected to a **digital-to-analog converter (DAC)**, which converts it to a proportional analog voltage or current. For example, the computer might produce a digital output ranging from 00000000 to 11111111, which the DAC converts to a voltage ranging from 0 to 10 V.
5. **Actuator.** The analog signal from the DAC is often connected to some device or circuit that serves as an actuator to control the physical variable. For our water temperature example, the actuator might be an electrically controlled valve that regulates the flow of hot water into the tank in accordance with the analog voltage from the DAC. The flow rate would vary in proportion to this analog voltage, with 0 V producing no flow and 10 V producing the maximum flow.

Thus, we see that ADCs and DACs function as *interfaces* between a completely digital system, such as a computer, and the analog world. This function has become increasingly more important as inexpensive microcomputers have moved into areas of process control where computer control was previously not feasible.

REVIEW QUESTIONS

1. What is the function of a transducer?
2. What is the function of an ADC?
3. What does a computer often do with the data that it receives from an ADC?
4. What function does a DAC perform?
5. What is the function of an actuator?

11-2 DIGITAL-TO-ANALOG CONVERSION

We will now begin our study of digital-to-analog (D/A) and analog-to-digital (A/D) conversion. Many A/D conversion methods utilize the D/A conversion process, so we will examine D/A conversion first.

Basically, D/A *conversion* is the process of taking a value represented in *digital* code (such as straight binary or BCD) and converting it to a voltage or current that is proportional to the digital value. Figure 11-2(a) shows the symbol for a typical four-bit D/A converter. We will not concern ourselves with the internal circuitry until later. For now, we will examine the various input/output relationships.

Notice that there is an input for a voltage reference, V_{ref}. This input is used to determine the **full-scale output** or maximum value that the D/A converter can produce. The digital inputs D, C, B, and A are usually derived from the output register of a digital system. The $2^4 = 16$ different binary numbers represented by these four bits are listed in Figure 11-2(b). For each input number, the D/A converter output voltage is a unique value. In fact, for this case, the analog output voltage V_{OUT} is equal in volts to the binary number. It

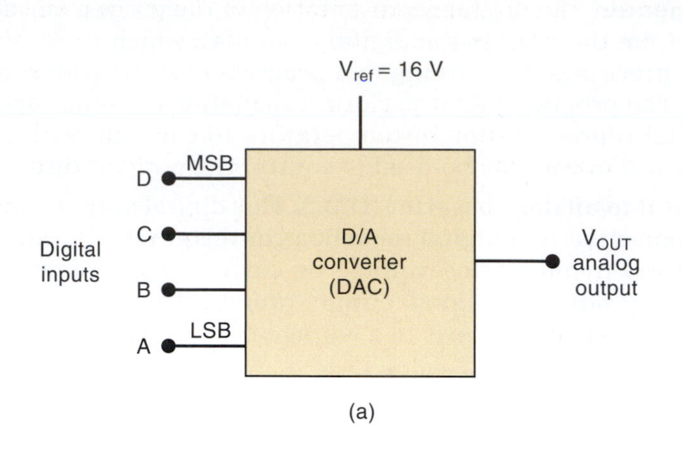

D	C	B	A	V_{OUT}
0	0	0	0	0 Volts
0	0	0	1	1
0	0	1	0	2
0	0	1	1	3
0	1	0	0	4
0	1	0	1	5
0	1	1	0	6
0	1	1	1	7
1	0	0	0	8
1	0	0	1	9
1	0	1	0	10
1	0	1	1	11
1	1	0	0	12
1	1	0	1	13
1	1	1	0	14
1	1	1	1	15 Volts

(b)

FIGURE 11-2 Four-bit DAC with voltage output.

could also have been twice the binary number or some other proportionality factor. The same idea would hold true if the D/A output were a current I_{OUT}.

In general,

$$\text{analog output} = K \times \text{digital input} \tag{11-1}$$

where K is the proportionality factor and is a constant value for a given DAC connected to a fixed reference voltage. The analog output can, of course, be a voltage or a current. When it is a voltage, K will be in voltage units, and when the output is a current, K will be in current units. For the DAC of Figure 11-2, $K = 1$ V, so that

$$V_{OUT} = (1 \text{ V}) \times \text{digital input}$$

We can use this to calculate V_{OUT} for any value of digital input. For example, with a digital input of $1100_2 = 12_{10}$, we obtain

$$V_{OUT} = 1 \text{ V} \times 12 = 12 \text{ V}$$

EXAMPLE 11-1A

A five-bit DAC has a current output. For a digital input of 10100, an output current of 10 mA is produced. What will I_{OUT} be for a digital input of 11101?

Solution

The digital input 10100_2 is equal to decimal 20. Because $I_{OUT} = 10$ mA for this case, the proportionality factor must be 0.5 mA. Thus, we can find I_{OUT} for any digital input such as $11101_2 = 29_{10}$ as follows:

$$\begin{aligned} I_{OUT} &= (0.5 \text{ mA}) \times 29 \\ &= 14.5 \text{ mA} \end{aligned}$$

Remember, the proportionality factor, K, varies from one DAC to another and depends on the reference voltage.

EXAMPLE 11-1B

What is the largest value of output voltage from an eight-bit DAC that produces 1.0 V for a digital input of 00110010?

Solution

$$00110010_2 = 50_{10}$$
$$1.0\text{ V} = K \times 50$$

Therefore,

$$K = 20\text{ mV}$$

The largest output will occur for an input of $11111111_2 = 255_{10}$.

$$V_{OUT}(\text{max}) = 20\text{ mV} \times 255 = 5.10\text{ V}$$

Analog Output

The output of a DAC is technically not an analog quantity because it can take on only specific values, such as the 16 possible voltage levels for V_{OUT} in Figure 11-2, as long as V_{ref} is constant. Thus, in that sense, it is actually digital. As we will see, however, the number of different possible output values can be increased and the difference between successive values decreased by increasing the number of input bits. This will allow us to produce an output that is more and more like an analog quantity that varies continuously over a range of values. In other words, the DAC output is a "pseudo-analog" quantity. We will continue to refer to it as analog, keeping in mind that it is an approximation to a pure analog quantity.

Input Weights

For the DAC of Figure 11-2, note that each digital input contributes a different amount to the analog output. This is easily seen if we examine the cases where only one input is HIGH (Table 11-1). The contributions of each digital input are *weighted* according to their position in the binary number. Thus, A, which is the LSB, has a *weight* of 1 V; B has a weight of 2 V; C has a weight of 4 V; and D, the MSB, has the largest weight, 8 V. The weights are successively doubled for each bit, beginning with the LSB. Thus, we can consider V_{OUT} to be the weighted sum of the digital inputs. For instance, to find V_{OUT} for the digital input 0111, we can add the weights of the C, B, and A bits to obtain 4 V + 2 V + 1 V = 7 V.

TABLE 11-1

D	C	B	A		V_{OUT} (V)
0	0	0	1	→	1
0	0	1	0	→	2
0	1	0	0	→	4
1	0	0	0	→	8

EXAMPLE 11-2

A five-bit D/A converter produces $V_{OUT} = 0.2\text{ V}$ for a digital input of 00001. Find the value of V_{OUT} for an input of 11111.

Solution

Obviously, 0.2 V is the weight of the LSB. Thus, the weights of the other bits must be 0.4 V, 0.8 V, 1.6 V, and 3.2 V, respectively. For a digital input of 11111, then, the value of V_{OUT} will be 3.2 V + 1.6 V + 0.8 V + 0.4 V + 0.2 V = 6.2 V.

Resolution (Step Size)

Resolution of a D/A converter is defined as the smallest change that can occur in the analog output as a result of a change in the digital input. Referring to the table in Figure 11-2, we can see that the resolution is 1 V because V_{OUT} can change by no less than 1 V when the digital input value is changed. The resolution is always equal to the weight of the LSB and is also referred to as the **step size** because it is the amount that V_{OUT} will change as the digital input value is changed from one step to the next. This is illustrated better in Figure 11-3, where the outputs from a four-bit binary counter provide the inputs to our DAC. As the counter is being continually cycled through its 16 states by the clock signal, the DAC output is a **staircase** waveform that goes up 1 V per step. When the counter is at 1111, the DAC output is at its maximum value of 15 V; this is its full-scale output. When the counter recycles to 0000, the DAC output returns to 0 V. The resolution (or step size) is the size of the jumps in the staircase waveform; in this case, each step is 1 V.

Note that the staircase has 16 levels corresponding to the 16 input states, but there are only 15 steps or jumps between the 0-V level and full-scale. In general, for an N-bit DAC the number of different levels will be 2^N, and the number of steps will be 2^N-1.

You may have already figured out that resolution (step size) is the same as the proportionality factor in the DAC input/output relationship:

$$\text{analog output} = K \times \text{digital input}$$

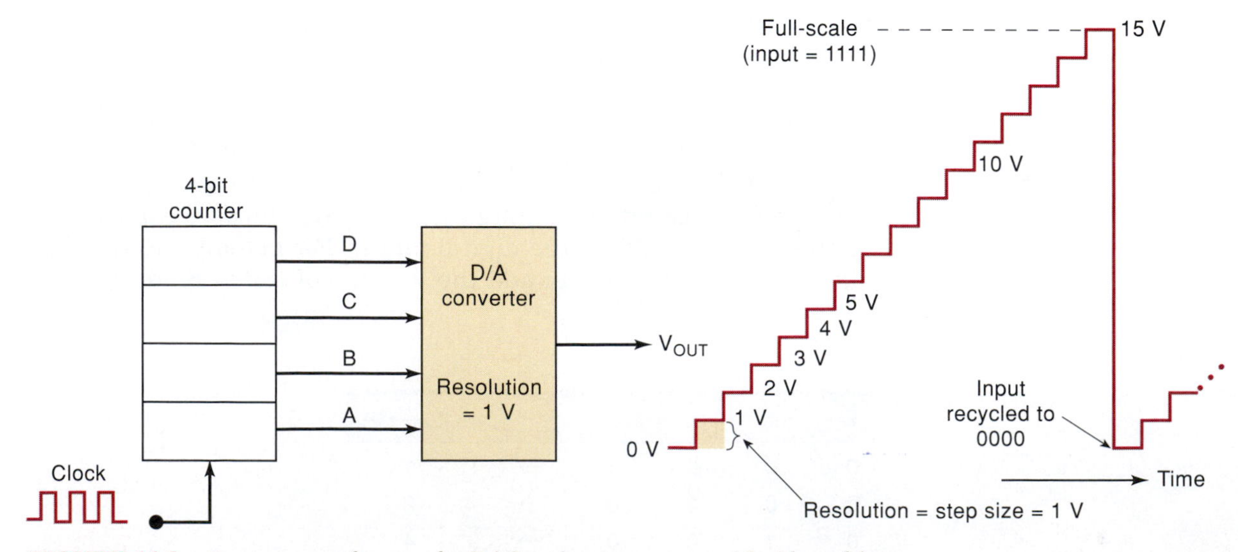

FIGURE 11-3 Output waveforms of a DAC as inputs are provided by a binary counter.

A new interpretation of this expression would be that the digital input is equal to the number of steps, K is the amount of voltage (or current) per step, and the analog output is the product of the two. We now have a convenient way of calculating the value of K for the D/A:

$$\text{resolution} = K = \frac{A_{\text{fs}}}{(2^n - 1)} \qquad \textbf{(11-2)}$$

where A_{fs} is the analog full-scale output and n is the number of bits.

EXAMPLE 11-3A

What is the resolution (step size) of the DAC of Example 11-2? Describe the staircase signal out of this DAC.

Solution

The LSB for this converter has a weight of 0.2 V. This is the resolution or step size. A staircase waveform can be generated by connecting a five-bit counter to the DAC inputs. The staircase will have 32 levels, from 0 V up to a full-scale output of 6.2 V, and 31 steps of 0.2 V each.

EXAMPLE 11-3B

For the DAC of Example 11-2, determine V_{OUT} for a digital input of 10001.

Solution

The step size is 0.2 V, which is the proportionality factor K. The digital input is $10001 = 17_{10}$. Thus, we have

$$\begin{aligned} V_{\text{OUT}} &= (0.2\text{ V}) \times 17 \\ &= 3.4\text{ V} \end{aligned}$$

Percentage Resolution

Although resolution can be expressed as the amount of voltage or current per step, it is also useful to express it as a percentage of the *full-scale output*. To illustrate, the DAC of Figure 11-3 has a maximum full-scale output of 15 V (when the digital input is 1111). The step size is 1 V, which gives a percentage resolution of

$$\%\text{ resolution} = \frac{\text{step size}}{\text{full scale (F.S.)}} \times 100\% \qquad \textbf{(11-3)}$$

$$= \frac{1\text{ V}}{15\text{ V}} \times 100\% = 6.67\%$$

EXAMPLE 11-4

A 10-bit DAC has a step size of 10 mV. Determine the full-scale output voltage and the percentage resolution.

Solution

With 10 bits, there will be $2^{10} - 1 = 1023$ steps of 10 mV each. The full-scale output will therefore be 10 mV × 1023 = 10.23 V, and

$$\% \text{ resolution} = \frac{10 \text{ mV}}{10.23 \text{ V}} \times 100\% \approx 0.1\%$$

Example 11-4 helps to illustrate the fact that the percentage resolution becomes smaller as the number of input bits is increased. In fact, the percentage resolution can also be calculated from

$$\% \text{ resolution} = \frac{1}{\text{total number of steps}} \times 100\% \tag{11-4}$$

For an N-bit binary input code, the total number of steps is $2^N - 1$. Thus, for the previous example,

$$\begin{aligned}\% \text{ resolution} &= \frac{1}{2^{10} - 1} \times 100\% \\ &= \frac{1}{1023} \times 100\% \\ &\approx 0.1\%\end{aligned}$$

This means that it is *only the number of bits* that determines the *percentage* resolution. Increasing the number of bits increases the number of steps to reach full scale, so that each step is a smaller part of the full-scale voltage. Most DAC manufacturers specify resolution as the number of bits.

What Does Resolution Mean?

A DAC cannot produce a continuous range of output values and so, strictly speaking, its output is not truly analog. A DAC produces a finite set of output values. In our water temperature example of Section 11-1, the computer generates a digital output to provide an analog voltage between 0 and 10 V to an electrically controlled valve. The DAC's resolution (number of bits) determines how many possible voltage values the computer can send to the valve. If a six-bit DAC is used, there will be 63 possible steps of 0.159 V between 0 and 10 V. When an eight-bit DAC is used, there will be 255 possible steps of 0.039 V between 0 and 10 V. The greater the number of bits, the finer the resolution (the smaller the step size).

The system designer must decide what resolution is needed on the basis of the required system performance. The resolution limits how close the DAC output can come to a given analog value. Generally, the cost of DACs increases with the number of bits, and so the designer will use only as many bits as necessary.

EXAMPLE 11-5

Figure 11-4 shows a computer controlling the speed of a motor. The 0- to 2-mA analog current from the DAC is amplified to produce motor speeds from 0 to 1000 rpm (revolutions per minute). How many bits should be used if the computer is to be able to produce a motor speed that is within 2 rpm of the desired speed?

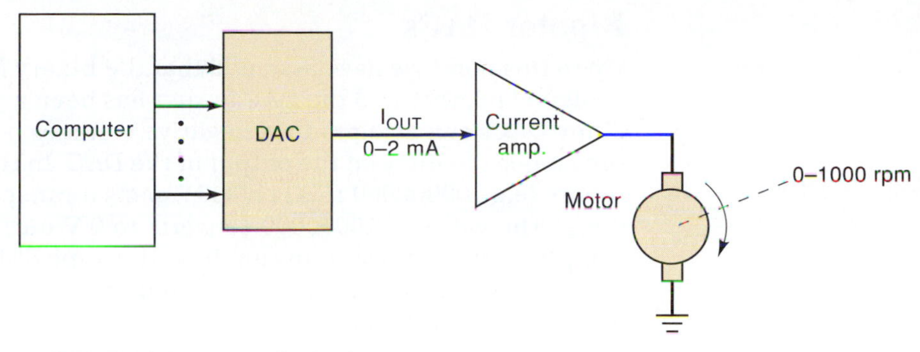

FIGURE 11-4 Example 11-5.

Solution

The motor speed will range from 0 to 1000 rpm as the DAC goes from zero to full scale. Each step in the DAC output will produce a step in the motor speed. We want the step size to be no greater than 2 rpm. Thus, we need at least 500 steps (1000/2). Now we must determine how many bits are required so that there are at least 500 steps from zero to full scale. We know that the number of steps is $2^N - 1$, and so we can say

$$2^N - 1 \geq 500$$

or

$$2^N \geq 501$$

Since $2^8 = 256$ and $2^9 = 512$, the smallest number of bits that will produce at least 500 steps is *nine*. We could use more than nine bits, but this might add to the cost of the DAC.

EXAMPLE 11-6

Using nine bits, how close to 326 rpm can the motor speed be adjusted?

Solution

With nine bits, there will be 511 steps ($2^9 - 1$). Thus, the motor speed will go up in steps of 1000 rpm/511 = 1.957 rpm. The number of steps needed to reach 326 rpm is 326/1.957 = 166.58. This is not a whole number of steps, and so we will round it to 167. The actual motor speed on the 167th step will be $167 \times 1.957 = 326.8$ rpm. Thus, the computer must output the nine-bit binary equivalent of 167_{10} to produce the desired motor speed within the resolution of the system.

In all of our examples, we have assumed that the DACs have been perfectly accurate in producing an analog output that is directly proportional to the binary input, and that the resolution is the only thing that limits how close we can come to a desired analog value. This, of course, is unrealistic because all devices contain inaccuracies. We will examine the causes and effects of DAC inaccuracy in Sections 11-3 and 11-4.

Bipolar DACs

Up to this point we have assumed that the binary input to a DAC has been an unsigned number and the DAC output has been a positive voltage or current. Many DACs can also produce negative voltages by making slight changes to the analog circuitry on the output of the DAC. In this case the range of binary inputs (e.g., 00000000 to 11111111) spans a range of $-V_{ref}$ to approximately $+V_{ref}$. The value of 10000000 converts to 0 V out. The output of a signed 2's complement digital system can drive this type of DAC by inverting the MSB, which converts the signed binary numbers to the proper values for the DAC as shown in Table 11-2.

TABLE 11-2

	Signed 2's Complement	DAC Inputs	DAC V_{out}
Most positive	01111111	11111111	$\sim+V_{ref}$
Zero	00000000	10000000	0 V
Most negative	10000000	00000000	$-V_{ref}$

Other DACs may have the extra circuitry built in and accept 2's complement signed numbers as inputs. For example, suppose that we have a six-bit bipolar DAC that uses the 2's-complement system and has a resolution of 0.2 V. The binary input values range from 100000 (−32) to 011111 (+31) to produce analog outputs in the range from −6.4 to +6.2 V. There are 63 steps ($2^6 - 1$) of 0.2 V between these negative and positive limits.

REVIEW QUESTIONS

1. An eight-bit DAC has an output of 3.92 mA for an input of 01100010. What are the DAC's resolution and full-scale output?
2. What is the weight of the MSB of the DAC of question 1?
3. What is the percentage resolution of an eight-bit DAC?
4. How many different output voltages can a 12-bit DAC produce?
5. For the system of Figure 11-4, how many bits should be used if the computer is to control the motor speed within 0.4 rpm?
6. *True or false:* The percentage resolution of a DAC depends *only* on the number of bits.
7. What is the advantage of a smaller (finer) resolution?

11-3 D/A-CONVERTER CIRCUITRY

There are several methods and circuits for producing the D/A operation that has been described. We shall examine several of the basic schemes to gain an insight into the ideas used. It is not important to be familiar with all of the various circuit schemes because D/A converters are available as ICs or as encapsulated packages that do not require any circuit knowledge. Instead, it is important to know the significant performance characteristics of DACs, in general, so that they can be used intelligently. These will be covered in Section 11-4.

Figure 11-5(a) shows the basic circuit for one type of four-bit DAC. The inputs A, B, C, and D are binary inputs that are assumed to have values of either 0 or 5 V. The *operational amplifier* is employed as a summing amplifier,

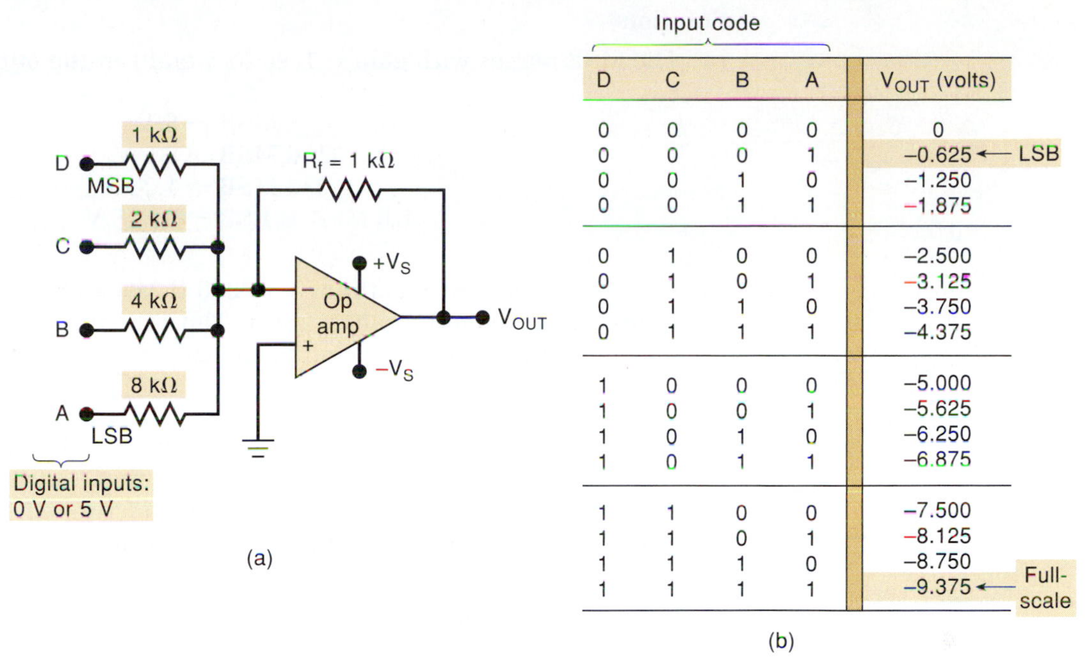

D	C	B	A	VOUT (volts)
0	0	0	0	0
0	0	0	1	−0.625 ← LSB
0	0	1	0	−1.250
0	0	1	1	−1.875
0	1	0	0	−2.500
0	1	0	1	−3.125
0	1	1	0	−3.750
0	1	1	1	−4.375
1	0	0	0	−5.000
1	0	0	1	−5.625
1	0	1	0	−6.250
1	0	1	1	−6.875
1	1	0	0	−7.500
1	1	0	1	−8.125
1	1	1	0	−8.750
1	1	1	1	−9.375 ← Full-scale

FIGURE 11-5 Simple DAC using an op-amp summing amplifier with binary-weighted resistors.

which produces the weighted sum of these input voltages. Recall that the summing amplifier multiplies each input voltage by the ratio of the feedback resistor R_F to the corresponding input resistor R_{IN}. In this circuit $R_F = 1\ \text{k}\Omega$, and the input resistors range from 1 to 8 kΩ. The *D* input has $R_{IN} = 1\ \text{k}\Omega$, so the summing amplifier passes the voltage at *D* with no attenuation. The *C* input has $R_{IN} = 2\ \text{k}\Omega$, so that it will be attenuated by $\frac{1}{2}$. Similarly, the *B* input will be attenuated by $\frac{1}{4}$, and the *A* input by $\frac{1}{8}$. The amplifier output can thus be expressed as

$$V_{OUT} = -(V_D + \tfrac{1}{2}V_C + \tfrac{1}{4}V_B + \tfrac{1}{8}V_A) \quad \textbf{(11-5)}$$

The negative sign is present because the summing amplifier is a polarity-inverting amplifier, but it will not concern us here.

Clearly, the summing amplifier output is an analog voltage that represents a weighted sum of the digital inputs, as shown by the table in Figure 11-5(b). This table lists all of the possible input conditions and the resultant amplifier output voltage. The output is evaluated for any input condition by setting the appropriate inputs to either 0 or 5 V. For example, if the digital input is 1010, then $V_D = V_B = 5\ \text{V}$ and $V_C = V_A = 0\ \text{V}$. Thus, using equation (11-5),

$$\begin{aligned} V_{OUT} &= -(5\ \text{V} + 0\ \text{V} + \tfrac{1}{4} \times 5\ \text{V} + 0\ \text{V}) \\ &= -6.25\ \text{V} \end{aligned}$$

The resolution of this D/A converter is equal to the weighting of the LSB, which is $\frac{1}{8} \times 5\ \text{V} = 0.625\ \text{V}$. As shown in the table, the analog output increases by 0.625 V as the binary input number advances one step.

EXAMPLE 11-7

(a) Determine the weight of each input bit of Figure 11-5(a).

(b) Change R_F to 250 Ω and determine the full-scale output.

Solution

(a) The MSB passes with gain = 1, so its weight in the output is 5 V. Thus,

$$\begin{aligned} \text{MSB} &\rightarrow 5\text{ V} \\ \text{2nd MSB} &\rightarrow 2.5\text{ V} \\ \text{3rd MSB} &\rightarrow 1.25\text{ V} \\ \text{4th MSB} = \text{LSB} &\rightarrow 0.625\text{ V} \end{aligned}$$

(b) If R_F is reduced by a factor of 4, to 250 Ω, each input weight will be four times *smaller* than the values above. Thus, the full-scale output will be reduced by this same factor and becomes $-9.375/4 = -2.344$ V.

If we look at the input resistor values in Figure 11-5, it should come as no surprise that they are *binarily weighted*. In other words, starting with the MSB resistor, the resistor values increase by a factor of 2. This, of course, produces the desired weighting in the voltage output.

Conversion Accuracy

The table in Figure 11-5(b) gives the *ideal* values of V_{OUT} for the various input cases. How close the circuit comes to producing these values depends primarily on two factors: (1) the precision of the input and feedback resistors and (2) the precision of the input voltage levels. The resistors can be made very accurate (within 0.01 percent of the desired values) by trimming, but the input voltage levels must be handled differently. It should be clear that the digital inputs cannot be taken directly from the outputs of FFs or logic gates because the output logic levels of these devices are not precise values like 0 V and 5 V but vary within given ranges. For this reason, it is necessary to add some more circuitry between each digital input and its input resistor to the summing amplifier, as shown in Figure 11-6.

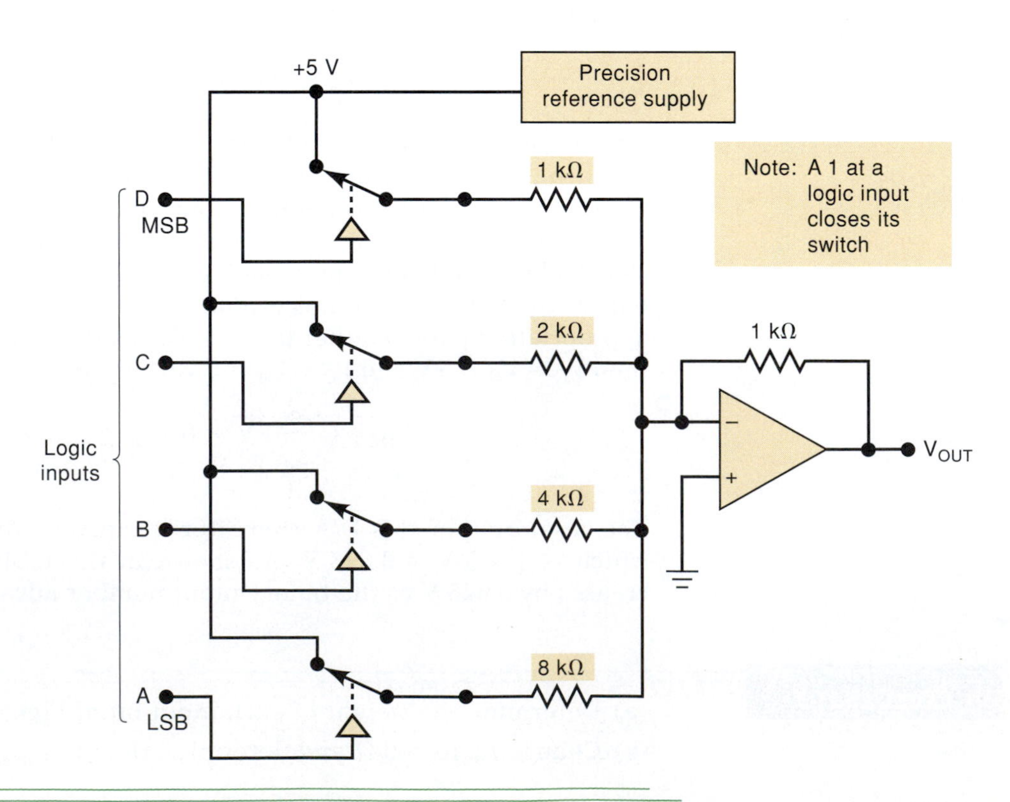

FIGURE 11-6 Complete four-bit DAC including a precision reference supply.

Each digital input controls a semiconductor switch such as the CMOS transmission gate we studied in Chapter 8. When the input is HIGH, the switch closes and connects a *precision reference supply* to the input resistor; when the input is LOW, the switch is open. The reference supply produces a very stable, precise voltage needed to generate an accurate analog output.

DAC with Current Output

Figure 11-7(a) shows one basic scheme for generating an analog output current proportional to a binary input. The circuit shown is a four-bit DAC using binarily weighted resistors. The circuit uses four parallel current paths, each controlled by a semiconductor switch such as the CMOS transmission gate. The state of each switch is controlled by logic levels at the binary inputs. The current through each path is determined by an accurate reference voltage, V_{REF}, and a precision resistor in the path. The resistors are binarily weighted so that the various currents will be binarily weighted, and the total current, I_{OUT}, will be the sum of the individual currents. The MSB path has the smallest resistor, R; the next path has a resistor of twice the value; and so on. The output current can be made to flow through a load R_L that is much smaller than R, so that it has no effect on the value of current. Ideally, R_L should be a short to ground.

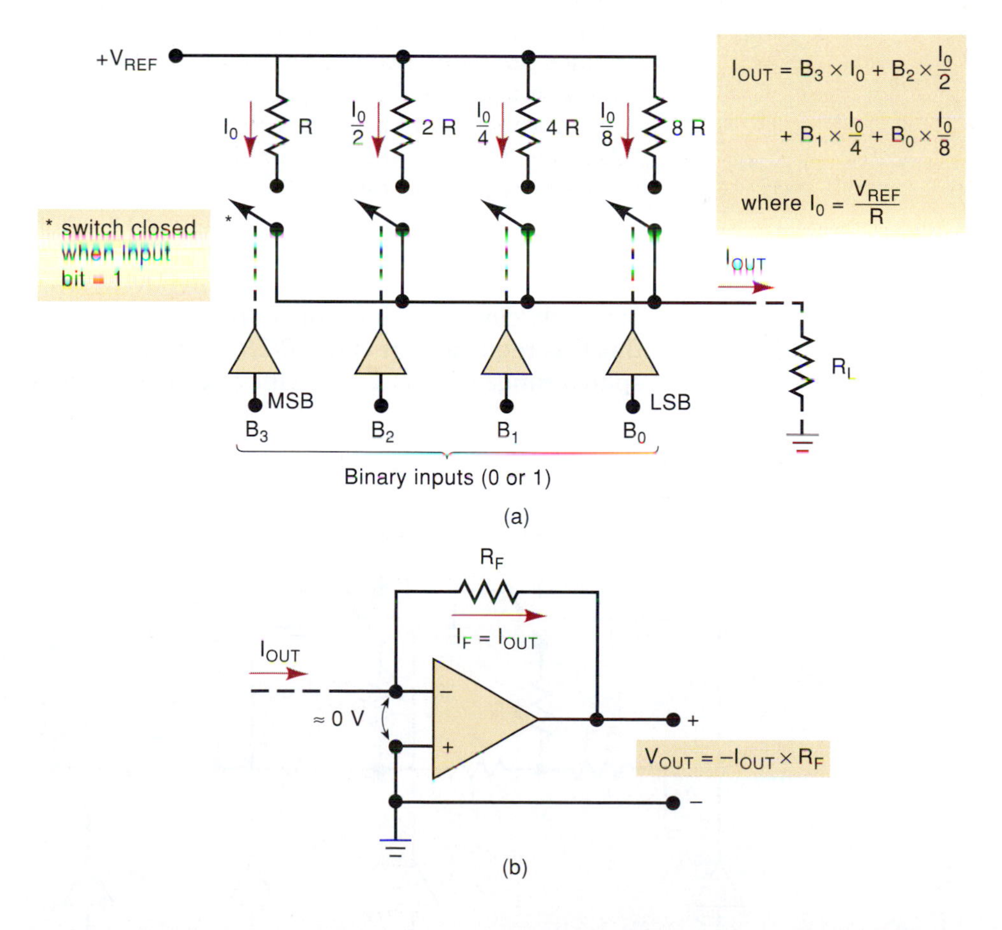

FIGURE 11-7 (a) Basic current-output DAC; (b) connected to an op-amp current-to-voltage converter.

EXAMPLE 11-8

Assume that $V_{REF} = 10$ V and $R = 10$ kΩ. Determine the resolution and the full-scale output for this DAC. Assume that R_L is much smaller than R.

Solution

$I_{OUT} = V_{REF}/R = 1$ mA. This is the weight of the MSB. The other three currents will be 0.5, 0.25, and 0.125 mA. The LSB is 0.125 mA, which is also the resolution.

The full-scale output will occur when the binary inputs are all HIGH so that each current switch is closed and

$$I_{OUT} = 1 + 0.5 + 0.25 + 0.125 = 1.875 \text{ mA}$$

Note that the output current is proportional to V_{REF}. If V_{REF} is increased or decreased, the resolution and the full-scale output will change proportionally.

For I_{OUT} to be accurate, R_L should be a short to ground. One common way to accomplish this is to use an op-amp as a current-to-voltage converter, as shown in Figure 11-7(b). Here, the I_{OUT} from the DAC is connected to the op-amp's "−" input, which is virtually at ground. The op-amp negative feedback forces a current equal to I_{OUT} to flow through R_F to produce $V_{OUT} = -I_{OUT} \times R_F$. Thus, V_{OUT} will be an analog voltage that is proportional to the binary input to the DAC. This analog output can drive a wide range of loads without being loaded down.

R/2*R* Ladder

The DAC circuits we have looked at thus far use binary-weighted resistors to produce the proper weighting of each bit. Whereas this method works in theory, it has some practical limitations. The biggest problem is the large difference in resistor values between the LSB and the MSB, especially in high-resolution DACs (i.e., many bits). For example, if the MSB resistor is 1 kΩ in a 12-bit DAC, the LSB resistor will be over 2 MΩ. With the current IC fabrication technology, it is very difficult to produce resistance values over a wide resistance range that maintain an accurate ratio, especially with variations in temperature.

For this reason, it is preferable to have a circuit that uses resistances that are fairly close in value. One of the most widely used DAC circuits that satisfies this requirement is the *R*/2*R ladder* network, where the resistance values span a range of only 2 to 1. One such DAC is shown in Figure 11-8.

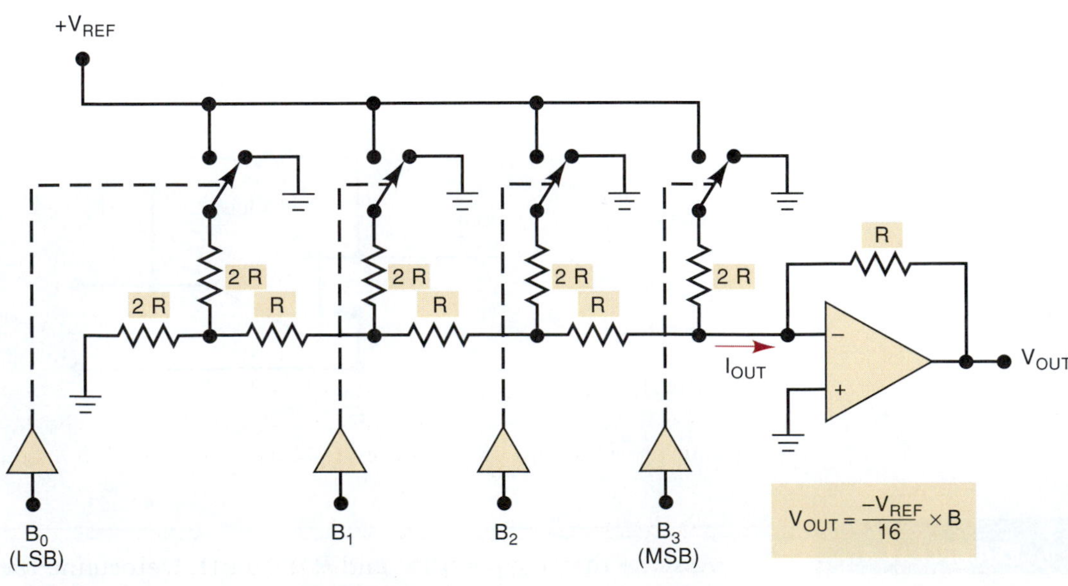

FIGURE 11-8 Basic *R*/2*R* ladder DAC.

Note how the resistors are arranged, and especially note that only two different values are used, R and $2R$. The current I_{OUT} depends on the positions of the four switches, and the binary inputs $B_3B_2B_1B_0$ control the states of the switches. This current is allowed to flow through an op-amp current-to-voltage converter to develop V_{OUT}. We will not perform a detailed analysis of this circuit here, but it can be shown that the value of V_{OUT} is given by the expression

$$V_{OUT} = \frac{-V_{REF}}{16} \times B \tag{11-6}$$

where B is the value of the binary input, which can range from 0000 (0) to 1111 (15).

EXAMPLE 11-9

Assume that $V_{REF} = 10$ V for the DAC in Figure 11-8. What are the resolution and full-scale output of this converter?

Solution

The resolution is equal to the weight of the LSB, which we can determine by setting $B = 0001 = 1$ in equation (11-6):

$$\text{resolution} = \frac{-10 \text{ V} \times 1}{16} = -0.625 \text{ V}$$

The full-scale output occurs for $B = 1111 = 15_{10}$. Again using equation (11-6),

$$\text{full-scale} = \frac{-10 \text{ V} \times 15}{16} = -9.375 \text{ V}$$

REVIEW QUESTIONS

1. What is the advantage of $R/2R$ ladder DACs over those that use binary-weighted resistors?
2. A certain six-bit DAC uses binary-weighted resistors. If the MSB resistor is 20 kΩ, what is the LSB resistor?
3. What will the resolution be if the value of R_F in Figure 11-5 is changed to 800 Ω?
4. What will happen to both resolution and full-scale output when V_{REF} is increased by 20 percent?

11-4 DAC SPECIFICATIONS

A wide variety of DACs are currently available as ICs or as self-contained, encapsulated packages. One should be familiar with the more important manufacturers' specifications in order to evaluate a DAC for a particular application.

Resolution

As mentioned earlier, the percentage resolution of a DAC depends solely on the number of bits. For this reason, manufacturers usually specify a DAC resolution as the number of bits. A 10-bit DAC has a finer (smaller) resolution than an eight-bit DAC.

Accuracy

DAC manufacturers have several ways of specifying accuracy. The two most common are called **full-scale error** and **linearity error**, which are normally expressed as a percentage of the converter's full-scale output (% F.S.).

Full-scale error is the maximum deviation of the DAC's output from its expected (ideal) value, expressed as a percentage of full scale. For example, assume that the DAC of Figure 11-5 has an accuracy of $\pm 0.01\%$ F.S. Because this converter has a full-scale output of 9.375 V, this percentage converts to

$$\pm 0.01\% \times 9.375 \text{ V} = \pm 0.9375 \text{ mV}$$

This means that the output of this DAC can, at any time, be off by as much as 0.9375 mV from its expected value.

Linearity error is the maximum deviation in step size from the ideal step size. For example, the DAC of Figure 11-5 has an expected step size of 0.625 V. If this converter has a linearity error of $\pm 0.01\%$F.S., this would mean that the actual *step size* could be off by as much as 0.9375 mV.

It is important to understand that accuracy and resolution of a DAC must be compatible. It is illogical to have a resolution of, say, 1 percent and an accuracy of 0.1 percent, or vice versa. To illustrate, a DAC with a resolution of 1 percent and an F.S. output of 10 V can produce an output analog voltage within 0.1 V of any desired value, assuming perfect accuracy. It makes no sense to have a costly accuracy of 0.01% F.S. (or 1 mV) if the resolution already limits the closeness of the desired value to 0.1 V. The same can be said for having a resolution that is very small (many bits) while the accuracy is poor; it is a waste of input bits.

EXAMPLE 11-10

A certain eight-bit DAC has a full-scale output of 2 mA and a full-scale error of $\pm 0.5\%$F.S. What is the range of possible outputs for an input of 10000000?

Solution

The step size is 2 mA/255 = 7.84 μA. Since $10000000 = 128_{10}$, the ideal output should be $128 \times 7.84\ \mu\text{A} = 1004\ \mu\text{A}$. The error can be as much as

$$\pm 0.5\% \times 2 \text{ mA} = \pm 10\ \mu\text{A}$$

Thus, the actual output can deviate by this amount from the ideal 1004 μA, so the actual output can be anywhere from 994 to 1014 μA.

Offset Error

Ideally, the output of a DAC will be zero volts when the binary input is all 0s. In practice, however, there will be a very small output voltage for this situation; this is called **offset error.** This offset error, if not corrected, will be added to the expected DAC output for *all* input cases. For example, let's say that a four-bit DAC has an offset error of +2 mV and a *perfect* step size of 100 mV. Table 11-3 shows the ideal and the actual DAC output for several input cases. Note that the actual output is 2 mV greater than expected; this is due to the offset error. Offset error can be negative as well as positive.

TABLE 11-3

Input Code	Ideal Output (mV)	Actual Output (mV)
0000	0	2
0001	100	102
1000	800	802
1111	1500	1502

Many DACs have an external offset adjustment that allows you to zero the offset. This is usually accomplished by applying all 0s to the DAC input and monitoring the output while an *offset adjustment potentiometer* is adjusted until the output is as close to 0 V as required.

Settling Time

The operating speed of a DAC is usually specified by giving its **settling time**, which is the time required for the DAC output to go from zero to full scale as the binary input is changed from all 0s to all 1s. Actually, the settling time is measured as the time for the DAC output to settle within $\pm\frac{1}{2}$ step size (resolution) of its final value. For example, if a DAC has a resolution of 10 mV, settling time is measured as the time it takes the output to settle within 5 mV of its full-scale value.

Typical values for settling time range from 50 ns to 10 μs. Generally speaking, DACs with a current output will have shorter settling times than those with voltage outputs. The main reason for this difference is the response time of the op-amp that is used as the current-to-voltage converter.

Monotonicity

A DAC is **monotonic** if its output increases as the binary input is incremented from one value to the next. Another way to describe this is that the staircase output will have no downward steps as the binary input is incremented from zero to full scale.

REVIEW QUESTIONS

1. Define *full-scale error.*
2. What is *settling time*?
3. Describe offset error and its effect on a DAC output.
4. Why are voltage DACs generally slower than current DACs?

11-5 AN INTEGRATED-CIRCUIT DAC

The AD7524, a CMOS IC available from several IC manufacturers, is an eight-bit D/A converter that uses an *R*/2*R* ladder network. Its block symbol is given in Figure 11-9(a). This DAC has an eight-bit input that can be latched internally under the control of the Chip Select ($\overline{CS}$) and WRITE ($\overline{WR}$) inputs. When both of these control inputs are LOW, the digital data inputs D_7–D_0 produce the analog output current *OUT 1* (the *OUT 2* terminal is normally grounded). When either control input goes HIGH, the digital input data are latched, and the analog output remains at the level corresponding to that latched digital data. Subsequent changes in the digital inputs will have no effect on *OUT 1* in this latched state.

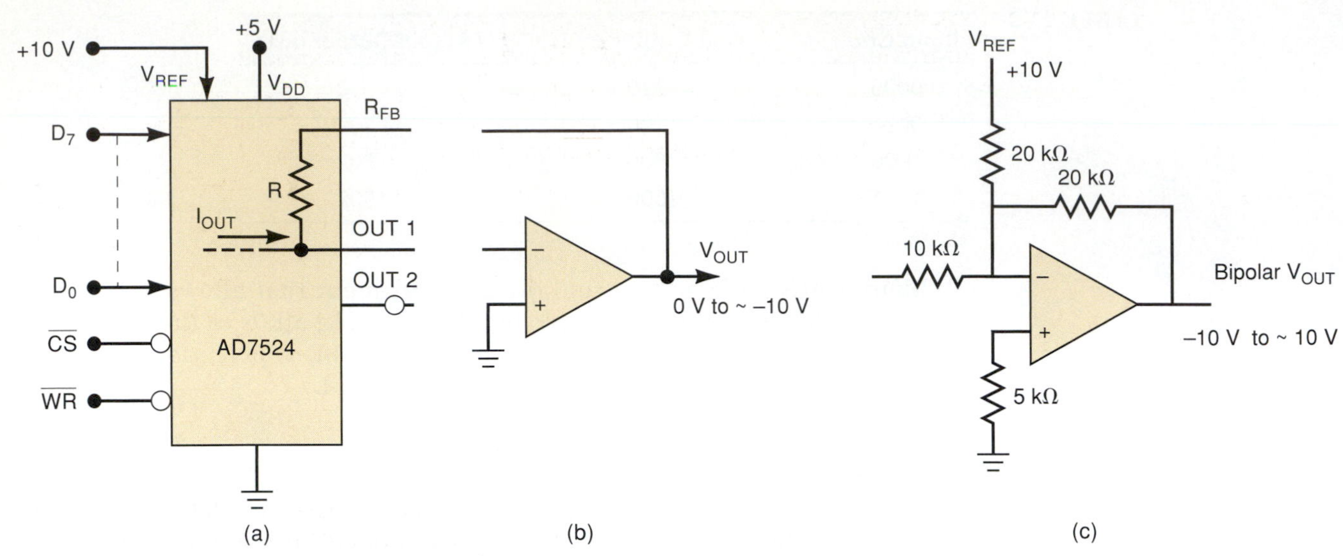

FIGURE 11-9 (a) AD7524 8-bit DAC with latched inputs; (b) op-amp current-to-voltage converter provides 0 to approximately 10 V out; (c) op-amp circuit to produce bipolar output from −10 V to approximately +10 V.

The maximum settling time for the AD7524 is typically 100 ns, and its full-range accuracy is rated at $\pm 0.2\%$F.S. The V_{REF} can range over both negative and positive voltages from 0 to 25 V, so that analog output currents of both polarities can be produced. The output current can be converted to a voltage using an op-amp connected as in Figure 11-9(b). Note that the op-amp's feedback resistor is already on the DAC chip. The op-amp circuit shown in Figure 11-9(c) can be added to produce a bipolar output that ranges from $-V_{ref}$ (when input = 00000000) to almost $+V_{ref}$ (when input = 11111111).

11-6 DAC APPLICATIONS

DACs are used whenever the output of a digital circuit must provide an analog voltage or current to drive an analog device. Some of the most common applications are described in the following paragraphs.

Control

The digital output from a computer can be converted to an analog control signal to adjust the speed of a motor or the temperature of a furnace, or to control almost any physical variable.

Automatic Testing

Computers can be programmed to generate the analog signals (through a DAC) needed to test analog circuitry. The test circuit's analog output response will normally be converted to a digital value by an ADC and fed into the computer to be stored, displayed, and sometimes analyzed.

Signal Reconstruction

In many applications, an analog signal is **digitized;** that is, successive points on the signal are converted to their digital equivalents and stored in memory. This conversion is performed by an analog-to-digital converter (ADC). A

DAC can then be used to convert the stored digitized data back to analog—one point at a time—thereby reconstructing the original signal. This combination of digitizing and reconstructing is used in digital storage oscilloscopes, audio compact disk systems, and digital audio and video recording. We will look at this further after we learn about ADCs.

A/D Conversion

Several types of ADCs use DACs as part of their circuitry, as we shall see in Section 11-8.

Digital Amplitude Control

DACs can also be used to reduce the amplitude of an analog signal by connecting the analog signal to the V_{REF} input as shown in Figure 11-10. The binary input scales the signal on V_{REF}: $V_{OUT} = V_{REF} \times \text{binary in}/2^N$. When the maximum binary input value is applied, the output is nearly the same as the V_{REF} input. However, when a value that represents about half of the maximum (e.g., 1000 000 for a unipolar eight-bit converter) is applied to the inputs, the output is about half of V_{REF}. If V_{REF} is a signal (e.g., a sine wave) that varies within the range of the reference voltage, the output will be the same fully analog wave shape whose amplitude depends on the digital number applied to the DAC. In this way a digital system can control things such as the volume of an audio system or the amplitude of a function generator.

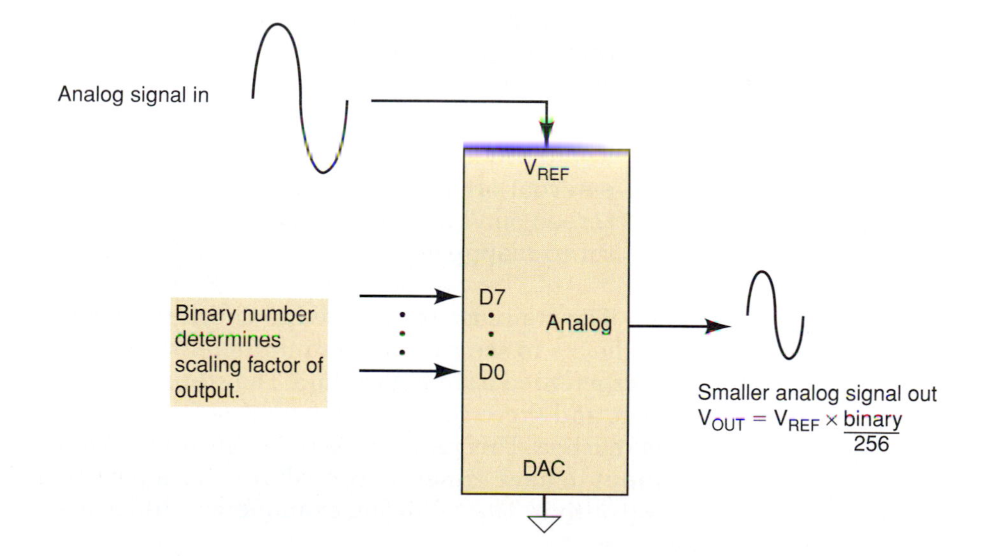

FIGURE 11-10 A DAC used to control the amplitude of an analog signal.

Serial DACs

Many of these DAC applications involve a microprocessor. The main problem with using the parallel-data DACs that have been described thus far is that they occupy so many port bits of the microcomputer. In cases where speed of data transfer is of little concern, a microprocessor can output the digital value to a DAC over a serial interface. Serial DACs are now readily available with a built-in serial in/parallel out shift register. Many of these devices have more than one DAC on the same chip. The digital data, along with a code that specifies which DAC you want, are sent to the chip, one bit at a time. As each bit is presented on the DAC input, a pulse is applied to the serial clock input to shift the bit in. After the proper number of clock pulses, the data value is latched and converted to its analog value.

11-7 TROUBLESHOOTING DACs

DACs are both digital and analog. Logic probes and pulsers can be used on the digital inputs, but a meter or an oscilloscope must be used on the analog output. There are basically two ways to test a DAC's operation: a *static accuracy test* and a *staircase test.*

The static test involves setting the binary input to a fixed value and measuring the analog output with a high-accuracy meter. This test is used to check that the output value falls within the expected range consistent with the DAC's specified accuracy. If it does not, there can be several possible causes. Here are some of them:

- Drift in the DAC's internal component values (e.g., resistor values) caused by temperature, aging, or some other factors. This condition can easily produce output values outside the expected accuracy range.
- Open connections or shorts in any of the binary inputs. This could either prevent an input from adding its weight to the analog output or cause its weight to be permanently present in the output. This situation is especially hard to detect when the fault is in the less significant inputs.
- A faulty voltage reference. Because the analog output depends directly on V_{REF}, this could produce results that are way off. For DACs that use external reference sources, the reference voltage can be checked easily with a digital voltmeter. Many DACs have internal reference voltages that cannot be checked, except on some DACs that bring the reference voltage out to a pin of the IC.
- Excessive offset error caused by component aging or temperature. This would produce outputs that are off by a fixed amount. If the DAC has an external offset adjustment capability, this type of error can initially be zeroed out, but changes in operating temperature can cause the offset error to reappear.

The staircase test is used to check the monotonicity of the DAC; that is, it checks to see that the output increases step by step as the binary input is incremented as in Figure 11-3. The steps on the staircase must be of the same size, and there should be no missing steps or downward steps until full scale is reached. This test can help detect internal or external faults that cause an input to have either no contribution or a permanent contribution to the analog output. The following example will illustrate.

EXAMPLE 11-11

How would the staircase waveform appear if the *C* input to the DAC of Figure 11-3 is open? Assume that the DAC inputs are TTL-compatible.

Solution

An open connection at *C* will be interpreted as a constant logic 1 by the DAC. Thus, this will contribute a constant 4 V to the DAC output so that the DAC output waveform will appear as shown in Figure 11-11. The dotted lines are the staircase as it would appear if the DAC were working correctly. Note that the faulty output waveform matches the correct one during those times when the bit *C* input would normally be HIGH.

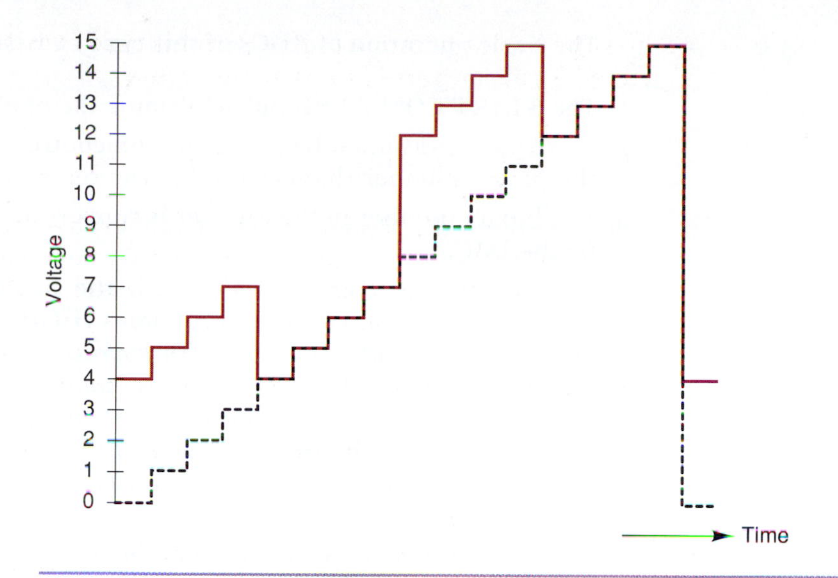

FIGURE 11-11 Example 11-11.

11-8 ANALOG-TO-DIGITAL CONVERSION

An analog-to-digital converter takes an analog input voltage and, after a certain amount of time, produces a digital output code that represents the analog input. The A/D conversion process is generally more complex and time-consuming than the D/A process, and many different methods have been developed and used. We shall examine several of these methods in detail, even though it may never be necessary to design or construct ADCs (they are available as completely packaged units). However, the techniques that are used provide an insight into what factors determine an ADC's performance.

Several important types of ADCs utilize a DAC as part of their circuitry. Figure 11-12 is a general block diagram for this class of ADC. The timing for the operation is provided by the input clock signal. The control unit contains the logic circuitry for generating the proper sequence of operations in response to the START COMMAND, which initiates the conversion process. The op-amp comparator has two *analog* inputs and a *digital* output that switches states, depending on which analog input is greater.

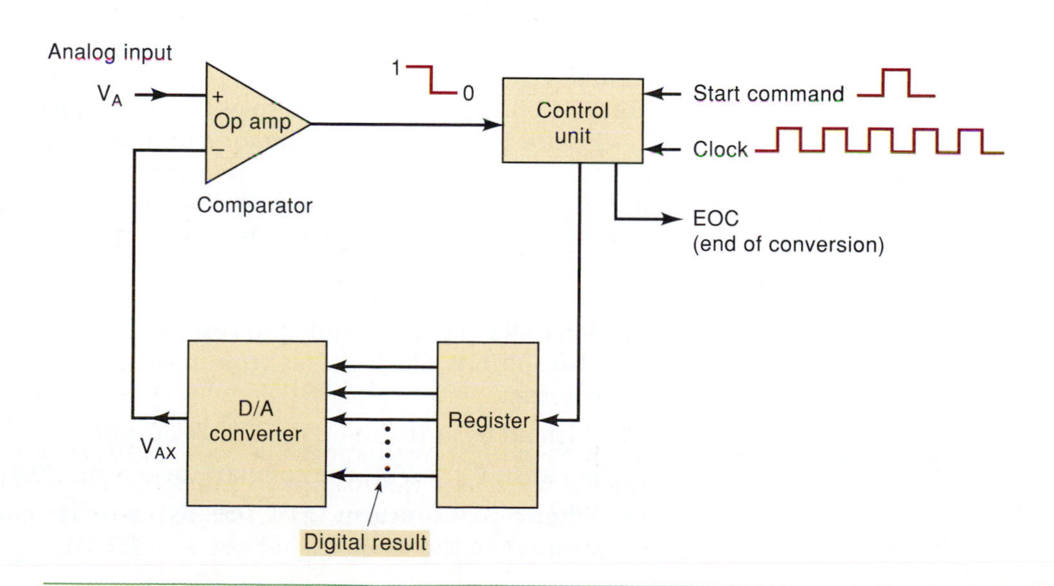

FIGURE 11-12 General diagram of one class of ADCs.

The basic operation of ADCs of this type consists of the following steps:

1. The START COMMAND pulse initiates the operation.
2. At a rate determined by the clock, the control unit continually modifies the binary number that is stored in the register.
3. The binary number in the register is converted to an analog voltage, V_{AX}, by the DAC.
4. The comparator compares V_{AX} with the analog input V_A. As long as $V_{AX} < V_A$, the comparator output stays HIGH. When V_{AX} exceeds V_A by at least an amount equal to V_T (threshold voltage), the comparator output goes LOW and stops the process of modifying the register number. At this point, V_{AX} is a close approximation to V_A. The digital number in the register, which is the digital equivalent of V_{AX}, is also the approximate digital equivalent of V_A, within the resolution and accuracy of the system.
5. The control logic activates the end-of-conversion signal, *EOC*, when the conversion is complete.

The several variations of this A/D conversion scheme differ mainly in the manner in which the control section continually modifies the numbers in the register. Otherwise, the basic idea is the same, with the register holding the required digital output when the conversion process is complete.

REVIEW QUESTIONS

1. What is the function of the comparator in the ADC?
2. Where is the approximate digital equivalent of V_A when the conversion is complete?
3. What is the function of the *EOC* signal?

11-9 DIGITAL-RAMP ADC

One of the simplest versions of the general ADC of Figure 11-12 uses a binary counter as the register and allows the clock to increment the counter one step at a time until $V_{AX} \geq V_A$. It is called a **digital-ramp ADC** because the waveform at V_{AX} is a step-by-step ramp (actually a staircase) like the one shown in Figure 11-3. It is also referred to as a *counter-type* ADC.

Figure 11-13 is the diagram for a digital-ramp ADC. It contains a counter, a DAC, an analog comparator, and a control AND gate. The comparator output serves as the active-LOW end-of-conversion signal $\overline{EOC}$. If we assume that V_A, the analog voltage to be converted, is positive, the operation proceeds as follows:

1. A START pulse is applied to reset the counter to 0. The HIGH at START also inhibits clock pulses from passing through the AND gate into the counter.
2. With all 0s at its input, the DAC's output will be $V_{AX} = 0$ V.
3. Because $V_A > V_{AX}$, the comparator output, $\overline{EOC}$, will be HIGH.
4. When START returns LOW, the AND gate is enabled and clock pulses get through to the counter.

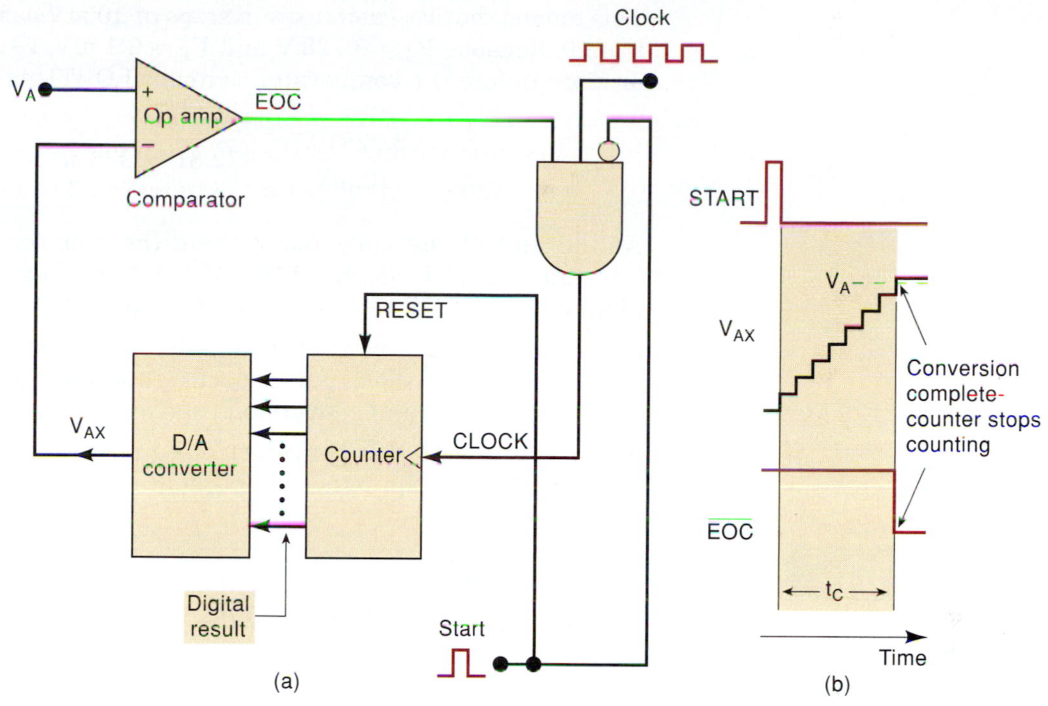

FIGURE 11-13 Digital-ramp ADC.

5. As the counter advances, the DAC output, V_{AX}, increases one step at a time, as shown in Figure 11-13(b).
6. This process continues until V_{AX} reaches a step that exceeds V_A by an amount equal to or greater than V_T (typically 10 to 100 μV). At this point, $\overline{EOC}$ will go LOW and inhibit the flow of pulses into the counter, and the counter will stop counting.
7. The conversion process is now complete, as signaled by the HIGH-to-LOW transition at $\overline{EOC}$, and the contents of the counter are the digital representation of V_A.
8. The counter will hold the digital value until the next START pulse initiates a new conversion.

EXAMPLE 11-12

Assume the following values for the ADC of Figure 11-13: clock frequency = 1 MHz; V_T = 0.1 mV; DAC has F.S. output = 10.23 V and a 10-bit input. Determine the following values.

(a) The digital equivalent obtained for $V_A = 3.728$ V
(b) The conversion time
(c) The resolution of this converter

Solution

(a) The DAC has a 10-bit input and a 10.23-V F.S. output. Thus, the number of total possible steps is $2^{10} - 1 = 1023$, and so the step size is

$$\frac{10.23\text{ V}}{1023} = 10\text{ mV}$$

This means that V_{AX} increases in steps of 10 mV as the counter counts up from 0. Because $V_A = 3.728$ V and $V_T = 0.1$ mV, V_{AX} must reach 3.7281 V or more before the comparator switches LOW. This will require

$$\frac{3.7281 \text{ V}}{10 \text{ mV}} = 372.81 = 373 \text{ steps}$$

At the end of the conversion, then, the counter will hold the binary equivalent of 373, which is 0101110101. This is the desired digital equivalent of $V_A = 3.728$ V, as produced by this ADC.

(b) Three hundred seventy-three steps were required to complete the conversion. Thus, 373 clock pulses occurred at the rate of one per microsecond. This gives a total conversion time of 373 μs.

(c) The resolution of this converter is equal to the step size of the DAC, which is 10 mV. Expressed as a percentage, it is $1/1023 \times 100\% \approx 0.1\%$.

EXAMPLE 11-13

For the same ADC of Example 11-12, determine the approximate range of analog input voltages that will produce the same digital result of $0101110101_2 = 373_{10}$.

Solution

TABLE 11-4

Step	V_{AX} (V)
371	3.71
372	3.72
373	3.73
374	3.74
375	3.75

Table 11-4 shows the ideal DAC output voltage, V_{AX}, for several of the steps on and around the 373rd. If V_A is slightly smaller than 3.72 V (by an amount $< V_T$), then $\overline{EOC}$ won't go LOW when V_{AX} reaches the 3.72-V step, but it will go LOW on the 3.73-V step. If V_A is slightly smaller than 3.73 V (by an amount $< V_T$), then $\overline{EOC}$ won't go LOW until V_{AX} reaches the 3.74-V step. Thus, as long as V_A is between approximately 3.72 and 3.73 V, $\overline{EOC}$ will go LOW when V_{AX} reaches the 3.73-V step. The exact range of V_A values is

$$3.72 \text{ V} - V_T \qquad \text{to} \qquad 3.73 \text{ V} - V_T$$

but because V_T is so small, we can simply say that the range is approximately 3.72 to 3.73 V—a range equal to 10 mV, the DAC's resolution. This is illustrated in Figure 11-14.

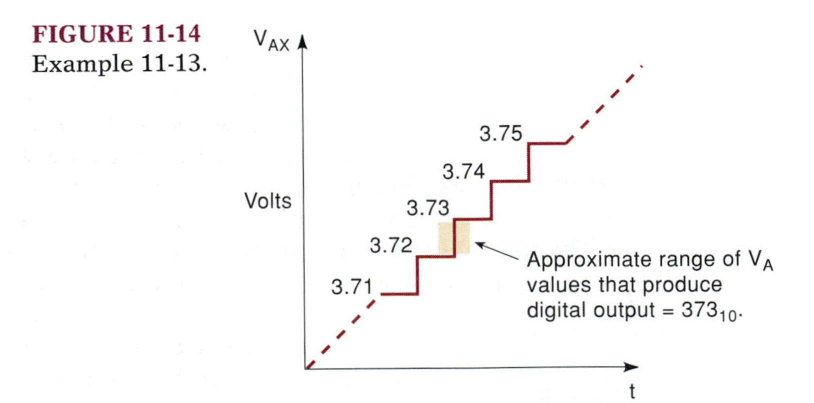

FIGURE 11-14 Example 11-13.

A/D Resolution and Accuracy

It is very important to understand the errors associated with making any kind of measurements. An unavoidable source of error in the digital-ramp method is that the step size or resolution of the internal DAC is the smallest unit of measure. Imagine trying to measure basketball players' heights by standing them next to a staircase with 12-in steps and assigning them the height of the first step higher than their head. Anyone over 6 ft would be measured as 7 ft tall! Likewise, the output voltage V_{AX} is a staircase waveform that goes up in discrete steps until it exceeds the input voltage, V_A. By making the step size smaller, we can reduce the potential error, but there will always be a difference between the actual (analog) quantity and the digital value assigned to it. This is called **quantization error**. Thus, V_{AX} is an approximation to the value of V_A, and the best we can expect is that V_{AX} is within 10 mV of V_A if the resolution (step size) is 10 mV. This quantization error, which can be reduced by increasing the number of bits in the counter and the DAC, is sometimes specified as an error of +1 LSB, indicating that the result could be off by as much as the weight of the LSB.

A more common practice is to make the quantization error symmetrical around an integer multiple of the resolution to make the quantization error $\pm\frac{1}{2}$ LSB. This is done by making sure the output changes at $\frac{1}{2}$ resolution unit below and above the nominal input voltage. For example, if the resolution is 10 mV, then the A/D output will ideally switch from 0 to 1 at 5 mV and from 1 to 2 at 15 mV. The nominal value (10 mV), which is represented by the digital value of 1, is ideally always within 5 mV ($\frac{1}{2}$ LSB) of the actual input voltage. Problem 11-28 explores a method to accomplish this. In any case, there is a small range of input voltages that will produce the same digital output.

The accuracy specification reflects the fact that the output of every ADC does not switch from one binary value to the next at the exact prescribed input voltage. Some change at slightly higher voltage than expected, and some at slightly lower. The inaccuracy and inconsistency is due to imperfect components such as precision resistors, comparators, current switches, and so on. Accuracy can be expressed as % full-scale, just as for the DAC, but it is more commonly specified as $\pm n$ LSB, where n is a fractional value or 1. For example, if the accuracy is specified as $\pm\frac{1}{4}$ LSB with a resolution of 10 mV, and assuming the output should ideally switch from 0 to 1 at 5 mV, then we know that the output could change from 0 to 1 at any input voltage between 2.5 and 7.5 mV. In this case we would be assured that any voltage between 7.5 and 12.5 mV would definitely produce the value 1. However, in the worst case, the output of binary 1 could be representing a nominal value of 10 mV with an actual applied voltage of 2.5 mV, an error of $\frac{3}{4}$ bit which is the sum of the quantization error and the accuracy.

EXAMPLE 11-14

A certain eight-bit ADC, similar to Figure 11-13, has a full-scale input of 2.55 V (i.e., $V_A = 2.55$ V produces a digital output of 11111111). It has a specified error of $\pm\frac{1}{4}$ LSB. Determine the maximum amount of error in the measurement.

Solution

The step size is 2.55 V/($2^8 - 1$), which is exactly 10 mV. This means that even if the DAC has no inaccuracies, the V_{AX} output could be off by as much as 10 mV because V_{AX} can change only in 10-mV steps; this is the quantization error. The specified error of $\pm\frac{1}{4}$ LSB is $\frac{1}{4} \times 10$ mV = 2.5 mV. This means that the V_{AX} value can be off by as much as 2.5 mV because of component inaccuracies. Thus, the total possible error could be as much as 10 mV + 2.5 mV = 12.5 mV.

For example, suppose that the analog input was 1.268 V. If the DAC output were perfectly accurate, the staircase would stop at the 127th step (1.27 V). But let's say that V_{AX} was off by −2 mV, so it was 1.268 V at the 127th step. This would not be large enough to stop the conversion; it would stop at the 128th step. Thus, the digital output would be $10000000_2 = 128_{10}$ (representing 12.8 V) for an analog input of 1.268 V, an error of 12 mV.

Conversion Time, t_C

The conversion time is shown in Figure 11-13(b) as the time interval between the end of the START pulse and the activation of the $\overline{EOC}$ output. The counter starts counting from 0 and counts up until V_{AX} exceeds V_A, at which point $\overline{EOC}$ goes LOW to end the conversion process. It should be clear that the value of the conversion time, t_C, depends on V_A. A larger value will require more steps before the staircase voltage exceeds V_A.

The maximum conversion time will occur when V_A is just below full scale so that V_{AX} must go to the last step to activate $\overline{EOC}$. For an N-bit converter, this will be

$$t_C(\text{max}) = (2^N - 1) \text{ clock cycles}$$

For example, the ADC in Example 11-12 would have a maximum conversion time of

$$t_C(\text{max}) = (2^{10} - 1) \times 1\ \mu\text{s} = 1023\ \mu\text{s}$$

Sometimes, average conversion time is specified; it is half of the maximum conversion time. For the digital-ramp converter, this would be

$$t_C(\text{avg}) = \frac{t_C(\text{max})}{2} \approx 2^{N-1} \text{ clock cycles}$$

The major disadvantage of the digital-ramp method is that conversion time essentially doubles for each bit that is added to the counter, so that resolution can be improved only at the cost of a longer t_C. This makes this type of ADC unsuitable for applications where repetitive A/D conversions of a fast-changing analog signal must be made. For low-speed applications, however, the relative simplicity of the digital-ramp converter is an advantage over the more complex, higher-speed ADCs.

EXAMPLE 11-15

What will happen to the operation of a digital-ramp ADC if the analog input V_A is greater than the full-scale value?

Solution

From Figure 11-13, it should be clear that the comparator output will never go LOW because the staircase voltage can never exceed V_A. Thus, pulses will be continually applied to the counter, so that the counter will repetitively count up from 0 to maximum, recycle back to 0, count up, and so on. This will produce repetitive staircase waveforms at V_{AX} going from 0 to full scale, and this will continue until V_A is decreased below full scale.

REVIEW QUESTIONS

1. Describe the basic operation of the digital-ramp ADC.
2. Explain *quantization error.*
3. Why does conversion time increase with the value of the analog input voltage?
4. *True or false:* Everything else being equal, a 10-bit digital-ramp ADC will have a better resolution, but a longer conversion time, than an eight-bit ADC.
5. Give one advantage and one disadvantage of a digital-ramp ADC.
6. For the converter of Example 11-12, determine the digital output for $V_A = 1.345$ V. Repeat for $V_A = 1.342$ V.

11-10 DATA ACQUISITION

There are many applications in which analog data must be *digitized* (converted to digital) and transferred into a computer's memory. The process by which the computer acquires these digitized analog data is referred to as *data acquisition.* Acquiring a single data point's value is referred to as **sampling** the analog signal, and that data point is often called a *sample.* The computer can do several different things with the data, depending on the application. In a storage application, such as digital audio recording, video recording, or a digital oscilloscope, the internal microcomputer will store the data and then transfer them to a DAC at a later time to reproduce the original analog signal. In a process control application, the computer can examine the data or perform computations on them to determine what control outputs to generate.

Figure 11-15(a) shows how a microcomputer is connected to a digital-ramp ADC for the purpose of data acquisition. The computer generates the START pulses that initiate each new A/D conversion. The EOC (end-of-conversion) signal from the ADC is fed to the computer. The computer monitors $\overline{EOC}$ to find out when the current A/D conversion is complete; then it transfers the digital data from the ADC output into its memory.

The waveforms in Figure 11-15(b) illustrate how the computer acquires a digital version of the analog signal, V_A. The V_{AX} staircase waveform that is generated internal to the ADC is shown superimposed on the V_A waveform for purposes of illustration. The process begins at t_0, when the computer generates a START pulse to start an A/D conversion cycle. The conversion is completed at t_1, when the staircase first exceeds V_A, and $\overline{EOC}$ goes LOW. This NGT at $\overline{EOC}$ signals the computer that the ADC has a digital output that now represents the value of V_A at point *a*, and the computer will load these data into its memory.

The computer generates a new START pulse shortly after t_1 to initiate a second conversion cycle. Note that this resets the staircase to 0 and $\overline{EOC}$ back HIGH because the START pulse resets the counter in the ADC. The second conversion ends at t_2 when the staircase again exceeds V_A. The computer then loads the digital data corresponding to point *b* into its memory. These steps are repeated at t_3, t_4, and so on.

The process whereby the computer generates a START pulse, monitors $\overline{EOC}$, and loads ADC data into memory is done under the control of a program that the computer is executing. This data acquisition program will determine how many data points from the analog signal will be stored in the computer memory.

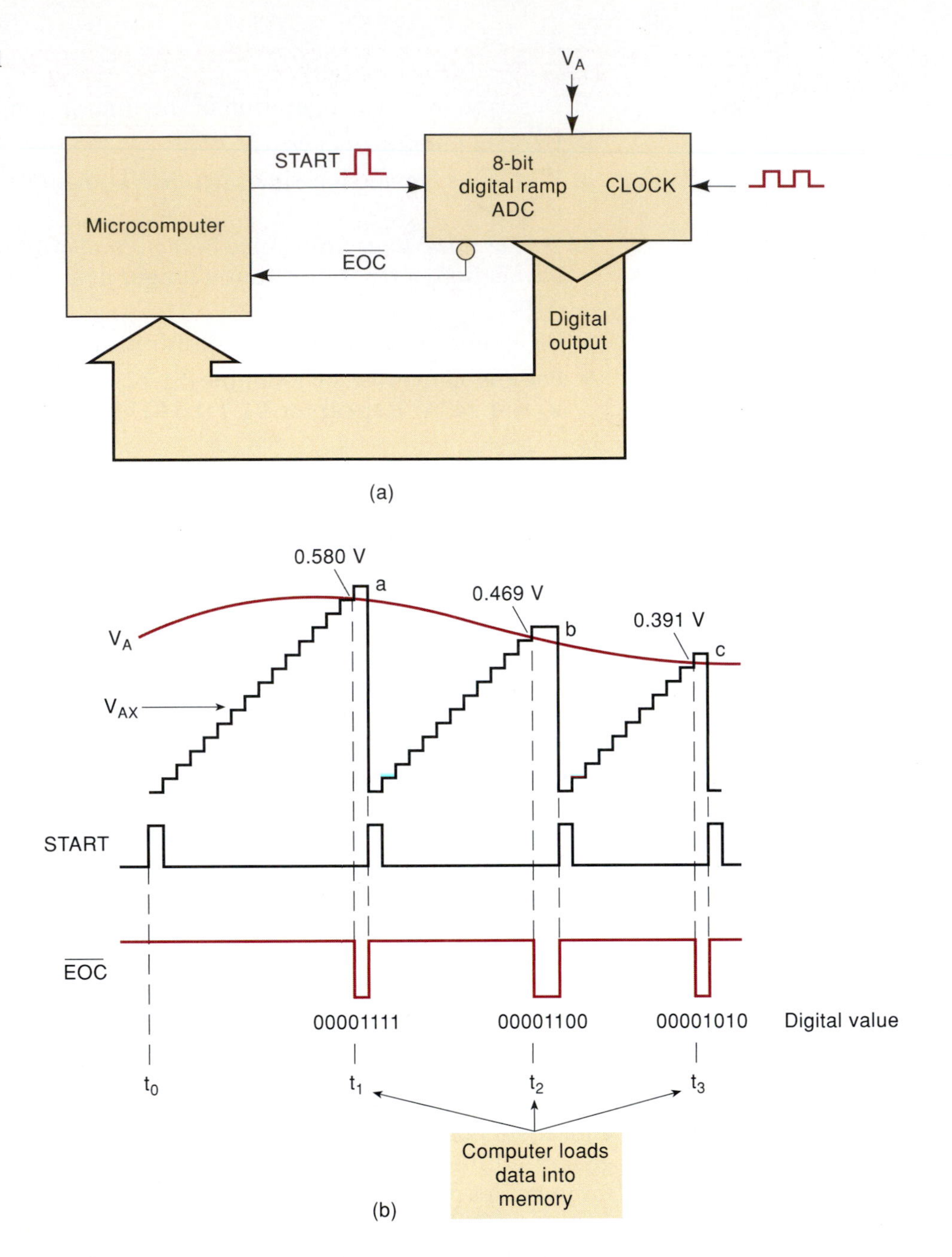

FIGURE 11-15 (a) Typical computer data acquisition system; (b) waveforms showing how the computer initiates each new conversion cycle and then loads the digital data into memory at the end of conversion.

Reconstructing a Digitized Signal

In Figure 11-15(b), the ADC is operating at its maximum speed because a new START pulse is generated immediately after the computer acquires the ADC output data from the previous conversion. Note that the conversion times are not constant because the analog input value is changing. The problem with this method of storing a waveform is that in order to reconstruct the waveform, we would need to know the point in time that each data value is to be plotted. Normally, when storing a digitized waveform, the samples are taken at fixed intervals at a rate that is at least two times greater than the highest frequency in the analog signal. The digital system will store the waveform as a list of sample data values. Table 11-5 shows the list of data that would be stored if the signal in Figure 11-16(a) were digitized.

TABLE 11-5 Digitized data samples.

Point	Actual Voltage (V)	Digital Equivalent
a	1.22	01111010
b	1.47	10010011
c	1.74	10101110
d	1.70	10101010
e	1.35	10000111
f	1.12	01110000
g	0.91	01011011
h	0.82	01010010

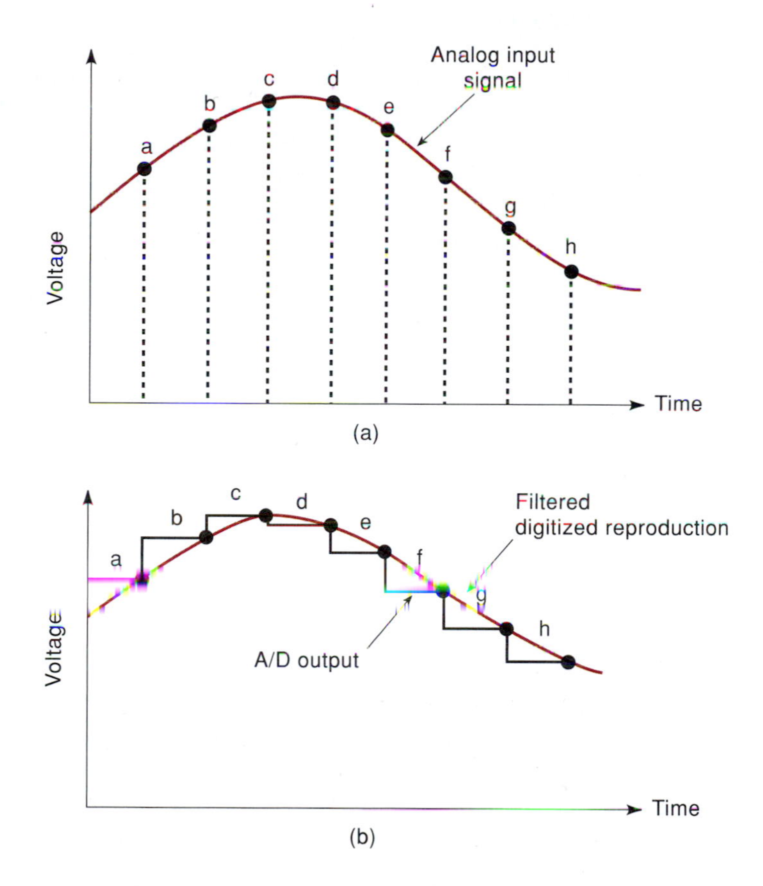

FIGURE 11-16 (a) Digitizing an analog signal; (b) reconstructing the signal from the digital data.

In Figure 11-16(a), we see how the ADC continually performs conversions to digitize the analog signal at points *a*, *b*, *c*, *d*, and so on. If these digital data are used to reconstruct the signal, the result will look like that in Figure 11-16(b). The black line represents the voltage waveform that would actually come out of the D/A converter. The red line would be the result of passing the signal through a simple low-pass RC filter. We can see that it is a fairly good reproduction of the original analog signal because the analog signal does not make any rapid changes between digitized points. If the analog signal contained higher-frequency variations, the ADC would not be able to follow the variations, and the reproduced version would be much less accurate.

Aliasing

The obvious goal in signal reconstruction is to make the reconstruction nearly identical to the original analog signal. In order to avoid loss of information, as has been proven by a man named Harry Nyquist, the incoming signal must be sampled at a rate greater than two times the highest-frequency component in the incoming signal. For example, if you are pretty sure that the highest frequency in an audio system will be less than 10 kHz, you must sample the audio signal at 20,000 samples per second in order to be able to reconstruct the signal. The frequency at which samples are taken is referred to as the **sampling frequency, F_S**. What do you think would happen if for some reason a 12-kHz tone is present in the input signal? Unfortunately, the system would *not* simply ignore it because it is too high! Rather, a phenomenon called *aliasing* would occur. A signal **alias** is produced by sampling the signal at a rate less than the minimum rate identified by Nyquist (twice the highest incoming frequency). In this case, any frequency over 10 kHz will produce an alias frequency. The alias frequency is always the difference between any integer multiple of the sampling frequency F_S (20 kHz) and the incoming frequency that is being digitized (12 kHz). Instead of hearing a 12-kHz tone in the reconstructed signal, you would hear an 8-kHz tone that was not in the original signal.

To see how aliasing can happen, consider the sine wave in Figure 11-17. Its frequency is 1.9 kHz. The dots show where the waveform is sampled every 500 μs ($F_S = 2$ kHz). If we connect the dots that make up the sampled waveform, we discover that they form a cosine wave that has a period of 10 ms and a frequency of 100 Hz. This demonstrates that the alias frequency is equal to the difference between the sample frequency and the incoming frequency. If we could hear the output that results from this data acquisition, it would not sound like 1.9 kHz; it would sound like 100 Hz.

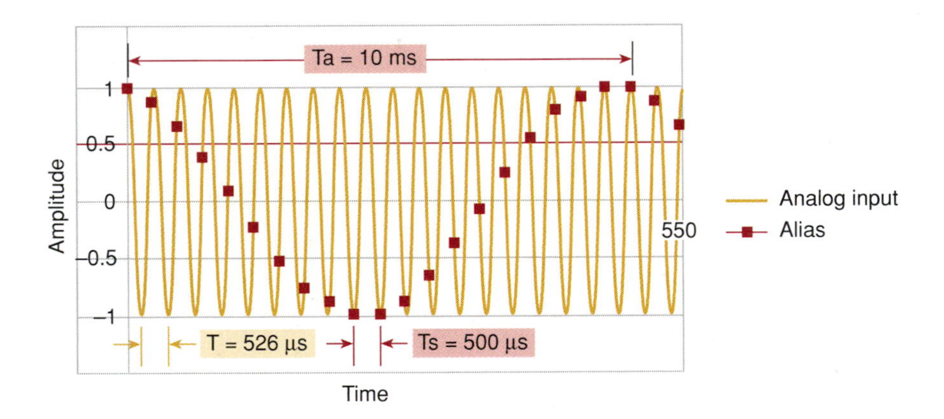

FIGURE 11-17 An alias signal due to undersampling.

The problem with **undersampling** ($F_S < 2\,F_{in}$ max) is that the digital system has no idea that there was actually a higher frequency at the input. It simply samples the input and stores the data. When it reconstructs the signal, the alias frequency (100 Hz) is present, the original 1.9 kHz is missing, and the reconstructed signal does not sound the same. This is why a data acquisition system must not allow frequencies greater than half of F_S to be placed on the input.

REVIEW QUESTIONS

1. What is *digitizing a signal*?
2. Describe the steps in a computer data acquisition process.
3. What is the minimum sample frequency needed to reconstruct an analog signal?
4. What occurs if the signal is sampled at less than the minimum frequency determined in question 3?

11-11 SUCCESSIVE-APPROXIMATION ADC

The **successive-approximation converter** is one of the most widely used types of ADC. It has more complex circuitry than the digital-ramp ADC but a much shorter conversion time. In addition, successive-approximation converters (SACs) have a fixed value of conversion time that is not dependent on the value of the analog input.

The basic arrangement, shown in Figure 11-18(a), is similar to that of the digital-ramp ADC. The SAC, however, does not use a counter to provide the input to the DAC block but uses a register instead. The control logic modifies the contents of the register bit by bit until the register data are the digital equivalent of the analog input V_A within the resolution of the converter. The basic sequence of operation is given by the flowchart in Figure 11-18(b). We will follow this flowchart as we go through the example illustrated in Figure 11-19.

For this example, we have chosen a simple four-bit converter with a step size of 1 V. Even though most practical SACs would have more bits and smaller resolution than our example, the operation will be exactly the same. At this point, you should be able to determine that the four register bits feeding the DAC have weights of 8, 4, 2, and 1 V, respectively.

Let's assume that the analog input is $V_A = 10.4$ V. The operation begins with the control logic clearing all of the register bits to 0 so that $Q_3 = Q_2 = Q_1 = Q_0 = 0$. We will express this as $[Q] = 0000$. This makes the DAC output $V_{AX} = 0$ V, as indicated at time t_0 on the timing diagram in Figure 11-19. With $V_{AX} < V_A$, the comparator output is HIGH.

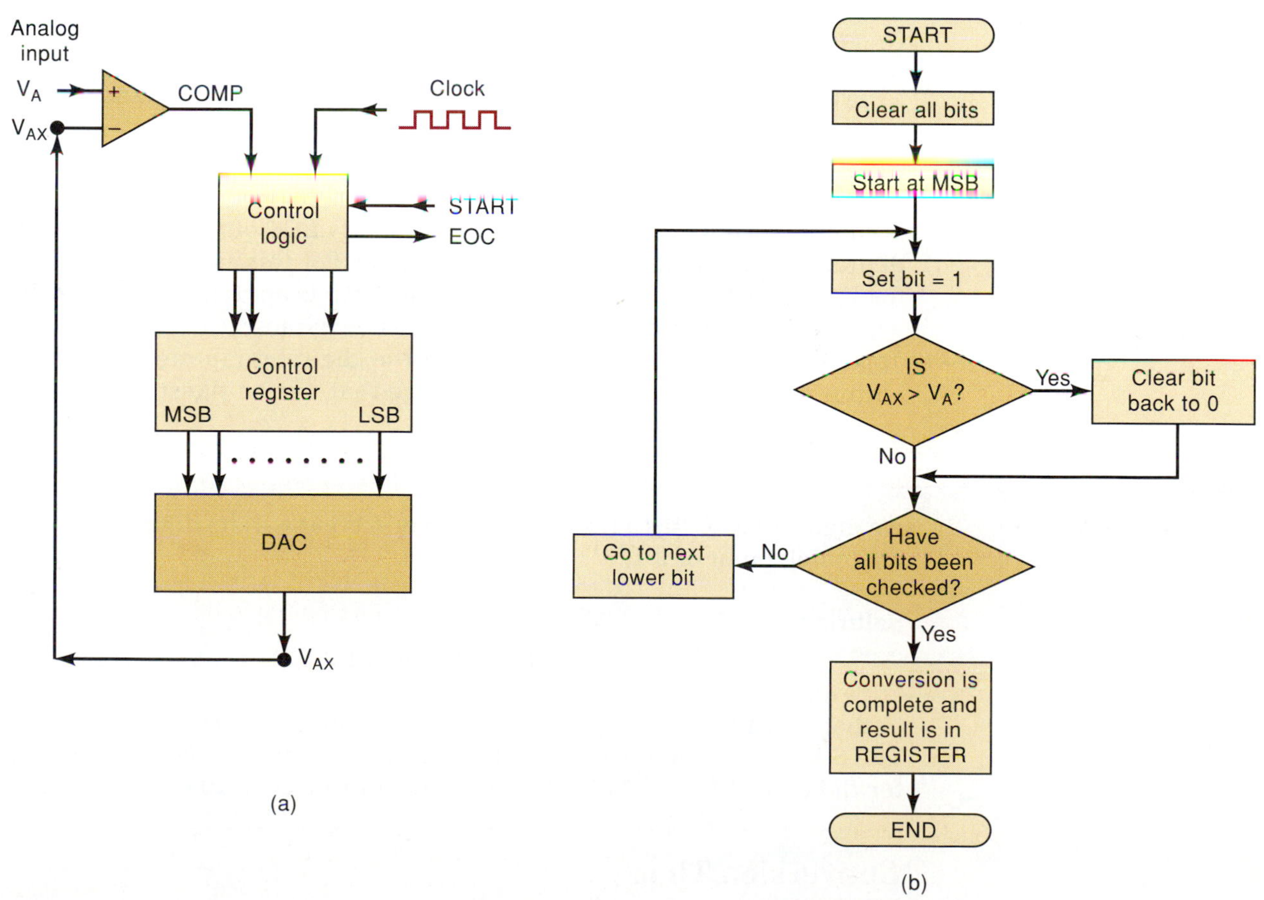

FIGURE 11-18 Successive-approximation ADC: (a) simplified block diagram; (b) flowchart of operation.

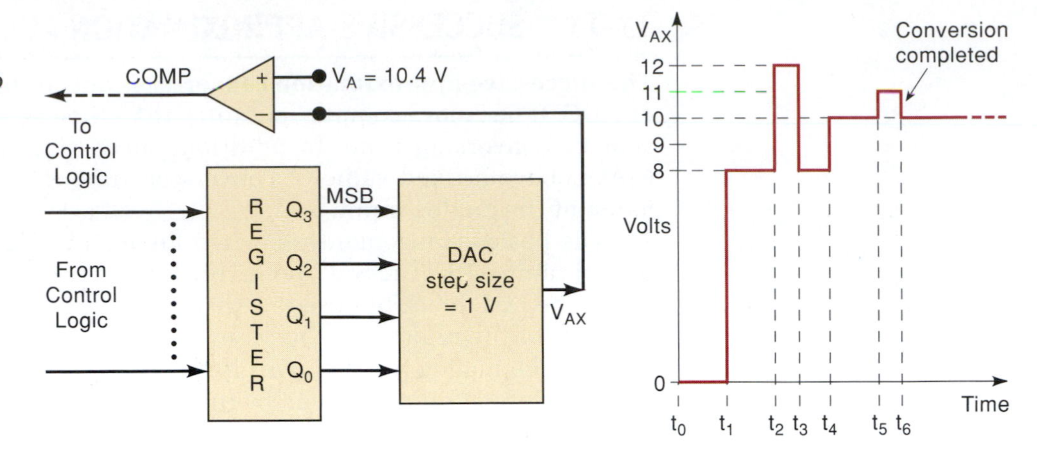

FIGURE 11-19 Illustration of four-bit SAC operation using a DAC step size of 1 V and $V_A = 10.4$ V.

At the next step (time t_1), the control logic sets the MSB of the register to 1 so that $[Q] = 1000$. This produces $V_{AX} = 8$ V. Because $V_{AX} < V_A$, the COMP output is still HIGH. This HIGH tells the control logic that the setting of the MSB did not make V_{AX} exceed V_A, so that the MSB is kept at 1.

The control logic now proceeds to the next lower bit, Q_2. It sets Q_2 to 1 to produce $[Q] = 1100$ and $V_{AX} = 12$ V at time t_2. Because $V_{AX} > V_A$, the COMP output goes LOW. This LOW signals the control logic that the value of V_{AX} is too large, and the control logic then clears Q_2 back to 0 at t_3. Thus, at t_3, the register contents are back to 1000 and V_{AX} is back to 8 V.

The next step occurs at t_4, where the control logic sets the next lower bit Q_1 so that $[Q] = 1010$ and $V_{AX} = 10$ V. With $V_{AX} < V_A$, COMP is HIGH and tells the control logic to keep Q_1 set at 1.

The final step occurs at t_5, where the control logic sets the next lower bit Q_0 so that $[Q] = 1011$ and $V_{AX} = 11$ V. Because $V_{AX} > V_A$, COMP goes LOW to signal that V_{AX} is too large, and the control logic clears Q_0 back to 0 at t_6.

At this point, all of the register bits have been processed, the conversion is complete, and the control logic activates its $\overline{EOC}$ output to signal that the digital equivalent of V_A is now in the register. For this example, digital output for $V_A = 10.4$ V is $[Q] = 1010$. Notice that 1010 is actually equivalent to 10 V, which is *less than* the analog input; this is a characteristic of the successive-approximation method. Recall that in the digital-ramp method, the digital output was always equivalent to a voltage that was on the step above V_A.

EXAMPLE 11-16

An eight-bit SAC has a resolution of 20 mV. What will its digital output be for an analog input of 2.17 V?

Solution

$$2.17\text{ V}/20\text{ mV} = 108.5$$

so that step 108 would produce $V_{AX} = 2.16$ V and step 109 would produce 2.18 V. The SAC always produces a final V_{AX} that is at the step *below* V_A. Therefore, for the case of $V_A = 2.17$ V, the digital result would be $108_{10} = 01101100_2$.

Conversion Time

In the operation just described, the control logic goes to each register bit, sets it to 1, decides whether or not to keep it at 1, and goes on to the next bit.

The processing of each bit takes one clock cycle, so that the total conversion time for an N-bit SAC will be N clock cycles. That is,

$$t_C \text{ for SAC} = N \times 1 \text{ clock cycle}$$

This conversion time will be the same *regardless of the value of* V_A because the control logic must process each bit to see whether or not a 1 is needed.

EXAMPLE 11-17

Compare the maximum conversion times of a 10-bit digital-ramp ADC and a 10-bit successive-approximation ADC if both utilize a 500-kHz clock frequency.

Solution

For the digital-ramp converter, the maximum conversion time is

$$(2^N - 1) \times (1 \text{ clock cycle}) = 1023 \times 2\ \mu\text{s} = 2046\ \mu\text{s}$$

For a 10-bit successive-approximation converter, the conversion time is always 10 clock periods or

$$10 \times 2\ \mu\text{s} = 20\ \mu\text{s}$$

Thus, it is about 100 times faster than the digital-ramp converter.

Because SACs have relatively fast conversion times, their use in data acquisition applications will permit more data values to be acquired in a given time interval. This feature can be very important when the analog data are changing at a relatively fast rate.

Because many SACs are available as ICs, it is rarely necessary to design the control logic circuitry, and so we will not cover it here. For those who are interested in the details of the control logic, many manufacturers' data books should provide sufficient detail.

An Actual IC: The ADC0804 Successive-Approximation ADC

ADCs are available from several IC manufacturers with a wide range of operating characteristics and features. We will take a look at one of the more popular devices to get an idea of what is actually used in system applications. Figure 11-20 is the pin layout for the ADC0804, which is a 20-pin CMOS IC that performs A/D conversion using the successive-approximation method. Some of its important characteristics are as follows:

- It has two analog inputs, $V_{IN}(+)$ and $V_{IN}(-)$, to allow **differential inputs.** In other words, the actual analog input, V_{IN}, is the difference in the voltages applied to these pins [analog $V_{IN} = V_{IN}(+) - V_{IN}(-)$]. In single-ended measurements, the analog input is applied to $V_{IN}(+)$, while $V_{IN}(-)$, is connected to analog ground. During normal operation, the converter uses $V_{CC} = +5$ V as its reference voltage, and the analog input can range from 0 to 5 V.
- It converts the differential analog input voltage to an eight-bit tristate buffered digital output. The internal circuitry is slightly more complex

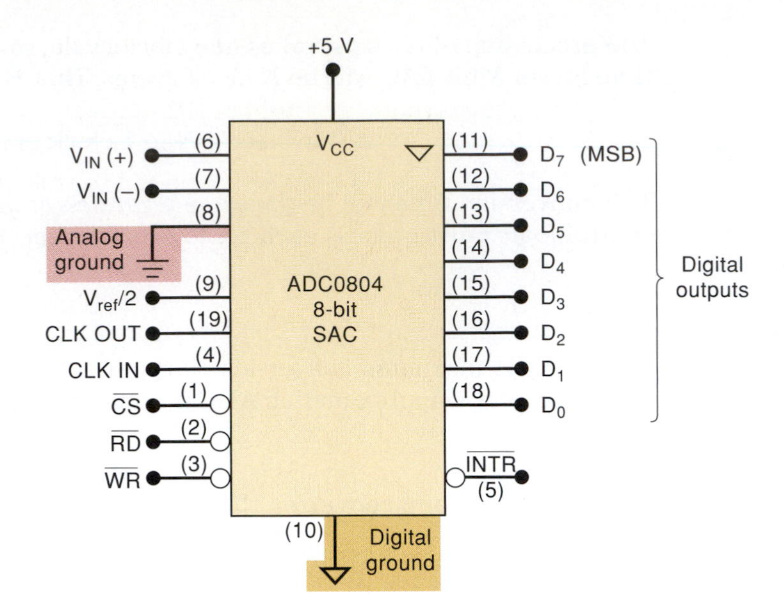

FIGURE 11-20 ADC0804 eight-bit successive-approximation ADC with tristate outputs. The numbers in parentheses are the IC's pin numbers.

than that described in Figure 11-19 in order to make transitions between output values occur at the nominal value $\pm\frac{1}{2}$ LSB. For example, with 10-mV resolution, the A/D output would switch from 0 to 1 at 5 mV, from 1 to 2 at 15 mV, and so on. For this converter the resolution is calculated as $V_{REF}/256$; with $V_{REF} = 5.00$ V, the resolution is 19.53 mV. The nominal full-scale input is $255 \times 19.53 = 4.98$ V, which should produce an output of 11111111. This converter will output 11111111 for any analog input between approximately 4.971 and 4.990 V.

- It has an internal clock generator circuit that produces a frequency of $f = 1/(1.1RC)$, where R and C are values of externally connected components. A typical clock frequency is 606 kHz using $R = 10$ kΩ and $C = 150$ pF. An external clock signal can be used, if desired, by connecting it to the CLK IN pin.
- Using a 606-kHz clock frequency, the conversion time is approximately 100 μs.
- It has separate ground connections for digital and analog voltages. Pin 8 is the analog ground that is connected to the common reference point of the analog circuit that is generating the analog voltage. Pin 10 is the digital ground that is the one used by all of the digital devices in the system. (Note the different symbols used for the different grounds.) The digital ground is inherently noisy because of the rapid current changes that occur as digital devices change states. Although it is not necessary to use a separate analog ground, doing so ensures that the noise from digital ground is prevented from causing premature switching of the analog comparator inside the ADC.

This IC is designed to be easily interfaced to a microprocessor data bus. For this reason, the names of some of the ADC0804 inputs and outputs are based on functions that are common to microprocessor-based systems. The functions of these inputs and outputs are defined as follows:

- $\overline{CS}$ (Chip Select). This input must be in its active-LOW state for $\overline{RD}$ or $\overline{WR}$ inputs to have any effect. With $\overline{CS}$ HIGH, the digital outputs are in the Hi-Z state, and no conversions can take place.

- $\overline{RD}$ (READ). This input is used to enable the digital output buffers. With $\overline{CS} = \overline{RD}$ = LOW, the digital output pins will have logic levels representing the results of the *last* A/D conversion. The microcomputer can then *read* (fetch) this digital data value over the system data bus.
- $\overline{WR}$ (WRITE). A LOW pulse is applied to this input to signal the start of a new conversion. This is actually a start conversion input. It is called a **WRITE** input because in a typical application, the microcomputer generates a WRITE pulse (similar to one used for writing to memory) that drives this input.
- $\overline{INTR}$ (INTERRUPT). This output signal will go HIGH at the start of a conversion and will return LOW to signal the end of conversion. This is actually an end-of-conversion output signal, but it is called INTERRUPT because in a typical situation, it is sent to a microprocessor's interrupt input to get the microprocessor's attention and let it know that the ADC's data are ready to be read.
- $V_{ref}/2$. This is an optional input that can be used to reduce the internal reference voltage and thereby change the analog input range that the converter can handle. When this input is unconnected, it sits at 2.5 V ($V_{CC}/2$) because V_{CC} is being used as the reference voltage. By connecting an external voltage to this pin, the internal reference is changed to twice that voltage, and the analog input range is changed accordingly (see Table 11-6).
- CLK OUT. A resistor is connected to this pin to use the internal clock. The clock signal appears on this pin.
- CLK IN. Used for external clock input, or for a capacitor connection when the internal clock is used.

TABLE 11-6

$V_{ref}/2$	Analog Input Range (V)	Resolution (mV)
Open	0–5	19.5
2.25	0–4.5	17.6
2.0	0–4	15.6
1.5	0–3	11.7

Figure 11-21(a) shows a typical connection of the ADC0804 to a microcomputer in a data acquisition application. The microcomputer controls when a conversion is to take place by generating the $\overline{CS}$ and $\overline{WR}$ signals. It then acquires the ADC output data by generating the $\overline{CS}$ and $\overline{RD}$ signals after detecting an NGT at $\overline{INTR}$, indicating the end of conversion. The waveforms in Figure 11-21(b) show the signal activity during the data acquisition process. Note that $\overline{INTR}$ goes HIGH when $\overline{CS}$ and $\overline{WR}$ are LOW, but the conversion process does not begin until $\overline{WR}$ returns HIGH. Also note that the ADC output data lines are in their Hi-Z state until the microcomputer activates $\overline{CS}$ and $\overline{RD}$; at that point the ADC's data buffers are enabled so that the ADC data are sent to the microcomputer over the data bus. The data lines return to the Hi-Z state when either $\overline{CS}$ or $\overline{RD}$ is returned HIGH.

In this application of the ADC0804, the input signal is varying over a range of 0.5 to 3.5 V. In order to make full use of the eight-bit resolution, the A/D must be matched to the analog signal specifications. In this case, the full-scale range is 3.0 V. However, it is offset from ground by 0.5 V. The offset of 0.5 V is applied to the negative input $V_{IN}(-)$, establishing this as the 0 value reference. The range of 3.0 V is set by applying 1.5 V to $V_{ref}/2$, which establishes V_{ref}

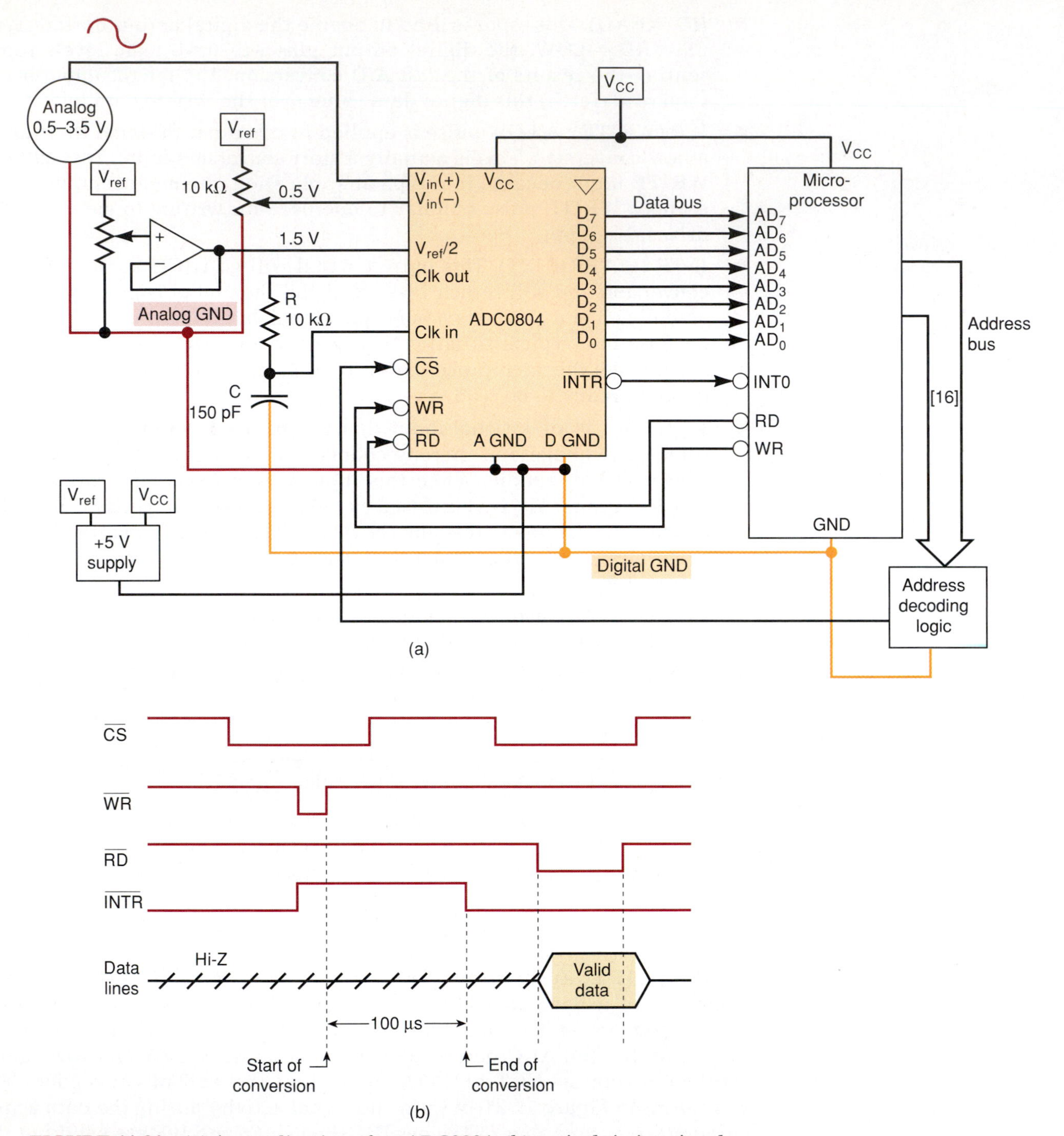

FIGURE 11-21 (a) An application of an ADC0804; (b) typical timing signals during data acquisition.

as 3.0 V. An input of 0.5 V will produce a digital value of 00000000, and an input of 3.5 V (or any value over 3.482) will produce 11111111.

Another major concern when interfacing digital and analog signals is *noise.* Notice that the digital and analog ground paths are separated. The two grounds are tied together at a point that is very close to the A/D converter. A very low-resistance path ties this point directly to the negative terminal of the power supply. It is also wise to route the positive supply lines separately to digital and analog devices and make extensive use of decoupling capacitors (0.01 μF) from very near each chip's supply connection to ground.

REVIEW QUESTIONS

1. What is the main advantage of a SAC over a digital-ramp ADC?
2. What is its principal disadvantage compared with the digital-ramp converter?
3. *True or false:* The conversion time for a SAC increases as the analog voltage increases.
4. Answer the following concerning the ADC0804.
   (a) What is its resolution in bits?
   (b) What is the normal analog input voltage range?
   (c) Describe the functions of the $\overline{CS}$, $\overline{WR}$, and $\overline{RD}$ inputs.
   (d) What is the function of the $\overline{INTR}$ output?
   (e) Why does it have two separate grounds?
   (f) What is the purpose of $V_{IN}(-)$?

11-12 FLASH ADCs

The **flash converter** is the highest-speed ADC available, but it requires much more circuitry than the other types. For example, a six-bit flash ADC requires 63 analog comparators, while an eight-bit unit requires 255 comparators, and a ten-bit converter requires 1023 comparators. The large number of comparators has limited the size of flash converters. IC flash converters are commonly available in two- to eight-bit units, and most manufacturers offer nine- and ten-bit units as well.

The principle of operation will be described for a three-bit flash converter in order to keep the circuitry at a workable level. Once the three-bit converter is understood, it should be easy to extend the basic idea to higher-bit flash converters.

The flash converter in Figure 11-22(a) has a three-bit resolution and a step size of 1 V. The voltage divider sets up reference levels for each comparator so that there are seven levels corresponding to 1 V (weight of LSB), 2V, 3V, . . . , and 7 V (full scale). The analog input, V_A, is connected to the other input of each comparator.

With $V_A < 1$ V, all of the comparator outputs C_1 through C_7 will be HIGH. With $V_A > 1$ V, one or more of the comparator outputs will be LOW. The comparator outputs are fed into an active-LOW priority encoder that generates a binary output corresponding to the highest-numbered comparator output that is LOW. For example, when V_A is between 3 and 4 V, outputs C_1, C_2, and C_3 will be LOW and all others will be HIGH. The priority encoder will respond only to the LOW at C_3 and will produce a binary output $CBA = 011$, which represents the digital equivalent of V_A, within the resolution of 1 V. When V_A is greater than 7 V, C_1 to C_7 will all be LOW, and the encoder will produce $CBA = 111$ as the digital equivalent of V_A. The table in Figure 11-22(b) shows the responses for all possible values of analog input.

The flash ADC of Figure 11-22 has a resolution of 1 V because the analog input must change by 1 V in order to bring the digital output to its next step. To achieve finer resolutions, we would have to increase the number of input voltage levels (i.e., use more voltage-divider resistors) and the number of

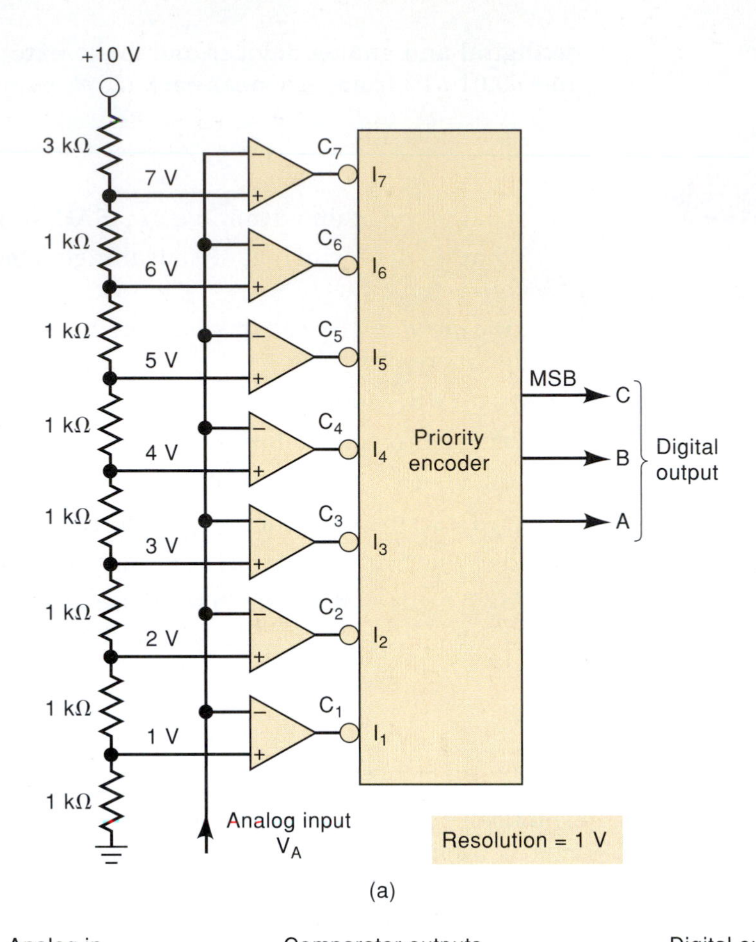

FIGURE 11-22 (a) Three-bit flash ADC; (b) truth table.

Analog in	Comparator outputs							Digital outputs		
V_A	C_1	C_2	C_3	C_4	C_5	C_6	C_7	C	B	A
0–1 V	1	1	1	1	1	1	1	0	0	0
1–2 V	0	1	1	1	1	1	1	0	0	1
2–3 V	0	0	1	1	1	1	1	0	1	0
3–4 V	0	0	0	1	1	1	1	0	1	1
4–5 V	0	0	0	0	1	1	1	1	0	0
5–6 V	0	0	0	0	0	1	1	1	0	1
6–7 V	0	0	0	0	0	0	1	1	1	0
> 7 V	0	0	0	0	0	0	0	1	1	1

(b)

comparators. For example, an eight-bit flash converter would require $2^8 = 256$ voltage levels, including 0 V. This would require 256 resistors and 255 comparators (there is no comparator for the 0-V level). The 255 comparator outputs would feed a priority encoder circuit that would produce an eight-bit code corresponding to the highest-order comparator output that is LOW. In general, an N-bit flash converter would require $2^N - 1$ comparators, 2^N resistors, and the necessary encoder logic.

Conversion Time

The flash converter uses no clock signal because no timing or sequencing is required. The conversion takes place continuously. When the value of analog input changes, the comparator outputs will change, thereby causing the encoder outputs to change. The conversion time is the time it takes for a new digital output to appear in response to a change in V_A, and it depends only on the propagation delays of the comparators and encoder logic. For this reason, flash converters have extremely short conversion times. For example, the Analog Devices AD9020 is a 10-bit flash converter with a conversion time under 17 ns.

REVIEW QUESTIONS

1. *True or false:* A flash ADC does not contain a DAC.
2. How many comparators would a 12-bit flash converter require? How many resistors?
3. State the major advantage and disadvantage of a flash converter.

11-13 OTHER A/D CONVERSION METHODS

Several other methods of A/D conversion have been in use for some time, each with its relative advantages and disadvantages. We will briefly describe some of them now.

Up/Down Digital-Ramp ADC (Tracking ADC)

As we have seen, the digital-ramp ADC is relatively slow because the counter is reset to 0 at the start of each new conversion. The staircase always begins at 0 V and steps its way up to the "switching point" where V_{AX} exceeds V_A and the comparator output switches LOW. The time it takes the staircase to reset to 0 and step back up to the new switching point is really wasted. The **up/down digital-ramp ADC** uses an up/down counter to reduce this wasted time.

The up/down counter replaces the up counter that feeds the DAC. It is designed to count up whenever the comparator output indicates that $V_{AX} < V_A$ and to count down whenever $V_{AX} > V_A$. Thus, the DAC output is always being stepped in the direction of the V_A value. Each time the comparator output switches states, it indicates that V_{AX} has "crossed" the V_A value, the digital equivalent of V_A is in the counter, and the conversion is complete.

When a new conversion is to begin, the counter is *not reset to 0* but begins counting either up or down from its last value, depending on the comparator output. It will count until the staircase crosses V_A again to end the conversion. The V_{AX} waveform, then, will contain both positive-going and negative-going staircase signals that "track" the V_A signal. For this reason, this ADC is often called a *tracking ADC.*

Clearly, the conversion times will generally be reduced because the counter does not start over from 0 each time but simply counts up or down from its previous value. Of course, the value of t_C will still depend on the value of V_A, and so it will not be constant.

Dual-Slope Integrating ADC

The **dual-slope converter** has one of the slowest conversion times (typically 10 to 100 ms) but has the advantage of relatively low cost because it does not require precision components such as a DAC or a VCO. The basic operation of this converter involves the *linear* charging and discharging of a capacitor using constant currents. First, the capacitor is charged up for a fixed time interval from a constant current derived from the analog input voltage, V_A. Thus, at the end of this fixed charging interval, the capacitor voltage will be proportional to V_A. At that point, the capacitor is linearly discharged from a constant current derived from a precise reference voltage, V_{ref}. When the capacitor voltage reaches 0, the linear discharging is terminated. During the discharge interval, a digital reference frequency is fed to a counter and counted. The duration of the discharge interval will be proportional to the initial capacitor voltage. Thus, at the end of the discharge interval, the counter will hold a count proportional to the initial capacitor voltage, which, as we said, is proportional to V_A.

In addition to its low cost, another advantage of the dual-slope ADC is its low sensitivity to noise and to variations in its component values caused by temperature changes. Because of its slow conversion times, the dual-slope ADC is not used in any data acquisition applications. The slow conversion times, however, are not a problem in applications such as digital voltmeters or multimeters, and this is where they find their major application.

Voltage-to-Frequency ADC

The **voltage-to-frequency ADC** is simpler than other ADCs because it does not use a DAC. Instead it uses a *linear voltage-controlled oscillator (VCO)* that produces an output frequency proportional to its input voltage. The analog voltage that is to be converted is applied to the VCO to generate an output frequency. This frequency is fed to a counter to be counted for a fixed time interval. The final count is proportional to the value of the analog voltage.

To illustrate, suppose that the VCO generates a frequency of 10 kHz for each volt of input (i.e., 1 V produces 10 kHz, 1.5 V produces 15 kHz, 2.73 V produces 27.3 kHz). If the analog input voltage is 4.54 V, then the VCO output will be a 45.4-kHz signal that clocks a counter for, say, 10 ms. After the 10-ms counting interval, the counter will hold the count of 454, which is the digital representation of 4.54 V.

Although this is a simple method of conversion, it is difficult to achieve a high degree of accuracy because of the difficulty in designing VCOs with accuracies of better than 0.1 percent.

One of the main applications of this type of converter is in noisy industrial environments where small analog signals must be transmitted from transducer circuits to a control computer. The small analog signals can be drastically affected by noise if they are directly transmitted to the control computer. A better approach is to feed the analog signal to a VCO, which generates a digital signal whose output frequency changes according to the analog input. This digital signal is transmitted to the computer and will be much less affected by noise. Circuitry in the control computer will count the digital pulses (i.e., perform a frequency-counting function) to produce a digital value equivalent to the original analog input.

Sigma/Delta Modulation*

Another approach to representing analog information in digital form is called **sigma/delta modulation**. A sigma/delta A/D converter is an oversampling device, which means that it effectively samples the analog information more often than the minimum sample rate. The minimum sample rate is two times higher than the highest frequency in the incoming analog wave. The sigma/delta approach, like the voltage-to-frequency approach, does not directly produce a multibit

*An excellent article published on the web by Jim Thompson, University of Washington, served as a basis for this description. Visit the *Digital Systems: Principles and Applications* Companion Web Site at http://www.prenhall.com/Tocci for the link to this article.

number for each sample. Instead, it represents the analog voltage by varying the density of logic 1s in a single-bit stream of serial data. To represent the positive portions of a waveform, a stream of bits with a high density of 1s is generated by the modulator (e.g., 0111110111110111110111). To represent the negative portions, a lower density of 1s (i.e., a higher density of 0s) is generated (e.g., 0001000100001000100 0).

Sigma/delta modulation is used in A/D as well as D/A conversion. One form of a sigma/delta modulator circuit is designed to convert a continuous analog signal into a modulated bit stream (A/D). The other form converts a sequence of digital samples into the modulated bit stream (D/A). We are coming from the perspective of digital systems, so it is easiest to understand the latter of these two circuits because it consists of all digital components that we have studied. Figure 11-23 shows a circuit that takes a five-bit signed digital value as its input and converts it into a sigma/delta bit stream. We will assume that the numbers that can be placed on this circuit's input range from −8 to +8. The first component is simply a subtractor (the delta section) similar to the one studied in Figure 6-14. The subtractor determines how far the input number is from its maximum or minimum value. This difference is often called the error signal. The second two components (the adder and the D register) form an accumulator very similar to the circuit in Figure 6-10 (the sigma section). For each sample that comes in, the accumulator adds the difference (error signal) to the running total. When the error is small, this running total (sigma) changes by small increments. When the error is large, the sigma changes by large increments. The last component compares the running total from the accumulator with a fixed threshold, which in this case is zero. In other words, it is simply determining if the total is positive or negative. This is accomplished by using the MSB (sign bit) of sigma. As soon as the total goes positive, the MSB goes LOW and feeds back to the delta section the maximum positive value (+8). When the MSB of sigma goes negative, it feeds back the maximum negative value (−8).

A spreadsheet is an excellent way to analyze a circuit like this. The tables in this section are taken from the spreadsheet that is included on the CD at

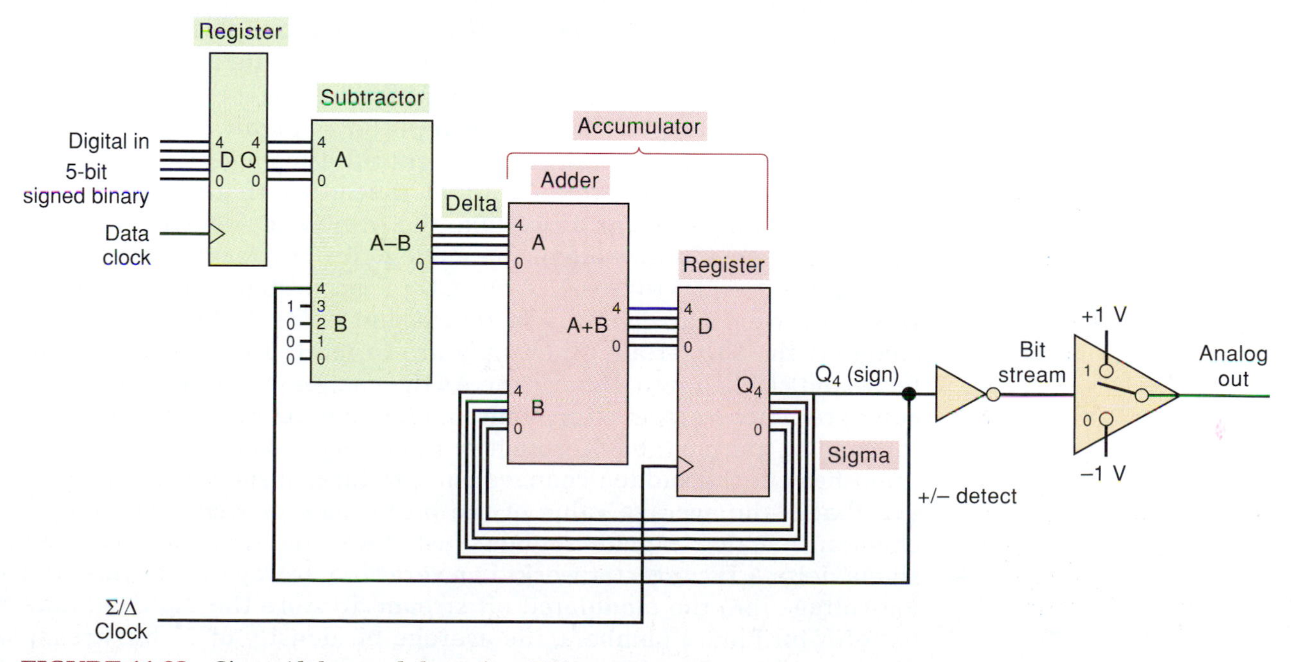

FIGURE 11-23 Sigma/delta modulator in a D/A converter.

TABLE 11-7 Sigma/delta modulator with an input of 0.

Sample (n)	Digital IN	Delta	Sigma	Bit Stream Out	Analog OUT	Feedback
1	0	−8	0	1	1	8
2	0	8	−8	0	−1	−8
3	0	−8	0	1	1	8
4	0	8	−8	0	−1	−8
5	0	−8	0	1	1	8
6	0	8	−8	0	−1	−8
7	0	−8	0	1	1	8
8	0	8	−8	0	−1	−8

TABLE 11-8 Sigma/delta modulator with an input of 4.

Sample (n)	Digital IN	Delta	Sigma	Bit Stream Out	Analog OUT	Feedback
1	4	−4	4	1	1	8
2	4	−4	0	1	1	8
3	4	12	−4	0	−1	−8
4	4	−4	8	1	1	8
5	4	−4	4	1	1	8
6	4	−4	0	1	1	8
7	4	12	−4	0	−1	−8
8	4	−4	8	1	1	8

the back of this book. Table 11-7 shows the operation of the converter when a value of zero is the input. Notice that the bit stream output alternates between 1 and 0, and the average value of the analog output is 0 volts. Table 11-8 shows what happens when the digital input is 4. If we assume that 8 is full scale, this represents $\frac{4}{8} = 0.5$. The output is HIGH for three samples and LOW for one sample, a pattern that repeats every four samples. The average value of the analog output is $(1 + 1 + 1 - 1)/4 = 0.5$ V.

As a final example, let's use an input of -5, which represents $-5/8 = -0.625$. Table 11-9 shows the resulting output. The pattern in the bit stream is not periodic. From the sigma column, we can see that it takes 16 samples for the pattern to repeat. If we take the overall bit density, however, and calculate the average value of the analog output over 16 samples, we will find that it is equal to -0.625. Your CD player probably uses a sigma/delta D/A converter that operates in this fashion. The 16-bit digital numbers come off the CD serially; then they are formatted into parallel data patterns and clocked into a converter. As the changing numbers come into the converter, the average value of the analog out changes accordingly. Next, the analog output goes through a circuit called a low-pass filter that smoothes out the sudden changes and produces a smoothly changing voltage that is the average value of the bit stream. In your headphones, this changing analog signal sounds just like the original recording. A sigma/delta A/D converter works in a very similar way but converts the analog voltage into the modulated bit stream. To store the digitized data as a list of N-bit binary numbers, the average bit density of 2^N bit-stream samples is calculated and stored.

TABLE 11-9 Sigma/delta modulator with an input of −5.

Sample (*n*)	Digital IN	Delta	Sigma	Bit Stream Out	Analog OUT	Feedback
1	−5	3	−5	0	−1	−8
2	−5	3	−2	0	−1	−8
3	−5	−13	1	1	1	8
4	−5	3	−12	0	−1	−8
5	−5	3	−9	0	−1	−8
6	−5	3	−6	0	−1	−8
7	−5	3	−3	0	−1	−8
8	−5	−13	0	1	1	8
9	−5	3	−13	0	−1	−8
10	−5	3	−10	0	−1	−8
11	−5	3	−7	0	−1	−8
12	−5	3	−4	0	−1	−8
13	−5	3	−1	0	−1	−8
14	−5	−13	2	1	1	8
15	−5	3	−11	0	−1	−8
16	−5	3	−8	0	−1	−8
17	−5	3	−5	0	−1	−8
18	−5	3	−2	0	−1	−8

REVIEW QUESTIONS

1. How does the up/down digital-ramp ADC improve on the digital-ramp ADC?
2. What is the main element of a voltage-to-frequency ADC?
3. Cite two advantages and one disadvantage of the dual-slope ADC.
4. Name three types of ADCs that do not use a DAC.
5. How many output data bits does a sigma/delta modulator use?

11-14 SAMPLE-AND-HOLD CIRCUITS

When an analog voltage is connected directly to the input of an ADC, the conversion process can be adversely affected if the analog voltage is changing during the conversion time. The stability of the conversion process can be improved by using a **sample-and-hold (S/H) circuit** to hold the analog voltage constant while the A/D conversion is taking place. A simplified diagram of a sample-and-hold (S/H) circuit is shown in Figure 11-24.

The S/H circuit contains a unity-gain buffer amplifier A_1 that presents a high impedance to the analog signal and has a low output impedance that can rapidly charge the hold capacitor, C_h. The capacitor will be connected to the output of A_1 when the digitally controlled switch is closed. This is called the *sample* operation. The switch will be closed long enough for C_h to charge to the present value of the analog input. For example, if the switch is closed at time t_0, the A_1 output will quickly charge C_h up to a voltage V_0. When the switch opens, C_h will *hold* this voltage so that the output of A_2 will apply this voltage to the ADC. The unity-gain buffer amplifier A_2 presents a high input impedance that will not discharge the capacitor voltage appreciably during

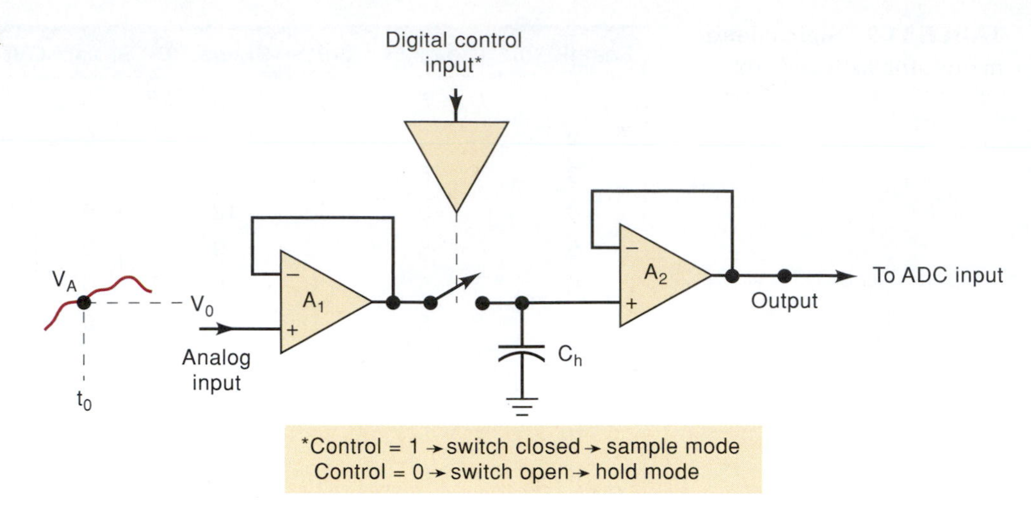

FIGURE 11-24 Simplified diagram of a sample-and-hold circuit.

the conversion time of the ADC, and so the ADC will essentially receive a dc input voltage V_0.

In a computer-controlled data acquisition system such as the one discussed earlier, the sample-and-hold switch would be controlled by a digital signal from the computer. The computer signal would close the switch in order to charge C_h to a new sample of the analog voltage; the amount of time the switch would have to remain closed is called the **acquisition time**, and it depends on the value of C_h and the characteristics of the S/H circuit. The computer signal would then open the switch to allow C_h to hold its value and provide a relatively constant analog voltage at the A_2 output.

The AD1154 is a sample-and-hold integrated circuit that has an internal hold capacitor with an acquisition time of 3.5 μs. During the hold time, the capacitor voltage will droop (discharge) at a rate of only 0.1 μV/μs. The voltage droop within the sampling interval should be less than the weight of the LSB. For example, a 10-bit converter with a full-scale range of 10 V would have an LSB weight of approximately 10 mV. It would take 100 ms before the capacitor droop would equal the weight of the ADC's LSB. It is not likely, however, that it would ever be necessary to hold the sample for such a long time in the conversion process.

REVIEW QUESTIONS

1. Describe the function of a sample-and-hold circuit.
2. *True or false:* The amplifiers in an S/H circuit are used to provide voltage amplification.

11-15 MULTIPLEXING

When analog inputs from several sources are to be converted, a multiplexing technique can be used so that one ADC may be time-shared. The basic scheme is illustrated in Figure 11-25 for a four-channel acquisition system. Rotary switch S is used to switch each analog signal to the input of the ADC, one at a time in sequence. The control circuitry controls the switch position according to the *select address* bits, A_1, A_0, from the MOD-4 counter. For example, with $A_1A_0 = 00$, the switch connects V_{A0} to the ADC input; $A_1A_0 = 01$ connects V_{A1} to the ADC input; and so on. Each input channel has a specific address code that, when present, connects that channel to the ADC.

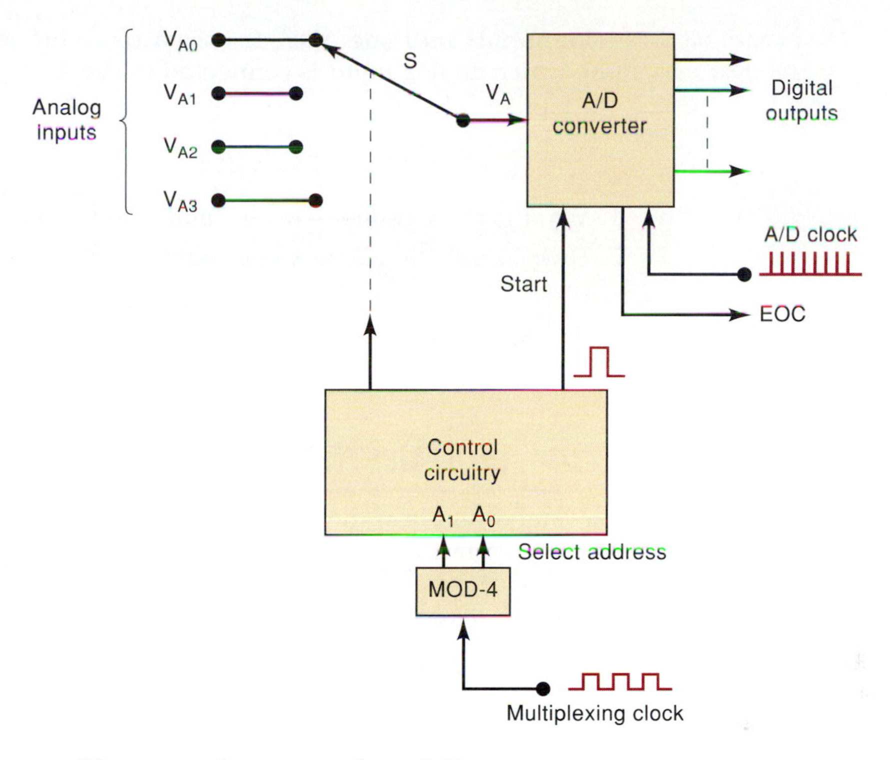

FIGURE 11-25 Conversion of four analog inputs by multiplexing through one ADC.

The operation proceeds as follows:

1. With select address = 00, V_{A0} is connected to the ADC input.
2. The control circuit generates a START pulse to initiate the conversion of V_{A0} to its digital equivalent.
3. When the conversion is complete, *EOC* signals that the ADC output data are ready. Typically, these data will be transferred to a computer over a data bus.
4. The multiplexing clock increments the select address to 01, which connects V_{A1} to the ADC.
5. Steps 2 and 3 are repeated with the digital equivalent of V_{A1} now present at the ADC outputs.
6. The multiplexing clock increments the select address to 10, and V_{A2} is connected to the ADC.
7. Steps 2 and 3 are repeated with the digital equivalent of V_{A2} now present at the ADC outputs.
8. The multiplexing clock increments the select address to 11, and V_{A3} is connected to the ADC.
9. Steps 2 and 3 are repeated with the digital equivalent of V_{A3} now present at the ADC outputs.

The multiplexing clock controls the rate at which the analog signals are sequentially switched into the ADC. The maximum rate is determined by the delay time of the switches and the conversion time of the ADC. The switch delay time can be minimized by using semiconductor switches such as the CMOS bilateral switch described in Chapter 8. It may be necessary to connect a sample-and-hold circuit at the input of the ADC if the analog inputs will change significantly during the ADC conversion time.

Many integrated ADCs contain the multiplexing circuitry on the same chip as the ADC. The ADC0808, for example, can multiplex eight different analog inputs into one ADC. It uses a three-bit select input code to determine which analog input is connected to the ADC.

REVIEW QUESTIONS

1. What is the advantage of this multiplexing scheme?
2. How would the address counter be changed if there were eight analog inputs?

11-16 DIGITAL STORAGE OSCILLOSCOPE

As a final example of the application of D/A and A/D converters, we will take a brief look at the digital storage oscilloscope (DSO). The DSO uses both of these devices to digitize, store, and display analog waveforms.

A block diagram of a DSO is shown in Figure 11-26. The overall operation is controlled and synchronized by the circuits in the CONTROL block, which usually contains a microprocessor executing a control program stored in ROM (read-only memory). The data acquisition portion of the system contains a sample-and-hold (S/H) and an ADC that repetitively samples and digitizes the input signal at a rate determined by the SAMPLE CLOCK and then sends the digitized data to memory for storage. The CONTROL block makes sure that successive data points are stored in successive memory locations by continually updating the memory's ADDRESS COUNTER.

When memory is full, the next data point from the ADC is stored in the first memory location, writing over the old data, and so on, for successive data points. This data acquisition and storage process continues until the CONTROL block receives a trigger signal from either the input waveform (INTERNAL trigger) or an EXTERNAL trigger source. When the trigger

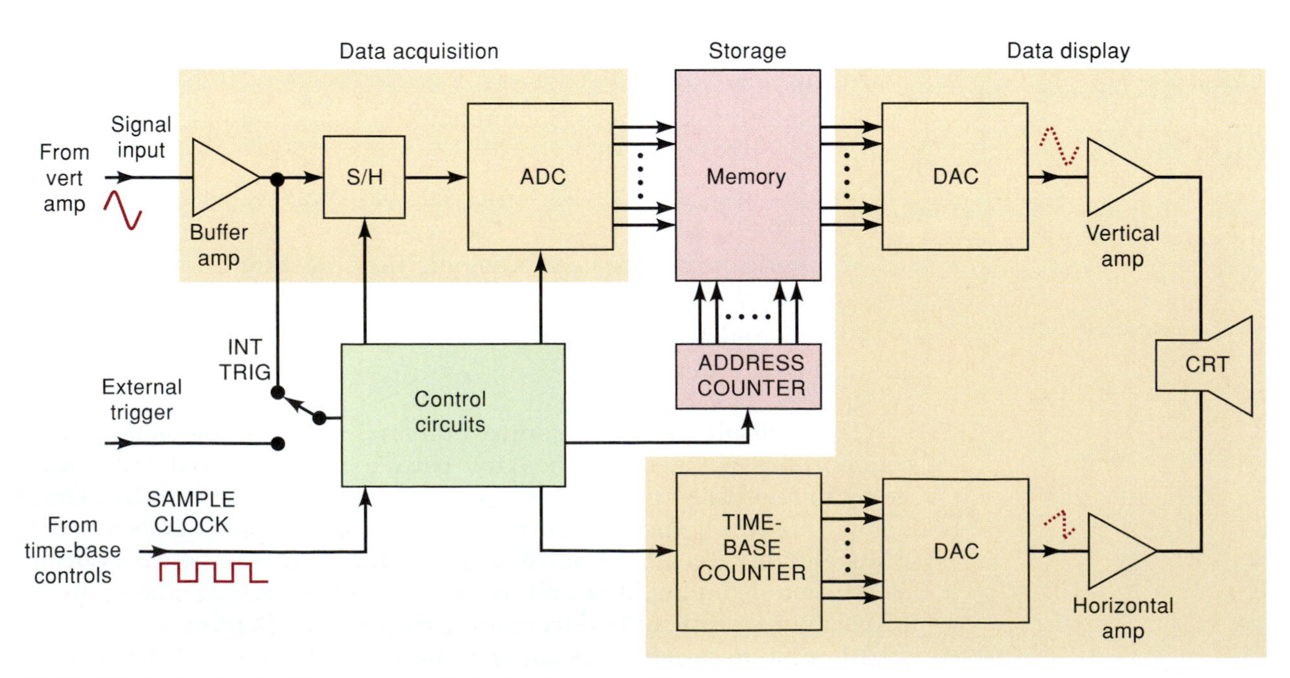

FIGURE 11-26 Block diagram of a digital storage oscilloscope.

occurs, the system stops acquiring new data and enters the display mode of operation, in which all or part of the memory data is repetitively displayed on the CRT.

The display operation uses two DACs to provide the vertical and horizontal deflection voltages for the CRT. Data from memory produce the vertical deflection of the electron beam, while the TIME-BASE COUNTER provides the horizontal deflection in the form of a staircase sweep signal. The CONTROL block synchronizes the display operation by incrementing the memory ADDRESS COUNTER and the TIME-BASE COUNTER at the same time so that each horizontal step of the electron beam is accompanied by a new data value from memory to the vertical DAC. The counters are continuously recycled so that the stored data points are repetitively replotted on the CRT screen. The screen display consists of discrete dots representing the various data points, but the number of dots is usually so large (typically 1000 or more) that they tend to blend together and appear to be a smooth, continuous waveform. The display operation is terminated when the operator presses a front-panel button that commands the DSO to begin a new data acquisition cycle.

Related Applications

The same sequence of operations performed in a DSO—data acquisition/digitizing/storage/data outputting—is used in other applications of DACs and ADCs. For example, heart monitors that can be found in any hospital are similar to DSOs but are constantly displaying a waveform showing the patient's heart activity over the past several seconds. As another example, digital video cameras digitize an image one picture element (pixel) at a time and store the information on magnetic tape or DVD. Digital still cameras digitize each pixel and store the data on a solid-state memory card. The data can later be transferred digitally and then output to a display device, where the data is converted to an analog "brightness" signal for each pixel and reassembled to form an image on the display.

REVIEW QUESTIONS

1. Look at Figure 11-26. How are waveforms "stored" in a DSO?
2. Describe the functions of the ADC and DACs that are part of the DSO.

11-17 DIGITAL SIGNAL PROCESSING (DSP)

One of the most dynamic areas of digital systems today is in the field of **digital signal processing (DSP)**. A DSP is a very specialized form of microprocessor that has been optimized to perform repetitive calculations on streams of digitized data. The digitized data are usually being fed to the DSP from an A/D converter. It is beyond the scope of this text to explain the mathematics that allow a DSP to process these data values, but suffice it to say that for each new data point that comes in, a calculation is performed (very quickly). This calculation involves the most recent data point as well as several of the preceding data samples. The result of the calculation produces a new output data point, which is usually sent to a D/A converter. A DSP system is similar to the block diagram shown in Figure 11-1. The main difference is in the specialized hardware contained in the computer section.

A major application for DSP is in filtering and conditioning of analog signals. As a very simple example, a DSP can be programmed to take in an analog waveform, such as the output from an audio preamplifier, and pass to the output only those frequency components that are below a certain frequency. All higher frequencies are attenuated by the filter. Perhaps you recall from your study of analog circuits that the same thing can be accomplished by a simple low-pass filter made from a resistor and capacitor. The advantage of DSP over resistors and capacitors is the flexibility of being able to change the critical frequency without switching any components. Instead, numbers are simply changed in the calculations to adapt the dynamic response of the filter. Have you ever been in an auditorium when the PA system started to squeal? This can be prevented if the degenerative feedback frequency can be filtered out. Unfortunately, the frequency that causes the squeal changes with the number of people in the room, the clothes they wear, and many other factors. With a DSP-based audio equalizer, the oscillation frequency can be detected and the filters dynamically adjusted to tune it out.

Digital Filtering

To help you understand digital filtering, imagine you are buying and selling stock. To decide when to buy and sell, you need to know what the market is doing. You want to ignore sudden, short-term (high-frequency) changes but respond to the overall trends (30-day averages). Every day you read the newspaper, take a sample of the closing price for your stock, and write it down. Then you use a formula to calculate the average of the last 30 days' prices. This average value is plotted as shown in Figure 11-27, and the resulting graph is used to make decisions. This is a way of filtering the digital signal (sequence of data samples) that represents the stock market activity.

Now imagine that instead of sampling stock prices, a digital system is sampling an audio (analog) signal from a microphone using an A/D converter. Instead of taking a sample once a day, it takes a sample 20,000 times each second (every 50 μs). For each sample, a weighted averaging calculation is performed using the last 256 data samples and produces a single output data point. A **weighted average** means that some of the data points are considered more important than others. Each of the samples is multiplied by a fractional number (between 0 and 1) before adding them together. This averaging calculation is processing (filtering) the audio signal.

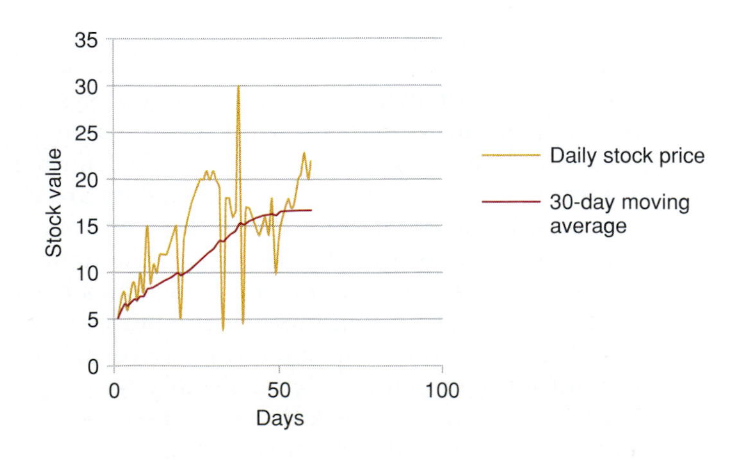

FIGURE 11-27 Digital filtering of stock market activity.

The most difficult part of this form of DSP is determining the correct weighting constants for the averaging calculation in order to achieve the desired filter characteristics. Fortunately, there is readily available software for PCs that makes this very easy. The special DSP hardware must perform the following operations:

Read the newest sample (one new number) from A/D.

Replace the oldest sample (of 256) with the new one from A/D.

Multiply each of the 256 samples by their corresponding weight constant.

Add all of these products.

Output the resulting sum of products (1 number) to the D/A.

Figure 11-28 shows the basic architecture of a DSP. The multiply and accumulate (**MAC**) section is central to all DSPs and is used in most applications. Special hardware, like you will study in Chapter 12, is used to implement the memory system that stores the data samples and weight values. The **arithmetic logic unit** and **barrel shifter** (shift register) provide the necessary support to deal with the binary number system while processing signals.

Another useful application of DSP is called **oversampling** or **interpolation filtering**. As you recall, the reconstructed waveform is always an approximation of the original due to quantization error. The sudden step changes from one data point to the next also introduce high-frequency noise into the reconstructed signal. A DSP can insert interpolated data points into the digital signal. Figure 11-29 shows how 4X oversampling interpolation filtering smoothes out the waveform and makes final filtering possible

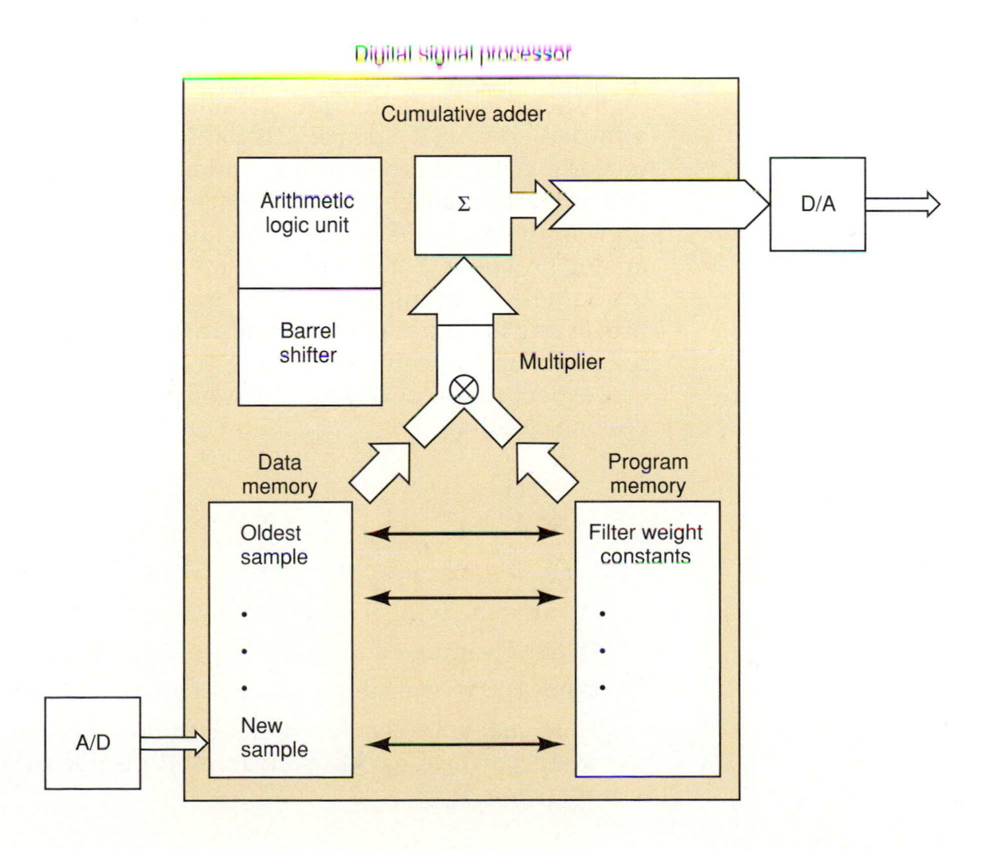

FIGURE 11-28 Digital signal processor architecture.

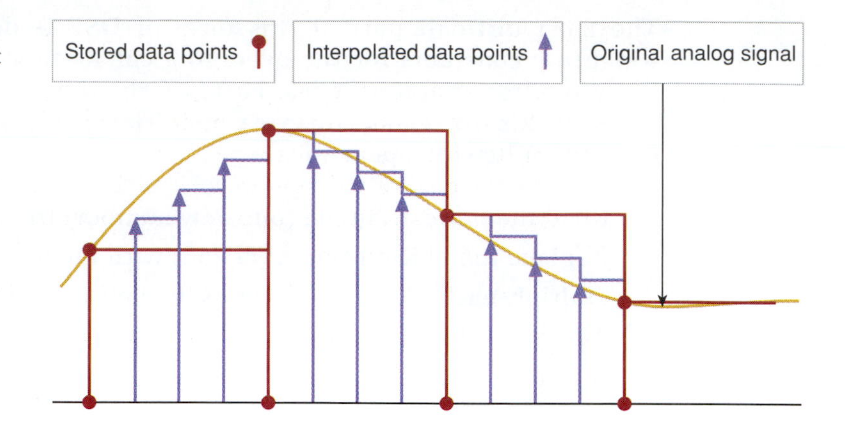

FIGURE 11-29 Inserting an interpolated data point into a digital signal to reduce noise.

with simpler analog circuitry. DSP performs this role in your CD player to provide an excellent audio reproduction. The round dots represent the digitally recorded data on your CD. The triangles represent the interpolated data points that the digital filter in your CD player inserts before the final analog output filter.

Many of the important concepts that you need to understand in order to move on to DSP have been presented in this and previous chapters. A/D and D/A conversion methods and hardware along with data acquisition and sampling concepts are vital. Topics such as signed binary number representations (including fractions), signed binary addition and multiplication (covered in Chapter 6), and shift registers (Chapter 7) are necessary to understand the hardware and programming of a DSP. Memory system concepts, which will be presented in the next chapter, will also be important.

DSP is being integrated into many common systems that you are familiar with. CD players use DSP to filter the digital data being read from the disk to minimize the quantization noise that is unavoidably caused by digitizing the music. Telephone systems use DSP to cancel echoes on the phone lines. The high-speed modems that are standard on PCs have been made possible and affordable by DSP. Special effects boxes for guitars and other instruments perform echo, reverb, phasing, and other effects using DSP. Applications of DSP are growing right now at the same rate that microprocessor applications grew in the early 1980s. They provide a digital solution to many traditionally analog problems. Some other examples of applications include speech recognition, telecommunications data encryption, fast Fourier transforms, image processing in digital television, ultrasonic beam forming in medical electronics, and noise cancellation in industrial controls. As this trend continues, you can expect to see nearly all electronic systems containing digital signal processing circuitry.

REVIEW QUESTIONS

1. What is a major application of DSP?
2. What is the typical source of digital data for a DSP to process?
3. What advantage does a DSP filter have over an analog filter circuit?
4. What is the central hardware feature of a DSP?
5. How many interpolated data points are inserted between samples when performing 4X oversampled digital filtering? How many for 8X oversampling?

SUMMARY

1. Physical variables that we want to measure, such as temperature, pressure, humidity, distance velocity, and so on, are continuously variable quantities. A transducer can be used to translate these quantities into an electrical signal of voltage or current that fluctuates in proportion to the physical variable. These continuously variable voltage or current signals are called *analog* signals.
2. To measure a physical variable, a digital system must assign a binary number to the analog value that is present at that instant. This is accomplished by an A/D converter. To generate variable voltages or current values that can control physical processes, a digital system must translate binary numbers into a voltage or current magnitude. This is accomplished by a D/A converter.
3. A D/A converter with n bits divides a range of analog values (voltage or current) into $2^n - 1$ pieces. The size or magnitude of each piece is the analog equivalent weight of the least significant bit. This is called the *resolution* or *step size.*
4. Most D/A converters use resistor networks that can cause weighted amounts of current to flow when any of its binary inputs are activated. The amount of current that flows is proportional to the binary weight of each input bit. These weighted currents are summed to create the analog signal out.
5. An A/D converter must assign a binary number to an analog (continuously variable) quantity. The precision with which an A/D converter can perform this conversion depends on how many different numbers it can assign and how wide the analog range is. The smallest change in analog value that an A/D can measure is called its *resolution,* the weight of its least significant bit.
6. By repeatedly sampling the incoming analog signal, converting it to digital, and storing the digital values in a memory device, an analog waveform can be captured. To reconstruct the signal, the digital values are read from the memory device at the same rate at which they were stored, and then they are fed into a D/A converter. The output of the D/A is then filtered to smooth the stair steps and re-create the original waveform. The bandwidth of sampled signals is limited to $\frac{1}{2} F_S$. Incoming frequencies greater than $\frac{1}{2} F_S$ create an *alias* that has a frequency equal to the difference between the nearest integer multiple of F_S and the incoming frequency. This difference will always be less than $\frac{1}{2} F_S$.
7. A digital-ramp A/D is the simplest to understand but it is not often used due to its variable conversion time. A successive-approximation converter has a constant conversion time and is probably the most common general-purpose converter.
8. Flash converters use analog comparators and a priority encoder to assign a digital value to the analog input. These are the fastest converters because the only delays involved are propagation delays.
9. Other popular methods of A/D include up/down tracking, integrating, voltage-to-frequency conversion, and sigma/delta conversion. Each type of converter has its own niche of applications.
10. Any D/A converter can be used with other circuitry such as analog multiplexers that select one of several analog signals to be converted, one at a time. Sample-and-hold circuits can be used to "freeze" a rapidly changing analog signal while the conversion is taking place.

11. Digital signal processing is an exciting new growth field in electronics. These devices allow calculations to be performed quickly in order to emulate the operation of many analog filter circuits digitally. The primary architectural feature of a DSP is a hardware multiplier and adder circuit that can multiply pairs of numbers together and accumulate the running total (sum) of these products. This circuitry is used to perform efficiently the weighted moving average calculations that are used to implement digital filters and other DSP functions. DSP is responsible for many of the recent advances in high-fidelity audio, high-definition TV, and telecommunications.

IMPORTANT TERMS

digital quantity
analog quantity
transducer
analog-to-digital converter (ADC)
digital-to-analog converter (DAC)
full-scale output
resolution
step size
staircase
full-scale error
linearity error
offset error
settling time
monotonicity
digitization
digital-ramp ADC
quantization error
sampling
sampling frequency, F_S
alias
undersampling
successive-approximation ADC
differential inputs
WRITE
flash ADC
up/down digital-ramp ADC
dual-slope ADC
voltage-to-frequency ADC
sigma/delta modulation
sample-and-hold (S/H) circuit
acquisition time
digital signal processing (DSP)
weighted average
MAC
arithmetic logic unit
barrel shifter
oversampling
interpolation filtering

PROBLEMS

SECTIONS 11-1 AND 11-2

B 11-1. **DRILL QUESTION**

(a) What is the expression relating the output and inputs of a DAC?

(b) Define *step size* of a DAC.

(c) Define *resolution* of a DAC.

(d) Define *full scale.*

(e) Define *percentage resolution.*

(f)* *True or false:* A 10-bit DAC will have a smaller resolution than a 12-bit DAC for the same full-scale output.

(g)* *True or false:* A 10-bit DAC with full-scale output of 10 V has a smaller percentage resolution than a 10-bit DAC with 12 V full scale.

B 11-2. An eight-bit DAC produces an output voltage of 2.0 V for an input code of 01100100. What will the value of V_{OUT} be for an input code of 10110011?

B 11-3.*Determine the weight of each input bit for the DAC of Problem 11-2.

*Answers to problems marked with an asterisk can be found in the back of the text.

B 11-4. What is the resolution of the DAC of Problem 11-2? Express it in volts and as a percentage.

B 11-5.*What is the resolution in volts of a 10-bit DAC whose F.S. output is 5 V?

B 11-6. How many bits are required for a DAC so that its F.S. output is 10 mA and its resolution is less than 40 μA?

B 11-7.*What is the percentage resolution of the DAC of Figure 11-30? What is the step size if the top step is 2 V?

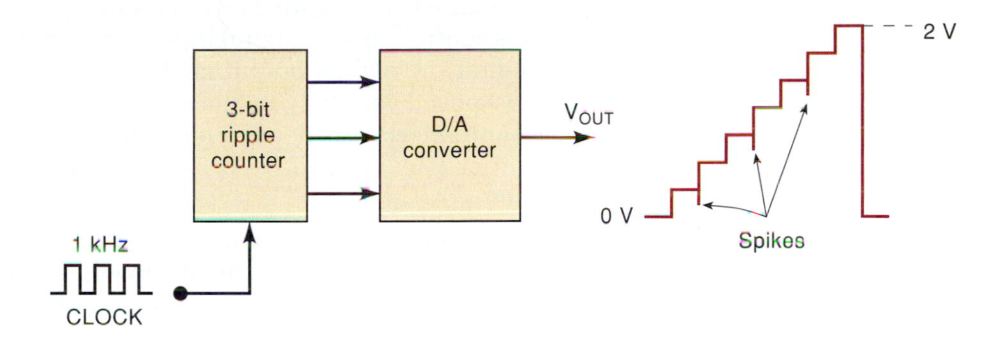

FIGURE 11-30 Problems 11-7 and 11-8.

C 11-8. What is the cause of the negative-going spikes on the V_{OUT} waveform of Figure 11-30? (*Hint:* Note that the counter is a ripple counter and that the spikes occur on every other step.)

B 11-9.*Assuming a 12-bit DAC with perfect accuracy, how close to 250 rpm can the motor speed be adjusted in Figure 11-4?

11-10. A 12-bit DAC has a full-scale output of 15.0 V. Determine the step size, the percentage resolution, and the value of V_{OUT} for an input code of 011010010101.

11-11.*A microcontroller has an eight-bit output port that is to be used to drive a DAC. The DAC that is available has 10 input bits and has a full-scale output of 10 V. The application requires a voltage that ranges between 0 and 10 V in steps of 50 mV or smaller. Which eight bits of the 10-bit DAC will be connected to the output port?

11-12. You need a DAC that can span 12 V with a resolution of 20 mV or less. How many bits are needed?

SECTION 11-3

D 11-13.*The step size of the DAC of Figure 11-5 can be changed by changing the value of R_F. Determine the required value of R_F for a step size of 0.5 V. Will the new value of R_F change the percentage resolution?

D 11-14. Assume that the output of the DAC in Figure 11-7(a) is connected to the op-amp of Figure 11-7(b).

(a) With $V_{REF} = 5$ V, $R = 20$ kΩ, and $R_F = 10$ kΩ, determine the step size and the full-scale voltage at V_{OUT}.

(b) Change the value of R_F so that the full-scale voltage at V_{OUT} is -2 V.

(c) Use this new value of R_F, and determine the proportionality factor, K, in the relationship $V_{OUT} = K(V_{REF} \times B)$.

11-15.*What is the advantage of the DAC of Figure 11-8 over that of Figure 11-7, especially for a larger number of input bits?

SECTIONS 11-4 TO 11-6

11-16. An eight-bit DAC has a full-scale error of 0.2% F.S. If the DAC has a full-scale output of 10 mA, what is the most that it can be in error for any digital input? If the D/A output reads 50 μA for a digital input of 00000001, is this within the specified range of accuracy? (Assume no offset error.)

C, N 11-17.*The control of a positioning device may be achieved using a *servomotor,* which is a motor designed to drive a mechanical device as long as an error signal exists. Figure 11-31 shows a simple servo-controlled system that is controlled by a digital input that could be coming directly from a computer or from an output medium such as magnetic tape. The lever arm is moved vertically by the servomotor. The motor rotates clockwise or counterclockwise, depending on whether the voltage from the power amplifier (P.A.) is positive or negative. The motor stops when the P.A. output is 0.

The mechanical position of the lever is converted to a dc voltage by the potentiometer arrangement shown. When the lever is at its 0 reference point, $V_P = 0$ V. The value of V_P increases at the rate of 1 V/inch until the lever is at its highest point (10 inches) and $V_P = 10$ V. The desired position of the lever is provided as a digital code from the computer and is then fed to a DAC, producing V_A. The *difference* between V_P and V_A (called *error*) is produced by the *differential* amplifier and is amplified by the P.A. to drive the motor in the direction that causes the error signal to decrease to 0—that is, moves the lever until $V_P = V_A$.

(a) If the lever must be positioned within a resolution of 0.1 in, what is the number of bits needed in the digital input code?

(b) In actual operation, the lever arm might oscillate slightly around the desired position, especially if a *wire-wound* potentiometer is used. Can you explain why?

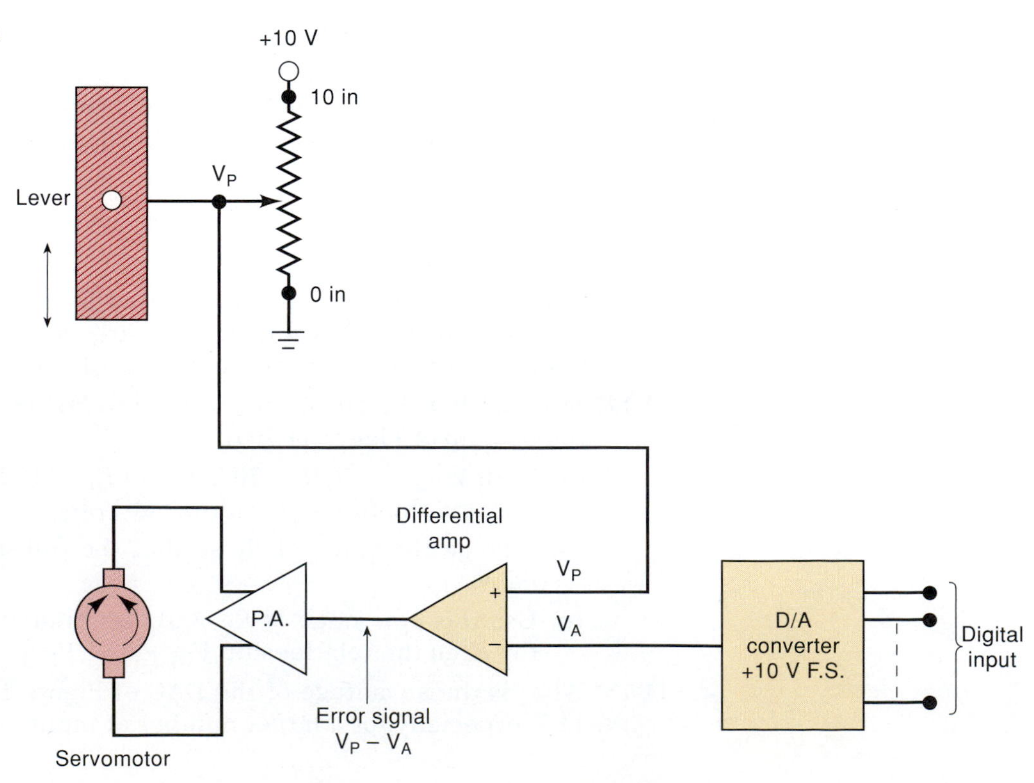

FIGURE 11-31 Problem 11-17.

B 11-18. **DRILL QUESTION**

(a) Define *binary-weighted resistor network.*

(b) Define *R/2R ladder network.*

(c) Define *DAC settling time.*

(d) Define *full-scale error.*

(e) Define *offset error.*

11-19.*A particular six-bit DAC has a full-scale output rated at 1.260 V. Its accuracy is specified as ±0.1% F.S., and it has an offset error of ±1 mV. Assume that the offset error has not been zeroed out. Consider the measurements made on this DAC (Table 11-10), and determine which of them are not within the device's specifications. (*Hint:* The offset error is added to the error caused by component inaccuracies.)

TABLE 11-10

Input Code	Output
000010	41.5 mV
000111	140.2 mV
001100	242.5 mV
111111	1.258 V

SECTION 11-7

T 11-20. A certain DAC has the following specifications: eight-bit resolution, full scale = 2.55 V, offset ≤ 2 mV; accuracy = ±0.1% F.S. A static test on this DAC produces the results shown in Table 11-11. What is the probable cause of the malfunction?

TABLE 11-11

Input Code	Output
00000000	8 mV
00000001	18.2 mV
00000010	28.5 mV
00000100	48.3 mV
00001111	158.3 mV
10000000	1.289 V

T 11-21.*Repeat Problem 11-20 using the measured data given in Table 11-12.

TABLE 11-12

Input Code	Output
00000000	20.5 mV
00000001	30.5 mV
00000010	20.5 mV
00000100	60.6 mV
00001111	150.6 mV
10000000	1.300 V

T 11-22.*A technician connects a counter to the DAC of Figure 11-3 to perform a staircase test using a 1-kHz clock. The result is shown in Figure 11-32. What is the probable cause of the incorrect staircase signal?

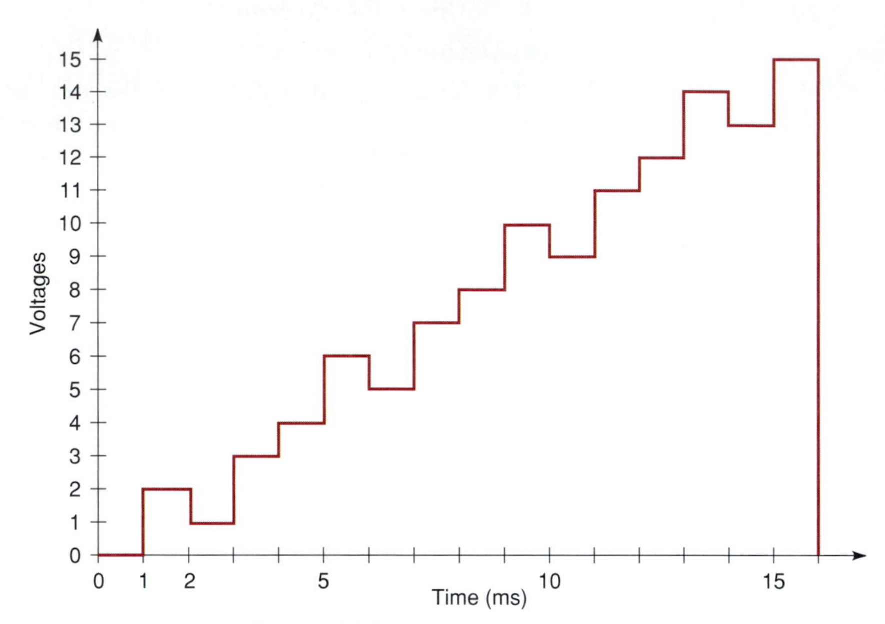

FIGURE 11-32 Problem 11-22.

SECTIONS 11-8 AND 11-9

11-23. **DRILL QUESTION**

Fill in the blanks in the following description of the ADC of Figure 11-13. Each blank may be one or more words.

A START pulse is applied to _____ the counter and to keep _____ from passing through the AND gate into the _____. At this point, the DAC output, V_{AX}, is _____ and $\overline{EOC}$ is _____.

When START returns _____, the AND gate is _____, and the counter is allowed to _____. The V_{AX} signal is increased one _____ at a time until it _____ V_A. At that point, _____ goes LOW to _____ further pulses from _____. This signals the end of conversion, and the digital equivalent of V_A is present at the _____.

B 11-24. An eight-bit digital-ramp ADC with a 40-mV resolution uses a clock frequency of 2.5 MHz and a comparator with $V_T = 1$ mV. Determine the following values.

(a)*The digital output for $V_A = 6.000$ V

(b) The digital output for 6.035 V

(c) The maximum and average conversion times for this ADC

B 11-25. Why were the digital outputs the same for parts (a) and (b) of Problem 11-24?

D 11-26. What would happen in the ADC of Problem 11-24 if an analog voltage of $V_A = 10.853$ V were applied to the input? What waveform would appear at the D/A output? Incorporate the necessary logic in this ADC so that an "overscale" indication will be generated whenever V_A is too large.

B 11-27.* An ADC has the following characteristics: resolution, 12 bits; full-scale error, 0.03% F.S.; full scale output, +5 V.

(a) What is the quantization error in volts?

(b) What is the total possible error in volts?

C, N 11-28. The quantization error of an ADC such as the one in Figure 11-13 is always positive because the V_{AX} value must exceed V_A in order for the comparator output to switch states. This means that the value of V_{AX} could be as much as 1 LSB greater than V_A. This quantization error can be modified so that V_{AX} would be within ±½ LSB of V_A. This can be done by adding a fixed voltage equal to ½ LSB (½ step) to the value of V_A. Figure 11-33 shows this symbolically for a converter that has a resolution of 10 mV/step. A fixed voltage of +5 mV is added to the D/A output in the summing amplifier, and the result, V_{AY}, is fed to the comparator, which has $V_T = 1$ mV.

For this modified converter, determine the digital output for the following V_A values.

(a)* $V_A = 5.022$ V

(b) $V_A = 50.28$ V

Determine the quantization error in each case by comparing V_{AX} and V_A. Note that the error is positive in one case and negative in the other.

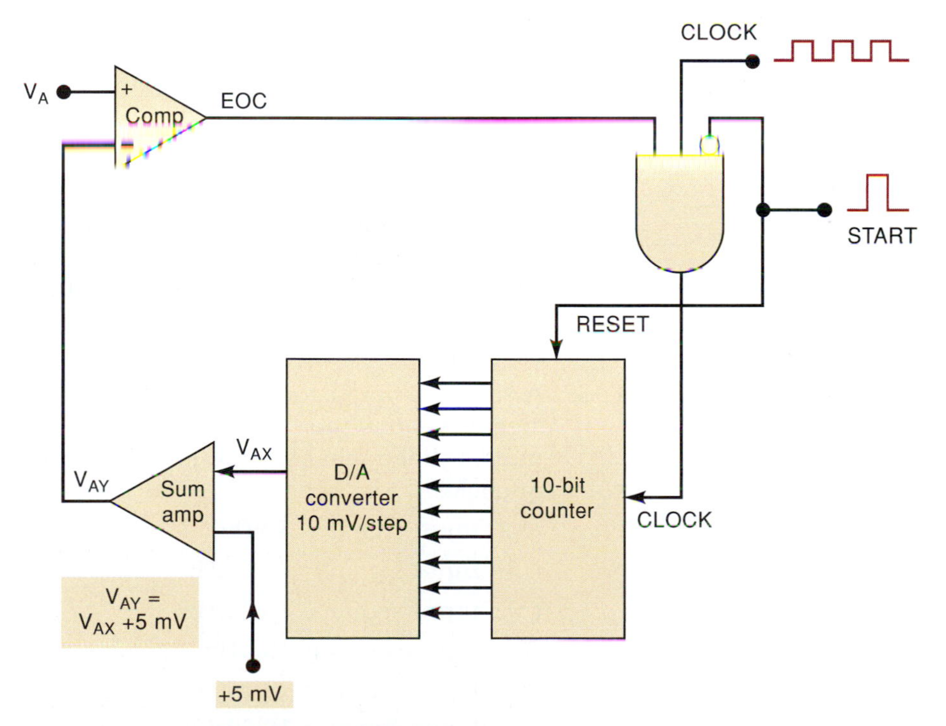

FIGURE 11-33 Problems 11-28 and 11-29.

C 11-29. For the ADC of Figure 11-33, determine the range of analog input values that will produce a digital output of 0100011100.

SECTION 11-10

N 11-30. Assume that the analog signal in Figure 11-34(a) is to be digitized by performing continuous A/D conversions using an eight-bit digital-ramp converter whose staircase rises at the rate of 1 V every 25 μs. Sketch the reconstructed signal using the data obtained during the digitizing process. Compare it with the original signal, and discuss what could be done to make it a more accurate representation.

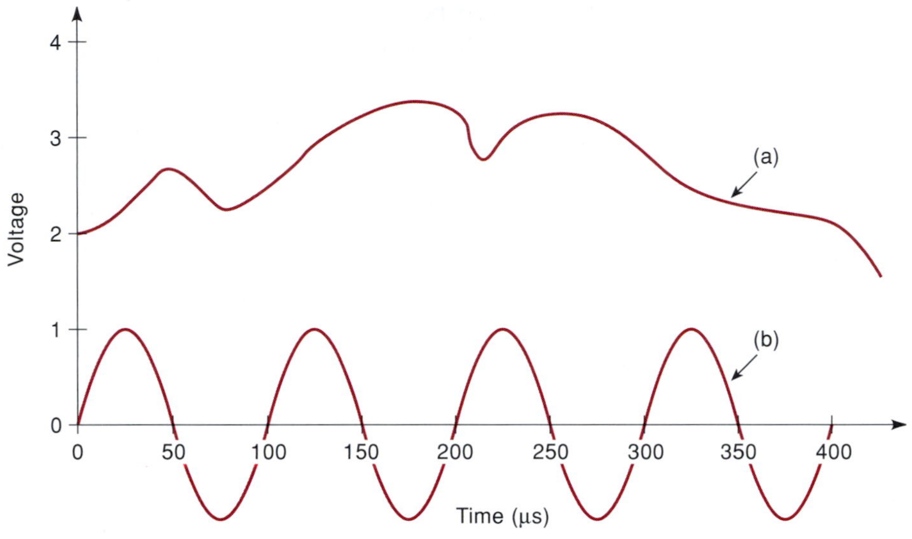

FIGURE 11-34 Problems 11-30, 11-31, and 11-41.

C 11-31.* On the sine wave of Figure 11-34(b), mark the points where samples are taken by a flash A/D converter at intervals of 75 μs (starting at the origin). Then draw the reconstructed output from the D/A converter (connect the sample points with straight lines to show filtering). Calculate the sample frequency, the sine input frequency, and the difference between them. Then compare to the resulting reconstructed waveform frequency.

11-32. A sampled data acquisition system is being used to digitize an audio signal. Assume the sample frequency F_S is 20 kHz. Determine the output frequency that will be heard for each of the following input frequencies.

(a)* Input signal = 5 kHz
(b)* Input signal = 10.1 kHz
(c) Input signal = 10.2 kHz
(d) Input signal = 15 kHz
(e) Input signal = 19.1 kHz
(f) Input signal = 19.2 kHz

SECTION 11-11

B 11-33.* **DRILL QUESTION**

Indicate whether each of the following statements refers to the digital-ramp ADC, the successive-approximation ADC, or both.

(a) Produces a staircase signal at its DAC output
(b) Has a constant conversion time independent of V_A
(c) Has a shorter average conversion time
(d) Uses an analog comparator
(e) Uses a DAC
(f) Uses a counter
(g) Has complex control logic
(h) Has an $\overline{EOC}$ output

11-34. Draw the waveform for V_{AX} as the SAC of Figure 11-19 converts $V_A = 6.7$ V.

11-35. Repeat Problem 11-34 for $V_A = 16$ V.

B 11-36.* A certain eight-bit successive-approximation converter has 2.55 V full scale. The conversion time for $V_A = 1$ V is 80 μs. What will be the conversion time for $V_A = 1.5$ V?

11-37. Figure 11-35 shows the waveform at V_{AX} for a six-bit SAC with a step size of 40 mV during a complete conversion cycle. Examine this waveform and describe what is occurring at times t_0 to t_5. Then determine the resultant digital output.

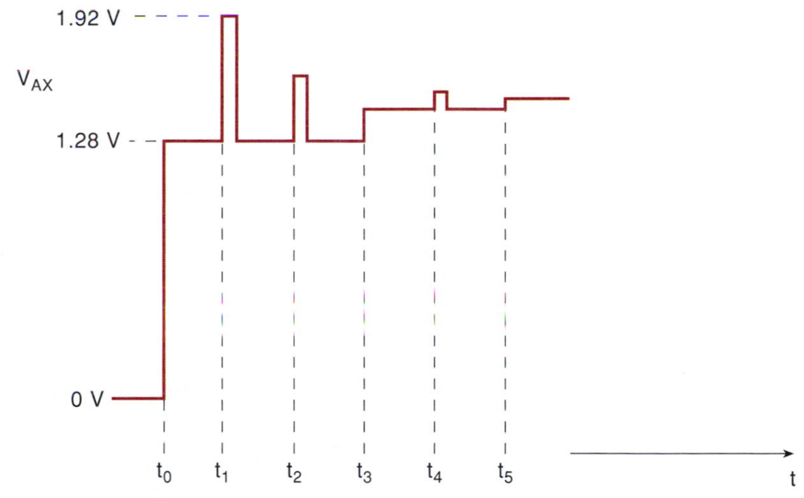

FIGURE 11-35 Problem 11-37.

B 11-38.* Refer to Figure 11-21. What is the approximate value of the analog input if the microcomputer's data bus is at 10010111 when $\overline{RD}$ is pulsed LOW?

D 11-39. Connect a 2.0-V reference source to $V_{ref}/2$, and repeat Problem 11-38.

C, D 11-40.* Design the ADC interface to a digital thermostat using an LM34 temperature sensor and the ADC0804. Your system must measure accurately (±0.2°F) from 50 to 101°F. The LM34 puts out 0.01 V per degree F (0°F = 0 V).

(a) What should the digital value for 50°F be for the best resolution?
(b) What voltage must be applied to $V_{in}(-)$?
(c) What is the full-scale range of voltage that will come in?
(d) What voltage must be applied to $V_{ref}/2$?
(e) What binary value will represent 72°F?
(f) What is the resolution in °F? In volts?

SECTION 11-12

B 11-41. Discuss how a flash ADC with a conversion time of 1 μs would work for the situation of Problem 11-30.

D 11-42. Draw the circuit diagram for a four-bit flash converter with BCD output and a resolution of 0.1 V. Assume that a +5 V precision supply voltage is available.

DRILL QUESTION

B 11-43. For each of the following statements, indicate which type of ADC—digital-ramp, SAC, or flash—is being described.

(a) Fastest method of conversion
(b) Needs a START pulse
(c) Requires the most circuitry
(d) Does not use a DAC
(e) Generates a staircase signal
(f) Uses an analog comparator
(g) Has a relatively fixed conversion time independent of V_A

SECTION 11-13

B 11-44. **DRILL QUESTION**

For each statement, indicate what type(s) of ADC is (are) being described.

(a) Uses a counter that is never reset to 0
(b) Uses a large number of comparators
(c) Uses a VCO
(d) Is used in noisy industrial environments
(e) Uses a capacitor
(f) Is relatively insensitive to temperature

SECTIONS 11-14 AND 11-15

T 11-45.* Refer to the sample-and-hold circuit of Figure 11-24. What circuit fault would result in V_{OUT} looking exactly like V_A? What fault would cause V_{OUT} to be stuck at 0?

C, D 11-46. Use the CMOS 4016 IC (Section 8-16) to implement the switching in Figure 11-25, and design the necessary control logic so that each analog input is converted to its digital equivalent in sequence. The ADC is a 10-bit, successive-approximation type using a 50-kHz clock signal, and it requires a 10-μs-duration start pulse to begin each conversion. The digital outputs are to remain stable for 100 μs after the conversion is complete before switching to the next analog input. Choose an appropriate multiplexing clock frequency.

MICROCOMPUTER APPLICATION

C, N, D 11-47.* Figure 11-21 shows how the ADC0804 is interfaced to a microcomputer. It shows three control signals, $\overline{CS}$, $\overline{RD}$, and $\overline{WR}$, that come from the microcomputer to the ADC. These signals are used to start each new A/D conversion and to read (transfer) the ADC data output into the microcomputer over the data bus.

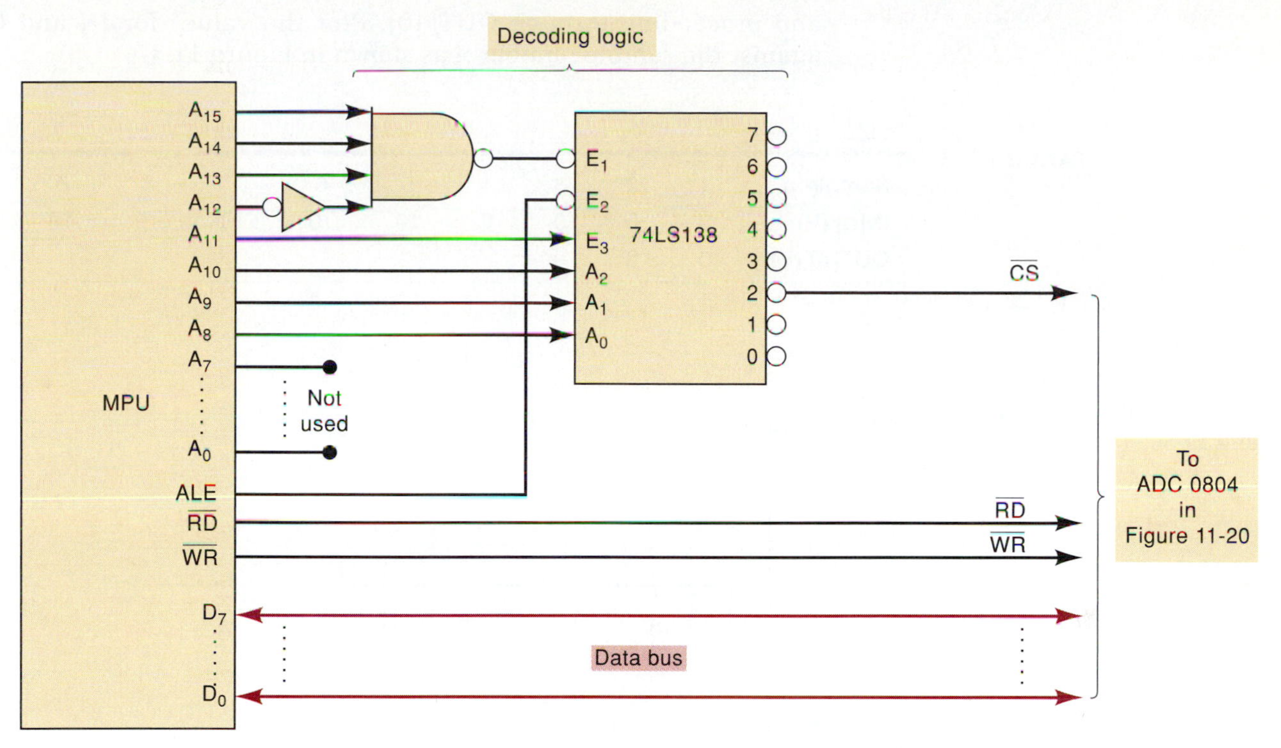

FIGURE 11-36 Problem 11-47: MPU interfaced to the ADC0804 of Figure 11-20.

Figure 11-36 shows one way the address decoding logic could be implemented. The $\overline{CS}$ signal that activates the ADC0804 is developed from the eight high-order address lines of the MPU address bus. Whenever the MPU wants to communicate with the ADC0804, it places the address of the ADC0804 onto the address bus, and the decoding logic drives the $\overline{CS}$ signal LOW. Notice that in addition to the address lines, a timing and control signal (*ALE*) is connected to the $\overline{E}_2$ enable input. Whenever *ALE* is HIGH, it means that the address is potentially in transition, so the decoder should be disabled until *ALE* goes LOW (at which time the address will be valid and stable). This serves a purpose for timing but has no effect on the address of the ADC.

(a) Determine the address of the ADC0804.

(b) Modify the diagram of Figure 11-36 to place the ADC0804 at address E8XX hex.

(c) Modify the diagram of Figure 11-36 to place the ADC0804 at address FFXX hex.

D 11-48. You have available a 10-bit SAC A/D converter (AD 573), but your system requires only eight bits of resolution and you have only eight port bits available on your microprocessor. Can you use this A/D converter, and if so, which of the 10 data lines will you attach to the port?

SECTION 11-17

11-49. The data in Table 11-13 are input samples taken by an A/D converter. Notice that if the input data were plotted, it would represent a simple step function like the rising edge of a digital signal. Calculate the simple average of the four most recent data points, starting with OUT[4] and proceeding through OUT[10]. Plot the values for IN and OUT against the sample number n as shown in Figure 11-37.

TABLE 11-13

Sample n	1	2	3	4	5	6	7	8	9	10
IN[n] (V)	0	0	0	0	10	10	10	10	10	10
OUT[n] (V)	0	0	0							

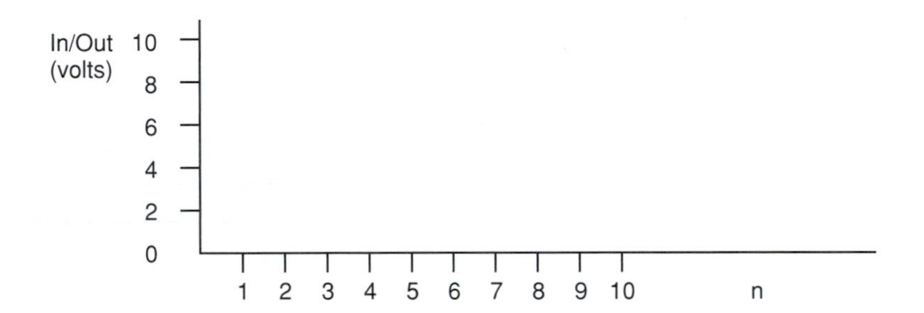

FIGURE 11-37 Graph format for Problems 11-49 and 11-50.

Sample calculations:

$$\begin{aligned}
OUT[n] &= (IN[n-3] + IN[n-2] + IN[n-1] + IN[n])/4 = 0 \\
OUT[4] &= (IN[1] + IN[2] + IN[3] + IN[4])/4 = 0 \\
OUT[5] &= (IN[2] + IN[3] + IN[4] + IN[5])/4 = 2.5
\end{aligned}$$

(Notice that this calculation is equivalent to multiplying each sample by ¼ and summing.)

11-50. Repeat the previous problem using a weighted average of the last four samples. The weights in this case are placing greater emphasis on recent samples and less emphasis on older samples. Use the weights 0.1, 0.2, 0.3, and 0.4.

$$\begin{aligned}
OUT[n] &= 0.1(IN[n-3]) + 0.2(IN[n-2]) + 0.3(IN[n-1]) + 0.4(IN[n]) \\
OUT[5] &= 0.1(IN[2]) + 0.2(IN[3] + 0.3(IN[4] + 0.4(IN[5]) = 4
\end{aligned}$$

11-51. What does the term MAC stand for?

11-52.* **DRILL QUESTIONS**

True or false:

(a) A digital signal is a continuously changing voltage.

(b) A digital signal is a sequence of numbers that represent an analog signal.

When processing an analog signal, the output may be distorted due to:

(a) Quantization error when converting analog to digital

(b) Not sampling the original signal frequently enough

(c) Temperature variation in the processor components

(d) The high-frequency components associated with sudden changes in voltage out of the DAC

(e) Electrical noise on the power supply

(f) Alias signals introduced by the digital system

ANSWERS TO SECTION REVIEW QUESTIONS

SECTION 11-1

1. Converts a nonelectrical physical quantity to an electrical quantity 2. Converts an analog voltage or current to a digital representation 3. Stores it; performs calculation or some other operation on it 4. Converts digital data to their analog representation 5. Controls a physical variable according to an electrical input signal

SECTION 11-2

1. 40 μA; 10.2 mA 2. 5.12 mA 3. 0.39 percent 4. 4096 5. 12 6. True 7. It produces a greater number of possible analog outputs between 0 and full scale.

SECTION 11-3

1. It uses only two different sizes of resistors. 2. 640 kΩ 3. 0.5 V 4. Increases by 20 percent

SECTION 11-4

1. Maximum deviation of DAC output from its ideal value, expressed as percentage of full scale 2. Time it takes DAC output to settle to within ½ step size of its full-scale value when the digital input changes from 0 to full scale 3. Offset error adds a small constant positive or negative value to the expected analog output for any digital input. 4. Because of the response time of the op-amp current-to-voltage converter

SECTION 11-8

1. Tells control logic when the DAC output exceeds the analog input 2. At outputs of register 3. Tells us when conversion is complete and digital equivalent of V_A is at register outputs

SECTION 11-9

1. The digital input to a DAC is incremented until the DAC staircase output exceeds the analog input. 2. The built-in error caused by the fact that V_{AX} does not continuously increase but goes up in steps equal to the DAC's resolution. The final V_{AX} can be different from V_A by as much as one step size. 3. If V_A increases, it will take more steps before V_{AX} can reach the step that first exceeds V_A. 4. True 5. Simple circuit; relatively long conversion time that changes with V_A 6. $0010000111_2 = 135_{10}$ for both cases

SECTION 11-10

1. Process of converting different points on an analog signal to digital and storing the digital data for later use 2. Computer generates START signal to begin an A/D conversion of the analog signal. When EOC goes LOW, it signals the computer that the conversion is complete. The computer then loads the ADC output into memory. The process is repeated for the next point on the analog signal. 3. Twice the highest frequency in the input signal 4. An alias frequency will be present in the output.

SECTION 11-11

1. The SAC has a shorter conversion time that doesn't change with V_A. 2. It has more complex control logic. 3. False 4. (a) 8 (b) 0–5 V (c) $\overline{CS}$ controls the effect of the $\overline{RD}$ and $\overline{WR}$ signals; $\overline{WR}$ is used to start a new conversion; $\overline{RD}$ enables the output buffers. (d) When LOW, it signals the end of a conversion. (e) It separates the usually noisy digital ground from the analog ground so as not to contaminate the analog input signal. (f) All analog voltages on $V_{in}(+)$ are measured with reference to this pin. This allows the input range to be offset from ground.

SECTION 11-12

1. True 2. 4095 comparators and 4096 resistors 3. Major advantage is its conversion speed; disadvantage is the number of required circuit components for a practical resolution.

SECTION 11-13

1. It reduces the conversion time by using an up/down counter that allows V_{AX} to track V_A without starting from 0 for each conversion. 2. A VCO
3. Advantages: low cost, temperature immunity; disadvantage: slow conversion time
4. Flash ADC, voltage-to-frequency ADC, and dual-slope ADC 5. One

SECTION 11-14

1. It takes a sample of an analog voltage signal and stores it on a capacitor.
2. False; they are unity-gain buffers with high input impedance and low output impedance.

SECTION 11-15

1. Uses a single ADC 2. It would become a MOD-8 counter.

SECTION 11-16

1. Digitized waveforms are stored in the memory block. 2. The ADC digitizes the points on the input waveform for storage in memory; the vertical DAC converts the stored data points back to analog voltages to produce the vertical deflection of the electron beam; the horizontal DAC produces a staircase sweep voltage that provides the horizontal deflection of the electron beam.

SECTION 11-17

1. Filtering analog signals 2. An A/D converter 3. To change their dynamic response, you simply change the numbers in the software program, not the hardware components. 4. The Multiply and Accumulate (MAC) unit 5. 3; 7

CHAPTER 12

MEMORY DEVICES

OUTLINE

■ OBJECTIVES

Upon completion of this chapter, you will be able to:

- Understand and correctly use the terminology associated with memory systems.
- Describe the difference between read/write memory and read-only memory.
- Discuss the difference between volatile and nonvolatile memory.
- Determine the capacity of a memory device from its inputs and outputs.
- Outline the steps that occur when the CPU reads from or writes to memory.
- Distinguish among the various types of ROMs and cite some common applications.
- Understand and describe the organization and operation of static and dynamic RAMs.
- Compare the relative advantages and disadvantages of EPROM, EEPROM, and flash memory.
- Combine memory ICs to form memory modules with larger word size and/or capacity.
- Use the test results on a RAM or ROM system to determine possible faults in the memory system.

■ INTRODUCTION

A major advantage of digital over analog systems is the ability to store easily large quantities of digital information and data for short or long periods. This memory capability is what makes digital systems so versatile and adaptable to many situations. For example, in a digital computer, the internal main memory stores instructions that tell the computer what to do under *all* possible circumstances so that the computer will do its job with a minimum amount of human intervention.

This chapter is devoted to a study of the most commonly used types of memory devices and systems. We have already become very familiar with the flip-flop, which is an electronic memory device. We have also seen how groups of FFs called *registers* can be used to store information and how this information can be transferred to other locations. FF registers are high-speed memory elements that are used extensively in the internal operations of a digital computer, where digital information is continually being moved from one location to another. Advances in LSI and VLSI technology have made it possible to obtain large numbers of

FFs on a single chip arranged in various memory-array formats. These bipolar and MOS semiconductor memories are the fastest memory devices available, and their cost has been continuously decreasing as LSI technology improves.

Digital data can also be stored as charges on capacitors, and a very important type of semiconductor memory uses this principle to obtain high-density storage at low power-requirement levels.

Semiconductor memories are used as the **main memory** of a computer (Figure 12-1), where fast operation is important. A computer's main memory—also called its *working memory*—is in constant communication with the central processing unit (CPU) as a program of instructions is being executed. A program and any data used by the program reside in the main memory while the computer is working on that program. RAM and ROM (to be defined shortly) make up main memory.

Another form of storage in a computer is performed by **auxiliary memory** (Figure 12-1), which is separate from the main working memory. Auxiliary memory—also called *mass storage*—has the capacity to store massive amounts of data without the need for electrical power. Auxiliary memory operates at a much slower speed than main memory, and it stores programs and data that are not currently being used by the CPU. This information is transferred to the main memory when the computer needs it. Common auxiliary memory devices are magnetic disk and compact disk (CD).

We will take a detailed look at the characteristics of the most common memory devices used as the internal memory of a computer. First, we define some of the common terms used in memory systems.

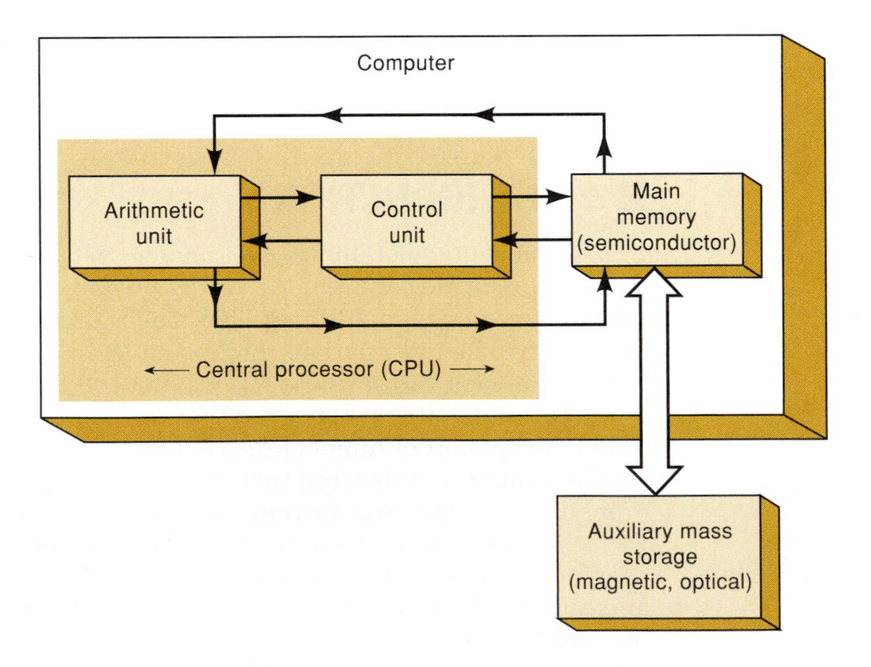

FIGURE 12-1 A computer system normally uses high-speed main memory and slower external auxiliary memory.

12-1 MEMORY TERMINOLOGY

The study of memory devices and systems is filled with terminology that can sometimes be overwhelming to a student. Before we get into any comprehensive discussion of memories, it would be helpful if you had the meaning of some of the more basic terms under your belt. Other new terms will be defined as they appear in the chapter.

- **Memory Cell.** A device or an electrical circuit used to store a single bit (0 or 1). Examples of memory cells include a flip-flop, a charged capacitor, and a single spot on magnetic tape or disk.
- **Memory Word.** A group of bits (cells) in a memory that represents instructions or data of some type. For example, a register consisting of eight FFs can be considered to be a memory that is storing an eight-bit word. Word sizes in modern computers typically range from 8 to 64 bits, depending on the size of the computer.
- **Byte.** A special term used for a group of eight bits. A byte always consists of eight bits. Word sizes can be expressed in bytes as well as in bits. For example, a word size of eight bits is also a word size of one byte, a word size of 16 bits is two bytes, and so on.
- **Capacity.** A way of specifying how many bits can be stored in a particular memory device or complete memory system. To illustrate, suppose that we have a memory that can store 4096 20-bit words. This represents a total capacity of 81,920 bits. We could also express this memory's capacity as 4096×20. When expressed this way, the first number (4096) is the number of words, and the second number (20) is the number of bits per word (word size). The number of words in a memory is often a multiple of 1024. It is common to use the designation "1K" to represent $1024 = 2^{10}$ when referring to memory capacity. Thus, a memory that has a storage capacity of $4K \times 20$ is actually a 4096×20 memory. The development of larger memories has brought about the designation "1M" or "1 meg" to represent $2^{20} = 1{,}048{,}576$. Thus, a memory that has a capacity of $2M \times 8$ is actually one with a capacity of $2{,}097{,}152 \times 8$. The designation "giga" refers to $2^{30} = 1{,}073{,}741{,}824$.

EXAMPLE 12-1A

A certain semiconductor memory chip is specified as $2\text{ K} \times 8$. How many words can be stored on this chip? What is the word size? How many total bits can this chip store?

Solution

$$2K = 2 \times 1024 = 2048 \text{ words}$$

Each word is eight bits (one byte). The total number of bits is therefore

$$2048 \times 8 = 16{,}384 \text{ bits}$$

EXAMPLE 12-1B

Which memory stores the most bits: a $5M \times 8$ memory or a memory that stores 1M words at a word size of 16 bits?

Solution

$$5M \times 8 = 5 \times 1{,}048{,}576 \times 8 = 41{,}943{,}040 \text{ bits}$$
$$1M \times 16 = 1{,}048{,}576 \times 16 = 16{,}777{,}216 \text{ bits}$$

The $5M \times 8$ memory stores more bits.

- **Density.** Another term for *capacity.* When we say that one memory device has a greater density than another, we mean that it can store more bits in the same amount of space. It is more dense.
- **Address.** A number that identifies the location of a word in memory. Each word stored in a memory device or system has a unique address. Addresses always exist in a digital system as a binary number, although octal, hexadecimal, and decimal numbers are often used to represent the address for convenience. Figure 12-2 illustrates a small memory consisting of eight words. Each of these eight words has a specific address represented as a three-bit number ranging from 000 to 111. Whenever we refer to a specific word location in memory, we use its address code to identify it.
- **Read Operation.** The operation whereby the binary word stored in a specific memory location (address) is sensed and then transferred to another device. For example, if we want to use word 4 of the memory of Figure 12-2 for some purpose, we must perform a read operation on address 100. The read operation is often called a *fetch* operation because a word is being fetched from memory. We will use both terms interchangeably.
- **Write Operation.** The operation whereby a new word is placed into a particular memory location. It is also referred to as a *store* operation. Whenever a new word is written into a memory location, it replaces the word that was previously stored there.
- **Access Time.** A measure of a memory device's operating speed. It is the amount of time required to perform a read operation. More specifically, it is the time between the memory receiving a new address input and the data becoming available at the memory output. The symbol t_{ACC} is used for access time.
- **Volatile Memory.** Any type of memory that requires the application of electrical power in order to store information. If the electrical power is removed, all information stored in the memory will be lost. Many semiconductor memories are volatile, while all magnetic memories are *nonvolatile,* which means that they can store information without electrical power.
- **Random-Access Memory (RAM).** Memory in which the actual physical location of a memory word has no effect on how long it takes to read

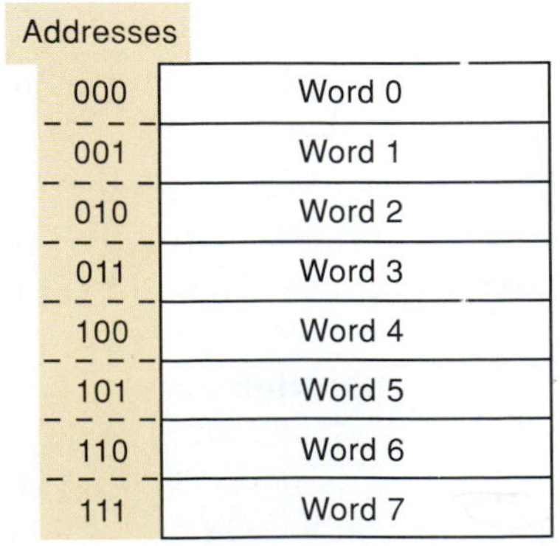

Addresses	
000	Word 0
001	Word 1
010	Word 2
011	Word 3
100	Word 4
101	Word 5
110	Word 6
111	Word 7

FIGURE 12-2 Each word location has a specific binary address.

from or write into that location. In other words, the access time is the same for any address in memory. Most semiconductor memories are RAMs.

- **Sequential-Access Memory (SAM).** A type of memory in which the access time is not constant but varies depending on the address location. A particular stored word is found by sequencing through all address locations until the desired address is reached. This produces access times that are much longer than those of random-access memories. An example of a sequential-access memory device is a magnetic tape backup. To illustrate the difference between SAM and RAM, consider the situation where you have recorded 60 minutes of songs on an audio tape cassette. When you want to get to a particular song, you have to rewind or fast-forward the tape until you find it. The process is relatively slow, and the amount of time required depends on where on the tape the desired song is recorded. This is SAM because you have to sequence through all intervening information until you find what you are looking for. The RAM counterpart to this would be an audio CD, where you can quickly select any song by punching in the appropriate code, and it takes approximately the same time no matter what song you select. Sequential-access memories are used where the data to be accessed will always come in a long sequence of successive words. Video memory, for example, must output its contents in the same order over and over again to keep the image refreshed on the CRT screen.
- **Read/Write Memory (RWM).** Any memory that can be read from or written into with equal ease.
- **Read-Only Memory (ROM).** A broad class of semiconductor memories designed for applications where the ratio of read operations to write operations is very high. Technically, a ROM can be written into (programmed) only once, and this operation is normally performed at the factory. Thereafter, information can only be read from the memory. Other types of ROM are actually read-mostly memories (RMM), which can be written into more than once; but the write operation is more complicated than the read operation, and it is not performed very often. The various types of ROM will be discussed later. *All ROM is nonvolatile* and will store data when electrical power is removed.
- **Static Memory Devices.** Semiconductor memory devices in which the stored data will remain permanently stored as long as power is applied, without the need for periodically rewriting the data into memory.
- **Dynamic Memory Devices.** Semiconductor memory devices in which the stored data will *not* remain permanently stored, even with power applied, unless the data are periodically rewritten into memory. The latter operation is called a *refresh* operation.
- **Main Memory.** Also referred to as the computer's *working memory.* It stores instructions and data the CPU is currently working on. It is the highest-speed memory in the computer and is always a semiconductor memory.
- **Auxiliary Memory.** Also referred to as *mass storage* because it stores massive amounts of information external to the main memory. It is slower in speed than main memory and is always nonvolatile. Magnetic disks and CDs are common auxiliary memory devices.

REVIEW QUESTION

1. Define the following terms.
   (a) *Memory cell*
   (b) *Memory word*
   (c) *Address*
   (d) *Byte*
   (e) *Access time*
2. A certain memory has a capacity of 8K × 16. How many bits are in each word? How many words are being stored? How many memory cells does this memory contain?
3. Explain the difference between the read (fetch) and write (store) operations.
4. *True or false:* A volatile memory will lose its stored data when electrical power is interrupted.
5. Explain the difference between SAM and RAM.
6. Explain the difference between RWM and ROM.
7. *True or false:* A dynamic memory will hold its data as long as electrical power is applied.

12-2 GENERAL MEMORY OPERATION

Although each type of memory is different in its internal operation, certain basic operating principles are the same for all memory systems. An understanding of these basic ideas will help in our study of individual memory devices.

Every memory system requires several different types of input and output lines to perform the following functions:

1. Select the address in memory that is being accessed for a read or write operation.
2. Select either a read or a write operation to be performed.
3. Supply the input data to be stored in memory during a write operation.
4. Hold the output data coming from memory during a read operation.
5. Enable (or disable) the memory so that it will (or will not) respond to the address inputs and read/write command.

Figure 12-3(a) illustrates these basic functions in a simplified diagram of a 32 × 4 memory that stores 32 four-bit words. Because the word size is four bits, there are four data input lines I_0 to I_3 and four data output lines O_0 to O_3. During a write operation, the data to be stored into memory must be applied to the data input lines. During a read operation, the word being read from memory appears at the data output lines.

Address Inputs

Because this memory stores 32 words, it has 32 different storage locations and therefore 32 different binary addresses ranging from 00000 to 11111 (0 to 31 in decimal). Thus, there are five address inputs, A_0 to A_4. To access one

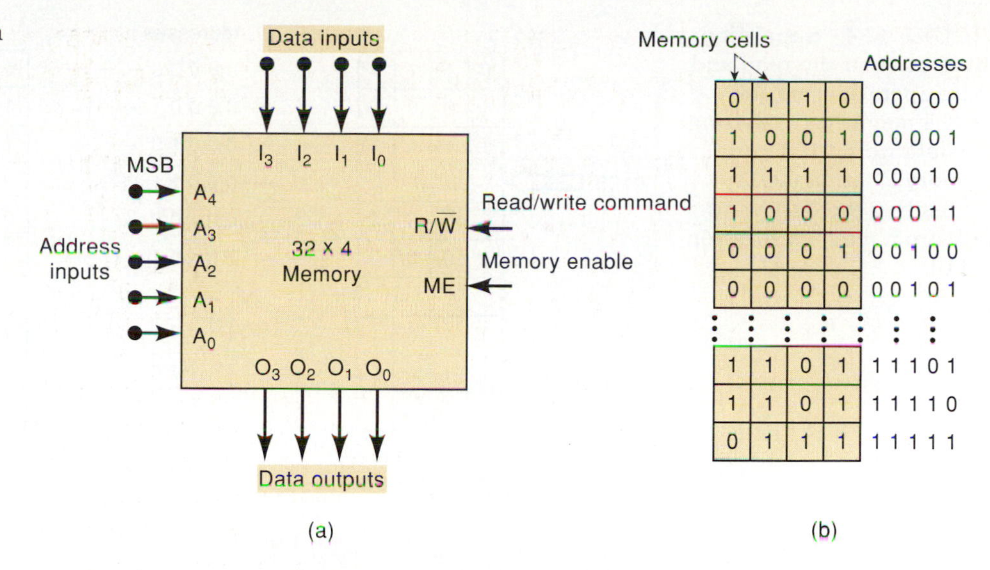

FIGURE 12-3 (a) Diagram of a 32 × 4 memory; (b) virtual arrangement of memory cells into 32 four-bit words.

of the memory locations for a read or a write operation, the five-bit address code for that particular location is applied to the address inputs. In general, N address inputs are required for a memory that has a capacity of 2^N words.

We can visualize the memory of Figure 12-3(a) as an arrangement of 32 registers, with each register holding a four-bit word, as illustrated in Figure 12-3(b). Each address location is shown containing four memory cells that hold 1s and 0s that make up the data word stored at that location. For example, the data word 0110 is stored at address 00000, the data word 1001 is stored at address 00001, and so on.

The $R/\overline{W}$ Input

This input controls which memory operation is to take place: read (R) or write (W). The input is labeled $R/\overline{W}$; there is no bar over the R, which indicates that the read operation occurs when $R/\overline{W} = 1$. The bar over the W indicates that the write operation takes place when $R/\overline{W} = 0$. Other labels are often used for this input. Two of the more common ones are $\overline{W}$ (write) and $\overline{WE}$ (write enable). Again, the bar indicates that the write operation occurs when the input is LOW. It is understood that the read operation occurs for a HIGH.

A simplified illustration of the read and write operations is shown in Figure 12-4. Figure 12-4(a) shows the data word 0100 being written into the memory register at address location 00011. This data word would have been applied to the memory's data input lines, and it replaces the data previously stored at address 00011. Figure 12-4(b) shows the data word 1101 being read from address 11110. This data word would appear at the memory's data output lines. After the read operation, the data word 1101 is still stored in address 11110. In other words, the read operation does not change the stored data.

Memory Enable

Many memory systems have some means for completely disabling all or part of the memory so that it will not respond to the other inputs. This is represented in Figure 12-3 as the MEMORY ENABLE input, although it can have different names in the various memory systems, such as chip enable (CE) or chip select (CS). Here, it is shown as an active-HIGH input that enables the memory to operate normally when it is kept HIGH. A LOW on this input disables the memory so that it will not respond to the address and $R/\overline{W}$ inputs.

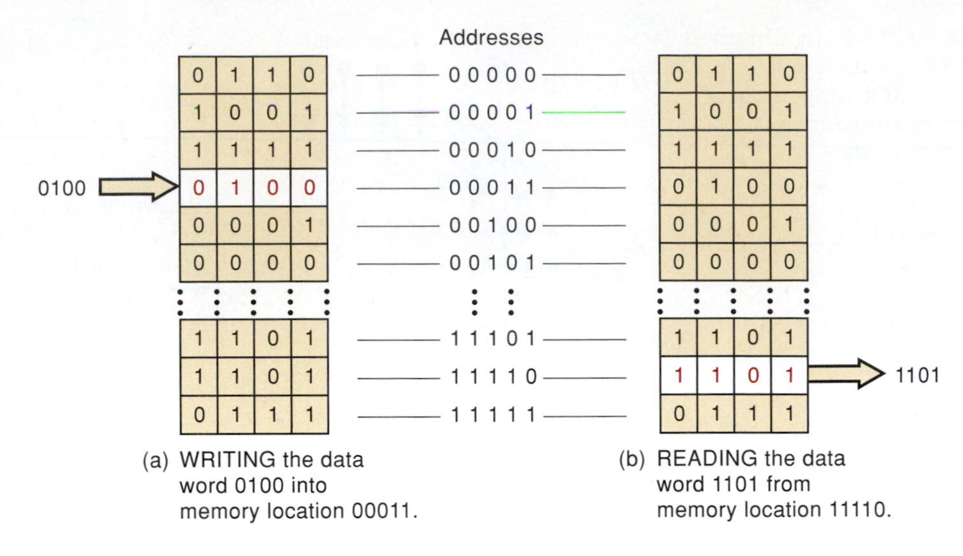

FIGURE 12-4 Simplified illustration of the read and write operations on the 32 × 4 memory: (a) writing the data word 0100 into memory location 00011; (b) reading the data word 1101 from memory location 11110.

This type of input is useful when several memory modules are combined to form a larger memory. We will examine this idea later.

EXAMPLE 12-2

Describe the conditions at each input and output when the contents of address location 00100 are to be read.

Solution

Address inputs: 00100

Data inputs: xxxx (not used)

$R/\overline{W}$: HIGH

MEMORY ENABLE: HIGH

Data outputs: 0001

EXAMPLE 12-3

Describe the conditions at each input and output when the data word 1110 is to be written into address location 01101.

Solution

Address inputs: 01101

Data inputs: 1110

$R/\overline{W}$: LOW

MEMORY ENABLE: HIGH

Data outputs: xxxx (not used; usually Hi-Z)

EXAMPLE 12-4

A certain memory has a capacity of 4K × 8.

(a) How many data input and data output lines does it have?

(b) How many address lines does it have?

(c) What is its capacity in bytes?

Solution

(a) Eight of each because the word size is eight.

(b) The memory stores 4K = 4 × 1024 = 4096 words. Thus, there are 4096 memory addresses. Because $4096 = 2^{12}$, it requires a 12-bit address code to specify one of 4096 addresses.

(c) A byte is eight bits. This memory has a capacity of 4096 bytes.

The example memory in Figure 12-3 illustrates the important input and output functions common to most memory systems. Of course, each type of memory may have other input and output lines that are peculiar to that memory. These will be described as we discuss the individual memory types.

REVIEW QUESTIONS

1. How many address inputs, data inputs, and data outputs are required for a 16K × 12 memory?
2. What is the function of the $R/\overline{W}$ input?
3. What is the function of the MEMORY ENABLE input?

12-3 CPU–MEMORY CONNECTIONS

A major part of this chapter is devoted to semiconductor memory, which, as pointed out earlier, makes up the main memory of most modern computers. Remember, this main memory is in constant communication with the central processing unit (CPU). It is not necessary to be familiar with the detailed operation of a CPU at this point, and so the following simplified treatment of the CPU–memory interface will provide the perspective needed to make our study of memory devices more meaningful.

A computer's main memory is made up of RAM and ROM ICs that are interfaced to the CPU over three groups of signal lines or buses. These are shown in Figure 12-5 as the address lines or address bus, the data lines or data bus, and the control lines or control bus. Each of these buses consists of several lines (note that they are represented by a single line with a slash), and the number of lines in each bus will vary from one computer to the next. The three buses play a necessary part in allowing the CPU to write data into memory and to read data from memory.

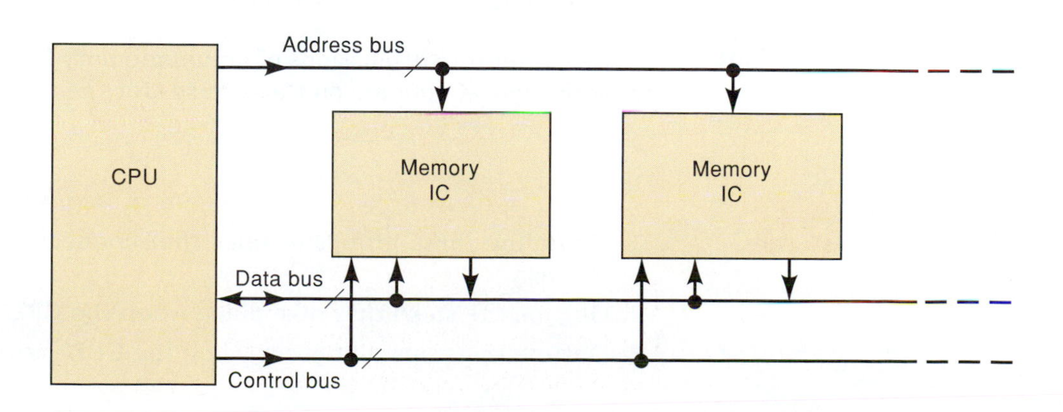

FIGURE 12-5 Three groups of lines (buses) connect the main memory ICs to the CPU.

When a computer is executing a program of instructions, the CPU continually fetches (reads) information from those locations in memory that contain (1) the program codes representing the operations to be performed and (2) the data to be operated upon. The CPU will also store (write) data into memory locations as dictated by the program instructions. Whenever the CPU wants to write data to a particular memory location, the following steps must occur:

Write Operation

1. The CPU supplies the binary address of the memory location where the data are to be stored. It places this address on the address bus lines.
2. The CPU places the data to be stored on the data bus lines.
3. The CPU activates the appropriate control signal lines for the memory write operation.
4. The memory ICs decode the binary address to determine which location is being selected for the store operation.
5. The data on the data bus are transferred to the selected memory location.

Whenever the CPU wants to read data from a specific memory location, the following steps must occur:

Read Operation

1. The CPU supplies the binary address of the memory location from which data are to be retrieved. It places this address on the address bus lines.
2. The CPU activates the appropriate control signal lines for the memory read operation.
3. The memory ICs decode the binary address to determine which location is being selected for the read operation.
4. The memory ICs place data from the selected memory location onto the data bus, from which they are transferred to the CPU.

The steps above should make clear the function of each of the system buses:

- **Address Bus.** This *unidirectional* bus carries the binary address outputs from the CPU to the memory ICs to select one memory location.
- **Data Bus.** This *bidirectional* bus carries data between the CPU and the memory ICs.
- **Control Bus.** This bus carries control signals (such as the $R/\overline{W}$ signal) from the CPU to the memory ICs.

As we get into discussions of actual memory ICs, we will examine the signal activity that appears on these buses for the read and write operations.

REVIEW QUESTIONS

1. Name the three groups of lines that connect the CPU and the internal memory.
2. Outline the steps that take place when the CPU reads from memory.
3. Outline the steps that occur when the CPU writes to memory.

12-4 READ-ONLY MEMORIES

The read-only memory is a type of semiconductor memory designed to hold data that either are permanent or will not change frequently. During normal operation, no new data can be written into a ROM, but data can be read from ROM. For some ROMs, the data that are stored must be built-in during the manufacturing process; for other ROMs, the data can be entered electrically. The process of entering data is called **programming** or *burning-in* the ROM. Some ROMs cannot have their data changed once they have been programmed; others can be *erased* and reprogrammed as often as desired. We will take a detailed look later at these various types of ROMs. For now, we will assume that the ROMs have been programmed and are holding data.

ROMs are used to store data and information that are not to change during the normal operation of a system. A major use for ROMs is in the storage of programs in microcomputers. Because all ROMs are *nonvolatile,* these programs are not lost when electrical power is turned off. When the microcomputer is turned on, it can immediately begin executing the program stored in ROM. ROMs are also used for program and data storage in microprocessor-controlled equipment such as electronic cash registers, appliances, and security systems.

ROM Block Diagram

A typical block diagram for a ROM is shown in Figure 12-6(a). It has three sets of signals: address inputs, control input(s), and data outputs. From our previous discussions, we can determine that this ROM is storing 16 words because it has $2^4 = 16$ possible addresses, and each word contains eight bits because there are eight data outputs. Thus, this is a 16×8 ROM. Another way to describe this ROM's capacity is to say that it stores 16 bytes of data.

The data outputs of most ROM ICs are tristate outputs, to permit the connection of many ROM chips to the same data bus for memory expansion. The most common numbers of data outputs for ROMs are four, eight, and 16 bits, with eight-bit words being the most common.

The control input $\overline{CS}$ stands for **chip select.** This is essentially an enable input that enables or disables the ROM outputs. Some manufacturers use different labels for the control input, such as *CE* (chip enable) or *OE* (output enable). Many ROMs have two or more control inputs that must be active in order to enable the data outputs so that data can be read from the selected address. In some ROM ICs, one of the control inputs (usually the *CE*) is used to place the ROM in a low-power standby mode when it is not being used. This reduces the current drain from the system power supply.

The $\overline{CS}$ input shown in Figure 12-6(a) is active-LOW; therefore, it must be in the LOW state to enable the ROM data to appear at the data outputs. Notice that there is no $R/\overline{W}$ (read/write) input because the ROM cannot be written into during normal operation.

The Read Operation

Let's assume that the ROM has been programmed with the data shown in the table of Figure 12-6(b). Sixteen different data words are stored at the 16 different address locations. For example, the data word stored at location 0011 is 10101111. Of course, the data are stored in binary inside the ROM, but very often we use hexadecimal notation to show the programmed data efficiently. This is done in Figure 12-6(c).

In order to read a data word from ROM, we need to do two things: (1) apply the appropriate address inputs and then (2) activate the control inputs. For

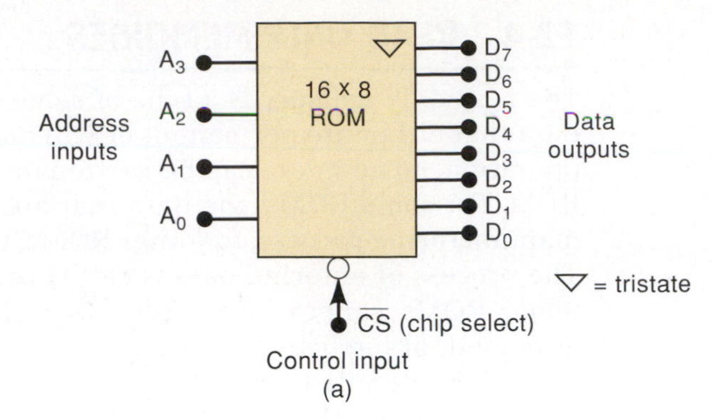

Word	Address A_3	A_2	A_1	A_0	Data D_7	D_6	D_5	D_4	D_3	D_2	D_1	D_0
0	0	0	0	0	1	1	0	1	1	1	1	0
1	0	0	0	1	0	0	1	1	1	0	1	0
2	0	0	1	0	1	0	0	0	0	1	0	1
3	0	0	1	1	1	0	1	0	1	1	1	1
4	0	1	0	0	0	0	0	1	1	0	0	1
5	0	1	0	1	0	1	1	1	1	0	1	1
6	0	1	1	0	0	0	0	0	0	0	0	0
7	0	1	1	1	1	1	1	0	1	1	0	1
8	1	0	0	0	0	0	1	1	1	1	0	0
9	1	0	0	1	1	1	1	1	1	1	1	1
10	1	0	1	0	1	0	1	1	1	0	0	0
11	1	0	1	1	1	1	0	0	0	1	1	1
12	1	1	0	0	0	0	1	0	0	1	1	1
13	1	1	0	1	0	1	1	0	1	0	1	0
14	1	1	1	0	1	1	0	1	0	0	1	0
15	1	1	1	1	0	1	0	1	1	0	1	1

(b)

Word	Address A_3 A_2 A_1 A_0	Data D_7–D_0
0	0	DE
1	1	3A
2	2	85
3	3	AF
4	4	19
5	5	7B
6	6	00
7	7	ED
8	8	3C
9	9	FF
10	A	B8
11	B	C7
12	C	27
13	D	6A
14	E	D2
15	F	5B

(c)

FIGURE 12-6 (a) Typical ROM block symbol; (b) table showing binary data at each address location; (c) the same table in hex.

example, if we want to read the data stored at location 0111 of the ROM in Figure 12-6, we must apply $A_3A_2A_1A_0 = 0111$ to the address inputs and then apply a LOW to $\overline{CS}$. The address inputs will be decoded inside the ROM to select the correct data word, 11101101, that will appear at outputs D_7 to D_0. If $\overline{CS}$ is kept HIGH, the ROM outputs will be disabled and will be in the Hi-Z state.

REVIEW QUESTIONS

1. *True or false:* All ROMs are nonvolatile.
2. Describe the procedure for reading from ROM.
3. What is *programming* or *burning-in* a ROM?

12-5 ROM ARCHITECTURE

The internal architecture (structure) of a ROM IC is very complex, and we need not be familiar with all of its detail. It is instructive, however, to look at a simplified diagram of the internal architecture, such as that shown in Figure 12-7, for the 16 × 8 ROM. There are four basic parts: *register array, row decoder, column decoder,* and *output buffers.*

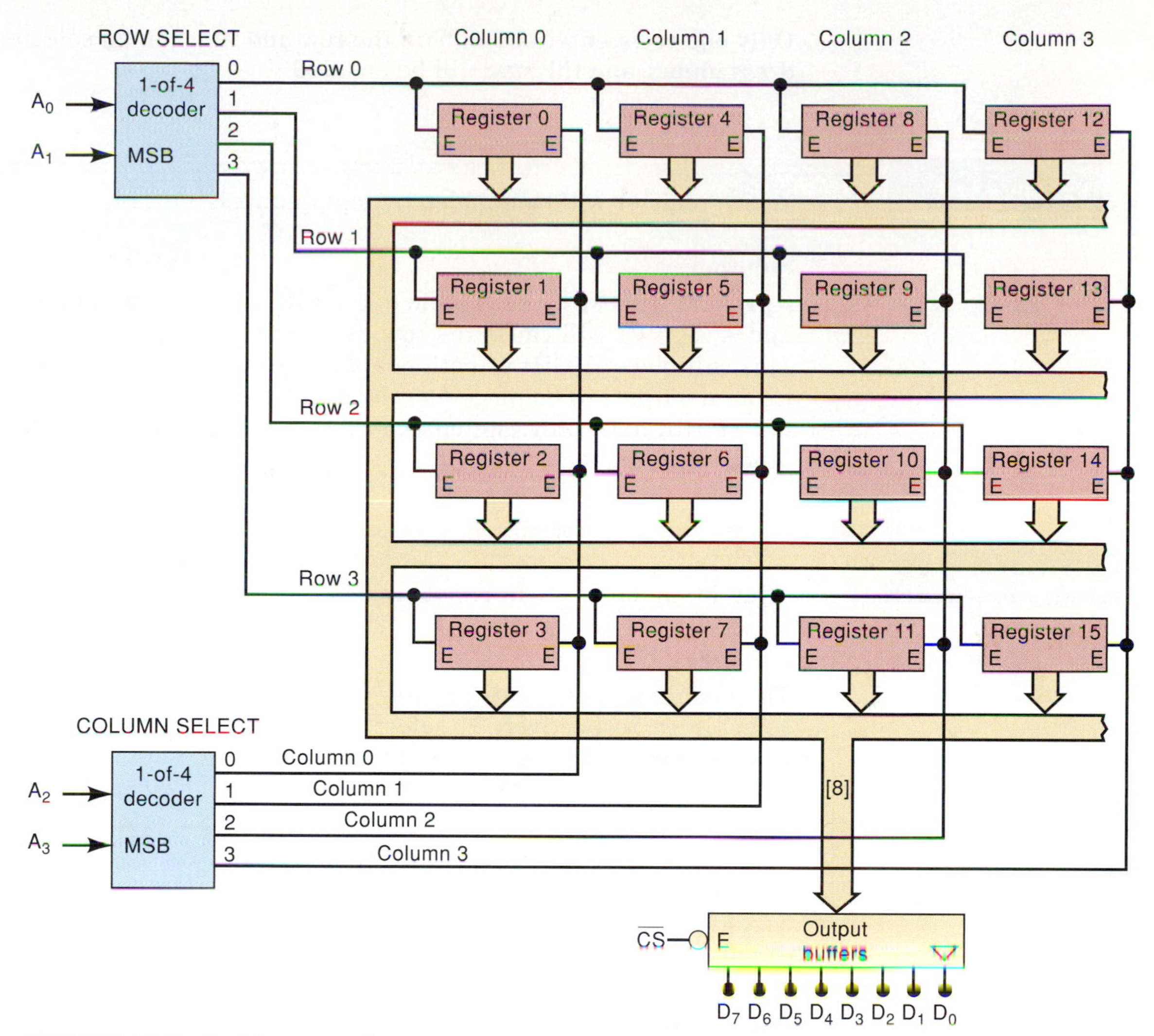

FIGURE 12-7 Architecture of a 16 × 8 ROM. Each register stores one eight-bit word.

Register Array

The register array stores the data that have been programmed into the ROM. Each register contains several memory cells equal to the word size. In this case, each register stores an eight-bit word. The registers are arranged in a square matrix array that is common to many semiconductor memory chips. We can specify the position of each register as being in a specific row and a specific column. For example, register 0 is in row 0, column 0, and register 9 is in row 1, column 2.

The eight data outputs of each register are connected to an internal data bus that runs through the entire circuit. Each register has two enable inputs (E); both must be HIGH in order for the register's data to be placed on the bus.

Address Decoders

The applied address code $A_3A_2A_1A_0$ determines which register in the array will be enabled to place its eight-bit data word onto the bus. Address bits A_1A_0 are fed to a 1-of-4 decoder that activates one row-select line, and address bits A_3A_2 are fed to a second 1-of-4 decoder that activates one column-select line.

Only one register will be in both the row and the column selected by the address inputs, and this one will be enabled.

EXAMPLE 12-5

Which register will be enabled by input address 1101?

Solution

$A_3A_2 = 11$ will cause the column decoder to activate the column 3 select line, and $A_1A_0 = 01$ will cause the row decoder to activate the row 1 select line. This will place HIGHs at both enable inputs of register 13, thereby causing its data outputs to be placed on the bus. Note that the other registers in column 3 will have only one enable input activated; the same is true for the other row 1 registers.

EXAMPLE 12-6

What input address will enable register 7?

Solution

The enable inputs of this register are connected to the row 3 and column 1 select lines, respectively. To select row 3, the A_1A_0 inputs must be at 11, and to select column 1, the A_3A_2 inputs must be at 01. Thus, the required address will be $A_3A_2A_1A_0 = 0111$.

Output Buffers

The register that is enabled by the address inputs will place its data on the data bus. These data feed into the output buffers, which will pass the data to the external data outputs, provided that $\overline{CS}$ is LOW. If $\overline{CS}$ is HIGH, the output buffers are in the Hi-Z state, and D_7 through D_0 will be floating.

The architecture shown in Figure 12-7 is similar to that of many IC ROMs. Depending on the number of stored data words, the registers in some ROMs will not be arranged in a square array. For example, the Intel 27C64 is a CMOS ROM that stores 8192 eight-bit words. Its 8192 registers are arranged in an array of 256 rows $\times$ 32 registers. ROM capacities range from 256 $\times$ 4 to 8M $\times$ 8.

EXAMPLE 12-7

Describe the internal architecture of a ROM that stores 4K bytes and uses a square register array.

Solution

4K is actually $4 \times 1024 = 4096$, and so this ROM holds 4096 eight-bit words. Each word can be thought of as being stored in an eight-bit register, and there are 4096 registers connected to a common data bus internal to the chip. Because $4096 = 64^2$, the registers are arranged in a 64×64 array; that is, there are 64 rows and 64 columns. This requires a 1-of-64 decoder to decode six address inputs for the row select, and a second 1-of-64 decoder to decode six other address inputs for the column select. Thus, a total of 12 address inputs is required. This makes sense because $2^{12} = 4096$, and there are 4096 different addresses.

REVIEW QUESTIONS

1. What input address code is required if we want to read the data from register 9 in Figure 12-7?
2. Describe the function of the row-select decoder, the column-select decoder, and the output buffers in the ROM architecture.

12-6 ROM TIMING

There will be a propagation delay between the application of a ROM's inputs and the appearance of the data outputs during a read operation. This time delay, called access time (t_{ACC}) is a measure of the ROM's operating speed. Access time is described graphically by the waveforms in Figure 12-8.

The top waveform represents the address inputs; the middle waveform is an active-LOW chip select, $\overline{CS}$; and the bottom waveform represents the data outputs. At time t_0 the address inputs are all at some specific level, some HIGH and some LOW. $\overline{CS}$ is HIGH, so that the ROM data outputs are in their Hi-Z state (represented by the hatched line).

Just prior to t_1, the address inputs are changing to a new address for a new read operation. At t_1, the new address is valid; that is, each address input is at a valid logic level. At this point, the internal ROM circuitry begins to decode the new address inputs to select the register that is to send its data to the output buffers. At t_2, the $\overline{CS}$ input is activated to enable the output buffers. Finally, at t_3, the outputs change from the Hi-Z state to the valid data that represent the data stored at the specified address.

The time delay between t_1, when the new address becomes valid, and t_3, when the data outputs become valid, is the access time t_{ACC}. Typical bipolar ROMs will have access times in the range from 30 to 90 ns; access times of NMOS devices will range from 35 to 500 ns. Improvements to CMOS technology have brought access times into the 20 to 60-ns range. Consequently, bipolar and NMOS devices are rarely produced in newer (larger) ROMs.

Another important timing parameter is the *output enable time* (t_{OE}), which is the delay between the $\overline{CS}$ input and the valid data output. Typical values for t_{OE} are 10 to 20 ns for bipolar, 25 to 100 ns for NMOS, and 12 to 50 ns for CMOS ROMs. This timing parameter is important in situations where

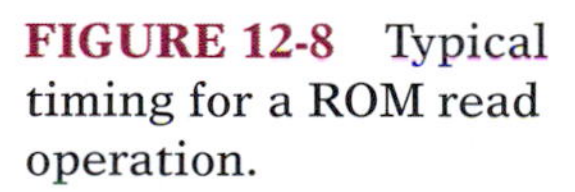

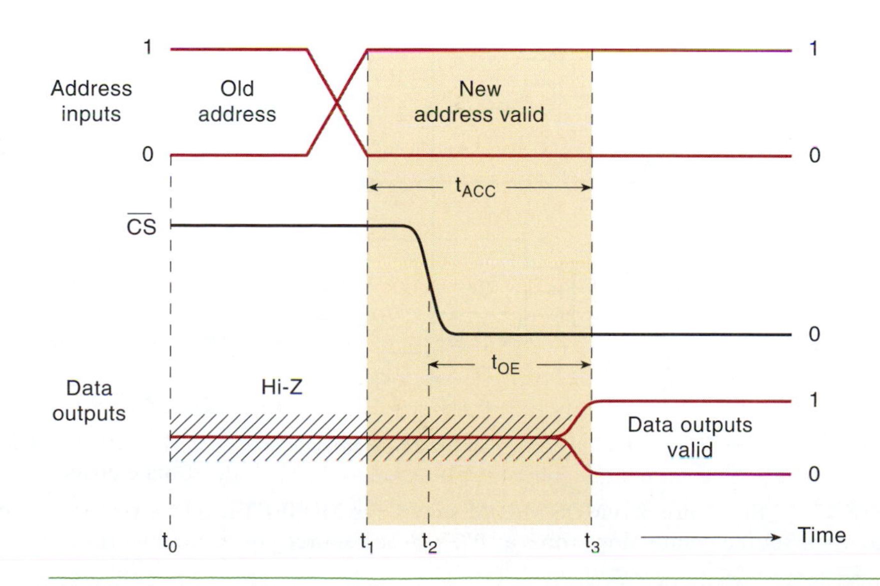

FIGURE 12-8 Typical timing for a ROM read operation.

the address inputs are already set to their new values, but the ROM outputs have not yet been enabled. When $\overline{CS}$ goes LOW to enable the outputs, the delay will be t_{OE}.

12-7 TYPES OF ROMs

Now that we have a general understanding of the internal architecture and external operation of ROM devices, we will look at the various types of ROMs to see how they differ in the way they are programmed, erased, and reprogrammed.

Mask-Programmed ROM

The mask-programmed ROM (MROM) has its information stored at the time the integrated circuit is manufactured. As you can see from Figure 12-9, ROMs are made up of a rectangular array of transistors. Information is stored by either connecting or disconnecting the source of a transistor to the output

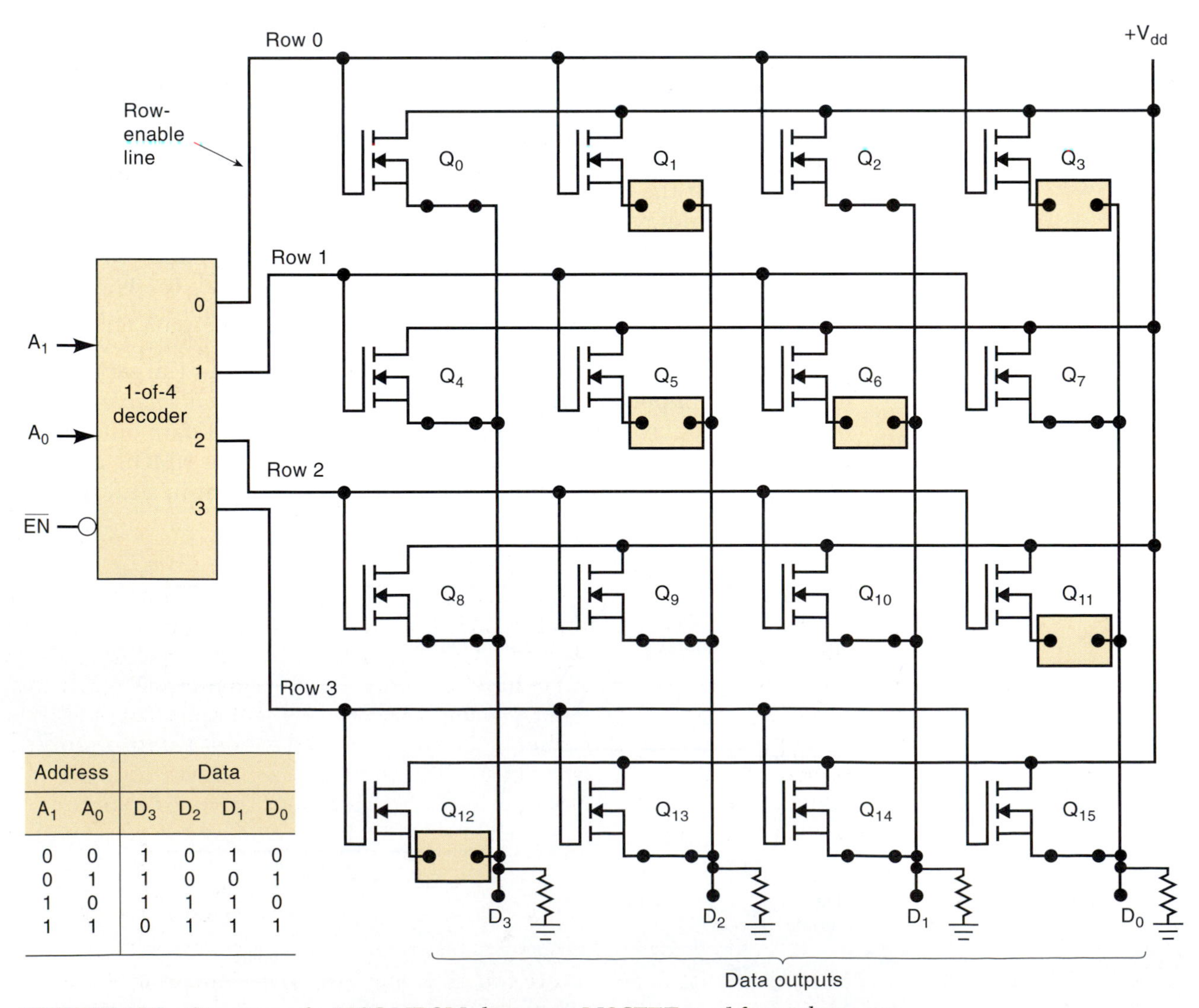

Address		Data			
A_1	A_0	D_3	D_2	D_1	D_0
0	0	1	0	1	0
0	1	1	0	0	1
1	0	1	1	1	0
1	1	0	1	1	1

FIGURE 12-9 Structure of a MOS MROM shows one MOSFET used for each memory cell. An open source connection stores a "0"; a closed source connection stores a "1."

column. The last step in the manufacturing process is to form all these conducting paths or connections. The process uses a "mask" to deposit metals on the silicon that determine where the connections form in a way similar to using stencils and spray paint but on a much smaller scale. The mask is very precise and expensive and must be made specifically for the customer, with the correct binary information. Consequently, this type of ROM is economical only when many ROMs are being made with exactly the same information.

Mask-programmed ROMs are commonly referred to as just ROMs, but this can be confusing because the term ROM actually represents the broad category of devices that, during normal operation, are only read from. We will use the abbreviation MROM whenever we refer to mask-programmed ROMs.

Figure 12-9 shows the structure of a small MOS MROM. It consists of 16 memory cells arranged in four rows of four cells. Each cell is an N-channel MOSFET transistor connected in the common-drain configuration (input at gate, output at source). The top row of cells (ROW 0) constitutes a four-bit register. Note that some of the transistors in this row (Q_0 and Q_2) have their source connected to the output column line, while others (Q_1 and Q_3) do not. The same is true of the cells in each of the other rows. The presence or absence of these source connections determines whether a cell is storing a 1 or a 0, respectively. The condition of each source connection is controlled during production by the photographic mask based on the customer-supplied data.

Notice that the data outputs are connected to column lines. Referring to output D_3, for instance, any transistor that has a connection from the source (such as Q_0, Q_4, and Q_8) to the output column can switch V_{dd} onto the column, making it a HIGH logic level. If V_{dd} is not connected to the column line, the output will be held at a LOW logic level by the pull-down resistor. At any given time, a maximum of one transistor in a column will ever be turned on due to the row decoder.

The 1-of-4 decoder is used to decode the address inputs A_1A_0 to select which row (register) is to have its data read. The decoder's active-HIGH outputs provide the ROW enable lines that are the gate inputs for the various rows of cells. If the decoder's enable input, $\overline{EN}$, is held HIGH, all of the decoder outputs will be in their inactive LOW state, and all of the transistors in the array will be off because of the absence of any gate voltage. For this situation, the data outputs will all be in the LOW state.

When $\overline{EN}$ is in its active-LOW state, the conditions at the address inputs determine which row (register) will be enabled so that its data can be read at the data outputs. For example, to read ROW 0, the A_1A_0 inputs are set to 00. This places HIGH at the ROW 0 line; all other row lines are at 0 V. This HIGH at ROW 0 turns on transistors Q_0, Q_1, Q_2, and Q_3. With all of the transistors in the row conducting, V_{dd} will be switched on to each transistor's source lead. Outputs D_3 and D_1 will go HIGH because Q_0 and Q_2 are connected to their respective columns. D_2 and D_0 will remain LOW because there is no path from the Q_1 and Q_3 source leads to their columns. In a similar manner, application of the other address codes will produce data outputs from the corresponding register. The table in Figure 12-9 shows the data for each address. You should verify how this correlates with the source connections to the various cells.

EXAMPLE 12-8

MROMs can be used to store tables of mathematical functions. Show how the MROM in Figure 12-9 can be used to store the function $y = x^2 + 3$, where the input address supplies the value for x, and the value of the output data is y.

TABLE 12-1

x		$y = x^2 + 3$			
A_1	A_0	D_3	D_2	D_1	D_0
0	0	0	0	1	1
0	1	0	1	0	0
1	0	0	1	1	1
1	1	1	1	0	0

Solution

The first step is to set up a table showing the desired output for each set of inputs. The input binary number, x, is represented by the address A_1A_0. The output binary number is the desired value of y. For example, when $x = A_1A_0 = 10_2 = 2_{10}$, the output should be $2^2 + 3 = 7_{10} = 0111_2$. The complete table is shown in Table 12-1. This table is supplied to the MROM manufacturer for developing the mask that will make the appropriate connections within the memory cells during the fabrication process. For instance, the first row in the table indicates that the connections to the source of Q_0 and Q_1 will be left unconnected, while the connections to Q_2 and Q_3 will be made.

MROMs typically have tristate outputs that allow them to be used in a bus system, as we discussed in Chapter 9. Consequently, there must be a control input to enable and disable the tristate outputs. This control input is usually labeled *OE* (for output enable). In order to distinguish this tristate enable input from the address decoder enable input, the latter is usually referred to as a chip enable (*CE*). The chip enable performs more than just enabling the address decoder. When *CE* is disabled, all functions of the chip are disabled, including the tristate outputs, and the entire circuit is placed in a **power-down** mode that draws much less current from the power supply. Figure 12-10 shows a 32K × 8 MROM. The 15 address lines (A0–A14) can identify 2^{15} memory locations (32, 767, or 32K). Each memory location holds an eight-bit data value that can be placed on the data lines D7–D0 when the chip is enabled and the outputs are enabled.

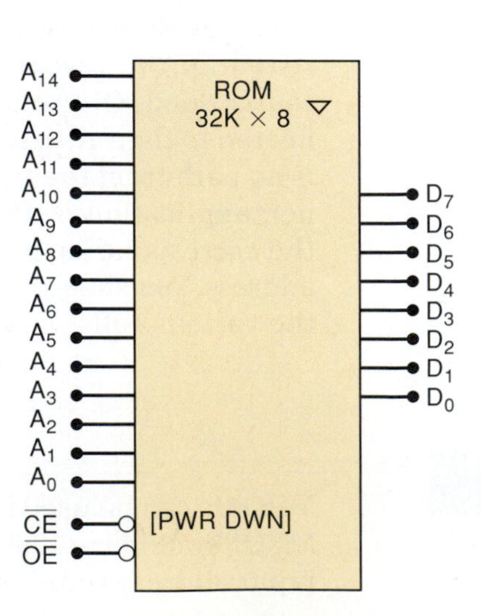

FIGURE 12-10 Logic symbol for a 32K × 8 MROM.

Programmable ROMs (PROMs)

A mask-programmable ROM is very expensive and would not be used except in high-volume applications, where the cost would be spread out over many units. For lower-volume applications, manufacturers have developed **fusible-link** PROMs that are user-programmable; that is, they are not programmed during the manufacturing process but are custom-programmed by the user. Once programmed, however, a PROM is like an MROM because it cannot be erased and reprogrammed. Thus, if the program in the PROM is faulty or must be changed, the PROM must be thrown away. For this reason, these devices are often referred to as "one-time programmable" (OTP) ROMs.

The fusible-link PROM structure is very similar to the MROM structure because certain connections either are left intact or are opened in order to program a memory cell as a 1 or a 0, respectively. A PROM comes from the manufacturer with a thin, fuse link connection in the source leg of every transistor. In this condition, every transistor stores a 1. The user can then "blow" the fuse for any transistor that needs to store a 0. Typically, data can be programmed or "burned into" a PROM by selecting a row by applying the desired address to the address inputs, placing the desired data on the data pins, and then applying a pulse to a special programming pin on the IC. Figure 12-11 shows the inner workings of how this is done.

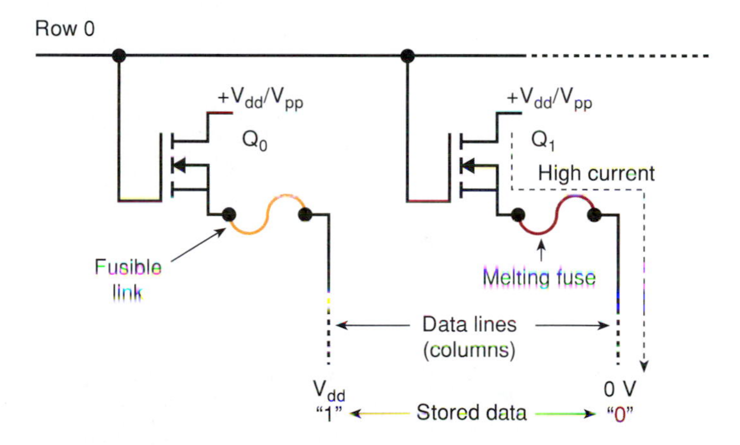

FIGURE 12-11 PROMs use fusible links that can be selectively blown open by the user to program a logic 0 into a cell.

All of the transistors in the selected row (row 0) are turned on, and V_{pp} is applied to their drain leads. Those columns (data lines) that have a logic 0 on them (e.g., Q_1) will provide a high-current path through the fusible link, burning it open and permanently storing a logic 0. Those columns that have a logic 1 (e.g., Q_0) have V_{pp} on one side of the fuse and V_{dd} on the other side, drawing much less current and leaving the fuse intact. Once all address locations have been programmed in this manner, the data are permanently stored in the PROM and can be read over and over again by accessing the appropriate address. The data will not change when power is removed from the PROM chip because nothing will cause an open fuse link to become closed again.

A PROM is programmed using the same equipment and process described in Chapter 4 for programming a PLD. The TMS27PC256 is a very popular CMOS PROM with a capacity of 32K × 8 and a standby power dissipation of only 1.4 mW. It is available with maximum access times ranging from 100 to 250 ns.

Erasable Programmable ROM (EPROM)

An EPROM can be programmed by the user, and it can also be *erased* and reprogrammed as often as desired. Once programmed, the EPROM is a *nonvolatile* memory that will hold its stored data indefinitely. The process for programming an EPROM is the same as that for a PROM.

The storage element of an EPROM is a MOS transistor with a silicon gate that has no electrical connection (i.e., a floating gate) but is very close to an electrode. In its normal state there is no charge stored on the floating gate and the transistor will produce a logic 1 whenever it is selected by the address decoder. To program a 0, a high-voltage pulse is used to leave a net charge on the floating gate. This charge causes the transistor to output a logic 0 when it is selected. Since the charge is trapped on the floating gate and has no discharge path, the 0 will be stored until it is erased. The data are erased by restoring all cells to a logic 1. To do this, the charge on the floating electrode is neutralized by exposing the silicon to high-intensity ultraviolet (UV) light for several minutes.

The 27C64 is an example of a small 8K × 8K memory IC that is available as a "one-time-programmable" (OTP) PROM or as an erasable UV EPROM. The obvious difference in the two ICs is the EPROM's clear quartz "window," shown in Figure 12-12(b), which allows the UV light to shine on the silicon. Both versions operate from a single +5-V power source during normal operation.

Figure 12-12(a) is the logic symbol for the 27C64. Note that it shows 13 address inputs (because $2^{13} = 8192$) and eight data outputs. It has four control inputs. $\overline{CE}$ is the chip enable input that is used to place the device in a standby mode where its power consumption is reduced. $\overline{OE}$ is the output enable and is used to control the device's data output tristate buffers so that the device can be connected to a microprocessor data bus without bus contention. V_{PP} is the special programming voltage required during the programming process. $\overline{PGM}$ is the program enable input that is activated to store data at the selected address.

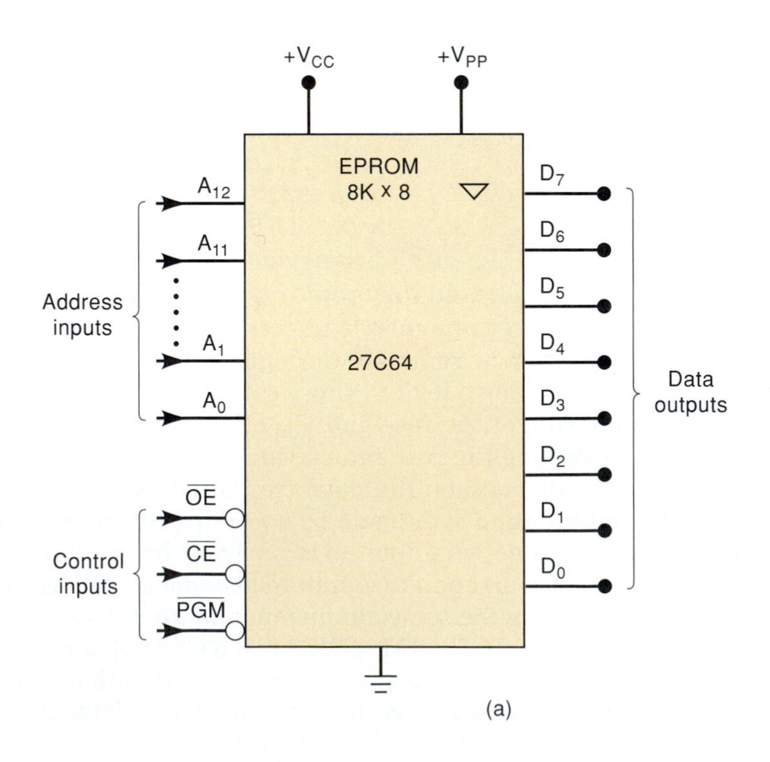

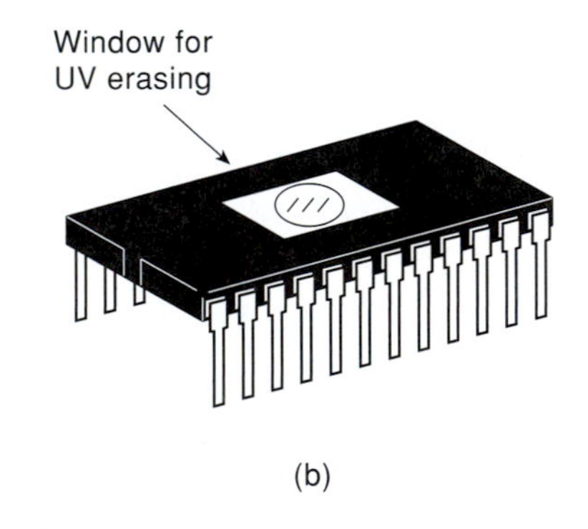

(b)

	Inputs				Outputs
Mode	$\overline{CE}$	$\overline{OE}$	$\overline{PGM}$	VPP	$D_7 - D_0$
Read	0	0	1	0–5V	$DATA_{out}$
Output Disable	0	1	1	0–5V	High Z
Standby	1	X	X	X	High Z
Program	0	1	0	12.75 V	$DATA_{in}$
PGM Verify	0	0	1	12.75 V	$DATA_{out}$

(c)

FIGURE 12-12 (a) Logic symbol for 27C64 EPROM; (b) typical EPROM package showing ultraviolet window; (c) 27C64 operating modes.

The 27C64 has several operating modes that are controlled by the $\overline{CE}$, $\overline{OE}$, V_{PP}, and PGM pins, as presented in Figure 12-12(c). The program mode is used to write new data into the EPROM cells. This is most often done on a "clean" EPROM, one that has previously been erased with UV light so that all cells are 1s. The programming process writes one eight-bit word into one address location at one time as follows: (1) the address is applied to the address pins; (2) the desired data are placed at the data pins, which function as inputs during the programming process; (3) a higher programming voltage of 12.75V is applied to V_{PP}; (4) $\overline{CE}$ is held LOW; (5) $\overline{PGM}$ is pulsed LOW for 100 μs and the data are read back. If the data were not successfully stored, another pulse is applied to $\overline{PGM}$. This is repeated at the same address until the data are successfully stored.

A clean EPROM can be programmed in less than a minute once the desired data have been entered, transferred, or downloaded into the EPROM programmer. The 27C512 is a common 64K $\times$ 8 EPROM that operates very much like the 27C64 but offers more storage capacity.

The major disadvantages of UVEPROMs are that they must be removed from the circuit to be programmed and erased, the erase operation erases the entire chip, and the erase operation takes up to 20 minutes.

Electrically Erasable PROM (EEPROM)

The disadvantages of the EPROM were overcome by the development of the **electrically erasable PROM (EEPROM)** as an improvement over the EPROM. The EEPROM retains the same floating-gate structure as the EPROM, but with the addition of a very thin oxide region above the drain of the MOSFET memory cell. This modification produces the EEPROM's major characteristic—its electrical erasability. By applying a high voltage (21 V) between the MOSFET's gate and drain, a charge can be induced onto the floating gate, where it will remain even when power is removed; reversal of the same voltage causes a removal of the trapped charges from the floating gate and erases the cell. Because this charge-transport mechanism requires very low currents, the erasing and programming of an EEPROM can be done *in circuit* (i.e., without a UV light source and a special PROM programmer unit).

Another advantage of the EEPROM over the EPROM is the ability to erase and rewrite *individual* bytes (eight-bit words) in the memory array electrically. During a write operation, internal circuitry automatically erases all of the cells at an address location prior to writing in the new data. This byte erasability makes it much easier to make changes in the data stored in an EEPROM.

The early EEPROMs, such as Intel's 2816, required appropriate support circuitry external to the memory chips. This support circuitry included the 21-V programming voltage (V_{PP}), usually generated from a +5 V supply through a dc-to-dc converter, and it included circuitry to control the timing and sequencing of the erase and programming operations. The newer devices, such as the Intel 2864, have integrated this support circuitry onto the same chip with the memory array, so that it requires only a single 5-V power pin. This makes the EEPROM as easy to use as the read/write memory we will be discussing shortly.

The byte erasability of the EEPROM and its high level of integration come with two penalties: density and cost. The memory cell complexity and the on-chip support circuitry place EEPROMs far behind an EPROM in bit capacity per square millimeter of silicon; a 1-Mbit EEPROM requires about twice as much silicon as a 1-Mbit EPROM. So despite its operational superiority, the EEPROM's shortcomings in density and cost-effectiveness have kept it from replacing the EPROM in applications where density and cost are paramount factors.

The logic symbol for the Intel 2864 is shown in Figure 12-13(a). It is organized as an 8K × 8 array with 13 address inputs ($2^{13} = 8192$) and eight data I/O pins. Three control inputs determine the operating mode according to the

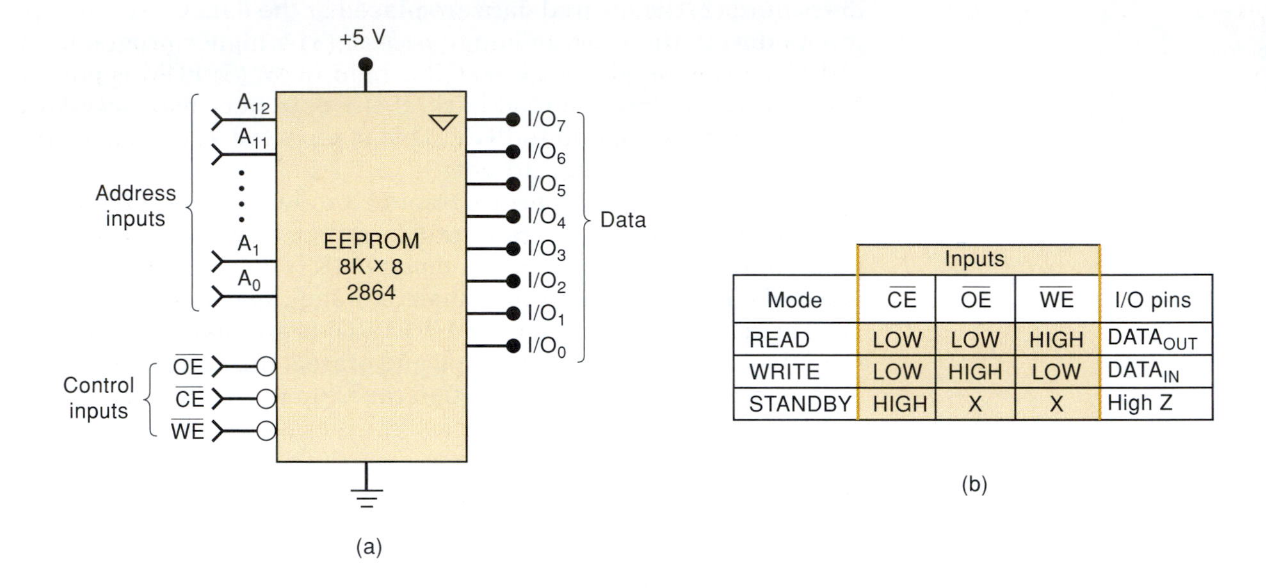

Mode	Inputs $\overline{CE}$	$\overline{OE}$	$\overline{WE}$	I/O pins
READ	LOW	LOW	HIGH	$DATA_{OUT}$
WRITE	LOW	HIGH	LOW	$DATA_{IN}$
STANDBY	HIGH	X	X	High Z

(b)

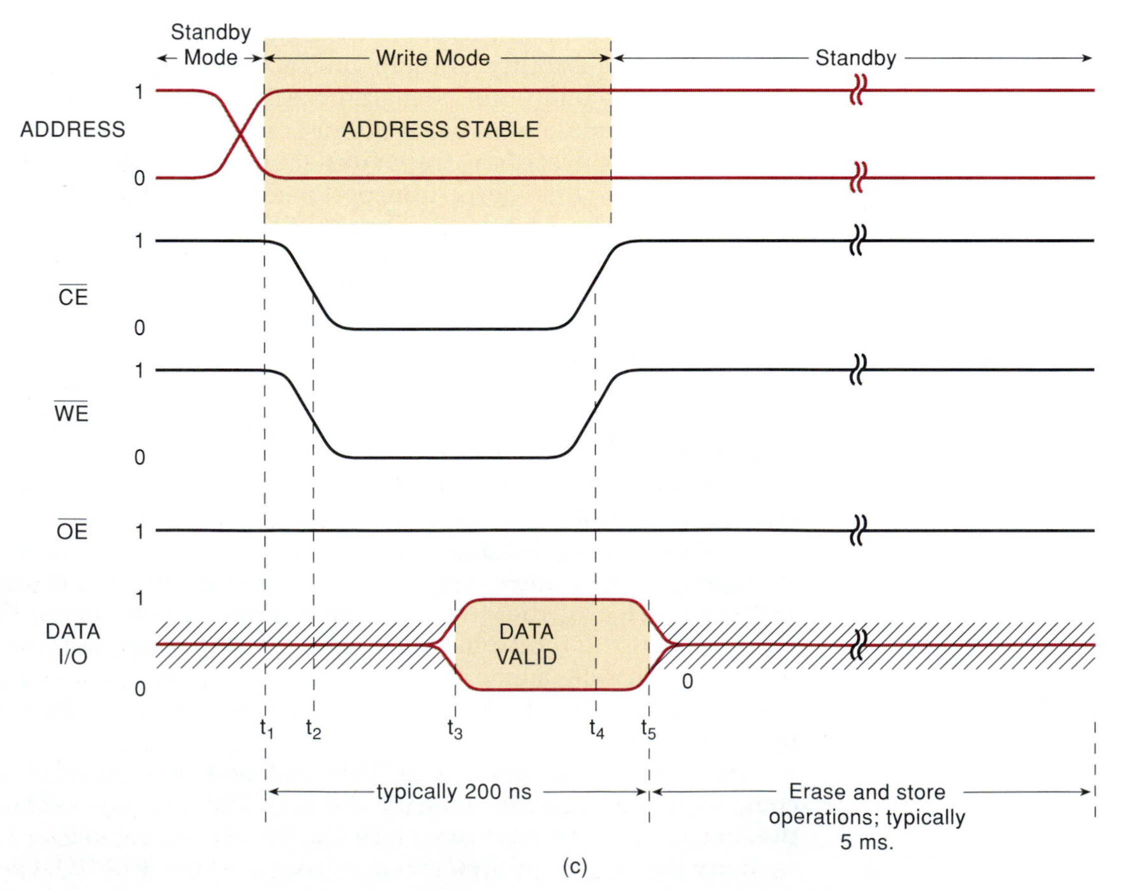

FIGURE 12-13 (a) Symbol for the 2864 EEPROM; (b) operating modes; (c) timing for the write operation.

table given in Figure 12-13(b). With $\overline{CE}$ = HIGH, the chip is in its low-power standby mode, in which no operations are being performed on any memory location and the data pins are in the Hi-Z state.

To read the contents of a memory location, the desired address is applied to the address pins; $\overline{CE}$ is driven LOW; and the output enable pin, $\overline{OE}$, is driven LOW to enable the chip's output data buffers. The write enable pin, $\overline{WE}$, is held HIGH during a read operation.

To write into (program) a memory location, the output buffers are disabled so that the data to be written can be applied as inputs to the I/O pins. The timing for the write operation is diagrammed in Figure 12-13(c). Prior to t_1, the device is in the standby mode. A new address is applied at that time. At t_2, the $\overline{CE}$ and $\overline{WE}$ inputs are driven LOW to begin the write operation; $\overline{OE}$ is HIGH so that the data pins will remain in the Hi-Z state. Data are applied to the I/O pins at t_3 and are written into the address location on the rising edge of $\overline{WE}$ at t_4. The data are removed at t_5. Actually, the data are first latched (on the rising edge of $\overline{WE}$) into a FF buffer memory that is part of the 2864 circuitry. The data are held there while other circuitry on the chip performs an erase operation on the selected address location in the EEPROM array, after which the data byte is transferred from the buffer to the EEPROM array and stored at that location. This erase and store operation typically takes 5 ms. With $\overline{CE}$ returned HIGH at t_4, the chip is back in the standby mode while the internal erase and store operations are completed.

The 2864 has an enhanced write mode that allows the user to write up to 16 bytes of data into the FF buffer memory, where it is held while the EEPROM circuitry erases the selected address locations. The 16 bytes of data are then transferred to the EEPROM array for storage at these locations. This process also takes about 5 ms.

Because the internal process of storing a data value in an EEPROM is quite slow, the speed of the data transfer operation can also be slower. Consequently, many manufacturers offer EEPROM devices in eight-pin packages that are interfaced to a two- or three-wire *serial* bus. This saves physical space on the system board as opposed to using a 2864 in a 28-pin, wide-DIP package. It also simplifies the hardware interface between the CPU and the EEPROM.

CD-ROM

A very prominent type of read-only storage used today in computer systems is the compact disk (CD). The disk technology and the hardware necessary to retrieve the information are the same as those used in audio systems. Only the format of the data is different. The disks are manufactured with a highly reflective surface. To store data on the disks, a very intense laser beam is focused on a *very* small point on the disk. This beam burns a light-diffracting pit at that point on the disk surface. Digital data (1s and 0s) are stored on the disk one bit at a time by burning or not burning a pit into the reflective coating. The digital information is arranged on the disk as a continuous spiral of data points. The precision of the laser beam allows very large quantities of data (over 550 Mbytes) to be stored on a small, 120-mm disk.

In order to read the data, a much less powerful laser beam is focused onto the surface of the disk. At any point, the reflected light is sensed as either a 1 or a 0. This optical system is mounted on a mechanical carriage that moves back and forth along the radius of the disk, following the spiral of data as the disk rotates. The data retrieved from the optical system come one bit at a time in a serial data stream. The angular rotation of the disk is controlled to maintain a constant rate of incoming data points. If the disk is being used for audio recording, this stream of data is converted into an analog waveform. If the disk is being used as ROM, the data are decoded into parallel bytes that the computer can use. The CD player technology, although very sophisticated, is relatively inexpensive and is becoming a standard way of loading large amounts of data into a personal computer. The major improvements that are occurring now in CD-ROM technology involve quicker access time in retrieving data.

REVIEW QUESTIONS

1. *True or false:* An MROM can be programmed by the user.
2. How does a PROM differ from an MROM? Can it be erased and reprogrammed?
3. *True or false:* A PROM stores a logic 1 when its fusible link is intact.
4. How is an EPROM erased?
5. *True or false:* There is no way to erase only a portion of an EPROM's memory.
6. What function is performed by PROM and EPROM programmers?
7. What EPROM shortcomings are overcome by EEPROMs?
8. What are the major drawbacks of EEPROM?
9. What type of ROM can erase one byte at a time?
10. How many bits are read from a CD-ROM disk at any point in time?

12-8 FLASH MEMORY

EPROMs are nonvolatile, offer fast read access times (typically 120 ns), and have high density and low cost per bit. They do, however, require removal from their circuit/system to be erased and reprogrammed. EEPROMs are nonvolatile, offer fast read access, and allow rapid in-circuit erasure and reprogramming of individual bytes. They suffer from lower density and much higher cost than EPROMs.

The challenge for semiconductor engineers was to fabricate a nonvolatile memory with the EEPROM's in-circuit electrical erasability, but with densities and costs much closer to those of EPROMs, while retaining the high-speed read access of both. The response to this challenge was the **flash memory.**

Structurally, a flash memory cell is like the simple single-transistor EPROM cell (and unlike the more complex two-transistor EEPROM cell), being only slightly larger. It has a thinner gate-oxide layer that allows electrical erasability but can be built with much higher densities than EEPROMs. The cost of flash memory is considerably less than for EEPROM. Figure 12-14 illustrates the trade-offs for the various semiconductor nonvolatile memories. As erase/programming flexibility increases (from base to apex of the triangle), so do device complexity and cost. MROM and PROM are the simplest and cheapest devices, but they cannot be erased and reprogrammed. EEPROM is the most complex and expensive because it can be erased and reprogrammed in circuit on a byte-by-byte basis.

Flash memories are so called because of their rapid erase and write times. Most flash chips use a *bulk erase* operation in which all cells on the chip are erased simultaneously; this bulk erase process typically requires hundreds of milliseconds compared to 20 minutes for UV EPROMs. Some newer flash memories offer a *sector erase* mode, where specific sectors of the memory array (e.g., 512 bytes) can be erased at one time. This prevents having to erase and reprogram all cells when only a portion of the memory needs to be updated. A typical flash memory has a write time of 10 μs per byte compared

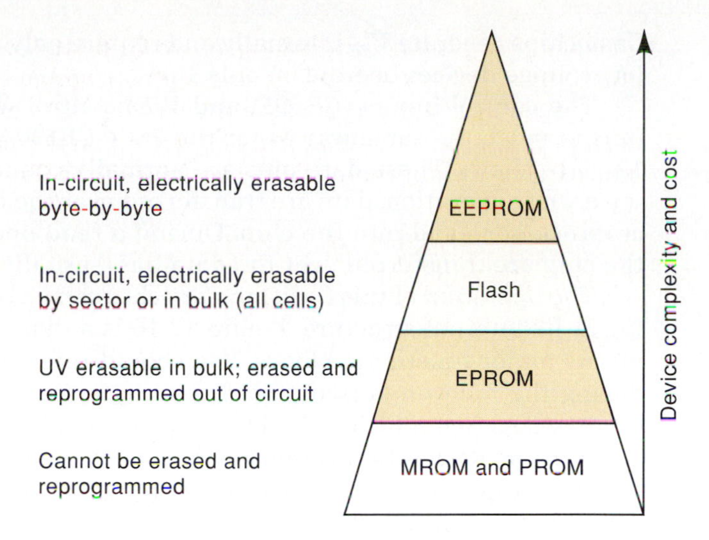

FIGURE 12-14 Trade-offs for semiconductor nonvolatile memories show that complexity and cost increase as erase and programming flexibility increases.

to 100 μs for the most advanced EPROM and 5 ms for EEPROM (which includes automatic byte erase time).

The 28F256A CMOS Flash Memory IC

Figure 12-15(a) shows the logic symbol for Intel Corporation's 28F256A CMOS flash memory chip, which has a capacity of 32K × 8. The diagram shows 15 address inputs (A_0–A_{14}) needed to select the different memory addresses; that is, 2^{15} = 32K = 32,768. The eight data input/output pins (DQ_0–DQ_7) are used as inputs during memory write operations and as outputs during memory read operations. These data pins float to the Hi-Z state when the chip is deselected ($\overline{CE}$ = HIGH) or when the outputs are disabled ($\overline{OE}$ = *HIGH*) The write enable input ($\overline{WE}$) is used to control memory write operations. Note that the chip requires two power-supply voltages: V_{CC} is the standard +5 V used for the logic circuitry; V_{PP} is the erase/programming power-supply voltage, nominally +12 V, which is needed for the erase and programming (write) operations. Newer

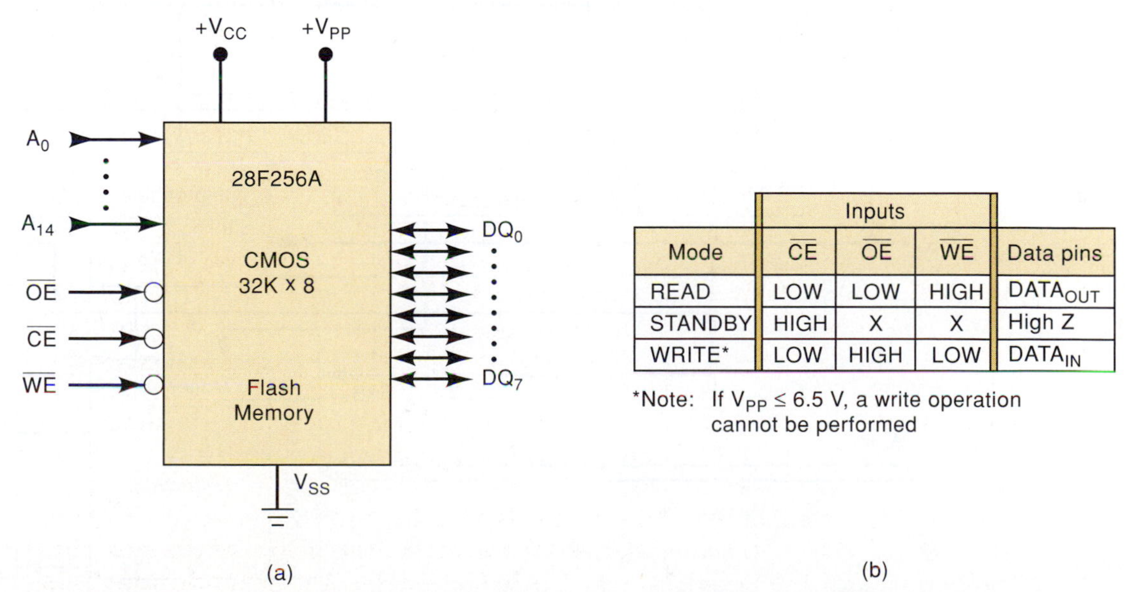

Mode	Inputs $\overline{CE}$	$\overline{OE}$	$\overline{WE}$	Data pins
READ	LOW	LOW	HIGH	$DATA_{OUT}$
STANDBY	HIGH	X	X	High Z
WRITE*	LOW	HIGH	LOW	$DATA_{IN}$

FIGURE 12-15 (a) Logic symbol for the 28F256A flash memory chip; (b) control inputs $\overline{CE}$, $\overline{WE}$, and $\overline{OE}$.

flash chips generate V_{PP} internally and require only a single supply. The latest low-voltage devices operate on only 1.8 V.

The control inputs ($\overline{CE}$, $\overline{OE}$, and $\overline{WE}$) control what happens at the data pins in much the same way as for the 2864 EEPROM, as the table in Figure 12-15(b) shows. These data pins are normally connected to a data bus. During a write operation, data are transferred over the bus—usually from the microprocessor—and into the chip. During a read operation, data from inside the chip are transferred over the data bus—usually to the microprocessor.

The operation of this flash memory chip can be better understood by looking at its internal structure. Figure 12-16 is a diagram of the 28F256A showing its major functional blocks. You should refer to this diagram as needed during the following discussion. The unique feature of this structure is the *command register,* which is used to manage all of the chip functions. Command codes are written into this register to control which operations take place inside the chip (e.g., erase, erase-verify, program, program-verify). These command codes usually come over the data bus from the microprocessor. State control logic examines the contents of the command register and generates logic and control signals to the rest of the chip's circuits to carry out the steps in the operation. Some examples of the types of commands that can be sent to the flash are shown here to give you an idea of why they are necessary. Each command is stored in the command register by using the same write cycle as described for the EEPROM in Figure 12-13(c).

Read Command. Writing a code of 00 hex into the command register prepares the memory IC for the read operation. After this, a normal read cycle can be used to access data stored at any address.

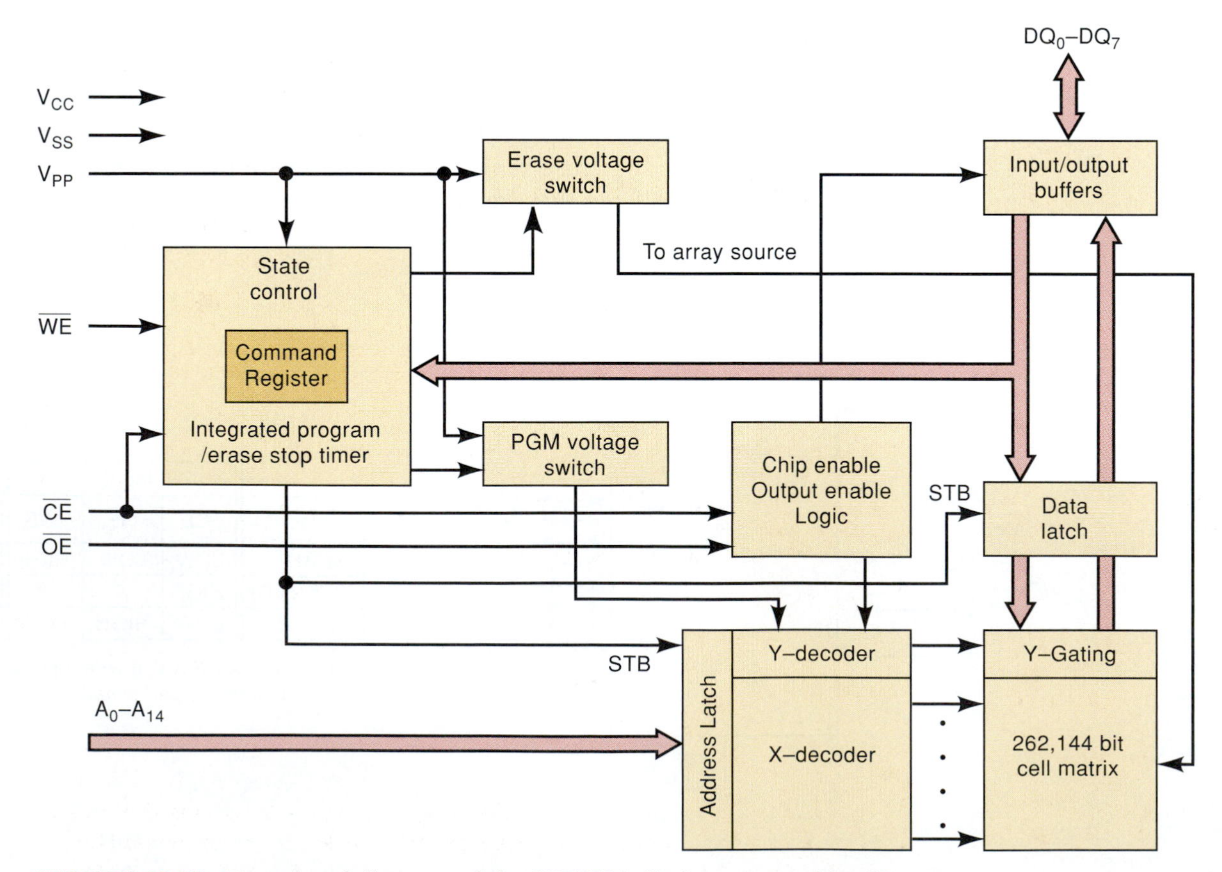

FIGURE 12-16 Functional diagram of the 28F256A flash memory chip. (Courtesy of Intel Corporation)

Set-Erase/Erase Command. The code of 20 hex must be written to the command register twice in a row to begin the internal erase sequence.

Erase Verify Command. This command (FF Hex) causes the memory IC to check all of its memory locations to verify that all bits are HIGH.

Set-Up Program/Program Command. This command (40 hex) puts the memory IC in a mode that allows subsequent write cycles to store data at a specified address, one byte at a time.

Program-Verify Command. This command (C0 hex) is used to verify that the correct data have been stored in the flash ROM. After this code is written to the command register, the next read operation will produce the contents of the last location that was written to, and these data can be compared with the intended value.

Improved Flash Memory

The core architecture of flash memory today and the basic set of command codes are very similar to those of the first-generation devices. The newest flash devices have new features, and new command codes to control these features, in addition to those common to earlier devices. Of course, the latest flash devices have much more capacity, run on much less power (and at lower voltages), come in smaller packages, and cost much less per bit than their predecessors. They also offer features such as the ability to read/write data while a block of memory is being erased. The V_{pp} programming voltage is generated internally, allowing it to use a single supply. The speed of operation can be enhanced by using a burst mode. This simply means that several addresses in a row can be accessed very rapidly, providing a burst of data transfer. A synchronous clock input is provided to control the burst operation. A base address is latched into the memory and then the contents of this location are transferred on the clock edge, which also increments the address to the next location. In this way, several sequential memory locations are accessed as fast as the system clock can oscillate, without the overhead of generating each address. All of these features have made flash memory the predominant solid-state nonvolatile memory technology in use today.

REVIEW QUESTIONS

1. What is the main advantage of flash memory over EPROMs?
2. What is the main advantage of flash memory over EEPROMs?
3. Where does the word *flash* come from?
4. What is V_{PP} needed for?
5. What is the function of the 28F256A's command register?
6. What is the purpose of an erase-verify command?
7. What is the purpose of the program-verify command?

12-9 ROM APPLICATIONS

With the exception of MROM and PROM, most ROM devices can be reprogrammed, so technically they are not *read-only* memories. However, the term *ROM* can still be used to include EPROMs, EEPROMs, and flash memory because, during normal operation, the stored contents of these devices is not changed nearly as often as it is read. So ROMs are taken to include all semiconductor, nonvolatile memory devices, and they are used in applications where nonvolatile storage of information, data, or program codes is needed and where the stored data rarely or never change. Here are some of the most common application areas.

Embedded Microcontroller Program Memory

Microcontrollers are prevalent in most consumer electronic products on the market today. Your car's automatic braking system and engine controller, your cell phone, your digital camcorder, your microwave oven, and many other products have a microcontroller for a brain. These little computers have their program instructions stored in nonvolatile memory—in other words, in a ROM. Most embedded microcontrollers today have flash ROM integrated into the same IC as the CPU. Many also have an area of EEPROM that offers the features of byte erasure and nonvolatile storage.

Data Transfer and Portability

The need to store and transfer large sets of binary information is a requirement of many low-power battery-operated systems today. Cell phones store photos and video clips. Digital cameras store many pictures on removable memory media. Flash drives connect to a computer's USB port and store gigabytes of information. Your MP-3 player is loaded up with music and runs all day on batteries. A PDA (personal digital assistant) stores appointment information, email, addresses, and even entire books. All of these common personal electronic gadgets require the low-power, low-cost, high-density, nonvolatile storage with in-circuit write capability that is available in flash memory.

Bootstrap Memory

Many microcomputers and most larger computers do not have their operating system programs stored in ROM. Instead, these programs are stored in external mass memory, usually magnetic disk. How, then, do these computers know what to do when they are powered on? A relatively small program, called a **bootstrap program**, is stored in ROM. (The term *bootstrap* comes from the idea of pulling oneself up by one's own bootstraps.) When the computer is powered on, it will execute the instructions that are in this bootstrap program. These instructions typically cause the CPU to initialize the system hardware. The bootstrap program then loads the operating system programs from mass storage (disk) into its main internal memory. At that point, the computer begins executing the operating system program and is ready to respond to the user commands. This startup process is often called "booting up the system."

Many of the digital signal processing chips load their internal program memory from an external bootstrap ROM when they are powered on. Some of the more advanced PLDs also load the programming information that configures their logic circuits from an external ROM into a RAM area inside the PLD. This is also done when power is applied. In this way, the PLD is reprogrammed by changing the bootstrap ROM, rather than changing the PLD chip itself.

Data Tables

ROMs are often used to store tables of data that do not change. Some examples are the trigonometric tables (i.e., sine, cosine, etc.) and code-conversion tables. The digital system can use these data tables to "look up" the correct value. For example, a ROM can be used to store the sine function for angles from 0° to 90°. It could be organized as a 128 × 8 with seven address inputs and eight data outputs. The address inputs represent the angle in increments of approximately 0.7°. For example, address 0000000 is 0°, address 0000001 is 0.7°, address 0000010 is 1.41°, and so on, up to address 1111111, which is 89.3°. When an address is applied to the ROM, the data outputs will represent the approximate sine of the angle. For example, with input address 1000000 (representing approximately 45°) the data outputs will be 10110101. Because the sine is less than or equal to 1, these data are interpreted as a fraction, that is, 0.10110101, which when converted to decimal equals 0.707 (the sine of 45°). It is vital that the user of this ROM understands the format in which the data are stored.

Standard look-up-table ROMs for functions such as these were at one time readily available TTL chips. Only a few are still in production. Today, most systems that need to look up equivalent values involve a microprocessor, and the "look-up" table data are stored in the same ROM that holds the program instructions.

Data Converter

The data-converter circuit takes data expressed in one type of code and produces an output expressed in another type. Code conversion is needed, for example, when a computer is outputting data in straight binary code and we want to convert it to BCD in order to display it on 7-segment LED readouts.

One of the easiest methods of code conversion uses a ROM programmed so that the application of a particular address (the old code) produces a data output that represents the equivalent in the new code. The 74185 is a TTL ROM that stores the binary-to-BCD code conversion for a six-bit binary input. To illustrate, a binary address input of 100110 (decimal 38) will produce a data output of 00111000, which is the BCD code for decimal 38.

Function Generator

The function generator is a circuit that produces waveforms such as sine waves, sawtooth waves, triangle waves, and square waves. Figure 12-17 shows how a ROM look-up table and a DAC are used to generate a sine-wave output signal.

The ROM stores 256 different eight-bit values, each one corresponding to a different waveform value (i.e., a different voltage point on the sine wave). The eight-bit counter is continuously pulsed by a clock signal to provide sequential address inputs to the ROM. As the counter cycles through the 256 different addresses, the ROM outputs the 256 data points to the DAC. The DAC output will be a waveform that steps through the 256 different analog voltage values corresponding to the data points. The low-pass filter smooths out the steps in the DAC output to produce a smooth waveform.

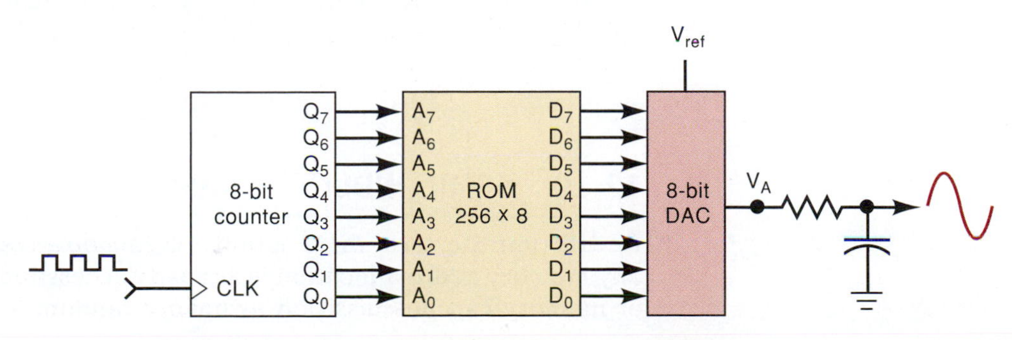

FIGURE 12-17 Function generator using a ROM and a DAC.

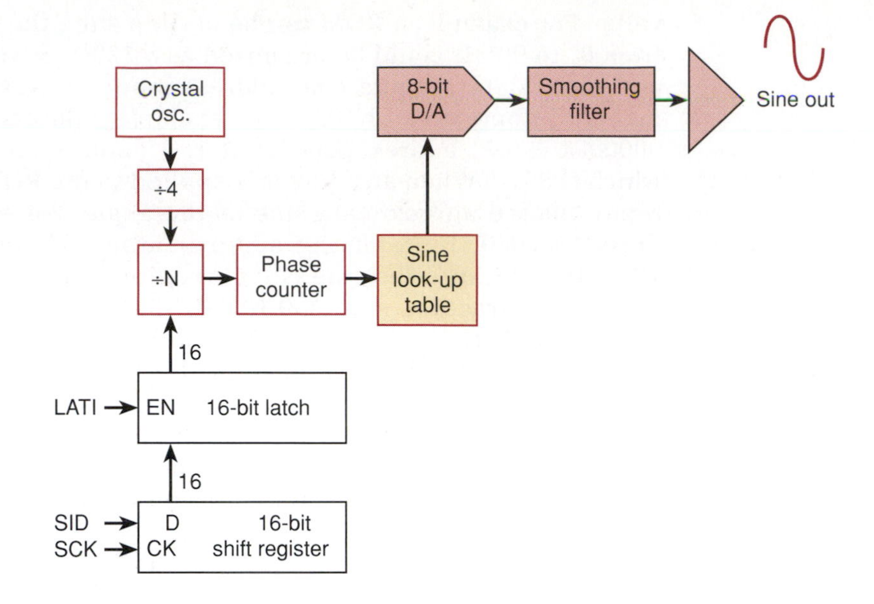

FIGURE 12-18 The ML2035 programmable sine-wave generator. (Courtesy of MicroLinear Corp.)

Circuits such as this are used in some commercial function generators. The same idea is employed in some speech synthesizers, where the digitized speech waveform values are stored in the ROM. The ML2035, illustrated in Figure 12-18, is a programmable sine-wave generator chip that incorporates this basic strategy to generate a sine wave of fixed amplitude and a frequency that can be selected from dc to 50 kHz. The number that is shifted into the 16-bit shift register is used to determine the clocking frequency for the counter that drives the address inputs on the ROM look-up table. The ML2035 is intended for telecommunications applications that require precise tones of various frequencies to be generated.

Auxiliary Storage

Because of their nonvolatility, high speed, low power requirements, and lack of moving parts, flash memory modules have become feasible alternatives to magnetic disk storage. This is especially true for lower capacities (5 Mbytes or less), where flash is cost-competitive with magnetic disk. The low power consumption of flash memory makes it particularly attractive for laptop and notebook computers that use battery power.

REVIEW QUESTIONS

1. Describe how a computer uses a bootstrap program.
2. What is a code converter?
3. What are the main elements of a function generator?
4. Why are flash memory modules a feasible alternative to auxiliary disk storage?

12-10 SEMICONDUCTOR RAM

Recall that the term *RAM* stands for *random-access memory,* meaning that any memory address location is as easily accessible as any other. Many types of memory can be classified as having random access, but when the term

RAM is used with semiconductor memories, it is usually taken to mean read/write memory (RWM) as opposed to ROM. Because it is common practice to use RAM to mean semiconductor RWM, we will do so throughout the following discussions.

RAM is used in computers for the *temporary* storage of programs and data. The contents of many RAM address locations will be read from and written to as the computer executes a program. This requires fast read and write cycle times for the RAM so as not to slow down the computer operation.

A major disadvantage of RAM is that it is volatile and will lose all stored information if power is interrupted or turned off. Some CMOS RAMs, however, use such small amounts of power in the standby mode (no read or write operations taking place) that they can be powered from batteries whenever the main power is interrupted. Of course, the main advantage of RAM is that it can be written into and read from rapidly with equal ease.

The following discussion of RAM will draw on some of the material covered in our treatment of ROM because many of the basic concepts are common to both types of memory.

12-11 RAM ARCHITECTURE

As with the ROM, it is helpful to think of the RAM as consisting of a number of registers, each storing a single data word, and each having a unique address. RAMs typically come with word capacities of 1K, 4K, 8K, 16K, 64K, 128K, 256K, and 1024K, and with word sizes of one, four, or eight bits. As we will see later, the word capacity and the word size can be expanded by combining memory chips.

Figure 12-19 shows the simplified architecture of a RAM that stores 64 words of four bits each (i.e., a 64 × 4 memory). These words have addresses

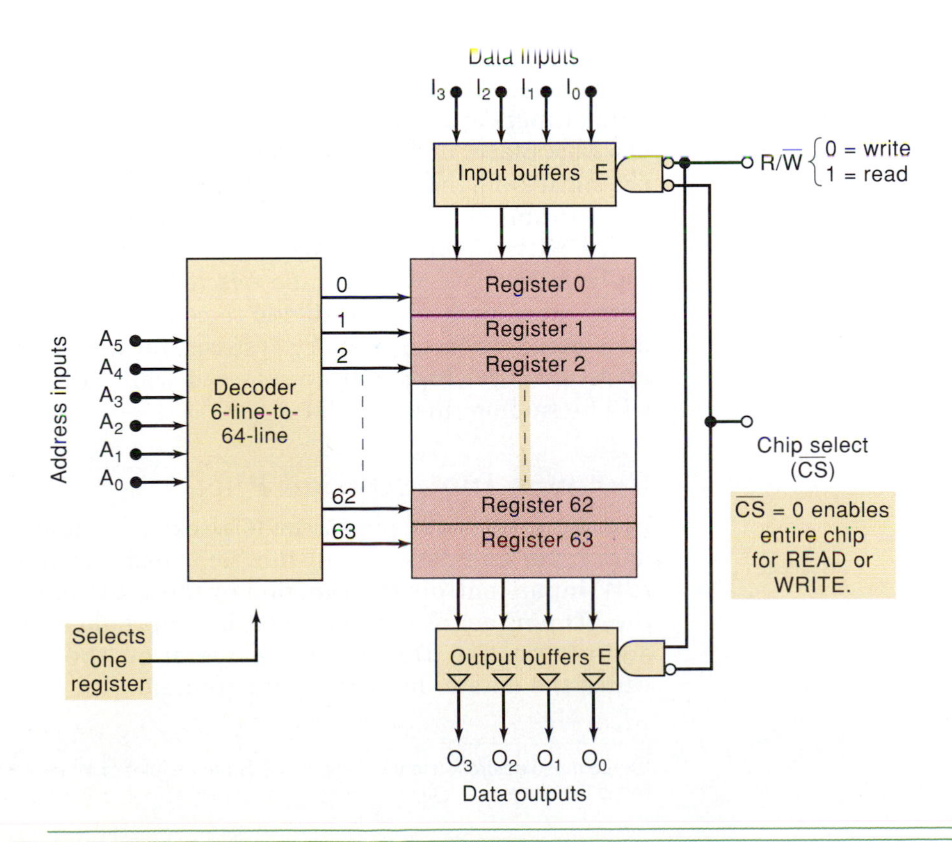

FIGURE 12-19 Internal organization of a 64 × 4 RAM.

ranging from 0 to 63_{10}. In order to select one of the 64 address locations for reading or writing, a binary address code is applied to a decoder circuit. Because $64 = 2^6$, the decoder requires a six-bit input code. Each address code activates one particular decoder output, which in turn enables its corresponding register. For example, assume an applied address code of

$$A_5A_4A_3A_2A_1A_0 = 011010$$

Because $011010_2 = 26_{10}$, decoder output 26 will go high, selecting register 26 for either a read or a write operation.

Read Operation

The address code picks out one register in the memory chip for reading or writing. In order to *read* the contents of the selected register, the READ/WRITE ($R/\overline{W}$)* input must be a 1. In addition, the CHIP SELECT ($\overline{CS}$) input must be activated (a 0 in this case). The combination of $R/\overline{W} = 1$ and $\overline{CS} = 0$ enables the output buffers so that the contents of the selected register will appear at the four data outputs. $R/\overline{W} = 1$ also *disables* the input buffers so that the data inputs do not affect the memory during a read operation.

Write Operation

To write a new four-bit word into the selected register requires $R/\overline{W} = 0$ and $\overline{CS} = 0$. This combination *enables* the input buffers so that the four-bit word applied to the data inputs will be loaded into the selected register. The $R/\overline{W} = 0$ also *disables* the output buffers, which are tristate, so that the data outputs are in their Hi-Z state during a write operation. The write operation, of course, destroys the word that was previously stored at that address.

Chip Select

Most memory chips have one or more *CS* inputs that are used to enable the entire chip or disable it completely. In the disabled mode, all data inputs and data outputs are disabled (Hi-Z) so that neither a read nor a write operation can take place. In this mode, the contents of the memory are unaffected. The reason for having *CS* inputs will become clear when we combine memory chips to obtain larger memories. Note that many manufacturers call these inputs CHIP ENABLE (*CE*). When the *CS* or *CE* inputs are in their active state, the memory chip is said to be *selected*; otherwise, it is said to be *deselected*. Many memory ICs are designed to consume much less power when they are deselected. In large memory systems, for a given memory operation, one or more memory chips will be selected while all others are deselected. More will be said on this topic later.

Common Input/Output Pins

In order to conserve pins on an IC package, manufacturers often combine the data input and data output functions using common input/output pins. The $R/\overline{W}$ input controls the function of these I/O pins. During a read operation, the I/O pins act as data outputs that reproduce the contents of the selected address location. During a write operation, the I/O pins act as data inputs to which the data to be written are applied.

---

*Some manufacturers use the symbol $\overline{WE}$ (write enable) or $\overline{W}$ instead of $R/\overline{W}$. In any case, the operation is the same.

We can see why this is done by considering the chip in Figure 12-19. With separate input and output pins, a total of 18 pins is required (including ground and power supply). With four common I/O pins, only 14 pins are required. The pin saving becomes even more significant for chips with larger word size.

EXAMPLE 12-9

The 2147H is an NMOS RAM that is organized as a 4K × 1 with separate data input and output and a single active-LOW chip select input. Draw the logic symbol for this chip, showing all pin functions.

Solution

The logic symbol is shown in Figure 12-20(a).

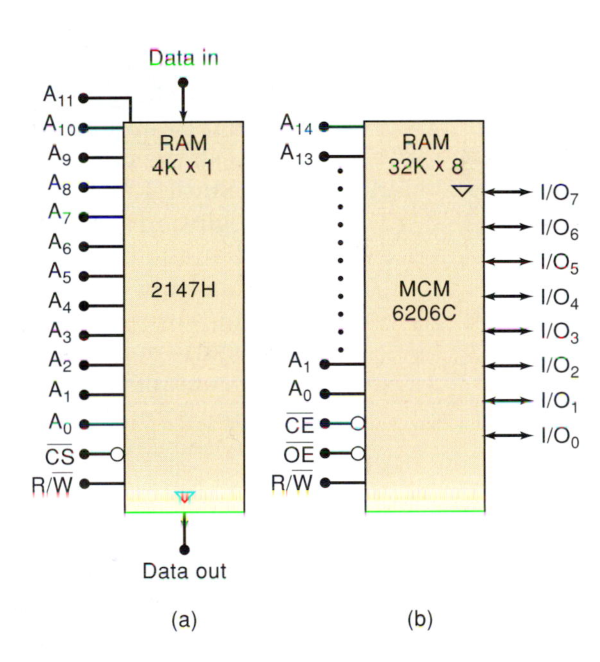

FIGURE 12-20 Logic symbols for (a) the 2147H RAM chip; (b) the MCM6206C RAM.

EXAMPLE 12-10

The MCM6206C is a CMOS RAM with 32K × 8 capacity, common I/O pins, an active-LOW chip enable, and an active-LOW output enable. Draw the logic symbol.

Solution

The logic symbol is shown in Figure 12-20(b).

In most applications, memory devices are used with a bidirectional data bus like we studied in Chapter 9. For this type of system, even if the memory chip had separate input and output pins, they would be connected together on the same data bus. A RAM having separate input and output pins is referred to as dual-port RAM. These are used in applications where speed is very important and the data in comes from a different device than the data out is going to. A good example is the video RAM on your PC. The RAM must be read repeatedly by the video card to refresh the screen and constantly filled with new updated information from the system bus.

REVIEW QUESTIONS

1. Describe the input conditions needed to read a word from a specific RAM address location.
2. Why do some RAM chips have common input/output pins?
3. How many pins are required for the MCM6208C 64K × 4 RAM with one *CS* input and common I/O?

12-12 STATIC RAM (SRAM)

The RAM operation that we have been discussing up to this point applies to a **static RAM**—one that can store data as long as power is applied to the chip. Static-RAM memory cells are essentially flip-flops that will stay in a given state (store a bit) indefinitely, provided that power to the circuit is not interrupted. In Section 12-13, we will describe **dynamic RAM**, which stores data as charges on capacitors. With dynamic RAMs, the stored data will gradually disappear because of capacitor discharge, so it is necessary to **refresh** the data periodically (i.e., recharge the capacitors).

Static RAMs (SRAMs) are available in bipolar, MOS, and BiCMOS technologies; the majority of applications use NMOS or CMOS RAMs. As stated earlier, the bipolars have the advantage in speed (although CMOS is gradually closing the gap), and MOS devices have much greater capacities and lower power consumption. Figure 12-21 shows for comparison a typical bipolar static memory cell and a typical NMOS static memory cell. The bipolar cell contains two bipolar transistors and two resistors, while the NMOS cell contains four N-channel MOSFETs. The bipolar cell requires more chip area than the MOS cell because a bipolar transistor is more complex than a MOSFET, and because the bipolar cell requires separate resistors while the MOS cell uses MOSFETs as resistors (Q_3 and Q_4). A CMOS memory cell would be similar to the NMOS cell except that it would use P-channel MOSFETs in place of Q_3 and Q_4. This results in the lowest power consumption but increases the chip complexity.

Static-RAM Timing

RAM ICs are most often used as the internal memory of a computer. The CPU (central processing unit) continually performs read and write operations on this memory at a very fast rate that is determined by the limitations

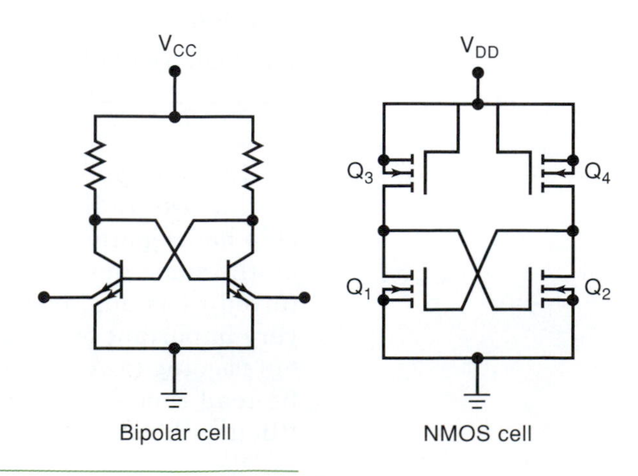

FIGURE 12-21 Typical bipolar and NMOS static-RAM cells.

of the CPU. The memory chips that are interfaced to the CPU must be fast enough to respond to the CPU read and write commands, and a computer designer must be concerned with the RAM's various timing characteristics.

Not all RAMs have the same timing characteristics, but most of them are similar, and so we will use a typical set of characteristics for illustrative purposes. The nomenclature for the different timing parameters will vary from one manufacturer to another, but the meaning of each parameter is usually easy to determine from the memory timing diagrams on the RAM data sheets. Figure 12-22 shows the timing diagrams for a complete read cycle and a complete write cycle for a typical RAM chip.

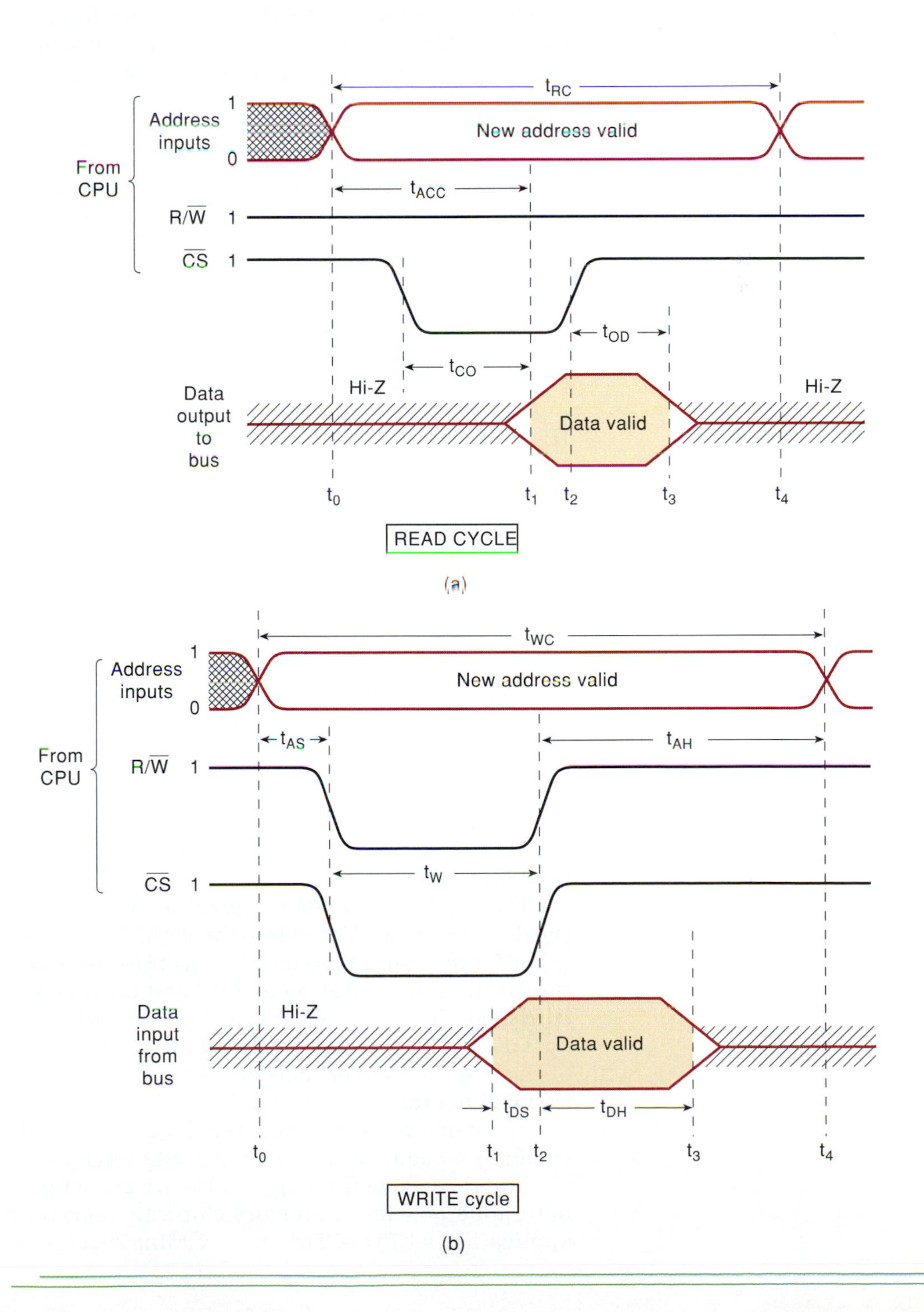

FIGURE 12-22 Typical timing for static RAM: (a) read cycle; (b) write cycle.

Read Cycle

The waveforms in Figure 12-22(a) show how the address, $R/\overline{W}$, and chip select inputs behave during a memory read cycle. As noted, the CPU supplies these input signals to the RAM when it wants to read data from a specific RAM address location. Although a RAM may have many address inputs coming from the CPU's address bus, for clarity the diagram shows only two. The RAM's data output is also shown; we will assume that this particular RAM has one data output. Recall that the RAM's data output is connected to the CPU data bus (Figure 12-5).

The read cycle begins at time t_0. Prior to that time, the address inputs will be whatever address is on the address bus from the preceding operation. Because the RAM's chip select is not active, it will not respond to its "old" address. Note that the $R/\overline{W}$ line is HIGH prior to t_0 and stays HIGH throughout the read cycle. In most memory systems, $R/\overline{W}$ is normally kept in the HIGH state except when it is driven LOW during a write cycle. The RAM's data output is in its Hi-Z state because $\overline{CS} = 1$.

At t_0, the CPU applies a new address to the RAM inputs; this is the address of the location to be read. After allowing time for the address signals to stabilize, the $\overline{CS}$ line is activated. The RAM responds by placing the data from the addressed location onto the data output line at t_1. The time between t_0 and t_1 is the RAM's access time, t_{ACC}, and is the time between the application of the new address and the appearance of valid output data. The timing parameter, t_{CO}, is the time it takes for the RAM output to go from Hi-Z to a valid data level once $\overline{CS}$ is activated.

At time t_2, the $\overline{CS}$ is returned HIGH, and the RAM output returns to its Hi-Z state after a time interval, t_{OD}. Thus, the RAM data will be on the data bus between t_1 and t_3. The CPU can take the data from the data bus at any point during this interval. In most computers, the CPU will use the PGT of the $\overline{CS}$ signal at t_2 to latch these data into one of its internal registers.

The complete read cycle time, t_{RC}, extends from t_0 to t_4, when the CPU changes the address inputs to a different address for the next read or write cycle.

Write Cycle

Figure 12-22(b) shows the signal activity for a write cycle that begins when the CPU supplies a new address to the RAM at a time t_0. The CPU drives the $R/\overline{W}$ and $\overline{CS}$ lines LOW after waiting for a time interval t_{AS}, called the *address setup time.* This gives the RAM's address decoders time to respond to the new address. $R/\overline{W}$ and $\overline{CS}$ are held LOW for a time interval t_W, called the write time interval.

During this write time interval, at time t_1, the CPU applies valid data to the data bus to be written into the RAM. These data must be held at the RAM input for at least a time interval t_{DS} prior to, and for at least a time interval t_{DH} after, the deactivation of $R/\overline{W}$ and $\overline{CS}$ at t_2. The t_{DS} interval is called the *data setup time,* and t_{DH} is called the *data hold time.* Similarly, the address inputs must remain stable for the address hold time interval, t_{AH}, after t_2. If any of these setup time or hold time requirements are not met, the write operation will not take place reliably.

The complete write-cycle time, t_{WC}, extends from t_0 to t_4, when the CPU changes the address lines to a new address for the next read or write cycle.

The read-cycle time, t_{RC}, and write-cycle time, t_{WC}, are what essentially determine how fast a memory chip can operate. For example, in an actual application, a CPU will often be reading successive data words from memory one right after the other. If the memory has a t_{RC} of 50 ns, the CPU can read one word every 50 ns, or 20 million words per second; with $t_{RC} = 10$ ns, the CPU can read 100 million words per second. Table 12-2 shows the minimum read-cycle and write-cycle times for some representative static-RAM chips.

TABLE 12-2

Device	t_{RC}(min) (ns)	t_{WC}(min) (ns)
CMOS MCM6206C, 32K × 8	15	15
NMOS 2147H, 4K × 1	35	35
BiCMOS MCM6708A, 64K × 4	8	8

Actual SRAM Chip

An example of an actual SRAM IC is the MCM6264C CMOS 8K × 8 RAM with read-cycle and write-cycle times of 12 ns and a standby power consumption of only 100 mW. The logic symbol for this IC is shown in Figure 12-23. Notice that it has 13 address inputs, because $2^{13} = 8192 = 8K$, and eight data I/O lines. The four control inputs determine the device's operating mode according to the accompanying mode table.

The $\overline{WE}$ input is the same as the $R/\overline{W}$ input that we have been using. A LOW at $\overline{WE}$ will write data into the RAM, provided that the device is selected—both chip select inputs are active. Note that the "&" symbol is used to denote that *both* must be active. A HIGH at $\overline{WE}$ will produce the read operation, provided that the device is selected and the output buffers are enabled by $\overline{OE}$ = LOW. When deselected, the device is in its low-power mode, and none of the other inputs have any effect.

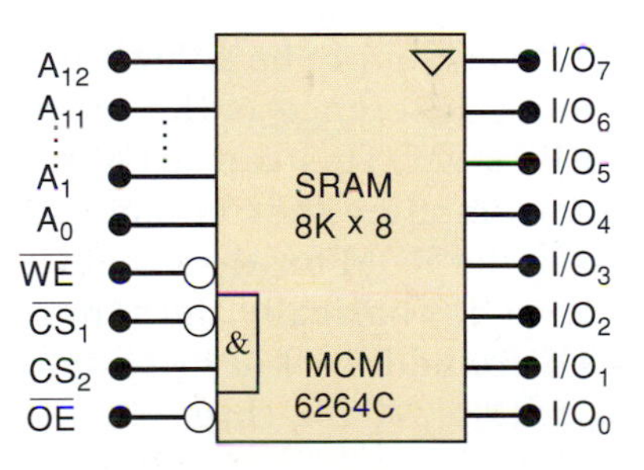

	Inputs				
Mode	$\overline{WE}$	$\overline{CS_1}$	CS_2	$\overline{OE}$	I/O pins
READ	1	0	1	0	$DATA_{OUT}$
WRITE	0	0	1	X	$DATA_{IN}$
Output disable	1	X	X	1	High Z
Not selected (power down)	X	1	X	X	High Z
	X	X	0	X	

X = don't care

FIGURE 12-23 Symbol and mode table for the CMOS MCM6264C.

Most of the devices that have been discussed in this chapter are available from several different manufacturers. Each manufacturer may offer different devices of the same dimension (e.g., 32K × 8) but with different specifications or features. There are also various types of packaging available such as DIP, PLCC, and various forms of gull-wing and surface-mount.

As you look at the various memory devices that have been described in this chapter, you will notice some similarities. For example, look at the chips in Figure 12-24 and take note of the pin assignments. The fact that the same function is assigned to the same pins on all of these diverse devices, manufactured

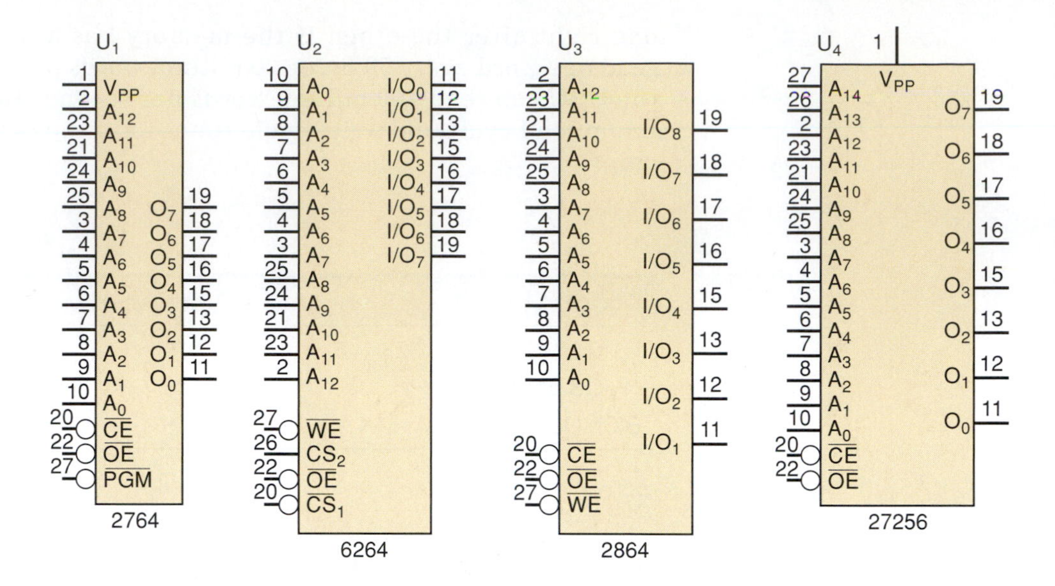

FIGURE 12-24 JEDEC standard memory packaging.

by different companies, is no coincidence. Industry standards created by the Joint Electronic Device Engineering Council **(JEDEC)** have led to memory devices that are interchangeable.

EXAMPLE 12-11

A system is wired for an 8K × 8 ROM chip (2764) and two 8K × 8 SRAM chips (6264). The entire 8K of ROM space is being used for storage of the microprocessor's instructions. You want to upgrade the system to have some nonvolatile read/write storage. Can the existing circuit be modified to accommodate the new revisions?

Solution

A 2864 EEPROM chip can simply be substituted into one of the RAM sockets. The only functional difference is the much longer write-cycle time requirements of the EEPROM. This can usually be handled by changing the program of the microcomputer that is using the memory device. Because there is no room left in the ROM for these changes, we need a larger ROM. A 32K × 8 ROM (27C256) has basically the same pin-out as a 2764. We simply need to connect two more address lines (A_{13} and A_{14}) to the ROM socket and replace the old chip with a 27C256 chip.

Many memory systems take advantage of the versatility that the JEDEC standards provide. The pins that are common for all of the devices are hardwired to the system buses. The few pins that are different among the various devices are connected to circuitry that can easily be modified to configure the system for the proper size and type of memory device. This allows the user to reconfigure the hardware without needing to cut or solder on the board. The configuration circuitry can be as simple as movable jumpers or DIP switches that the user sets up and as complicated as an in-circuit programmable logic device that the computer can set up or modify to meet the system requirements.

REVIEW QUESTIONS

1. How does a static-RAM cell differ from a dynamic-RAM cell?
2. Which memory technology generally uses the least power?
3. What device places data on the data bus during a read cycle?
4. What device places data on the data bus during a write cycle?
5. What RAM timing parameters determine its operating speed?
6. *True or false:* A LOW at $\overline{OE}$ will enable the output buffers of an MCM6264C provided that both chip select inputs are active.
7. What must be done with pin 26 and pin 27 if a 27256 is replaced with a 2764?

12-13 DYNAMIC RAM (DRAM)

Dynamic RAMs are fabricated using MOS technology and are noted for their high capacity, low power requirement, and moderate operating speed. As we stated earlier, unlike static RAMs, which store information in FFs, dynamic RAMs store 1s and 0s as charges on a small MOS capacitor (typically a few picofarads). Because of the tendency for these charges to leak off after a period of time, dynamic RAMs require periodic recharging of the memory cells; this is called refreshing the dynamic RAM. In modern DRAM chips, each memory cell must be refreshed typically every 2, 4, or 8 ms, or its data will be lost.

The need for refreshing is a drawback of dynamic RAM compared to static RAM because it may require external support circuitry. Some DRAM chips have built-in refresh control circuitry that does not require extra external hardware but does require special timing of the chip's input control signals. Additionally, as we shall see, the address inputs to a DRAM must be handled in a less straightforward way than SRAM. So, all in all, designing with and using DRAM in a system is more complex than with SRAM. However, their much larger capacities and much lower power consumption make DRAMs the memory of choice in systems where the most important design considerations are keeping down size, cost, and power.

For applications where speed and reduced complexity are more critical than cost, space, and power considerations, static RAMs are still the best. They are generally faster than dynamic RAMs and require no refresh operation. They are simpler to design with, but they cannot compete with the higher capacity and lower power requirement of dynamic RAMs.

Because of their simple cell structure, DRAMs typically have four times the density of SRAMs. This increased density allows four times as much memory capacity to be placed on a single board; alternatively, it requires one-fourth as much board space for the same amount of memory. The cost per bit of dynamic RAM storage is typically one-fifth to one-fourth that of static RAMs. An additional cost saving is realized because the lower power requirements of a dynamic RAM, typically one-sixth to one-half those of a static RAM, allow the use of smaller, less expensive power supplies.

The main applications of SRAMs are in areas where only small amounts of memory are needed or where high speed is required. Many microprocessor-controlled instruments and appliances have very small memory capacity requirements. Some instruments, such as digital storage oscilloscopes and logic analyzers, require very high-speed memory. For applications such as these, SRAM is normally used.

The main internal memory of most personal microcomputers (e.g., Windows-based PCs or Macs) uses DRAM because of its high capacity and low power consumption. These computers, however, sometimes use some small amounts of SRAM for functions requiring maximum speed, such as video graphics, look-up tables, and cache memory.

REVIEW QUESTIONS

1. What are the main drawbacks of dynamic RAM compared with static?
2. List the advantages of dynamic RAM compared with static RAM.
3. Which type of RAM would you expect to find on the main memory modules of your PC?

12-14 DYNAMIC RAM STRUCTURE AND OPERATION

The dynamic RAM's internal architecture can be visualized as an array of single-bit cells, as illustrated in Figure 12-25. Here, 16,384 cells are arranged in a 128 × 128 array. Each cell occupies a unique row and column position within the array. Fourteen address inputs are needed to select one of the cells ($2^{14} = 16,384$); the lower address bits, A_0 to A_6, select the column, and the higher-order bits, A_7 to A_{13}, select the row. Each 14-bit address selects a unique cell to be written into or read from. The structure in Figure 12-25 is a 16K × 1 DRAM chip. DRAM chips are currently available in various configurations. DRAMs with a four-bit (or greater) word size have a cell arrangement similar to that of Figure 12-25 except that each position in the array contains four cells, and each applied address selects a group of four cells for a read or a write operation. As we will see later, larger word sizes can also be attained by combining several chips in the appropriate arrangement.

Figure 12-26 is a symbolic representation of a dynamic memory cell and its associated circuitry. Many of the circuit details are not shown, but this simplified diagram can be used to describe the essential ideas involved in

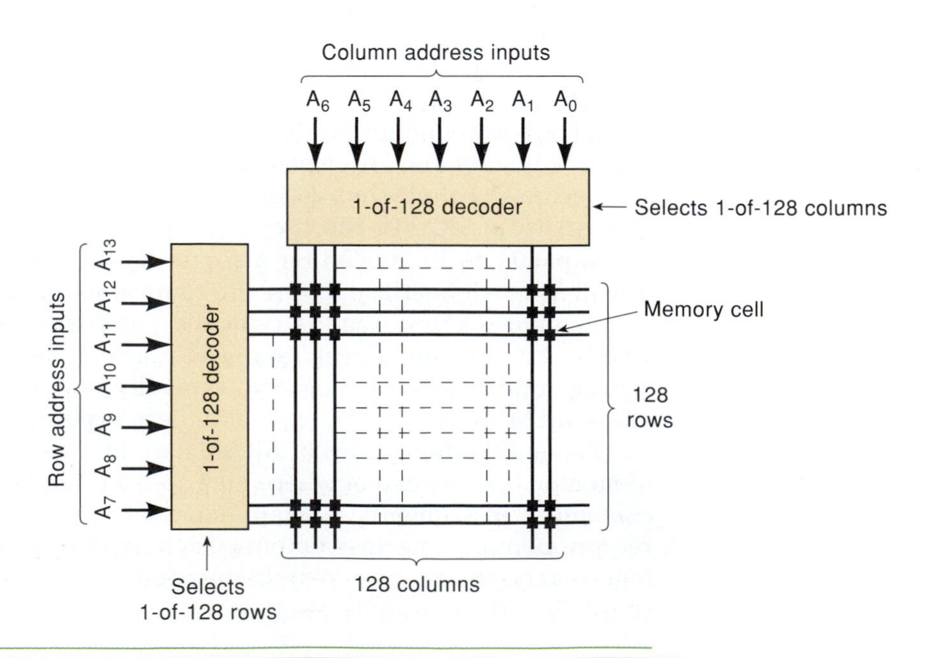

FIGURE 12-25 Cell arrangement in a 16K × 1 dynamic RAM.

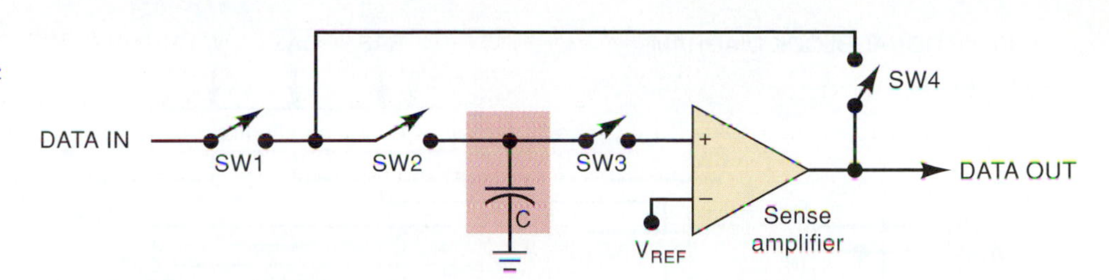

FIGURE 12-26 Symbolic representation of a dynamic memory cell. During a WRITE operation, semiconductor switches SW1 and SW2 are closed. During a read operation, all switches are closed except SW1.

writing to and reading from a DRAM. The switches SW1 through SW4 are actually MOSFETs that are controlled by various address decoder outputs and the $R/\overline{W}$ signal. The capacitor, of course, is the actual storage cell. One sense amplifier would serve an entire column of memory cells, but operate only on the bit in the selected row.

To write data to the cell, signals from the address decoding and read/write logic will close switches SW1 and SW2, while keeping SW3 and SW4 open. This connects the input data to *C*. A logic 1 at the data input charges *C*, and a logic 0 discharges it. Then the switches are open so that *C* is disconnected from the rest of the circuit. Ideally, *C* would retain its charge indefinitely, but there is always some leakage path through the off switches, so that *C* will gradually lose its charge.

To read data from the cell, switches SW2, SW3, and SW4 are closed, and SW1 is kept open. This connects the stored capacitor voltage to the *sense amplifier*. The sense amplifier compares the voltage with some reference value to determine if it is a logic 0 or 1, and it produces a solid 0 V or 5 V for the data output. This data output is also connected to *C* (SW2 and SW4 are closed) and refreshes the capacitor voltage by recharging or discharging. In other words, the data bit in a memory cell is refreshed each time it is read.

Address Multiplexing

The 16K × 1 DRAM array depicted in Figure 12-25 is obsolete and nearly unavailable. It has 14 address inputs; a 64K × 1 DRAM array would have 16 address inputs. A 1M × 4 DRAM needs 20 address inputs; a 4M × 1 needs 22 address inputs. High-capacity memory chips such as these would require many pins if each address input required a separate pin. In order to reduce the number of pins on their high-capacity DRAM chips, manufacturers utilize **address multiplexing** whereby each address input pin can accommodate two different address bits. The saving in pin count translates to a significant decrease in the size of the IC packages. This is very important in large-capacity memory boards, where you want to maximize the amount of memory that can fit on one board.

In the discussions that follow, we will be describing the order in which the address is multiplexed into the DRAM chips. It should be noted that in older, small-capacity DRAMs, the convention was to present the low-order address first specifying the row, followed by the high-order address specifying the column. The newer DRAMs and the controllers that perform the multiplexing use the opposite convention of applying the high-order bits as the row address and then the low-order bits as the column address. We will describe the more recent convention, but you should be aware of this change as you investigate older systems.

We will use the TMS44100 4M × 1 DRAM from Texas Instruments to illustrate the operation of DRAM chips today. The functional block diagram of this chip's internal architecture (shown in Figure 12-27) is typical of

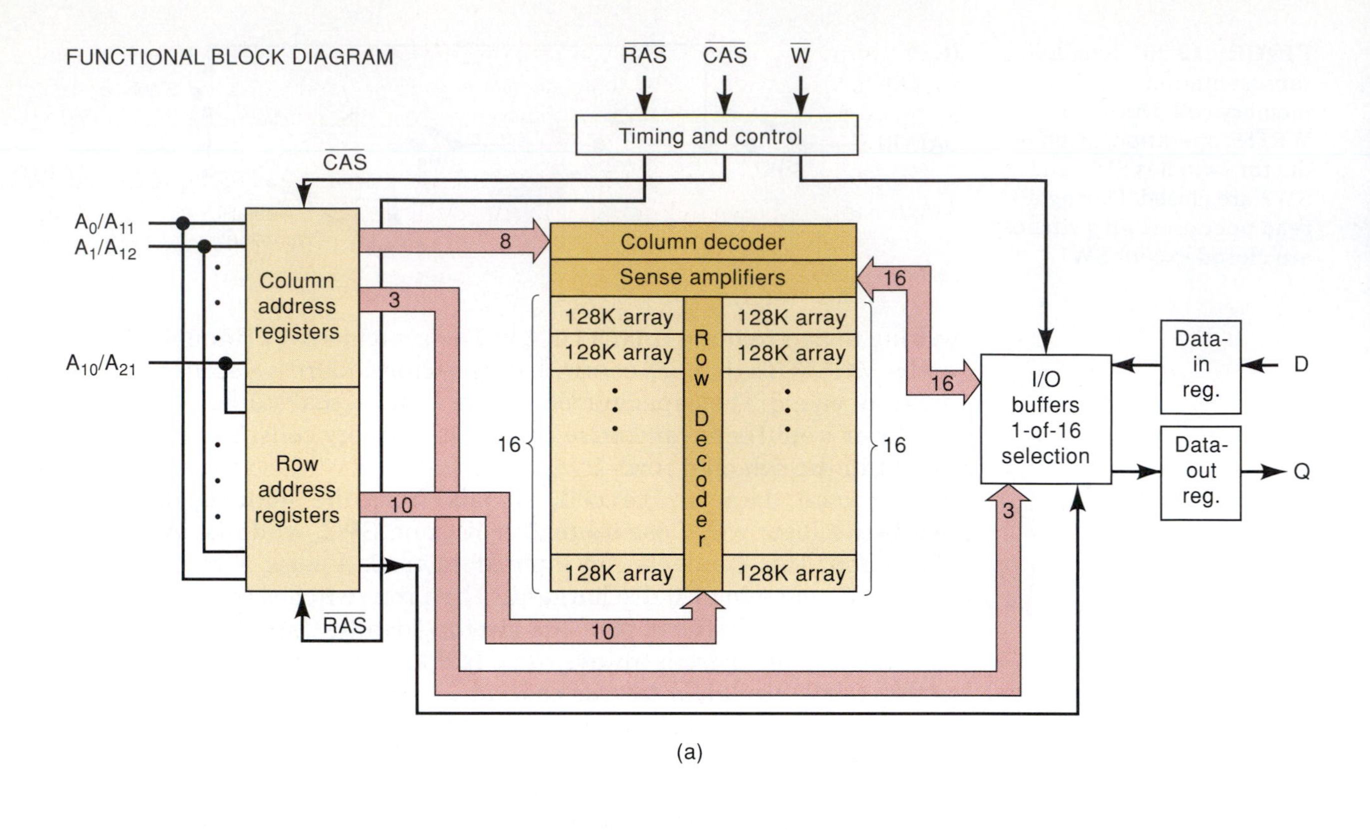

(a)

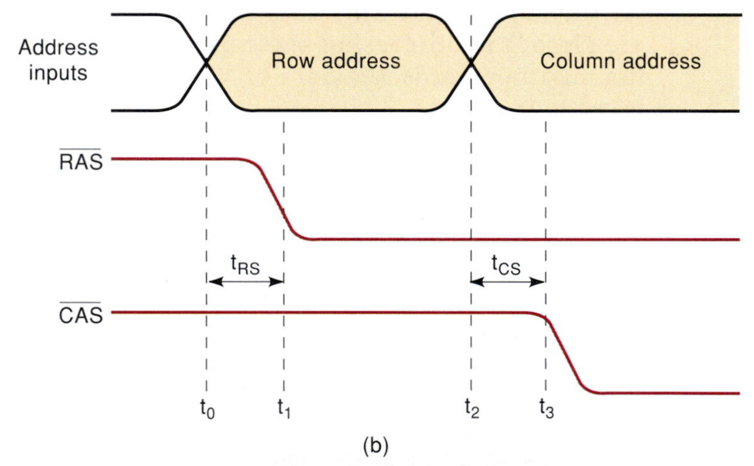

(b)

FIGURE 12-27 (a) Simplified architecture of the TMS44100 4M × 1 DRAM; (b) $\overline{RAS}/\overline{CAS}$ timing. (Reprinted by permission of Texas Instruments)

diagrams you will find in data books. The layout of the memory array in this diagram may appear complicated at first glance, but it can be thought of as just a bigger version of the 16K × 1 DRAM in Figure 12-25. Functionally, it is an array of cells arranged as 2048 rows by 2048 columns. A single row is selected by address decoder circuitry that can be thought of as a 1-of-2048 decoder. Likewise, a single column is selected by what is effectively a 1-of-2048 decoder. Because the address lines are multiplexed, the entire 22-bit address cannot be presented simultaneously. Notice that there are only 11 address lines and that they go to both the row and the column address registers. Each of the two address registers stores half of the 22-bit address. The row register stores the upper half, and the column register stores the lower half. Two very important **strobe** inputs control when the address information is latched. The **row address strobe ($\overline{RAS}$)** clocks the 11-bit row address register. The **column address strobe ($\overline{CAS}$)** clocks the 11-bit column address register.

A 22-bit address is applied to this DRAM in two steps using $\overline{RAS}$ and $\overline{CAS}$. The timing is shown in Figure 12-27(b). Initially, $\overline{RAS}$ and $\overline{CAS}$ are both HIGH. At time t_0, the 11-bit row address (A_{11} to A_{21}) is applied to the address inputs. After allowing time for the setup time requirement (t_{RS}) of the row address register, the $\overline{RAS}$ input is driven LOW at t_1. This NGT loads the row address into the row address register so that A_{11} to A_{21} now appear at the row decoder inputs. The LOW at $\overline{RAS}$ also enables this decoder so that it can decode the row address and select one row of the array.

At time t_2, the 11-bit column address (A_0 to A_{10}) is applied to the address inputs. At t_3, the $\overline{CAS}$ input is driven LOW to load the column address into the column address register. $\overline{CAS}$ also enables the column decoder so that it can decode the column address and select one column of the array.

At this point the two parts of the address are in their respective registers, the decoders have decoded them to select the one cell corresponding to the row and column address, and a read or a write operation can be performed on that cell just as in a static RAM.

You may have noticed that this DRAM does not have a chip select (CS) input. The $\overline{RAS}$ and $\overline{CAS}$ signals perform the chip select function because they must both be LOW for the decoders to select a cell for reading or writing.

As you can see, there are several operations that must be performed before the data that is stored in the DRAM can actually appear on the outputs. The term **latency** is often used to describe the time required to perform these operations. Each operation takes a certain amount of time, and this amount of time determines the maximum rate at which we can access data in the memory.

EXAMPLE 12-12

How many pins are saved by using address multiplexing for a 16M × 1 DRAM?

Solution

Twelve address inputs are used instead of 24; $\overline{RAS}$ and $\overline{CAS}$ are added; no $\overline{CS}$ is required. Thus, there is a net saving of *eleven* pins.

In a simple computer system, the address inputs to the memory system come from the central processing unit (CPU). When the CPU wants to access a particular memory location, it generates the complete address and places it on address lines that make up an address bus. Figure 12-28(a) shows this for a small computer memory that has a capacity of 64K words and therefore requires a 16-line address bus going directly from the CPU to the memory.

This arrangement works for ROM or for static RAM, but it must be modified for DRAM that uses multiplexed addressing. If all 64K of the memory is DRAM, it will have only eight address inputs. This means that the 16 address lines from the CPU address bus must be fed into a

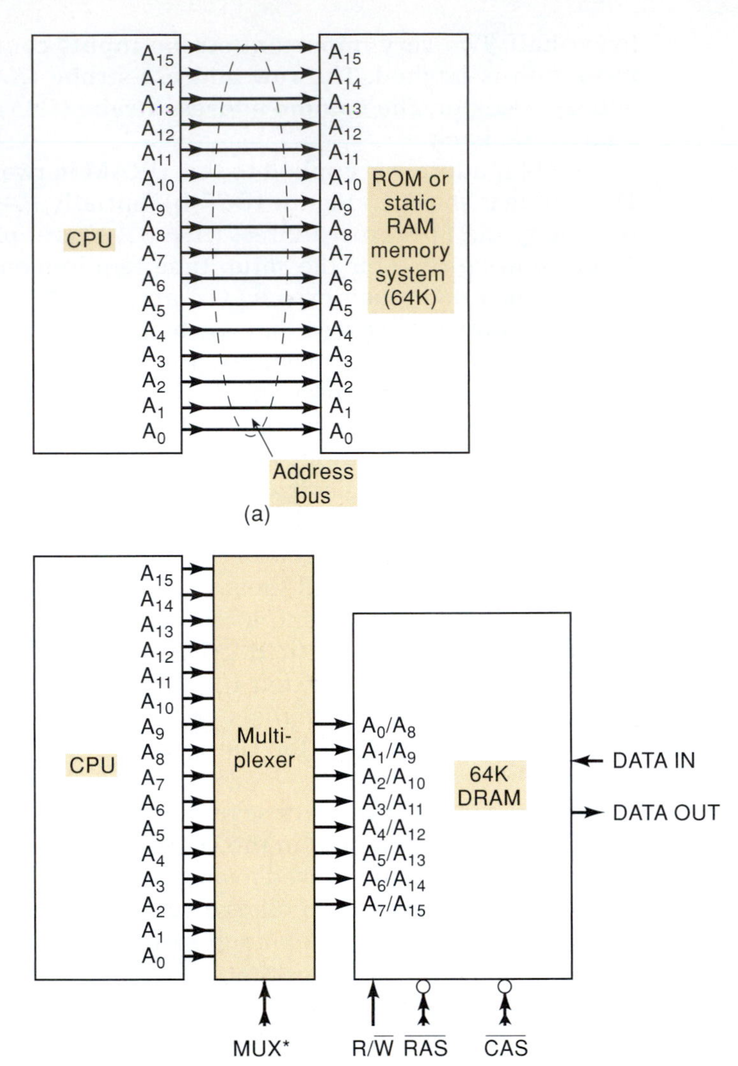

FIGURE 12-28 (a) CPU address bus driving ROM or static-RAM memory; (b) CPU addresses driving a multiplexer that is used to multiplex the CPU address lines into the DRAM.

multiplexer circuit that will transmit eight address bits at a time to the memory address inputs. This is shown symbolically in Figure 12-28(b). The multiplexer select input, labeled *MUX*, controls whether CPU address lines A_0 to A_7 or address lines A_8 to A_{15} will be present at the DRAM address inputs.

The timing of the *MUX* signal must be synchronized with the $\overline{RAS}$ and $\overline{CAS}$ signals that clock the addresses into the DRAM. This is shown in Figure 12-29. *MUX* must be LOW when $\overline{RAS}$ is pulsed LOW so that address lines A_8 to A_{15} from the CPU will reach the DRAM address inputs to be loaded on the NGT of $\overline{RAS}$. Likewise, *MUX* must be HIGH when $\overline{CAS}$ is pulsed LOW so that A_0 to A_7 from the CPU will be present at the DRAM inputs to be loaded on the NGT of $\overline{CAS}$.

The actual multiplexing and timing circuitry will not be shown here but will be left to the end-of-chapter problems (Problems 12-26 and 12-27).

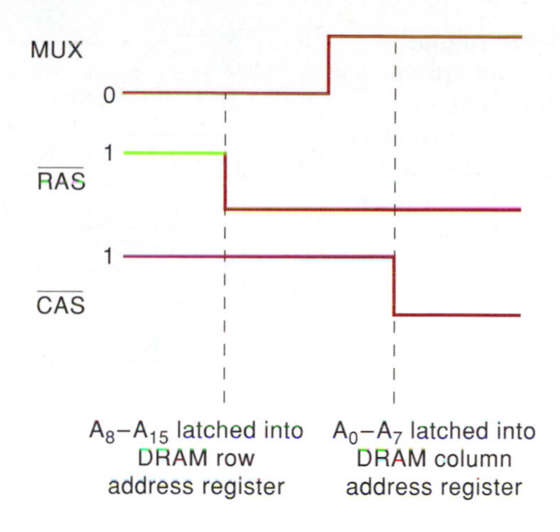

FIGURE 12-29 Timing required for address multiplexing.

REVIEW QUESTIONS

1. Describe the array structure of a 64K × 1 DRAM.
2. What is the benefit of address multiplexing?
3. How many address inputs would there be on a 1M × 1 DRAM chip?
4. What are the functions of the $\overline{RAS}$ and $\overline{CAS}$ signals?
5. What is the function of the *MUX* signal?

12-15 DRAM READ/WRITE CYCLES

The timing of the read and write operations of a DRAM is much more complex than for a static RAM, and there are many critical timing requirements that the DRAM memory designer must consider. At this point, a detailed discussion of these requirements would probably cause more confusion than enlightenment. We will concentrate on the basic timing sequence for the read and write operations for a small DRAM system like that of Figure 12-28(b).

Dram Read Cycle

Figure 12-30 shows typical signal activity during the read operation. It is assumed that $R/\overline{W}$ is in its HIGH state throughout the operation. The following is a step-by-step description of the events that occur at the times indicated on the diagram.

- t_0: *MUX* is driven LOW to apply the row address bits (A_8 to A_{15}) to the DRAM address inputs.
- t_1: $\overline{RAS}$ is driven LOW to load the row address into the DRAM.
- t_2: *MUX* goes HIGH to place the column address (A_0 to A_7) at the DRAM address inputs.
- t_3: $\overline{CAS}$ goes LOW to load the column address into the DRAM.
- t_4: The DRAM responds by placing valid data from the selected memory cell onto the DATA OUT line.
- t_5: *MUX*, $\overline{RAS}$, $\overline{CAS}$, and DATA OUT return to their initial states.

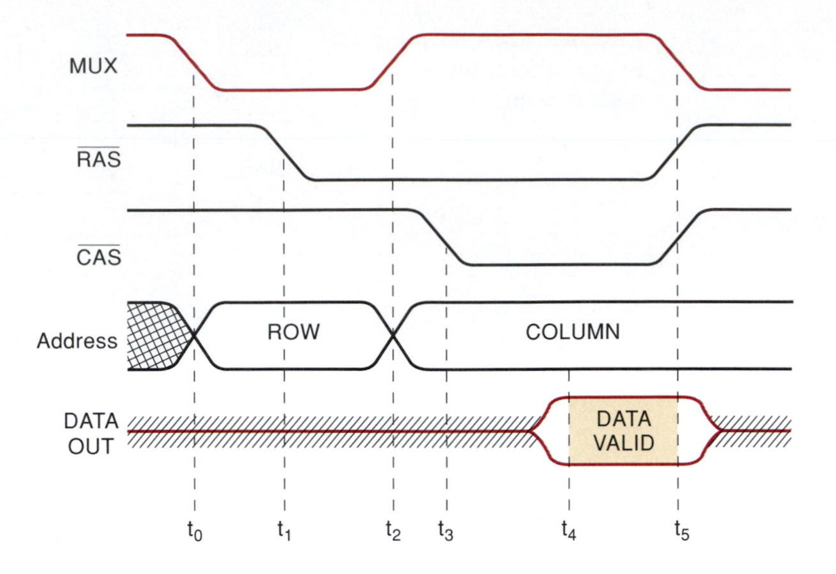

FIGURE 12-30 Signal activity for a read operation on a dynamic RAM. The $R/\overline{W}$ input (not shown) is assumed to be HIGH.

Dram Write Cycle

Figure 12-31 shows typical signal activity during a DRAM write operation. Here is a description of the sequence of events.

- t_0: The LOW at *MUX* places the row address at the DRAM inputs.
- t_1: The NGT at $\overline{RAS}$ loads the row address into the DRAM.
- t_2: *MUX* goes HIGH to place the column address at the DRAM inputs.
- t_3: The NGT at $\overline{CAS}$ loads the column address into the DRAM.
- t_4: Data to be written are placed on the DATA IN line.
- t_5: $R/\overline{W}$ is pulsed LOW to write the data into the selected cell.
- t_6: Input data are removed from DATA IN.
- t_7: *MUX*, $\overline{RAS}$, $\overline{CAS}$, and $R/\overline{W}$ are returned to their initial states.

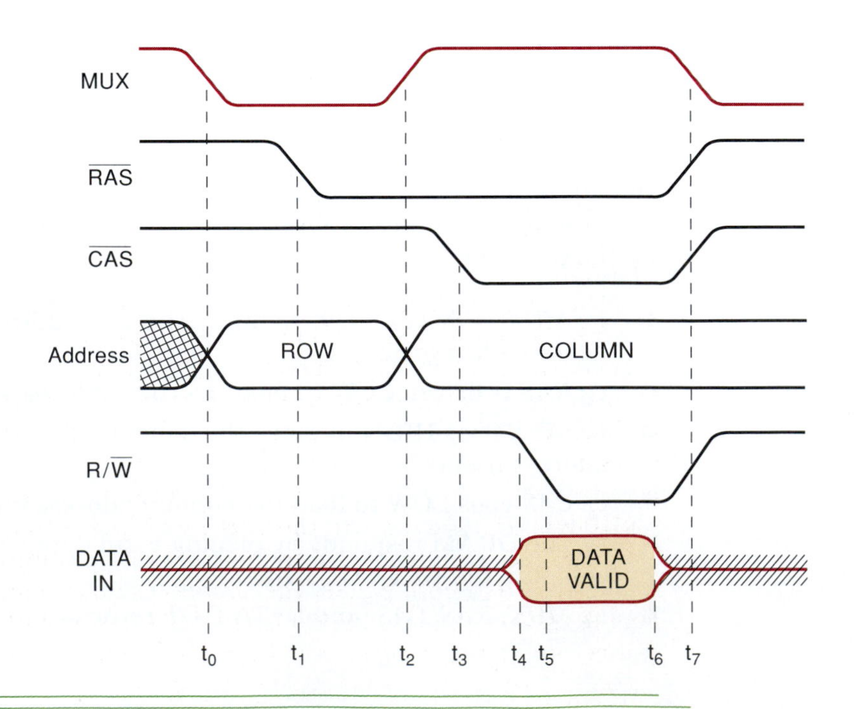

FIGURE 12-31 Signal activity for a write operation on a dynamic RAM.

REVIEW QUESTIONS

1. *True or false:*
   (a) During a read cycle, the $\overline{RAS}$ signal is activated before the $\overline{CAS}$ signal.
   (b) During a write operation, $\overline{CAS}$ is activated before $\overline{RAS}$.
   (c) $R/\overline{W}$ is held LOW for the entire write operation.
   (d) The address inputs to a DRAM will change twice during a read or a write operation.
2. Which signal in Figure 12-28(b) makes sure that the correct portion of the complete address appears at the DRAM inputs?

12-16 DRAM REFRESHING

A DRAM cell is refreshed each time a read operation is performed on that cell. Each memory cell must be refreshed periodically (typically, every 4 to 16 ms, depending on the device) or its data will be lost. This requirement would appear to be extremely difficult, if not impossible, to meet for large-capacity DRAMs. For example, a 1M $\times$ 1 DRAM has 10^{20} = 1,048,576 cells. To ensure that each cell is refreshed within 4 ms, it would require that read operations be performed on successive addresses at the rate of one every 4 ns (4 ms/1,048,576 $\approx$ 4 ns). This is much too fast for any DRAM chip. Fortunately, manufacturers have designed DRAM chips so that

> **whenever a read operation is performed on a cell, all of the cells in that row will be refreshed.**

Thus, it is necessary to do a read operation only on each *row* of a DRAM array once every 4 ms to guarantee that each *cell* of the array is refreshed. Referring to the 4M $\times$ 1 DRAM of Figure 12-27(a), if any address is strobed into the row address register, all 2048 cells in that row are automatically refreshed.

Clearly, this row-refreshing feature makes it easier to keep all DRAM cells refreshed. However, during the normal operation of the system in which a DRAM is functioning, it is unlikely that a read operation will be performed on each row of the DRAM within the required refresh time limit. Therefore, some kind of refresh control logic is needed either external to the DRAM chip or as part of its internal circuitry. In either case, there are two refresh modes: a *burst* refresh and a *distributed* refresh.

In a burst refresh mode, the normal memory operation is suspended, and each row of the DRAM is refreshed in succession until all rows have been refreshed. In a distributed refresh mode, the row refreshing is interspersed with the normal operations of the memory.

The most universal method for refreshing a DRAM is the **$\overline{RAS}$-only refresh.** It is performed by strobing in a row address with $\overline{RAS}$ while $\overline{CAS}$ and $R/\overline{W}$ remain HIGH. Figure 12-32 illustrates how $\overline{RAS}$-only refresh is used for a burst refresh of the TMS44100. Some of the complexity of the memory array in this chip is there to make refresh operations simpler. Because two banks are lined up in the same row, both banks can be refreshed at the same time, effectively making it the same as if there were only 1024 rows. A **refresh counter** is used to supply 10-bit row addresses to the DRAM address inputs starting at 0000000000 (row 0). $\overline{RAS}$ is pulsed LOW to load this address into the DRAM, and this refreshes row 0 in both banks. The counter is incremented and the process is repeated up to address 1111111111 (row 1023). For the TMS44100, a burst refresh can be completed in just over 113 μs and must be repeated every 16 ms or less.

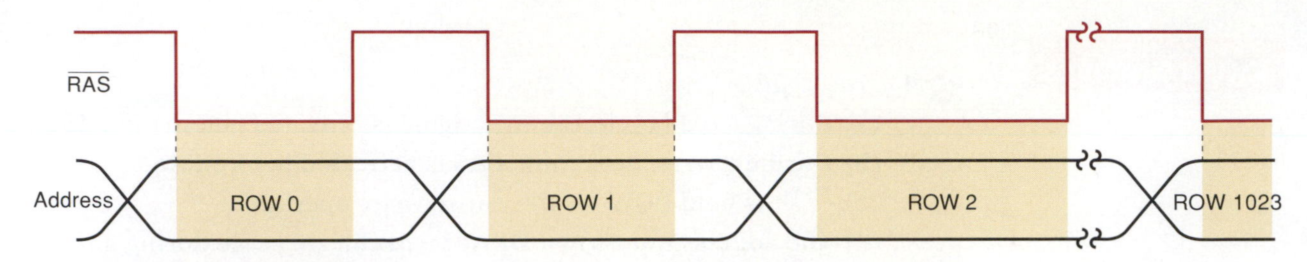

* $R/\overline{W}$ and $\overline{CAS}$ lines held HIGH

FIGURE 12-32 The $\overline{RAS}$-only refresh method uses only the $\overline{RAS}$ signal to load the row address into the DRAM to refresh all cells in that row. The $\overline{RAS}$-only refresh can be used to perform a burst refresh as shown. A refresh counter supplies the sequential row addresses from row 0 to row 1023 (for a 4M × 1 DRAM).

While the refresh counter idea seems easy enough, we must realize that the row addresses from the refresh counter cannot interfere with the addresses coming from the CPU during normal read/write operations. For this reason, the refresh counter addresses must be multiplexed with the CPU addresses, so that the proper source of DRAM addresses is activated at the proper times.

In order to relieve the computer's CPU from some of these burdens, a special chip called a **dynamic RAM (DRAM) controller** is often used. At a minimum, this chip will perform address multiplexing and refresh count sequence generation, leaving the generation of the timing for $\overline{RAS}$, $\overline{CAS}$, and *MUX* signals up to some other logic circuitry and the person who programs the computer. Other DRAM controllers are fully automatic. Their inputs look very much like a static RAM or ROM. They automatically generate the refresh sequence often enough to maintain the memory, multiplex the address bus, generate the $\overline{RAS}$ and $\overline{CAS}$ signals, and arbitrate control of the DRAM between the CPU read/write cycles and local refresh operations. In current personal computers, the DRAM controller and other high-level controller circuits are integrated into a set of VLSI circuits that are referred to as a "chip set." As newer DRAM technologies are developed, new chip sets are designed to take advantage of the latest advances. In many cases, the number of existing (or anticipated) chip sets supporting a certain technology in the market determines the DRAM technology in which manufacturers invest.

Most of the DRAM chips in production today have on-chip refreshing capability that eliminates the need to supply external refresh addresses. One of these methods, shown in Figure 12-33(a), is called $\overline{CAS}$-before-$\overline{RAS}$ *refresh.* In this method, the $\overline{CAS}$ signal is driven LOW first and is held LOW until after $\overline{RAS}$ goes LOW. This sequence will refresh one row of the memory array and increment an internal counter that generates the row addresses. To perform a burst refresh using this feature, $\overline{CAS}$ can be held LOW while $\overline{RAS}$ is pulsed once for each row until all are refreshed. During this refresh cycle, all external addresses are ignored. The TMS44100 also offers "hidden refresh," which allows a row to be refreshed while holding data on the output. This is done by holding $\overline{CAS}$ LOW after a read cycle and then pulsing $\overline{RAS}$ as in Figure 12-33(b).

The self-refresh mode of Figure 12-33(c) fully automates the process. By forcing $\overline{CAS}$ LOW before $\overline{RAS}$ and then holding them both LOW for at least 100 μs, an internal oscillator clocks the row address counter until all cells are refreshed. The mode that a system designer chooses depends on how busy the computer's CPU is. If it can spare 100 μs without accessing its memory, and if it can do this every 16 ms, then the self-refresh is the way to go.

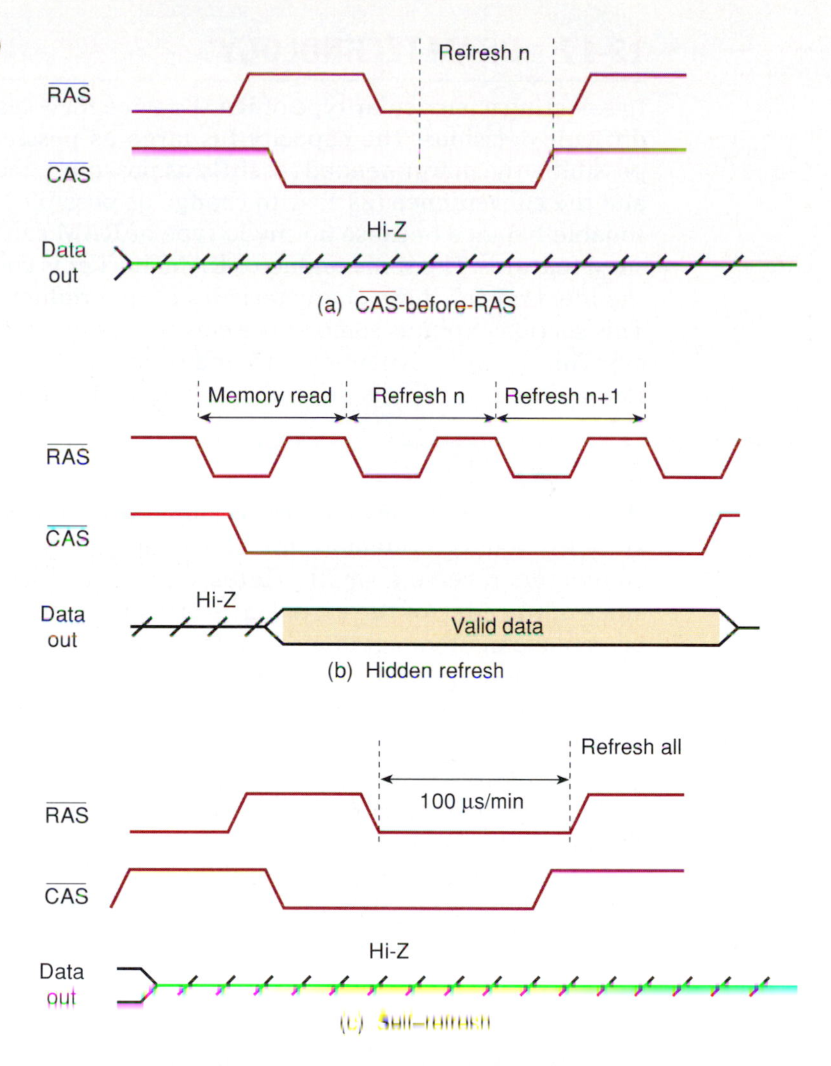

FIGURE 12-33 TMS44100 refresh modes.

However, if this will slow the program execution down too much, it may require some distributed refreshing using $\overline{CAS}$-before-$\overline{RAS}$ or hidden refresh cycles. In any case, all cells must be refreshed within the allotted time or data will be lost.

REVIEW QUESTIONS

1. *True or false:*
   (a) In most DRAMs, it is necessary to read only from one cell in each row in order to refresh all cells in that row.
   (b) In the burst refresh mode, the entire array is refreshed by one $\overline{RAS}$ pulse.
2. What is the function of a refresh counter?
3. What functions does a DRAM controller perform?
4. *True or false:*
   (a) In the $\overline{RAS}$-only refresh method, the $\overline{CAS}$ signal is held LOW.
   (b) $\overline{CAS}$-before-$\overline{RAS}$ refresh can be used only by DRAMs with on-chip refresh control circuitry.

12-17 DRAM TECHNOLOGY*

In selecting a particular type of RAM device for a system, a designer has some difficult decisions. The capacity (as large as possible), the speed (as fast as possible), the power needed (as little as possible), the cost (as low as possible), and the convenience (as easy to change as possible) must all be kept in a reasonable balance because no single type of RAM can maximize all of these desired features. The semiconductor RAM market is constantly trying to produce the ideal mix of these characteristics in its products for various applications. This section explains some of the current terms used regarding RAM technology. This is a very dynamic topic, and perhaps some of these terms will be history before this book is printed, but here is the state of the art today.

Memory Modules

With many companies manufacturing motherboards for personal computer systems, standard memory interface connectors have been adopted. These connectors receive a small printed circuit card with contact points on both sides of the edge of the card. These modular cards allow for easy installation or replacement of memory components in a computer. The single-in-line memory module (SIMM) is a circuit card with 72 functionally equivalent contacts on both sides of the card. A redundant contact point on each side of the board offers some assurance that a good, reliable contact is made. These modules use 5-V-only DRAM chips that vary in capacity from 1 to 16 Mbits in surface-mount gull-wing or J-lead packages. The memory modules vary in capacity from 1 to 32 Mbytes.

The newer, 168-pin, dual-in-line memory module (DIMM) has 84 functionally unique contacts on each side of the card. The extra pins are necessary because DIMMs are connected to 64-bit data buses such as those found in modern PCs. Both 3.3-V and 5-V versions are available. They also come in buffered and unbuffered versions. The capacity of the module depends on the DRAM chips that are mounted on it; and as DRAM capacity increases, the capacity of the DIMMs will increase. The chip set and motherboard design that is used in any given system determines which type of DIMM can be used. For compact applications, such as laptop computers, a small-outline, dual-in-line memory module is available (SODIMM).

The primary problem in the personal computer industry is providing a memory system that is fast enough to keep up with the ever-increasing clock speeds of the microprocessors while keeping the cost at an affordable level. Special features are being added to the basic DRAM devices to enhance their total bandwidth. A new type of package called the RIMM has entered the market. RIMM stands for Rambus In-line Memory Module. Rambus is a company that has invented some revolutionary new approaches to memory technology. The RIMM is their proprietary package that holds their proprietary memory chips called Direct Rambus DRAM (DRDRAM) chips. Although these methods of improving performance are constantly changing, the technologies described in the following sections are currently being referred to extensively in memory-related literature.

FPM DRAM

Fast page mode (FPM) allows quicker access to random memory locations within the current "page." A page is simply a range of memory addresses

*This topic may be omitted without affecting the continuity of the remainder of the book.

that have identical upper address bit values. In order to access data on the current page, only the lower address lines must be changed.

EDO DRAM

Extended data output (EDO) DRAMs offer a minor improvement to FPM DRAMs. For accesses on a given page, the data value at the current memory location is sensed and latched onto the output pins. In the FPM DRAMs, the sense amplifier drives the output without a latch, requiring $\overline{CAS}$ to remain low until data values become valid. With EDO, while these data are present on the outputs, $\overline{CAS}$ can complete its cycle, a new address on the current page can be decoded, and the data path circuitry can be reset for the next access. This allows the memory controller to be outputting the next address at the same time that the current word is being read.

SDRAM

Synchronous DRAM is designed to transfer data in rapid-fire *bursts* of several sequential memory locations. The first location to be accessed is the slowest due to the overhead (latency) of latching the row and column address. Thereafter, the data values are clocked out by the bus system clock (instead of the $\overline{CAS}$ control line) in bursts of memory locations within the same page. Internally, SDRAMs are organized in two banks. This allows data to be read out at a very fast rate by alternately accessing each of the two banks. In order to provide all of the features and the flexibility needed for this type of DRAM to work with a wide variety of system requirements, the circuitry within the SDRAM has become more complex. A command sequence is necessary to tell the SDRAM which options are needed, such as burst length, sequential or interleaved data, and $\overline{CAS}$-before-$\overline{RAS}$ or self-refresh modes. Self-refresh mode allows the memory device to perform all of the necessary functions to keep its cells refreshed.

DDRSDRAM

Double Data Rate SDRAM offers an improvement of SDRAM. In order to speed up the operation of SDRAM, while operating from a synchronous system clock, this technology transfers data on the rising and falling edges of the system clock, effectively doubling the potential rate of data transfer.

SLDRAM

Synchronous-Link DRAM is an evolutionary improvement over DDRSDRAM. It can operate at bus speeds up to 200 MHz and clocks data synchronously on the rising and falling edges of the system clock. A consortium of several DRAM manufacturers is developing it as an open standard. If chip sets are developed that can take advantage of these memory devices and enough system designers adopt this technology, it is likely to become a widely used form of DRAM.

DRDRAM

Direct Rambus DRAM is a proprietary device developed and marketed by Rambus, Inc. It uses a revolutionary new approach to DRAM system architecture with much more control integrated into the memory device. This technology is still battling with the other standards to find its niche in the market.

REVIEW QUESTIONS

1. Are SIMMs and DIMMs interchangeable?
2. What is a "page" of memory?
3. Why is "page mode" faster?
4. What does *EDO* stand for?
5. What term is used for accessing several consecutive memory locations?
6. What is an SDRAM synchronized to?

12-18 EXPANDING WORD SIZE AND CAPACITY

In many memory applications, the required RAM or ROM memory capacity or word size cannot be satisfied by one memory chip. Instead, several memory chips must be combined to provide the capacity and/or the word size. We will see how this is done through several examples that illustrate the important ideas that are used when memory chips are interfaced to a microprocessor. The examples that follow are intended to be instructive, and the memory chip sizes that are used were chosen to conserve space. The techniques that are presented can be extended to larger memory chips.

Expanding Word Size

Suppose that we need a memory that can store 16 eight-bit words and all we have are RAM chips that are arranged as 16×4 memories with common I/O lines. We can combine two of these 16×4 chips to produce the desired memory. The configuration for doing so is shown in Figure 12-34. Examine this diagram carefully and see what you can find out from it before reading on.

Because each chip can store 16 four-bit words and we want to store 16 eight-bit words, we are using each chip to store *half* of each word. In other words, RAM-0 stores the four *higher*-order bits of each of the 16 words, and RAM-1 stores the four *lower*-order bits of each of the 16 words. A full eight-bit word is available at the RAM outputs connected to the data bus.

Any one of the 16 words is selected by applying the appropriate address code to the four-line *address bus* (A_3, A_2, A_1, A_0). The address lines typically originate at the CPU. Note that each address bus line is connected to the corresponding address input of each chip. This means that once an address code is placed on the address bus, this same address code is applied to both chips so that the same location in each chip is accessed at the same time.

Once the address is selected, we can read or write at this address under control of the common $R/\overline{W}$ and $\overline{CS}$ line. To read, $R/\overline{W}$ must be high and $\overline{CS}$ must be low. This causes the RAM I/O lines to act as *outputs*. RAM-0 places its selected four-bit word on the upper four data bus lines, and RAM-1 places its selected four-bit word on the lower four data bus lines. The data bus then contains the full selected eight-bit word, which can now be transmitted to some other device (usually a register in the CPU).

To write, $R/\overline{W} = 0$ and $\overline{CS} = 0$ causes the RAM I/O lines to act as *inputs*. The eight-bit word to be written is placed on the data bus (usually by the CPU). The higher four bits will be written into the selected location of RAM-0, and the lower four bits will be written into RAM-1.

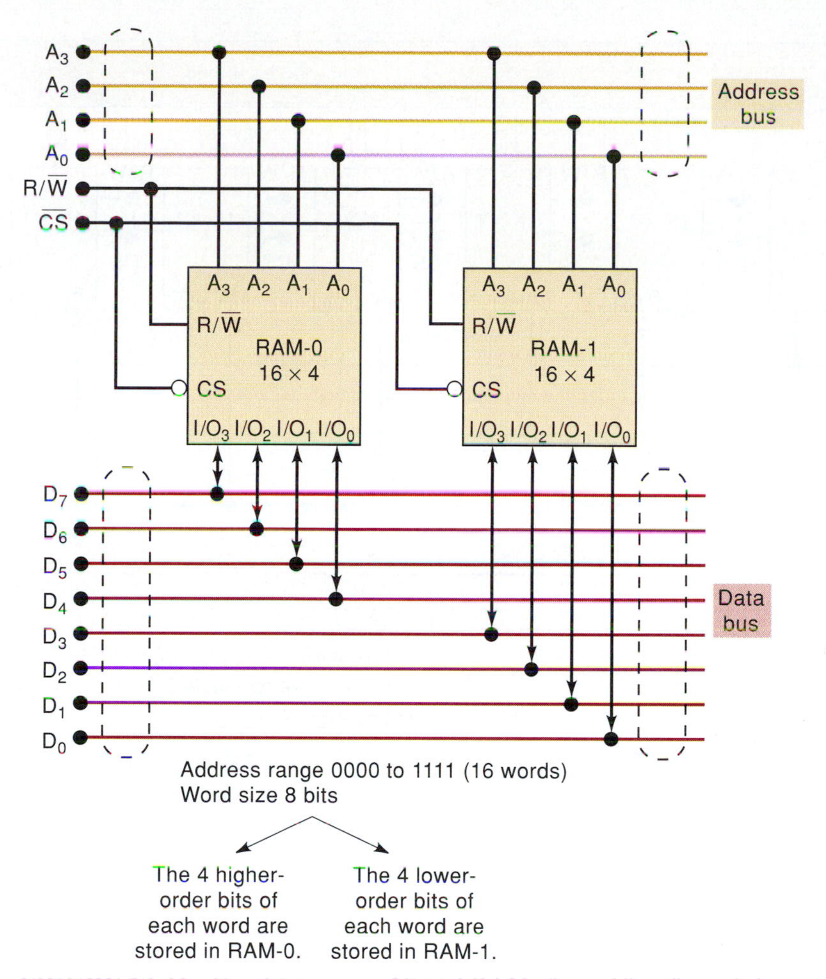

FIGURE 12-34 Combining two 16 × 4 RAMs for a 16 × 8 module.

In essence, the combination of the two RAM chips acts like a single 16 × 8 memory chip. We refer to this combination as a 16 × 8 *memory module.*

The same basic idea for expanding word size will work for many different situations. Read the following example and draw a rough diagram for what the system will look like before looking at the solution.

EXAMPLE 12-13

The 2125A is a static-RAM IC that has a capacity of 1K × 1, one active-LOW chip select input, and separate data input and output. Show how to combine several 2125A ICs to form a 1K × 8 module.

Solution

The arrangement is shown in Figure 12-35, where eight 2125A chips are used for a 1K × 8 module. Each chip stores one of the bits of each of the 1024 eight-bit words. Note that all of the $R/\overline{W}$ and $\overline{CS}$ inputs are wired together, and the 10-line address bus is connected to the address inputs of each chip. Also note that because the 2125A has separate data in and data out pins, both of these pins of each chip are tied to the same data bus line.

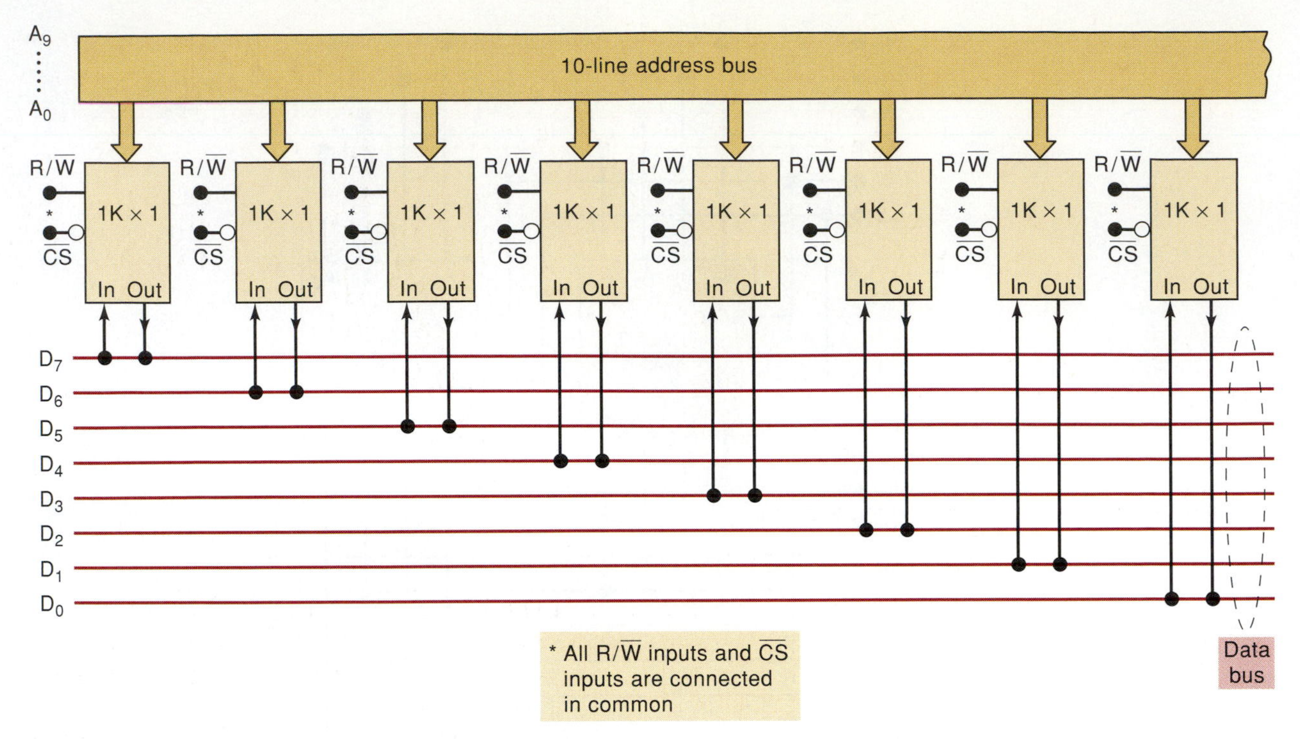

FIGURE 12-35 Eight 2125A 1K × 1 chips arranged as a 1K × 8 memory.

Expanding Capacity

Suppose that we need a memory that can store 32 four-bit words and all we have are the 16 × 4 chips. By combining two 16 × 4 chips as shown in Figure 12-36, we can produce the desired memory. Once again, examine this diagram and see what you can determine from it before reading on.

Each RAM is used to store 16 four-bit words. The four data I/O pins of each RAM are connected to a common four-line data bus. Only one of the RAM chips can be selected (enabled) at one time so that there will be no bus-contention problems. This is ensured by driving the respective $\overline{CS}$ inputs from different logic signals.

The total capacity of this memory module is 32 × 4, so there must be 32 different addresses. This requires *five* address bus lines. The upper address line A_4 is used to select one RAM or the other (via the $\overline{CS}$ inputs) as the one that will be read from or written into. The other four address lines A_0 to A_3 are used to select the one memory location out of 16 from the selected RAM chip.

To illustrate, when $A_4 = 0$, the $\overline{CS}$ of RAM-0 enables this chip for read or write. Then any address location in RAM-0 can be accessed by A_3 through A_0. The latter four address lines can range from 0000 to 1111 to select the desired location. Thus, the range of addresses representing locations in RAM-0 is

$$A_4A_3A_2A_1A_0 = 00000 \text{ to } 01111$$

Note that when $A_4 = 0$, the $\overline{CS}$ of RAM-1 is high, so that its I/O lines are disabled (Hi-Z) and cannot communicate with (give data to or take data from) the data bus.

It should be clear that when $A_4 = 1$, the roles of RAM-0 and RAM-1 are reversed. RAM-1 is now enabled, and the lines A_3 to A_0 select one of its locations. Thus, the range of addresses located in RAM-1 is

$$A_4A_3A_2A_1A_0 = 10000 \text{ to } 11111$$

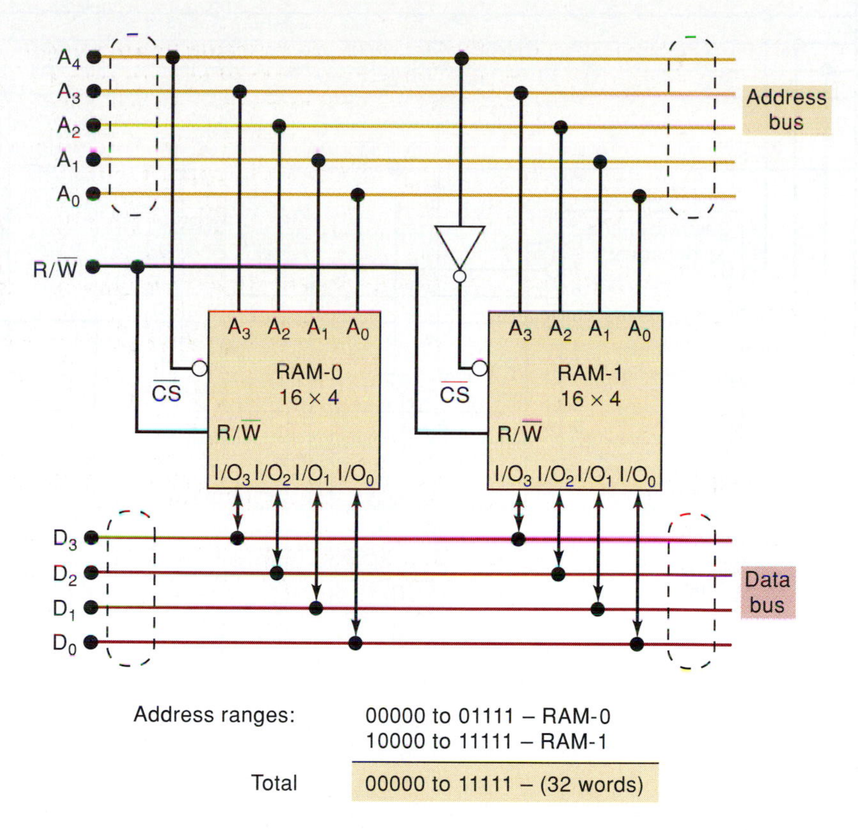

FIGURE 12-36 Combining two 16 × 4 chips for a 32 × 4 memory.

EXAMPLE 12-14

We want to combine several 2K × 8 PROMs to produce a total capacity of 8K × 8. How many PROM chips are needed? How many address bus lines are required?

Solution

Four PROM chips are required, with each one storing 2K of the 8K words. Because $8K = 8 \times 1024 = 8192 = 2^{13}$, *thirteen* address lines are needed.

The configuration for the memory of Example 12-14 is similar to the 32 × 4 memory of Figure 12-36. It is slightly more complex, however, because it requires a decoder circuit for generating the $\overline{CS}$ input signals. The complete diagram for this 8192 × 8 memory is shown in Figure 12-37(a).

The total capacity of the block of ROM is 8192 bytes. This system containing the block of memory has an address bus of 16 bits, which is typical of a small microcontroller-based system. The decoder in this system can only be enabled when A_{15} and A_{14} are LOW and E is HIGH. This means that it can only decode addresses less than 4000 hex. It is easier to understand this by looking at the memory map of Figure 12-37(b). You can see that the top two MSBs (in red) are always LOW for addresses under 4000 hex. Address lines A_{13}–A_{11} (blue font) are connected to decoder inputs C–A, respectively. These three bits are decoded and used to select one of the memory ICs. Notice in the bit map of Figure 12-37(b) that all the addresses that are contained in PROM-0 have $A_{13}, A_{12}, A_{11} = 0, 0, 0$; PROM-1 is selected when these bits have a value of 0, 0, 1; PROM-2 when 0, 1, 0; and PROM-3 when 0, 1, 1. When any PROM is selected, the address lines A_{10}–A_0 can range from all 0s to all 1s. To summarize the address scheme of this system, the top two bits are used to select this decoder,

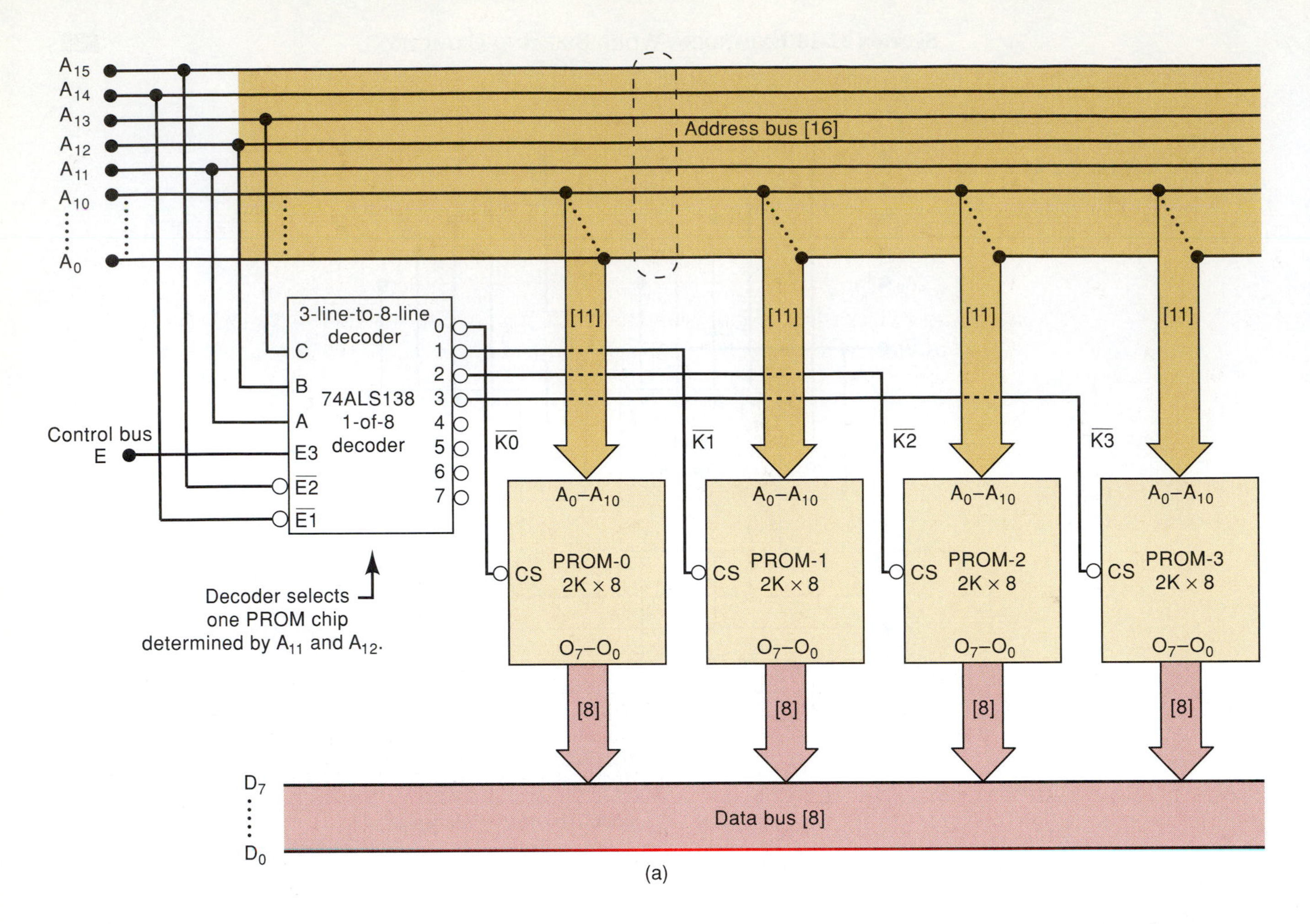

A_{15}	A_{14}	A_{13}	A_{12}	A_{11}	A_{10}	A_9	A_8	A_7	A_6	A_5	A_4	A_3	A_2	A_1	A_0	Address	System Map	
0	0	0	0	0	0	0	0	0	0	0	0	0	0	0	0	0000	PROM-0	2K
0	0	0	0	0	1	1	1	1	1	1	1	1	1	1	1	07FF		
0	0	0	0	1	0	0	0	0	0	0	0	0	0	0	0	0800	PROM-1	2K
0	0	0	0	1	1	1	1	1	1	1	1	1	1	1	1	0FFF		
0	0	0	1	0	0	0	0	0	0	0	0	0	0	0	0	1000	PROM-2	2K
0	0	0	1	0	1	1	1	1	1	1	1	1	1	1	1	17FF		
0	0	0	1	1	0	0	0	0	0	0	0	0	0	0	0	1800	PROM-3	2K
0	0	0	1	1	1	1	1	1	1	1	1	1	1	1	1	1FFF		
0	0	1	0	0	0	0	0	0	0	0	0	0	0	0	0	2000	O_4 Decoded Expansion	8K
		1	0	1													O_5	
		1	1	0													O_6	
0	0	1	1	1	1	1	1	1	1	1	1	1	1	1	1	3FFF	O_7	
0	1	0	0	0	0	0	0	0	0	0	0	0	0	0	0	4000	Available	48K
1	1	1	1	1	1	1	1	1	1	1	1	1	1	1	1	FFFF		

(b)

FIGURE 12-37 (a) Four 2K × 8 PROMs arranged to form a total capacity of 8K × 8. (b) Memory map of the full system.

the next three bits (A_{13}–A_{11}) are used to select one out of four PROM chips, and the lower 11 address lines are used to select one out of 2048 byte-sized memory locations in the enabled PROM.

When the system address of 4000 or more is on the address bus, none of the PROMs will be enabled. However, decoder outputs 4–7 can be used to enable more memory chips if we wish to expand the capacity of the memory system. The memory map on the right side of Figure 12-37(b) shows a 48K area of the system's space that is not occupied by this memory block. In order to expand into this area of the memory map, more decoding logic would be needed.

EXAMPLE 12-15

What would be needed to expand the memory of Figure 12-37 to 32K × 8? Describe what address lines are used.

Solution

A 32K capacity will require 16 of the 2K PROM chips. Four are already shown and four more can be connected to the O_4–O_7 of the existing decoder outputs. This accounts for half of the system. The other eight PROM chips can be selected by adding another 74ALS138 decoder and enabling it only when $A_{15} = 0$ and $A_{14} = 1$. This can be accomplished by connecting an inverter between A_{14} and $\overline{E_1}$ while connecting A_{15} directly to $\overline{E_2}$. The other connections are the same as in the existing decoder.

Incomplete Address Decoding

In many instances, it is necessary to use various memory devices in the same memory system. For example, consider the requirements of a digital dashboard system on an automobile. Such a system is typically implemented using a microprocessor. Consequently, we need some nonvolatile ROM to store the program instructions. We need some read/write memory to store the digits that represent the speed, RPM, gallons of fuel, and so on. Other digitized values must be stored to represent oil pressure, engine temperature, battery voltage, and so on. We also need some nonvolatile read/write storage (EEPROM) for the odometer readout because it would not be good to have this number reset to 0 or assume a random value whenever the car battery is disconnected.

Figure 12-38 shows a memory system that could be used in a microcomputer system. Notice that the ROM portion is made up of two 8K × 8 devices (PROM-0 and PROM-1). The RAM section requires a single 8K × 8 device. The EEPROM available is only a 2K × 8 device. The memory system requires a decoder to select only one device at a time. This decoder divides the entire memory space (assuming 16 address bits) into 8K address blocks. In other words, each decoder output is activated by 8192 (8K) different addresses. Notice that the upper three address lines control the decoder. The 13 lower-order address lines are tied directly to the address inputs on the memory chips. The only exception to this is the EEPROM, which has only 11 address lines for its 2-Kbyte capacity. If the address (in hex) of this EEPROM is intended to range from 6000 to 67FF, it will respond to these addresses as intended. However, the two address lines, A_{11} and A_{12}, are not involved in the decoding scheme for this chip. The decoder output ($K3$) is active for 8K addresses, but the chip that it is connected to contains only 2K locations. As a result, the EEPROM will also respond to the other 6K of addresses in this

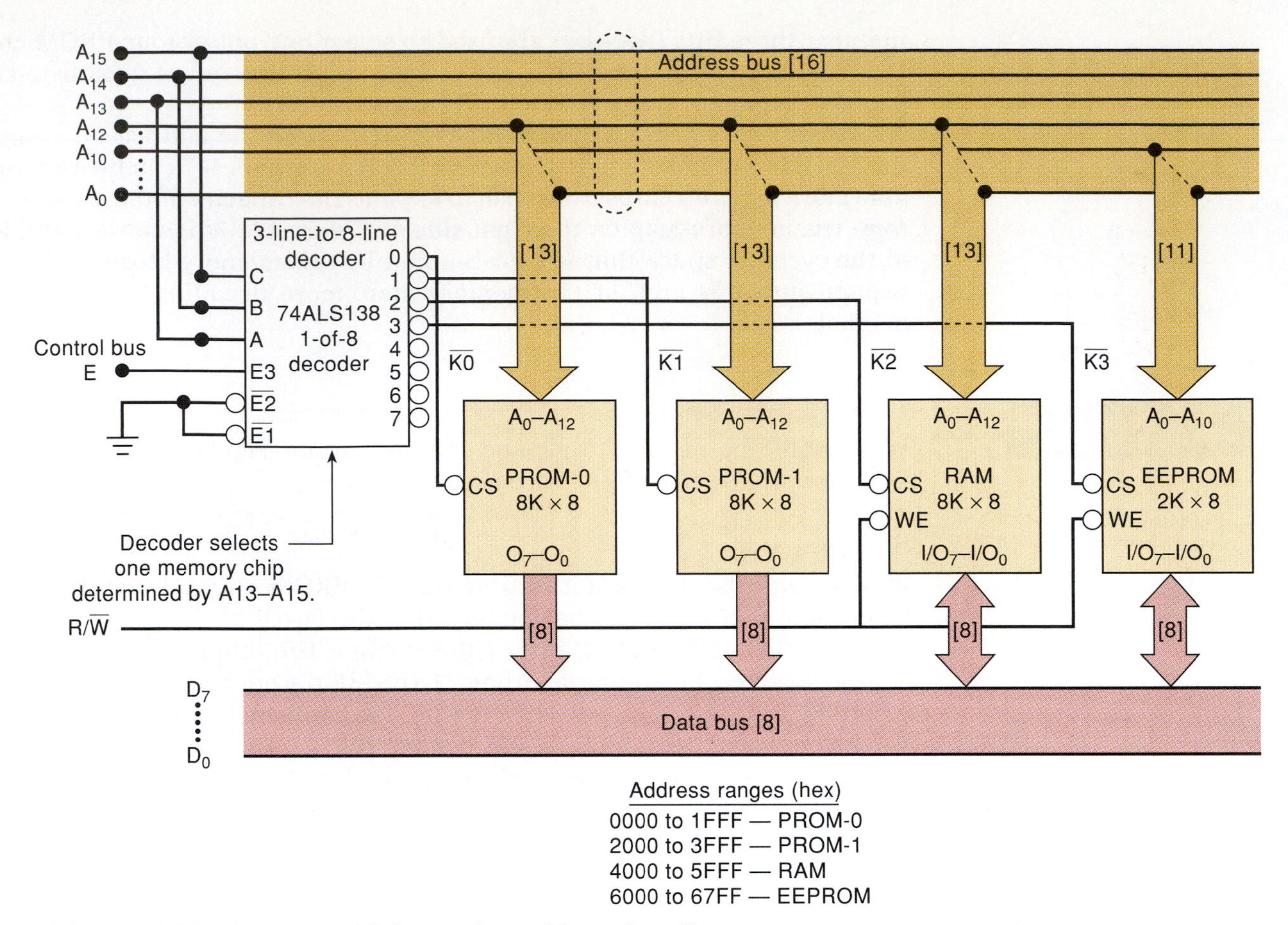

FIGURE 12-38 A system with incomplete address decoding.

decoded block of memory. The same contents of the EEPROM will also appear at addresses 6800–6FFF, 7000–77FF, and 7800–7FFF. These areas of memory that are redundantly occupied by a device due to incomplete address decoding are referred to as **memory foldback** areas. This occurs frequently in systems where there is an abundance of address space and a need to minimize decoding logic. A **memory map** of this system (see Figure 12-39)

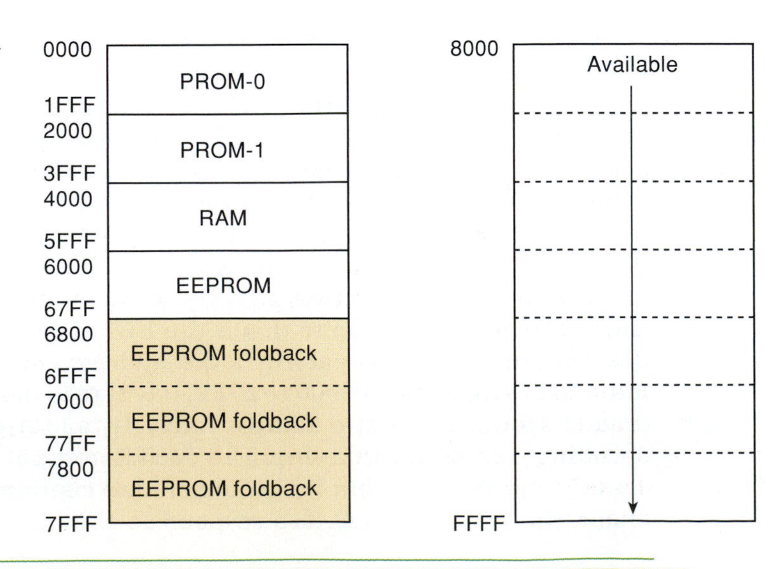

FIGURE 12-39 A memory map of a digital dashboard system.

clearly shows the addresses that each device is assigned to as well as the memory space that is available for expansion.

Combining DRAM Chips

DRAM ICs often come with word sizes of one or four bits, so it is necessary to combine several of them to form larger word size modules. Figure 12-40 shows how to combine eight TMS44100 DRAM chips to form a 4M × 8 module. Each chip has a 4M × 1 capacity.

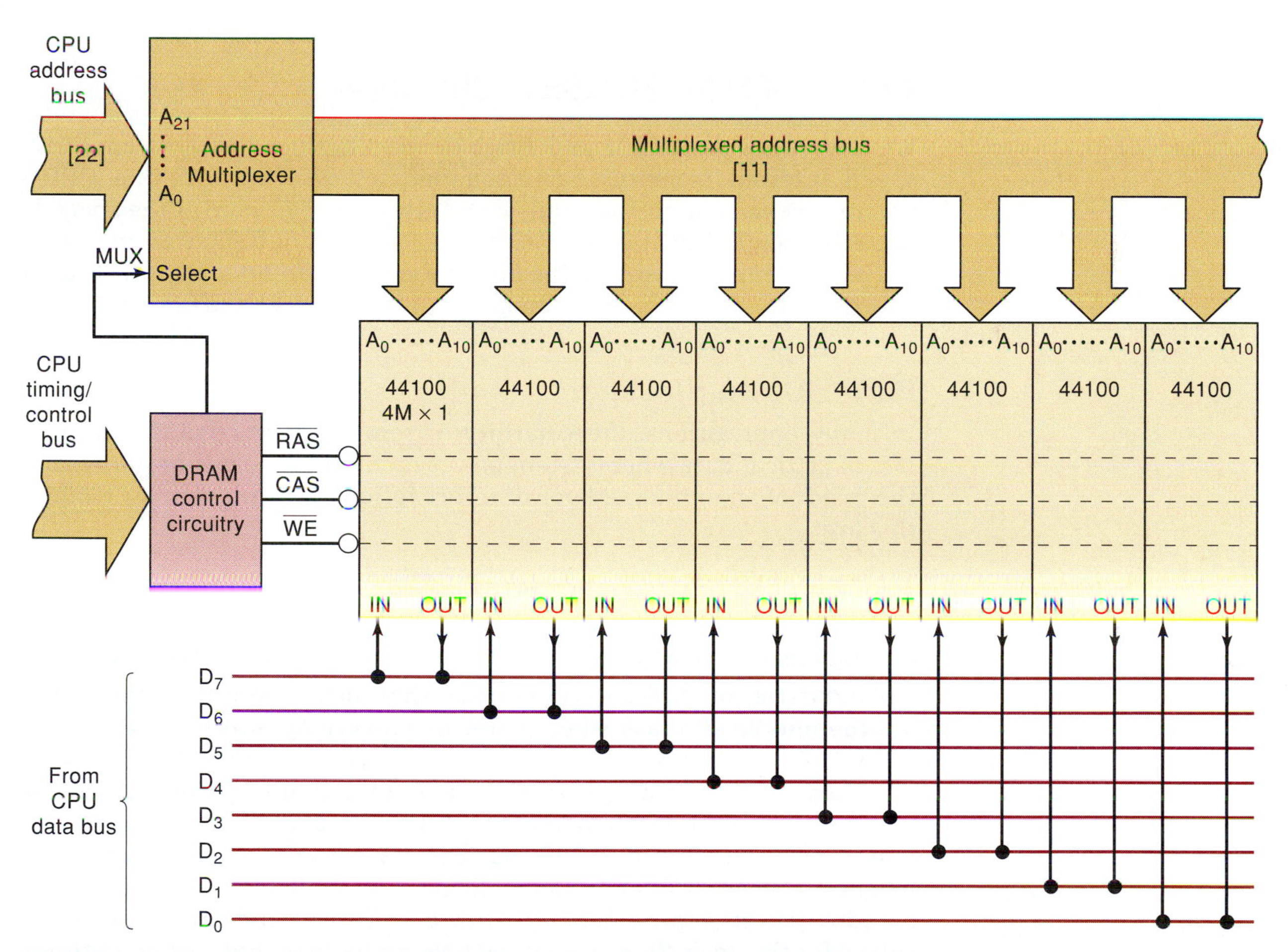

FIGURE 12-40 Eight 4M × 1 DRAM chips combined to form a 4M × 8 memory module.

There are several important points to note. First, because $4M = 2^{22}$, the TMS44100 chip has *eleven* address inputs; remember, DRAMs use multiplexed address inputs. The address multiplexer takes the 22-line CPU address bus and changes it to an 11-line address bus for the DRAM chips. Second, the $\overline{RAS}$, $\overline{CAS}$, and $\overline{WE}$ inputs of all eight chips are connected together so that all chips are activated simultaneously for each memory operation. Finally, recall that the TMS44100 has on-chip refresh control circuitry, so there is no need for an external refresh counter.

REVIEW QUESTIONS

1. The MCM6209C is a 64K × 4 static-RAM chip. How many of these chips are needed to form a 1M × 4 module?
2. How many are needed for a 64K × 16 module?
3. *True or false:* When memory chips are combined to form a module with a larger word size or capacity, the *CS* inputs of each chip are always connected together.
4. *True or false:* When memory chips are combined for a larger capacity, each chip is connected to the same data bus lines.

12-19 SPECIAL MEMORY FUNCTIONS

We have seen that RAM and ROM devices are used as a computer's high-speed internal memory that communicates directly with the CPU (e.g., microprocessor). In this section, we briefly describe some of the special functions that semiconductor memory devices perform in computers and other digital equipment and systems. The discussion is not intended to provide details of how these functions are implemented but to introduce the basic ideas.

Power-Down Storage

In many applications, the volatility of semiconductor RAM can mean the loss of critical data when system power is shut down—either purposely or as the result of an unplanned power interruption. Just two of many examples of this are:

1. Critical operating parameters for graphics terminals, intelligent terminals, and printers. These changeable parameters determine the operating modes and attributes that will be in effect upon power-up.
2. Industrial process control systems that must never "lose their place" in the middle of a task when power unexpectedly fails.

There are several approaches to providing storage of critical data in power-down situations. In one method, all critical data during normal system operation are stored in RAM that can be powered from backup batteries whenever power is interrupted. Some CMOS RAM chips have very low standby power requirements (as low as 0.5 mW) and are particularly well suited for this task. Some CMOS SRAMs actually include a small lithium battery right on the chip. Of course, even with their low power consumption, these CMOS RAMs will eventually drain the batteries if power is out for prolonged periods, and data will be lost.

Another approach stores all critical system data in nonvolatile flash memory. This approach has the advantage of not requiring backup battery power, and so it presents no risk of data loss even for long power outages. Flash memory, however, cannot have its data changed as easily as static RAM. Recall that with a flash chip, we cannot erase and write to one or two bytes; it must be erased a sector at a time. This requires the CPU to have to rewrite a large block of data even when only a few bytes need to be changed.

In a third approach, the CPU stores all data in high-speed, volatile RAM during normal system operation. On power-down, the CPU executes a short power-down program (from ROM) that transfers critical data from the system

RAM into either battery-backup CMOS RAM or nonvolatile flash memory. This requires a special circuit that senses the onset of a power interruption and sends a signal to the CPU to tell it to begin executing the power-down sequence.

In any case, when power is turned back on, the CPU executes a power-up program (from ROM) that transfers the critical data from the backup storage memory to the system RAM so that the system can resume operation where it left off when power was interrupted.

Cache Memory

Computers and other digital systems may have thousands or millions of bytes of internal memory (RAM and ROM) that store programs and data that the CPU needs during normal operation. Normally, this would require that all of the internal memory have an operating speed comparable to that of the CPU in order to achieve maximum system operation. In many systems, it is not economical to use high-speed memory devices for all of the internal memory. Instead, system designers use a block of high-speed **cache** memory. This cache memory block is the only block that communicates directly with the CPU at high speed; program instructions and data are transferred from the slower, cheaper internal memory to the cache memory when they are needed by the CPU. The success of cache memory depends on many complex factors, and some systems will not benefit from using cache memory.

Modern PC CPUs have a small (8–64 Kbytes) internal memory cache. This is referred to as a level 1 or L1 cache. The chip set of most computer systems also controls an external bank of static RAM (SRAM) that implements a level 2 or L2 cache (64 Kbytes to 2 Mbytes). The cache memory is filled with a sequence of instruction words from the system memory. The CPU (many operating at over 2 GHZ clock rates) can access the cache contents at very high speed. However, when the CPU needs a piece of information that is not currently in either the L1 or the L2 cache (i.e., a cache miss), it must go out to the slow system DRAM to get it. This transfer must occur at the much slower *bus clock rate*, which may be from 66 MHz to 800 MHz depending on your system. In addition to the slower clock rate, the DRAM access time (latency) is much greater.

The specification of 7-2-2-2 or 5-1-1-1 for a memory system refers to the number of *bus* clock cycles necessary to transfer a burst of four 64-bit words from DRAM to the L2 cache. The first access takes the most time due to latency associated with RAS/CAS cycles. Subsequent data are clocked out in a burst that takes much less time. For example, the 7-2-2-2 system would require 7 clocks to obtain the first 64-bit word, and each of the next three 64-bit words would require 2 clock cycles each. A total of 13 clock cycles are necessary to get the four words out of memory.

First-In, First-Out Memory (FIFO)

In **FIFO** memory systems, data that are written into the RAM storage area are read out in the same order that they were written in. In other words, the first word written into the memory block is the first word that is read out of the memory block: hence the name FIFO. This idea is illustrated in Figure 12-41.

Figure 12-41(a) shows the sequence of writing three data bytes into the memory block. Note that as each new byte is written into location 1, the other bytes move to the next location. Figure 12-41(b) shows the sequence of reading the data out of the FIFO block. The first byte read is the same as the first byte that was written, and so on. The FIFO operation is controlled by special *address pointer registers* that keep track of where data are to be written and the location from which they are to be read.

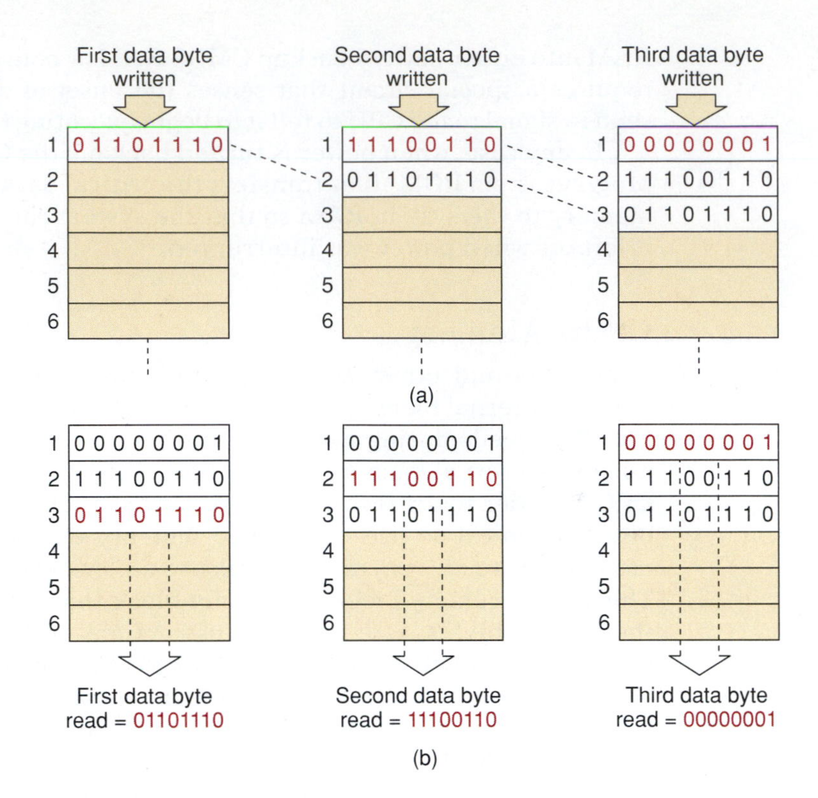

FIGURE 12-41 In FIFO, data values are read out of memory (b) in the same order that they were written into memory (a).

A FIFO is useful as a **data-rate buffer** between systems that transfer data at widely different rates. One example is the transfer of data from a computer to a printer. The computer sends character data to the printer at a very high rate, say, one byte every 10 μs. These data fill up a FIFO memory in the printer. The printer then reads out the data from the FIFO at a much slower rate, say, one byte every 5 ms, and prints out the corresponding characters in the same order as sent by the computer.

A FIFO can also be used as a data-rate buffer between a slow device, such as a keyboard, and a high-speed computer. Here, the FIFO accepts keyboard data at the slow and asynchronous rate of human fingers and stores them. The computer can then read all of the recently stored keystrokes very quickly at a convenient point in its program. In this way, the computer can perform other tasks while the FIFO is slowly being filled with data.

Circular Buffers

Data rate buffers (FIFOs) are often referred to as **linear buffers.** As soon as all the locations in the buffer are full, no more entries are made until the buffer is emptied. This way, none of the "old" information is lost. A similar memory system is called a **circular buffer.** These memory systems are used to store the last n values entered, where n is the number of memory locations in the buffer. Each time a new value is written to a circular buffer, it overwrites (replaces) the oldest value. Circular buffers are addressed by a MOD-n address counter. Consequently, when the highest address is reached, the address counter will "wrap around" and the next location will be the lowest address. As you recall from Chapter 11, digital filtering and other DSP operations perform calculations using a group of recent samples. Special hardware included in a DSP allows easy implementation of circular buffers in memory.

REVIEW QUESTIONS

1. What are the various ways to handle the possible loss of critical data when power is interrupted?
2. What is the principal reason for using a cache memory?
3. What does *FIFO* mean?
4. What is a data-rate buffer?
5. How does a circular buffer differ from a linear buffer?

12-20 TROUBLESHOOTING RAM SYSTEMS

All computers use RAM. Many general-purpose computers and most special-purpose computers (such as microprocessor-based controllers and process-control computers) also use some form of ROM. Each RAM and ROM IC that is part of a computer's internal memory typically contains thousands of memory cells. A single faulty memory cell can cause a complete system failure (commonly referred to as a "system crash") or, at the least, unreliable system operation. The testing and troubleshooting of memory systems involves the use of techniques that are not often used on other parts of the digital system. Because memory consists of thousands of identical circuits acting as storage locations, any tests of its operation must involve checking to see exactly which locations are working and which are not. Then, by looking at the pattern of good and bad locations along with the organization of the memory circuitry, one can determine the possible causes of the memory malfunction. The problem typically can be traced to a bad memory IC; a bad decoder IC, logic gate, or signal buffer; or a problem in the circuit connections (i.e., shorts or open connections).

Because RAM must be written to and read from, testing RAM is generally more complex than testing ROM. In this section, we will look at some common procedures for testing the RAM portion of memory and interpreting the test results. We will examine ROM testing in the next section.

Know the Operation

The RAM memory system shown in Figure 12-42 will be used in our examples. As we emphasized in earlier discussions, successful troubleshooting of a relatively complex circuit or system begins with a thorough knowledge of its operation. Before we can discuss the testing of this RAM system, we should first analyze it carefully so that we fully understand its operation.

The total capacity is $4K \times 8$ and is made up of four $1K \times 8$ RAM modules. A module may be just a single IC, or it may consist of several ICs (e.g., two $1K \times 4$ chips). Each module is connected to the CPU through the address and data buses and through the $R/\overline{W}$ control line. The modules have common I/O data lines. During a read operation, these lines become data output lines through which the selected module places its data on the bus for the CPU to read. During a write operation, these lines act as input lines for the memory to accept CPU-generated data from the data bus for writing into the selected location.

The 74ALS138 decoder and the four-input OR gate combine to decode the six high-order address lines to generate the chip select signals $\overline{K0}$, $\overline{K1}$, $\overline{K2}$, and $\overline{K3}$. These signals enable a specific RAM module for a read or a write operation. The INVERTER is used to invert the CPU-generated Enable signal (E)

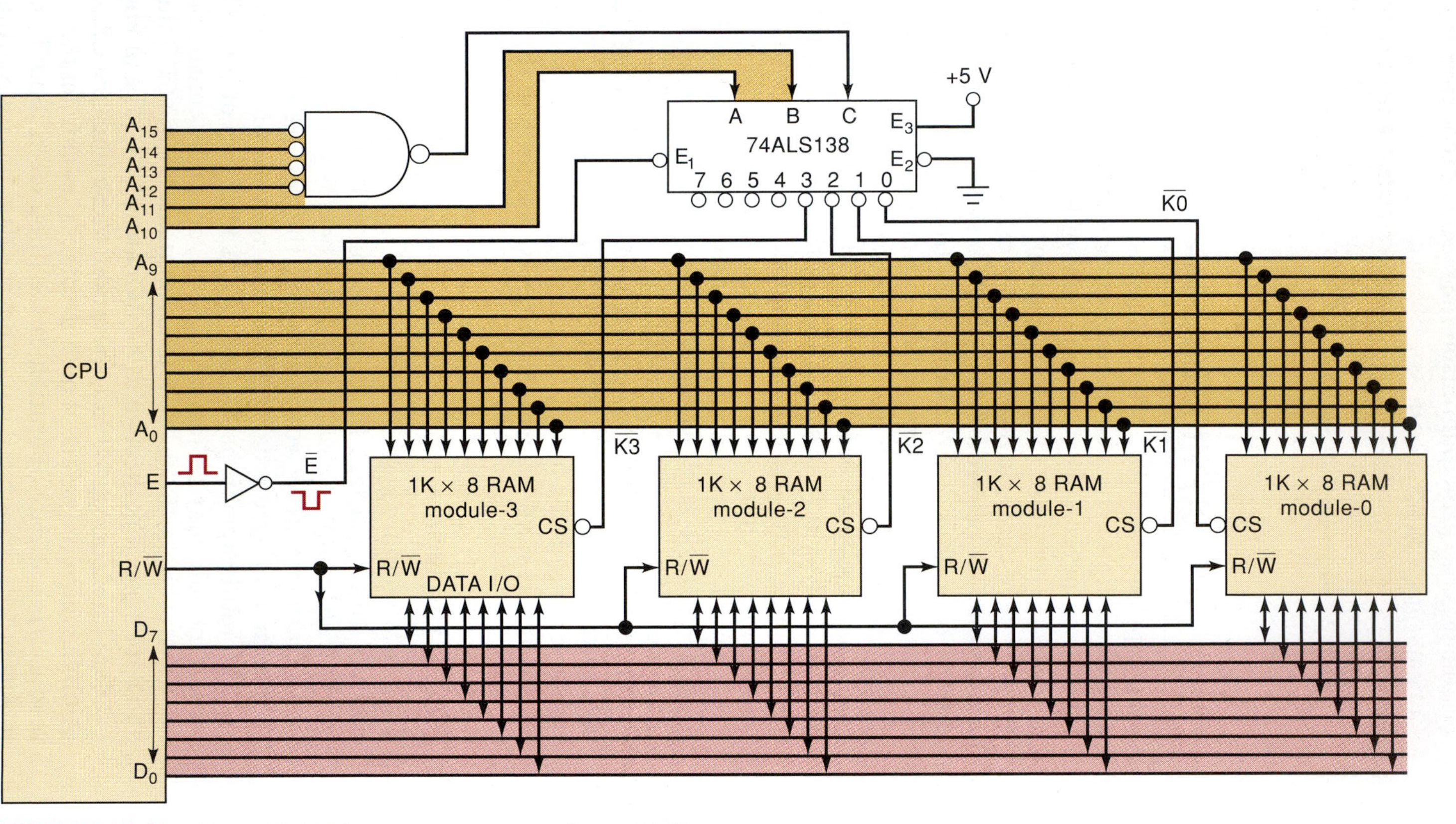

FIGURE 12-42 4K × 8 RAM memory connected to a CPU.

so that the decoder is enabled only while E is HIGH. The E pulse occurs only after allowing enough time for the address lines to stabilize following the application of a new address on the address bus. E will be LOW while the address and $R/\overline{W}$ lines are changing; this prevents the occurrence of decoder output glitches that could erroneously activate a memory chip and possibly cause invalid data to be stored.

Each RAM module has its address inputs connected to the CPU address bus lines A_0 through A_9. The high-order address lines A_{10} through A_{15} select one of the RAM modules. The selected module decodes address lines A_0 through A_9 to find the word location that is being addressed. The following examples will show how to determine the addresses that correspond to each module.

EXAMPLE 12-16

Assume that the CPU is performing a read operation from address 06A3 (hex). Which RAM module, if any, is being read from?

Solution

First write out the address in binary.

A_{15}	A_{14}	A_{13}	A_{12}	A_{11}	A_{10}	A_9	A_8	A_7	A_6	A_5	A_4	A_3	A_2	A_1	A_0
0	0	0	0	0	1	1	0	1	0	1	0	0	0	1	1

You should be able to verify that levels A_{15} to A_{10} will activate decoder output $\overline{K1}$ to select RAM module-1. This module internally decodes the address lines A_9 to A_0 to select the location whose data are to be placed on the data bus.

EXAMPLE 12-17

Which RAM module will have data written into it when the CPU executes a write operation to address 1C65?

Solution

Writing out the address in binary, we can see that $A_{12} = 1$. This produces a HIGH out of the OR gate and at the C input of the decoder. With $A_{11} = A_{10} = 1$, the decoder inputs are 111, which activates output 7. Outputs $\overline{K0}$ to $\overline{K3}$ will be inactive, and so none of the RAM modules will be enabled. In other words, the data placed on the data bus by the CPU will not be accepted by any of the RAMs.

EXAMPLE 12-18

Determine the range of addresses for each module in Figure 12-42.

Solution

Each module stores 1024 eight-bit words. To determine the addresses of the words stored in any module, we start by determining the address bus conditions that activate that module's chip select input. For example, module-3 will be selected when decoder input $\overline{K3}$ is LOW (Figure 12-43). $\overline{K3}$ will be LOW for CBA = 011. Working back to the CPU address lines A_{15} to A_{10}, we see that module-3 will be enabled when the following address is placed on the address bus:

A_{15}	A_{14}	A_{13}	A_{12}	A_{11}	A_{10}	A_9	A_8	A_7	A_6	A_5	A_4	A_3	A_2	A_1	A_0
0	0	0	0	1	1	x	x	x	x	x	x	x	x	x	x

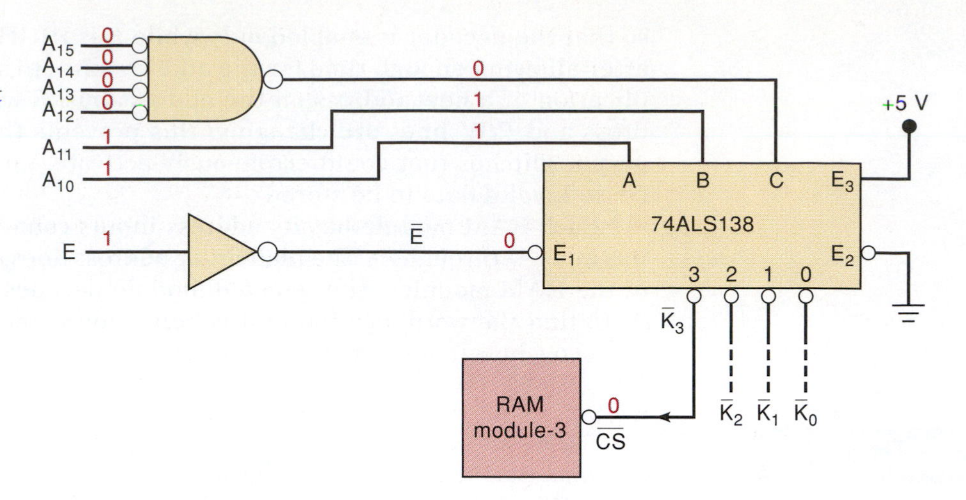

FIGURE 12-43 Example 12-18, showing address bus conditions needed to select RAM module-3.

The *x*'s under A_9 through A_0 indicate don't care because these address lines are not used by the decoder to select module-3. A_9 to A_0 can be any combination ranging from 0000000000 to 1111111111, depending on which word in module-3 is being accessed. Thus, the complete range of addresses for module-3 is determined by using all 0s, and then all 1s for the *x*'s.

A_{15}	A_{14}	A_{13}	A_{12}	A_{11}	A_{10}	A_9	A_8	A_7	A_6	A_5	A_4	A_3	A_2	A_1	A_0		
0	0	0	0	1	1	0	0	0	0	0	0	0	0	0	0	→	$0C00_{16}$
0	0	0	0	1	1	1	1	1	1	1	1	1	1	1	1	→	$0FFF_{16}$

Finally, this gives us 0C00 to 0FFF as the range of hex addresses stored in module-3. When the CPU places any address in this range onto the address bus, only module-3 will be enabled for either a read or a write, depending on the state of $R/\overline{W}$.

A similar analysis can be used to determine the address ranges for each of the other RAM modules. The results are as follows:

- Module-0: 0000–03FF
- Module-1: 0400–07FF
- Module-2: 0800–0BFF
- Module-3: 0C00–0FFF

Note that the four modules combine for a total address range of 0000 to 0FFF.

Testing the Decoding Logic

In some situations, the decoding logic portion of the RAM circuit (Figure 12-43) can be tested using the various techniques that we have applied to combinatorial circuits. It can be tested by applying signals to the six most significant address lines and E and by monitoring the decoder outputs. To do this, it must be possible to disconnect the CPU easily from these signal lines. If the CPU is a microprocessor chip in a socket, it can simply be removed from its socket.

Once the CPU is disconnected, you can supply the A_{10}–A_{15} and E signals from an external test circuit to perform a static test, using manually operated switches for each signal, or a dynamic test, using some type of counter to cycle through the various address codes. With these test signals applied, the decoder output lines can be checked for the proper response. Standard signal-tracing techniques can be used to isolate any faults in the decoding logic.

If you do not have access to the system address lines or do not have a convenient way of generating the static logic signals, it is often possible to force the system to generate a sequence of addresses. Most computer systems used for development have a program stored in a ROM that allows the user to display and change the contents of any memory location. Whenever the computer accesses a memory location, the proper address must be placed on the bus, which should cause the decoder output to go low, even if it is for a short time. Enter the following command to the computer:

```
Display from 0400 to 07FF
```

Then place the logic probe on the $\overline{K1}$ output. The logic probe should show pulses during the time the data values are being displayed.

EXAMPLE 12-19

A dynamic test is performed on the decoding logic of Figure 12-43 by keeping $E = 1$ and connecting the outputs of a six-bit counter to the address inputs A_{10} through A_{15}. The decoder outputs are monitored as the counter repetitively cycles through all six-bit codes. A logic probe check on the decoder outputs shows pulses at $\overline{K1}$ and $\overline{K3}$, but shows $\overline{K0}$ and $\overline{K2}$ remaining HIGH. What are the most probable faults?

Solution

It is possible, but highly unlikely, that $\overline{K0}$ and $\overline{K2}$ could both be stuck HIGH due to either an internal or an external short to V_{CC}. A more likely fault is an open between A_{10} and the A input of the decoder because this would act as a logic HIGH and prevent any even-numbered decoder output from being activated. It is also possible that the decoder's A input is shorted to V_{CC}, but this is also unlikely because this short would have probably affected the operation of the counter that is supplying the address inputs.

Testing the Complete RAM System

Testing and troubleshooting the decoding logic will not reveal problems with the memory chips and their connections to the CPU buses. The most common methods for testing the operation of the *complete* RAM system involve writing known patterns of 1s and 0s to each memory location and reading them back to verify that the location has stored the pattern properly. While many different patterns can be used, one of the most widely used is the "checkerboard pattern." In this pattern, 1s and 0s are alternated as in 01010101. Once all locations have been tested using this pattern, the pattern is reversed (i.e., 10101010), and each location is tested again. Note that this sequence of tests will check each cell for the ability to store and read both a 1 and a 0. Because it alternates 1s and 0s, the checkerboard pattern will also detect any interactions or shorts between adjacent cells. Many other patterns can be used to detect various failure modes within RAM chips.

No memory test can catch all possible RAM faults with 100 percent accuracy, even though it may show that each cell can store and read a 0 or a 1. Some faulty RAMs can be pattern-sensitive. For instance, a RAM may be able to store and read 01010101 and 10101010, but it may fail when 11100011 is stored. Even for a small RAM system, it would take a prohibitively long time to try storing and reading every possible pattern in each location. For this reason, if a RAM system passes the checkerboard test, you can conclude that it is *probably* good; if it fails the test, then it *definitely* contains a fault.

Manually testing thousands of RAM locations by storing and reading checkerboard patterns would take hundreds of hours and is obviously not feasible. RAM pattern testing is usually done automatically either by having the CPU run a memory test program or by connecting a special test instrument to the RAM system buses in place of the CPU. In fact, in many computers and microprocessor-based equipment, the CPU will automatically run a memory test program every time it is powered up; this is called a **power-up self-test.** The self-test routine (we will call it SELF-TEST) is stored in ROM, and it is executed whenever the system is turned on or when the operator requests it from the keyboard. As the CPU executes SELF-TEST, it will write test patterns to and read test patterns from each RAM location and will display some type of message to the user. It may be something as simple as an LED to indicate faulty memory, or it may be a descriptive message printed on the screen or printer. Typical messages might be:

```
RAM module-3 test OK
ALL RAM working properly
Location 027F faulty in bit positions 6 and 7
```

With messages like these and a knowledge of the RAM system operation, the troubleshooter can determine what additional action is needed to isolate the fault.

REVIEW QUESTIONS

1. What is *E*'s function in the RAM circuit of Figure 12-42?
2. What is the checkerboard test? Why is it used?
3. What is a power-up self-test?

12-21 TESTING ROM

The ROM circuitry in a computer is very similar to the RAM circuitry (compare Figures 12-37 and 12-42). The ROM decoding logic can be tested in the same manner described in the preceding section for the RAM system. The ROM chips, however, must be tested differently from RAM chips because we cannot write patterns into ROM and read them back as we can for RAM. Several methods are used to check the contents of a ROM IC.

In one approach, the ROM is placed in a socket in a special test instrument that is typically microprocessor-controlled. The special test instrument can be programmed to read every location in the test ROM and print out a listing of the contents of each location. The listing can then be compared with what the ROM is supposed to contain. Except for low-capacity ROM chips, this process can be very time-consuming.

In a more efficient approach, the test instrument has the correct data stored in its own *reference* ROM chip. The test instrument is then programmed to read the contents of each location of the test ROM and compare it with the contents of the reference ROM. This approach, of course, requires the availability of a preprogrammed reference ROM.

A third approach uses a **checksum,** a special code placed in the last one or two locations of the ROM chip when it is programmed. This code is derived by adding up the data words to be stored in all of the ROM locations (excluding

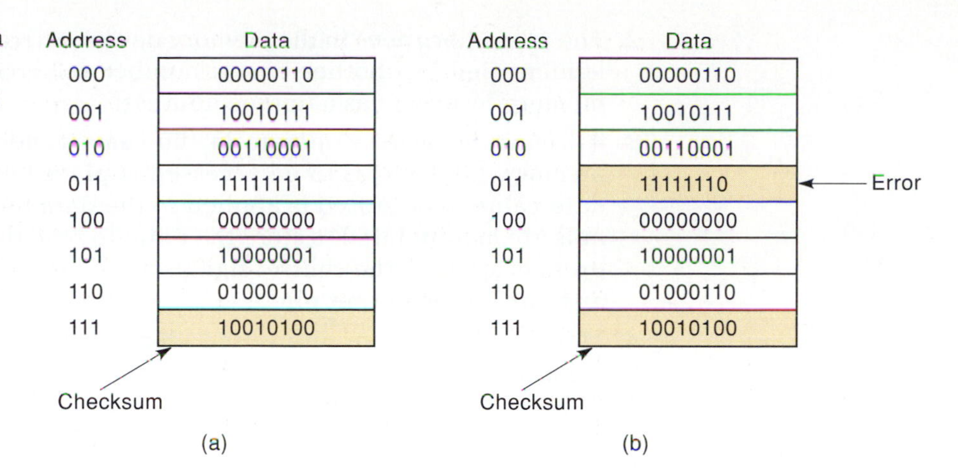

FIGURE 12-44 Checksum method for an 8 × 8 ROM: (a) ROM with correct data; (b) ROM with error in its data.

those containing the checksum). As the test instrument reads the data from each test ROM location, it will add them up and develop its own checksum. It then compares its calculated checksum with that stored in the last ROM locations, and they should agree. If so, there is a high probability that the ROM is good (there is a very small chance that a combination of errors in the test ROM data could still produce the same checksum value). If they do not agree, then there is a definite problem in the test ROM.

The checksum idea is illustrated in Figure 12-44(a) for a very small ROM. The data word stored in the last address is the eight-bit sum of the other seven data words (ignoring carries from the MSB). When this ROM is programmed, the checksum is placed in the last location. Figure 12-44(b) shows the data that might actually be read from a faulty ROM that was originally programmed with the data in Figure 12-44(a). Note the error in the word at address 011. As the test instrument reads the data from each location of the faulty ROM, it calculates its own checksum from those data. Because of the error, the calculated checksum will be 10010011. When the test instrument compares this with the checksum value stored at ROM location 111, it sees that they disagree, and a ROM error is indicated. Of course, the exact location of the error cannot be determined.

The checksum method can also be used by a computer or microprocessor-based equipment during an automatic power-up self-test to check out the contents of the system ROMs. Again, as in the self-test used for RAM, the CPU would execute a program on power-up that would do a checksum test on each ROM chip and would print out some type of status message. The self-test program itself will be located in a ROM, and so any error in that ROM could prevent the successful execution of the checksum tests.

REVIEW QUESTIONS

1. What is a checksum? What is its purpose?

SUMMARY

1. All memory devices store binary logic levels (1s and 0s) in an array structure. The size of each binary word (number of bits) that is stored varies depending on the memory device. These binary values are referred to as *data*.

2. The place (location) in the memory device where any data value is stored is identified by another binary number referred to as an *address*. Each memory location has a unique address.
3. All memory devices operate in the same general way. To write data in memory, the address to be accessed is placed on the address input, the data value to be stored is applied to the data inputs, and the control signals are manipulated to store the data. To read data from memory, the address is applied, the control signals are manipulated, and the data value appears on the output pins.
4. Memory devices are often used along with a microprocessor CPU that generates the addresses and control signals and either provides the data to be stored or uses the data from the memory. Reading and writing are *always* done from the CPU's perspective. Writing puts data into the memory, and reading gets data out of the memory.
5. Most read-only memories (ROMs) have data entered one time, and from then on their contents do not change. This storage process is called *programming*. They do not lose their data when power is removed from the device. MROMs are programmed during the manufacturing process. PROMs are programmed one time by the user. EPROMs are just like PROMs but can be erased using UV light. EEPROMs and flash memory devices are electrically erasable and can have their contents altered after programming. CD ROMs are used for mass storage of information that does not need to change.
6. Random access memory (RAM) is a generic term given to devices that can have data easily stored and retrieved. Data are retained in a RAM device only as long as power is applied.
7. Static RAM (SRAM) uses storage elements that are basically latch circuits. Once the data are stored, they will remain unchanged as long as power is applied to the chip. Static RAM is easier to use but more expensive per bit and consumes more power than dynamic RAM.
8. Dynamic RAM (DRAM) uses capacitors to store data by charging or discharging them. The simplicity of the storage cell allows DRAMs to store a great deal of data. Because the charge on the capacitors must be refreshed regularly, DRAMs are more complicated to use than SRAMs. Extra circuitry is often added to DRAM systems to control the reading, writing, and refreshing cycles. On many new devices, these features are being integrated into the DRAM chip itself. The goal of DRAM technology is to put more bits on a smaller piece of silicon so that it consumes less power and responds faster.
9. Memory systems require a wide variety of different configurations. Memory chips can be combined to implement any desired configuration, whether your system needs more bits per location or more total word capacity. All of the various types of ROM and RAM can be combined within the same memory system.

IMPORTANT TERMS

main memory
auxiliary memory
memory cell
memory word
byte
capacity
density
address
read operation
write operation
access time
volatile memory
random-access memory (RAM)

sequential-access memory (SAM)
read/write memory (RWM)
read-only memory (ROM)
static RAM (SRAM)
dynamic RAM
address bus
data bus
control bus
programming
chip select
power-down
fusible link
electrically erasable PROM (EEPROM)
flash memory
bootstrap program
refresh
JEDEC
address multiplexing
strobing
row address strobe (*RAS*)
column address strobe (*CAS*)
latency
$\overline{RAS}$-only refresh
refresh counter
DRAM controller
memory foldback
memory map
cache
FIFO
data-rate buffer
linear buffer
circular buffer
power-up self-test
checksum

PROBLEMS

SECTIONS 12-1 TO 12-3

B 12-1.* A certain memory has a capacity of 16K × 32. How many words does it store? What is the number of bits per word? How many memory cells does it contain?

B 12-2. How many different addresses are required by the memory of Problem 12-1?

B 12-3.* What is the capacity of a memory that has 16 address inputs, four data inputs, and four data outputs?

B 12-4. A certain memory stores 8K 16-bit words. How many data input and data output lines does it have? How many address lines does it have? What is its capacity in bytes?

DRILL QUESTIONS

12-5. Define each of the following terms.

B
(a) RAM
(b) RWM
(c) ROM
(d) Internal memory
(e) Auxiliary memory
(f) Capacity
(g) Volatile
(h) Density
(i) Read
(j) Write

12-6. (a) What are the three buses in a computer memory system?

B
(b) Which bus is used by the CPU to select the memory location?
(c) Which bus is used to carry data from memory to the CPU during a read operation?
(d) What is the source of data on the data bus during a write operation?

*Answers to problems marked with an asterisk can be found in the back of the text.

SECTIONS 12-4 AND 12-5

12-7.* Refer to Figure 12-6. Determine the data outputs for each of the following input conditions.

B (a) $[A] = 1011$; $CS = 1$

(b) $[A] = 0111$; $CS = 0$

B 12-8. Refer to Figure 12-7.

(a) What register is enabled by input address 1011?

(b) What input address code selects register 4?

B 12-9.* A certain ROM has a capacity of 16K × 4 and an internal structure like that shown in Figure 12-7.

(a) How many registers are in the array?

(b) How many bits are there per register?

(c) What size decoders does it require?

DRILL QUESTION

B 12-10. (a) *True or false:* ROMs cannot be erased.

(b) What is meant by *programming* or *burning* a ROM?

(c) Define a ROM's access time.

(d) How many data inputs, data outputs, and address inputs are needed for a 1024 × 4 ROM?

(e) What is the function of the decoders on a ROM chip?

SECTION 12-6

C, D 12-11.* Figure 12-45 shows how data from a ROM can be transferred to an external register. The ROM has the following timing parameters: $t_{ACC} =$ 250 ns and t_{OE} = 120 ns. Assume that the new address inputs have been applied to the ROM 500 ns before the occurrence of the TRANSFER pulse. Determine the minimum duration of the TRANSFER pulse for reliable transfer of data.

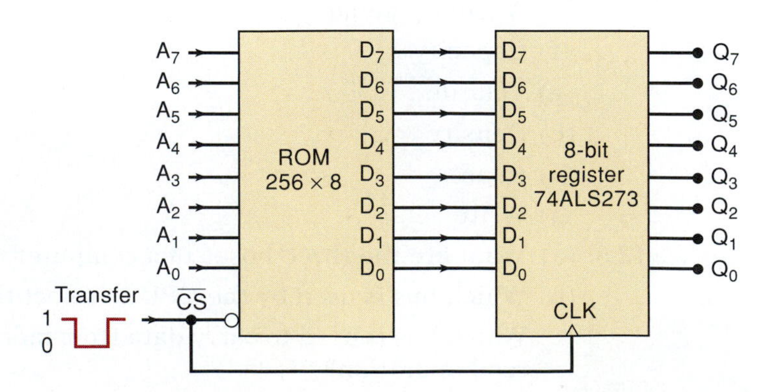

FIGURE 12-45 Problem 12-11.

C, D 12-12. Repeat Problem 12-11 if the address inputs are changed 70 ns prior to the TRANSFER pulse.

SECTIONS 12-7 AND 12-8

B 12-13. **DRILL QUESTION**

For each item below, indicate the type of memory being described: MROM, PROM, EPROM, EEPROM, flash. Some items will correspond to more than one memory type.

(a) Can be programmed by the user but cannot be erased.

(b) Is programmed by the manufacturer.

(c) Is volatile.

(d) Can be erased and reprogrammed over and over.

(e) Individual words can be erased and rewritten.

(f) Is erased with UV light.

(g) Is erased electrically.

(h) Uses fusible links.

(i) Can be erased in bulk or in sectors of 512 bytes.

(j) Does not have to be removed from the system to be erased and reprogrammed.

(k) Requires a special supply voltage for reprogramming.

(l) Erase time is about 15 to 20 min.

B 12-14. Which transistors in Figure 12-9 will be conducting when $A_1 = A_0 = 1$ and $\overline{EN} = 0$?

12-15.* Change the MROM connections in Figure 12-9 so that the MROM stores the function $y = 3x + 5$.

D 12-16. Figure 12-46 shows a simple circuit for manually programming a 2732 EPROM. Each EPROM data pin is connected to a switch that can be set at a 1 or a 0 level. The address inputs are driven by a 12-bit counter. The 50-ms programming pulse comes from a one-shot each time the PROGRAM push button is actuated.

(a) Explain how this circuit can be used to program the EPROM memory locations sequentially with the desired data.

FIGURE 12-46 Problem 12-16.

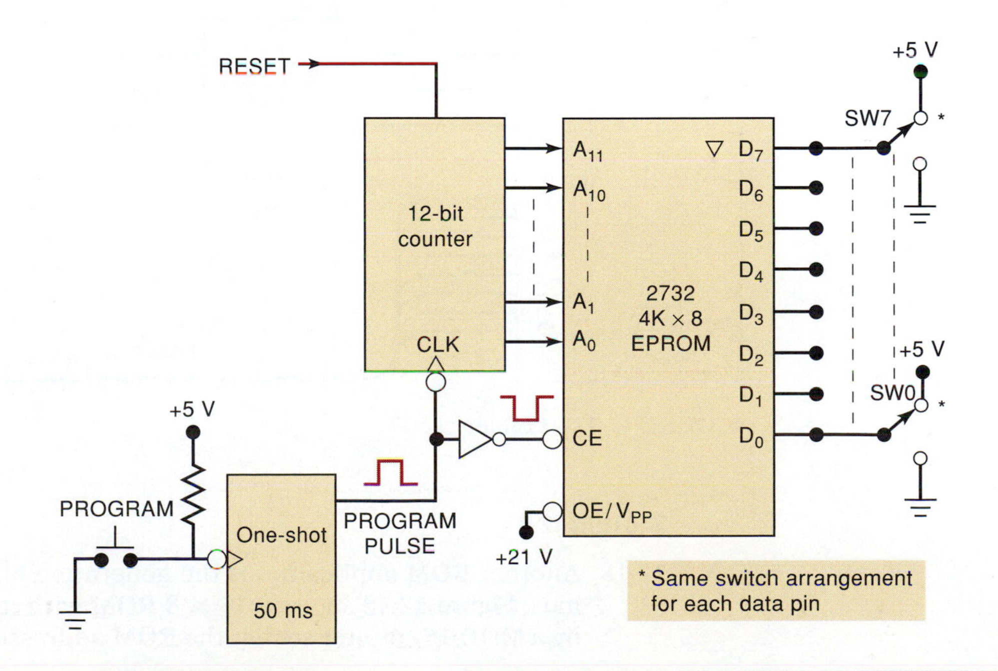

(b) Show how 74293s and a 74121 can be used to implement this circuit.

(c) Should switch bounce have any effect on the circuit operation?

N 12-17.* Figure 12-47 shows a 28F256A flash memory chip connected to a CPU over a data bus and an address bus. The CPU can write to or read from the flash memory array by sending the desired memory address and generating the appropriate control signals to the chip [Figure 12-15(b)]. The CPU can also write to the chip's command register (Figure 12-16) by generating the appropriate control signals and sending the desired command code over the data bus. For this latter operation, the CPU does not have to send a specific memory address to the chip; in other words, the address lines are don't-cares.

(a) Consider the following sequence of CPU operations. Determine what will have happened to the flash memory at the completion of the sequence. Assume that the command register is holding 00_{16}.

1. The CPU places 20_{16} on the data bus and pulses $\overline{CE}$ and $\overline{WE}$ LOW while holding $\overline{OE}$ HIGH. The address bus is at 0000_{16}.
2. The CPU repeats step 1.

(b) After the sequence above has been executed, the CPU executes the following sequence. Determine what this does to the flash memory chip.

1. The CPU places 40_{16} on the data bus and pulses $\overline{CE}$ and $\overline{WE}$ LOW while holding $\overline{OE}$ HIGH. The address bus is at 0000_{16}.
2. The CPU places $3C_{16}$ on the data bus and 2300_{16} onto the address bus, and it pulses $\overline{CE}$ and $\overline{WE}$ LOW while holding $\overline{OE}$ HIGH.

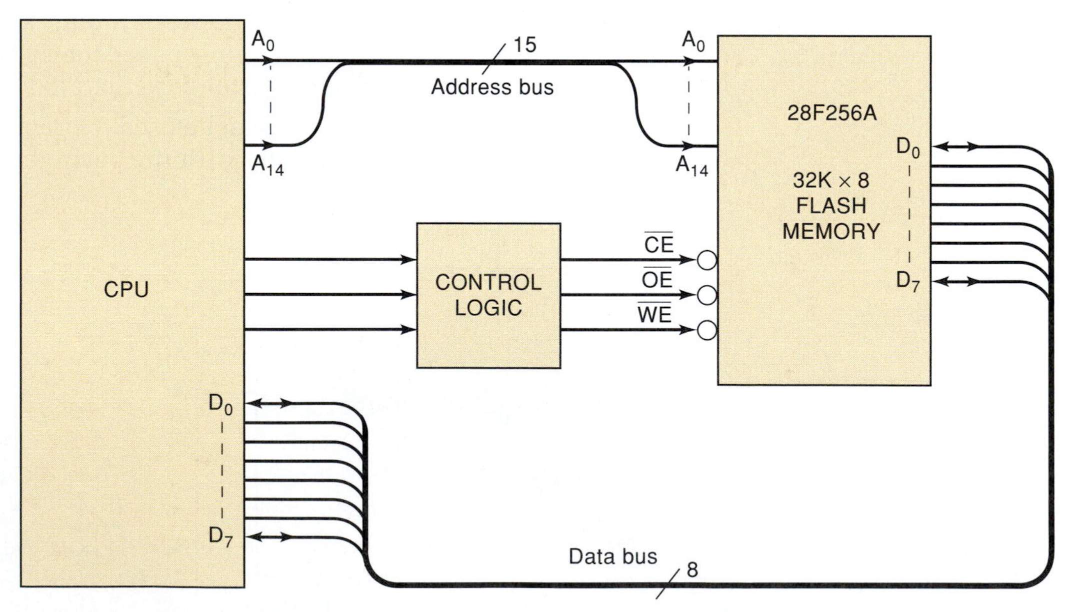

FIGURE 12-47 Problem 12-17.

SECTION 12-9

N 12-18. Another ROM application is the generation of timing and control signals. Figure 12-48 shows a 16 × 8 ROM with its address inputs driven by a MOD-16 counter so that the ROM addresses are incremented with

FIGURE 12-48 Problem 12-18.

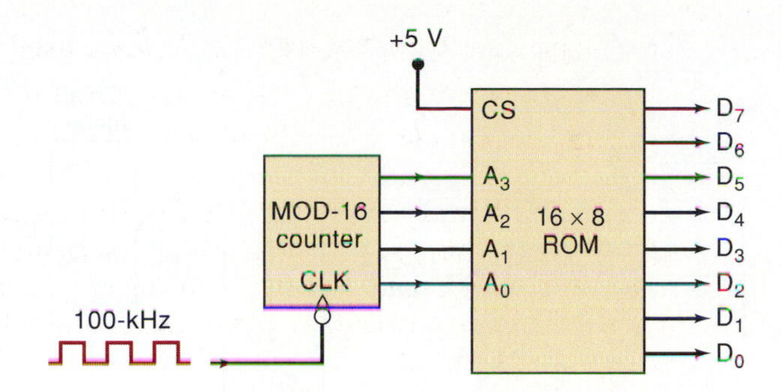

each input pulse. Assume that the ROM is programmed as in Figure 12-6, and sketch the waveforms at each ROM output as the pulses are applied. Ignore ROM delay times. Assume that the counter starts at 0000.

D 12-19.* Change the program stored in the ROM of Problem 12-18 to generate the D_7 waveform of Figure 12-49.

FIGURE 12-49 Problem 12-19.

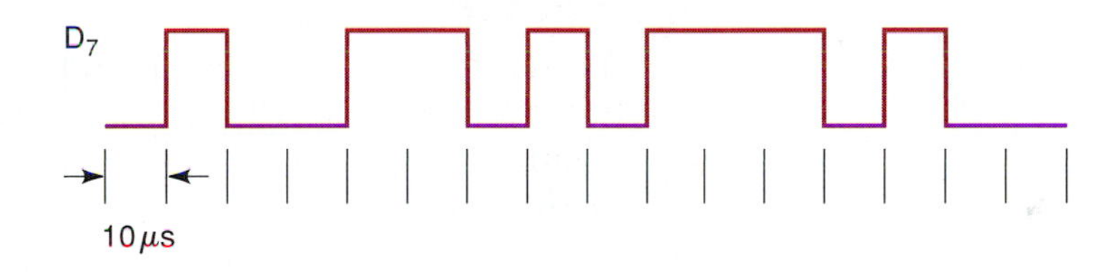

D 12-20.* Refer to the function generator of Figure 12-17.

(a) What clock frequency will result in a 100-Hz sine wave at the output?

(b) What method could be used to vary the peak-to-peak amplitude of the sine wave?

C 12-21. For the ML2035 of Figure 12-18, assume that a value of 038E (hex) in the latch will produce the desired frequency. Draw the timing diagram for the *LATI*, *SID*, and *SCK* inputs, and assume that the LSB is shifted in first.

N, C 12-22.* The system shown in Figure 12-50 is a waveform (function) generator. It uses four 256-point look-up tables in a 1-Kbyte ROM to store one cycle each of a sine wave (address 000–0FF), a positive slope ramp (address 100–1FF), a negative slope ramp (200–2FF), and a triangle wave (300–3FF). The phase relationship among the three output channels is controlled by the values initially loaded into the three counters. The critical timing parameters are $t_{pd(\text{ck-Q and OE–Q max})}$, counters = 10 ns, latches = 5 ns, and t_{ACC} ROM = 20 ns. Study the diagram until you understand how it operates and then answer the following:

(a) If counter A is initially loaded with 0, what values must be loaded into counters B and C so that A lags B by 90° and A lags C by 180°?

(b) If counter A is initially loaded with 0, what values must be loaded into counters B and C to generate a three-phase sine wave with 120° shift between each output?

(c) What must the frequency of pulses on DAC_OUT be in order to generate a 60-Hz sine wave output?

(d) What is the maximum frequency of the CLK input?

(e) What is the maximum frequency of the output waveforms?

(f) What is the purpose of the function select counter?

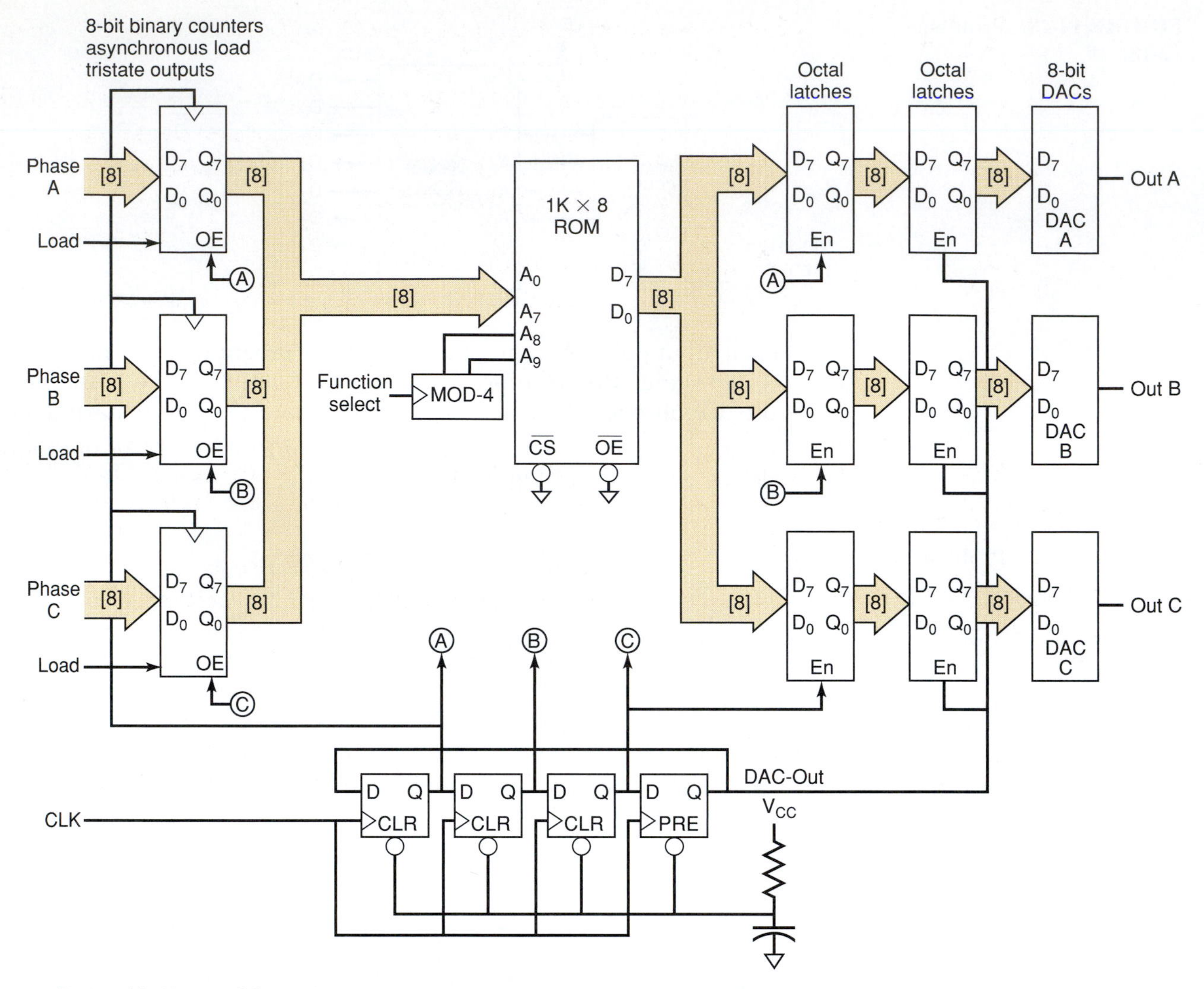

FIGURE 12-50 Problem 12-22.

SECTION 12-11

12-23. (a) Draw the logic symbol for an MCM101514, a CMOS static RAM organized as a 256K × 4 with separate data in and data out, and an active-LOW chip enable.

(b) Draw the logic symbol for an MCM6249, a CMOS static RAM organized as a 1M × 4 with common I/O, an active-LOW chip enable, and an active-LOW output enable.

SECTION 12-12

12-24.* A certain static RAM has the following timing parameters (in nanoseconds):

$$t_{RC} = 100 \qquad t_{AS} = 20$$
$$t_{ACC} = 100 \qquad t_{AH} = \text{not given}$$
$$t_{CO} = 70 \qquad t_{W} = 40$$
$$t_{OD} = 30 \qquad t_{DS} = 10$$
$$t_{WC} = 100 \qquad t_{DH} = 20$$

(a) How long after the address lines stabilize will valid data appear at the outputs during a read cycle?

(b) How long will output data remain valid after $\overline{CS}$ returns HIGH?

(c) How many read operations can be performed per second?

(d) How long should $R/\overline{W}$ and $\overline{CS}$ be kept HIGH after the new address stabilizes during a write cycle?

(e) What is the minimum time that input data must remain valid for a reliable write operation to occur?

(f) How long must the address inputs remain stable after $R/\overline{W}$ and $\overline{CS}$ return HIGH?

(g) How many write operations can be performed per second?

SECTIONS 12-13 TO 12-17

12-25. Draw the logic symbol for the TMS4256, which is a 256K × 1 DRAM. How many pins are saved by using address multiplexing for this DRAM?

D 12-26. Figure 12-51(a) shows a circuit that generates the $\overline{RAS}$, $\overline{CAS}$, and MUX signals needed for proper operation of the circuit of Figure 12-28(b). The 10-MHz master clock signal provides the basic timing for the computer. The memory request signal ($MEMR$) is generated by the CPU in synchronism with the master clock, as shown in part (b) of the figure. $MEMR$ is normally LOW and is driven HIGH whenever the CPU wants to access memory for a read or a write operation. Determine the waveforms at Q_0, $\overline{Q}_1$, and Q_2, and compare them with the desired waveforms of Figure 12-29.

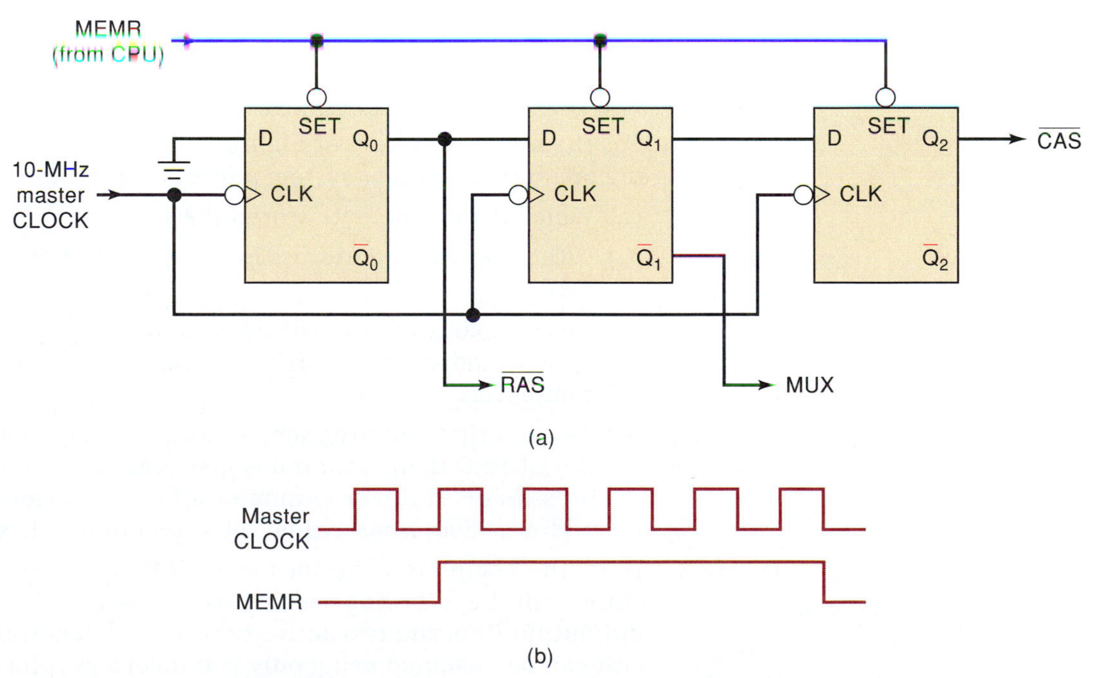

FIGURE 12-51 Problem 12-26.

D 12-27. Show how to connect two 74157 multiplexers (Section 9-6) to provide the multiplexing function required in Figure 12-28(b).

12-28. Refer to the signals in Figure 12-30. Describe what occurs at each of the labeled time points.

12-29. Repeat Problem 12-28 for Figure 12-31.

C 12-30.* The 21256 is a 256K × 1 DRAM that consists of a 512 × 512 array of cells. The cells must be refreshed within 4 ms for data to be retained. Figure 12-33(a) shows the signals used to execute a $\overline{CAS}$-before-$\overline{RAS}$ refresh cycle. Each time a cycle such as this occurs, the on-chip refresh circuitry will refresh a row of the array at the row address specified by a refresh counter. The counter is incremented after each refresh. How often should $\overline{CAS}$-before-$\overline{RAS}$ cycles be applied in order for all of the data to be retained?

12-31.* Study the functional block diagram of the TMS44100 DRAM in Figure 12-27.

(a) What are the actual dimensions of the DRAM cell array?

(b) If the cell array were actually square, how many rows would there be?

(c) How would this affect the refresh time?

SECTION 12-18

D 12-32. Show how to combine two 6206 RAM chips (Figure 12-20) to produce a 32K × 16 module.

D 12-33. Show how to connect two of the 6264 RAM chips symbolized in Figure 12-23 to produce a 16K × 8 RAM module. The circuit should not require any additional logic. Draw a memory map showing the address range of each RAM chip.

D 12-34.* Describe how to modify the circuit of Figure 12-37 so that it has a total capacity of 16K × 8. Use the same type of PROM chips.

D 12-35. Modify the decoding circuit of Figure 12-37 to operate from a 16-line address bus (i.e., add A_{13}, A_{14}, and A_{15}). The four PROMs are to maintain the same hex address ranges.

C 12-36. For the memory system of Figure 12-38, assume that the CPU is storing one byte of data at system address 4000 (hex).

(a) Which chip is the byte stored in?

(b) Is there any other address in this system that can access this data byte?

(c) Answer parts (a) and (b) by assuming that the CPU has stored a byte at address 6007. (*Hint:* Remember that the EEPROM is not completely decoded.)

(d) Assume that the program is storing a sequence of data bytes in the EEPROM and that it has just completed the 2048th byte at address 67FF. If the programmer allows it to store one more byte at address 6800, what will be the effect on the first 2048 bytes?

D 12-37. Draw the complete diagram for a 256K × 8 memory that uses RAM chips with the following specifications: 64K × 4 capacity, common input/output line, and two active-LOW chip select inputs. [*Hint:* The circuit can be designed using only two inverters (plus memory chips).]

SECTION 12-20

12-38.* Modify the RAM circuit of Figure 12-42 as follows: change the OR gate to an AND gate and disconnect its output from *C*; connect the

AND output to E_3; connect C to ground. Determine the address range for each RAM module.

C, D 12-39. Show how to expand the system of Figure 12-42 to an 8K × 8 with addresses ranging from 0000 to 1FFF. (*Hint:* This can be done by adding the necessary memory modules and modifying the existing decoding logic.)

T 12-40.* A dynamic test is performed on the decoding logic of Figure 12-42 by keeping $E = 1$ and connecting the outputs of a six-bit counter to address inputs A_{10} to A_{15}. The decoder outputs are monitored with an oscilloscope (or a logic analyzer) as the counter is continuously pulsed by a 1-MHz clock. Figure 12-52(a) shows the displayed signals. What are the most probable faults?

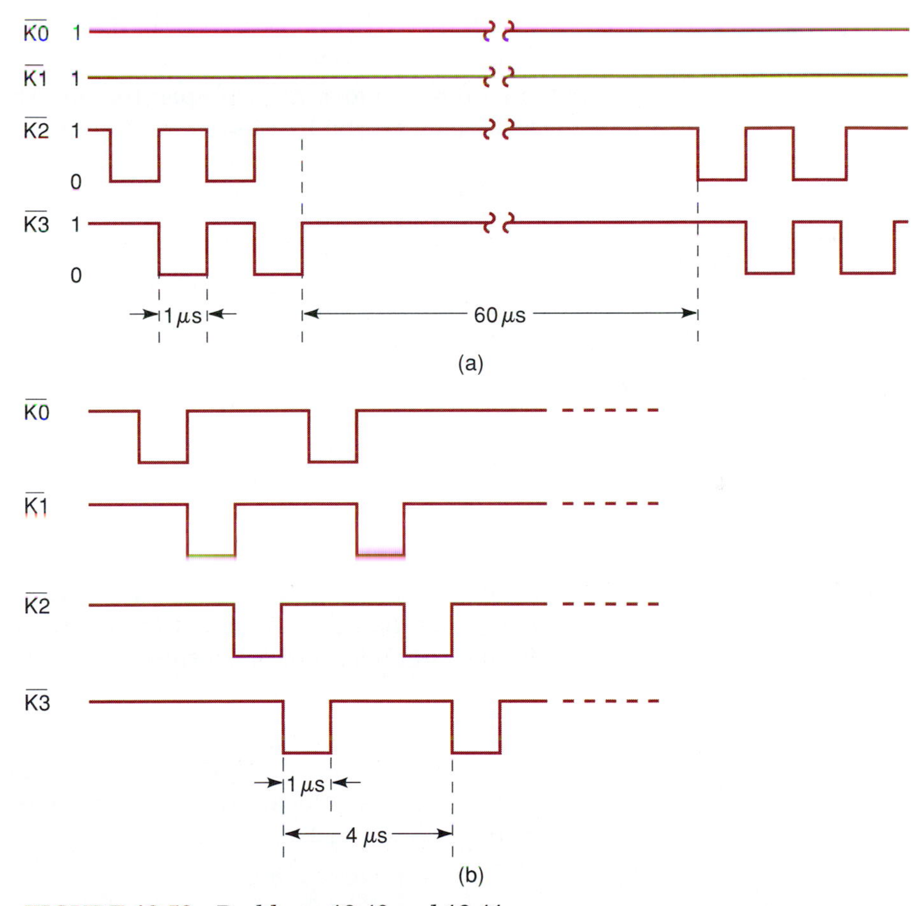

FIGURE 12-52 Problems 12-40 and 12-41.

C, T 12-41. Repeat Problem 12-40 for the decoder outputs shown in Figure 12-52(b).

C, D 12-42.* Consider the RAM system of Figure 12-42. The checkerboard pattern test will not be able to detect certain types of faults. For instance, assume that there is a break in the connection to the A input to the decoder. If a checkerboard pattern SELF-TEST is performed on this circuit, the displayed messages will state that the memory is OK.

(a) Explain why the circuit fault was not detected.

(b) How would you modify the SELF-TEST so that faults such as this will be detected?

T 12-43.* Assume that the 1K × 8 modules used in Figure 12-42 are formed from two 1K × 4 RAM chips. The following messages are printed out when the power-up self-test is performed on this RAM system:

```
module-0 test OK
module-1 test OK
address 0800 faulty at bits 4-7
address 0801 faulty at bits 4-7
address 0802 faulty at bits 4-7
   .  .  .  .  .  .  .
   .  .  .  .  .  .  .
   .  .  .  .  .  .  .
address 0BFE faulty at bits 4-7
address 0BFF faulty at bits 4-7
module-3 test OK
```

Examine these messages and list the possible faults.

T 12-44.* The following messages are printed out when the power-up self-test is performed on the RAM system of Figure 12-42.

```
module-0 test OK
module-1 test OK
module-2 test OK
address 0C00 faulty at bit 7
address 0C01 faulty at bit 7
address 0C02 faulty at bit 7
  .     .  .     .   .
  .     .  .     .   .
  .     .  .     .   .
address 0FFE faulty at bit 7
address 0FFF faulty at bit 7
```

Examine these messages and list the possible faults.

T 12-45. What messages would be printed out when a power-up self-test is performed on the RAM system of Figure 12-42 if there is a short between the decoder's $\overline{K2}$ and $\overline{K3}$ outputs?

SECTION 12-21

T 12-46.* Consider the 16 × 8 ROM in Figure 12-6. Replace the data word stored at address location 1111 with a checksum calculated from the other 15 data words.

ANSWERS TO SECTION REVIEW QUESTIONS

SECTION 12-1

1. See text. 2. 16 bits per word; 8192 words; 131,072 bits or cells 3. In a read operation, a word is taken from a memory location and is transferred to another device. In a write operation, a new word is placed in a memory location and replaces the one previously stored there. 4. True 5. SAM: Access time is not constant but depends on the physical location of the word being accessed. RAM: Access time is the same for any address location. 6. RWM is memory that can be read from or written to with equal ease. ROM is memory that is mainly read from and is written into very infrequently. 7. False; its data must be periodically refreshed.

SECTION 12-2

1. 14, 12, 12 2. Commands the memory to perform either a read operation or a write operation 3. When in its active state, this input enables the memory to perform the read or the write operation selected by the $R/\overline{W}$ input. When in its inactive state, this input disables the memory so that it cannot perform the read or the write function.

SECTION 12-3

1. Address lines, data lines, control lines 2. See text. 3. See text.

SECTION 12-4

1. True 2. Apply desired address inputs; activate control input(s); data appear at data outputs. 3. Process of entering data into ROM

SECTION 12-5

1. $A_3A_2A_1A_0 = 1001$ 2. The row-select decoder activates one of the enable inputs of all registers in the selected row. The column-select decoder activates one of the enable inputs of all registers in the selected column. The output buffers pass the data from the internal data bus to the ROM output pins when the *CS* input is activated.

SECTION 12-7

1. False; by the manufacturer 2. A PROM can be programmed once by the user. It cannot be erased and reprogrammed. 3. True 4. By exposure to UV light 5. True 6. Automatically programs data into memory cells one address at a time 7. An EEPROM can be electrically erased and reprogrammed without removal from its circuit, and it is byte erasable. 8. Low density; high cost 9. EEPROM 10. One

SECTION 12-8

1. Electrically erasable and programmable in circuit 2. Higher density; lower cost 3. Short erase and programming times 4. For the erase and programming operations 5. The contents of this register control all internal chip functions. 6. To confirm that a memory address has been successfully erased (i.e., data = all 1s) 7. To confirm that a memory address has been programmed with the correct data

SECTION 12-9

1. On power-up, the computer executes a small bootstrap program from ROM to initialize the system hardware and to load the operating system from mass storage (disk). 2. Circuit that takes data represented in one type of code and converts it to another type of code 3. Counter, ROM, DAC, low-pass filter 4. They are nonvolatile, fast, reliable, small, and low-power.

SECTION 12-11

1. Desired address applied to address inputs; $R/\overline{W} = 1$; *CS* or *CE* activated 2. To reduce pin count 3. 24, including V_{CC} and ground

SECTION 12-12

1. SRAM cells are flip-flops; DRAM cells use capacitors. 2. CMOS 3. Memory 4. CPU 5. Read- and write-cycle times 6. False; when $\overline{WE}$ is LOW, the I/O pins act as data inputs regardless of the state of $\overline{OE}$ (second entry in mode table). 7. A_{13} can remain connected to pin 26. A_{14} must be removed, and pin 27 must be connected to +5 V.

SECTION 12-13

1. Generally slower speed; need to be refreshed 2. Low power; high capacity; lower cost per bit 3. DRAM

SECTION 12-14

1. 256 rows × 256 columns 2. It saves pins on the chip. 3. 1M = 1024K = 1024 × 1024. Thus, there are 1024 rows by 1024 columns. Because $1024 = 2^{10}$, the chip needs 10 address inputs. 4. $\overline{RAS}$ is used to latch the row address into the DRAM's row address register. $\overline{CAS}$ is used to latch the column address into the column address register. 5. *MUX* multiplexes the full address into the row and column addresses for input to the DRAM.

SECTION 12-15

1. (a) True (b) False (c) False (d) True 2. *MUX*

SECTION 12-16

1. (a) True (b) False 2. It provides row addresses to the DRAM during refresh cycles. 3. Address multiplexing and the refresh operation 4. (a) False (b) True

SECTION 12-17

1. No 2. Memory locations with same upper address (same row) 3. Only the column address must be latched. 4. Extended data output 5. *Burst* 6. The system clock

SECTION 12-18

1. Sixteen 2. Four 3. False; when expanding memory capacity, each chip is selected by a different decoder output (see Figure 12-43). 4. True

SECTION 12-19

1. Battery backup for CMOS RAM; flash memory 2. Economics 3. Data are read out of memory in the same order they were written in. 4. A FIFO used to transfer data between devices with widely different operating speeds 5. Circular buffers "wrap around" from the highest address to the lowest, and the newest datum always overwrites the oldest.

SECTION 12-20

1. Prevents decoding glitches by disabling the decoder while the address lines are changing 2. A way to test RAM by writing a checkerboard pattern (first 01010101, then 10101010) into each memory location and then reading it. It is used because it will detect any shorts or interactions between adjacent cells. 3. An automatic test of RAM performed by a computer on power-up

SECTION 12-21

1. A code placed in the last one or two ROM locations that represents the sum of the expected ROM data from all other locations. It is used as a means to test for errors in one or more ROM locations.

CHAPTER 13

PROGRAMMABLE LOGIC DEVICE ARCHITECTURES*†

OUTLINE

---

*Diagrams of the GAL 16V8 device presented in this chapter have been reproduced through the courtesy of Lattice Semiconductor Corporation, Hillsboro, Oregon.

†Diagrams of the MAX7000S and FLEX10K family devices presented in this chapter have been reproduced through the courtesy of Altera Corporation, San Jose, California.

OBJECTIVES

Upon completion of this chapter, you will be able to:

- Describe the different categories of digital system devices.
- Describe the different types of PLDs.
- Interpret PLD data book information.
- Define PLD terminology.
- Compare the different programming technologies used in PLDs.
- Compare the architectures of different types of PLDs.
- Compare the features of the Altera MAX7000S and FLEX10K families of PLDs.

INTRODUCTION

Throughout the chapters of this book you have been introduced to a wide variety of digital circuits. You now know how the building blocks of digital systems work and can combine them to solve a wide variety of digital problems. More complicated digital systems, such as microcomputers and digital signal processors, have also been briefly described. The defining difference between microcomputer/DSP systems and other digital systems is that the former follow a programmed sequence of instructions that the designer specifies. Many applications require faster response than a microcomputer/DSP architecture can accommodate and in these cases, a conventional digital circuit must be used. In today's rapidly advancing technology market, most conventional digital systems are not being implemented with standard logic device chips containing only simple gates or MSI-type functions. Instead, programmable logic devices, which contain the circuitry necessary to create logic functions, are being used to implement digital systems. These devices are not programmed with a list of instructions, like a computer or DSP. Instead, their internal hardware is configured by electronically connecting and disconnecting points in the circuit.

Why have PLDs taken over so much of the market? With programmable devices, the same functionality can be obtained with one IC rather than using several individual logic chips. This characteristic means less board space, less power required, greater reliability, less inventory, and overall lower cost in manufacturing.

In the previous chapters you have become familiar with the process of programming a PLD using either AHDL or VHDL. At the same time, you have learned about all the building blocks of digital systems. The PLD implementations of digital circuits up to this point have been presented as a "black box." We have not been concerned with what was going on inside the PLD to make it work. Now that you understand all the circuitry inside the black box, it is time to turn the lights on in there and look at how it works. This will allow you to make the best decisions when selecting and applying a PLD to solve a problem. This chapter will take a look at the various types of hardware available to design digital systems. We will then introduce you to the architectures of various families of PLDs.

13-1 DIGITAL SYSTEMS FAMILY TREE

While the major goal of this chapter is to investigate PLD architectures, it is also useful to look at the various hardware choices available to digital system designers because it should give us a little better perception of today's digital hardware alternatives. The desired circuit functionality can generally be achieved by using several different types of digital hardware. Throughout this book, we have described both standard logic devices as well as how programmable logic devices can be used to create the same functional blocks. Microcomputers and DSP systems can also often be applied with the necessary sequence of instructions (i.e., the application's program) to produce the desired circuit function. The design engineering decisions must take into account many factors, including the necessary speed of operation for the circuit, cost of manufacturing, system power consumption, system size, amount of time available to design the product, etc. In fact, most complex digital designs include a mix of different hardware categories. Many trade-offs between the various types of hardware have to be weighed to design a digital system.

A digital system family tree (see Figure 13-1) showing most of the hardware choices that are currently available can be useful in sorting out the many categories of digital devices. The graphical representation in the figure does not show all the details—some of the more complex device types have many additional subcategories, and older, obsolete device types have been omitted for clarity. The major digital system categories include standard logic, application-specific integrated circuits (ASICs) and microprocessor/digital signal processing (DSP) devices.

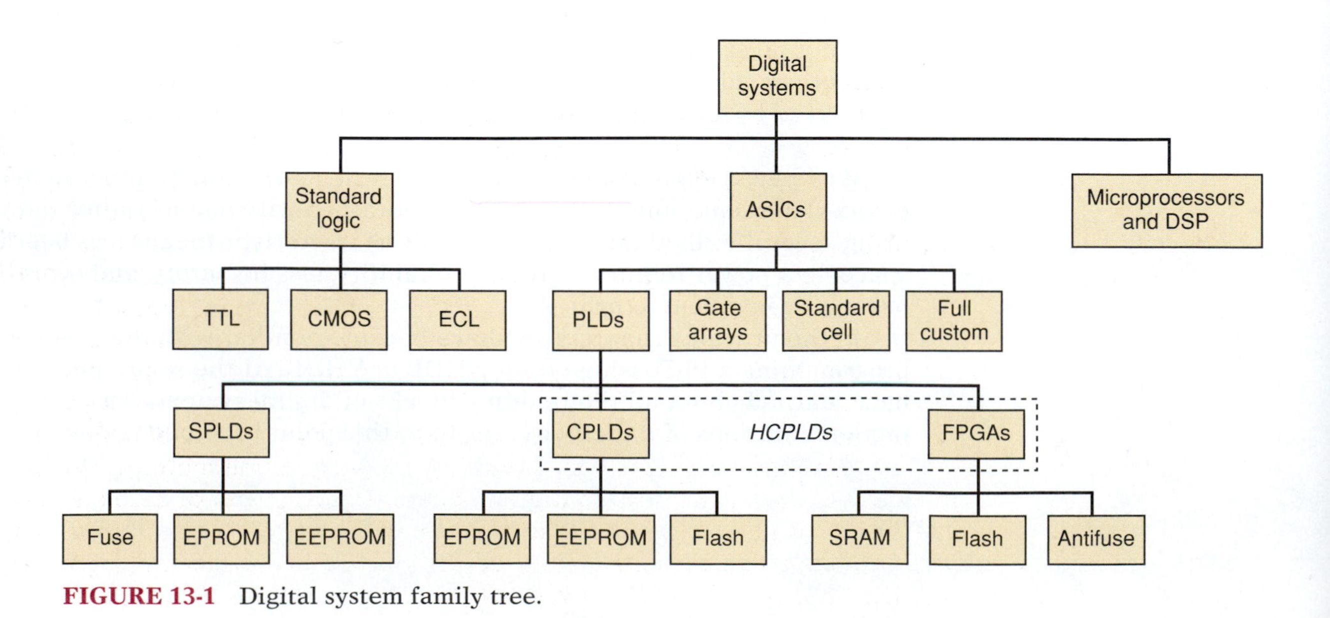

FIGURE 13-1 Digital system family tree.

The first category of **standard logic** devices refers to the basic functional digital components (gates, flip-flops, decoders, multiplexers, registers, counters, etc.) that are available as SSI and MSI chips. These devices have been used for many years (some more than 30 years) to design complex digital systems. An obvious drawback is that the system may literally consist of hundreds of such chips. These inexpensive devices can still be useful if our design is not very complex. As discussed in Chapter 8, there are three major families of standard logic devices: TTL, CMOS, and ECL. TTL is a mature technology consisting of numerous subfamilies that have been developed over many years of use. Very few new designs apply TTL logic, but many, many digital systems still contain TTL devices. CMOS is the most popular standard logic device family today, primarily due to its low power consumption. ECL technology, of course, is applied for higher-speed designs. Standard logic devices are still available to the digital designer, but if the application is very complex, a lot of SSI/MSI chips will be needed. That solution is not very attractive for our design needs today.

The **microprocessor/digital signal processing (DSP)** category is a much different approach to digital system design. These devices actually contain the various types of functional blocks that have been discussed throughout this text. With microcomputer/DSP systems, devices can be controlled electronically, and data can be manipulated by executing a program of instructions that has been written for the application. A great deal of flexibility can be achieved with microcomputer/DSP systems because all you have to do is change the program. The major downfall with this digital system category is speed. *Using a hardware solution for your digital system design is always faster than a software solution.*

The third major digital system category is called **application-specific integrated circuits (ASICs)**. This broad category represents the modern hardware design solution for digital systems. As the acronym implies, an integrated circuit is designed to implement a specific desired application. Four subcategories of ASIC devices are available to create digital systems: programmable logic devices, gate arrays, standard-cell, and full-custom.

Programmable logic devices (PLDs), sometimes referred to as field-programmable logic devices (FPLDs), can be custom-configured to create any desired digital circuit, from simple logic gates to complex digital systems. Many examples of PLD designs have been given in earlier chapters. This ASIC choice for the designer is very different from the other three subcategories. With a relatively small capital investment, any company can purchase the necessary development software and hardware to program PLDs for their digital designs. On the other hand, to obtain a gate array, standard-cell or full-custom ASIC requires that most companies contract with an IC foundry to fabricate the desired IC chip. This option can be extremely expensive and usually requires that your company purchase a large volume of parts to be cost effective.

Gate arrays are ULSI circuits that offer hundreds of thousands of gates. The desired logic functions are created by the interconnections of these prefabricated gates. A custom-designed mask for the specific application determines the gate interconnections, much like the stored data in a mask-programmed ROM. For this reason, they are often referred to as mask-programmed gate arrays (MPGAs). Individually, these devices are less expensive than PLDs of comparable gate count, but the custom programming process by the chip manufacturer is very expensive and requires a great deal of lead time.

Standard-cell ASICs use predefined logic function building blocks called cells to create the desired digital system. The IC layout of each cell has been designed previously, and a library of available cells is stored in a computer database. The needed cells are laid out for the desired application, and the interconnections between the cells are determined. Design costs for standard-cell ASICs are even higher than for MPGAs because all IC fabrication masks that define the components and interconnections must be custom designed. Greater lead time is also needed for the creation of the additional masks. Standard cells do have a significant advantage over gate arrays. The cell-based functions have been designed to be much smaller than equivalent functions in gate arrays, which allows for generally higher-speed operation and cheaper manufacturing costs.

Full-custom ASICs are considered the ultimate ASIC choice. As the name implies, all components (transistors, resistors, and capacitors) and the interconnections between them are custom-designed by the IC designer. This design effort requires a significant amount of time and expense, but it can result in ICs that can operate at the highest possible speed and require the smallest die (individual IC chip) area. Smaller IC die sizes allow for many more die to fit on a silicon wafer, which significantly lowers the manufacturing cost for each IC.

More on PLDs

This chapter is mainly about PLDs, so we will look a little further down that branch of the family tree. The development of PLD technology has advanced continuously since the first PLDs appeared more than 30 years ago. The early devices contained the equivalent of a few hundred gates, and now we have parts available that contain a few million gates. The old devices could handle a few inputs and a few outputs with limited logic capabilities. Now there are PLDs that can handle hundreds of inputs and outputs. Original devices could be programmed only once and, if the design changed, the old PLD would have to be removed from the circuit and a new one, programmed with the updated design, would have to be inserted in its place. With newer devices, the internal logic design can be changed on the fly, while the chip is still connected to a printed circuit board in an electronic system.

Generally, PLDs can be described as being one of three different types: **simple programmable logic devices (SPLDs), complex programmable logic devices (CPLDs)**, or **field programmable gate arrays (FPGAs)**. There are several manufacturers with many different families of PLD devices, so there are many variations in architecture. We will attempt to discuss the general characteristics for each of the types, but be forewarned: the differences are not always clear-cut. The distinction between CPLDs and FPGAs is often a little fuzzy, with the manufacturers constantly designing new, improved architectures and frequently muddying the waters for marketing purposes. Together, CPLDs and FPGAs are often referred to as **high-capacity programmable logic devices (HCPLDs)**. The programming technologies for PLD devices are actually based on the various types of semiconductor memory. As new types of memory have been developed, the same technology has been applied to the creation of new types of PLD devices.

The amount of logic resources available is the major distinguishing feature between SPLDs and HCPLDs. Today, SPLDs are devices that typically contain the equivalent of 600 or fewer gates, while HCPLDs have thousands and hundreds of thousands of gates available. Internal programmable signal interconnect resources are much more limited with SPLDs. SPLDs are generally much less complicated and much cheaper than HCPLDs. Many small digital applications need only the resources of an SPLD. On the other hand, HCPLDs are capable of providing the circuit resources for complete complex digital systems, and larger, more sophisticated HCPLD devices are designed every year.

The SPLD classification includes the earliest PLD devices. The amount of logic resources contained in the early PLDs may be relatively small by today's standards, but they represented a significant technological step in their ability to create easily a custom IC that can replace several standard logic devices. Over the years, numerous semiconductor advances have created different SPLD types. The first PLD type to gain the interest of circuit designers was programmed by literally burning open selected fuses in the programming matrix. The fuses that were left intact in these **one-time programmable (OTP)** devices provided the electrical connections for the AND/OR circuits to produce the desired functions. This logic device was based on the fuse links in PROM memory technology (see Section 12-7) and was most commonly referred to as a programmable logic array (PLA). PLDs didn't really gain widespread acceptance with digital designers until the late 1970s, when a device called a **programmable array logic (PAL)** was introduced. The programmable fuse links in a PAL are used to determine the input connections to a set of AND gates that are wired to fixed OR gates. With the development of the ultraviolet erasable PROM came the EPROM-based PLDs in the mid 1980s, followed soon by PLDs using electrically erasable (EEPROM) technology.

CPLDs are devices that typically combine an array of PAL-type devices on the same chip. The logic blocks themselves are programmable AND/fixed-OR logic circuits with fewer product terms available than most PAL devices. Each logic block (often called a **macrocell**) can typically handle many input variables, and the internal programmable logic signal routing resources tend to be very uniform throughout the chip, producing consistent signal delays. When more product terms are needed, gates may be shared between logic blocks, or several logic blocks can be combined to implement the expression. The flip-flop used to implement the register in the macrocell can often be configured for D, JK, T (toggle), or SR operation. Input and output pins for some CPLD architectures are associated with a specific macrocell, and typically additional macrocells are buried (that is, not connected to a pin). Other CPLD architectures may have independent I/O blocks with built-in registers that can be used to latch incoming or outgoing data. The programming technologies used in CPLD devices are all nonvolatile and include EPROM, EEPROM, and flash, with EEPROM being the most common. All three technologies are erasable and reprogrammable.

FPGAs also have a few fundamental characteristics that are shared. They typically consist of many relatively small and independent programmable logic modules that can be interconnected to create larger functions. Each module can usually handle only up to four or five input variables. Most FPGA logic modules utilize a **look-up table (LUT)** approach to create the desired logic functions. A look-up table functions just like a truth table in which the output can be programmed to create the desired combinational function by storing the appropriate 0 or 1 for each input combination. The programmable signal routing resources within the chip tend to be quite varied, with many different path lengths available. The signal delays produced for a design depend on the actual signal routing selected by the programming software. The logic modules also contain programmable registers. The logic modules are not associated with any I/O pin. Instead, each I/O pin is connected to a programmable input/output block that, in turn, is connected to the logic modules with selected routing lines. The I/O blocks can be configured to provide input, output, or bidirectional capability, and built-in registers can be used to latch incoming or outgoing data. A general architecture of FPGAs is shown in Figure 13-2. All of the logic blocks and input/output blocks can be programmed to implement almost any logic circuit. The programmable interconnections are accomplished via lines that run through the rows and columns in the channels between the logic blocks. Some FPGAs include large blocks of RAM memory; others do not.

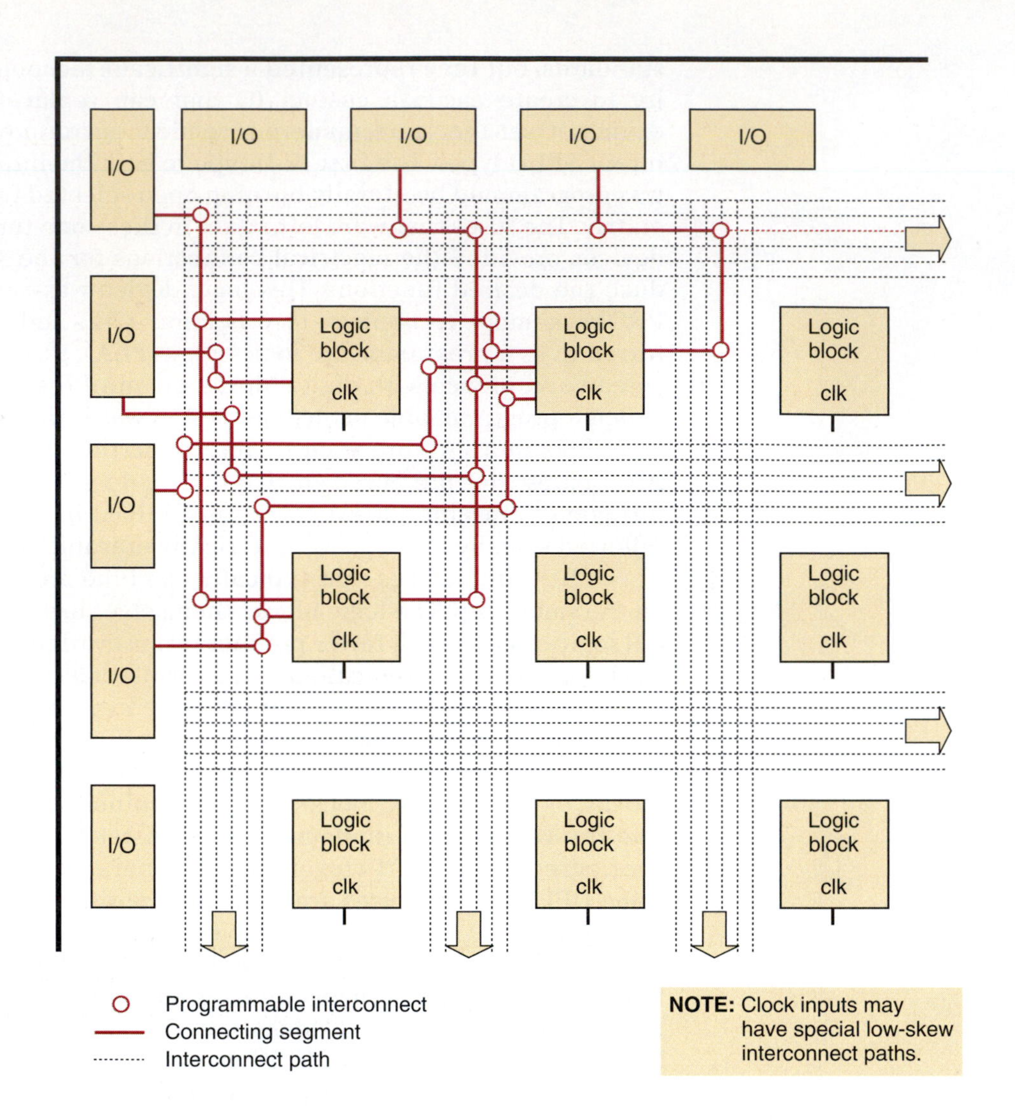

FIGURE 13-2 FPGA architecture.

The programming technologies used in FPGA devices include SRAM, flash, and antifuse, with SRAM being the most common. SRAM-based devices are volatile and therefore require the FPGA to be reconfigured (programmed) when it is powered-up. The programming information that defines how each logic block functions, which I/O blocks are inputs and outputs, and how the blocks are interconnected is stored in some type of external memory that is downloaded to the SRAM-based FPGA when power is applied. Antifuse devices are one-time programmable and are therefore nonvolatile. Antifuse memory technology is not currently used for memory devices but, as its name implies, it is the opposite of fuse technology. Instead of opening a fuse link to prevent a signal connection, an insulator layer between interconnects has an electrical short created to produce a signal connection. Antifuse devices are programmed in a device programmer either by the end-user or by the factory or distributor.

Differences in architecture between CPLDs and FPGAs, among different HCPLD manufacturers, and among different families of devices from a single manufacturer can affect the efficiency of design implementation for a particular application. You may ask, "Does the architecture of this PLD family provide the best fit for my application?" It is very difficult, however, to predict which architecture may be the best choice to use for a complex digital system. Only a portion of the available gates can be utilized. Who knows how many equivalent gates will be needed for a large design? The basic design of the signal routing resources can affect how much of the PLD's logic resources can be utilized. The segmented interconnects often found in FPGAs can produce shorter delays between adjacent logic blocks, but they may also produce longer delays between the blocks that are further apart than would be produced by the continuous type of interconnect found in most CPLDs. There is no easy answer to your question, but every HCPLD manufacturer will give you an answer anyway: their product is best!

As you can see, the field of PLDs is quite diverse and it is constantly changing. You should now have the basic knowledge of the various types and technologies necessary to interpret PLD data sheets and learn more about them.

REVIEW QUESTIONS

1. What are the three major categories of digital systems?
2. What is the major disadvantage of a microprocessor/DSP design?
3. What does ASIC stand for?
4. What are the four types of ASICs?
5. What are HCPLDs?
6. What are two major differences between CPLDs and FPGAs?
7. What does volatility refer to?

13-2 FUNDAMENTALS OF PLD CIRCUITRY

A simple PLD device is shown in Figure 13-3. Each of the four OR gates can produce an output that is a function of the two input variables, A and B. Each output function is programmed with the fuses located between the AND gates and each of the OR gates.

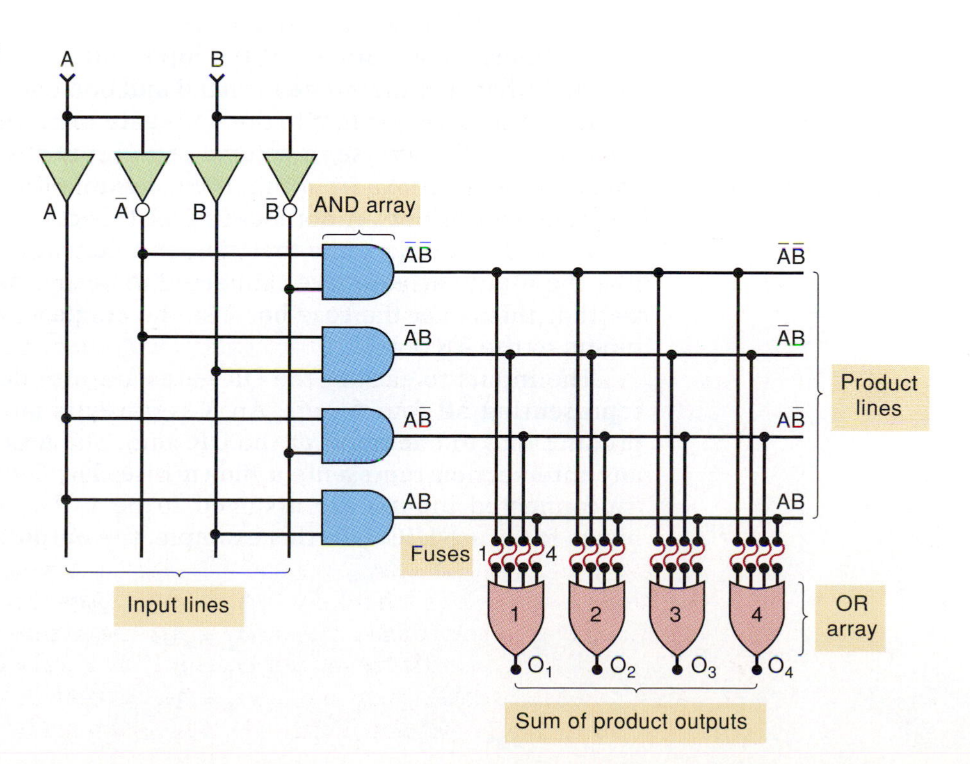

FIGURE 13-3 Example of a programmable logic device.

Each of the inputs A and B feed both a noninverting buffer and an inverting buffer to produce the true and inverted forms of each variable. These are the *input lines* to the AND gate array. Each AND gate is connected to two different input lines to generate a unique product of the input variables. The AND outputs are called the *product lines.*

Each of the product lines is connected to one of the four inputs of each OR gate through a fusible link. With all of the links initially intact, each OR output will be a constant 1. Here's the proof:

$$\begin{aligned} O_1 &= \overline{A}\,\overline{B} + \overline{A}B + A\overline{B} + AB \\ &= \overline{A}(\overline{B} + B) + A(\overline{B} + B) \\ &= \overline{A} + A = 1 \end{aligned}$$

Each of the four outputs O_1, O_2, O_3, and O_4 can be *programmed* to be any function of A and B by selectively blowing the appropriate fuses. PLDs are designed so that a blown OR input acts as a logic 0. For example, if we blow fuses 1 and 4 at OR gate 1, the O_1 output becomes

$$O_1 = 0 + \overline{A}B + A\overline{B} + 0 = \overline{A}B + A\overline{B}$$

We can program each of the OR outputs to any desired function in a similar manner. Once all of the outputs have been programmed, the device will permanently generate the selected output functions.

PLD Symbology

The example in Figure 13-3 has only two input variables and the circuit diagram is already quite cluttered. You can imagine how messy the diagram would be for PLDs with many more inputs. For this reason, PLD manufacturers have adopted a simplified symbolic representation of the internal circuitry of these devices.

Figure 13-4 shows the same PLD circuit as Figure 13-3 using the simplified symbols. First, notice that the input buffers are represented as a single buffer with two outputs, one inverted and one noninverted. Next, note that a *single line* is shown going into the AND gate to represent all four inputs. Each time the row line crosses a column represents a separate input to the AND gate. The connections from the input variable lines to the AND gate inputs are indicated as dots. A dot means that this connection to the AND gate input is hard-wired (i.e., one that cannot be changed). At first glance, it looks like the input variables are connected to each other. It is important to realize that this is *not* the case because the single row line represents *multiple* inputs to the AND gate.

The inputs to each of the OR gates are also designated by a single line representing all four inputs. An X represents an intact fuse connecting a product line to one input of the OR gate. The absence of an X (or a dot) at any intersection represents a blown fuse. For OR gate inputs, blown fuses (unconnected inputs) are assumed to be LOW, and for AND gate inputs, blown fuses are HIGH. In this example, the outputs are programmed as

$$\begin{aligned} O_1 &= \overline{A}B + A\overline{B} \\ O_2 &= AB \\ O_3 &= 0 \\ O_4 &= 1 \end{aligned}$$

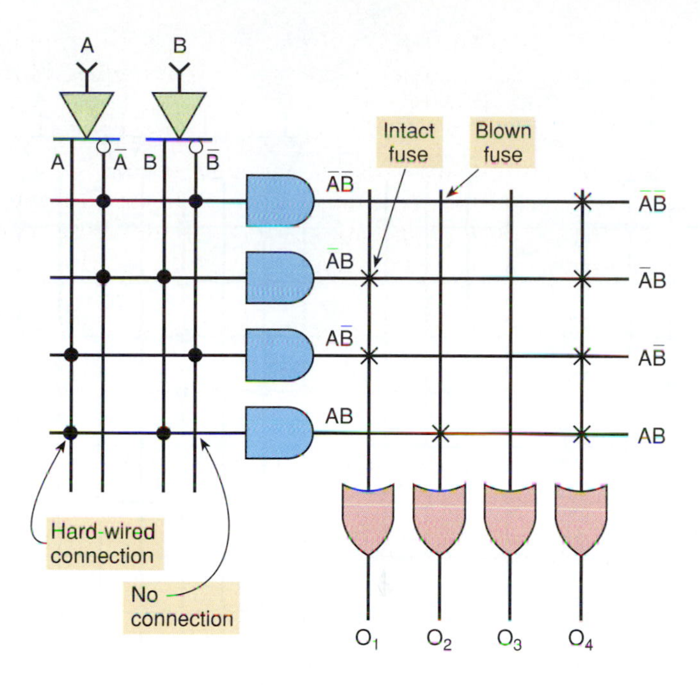

FIGURE 13-4 Simplified PLD symbology.

REVIEW QUESTIONS

1. What is a PLD?
2. What would output O_1 be in Figure 13-3 if fuses 1 and 2 were blown?
3. What does an X represent on a PLD diagram?
4. What does a dot represent on a PLD diagram?

13-3 PLD ARCHITECTURES

The concept of PLDs has led to many different architectural designs of the inner circuitry of these devices. In this section, we will explore some of the basic differences in architecture.

PROMs

The architecture of the programmable circuits in the previous section involves programming the connections to the OR gate. The AND gates are used to decode all the possible combinations of the input variables, as shown in Figure 13-5(a). For any given input combination, the corresponding row is activated (goes HIGH). If the OR input is connected to that row, a HIGH appears at the OR output. If the input is not connected, a LOW appears at the OR output. Does this sound familiar? Refer back to Figure 12-9. If you think of the input variables as address inputs and the intact/blown fuses as stored 1s and 0s, you should recognize the architecture of a PROM.

Figure 13-5(b) shows how the PROM would be programmed to generate four specified logic functions. Let's follow the procedure for output $O_3 = AB + \overline{C}\overline{D}$.

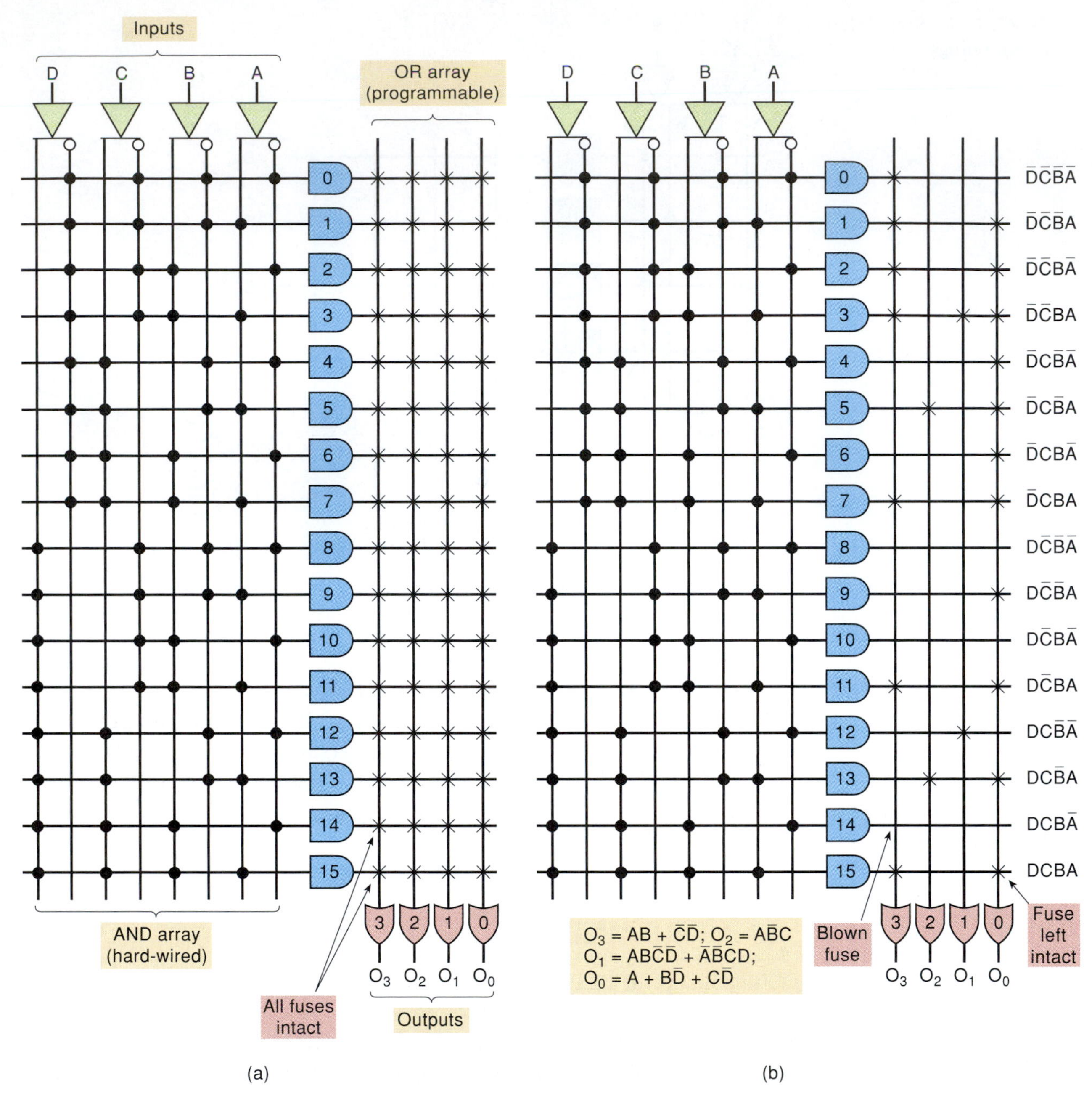

FIGURE 13-5 (a) PROM architecture makes it suitable for PLDs; (b) fuses are blown to program outputs for given functions.

The first step is to construct a truth table showing the desired O_3 output level for all possible input combinations (Table 13-1).

Next, write down the AND products for those cases where the output is to be a 1. The O_3 output is to be the OR sum of these products. Thus, only the fuses that connect these product terms to the inputs of OR gate 3 are to be left intact. All others are to be blown, as indicated in Figure 13-5(b). This same procedure is followed to determine the status of the fuses at the other OR gate inputs.

The PROM can generate any possible logic function of the input variables because it generates every possible AND product term. In general, any application that requires every input combination to be available is a good candidate for a PROM. However, PROMs become impractical when a large

TABLE 13-1

D	C	B	A	O_3		
0	0	0	0	1	→	$\overline{D}\,\overline{C}\,\overline{B}\,\overline{A}$
0	0	0	1	1	→	$\overline{D}\,\overline{C}\,\overline{B}A$
0	0	1	0	1	→	$\overline{D}\,\overline{C}B\overline{A}$
0	0	1	1	1	→	$\overline{D}\,\overline{C}BA$
0	1	0	0	0		
0	1	0	1	0		
0	1	1	0	0		
0	1	1	1	1	→	$\overline{D}CBA$
1	0	0	0	0		
1	0	0	1	0		
1	0	1	0	0		
1	0	1	1	1	→	$D\overline{C}BA$
1	1	0	0	0		
1	1	0	1	0		
1	1	1	0	0		
1	1	1	1	1	→	$DCBA$

number of input variables must be accommodated because the number of fuses doubles for each added input variable.

Calling a PROM a PLD is really just a semantics issue. You already knew that a PROM is programmable and it is a logic device. This is just a way of using a PROM and thinking of its purpose as implementing SOP logic expressions rather than storing data values in memory locations. The real problem is translating the logic equations into the fuse map for a given PROM. A general-purpose logic compiler designed to program SPLDs has a list of PROM devices that it can support. If you choose to use any old scavenged EPROM as a PLD, you may need to generate your own bit map (like they used to do it), which is very tedious.

Programmable Array Logic (PAL)

The PROM architecture is well suited for those applications where every possible input combination is required to generate the output functions. Examples are code converters and data storage (look-up) tables that we examined in Chapter 12. When implementing SOP expressions, however, they do not make very efficient use of circuitry. Each combination of address inputs must be fully decoded, and each expanded product term has an associated fuse that is used to OR them together. For example, notice how many fuses were required in Figure 13-5 to program the simple SOP expressions and how many product terms are often not used. This has led to the development of a class of PLDs called programmable array logic (PAL). The architecture of a PAL differs slightly from that of a PROM, as shown in Figure 13-6(a).

The PAL has an AND and OR structure similar to a PROM but in the PAL, inputs to the AND gates are programmable, whereas the inputs to the OR gates are hard-wired. This means that every AND gate can be programmed to generate any desired product of the four input variables and their complements. Each OR gate is hard-wired to only four AND outputs. This limits each output function to four product terms. If a function requires

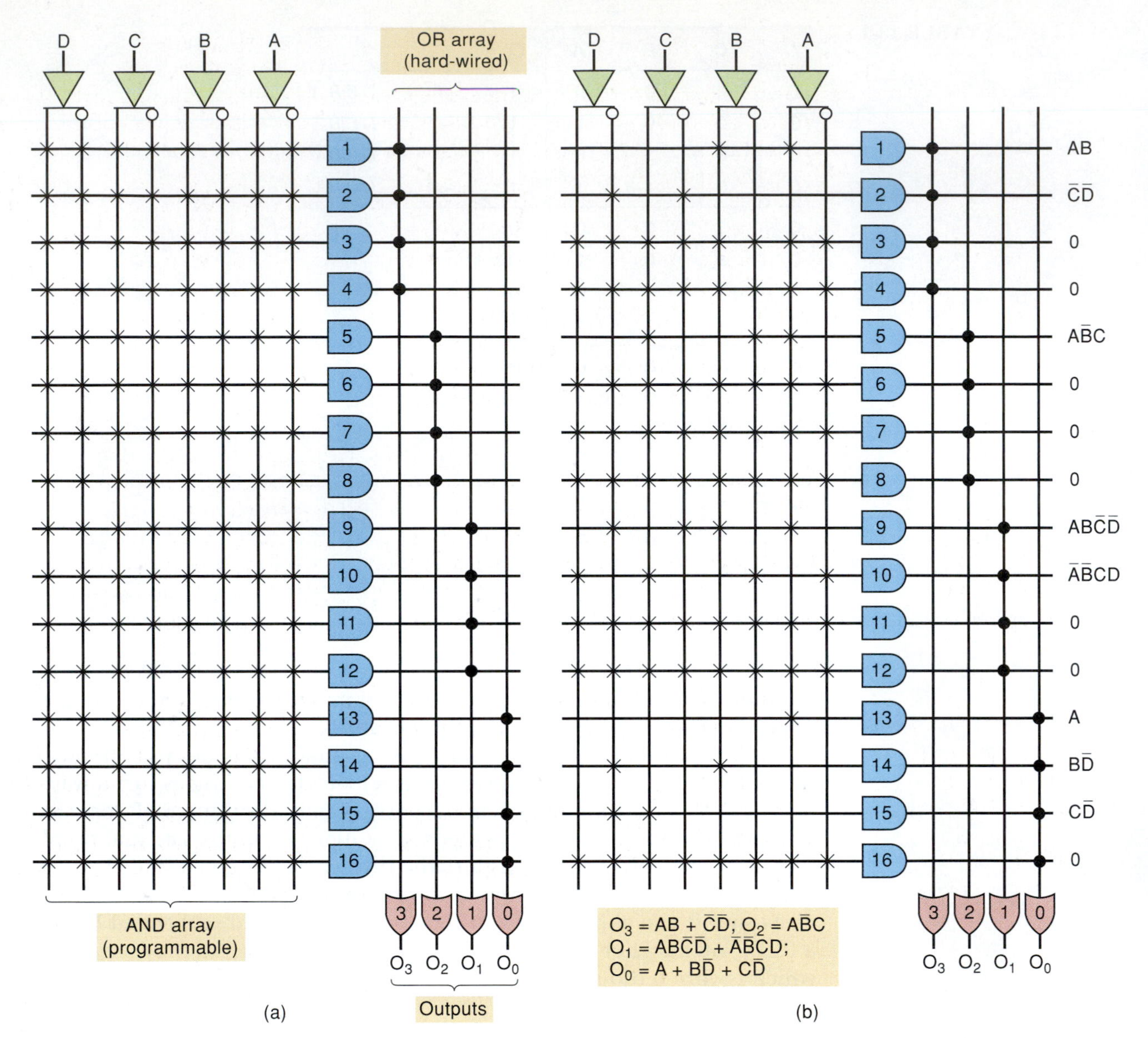

FIGURE 13-6 (a) Typical PAL architecture; (b) the same PAL programmed for the given functions.

more than four product terms, it cannot be implemented with this PAL; one having more OR inputs would have to be used. If fewer than four product terms are required, the unneeded ones can be made 0.

Figure 13-6(b) shows how this PAL is programmed to generate four specified logic functions. Let's follow the procedure for output $O_3 = AB + \overline{C}\overline{D}$. First, we must express this output as the OR sum of four terms because the OR gates have four inputs. We do this by putting in 0s. Thus, we have

$$O_3 = AB + \overline{C}\overline{D} + 0 + 0$$

Next, we must determine how to program the inputs to AND gates 1, 2, 3, and 4 so that they provide the correct product terms to OR gate 3. We do this term by term. The first term, AB, is obtained by leaving intact the fuses that connect inputs A and B to AND gate 1 and by blowing all other fuses on that line.

Likewise, the second term, $\overline{C}\overline{D}$, is obtained by leaving intact only the fuses that connect inputs $\overline{C}$ and $\overline{D}$ to AND gate 2. The third term is a 0. A constant 0 is produced at the output of AND gate 3 by leaving all of its input fuses intact. This would produce an output of $A\overline{A}B\overline{B}C\overline{C}D\overline{D}$, which, as we know, is 0. The fourth term is also 0, so the input fuses to AND gate 4 are also left intact.

The inputs to the other AND gates are programmed similarly to generate the other output functions. Note especially that many of the AND gates have all of their input fuses intact because they need to generate 0s.

An example of an actual PAL integrated circuit is the PAL16L8, which has 10 logic inputs and eight output functions. Each output OR gate is hard-wired to seven AND gate outputs, and so it can generate functions that include up to seven terms. An added feature of this particular PAL is that six of the eight outputs are fed back into the AND array, where they can be connected as inputs to any AND gate. This makes it very useful in generating all sorts of combinational logic.

The PAL family also contains devices with variations of the basic SOP circuitry we have described. For example, most PAL devices have a tristate buffer driving the output pin. Others channel the SOP logic circuit to a D FF input and use one of the pins as a clock input to clock all of the output flip-flops synchronously. These devices are referred to as *registered PLDs* because the outputs pass through a register. An example is the PAL16R8, which has up to eight registered outputs (which can also serve as inputs) plus eight dedicated inputs.

Field Programmable Logic Array (FPLA)

The field programmable logic array (FPLA) was developed in the mid-1970s as the first nonmemory programmable logic device. It used a programmable AND array as well as a programmable OR array. Although the FPLA is more flexible than the PAL architecture, it has not been as widely accepted by engineers. FPLAs are used mostly in state-machine design where a large number of product terms are needed in each SOP expression.

REVIEW QUESTIONS

1. Verify that the correct fuses are blown for the O_2, O_1, and O_0 functions in Figure 13-5(b).
2. A PAL has a hard-wired _____ array and a programmable _____ array.
3. A PROM has a hard-wired _____ array and a programmable _____ array.
4. How would the equation for the output of O_1 in Figure 13-5(b) change if all the fuses from AND gate 14 were left intact?

13-4 THE GAL 16V8 (GENERIC ARRAY LOGIC)

The GAL 16V8, introduced by Lattice Semiconductor, has an architecture that is very similar to the PAL devices described in the previous section. Standard, low-density PALs are one-time programmable. The GAL chip, on the other hand, uses an EEPROM array (located at row and column intersections in Figure 13-7) to control the programmable connections to the AND matrix, allowing them to be erased and reprogrammed at least 100 times. In addition to the AND and OR gates used to produce the sum of product functions, the GAL 16V8 contains optional flip-flops for register and counter applications, tristate buffers for the outputs, and control multiplexers used

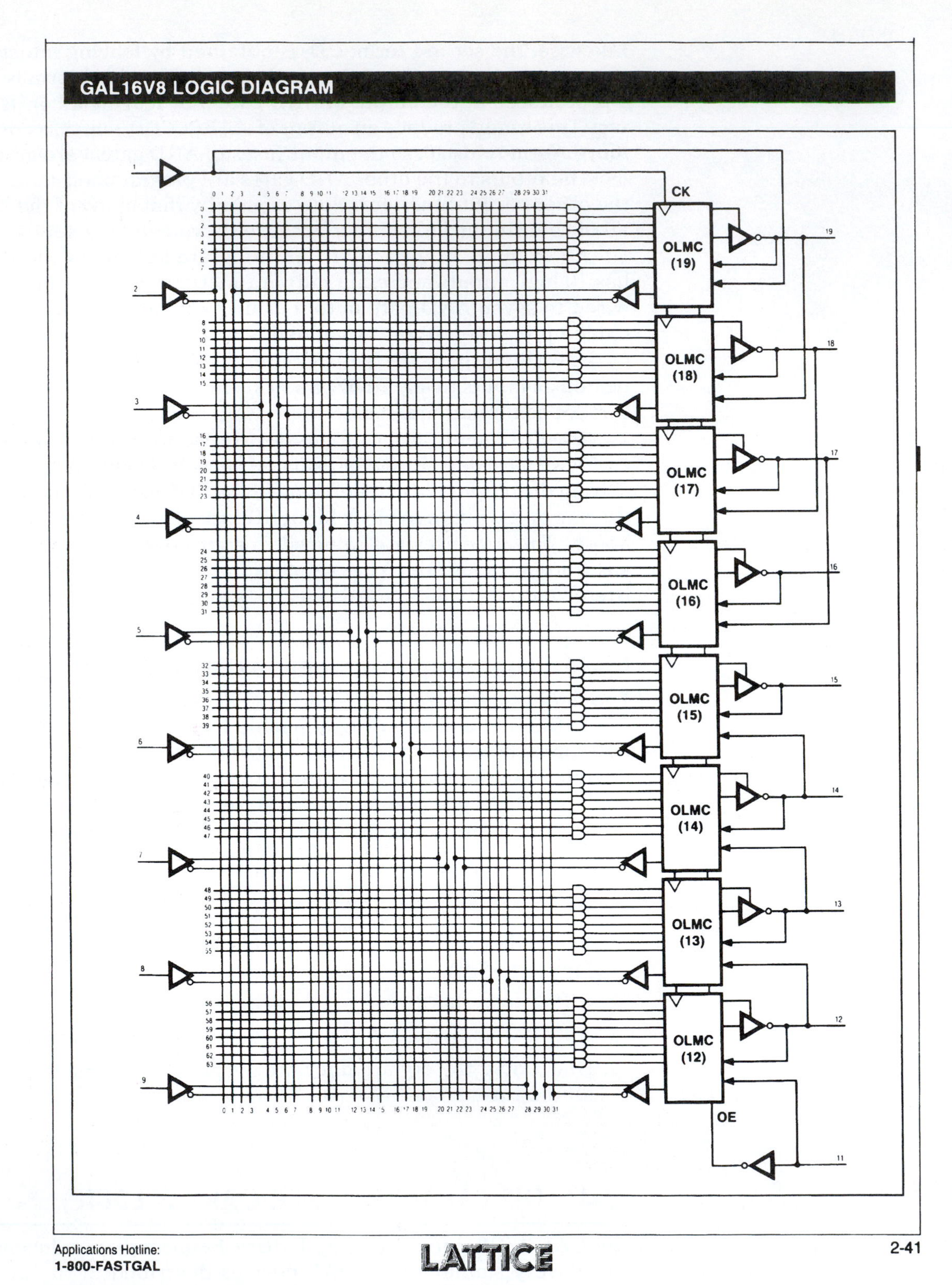

FIGURE 13-7 GAL 16V8 logic diagram. (Reprinted with permission of Lattice Semiconductor.)

to select the various modes of operation. Consequently, it can be used as a generic, pin-compatible replacement for most PAL devices. Specific locations in the memory array are designated to control the various programmable connections in the chip. Fortunately, it is not necessary to delve into the addresses of each bit location in the matrix. The programming software takes care of these details in a user-friendly manner.

The complete logic diagram of the GAL 16V8 is shown in Figure 13-7. This device has eight dedicated input pins (pins 2–9), two special function inputs (pins 1 and 11), and eight pins (12–19) that can be used as inputs or outputs. The major components of the GAL devices are the input term matrix; the AND gates, which generate the products of input terms; and the output logic macrocells (OLMCs). Notice that the eight inputs (pins 2–9) are each connected directly to a column of the input term matrix. The complement of each of these inputs is also connected to a column of the matrix. These pins must always be specified as inputs when programming the 16V8. A logic level and its complement are also fed from each OLMC back to a column of the input matrix. This accounts for the 32 input variables (columns in the input matrix) that can be programmed as connections to the 64 multiple-input AND gates.

The flexibility of the GAL 16V8 lies in its programmable output logic macrocell. Eight different products (outputs of AND gates) are applied as inputs to each of the eight output logic macrocells. Within each OLMC, the products are ORed together to generate the sum of products (SOP). Recall from Chapter 4 that any logic function can be expressed in SOP form. Within the OLMC, the SOP output may be routed to the output pin to implement a combinational circuit, or it may be clocked into a D flip-flop to implement a registered output circuit.

To understand the detailed operation of the OLMC, refer to Figure 13-8. The figure shows the structure of OLMC(n), where n is a number from 12 to 19. Notice that seven of the products are unconditionally connected to the OR

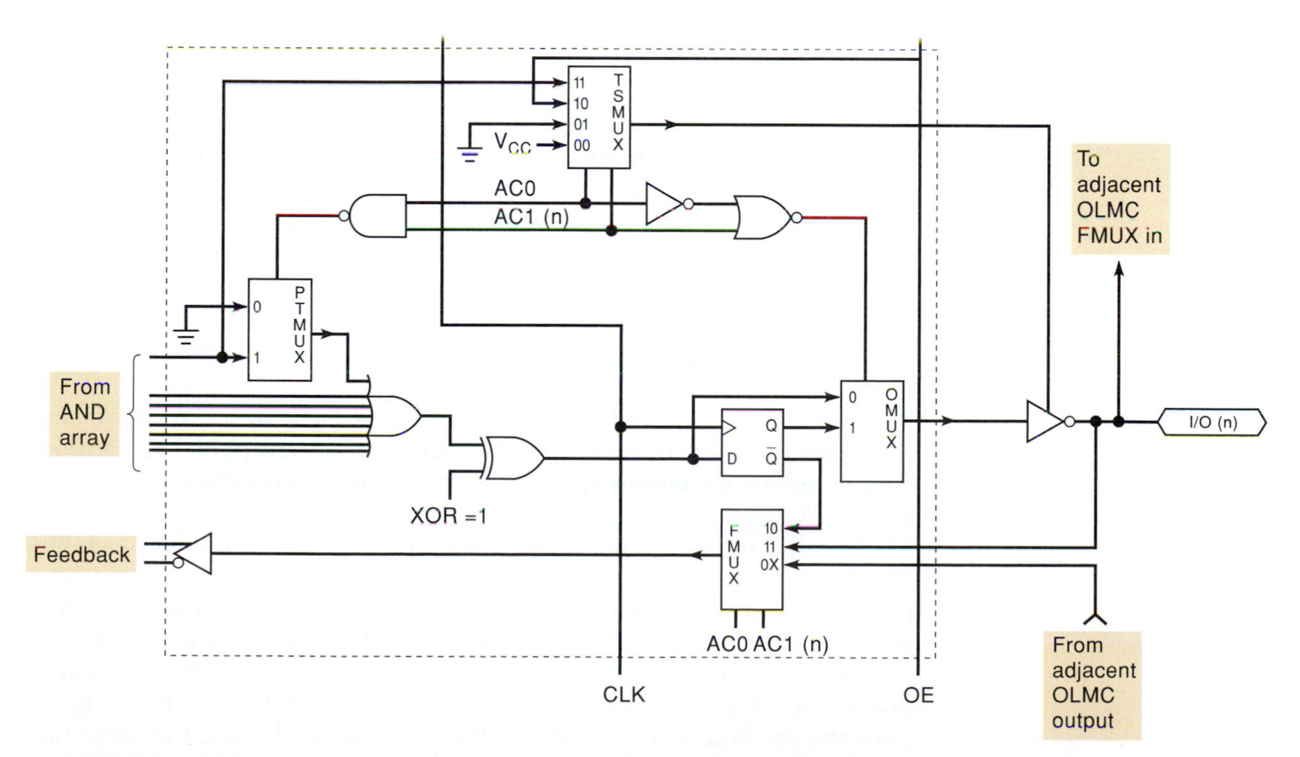

FIGURE 13-8 Output logic macrocell for the GAL 16V8. (Reprinted with permission of Lattice Semiconductor.)

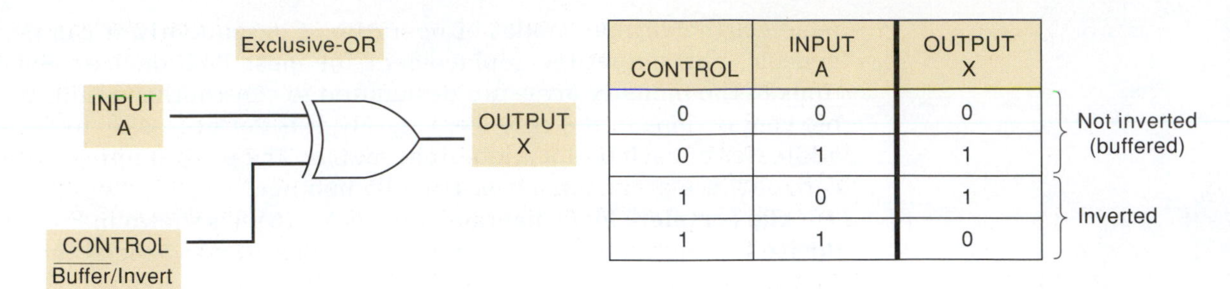

CONTROL	INPUT A	OUTPUT X
0	0	0
0	1	1
1	0	1
1	1	0

FIGURE 13-9 Using XOR to complement selectively.

gate inputs. The eighth product term is connected to a two-input product term multiplexer (PTMUX), which drives the eighth input to the OR gate. The eighth product term also connects to one input of a four-input multiplexer (TSMUX). The output of TSMUX enables the tristate inverter that drives the output pin [I/O(n)]. The output multiplexer (OMUX) is a two-input MUX that selects between the combinational output (OR gate) and the registered output (the D flip-flop). A fourth MUX selects the logic signal that is fed back to the input matrix. This is called the *feedback* multiplexer (FMUX).

Each of these multiplexers is controlled by programmable bits (AC1 and AC0) in the EEPROM matrix. This is the way that the OLMC configuration can be altered by the programmer. Another programmable bit is the input to the XOR gate. This provides the programmable output polarity feature. Recall that an XOR gate can be used to complement a logic signal selectively, as shown in Figure 13-9. When the control line is a logic 0, the XOR will pass the logic level at input A with no inversion. When the control bit is a logic 1, the XOR will invert the signal so that $X = \overline{A}$. In Figure 13-8, the programmable bit (labeled XOR) is a logic 1 under normal positive logic conditions. This inverts the output of the OR gate, which is inverted again when it passes through the tristate inverting buffer on the output.

We can understand the various configuration options by studying the possible inputs to each multiplexer. The TSMUX controls the tristate buffer's enable input. If the V_{CC} input is selected, the output is always enabled, like a standard combinational logic gate. If the grounded input is selected, the tristate output of the inverter is always in its high-impedance state (allowing the I/O pin to be used as an input). Another input to the MUX that may be selected comes from the *OE* input, which is pin 11. This allows the output to be enabled or disabled by an external logic signal applied to pin 11. The last possible input selection is a product term from the eighth AND gate. This allows an AND combination of terms from the input matrix to enable or disable the output.

The FMUX selects the signal that is fed back into the input matrix. In this case, there are three possible selections. Selecting the *MUX* input that is connected to an adjacent stage or the *MUX* input connected to its own OLMC I/O pin allows an existing output state to be fed back to the input matrix in some of the modes of operation. This feature gives the GAL 16V8 the ability to implement sequential circuits such as the cross-coupled NAND gate latch circuit described in Chapter 5. This feedback option also allows an I/O pin to be used as a dedicated input as opposed to an output. One of these two feedback paths is chosen, depending on the MODE that the chip is programmed for. The third option, selecting the output from the D flip-flop, allows the present state of the flip-flop (which can be used to determine the next state) to be fed back to the input matrix. This allows synchronous sequential circuits, such as counters and shift registers, to be implemented.

With all of these options, it would seem that there must be a long list of possible configurations. In actual practice, all these configuration decisions are made by the software. Actually, the GAL 16V8 has only three different modes: (1) *simple mode,* which is used to implement simple SOP combinational logic without tristate outputs; (2) *complex mode,* which implements SOP combinational logic with tristate outputs that are enabled by an AND product expression; and (3) *registered mode,* which allows individual OLMCs to operate in a combinational configuration with tristate outputs (similar to the complex mode) or in a synchronous mode with clocked D FFs synchronized to a common clock signal.

The GAL 16V8 is an inexpensive and versatile PLD chip, but what if a design requires more hardware resources than is contained in the 16V8? It may be possible to split the design into smaller blocks that can be implemented in several 16V8 chips. Fortunately, there are other members of the GAL family to choose from. Another popular, general-purpose PLD is the GAL 22V10. This device has 10 output pins and 12 input pins in an architecture that is similar but not identical to the GAL 16V8. Groups of product terms are logically summed with an OR gate, which feeds an OLMC. Unlike the 16V8, however, each OR gate in the 22V10 does not combine the same number of product terms. The number of terms ranges from eight all the way up to 16. To take advantage of the extra terms, you must assign the larger Boolean expressions to the correct output pin. The D flip-flops contained in the OLMCs also have asynchronous reset and synchronous preset capabilities. A newer version of the 22V10—the ispGAL 22V10—is now available. This device is said to be in-system programmable (ISP). Instead of requiring a programmer, as is needed to program PALs and standard GAL chips, a cable from the PC is connected directly to a special set of pins on the ISP device to do the programming.

REVIEW QUESTIONS

1. Name two advantages of GAL devices over PAL devices.
2. Name the three modes of operation for a GAL 16V8.

13-5 THE ALTERA EPM7128S CPLD

We will investigate the architecture of the EPM7128S, an EEPROM-based device in the Altera MAX7000S CPLD family. This device is found on several educational development boards, including the Altera UP2, DeVry eSOC, and RSR PLDT-2. The block diagram for this family is shown in Figure 13-10. The major structures in the MAX7000S are the **logic array blocks (LABs)** and the **programmable interconnect array (PIA)**. A LAB contains a set of 16 macrocells and looks very similar to a single SPLD device. Each macrocell consists of a programmable AND/OR circuit and a programmable register (flip-flop). The macrocells in a single LAB can share logic resources such as common product terms or unused AND gates. The number of macrocells contained in one of the MAX7000S family devices depends on the part number. As shown in Table 13-2, the EPM7128S has 128 macrocells arranged in eight LABs.

Logic signals are routed between LABs via the PIA. The PIA is a global bus that connects any signal source to any destination within the device. All inputs to the MAX7000S device and all macrocell outputs feed the PIA. Up to 36 signals can feed each LAB from the PIA. Only signals needed to produce the required functions for any LAB are fed into that LAB.

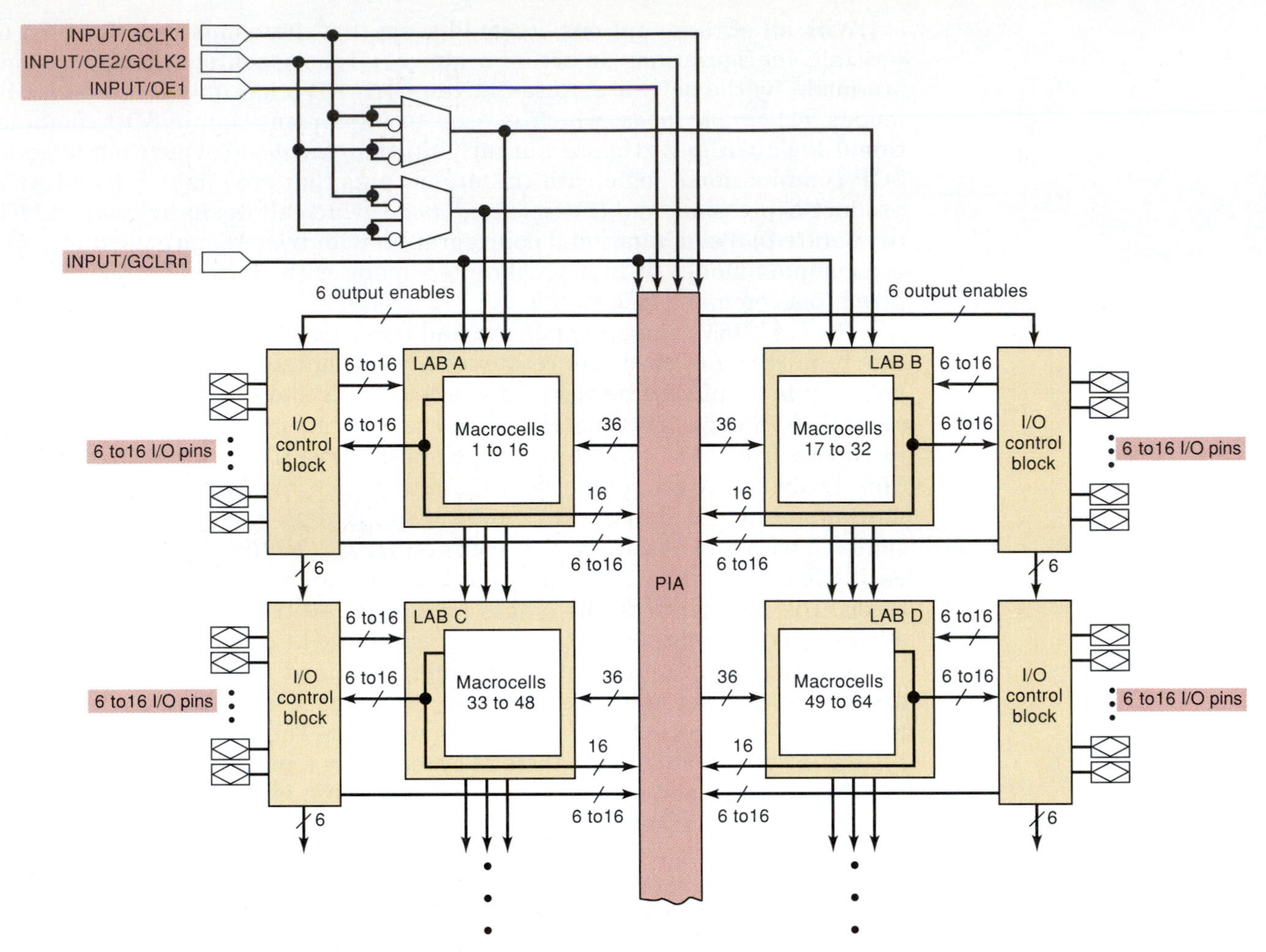

FIGURE 13-10 MAX7000S family block diagram. (Courtesy of Altera Corporation.)

I/O pins in the MAX7000S family are connected to specific macrocells. The number of I/O pins available to the user depends on the device package. An EPM7128S in a 160-pin PQFP package has 12 I/Os per LAB plus four additional input-only pins, for a total of 100 pins. On the other hand, in an 84-pin PLCC package, which is included on the above-mentioned development boards, there are eight I/Os per LAB plus the four extras, for a total of 68 I/O pins. The EPM7128S is an in-system programmable (ISP) device. The ISP feature utilizes a joint test action group (JTAG) interface that requires four specific pins to be dedicated to the programming interface and are therefore not available for general user I/O. The target PLD can be programmed in-system via the JTAG pins by connecting them to the parallel port of a PC with driver gates, as shown

TABLE 13-2 Altera MAX7000S family device features.

Feature	EPM7032S	EPM7064S	EPM7128S	EPM7160S	EPM7192S	EPM7256S
Usable gates	600	1250	2500	3200	3750	5000
Macrocells	32	64	128	160	192	256
LABs	2	4	8	10	12	16
Maximum number of user I/O pins	36	68	100	104	124	164

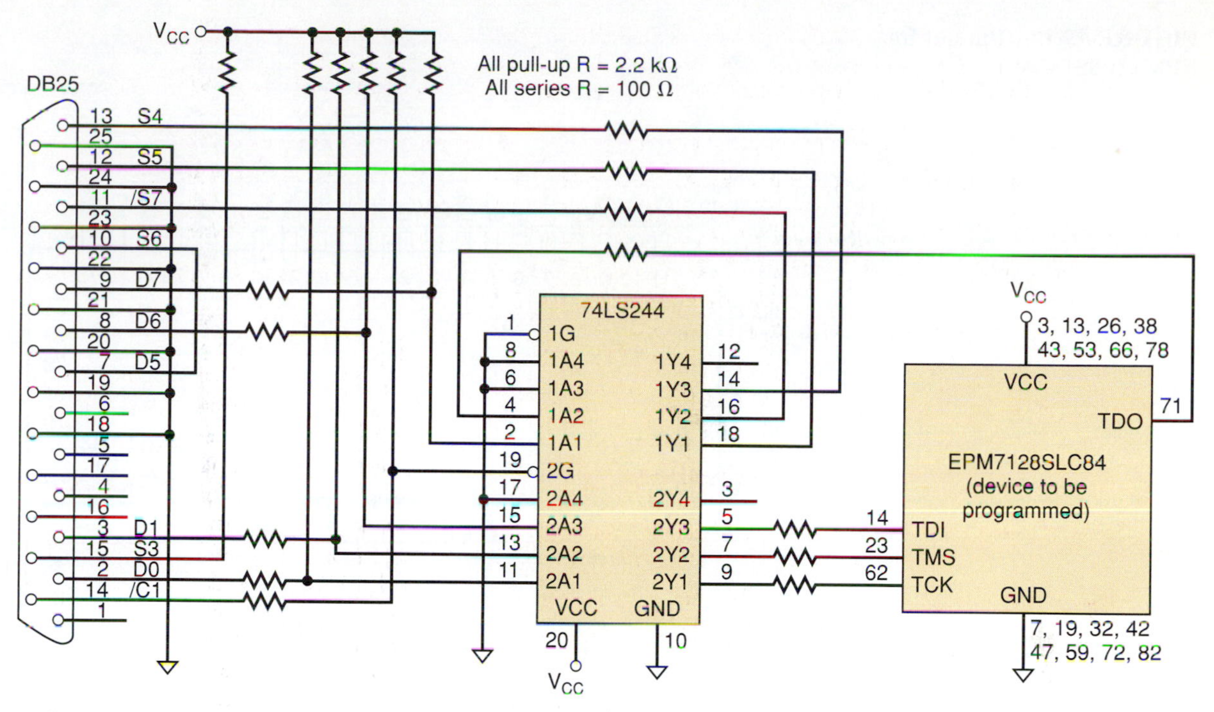

FIGURE 13-11 JTAG interface between PC parallel port and EPM7128SLC84.

in Figure 13-11. The JTAG signals are named TDI (test data in), TDO (test data out), TMS (test mode select), and TCK (test clock). This brings the user I/O pin total for an EPM7128SLC84 (an EPM7128S in an 84-pin PLCC package) down to 64 pins. All 68 pins, however, can be used for user I/O if the EPM7128SLC84 is programmed in a PLD programmer instead of in-system. When the design is compiled, you must indicate whether or not the device will use a JTAG interface. In either case, you can see that some macrocells will not be connected directly to user I/O pins. These macrocells can be utilized by the compiler for internal (buried) logic.

The four input-only pins found on devices in the MAX7000S family can be configured as specific high-speed control signals or as general user inputs. GCLK1 is the primary global clock input for all macrocells in the device. It is used to clock all registers synchronously in a design. It is located on pin 83 on an EPM7128SLC84 (see Figure 13-12). Pin 2 on this device is GCLK2 (secondary global clock). As an alternative, this pin may be used as a secondary global output enable (OE2) for any macrocells designated to have a tristate output. The primary tristate enable, OE1, is located on pin 84. The last of the four global control signals is GCLRn on pin 1. This active-LOW input can control the asynchronous clear on any macrocell register. How these pins are to be used for a specific application is assigned in MAX+PLUS II or Quartus II either automatically by the compiler or manually by the designer during the design process.

The I/O control blocks (see Figure 13-10) configure each I/O pin for input, output, or bidirectional operation. All I/O pins in the MAX7000S family have a tristate output buffer that is (1) permanently enabled or disabled, (2) controlled by one of the two global output enable pins, or (3) controlled by other inputs or functions generated by other macrocells. When an I/O pin is configured as an input, the associated macrocell can be used for buried logic. During in-system programming, the I/O pins will be made tristate and internally pulled up to eliminate board conflicts.

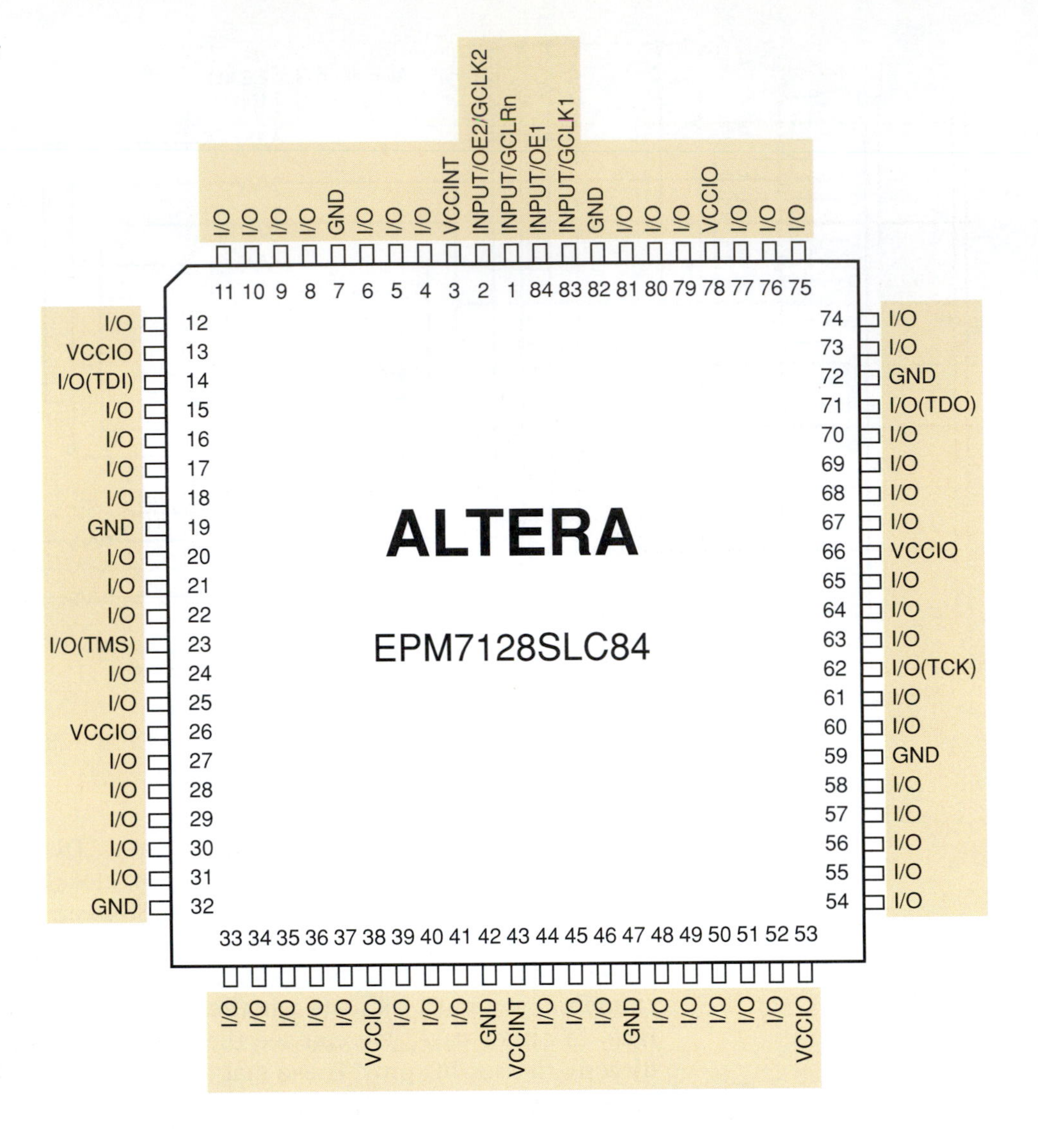

FIGURE 13-12 Pin-out for EPM7128SLC84.

Figure 13-13 shows the block diagram for a MAX7000S macrocell. Each macrocell can produce either a combinational or a registered output. The register (flip-flop) contained in a macrocell will be bypassed to produce a combinational output. The programmable sum of product circuit looks very much like that found in a GAL chip. Each macrocell can produce five product terms. While this is fewer than was found in the simpler GAL chips discussed earlier, it is often sufficient for most logic functions. If more product terms are needed, the compiler will automatically program a macrocell to borrow up to five product terms from each of three adjacent macrocells in the same LAB. This parallel logic expander option can provide a total of 20 product terms. The borrowed gates are no longer usable by the macrocell from which they are borrowed. Another expansion option, available in each LAB, is called shared logic expanders. Instead of adding more product terms, this option allows a common product term to be produced once and then used by several macrocells within the LAB. Only one product term per macrocell can be used in this fashion, but with 16 macrocells per LAB, this makes up to 16 common product terms available. The compiler automatically optimizes the allocation of available product terms within a LAB according to the logic requirements of the design. Using either expander option does incur a small amount of additional propagation delay.

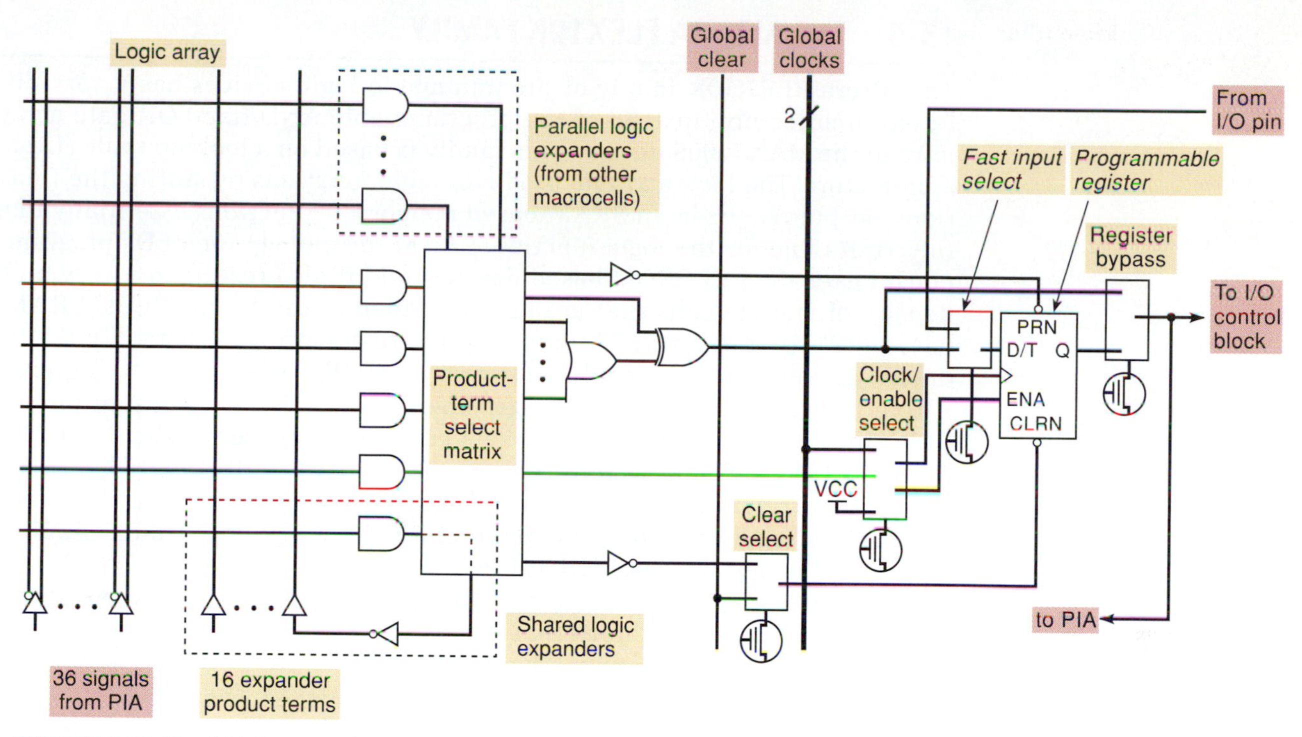

FIGURE 13-13 MAX7000S family macrocell. (Courtesy of Altera Corporation.)

For registered functions, each macrocell flip-flop can be programmed individually to implement D, T, JK, or SR operation. Each programmable register can be clocked in three different modes: (1) with a global clock signal, (2) with a global clock signal when the flip-flop is enabled, or (3) with an array clock signal produced by a buried macrocell or a (nonglobal) input pin. In the EPM7128S, either of the two global clock pins (GCLK1 or GCLK2) can be used to produce the fastest clock-to-Q performance. Either clock edge can be programmed to trigger the flip-flops. Each register can be preset asynchronously or cleared with an active-HIGH or active-LOW product term. Each register may also be cleared with the active-LOW global clear pin (GCLRn). A fast data input path from an I/O pin to the registers, bypassing the PIA, is also available. All registers in the device will be reset automatically at power-up.

MAX7000S devices have a power-saving option that allows the designer to program each individual macrocell for either high-speed (turbo bit turned on) or low-power (turbo bit turned off) operation. Because most logic applications require only a small fraction of all gates to operate at maximum frequency, this feature may produce a significant savings in total system power consumption. Speed-critical paths in the design can run at maximum speed, while the remaining signal paths can operate at reduced power.

REVIEW QUESTIONS

1. What is a macrocell?
2. What is an ISP device?
3. What special control functions are provided with the four input-only pins on a MAX7000S device?
4. What system advantage is achieved by slowing down selected macrocells in a MAX7000S device?

13-6 THE ALTERA FLEX10K FAMILY

The Altera FLEX10K family of programmable logic devices has a very different architecture. Instead of the programmable AND/fixed-OR gate array used in the MAX7000S devices, this family is based on a look-up table (LUT) architecture. The look-up table produces logic functions by storing the function's output results in an SRAM-based memory. It functions essentially like the truth table for the logic function. SRAM technology for PLDs programs much faster than EEPROM-based devices, and it also results in a very high density of storage cells that are used to program the larger PLDs. SRAM-based PLDs that use the LUT architecture are generally classified in the industry as field programmable gate arrays. Unlike most FPGAs, however, Altera has chosen to utilize a programmable signal routing design for the FLEX10K family that looks more like an enhanced version of the PIA found in the CPLD MAX7000S family. As a result, the FLEX10K family has architectural characteristics that are a combination of the two HCPLD classifications. Based on the high-density architecture of the logic cells, the FLEX10K devices are generally classified as FPGAs.

Let us examine the concept of a look-up table. The LUT is the portion of the programmable logic block that produces a combinational function (see Figure 13-14). This function can be used as the output of the logic block or it may be registered (controlled by the internal MUX). The look-up table itself consists of a set of flip-flops that store the desired truth table for our function. LUTs are usually rather small, typically handling four input variables, and so our truth table would have a total of 16 combinations. We will need a flip-flop to store each of the 16 function values (see Figure 13-15). Up to four input variables in our example LUT will be connected to the data inputs on the decoder block using programmable interconnects. The input combination that is applied will determine which of the 16 flip-flops will be selected to feed the output via the tristate buffers. The look-up table is basically a 16×1 SRAM memory block. All we have to do to create any desired function (of up to four input variables) is to store the appropriate set of 0s and 1s in the LUT's flip-flops. That is essentially what is done to program this type of PLD. Because the flip-flops are volatile (they are SRAM), we need to load the LUT memory for the desired functions whenever the PLD is powered-up. This process is called configuring the PLD. Other portions of the device are also programmed in the same fashion using other SRAM memory bits to store the programming information. This is the basic programming technique for the logic blocks, called **logic elements (LEs)**, found in the FLEX10K devices.

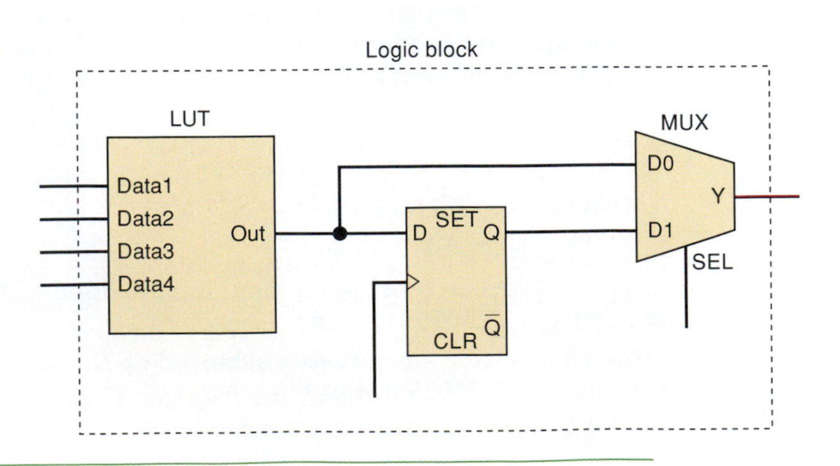

FIGURE 13-14 Simplified logic block diagram for FLEX10K device.

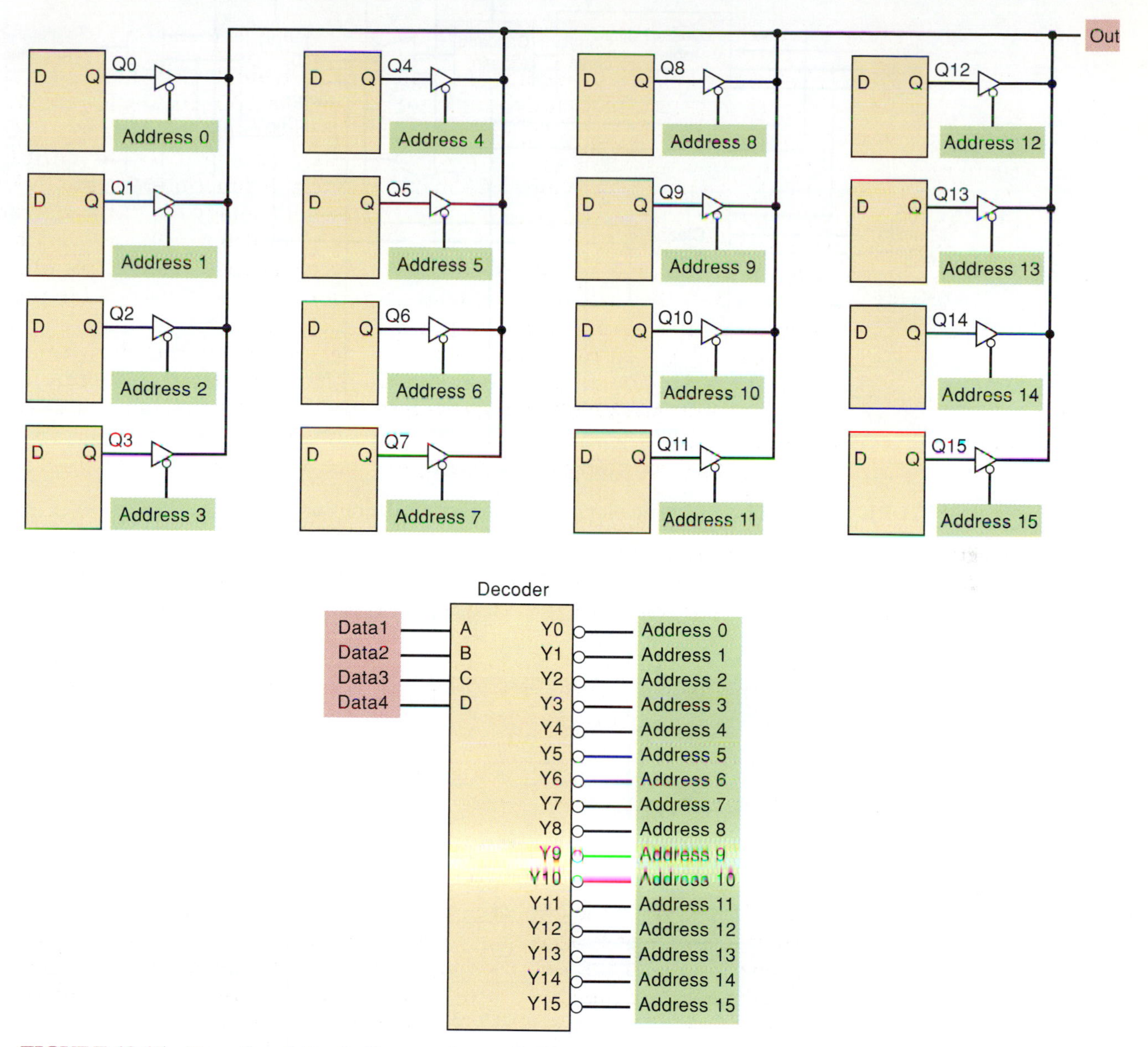

FIGURE 13-15 Functional block diagram for an LUT.

Figure 13-16 shows the block diagram for a FLEX10K logic element. It contains the LUT and programmable register, as well as cascade- and carry-expansion circuitry, programmable control functions, and local and global bus interconnections. The programmable flip-flop can be configured for D, T, JK, or SR operation and will be bypassed for combinational functions. The flip-flop control signals (clock, clear, and preset) can be driven selectively by global inputs, general-purpose I/O pins, or any internally created functions. The LE can produce two outputs to drive local (LAB) and global (FastTrack) interconnects on the chip. This allows the LUT and the register in one LE to be used for unrelated functions. Two types of high-speed data paths—cascade chains and carry chains—connect adjacent LEs without using local interconnects. The cascade-chain expansion allows the FLEX10K architecture to create functions with more than four input variables. Adjacent LUTs can be paralleled together, with each additional LUT providing four more input variables. The carry chain provides a fast carry-forward function between

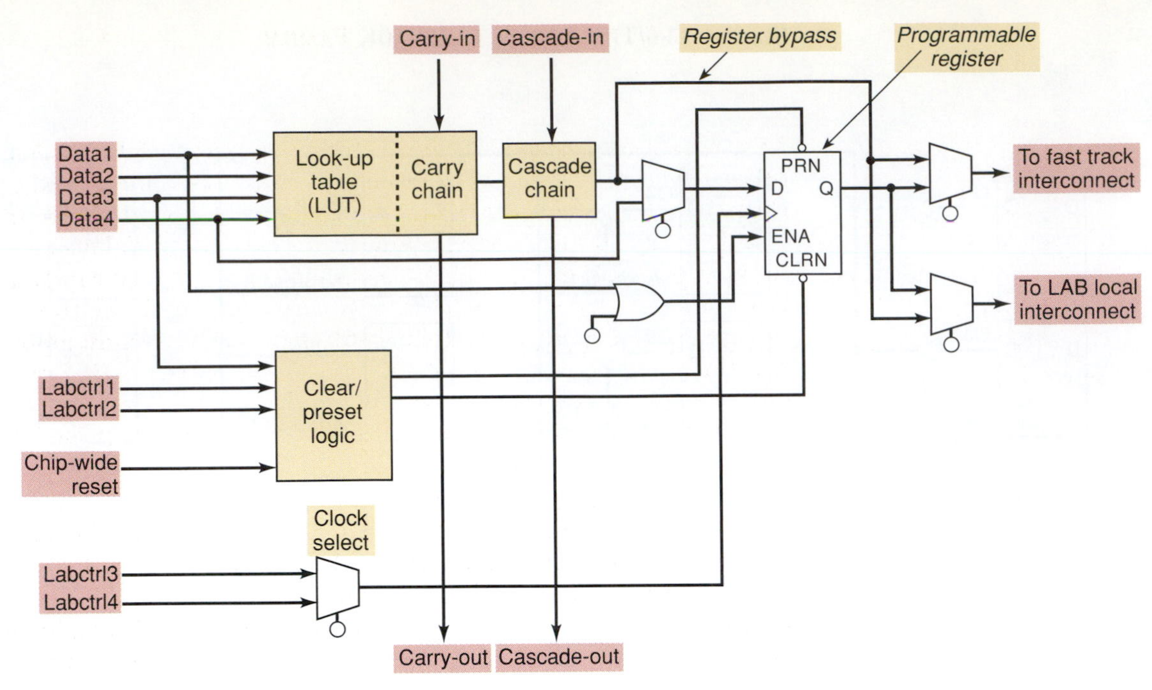

FIGURE 13-16 FLEX10K logic element. (Courtesy of Altera Corporation.)

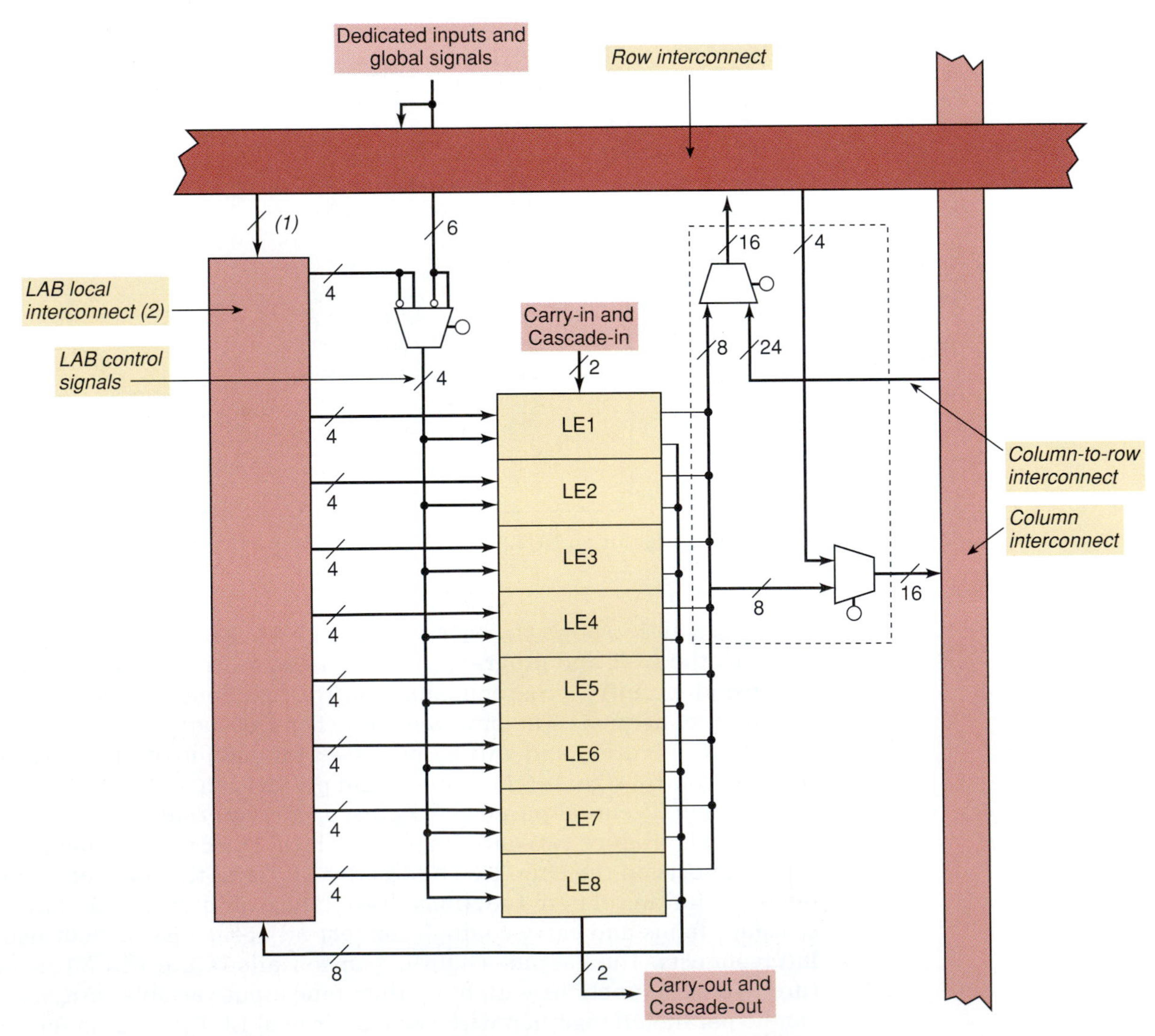

FIGURE 13-17 FLEX10K logic array block. (Courtesy of Altera Corporation.)

LEs, which allows for efficient implementation of functions that build on other functions such as those found in counters, adders, and comparators. In these functions, the upper bits depend on the lower bits. Without an expansion feature like the carry chain, the propagation delays can become quite long for larger circuits. Cascade-chain and carry-chain logic can be created automatically by the compiler software or manually by the designer during design entry. Propagation delays will increase by a small amount when using the expansion options. The MAX+PLUS II or Quartus II Timing Analyzer calculates these added delays for a given design. Intensive use of carry and cascade chains can reduce routing flexibility and should therefore be limited to speed-critical portions of a design.

The logic array block for the FLEX10K family contains eight logic elements and the local interconnect for that LAB (see Figure 13-17). Signals from one LE to another within an LAB are routed with the local interconnect. The row and column interconnects, which Altera has named a FastTrack interconnect, provide the signal pathways between LABs. Each LAB has four control signals available to all eight LEs. Two can be used for register clocks and the other two are for preset or clear.

The overall block diagram for a FLEX10K device is shown in Figure 13-18. In addition to the logic array blocks and FastTrack interconnects that we have already described, the devices contain I/O elements (IOEs) and embedded array blocks (EABs). The IOEs each contain a bidirectional I/O buffer and a register that can be used for either input or output data storage. Each EAB

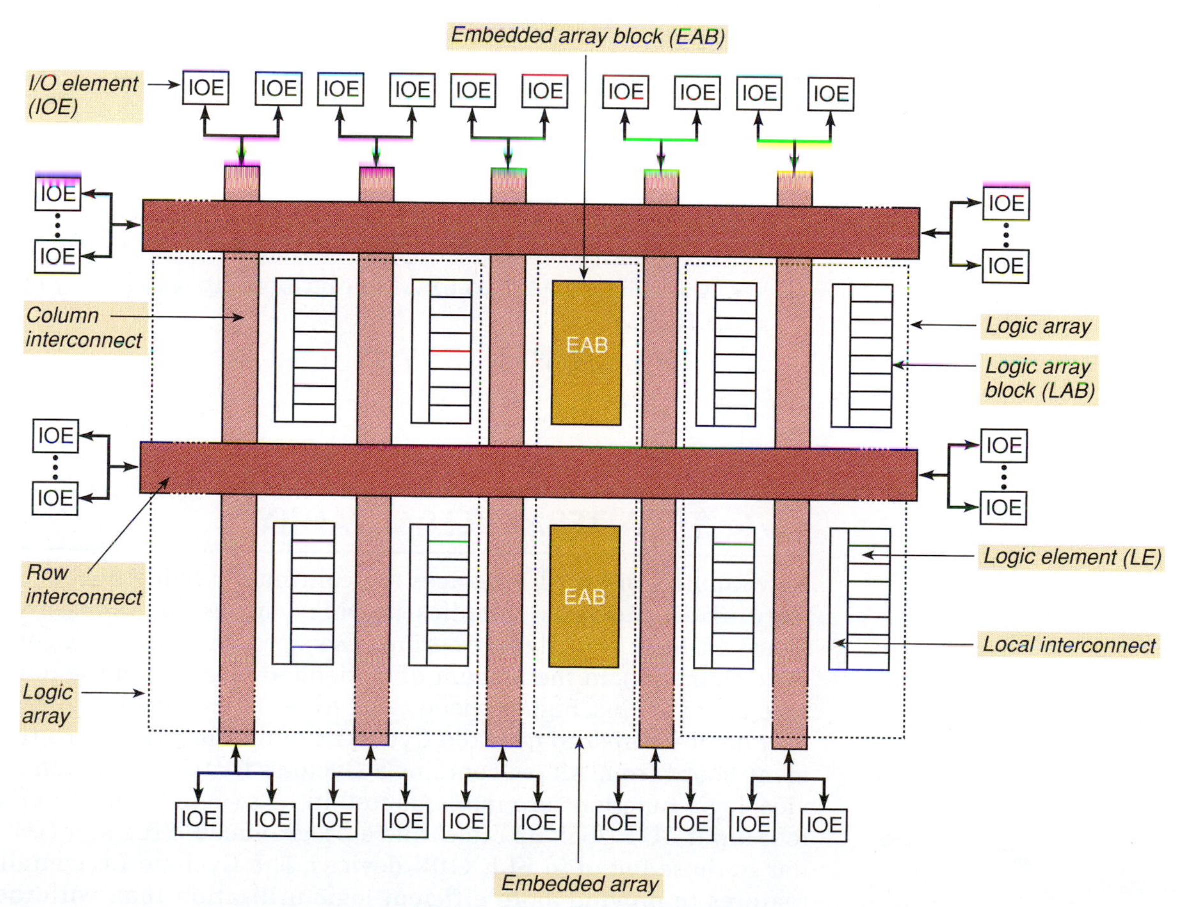

FIGURE 13-18 FLEX10K device block diagram. (Courtesy of Altera Corporation.)

TABLE 13-3 Altera FLEX10K family device features.

Feature	EPF10K10	EPF10K20	EPF10K30	EPF10K40	EPF10K50	EPF10K70	EPF10K100	EPF10K120	EPF10K250
Typical number of gates	10,000	20,000	30,000	40,000	50,000	70,000	100,000	120,000	250,000
Maximum number of gates	31,000	63,000	69,000	93,000	116,000	118,000	158,000	211,000	310,000
LEs	576	1,152	1,728	2,304	2,880	3,744	4,992	6,656	12,160
LABs	72	144	216	288	360	468	624	832	1,520
EABs	3	6	6	8	10	9	12	16	20
Maximum number of I/O pins	150	189	246	189	310	358	406	470	470

provides a flexible block of 2048 bits of RAM storage for various internal memory applications. Combining multiple EABs on one chip can create larger blocks of RAM. An EAB can also be used to create large combinational functions by implementing an LUT.

The FLEX10K family contains several different sizes of parts, as shown in Table 13-3. The Altera UP2 educational development board also contains an EPF10K70 device in a 240-pin package. As you can see in the table, this device has a lot of logic resources available!

REVIEW QUESTIONS

1. What is a look-up table?
2. What advantage does SRAM programming technology have over EEPROM?
3. What disadvantage does SRAM programming technology have compared to EEPROM?
4. What are EABs? What can they be used for?

13-7 THE ALTERA CYCLONE FAMILY

New families of HCPLD devices are continually being developed. The architectures of these new families provide various combinations of enhancements in logic and signal routing resources, in density (higher number of logic elements), in the amount of embedded memory, in the number of available user I/O pins, higher speeds, and lower costs. Another Altera family that may be of interest to us is the Cyclone family. The UP3 educational development board from Altera contains a Cyclone EP1C6 device. In a Cyclone device, logic functions are implemented in LEs (logic elements) that contain a four-input LUT (look-up table) and a programmable register (D flip-flop) similar to those found in FLEX10K devices. The Cyclone LE contains advanced features to provide more efficient logic utilization than with the FLEX10K. The Cyclone LE, for example, has been enhanced to more efficiently create

TABLE 13-4 Altera Cyclone family device features.

Feature	EP1C3	EP1C4	EP1C6	EP1C12	EP1C20
LEs	2,910	4,000	5,980	12,060	20,060
M4K RAM blocks	13	17	20	52	64
Total RAM bits	59,904	78,336	92,160	239,616	294,912
PLLs	1	2	2	2	2
Maximum number of I/O pins	104	301	185	249	301

digital applications that use adder/subtractors, asynchronous loading of the programmable register, and shift registers. The logic array blocks in Cyclone devices consist of 10 LEs and a local interconnect. This family also contains blocks of 4K bits of RAM memory that can be configured as dual-port or single-port memory with words up to 36 bits wide. A global clock network with eight global clock lines provides clocks for all I/O elements, LEs, and memory blocks. Internal phase-lock loops (PLLs) provide clock frequency multiplication and division and clock signal phase shifting. The features of the Cyclone family devices are compared in Table 13-4. Cyclone devices have the capability to interface with other digital circuits using multiple I/O standards, but they do not support 5-V I/O. Cyclone family devices are not supported by MAX+PLUS II design software.

SUMMARY

1. Programmable logic devices (PLDs) are the key technology in the future of digital systems.
2. PLDs can reduce parts inventory, simplify prototype circuitry, shorten the development cycle, reduce the size and power requirements of the product, and allow the hardware of a circuit to be upgraded easily.
3. The major digital system categories are standard logic, application-specific integrated circuits (ASICs), and microprocessor/digital signal processing (DSP) devices.
4. ASIC devices may be programmable logic devices (PLDs), gate arrays, standard cells, or full-custom devices.
5. PLDs are the least expensive type of ASIC to develop.
6. Simple PLDs (SPLDs) contain the equivalent of 600 or fewer gates and are programmed with fuse, EPROM, or EEPROM technology.
7. High-capacity PLDs (HCPLDs) have two major architectural categories: complex programmable logic devices (CPLDs) and field programmable gate arrays (FPGAs).
8. The most common CPLD programming technologies are EEPROM and flash, both of which are nonvolatile.
9. The most common FPGA programming technology is SRAM, which is volatile.
10. The GAL 16V8 is one of the simplest PLDs available but is still widely used and demonstrates the basic principles behind all PLDs.
11. The Altera EPM7128S CPLD contains 128 macrocells, each of which contains a programmable AND/OR circuit and a programmable register.
12. The EPM7128SLC84 can have up to 68 inputs and outputs.

13. The MAX7000S family of CPLDs is in-system programmable (ISP).
14. The Altera FLEX10K and Cyclone families of devices use a look-up table (LUT) architecture in an SRAM technology.
15. SRAM programming technology is volatile, meaning that the devices must be reconfigured at power-up.

IMPORTANT TERMS

standard logic
microprocessor
digital signal processing (DSP)
application-specific integrated circuit (ASIC)
programmable logic device (PLD)
gate array
standard-cell ASIC
full-custom ASIC
simple PLD (SPLD)
complex PLD (CPLD)
field programmable gate array (FPGA)
high-capacity PLD (HCPLD)
one-time programmable (OTP)
programmable array logic (PAL)
macrocell
look-up table (LUT)
logic array block (LAB)
programmable interconnect array (PIA)
logic element (LE)

PROBLEMS

SECTION 13-1

13-1. Describe each of the following major digital system categories:
- (a) Standard logic
- (b) ASICs
- (c) Microprocessor/DSP

13-2.*Name three factors that are generally considered when making design engineering decisions.

13-3. Why is a microprocessor/DSP system called a software solution for a design?

13-4.*What major advantage does a hardware design solution have over a software solution?

13-5. Describe each of the following four ASIC subcategories:
- (a) PLDs
- (b) Gate arrays
- (c) Standard-cell
- (d) Full-custom

13-6.*What are the major advantages and disadvantages of a full-custom ASIC?

13-7. Name the six PLD programming technologies. Which is one-time programmable? Which is volatile?

13-8.*How is the programming of SRAM-based PLDs different from other programming technologies?

SECTION 13-5

13-9. Describe the functions of each of the following architectural structures found in the Altera MAX7000S family:

*Answers to problems marked with an asterisk can be found in the back of the text.

(a) LAB
(b) PIA
(c) Macrocell

13-10.* What two ways can be used to program the MAX7000S family devices?

13-11. What standard device interface is used for in-system programming in the MAX7000S family?

13-12.* What are the four input-only pins on the EPM7128SLC84 (by pin number and function)?

13-13. What is the advantage of using one of the global clock inputs for registered operation?

SECTION 13-6

13-14.* What is the fundamental architectural difference between the MAX 7000S and FLEX10K families? What is the programming technology used by each family? Which family is nonvolatile? Which family contains more logic resources?

ANSWERS TO SECTION REVIEW QUESTIONS

SECTION 13-1

1. Standard logic, ASICs, microprocessor 2. Speed 3. Application-specific integrated circuit 4. Programmable logic devices, gate arrays, standard cells, full custom 5. High-capacity programmable logic device 6. (1) Logic blocks: programmable AND/fixed-OR CPLD versus look-up table FPGA (2) Signal routing resources: uniform CPLD versus varied FPGA 7. Volatility refers to whether a PLD (or memory device) loses stored information when it is powered-down.

SECTION 13-2

1. An IC that contains a large number of gates whose interconnections can be modified by the user to perform a specific function. 2. $O_1 = A$ 3. An intact fuse 4. A hard-wired connection

SECTION 13-3

2. Hard-wired OR; programmable AND 3. Hard-wired AND; programmable OR 4. $O_1 = ABC\overline{D} + \overline{A}\overline{B}CD + \overline{A}BCD = AB\overline{C}\overline{D} + \overline{A}CD$

SECTION 13-4

1. Erasable and reprogrammable; has an OLMC 2. Simple, complex, registered

SECTION 13-5

1. A macrocell is the programmable logic block in MAX7000S CPLDs consisting of a programmable AND/OR circuit and a programmable register (flip-flop). 2. An ISP PLD device is in-system programmable, which means that it can be programmed while connected in the circuit. 3. Global clocks, tristate output enables, asynchronous clear 4. Power consumption may be decreased by slowing down macrocells.

SECTION 13-6

1. A look-up table is typically a 16-word by 1-bit SRAM array used to store the desired output logic levels for a simple logic function. 2. SRAM programs faster and has a higher logic cell density than EEPROM. 3. SRAM is volatile and must be reconfigured upon power-up of the device. 4. Embedded array blocks provide RAM storage on the PLD.

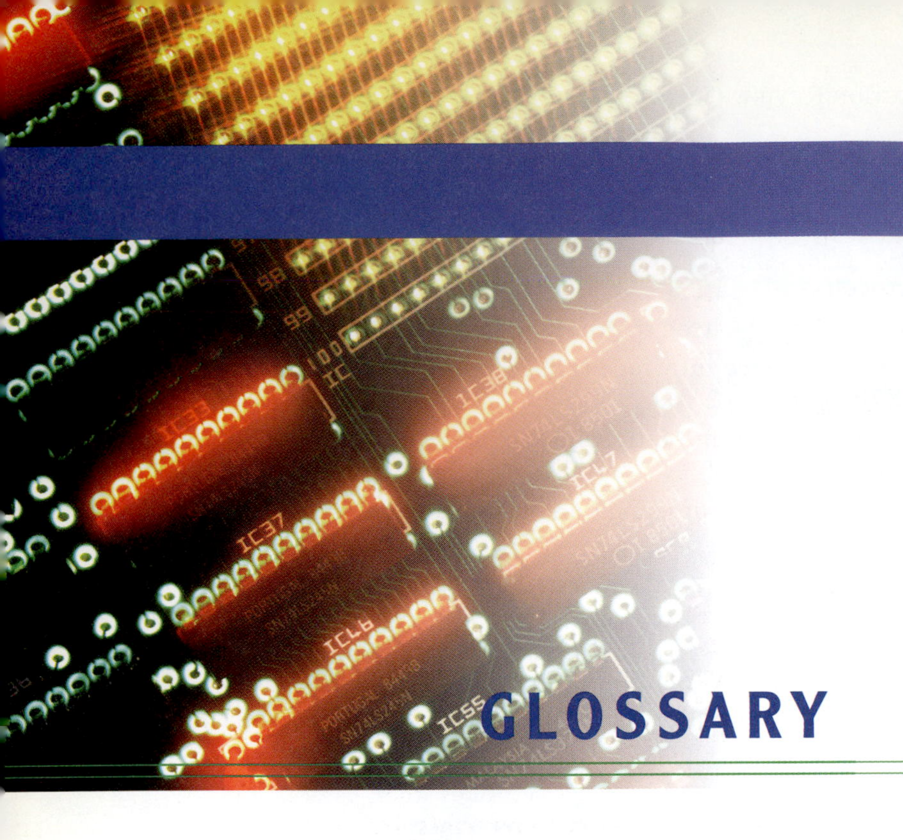

GLOSSARY

Access Time Time between the memory's receiving a new input address and the output data's becoming available in a read operation.

Accumulator Principal register of an arithmetic/logic unit (ALU).

Acquisition Time Time required for a sample-and-hold circuit to capture the analog value that is present on its input.

Active-HIGH (LOW) Decoder Decoder that produces a logic HIGH (LOW) at the output when detection occurs.

Active Logic Level Logic level at which a circuit is considered active. If the symbol for the circuit includes a bubble, the circuit is active-LOW. On the other hand, if it doesn't have a bubble, then the circuit is active-HIGH.

Actuator Electrically controlled device that controls a physical variable.

Addend Number to be added to another.

Adder/Subtractor An adder circuit that can subtract by complementing (negating) one of the operands. *See also* Parallel/Adder.

Address Number that uniquely identifies the location of a word in memory.

Address Bus Unidirectional lines that carry the address code from the CPU to memory and I/O devices.

Address Multiplexing Multiplexing used in dynamic RAMs to save IC pins. It involves latching the two halves of a complete address into the IC in separate steps.

Alias A digital signal that results from sampling an incoming signal at a rate less than twice the highest frequency contained in the incoming signal.

Alphanumeric Codes Codes that represent numbers, letters, punctuation marks, and special characters.

Altera Hardware Description Language (AHDL) A proprietary HDL developed by Altera Corporation for programming their programmable logic devices.

Alternate Logic Symbol A logically equivalent symbol that indicates the active level of the inputs and outputs.

Analog Representation Representation of a quantity that varies over a continuous range of values.

Analog System Combination of devices designed to manipulate physical quantities that are represented in analog form.

Analog-to-Digital Converter (ADC) Circuit that converts an analog input to a corresponding digital output.

Analog Voltage Comparator Circuit that compares two analog input voltages and produces an output that indicates which input is greater.

& When used inside an IEEE/ANSI symbol, an indication of an AND gate or AND function.

AND Gate Digital circuit that implements the AND operation. The output of this circuit is HIGH (logic level 1) only if all of its inputs are HIGH.

AND Operation Boolean algebra operation in which the symbol is used to indicate the ANDing of two or more logic variables. The result of the AND operation will be HIGH (logic level 1) only if all variables are HIGH.

Application-Specific Integrated Circuit (ASIC) An IC that has been specifically designed to meet the requirements of an application. Subcategories include PLDs, gate arrays, standard cells, and full-custom ICs.

ARCHITECTURE Keyword in VHDL used to begin a section of code that defines the operation of a circuit block (ENTITY).

Arithmetic/Logic Unit (ALU) Digital circuit used in computers to perform various arithmetic and logic operations.

ASCII Code (American Standard Code for Information Interchange) Seven-bit alphanumeric code used by most computer manufacturers.

Asserted Term used to describe the state of a logic signal; synonymous with "active."

Astable Multivibrator Digital circuit that oscillates between two unstable output states.

Asynchronous Counter Type of counter in which each flip-flop output serves as the clock input signal for the next flip-flop in the chain.

Asynchronous Inputs Flip-flop inputs that can affect the operation of the flip-flop independent of the synchronous and clock inputs.

Asynchronous Transfer Data transfer performed without the aid of the clock.

Augend Number to which an addend is added.

Auxiliary Memory The part of a computer's memory that is separate from the computer's main working memory. Generally has high density and high capacity, such as magnetic disk.

Backplane Electrical connection common to all segments of an LCD.

Barrel Shifter A shift register that can very efficiently shift a binary number left or right by any number of bit positions.

BCD Counter Binary counter that counts from 0000_2 to 1001_2 before it recycles.

BCD-to-Decimal Decoder Decoder that converts a BCD input into a single decimal output equivalence.

BCD-to-7-Segment Decoder/Driver Digital circuit that takes a four-bit BCD input and activates the required outputs to display the equivalent decimal digit on a 7-segment display.

Behavioral Level of Abstraction A technique of describing a digital circuit that focuses on how the circuit reacts to its inputs.

Bidirectional Data Line Term used when a data line functions as either an input or an output line depending on the states of the enable inputs.

Bilateral Switch CMOS circuit that acts like a single-pole, single-throw (SPST) switch controlled by an input logic level.

Binary-Coded-Decimal Code (BCD Code) Four-bit code used to represent each digit of a decimal number by its four-bit binary equivalent.

Binary Counter Group of flip-flops connected in a special arrangement in which the states of the flip-flops represent the binary number equivalent to the number of pulses that have occurred at the input of the counter.

Binary Digit Bit.

Binary Point Mark that separates the integer from the fractional portion of a binary quantity.

Binary System Number system in which there are only two possible digit values, 0 and 1.

Bipolar DAC Digital-to-analog converter that accepts signed binary numbers as input and produces the corresponding positive or negative analog output value.

Bipolar ICs Integrated digital circuits in which NPN and PNP transistors are the main circuit elements.

BIT In VHDL, the data object type representing a single binary digit (bit).

Bit Digit in the binary system.

Bit Array A way to represent a group of bits by giving it a name and assigning an element number to each bit's position. This same structure is sometimes called a bit vector.

BIT_VECTOR In VHDL, the data object type representing a bit array. *See also* Bit Array.

Boolean Algebra Algebraic process used as a tool in the design and analysis of digital systems. In Boolean algebra only two values are possible, 0 and 1.

Boolean Theorems Rules that can be applied to Boolean algebra to simplify logic expressions.

Bootstrap Program Program, stored in ROM, that a computer executes on power-up.

Bubbles Small circles on the input or output lines of logic-circuit symbols that represent inversion of a particular signal. If a bubble is present, the input or output is said to be active-LOW.

Buffer/Driver Circuit designed to have a greater output current and/or voltage capability than an ordinary logic circuit.

Buffer Register Register that holds digital data temporarily.

Buried Node A defined point in a circuit that is not accessible from outside that circuit.

Bus Group of wires that carry related bits of information.

Bus Contention Situation in which the outputs of two or more active devices are placed on the same bus line at the same time.

Bus Drivers Circuits that buffer the outputs of devices connected to a common bus; used when a large number of devices share a common bus.

Byte Group of eight bits.

Cache A high-speed memory system that can be loaded from the slower system DRAM and accessed quickly by the high-speed CPU.

Capacity Amount of storage space in a memory expressed as the number of bits or number of words.

Carry Digit or bit that is generated when two numbers are added and the result is greater than the base for the number system being used.

Carry Propagation Intrinsic circuit delay of some parallel adders that prevents the carry bit (C_{OUT}) and the result of the addition from appearing at the output simultaneously.

Carry Ripple *See* Carry Propagation.

CAS (Column Address Strobe) Signal used to latch the column address into a DRAM.

CAS-before-RAS Method for refreshing DRAMs that have built-in refresh counters. When the *CAS* input is driven LOW and held there as *RAS* is pulsed LOW, an internal refresh operation is performed at the row address given by the on-chip refresh counter.

Cascading Connecting logic circuits in a serial fashion with the output of one circuit driving the input of the next, and so on.

CASE A control structure that selects one of several options when describing a circuit's operation based on the value of a data object.

Central Processing Unit (CPU) Part of a computer that is composed of the arithmetic/logic unit (ALU) and the control unit.

Checksum Special data word stored in the last ROM location. It is derived from the addition of all other data words in the ROM, and it is used for error-checking purposes.

Chip Select Input to a digital device that controls whether or not the device will perform its function. Also called *chip enable.*

Circuit Excitation Table Table showing a circuit's possible PRESENT-to-NEXT state transitions and the required J and K levels at each flip-flop.

Circular Buffer A memory system that always contains the last n data values that have been written. Whenever a new data value is stored, it overwrites the oldest value in the buffer.

Circulating Shift Register Shift register in which one of the outputs of the last flip-flop is connected to the input of the first flip-flop.

CLEAR An input to a latch or FF used to make $Q = 0$.

CLEAR State The $Q = 0$ state of a flip-flop.

Clock Digital signal in the form of a rectangular pulse train or a square wave.

Clock Skew Arrival of a clock signal at the clock inputs of different flip-flops at different times as a result of propagation delays.

Clock Transition Times Minimum rise and fall times for the clock signal transitions used by a particular IC, specified by the IC manufacturer.

Clocked D Flip-Flop Type of flip-flop in which the D (data) input is the synchronous input.

Clocked Flip-Flops Flip-flops that have a clock input.

Clocked J-K FLip-Flop Type of flip-flop in which inputs J and K are the synchronous inputs.

Clocked S-R Flip-Flop Type of flip-flop in which the inputs SET and RESET are the synchronous inputs.

CMOS (Complementary Metal-Oxide-Semiconductor) Integrated-circuit technology that uses MOSFETs as the principal circuit element. This logic family belongs to the category of unipolar digital ICs.

Combinational Logic Circuits Circuits made up of combinations of logic gates, with no feedback from outputs to inputs.

Comments Text added to any HDL design file or computer program to describe the purpose and operation of the code in general or of individual statements in the code. Documentation regarding author, date, revision, etc., may also be contained in the comments.

Common Anode LED display that has the anodes of all of the segment LEDs tied together.

Common Cathode LED display that has the cathodes of all of the segment LEDs tied together.

Common-Control Block Symbol used by the IEEE/ANSI standard to describe when one or more inputs are common to more than one of the circuits in an IC.

Compiler A program that translates a text file written in a high-level language into a binary file that can be loaded into a programmable device such as a PLD or a computer's memory.

Complement *See* Invert.

Complex PLD (CPLD) Class of PLDs that contain an array of PAL-type blocks that can be interconnected.

COMPONENT A VHDL keyword used at the top of a design file to provide information about a library component.

Computer Word Group of binary bits that form the primary unit of information in a computer.

Concatenate A term used to describe the arrangement or linking of two or more data objects into ordered sets.

Concurrent Events that occur simultaneously (at the same time). In HDL, the circuits generated by concurrent statements are not affected by the order or sequence of the statements in the code.

Concurrent Assignment Statement A statement in AHDL or VHDL that describes a circuit that works concurrently with all other circuits that are described by concurrent statements.

Conditional Signal Assignment A VHDL concurrent construct that evaluates a series of conditions sequentially to determine the appropriate value to assign to a signal. The first true condition evaluated determines the assigned value.

Constants Symbolic names that can be used to represent fixed numeric (scalar) values.

Contact Bounce The tendency of all mechanical switches to vibrate when forced to a new position. The vibrations cause the circuit to make contact and break contact repeatedly until the vibrations settle out.

Contention Two (or more) output signals connected together trying to drive a common point to different voltage levels. *See also* Bus Contention.

Control Bus Set of signal lines that are used to synchronize the activities of the CPU and the separate μC elements.

Control Inputs Input signals synchronized with the active clock transition that determine the output state of a flip-flop.

Control Unit Part of a computer that provides decoding of program instructions and the necessary timing and control signals for the execution of such instructions.

Count Enable An input on a synchronous counter that controls whether the outputs respond to or ignore an active clock transition.

Crystal-Controlled Clock Generator Circuit that uses a quartz crystal to generate a clock signal at a precise frequency.

Current-Sinking Logic Logic family in which the output of a logic circuit sinks current from the input of the logic circuit that it is driving.

Current-Sinking Transistor Name given to the output transistor (Q_4) of a TTL circuit. This transistor is turned on when the output logic level is LOW.

Current-Sourcing Logic Logic family in which the output of a logic circuit sources, or supplies, current to the input of the logic circuit that it is driving.

Current-Sourcing Transistor Name given to the output transistor (Q_3) of most TTL circuits. This transistor is conducting when the output logic level is HIGH.

Current Transients Current spikes generated by the totem-pole output structure of a TTL circuit and caused when both transistors are simultaneously turned on.

D Flip-Flop *See* Clocked D Flip-Flop.

D Latch Circuit that contains a NAND gate latch and two steering NAND gates.

Data Binary representations of numerical values or nonnumerical information in a digital system. Data are used and often modified by a computer program.

Data Acquisition Process by which a computer acquires digitized analog data.

Data Bus Bidirectional lines that carry data between the CPU and the memory, or between the CPU and the I/O devices.

Data Distributors *See* Demultiplexer.

Data-Rate Buffer Application of FIFOs in which sequential data are written into the FIFO at one rate and read out at a different rate.

Data Selectors *See* Multiplexer.

Data Transfer *See* Parallel Data Transfer *or* Serial Data Transfer.

Decade Counter Any counter capable of going through 10 different logic states.

Decimal System Number system that uses 10 different digits or symbols to represent a quantity.

Decision Control Structures The statements and syntax that describe how to choose between two or more options in the code.

Decoder Digital circuit that converts an input binary code into a corresponding single active output.

Decoding Act of identifying a particular binary combination (code) in order to display its value or recognize its presence.

DEFAULTS An AHDL keyword used to establish a default value for a combinational signal for instances when the code does not explicitly specify a value.

DeMorgan's Theorems (1) Theorem stating that the complement of a sum (OR operation) equals the product (AND operation) of the complements, and (2) theorem stating that the complement of a product (AND operation) equals the sum (OR operation) of the complements.

Demultiplexer (DEMUX) Logic circuit that, depending on the status of its select inputs, will channel its data input to one of several data outputs.

Density A relative measure of capacity to store bits in a given amount of space.

Dependency Notation Method used to represent symbolically the relationship between inputs and outputs of logic circuits. This method employs the use of qualifying symbols embedded near the top center or geometric center of a symbol element.

Differential Inputs Method of connecting an analog signal to an analog circuit's + and − inputs, neither of which is ground, such that the analog circuit acts upon the voltage difference between the two inputs.

Digital Computer System of hardware that performs arithmetic and logic operations, manipulates data, and makes decisions.

Digital Integrated Circuits Self-contained digital circuits made by using one of several integrated-circuit fabrication technologies.

Digital One-Shot A one-shot that uses a counter and clock rather than an RC circuit as a time base.

Digital-Ramp ADC Type of analog-to-digital converter in which an internal staircase waveform is generated and utilized for the purpose of accomplishing the conversion. The conversion time for this type of analog-to-digital converter varies depending on the value of the input analog signal.

Digital Representation Representation of a quantity that varies in discrete steps over a range of values.

Digital Signal Processing (DSP) Method of performing repetitive calculations on an incoming stream of digital data words to accomplish some form of signal conditioning. The data are typically digitized samples of an analog signal.

Digital Storage Oscilloscope Instrument that samples, digitizes, stores, and displays analog voltage waveforms.

Digital System Combination of devices designed to manipulate physical quantities that are represented in digital form.

Digital-to-Analog Converter (DAC) Circuit that converts a digital input to a corresponding analog output.

Digitization Process by which an analog signal is converted to digital data.

Disable Action in which a circuit is prevented from performing its normal function, such as passing an input signal through to its output.

Divide-and-Conquer Troubleshooting technique whereby tests are performed that will eliminate half of all possible remaining causes of the malfunction.

Don't-Care Situation when a circuit's output level for a given set of input conditions can be assigned as either a 1 or a 0.

Down Counter Counter that counts from a maximum count downward to 0.

Downloading Process of transferring output files to a programming fixture.

DRAM Controller IC used to handle refresh and address multiplexing operations needed by DRAM systems.

Driver Technical term sometimes added to an IC's description to indicate that the IC's outputs can operate at higher current and/or voltage limits than a normal standard IC.

Dual-in-Line Package (DIP) A very common IC package with two parallel rows of pins intended to be inserted into a socket or through holes drilled in a printed circuit board.

Dual-Slope Analog-to-Digital Converter Type of analog-to-digital converter that linearly charges a capacitor from a current proportional to V_A for a fixed time interval and then increments a counter as the capacitor is linearly discharged to 0.

Dynamic RAM (DRAM) Type of semiconductor memory that stores data as capacitor charges that need to be refreshed periodically.

ECL Emitter-coupled logic; also referred to as *current-mode logic.*

Edge-Detector Circuit Circuit that produces a narrow positive spike that occurs coincident with the active transition of a clock input pulse.

Edge-Triggered Manner in which a flip-flop is activated by a signal transition. A flip-flop may be either a positive- or a negative-edge-triggered flip-flop.

Electrically Compatible When two ICs from different logic series can be connected directly without any special measures taken to ensure proper operation.

Electrically Erasable Programmable ROM (EEPROM) ROM that can be electrically programmed, erased, and reprogrammed.

Electrostatic Discharge (ESD) The often detrimental act of the transfer of static electricity (i.e., an electrostatic charge) from one surface to another. This impulse of current can destroy electronic devices.

ELSE A control structure used in conjunction with IF/THEN to perform an alternate action in the case that the condition is false. An IF/THEN/ELSE always performs one of two actions.

ELSIF A control structure that can be used multiple times following an IF statement to select one of several options in describing a circuit's operation based on whether the associated expressions are true or false.

Embedded Microcontroller Microcontroller that is embedded in a marketable product such as a VCR or an appliance.

Emitter-Coupled Logic *See* ECL.

Enable Action in which a circuit is allowed to perform its normal function, such as passing an input signal through to its output.

Encoder Digital circuit that produces an output code depending on which of its inputs is activated.

Encoding Use of a group of symbols to represent numbers, letters, or words.

ENTITY Keyword in VHDL used to define the basic block structure of a circuit. This word is followed by a name for the block and the definitions of its input/output ports.

Enumerated Type A VHDL user-defined type for a signal or variable.

Erasable Programmable ROM (EPROM) ROM that can be electrically programmed by the user. It can be erased (usually with ultraviolet light) and reprogrammed as often as desired.

EVENT A VHDL keyword used as an attribute attached to a signal to detect a transition of that signal. Generally, an event means a signal changed state.

Exclusive-NOR (XNOR) Circuit Two-input logic circuit that produces a HIGH output only when the inputs are equal.

Exclusive-OR (XOR) Circuit Two-input logic circuit that produces a HIGH output only when the inputs are different.

Fan-Out Maximum number of standard logic inputs that the output of a digital circuit can reliably drive.

Field Programmable Gate Array (FPGA) Class of PLDs that contain an array of more complex logic cells that can be very flexibly interconnected to implement high-level logic circuits.

Field Programmable Logic Array (FPLA) A PLD that uses both a programmable AND array and a programmable OR array.

Firmware Computer programs stored in ROM.

First-In, First-Out (FIFO) Memory Semiconductor sequential-access memory in which data words are read out in the same order in which they were written in.

555 Timer TTL-compatible IC that can be wired to operate in several different modes, such as a one-shot and an astable multivibrator.

Flash ADC Type of analog-to-digital converter that has the highest operating speed available.

Flash Memory Nonvolatile memory IC that has the high-speed access and in-circuit erasability of EEPROMs but with higher densities and lower cost.

Flip-Flop Memory device capable of storing a logic level.

Floating Bus When all outputs connected to a data bus are in the Hi-Z state.

Floating Input Input signal that is left disconnected in a logic circuit.

FOR Loop *See* Iterative Loop.

4-to-10 Decoder *See* BCD-to-Decimal Decoder.

Frequency The number of cycles per unit time of a periodic waveform.

Frequency Counter Circuit that can measure and display a signal's frequency.

Frequency Division The use of flip-flop circuits to produce an output waveform whose frequency is equal to the input clock frequency divided by some integer value.

Full Adder Logic circuit with three inputs and two outputs. The inputs are a carry bit (C_{IN}) from a previous stage, a bit from the augend, and a bit from the addend, respectively. The outputs are the sum bit and the carry-out bit (C_{OUT}) produced by the addition of the bit from the addend with the bit from the augend and C_{IN}.

Full-Custom An application-specific integrated circuit (ASIC) that is completely designed and fabricated from fundamental elements of electronic devices such as transistors, diodes, resistors, and capacitors.

Full-Scale Error Term used by some digital-to-analog converter manufacturers to specify the accuracy of a digital-to-analog converter. It is defined as the maximum deviation of a digital-to-analog converter's output from its expected ideal value.

Full-Scale Output Maximum possible output value of a digital-to-analog converter.

Function Generator Circuit that produces different waveforms. It can be constructed using a ROM, a DAC, and a counter.

Function Prototype A text description that contains all the essential defining attributes of a library function or module.

Functionally Equivalent When the logic functions performed by two different ICs are exactly the same.

Fusible Link Conducting material that can be made nonconducting (i.e., open) by passing too much current through it.

Gate Array An application-specific integrated circuit (ASIC) made up of hundreds of thousands of prefabricated basic gates that can be custom interconnected in the last stages of manufacture to form the desired digital circuit.

GENERATE A VHDL keyword used with the FOR construct to iteratively define multiple similar components and to interconnect them.

Glitch Momentary, narrow, spurious, and sharply defined change in voltage.

Gray Code A code that never has more than one bit changing when going from one state to another,

GSI Giga-scale integration (1,000,000 gates or more).

Half Adder Logic circuit with two inputs and two outputs. The inputs are a bit from the augend and a bit from the addend. The outputs are the sum bit produced by the addition of the bit from the addend with the bit from the augend and the resulting carry (C_{OUT}) bit, which will be added to the next stage.

Hard Disk Rigid metal magnetic disk used for mass storage.

Hardware Description Language (HDL) A text-based method of describing digital hardware that follows a rigid syntax for representing data objects and control structures.

Hexadecimal Number System Number system that has a base of 16. Digits 0 through 9 plus letters A through F are used to express a hexadecimal number.

Hierarchical Design A method of designing a project by breaking it into constituent modules, each of which can be broken further into simpler constituent modules.

Hierarchy A group of tasks arranged in rank order of magnitude, importance, or complexity.

High-Capacity PLD (HCPLD) A PLD with thousands of logic gates and many programmable macrocell resources, along with very flexible interconnection resources.

Hold Time (t_H) Time interval immediately following the active transition of the clock signal during which the control input must be maintained at the proper level.

Hybrid System System that employs both analog and digital techniques.

IEEE/ANSI Institute of Electrical and Electronics Engineers/American National Standards Institute, both professional organizations that establish standards.

IF/THEN A control structure that evaluates a condition and performs an action if the condition is true or bypasses the action and continues on if the condition is false.

Indeterminate Of a logic voltage level, outside the required range of voltages for either logic 0 or logic 1.

Index Another name for the element number of any given bit in a bit array.

Inhibit Circuits Logic circuits that control the passage of an input signal through to the output.

Input Term Matrix Part of a programmable logic device that allows inputs to be selectively connected to or disconnected from internal logic circuitry.

Input Unit Part of a computer that facilitates the feeding of information into the computer's memory unit or ALU.

Instructions Binary codes that tell a computer what operation to perform. A program is made up of an orderly sequence of instructions.

INTEGER In VHDL, the data object type representing a numeric value.

Interfacing Joining of dissimilar devices in such a way that they are able to function in a compatible and coordinated manner; connection of the output of a system to the input of a different system with different electrical characteristics.

Interpolation Filtering Another name for oversampling. Interpolation refers to intermediate values inserted into the digital signal to smooth out the waveform.

Invert Cause a logic level to go to the opposite state.

INVERTER Also referred to as the NOT circuit; logic circuit that implements the NOT operation. An INVERTER has only one input, and its output logic level is always the opposite of this input's logic level.

Iterative Loop A control structure that implies a repetitive operation and a stated number of iterations.

Jam Transfer *See* Asynchronous Transfer.

JEDEC Joint Electronic Device Engineering Council, which established standards for IC pin assignments and PLD file format.

J-K Excitation Table Table showing the required J and K input conditions for each possible state transition for a single J-K flip-flop.

Johnson Counter Shift register in which the inverted output of the last flip-flop is connected to the input of the first flip-flop.

JTAG Joint Test Action Group, which created a standard interface that allows access to the inner workings of an IC for testing, controlling, and programming purposes.

Karnaugh Map (K Map) Two-dimensional form of a truth table used to simplify a sum-of-products expression.

Latch Type of flip-flop; also, the action by which a logic circuit output captures and holds the value of an input.

Latch-Up Condition of dangerously high current in a CMOS IC caused by high-voltage spikes or ringing at device input and output pins.

Latency The inherent delay associated with reading data from a DRAM. It is caused by the timing requirements of supplying the row and column addresses, and the time for the data outputs to settle.

LCD Liquid-crystal display.

Lead Pitch The distance between the centers of adjacent pins on an IC.

Least Significant Bit (LSB) Rightmost bit (smallest weight) of a binary expressed quantity.

Least Significant Digit (LSD) Digit that carries the least weight in a particular number.

LED Light-emitting diode.

Libraries A collection of descriptions of commonly used hardware circuits that can be used as modules in a design file.

Library of Parameterized Modules (LPM) A set of generic library functions designed to be very flexible in allowing the user to specify the number of bits, mod number, control options, etc.

Linear Buffer A first-in, first-out memory system that fills at one rate and empties at another rate. After it is full, no data can be stored until data is read from the buffer. *See also* First In, First-Out (FIFO) Memory.

Linearity Error Term used by some digital-to-analog converter manufacturers to specify the device's accuracy. It is defined as the maximum deviation in step size from the ideal step size.

Literals In VHDL, a scalar value or bit pattern that is to be assigned to a data object.

Load Operation Transfer of data into a flip-flop, a register, a counter, or a memory location.

Local Signal *See* Buried Node.

Logic Array Block (LAB) A term Altera Corporation uses to describe building blocks of their CPLDs. Each LAB is similar in complexity to an SPLD.

Logic Circuit Any circuit that behaves according to a set of logic rules.

Logic Elements A term Altera Corporation uses to describe the building blocks of their FLEX10K family of PLDs. The logic elements are programmed as a ram-based look-up table.

Logic Function Generation Implementation of a logic function directly from a truth table by means of a digital IC such as a multiplexer.

Logic Level State of a voltage variable. The states 1 (HIGH) and 0 (LOW) correspond to the two usable voltage ranges of a digital device.

Logic Primitive A circuit description of a fundamental component that is built into the MAX+PLUS II system of libraries.

Logic Probe Digital troubleshooting tool that senses and indicates the logic level at a particular point in a circuit.

Logic Pulser Testing tool that generates a short-duration pulse when actuated manually.

Look-Ahead Carry Ability of some parallel adders to predict, without having to wait for the carry to propagate through the full adders, whether or not a carry bit (C_{OUT}) will be generated as a result of the addition, thus reducing the overall propagation delays.

Look-Up Table (LUT) A way to implement a single logic function by storing the correct output logic state in a memory location that corresponds to each particular combination of input variables.

Looping Combining of adjacent squares in a Karnaugh map containing 1s for the purpose of simplification of a sum-of-products expression.

Low-Power Schottky TTL (LS-TTL) TTL subfamily that uses the identical Schottky TTL circuit but with larger resistor values.

Low-Voltage Differential Signaling (LVDS) A technology for driving high-speed data lines in low-voltage systems that uses two conductors and reverses the polarity to distinguish between HIGH and LOW.

Low-Voltage Technology New line of logic devices that operate from a nominal supply voltage of 3.3 V or less.

LSI Large-scale integration (100 to 9999 gates).

MAC An abbreviation for Multiply Accumulate Unit, the hardware section of a DSP that multiplies a sample with a coefficient and then accumulates (sums) a running total of these products.

MACHINE An AHDL keyword used to create a state machine in a design file.

Macrocell A circuit made up of a group of basic digital components such as AND gates, OR gates, registers, and tristate control circuits that can be interconnected within a PLD via a program.

Macrofunctions A term used by Altera Corporation to describe the predefined hardware descriptions in their libraries that represent standard IC parts.

Magnetic Disk Memory Mass storage memory that stores data as magnetized spots on a rotating, flat disk surface.

Magnetic Tape Memory Mass storage memory that stores data as magnetized spots on a magnetically coated plastic tape.

Magnitude Comparator Digital circuit that compares two input binary quantities and generates outputs to indicate whether the inputs are equal or, if not, which is greater.

Main Memory High-speed portion of a computer's memory that holds the program and data the computer is currently working on. Also called *working memory*.

Mask-Programmed ROM (MROM) ROM that is programmed by the manufacturer according to the customer's specifications. It cannot be erased or reprogrammed.

Mass Storage Storage of large amounts of data; not part of a computer's internal memory.

Maximum Clocking Frequency (f_{MAX}) Highest frequency that may be applied to the clock input of a flip-flop and still have it trigger reliably.

Mealy Model A state-machine model in which the output signals are controlled by combinational inputs as well as the state of the sequential circuit.

Megafunctions A complex or high-level building block available in the Altera library.

Memory Ability of a circuit's output to remain at one state even after the input condition that caused that state is removed.

Memory Cell Device that stores a single bit.

Memory Foldback Redundant enabling of a memory device at more than one address range as a result of incomplete address decoding.

Memory Map Diagram of a memory system that shows the address range of all existing memory devices as well as available memory space for expansion.

Memory Unit Part of a computer that stores instructions and data received from the input unit, as well as results from the arithmetic/logic unit.

Memory Word Group of bits in memory that represents instructions or data of some type.

Microcomputer Newest member of the computer family, consisting of microprocessor chip, memory chips, and I/O interface chips. In some cases, all of the aforementioned are in one single IC.

Microcontroller Small microcomputer used as a dedicated controller for a machine, a piece of equipment, or a process.

Microprocessor (MPU) LSI chip that contains the central processing unit (CPU).

Minuend Number from which the subtrahend is to be subtracted.

MOD Number Number of different states that a counter can sequence through; the counter's frequency division ratio.

Mode The attribute of a port in a digital circuit that defines it as input, output, or bidirectional.

Monostable Multivibrator *See* One-Shot.

Monotonicity Property whereby the output of a digital-to-analog converter increases as the binary input is increased.

Moore Model A state-machine model in which the output signals are controlled only by the sequential circuit outputs.

MOSFET Metal-oxide-semiconductor field-effect transistor.

Most Significant Bit (MSB) Leftmost binary bit (largest weight) of a binary expressed quantity.

Most Significant Digit (MSD) Digit that carries the most weight in a particular number.

MSI Medium-scale integration (12 to 99 gates).

Multiplexer (MUX) Logic circuit that, depending on the status of its select inputs, will channel one of several data inputs to its output.

Multiplexing Process of selecting one of several input data sources and transmitting the selected data to a single output channel.

Multistage Counter Counter in which several counter stages are connected so that the output of one stage serves as the clock input of the next stage to achieve greater counting range or frequency division.

NAND Gate Logic circuit that operates like an AND gate followed by an INVERTER. The output of a NAND gate is LOW (logic level 0) only if all inputs are HIGH (logic level 1).

NAND Gate Latch Flip-flop constructed from two cross-coupled NAND gates.

Negation Operation of converting a positive number to its negative equivalent, or vice versa. A signed binary number is negated by the 2's-complement operation.

Negative-Going Transition When a clock goes from 1 to 0.

Nested To have one control structure embedded within another control structure.

Nibble A group of four bits.

N-MOS (N-Channel Metal-Oxide-Semiconductor) Integrated-circuit technology that uses N-channel MOSFETs as the principal circuit element.

NODE A keyword in AHDL used to declare an intermediate variable (data object) that is local to that subdesign.

Noise Spurious voltage fluctuations that may be present in the environment and cause digital circuits to malfunction.

Noise Immunity Circuit's ability to tolerate noise voltages on its inputs.

Noise Margin Quantitative measure of noise immunity.

Nonretriggerable One-Shot Type of one-shot that will not respond to a trigger input signal while in its quasi-stable state.

Nonvolatile Memory Memory that will keep storing its information without the need for electrical power.

Nonvolatile RAM Combination of a RAM array and an EEPROM or flash on the same IC. The EEPROM serves as a nonvolatile backup to the RAM.

NOR Gate Logic circuit that operates like an OR gate followed by an INVERTER. The output of a NOR gate is LOW (logic level 0) when any or all inputs are HIGH (logic level 1).

NOR Gate Latch Flip-flop constructed from two cross-coupled NOR gates.

NOT Circuit *See* INVERTER.

NOT Operation Boolean algebra operation in which the overbar (¯) or the prime (′) symbol is used to indicate the inversion of one or more logic variables.

Objects Various ways of representing data in the code of any HDL.

Observation/Analysis Process used to troubleshoot circuits or systems in order to predict the possible faults before ever picking up a troubleshooting instrument. When this process is used, the troubleshooter must understand the circuit operation, observe the symptoms of the failure, and then reason through the operation.

Octal Number System Number system that has a base of 8; digits from 0 to 7 are used to express an octal number.

Octets Groups of eight 1s that are adjacent to each other within a Karnaugh map.

Offset Error Deviation from the ideal 0 V at the output of a digital-to-analog converter when the input is all 0s. In reality, there is a very small output voltage for this situation.

1-of-10 Decoder *See* BCD-to-Decimal Decoder.

1's-Complement Form Result obtained when each bit of a binary number is complemented.

One-Shot Circuit that belongs to the flip-flop family but that has only one stable state (normally $Q = 0$).

One-Time Programmable (OTP) A broad category of programmable components that are programmed by permanently altering the connections (e.g., melting a fuse element).

Open-Collector Output Type of output structure of some TTL circuits in which only one transistor with a floating collector is used.

Optical Disk Memory Class of mass memory devices that uses a laser beam to write onto and read from a specially coated disk.

OR Gate Digital circuit that implements the OR operation. The output of this circuit is HIGH (logic level 1) if any or all of its inputs are HIGH.

OR Operation Boolean algebra operation in which the symbol + is used to indicate the ORing of two or more logic variables. The result of the OR operation will be HIGH (logic level 1) if one or more variables are HIGH.

Output Logic Macrocell (OLMC) A group of logic elements (gates, multiplexers, flip-flops, buffers) in a PLD that can be configured in various ways.

Output Unit Part of a computer that receives data from the memory unit or ALU and presents it to the outside world.

Overflow When in the process of adding signed binary numbers, a carry of 1 is generated from the MSB position of the number into the sign bit position.

Override Inputs Synonymous with "asynchronous inputs."

Oversampling Inserting data points between sampled data in a digital signal to make it easier to filter out the rough edges of the waveform coming out of the DAC.

PACKAGE A VHDL keyword used to define a set of global elements that are available to other modules.

Parallel Adder Digital circuit made from full adders and used to add all of the bits from the addend and the augend together simultaneously.

Parallel Counter *See* Synchronous Counter.

Parallel Data Transfer Operation by which several bits of data are transferred simultaneously into a counter or a register.

Parallel In/Parallel Out Register Type of register that can be loaded with parallel data and has parallel outputs available.

Parallel In/Serial Out Register Type of register that can be loaded with parallel data and has only one serial output.

Parallel Load *See* Parallel Data Transfer.

Parallel-to-Serial Conversion Process by which all data bits are presented simultaneously to a circuit's input and then transmitted one bit at a time to its output.

Parallel Transmission Simultaneous transfer of all bits of a binary number from one place to another.

Parity Bit Additional bit that is attached to each code group so that the total number of 1s being transmitted is always even (or always odd).

Parity Checker Circuit that takes a set of data bits (including the parity bit) and checks to see if it has the correct parity.

Parity Generator Circuit that takes a set of data bits and produces the correct parity bit for the data.

Parity Method Scheme used for error detection during the transmission of data.

Percentage Resolution Ratio of the step size to the full-scale value of a digital-to-analog converter. Percentage resolution can also be defined as the reciprocal of the maximum number of steps of a digital-to-analog converter.

Period The amount of time required for one complete cycle of a periodic event or waveform.

Periodic A cycle that repreats itself regularly in time and form.

Pin-Compatible When the corresponding pins on two different ICs have the same functions.

Pixel Small dots of light that make up a graphical image on a display.

P-MOS (P-channel Metal Oxide Semiconductor) Integrated-circuit technology that uses P-channel MOSFETs as the principal circuit element.

PORT MAP A VHDL keyword that precedes the list of connections specified between components.

Positional-Value System System in which the value of a digit depends on its relative position.

Positive-Going Transition (PGT) When a clock signal changes from a logic 0 to a logic 1.

Power-Down Operating mode in which a chip is disabled and draws much less power than when it is fully enabled.

Power-Supply Decoupling Connection of a small RF capacitor between ground and V_{CC} near each TTL integrated circuit on a circuit board.

Power-Up Self-Test Program stored in ROM and executed by the CPU on power-up to test RAM and/or ROM portions of the computer circuitry.

Preprocessor Commands Compiler commands that are processed before the main program code in order to control how the code is interpreted.

Prescaler A counter circuit that takes base reference frequency and scales it by dividing the frequency down to a rate required by the system.

Present State–Next State Table A table which lists each possible present state of a sequential (counter) circuit and identifies the corresponding next state.

PRESET Asynchronous input used to set $Q = 1$ immediately.

Presettable Counter Counter that can be preset to any starting count either synchronously or asynchronously.

Priority Encoder Special type of encoder that senses when two or more inputs are activated simultaneously and then generates a code corresponding to the highest-numbered input.

PROCESS A VHDL keyword that defines the beginning of a block of code that describes a circuit that must respond whenever certain signals (in the sensitivity list) change state. All sequential statements must occur inside a process.

Product-of-Sums Form Logic expression consisting of two or more OR terms (sums) that are ANDed together.

Program Sequence of binary-coded instructions designed to accomplish a particular task by a computer.

Programmable Array Logic (PAL) Class of programmable logic devices. Its AND array is programmable, whereas its OR array is hard-wired.

Programmable Interconnect Array (PIA) A term Altera Corporation uses to describe the resources used to connect the LABs with each other and also with the input/output modules.

Programmable Logic Array (PLA) Class of programmable logic devices. Both its AND and its OR arrays are programmable. Also called a *field programmable logic array (FPLA).*

Programmable Logic Device (PLD) IC that contains a large number of interconnected logic functions. The user can program the IC for a specific function by selectively breaking the appropriate interconnections.

Programmable Output Polarity Feature of many PLDs whereby an XOR gate with a polarity fuse gives the designer the option of inverting or not inverting a device output.

Programmable ROM (PROM) ROM that can be electrically programmed by the user. It cannot be erased and reprogrammed.

Programmer A fixture used to apply the proper voltages to PLD and PROM chips in order to program them.

Programming The act of storing 1s and 0s in a programmable logic device to configure its behavioral characteristics.

Propagation Delays (t_{PLH}/t_{PHL}) Delay from the time a signal is applied to the time when the output makes its change.

Pull-Down Transistor *See* Current-Sinking Transistor.

Pull-Up Transistor *See* Current-Sourcing Transistor.

Pulse A momentary change of logic state that represents an event to a digital system.

Pulse-Steering Circuit A logic circuit that can be used to select the destination of an input pulse, depending on the logic levels present at the circuit's inputs.

Quantization Error Error caused by the nonzero resolution of an analog-to-digital converter. It is an inherent error of the device.

Quasi-Stable State State to which a one-shot is temporarily triggered (normally $Q = 1$) before returning to its stable state (normally $Q = 0$).

R/2R Ladder DAC Type of digital-to-analog converter whose internal resistance values span a range of only 2 to 1.

Random-Access Memory (RAM) Memory in which the access time is the same for any location.

RAS (Row Address Strobe) Signal used to latch the row address into a DRAM chip.

RAS-Only Refresh Method for refreshing DRAM in which only row addresses are strobed into the DRAM using the *RAS* input.

Read Term used to describe the condition when the CPU is receiving data from another element.

Read-Only Memory (ROM) Memory device designed for applications where the ratio of read operations to write operations is very high.

Read Operation Operation in which a word in a specific memory location is sensed and possibly transferred to another device.

Read/Write Memory (RWM) Any memory that can be read from and written into with equal ease.

Refresh Counter Counter that keeps track of row addresses during a DRAM refresh operation.

Refreshing Process of recharging the cells of a dynamic memory.

Register Group of flip-flops capable of storing data.

RESET Term synonymous with "CLEAR."

RESET State The $Q = 0$ state of a flip-flop.

Resolution In a digital-to-analog converter, smallest change that can occur in the output for a change in digital input; also called *step size.* In an analog-to-digital converter, smallest amount by which the analog input must change to produce a change in the digital output.

Retriggerable One-Shot Type of one-shot that will respond to a trigger input signal while in its quasi-stable state.

Ring Counter Shift register in which the output of the last flip-flop is connected to the input of the first flip-flop.

Ripple Counter *See* Asynchronous Counter.

Sample-and-Hold Circuit Type of circuit that utilizes a unity-gain buffer amplifier in conjunction with a capacitor to keep the input stable during an analog-to-digital conversion process.

Sampling Acquiring and digitizing a data point from an analog signal at a given instant of time.

Sampling Frequency The rate at which an analog signal is digitized (samples per second).

Sampling Interval Time window during which a frequency counter samples and thereby determines the unknown frequency of a signal.

SBD Schottky barrier diode used in all Schottky TTL series.

Schematic Capture A computer program that can interpret graphic symbols and signal connections and translate them into logical relationships.

Schmitt Trigger Digital circuit that accepts a slow-changing input signal and produces a rapid, oscillation-free transition at the output.

Schottky TTL TTL subfamily that uses the basic TTL standard circuit except that it uses a Schottky barrier diode (SBD) connected between the base and the collector of each transistor for faster switching.

Selected Signal Assignment A VHDL statement that allows a data object to be assigned a value from one of several signal sources depending on the value of an expression.

Self-Correcting Counter A counter that always progresses to its intended sequence, regardless of its initial state.

Sensitivity List The list of signals used to invoke the sequence of statements in a PROCESS.

Sequential Occuring one at a time in a certain order. In HDL, the circuits that are generated by sequential statements behave differently, depending on the order of the statements in the code.

Sequential-Access Memory (SAM) Memory in which the access time will vary depending on the storage location of the data.

Sequential Circuit A logic circuit whose outputs can change states in synchronism with a periodic clock signal. The new state of an output may depend on its current state as well as the current states of other outputs.

Serial Data Transfer Transfer of data from one place to another one bit at a time.

Serial In/Parallel Out Type of register that can be loaded with data serially and has parallel outputs available.

Serial In/Serial Out Type of register that can be loaded with data serially and has only one serial output.

Serial Transmission Transfer of binary information from one place to another a bit at a time.

SET An input to a latch or FF used to make $Q = 1$.

Set A grouping of concatenated variables or signals.

SET State The $Q = 1$ state of a flip-flop.

Settling Time Amount of time that it takes for the output of a digital-to-analog converter to go from 0 to within one-half step size of its full-scale value as the input is changed from all 0s to all 1s.

Setup Time (t_S) Time interval immediately preceding the active transition of the clock signal during which the control input must be maintained at the proper level.

Shift Register Digital circuit that accepts binary data from some input source and then shifts these data through a chain of flip-flops one bit at a time.

Sigma (Σ) Greek letter that represents addition and is often used to label the sum output bits of a parallel adder.

Sigma/Delta Modulation Method of sampling ananalog signal and converting its data points into a bit stream of serial data.

Sign Bit Binary bit that is added to the leftmost position of a binary number to indicate whether that number represents a positive or a negative quantity.

Sign-Magnitude System A system for representing signed binary numbers where the most significant bit represents the sign of the number and the remaining bits represent the true binary value (magnitude).

Simple PLD (SPLD) A PLD with a few hundred logic gates and possibly a few programmable macrocells available.

Simulator Computer program that calculates the correct output states of a logic circuit based on a description of the logic circuit and on the current inputs.

Spike *See* Glitch.

SSI Small-scale integration (fewer than 12 gates).

Staircase Test Process by which a digital-to-analog converter's digital input is incremented and its output monitored to determine whether or not it exhibits a staircase format.

Staircase Waveform Type of waveform generated at the output of a digital-to-analog converter as its digital input signal is incrementally changed.

Standard Cell An application-specific integrated circuit (ASIC) made of predesigned logic blocks from a library of standard cell designs that are interconnected during the system design stage and then fabricated on a single IC.

Standard Logic The large assortment of basic digital IC components available in various technologies as MSI, SSI chips.

State Machines A sequential circuit that advances through several defined states.

State Table A table whose entries represent the sequence of individual FF states (i.e., 0 or 1) for a sequential binary circuit.

State Transition Diagram A graphic representation of the operation of a sequential binary circuit, showing the sequence of individual FF states and conditions needed for transitions from one state to the next.

Static Accuracy Test Test in which a fixed binary value is applied to the input of a digital-to-analog converter and the analog output is accurately measured. The measured result should fall within the expected range specified by the digital-to-analog converter's manufacturer.

Static RAM (SRAM) Semiconductor RAM that stores information in flip-flop cells that do not have to be periodically refreshed.

STD_LOGIC In VHDL, a data type defined as an IEEE standard. It is similar to the BIT type, but it offers more possible values than just 1 or 0.

STD_LOGIC_VECTOR In VHDL, a data type defined as an IEEE standard. It is similar to the BIT_VECTOR type, but it offers more possible values than just 1 or 0 for each element.

Step Size *See* Resolution.

Straight Binary Coding Representation of a decimal number by its equivalent binary number.

Strobe Another name for an enable input usually used to latch a value into a register.

Strobing Technique often used to eliminate decoding spikes.

Structural Level of Abstraction A technique for describing a digital circuit that focuses on connecting ports of modules with signals.

SUBDESIGN Keyword in AHDL used to begin a circuit description.

Substrate Piece of semiconductor material that is part of the building block of any digital IC.

Subtrahend Number that is to be subtracted from a minuend.

Successive-Approximation ADC Type of analog-to-digital converter in which an internal parallel register and complex control logic are used to perform the conversion. The conversion time for this type of analog-to-digital converter is always the same regardless of the value of the input analog signal.

Sum-of-Products Form Logic expression consisting of two or more AND terms (products) that are ORed together.

Supercomputers Computers with the greatest speed and computational power.

Surface Mount A method of manufacturing circuit boards whereby ICs are soldered to conductive pads on the surface of the board.

Synchronous Control Inputs *See* Control Inputs.

Synchronous Counter Counter in which all of the flip-flops are clocked simultaneously.

Synchronous Systems Systems in which the circuit outputs can change states only on the transitions of a clock.

Synchronous Transfer Data transfer performed by using the synchronous and clock inputs of a flip-flop.

Syntax The rules defining keywords and their arrangement, usage, punctuation, and format for a given language.

Test Vector Sets of inputs used to test a PLD design before the PLD is programmed.

Timing Diagram Depiction of logic levels as related to time.

Toggle Mode Mode in which a flip-flop changes states for each clock pulse.

Toggling Process of changing from one binary state to the other.

Top-Down A design method that starts at the overall system level and then defines a hierarchy of modules.

Totem-Pole Output Term used to describe the way in which two bipolar transistors are arranged at the output of most TTL circuits.

Transducer Device that converts a physical variable to an electrical variable (for example, a photocell or a thermocouple).

Transmission Gate *See* Bilateral Switch.

Transparent Of a *D* latch, operating so that the *Q* output follows the *D* input.

Trigger Input signal to a flip-flop or one-shot that causes the output to change states depending on the conditions of the control signals.

Tristate Type of output structure that allows three types of output states: HIGH, LOW, and high-impedance (Hi-Z).

Truth Table Logic table that depicts a circuit's output response to the various combinations of the logic levels at its inputs.

TTL (Transistor/Transistor Logic) Integrated-circuit technology that uses the bipolar transistor as the principal circuit element.

2's-Complement Form Result obtained when a 1 is added to the least significant bit position of a binary number in the 1's-complement form.

Type The attribute of a variable in a computer-based language that defines its size and how it can be used.

ULSI Ultra-large-scale integration (100,000 or more gates).

Unasserted Term used to describe the state of a logic signal; synonymous with "inactive."

Undersampling Acquiring samples of a signal at a rate less than twice the highest frequency contained in the signal.

Unipolar ICs Integrated digital circuits in which unipolar field-effect transistors (MOSFETs) are the main circuit elements.

Up Counter Counter that counts upward from 0 to a maximum count.

Up/Down Counter Counter that can count up or down depending on how its inputs are activated.

Up/Down Digital-Ramp ADC Type of analog-to-digital converter that uses an up/down counter to step up or step down the voltage from a digital-to-analog converter until it intersects the analog input.

VARIABLE A keyword in AHDL used to begin a section of the code that defines the names and types of data objects and library primitives. A keyword used in VHDL to declare a local data object within a PROCESS.

Very High Speed Integrated Circuit (VHSIC) Hardware Description Language (VHDL) A hardware description language developed by the Department of Defense to document, simulate, and synthesize complex digital systems.

VLSI Very large-scale integration (10,000 to 99,999 gates).

Volatile Memory Memory requiring electrical power to keep information stored.

Voltage-Controlled Oscillator (VCO) Circuit that produces an output signal with a frequency proportional to the voltage applied to its input.

Voltage-Level Translator Circuit that takes one set of input voltage levels and translates it to a different set of output levels.

Voltage-to-Frequency ADC Type of analog-to-digital converter that converts the analog voltage to a pulse frequency that is then counted to produce a digital output.

Weighted Average An average calculation of a group of samples that assigns a different weight (between 0.0 and 1.0) to each sample.

Wired-AND Term used to describe the logic function created when open-collector outputs are tied together.

Word Group of bits that represent a certain unit of information.

Word Size Number of bits in the binary words that a digital system operates on.

WRITE Term used to describe the condition when the CPU is sending data to another element.

Write Operation Operation in which a new word is placed into a specific memory location.

ZIF Zero-insertion-force IC socket.

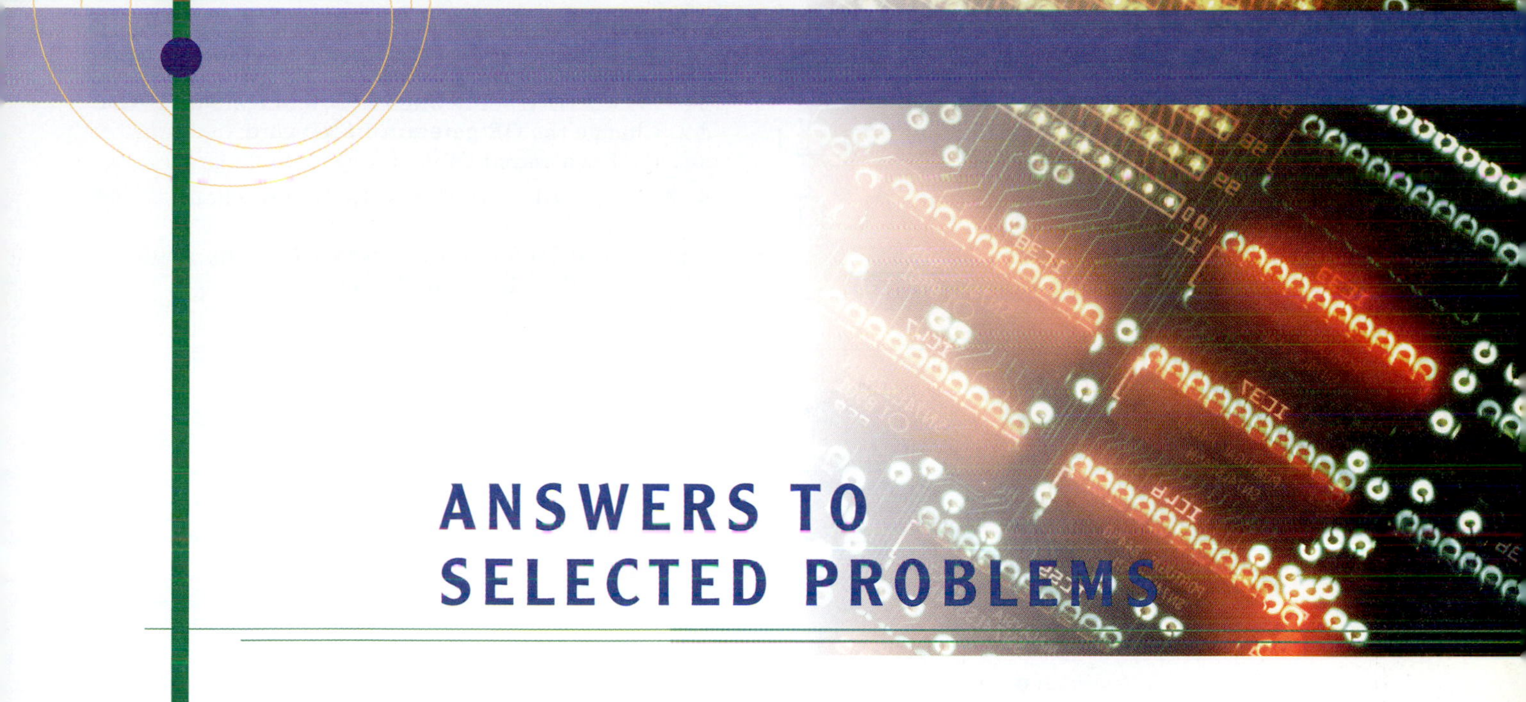

ANSWERS TO SELECTED PROBLEMS

CHAPTER 1

1-1. (a) and (e) are digital; (b), (c) and (d) are analog
1-3. (a) 25 (b) 9.5625 (c) 1241.6875
1-5. 000, 001, 010, 011, 100, 101, 110, 111
1-7. 1023
1-9. Nine bits
1-11.

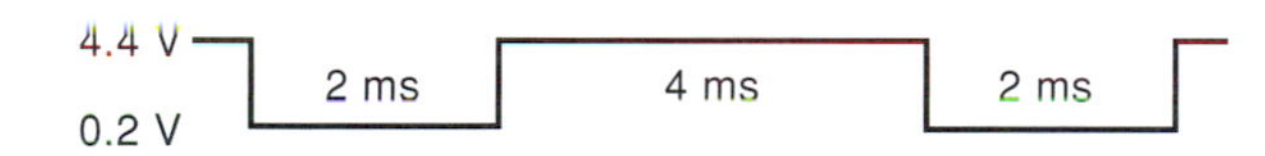

1-13. (a) $2^N - 1 = 15$ and $N = 4$; therefore, four lines are required for parallel transmission. (b) Only one line is required for serial transmission.

CHAPTER 2

2-1. (a) 22 (c) 2313 (e) 255 (g) 983
2-2. (a) 100101 (c) 10111101 (e) 1001101 (g) 11001101 (i) 111111111
2-3. (a) 255
2-4. (a) 1859 (c) 14333 (e) 357 (g) 2047
2-5. (a) 3B (c) 397 (e) 303 (g) 10000
2-6. (a) 11101000011 (c) 11011111111101 (e) 101100101 (g) 011111111111
2-7. (a) 16 (c) 909 (e) FF (g) 3D7
2-9. $2133_{10} = 855_{16} = 100001010101_2$
2-11. (a) 146 (c) 14,333 (e) 15 (g) 704
2-12. (a) 4B (c) 800 (e) 1C4D (g) 6413
2-15. (a) 16 (c) 909 (e) FF (g) 3D7
2-16. (a) 10010010 (c) 0011011111111101 (e) 1111 (g) 1011000000
2-17. 280, 281, 282, 283, 284, 285, 286, 287, 288, 289, 28A, 28B, 28C, 28D, 28E, 28F, 290, 291, 292, 293, 294, 295, 296, 297, 298, 299, 29A, 29B, 29C, 29D, 29E, 29F, 2A0
2-19. (a) 01000111 (c) 000110000111 (e) 00010011 (g) 10001001011000100111
2-21. (a) 9752 (c) 695 (e) 492
2-22. (a) 64 (b) FFFFFFFF (c) 999,999
2-25. 78, A0, BD, A0, 33, AA, F9
2-26. (a) BEN SMITH
2-27. (a) 101110100 (parity bit on the left) (c) 11000100010000100 (e) 0000101100101
2-28. (a) No single-bit error (b) Single-bit error (c) Double error (d) No single-bit error
2-30. (a) 10110001001 (b) 11111111 (c) 209 (d) 59,943 (e) 9C1 (f) 010100010001 (g) 565 (h) 10DC (i) 1961 (j) 15,900 (k) 640 (l) 952B (m) 100001100101 (n) 947 (o) 10001100101 (p) 101100110100 (q) 1001010 (r) 01011000 (BCD)
2-31. (a) 100101 (b) 00110111 (c) 25 (d) 0110011 0110111 (e) 45
2-32. (a) Hex (b) 2 (c) Digit (d) Gray (e) Parity; single-bit errors (f) ASCII (g) Hex (h) byte
2-33. (a) 1000
2-34. (a) 0110
2-35. (a) 777A (c) 1000 (e) A00
2-36. (a) 7778 (c) OFFE (e) 9FE
2-37. (a) 1,048,576 (b) Five (c) 000FF
2-39. Eight

CHAPTER 3

3-1.

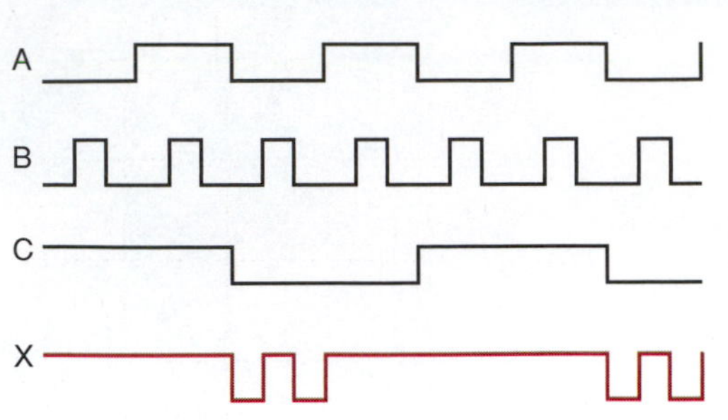

3-3. x will be a constant HIGH.
3-6. (a) x is HIGH only when A, B, and C are all HIGH.
3-7. Change the OR gate to an AND gate.
3-8. OUT is always LOW.
3-12. (a) $x = (\overline{\overline{A} + \overline{B}})BC$. x is HIGH only when ABC = 111
3-13. X is HIGH for all cases where $E = 1$ except for $EDCBA$ = 10101, 10110, and 10111.
3-14. (a) $x = D \cdot (\overline{AB + C}) + E$
3-16.

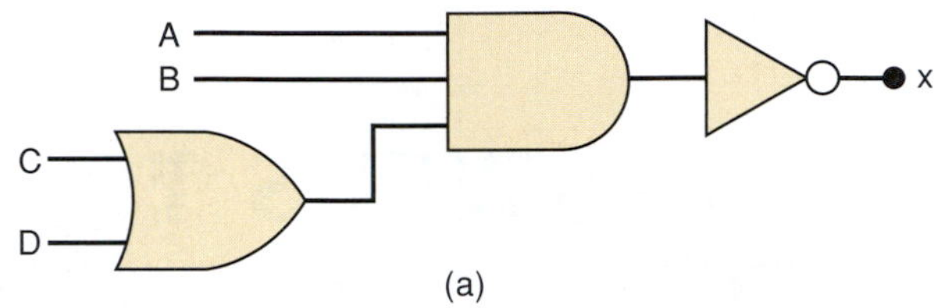

(a)

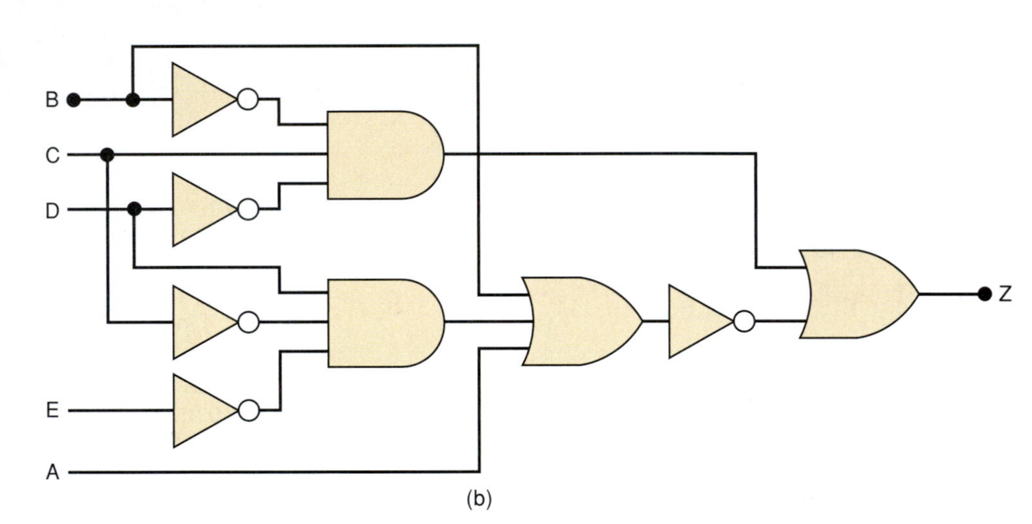

(b)

3-17.–3-18.

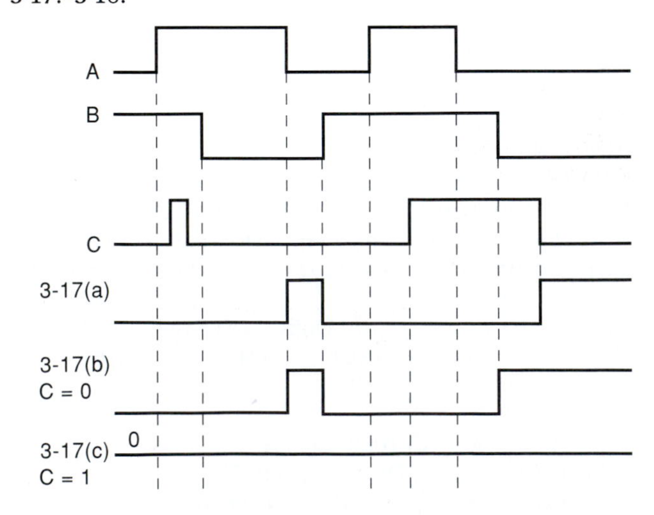

3-19. $x = \overline{(\overline{A + B}) \cdot (\overline{B + \overline{C}})}$
$x = 0$ only when $A = B = 0$, $C = 1$.
3-23. (a) 1 (b) A (c) 0 (d) C (e) 0 (f) D (g) D (h) 1 (i) G (j) y
3-24. (a) $MP\overline{N} + \overline{M}PN$
3-26. (a) $A + \overline{B} + C$ (c) $\overline{A} + \overline{B} + CD$ (e) $A + B$ (g) $\overline{A} + B + \overline{C} + \overline{D}$
3-27. $A + B + \overline{C}$
3-32. (a) $W = 1$ when $T = 1$ and either $P = 1$ or $R = 0$.
3-33. (a) NOR (b) AND (c) NAND
3-35. (a)

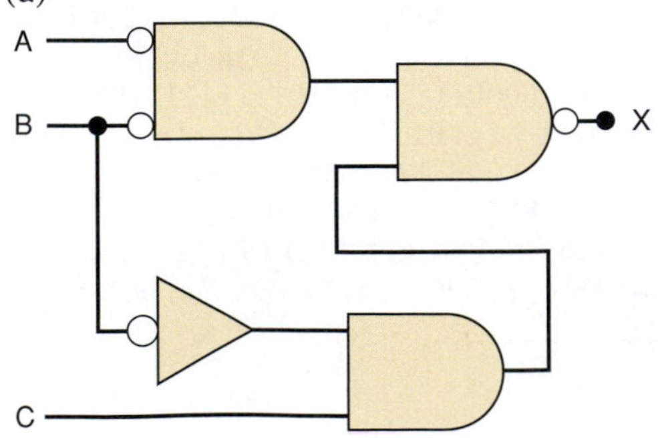

3-38. X will go HIGH when $E = 1$, or $D = 0$, or $B = C = 0$, or when $B = 1$ and $A = 0$.
3-39. (a) HIGH (b) LOW
3-41. $\overline{\text{LIGHT}} = 0$ when $A = B = 0$ or $A = B = 1$.
3-42. (a)

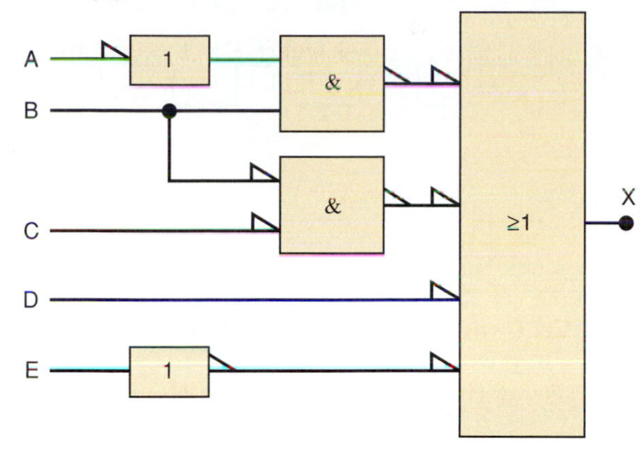

3-43. (a) False (b) True (c) False (d) True (e) False (f) False (g) True (h) False (i) True (j) True
3-45. AHDL and VHDL solutions are on the enclosed CD.
3-47. Put INVERTERs on the A_7, A_5, A_4, A_2 inputs to the 74HC30.
3-49. Requires six 2-input NAND gates.

CHAPTER 4

4-1. (a) $C\overline{A} + CB$ (b) $\overline{Q}R + Q\overline{R}$ (c) $C + \overline{A}$ (d) $\overline{R}\,\overline{S}\,\overline{T}$
(e) $BC + \overline{B}(\overline{C} + A)$
(f) $BC + \overline{B}(\overline{C} + A)$ or $BC + \overline{B}\,\overline{C} + AC$
(g) $\overline{D} + A\overline{B}\,\overline{C} + \overline{A}B\overline{C}$
(h) $x = ABC + AB\overline{D} + \overline{A}BD + \overline{B}\,\overline{C}\,\overline{D}$
4-3. $MN + Q$
4-4. One solution: $\overline{x} = \overline{B}C + AB\overline{C}$. Another: $x = \overline{A}B + \overline{B}\,\overline{C} + BC$. Another: $BC + \overline{B}\,\overline{C} + \overline{A}\,\overline{C}$
4-7. $x = \overline{A}_3(A_2 + A_1A_0)$
4-9.

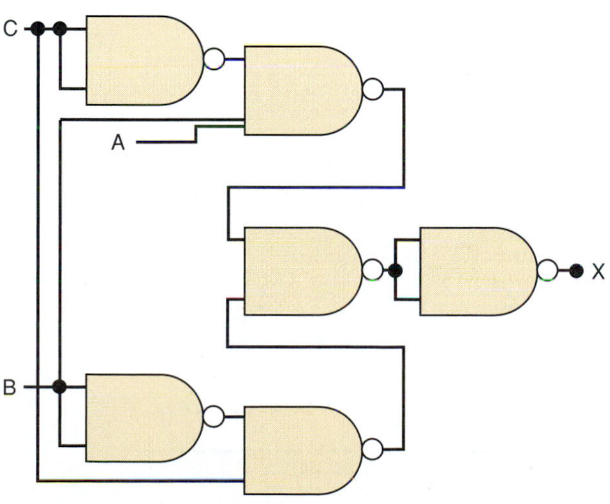

4-11. (a) $x = \overline{A}\,\overline{C} + \overline{B}C + AC\overline{D}$

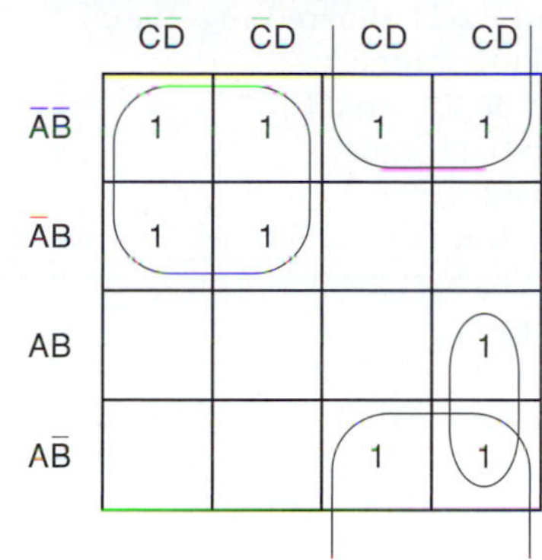

4-14. (a) $x = BC + \overline{B}\,\overline{C} + AC$; or $x = BC + \overline{B}\,\overline{C} + A\overline{B}$
(c) One possible looping:
$x = \overline{A}BD + ABC + AB\overline{D} + \overline{B}\,\overline{C}\,\overline{D}$; another one is:
$x = ABC + \overline{A}BD + A\overline{C}\,\overline{D} + \overline{B}\,\overline{C}\,\overline{D}$
4-15. $x = \overline{A}_3A_2 + \overline{A}_3A_1A_0$
4-16. (a) Best solution: $x = B\overline{C} + AD$
4-17. $x = \overline{S}_1\overline{S}_2 + \overline{S}_1\overline{S}_3 + \overline{S}_3\overline{S}_4 + \overline{S}_2\overline{S}_3 + \overline{S}_2\overline{S}_4$
4-18. $z = \overline{B}C + \overline{A}B\overline{D}$
4-21. $A = 0, B = C = 1$
4-23. One possibility is shown below.

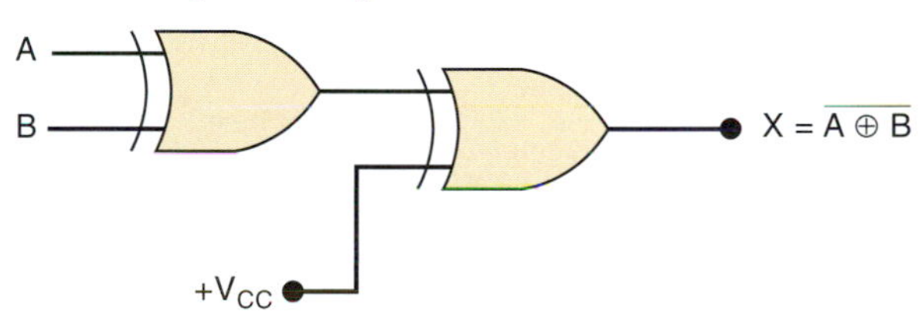

4-24. Four XNORs feeding an AND gate
4-26. Four outputs where z_3 is the MSB
$z_3 = y_1y_0x_1x_0$
$z_2 = y_1x_1(\overline{y}_0 + \overline{x}_0)$
$z_1 = y_0x_1(\overline{y}_1 + \overline{x}_0) + y_1x_0(\overline{y}_0 + \overline{x}_1)$
$z_0 = y_0x_0$
4-28. $x = AB\overline{(C \oplus D)}$
4-30. $\text{N-S} = \overline{C}\,\overline{D}(A + B) + AB(\overline{C} + \overline{D})$; $\text{E-W} = \overline{\text{N-S}}$
4-33. (a) No (b) No
4-35. $x = A + BCD$
4-38. $z = x_1x_0y_1y_0 + x_1\overline{x}_0y_1\overline{y}_0 + \overline{x}_1x_0\overline{y}_1y_0 + \overline{x}_1\overline{x}_0\overline{y}_1\overline{y}_0$
No pairs, quads, or octets
4-40. (a) Indeterminate (b) 1.4–1.8 V (c) See below.

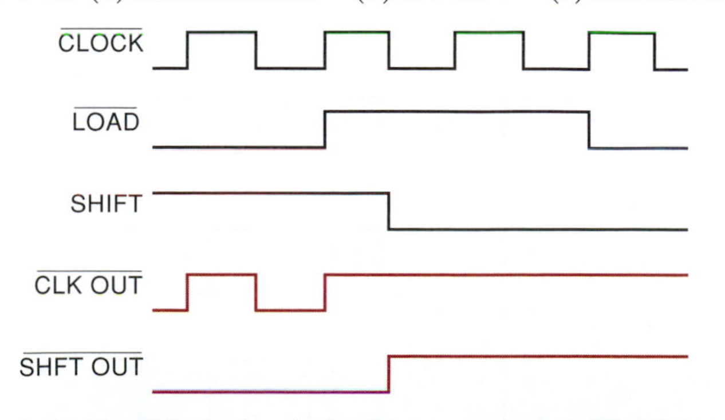

4-43. Possible faults: faulty V_{CC} or ground on Z2; Z2-1 or Z2-2 open internally or externally; Z2-3 internally open

4-44. Yes: (c), (e), (f). No: (a), (b), (d), (g).
4-46. Z2-6 and Z2-11 shorted together
4-48. Most likely faults:
faulty ground or V_{CC} on Z1;
Z1 plugged in backwards;
Z1 internally damaged
4-49. Possible faults:
Z2-13 shorted to V_{CC};
Z2-8 shorted to V_{CC};
broken connection to Z2-13;
Z2-3, Z2-6, Z2-9, or Z2-10 shorted to ground
4-50. (a) T, (b) T, (c) F, (d) F, (e) T
4-54. Boolean equation; truth table; schematic diagram
4-56. (a) AHDL: gadgets[7..0] :OUTPUT;
VHDL: gadgets :OUT BIT_VECTOR (7 DOWNTO 0);
4-57. (a) AHDL: H"98" B"10011000" 152
VHDL: X"98" B"10011000" 152
4-58.

```
AHDL: outbits[3]  =   inbits[1];
      outbits[2]  =   inbits[3];
      outbits[1]  =   inbits[0];
      outbits[0]  =   inbits[2];
VHDL: outbits(3)  <=  inbits(1);
      outbits(2)  <=  inbits(3);
      outbits(1)  <=  inbits(0);
      outbits(0)  <=  inbits(2);
```

4-60.

```
BEGIN
       IF digital_value[] < 10 THEN
            z = VCC; --output a 1
       ELSE z = GND; --output a 0
       END IF;
END;
```

4-62.

```
PROCESS (digital_value)
     BEGIN
        IF (digital_value < 10) THEN
               z  < =  '1';
        ELSE
               z  < =  '0';
        END IF;
END PROCESS
```

4-65. S=!P#Q&R
4-68. (a) 00 to EF

CHAPTER 5

5-1.

5-3.

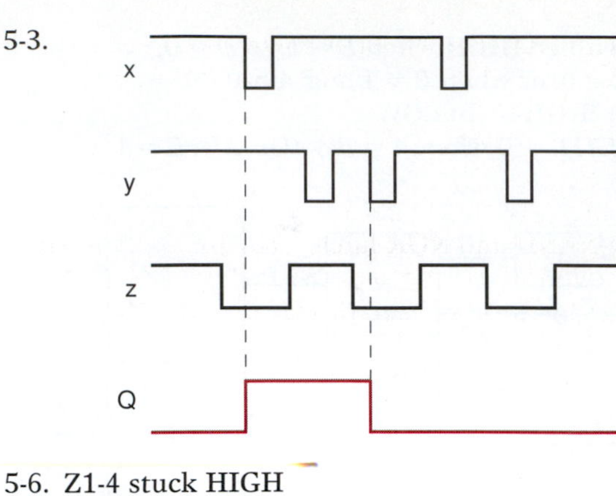

5-6. Z1-4 stuck HIGH
5-9. Assume $Q = 0$ initially.
For PGT FF: Q will go HIGH on first PGT of CLK.
For NGT FF: Q will go HIGH on first NGT of CLK, LOW on second NGT, and HIGH again on fourth NGT.
5-11.

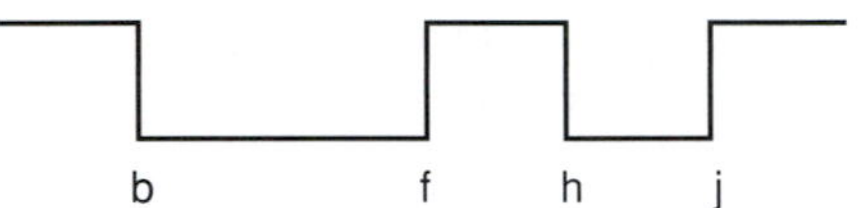

5-12. (a) 5-kHz square wave
5-14.

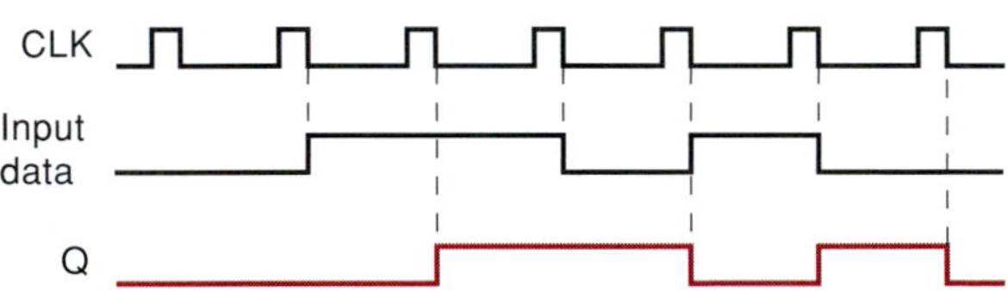

5-16. 500-Hz square wave
5-21.

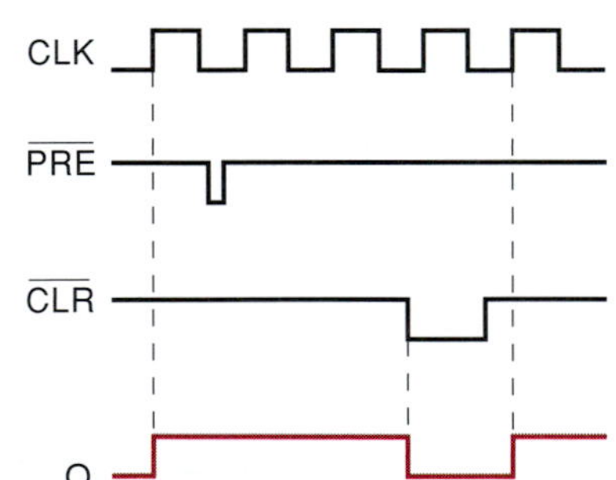

5-23. (a) 200 ns (b) 7474; 74C74
5-25. Connect A to J, $\overline{A}$ to K.
5-27. (a) Connect X to J, $\overline{X}$ to K. (b) Use arrangement of Figure 5-41.
5-29. Connect X_0 to D input of X_2.
5-30. (a) 101;011;000
5-33. (a) 10 (b) 1953 Hz (c) 1024 (d) 12
5-36. Put INVERTERs on A_8, A_{11}, and A_{14}.
5-41.

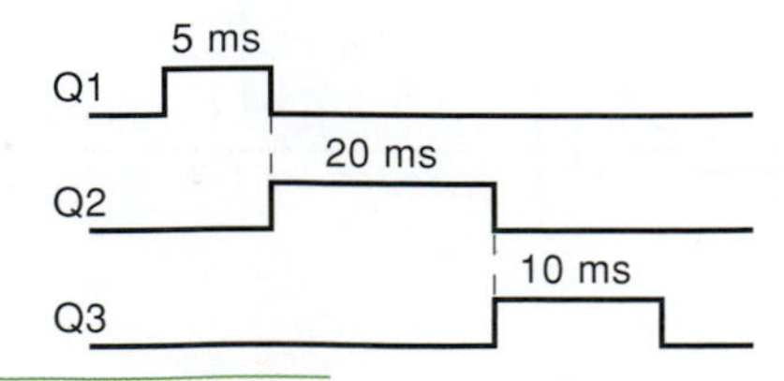

5-43. (a) A_1 or A_2 must be LOW when a PGT occurs at B.
5-45. One possibility is $R = 1\ k\Omega$ and $C = 80$ nF.
5-50. (a) No (b) Yes
5-51. (a) Yes
5-53. (a) No (b) No
5-55. (a) No (b) No (c) Yes
5-56. (a) NAND and NOR latch (b) J-K (c) D latch (d) D flip-flop
5-59. See Prob5_59.tdf and prob5_59.vhd on the enclosed CD.
5-61. See Prob5_61.tdf and prob5_61.vhd on the enclosed CD.
5-66. (a) See Prob5_66a.tdf on the enclosed CD. (b) See Prob5_66b.vhd on the enclosed CD.

CHAPTER 6

6-1. (a) 10101 (b) 10010 (c) 1111.0101
6-2. (a) 00100000 (including sign bit) (b) 11110010 (c) 00111111 (d) 10011000 (e) 01111111 (f) 10000001 (g) 01011001 (h) 11001001
6-3. (a) +13 (b) −3 (c) +123 (d) −103 (e) +127
6-5. -16_{10} to 15_{10}
6-6. (a) 01001001; 10110111 (b) 11110100; 00001100
6-7. 0 to 1023; −512 to +511
6-9. (a) 00001111 (b) 11111101 (c) 11111011 (d) 10000000 (e) 00000001
6-11. (a) 100011 (b) 1111001
6-12. (a) 11 (b) 111
6-13. (a) 10010111 (BCD) (b) 10010101 (BCD) (c) 010100100111 (BCD)
6-14. (a) 6E24 (b) 100D (c) 18AB
6-15. (a) 0EFE (b) 229 (c) 02A6
6-17. (a) 119 (b) +119
6-19. SUM = $A \oplus B$; CARRY = AB
6-21. $[A]$ = 1111, or $[A]$ = 000 (if $C_0 = 1$)
6-25. $C_3 = A_2B_2 + (A_2 + B_2)\{A_1B_1 + (A_1 + B_1)[A_0B_0 + A_0C_0 + B_0C_0]\}$
6-27. (a) SUM = 0111
6-32.

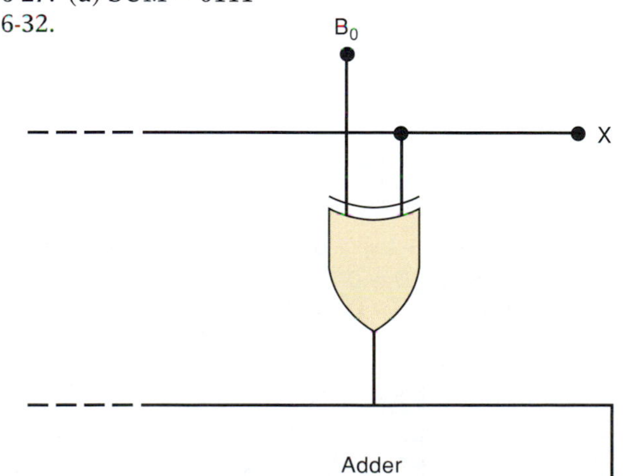

6-33.

	[F]	C_{N+4}	OVR
(a)	1001	0	1

6-35. (a) 00001100
6-37. (a) 0001 (b) 1010
6-39. (a) 1111 (b) HIGH (c) No change (d) HIGH
6-41. (a) 00000100 (b) 10111111
6-43. (a) 0 (b) 1 (c) 0010110
6-44. **AHDL**

```
z[6..0] = a[7..1];
z[7] = a[0];
```

VHDL

```
z(6..0) < = a(7..1);
z(7) < = a(0);
```

6-47. AHDL: ovr <= c[4] $ c[3)];
VHDL: ovr <= c(4) XOR c(3);
6-48. See Prob6_48.tdf and Prob6_48.vhd on the enclosed CD.
6-53. Use D flip-flops. Connect $\overline{(S_3 + S_2 + S_1 + S_0)}$ to the D input of the 0 FF; C_4 to the D input of the carry FF; and S_3 to the D input of the sign FF.
6-54. 0000000001001001; 1111111110101110

CHAPTER 7

Note: Solutions to some problems in Chapter 7 are provided in a document file (*Chapter 7 solutions.doc*) on the enclosed CD. Please see this file as indicated below.

7-1. (a) 250 kHz; 50% (b) Same as (a) (c) 1 MHz (d) 32
7-3. 10000_2
7-5. 1000 and 0000 states never occur
7-7. (a) See schematic on CD. (b) 33 MHz
7-9. Frequency at D = 100 Hz (see diagram on CD)
7-11. Replace four-input NAND with a three-input NAND driving all FF CLRs whose inputs are Q5, Q4, and Q1
7-13. See diagram on CD.
7-15. Counter switches states between 000 and 111 on each clock pulse
7-17. See timing on CD.
7-19. See timing on CD.
7-21. (a) 0000, 0001, 0010, 0011, 0100, 0101, 0110, 0111, 1000, 1001, 1010, 1011, & repeat (b) MOD-12 (c) Frequency at QD (MSB) is 1/12 of CLDK frequency (d) 33.3%
7-23. (a) see timing on CD (b) MOD-10 (c) 10 down to 1 (d) Can produce MOD-10, but not same sequence
7-25. (a), (b) See diagrams on CD.
7-27. See diagrams on CD.
7-29.

Output:	QA	QB	QC	QD	RCO
Frequency:	3 MHz	1.5 MHz	750 kHz	375 kHz	375 kHz
Duty cycle:	50%	50%	50%	50%	6.25%

7-31. Frequency at f_{out1} = 500 kHz, at f_{out2} = 100 kHz
7-33. 12M/8 = 1.5M 1.5M/10 = 150k 1.5M/15 = 100k See diagram on CD
7-35. See gate symbols on CD.
7-37. See simulation on CD.
7-39. See simulation on CD.
7-41. See diagram on CD.
7-43. (a) $J_A = B\ \overline{C}$, $K_A = 1$, $J_B = C\ A + \overline{C}\ \overline{A}$, $K_B = 1$, $J_C = B\ \overline{A}$, $K_C = B + \overline{A}$
(b) $J_A = B\ \overline{C}$, $K_A = 1$, $J_B = K_B = 1$, $J_C = K_C = B$
7-45. $J_A = K_A = 1$, $J_B = C\ \overline{A} + D\ \overline{A}$, $K_B = \overline{A}$, $J_C = D\ \overline{A}$, $K_C = \overline{B}\ \overline{A}$, $J_D = \overline{C}\ \overline{B}\ \overline{A}$, $K_D = \overline{A}$
7-47. $D_A = \overline{A}$, $D_B = B\ A + \overline{B}\ \overline{A}$, $D_C = C\ A + C\ B + \overline{C}\ \overline{B}\ \overline{A}$
7-49. See HDL files on CD. mod13_ahdl mod13_vhdl
7-51. See HDL files on CD. gray_ahdl gray_vhdl
7-53. See HDL files on CD. divide_by50_ahdl divide_by50_vhdl
7-55. See HDL files on CD. mod256_ahdl mod256_vhdl
7-57. See HDL files on CD. mod16_ahdl mod16_vhdl
7-59. See diagram on CD.
7-61. See HDL files on CD. mod10_ahdl mod5_ahdl mod50_vhdl mod10_vhdl mod5_vhdl
7-63. See HDL files on CD. wash_mach_delux wash_mach_delux
7-65. See table on CD.
7-67. Eight clock pulses are needed to serially load a 74166, since there are eight FFs in the chip.
7-69. See timing on CD.
7-71. See answer on CD.
7-73. See diagram on CD.
7-75. See diagram on CD.
7-77. Output of 3-in AND or J, K inputs to FF D shorted to ground, FF D output shorted to ground, CLK input on FF D open, B input to NAND is open
7-79. See HDL files on CD. siso8_ahdl siso8_vhdl
7-81. See HDL files on CD. piso8_ahdl piso8_vhdl
7-83. See simulation on CD.
7-85. See HDL files on CD. johnson_ahdl johnson_vhdl
7-87. See simulation on CD.
7-89. (a) Parallel (b) Binary (c) MOD-8 down (d) MOD-10, BCD, decade (e) Asynchronous, ripple (f) Ring (g) Johnson (h) All (i) Presettable (j) Up/down (k) Asynchronous, ripple (l) MOD-10, BCD, decade (m) Synchronous, parallel

CHAPTER 8

8-1. (a) A; B (b) A (c) A
8-2. (a) 39.4 mW, 18.5 ns (b) 65.6 mW, 7.0 ns
8-3. (a) 0.9 V
8-4. (a) I_{IH} (b) I_{CCL} (c) t_{PHL} (d) V_{NH} (e) Surface-mount (f) Current sinking (g) Fan-out (h) Totem-pole (i) Sinking transistor (j) 4.75 to 5.25 V (k) 2.5 V; 2.0 V (l) 0.8 V; 0.5 V (m) Sourcing
8-5. (a) 0.7 V; 0.3 V (b) 0.5 V; 0.4 V (c) 0.5 V; 0.3 V
8-6. (b) AND, NAND (c) Unconnected inputs
8-7. (a) 40 (b) 33
8-8. (a) 20 μA/0.4 mA
8-9. (a) 30/15 (b) 24 mA
8-11. Fan-out is not exceeded in either case.
8-13. 60 ns; 38 ns
8-14. (a) 2 kΩ
8-15. (b) 4.7-kΩ resistor is too large.
8-19. *a*, *c*, *e*, *f*, *g*, *h*
8-21. 12.6 mW
8-27. $AB + CD + FG$
8-29. (a) 5 V (b) R_S = 110 Ω for LED current of 20 mA
8-30. (a) 12 V (b) 40 mA
8-33. Ring counter
8-36. 1.22 V; 0 V
8-37.

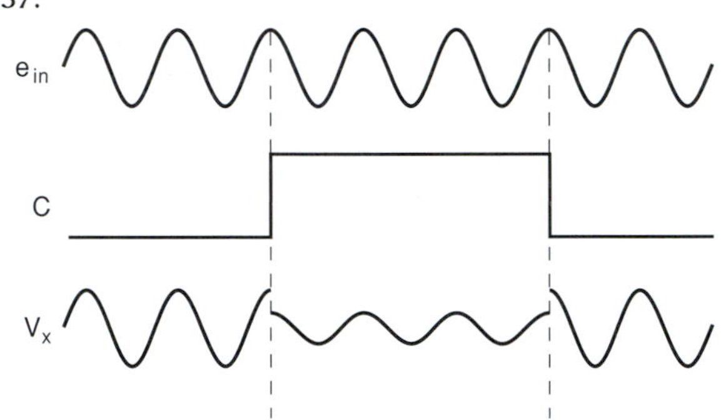

8-38. −1 and −2
8-39. (a) 74HCT (b) Converts logic voltages (c) CMOS cannot sink TTL current. (d) False
8-41. (a) None
8-44. Fan-out of 74HC00 is exceeded; disconnect pin 3 of 7402 and tie it to ground.
8-46. R_2 = 1.5 kΩ, R_1 = 18 kΩ
8-49. (b) is a possible fault.
8-50. 0 V to −11.25 V and back up to −6 V

CHAPTER 9

9-1. (a) All HIGH (b) $\overline{O}_0$ = LOW
9-2. Six inputs, 64 outputs
9-3. (a) $E_3E_2E_1$ = 100; [A] = 110 (b) $E_3E_2E_1$ = 100; [A] = 011
9-5.

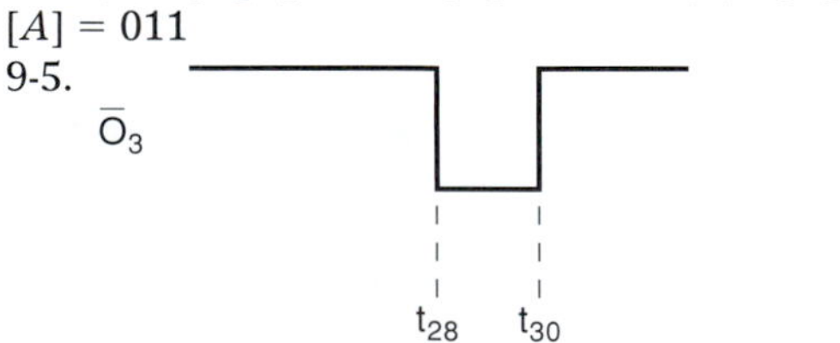

9-7. Enabled when D = 0
9-10. Resistors are 250 Ω.
9-12.

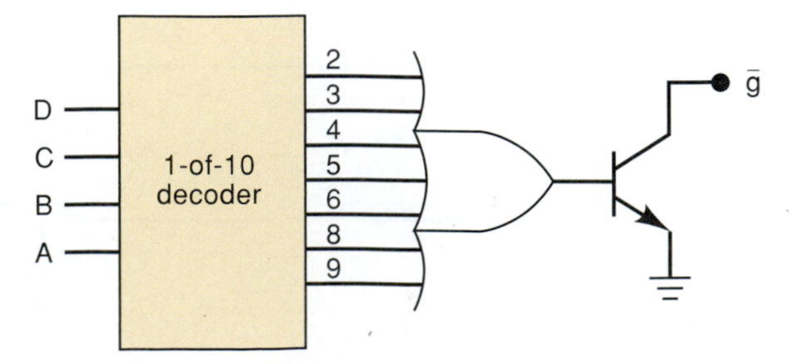

9-13. (a), (b) Encoder (c), (d), (e) Decoder
9-17. The fourth key actuation would be entered into the MSD register.
9-18. Choice (b)
9-20. (a) Yes (b) No (c) No
9-21. A_2 bus line is open between Z2 and Z3.
9-23. *g* segment or decoder output transistor would burn out.
9-25. Decoder outputs: *a* and *b* are shorted together.
9-26. Connection 'f' from decoder/driver to XOR gate is open.
9-29. A 4-to-1 MUX
9-31.

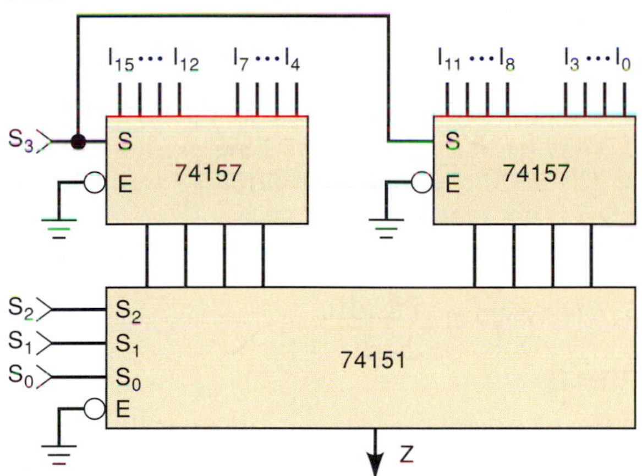

9-32. (b) The total number of connections in the circuit using MUXes is 63, not including V_{CC} and GND, and not including the connections to counter clock inputs. The total number for the circuit using separate decoder/drivers is 66.
9-33.

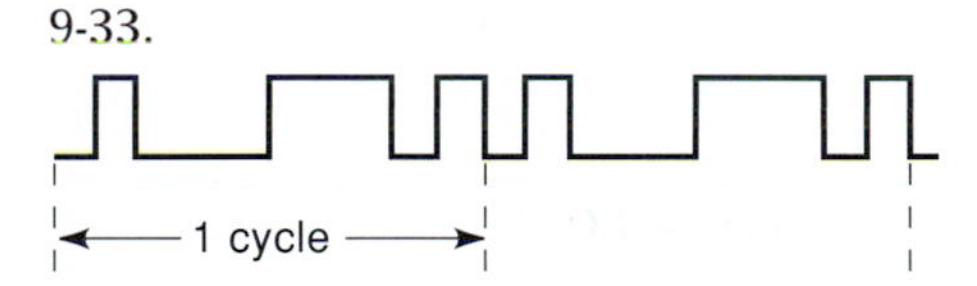

9-35.

A	B	C	
0	0	0	$0 \Rightarrow I_0$
0	0	1	$0 \Rightarrow I_1$
0	1	0	$0 \Rightarrow I_2$
0	1	1	$1 \Rightarrow I_3$
1	0	0	$0 \Rightarrow I_4$
1	0	1	$1 \Rightarrow I_5$
1	1	0	$1 \Rightarrow I_6$
1	1	1	$1 \Rightarrow I_7$

9-37. Z = HIGH for *DCBA* = 0010, 0100, 1001, 1010.
9-39. (a) Encoder, MUX (b) MUX, DEMUX (c) MUX (d) Encoder (e) Decoder, DEMUX (f) DEMUX (g) MUX
9-41. Each DEMUX output goes LOW, one at a time in sequence.
9-43. Five lines
9-46. (a) Sequencing stops after actuator 3 is activated.
9-47. Probable fault is short to ground at MSB of tens MUX.
9-48. Q_0 and Q_1 are probably reversed.
9-49. Inputs 6 and 7 of MUX are probably shorted together.
9-50. S_1 stuck LOW
9-53. Use three 74HC85s
9-55. A_0 and B_0 are probably reversed.
9-57. $\overline{OE_C} = 0$, $IE_C = 1$; $\overline{OE_B} = \overline{OE_A} = 1$; $\overline{IE_B} = \overline{IE_A} = 0$; apply a clock pulse.
9-61. (a) At t_3, each register holds 1001.
9-63. (a) 57FA (b) 5000 to 57FF (c) 9000 to 97FF (d) no
9-65. See Prob9_65.tdf and Prob9_65.vhd on the enclosed CD.

CHAPTER 10

10-1. (d) 20 Hz (e) Only one LED will be lit at any time.
10-2. 24
10-3. Four states = four steps * 15°/step = 60° of rotation
10-5. Three state transitions * 15°/step = 45° of rotation
10-10. 1111
10-12. (a) 1011
10-13. No
10-15. The data go away (hi-Z) before the DAV goes LOW. The hi-Z state is latched.
10-16.

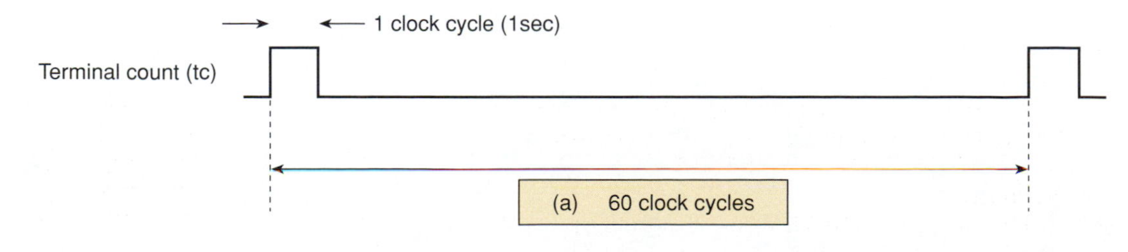

10-17. 60 cycles/sec * 60 sec/min * 60 min/hr * 24 hr/day = 5,184,000 cycles/day. This takes a long time to generate a simulation file.
10-18. When the set input is active, bypass the prescaler and feed the 60-Hz clock directly into the units of seconds counter.
10-22. See Prob10_22.tdf and Prob10_22.vhd on the enclosed CD.

CHAPTER 11

11-1. (f), (g) False
11-3. LSB = 20 mV
11-5. Approximately 5 mV
11-7. 14.3 percent, 0.286 V
11-9. 250.06 rpm
11-11. The eight MSBs: PORT[7..0] ⇒ DAC[9..2]
11-13. 800 Ω; no
11-15. Uses fewer different *R* values
11-17. (a) Seven
11-19. 242.5 mV is not within specifications.
11-21. Bit 1 of DAC is open or stuck HIGH.
11-22. Bits 0 and 1 are reversed.
11-24. (a) 10010111
11-27. (a) 1.2 mV (b) 2.7 mV
11-28. (a) 0111110110
11-31. Reconstructed waveform frequency is 3.33 kHz.
11-32. (a) 5 kHz (b) 9.9 kHz
11-33. Digital ramp: *a*, *d*, *e*, *f*, *h*. SAC: *b*, *c*, *d*, *e*, *g*, *h*
11-36. 80 μs
11-38. 2.276 V
11-40. (a) 00000000 (b) 500 mV (c) 510 mV (d) 255 mV (e) 01101110 (f) 0.2°F; 2 mV
11-45. Switch is stuck closed; switch is stuck open, or capacitor is shorted.
11-47. (a) Address is EA*xx*.
11-52. False: a, e, g; True: b, c, d, f, h

CHAPTER 12

12-1. 16,384; 32; 524,288
12-3. 64K × 4
12-7. (a) Hi-Z (b) 11101101
12-9. (a) 16,384 (b) Four (c) Two 1-of-128 decoders
12-11. 120 ns
12-15. The following transistors will have open source connections: Q_0, Q_2, Q_5, Q_6, Q_7, Q_9, Q_{15}.
12-17. (a) Erases all memory locations to hold FF_{16} (b) Writes $3C_{16}$ into address 2300_{16}
12-19. Hex data: 5E, BA, 05, 2F, 99, FB, 00, ED, 3C, FF, B8, C7, 27, EA, 52, 5B
12-20. (a) 25.6 kHz (b) Adjust V_{ref}.
12-22. (a) [*B*] = 40 (hex); [*C*] = 80 (hex) (b) [*B*] = 55 (hex); [*C*] = AA (hex) (c) 15,360 Hz (d) 28.6 MHz (e) 27.9 kHz
12-24. (a) 100 ns (b) 30 ns (c) 10 million (d) 20 ns (e) 30 ns (f) 40 ns (g) 10 million
12-30. Every 7.8 μs
12-31. (a) 4096 columns, 1024 rows (b) 2048 (c) It would double.
12-34. Add four more PROMs (PROM-4 through PROM-7) to the circuit. Connect their data outputs and address inputs to data and address bus, respectively. Connect AB_{13} to *C* input of decoder, and connect decoder outputs 4 through 7 to *CS* inputs of PROMs 4 through 7, respectively.
12-38. F000–F3FF; F400–F7FF; F800–FBFF; FC00–FFFF
12-40. *B* input of decoder is open or stuck HIGH.
12-42. Only RAM modules 1 and 3 are getting tested.
12-43. The RAM chip with data outputs 4 through 7 in module 2 is not functioning properly.
12-44. RAM module 3, output 7 is open or stuck HIGH.
12-46. Checksum = 11101010.

CHAPTER 13

13-2. The necessary speed of operation for the circuit, cost of manufacturing, system power consumption, system size, amount of time available to design the product, etc.
13-4. Speed of operation
13-6. Advantages: highest speed and smallest die area; Disadvantages: design/development time and expense
13-8. SRAM-based PLDs must be configured (programmed) upon power-up.
13-10. In a PLD programmer or in-system (via JTAG interface)
13-12. pin 1—GCLRn (Global Clear)
pin 2—OE2/GCLK2 (Output Enable 2/Global Clock 2)
pin 83—GCLK1 (Global Clock 1)
pin 84—OE1 (Output Enable 1)
13-14. Logic cell in MAX7000S is AND/OR circuit versus look-up table in FLEX10K; EEPROM (MAX7000S) and SRAM (FLEX10K); MAX7000S is nonvolatile; FLEX10K has greater logic resources.

INDEX OF ICs

OTHER ICs

INDEX

D

E

L

M

N

Q

R

T

U

V

W

Z